R a Riddle

AF335096

HYDROGEN EFFECTS IN METALS

Hydrogen Effects in Metals

Proceedings of the Third International Conference on Effect of Hydrogen on Behavior of Materials sponsored by the Physical Metallurgy and Mechanical Metallurgy Committees of The Metallurgical Society of AIME, the National Science Foundation, and the U.S. Air Force Office of Scientific Research, Moran, Wyoming, August 26-31, 1980.

Edited by

I.M. BERNSTEIN and ANTHONY W. THOMPSON
Carnegie-Mellon University
Pittsburgh, Pennsylvania

A Publication of

A Publication of The Metallurgical Society of AIME
P.O. Box 430
420 Commonwealth Drive
Warrendale, Pa. 15086
(412) 776-9000

Printed in the United States of America.
Library of Congress Card Catalogue Number 81-82943
ISBN Number 0-89520-378-2

FOREWORD

More than five years since the last major hydrogen conference in the
United States, 1980 appeared appropriate to assess once again the state of
our progress in understanding the multi-complex set of phenomena that
characterizes hydrogen's interaction with metals.

As with the period between 1972 and 1975, between the first and second
of these conferences, this period has been exemplified by two seemingly contradictory trends: As the breadth of our understanding improves about hydrogen interactions with the atomic structure, microstructure, and defects
in a number of metals and alloys, the mechanism(s) underlying such behavior
has (have) still not been satisfactorily identified. Among the reasons for
this have been the apparent diversity of susceptible metals and alloys, and
the almost dazzling array of failure modes that fall under the rubric of
hydrogen embrittlement.

The results and discussions stemming from this conference seem to suggest, however, that the situation may have changed and that a coherent link
is emerging between the various forms of hydrogen embrittlement. Evidence
is rapidly accumulating to support hydrogens major role in modifying the
response to plastic flow in ferrous alloys and further that the failure process itself is closely connected to such hydrogen-modified plasticity effects
as strain localization. If such patterns can be shown to occur in the hydrogen embrittlement of nonferrous alloys, then the real possibility exists
that experimenters can present to the theorists a coherent set of results to
test current or future theories.

In nonferrous metals and alloys, there is now little doubt that hydrogen is capable of dominating environmentally induced embrittlement in nickel,
austenitic stainless steels, and several aluminum alloys. A number of papers
in this volume describe detailed studies of such behavior, which provide a
useful parallel to the considerable work continuing in ferrous alloys. In
particular, the contrast in the relative mobility of hydrogen in these different material classes has provoked a number of important studies on the
possible occurrence of enhanced transport of hydrogen by dislocations, and
the subsequent importance of such transport on the fracture process. The
question has been raised as to whether dislocation transport can provoke a
transient build-up of internal hydrogen pressure far in excess of that predicted from equilibrium considerations, thus promoting premature failure by
a modified "pressure" mechanism. This suggestion has generated considerable
and important debate and the issue is by no means settled, as will be
obvious to the readers of this volume.

Have we made real progress in the field? We think the answer is a
clear "yes." Not only has understanding been broadened of how hydrogen
modifies mechanical properties but there are now suggestive clues pointing
to the mechanisms by which such modifications occur. The real test for such

optimism will come from future research results, which hopefully will be stimulated by the papers and discussions presented herein.

The papers in this book have been reproduced from manuscripts prepared by the authors. It was possible to correct some grammatical and typographical errors, but corrections had to be minimized in the interest of quick and economical publication. It is hoped that the reader will view any errors which are discovered in this light.

We want to thank a number of people and agencies who helped make the Conference and subsequent proceedings a success. Dr. Clyde Briant of the General Electric R&D Laboratory was of great help in the initial planning and in helping us with many of the details. A number of our colleagues at Carnegie-Mellon University gave unstintingly of their time before and during the Conference, and we want to gratefully thank Drs. Dallis Hardwick and Leo Christodoulou, and Mary Juhas, Mike Stevens, John Lewandowski, Bob Swanson, and Cherngye Hwang. Our wives Kathie Bernstein and Mary Thompson helped ensure that activities ran smoothly and as with all things, provided support and encouragement. Annette Crelli and Maxine Patterson did the myriad secretarial chores and typing so well that we were never aware of any problems or time pressures.

Financial support was generously provided by the National Science Foundation and the Air Force Office of Scientific Research and we would like to thank Drs. R. J. Reynik and A. F. Rosenstein of these respective organizations. General Electric and the Sandia Laboratories donated printing and mailing facilities for which we are grateful.

I. M. Bernstein
A. W. Thompson
Carnegie-Mellon University
Pittsburgh, Pennsylvania

March, 1981

TABLE OF CONTENTS

EFFECTS ON DEFORMATION AND MECHANICAL PROPERTIES

EFFECTS OF MICROSTRUCTURE

PERMEATION AND TRAPPING

HYDROGEN AND DEUTERIUM TRAPPING IN IRON[*]

Herbert H. Johnson and Ruey Way Lin

Department of Materials Science and Engineering

Cornell University

Ithaca, New York 14853

Hydrogen and deuterium trapping in annealed and deformed iron were studied with a gas phase permeation technique over a range of lattice hydrogen concentrations. Transient, steady state, and time lag measurements were made. The McNabb-Foster formalism provided a self-consistent interpretation of the transient and time lag data.

In both annealed and deformed iron the trap binding energy for hydrogen and deuterium was 48-58 kJ/mol, while trap densities varied from about 10^{20} m^{-3} for annealed iron to over 10^{23} m^{-3} for heavily deformed iron. The smaller binding energy, about 27 kJ/mol, reported by earlier investigators, could not be reconciled with transient and time lag measurements for different lattice hydrogen concentrations.

A discussion is given of the conditions under which permeation methods are sensitive to trapping phenomena.

[*] This research was supported by the Materials Sciences Program of the U. S. Department of Energy; it benefited from facilities provided by the National Science Foundation through the MRL Program.

3

Introduction

Hydrogen embrittlement shows a characteristic dependence upon temperature and strain rate, which implies strongly that hydrogen transport is an essential component of the embrittlement mechanism. In the temperature range (around 300 K) where embrittlement is encountered, there is ample but qualitative evidence that hydrogen transport, although still rapid, is hindered by interactions with heterogeneities such as dislocations, impurity atoms, interfaces and voids, which have been subsumed as "traps". There is also substantial evidence for hydrogen transport during plastic deformation by moving dislocations. For a comprehensive and critical review of these and related topics see Hirth's 1980 Institute of Metals Lecture (1).

Substantial evidence for trapping is found in compilations (2,3) of effective hydrogen diffusivities in iron. There is much scatter, but diffusivities measured near 300 K fall well below the values estimated by extrapolation of high temperature data. Since most of the experiments were based on permeability techniques, some of the low values and scatter may be caused by surface impedences to hydrogen entry, but there seems little doubt that the smaller diffusivities are at least in part caused by trapping.

Darken and Smith (4) were apparently the first investigators to suggest that the hindered transport of hydrogen through cold worked steels, as shown by permeation transients, was caused by trapping. Some twenty years later, Oriani (5) in a seminal paper analyzed the then available data for steel, obtained from various absorption, permeation and outgassing experiments, in terms of trapping concepts. His calculations suggested a surprising uniformity of the trap binding energy at a value of 27.2 kJ/mol H. The trap density for annealed steels was estimated to be of the order of $10^{25} m^{-3}$. Oriani (5) suggested the possibility of multiple trapping sites for hydrogen in steel, including but not necessarily limited to dislocations, internal interfaces, and microcracks. A trap binding energy of 27.2 kJ/mol H was also estimated from subsequent permeation experiments (6,7) using electrolytic charging of iron.

The binding energy between hydrogen and dislocations has also been inferred to be about 27.2 kJ/mol H from internal friction measurements on deformed and charged iron (see 8 and 9 for reviews of earlier work, 10 for a recent study, and 1 for a critical review). These experiments were interpreted in terms of the Schoeck (11,12) model, wherein the relaxation mechanism is assumed to be the viscous dragging of solutes, such as hydrogen, by dislocations moving between fixed pinning points. The parameters inferred from the measurements are strongly model dependent, and it does not seem possible to obtain trap densities.

Two quite different approaches have therefore led to the same value for the trap binding energy, but not to a consensus identification of the trap site. Quite remarkably, the value obtained, 27.2 kJ/mol H, is the negative of the energy of solution for the hydrogen atom in iron, with the gas molecule as the reference state. Thus, the results of these earlier studies imply that the binding energy for microstructural trapping is equivalent to the energy of desorption of a hydrogen atom to the gas phase.

In this paper we summarize recent research at Cornell on hydrogen and deuterium trapping in iron and steel. This investigation, and a parallel study at Sandia Laboratories (13), has led to the conclusion that the energy with which hydrogen is bound to traps is greater, and the trap density lower, than had been inferred from the earlier research just summarized.

There are three components to our program which in the aggregate set
it somewhat apart from earlier and most concurrent efforts:

1) The extraction of trapping parameters, specifically the binding
    energy and trap density, from measured permeation time lags and
    analytic solutions for the time lag developed by the methods of
    McNabb and Foster (2) and Frisch (14).

2) The coordinated use of vacuum and electrochemical techniques for
    hydrogen flux measurements at high and low temperatures, respec-
    tively.

3) The development of gas phase charging techniques (15) utilizing
    thin palladium layers on both surfaces of the permeation membrane,
    for use at temperatures down to at least 278K and pressures well
    below one atmosphase.

Hydrogen Entry

Permeation techniques can be used to study bulk transport processes
only if it is established that the measured fluxes are not influenced by
surface impedances, and that the boundary conditions at both surfaces are
well characterized. It has long been the custom to assume local equilibrium
at the input and output surfaces. This leads immediately to the requirement
that the steady state flux (J_∞) be linearly dependent upon the inverse
(a^{-1}) of the specimen thickness, and for gas phase charging linearly dependent upon the square root of the input pressure ($P^{1/2}$) (e.g., 16). The
advantage of gas phase charging is that through Sievert's law the lattice
hydrogen concentration at the input surface of the membrane can be obtained.
With electrochemical charging it is not possible to get a good estimate of
the hydrogen concentration.

The realization of these criteria is shown in Figures 1 and 2 (15),
which depict the membrane thickness and input pressure dependence respectively, of the steady state flux of hydrogen through zone refined and annealed Ferrovac E iron. Similar results have been obtained for deuterium,
(17, 18). The agreement with the criterion for volume transport control is
quite good; there is little scatter and the data can be linearly extrapolated through the origin in an unforced way. The scatter is much less with
the present gas phase charging technique than is commonly encountered with
electrochemical charging. Gas phase charging in the present temperature
and pressure regime is possible only with the palladium layer technique.

Permeability: Lattice Diffusivity and Solubility

Permeability

Permeability measurements suggest that there are no anomalies in the
lattice diffusivity and solubility over the temperature range from 278 K to
1173 K. Early investigations were anlyzed by Gonzalez (19), who developed
a best fit equation for 399 K to 1173 K. More recent investigations (15,
17, 18, 20) are in agreement and have extended the temperature range to
278 K, as is shown in Table I. Good agreement between vacuum RGA and electrochemical measurement techniques is evident.

Permeability values are not influenced by trapping, since the permeability is a steady state characteristic of the lattice. The definition of
the steady state requires that the trapped and lattice hydrogen populations

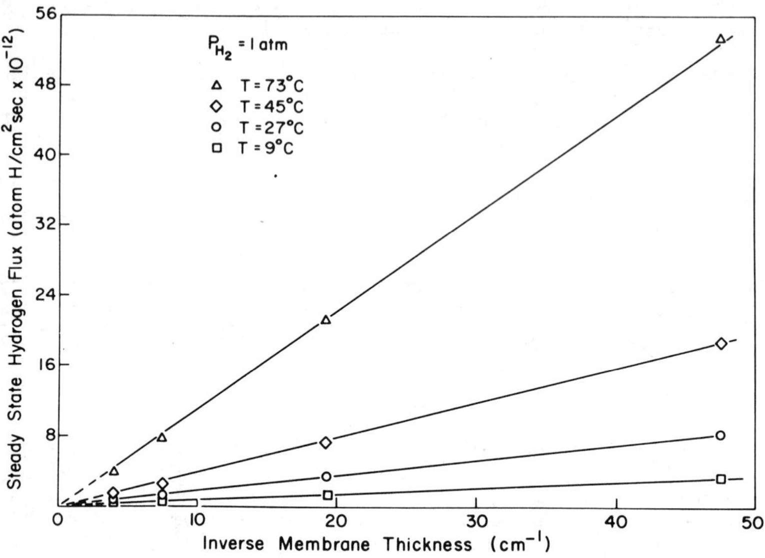

Fig. 1 – Variation of steady state flux with specimen thickness.

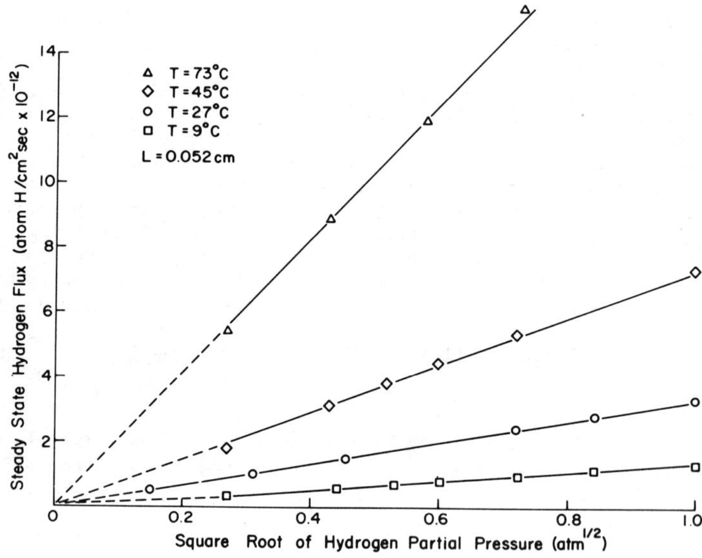

Fig. 2 – Variation of steady state flux with input pressure.

be in local equilibrium. Transport is then governed exclusively by lattice
parameters, as has been demonstrated (15) for annealed and deformed iron.
The deformed iron has a much higher trap density, and a correspondingly
longer time lag in permeation transient measurements, but the steady state
fluxes, and therefore permeabilities are identical because the local equilibrium condition is satisfied (15).

Table I. Permeability Preexponentials (Φ_o) and Activation

Energies (E_p), Hydrogen in Iron

Investigator	Temperature Range	Φ_o	E_p
Gonzalez (19)	399–173 K	$1.6\pm.3 \times 10 \dfrac{17 \text{ atom H}}{\text{cm-s-atm}^{\frac{1}{2}}}$	$8400\pm500 \ \dfrac{\text{cal}}{\text{mol}}$
Kummick & Johnson (15)	273–333 K	$2.6 \times 10 \dfrac{17 \text{ atom H}}{"}$	$8500\pm600 \ \dfrac{\text{cal}}{\text{mol}}$
Kummick & Johnson (17)	232–346 K	$(1.9 {\pm}^{12.2}_{\ 1.65}) \times 10 \dfrac{17 \text{atomH}}{"}$	$8380\pm610 \ \dfrac{\text{cal}}{\text{mol}}$
Quick & Johnson (18)	322–779 K	$(2.53 {\pm}^{7.97}_{\ 1.62}) \times 10 \dfrac{17 \text{atomH}}{"}$	$8520\pm670 \ \dfrac{\text{cal}}{\text{mol}}$
Nelson & Stein (20)	340–770 K	1×10^{17}	$8200 \ \dfrac{\text{cal}}{\text{mol}}$

Lattice Diffusivity

Compilations (2, 3) of diffusivities of hydrogen on iron display rather
more scatter at high temperatures, where lattice diffusivity is controlling,
than might have been expected. Since surface effects are important only at
lower temperatures, the higher temperature scatter can, most probably, be
attributed to the evolution in specimen preparation and measurement techniques over several decades.

It is expected that the highest values of diffusivity will be the most
accurate values. Table II gives the preexponential factors and activation

Table II. Lattice Diffusion Preexponential (D_o) Factors and

Activation Energies (E_d) Hydrogen in Iron

Investigator	Temperature Range	D_o	E_d
Quick & Johnson (18)	322–779 K	$(1.6 {+}^{0.94}_{-0.59}) \times 10^3 \dfrac{\text{cm}^2}{\text{s}}$	$1690\pm30 \ \dfrac{\text{cal}}{\text{mol}}$
Nelson & Stein (20)	327–770 K	$2.33 \times 10^3 \dfrac{\text{cm}^2}{\text{s}}$	$1600 \ \dfrac{\text{cal}}{\text{mol}}$

energies from two relatively recent investigations (18, 20), which satisfy
that condition and are in close agreement.

Lattice Solubility

The majority of the solubility equations reported in the literature are
based on permeability and diffusivity measurements, as the low solubility
makes direct measurements rather difficult except perhaps at higher temperatures. Older results summarized by Oriani (21) are in reasonable agreement

with more recent measurements based upon both direct analytic and permeation techniques (18, 22-24).

Isotope Effect, Hydrogen and Deuterium

Recent investigations (17, 18) have established that the activation energies for permeation and diffusion are statistically independent of isotopic mass for hydrogen and deuterium. Consequently, the energies of solution for hydrogen and deuterium are equal, and within experimental error their solubilities are equal.

The effect of isotopic mass is found in the transport behavior only, more particularly in the preexponential factors of the permeability and diffusivity equations. However, the statistical errors, which are inevitable for even very precise data because of the mathematical form of the permeability and diffusivity equations, make the preexponential factors unsuitable for evaluating isotope effects. A more satisfactory procedure is, at a fixed temperature, to calculate the ratios of the steady state fluxes and/or diffusivities as a function of pressure. It is found (17, 18) that the isotope effect shows small but measureable deviations from the classical square root of mass ratio. It is somewhat larger than 1.41 at 300 K, and decreases with temperature to a value just slightly less than 1.41 at 780 K.

Effective Diffusivity and Time Lag (25)

Trapping phenomena may be well characterized by transport measurements carried out as a function of lattice concentration of the diffusing species. This concept originated with Darken and Smith (4), who studied the permeation of hydrogen through steel after discharge from an acid solution, and found that the effective diffusivity was greater at higher lattice concentrations than lower. They argued that this could not be accounted for on the basis of molecular trapping in voids since, by Sievert's Law, the trapped hydrogen should be an increasing fraction of the total hydrogen as the lattice concentration was increased. This would require that the effective diffusivity decrease with increasing lattice concentration, in contrast to their experimental observations. This concept, that the change of effective diffusivity with lattice concentration can be used to characterize trapping behavior, has lain dormant in the literature for many years.

Effective Diffusivity

The diffusion equation for a system with trapping may be written (e.g., 2,26) quite generally as

$$\frac{\partial c_L}{\partial t} + \frac{\partial c_T}{\partial t} = D_L \nabla^2 c_L \qquad (1)$$

where c_L is the lattice hydrogen concentration, c_T the trapped hydrogen contration, and D_L the (concentration independent) lattice diffusivity. It is implicit in Eq. 1 that the density of trap sites is much less than the density of lattice sites.

Following earlier treatments (5, 26), if a relation between the lattice and trapped hydrogen concentrations is known, say $c_T(c_L)$, then equation (1) may be written as

$$\frac{\partial c_L}{\partial t} = \frac{D_L}{[1+c_T'(c_L)]} \nabla^2 c_L, \text{ where } c_T'(c_L) \equiv \frac{dc_T}{dc_L} \qquad (2)$$

an effective diffusivity may then be defined as

$$D_{eff} \equiv D_L[1+c_T'(c_L)]^{-1} \qquad (3)$$

D_{eff} is in general concentration dependent, which tends to limit its usefulness, as explicit solutions of the diffusion equation are then unlikely.
However, useful insights are obtained when D_{eff} is examined for $c_T(c_L)$ relations which represent different trapping models.

Non-Saturable Traps

An important class of trapping models is characterized by the equilibrium relation,

$$c_T = \alpha c_L^{m}; \quad m = 1,2,3----- \qquad (4)$$

These traps are non-saturable since the trapped hydrogen concentration
increases without limit as c_L increases. α will contain parameters for the
particular trapping model under consideration. An m of two characterizes
the oft- discussed case of hydrogen trapped as molecules in internal voids.
m values greater than two would apply to trapping at fixed sites in m-order
clusters, a perhaps physically unlikely event. m of one corresponds to
linear trapping. For this broad group of models the effective diffusivity
is

$$D_{eff} = D_L(1+mac_L^{m-1})^{-1} = D_L(1+m\frac{c_T}{c_L})^{-1}, \qquad (5)$$

which specifies the influence of c_L upon D_{eff}.

For m>1, which includes the important case of molecular trapping, D_{eff}
decreases as c_L increases, and more rapidly with larger m, in accord with
reasons first given by Darken and Smith (4), and summarized earlier. The
lattice diffusivity may be obtained as the low lattice concentration limit
of the effective diffusivity. For linear trapping, an m of one, D_{eff} is
independent of c_L but less than D_L by the factor $(1+\alpha)^{-1}$.

The total hydrogen concentration, c_{tot}, is given by

$$c_{tot} = c_T+c_L = c_L(1+\alpha c_L^{m-1}) = c_L(1+\frac{c_T}{c_L}). \qquad (6)$$

from which it is evident that the total hydrogen concentration may vary more
strongly with temperature than does the lattice solubility. Trapping data
for hydrogen in steel have been interpreted (27-31) in terms of molecular
trapping in voids, using Eq. 5 with m = 2. However, the measurements did
not allow an examination of the dependence of D_{eff} on c_L, which is the most
discriminating test of the void trapping model.

Saturable Traps

There is an equally important class of potential trapping sites, such
as dislocations, impurity atoms, and internal interfaces, where plausible
physical arguments suggest that the capacity of the traps for hydrogen is
finite. In these models the trapped hydrogen concentration then saturates
as the lattice hydrogen concentration is increased. If each trap can accomodate one gas atom, the saturation concentration is equal to the trap density. At local equilibrium the trapped and lattice concentrations are related by (5),

$$H_L = H_T: \quad K = \frac{n}{\dfrac{c_L}{N_L}(1-n)} = \exp(E_b/RT) \tag{7}$$

where K is the equilibrium constant for the trap reaction, E_b the trap binding energy, n the fractional occupancy $(0 < n < 1)$ of the traps, and N_L the number of (interstitial) lattice sites per unit volume, c_T is then the product of n and N_T, the structurally fixed trap density,

$$C_T = \frac{\dfrac{N_T}{N_L}Kc_L}{1+\dfrac{K}{N_L}c_L} \tag{8}$$

The temperature dependence of c_T is controlled by the parameter $\frac{K}{N_L}c_L$; for $\frac{K}{N_L}c_L \gg 1$, c_T saturates and is equal to the trap density N_T. This may be simply interpreted in terms of an energy diagram, Fig. 3. Temperature

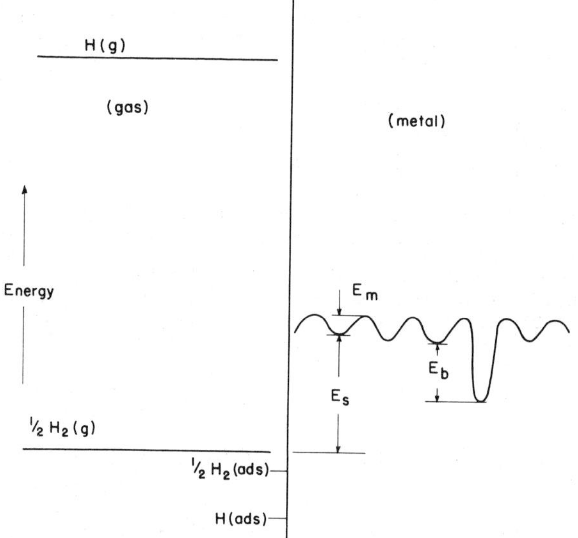

Fig. 3 - Schematic view of energy relations in gas—metal equilibria, including trapping.

is contained in Kc_L which, when experiments are carried out in equilibrium with hydrogen gas, is proportional to $\exp(\frac{E_b-E_s}{RT})$, where E_s is the energy of solution. Thus, it is apparent that c_T will saturate at high temperatures when $E_b < E_s$, and at low temperatures when $E_b > E_s$.

From equation 3 the effective diffusivity is (5)

$$D_{eff} = D_L \left[1 + \frac{\frac{N_T}{N_L} K}{(1 + \frac{K}{N_L} c_L)^2} \right]^{-1} \tag{9}$$

In striking contrast to the results obtained for non-saturable traps, Eq. 5, D_{eff} now increases as c_L increases, and approaches the lattice diffusivity at high, rather than low, values of c_L. This is to be expected for saturable traps.

In the region of low trap coverage ($n \ll 1$), saturable traps are equivalent to linear traps (non-saturable traps with m equal to one). In agreement, for $n \ll 1$ equation 9 reduces to

$$D_{eff} = D_L (1+\alpha)^{-1}, \quad \alpha \equiv \frac{N_T}{N_L} K \tag{10}$$

and equation 8 to

$$c_T = \alpha c_L \tag{11}$$

These agree, as they should, with the non-saturable trap analysis, equations 5 and 6, for the case of m equal to one.

These relationships, which illustrate the important effect of lattice concentration upon effective diffusivity, have not been exploited in previous studies. Perhaps this is because electrochemical charging of hydrogen has almost always been used; the relation between electrochemical charging parameters and lattice hydrogen concentration is often uncertain. Nevertheless, it should be feasible with electrochemical charging to discriminate between void trapping and saturable traps. In experiments where the current density has been varied (32, 33), the results indicate qualitatively an increase in effective diffusivity with concentration, and therefore saturable traps.

For these models the calculated variation of the effective diffusivity with temperatures is of little value in defining trap characteristics. A comparison is given in Fig. 4 for void trapping as molecules, for two void densities and external hydrogen pressures. Figure 5 is for linear traps, with two sets of binding energies and trap densities. Trade-offs in trap parameters are evident in both figures. Both models show reasonable agreement with the broad spread of literature values.

The relations just discussed are not easily adapted to quantitative studies, in the absence of closed form solutions of the diffusion equation with trapping. Some numerical analyses of the diffusion equation with trapping have appeared (34-36), and we will use numerical procedures in a later section to interpret experimental results. The main thrust of our program, however, has been to develop experimental techniques based upon time-lag measurements which are amenable to exact interpretation.

<u>Time Lag</u>

Time-lag measurements on trapping systems are susceptible to exact interpretations by the equivalent analyses developed separately by McNabb and Foster (2) and by Frisch (14). These analyses provide a mathematical prescription for quantitatively relating measured time lags to trap parameters, even though a full solution to the diffusion equation cannot be obtained.

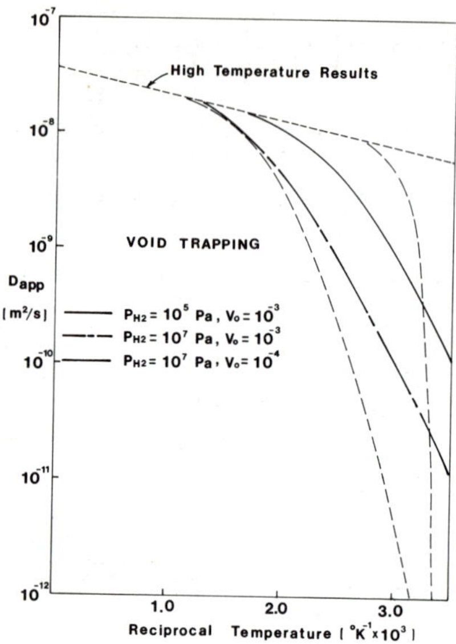

Fig. 4 – Variation of apparent diffusivity with temperature for void trapping.

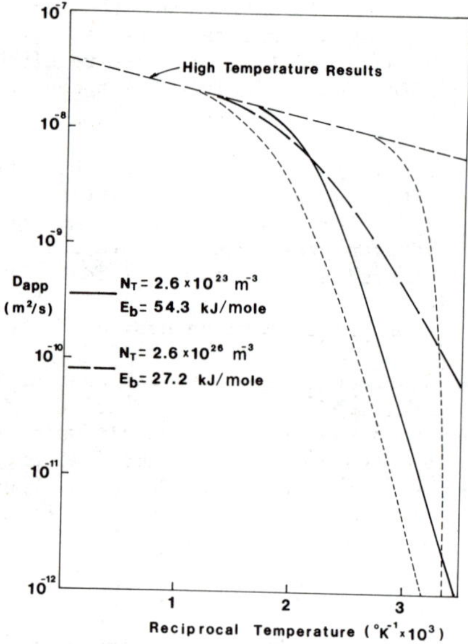

Fig. 5 – Variation of apparent diffusivity with temperature for linear trapping.

The derived time lag formulas require for application only a knowledge of the position dependence of the steady state distributions of both the lattice and trapped hydrogen concentrations.

McNabb and Foster (2) start by writing the diffusion equation with trapping as

$$\frac{\partial c_{tot}}{\partial t} = \frac{\partial c_L}{\partial T} + \frac{\partial c_T}{\partial T} = \frac{\partial c_L}{\partial T} + N_T \frac{\partial n}{\partial T} = D_L \nabla^2 c_L \qquad (12)$$

while Frisch (14) starts with

$$\frac{\partial c_{tot}}{\partial x} = \frac{\partial}{\partial x} \left(D(c_{tot}) \frac{\partial c_{tot}}{\partial x} \right) \qquad (13)$$

These may be shown to be equivalent, however, since from our previous discussion

$$D(c_{tot}) = D_{eff} = \frac{D_L}{1+c_T'} \text{ and } \frac{\partial c_{tot}}{\partial x} = (1+c_T') \frac{\partial c_L}{\partial x} \qquad (14)$$

When these relations are put into Equation 13, it becomes equal to Equation 12. Thus, the McNabb-Foster and Frisch analyses have the same starting point and, not surprisingly, they lead to equivalent formulas for the time lag.

Non-Saturable Traps

When this method is applied to the broad class of non-saturable traps defined by Equation 4 the following equation is obtained for the time lag,

$$t_T = \frac{a^2}{6D_L} \left[1 + \frac{6\alpha c_o^{m-1}}{(m+1)(m+2)} \right] \qquad (15)$$

where a is the membrane thickness and c_o the lattice concentration at the input surface of the permeation membrane. In the absence of trapping, this reduces to the well known (37) relation relating lattice diffusion and time lag,

$$t_L = \frac{a^2}{6D_L} \qquad (16)$$

It is evident from Eq. 15 that t_T increases with lattice concentration, and approaches the lattice time lag, Eq. 16, in the limit of low lattice concentration. Thus time lag measurements as a function of lattice concentration can be used to discriminate between saturable and non-saturable traps. These measurements are best carried out by gas phase charging, where as pointed out earlier the lattice concentration is quantitatively controlled by the input pressure, and the concentration at the input surface is proportional to the square root of pressure through Sievert's Law.

Saturable Traps

McNabb and Foster (2) wrote the kinetic equation for saturable traps from straight forward probabilistic considerations as

$$\frac{\partial n}{\partial t} = kc_T (1-n)-pn \qquad (17)$$

   k = transition probability for hydrogen transport from lattice into trap
   p = transition probability for hydrogen escape from trap to lattice site.

This model is broadly applicable in that it does not prescribe a specific structural feature as the trap site, and thus places few constraints on possible trapping mechanisms. The trap and escape parameters, k and p, are related to the equilibrium constant K of Equation 7 by

$$K = N_L \frac{k}{p} = \exp \frac{E_b}{RT} \tag{18}$$

For permeation boundary conditions the time lag was shown (4) to be

$$t_T = \frac{a^2}{6D_L}[1 + \frac{3\alpha}{\beta} + \frac{6\alpha}{\beta^2} - \frac{6\alpha}{\beta^3} (1+\beta) \ln (1+\beta)], \tag{19}$$

$$\alpha = N_T \frac{k}{p} \text{ (as before)}$$

$$\beta = c_o \frac{k}{p} = \frac{n}{1-n} = \text{hydrogen activity in traps just beneath input surface}$$

The parameter β is therefore a quantitative measure of the trap coverage.

This equation has interesting, simple, and useful forms in the limits of both low and high trap coverage. For low trap coverage, $\beta, n \ll 1$, the equation becomes

$$t_T = t_L (1+\alpha), \tag{20}$$

in agreement with the earlier analysis for linear trapping, Equation 10 and 11. In this limit the time lag does not depend upon lattice concentration, and therefore should be insensitive to input pressure in a gas phase permeation experiment. The time lag with trapping is simply the lattice time lag multiplied by the factor $(1+\alpha)$, where α contains the trap density and binding energy. This behavior is expected at low concentrations and/or at temperatures rather high in the trapping regime, for here thermal emptying of the traps should produce low and linear coverage.

For endothermic solutes such as hydrogen in iron, as the temperature is decreased the fractional occupancy will increase and eventually β will become much larger than one, while n will become very close to one. In this near saturation limit the time lag becomes

$$t_T = t_L (1 + \frac{3N_T}{c_o}) \tag{21}$$

The time lag now decreases with increasing concentration, or input pressure, and approaches the lattice time lag at high, rather than low, lattice concentrations i.e., when $c_o/3N_T \gg 1$. Therefore, experimental time lag measurements as a function of input pressure provides basic information on trap characteristics, i.e. whether or not the traps are saturable.

<u>Hydrogen and Deuterium Trapping in Iron</u>

<u>Saturable or Non-Saturable Traps?</u>

The traps for hydrogen and deuterium in iron are saturable, as may be seen from a representative plot of measured time lag vs. input pressure, Fig. 6. Although the data presented are for hydrogen and deformed iron, plots of similar shape are observed for hydrogen in annealed iron, and for hydrogen and deuterium in deformed and annealed iron. Since the time lag always decreases with increasing input pressure, the traps must be saturable as shown by the analysis given earlier.

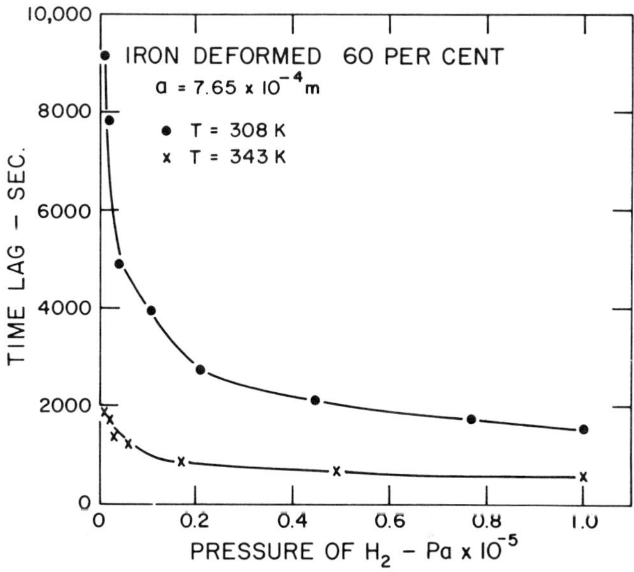

Fig. 6 — Variation of time lag with input pressure.

Methods of Data Analysis

The McNabb–Foster formalism described by Equations 17 and 19 will be used, as it is the simplest model that can be applied to data which suggest saturable traps. There are two methods that can be used to extract the trap parameters from measured permeation transients and time lags. The first is essentially a graphical technique applied to the time lags, while the second is a direct numerical least squares fitting of the permeation transient.

In the graphical method (38), we first rewrite Equation 21 by substituting for c_o through Sievert's Law, whereupon Equation 21 becomes

$$\frac{t_T}{t_L} = 1 + \frac{3N_T}{S(T)P^{1/2}} \tag{22}$$

where $S(T)$ = Sievert's Law coefficient. The forms of Equation 19 and 22 suggest that the experimental data is best analyzed for binding energies and trap densities through plots of $t_T/t_L - 1$ vs $P^{-\frac{1}{2}}$. A schematic representation is given in Fig. 7. In the saturation region A, described by Equation 22, a straight line of slope $3N_T/S(T)$ is anticipated, from which the trap density may be calculated directly. In the dilute occupation region C, described by Equation 20, the data should conform to a horizontal line at a level controlled by N_T, E_b, and T, through the parameter α. In the intermediate region B, all terms of Equation 19 contribute to the schematic curve. Thus, if N_T is known from measurements in the saturation region, E_b may be attained from a single parameter fit to the intermediate region B or the dilute occupation region C. It is probably unlikely that a given set of

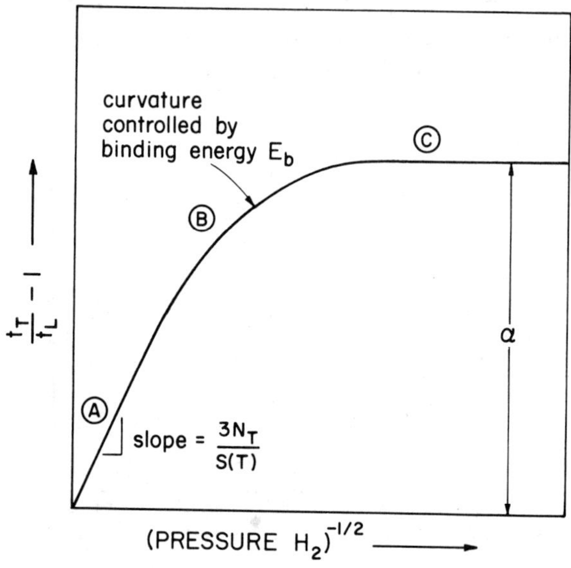

Fig. 7 – Schematic view of time lag as a function of input
pressure for saturable traps.

measurements can span the full range from dilute occupation to trap saturation. Nevertheless, this method can be used with confidence (38) to evaluate trap parameters.

A second method of evaluating the trap parameters is to, by computer
calculation, determine the best fit of the measured permeation transient to
numerical solutions of the full non-linear trapping equation. Caskey and
Pillinger (36) have developed techniques for numerical solutions. Standard
computer programs for least squares fitting may be combined with the Caskey-
Pillinger (36) method to obtain best fits to measured transients, and hence
best values for the trapping parameters. This is essentially a three parameter least-squares fitting procedure, since in the transient region the un-
knowns are the capture parameter k, the escape parameter p, and the trap
density. E_b may be obtained from the ratio of k to p, Equation 18.

Sensitivity of Permeation Measurements to Trapping

It is convenient at this point, in the context of saturable traps and
Fig. 7, to discuss the conditions under which time lag measurements can give
information on trapping. There are three conditions that must be met; two
may be developed from Fig. 7, while the third is simply that the steady
state flux, the signal, must equal or exceed the sensitivity of the measuring system.

It is evident from Fig. 7 that the measured time lag will show the
largest deviation from the value expected for purely lattice transport in
the dilute occupation region (C). If, for example, the parameter α is much
less than one, time lag measurements cannot detect trapping, since then

$t_T \sim t_L$. There is clearly a steep trade off between trap density and binding energy, for a given α. To give a concrete example, we may stipulate that for adequate sensitivity at 300° K, α must equal, say, three. Then if the binding energy is 27.2 K J/mol H, a trap density of about $4.4 \times 10^{24} m^{-3}$ is required, while if the binding energy is doubled, to 54.4 kJ/mol H, the trap density need only be $2.6 \times 10^{20} m^{-3}$. The trade-off between trap density and binding is also evident in Fig. 5.

The second sensitivity condition involves the relative lattice hydrogen concentrations and trap densities; it may be developed from region A of Fig. 7 and Equations 21 and 22. It is evident that c_0 must not be much larger than N_T, for if $c_0 >> N_T$, then again $t_T \simeq t_L$, and the time lag method is insensitive. Physically, if $c_0 >> N_T$ the trapped and lattice hydrogen concentrations equilibriate so rapidly that nearly all of the hydrogen atoms do not encounter a trap upon passage through the membrane, and $t_T \simeq t_L$.

The third condition is simply that the steady state flux, the signal in the experiment, must equal or exceed the sensitivity of the measuring system, else there is nothing that can be measured. This condition combines lattice hydrogen characteristics and specimen geometry. Since the electrochemical measuring technique has at present a threshold sensitivity of about 6×10^{14} atom H/m^2-s, it follows that the condition may be expressed as $D_L c_0 /a \geq 6 \times 10^{14}$ atom H/m^2-s. This condition contains a tradeoff between lattice solubility and diffusivity.

The second and third conditions do to some extent work against one another, which explains why trapping is readily observed in iron, but not in nickel or stainless steel. The lattice diffusivities of hydrogen in nickel and stainless steel are much lower than in iron, so that, by condition three, for a measurable steady state flux the lattice hydrogen concentrations must be much higher. This, however, leads to a violation of condition two, for the lattice concentration may now greatly exceed the trap density. Consequently, although trapping may well be present in lower diffusivity metals and alloys, it may not be seen in time lag or permeation transient measurements, except possibly at much lower temperatures.

<u>Trap Parameters for Deformed Iron (38)</u>

Evaluation of the trap density and binding energy by the graphical method is shown in Fig. 8, a representative plot which compares measured time lags with Equation 19 for one trap density and three different, but not widely spaced, binding energies. The goodness of fit by the graphical method is seen to be quite sensitive to the binding energy. The curvature exhibited in the data establishes that the measurements were not carried out in the dilute occupation region, and that the traps were substantially saturated. Indeed, using a trap density of 6×10^{22} m^{-3} and a binding energy of 58.6 kJ/mol H, for the data in Figure 8 α is 550, well outside the pressure range of feasible measurements at 329 K. Correspondingly, for a hydrogen pressure of 1 atm β is 19.5 and n is .95, while at .01 atm pressure β is 1.95 and n is .66.

Measurements at several deformation levels, interpreted by the graphical method, supplemented in some instances by least squares analysis of the measured time lags, established (38) a self consistent picture for the trap parameters. The binding energy was found to be independent of deformation level and measurement temperature, suggesting that a single trap site was under study. The trap densities were, as would be expected, independent of temperature, but increased sharply with deformation at low deformation levels, and then much more gradually with a further increase in deformation.

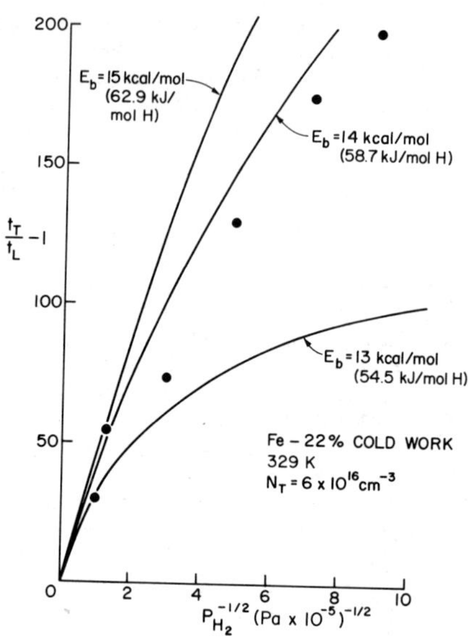

Fig. 8 - Time lag as a function of input pressure for deformed iron.

A summary of this data is given in Table III.

Table III.   Summary of Trap Parameters of Deformed Iron
Determined at Different Deformation Levels and Temperatures

% Cold Work	Trap Density N_T	Binding Energy E_b
0 (annealed)	8.5×10^{20}	
15	5.9×10^{22}	
30	5×10^{23}	14.3 ± 1.1 k cal/mol H
40	7×10^{22}	$(59.9 \pm 4.6$ k J/mol H)
60	1.5×10^{23}	
80	1.8×10^{23}	

The physical picture that may be inferred from these observations is
one of deep isolated traps with a single binding energy and a concentration
which is dependent upon deformation level. The variation of trap density
with deformation is similar to the variation of quantities such as yield
strength, hardness, and dislocation density with deformation level. This
implies that the traps are associated with the imperfection structures,
probably dislocations but perhaps point defects or point defect aggregates,
developed during plastic deformation. Hirth (1) has analyzed critically
the implications of the high binding energy, and suggested a new model to
account for it.

Hydrogen and Deuterium in Annealed Iron

The trap binding energy and density are the same for hydrogen and deuterium in annealed iron, as is shown by representative data in Fig. 9.

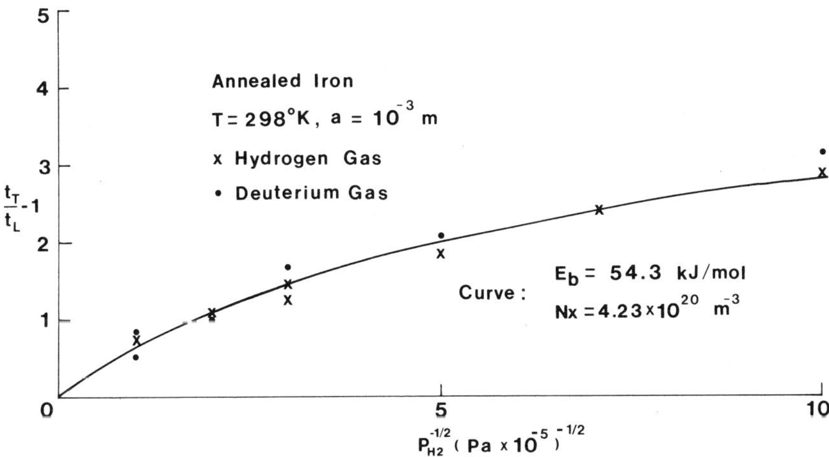

Fig. 9 - Time lag as a function of input pressure for annealed iron.

The same curve fits the data for both isotopes; in the calculation of t_L, necessary to construct the ordinate, the separate lattice time lag determinations (18) for hydrogen and deuterium were of course used. The binding energy is slightly lower than the average value quoted in the previous section, but is consistent with values supported earlier (38) for annealed iron. Further investigation will be required to establish whether or not the trap binding energy for annealed iron is statistically smaller in value than for deformed iron.

In constrast to the α value of 550 for deformed iron presented in the last section, α for the annealed iron data in Fig. 9 is about 5.5. The differences arise primarily from the trap densities, and to a lesser degree from the different temperatures. The parameter β and the fractional occupancy n are of course independent of trap density for an open system

With annealed iron the permeation transient times are sufficiently short that a least squares analysis for the trap parameters may be easily carried out by computer, as described earlier. Examples are given in Figures 10 and 11 for gas phase and electrochemical charging (33), respectively. To fit the transient obtained by electrochemical charging, c_o was first determined from Sievert's Law after a comparison of the steady state fluxes for the two charging methods, which gave an effective pressure for electrochemical charging. The same fitting procedure was then used for both gas phase and electrochemical charging.

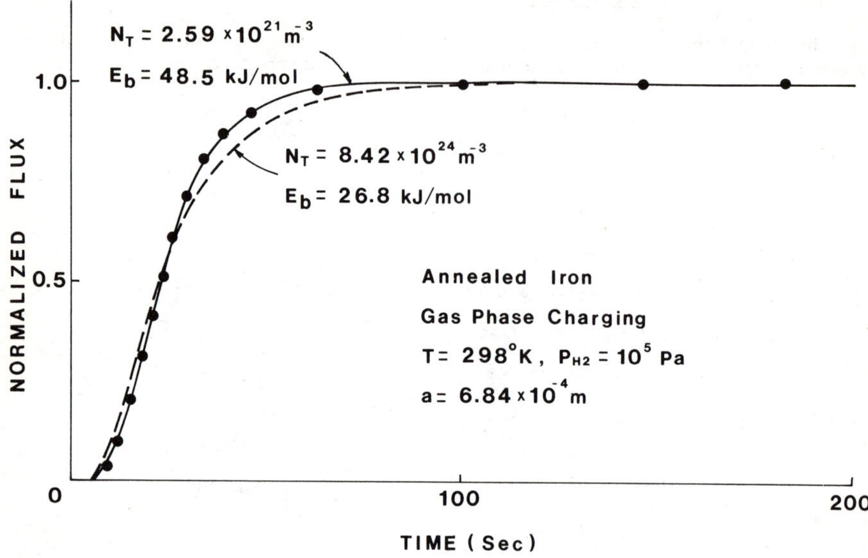

Fig. 10 – Permeation transient for annealed iron, gas phase charging.

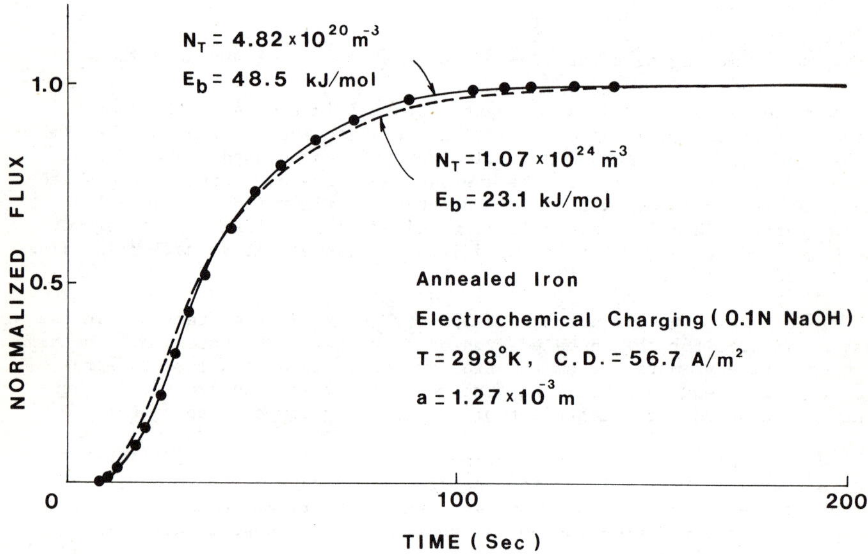

Fig. 11 – Permeation transient for annealed iron, electrochemical charging.

In reasonable agreement with the values determined by the graphical method, for both charging methods the best fit to the permeation transients is given by the high binding energy and low trap density. The strong tradeoff between binding energy and trap density is shown by the dashed lines, which assumed the lower binding energy reported by earlier investigators. The dashed lines seem to correspond to a local minimum in the errors, which are substantially larger than the errors associated with the solid lines.

These results demonstrate that the sharp tradeoff between binding energy and trap density can result in two quite different sets of parameters providing a good visual fit to the same permeation transient. The importance of very accurate transient measurements is also apparent. It is easy to see why earlier investigators reported the smaller binding energy.

The validity of the higher binding energy is further buttressed by the comparisons presented in Figures 12 and 13. Figure 12 compares the graphical and least squares method of extracting the trap parameters. The solid line is the best fit to the experimental time-lag data given as crosses. The filled in circles are the time lags computed from the experimental transients. There is reasonable agreement.

In Figure 13 we return to the theme of this paper, which is that time-lag measurements as a function of lattice hydrogen concentration, or input hydrogen pressure, are a powerful tool for characterizing trap parameters. It is evident that the higher binding energy and lower trap density provide a self-consistent fit over the range of experimental input pressures. On the other hand, the lower binding energy and higher trap density do not lead to a consistent fit.

Conclusions

A study of hydrogen and deuterium transport in iron, with a focus upon trapping, was presented. The experimental approach was based upon measurement of permeation transients and time lags, utilizing a gas phase charging proceedure useful down to 273 K and .01 atm H_2, and an electrochemical flux-measuring technique. The lattice hydrogen concentrations are in the range of one part in 10^8.

It was shown that the variation of the time lag with lattice hydrogen concentration, or input pressure, may be used to classify traps as either non-saturable or saturable . The traps for hydrogen and deuterium in annealed and deformed iron were shown to be saturable.

Two methods, one graphical and the other computer least squares fitting, were described for extracting trap parameters from time lag and transient measurements, respectively. The methods were shown to give comparable results.

The measurements were self-consistently interpreted in terms of the McNabb-Foster formalism for saturable traps. The binding energies for both hydrogen and deuterium in annealed and deformed iron were 48-58 kJ/mol H; the trap densities were about $10^{20}m^{-3}$ for annealed iron and over $10^{23}m^{-3}$ for heavily deformed iron. The smaller binding energies and larger trap densities reported by earlier investigators were shown not to be consistent with time lag and transient measurements over a range of lattice hydrogen concentrations.

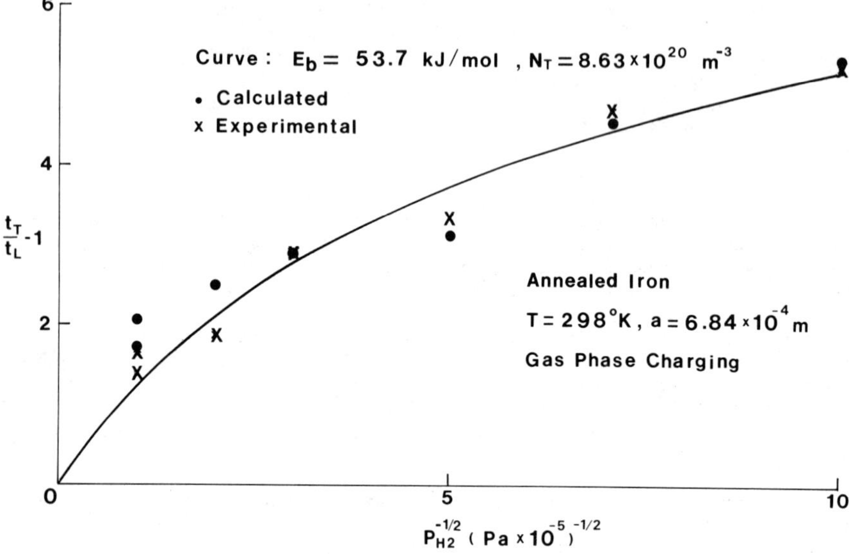

Fig. 12 – Comparison of time lag and transient methods for determining trap parameters.

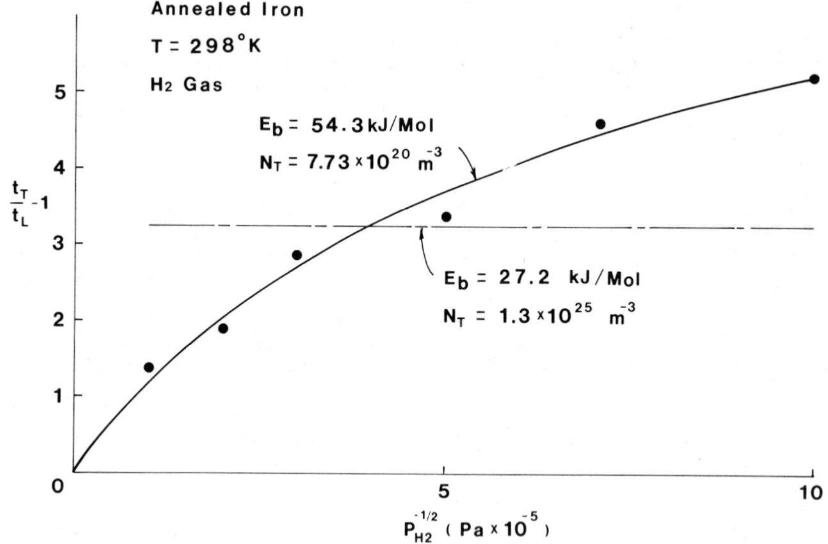

Fig. 13 – Comparison of high and low binding energy fits to experimental data.

References

1. John P. Hirth, Met. Trans. A 11A, 861-870 (1980).
2. A. McNabb and P. K. Foster, Trans. Met. Soc. AIME, 227, 618-627 (1963).
3. J. Volkl and G. Alefeld, Topics in Applied Physics. Vol. 28, 321-348, J. Volkl and G. Alefeld, Eds. (1978).
4. L. S. Darken and R. P. Smith, Corrosion, 5, 1 (1949).
5. R. A. Oriani, Acta Met. 18, 147-157 (1970).
6. Y. Sakamoto and J. Eguchi, Proc. Japan Congress on Materials Research, 19, 91 (1976).
7. E. M. Riecke, International Congress on Hydrogen in Metals, Paper 2B8. Paris (1977); also Arch. Eisenhütten, 97, 247 (1976).
8. R. Gibala, Stress Corrosion Cracking and Hydrogen Embrittlement of Iron Base Alloys, NACE, Houston, Texas (1977).
9. A. P. Miodownik, Stress Corrosion Cracking and Hydrogen Embrittlement of Iron Base Alloys, NACE, Houston, Texas (1977).
10. A. Zielinski, E. Lunarska and M. Smialowska, Acta Met 25, 551 (1977).
11. G. Schoeck, Acta Met. 11, 617 (1963).
12. R. E. Miner, R. Gibala and F. A. Hultgren, Acta Met. 24, 233 (1976).
13. S. M. Myers, S. T. Picraux and R. E. Stoltz, J. Appl. Phys. 50, 5710 (1979).
14. H. L. Frisch, J. Phys. Chem. 61, 93 (1957).
15. A. J. Kumnick and H. H. Johnson, Met. Trans. A 6A, 1087-91 (1975).
16. D. S. Shupe, J. Chem. Phys. 68(6), 2612-2620 (1978).
17. A. J. Kumnick and H. H. Johnson, Acta Met. 25, 894-895 (1977).
18. N. R. Quick and H. H. Johnson, Acta Met. 26, 903-907 (1978).
19. O. D. Gonzalez, Trans. TMS-AIME, 245, 607 (1969).
20. H. G. Nelson and J. E. Stein, NASA Report TND-7265, NASA Ames Research Center, Moffett Field, California (1973).
21. R. A. Oriani, Fundamental Aspects of Stress Corrosion Cracking, NACE, Houston, Texas (1969).
22. J. R. G. DaSilva, S. W. Stafford and R. B. McLellan, J. Less-Common Metals, 49, 407 (1976).
23. J. R. G. DaSilva, S. W. Stafford and R. B. McLellan, J. Less-Common Metals, 50, 1 (1977).
24. M. R. Louthan, R. G. Derrick, J. A. Donovan and G. R. Caskey, Jr., J. Effect of Hydrogen on Behavior of Materials, TMS-AIME, A. W. Thompson and I. M. Bernstein, Eds. New York, 337 (1976).
25. H. H. Johnson, N. R. Quick and A. J. Kumnick, Scripta Met. 13, 67-72 (1979).
26. J. Crank, Mathematics of Diffusion, Clarendon Press, Oxford (1957).
27. G. M. Evans and E. C. Rollason, JISI, 207, 1484 (1969).
28. G. M. Evans and E. C. Rollason, JISI, 207, 1591 (1969).
29. B. Chew, Metal Science Journal, 5, 195 (1971).
30. M. R. Louthan, A. H. Dexter and J. A. Donovan, JISI 210, 58 (1972).
31. von H.-G. Ellerbrock, G. Vibrans and H. P. Stuwe, Acta Met. 20, 53 (1972).
32. L. N. Nanis and T. K. Govindam Namboodhiri, Acta Met. 21, 663 (1973).
33. H.-M. Shih, M. S. Thesis, Cornell University (1975).
34. D. M. Allen-Booth and J. Hewitt, Acta Met. 22, 171 (1974).
35. D. M. Allen-Booth and J. Hewitt, Acta Met. 23, 371 (1975).
36. G. R. Caskey, Jr. and W. L. Pillinger, Met. Trans A, 6A, 467 (1975).
37. R. M. Barrer, Diffusion in and through Solids, Cambridge University Press, England (1951).
38. A. J. Kumnick and H. H. Johnson, Acta Met. 28, 33-39 (1980).

DISCUSSION

R. A. Oriani, University of Minnesota, Minneapolis: This is beautiful work and one must believe in its reliability. However, accepting a trap depth of about 14 kcal/mol and a trap density of 10^{15} to $10^{17} cm^{-3}$ which increases with increasing deformation of iron leads to the problem of their identity. If these are dislocation core sites, the N_T should be larger, especially since tritium autoradiography shows that dislocations abundantly attract tritium. Is it possible that your deep traps are special dislocation core sites, perhaps having other adsorbed sites, or perhaps microvoid interfaces produced by cold work, and that the permeation measurements at low fugacity are sensitive mostly to the deep traps whereas techniques such as internal friction respond to the shallower core sites? To shed light on this problem I would suggest permeation experiments over a large range of input fugacity, since the presumably more numerous, shallower traps should make a larger contribution at the higher fugacities.

H. H. Johnson: The physical nature of the deep traps has not yet been established. We (38) have suggested special core sites as a possibility, while Myers et al (13) have suggested vacancies as the trapping site. For permeation transients to be sensitive to trapping, the McNabb-Foster parameter α must, as discussed in the text, be larger than, say, three. Since α contains the trap density, trap binding energies and temperatures, it may be more promising, for the study of shallower traps, to pursue measurements at lower temperatures, rather than higher input fugacities.

 C. Altstetter, University of Illinois, Urbana: In another paper in this conference we present permeation results for stainless steel. We find that the steady states permeation flux increases with stress. This appears to be due to an increased diffusivity rather than a change in solubility. Would you care to comment on the differences observed for iron and stainless steel?

H. H. Johnson: Professor Altstetter's comment is quite interesting, for it raises the possibility that elastic stress may influence hydrogen transport in iron and stainless steel in fundamentally different ways. Although this seems to us to be improbable, we have no explanation to offer at the present time.

G. R. Pressouyre, Creusot-Loire Research Center, France: I find the term "saturable traps" confusing. If the trapping reaction is: $dn/dt = kC_o(1-n) - pn$, then at steady state: $dn/dt = o \Rightarrow n = C_o/(1 + \frac{k}{p} C_o)$. Clearly, this corresponds to an __equilibrium__ value (n depends on C_o), not a saturation value. The best term in that case is a __reversible__ trap. If the trap is __irreversible__ ($p \simeq o$), then at steady state, $n = 1$, and this now is a __saturation__ value.

H. H. Johnson: It appears to us that Dr. Pressouyre's comment is one of semantics. In our usage, "saturable trap" is a generic term for traps whose physical nature is such that they can only accommodate, under any conditions, a finite number of hydrogen atoms. The McNabb-Foster trap, discussed at length by us and cited by Dr. Pressouyre, is an example of a "saturable trap". We find it useful to distinguish this from, say, void trapping, where by Sieverts Law the amount of hydrogen in a void should increase indefinitely as the lattice concentration is increased. It may be of interest to note that for irreversible traps, as defined by $p = o$ in Dr. Pressouyre's comment, the McNabb-Foster formalism yields for the time lag $t_T = t_L(1 + \frac{3N_T}{C_o})$, which is the same as equation 21 of the manuscript. Therefore, with respect to permeation absorption transients, irreversible traps behave like saturable traps in the saturation limit.

<u>R. Gibala, Case Western Reserve University, Cleveland</u>: If you plot your per-
meation data in the form of ℓnD (apparent) as a function of reciprocal temperature, what values of apparent activation energies are obtained in the lower
temperature range?

<u>H. H. Johnson</u>: Figures 4 and 5 of the text present ℓnD_{eff} vs $1/T$ as calculated
for void trapping and saturable traps. It seems that the broad scatter of data
in the literature limits the usefulness of this form of data analysis. The
figure accompanying this discussion shows how our data fits the McNabb-Foster
theory from the viewpoint of temperature dependence.

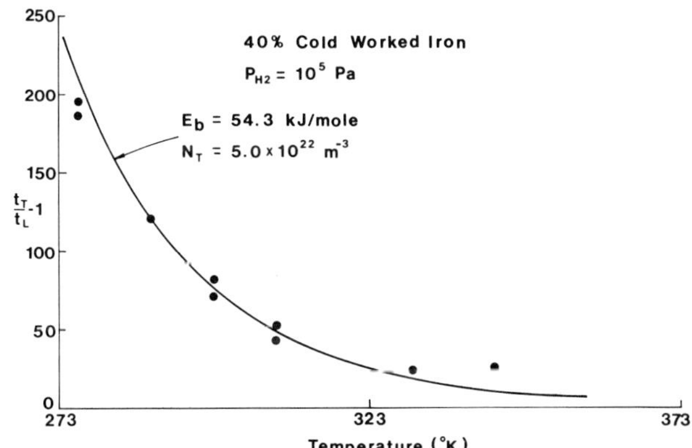

TRAP THEORY OF HYDROGEN EMBRITTLEMENT : EXPERIMENTAL INVESTIGATIONS

G.M. PRESSOUYRE
Creusot-Loire Research Center, 71208 Le Creusot (F)
J.P. FIDELLE and R. ARNOULD-LAURENT
CEA. Service Métallurgie BP.511, 75752 Paris Cedex 15   France (F)

Introduction

In spite of the considerable amount of research performed on hydrogen embrittlement (HE) of metals, many enigma of this phenomenon still remain as such. For example, (e.g. Ref.1), it is a well accepted, but puzzling fact, that HE can display a number of dissimilarities, whether H has been first introduced into the material ("internal hydrogen embrittlement", IHE) or is present as an external atmosphere ("external hydrogen embrittlement", EHE). This paper is an attempt at elucidating such enigma by experimentally verifying recent ideas on HE, particularly on the interaction between traps and hydrogen (Ref.2).

Trap Theory

A basic assumption of the trap theory of HE (Ref.2) is that concerning the "critical concentration", C_K. As shown in figure 1, a crack will initiate on a given defect if the quantity of trapped hydrogen, C_H (t), goes over the critical concentration, C_K. Both C_K and C_H (t) are dependent on various parameters, but C_K is specific to the defect considered. Thus, the critical concentration, C_K, of a given defect depends on its shape (e.g. : an elongated manganese sulfide is more susceptible to initiate a crack than a round one), on the neighbouring structure (e.g. : C_K for an elongated MnS in martensite is lower than in ferrite), on its coherency (e.g. : very incoherent grain boundaries are more susceptible than more coherent ones), on the applied stress and its direction (e.g. : a tensile stress perpendicular to the main axis of an elongated MnS lowers C_K), on adsorbed impurities at the interface defect-matrix, on temperature, etc... In the same way, C_H (t) is influenced by numerous parameters, such as time, stress, temperature, amount of available hydrogen (itself dependent on external fugacity) (Ref..3), etc... The trap theory of HE (Ref.2) quantitatively describes the role of three groups of C_H (t) parameters : trap characteristics, hydrogen transport mode, and localisation of hydrogen before the test (external versus, internal hydrogen). By trap characteristics, we mean whether or not a trap is reversible, in the existing conditions (temperature, stress), i.e. whether or not it may exchange hydrogen : classifications of traps according to reversibility exist in the literature (e.g. Ref.4). For transport mode, we consider that hydrogen may diffuse by interstitial jumps (normal diffusion), or as dislocations atmospheres, or along short circuit paths ; the differences between these various means by which hydrogen diffuses are fundamental : for instance, interstitial diffusion is slow and little energetic (0.08 eV) while transport by dislocations is faster and the interaction dislocation-hydrogen is more energetic (0.25 eV ; Ref.5). With these differences in mind, it can easily be shown that the trap population of a given material will respond differently depending on test conditions. An example is given in figure 2 (from Ref.2), where hydrogen is assumed to move as dislocations atmospheres. In one case, hydrogen is external to the material : all traps (both reversible and irreversible) act as sinks and little hydrogen is dumped into the potential flaw F ; in the other case, the material has been precharged, and the reversible trap may act as a source that will replenish the dislocation before it encounters the potential flaw. Thus, in the first case, it will take a longer time than in the second case before C_H (t) goes over C_K, i.e. before cracking initiates, and this is not solely due to hydrogen being already present on the flaw. Another example would be to consider two materials already containing hydrogen (iHE), that are being plastically deformed (dislocation transport), but that exhibit the same types of traps in different quantities : the one containing the largest number of irreversible traps, provided they have a high C_K value, should be less sensitive to iHE. Besides these considerations, things would again be different if hydrogen diffused by interstitial jumps : because hydrogen is more strongly bound to a dislocation than to an interstitial site, the sink character of all traps will be more predominant when hydrogen normally diffuses. Therefore, transport mode also affects EHE and iHE susceptibility, depending on traps characters.

Such differences have been modelled (for more complete details, see Ref.2) and allow to calculate hydrogen concentration profiles that would be obtained according to various test conditions. Examples of such calculations are given in fig.3. In these figures, $n_H^{F,p}$ is the number of hydrogen atoms located on a potential flaw F (that is also an irreversible trap) as a function of dislocation number, p, emitted from a source ; $n_o^{\perp,r}$ is the initial number of hydrogen atoms on dislocations, $\perp$, or reversible traps, r ; k is a measure of trap strength ; $d_{i,r,f}$ is the diameter of an irreversible, reversible trap, potential flaw, respectively ; x is the average distance between traps. As seen, hydrogen concentration profiles are wholly different, depending on test conditions and trap characteristics. In figure 3a, for instance, the critical concentration, that corresponds to a critical value of $n_H^{F,p}$, is first reached in A (subsurface crack). In fig.3b, the profile may or may not exhibit a maximum depending on the presence of irreversible traps ($d_i \neq o$ or $d_i = o$) ; cracking will happen in the interior, where a suitable crack site is close to a dislocation source. Obviously, other test parameters also influence the time at which C_K is reached ; this is the case of the stress state around potential flaws, that may increase or decrease C_K as time goes on. What the trap theory says is that trap characteristics also play an important role and should be taken into account. The procedure chosen to verify these ideas is the following : take materials with well-known trap parameters, and try to explain experimental results with the help of the theory ; then take a material with unknown trap characteristics, and from the obtained results, try to understand its trap parameters.

Experimental

Materials

Two groups of materials were used here : the first consists of pure iron alloys containing titanium and carbon and having well-known trap characteristics (Ref.5 : Ti atoms are reversible traps and TiC particles are irreversible traps), and the second is a 35 NCD 16 steel with unknown trap characteristics. Chemical compositions are given below.

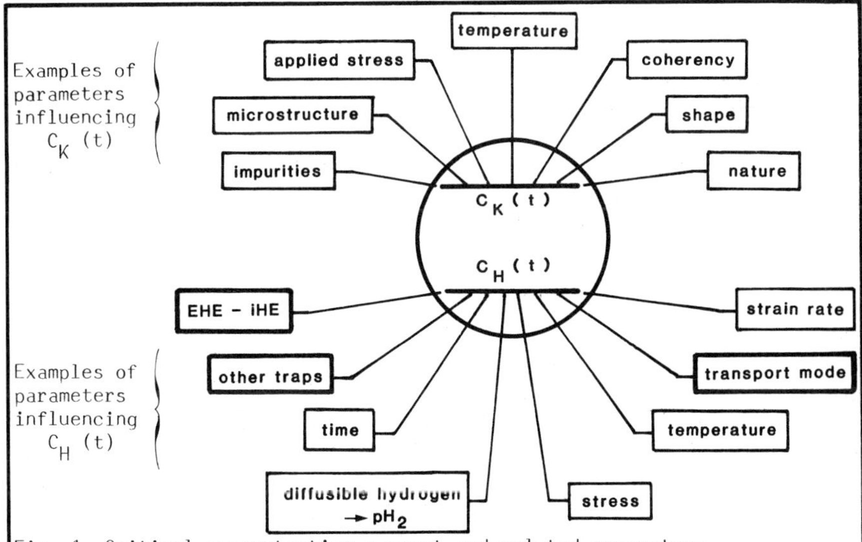

Fig. 1. Critical concentration concept and related parameters. C_K is the critical concentration and C_H the hydrogen concentration trapped on the defect considered.

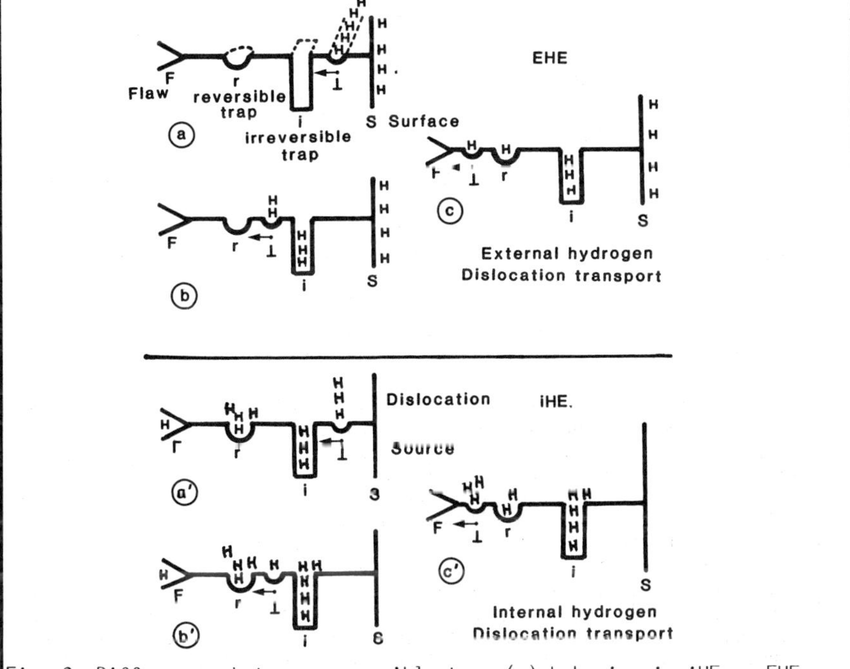

Fig. 2. Differences between reversible trap (r) behavior in iHE or EHE. r trap is a sink in EHE, and a source in iHE. (from Ref.2).

Table I. Typical chemical compositions

Material	C	Si	Mn	Ni	Cr	Ti	Mo	UTS (MPa)
A	320 ppm	-	-	-	-	.09	-	220
B	50 ppm	-	-	-	-	.15	-	206
C	650 ppm	-	-	-	-	.50	-	230
35 NCD 16	4100 ppm	.36	.26	4.3	1.9	-	0.5	1862
2 x 2 hrs at 230°C				Wt %				

Test techniques

According to our premises that hydrogen embrittlement depends on hydrogen location at the start of the test (EHE versus iHE), on transport mode and on traps reversibility character, all test techniques may then be classified according to these parameters. This classification is given below :

Table II. Classification of test techniques according
to trap theory parameters

Group of parameters	Examples of corresponding test techniques		
EH,⊥	Normal disk pressure test (DPT)	tensile test in hydrogen	fracture mechanics in hydrogen (K_{iH})
EH, ij	Delayed failure tests below σ_y, in hydrogen (DPT, tensile test, etc...)		
iH,⊥	DPT and tensile test on precharged specimens		
iH, ij	Delayed failure tests below σ_y (DPT, tensile) on precharged specimens.		

Code : EH : external hydrogen ; iH : internal hydrogen ; ⊥ : transport by dislocations; ij : interstitial jumps (normal diffusion).

In order to test the influence of various groups of parameters on HE, the simplest test technique is the disk pressure test (DPT). This test is well documented in the literature (e.g. Refs. 7,9). Disk shaped specimens are clamped between two flasks of a high pressure cell and strained to rupture under increasing helium or hydrogen pressure ; the effect of helium is only mechanical, while that of hydrogen adds embrittlement. Thus, the ratio pH_e/pH_2 at rupture is a good indication of hydrogen induced embrittlement EHE may easily be studied by straining in hydrogen uncharged specimens, while iHE is obtained by straining in helium previously charged specimens. Another experiment performed here was to plastically predeform the specimens before hydrogen testing ; this adds two differences to the usual procedure : for one thing, traps have been added (dislocations) and dislocation transport is not as predominant as before. These four possibilities are described in fig.4 in greater details. Disks dimensions were either Ø 1 cm x thickness 0.025 cm for alloys A, B, C, or Ø 5.8 cm x thickness 0.075 cm for 35 NCD 16.

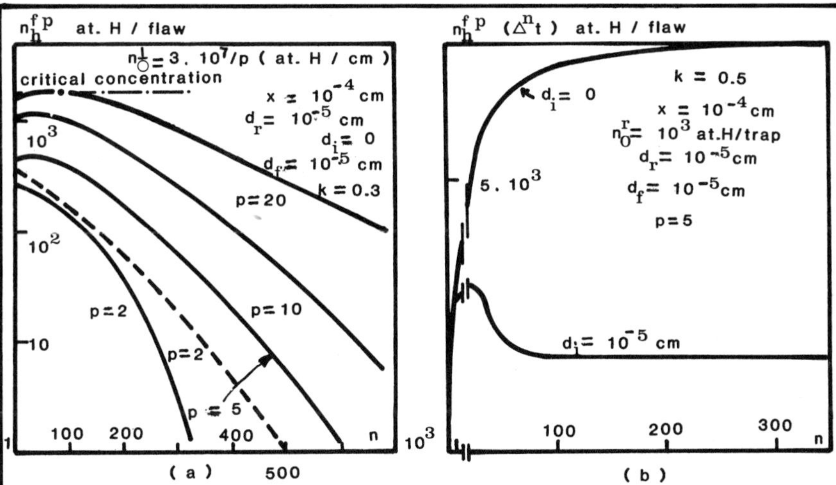

Fig. 3. Examples of concentration profiles obtained when hydrogen is transported by dislocations ; 3a : external hydrogen ; 3b : internal hydrogen.

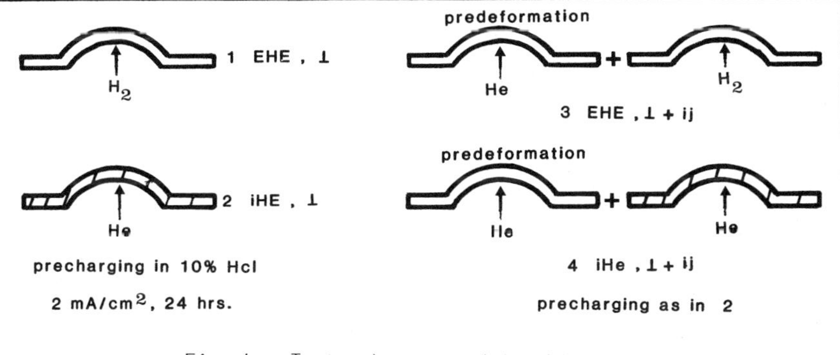

Fig. 4. Test schemes used in this study

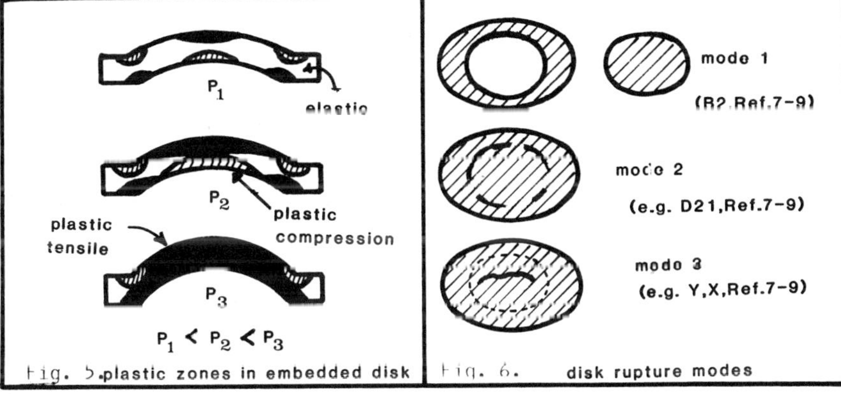

Fig. 5. plastic zones in embedded disk   Fig. 6.   disk rupture modes

All tests were performed at room temperature, using variable pressure increase rates ($\Delta p/\Delta t$) and high purity hydrogen ($O_2 < 2$ wt ppm).

Results

Preliminary

. The DPT involves inhomogeneous triaxial stresses (Ref.9) ; at first, tensile stresses are generated at the upstream anchorage and at the downstream pole. When yield stress is reached, a notch like plastic enclave appears at the upstream anchorage and grows (see fig. 5). This state of stress may give rise to various rupture modes of the disks, three of which were observed here (fig. 6) : complete shearing-off of the cupola (aspect 1),partial cracking of the anchorage (case 2), partial cracking of the cupola (case 3).

. Iron-titanium-carbon alloys have been shown (Ref.4) to possess both irreversible traps (TiC particles) and reversible traps (Ti atoms). Important trap characteristics are given below :

Table III.   Trap characteristics of alloys A, B, C

Alloy	N_{Ti} at./cm^3	N^1_{TiC}* sites./cm^3	D^1_{TiC} μ	N^2_{TiC}* sites./cm^3	D^2_{TiC} μ
A	8×10^{19}	7.4×10^{18}	.04	1.3×10^{16}	1
B	1.5×10^{20}	1.9×10^{17}	.12	1.6×10^{16}	1.3
C	5×10^{20}	10^{19}	.2	6.6×10^{16}	1

* Note : $N^{1,2}_{TiC}$ is the total number of sites at interfaces TiC-matrix, for the two populations of diameters $D1,2_{TiC}$.

. For a better comparison between EH.I and IH.I tests, the total quantity of hydrogen was kept the same. Thus, the measure of the quantity of hydrogen introduced at the anchorage in a EH.I test gave an average value of 2 ppm for 35 NCD 16 and 1.5 ppm for other alloys. The charging current density for IH.I tests was thus chosen around 2 cm A/cm^2 to give an identical initial quantity of hydrogen. Obviously, if the total quantity of hydrogen is approximatly the same in both tests, its distribution is not.

. pHe/pH$_2$ or pHe/pHe$_c$ ratios : these ratios are given below ; pHe$_c$ is the He rupture pressure of precharged specimens ; number between parenthesis correspond to the rupture mode (fig.6).

Table IV.   Average embrittlement ratios

Alloy	pHe/pH$_2$ (I)	pHe/pHe$_c$ (I)	pHe/pH$_2$ (2)	pHe/pHe$_c$ (2)	pHe/pH$_2$ (3)	pHe/pHe$_c$ (3)
A	1.0	1.1	-	-	-	1.5
B	1.0	1.0	-	-	1.6	1.3
C	1.4	1.2	1.2	-	1.2	-
35 NCD 16	-	1.5	4.1	3.1	-	5.6

All ratios have been computed with the helium rupture pressure pHe (I) ; however, some ruptures in helium alone were of type (2) or (3). This points out to the fact that the DPT is sensitive to the location of large defects, HE being absent (e.g. a large TiC in the cupola may give rise to a mode (3) rupture). Thus, HE susceptibility can be compared between disks having the same rupture mode. In particular, mode (I) ruptures are best for trap considerations, since they are relevant of a global response of the material and not of a localized defect. All results are shown on fig.7, as a function of pressure increase rate, $\Delta p/\Delta t$.

. Results on predeformed specimens are offered on fig.8 ; again, all disks exhibited various rupture modes, and comparisons can be made between similarly ruptured disks. In agreement with the assumption of a localized defect in mode (2) and (3), all rupture pressures for these modes are lower than for mode (I) ; this was also the case in fig.7.

Discussion

. IH-I :for mode (I) and (3) ruptures, the ranking of alloys A, B, C, according to increasing embrittlement, is : B > A > C ; also, pressure ratios are close to unity, meaning little susceptibility to embrittlement. For these tests conditions, the Trap Theory (Ref. 2) predicts a beneficial role of finely distributed irreversible traps (N_{TiC}) and a deleterious influence (sources) of reversible traps (N_{Ti}). These predictions are verified here for reversible traps, since the increasing N_{Ti} ranking is : A,B > C ; as for the low embrittlement ratio, it may be due to the beneficial role of N'_{TiC}. Charging the disks at higher current densities (up to 30 mA/cm^2) may considerably increase all ratios (see obtained lower limits on fig.7) ; however, reasoning on traps is now difficult since cracks are induced prior to testing. In the same way, the embrittlement ratio of 35 NCD 16 charged with 2 ppm hydrogen is relatively low (I.5). Since traps characteristics in this material are unknown, the reverse reasoning on the basis of the Trap Theory would indicate that beneficial irreversible traps or strong reversible traps are present ; again however, charging at higher densities changes the rupture mode and increases the ratio pHe/pHe$_c$ up to 5. In this last case, hydrogen induced cracks were not involved ; the most probable interpretation then is that various potential flaws with a wide range of C_K values coexist, and that charging with 2 ppm only helped to go over the lowest C_K's while testing. Also, one should note that mode (3) ruptures for all alloys are more frequent in IH-I than in EH-I. This has a simple explanation : when testing with EH-I parameters, the pole, that works in compression, will not admit much H entry ; on the contrary, in IH-I testing, H being already there will accumulate in the tensile stressed downstream pole.

. EH-I : for Fe-Ti-C alloys, the ranking is approximately the same as before: A, B > C. Alloys A and B are unaffected (pHe/pH$_2$ = 1.0), which is in agreement with the fact that in EH-I, beneficial irreversible traps should be even more so than in IH-I (Table IV). However, this is not the case for alloy C and for 35 NCD 16 (mode (2) rupture) for which embrittlement in EH-I tests is not less than in IH-I tests. A possible explanation for alloy C, according to the Trap Theory, is the following : if alloy C is bad, since it has the highest number of reversible traps, it is because these traps are forced to act as sources. The only case, in EH-I conditions, for which this may happen is illustrated in fig.3 a : incoming dislocations carry less hydrogen than previous ones, but still transport appreciable amounts of hydrogen into the lattice because they keep recharging on reversible traps. Why $n_0^{\pm}$ should decrease with number of dislocations may stem from the particular strain state at the anchorage : the rate at which dislocations are nucleated increases with test duration (inhomogeneous strain increments) ; thus, at any time

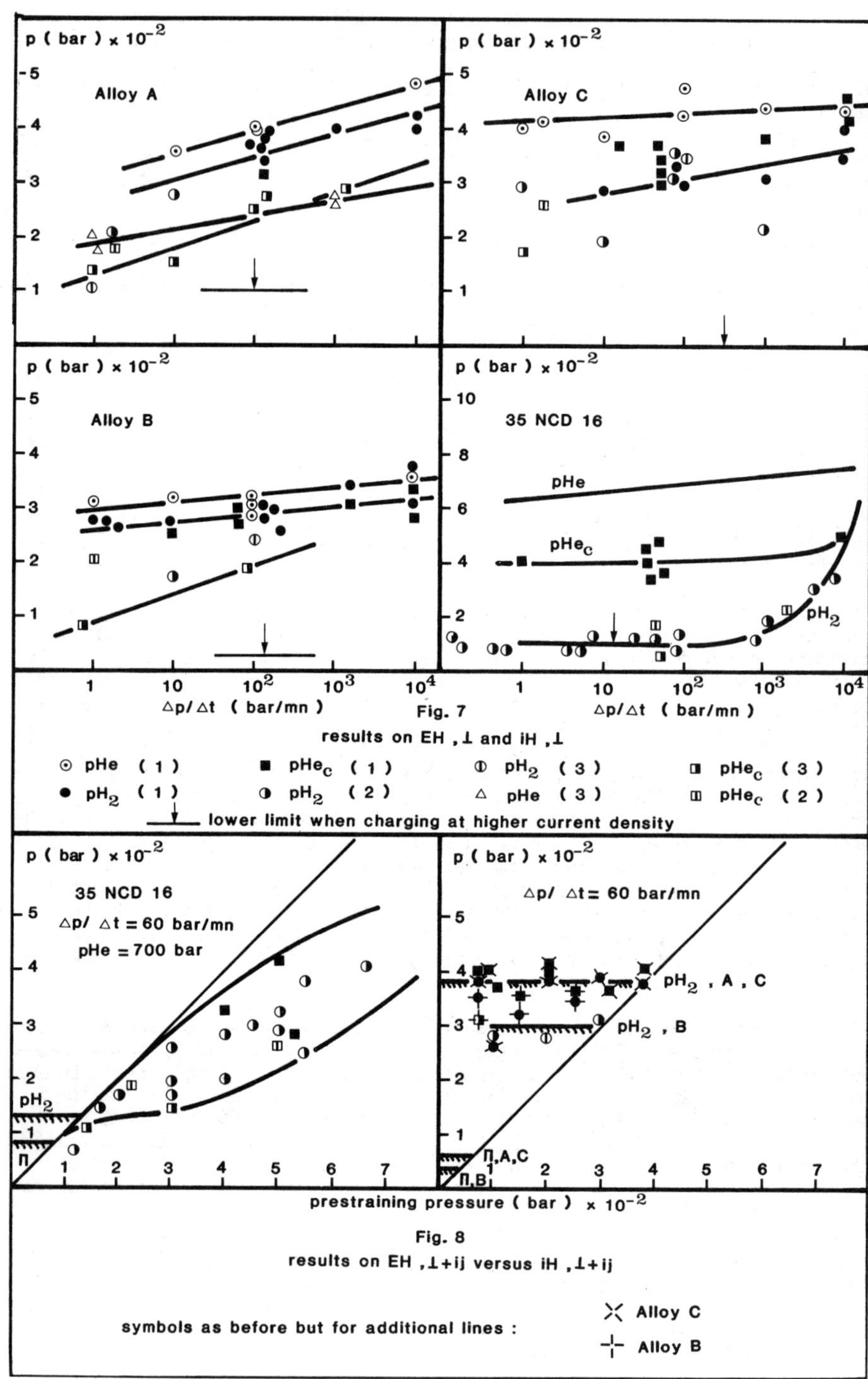

p (bar) x 10^-2
Alloy A
Alloy C
Alloy B
35 NCD 16
pHe
pHe_c
pH_2
1   10   10^2   10^3   10^4
Δp/Δt (bar/mn)
Fig. 7
results on EH ,⊥ and iH ,⊥
⊙ pHe (1)
■ pHe_c (1)
pH_2 (3)
pHe_c (3)
● pH_2 (1)
pH_2 (2)
△ pHe (3)
pHe_c (2)
lower limit when charging at higher current density
35 NCD 16
Δp/ Δt = 60 bar/mn
pHe = 700 bar
pH_2
π
Δp/ Δt = 60 bar/mn
pH_2 , A , C
pH_2 , B
π,A,C
π,B
1   2   3   4   5   6   7
prestraining pressure (bar) x 10^-2
Fig. 8
results on EH ,⊥+ij versus iH ,⊥+ij
symbols as before but for additional lines :
Alloy C
Alloy B

interval, more and more dislocations have to share hydrogen. If however necessary, this reason alone is not sufficient : even if reversible traps act as sources in EH-$\perp$, dislocations may still carry less hydrogen than in IH-$\perp$. This is not true however in our case : the initial H content in IH-$\perp$ and the final one in EH-$\perp$ may be the same, but in one case (EH-$\perp$) the hydrogen source is infinite (external source), and in the other (IH-$\perp$) it is not (reversible traps) ; thus the amount of hydrogen, at any time, that is transported by dislocations may be higher in EHE than in IHE, making reversible traps more noxious.

EH , $\perp$ + ij : a first striking result of fig.8 is that the rupture pressure is always higher than the predeformation pressure for alloys A, B and C, while the reverse happens for 35 NCD 16. Thus, for alloys A, B and C, dislocation transport will prevail, while it does not necessarily for 35 NCD 16. In all cases however, predeformation above the yield strength (pressure 7 on figures) introduces dislocations (additionnal reversible traps). Another constatation is that the rupture pressure after predeformation is always higher, for a same rupture mode, than the rupture pressure without prestraining ; in other words, predeforming is beneficial. However, if the rupture pressure after prestraining increases with prestraining for 35 NCD 16, it does not for other alloys. Finally, the ranking of alloys A, B and C differs from preceding rankings : alloy A seems the most embrittled, with modes (2) and (3) ruptures.The first remark has a simple explanation (fig.8) : for 35 NCD 16, the rupture pressure is lower than the predeformation pressure (PP) once PP is sufficiently above the rupture pressure without prestraining ; since this pressure is close to pHe for alloys A, B and C, predeformation pressures could not be chosen much higher. Thus, the difference between alloys is only apparent. The second remark concerning the beneficial effect of prestraining is mainly valid for 35 NCD 16, and may receive an explanation in the light of the Trap Theory : the additional dislocations created before hydrogen is introduced (as dislocation atmospheres and as diffusing interstitials) will act as many sinks for hydrogen, allowing greater deformation before C_K is reached anywhere. Thus, predeformation is beneficial.

. IH , $\perp$ + ij : little differences exist between results of this test and of the previous one, but for alloy A where IHE is less than EHE ; but then, comparison is difficult because rupture modes are different. The fact that IHE is similar to EHE is noticeable for 35 NCD 16, since the difference was important for EH-$\perp$ and IH-$\perp$ tests. However, we had concluded before that strong trapping (irreversible or reversible) was predominant in this alloy, and explained differences between EH-$\perp$ and IH-$\perp$ in this way : dislocations could not carry much hydrogen in IH-$\perp$ because it was strongly trapped on possible sources, while this was not the case in EH-$\perp$ with an infinite surface source. The difference now is that average reversible traps (dislocations) have been created by prestraining, and that they act as sources in IH-$\perp$ + ij test, as does the surface in EH-$\perp$ + ij. Thus, results are the same.

Conclusions

Several experiments have been devised to verify recent ideas on trapping (Réf.4), using the DPT test. The most important result concerns the different response of various alloys, depending on test procedure and Trap Theory parameters. This is summarized below and compared to other literature results (Réf.10).

Table V. Ranking of alloys A, B, C according to test procedure

Test	Ranking : HE increases —>
iH, $\perp$	A,B ————————> C
EH, $\perp$	A,B ————————> C
EH, ij (Ref.10)	C ——> B ——> A
EH, ij + $\perp$	B,C ————————> A
iH, ij + $\perp$	A,B,C
N_{Ti}	A ——> B ——> C
N_{Tic}^{1}	B ——> A ——> C
N_{Tic}^{2}	A ——> B ——> C

  Most of this study's results can simply be accounted for using Trap Theory considerations (e.g. iH, $\perp$). Others require very particular cases of Trap Theory parameters combination (e.g. EH, $\perp$). Perhaps the best agreement between theoretical predictions and experimental data is that obtained when comparing iH,$\perp$ and EH,ij rankings : iH,$\perp$ follows the ranking of N_{Ti} (Ti atoms act as sources), while EH,ij (Ref.10 : hydrogen induced cracking while cathodically charging) has the opposite ranking (Ti atoms now act as beneficial sinks).

  Reasoning in the same way on complex materials may give indications as to their trap characteristics. However, interpretations are far from being straight forward and necessitate further developments and experiments on the basis of this model.

References.

1. J.P. Fidelle ; H.G. Nelson ; ASTP 543, p.267 and p.273, 1972.
2. G.M. Pressouyre, Acta Met., 1980, 28, p. 895.
3. G.M. Pressouyre, ISS-AIME, Toronto, Canada, 1980.
4. G.M. Pressouyre, Met. Trans., 1979, 10A, p. 1571.
5. J.K. Tien, A.W. Thompson, I.M. Bernstein and R.J. Richards, Met. Trans. A, 1976, 7A, p. 821
6. G.M. Pressouyre and I.M Bernstein, Met. Trans., 1978, 9A, p. 1571.
7. J.P. Fidelle, Ivry Conference, France, 1975.
8. J.P. Fidelle et al., ibid Ref. 1, pp. 34 and 221.
9. J.P. Fidelle et al., in. Proc. Int. Conf. on Effect of Hydrogen on Behavior of Materials, ed. by A.W. Thompson and I.M Bernstein, 1975.
10. G.M. Pressouyre and I.M. Bernstein, Acta Met., 1979, 27, p. 89.

DISLOCATION SWEEPING OF HYDROGEN AND HYDROGEN EMBRITTLEMENT*

37

J.K. Tien, S.V. Nair and R.R. Jensen
Henry Krumb School of Mines
Columbia University
New York, NY 10027

The kinetics of embrittlement of materials by hydrogen may be closely related to the transport of hydrogen to critical failure sites within the material. Such sites can be, for example, internal hydrogen traps in the material, such as voids, inclusions or precipitate particle interfaces or grain boundaries, which have a negative binding energy for hydrogen atoms.

Enrichment of hydrogen at or near critical sites, whether or not leading to large supersaturations of hydrogen, have been treated quantitatively in the recent past as occurring, for example, by the stripping of solutes from moving dislocations by traps and by the annihilation of hydrogen containing dislocations. This paper will thoroughly and critically review the recent theoretical and experimental efforts into hydrogen transport and trapping in heterogeneous alloys. Kinetic models will be presented whereby internal hydrogen supersaturations are achieved by dislocation sweeping processes.

* Invited Lecture

Introduction

Pivoting about the sweeping of hydrogen as Cottrell atmospheres by mobile dislocations,[1] a kinetic model for plastically driven movement of hydrogen atoms in solids took shape four years ago.[2-5]   In the interim, experimental studies have been completed[6-31] that sought to verify the very long transport distances,[6-11] the strain rate dependence,[12,13] the temperature dependence,[14-19] the inclusion or trap microstructural dependence[20-25] the slip mode dependence,[12,26-28] and the fracture mode or embrittlement predictions[21,22,29-31] of the model. In what follows, we critically review these experimental results in terms of the sweeping model. We also review the model and recent constructive criticisms of the model.[32]   Recent extension of the model to predict the kinetics of plastically driven subcritical crack growth[33,34] of high strength steels in hydrogen or aqueous environments are also discussed. It will be shown that microstructure, plastic flow and hydrogen often combine to result in transport of hydrogen to traps, the interfaces of which in turn prematurely decohere and initiate fracture.

The Dislocation-Hydrogen Interaction and Sweeping Velocity Model

The carrying capacity of mobile dislocations is dependent upon the equilibrium segregation tendency of hydrogen to dislocation cores or to the elastically stressed regions around dislocations. This tendency is expressed by E_B, the binding energy. According to Cottrell[35] the hydrogen concentration at the dislocation core, $C_\perp$, can be related to the lattice atom fraction of hydrogen near the dislocation, C_0, by the Boltzman equation when $C_\perp$ is low

$$C_\perp = C_0 \exp(-E_B/kT) \tag{1}$$

The Fermi-Dirac relationship can be used[36] when $C_\perp$ approaches unity

$$C_\perp = Co \exp(-E_B/kT) \, / [1 + C_0 \exp(-E_B/kT)] \tag{2}$$

The concentration of hydrogen in the dislocation's elastic field can be calculated using these same equations with the appropriate E_B. This elastic contribution to E_B is believed to be small.[32]

E_B values for dislocations, as well as for such traps as grain boundaries and interface boundaries, are given in Table II of Ref. 4. Johnson, in a preceding paper in this conference proceedings, offers newer experimental values of E_B and discusses the intricacies of the interpretation of experimental results.[37] With a typical E_B value of about -0.3eV, dislocations like grain boundaries behave like reversible traps, whereas, for example, a TiC-Fe interface, with a high negative binding energy of about -0.9eV, behaves like an irreversible trap.[38-40] Voids, associated with decohered interfaces where H can enter and form H_2, can be viewed as unsaturable traps.[37]

Accordingly, dislocations can deliver hydrogen to traps such as grain boundaries and other interfaces, thus resulting in the much-discussed lowering of cohesive strength. Under the appropriate conditions where leakage out of voids is slow, hydrogen carrying dislocations can pressurize voids and, hence, aid in faster dimple growth.[2-5,32]

The carrying capacity, $C_\perp$, of dislocations, assuming E_B of -0.3eV and C_0 of only 5-6ppm, is about unity at ambient temperature. The velocity, $\bar{v}$, with which this unity hydrogen cloud is swept along by dislocations can be derived[1-4] through the Einstein-Stokes relationship and is

$$\bar{v} = (D_H/kT)F \tag{3}$$

where D_H is the effective diffusivity of hydrogen in the actual matrix and F is the effective driving force per atom of hydrogen on the dislocation. An upper bound for F can be estimated to be E_B graded over the maximum interaction distance of dislocation and solute of 30 times Burgers vector, in which case a critical velocity can be expressed as

$$\bar{v}_c = -(D_H/kT)(E_B/30b) \tag{4}$$

Above $\bar{v}_c$, or at a strain rate greater than

$$\dot{\varepsilon}_c = \rho b \bar{v}_c \tag{5}$$

the dislocation line can be expected to break away from its H cloud so that dislocation sweeping is no longer effective.[3-5] Equation (5), with $\bar{v} < \bar{v}_c$, also predicts that sweeping velocity, or penetration distance, or embrittlement, can be directly proportional to the applied strain rate. In the actual case, of course, the material would have a distribution of dislocation velocities at any given strain rate so that, as the strain rate increases the fraction of dislocations moving at velocities below $\bar{v}_c$ decreases. An upper bound of the penetration distance due to dislocation sweeping can then be given by

$$\bar{x}_c = \bar{v}_c t_p \tag{6}$$

where t_p is the duration of plastic deformation. The corresponding penetration distance by lattice diffusion alone is

$$\bar{x}_D = 4\sqrt{D_H t_p} \tag{7}$$

Table I compares values $\bar{x}_c/\bar{x}_D$ for Monel and iron and nickel-base alloys with D_H values ranging from $10^{-6}\text{cm}^2/\text{s}$ for ferritic steels to $10^{-10}\text{cm}^2/\text{s}$ for austenitic steels or Ni-base alloys. It is apparent from the table that, except at extremely short times, dislocations can carry hydrogen much more deeply into the material than lattice diffusion can. Correspondingly, mobile dislocations can carry hydrogen to material weak links much more rapidly than lattice diffusion.

Table I.   Penetration Depth Ratios for $\sqrt{Dt}$ vs. Dislocation Sweep-in Transport of Hydrogen in Fe and Ni Base Alloys*

t(sec)	f	$\bar{x}_C/\bar{x}_D$ Fe-Base	$\bar{x}_C/\bar{x}_D$ Ni-Base	$\bar{x}_C/\bar{x}_D$ Monel
1 hr	1 CPH	2.0×10^5	2.0×10^3	6.3×10^3
60	1 CPM	2.6×10^4	2.6×10^2	8.2×10^2
1	1 Hz	3.3×10^3	3.3×10^1	1.1×10^2
2×10^{-2}	60 Hz	4.3×10^2	4.3×10^0	1.4×10^1
1×10^{-3}	1 kHz	1.1×10^2	1.1×10^0	3.3×10^0
1×10^{-4}	10 kHz	3.3×10^1	3.3×10^{-1}	1.1×10^0
		D 10^{-6}cm^2/sec	D 10^{-10}cm^2/sec	D 10^{-9}cm^2/sec

*from Ref. 4 (Note values in Ref. 4 are in slight error).

The most convincing evidence for rapid dislocation sweeping of hydrogen with large penetration distances, and the predicted strain rate dependence, is the very recent work by Kurkela and Latanision.[6] They measured hydrogen permeation in thin membranes of Ni270 (99.97% Ni) at different plastic strain rates and found that the steady state permeation current increased with strain rate. Using their data, we found, Fig. 1, that this increase was linear in strain rate, as predicted by a dislocation sweeping model.[3] Had their data been carried over to larger strain rates, a drop in the permeation current could have been expected due to dislocation breakaway effects mentioned. In addition, Kurkela and Latanision[6] found that the time taken for hydrogen to permeate through the specimen thickness was only about 5 secs, in full agreement with that calculated using Eq. (6), whereas it is impossible to explain this permeation time using lattice diffusion kinetics.

These results are in agreement with the earlier results, already reviewed,[4] of the tritium release rate experiments[8] in Armco iron, type 304 stainless steel, commercially pure nickel, 5086 aluminum and Inconel 718, as well as the tritium penetration experiments[9] in type 304L stainless steel. However, the penetration results from a nickel system[9] appear inconsistent with dislocation sweeping. It was found that the tritium penetration after deformation was less in the gauge length as compared to the elastically deformed shoulder region. A close examination of his results show that this can be a consequence of the existence of a concentration profile of tritium in the material prior to deformation. The concentration profile prior to and after deformation given in Figs. 4 and 2, respectively, in Ref. 9 is replotted in our Fig. 2. It appears that plastic deformation has caused a redistribution of hydrogen towards the external surface. Such a redistribution is not possible by lattice diffusion since the concentration gradient favors diffusion into the material; but, the results are fully consistent with enhanced dislocation sweeping of hydrogen from inside the material out to the surface.

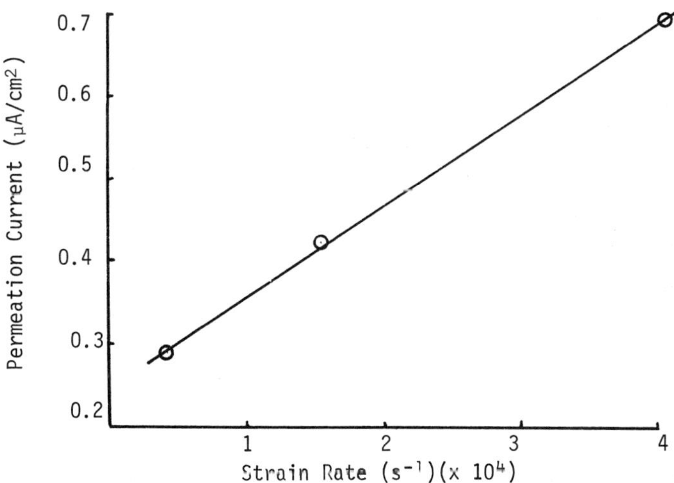

Fig. 1.   Permeation Current as a Function of Strain Rate
for Ni270 (99.97% Ni) Specimens.   (Data points
from Ref. 6)

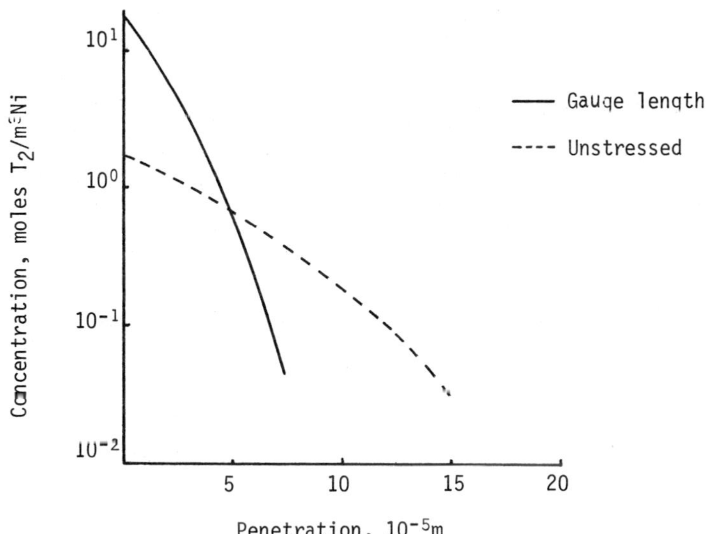

Fig. 2.   Tritium Penetration into Nickel.   The
Data Points from Donovan[9] Have Been
Omitted.

<u>Models for Matrix Enrichment by Dislocation Sweeping</u>

Lattices can be enriched with hydrogen by the dislocation sweeping mechanism through two major mutually complementary means.[3-5] One is by the annihilation of hydrogen carrying mobile dislocations with each other, and the other is by the transfer of hydrogen from the dislocations to deeper traps such as grain boundaries, inclusion or precipitate interfaces and voids as the mobile dislocations encounter these obstacles. The first mechanism, known as the dislocation annihilation model, was first analyzed by Johnson and Hirth[42] and found to be non-operative as a lattice enrichment mechanism. Nair <u>et al.</u> very recently re-examined this mechanism and found that it can indeed enrich when traps are present.[43] The second mechanism, known as the stripping model, was derived by Tien <u>et al.</u>,[3-5] and up to the Nair <u>et al.</u> work[43] was the only quantitative model which predicts lattice enrichment. Based on the stripping model,Pressouyre[44] also recently developed a hydrogen embrittlement model specifically to discuss in detail the influence of trap type and trap distribution. This model is not a kinetic one and is further discussed in Pressouyre <u>et al</u>'s paper[45] in this conference proceedings.

In the model of Johnson and Hirth,[42] local kinetic supersaturations of hydrogen are assumed to be possible by dislocation annihilation in the material. A quantitative estimate of the supersaturation was then obtained by balancing the arrival rate at annihilation sites by a departure rate out to the surface of the material. It was concluded that the supersaturation values were extremely small. It was further concluded on the basis of this calculation that dislocation annihilation could not contribute to any embrittlement. However, Johnson and Hirth[42] did not take into account that, besides the external surfaces, hydrogen can proceed from the annihilation sites to local traps, i.e., to the weak links.

In the model of Nair <u>et al.</u>,[43] a trap included solution was obtained for a one-dimensional case with the assumption of a planar trap interface. It was also assumed that once annihilated the dumped hydrogen act as planar sources. Schematically, these assumptions are embodied in Fig. 3, which shows the situation at the instant dislocations annihilate at a distance, s, from the trap, resulting in a supersaturated concentration value, C_d, a distance away from the trap.

Under these assumptions, it was shown that the fraction $F_r(t)$ of the initial deposited amount retained in the trap after time t is given by

$$F_r(t) = \alpha \{ \text{erfc}(s/2\sqrt{D_H t}) - \text{erfc}(r/2\sqrt{D_H t})$$

$$- \exp(-s^2/4D_H t)\exp Z^2(s)\ \text{erfc}\ Z(y)$$

$$+ \exp(-r^2/4D_H t)\exp Z^2(r)\ \text{erfc}\ Z(r)\ \} \qquad (8)$$

where,

$$Z(z) = [z/2\sqrt{D_H t} + D_H\sqrt{t/D_H}]/\alpha\delta \qquad (9)$$

$$\alpha = \exp(-E_B/kT) \qquad (10)$$

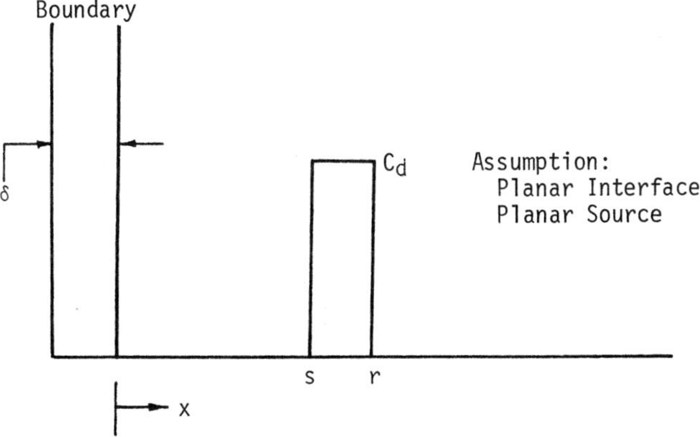

Fig. 3.   Schematic showing excess concentration, C_d, ahead of the trap-matrix boundary as a result of dislocation annihilation.

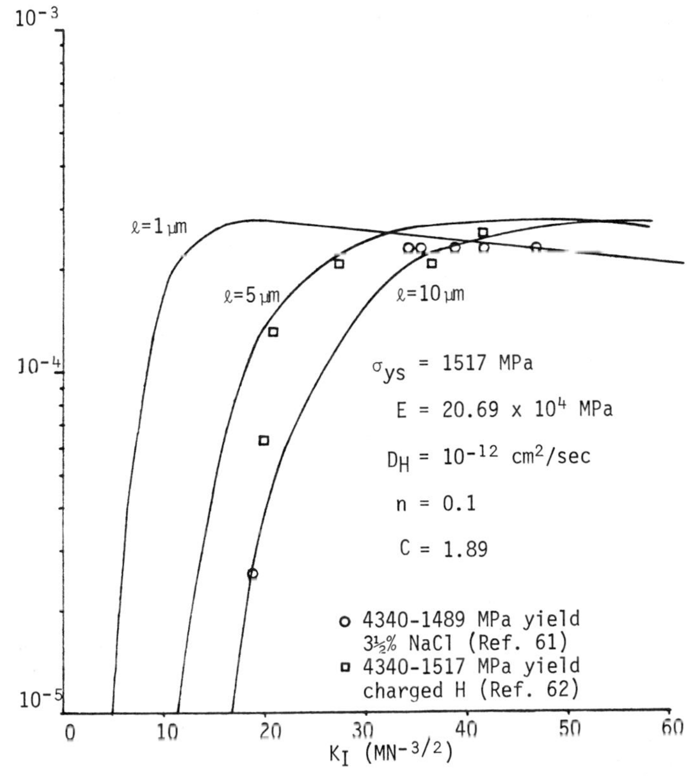

Fig. 4.   Crack growth rates for different ℓ values calculated from Eq. (19). Also shown are experimental data points.

E_B is the binding energy of hydrogen to the trap, D_H the diffusion coefficient of hydrogen and δ the interface thickness.

Table II shows $F_R(t)$ for different values of s and t for an E_B value of -0.4eV. It can be seen that, even after long times (long relative to the time interval between annihilation events), almost all the hydrogen deposited by annihilation is retained by the trap and the leakage rate out to the surface is insignificant. This is true even if the annihilation event occurs relatively far from the traps, that is, of the order of a grain size ($\sim$100μm). Consequently, contrary to the conclusion of Johnson and Hirth,[42] annihilation can result in enrichment at interfaces. The relative contributions of the annihilation model and the stripping model depend on the trap density, annihilation rate and strain rate and are discussed in the original paper.[43]

Table II: Values of $F_R(t)$ as a Function of s and t (for E_B =-0.4eV)

s (μm) \ t(secs)	1	10^3	10^4
0 (annihilation at the interface)	0.99	0.94	0.82
1	0.89	0.94	0.82
100	--	0.58	0.82

The dislocation stripping model[3-6] is based on dislocation and barrier collision rate, the stripping of the hydrogen from the dislocation along the encounter length, and the leakage rate from the trap. The details of the thermodynamics of hydrogen exchange between dislocation and reversible traps, irreversible traps and unsaturable traps are discussed elsewhere.[3,38-40] The kinetic equations of interface enrichment are as follows:[3-5]

$$\dot{C}_b = (gD^2/b^2f^{1/3})\dot{\varepsilon} \tag{11}$$

where $\dot{C}_b$ is the arrival rate of hydrogen atoms to the interface or boundary, g is a geometric factor, D is the particle dimension intercepting the dislocation, f is volume fraction of the particles, b is Burgers vector and $\dot{\varepsilon}$ is the test or application strain rate. $\dot{C}_b$ drops to zero as $\dot{\varepsilon}$ exceeds $\dot{\varepsilon}_c$. The above expression basically provides the kinetics of hydrogen embrittlement when embrittlement is controlled by C_b and subsequent decohesion of the interface resulting in crack initiation.

When the interface decoheres, forming a void, or in the case of pre-existing voids attached to, say, inclusions, $\dot{\varepsilon}_b$ can result in the pressurization of the void if H_2 formation is rapid in the void, the H_2 dissociation reaction is slow, and when leakage from the void or inclusion interface is low.[3-5,32] Such pressurization may further contribute to speedy embrittlement through enhancement of crack growth or dimple growth in ductile materials. Equation (11) processed through the gaseous equation of state relationship results in an equation which relates pressure in the void, P_i, to the plastic strain, ε, and the ratio of inclusion size D and void size D_V, where[3]

$$P_i = g(kT/2b^2)(1/f^{1/3})(D^2/D_V^3) \tag{12}$$

The geometric factor g can take the value unity when inclusion size is comparable to inclusion spacing[3-5] and the value $f^{1/3}$ [32] when the inclusion size is small. Roughly, g varies between unity and six for most cases. For g = $f^{1/3}$, $\varepsilon \simeq 10\%$ and $D_V = 10^{-2}\mu m$ (i.e., freshly initiated small void), Eq. (12) reduces to

$$P_i(atm) = 3.3 \times 10^4 D^2(\mu m) \tag{13}$$

so that even for a particle or inclusion size of 1 μm, pressures on the order of 10^4 atmospheres can develop inside the voids.

Leakage from the void or inclusion interface can, of course, prevent significant pressurization. Such leakage will not occur if H_2 dissociation or H adsorption poisons such as oxygen, segregates to the interface either through rejection by the matrix or by the inclusions.[23,46] It can occur if dissociation catalysts segregate to the interface.

Tien et al.[3] concluded through a diffusion calculation (Eq. (14) in Ref. 3) that leakage from voids on large flat inclusions can be considered negligible at realistic test times. Using a one-dimensional analysis, Tien et al.[3] showed that the diffusion current away from the interface can be given by

$$I_t = (D^2 c_s/4) \cdot \sqrt{D_H/t_p} \tag{14}$$

where c_s is the concentration in the matrix at the interface. Equation (14) is a reasonable assumption only for large flat particles, and is not valid at the particle extremeties. These limitations notwithstanding, it was shown, on the basis of Eq. (14), that I_t was greater than the arrival rate via dislocations, $\dot{C}_b$, only at times $t_p < {\sim}100$secs. Thus, it was rationalized that, for most practical situations, I_t can be neglected in the calculation of void pressurization.

Recently Hirth[32] calculated a modified leakage current, I_H, for the case of small spherical inclusions using the form at steady state given by Jost,[47] namely,

$$I_H = 2\pi DD_H c_s \tag{15}$$

According to Hirth, I_H/I_t, given by

$$I_H/I_t = (4.3/D)\sqrt{6D_H t_p} \tag{16}$$

could be as high as 10^8 for the case of ferritic steels with $D_H = 10^{-6} cm^2/s$. It is not clear to us how a factor as large as 10^8 can be obtained using Eq. (16). In Table III this ratio, I_H/I_t, is recalculated for inclusion sizes typical of those preferentially nucleating voids at their interfaces and for a range of hydrogen diffusivities. A value of $D_H = 10^{-12} cm^2/s$ is applicable for aluminum alloys and is also said to be representative of austenitic alloys[41] although, for the latter, D_H of $10^{-10} cm^2/s$ is more generally used. Table III shows that except, perhaps, for the case of ferritic alloys in certain inclusion size ranges Eq. (14) is not an underestimate of the leakage current via lattice diffusion. Even in the ferritic case, I_H/I_t is no more than 10 to 10^2.

Table III. Values of Leakage Ratio, I_H/I_T, for Different Values of Hydrogen Diffusivity and Particle Size*

$D_H (cm^2/s)$ \ $D (\mu m)$	10	50	100
10^{-6}	10^2	21	11
10^{-10}	1	0.21	0.11
10^{-12}	0.1	0.021	0.011

*Value of $t_p = 10^2$ s

In both the above treatments of diffusion leakage, the supersaturated concentration c_s in the matrix at the interface does not recognize the presence of the interface since the H-interface binding energy does not enter into the solutions of the diffusion euqations.

Recently we treated the leakage rate allowing for the interface to have a finite binding energy for hydrogen.[43] The solution attempted again assumed a planar interface and a semi-infinite matrix. It can be shown[43] by solving Fick's second law under the constraints of local equilibrium and mass conservation that the ratio of leakage rate, I_L in this case, and the hydrogen arrival rate, $\dot{C}_B$, is

$$I_L/\dot{C}_B = 1 - \exp(t/\tau)\, erfc[(t/\tau)^{1/2}] \tag{17}$$

where

$$\tau = \alpha^2\delta^2/D_H \tag{18}$$

α and δ mean the same as in Eqs. (8) through (10).

When $t \ll \tau$, then $I_L \ll I_A$, or leakage becomes negligible compared to arrival rate. Table IV shows the calculated values of τ for the hydrogen diffusivity range of interest in ferritic, austenitic and nickel alloys and for E_B from -0.3 to -0.8eV. As can be seen, τ is very, very large. Consequently, at all reasonable test or embrittlement times, leakage can be ignored.

Table IV: Values of Time Constant, τ, in Seconds

D_H (cm^2/s) \ $-E_B(eV)$	0.3	0.4	0.5	0.8
10^{-6}	265	8×10^5	2.4×10^9 (76 years)	6.2×10^{19}
10^{-10}	2.65×10^6	8×10^9	2.4×10^{13}	6.2×10^{23}
10^{-12}	2.65×10^8	8×10^{11}	2.4×10^{15}	6.2×10^{25}

Discussion

Assuming that the dislocation stripping model for matrix enrichment is the operative and controlling sweeping mechanism, then the predictions for embrittlement can be obtained from Eqs. (11) and (12) and are as follows:

1. The preferred initiation sites for voids or cracks in the presence of hydrogen are plastically deformed areas in the alloy.

2. The kinetics of embrittlement is faster than hydrogen lattice diffusion kinetics.

3. Embrittlement increases linearly with strain rate reaching a peak at a calculable critical value.

4. Embrittlement shows a maximum at some intermediate temperature.

5. Larger and flatter particles cause more embrittlement since trapping cross-section is increased.

6. Embrittlement increased by greater degree of coplanar slip due to greater penetration and concentration on a localized scale.

7. With regard to fracture surface characteristics, a dimple size ratio greater than unity is predicted in cases where hydrogen enhanced ductile fracture is due to pressurization induced void growth. When hydrogen predominantly influences void nucleation dimple size ratio is expected to be less than unity. Here dimple size ratio is the ratio of the dimple size in hydrogen to that in the absence of hydrogen.

The most basic hypothesis of the model is the requirement of plastic deformation for hydrogen assisted failure. A particularly convincing demonstration of this is Lee et al.'s[21,22] finding, using notched 1095 and 4340 steels, that void nucleations for the case of a hydrogen precharged material occurs preferentially in the plastically deformed region along characteristic slip lines.

During subcritical crack growth of high strength steels, Chu et al.[48,49] and Nair *et al.*[50] observed that hydrogen induced a slow growth of plastic wings ahead of the crack tip in precracked WOL type specimens. When the plastic wings at the surface collide, a new crack initiates at the point of collision and then travels back to meet the main crack along the plastically deformed regions. The cycle repeats itself until the K_I values decrease to below the threshold K_I. This sequence of events contradicts the hydrogen induced subcritical crack growth model of Gerberich et al.[51] They allow the site of crack initiation to be the region of maximum hydrostatic tensile stresses which is very near the main crack tip in work hardened material and not at the observed collision point of the plastic wings. They also assume that the kinetics is controlled by lattice diffusion from the through-crack tip to the maximum dilation region. This is inconsistent with the observation that the average subcritical crack growth rate is basically the rate of plastic zone growth. Recently Tien *et al.*[33,34] modelled the actual situation and concluded that subcritical crack growth in high strength steels is rate controlled by the kinetics of dislocation sweep-in of hydrogen at the crack tip. Beginning with Eq. (3), and identifying F to be the force on the dislocation due to the shear stress resolved from the maximum local tensile stress in the plastic zone, they derived a subcritical crack growth rate, $\bar{v}_{sc}$, of the form

$$\bar{v}_{sc} = (D_H/\sqrt{3}kT)b^2\sigma_{ys}\exp(-2Cn\ell\sigma_{ys}E/K_I^2)\{1 + \ln[1 + (2\ell\sigma_{ys}E/K_I^2)]\} \qquad (19)$$

where n is the work hardening exponent, K_I is the stress intensity factor, σ_{ys} is yield strength, E is Young's modulus, and ℓ is the distance ahead of the crack tip. Figure 4 shows plots of $\bar{v}_{sc}$ vs. K_I for ℓ values on the order of 5-10 times the crack opening displacement. Although there remains some question as to the applicable value of D_H in a work hardened and tempered martensitic microstructure, Fig. 4 shows good qualitative agreement with experimental observations. Hence, it appears that subcritical crack growth in high strength steels, heretofore believed to be independent of plasticity, is microplastically driven and may be rate controlled by dislocation sweeping.

With respect to kinetic predictions, a dislocation sweeping model is possibly the only realistic alternative transport mode in the often observed cases[11,12,28,48-50] of rapid embrittlement kinetics that cannot be explained in terms of lattice diffusion. Swanson and Marcus,[10] and Zurek and Marcus[11] in a paper in these Proceedings, using O^{18} and deuterium in fatigue crack growth experiments, found penetration of oxygen and deuterium ahead of the crack tip during fatigue in excess of the penetration due to diffusion alone. This increased penetration also explained the enhanced crack growth rates in oxygen or hydrogen.

It may be argued that the above observations result from short circuit diffusion of hydrogen. However Latanision[7] found no difference in hydrogen permeation currents between polycrystalline and amorphous nickel. Also in the same study,[7] it was found that, at any given strain rate, the hydrogen permeation current decreased with decreasing grain size quite opposite to what would be predicted had short circuit diffusion played an accelerating role. Similar conclusions were arrived at by Robertson.[52]

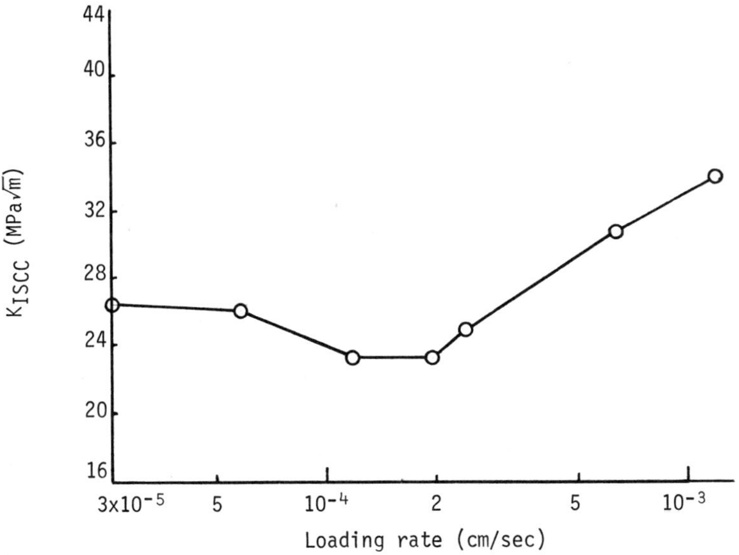

Fig. 5.   Dependence of K_{ISCC} on loading rate for a Ti-6V-6Al-2Sn alloy in 3.5% NaCl (from Ref. 13)

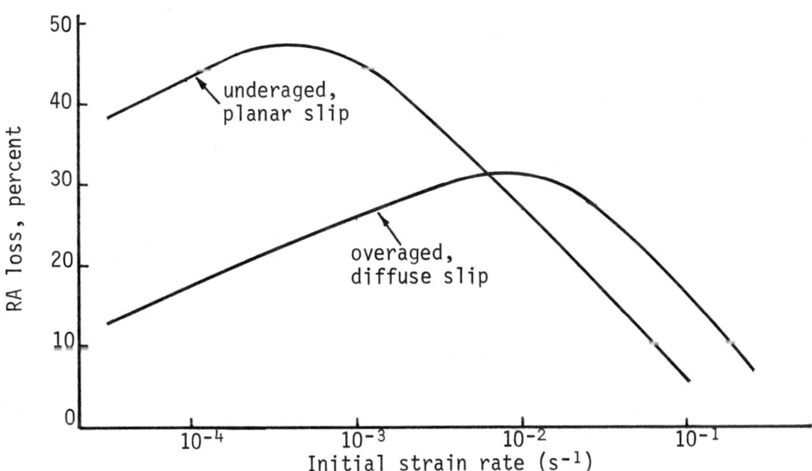

Fig. 6.   Dependence of susceptibility to hydrogen embrittlement on slip planarity in 7075 Aluminum alloy (from Ref. 12).

The sweeping theory predicts a peak embrittlement at an intermediate temperature, following from Eqs. (1) and (3).   At high temperatures C is too low whereas at low temperatures, the mobility term, in Eq. (3), given by D_H/kT is sharply reduced. Such a temperature dependence is widely observed for hydrogen embrittlement.[17-19] It must be noted, however, that such a temperature window behavior may not be unique to dislocation sweeping controlled embrittlement. For example, Ficalora and co-workers[14-16] have shown that chemisorbtion of hydrogen due to surface entry slow steps may also show a similar temperature dependence.

The predicted effect of strain rate follows from Eq. (11). We have already discussed permeation rates and strain rate effects. Embrittlement should increase linearly with strain rate but, as the critical strain rate, $\dot{\varepsilon}_c$, Eq. (5), is approached dislocation sweeping of H would be less effective. Thus, at large enough strain rates, embrittlement would not be observed. As predicted, embrittlement at intermediate strain rates has been observed in 0.5 pct Cr steel, in 304L and an $\alpha+\beta$ Ti alloy[53,54] and very recently in 7075 Aluminum[12] and Ti-6Al-6V-2Sn alloy.[13] In the latter work on the Ti-alloy, Moskovitz and Pelloux[13] determined K_{ISCC} values in a rising load test for the Ti alloy in a 3.5% NaCl solution. As shown in Fig. 5, K_{ISCC} values show a minimum at some intermediate loading rates. These authors converted the minima in terms of strain rates[3-5] and found them coinciding with the critical strain rate for dislocation sweeping.

The existence of a critical strain rate above which embrittlement susceptibility decreases was also observed by Taheri et al.,[12] Fig. 6.   The information in this figure also shows the predicted influence of planar slip in accelerating embrittlement. Similar results on the effect of planar slip were obtained by Albrecht et al.[28] on the same material. That planar slip enhances embrittlement was also reported in many papers in these proceedings, including those of Wilcox and Koss[26] and McMahon.[27]

Finally, as to the predictions with respect to fracture surface dimple sizes, experimental evidence indicates that dimple size can be increased[29,55] or decreased[6] by hydrogen. In a significant work, Garber et al.,[29] using a spherodized 1080 steel, found that the dimple size ratio was greater than unity and greater than that predicted by a purely geometric contribution[57] due to hydrogen. In addition, Garber et al.[29] used the technique of Argon and Im[58-60] to show that the critical particle-matrix interfacial stress for void nucleation was not affected by hydrogen. On the basis of this additional confirmation, Garber et al.[29] concluded that enhanced void growth due to hydrogen pressure in these voids was the cause of the observed reduction in area loss. On the other hand, Lee et al.,[21,22] in a notched 1095 and 4340 steel, metallographically determined the number density of voids at particle interfaces in plastically deformed areas ahead of the notch tip. They observed a higher void number density in the hydrogen precharged steel as compared to the uncharged case and they concluded that hydrogen enhanced void nucleation in these steels.

It appears then, on the basis of the available evidence, that both void nucleation and void pressurization effects need to be given important consideration as failure mechanisms in the hydrogen embrittlement of materials.

Concluding Remarks

We conclude that knowledge of hydrogen sweeping has broadened since the Second International Conference on Effect of Hydrogen on Behavior of Materials. Although there are still very few groups concentrating on the theoretical aspects, these groups are beginning to interact and to reach equilibrium with respect to differing opinions over the assumptions and boundary conditions which are critical to any theoretical development.

The situation is perhaps most settled in terms of mobile dislocations acting as carriers of hydrogen, in ferritic as well as austenitic alloys, and providing a much faster transport mechanism than lattice diffusion. Equations like Eqs. (3) through (6) are experimentally confirmed to be adequate predictors of hydrogen penetration distances and strain rate dependences of hydrogen transport.

The situation with respect to lattice enrichment by dislocation sweep-in, or sweeping, is somewhat less settled. As discussed in this paper, the solute stripping model[3] for enrichment of such system weak-links as grain boundaries and inclusion or precipitate boundaries has been experimentally examined and found to be reasonably adequate. The dislocation annihilation model was first treated and judged non-operable,[42] but later proposed to be a viable mechanism[43] that will occur in conjunction with the stripping mechanism.

The situation is still reasonably unsettled with respect to the pressurization of small voids associated with inclusions and other heterogeneities. The problem lies in the calculation of leakage rates out of the void. The initial calculations[3] show low leakage rates; a recent argument concluded much higher leakage rates[32]; and, as presented in this paper, an even more recent treatment shows very low leakage rates.[43] The resolution of this aspect awaits experimental efforts and input.

Current mechanisms are basically single dislocation mechanisms. The translation from single dislocation kinetic models to more comprehensive models of enrichment requires such unavailable knowledge as the velocity distribution of dislocations in an alloy undergoing work hardening.

In this paper we have reviewed--we hope in an objective manner--the experimental results of hydrogen penetration and hydrogen embrittlement which support, and in many cases have been explained by using, the dislocation sweeping mechanism (the stripping model, in particular) as the only viable explanation for the enhanced transport kinetics, embrittlement kinetics, strain rate effects, slip mode effects and fracture surface appearances.

Certainly, more experiments are necessary to confirm, deny or moderate the various aspects of the proposed mechanisms. The theories, although still maturing, are sufficiently quantitative to dictate single variable experiments.

Acknowledgments

We wish to thank Professors I. Mel Bernstein, Anthony Thompson, Harris Marcus and Otto Buck for providing us with information and for valuable discussions.

References

1. P. Bastien and P. Azou: C.R. Acad. Sci. Paris, 1951, Vol. 232, pp. 1845-48.
2. J.K. Tien, R.J. Richards, O. Buck and H.L. Marcus: Scripta Met., 1975, Vol. 9, pp. 1097-1101.
3. J.K. Tien, A.W. Thompson, I.M. Bernstein and R.J. Richards, Met. Trans., 1976, Vol. 7A, pp. 821-29.
4. J.K. Tien: "Effect of Hydrogen on the Behavior of Materials," eds. A.W. Thompson and I.M. Bernstein, TMS-AIME, NY, 1976, pp. 309-26.
5. J.K. Tien, N.F. Panayotou and R.J. Richards: Proceedings of the Second International Congress on Hydrogen in Metals, Paris, France, 1977, pp. 2B11.
6. M. Kurkela and R.M. Latanision: Scripta Met., 1979, Vol. 13, pp. 927-32.
7. R.M. Latanision, M. Kurkela and F. Lee, this Conference Proceedings.
8. J.A. Donovan: Met. Trans., 1976, Vol. 7A, pp. 1677-83.
9. J.A. Donovan: Met. Trans., 1976, Vol. 7A, pp. 145.
10. J.W. Swanson and H.L. Marcus, Met. Trans., 1978, Vol. 9A, pp. 291-93.
11. A.D. Zurek and H.L. Marcus, this Conference Proceedings.
12. M. Taheri, J. Albrecht, I.M. Bernstein and A.W. Thompson: Scripta Met., 1979, Vol. 13, pp. 871-75.
13. J.A. Moskovitz and R.M. Pelloux: Corrosion, 1979, Vol. 35, pp. 509-14.
14. K. Sieradzki and P.J. Ficalora: J. of Mat. Sc., 1979, Vol. 14, pp. 2703-08.
15. L. Pecora and P.J. Ficalora: Met. Trans., 1980, Vol. 11A, pp. 1841-48.
16. C.M. Ransom and P.J. Ficalora: Met. Trans., 1980, Vol. 11A, pp. 801-07.
17. D.P. Williams and H.G. Nelson: Met. Trans., 1972, Vol. 3, pp. 2107-13.
18. H.G. Nelson and D.P. Williams: "Stress Corrosion Cracking and Hydrogen Embrittlement of Iron Base Alloys," eds. R.W. Staehle, J. Hochmann, R.D. McCright and J.E. Slater, NACE-5, NACE, Houston, 1977, pp. 390-404.
19. R.P. Gangloff and R.P. Wei: Met. Trans., 1977, Vol. 8A, pp. 1043.
20. G.R. Caskey, Jr.: Scripta Met., 1977, Vol. 11, pp. 1077-83.
21. T.D. Lee, T. Goldenberg and J.P. Hirth: Met. Trans., 1979, Vol. 10A, pp. 199-208.
22. T.D. Lee, T. Goldenberg and J.P. Hirth: Met. Trans., 1979, Vol. 10A, pp. 439-47.
23. I.E. French, P.F. Weinrich and C.W. Weaver: Scripta Met., 1979, Vol. 13, pp. 285-88.
24. J.E. Smugeresky: Met. Trans., 1977, Vol. 8A, pp. 1283-89.
25. G.M. Pressouyre: Ph.D. Thesis, Carnegie-Mellon University, 1977. Also issued on ONR Technical Report NR036-099-7.
26. J.R. Wilcox and D.A. Koss: this Conference Proceedings.
27. C.J. McMahon: this Conference Proceedings.
28. J. Albrecht, A.W. Thompson and I.M. Bernstein: Met. Trans., 1979, Vol. 10A, pp. 1759-66.
29. R. Garber, I.M. Bernstein and A.W. Thompson: Scripta Met., 1976, Vol. 10, pp. 341-45.
30. A.J. West and M.R. Louthan, Jr.: Met. Trans., 1979, Vol. 10A, pp. 1675-82.
31. N. Sridhar, J.A. Kargol and N.F. Fiore: Unpublished report.
32. J.P. Hirth: 1980 Institute of Metals Lecture, Met. Trans., 1980, Vol. 11A, pp. 861-90.
33. J.K. Tien, O. Buck, R.C. Bates and S. Nair: Scripta Met., 1980, Vol. 14, pp. 583-90.
34. J.K. Tien, S. Nair, R.C. Bates and O. Buck: Scripta Met., 1980, Vol. 14, pp. 591-94.

35. A.H. Cottrell: "Report of a Conference on the Strength of Solids,"
    30; 1948, London Physical Society.
36. R. Gibala: Ref. 18, pp. 244-68.
37. H.H. Johnson: this Conference Proceedings.
38. G.M. Pressouyre and I.M. Bernstein: Met. Trans., 1978, Vol. 9A,
    pp. 1571-80.
39. G.M. Pressouyre and I.M. Bernstein: Acta Met., 1979, Vol. 27,
    pp. 89.
40. G.M. Pressouyre: Met. Trans., 1979, Vol. 10A, pp. 1571.
41. J. Volkl and G. Alefeld: "Diffusion in Solids--Recent Developments,"
    A.S. Nowick and J.J. Burton, eds., Academic Press, NY, 1975, pp. 272-
    302.
42. H.H. Johnson and J.P. Hirth: Met. Trans., 1976, Vol. 7A, pp. 1543-48.
43. S.V. Nair, R.R. Jensen and J.K. Tien: to be submitted to Met. Trans.
44. G.M. Pressouyre: Acta Met., 1980, Vol. 28, pp. 895-911.
45. G.M. Pressouyre, J.P. Fidelle and R.A. Laurent: this Conference
    Proceedings.
46. A.W. Thompson and I.M. Bernstein: "Advances in Corrosion Science and
    Technology," Vol. 7, M.G. Fontana and R.W. Staehle, eds., Plenum Press,
    NY, 1980, pp. 53-175.
47. W. Jost: "Diffusion," Academic Press, NY, 1955, p. 12.
48. W-Y. Chu, C-M, Hsiao and S-Q Li: Scripta Met., 1979, Vol. 13,
    pp. 1063.
49. C-M. Hsiao and W-Y. Chu: this Conference Proceedings.
50. S.V. Nair and J.K. Tien: Unpublished research.
51. W.W. Gerberich, Y.T. Chen and C. St. John: Met. Trans., 1975, Vol. 6,
    pp. 1485.
52. W.D. Robertson: Z. Metallkunde, 1973, Vol. 64, pp. 436.
53. T. Toh and W.M. Baldwin: "Stress Corrosion Cracking and Embrittlement,"
    W.D. Robertson, ed., Wiley, NY, 1956, pp. 176-86.
54. I.M. Bernstein: Mat. Sc. Eng., 1970, Vol. 6, pp. 1-19.
55. A.W. Thompson: Carnegie-Mellon University, Pittsburgh, Unpublished
    research, 1975.
56. A.W. Thompson and J.A. Brocks: Met. Trans., 1975, Vol. 6A, pp. 1431-
    42.
57. A.W. Thompson: Met. Trans., 1979, Vol. 10A, pp. 727-31.
58. A.S. Argon, J. Im and A. Needleman: Met. Trans., 1975, Vol. 6A,
    pp. 815-24.
59. A.S. Argon, J. Im and R. Safoglu: Met. Trans., 1975, Vol. 6A,
    pp. 825-37.
60. A.S. Argon and J. Im: Met. Trans., 1975, Vol. 6A, pp. 839-51.
61. C.S. Carter: Eng. Fract. Mech., 1971, Vol. 3, pp. 1.
62. H.L. Dunegan and A.S. Tetelman: Eng. Fract. Mech., 1971, Vol. 2,
    pp. 387.

DISCUSSION

J. P. Hirth, Ohio State University, Columbus: Aside from the geometric factor g referred to in the paper by Tien et al[1], there has been no question about the correctness of the theory by Tien et al[2]. Yet Johnson and myself[3,4] cite very small supersaturations of hydrogen in voids, while Tien and coworkers[1,2] cite large supersaturations. The difference lies in the equilibrium boundary condition for the leakage rate from the voids. Other differences, such as the greater or lower leakage rate from disc-shaped voids[1] compared to spherical voids[4] are reasonable consequences of the different geometries. Thus, we consider only the leakage boundary condition.

The key to this condition is whether the binding energy must be attained in the first jump of a hydrogen atom away from the trap or whether it is achieved over a number, say ten, of such distances, both situations illustrated, for example, in the figures in the work of Pressouyre[5]. In the first case[2], the jump rate out of the trap will be small and it is likely that "interface control" will intrude into the leakage rate, i.e. the local equilibrium boundary condition will not be attained. In the second case, the jump rates out of the trap will be larger (as will the reverse jumps locally) and it is more likely that a local equilibrium boundary condition will be maintained as assumed by myself[4]. The difference in these assumed boundary conditions gives rise to the factor of 10^8 difference between estimated supersaturations from hydrogen in iron, in references 1 and 2 on the one hand and 3 and 4 on the other. Thus, again, there is no question that dislocations can carry hydrogen, particularly in their <u>core</u> atmospheres, and that they can "dump" the hydrogen when they annihilate or when they encounter a trap with larger binding energy. The issue is what supersaturation can be attained.

The binding energy must be long range for dislocations on coherent precipitates because of the elastic interaction of these defects with the internal dilatation field of the hydrogen (~ 0.2 atomic volumes in iron). Whether the binding energy is short or long range for the void could only be decided by atomic calculations and these are not yet reliable enough to decide the issue. Thus, the above estimates of the leakage rate represent upper[4] or lower[2] bounds. I chose the upper bound condition[4] on the basis that a void surface where permeation experiments suggest some surface impediment (interface control) but not a large one. It is true that internal voids are likely to nucleate on defects, particularly internal interfaces, and the void surfaces can inherit impurities segregated on these interfaces. In the case that these impurities are Group V-VI metalloids as occur in temper-embrittled steels, they would act as catalysts and enhance hydrogen entry from the void.

I hope that these remarks put any differences in the theories[1-4] in context. More accurate theories would have to consider whether hydrogen was "dumped" into a void in atomic or molecular form and consider competition between entry of atomic hydrogen with recombination to form molecules. Also, binding entropies and local enthalpies and entropies of activation for diffusion in the vicinity of a trap would be important to consider. Finally, all of the diffusion in the vicinity of trap would be important to consider. Finally, all of the diffusion models for dislocation transport of hydrogen[1-4] are somewhat simplified in that they consider noninteracting moving sources, so more elaborate models are possible. Thus, there is evidently room for further work on the problem.

1.   J. K. Tien, S. V. Nair and R. R. Jensen, preceding article.
2.   J. K. Tien, A. W. Thompson, I. M. Bernstein and R. Richards, Met. Trans., <u>7A</u>, pp 821-829 (1976).
3.   H. H. Johnson and J. P. Hirth, Met. Trans.,<u>7A</u>, pp 1543-1548 (1976).

4.   J. P. Hirth, Met. Trans., 11A, pp 861-890 (1980).
5.   G. M. Pressouyre, Met. Trans., 10A, pp 1571-1573 (1979).

J. K. Tien: Dr. Hirth's discussion and comments further provide us with confidence that dislocation sweeping can indeed result in matrix enrichment, especially at the traps which are often the weak links in an alloy. The origin of the geometric factor has been discussed in the text as also in Dr. Hirth's Mehl Lecture in Met. Trans. 1980.

It is not at all clear to us how Dr. Hirth can realistically obtain the factor of 10^8 mentioned in Dr. Hirth's comments above and in the Mehl Lecture. As mentioned in the text and as shown in Table III, our calculations show that this factor for realistic cases is no larger than about 10^2.

We disagree with Dr. Hirth that this factor, whatever its exact magnitude, arises from any consideration of local equilibrium at the trap boundary. This factor is a consequence only of the different geometries, namely, the difference in leakage currents between large flat particles and small spherical particles. The local equilibrium boundary condition at the trap, including the long- or short-range nature of binding energy, does not appear explicitly in the solutions of the diffusion equations either in the dislocation annihilation model of Johnson and Hirth, published in 1976, or in the diffusion leakage calculations recently performed by Dr. Hirth in the Mehl Lecture paper for spherical particles, or by Tien et al. for large flat particles in their 1976 Met. Trans. paper.

We agree with Dr. Hirth that the internal void surface can be very different from that of an external surface. The surface can contain not only Group V-VI impurities as suggested by Dr. Hirth, but also impurities such as oxygen, very commonly found in typical inclusion surfaces which can serve to preserve hydrogen gas in the voids.

Certainly the next step in theory advancement is the development of a multi-dislocation model for hydrogen sweeping involving interacting moving sources. We enthusiastically look forward to such developments from Dr. Hirth.

M. R. Louthan, Jr., Virginia Polytechnic Institute and State University, Blacksburg: You propose that dislocation transport carries hydrogen large distances into many metal lattices; yet, Donovan's work has shown that plastic deformation decreases the effective diffusivity and depth of penetration of hydrogen on nickel; Johnson showed that plastic deformation during charging decreased the effective diffusivity of hydrogen on iron. These results contrast the tritium release data and the tritium uptake data for austenitic steels. Does your model contain parameters which provide for such diverse experimental results?

J. K. Tien: The Donovan work that Dr. Louthan refers to is discussed in the text and we concluded that his results can be the consequence of the samples being tritium charged prior to deformation. Indeed, if anything, his results support dislocation sweeping of hydrogen. No, our model does not contain parameters that distinguish between an initially hydrogen-free specimen and hydrogen-containing specimen. Some work has been initiated in this direction recently by Pressouyre, which is reported in the July 1980 issue of Acta. Met.

G. M. Pressouyre, Creusot-Loire Research Center, France: Leakage should not be important for at least three reasons:
1) Most carbide particles have high trapping energies; they are then irrevers-

<u>ible</u>, hence no leakage.
2) If a <u>reversible</u> trap is met by a dislocation, some equilibrium may be
assumed to be reached at the time of the encounter; now, lattice sites may
be considered as an extreme case of weak reversible traps. Hence, at the
time of the encounter, one has a triple-body equilibrium: H on $\perp$ $\rightleftharpoons$ H on r trap
$\rightleftharpoons$ H in lattice. Since equilibrium was reached, there is really not much in-
centive (driving force) for H to leave the r trap for the lattice.
3) Leakage occurs by <u>diffusion</u>. It thus involves low H quantities that move
slowly; on the contrary, hydrogen is dumped in high quantities at a high rate
(high encounter frequency). Thus, leakage could be neglected in external H
embrittlement

Simple evidence for H transport by dislocations and concurrent related embrittlement may be found every day. Example: the disk pressure results
presented in my talk show that in steel, rupture in helium occurs in less
than 2 minutes, while in hydrogen, the rupture pressure is much lower and
occurs in a few seconds (i.e. there is dislocation transport).

J. K. Tien: Dr. Pressouyre provides additional evidence in support of dislocation sweeping playing an important role in hydrogen embrittlement. With
regard to his comments on leakage, we would like to add that the effect of
trap type and hydrogen diffusivity on leakage rates are all embodied in the
time constant τ, in our Eq. (18) in the text, the values of which are shown
in Table IV, in which we show diffusion leakage is not significant.

STUDIES OF THE DEUTERIUM TRANSPORT MECHANISM

DURING FATIGUE OF ALUMINUM ALLOYS

A.K. Zurek and H.L. Marcus
Mechanical Engineering
and
Materials Science and Engineering
The University of Texas at Austin
Austin, Texas 78712

Diffusion and transport mechanisms of deuterium and oxygen during
fatigue crack growth of 2000 and 7000 series aluminum alloys have been
studied with SIMS and AES techniques for samples fatigued in vacuum, D_2O,
H_2O^{18}, O_2^{18}, and D_2. The dependence of the presence of different trapping
sites on the deuterium diffusion was studied using a variety of deuterium-
ion implanted surfaces. The deuterium profiles of D_2O vapor fatigue fractured specimens were closely duplicated in the deuterium-ion implantation
experiment. It is suggested that surface oxide serves as a strong trap and
a barrier for the outdiffusing deuterium. An image model is used to simulate the observed deuterium time and trapping dependent distribution.

Sputtering rates of the 2219-T851 and 7075-T6 aluminum alloys were
determined. The sputtering rate for the metal matrix is about 5 to 6
times greater than for the surface aluminum oxide.

I.   Introduction

The environmental influence on the mechanical properties of structural materials has posed a problem of extraordinary importance for material science. That hydrogen can severely embrittle metals is well known (1-11) and one of the least understood aspects of this embrittlement is the mechanism by which it happens. The transport of hydrogen is involved in the embrittlement process. A quantitative analysis of hydrogen transport was proposed by A. McNabb and P.K. Foster (12) followed by others (13,14,15). Their theory is based on a physical model which supposes that hydrogen atoms wander in a random manner through the crystal lattice, but tend to get trapped or delayed at certain fixed sites distributed throughout the metal. Quantitative analysis of trapping the hydrogen at different sites for specific alloys was discussed in (15,20,21). Tien and co-authors (16-19) discuss the dislocation sweeping-in model, suggesting that large excess of hydrogen in front of the propagating crack can be made available by mobile dislocation transport during cyclic loading. Unfortunately, not much data is available for hydrogen interactions in aluminum alloys. As in other alloy systems, aluminum alloys are apparently degraded by hydrogen (7,8,10, 11,21).

In this paper we report data obtained from the fatigue crack propagation tests of aluminum alloys fatigued in vacuum and gaseous environments. In addition, transport and trapping experiments of deuterium ion-implanted aluminum alloys are used to model data obtained in the fatigue crack growth experiments. A simplified model describing the redistribution and trapping for the ion-implanted deuterium is discussed.

II.   Experimental Procedure

The chemical composition of 2219-T851 and 7075-T6 age hardenable aluminum alloys used in this study and conditions of fatigue crack growth experiments were described in detail elsewhere (22,23). Fatigue crack growth tests were performed in a variety of environments, such as deuterium gas (D_2), oxygen (18) gas, H_2O^{18}, and D_2O water vapor, and compared to the fatigue crack growth in the 1 µPa vacuum. All fatigue fractured experiments can be divided into three categories:

1. Samples in which the fatigue crack propagation was first in vacuum followed by crack growth in D_2 and the last segment fatigued in D_2O vapor.

2. Samples in which fatigue crack growth occurred first in vacuum followed by further crack propagation in O^{18} gas or D_2.

3. Samples fatigue fractured in vacuum and then followed by H_2O^{18}.

SIMS and AES profiles of deuterium and oxygen penetration into the fractured surfaces were taken after fatigue tests.

Ion-implanted experiments were performed on specimens in the as-received condition, polished plate, 7-10% cold rolled plate, ion radiation damaged plate, and the fracture surfaces of samples fatigued in 50% relative humidity (H_2O) laboratory air. The radiation damaged samples were subjected to the neon irradiation with a beam 130 KeV and a dose 8 x 10^{15} ions/cm^2 at a projected range 0.23 µm. Deuterium implantation was done with a beam of 25 KeV, and a dose 6 x 10^{15} ions/cm^2 at a projected range of 0.24 µm. Time dependent diffusion profiles of deuterium (D^-) and oxygen were obtained using SIMS and Auger Electron Spectroscopy, respectively.

The inert ion sputtering rates for 2219-T851 and 7075-T6 aluminum alloys were determined by argon-gas-ion sputtering of metallographically polished samples. The measurement of the crater depth of the sputtered area was done using a Sloan profilometer. The pressure of argon in the ion gun was fixed at a 1 x 10^{-2} Pa, with the target chamber pressure 1.6 x 10^{-5} Pa and a rastered ion current density of 140 µA/cm^2. The accelerating voltage of 3 KeV was used.

III. Results and Discussion

The observation of the fatigue crack growth of 2219-T851 and 7075-T6 aluminum alloys showed the ratio of the crack growth rate in water vapor (D_2O) to that in vacuum varied from 1.8 to 3.6, depending on the ΔK value. The effect was more pronounced for intermediate ΔK levels (22). No change in growth rate in D_2 gas was noticed. A distinct difference in the depth of penetration of deuterium and oxygen was noted for the case of the surface first fatigued in a vacuum of 1 µPa or D_2 gas and then exposed to the D_2O vapor, when compared to fatiguing in the environment of the D_2O vapor itself. In all cases a higher concentration of deuterium that more deeply penetrated into the fracture surface and had a longer lasting concentration was observed for the segment exposed to D_2O during the fatigue cycle. No major effect of D_2 gas itself was observed [Fig. 1].

Figure 2 represents combined data of deuterium and oxygen concentrations obtained using SIMS and AES, respectively, for the specimens fatigue fractured in D_2O water vapor and compared to the oxygen concentration profile for vacuum fatigue samples and subsequently exposed to D_2O vapor. Significantly thicker oxide, formed on the D_2O fatigue surfaces with the deuterium penetration extending much beyond the oxide layer was observed.

An insignificant increase in the fatigue crack propagation rate was observed for the samples fatigued in $O_2{}^{18}$ gas in comparison to the vacuum fatigue crack growth tests (23). However, the oxide layer formed on the $O_2{}^{18}$ fatigue fracture surfaces was twice the depth of the oxide formed for the vacuum fatigue fracture surfaces [Fig. 3]. Even thicker oxide formation was observed for the H_2O^{18} fatigue fracture surface segments.

Striking similarities in the concentrations and the depth of presence of deuterium were observed (24) for the deuterium ion-implanted specimens [Fig. 4]. The high energy of the implanted deuterium ions compared to the binding energies of trapping sites (15,20,21,25) allowed for the deep internal charging of the samples. Subsequent redistribution by internal diffusion and the trapping of the deuterium at the trapping sites discussed below are presumably responsible for the observed deuterium profiles. Fatigue fracture surfaces are expected to have the highest density of microvoids, microcracks, and dislocations, with an increasing density of these traps toward the surface. The accumulation of these traps close to the surface is expected because of the large strain fields formed in the vicinity of the propagating crack due to the crack opening displacement and reverse yielding. This zone should extend several microns from the surface, depending on the value of ΔK. The largest, deepest, and longest lasting concentration of ion-implanted deuterium (up to the 14th day after implantation) is observed in the fatigue fracture surfaces [Fig. 4]. In the case of the cold rolled specimens, near surface accumulation of microcracks, damaged boundaries, and high dislocation density are known to be sinks for hydrogen. The as-received unrolled specimens with fewer but uniformly distributed trapping sites, such as second phase particles, dislocations, and very reversible diffusion lattice sites, fall in the third category of

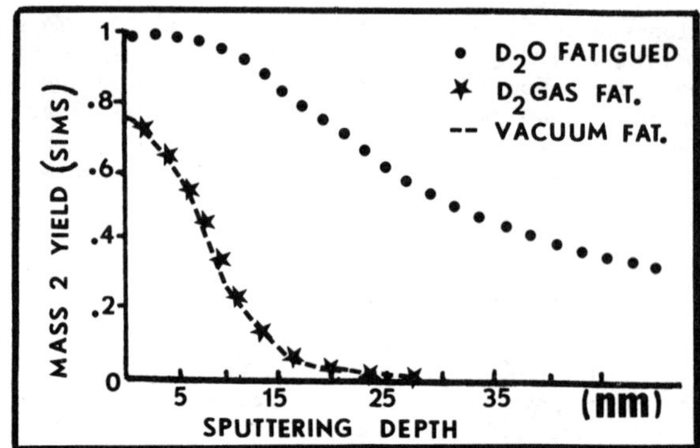

Fig. 1   Deuterium concentration profiles (negative SIMS) for the
2219-T851 aluminum alloy, fatigue fractured surfaces.

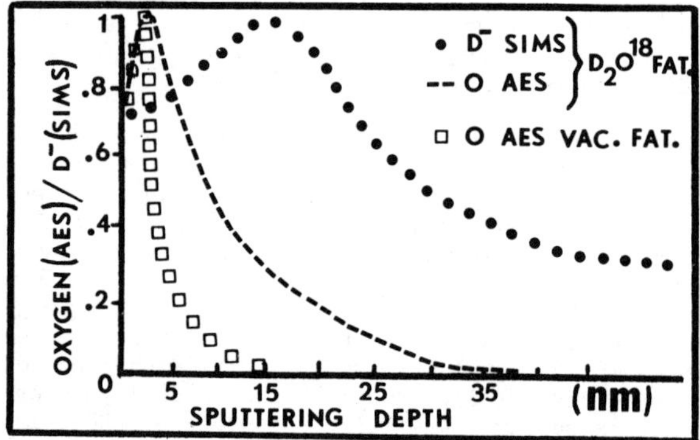

Fig. 2   Oxygen yield obtained by AES and deuterium yield obtained
for SIMS from the 7075-T6 aluminum alloy, fatigue fractured
surfaces.

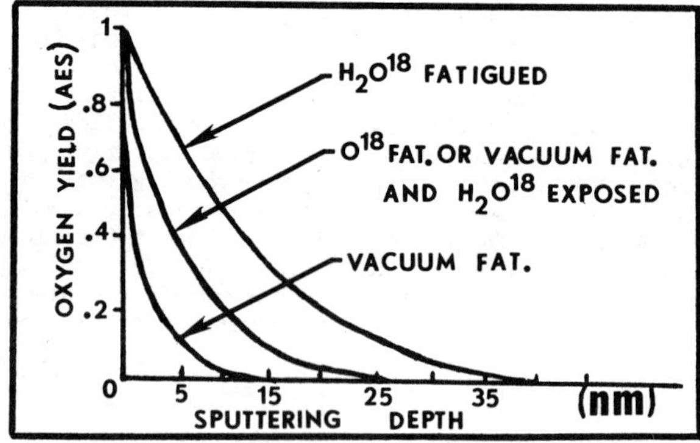

Fig. 3   Oxygen concentration profiles obtained by AES for the
2219-T851, fatigue fractured surfaces.

the surfaces capable of trapping implanted deuterium. The unrolled specimens and the neon radiation predamaged specimens seem to behave identically
in terms of the observed near surface concentration profile for deuterium.
The light mechanical polishing of the unrolled specimens with and without
neon predamage would be expected to introduce limited amounts of near
surface trapping sites.

To explain the results shown in Figure 4 we have considered a simple
diffusion model, based on the approximation that the oxide layer formed at
the surface of the ion-implanted sample acts as a reflecting (nonpermeable)
barrier for the outdiffusing deuterium (24). In order to solve that
problem, we have used the "method of images" (26) discussed below. This
is shown schematically in Figure 5.

The initial distribution of deuterium at a time of implantation can be
represented by:

$$(x, t = 0) - \lim_{t \to 0} \frac{1}{\sqrt{\pi Dt}} \exp \left(\frac{-(x - x_o)^2}{4Dt} \right) = \delta(x - x_o) \tag{1}$$

We seek a solution to the Fick's equation with the initial condition (1) and
a reflecting barrier at x = 0

$$\frac{\partial \rho(x,t)}{\partial x} \Big|_{x=0} = 0 \tag{2}$$

In order to simplify the difficult boundary problem, the "mirror image" of
the initial distribution is introduced. This yields as a solution the sum
of the Gaussians:

$$\rho(x,t) = \frac{1}{\sqrt{\pi Dt}} \left\{ \exp \left(-\frac{(x - x_o)^2}{4Dt} \right) + \exp \left(-\frac{(x + x_o)^2}{4Dt} \right) \right\} \tag{3}$$

which satisfies boundary condition (2).

In the region x > 0, $\rho(x,t)$ evolves with time in a manner shown in
Figure 6. Solutions for times t_3, t_4 and later are in good qualitative
agreement with the experimental observations shown in Figure 4. However,
a more realistic model would be to permit the oxide layer to be a partially
permeable barrier for the outdiffusing deuterium. Mathematical representation of such boundary condition using the same "method of images" as well
as the influence of trapping sites within the matrix will be considered in
the near future.

In order to calibrate the sputtering rate of aluminum alloys used in
our experiments, we have performed argon-ion sputtering of polished samples
over a range of times. For the conditions described in the experimental
procedure section, the measured sputtering rates increased with time due to
the difference in sputtering rates of the oxide and the metal. Calculations
based on this data result in a sputtering rate for the metal matrix of
approximately 11 nm/min and of the oxide layer of approximately 2 nm/min.
Based on these sputtering rates, the oxide thicknesses are about 16 nm for
the cases of fatigue cracks grown in the D_2O vapor and about 8 nm for the
case of fatigue crack growth in vacuum and D_2 dry-gas subsequently exposed
to D_2O.

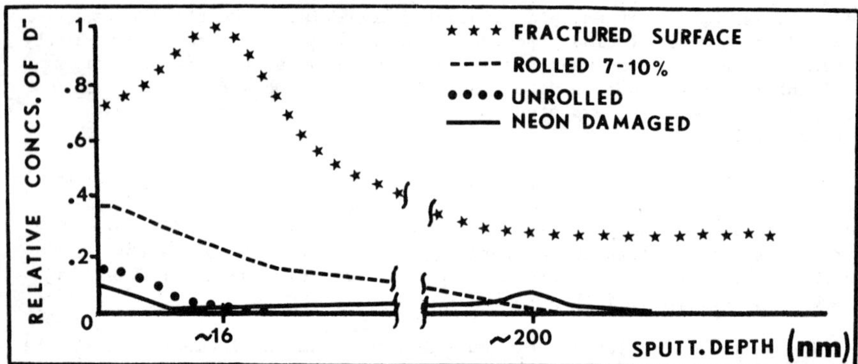

Fig. 4   Deuterium concentration profiles obtained by SIMS for 7075-T6 aluminum alloy, one day after the deuterium implantation.

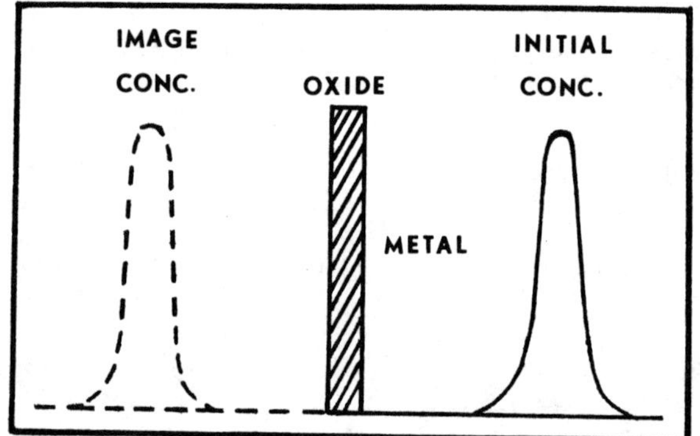

Fig. 5   Schematic representation of an "image model."

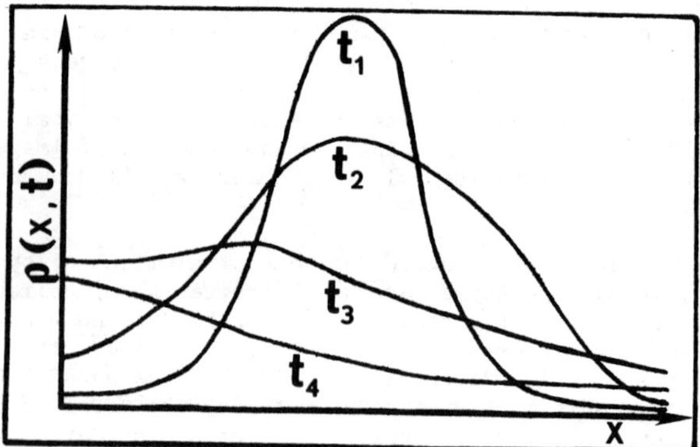

Fig. 6   Concentration profiles evolving in time according to equation 4 (24).

The results of these experiments combined with those reported previously (22-24) for the 2219 and 7075 aluminum alloys lead to the following observations. Fatigue crack growth rate in O_2^{18} and D_2 gases are the same as in the baseline vacuum. Fatigue crack growth rate is significantly higher in water vapor. Oxygen is transported deeper into the fracture surface when the sample is fatigued in the O_2^{18} or water vapor than if only exposed to the O_2^{18} or water vapor following fatigue. No indication of deuterium in the fracture surface is found for samples fatigued in deuterium gas or fatigued in vacuum and then exposed to deuterium gas. When the fatigue crack growth takes place in D_2O, deuterium is found at great depth into the fracture surface. Sputtering deeper than a µm still shows the presence of deuterium.

From the above the following model can be suggested. The D_2 gas does not dissociate significantly on the aluminum matrix. Therefore, no deuterium is available to be transported during fatigue. When the D_2O gas is present during fatigue, it dissociates into D and OD on the fresh metal surface, making available deuterium atoms at the crack tip. These atoms are then transported deeply into the aluminum, most likely by a dislocation transport mechanism. The above, however, does not prove that the enhanced crack growth is strictly a hydrogen embrittlement phenomenon. In an attempt to establish this, a series of experiments is planned where ionized atomic deuterium will be introduced at the advancing crack front.

Acknowledgements

This research was sponsored by AFOSR Grant Number 76-2955. Assistance of Dr. N. Hertel and J. Cecil in the specimen preparation is greatly appreciated. Dr. W.H. Zurek, Professor J.P. Stark, and Dr. D. Bourell provided helpful criticism of the transport and trapping model.

References

1. A.R. Troiano, _Trans. of the ASM_, _52_ (1960).

2. R.A. Oriani, _Berichte der Bunsen-Gesellschaft fur Physikalische Chemie_, _76_ (1972).

3. I.M. Bernstein, R. Garber, and G.M. Pressouyre, _Effect of Hydrogen on Behavior of Metals_, p. 37, TMS-AIME, Pittsburgh, PA, 1975.

4. J.D. Frandsen and H.L. Marcus, _Scripta Met._, _9_ (1975) p. 1089.

5. A.S. Tetelman and W. Robertson, _Trans. of the Met. Society of AIME_, _24_ (1962).

6. C.D. Beachem, _Met. Trans._, _3_ (1972).

7. M.O. Speidel, _Hydrogen in Metals_, ASM, 1974.

8. E.F. Smith, R. Jacko, and D.J. Duguette, _Effect of Hydrogen on Behavior of Metals_, p. 218, TMS-AIME, Pittsburgh, PA, 1975.

9. J.D. Frandsen and H.L. Marcus, _Effect of Hydrogen on Behavior of Metals_, p. 233, TMS-AIME, Pittsburgh, PA, 1975.

10. R. Alani and P.R. Swann, _Br. Corrosion Journal_, _12_ (2) (1977) p. 80.

11.   H.L. Marcus, "Environmental Degradation of Engineering Materials,"
      paper presented at Environmental Degradation of Engineering Materials
      Conference, Virginia Polytechnic Institute and State University,
      Blacksburg, VA, 1977.

12.   A. McNabb and P.K. Foster, Trans. TMS-AIME, 227 (1963) p. 618.

13.   G.R. Caskey, Jr. and W.L. Pillinger, Met. Trans. A, 6A (1975) p. 467.

14.   R.C. Frank, C.W. Wert, and H.K. Birnbaum, Met. Trans. A, 10A (1979)
      p. 1627.

15.   C.A. Wert, Hydrogen in Metals II, p. 305, Springer-Verlag, New York,
      NY, 1978.

16.   J.K. Tien, R.J. Richards, O. Buck, and H.L. Marcus, Scripta Met., 9
      (1975) p. 1097.

17.   J.K. Tien, Effect of Hydrogen on Behavior of Metals, p. 309,
      TMS-AIME, Pittsburgh, PA, 1975.

18.   J.K. Tien, O. Buck, R.C. Bates, and S. Nair, Scripta Met., 14 (1980)
      p. 583.

19.   J.K. Tien, S. Nair, R.C. Bates, and O. Buck, Scripta Met., 14 (1980)
      p. 591.

20.   G.M. Pressouyre and I.M. Bernstein, Acta Metallurgica, 27 (1979)
      p. 89.

21.   G.M. Pressouyre and I.M. Bernstein, Met. Trans. A, 9A (1978) p. 1571.

22.   A.K. Zurek and H.L. Marcus, Chemical Physics: Secondary Ion Mass
      Spectrometry SIMS II, Vol. 9, p. 163, Springer-Verlag, New York,
      NY, 1979.

23.   J.W. Swanson and H.L. Marcus, Met. Trans. A, 9A (1977) p. 291.

24.   A.K. Zurek, H.L. Marcus, T.N. Cecil, and R. Powers, Met. Trans. A,
      in press (1980).

25.   A.T. West and MacIntyre, R. Louthan, Jr., Met. Trans. A, 10A (1979)
      p. 1723.

26.   P.M. Morse and H. Feshback, Methods of Theoretical Physics, 1st ed.,
      p. 860, McGraw-Hill, New York, NY, 1953.

DISCUSSION

R. A. Oriani, University of Minnesota, Minneapolis: One need not be surprised that H_2O vapor can manifest embrittlement of the aluminum alloys, whereas a low pressure of hydrogen gas does not, because the input fugacity of the hydrogen generated from the H_2O+Al reaction can be extremely large, much larger than that of the hydrogen gas employed.

A. K. Zurek: We fully agree with Dr. Oriani's comment. In our experiment, however, we did not try to compare the influence of equal atomic hydrogen activities from H_2O and H_2 on the penetration of free hydrogen into the aluminum and related embrittlement.

We have focused on the fact that the penetration and the deuterium concentration present in the specimens fatigued in D_2O water vapor was significantly higher than for segments just exposed to D_2O after fatigue in vacuum.

In order to be able to compare equal activities experiments, this required either extremely high pressure of hydrogen gas or some other source of surface available atomic hydrogen.

At present we are in the preparation stage of an experiment where ionized hydrogen will be focused in the vicinity of the advancing crack front. This should overcome the dissociation problem of the pure gas on the Al metal, significantly increasing the atomic hydrogen activity.

C. Altstetter, University of Illinois, Urbana: Your table shows that in D_2O vapor deuterium was detected very far from the fatigue cracked surface. How does this distance compare with the crack advance per cycle or the plastic zone size?

A. K. Zurek: We sputtered normal to the fracture surface for approximately 1μm. The crack growth per cycle under our loading conditions was approximately 0.8μm. The plane strain forward plastic zone radius is approximately 150μm. The main point of our result is that the deuterium is present deeper than 1μm into the fracture surface when the sample is fatigued in D_2O but the deuterium is localized at the oxide interface when the sample is fatigued in vacuum and subsequently exposed to D_2O. It should be remembered that the measured deuterium profiles are after diffusion has taken place for six to twelve hours due to the transfer of the specimen from the fracture chamber to the AES/SIMS system.

HYBRID CALCULATIONS OF THE PROPERTIES OF HYDROGEN IN METALS

M. I. Baskes, C. F. Melius, and W. D. Wilson
Sandia National Laboratories
Livermore, California 94550

A number of recent fundamental calculations which have been corroborated by experiment may provide a framework from which to view hydrogen embrittlement models.

The calculations are based upon a hybrid approach which allows electronic rearrangement in the vicinity of a hydrogen defect, as well as lattice atom relaxation. The far-region from the defect can be treated using a self-consistent finite-element technique or analytic continuum theory. The calculations have been applied to a number of point defects such as vacancies and interstitials, as well as to extended defects such as dislocations. The hydrogen solubility and diffusivity in the vicinity of a crack tip have been studied. This hybrid method has been applied to both f.c.c. and b.c.c metals.

The results are in good agreement with recent experiments: The activation energy for hydrogen migration in Ni is calculated to be 0.45 eV as compared with the 0.41 eV experimental value. Channeling measurements indicate a H location off center in a vacancy for both f.c.c. and b.c.c. materials (by 0.104 nm along a $\langle 100 \rangle$ direction in Fe) to be compared with the calculated values of 0.090 nm and 0.096 nm along the $\langle 100 \rangle$ direction in f.c.c. and b.c.c. lattices, respectively. Hydrogen binding to a vacancy in an f.c.c. lattice is calculated to be small ($\sim .05$ eV) consistent with desorption experiments of implanted stainless steel (~ 0.3 eV).
Deuterium desorption measurements of charged, cold-worked nickel indicate a small (0.09 ev) binding of H to a dislocation compared to our calculated value of ~ 0.05 eV. Our calculated values for binding of hydrogen to a dislocation in a b.c.c. lattice (0.26 eV) agree with the experimental value of 0.26 eV for Fe. Calculations of the hydrogen diffusivity and solubility in the vicinity of a crack tip in an f.c.c. metal show a surprisingly small (less than a factor of 2) effect on these properties due to the presence of such stresses. Autoradiography shows no enhancement in the tritium concentration at grain boundaries in Ni even in the presence of an active impurity such as sulfur. These are manifestations of the very small lattice expansion due to a hydrogen interstitial atom in the f.c.c. nickel lattice. This expansion is calculated to be 0.035 atomic volumes and has been quantitatively confirmed by laser interferometry. These results illustrate the difference between f.c.c. and b.c.c. lattices with respect to hydrogen embrittlement models.

General Method

The mechanisms of degradation of materials by hydrogen may depend upon a large number of parameters; e.g., heat treatment, mechanical working, impurity content, phase segregation, etc. It is difficult to sort out a fundamental failure mechanism with so many variables clouding the issue. Furthermore, the manifestation of degradation; for example, as a loss in the reduction of area at fracture, occurs on the scale of centimeters while the processes responsible for the effect may occur on the scale of nanometers. The problem is therefore not amenable to solution by any single theoretical approach, and any theoretical approach which is taken should be carefully checked by appropriate experiments.

Our general approach has been to apply a variety of calculational techniques, each aimed at a specific aspect or dimension of the problem, and then combine these techniques in a self-consistent manner. Figure 1 is an illustration of the manner in which we have coupled together these calculational techniques. First principles (ab initio) calculations (1) have been developed (beginning in the 1930's) for treating the electronic structure of atoms and small molecules (Box A in Fig. 1). More recently, psuedopotential formalisms (2) (Box B) have arisen which allow the core electrons of atoms to be replaced with an effective potential, allowing high Z atoms and larger clusters of atoms (3) (Box C) to be studied. In our calculations, a cluster of up to $\sim$ 40 Ni atoms (with 20 Ni atoms being commonly employed) containing a defect (H or He atom) has been investigated from an electronic structure point of view. We recognize, however, that when a hydrogen or

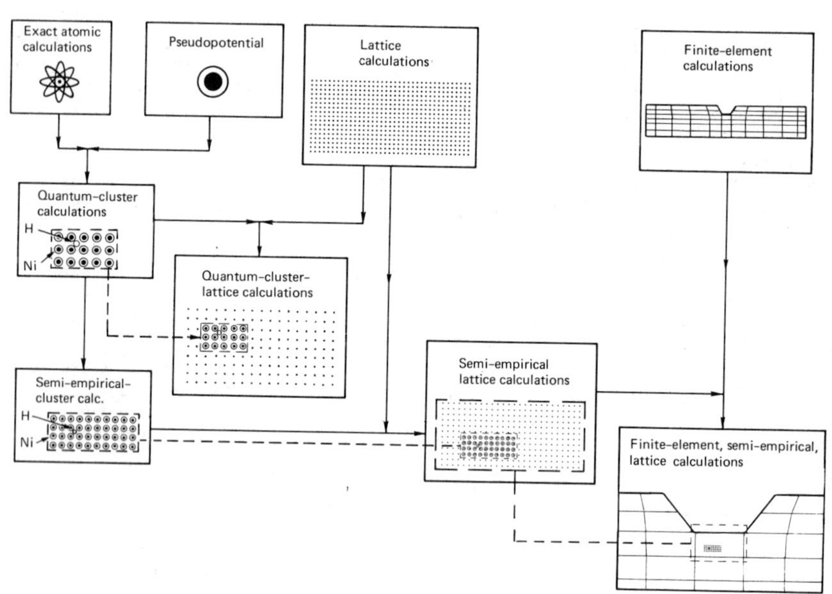

Fig. 1 - Summary of the calculational procedure.

helium atom moves about (diffuses) in a metal lattice, it can cause displacements of the lattice atoms, and that these displacements may have a large effect on the migration energy (factors of, perhaps, five). Such atomic relaxations are usually taken into account by minimizing the potential energy of a lattice of atoms interacting via two-body forces (Box D). These methods (4) have had great success in the field of radiation damage theory where the defect treated is of the same type as the host lattice. When an impurity such as hydrogen is present, however, the electronic rearrangement in the vicinity of the defect needs to be explicitly included and often cannot be treated using a two-body approximation. By imbedding our quantum cluster calculation (Box C) into a two-body lattice (hundreds or thousands of atoms) calculation (Box D), we are able to minimize the total energy including electronic rearrangement and atomic relaxation (3).

Despite the elaborate and sophisticated computational level of this method, care must be taken in applying the results. For example, when we place a H in our cluster and move it from point to point, we have some degree of confidence in the energy difference, (and hence activation energies) but have a great deal less confidence in the absolute value of any one energy (formation energy or solubility).

In order to investigate more complex defects, such as hydrogen interacting with dislocations, it is necessary to employ a method taking more than 20 to 40 metal atoms into account. To this end, we have developed a method along the lines of the extended Hückel techniques (5, 6) which parameterizes certain matrix elements. The parameters are obtained from the ab initio quantum mechanical treatment (Box C) for small clusters. The method is labelled "semi-empirical" (Box F) and allows $\sim$ 300 Ni atoms to be treated directly in the vicinity of the defect hydrogen. The method furthermore allows the analytic determination of the forces on each atom and is, therefore, very simply coupled to the two-body lattice defect calculation to form Box G of Figure 1.

Clearly, even though thousands of atoms can be explicitly included in such a calculation, a method is needed to couple this atomistic scale to the scale of observation of, for example, cracking. The finite element method (7) (Box H of Figure 1), briefly, involves mathematically breaking up a sample into elements and solving the equations of elasticity and plasticity within each element subject to the boundary conditions imposed by adjacent elements. Such a method allows the treatment of the vicinity of a crack tip in a notched tensile specimen from a continuous elastic/plastic point of view. By zoning down one of the elements by six orders of magnitude, we were able to include our semi-empirical lattice defect calculation within the finite element scheme (Box 1 of Figure 1). That is, the boundaries of one of the elements become the boundaries of the atomistic calculation. Changes in those boundaries resulting from external stress and crack-tip elastic/plastic effects, as well as forces from the atomistic region, can be self-consistently included (8).

This calculational method, therefore, allows electronic structure and atomic relaxation effects to be treated on the scale of angstroms and continuum effects to be included on the scale of centimeters. An alternative method we have also used (8) which replaces the finite element method is application of an analytic continuum elastic-plastic model (9) to obtain the local stress states used in semi empirical lattice calculations (Box C).

Results

As stated above, one must restrict the calculations to simple, well-defined problems for which the results can be understood by themselves and can be checked by experiment. Our effort has been to concentrate on those aspects of hydrogen embrittlement which are affected by the intrinsic lattice structure (f.c.c. vs. b.c.c.) of the metal. Nickel was chosen for computation ease. Studying nickel allows us to separate out the effects of multiple phases (e.g., austenite, martensite, carbides and sulfides) and alloy constituents in a stainless steel. Nickel has the same face-centered cubic structure as austenitic stainless steel, exhibits a similar ductility in hydrogen, and has approximately the same activation energy for hydrogen migration and trapping to defects such as vacancies and dislocations. Pseudopotentials Box B, Figure 1) for nickel (3) have also been extensively investigated and allow us to use existing theoretical understanding.

The two-body nickel-nickel potential chosen for the lattice defect region was the long-range form derived by Baskes and Melius (10). This potential has the advantage of being fitted to a variety of experimental data: Lattice parameter, cohesive energy, vacancy formation and migration energy, elastic constants and stacking fault energy. Hence, the potential allows us to perform calculations of the splitting of the partials of an edge dislocation in an f.c.c. material as a precursor to obtaining the binding energy of hydrogen to such defects (11). Furthermore, when this two-body potential is applied to a b.c.c. structure, a model b.c.c. "nickel" lattice can be formed having the properties of a solid somewhat similar to α-Fe as shown in Table I. The lattice parameter is close to that of α-Fe and the elastic constants are of the same size as those of iron and nickel. Note that the atomic volume decreases slightly from f.c.c. to b.c.c. nickel as occurs (experimentally) in iron. This b.c.c. lattice is a metastable structure having an energy of $\sim$.05 eV per atom above its f.c.c. phase. It is held in the b.c.c. structural configuration by the boundary atoms. By comparing hydrogen effects in the resultant f.c.c. and b.c.c. lattices generated by the same potential, any differences may be attributed to crystal structure alone.

We begin with the treatment of simple defects such as interstitial hydrogen in a perfect lattice and then investigate hydrogen behavior near vacancies, dislocations, and the region of a crack tip. The results are summarized in Table II. It should first be noted that the calculated hy-

Table I.

Atom	Structure	Lattice Parameter (Å)	C_{11}	C_{12}	C_{44} $\left(10^{12} \text{ dynes/cm}^2 \right)$	Atomic Volume (Å^3)
Ni	f.c.c.	3.52	2.5	1.5	1.2	10.90
Ni	b.c.c.	2.76	2.3	2.1	1.8	10.55
Fe	f.c.c.	3.65	1.9	1.2	1.4	12.12
Fe	b.c.c.	2.87	2.4	1.4	1.2	11.78

Table II.

	F.C.C. Lattice		B.C.C. Lattice	
	Calc	Exp	Calc	Exp
Lattice Location	O_h	O_h [a]	T_d	Inter-stitial [b]
Interstitial Migration Energy (eV)	0.45	0.41 [c]	0.05	0.07 [c]
Interstitial Migration Path	$\langle 111\rangle$ $O_h{\to}T_d{\to}O_h$		$\langle 110\rangle$ $T_d{\to}T_d$	
Volume Expansion at Interstitial Site (atomic volumes)	0.04	$\sim$0.05 [d]	0.13	$\lesssim$0.17 [e]
Binding to a Vacancy (eV)	$\sim$0.05	$\sim$ 0.3 [f]	0.30	0.48 [g]
Location in the Vacancy	0.9 Å off center along$\langle 100\rangle$	off vacancy site [a]	0.96 Å off center along$\langle 100\rangle$	1.04 Å [g]
Binding to an Edge Dislocation (eV)	0.05	0.09 [h]	0.26	0.26 [i]
Location at an Edge Dislocation	Below slip plane (expansion)		Below slip plane (expansion)	

```
a.  Ref. 12 (Cu)          b.  Ref. 14 (Fe)
c.  Ref. 13 (Ni/Fe)       d.  Ref. 16 (Ni)
e.  Ref. 17 (Fc)          f.  Ref. 21 (SS)
g.  Ref. 20 (Fe)          h.  Ref. 22 (Ni)
i.  Ref. 17 (Fe); calc. from a detrapping energy of 0.33 eV
                 and a migration energy of 0.07 eV.
```

drogen interstitial location and activation energy for motion in f.c.c.
nickel is in good agreement with experiment (12, 13), and that the calculated 0.05 eV value for motion in our b.c.c. nickel lattice is in agreement with measurements on α-Fe (13, 14). The calculated path of motion
in the f.c.c. lattice is through the tetrahedral site; with the saddle
point occurring in the plane of the three nickel atoms, it must cross in
order to migrate. For the b.c.c. case, the migration energy is small due
to the more open nature of the b.c.c. lattice. Of special importance in
Table II is the volume expansion caused by a hydrogen atom. It is small
in the f.c.c. material (calculated: using the method of Ref. 15 to be
0.035 Ω; experimental (16): approximately < 0.05 Ω where Ω equals one Ni
atomic volume), while it is large in the b.c.c. lattice (calculated:
0.13 Ω; experimental (17) approximately < 0.17 Ω). This smaller calculated volume expansion for f.c.c. Ni (only recently measured and confirmed by
laser interferometry) is contrary to previous beliefs (18), which assumed a
large volume expansion for hydrogen in nickel and has far reaching implications as discussed below. For iron, the experimental volume should be
considered an upper limit as the measured expansion has been shown to decrease with decreasing hydrogen content.

When a hydrogen atom is introduced into a vacancy, the atom in both
the f.c.c. and b.c.c. lattice lies considerably off the vacant site. It
is as if the vacancy is large enough to be considered a "void" in the lattice, with the hydrogen chemically binding itself to the void surface.
(This is in sharp contrast to the central position which a helium atom takes
in a vacancy (19). In Figure 2, the electron distribution in the neighborhood of a vacancy in a (110) plane of our f.c.c. nickel lattice containing
a hydrogen is given and the off-center configuration is clearly seen. This
configuration is consistent with off-vacancy location of hydrogen in channeling measurements (12, 20) in both f.c.c. and b.c.c. materials (channeling
measurements do not "see" the vacancies - only the hydrogen interstitial -
and so, the agreement is along the lines of being consistent with experiments,
rather than in direct agreement). The calculated binding energy of hydrogen
to the vacancy is somewhat less than the experimentally measured binding
energy of hydrogen to radiation-produced effects (20, 21). However, the
calculated results do show, in agreement with experiment, that the trapping
to vacancies is greater in b.c.c. lattices than in f.c.c. lattices.

The binding energies of hydrogen to edge dislocations in both f.c.c.
and b.c.c. materials are seen from Table II to be in good agreement with
experiment (17, 22). As mentioned earlier, the edge dislocation in the

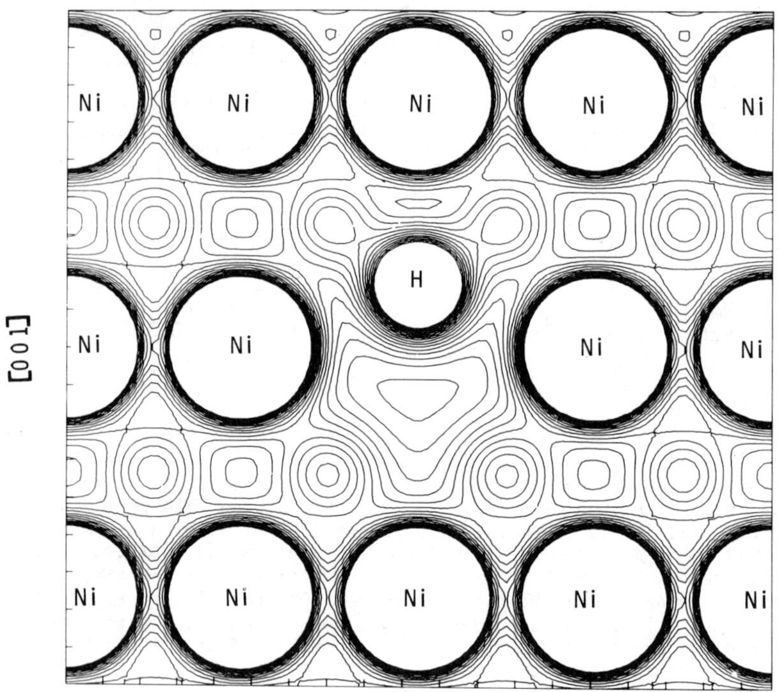

Fig. 2 - Electronic charge density contour plot of a hydrogen atom in a
vacancy in nickel. The (110) plane of the f.c.c. lattice passing through
the vacancy site is plotted. Note that the hydrogen atom in its minimum
energy site does not reside at the vacant atom site, but rather about
half-way between the vacant atom site and a neighboring octahedral site.

f.c.c. lattice was allowed to split into partials before the hydrogen was introduced (11). Again, it should be noted that the binding energy is smaller in the f.c.c. lattice than in the b.c.c. lattice.

Using the continuum elastic-plastic model (9) to determine the stress state around the crack tip, the ratio of the room temperature solubility in the stressed state to that in the stress-free state is shown as a function of reduced distance (Figure 3). The mechanical properties of the continuum material have been taken as those of a modified A-286 stainless steel with a yield stress, σ_0, of 100 ksi. In the f.c.c. lattice, the hydrogen solubility is essentially unaffected by the stress state near the crack tip; while in the b.c.c. lattice, an enhancement of more than a factor of five is seen. It is thus clear that the b.c.c. lattice structure has a greater effect than the f.c.c. lattice on the trapping of hydrogen in the stress field of a crack.

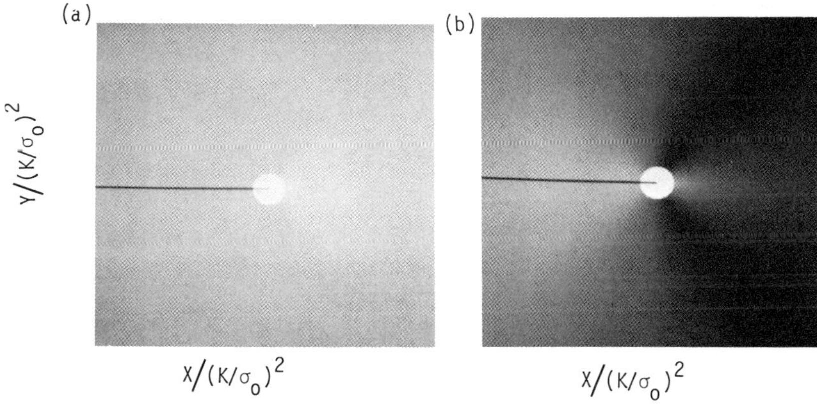

Fig. 3 - Hydrogen solubility ratio near a crack (black line) tip for a) an f.c.c. and b) a b.c.c. material. Black represents about a factor of five enhancement in solubility while white represents about a factor of two decrease. The white circle surrounding the crack tip represents the region in which the continuum stress calculation is invalid. Note that the figure is presented in terms of a reduced spatial variable; i.e., the actual distances are determined only after a stress intensity K and yield stress σ_0 are specified. The dramatic effect of crystal structure on solubility is evident.

Conclusions

From our results, a pattern emerges which may have fundamental importance to hydrogen embrittlement. Whereas in f.c.c. materials, the binding energy of hydrogen to defects and its attraction to a stressed crack tip is small compared to its interstitial activation energy; in a b.c.c. material, the defects and stress fields trap hydrogen by energies as high as an order of magnitude above this fundamental migration energy. This implies that defects - perhaps grain boundaries, cracks, flaws, and precipitates - will dominate the hydrogen transport and location in a material in b.c.c.'s, but will be of much less importance in f.c.c.'s. Autoradiographic evidence (23-25) supports this conclusion, as does the observation that b.c.c. steels are more susceptible to hydrogen degradation than are f.c.c. steels. The direct connection between the hydrogen embrittlement processes and the enhanced hydrogen concentrations which we have predicted is historically made through a decohesion based model. Atomistic investigations are now proceeding to evaluate the validity of such models and the critical hydrogen concentrations necessary to cause such embrittlement.

Acknowledgments

We would like to thank G. J. Thomas for communicating his experimental results prior to publication and to C. L. Bisson for his help with the calculations. This work was supported by the Division of Basic Energy Sciences of the U. S. Department of Energy.

References

1. Methods of Electronic Structure Theory, H. F. Schaefer III, ed., Plenum Press, NY, 1978; Quantum Theory of Molecules and Solids IV, J. C. Slater, McGraw Hill, NY, 1974.

2. J. D. Weeks, A. Hazi, and S. A. Rice: Adv. Chem. Phys., 1969, vol. 16, p. 283.

3. C. F. Melius, C. L. Bisson, and W. D. Wilson: Phys. Rev., 1978, vol. B18, p. 1647.

4. R. A. Johnson: J.Phys. F: Metal Physics, 1973, vol. 3, p. 295.

5. R. Hoffman: J. Chem. Phys., 1963, vol. 39, p. 1397; R. Hoffman and W. N. Lipscomb, ibid, 1962, vol. 36, pp. 2179, 3479; 1962, vol. 37, p. 2872.

6. C. F. Melius, W. D. Wilson, and C. L. Bisson: Proceedings of the Harwell Consultants Symposium on Inert (Rare) Gases in Metals and Ionic Solids, AERE Harwell, 1980; to be publ. in Rad. Effects.

7. O. C. Zienkiewicz: The Finite Element Method, McGraw Hill, London, 1977.

8. M. I. Baskes, C. F. Melius, and W. D. Wilson: Proceedings of Symposium on Interatomic Potentials and Computer Simulation of Defects of Metals, Pittsburgh, PA, 1980.

9. F. A. McClintock: Fracture III, H. Leibowitz, ed., 162, Academic Press, NY, 1971.

10. M. I. Baskes and C. F. Melius: Phys. Rev., 1979, vol. B20, p. 3197.

11. M. I. Baskes and C. F. Melius: Zeitschrift für Phys. Chem. Neue Folge, 1979, vol. 116, p. 289.

12. J. P. Bugeat and F. Ligeon: Phys. Lett., 1979, vol. 71A, p. 93.

13. J. Völkl and J. G. Alefeld: Diffusion in Solids, A. S. Nowick and J. J. Burton, eds., 231, Academic Press, NY, 1975.

14. L. E. Brainin, V. A. Karchenko, and B. A. Brusilowskii: Izv. Akad. Nauk. SSSR, Otd. Tekhn. Nauk, Met. i Toplivo, 1961, vol. 6, p. 115.

15. M. I. Baskes and J. H. Holbrook: Phys. Rev., 1978, vol. B17, p. 422.

16. G. J. Thomas and W. Drotning, private communication.

17. J. P. Hirth: Met. Trans., 1980, vol. 11A, p. 861.

18. E. O. Wollan, J. W. Cable, and W. C. Koehler: J. Phys. Chem. Solids, 1963, vol. 24, p. 1141.

19. W. D. Wilson, M. I. Baskes, C. L. Bisson: Phys. Rev., 1976, vol. B13, p. 2470.

20. S. M. Myers, S. T. Picraux, and R. E. Stoltz: J. Appl. Phys., 1979, vol. 50, p. 5710.

21. K. L. Wilson and M. I. Baskes: J. Nucl. Mater., 1978, vols. 76 & 77, p. 291.

22. G. J. Thomas: this conf.

23. G. J. Thomas, private communication.

24. I. Taguchi: Proceedings JIMIS-2, Hydrogen in Metals, supp. to Trans. JIM, 1980, vol. 21, p. 225.

25. R. Clermont, J. Ovejero-Garcia, J. Chêne, M. Aucouturier, C. Pirrovani, P. Tison, and J-P. Fidelle, Proceedings JIMIS-2, Hydrogen in Metals, supp. to Trans. JIM, 1980 vol. 21, p. 233.

DISCUSSION

<u>A. W. Thompson, Carnegie-Mellon University, Pittsburgh</u>: It seems to me that if your various interatomic parameters were really "accurate", the calculations for BCC nickel wouldn't work in the sense that the calculated lattice would try to be FCC. Can you comment?

<u>W. D. Wilson</u>: Our b.c.c. lattice is metastable having an energy higher (0.05 eV/atm) than our f.c.c. lattice. The imposition of fixed boundaries of b.c.c. crystal structure maintains stability of the b.c.c. phase. Even the imposition of stress or the creation of a dislocation does not cause recrystallization.

<u>I. M. Bernstein, Carnegie-Mellon University, Pittsburgh</u>: Your results show only weak traps in fcc nickel. However, is this simply because you have not yet considered how hydrogen will interact with traps that may be quite strong, e.g. precipitates?

<u>W. D. Wilson</u>: Let us reemphasize again that the main reason that the trapping of hydrogen to vacancies, dislocations, and the stress field at a crack tip in an f.c.c. lattice is very weak is because the volume expansion caused by an interstitial hydrogen is quite small. The atomic rearrangement at a precipitate-matrix interface is certainly not much more severe than the rearrangement at a dislocation, thus we would not expect a precipitate to have a large strain-induced trapping energy for hydrogen in an f.c.c. material. On the other hand, since hydrogen is a chemically active element, it is certainly possible that trapping of hydrogen to specific impurities in an f.c.c. material could be large. Calculations are now proceeding in this direction, looking for trapping to such impurities as sulfur, oxygen, and carbon.

<u>M. R. Shanabarger, University of California, Santa Barbara</u>: From your calculation the hydrogen proton appears to sit in an off-center potential minima center at the vacancy. I would expect that the proton must tunnel to the other potential minima. Such a process may be observable in a specific heat measurement. Have you tried to estimate the magnitude of such a specific heat contribution and would it be measurable.

<u>W. D. Wilson</u>: Indeed our calculations point to a very flat potential energy surface for the off-center hydrogen in a vacancy. We do not feel the calculations are accurate enough (e.g. zero point energies) to predict the magnitude of possible tunnel splittings between, for example, pocket states separated by $\gtrsim$ 0.01 which the hydrogen is probably tunneling between. Such effects have been observed years ago for small, off-center Li^+ in KCl (and other such impurities in alkali halides) and do give rise to Shottky anomalies in the specific heat at very low temperatures ($\sim 1°K$). Whether skin depth problems (in the microwave region) make such measurements intractable in metals is worthy of attention.

HYDROGEN TRAPPING IN FCC METALS[*]

G. J. Thomas

Sandia National Laboratory
Livermore, CA 94550

Desorption measurements have been employed to study trapping in
cold-worked and impurity-doped nickel. The data was analyzed using numerical solutions to diffusion models to obtain trap binding energies. The
present results, as well as literature data, indicate low binding energies
relative to bulk diffusional activation energies in nickel and austenitic
steels, in contrast to the ferritic case. This suggests that enhancement
of hydrogen concentrations at trap sites may not be very large. The small
binding energies to defect strain fields indicate that hydrogen in solution
induces a small lattice relaxation in these metals.

[*]This work was supported by U. S. Department of Energy, Division of Basic
Energy Sciences.

I.   Introduction

A knowledge of hydrogen trapping to defects is important to an understanding of hydrogen degradation of material properties. In b.c.c iron and ferritic alloys, trapping effects have been recognized for some time. Trapping has been found during permeation measurements[1] and autoradiography has shown[2] enhancements of hydrogen concentration (over the lattice concentration) at grain boundaries and other sites. This enhancement could be a crucial factor in brittle failure of these metals and, hence, are central to embrittlement models.

The situation is not as clear, however, in the case of f.c.c. metals such as austenitic steels and nickel. Although it is known that these metals suffer degradation upon hydrogen exposure,[3,4] there is little data on the magnitude of trapping effects. In this paper, thermal desorption has been applied to measure binding energies of hydrogen to specific defects in nickel. These results, as well as literature data, indicate that there is generally a weak interaction between dissolved hydrogen and localized defect strain fields. This is consistent with a small partial molal volume for hydrogen and further suggests that, in the absence of deformation, significant concentration enhancement may not be critical to embrittlement of f.c.c. metals.

II.   Experimental

Thermal desorption techniques have been applied to the measurement of bulk hydrogen transport properties for many years. In the present case, this technique is employed under conditions which should enhance the observation of trapping phenomena--transient or non-steady-state conditions, high defect concentrations, low hydrogen concentrations, and low temperatures. A numerical solution to the governing diffusion equations, including trapping terms, is fitted by a least squares method to the experimental results for the exact time-temperature schedule employed in the measurements. The parameters in the analysis are the bulk transport properties (lattice activation energy, E_D, pre-exponential diffusivity factor, D_0, and initial concentration) and trap parameters (binding energies, E_T, and trap densities, N_T). Thus, for a one-dimensional case:[5]

$$\frac{dc(x,t)}{dt} = \frac{d}{dx}\left[D\,\frac{dc(x,t)}{dt}\right] - \frac{dc_T}{dt} \tag{1}$$

the trapping term $\dfrac{dc_T}{dt}$ is

$$\frac{dc_T}{dt} = \text{trap rate} - \text{detrap rate}$$

$$= \frac{Dc(x,t)}{\lambda^2}\,[N_T - fc_T] - c_T\,\nu\exp\left[-(E_D + E_T)/_{kT}\right] \tag{2}$$

where: $c(x,t)$ = lattice concentration

D = diffusivity = $D_0\,\exp\left[-E_D/kT\right]$

c_T = trapped concentration

N_T = density of traps

ν = attempt frequency = 10^{13} sec^{-1}

λ = jump distance

The computer code for solving this problem with up to two types of traps has been developed by Baskes.[5]

The experimental procedure is schematically shown in Figure 1. Samples were deuterium charged to saturation, then rapidly cooled by dropping the charging vessel into ice water. The cool-down time to room temperature was roughly 40 seconds. Charged samples were then placed in the desorption apparatus after a short air exposure at room temperature. This time, as well as the charging time, temperature and cool-down, are included in the thermal history when solving equation 1.

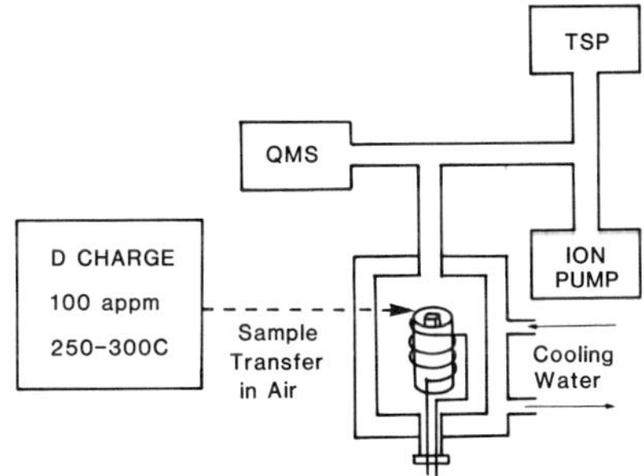

Fig. 1 - Schematic diagram of the desorption apparatus. "QMS" indicates the quadrupole mass spectrometer.

The desorption apparatus consists of an all-metal high vacuum system. Both the ion pump and titanium sublimation pump were used to maintain a low background ($\sim$ 5 x 10^{-8} torr) and relatively constant pumping speed over a large range of release rates. The linearity of the response of the quadrupole mass spectrometer (UTI 100C) was verified using a series of calibrated deuterium leaks supplied by Teledyne-Hastings.

Trapping to dislocations in nickel was studied by measuring cold-worked material and comparing the results with annealed samples. MARZ grade (Materials Research Corp.) Ni sheet 1mm thick was annealed and subsequently cold-rolled to 50% reduction in thickness. A deuterium charging temperature of 425K was used to prevent loss of the cold work structure. Defect-free samples were prepared from the cold-rolled material by subsequent annealing at 1000K. Samples of Ni rods (prepared at the Max Planck Institut, Stuttgart, West Germany) doped with atomically dispersed substitutional impurities of Fe or Ti were also studied. These samples were initially solutionized at $\sim$ 1100K, then deuterium charged at 573K. The impurity concentrations were determined by electron microprobe analysis to be nominally 2 at. % in both the Fe doped material and Ti doped material. MARZ grade Ni rod which was swaged down to the same diameter as the impurity doped rods and subsequently annealed was used as a defect-free standard in order to eliminate uncertainties in comparing different geometries. The diffusion equation, eq. 1, was changed to a cylindrical coordinate system for fitting these data.

III.   Results

The deuterium release rate for an annealed sample as a function of
time during a series of isothermal anneals is shown in Figure 2. During
each of the plateau regions the sample was held at the indicated temperature.
The squares are the experimentally measured release rates (the vertical
scale is related to the QMS current) and the solid line is a calculated
rate determined by the exact numerical solution of equation 1 for parameters
which minimize the difference between experiment and calculation. One can
see that the overall fit is reasonably good and, further, that all data
points, including the ramp intervals were taken into account. The diffusivity parameters determined by this procedure are summarized in Table
I.   Several runs were made on each sample condition. The diffusivity
values are in excellent agreement with literature values.[6] A sample
from cold-worked material is shown in Figure 3 with the solid curves
calculated for two different conditions. The lower one indicates the
release rate when the defect-free case is assumed (i.e. no traps and
diffusivity values of Figure 2). Much better agreement was obtained with
the upper curve which includes trapping. The parameters determined from
fitting the cold-worked data are also given in Table I. The density of
traps, N_T, corresponds to dislocation densities of $\sim 10^{11}$-10^{12} cm/cm^3
for trapping radii of 1-2 lattice spacings.

Table I. Desorption Summary Cold-Worked Nickel

| | Range of Values from Fitting Proc. | | |
	Annealed	Cold-worked	Literature Values
E_D	$0.40 - 0.43$ eV	0.42 eV (held fixed)	0.415 eV
D_0	$3\text{-}6 \times 10^{-3}$ cm^2/s	4.5×10^{-3} (held fixed)	4×10^{-3} cm^2/s
Sol.	100 appm	100 appm	110 appm (calc.)
N_T	--	$1\text{-}2 \times 10^{-3}$ at fract.	--
E_T	--	$0.1 - 0.15$ eV	--

Similar results were obtained with the impurity doped material. An
example is shown in Figure 4. As one can see, a different temperature
schedule was employed for these samples. However, as before, good agreement
was obtained with published values for the bulk parameters. The best fit
results for the impurity doped cases are summarized in Table II.

Table II. Desorption Summary Impurity Doped Nickel

| | Range of Values from Fitting Proc. | | |
	Annealed	Ti impurity	Fe Impurity
E_D	$0.40 - 0.44$ eV	0.42 eV (held fixed)	0.42 eV (held fixed)
D_0	$3\text{-}6 \times 10^{-3}$ cm^2/s	4.5×10^{-3} cm^2/s	4.5×10^{-3} cm^2/s
N_T	--	0.02 at fract.	0.02 at fract.
E_T	--	$0.05 - 0.1$ eV	$0.07 - 0.12$ eV

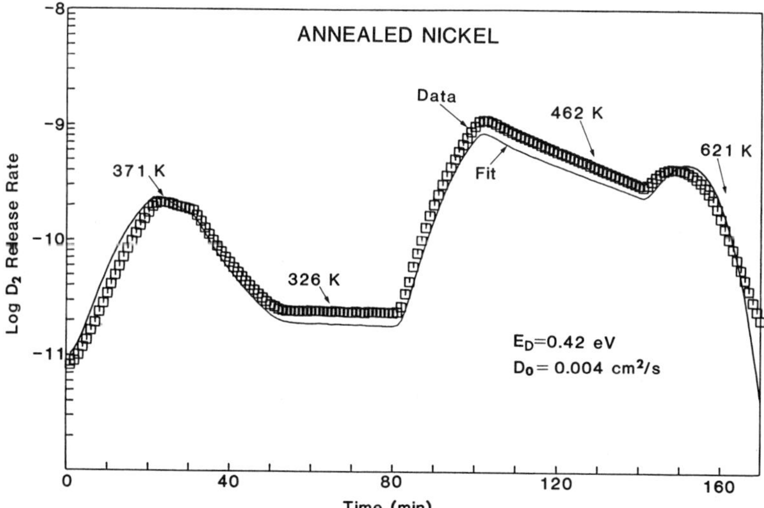

Fig. 2 - Plot of deuterium release rate versus time for annealed Ni during
a series of isothermal anneals. The solid line is fit to the
data by a numerical solution of equation 1, resulting in determination of the indicated parameters.

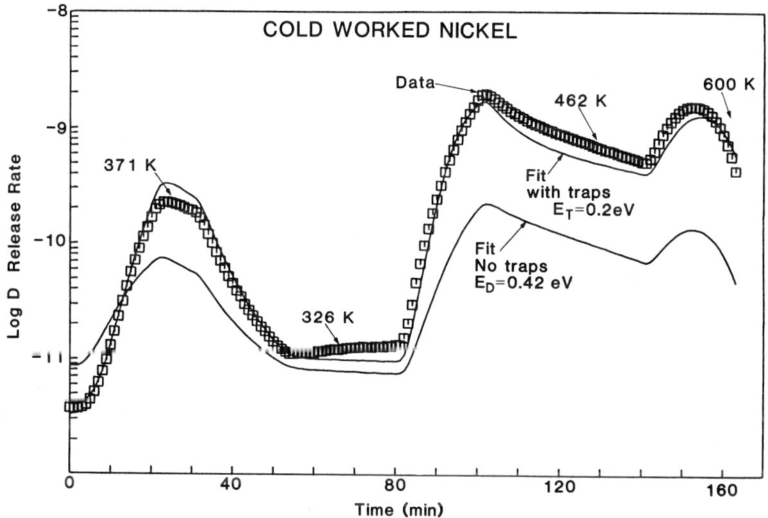

Fig. 3 - Plot of deuterium release rate versus time for a cold worked Ni
sample. The lower solid curve was calculated assuming no traps
and using the parameters shown in Fig. 2.

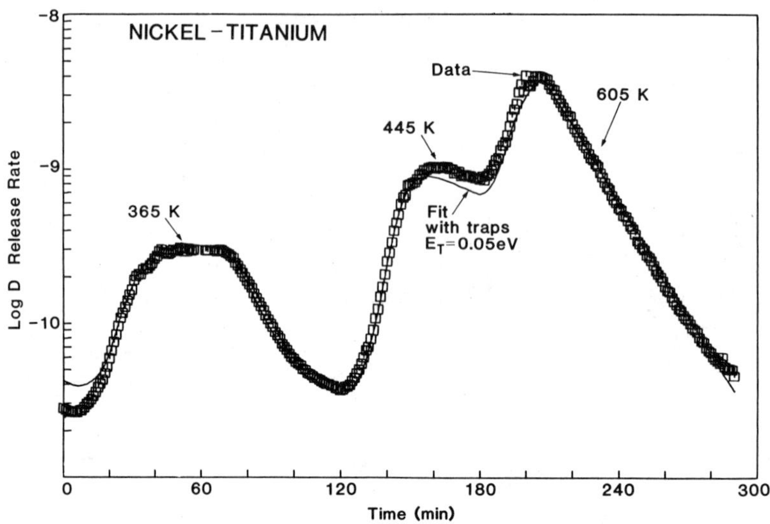

Fig. 4 - Plot of deuterium release rate versus time for a Ni sample doped
with 2 at.% of titanium. The bulk diffusion parameters shown in
Fig. 2 were used during the fitting procedure to obtain E_T.

IV. Discussion

Our results show that hydrogen binding to dislocations (Table I) and
to substitutional Fe or Ti impurities (Table II) in Ni is only of the
order of 0.1 eV. It is interesting to note that binding to the clearly
elastic defect, the dislocation, is slightly higher than to the highly
reactive Fe or Ti atoms. This suggests that the affinity of hydrogen for
these impurities is essentially due to elastic interactions. The value
for the binding energy to the dislocation is in agreement with the theoretical calculations of Baskes, et al.[7]

There have been other measurements of hydrogen binding to various
defects in nickel and austenitic stainless steels. Some of these data,
along with the present results, are summarized in Table III. (No attempt
was made to be all inclusive in this compilation) It is readily apparent
that hydrogen trapping is generally weak in these metals relative to the
bulk activation energy. Indeed, the autoradiographic results indicate no
enhancement of hydrogen at grain boundaries in stainless steel at room
temperature.

The low binding energy of hydrogen to the strain fields of defects
such as dislocations, impurities and precipitates implies a small volume
lattice expansion. That is, for a hydrostatic stress σ, the interaction
enthalpy for an elastic misfit is given by[15]

$$\Delta H = \sigma \, \Delta v$$

where Δv is the excess volume of the hydrogen. For an edge dislocation[15]

$$\Delta H = 4 \, \mu b \, \frac{\Delta r}{r_0} \, r_0^{\,3} \, \sin \theta / r$$

Hence the volume excess in terms of the binding energy is

$$\Delta v \sim \parallel \frac{E_T}{\mu}$$

for nickel this gives $\Delta v \sim 1$ Å^3 or roughly 30% of the value of the "universal" volume excess of 2.9 Å^3.[16] Length change measurements on Ni during hydrogen evolution are currently in progress to re-examine this quantity. Preliminary results indicate agreement with the lower value.

Table III. Experimental Values of Hydrogen-Defect Binding Energies

Exp. Technique	Defect	E_T(eV)	Material
Int. Frict.[8]	disl.	0.14	SS
Int. Frict.[9]	disl.	<0.1	SS
Serrated[3] Yielding	disl.	0.09	Ni
Release[10] During Def.	disl.	0.14	Ni
Desorption	disl.	0.1 - 0.2	Ni
Mag. Rel.[11]	impurities	0.05 - 0.12	Ni
Desorption	Ti, Fe impurities	< 0.1	Ni
Autorad.[12]	g.b.	0	SS, Ni
Desorption[13]	coherent precipitates	0.1 - 0.15	SS
Permeation[14]	incoherent oxide	0.3	Ni

The weak binding relative to the bulk activation energy has important implications to accumulation of hydrogen at trap sites. Equation 2 shows that the trapping rate is proportional to the local concentration and the bulk diffusivity. Hence, the trap rate is proportional to c exp $(- E_D/kT)$ whereas the detrapping rate from equation 2 is proportional to exp $[- (E_D + E_T)/kT]$. In b.c.c. iron, $E_D < 0.1$ eV and $E_T \sim 0.2$-0.5 eV. Consequently, traps can fill quickly and the residence time in traps can be relatively long. In contrast $E_D \sim 0.4$ eV in Ni and ~ 0.5 eV in eustenitic steel[5], while $E_T \sim 0.1$ - 0.2 eV in both materials. Thus, traps can be expected to fill relatively slower and hydrogen more easily de-trapped.

Enhanced hydrogen concentrations at defect sites are therefore more likely in b.c.c. metals, while in f.c.c. metals the concentration at defects is roughly equivalent to the lattice concentration. This effect should be taken into consideration in any model of hydrogen embrittlement pertaining to f.c.c. metals.

V.   Acknowledgements

The author is indebted to M. I. Baskes for development of the DIFFUSE code, and to B. Hohler of the Max Planck Institut, Stuttgart, for the impurity doped samples. The experimental assistance of L. A. Brown and Diane Look are gratefully acknowledged. The author wishes to thank W. D. Wilson, R. E. Stoltz and M. I. Baskes for many useful discussions.

REFERENCES

1.   See, for example, R. A. Oriani: Acta. Met., 1970, Vol. 18, p. 147.
2.   T. Asaoka, G. Lapasset, M. Aucouturier and P. Lacombe: Corrosion, 1978, Vol. 34, p. 39.
3.   T. Boniszewski and G. C. Smith: Acta Met, 1963, Vol. 11, p. 165.
4.   M. R. Louthan, G. R. Caskey, J. A. Donovan and D. E Rawl: Mat. Science and Eng., 1972, Vol. 10, p.
5.   M. I. Baskes: Sandia National Laboratories Report, SAND80-8201.
6.   W. M. Robertson: Z. Metallk, 1963, Vol. 64, p. 436.
7.   M. I. Baskes, C. F. Melius and W. D. Wilson: this conference.
8.   A. Atrens, N. F. Fiore and K. Miura: J. Appl. Phys. 1977, Vol. 48, p. 4247.
9.   J. A. Peterson, R. Gibala and A. R. Troiano: J. Iron Steel Inst., 1969, Vol. 207, p. 86.
10. J. A. Donovan: Met. Trans., 7A, 1677 (1976), 1976, Vol. 7A, p. 1677.
11. B. Hohler and H. Krönmiller: Proc HYDROGEN IN METALS, Munster, 1979, p. 301.
12. J. Chene, J. Ovejero - Garcia, C. Paes de Oliveira, M. Aucouturier, P. Lacombe: J. Micros. Sp. Electr., 1979, Vol. 4, p. 37.
13. G. J. Thomas, M. I. Baskes and R. E. Stoltz: Proc HYDROGEN IN METALS, Munster, 1979, p. 727 (Abstract only).
14. W. M. Robertson: Met. Trans, 1979, Vol. 10A, p. 489.
15. N. F. Fiore and C. L. Bauer: Progr. Mater. Sci, 1967, Vol. 13, p. 1.
16. H. Peisl: HYDROGEN IN METALS I, p. 53 Ed. by G. Alefeld and J. Völkl Springer-Verlag, Berlin, 1978.

DISCUSSION

<u>A. Atrens, Brown Boveri Research Center, Baden, Switzerland</u>: (1) Are you
going to be doing any work on austenitic stainless steels? (2) Have you
measured any hydrogen-grain boundary binding energies?

<u>G. J. Thomas</u>: Measurements of deuterium trapping at oxide particles in 316
stainless steel are currently in progress (in collaboration with R. Stoltz).
Grain boundary trapping measurements would likely be difficult with the desorption technique. Autoradiography results indicate little, if any, segregation of tritium at grain boundaries in nickel and stainless steel.

<u>M. R. Shanabarger, University of California, Santa Barbara</u>: As I understand
your experimental technique, you are observing gaseous hydrogen evolved from
a charged specimen as the specimen is heated in a programmed fashion. The
dissolved hydrogen desorbs by recombination at the surface of the specimen.
The kinetics of desorption are changing significantly over the temperature
range of your measurements ($100^\circ K$-$500^\circ K$) (1). How do you justify neglecting
the desorption process in interpreting your measurements.

(1) See M. R. Shanabarger, Phys. Rev. Letters, Dec. 24, 1979. (I think).

<u>G. J. Thomas</u>: The rate-limiting process during these measurements, as in all
bulk transport measurements, is the rate of gas arrival to the surface. Hence,
although the time for desorbing a monolayer of adsorbed gas may be changing
over the temperature range of the experiment, the present results will not
be significantly affected.

TRAPPING OF DEUTERIUM IN ION-IMPLANTED Fe*

S. M. Myers and S. T. Picraux
Sandia National Laboratories†
Albuquerque, NM 87185

R. E. Stoltz
Sandia National Laboratories†
Livermore, CA 94550

Ion-beam techniques have been used to investigate hydrogen-trapping mechanisms in Fe. Lattice defects and the solutes Ti, Kr, and Y were initially introduced by ion implantation. The deuterium isotope (D) was then implanted, and its migration during subsequent heating was monitored by ion beam analysis. Trap binding enthalpies were extracted from the resulting data. Microstructural information was obtained by ion channeling and transmission electron microscopy.

Traps having strengths of 0.5 and 0.8 eV were observed in irradiated pure Fe and attributed respectively to monovacancies and vacancy clusters. Addition of Ti increased both these enthalpies by ≈0.05 eV. This is believed due to a Ti D chemical affinity which augments the interactions with the structural defects. Ion implantation of Kr and Y produced substantially stronger binding, 1.1 and 1.3 eV respectively. It is proposed that these greatly oversized solutes stabilize an impurity-vacancy complex which is a strong trap for hydrogen, and that the coupling is further strengthened in the case of Y by a Y-D chemical affinity.

* This work was supported by the U.S. Department of Energy under Contract DE-AC04-76-DP00789.

† A U.S. Department of Energy Facility.

Trapping of hydrogen in ferritic steels is believed to reduce embrittlement by this impurity, and the resulting prospect for improved hydrogen compatibility has motivated numerous investigations of trapping mechanisms. As an example, heavily cold-worked Fe was found to contain 0.6 eV traps at a concentration of 10^{-4} at.% (1). (Trap depths herein are referenced to an equilibrium site in solution.) In another study, solutionized Ti was reported to bind hydrogen with an enthalpy of $\simeq 0.2$ eV (2). Finally, incoherent TiC precipitates are the deepest traps previously observed in Fe, with an estimated binding enthalpy of 1 eV (2).

The investigation reported in this paper was concerned with trapping of hydrogen in bcc Fe by ion-irradiation defects and ion-implanted solutes. The isotope deuterium (D) was implanted into Fe at a temperature of 90 K, and during subsequent linear ramping of temperature, the quantity of D remaining within the implanted region was monitored by ion beam analysis. The resulting retention-versus-temperature profiles provided information on the number and strength of traps. In some cases the injection of D was preceeded by room-temperature implantation of Fe, Ti, Kr, or Y ions. Deuterium trapping observed in the highly metastable implanted alloys is attributed to monovacancies, vacancy clusters, and impurity-vacancy complexes.

Our results provide new insight into the factors affecting trap depths, and they demonstrate that hydrogen can be bound strongly (>1 eV) to entities other than precipitates. Furthermore, it is conceivable that certain of the deep traps identified in these studies can be introduced into commercial steels by means less novel than ion implantation.

Experimental Procedure

Iron discs of specified purity 99.99 wt.% were implanted at room temperature with 190 keV ions of Fe, Ti, Kr, or Y, producing an alloyed surface layer extending to ~0.1 μm. The depth profiles of Ti, Kr, and Y were measured by Rutherford backscattering of 2.5 MeV ^{4}He. These samples and others of virgin Fe were then implanted with 15 keV deuterium (D) at a temperature of 90 K, so that the maximum concentration occurred at $\simeq 0.1$ μm. Subsequently the temperature was linearly ramped at 2 K/min., and at 10 K intervals the amount of D remaining within the implanted region was measured using the nuclear reaction ^{2}D(^{3}He, ^{1}H)^{4}He. The number of protons detected during bombardment by 0.7 MeV ^{3}He was proportional to the total areal density of D at depths $\lesssim 0.4$ μm.

Ion channeling of 0.7 MeV ^{3}He was employed in selected cases to determine the lattice position of implanted D. The over-all microstructure was characterized by transmission electron microscopy, using bright and dark-field imaging and selected-area diffraction at 100 keV. The microscopy was performed on 50 μm Fe foils which were ion-implanted and then thinned from the unimplanted side.

Additional procedural detail has been given elsewhere (3).

Results and Interpretation

Trapping of ion-implanted deuterium (D) at irradiation defects in pure Fe has been discussed in detail previously (3). Here we summarize the principal features of that work and further develop the interpretation. More extended consideration is given to D trapping in Fe alloyed by implantation with Ti, Kr, and Y.

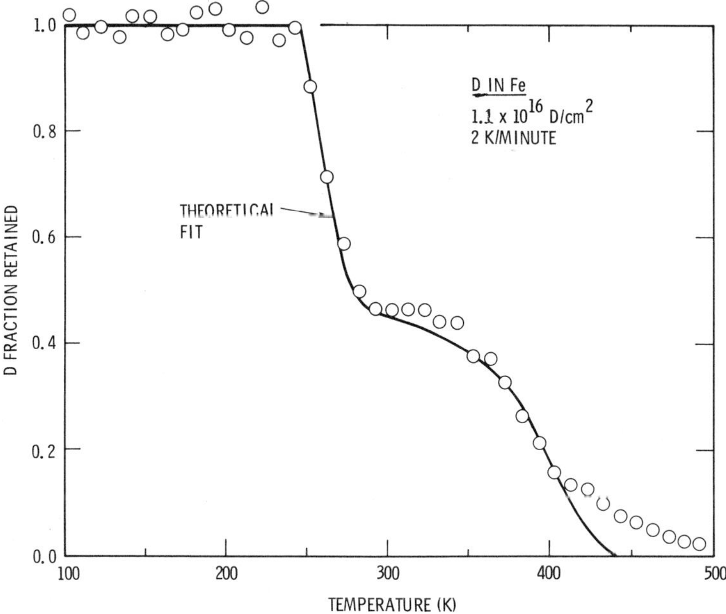

Fig. 1 - Retention of ion-implanted D in Fe during linear
ramping of temperature.

D in Unalloyed Fe

Figure 1 shows the fraction of initially injected D remaining within
the implanted region as a function of increasing temperature. Here the D
dose of 1.1 X 10^{16} at./cm produced a peak concentration of ~1 at.% at ≃0.1 μm.
In this case no other ions were implanted, so that all lattice damage occurred
during the introduction of D; it is estimated theoretically that each incident
particle produced ~10 atomic displacements within the Fe matrix. Measurement
conditions were such that ≃2400 protons from the ^{2}D(^{3}He,^{1}H)^{4}He nuclear reaction were detected for each plotted datum prior to the onset of D release.

The two release stages seen in Figure 1 occur far too high in temperature for perfect-lattice diffusion, for which the activation energy is only
≃0.05 eV (4). Consequently, these stages are attributed to trapping of D at
irradiation defects. The associated binding enthalpies, expressed with
reference to an untrapped site in solution, were obtained by fitting to a
theoretical model of the system. The McNabb-Foster formalism for diffusion
in a field of traps (5) was solved numerically, assuming two types of defect
trap within the irradiated layer. The respective binding enthalpies were
then adjusted to produce agreement with the data. This procedure, described
in detail elsewhere (3), yielded the solid curve in Figure 1, which is seen
to accurately represent the experimental results. The trap depths so obtained
are 0.5 and 0.8 eV.

Ion channeling measurements, performed as a function of temperature,
indicate that D in the 0.5 eV trap occupies the lattice position shown in
Figure 2. Open circles represent Fe atoms about an octahedral cell in the

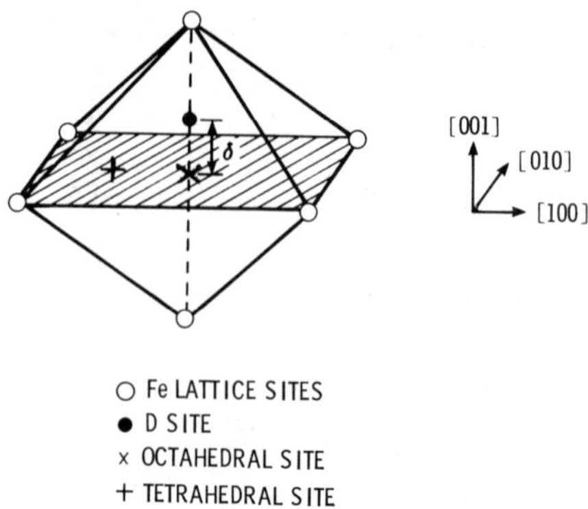

Fig. 2 - Lattice location of D in the 0.5 eV trap.

bcc matrix, while the crosses indicate octahedral and tetrahedral interstitial
sites. Trapped D, shown as a solid circle, is displaced by δ = 0.04 nm from
the octahedral position toward a nearest-neighbor site. In the case of D
occupying the 0.8 eV traps, the channeling behavior is more complex and is
difficult to reconcile with a unique lattice position; this suggests more
than one atomic configuration within the traps. Transmission-electron micrographs of Fe foils implanted with D at room temperature show a high density
of dislocation loops but no resolvable bubbles.

We previously hypothesized that the 0.5 eV D trap is a monovacancy (3).
This was prompted in part by the sharpness of the associated release stage
and the uniqueness of the D lattice site, both suggestive of a point defect.
This interpretation has subsequently been reinforced by the theoretical work
of M. Baskes et. al. (6), who calculated the minimum-energy position for
hydrogen associated with a vacancy in a bcc metal; they located the solute
0.096 nm from the center of the vacancy along a [100] direction. If, in
Figure 2, a vacancy is assumed to exist at the top Fe site, then the D is
0.10 nm from the vacancy along [100]. Although the theoretical calculations
were performed for the bcc lattice using Ni potentials rather than Fe, we
nevertheless feel that such consistency between theory and experiment
supports our identification of the 0.5 eV trap as a monovacancy.

The 0.8 eV traps are presumed to be vacancy clusters. This interpretation is tentative, but it is consistent with the following consideration:
the binding enthalpy of hydrogen to a vacancy cluster probably lies between
the value for a monovacancy, 0.5 eV, and that for a free surface, 0.7 to 1.0
eV depending upon the experimental results used (7).

D in Alloyed Fe

The influence of implanted alloying additions on D trapping is seen in
Figure 3. In these experiments, 190 keV ions of Fe, Ti, Kr, or Y were
implanted into Fe at room temperature, the dose being about 4 X 10^{16} at./cm^2
in each case. Penetration decreased somewhat with increasing atomic number:
for Y, a peak concentration of 8 at.% occurred at $\simeq$0.04 µm. The specimens
were then implanted at 90 K with $\simeq$1.2 X 10^{16} at./cm^2 of 15 keV D, and retention-versus-temperature profiles were measured as discussed in the preceeding
subsection. The data for self-ion-irradiated Fe in Figure 3 closely reproduce

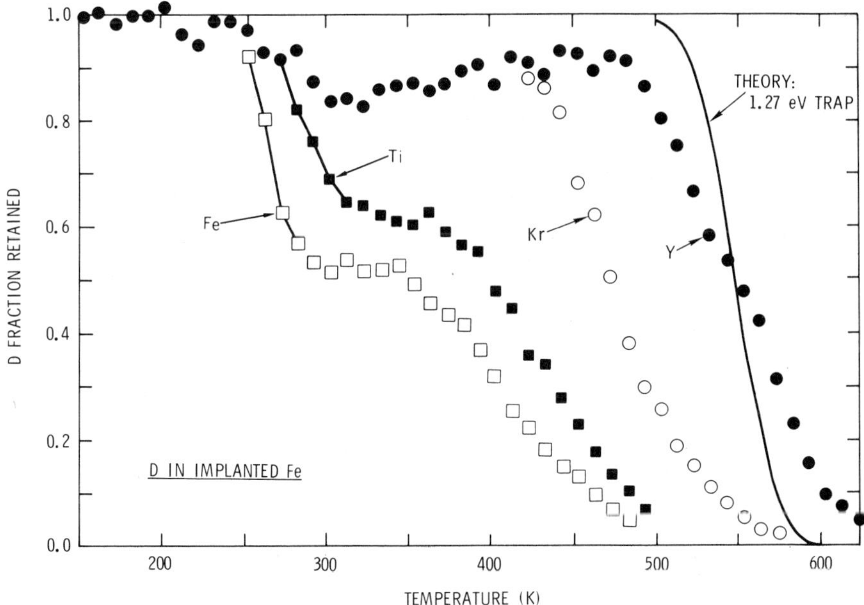

Fig. 3 – Deuterium retention during linear ramping of temperature
for Fe implanted with Fe, Ti, Kr, or Y ions.

those in Figure 1 where only D was implanted, indicating similar trapping
conditions in the two cases. However, D binding was strengthened in the
presence of Ti, Kr, and Y. Titanium shifted the two pure-Fe stages upward
in temperature by ≈30 K, corresponding to an increase in both trapping
enthalpies of ≈0.05 eV. Much larger changes were produced by Kr and Y: the
theoretical curve through the Fe-Y data in Figure 3, calculated using the
model discussed earlier, is for a single trap depth of 1.3 eV, whereas in
the case of Fe-Kr, a value of 1.1 eV is obtained.

We interpret the data of Figure 3 as follows. Both Y and Kr are greatly
oversized solutes, having larger atomic radii than Fe by 41 and ≈60% respectively. As a result, these atoms collect vacancies produced during implantation, forming impurity-vacancy complexes which strongly bind hydrogen. In
the case of Y, this binding is augmented by a Y-D chemical interaction, so
that the trap depth is greater than for Kr. Titanium, only 17% oversized in
Fe, does not stabilize the impurity-vacancy complex. Instead, it simply
increases the coupling to pure-Fe defects via a Ti-D chemical interaction.
The strength of the chemical coupling between D and Ti, 0.05 eV, is less
than that for Y, 1.3 - 1.1 = 0.2 eV, consistent with the larger heat of
hydride formation for Y. Evidence relating to the validity of this picture
will be considered in the remainder of the present section.

Mössbauer (8-10) and ion-channeling (10) results strongly indicate that
a sufficiently oversized atom in a bcc metal can collect as many as three
vacancies, forming a stable open region within the host lattice. In particu
lar, Yb and Xe have both been reported to produce this effect in Fe (8,10).
Since these two elements are respectively similar in size and chemistry to Y
and Kr, this supports the proposed atomic configuration of the Y and Kr
traps. Furthermore, ion-channeling measurements on a number of

implanted elements in Fe suggest that the solute must be ≳20% oversized for
the impurity-vacancy complex to form (10). Since Ti is only 17% oversized,
the relatively small effect of this atom on D trapping is to be expected.

The D trapping mechanisms discussed above imply that Ti, Kr, and Y are
in solid solution in the Fe host rather than being precipitated. This is
borne out by transmission electron microscopy of Fe foils implanted at room
temperature with Ti or Y. Micrographs and diffraction patterns show bcc Fe
containing a dense array of dislocations, and there is no evidence of precipitation. Such a highly supersaturated state persisted because of the small
atomic mobilities at room temperature. A similar condition was observed after
annealing Y-implanted Fe for one hour at 673 K, which is above the temperature
of the highest D-release stage in Figure 3. However, after one hour at 973 K,
both the microstructure and the D-trapping behavior were qualitatively different: there was precipitation of YFe_9, and the D-retention profile reverted
almost to that of unalloyed Fe. This reinforces the view that dispersed
implanted atoms produce the strong trapping observed in Fe-Kr and Fe-Y.

Figure 4 shows D-retention-versus-temperature profiles for Fe implanted
with two different Y doses and, for comparison, data from self-ion-irradiated
material. The amount of D has been increased from Figure 3 to more fully
occupy the traps. These results exhibit two additional features of the Y-
associated traps. First, the gradual D release prior to the final, abrupt

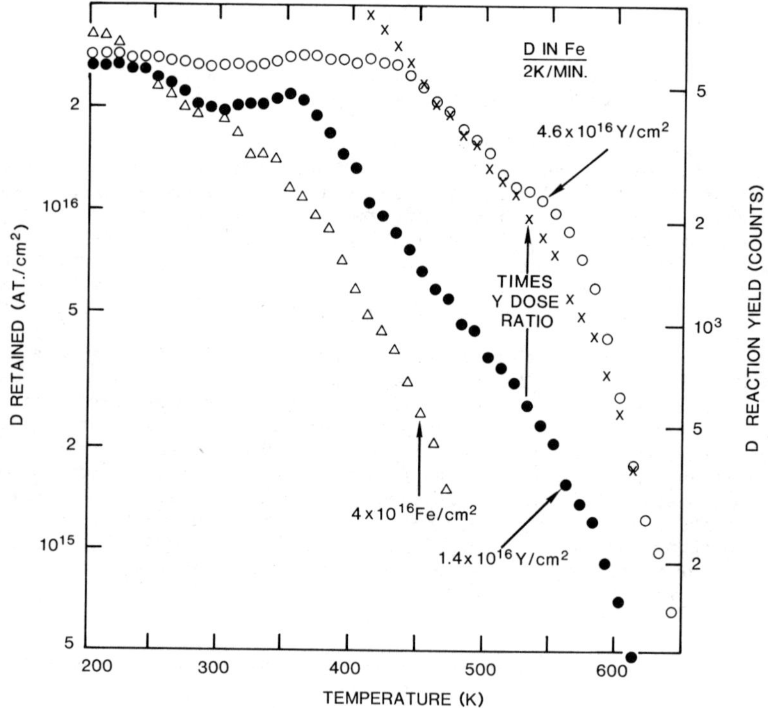

Fig. 4 – Deuterium retention profiles for Fe implanted with two
      Y doses or self-ions. Note the logarithmic vertical scale.

drop indicates a range of binding enthalpies. This is not surprising since
the impurity-vacancy complexes may contain differing numbers of vacancies
(8,9). (The enthalpy range is less apparent in Figure 3 because there are
enough of the strongest traps to bind the smaller quantity of D.)

The second feature to be noted in Figure 4 is that, at temperatures
above the trapping range for pure Fe, the amount of trapped D is nearly proportional to the amount of Y. This is demonstrated by multiplying the
retained-D quantities in the lower Fe-Y profile by the ratio of Y doses,
giving the crosses in the figure. Such a proportionality supports the proposal that the D traps are associated with individual Y atoms.

Conclusion

Table I lists the binding enthalpies observed in this work, referenced
to an untrapped site in solution, together with the proposed trapping entities. The basic elements of the interpretation are a strong coupling of
hydrogen to vacancy-type defects, combined with a superimposed chemical
binding in the presence of hydride-forming elements such as Ti and Y. The
impurity-vacancy complex is believed to be stabilized by solutes which are
> 20% oversized.

Table I. Deuterium Traps in Ion-Implanted Fe

Monovacancy	0.5 eV (50 kJ/mole)
with Ti	+ 0.05 eV (5 kJ/mole)
Vacancy cluster	0.8 eV (80 kJ/mole)
with Ti	+ 0.05 eV (5 kJ/mole)
Kr-vacancy complex	1.1 eV (110 kJ/mole)
Y-vacancy complex	1.3 eV* (130 kJ/mole)

(*A substantial range of binding enthalpies
was observed.)

Since the Y-associated trap is the deepest yet reported in Fe, the
responsible mechanism may prove useful in improving the hydrogen compatibility of ferritic steels. This application will depend upon developing an
economic means of introducing the trapping entities into a macroscopic volume.
One conceivable procedure is to form a solid solution of an oversized,
strongly hydriding element such as Sc, Y, or a rare earth, and then introduce the necessary vacancies by techniques such as cold-working.

Acknowledgements

The authors gratefully acknowledge valuable discussions with C. F.
Melius and W. R. Wampler, transmission-electron-microscopy measurements by
D. M. Follstaedt and F. A. Greulich, and the technical assistance of C. T.
Fuller.

References

1. A. J. Kumnick and H. H. Johnson, "Deep Trapping States for Hydrogen in Deformed Iron," Acta Metall., 28 (1980), pp. 33–39.

2. G. M. Pressouyre and I. M. Bernstein, "A Quantitative Analysis of Hydrogen Trapping," Metall. Trans., 9A (1978) pp. 1571–1580.

3. S. M. Myers, S. T. Picraux, and R. E. Stoltz, "Defect Trapping of Ion-Implanted Deuterium in Fe," J. Appl. Phys., 50 (1979) pp. 5710–5719.

4. J. Völkl and G. Alefeld, "Hydrogen Diffusion in Metals," pp. 250–252 in Diffusion in Solids: Recent Developments, A. S. Nowick and J. J. Burton eds.; Academic, N.Y., 1975.

5. A. McNabb and P. K. Foster, "A New Analysis of the Diffusion of Hydrogen in Iron and Ferritic Steels," Trans. AIME, 227 (1963) pp. 618–627.

6. M. Baskes, C. Melius, and W. Wilson, "Hybrid Calculations of the Properties of Hydrogen in Metals," these proceedings.

7. J. P. Hirth, "Effects of Hydrogen on the Properties of Iron and Steel," Metall. Trans., 11A (1980) pp. 861–890.

8. S. R. Reintsema, S. A. Drentje, P. Schurer, and H. DeWaard, "Lattice Location of Xenon Impurities Implanted in Iron Derived from Mössbauer Effect Measurements," Radiat. Eff., 24 (1975) pp. 145–154.

9. S. R. Reintsema, E. Verbiest, J. Odeurs, and H. Pattyn, "Vacancy Recovery in Heavy Ion-Implanted Refractory BCC Metals Studied by Mössbauer Spectroscopy," J. Phys. F: Metal Phys., 9 (1979) pp. 1511–1527.

10. L. Thomé, H. Bernas, and C. Cohen, "Relation Between Hyperfine-Field and Lattice-Location Measurements for Heavy Impurities in Iron: Influence of Radiation Damage," Phys. Rev. B, 20 (1979) pp. 1789–1798.

DISCUSSION

<u>C. Altstetter, University of Illinois, Urbana</u>: You suggested that Y or Ce
in steel may be a useful way to trap H. Yet you say that the H-trap is a
Y-vacancy complex. How do you propose to get these complexes in a Y-steel?

<u>S. M. Myers</u>: In the text we suggest that the necessary vacancies might be
introduced by cold working. Future experiments with bulk Fe-Ce alloys are
planned to determine whether such a procedure produces strong H trapping.

<u>M. B. Lewis, ORNL, Tennessee</u>: We have made similar measurements of D-vacancy
binding in type 316 steel (FCC). We find a somewhat smaller value (0.33eV)
for the binding energy. Can you rationalize why there should be a significant
difference between BCC iron and FCC steel?

<u>S. M. Myers</u>: The papers by Thomas and by Baskes et.al. in these proceedings
discuss the weaker trapping observed in fcc lattices and the reasons for it.

<u>B. O. Hall, Westinghouse Research, Pittsburgh</u>: Have you considered inter-
actions between deuterium and interstitials/interstitial clusters produced
by ion damage?

<u>S. M. Myers</u>: Since self-interstitials are mobile in Fe above about 120K,
we expect these point defects to agglomerate into dislocations before the
onset of D detrapping at about 260K. In fact, room-temperature TEM of D-
implanted Fe shows a dense array of dislocation loops (Ref. 3). We nevertheless attribute the observed trapping in unalloyed Fe to vacancies and vacancy clusters. One reason is that D in the 0.5eV trap occupies the lattice
position predicted theoretically for a monovacancy, as discussed. Additionally, the observed array of dislocation loops seems inadequate to accommodate
the high concentration of trapped D, about 1 at.%.

<u>P. W. Adler, Grumman Aerospace Corp. NY</u>: What do you feel is responsible for
the high temperature tail indicated in your results for unalloyed and implanted
specimens?

<u>S. M. Myers</u>: These tails involve less than 10% of the implanted D. We have
not identified the responsible mechanism.

<u>G. Pressouyre, Creusot-Loire Res. Center, France</u>: Part of the chemical inter-
action Ti-H is due to the electronic nature of Ti in α Fe: Ti has <u>less</u> electrons than Fe in the 3d band, and H carries an <u>excess</u> electron. Therefore,
all elements on the left of iron in the periodic table should have such an
electronic attractive force for H. Example: Sc, Nd, etc... See: G. M.
Pressouyre, Met. Trans., 1979, 10A, p. 1571.

<u>S. M. Myers</u>: Our interpretation includes a chemical interaction of H with
implanted Ti and Y, but none with Ki, consistent with your criterion. Alternatively, in the text we note a qualitative correlation of the chemical bind-
ings with the heats of hydriding for the implanted elements.

INFLUENCE OF MICROSTRUCTURE OF IRON AND STEELS

ON STEADY STATE HYDROGEN PERMEATION

E. Riecke, F. Schambil and K. Bohnenkamp

Max-Planck-Institut für Eisenforschung GmbH,

Düsseldorf, FRG

ABSTRACT

The effect of microstructural heterogeneities in iron and two low
alloyed steels on steady state hydrogen permeation was studied in the
temperature range of 15 to 80°C by means of the electrochemical permeation
method using the hydrogen gas phase charging technique at f_{H_2} = 1bar. The
permeation coefficient for annealed iron was found to be ϕ_H = 3,04 10^{-7}
exp(-34.7kJ/mole·RT) $[$moleH/cm·s·bar$^{1/2}]$. In iron, there was no visible
influence of impurities, of oxide inclusions or of a high density of lattice
defects introduced by cold rolling on the steady state permeation rate. In
steel specimens of different martensitic, bainitic or pearlitic structures,
the hydrogen permeability is more or less decreased. Additional cold
deformation by rolling of about 40 to 70% causes further decrease in the
permeability. In all cases, however, no change in the temperature dependence
was observed.

INTRODUCTION

The main effect of microstructural heterogeneities in iron and steels on instationary hydrogen diffusion at temperatures lower than 100°C is due to the attractive interactions of dissolved hydrogen with the various types of lattice imperfections (1-8). In case of steady state hydrogen permeation, local equilibrium between the populations of dissolved and trapped hydrogen is established. Hydrogen transport is carried on predominately by the mobile hydrogen atoms on normal interstitial positions and is hindered by impurities and substitute elements blocking interstitial sites, by precipitates decreasing the diffusion cross-section and by compressive elastic stress fields for which repellent interactions with hydrogen have been demonstrated (9). At high hydrogen fugacities in steels, trapped hydrogen itself may decrease the permeability of the material (10, 11).

In this paper, hydrogen permeation behavior of two low alloyed steels of different microstructure is studied at low hydrogen fugacities and is compared with the behavior of more or less pure iron.

EXPERIMENTAL

The materials used for this work and their chemical compositions are shown in Table I. Vacuum molten pure iron was used to study the influence

TableI: Iron and steel compositions (wt%)

material	C	Si	Mn	P	S	N	Al	Cu	Cr	Ni	Mo	O
pure iron	0.006	0.007	0.001	0.005	0.004	0.001	0.002	—	—	—	—	0.002
techn. iron	0.02	0.01	0.01	0.002	0.009	0.003	0.001	—	<0.01	—	—	0.11
steel A	0.46	1.58	0.75	0.03	0.018	—	0.018	—	0.38	—	—	—
steel B	0.34	0.22	0.60	0.032	0.010	0.008	0.008	0.01	0.97	0.03	0.19	—

of high densities of crystal defects without precipitates. The less pure technical iron is characterized by a high content of spherical oxide inclusions from the metallurgical treatment. Steel A represents a commercial type of quenched and·tempered high strength steel. Steel B is a usual high pressure hydrogen vessel steel.

Sheet specimens about 0.1 to 1.2 mm thick were used. The iron specimens were either annealed or cold rolled to about 80%. Steel specimens were given the specific heat treatment to obtain different microstructures. A martensitic structure for steel A or B was obtained by austenizing the sheets at 880°C for 15 min or 850°C for 30 min and subsequent quenching in water or oil at ambient temperature, respectively. Some martensite specimens were tempered at various temperatures, as indicated in Figs.1, 3 and 4.

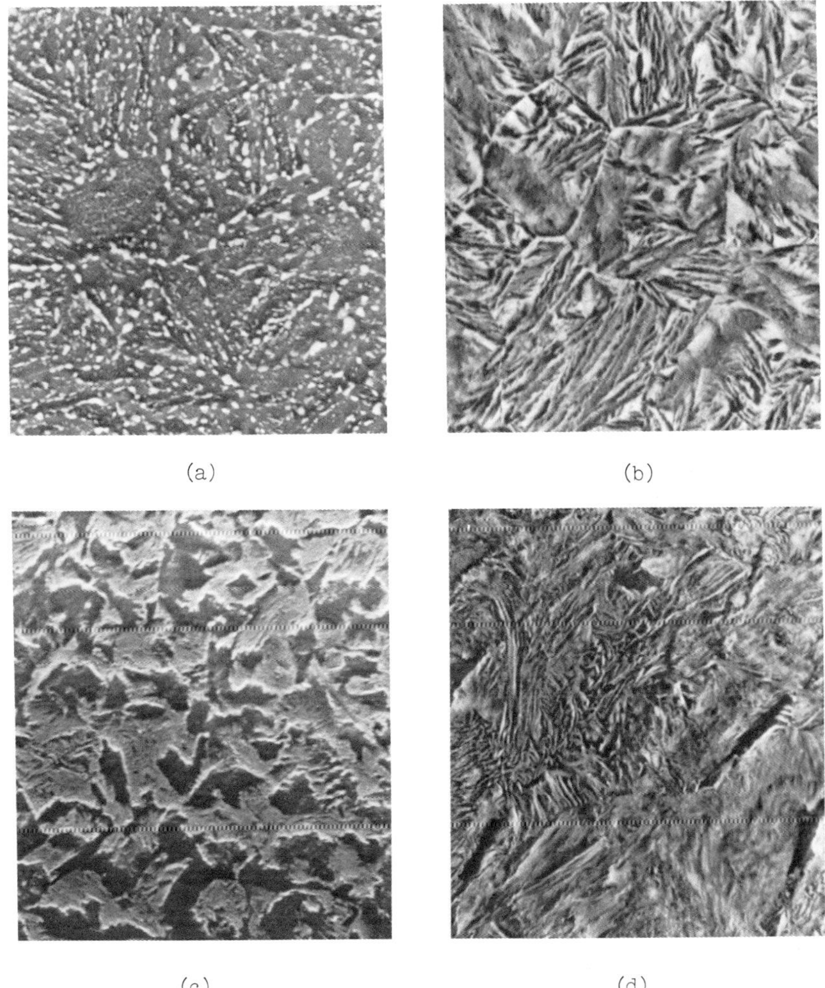

(a)   (b)

(c)   (d)

Fig.1-Microstructures of steel A (carbides light, ferrite dark; X3000);
(a) tempered martensite (2h/700°C), (b) lower bainite,
(c) fine lamellar pearlite, (d) pearlite 70% cold rolled

A fine lamellar pearlitic structure in steel A was obtained isothermally at 550°C for 5 min after the austeniting treatment and a bainitic structure similar by heating at 350°C for 17 min, Fig.1. All the structures were air cooled.

Hydrogen permeation measurements were carried out using the sensitive electrochemical methode (12) adapted to gas phase charging. The entry side of the iron and steel membranes was exposed to a hydrogen fugacity f_{H_2}= 1bar. The exit side was subjected to a constant anodic potential of +200 mV (NHE) in a 0.1N sodium hydroxide solution. Both the entry and exit surfaces were palladium plated to minimize influences of disturbing surface reactions. For the most part the membrane thickness L was between 0.25 and 0.4 mm. The surface area A was about 12 cm^2.

The hydrogen flux $j_H = I_a/A$ F ⌊moleH/$cm^2\cdot$s⌋ is given by the anodic current I_a ⌊A⌋ as measured at the exit side of the membrane and related to surface area and Faraday's constant F ⌊A·s/mole⌋. In case of the stationary state of hydrogen permeation and provided that hydrogen bulk diffusion is the rate determining step in the overall process, Fick's 1.Law can be applied and the steady state flux of hydrogen becomes

$$j_H^\infty = \phi_H \frac{\Delta \sqrt{f_{H_2}}}{L} \qquad (1).$$

$\phi_H(f_{H_2},T)$ ⌊moleH/cm s $bar^{1/2}$⌋ is the hydrogen permeation coefficient and can be studied as a function of hydrogen fugacity, temperature, membrane thickness, material and microstructure. In this investigations, $\phi_H = j_H^\infty L$ because input and output fugacities were $f_{H_2}(x=0)$ = 1bar and $f_{H_2}(x=L) \approx 0$, respectively.

<u>RESULTS AND DISCUSSION</u>

The permeation coefficients ϕ_H obtained under various experimental conditions are presented in Figs.2-5 as a function of temperature in the range of 15 to 80°C.

First, the dependence of ϕ_H on membrane thickness was studied for pure annealed iron, for steel A with a bainitic structure and for steel B in the state of tempered (1h 600°C) martensite. The results in Fig.2 prove the permeation coefficient for these materials to be independent of thickness L and confirm the assumption that the permeation experiments have been diffusion controlled.

The temperature dependence of the permeation coefficient at f_{H_2} = 1bar has been evaluated for pure iron and is given by

$$\phi_H(T) = 3.04 \; 10^{-7} \exp(-34.7kJ/mole\cdot RT) \; \text{⌊moleH/cm·s } bar^{1/2}\text{⌋}$$

This is in good agreement with earlier data (13). In case of steels A and B the permeation coefficient is markedly decreased but essentially unchanged in the temperature dependence. This implies that hydrogen transport in carbon steels is merely impeded by carbides and other obstructive structure elements.

The specific effect of material and microstructure on hydrogen permeability after various heat treatments is described in more detail in

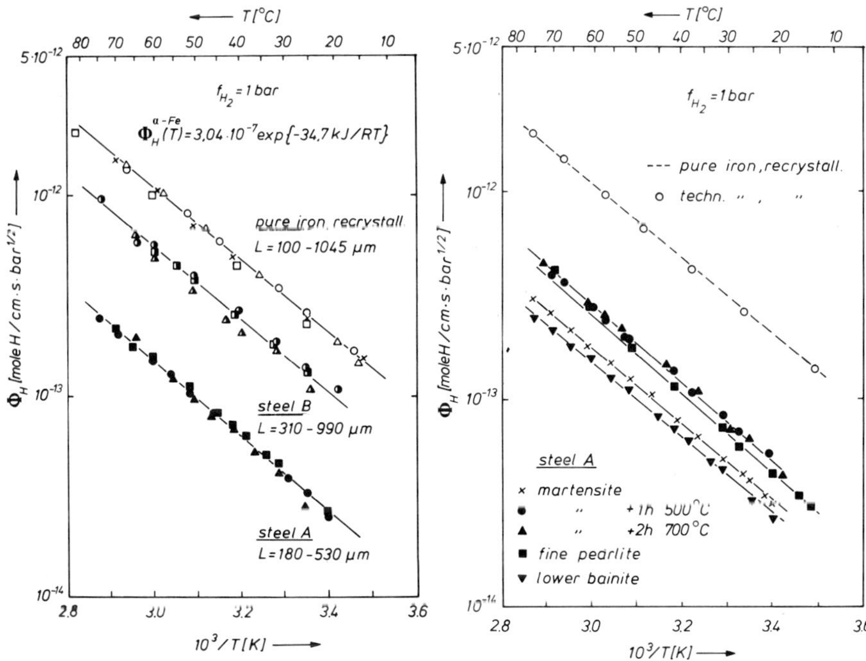

Fig.2-Permeation coefficient as a function of temperature and membrane thickness for iron and two alloyed steels.

Fig.3-Permeation coefficient as a function of temperature and microstructure for iron and steel A.

Figs.3 and 4. As shown in Fig.3, Φ_H-values obtained with less pure iron coincide with those from pure iron; that is a high content of uniformly distributed spherical oxide inclusions and an increased concentration of impurities as given with the technical iron do not change the steady state rate of hydrogen permeation remarkably. On the other hand, it is shown that in steel A steady state permeation is decreased by at least a factor 4. The lowest Φ_H-values were obtained from specimens with a martensitic and a bainitic structure. Both of these structures are characterized by a highly distorted lattice with increased internal stresses, by a high density of lattice defects and a supersaturation in carbon due to the diffusionless transformation of the prior austenitic structure at low temperatures (martensite) or at a sufficient low carbon content with carbide precipitation (bainite). In case of the annealed structures of fine lamellar pearlite or of tempered martensite the internal stresses are decreased, lattice distortions and defects are widely removed and the concentration of carbon atoms is reduced substantially by carbide precipitation. The residual hydrogen traps in the matrix are shallow and do not hinder the

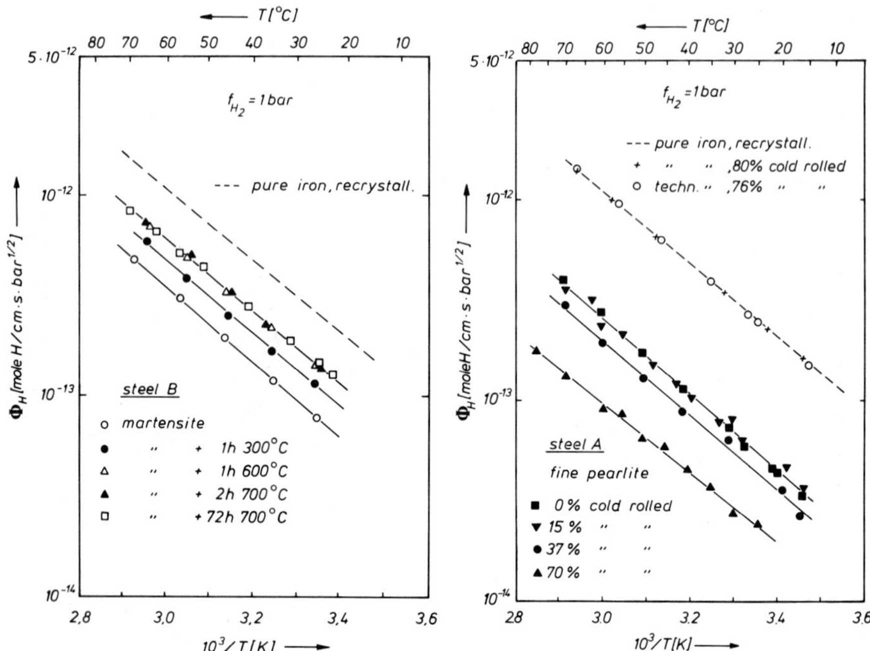

Fig.4-Permeation coefficient as a function of temperature and microstructure for steel B.

Fig.5-Permeation coefficient as a function of temperature and degree of cold rolling for iron and steel A.

steady state hydrogen transport essentially as shown with iron in Fig.5. There was no size effect of the carbide precipitates.

Approximately the same results were obtained from steel B, Fig.4, the Φ_H-values lying about a factor 2 higher than those from steel A. As before, the lowest values were obtained from specimens of martensitic microstructure. An additional heat treatment at temperatures between 300 and 700°C increased the permeation coefficient reaching at 600°C final values. No size effect of the carbide particles on steady state permeation was observed, not even after a 700°C-treatment over 72 h. There was also no effect of the grain size of the prior austenite on steady state permeation.

Recent results from steel B specimens decarburized to ≤ 5 ppm C yielded nearly the same Φ_H-values as obtained from martensite tempered at ≥ 600°C. This implies that carbide precipitates do not play a dominant role in tempered martensite. The observed decrease in steady state hydrogen permeation in comparison with pure iron seems also to be due to elements as manganese, silicon and other. In martensite and bainite structures also supersaturated carbon-atoms may affect steady state permeation.

Finally, the influence of cold rolling on hydrogen transport at steady state permeation through iron and steel specimens is shown in Fig.5. There is evidence, in pure iron as well as in less pure technical iron steady state hydrogen permeation at low fugacities is not affected by a high density of dislocations and other defects introduced by cold rolling up to 80%. The same behavior is indicated by the results from steel A specimens with a fine lamellar pearlitic structure additionally cold rolled to about 15%. The corresponding ϕ_H-values are unchanged compared with the results from undeformed specimens. A remarkable decrease of the permeation coeffecient was found only at higher degrees of cold deformation (>30%) (10, 11). After cold rolling to about 70% steady state permeation was decreased by nearly a factor 3. This may be due to the numerous tiny carbide plates which have been cracked and distributed by rolling throughout the ferrite volume turning into the rolling plane almost perpendicular to the direction of hydrogen permeation. The cementite plates can be assumed to be much less permeable by hydrogen than ferrite zones. Thus the mean diffusion way is increased.

The temperature dependence of the permeation coefficient was found to be essentially unchanged for all the microstructures investigated, Figs.2-5. This is reasonable for tempered martensitic and pearlitic microstructures. In structures without precipitates, carbon-atoms and alloying elements in solution will change the solubility as well as the mobility cf hydrogen. Further investigations are needed.

<u>CONCLUSIONS</u>

1) Steady state hydrogen permeation through more or less pure iron at low hydrogen fugacity (f_{H_2}= 1bar) is independent of usual amounts of impurities, of elevated contents of spherical oxide inclusions and of lattice defects introduced by cold rolling up to 80% reduction in thickness.

2) The permeation coefficient for hydrogen in iron is found to be $\phi_H(T)$ = $3.04 \cdot 10^{-7} \exp(-34.7 kJ/mole\ RT)$.

3) In low alloyed steels of different microstructure the permeation coefficient is decreased by a factor 2 to 10. Martensite, lower bainite, and **heavily** deformed pearlite showed the lowest ϕ_H-values.

4) The temperature dependence of the permeation coefficient for iron and steels investigated is nearly independent of microstructure.

5) The ϕ_H-values for f_{H_2}= 1bar can be used to calibrate hydrogen charging experiments in terms of hydrogen fugacities or activities.

REFERENCES

(1) A. McNabb and P.K.Foster: Trans. Met. Soc. AIME, 1963, vol.227, p.618

(2) J.F. Newman and L.L.Sheir: J. Iron Steel Inst., 1969, vol.207, p.1369

(3) R.A. Oriani: Acta Met., 1970, vol.18, p.147

(4) A.J. Kumnick and H.H.Johnson: Met. Trans., 1974, vol.5, p.1199

(5) E.Riecke: Arch. Eisenhüttenwes., 1976, vol.47, p.247
presented at the 2nd Intern. Congress on Hydrogen in Metals, Paris,
France, 6-10 June, 1977; Proceedings, vol.5, Sect.2B8, Pergamon Press
Oxford, 1977

(6) A.J. Kumnick and H.H.Johnson: Acta Met., 1979, vol.28, p.33

(7) W.W. Robertson and A.W.Thompson: Met. Trans. A, 1980, vol.11A, p.553

(8) R.B. McLellan: Scripta Met., 1980, vol.14, p.513

(9) H.A. Wriedt and R.A.Oriani: Acta Met., 1970, vol.18, p.753

(10) H.G. Schumann and F.Erdmann-Jesnitzer: Arch. Eisenhüttenwes., 1953,
vol.24, p.353

(11) R. Pöpperling and W.Schwenck: Werkstoffe und Korrosion, 1979,
vol.30, p.603

(12) M.A.V. Devanathan and Z.Stachurski: Proc. Roy. Soc., 1962,
vol.A270, p.90

(13) A.J. Kumnick and H.H.Johnson: Met. Trans. A, 1975, vol.6A, p.1087

THE EFFECT OF SULPHIDE INCLUSIONS ON THE

DIFFUSION OF HYDROGEN IN STEELS*

P.H. Pumphrey
Central Electricity Generating Board
Central Electricity Research Laboratories
Leatherhead, Surrey, England.

Internal surfaces may affect the permeation of hydrogen through a steel membrane both by acting as traps for the accumulation of hydrogen and by facilitating the local, short circuit, diffusion of hydrogen. Immediately following the initial exposure, trapping by grain boundaries and particle/ matrix interfaces contributes to the well known, anomalously low diffusion coefficient of hydrogen in iron and ferritic steels at temperatures $\lesssim 200^{\circ}C$.[1] Once the traps have been filled, steady state permeation independent of the trap density is achieved[1-3]. It is at this stage that any short circuit diffusion along internal interfaces can be separated from trapping effects.

Although there have been a number of experimental investigations of possible short circuit diffusion of hydrogen along grain boundaries in iron [4-6], in the case of particle/matrix interfaces, work seems to have been directed exclusively towards investigating trapping at interfaces[7] rather than to possible short circuit diffusion. However, steels containing elongated Type II (FeMn)S inclusions have been shown to have lower steady state permeation rates where inclusions lie parallel to the exposed surface rather than perpendicular to it[8,9]. This has been shown to result from either a higher degree of internal cracking and surface blistering[8] or, in the absence of promoter added to an acid electrolyte, to a lower concentration at the metal surface of H_2S produced by the dissolution of the inclusions[9]. The object of the present work was to investigate the contribution of short circuit diffusion where inclusions lie perpendicular to the exposed surface.

First, expressions will be derived theoretically for the effective diffusion coefficient at steady state of a steel containing an array of elongated inclusions. By substituting parameters appropriate to typical steels, the conditions under which the effects of interfacial diffusion might be detected by permeation experiments at temperatures close to ambient will be deduced. The results of these experiments will be presented and discussed.

* This work was carried out at the Central Electricity Research Laboratories
and is published by permission of the Central Electricity Generating Board.

Calculation of the Effective Diffusion Coefficient

Consider a steel containing inclusion stringers to be made up of an assembly of elements identical to that in Fig. 1(a). Each element side y and length (ℓ + x) contains one inclusion of square cross section, side a and length ℓ. The interface is treated as a narrow channel of width δ.

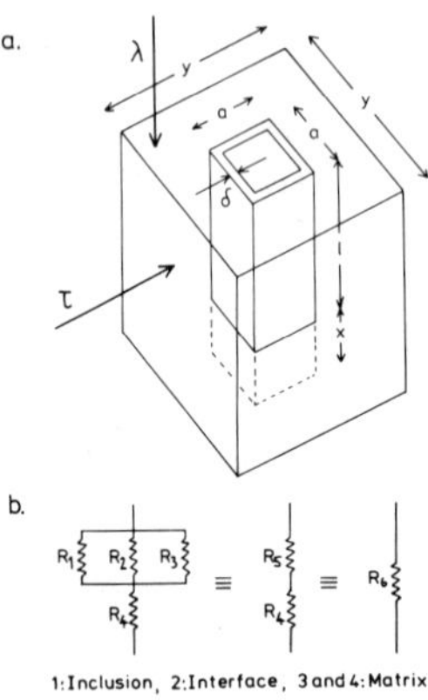

Fig. 1 - (a) Model of element in steel containing inclusion.

     (b) The electrical resistance analogue

The time dependent permeation through such a composite may be derived using the technique employed by Ash, Barrer and Palmer[10]. However, in the present context only steady state permeation is of interest and the effective diffusion coefficient D* may be determined more simply using the electrical analogue[11] in Fig. 1b. By comparing Ficks law for steady state permeation and Ohms Law for current flow, it follows that if concentration differences between adjacent phases are neglected, the resistance to permeation R is given by

$$R = L/DA$$

L = thickness, D = diffusion coefficient, A = area.

Therefore, for diffusion in the longitudinal direction λ

$$R_1 \simeq \frac{\ell}{D_1 a^2} \qquad R_2 \simeq \frac{\ell}{D_2 \cdot 4a\delta}$$

Both are good approximations as $\delta \ll a$

$$R_3 = \frac{\ell}{D_1 (y^2 - a^2)} \qquad R_4 = \frac{x}{D_1 y^2} \qquad R_6 = \frac{(\ell + x)}{D^* y^2}$$

D_1 = diffusion coefficient for the matrix and the inclusion which is assumed to be the same in the absence of better information

D_2 = diffusion coefficient for the interface

After some manipulation it may be shown that:

$$D^*_\lambda = \frac{\left(\dfrac{1}{4D_2\delta(y^2-a^2)} + \dfrac{1}{D_1a(y^2-a^2)} + \dfrac{1}{4D_2a^2\delta}\right)\left(\dfrac{\ell+x}{y^2}\right)}{\left(\dfrac{\ell}{4D_1D_2a^2\delta(y^2-a^2)} + \dfrac{x}{4D_1D_2y^2\delta(y^2-a^2)} + \dfrac{x}{D_1^2ay^2(y^2-a^2)} + \dfrac{x}{4D_1D_2\delta a^2y^2}\right)}$$

An expression for the effective diffusion coefficient D^*_τ in the transverse direction τ may be derived in the same way.

The only completely unknown parameter in these expressions is the interfacial diffusion coefficient D_2. Therefore, D^* is plotted in Fig. 2 for a range of D_2 as the other parameters are varied. The curves are for diffusion in the longitudinal direction, except where indicated in Fig. 2(d). The volume fraction of particles is controlled by adjusting the cell size, y. The width of the interface δ is taken to be 0.5nm, a value commonly assumed in grain boundary studies. The values of the other parameters are as indicated in Fig. 2.

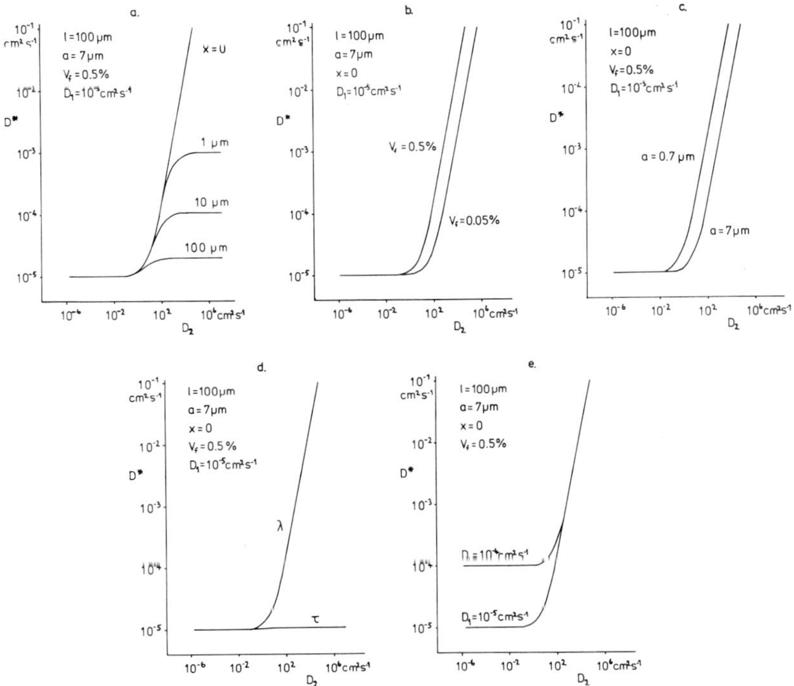

Fig. 2 - Effective diffusion coefficient D^* as function of interfacial diffusion coefficient D_2 showing the effects of (a) particle separation, x (b) particle volume fraction, V_f (c) particle cross section, a (d) longitudinal λ and transverse τ orientation (e) matrix diffusion coefficient D_1.

The separation x between the inclusions is a measure of the continuity and alignment of the inclusions in the stringers. It is clear from Fig. 2(a) that at high D_2, D* is considerably higher for continuous inclusions (x=O) than for even small separations (say, x = 1Oµm). This is because permeation rates are limited by diffusion through the matrix between the inclusions. In considering the other variables, the effect of the inclusions is maximized by putting x = O. This is also the appropriate value of x in assessing the local effect where inclusions do not penetrate the membrane.

The variations of D* with inclusion volume fraction and particle size commonly encountered in steels are shown in Fig. 2(b) and (c) respectively. At high D_2 where interfacial diffusion dominates, D* is proportional to the total inclusion perimeter. Hence D* varies directly as the volume fraction and inversely as the particle size at constant volume fraction.

The comparison between D* in the longitudinal and transverse directions in Fig. 2(d) shows that even for a relatively high volume fraction of 0.5%, short circuit diffusion has virtually no effect on diffusion in the transverse direction even at high D_2. Permeation in this orientation is limited by diffusion through the matrix between the inclusions.

To cover the range of diffusion coefficient most commonly quoted for hydrogen in α-iron[2,12-16] the effect of increasing D_1 from 10^{-5} to $10^{-4} cm^{-2} s^{-1}$ is shown in Fig. 2(e). As might be expected, D* is increased only at relatively small D_2 where matrix diffusion dominates.

On the basis of Fig. 2, it is clear that the best chance of detecting experimentally the effect of interfacial short circuit diffusion on steady state permeation will be where the number of inclusions penetrating the membrane is maximized. This will occur for thin membranes of a high sulphur steel in the longitudinal orientation.

<u>Experimental</u>

The materials used were the hot rolled bars of En3a and Enla mild steel employed in the earlier work[9]. Both alloys contained sulphide inclusions elongated parallel to the rolling direction (Fig. 3) and had sulphur levels of 0.046 and 0.24% respectively*.

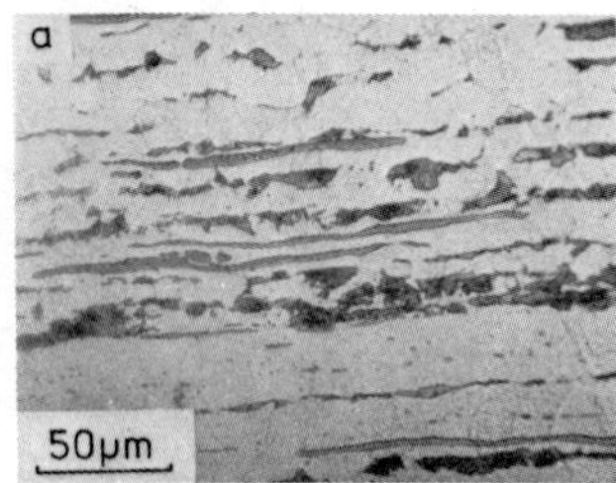

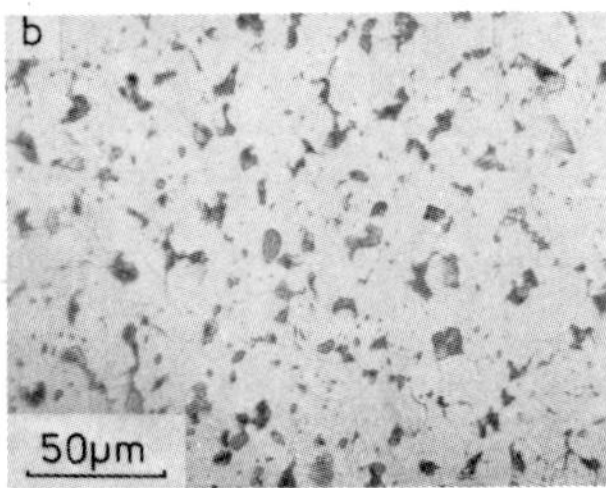

Fig. 3 –

High sulphur Enla steel sectioned (a) parallel (b) perpendicular to rolling direction

*The full analyses were as follows:
<u>En3a</u> 0.16C, 0.1Si, 0.75Mn, 0.046S, 0.014P, 0.50Ni, 0.20Cu, 0.80Cr, 0.09Mo, <0.02V, <0.01Ti, <0.07Co, <0.09Al

<u>Enla</u> 0.12C, <0.1Si, 1.5Mn, 0.24S, 0.061P, <0.06Ni, 0.30Cu, 0.08Cr, <0.03Mo, <0.02V, <0.01Ti, <0.07Co, <0.09Al

Samples $\sim$13mm square and $\sim$1mm thick were cut with a fine saw, both perpendicular and parallel to the rolling direction for permeation in the longitudinal and transverse directions respectively. Membranes 0.03 to 0.30mm thick were prepared by grinding on successively finer carborundum papers with a final finish of 1200 grit.

Membranes were glued to a glass specimen holder so that an area of 25mm^2 on one side could be cathodically charged at 300μA mm^{-2} in sulphuric acid $_{pH\sim2+}$ $\sim$10ppm H_2S at 2°C and 50°C in the cell shown in Fig. 4. The volume of hydrogen permeated was monitored by the displacement of a water droplet in the capillary tube.

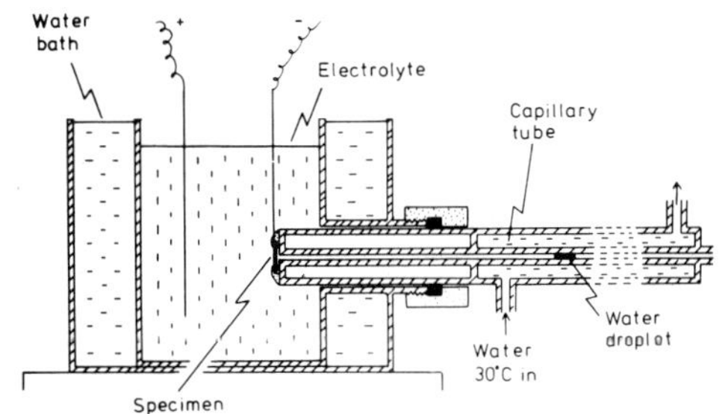

Fig. 4 - Specimen holder and
cell for permeation
measurements

Following the permeation experiments, the numbers of inclusions in the longitudinal orientation penetrating the membranes were determined by making use of the earlier observation that inclusions are selectively dissolved by 1N H_2SO_4 at 2°C[9]. Inclusions removed in this way were counted by viewing the membranes in transmitted light.

<u>Results</u>

It is clear from Fig. 5 that within $\pm$10% there are linear relationships between the inverse of the steady state permeation rate (p^{-1}) and the membrane thickness (L) at both 2°C and 50°C. There is no evidence that thin membranes of the high sulphur steel with inclusions in the longitudinal orientation show anomalously high permeation rates.

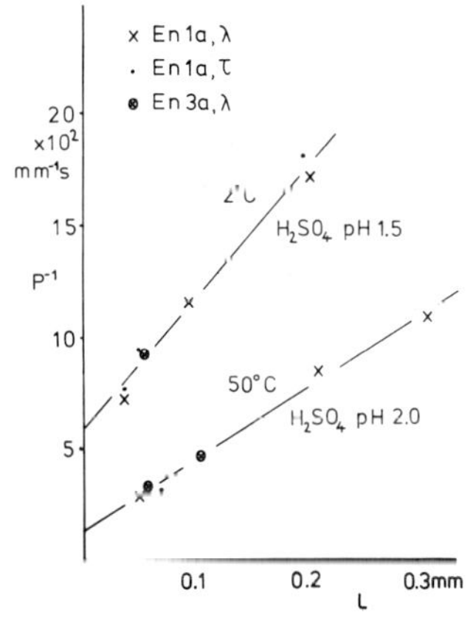

Fig. 5 - Inverse of steady state
permeation rate p^{-1} as
function of membrane
thickness L for cathodic
charging at 300μA mm^{-2}

The transmitted light micrograph in Fig. 6 shows the tunnels resulting from the dissolution of inclusions in the longitudinal orientation penetrating a membrane of Enla, following exposure to 1N H_2SO_4 at $2^{\circ}C$. The number of penetrating inclusions as a function of membrane thickness (Fig. 7) was determined by counting the tunnels in micrographs such as that in Fig. 6. The mean inclusion diameter was $\sim 2\mu m$.

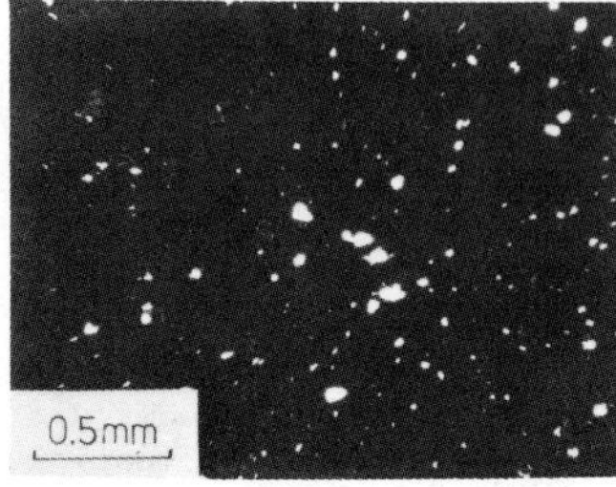

Fig. 6 - Micrograph in transmitted light of Enla membrane (L = 0.047mm) in longitudinal orientation λ, etched 6m in 1N H_2SO_4 at $2^{\circ}C$

Fig. 7 - Number of inclusions N in longitudinal orientation λ penetrating unit area of Enla and corresponding volume fraction V_f as function of membrane thickness L.

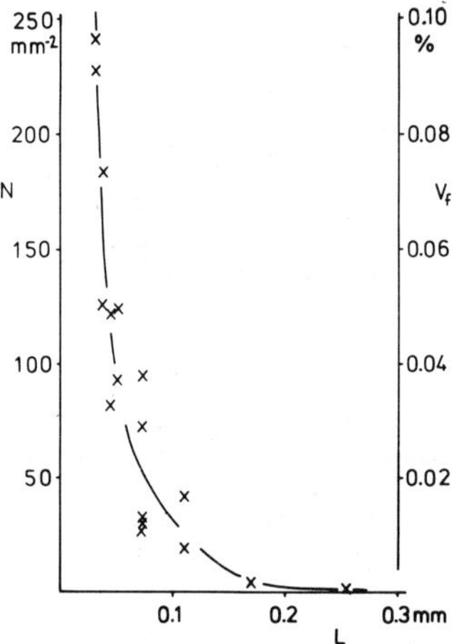

Using the above information, an upper limit estimate of the interfacial diffusion coefficient (D_2) can be made from calculated plots of effective diffusion coefficient (D^*) against membrane thickness (L). As an example, the plots comparing D^*_λ and D^*_τ for the high sulphur Enla steel are shown in Fig. 8 for a lattice diffusion coefficient D_1 of $10^{-4}cm^2s^{-1}$.

For the purpose of calculating D^*_λ, two populations of inclusions were assumed. Those in the first are continuous (i.e. x = 0) and decrease in volume fraction with increasing membrane thickness (Fig. 7). Those in the second, larger, population are discontinuous ($\ell=20\mu m$, $x=50\mu m$) and have a constant volume fraction ($V_f=0.9$%). As expected, Fig. 8 shows that D^*_λ increases with increasing D_2 and decreasing L. Because of the lack of penetrating inclusions, membranes with inclusions in the transverse orientation were taken to contain a single population of discontinuous inclusions ($\ell=20\mu m$, $x=50\mu m$) with $V_f=1.0$% independent of L. Taking the maximum difference of 20% in D^* for thin membranes, an upper limit of $D_2 \sim 10^1 cm^2s^{-1}$ may be deduced. A similar result is obtained by comparing D^*_λ for thin membranes of Enla and En3a. The upper limit for D_2 is reduced to $\sim 10^0 cm^2s^{-1}$ for a lattice diffusion coefficient of $10^{-5}cm^2s^{-1}$.

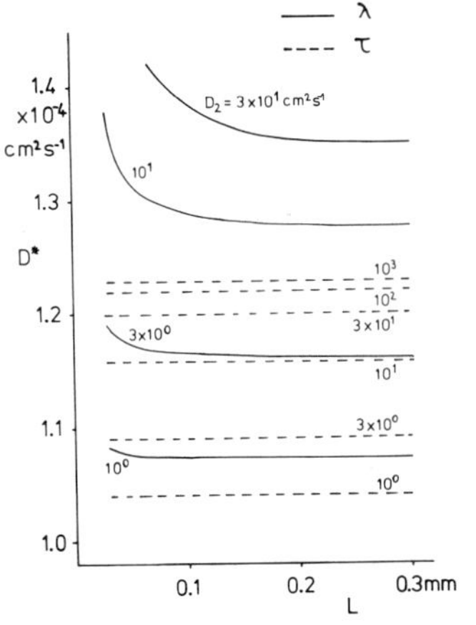

Fig. 8 - Calculated effective diffusion coefficient D* as function of thickness L of Enla membranes in the longitudinal λ and transverse orientations for a range of values of the interfacial diffusion coefficient D_2. Matrix diffusion coefficient D_1, 10^{-4}cm^{2}s^{-1}.

Discussion

The relationships between P^{-1} and L in Fig. 5 are not of the simple form $P^{-1} \propto L$ predicted directly by Ficks Law. They are, however consistent with the relationship, shown in earlier work to be valid for iron[17,18], which takes account of the finite rate constants for the transfer of hydrogen between the adsorbed and absorbed states at the metal surface.

The present results amplify the conclusions of previous work on grain boundaries[4-6] that, at steady state, interfaces in ferritic steels are not significant short circuit diffusion paths for hydrogen at ambient temperatures.

It has been argued[19] that if grain boundaries may be treated as liquid films, short circuit diffusion of hydrogen in α-iron should not be significant, as the activation energy for diffusion in the liquid is higher than in the solid. However, it is well known[20] that by virtue of a lower activation energy, grain boundary dominates matrix diffusion of metallic species in polycrystals at low temperatures. If this applies to the diffusion of hydrogen in α-iron, short circuit diffusion should become significant at temperatures below ambient. However, the absence of any significant effect in the present experiments precludes a distinction between the opposing viewpoints.

Conclusions

1. The effective diffusion coefficient D* of a composite material at steady state may be estimated using an electrical analogue, treating the components as resistors.

2. Calculations for a steel containing elongated inclusions show that where the interfacial diffusion coefficient is sufficiently high:

(a) D* is higher for diffusion in the longitudinal than in the transverse direction

(b) D* in the longitudinal direction is considerably reduced where inclusions are not continuous, even for relatively small separations.

(c)   D* increases as the particle size decreases

(d)   D* increases as the volume fraction of particles increases.

3.   Experimental measurements of steady state permeation rates through membranes of mild steels containing 0.046 and 0.24%S failed to reveal any significant effect of interfacial diffusion on permeation.

4.   On the basis of a $\pm$10% scatter in experimentally measured permeation rates, upper limits for the interfacial diffusion coefficient of $\sim10^1$ and $\sim10^0 \mathrm{cm}^2\mathrm{s}^{-1}$ were deduced for assumed matrix diffusion coefficients of 10^{-4} and $10^{-5}\mathrm{cm}^2\mathrm{s}^{-1}$.

References

1.   A. McNabb and P.K. Foster, Trans, Met. Soc. AIME, 1963, vol.227, p.618.

2.   A.J. Kumnick and H.H. Johnson, Met. Trans., 1974, vol. 5, p.1199.

3.   G.R. Caskey and W.L. Pillinger, Met. Trans. 1975, vol. 6A, p.467.

4.   U.V. Bhat and H.K. Lloyd, J.I.S.I., 1950, vol. 165, p.382.

5.   J.O'M. Bockris, J. McBreen and L. Nanis, J. Electrochem. Soc., 1965, vol. 112, p. 1025.

6.   T.P. Radhakrishnan and L.L. Shreir, Electrochim, Acta, 1967, vol. 12, p.889.

7.   J.F. Newman and L.L. Shreir, J.I.S.I., 1969, vol. 207, p.1369.

8.   M. Smialowski, 'Stress Corrosion Cracking and Hydrogen Embrittlement of Iron Base Alloys', NACE 5, Unieux Firminy, France, 1973, p.405.

9.   P.H. Pumphrey, Corrosion NACE, 1980, in press.

10.   R. Ash, R.M. Barrer and D.G. Palmer, Brit. J. Appl. Phys., 1965, vol. 16, p.873.

11.   W. Beck, J. O'M. Bockris, M.A. Genshaw and P.K. Subramanyan. Met. Trans. 1971, vol. 2, p.883.

12.   J. McBreen, Ph. D. Thesis, University of Pennsylvania 1965.

13.   P.K. Subramanyan, Ph. D. Thesis, University of Pennsylvania 1970.

14.   H.G. Nelson and J.E. Stein, NASA TND-7265, 1972.

15.   S. Yoshizawa and K. Yamakawa, 5th Int. Cong. on Metallic Corrosion, NACE, Tokyo, 1972, p. 991.

16.   D.L. Dull, Ph.D. Thesis, University of California, Los Angeles, 1976.

17.   M.A.V. Devanathan and Z. Stachurski, J. Electrochem. Soc., 1964, vol. 111, p. 619.

18.   C.D. Kim and B.E. Wilde, J. Electrochem. Soc., 1971, vol. 118, p. 202.

19.   A.P. Miodownik, 'Stress Corrosion Cracking and Hydrogen Embrittlement of Iron Base Alloys', NACE 5, Unieux-Firminy, France, 1973, p.272.

20.   D. McLean, 'Grain Boundaries in Metals', Clarendon, Oxford, 1957, p.219.

HYDROGEN IN IRON AND STEELS: INTERACTIONS, TRAPS AND CRACK PATHS*

113

R. Gibala and D.S. DeMiglio[x]

Department of Metallurgy and Materials Science
Case Western Reserve University
Cleveland, Ohio 44106

A number of defect sites interact significantly enough with hydrogen in
ferrous alloys to act as traps under circumstances in which the material
exhibits hydrogen-induced loss of ductility. These sites include point defects (solute atoms, vacancies), dislocations (edge and screw), interfaces
and surfaces (grain boundaries, particle-matrix interfaces, cracks, external
surfaces), and volume defects (voids, second phase particles). Experiments
performed under controlled conditions involving careful microstructural
characterization have generated quantitative estimates of number densities,
binding energies and saturabilities of defects, which in turn can be used
to imply probable susceptibilities to hydrogen embrittlement. Current information on hydrogen-defect interactions is reviewed and applied to results
on the nature of hydrogen-assisted crack paths in a high strength steel. It
is shown that both the primary and secondary crack paths can be explained in
terms of the types of microstructural traps that predominate in a given
material.

*Research sponsored by the U.S. Department of Energy
[x]Now at: Babcock and Wilcox, Alliance, Ohio 44601

Introduction

It is now well-established that hydrogen embrittlement of steels correlates closely with hydrogen-defect interaction processes (trapping) which
have been examined by a number of experimental techniques (1-3). Figure 1
illustrates this correlation in an ideal manner with diffusion data as normally obtained in many investigations (4). A steel can be embrittled by
hydrogen if it is stressed at temperatures high enough for diffusion of hydrogen to occur to potential embrittlement sites (T_A in Figure 1), but also
low enough that the hydrogen will not be depleted in some way from embrittle-
ment-producing traps (T_B in Figure 1). In the range of increasing temperatures from $T<T_A$ to $T>T_B$, a given material is expected to undergo a ductile-
brittle-ductile transition determined largely by the nature of hydrogen transport among the populations of various hydrogen traps.

A defect trap for hydrogen is well-characterized if its binding energy
E_B, its number density N_x and its specific saturability (proportional to the
number of hydrogen atoms trapped per defect or defect site) are known. Research in the past decade has resulted in a significant increase in knowledge
of the hydrogen trapping ability of various imperfections in steels. Hirth
(3) has summarized most of the current literature. Table I gives value of
E_B and N_x for a number of traps, along with a few other recent results not
included in his compilations.

Examination of Table I shows that most traps can be classified into one
of three categories relative to E_s, the heat of solution of hydrogen in iron,
which is $\sim$ 29 kJ/mol: weak traps with $E_B < E_s$, moderate traps with $E_B \sim E_s$,
and strong traps with $E_B > E_s$. Each category of trap can affect or control
the mechanical behavior of the hydrogenated material. Strong traps behave
irreversibly in the sense that the rate of escape of hydrogen is relatively
small. Pressouyre and Bernstein (5) have argued that $E_B > 60$ kJ/mol dictates
irreversibility in hydrogen embrittlement situations. Moderate or weak traps
have appreciable escape rates by comparison and act reversibly under the same
experimental conditions. Strong traps with large specific saturabilities
for hydrogen, e.g. incoherent precipitates and other interfaces (5), appear
to have the largest detrimental effects, although the detailed kinetic behavior can be complicated and confusing when the various categories of traps
are present in differing densities (5,6).

By the previous arguments, the nature of the crack path in hydrogen embrittlement should be determined by the extent to which strong traps present
in adequate numbers can collect hydrogen from external sources or reversible
traps. Table II summarizes this statement in a general way. Briefly, one
would expect materials with relatively large densities of reversible traps
to be resistant to well-defined interfacial cracking. The same result could
be obtained with a high density of well-dispersed strong traps characterized
by small specific saturabilities (5,6). In the present paper, we have examined the nature of the crack path in a commercial 4340 steel as a function
of strength level by scanning and transmission electron microscopy. It is
shown that the classification scheme given in Tables I and II can be used to
explain the nature of both primary and secondary crack paths.

Experimental

The 4340 steel used in this investigation was obtained from The Timken
Corporation as 38 mm diameter hot-rolled bars approximately 1 m long. The
composition in wt. pct. was: C-0.40, Mn-0.73, Si-0.30, Ni-1.72, Cr-0.75,
Mo-0.28, S-0.013, P-0.010. Sharp-notched stress rupture specimens were machined

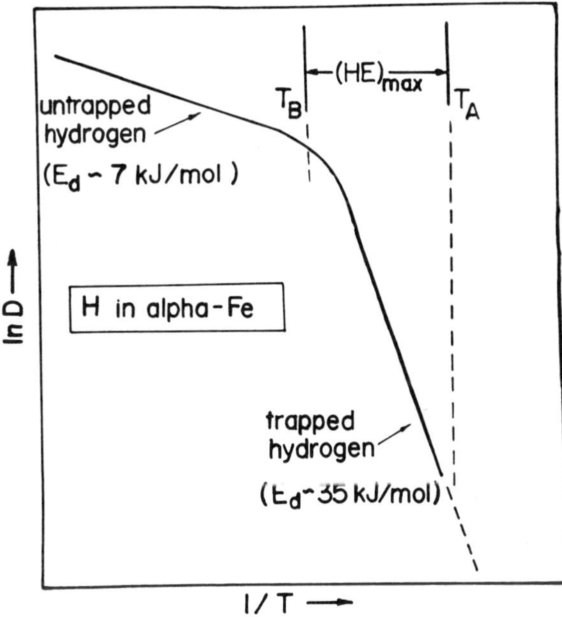

Fig. 1. Schematic illustration of the diffusivity of hydrogen in α-iron.
Room temperature typically falls in the range between T_A and T_B.

Table I. Hydrogen-Trap Interactions

Trap	E_B (kJ/mol)	N_x (no/m^3)
Interstitial solutes (N,C)	$\sim$3–15	10^{25}
Si atom	>20	10^{27}
Ti atom	26	10^{27}
Vacancy	46	$<10^{23}$
Y-vacancy	126	$\sim 10^{23}$
– elastic stress field	20	10^{19}–10^{26}
$\perp$ core (screw)	20–30	10^{19}–10^{26}
$\perp$ core (mixed)	59	10^{19}–10^{26}
$\frac{1}{2}$ H$_2$ (vapor/void)	29	--
Grain boundary	$\sim$59	10^{19}–10^{23}
Free surface	70–95	10^{21}
AlN interface	65	10^{24}–10^{25}
Fe$_3$C interface	84	10^{24}–10^{25}
TiC interface	96	10^{24}–10^{25}

Table II. Trap Classification for Hydrogen-Assisted Cracking

Principal Microstructural Traps	Nature of Trap	H.-A.C. Susceptibility
Solute atoms Microvoids Low angle boundaries Dislocations Coherent precipitates	Weak/moderate Reversible $E_B \lesssim 60$ kJ/mol	Cracking along well-defined interfaces is not favored.
Pre-existing⎱ microcracks Propagating ⎰ High angle boundaries Incoherent ⎰precipitates ⎱interfaces	Strong Irreversible $E_B \gtrsim 60$ kJ/mol	Cracking along interfaces that are irreversible traps is favored.

from these bars, austenitized for 1 h at 870°C, oil-quenched to room temperature, and tempered for 1 h at various temperatures in the range 200°C–600°C or left untempered.

Hydrogen charging was done cathodically in 4% H_2SO_4 at 30 A/m^2 for 300 s with no catalytic poison and was followed immediately by cadmium plating. Specimens were baked for 0.5-3.0 h in air at 150°C to homogenize the hydrogen distribution and then tested to failure (usually) in static loading at stresses below the yield stresses of the unnotched, unhydrogenated materials. Table III gives some of the typical results.

Scanning electron microscopy (SEM) was used to characterize the primary crack paths. Secondary crack paths were determined by optical microscopy and transmission electron microscopy (TEM). TEM specimens were back-thinned from sections cut $\leq$ 1 mm below the fracture surface (7). Conventional bright field and dark field TEM was done on either a 125 kV Siemens 102 or 650 kV Hitachi 650B microscope. Mass thickness imaging was also done to locate fine detail of the crack paths.

Table III. Results of Hydrogen-Assisted Cracking Experiments

Tempering Treatment	Unnotched Yield Stress (MN/m^2)	Notched Tensile Stress (MN/m^2)	Baking Time at 150°C (h)	Applied Stress (MN/m^2)	Time to Fracture (h)
600°C	950	-----	1	860	no failure
500°C	1035	1930	1	860	no failure
400°C	1205	2140	1	860	0.75
320°C	1380	2220	3	690	0.75
200°C	1550	2290	3	515	0.75
As-quenched	>1550	>2290	0.5	240	0.25

Results

Detailed microscopical documentation of the results on the crack path determinations in the form of SEM and TEM micrographs is presented elsewhere (8,9). Table IV summarizes the findings for the various materials, which fall into three categories according to the tempering temperatures.

Tempering Temperatures 500°C and 600°C

These materials did not undergo hydrogen-assisted fracture at stresses below the yield stress. The fracture surfaces of the uncharged and the charged materials that were deformed to fracture consisted primarily of ductile tearing as observed by SEM. The charged materials exhibited a decreased reduction in area in tensile tests and a larger dimple size than their uncharged counterparts, consistent with observations by other investigators (10). TEM observations of the secondary crack path in these materials were limited, but appeared to consist of predominately random translath cracks that often intersected cementite particles. Occasional cracking occurred along lath and lath colony boundaries, but mainly as transition paths between larger translath segments.

Table IV. Crack Path Data in 4340 Steel

Tempering Treatment	SEM Fracture Characteristics of the Uncharged Material	SEM Fracture Characteristics of the Primary Hydrogen-Assisted Crack Path	TEM Fracture Characteristics of the Secondary Hydrogen-Assisted Crack Path
600°C	Ductile tearing	Ductile tearing	Translath cracking Interlath cracking
500°C	Ductile tearing	Ductile tearing	Translath cracking Interlath cracking

400°C	Ductile tearing
	Transgranular
	  cleavage
		Ductile tearing
		Transgranular
		  cleavage
			Translath cracking
			Interlath cracking

320°C	Ductile tearing
	Transgranular
	  cleavage
		Ductile tearing
		Transgranular
		  cleavage
			— — — —

200°C	Ductile tearing
	Transgranular
	  cleavage
		Cracking along
		  prior γ gb's
		Transgranular
		  cleavage
			Translath cracking
			Interlath cracking
			Intercolony cracking
			Cracking along
			  prior γ gb's

As-quenched	Ductile tearing
	Transgranular
	  cleavage
	Cracking along
	  prior γ gb's
		Cracking along
		  prior γ gb's
		Ductile tearing
			Translath cracking
			Interlath cracking
			Intercolony cracking
			Cracking along prior
			  γ gb's

Tempering Temperatures 320°C and 400°

Table III illustrates that these materials exhibited delayed hydrogen-
induced failure at stresses below the yield stress. The primary crack path
observed by SEM is transgranular such that the fracture surface consists of
approximately equal amounts of ductile tearing and transgranular cleavage in
the hydrogen-charged material, similar to results by Beachem (11). Both
fracture modes exist in the uncharged material, but the proportion of transgranular cleavage is smaller. The secondary crack path observed by TEM is
predominately translath, with some tendency toward fracture on {110} translath slip planes. Some interlath fracture is also observed, more so than in
materials tempered at 500°C and 600°C.

As-quenched Steels and Steels Tempered at 200°C

These materials, which failed at stresses well below the yield stress,
exhibited primary crack paths that were more than 50% intergranular fracture
along prior austenite grain boundaries, similar to other observations (11-13).
Transgranular cleavage constituted another 20-30%, with the balance, often
10% or less, being ductile tearing. The secondary crack paths were varied
in these materials. Translath, interlath and intercolony cracks were observed in most foils, with higher percentages of the latter two modes than
in materials tempered at higher temperatures. In addition, secondary intergranular cracks along prior austenite grain boundaries were often observed.
The general trend was toward more interfacially-oriented primary and secondary
crack paths in the stronger materials. It should be noted that the as-
quenched material was mildly tempered at 150°C as part of the hydrogen baking
treatment. Thus the similarity in results of these two materials is expected.

Discussion

It has become widely accepted in recent years that manipulation and
control of microstructure is an effective method of obtaining resistance to
hydrogen-assisted cracking in steels (14-16). In view of the nature of
hydrogen trapping by many types of microstructural imperfections given in
Tables I and II, it is clear that the effect of microstructure on hydrogen
embrittlement is an effect of hydrogen trapping on the nature of the hydrogen-assisted crack path. The following discussion is intended to illustrate
that the nature of microstructural traps in 4340 steel as a function of tempering temperature is consistent with the trap classification scheme given in
Table II and the crack path expected from the classification.

Table V lists the major microstructural traps for hydrogen in 4340
steel and attempts to identify which traps are reversible or irreversible
in nature. The list and categorization are not intended to be complete or
absolute, but they do indicate a simple generalization: materials tempered
at the higher temperatures tend to have an abundance of reversible types of
traps, whereas materials tempered at the lower temperatures tend to have an
abundance of irreversible traps. This observation, plus the recognition in
Table II that prediction of the hydrogen-assisted crack path relies mainly
on identification of the irreversible trapping sites, allows one to understand the results on the crack path determinations given in Table IV and
summarized schematically in Figure 2. Figure 2 illustrates the crack paths
only for the materials examined extensively: those tempered at 320°C or 400°C
(in(i)) and the as-quenched materials or those tempered at 200°C (in(ii)).
The crack path in materials tempered at 500°C and 600°C resembles that in
(i), but with relatively little inter-boundary cracking.

Table V. Nature of Microstructural Traps for Hydrogen in 4340 Steel

Tempering Treatment	Description of the Microstructure	Reversibile Traps for Hydrogen	Irreversible Traps for Hydrogen
600°C 500°C	Spheroidized, incoherent Fe_3C at low angle lath bdies Dislocation recovery	Low angle lath bdies and lath colony bdies Dislocations Solute atoms Incoherent and coherent Fe_3C	Microcracks Incoherent Fe_3C
400°C	Early-stage Fe_3C spheroidization, dislocation recovery, and high-to-low angle lath bdy transformation		
320°C	Very little dislocation recovery Coherent Fe_3C at high angle lath bdies		Microcracks High angle lath bdies and lath colony bdies Prior γ gb's
200°C	No significant dislocation recovery Coherent ϵ-carbide at high angle lath bdies	Dislocations Solute atoms ϵ-carbide	
As-quenched	Precipitate-free high angle lath bdies Carbon in solution in α'	Dislocations Solute atoms	

The prediction of the crack path in the classification scheme in Table II relies on the fact that <u>all</u> irreversible traps bind hydrogen effectively by virtue of the large E_B and N_x and that <u>most</u> irreversible traps are also characterized by large specific saturabilities. A large saturability in general can permit accumulation of hydrogen to a critical concentration necessary to initiate a hydrogen-assisted crack (17). The classification scheme in Table II works because most of the irreversible traps are either extended interfaces of various types or second phase particles that precipitate at such interfaces. The yttrium-vacancy point defect trap with $E_B \sim 126$ kJ/mol and $N_x \sim 10^{23}$ per m^3 reported by Myers et al. (18) is a probable example of an irreversible trap with a low specific saturability. So are other point defect traps reported to have large E_B's (6) and small, finely dispersed incoherent precipitates.

The presence of reversible traps can greatly complicate the sensitivity to hydrogen embrittlement (6). Reversible interfacial traps, which predominate in the materials tempered at 500°C and 600°C, do not trap hydrogen for

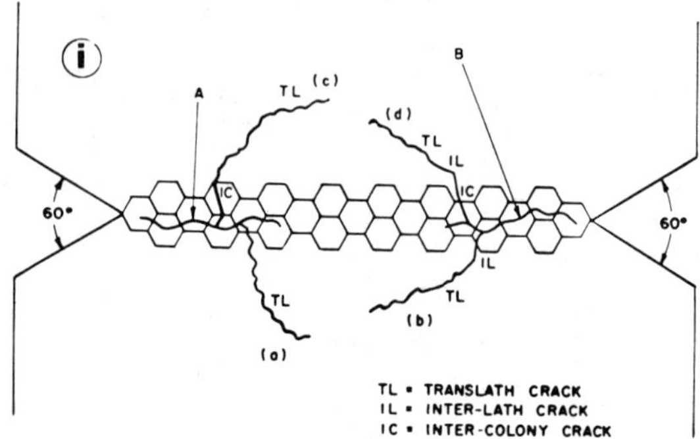

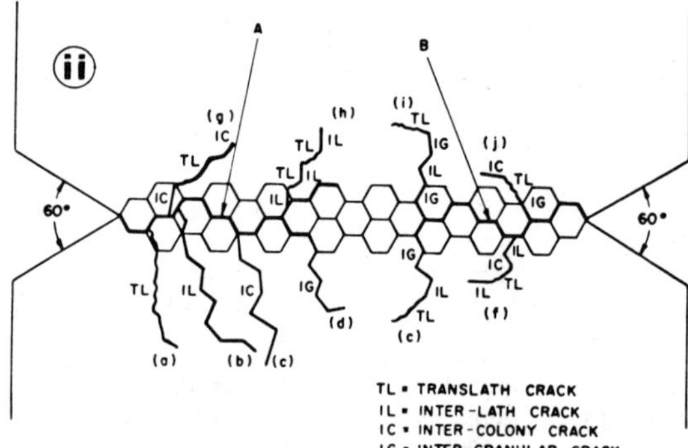

Fig. 2. Schematic illustration of hydrogen-assisted primary (A,B) and secondary (a,b,......) cracking in sharp-notched 4340 steel as observed
by SEM and TEM respectively. (i) Summary of results for materials
tempered at 400°C and 320°C. (ii) Summary of results for materials
tempered at 200°C and as-quenched materials.

long enough times to allow a large hydrogen accumulation. Thus hydrogen-
assisted cracking along reversible interfacial traps is not expected. However, these and other reversible traps can help deter irreversible interfaces
from trapping adequate hydrogen to become crack initiation sites. On the
other hand, mobile reversible traps (dislocations) might actually transport
hydrogen to pre-existing or propagating cracks or other irreversible microstructural traps (19) and initiate and accelerate hydrogen cracking at those
sites. The same argument applies to the materials tempered at 400°C, except
that incomplete recovery and carbide spheroidization leaves these materials
with a smaller relative density of reversible traps and a correspondingly
higher remnant irreversible trap population in the incompletely recovered
martensite substructure. None of the materials tempered at 400°C and above
contain a high enough density of major irreversible traps with high specific
supersaturabilities (apart from pre-existing or propagating cracks) to promote a large fraction of interfacially-aligned fracture paths.

The untempered materials or ones tempered at relatively low temperatures
contain a large density of irreversible interfacial traps in the form of
high angle austenite grain boundaries and unrecovered lath and lath colony

boundaries. The presence in the microstructure of these major irreversible traps with relatively high supersaturabilities results in predominately interfacial hydrogen-assisted primary and secondary cracking. Evidently, the presence of coherent ε or Fe_3C as immobile reversible traps is not adequate to prevent either interfacial primary or secondary cracking at irreversible traps.

It might also be argued that the nature of the primary and secondary hydrogen-assisted crack path correlates with the strength of the material. Certainly the primary crack path undergoes a transition from intergranular to transgranular cracking with decreasing strength, at the same time that secondary cracking path undergoes a transition from predominately interfacial to random trans-boundary cracking. However, in the same manner that it has been shown that susceptibility to hydrogen embrittlement does not correlate with strength, but is more directly an effect of microstructure (15,16), we would argue that the correlation of crack path with strength is an effect of microstructure in the manner outlined in Tables IV and V.

Summary and Conclusions

The nature of the hydrogen assisted crack path in quenched and tempered 4340 steels can be correlated with the trapping ability of the major microstructural features of the steels. Materials tempered at lower temperatures contain a relatively high density of irreversible traps (high binding energy, high number density, large specific saturability) and tend to exhibit primary and secondary crack paths along irreversible interfacial traps. Materials tempered at higher temperatures contain a relatively high density of reversible traps and exhibit more random trans-boundary primary and secondary crack paths.

Acknowledgements

The authors are indebted to Drs. Terence Mohr and Victor Lou for their help and advice in the experimental work. The research was supported by the United States Department of Energy.

References

1. R. Gibala, pp. 244-268 in _Stress Corrosion Cracking and Hydrogen Embrittlement of Iron Base Alloys_, R.W. Staehle et al., eds., NACE, Houston, 1977.

2. A. Miodownik, pp. 272-285, ibid.

3. J.P. Hirth, _Met. Trans. A_, _11A_ (1980) pp. 861-890.

4. J. Völkl and G. Alefeld, pp. 321-348 in _Hydrogen in Metals_, _Topics in Applied Physics_, Vol. 28, G. Alefeld and J. Völkl, eds., Springer-Verlag, New York, 1978.

5. G.M. Pressouyre and I.M. Bernstein, _Met. Trans. A_, _9A_ (1978) pp. 1571-1580; _Acta Met._, _27_ (1979) pp. 89-100.

6. G.M. Pressouyre, _Met. Trans. A_, _10A_ (1979) pp. 1571-1573; _Acta Met._, _28_ (1980) pp. 895-911.

7. J. Awatani, K. Katagiri and H. Nakai, <u>Met. Trans. A</u>, <u>9A</u> (1978) pp. 111-116.

8. D.S. DeMiglio, M.S. Thesis, Case Western Reserve University, Cleveland, Ohio, 1979.

9. D.S. DeMiglio and R. Gibala, submitted to <u>Met. Trans.A</u>.

10. H. Cialone and R.J. Asaro, <u>Met. Trans. A</u>, <u>10A</u> (1979) pp. 367-375. See also these proceedings.

11. C.D. Beachem, <u>Met. Trans. A</u>, <u>3A</u> (1972) pp. 437-451.

12. S.K. Banerji, C.J. McMahon, Jr. and H.C. Feng, <u>Met. Trans. A</u>, <u>9A</u> (1978) pp. 237-247.

13. G.W. Simmons, P.S. Pao and R.P. Wei, <u>Met. Trans. A</u>, <u>9A</u> (1978) pp. 1147-1158.

14. J.P. Hirth and H.H. Johnson, <u>Corrosion</u>, <u>32</u> (1976) pp. 3-26.

15. I.M. Bernstein and A.W. Thompson, <u>Int. Met. Rev.</u>, <u>21</u> (1976) pp. 269-297.

16. A.W. Thompson and I.M. Bernstein, pp. 53-175 in <u>Advances in Corrosion Science and Technology</u>, Vol. 7, R.W. Staehle and M. Fontana, eds. Plenum Press, New York, 1979.

17. A.R. Troiano, <u>Trans. ASM</u>, <u>52</u> (1960) pp. 54-80.

18. S.M. Myers, S.T. Picraux and R.E. Stoltz, <u>Appl. Phys. Letters</u>, in press; see also these proceedings.

19. J.K. Tien, A.W. Thompson, I.M. Bernstein and R.J. Richards, <u>Met. Trans. A</u>, <u>7A</u> (1976) pp. 821-829.

DISCUSSION

<u>Paul Shewmon, Ohio State University, Columbus</u>: Do these cracks ever run along inclusions?

<u>R. Gibala</u>: We've seen some secondary cracks intersect and run along inclusions in optical microscopy observations. However, it is not a major observation.

<u>J. W. Morris, Jr., University of California, Berkeley</u>: We've had two reports so far in the conference by McMahon and Araki that the trans-lath fracture is on the (110) slip plane. Is that your observation as well?

<u>R. Gibala</u>: We have seen (110) translath fracture, but are not certain if it is a general observation.

A HIGH-PERFORMANCE ATOM PROBE FIELD ION MICROSCOPE STUDY OF

SEGREGATION AND HYDROGEN CRACKING IN Fe-0.29 Ti[*]

Y. Kuk, H. W. Pickering and T. Sakurai
The Pennsylvania State University
University Park, PA 16802

With the greatly improved resolution now available in energy focused
atom probes, hydrogen can be readily resolved even when combined with metals
having several isotopes. In addition to finding that H, H_2, FeH and TiH_2
accumulate at segregated grain boundaries in Fe-0.29 wt. % Ti, a striking
observation was made – the formation and propagation of a microcrack when
the (field ion microscope) tip was exposed to hydrogen gas at elevated
temperature. A small crack (~200Å in length) was first noticed at a grain-
boundary intersection during field ion imaging. This was an open crack,
formed by detachment of metal between the intersecting grain boundaries,
which was observed to be much larger after the tip was reheated to 1300K for
10 min. in the presence of 10^2 Pa (1 Torr) H_2. This crack could be easily
reduced in size by gradually field evaporating the surface. Its propagation
was repeated several times and reproducible results were obtained. Hydrogen
was identified in quantity in the crack surface, though not elsewhere.

The observation of H_2 is unusual and is taken to mean that H_2 gas was
trapped in the grain boundary. The grain boundary was also observed to be
enriched in Ti, O, C and S, in agreement with earlier results for Fe-Ti.

[*]This work was supported by the U.S. Department of Energy, Contract No.
DE AS02 79ER10430.

Introduction

This paper presents some of our current progress on an investigation of hydrogen behavior in metals using the recently developed (1,2) high resolution atom-probe field ion microscope (FIM). In spite of the need and importance of the study of hydrogen-metal interactions there have not been, to our knowledge, any systematic studies of hydrogen-metal systems using atom-probes. This, of course, is not without reason. Because hydrogen interacts very strongly with various metals, diffuses very rapidly in iron and certain other metals and cannot be removed easily from the vacuum system, controlled experiments are difficult and extremely time-consuming. Also resolving metal from metal hydrides requires rather high mass resolution not available in conventional straight time-of-flight (ToF) atom-probes.

Since undertaking this investigation, we have obtained a number of encouraging and exciting results: Cluster-formation consisting of Ti, C, N and O in Fe-Ti alloys (3,4); segregation of Ti and P to grain boundaries in iron alloys (3,5); analysis of hydrogen in iron (6); and the direct observation of a microcrack and its propagation when an Fe-Ti tip was exposed to hydrogen (6). The present paper reports on our results obtained for hydrogen charged Fe-Ti alloys. In addition, test data from conventional straight and the Penn State energy focused ToF atom probes are presented in order to illustrate that the latter, with its order-of-magnitude increase in resolution, is capable of analyzing hydrogen and the hydrides of metals.

Experimental Procedure

The Penn State high performance atom probe FIM is shown in Figure 1. Mass resolution and the percent of artifact signals were determined using several different metal and alloy tips. Figures 2 and 3 show the order-of-magnitude improvement in resolution of the Penn State energy focased atom probe over the conventional straight ToF atom probes. Figure 3 also illustrates the elimination of artifact signals frequently observed with straight ToF atom probes.

Fe-0.29 wt. % Ti tips were prepared by wire drawing an homogenized alloy ingot to a diameter of 0.025 cm from a starting diameter of 0.2 cm, annealing the wires in a vacuum chamber at 1090K for 48 hrs. at 10^{-5} Pa (10^{-7} Torr), and preparing in the shape of sharp needles by electrochemical etching in a HNO_3, HCl, and H_2O (1:1:2) solution (8). The alloy was reported by the supplier to contain 0.29 wt. % Ti, 0.52 wt. % Mn and C, P, S, Si, O, N, and Al in the ppm range.

After installing a tip in the FIM vacuum chamber and a brief bake out at 100°C, the chamber pressure was about 10^{-7} Pa (10^{-9} Torr). The atom probe analyses reported in this paper were done at 10^{-7} Pa and a tip temperature of 78 K. Field ion microscopy imaging was done using hydrogen gas at about 10^{-4} Pa (10^{-6} Torr) and 78 K. The Fe-Ti tips were previously characterized in terms of clustering and surface and grain boundary segregation; these phenomena were observed to occur at certain temperatures in the approximate range 710-1270 K (3-5). The Fe-0.29 Ti tip described in this paper contained a grain boundary. Following atom probe analysis it was exposed in situ to hydrogen at elevated temperatures in the following sequence: 10^{-1} Pa (10^{-3} Torr) at 1220 K for 10 min., 10^{-1} Pa at 870 K for 10 min., 1 Pa at 1220 K for 10 min., and 1 Pa at room temperature for 1 day. This was followed by a second sequence with pressures and temperatures in the ranges 1 to 10^2 Pa and 870 to 1300 K. The temperature was measured using an optical pyrometer with an accuracy estimated to be within ± 20 K. After the

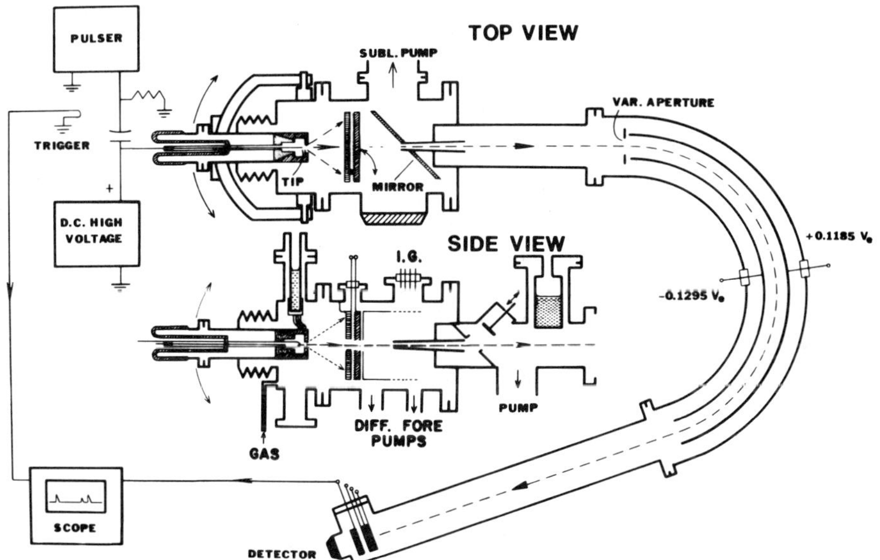

Fig. 1. A schematic of the high performance time-of-flight atom-probe FIM equipped with a curved electrostatic lens(2), an advanced version of conventional straight-type ToF atom-probes. The unique features of the high performance ToF atom probe are (1) high mass resolution of close to $m/\Delta m \sim 5000$ and (2) elimination of noise or artifact signals. Those are very essential to the study of the surface or interface.

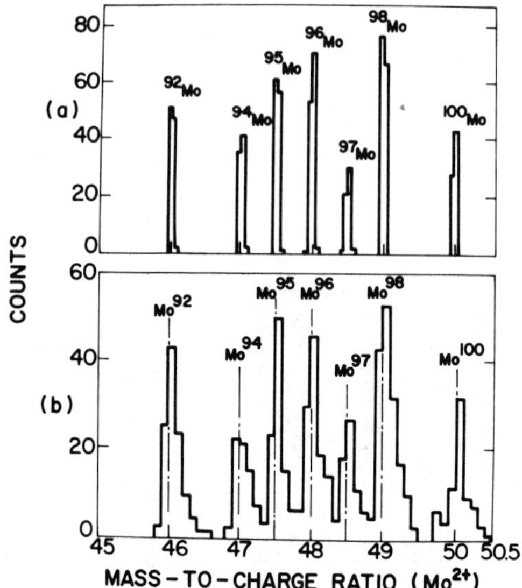

Fig. 2. Mass spectra of field evaporated Mo obtained by using (a) the high performance TOF atom-probe(1) and (b) a conventional straight-type ToF atom-probe(7).

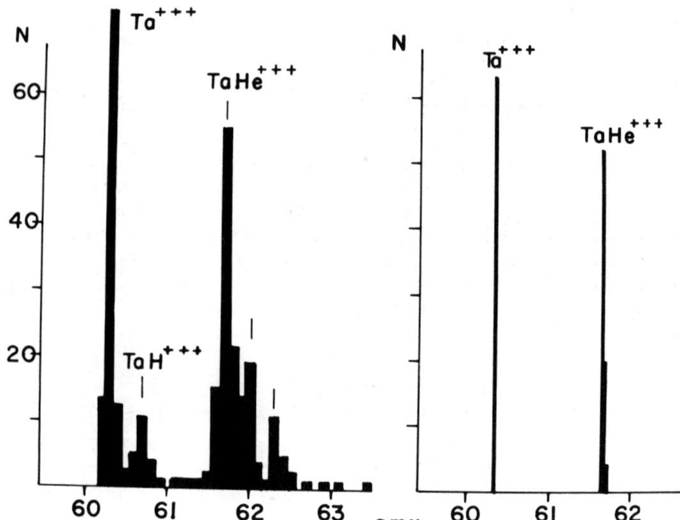

Fig. 3. Mass spectra obtained by field evaporation ofTa in the He imaging gas(2). Left: conventional straight-type ToF atom-probe. Right: the high performance ToF atom-probe. The signal identified as TaH at left turned out to be the artifact signal by the recent study using the high performance ToF atom-probe.

exposures, compositional analyses were performed along the grain boundary by slow (~1 species per pulse) field evaporation. A 17 ns pulse of 3600 volts was superimposed on the 8000–16000 volt field for evaporation.

Results

Atom probe analysis taken after insertion of the tip in the FIM vacuum chamber and prior to elevated temperature hydrogen exposure is shown on the left in Table I.

Table I.   Atom Probe Analyses

Surface/Subsurface* Species	at. %	Grain Boundary† Species	at. %	Fracture Surface Species	at. %
Ti,TiH,TiO	0.3	$TiC,TiO,TiO_2,TiH_2,Ti_2C$	12	$Ti^{\ddagger}$	21
		Ti	3		
FeH,FeH_2	1.4	FeH,FeH_2	11	FeH,FeH_2	13
H	6	H,H_2	8	H	5
FeO	0.2	FeO,FeO_2,FeC_2,FeC	12	$O^{\ddagger}$	20
		O,S,C	4	$C^{\ddagger}$	6
Fe	bal.	Fe	51	Fe	35

*prior to hydrogen exposure at elevated temperature; †after 1st elevated temperature hydrogen exposure; ‡total: atomic and combined forms

The elevated temperature hydrogen exposure causes hydrogen, along with titanium and other elements in the alloy, to accumulate at the grain boundary, Table I (middle column) and Figure 4. A very large number of hydrogen or hydrogen associated signals were detected. These, indeed, amount to > 19% of the total signals, 8% as pure hydrogen (H and H_2) and > 11% as iron and titanium hydrides (FeH, FeH_2 and TiH_2). The Ti content in the grain boundary was 15%, mostly as compound forms with C, O and H ($\gtrsim$ 12%). Total O and C levels detected in the grain boundary were even much higher, with an additional 12% combined with Fe and 4% in the atomic state. Figure 5a shows the grain boundary which crosses the tip and another grain boundary which intersects the main boundary at the lower right of the FIM image.

Toward the end of the series of hydrogen exposures, a small micro-crack was observed at the grain boundary intersection, Figure 5c, though its origin could not be traced back because of the limited view of a FIM image. This was an open crack formed by the detachment of metal bounded by the two intersecting grain boundaries. Its depth was of the order of several nm, since it shrunk readily by evaporating the surface to a depth of a few tens of atomic layers. The crack, once shrunk, could be grown again by exposing the tip to hydrogen ($\gtrsim$ 1 Pa for 10 min. to one hour at 1220–1300K). More and more of the tip surface was dissected in this process (Figure 5d and 5e).

Composition analyses were attempted from one of the fracture surfaces, along the line A to D (Figure 5e). The results are summarized in Table I

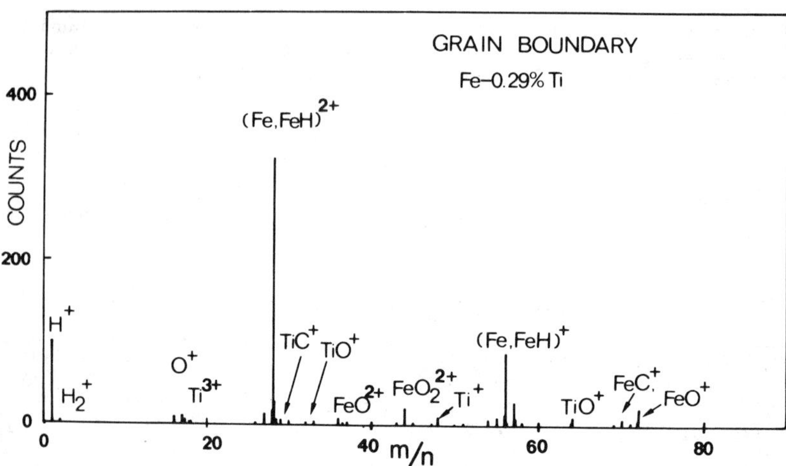

Fig. 4. Grain boundary composition after exposure to hydrogen at 10^{-1} Pa (10^{-3} Torr) for 10 min at 1220 K.

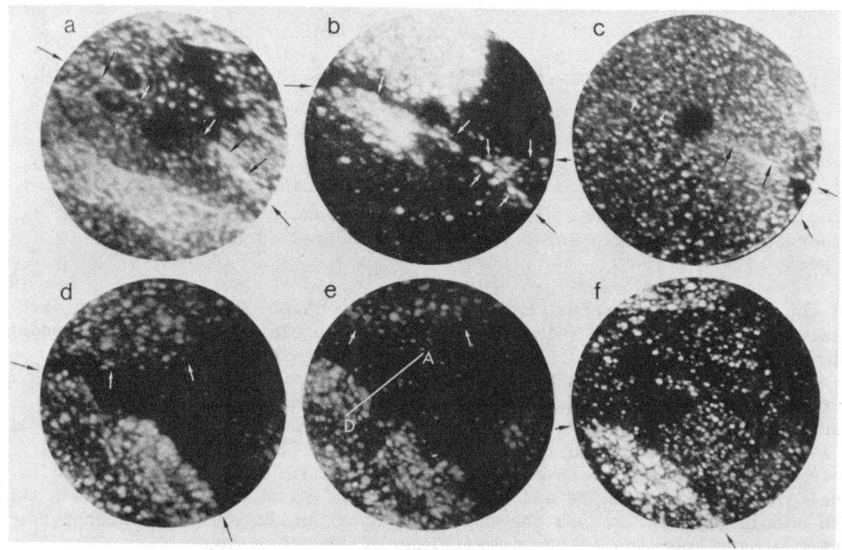

Fig. 5. FIM micrographs of the Fe-0.29% Ti tip showing the grain boundary network (a and b) and crack propagation (c – f). (a) prior to hydrogen charging and heating, (b) after the first elevated temperature hydrogen charging (10^{-1} Pa at 1220 K for 10 min), (c) after 1 Pa at 1220 K for 10 min, (d) after 10^2 Pa at 1300 K for 10 min, (e) after another dose under condition in (d), and (f) after removal of tens of atom layers by field evaporating following (e).

and Figure 6. At the fracture surface, large amounts of titanium (~20%),
hydrogen (~13%), and carbon (~6%) were detected while no carbon, and only
~7% titanium and ~9% oxygen were found in the normal tip surface. No
attempt was made to separate metal elements and their hydrides since the
fracture surface cannot be reproduced to accumulate enough data, which is
necessary to decouple hydrides from metal isotopes. However, it is not
unreasonable to assume that the ratios of metal elements and their hydrides,
are nearly the same as those observed above in the study of grain boundary
segregation. Table I therefore, reflects this assumption when separately
listing abundance of pure iron and iron hydride.

Discussion

Comparisons in Figures 2 and 3 illustrate the advantages of the high
performance atom probe in that (i) a very much greater mass resolution is
obtained, (ii) mass differences of one, e.g., FeH vs. Fe, and H vs. H_2, are
readily resolved, and (iii) artifact signals are no longer observed. Thus,
the high performance atom probe is capable of noise-free analysis of solute
atoms including H, with a mass resolution $(m/\Delta m)$ of over 1,000. This resolution is to be compared with that of the conventional (straight drift tube)
ToF atom probes for which $m/\Delta m \cong 100$. The factor accounting for the improved
resolution is the incorporation of a curved (Poschenrieder) electrostatic
focusing lens (9). In addition, atom probes provide analyses on an absolute
scale and, thus, can yield useful information which can only be obtained
indirectly, and usually with difficulty, by other techniques.

The atom probe data from the surface and subsurface layers taken prior
to heating (Table I) can be regarded as typical of the bulk composition
with respect to Ti. The hydrogen and oxygen are virtually entirely from the
gas phase; hydrogen was used as the imaging gas. The 6%H figure represents,
therefore, that hydrogen which was adsorbed on the surface from the gas phase.
Thus, except for a small amount of carbon (detected subsequently in the
fracture surface), the Fe-Ti tip appears to be quite pure, in agreement with
the reported analysis.

Detection of the H_2 molecule among the large amounts of atomic hydrogen
at the grain boundary following hydrogen exposure, is a new result. Hydrogen
previously has been detected as atomic hydrogen, which is not surprising
since hydrogen adsorbs in the atomic state. Furthermore, H_2 molecules have
not previously been observed when lower pressures of hydrogen have been used
as the imaging gas. The appearance of signals with mass 2 also is not a
field effect since the field tends to dissociate rather than combine species,
i.e., there is likely to have been more H_2 present than was actually detected.
Nor are they expected to be artifact signals in view of the elimination of
same by use of the energy focasing Poschenrieder lens discussed above. Thus,
we conclude that H_2 gas formed at suitable locations in the grain boundary
during or following hydrogen exposure and was then released when field
evaporation opened these internal cavities to the vacuum.

The hydride species evaporated on a one-per-pulse basis under normal
operating conditions as did other species in the grain boundary. On the
other hand, when a cluster appears at the surface it evaporates as a unit
for the same operating conditions. The latter has been observed to occur
for clusters of Ti, O, C and N in an Fe-Ti tip (3,4). Precursors to precipitate formation may be similarly identified. Since neither clusters nor
precipitate precursors of Fe_xH were indicated, we conclude at this time that
the identified iron hydride compound form was the result of the evaporation
of favorably situated (kink) iron atoms to which hydrogen atom(s) were bound.

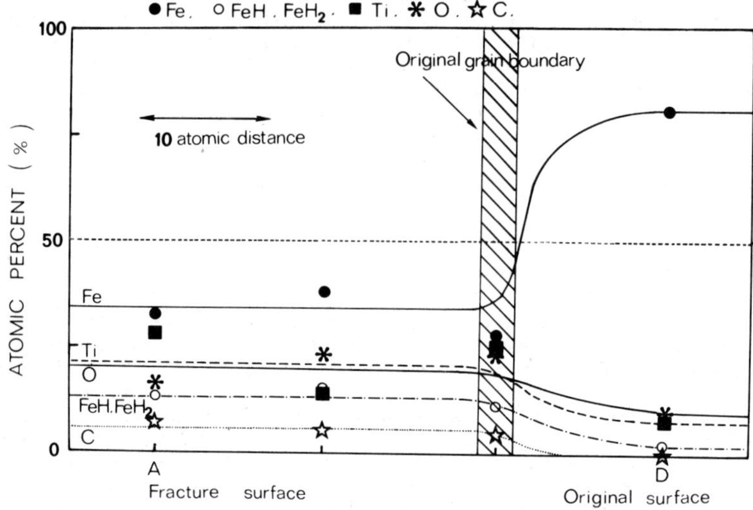

Fig. 6. Composition change across the interface between the fracture surface and the original tip surface. Regions to the left and right of the original grain boundary are regions A and D in Fig. 5(e). Percentages do not include a constant low level of H$^+$, which was concluded to be from the residual gas in the chamber.

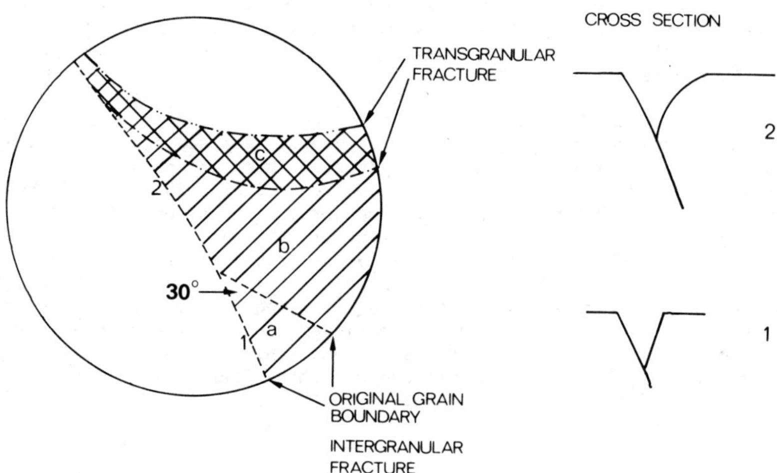

Fig. 7. Schematic of the FIM tip showing the location of the two intersecting grain boundaries and the regions of inter and transgranular fracture.

Examining all of the results before and after the initial hydrogen exposure, e.g., Table I and Figure 4 the most reasonable conclusions are that (1) hydrogen readily absorbed in the Fe-Ti tip with a major disproportinate accumulation occurring at the grain boundaries, (2) most of the hydrogen atoms are bound with Fe and Ti atoms and evaporate as metal hydride species, while lesser numbers form H_2 molecules and reside as gas in cavities mainly within the grain boundary region, and (3) strong segregation of Ti, O and C occurs to the grain boundary during the elevated temperature hydrogen exposure. Because of the tendency of the field to dissociate compound forms, the ratio of combined (hydride and H_2) to atomic hydrogen was probably greater than indicated by the data in Table I. With regard to the strong grain boundary segregation indicated for Ti, O and C, basically the same results were previously found in the case of another Fe-Ti alloy which had been annealed at 1270 K and 10^{-7} Pa (3,4).

The analysis of the fracture surface in Table I and Figure 6 shows that it is much different in chemical composition from the normal surface, and rather similar to that of the grain boundary. This is consistent with the fact that the fracture surface is along the original main grain boundary. Since crack propagation in this case involved dissection of the tip rather than merely an opening along the grain boundary, fracture necessarily occurred also through the grain (to the upper right of the main grain boundary, Figures 5d and 5e). A schematic drawing illustrating this fracture situation is shown in Figure 7.

It is not clear how a micro-crack formed under the hydrogen ambient. The usually rapid quenching of an FIM tip from the γ-phase region may result in the super saturation of hydrogen in α-phase iron, residual stresses from cooling and from the reduced hydrogen solubility, and the formation of a crack at the atomic level. The crack may also have formed and propagated along the hydrogen-rich boundary because of stresses introduced by the field during imaging. Cracks have never previously been seen in tips in the absence of rather high H_2 pressures. Also unclear is the retention of the original grain boundary network following the expected α-to-γ and γ-to-α transformations during heating to and cooling from temperatures in the austenite region.

Concluding Remarks

Since the data are preliminary, further analysis is needed. What can be stated at this time is that the energy focused atom probe has adequate resolution to separate H, H_2, hydride and metal isotopes. Secondly, these results show for the first time that the atom probe FIM can successfully study reactions involving gases at relatively high pressures. Thirdly, the evidence for the presence of the H_2 and hydride molecules and hydrogen, Ti, O and C accumulation in the grain boundary is rather convincing. Finally, the formation of a crack in the tip and a primary role of hydrogen in that process also seems clear. Whether or not hydrogen cracking can be induced in other tips remains to be seen[10] and additional analysis is needed to quantitatively establish the segregation patterns of solute and hydrogen as a function of temperature and composition.

References

1. E. W. Müller and S. U. Kirshnaswamy: Rev. Sci. Instrum. 1974, vol. 45, pp 1053–1059.
2. E. W. Muller and S. V. Kirshnaswamy: Characterization of Metal and Polymer Surfaces, vol. 1, pp. 21–47, Academic Press, New York, 1977.

3. H. W. Pickering, Y. Kuk and T. Sakurai: Appl. Phys. Lett. 1980, vol. 36, pp. 902–904.

4. H. W. Pickering, Y. Kuk and T. Sakurai: 27th International Field Emission Symposium, pp. 260–266, ed. Y. Yashiro and N. Igata, The University of Tokyo, Tokyo, 1980.

5. T. Sakurai, Y. Kuk, H. J. Grabke, A. K. Birchenall and H. W. Pickering: 27th International Field Emission Symposium, pp. 334–338, ed. Y. Yashiro and N. Igata, The University of Tokyo, Tokyo, 1980.

6. Y. Kuk, T. Sakurai and H. W. Pickering: 27th International Field Emission Symposium, pp. 339–343, ed. Y. Yashiro and N. Igata, The University of Tokyo, Tokyo, 1980.

7. A. Wagner, T. M. Hall and D. N. Seidman: J. Nuclear Mat. 1978, vol. 69, p. 413.

8. A. J. Melmed, in Abstracts of the 26th Field Emission Symposium, Berlin, 1979 (available from author, NBS, Washington, D.C.).

9. W. P. Poschenrieder: Int. J. Mass Spectrom. Ion Phys., 1972, vol. 9, pp. 357–373.

10. A similar microcrack induced by hydrogen has been observed recently by the authors in a Fe-0.29 wt. % Ti tip, giving strong encouragement in this research using the atom-probe.   (To be published).

TRANSPORT MODES

A COMPARISON OF ADSORPTION KINETICS

ON IRON OF H_2 AND H_2S*

M. R. Shanabarger

Quantum Institute, University of California
Santa Barbara, California (USA) and

NASA-Ames Research Center
Moffett Field, California (USA)

1. Introduction

A gaseous environment containing H_2S is known to be significantly more
aggressive in embrittling ferrous alloys than is gaseous H_2. The adsorption
of gaseous H_2 has been suggested to play an important role in determining
embrittlement kinetics(1). More rapid adsorption kinetics, therefore, may
be the explanation for the enhanced embrittlement by H_2S. Little has been
known, however, of the isothermal adsorption kinetics of H_2 and H_2S on ferrous alloys or even pure Fe to attempt to make a correlation between the adsorption kinetics and the embrittlement behavior.

Reported here is a summary of the results of studies to elucidate the
isothermal adsorption kinetics of H_2 and H_2S chemisorbed onto pure Fe surfaces.
Portions of the H_2 adsorption work have been reported earlier (2,3). A more
detailed discussion of the studies will be published elsewhere. Adsorption
of either H_2 or H_2S was found to occur via a molecular precursor. In both
cases, the reaction step from the molecular precursor to the dissociated adsorbed (chemisorbed) state is a rate-limiting step for adsorption at high
pressures. In this paper, it is shown that for adsorption onto pure Fe at
high pressure the adsorption rates for H_2 and H_2S are nearly equal at 300K.

2. General Experimental Details

The surfaces of evaporated thin iron films were used for the isothermal
adsorption studies. The films were made and the measurements performed in an
ultra-high-vacuum (UHV) apparatus (base pressure was about $2.7 \times 10^{-8} N/m^2$).
Films were made by sublimation from high purity iron wire (99.998%) filaments
onto either glass substrates for polycrystalline films (H_2 and H_2S studies)
or onto the (111) surfaces of polished Si monocrystals for epitaxially-
oriented [111] Fe films (H_2S studies only). Substrate temperature during
evaporation was about 300K. Films used at temperatures above 300K were
annealed at temperatures up to 450K, otherwise, the films were not annealed.
Annealing did not significantly affect the results and was necessary for
measurement purposes. The purity of the surfaces of the films was checked

*This work supported by the National Aeronautics and Space Administration,
Ames Research Center.

with Auger Electron Spectroscopy (AES) whenever possible. The only impurity
observed was oxygen which was estimated to be no greater than 5% of a monolayer. The oxygen was not the result of gas-phase adsorption, but was most
likely the result of oxides and oxygen present in the evaporation source.

The isothermal kinetics were determined by monitoring physical parameters
proportional to the coverage, Θ, of adsorbed (chemisorbed) adatoms as a function either of time at fixed pressure, or as a function of exposure. Coverage
is defined as the ratio of adatoms on the surface to the number of possible
adsorption sites. Exposure is the time integral of the pressure of the adsorbate gas. The coverage of the chemisorbed adatoms was determined by
either an electrical resistance-change technique or by AES. In the former
technique the magnitude of the electrical resistance-change is known to be
proportional to the coverage of chemisorbed adatoms and inversely proportional
to film thickness (4). The H_2 studies were done exclusively with the resistance change technique.

The AES technique is insensitive to hydrogen and was useful only for the
H_2S studies where the sulfur coverage in the chemisorbed state could be monitored. It has been assumed in the Auger measurements that the amplitude of
the peak-to-peak derivative signal for the 150eV sulfur Auger transition is
directly proportional to the coverage of chemisorbed sulfur. Extreme care was
taken to minimize the influence of the electron beam from the Auger spec-
tromer on the H_2S adsorption measurements.

Several procedures were followed to minimize or eliminate spurious impurity gas generation in the UHV apparatus which could potentially influence
the adsorption measurements. The vacuum system after exposure to the atmosphere was always baked at about 520K for several hours. To reduce the
effects of impact desorption and surface reactions occurring on the chamber
surfaces, the vacuum system was continuously purged with the adsorbate gas
for periods up to 7-8 hours at pressures up to $1.3 \times 10^{-3} N/m^2$. If this is not
done, for example, H_2S at $1.3 \times 10^{-4} N/m^2$ will react with the chamber surfaces
to produce a partial pressure of H_2 and H_2O which individually may exceed
$1.3 \times 10^{-4} N/m^2$. The adsorbate gas was admitted under flowing conditions to
further minimize contamination by continuous pumping. In addition, the ion
pump was isolated from the experimental chamber by a titanium sublimation
pump and a liquid nitrogen cooled cryopanel to essentially eliminate regurgi-
tation of reactive gases from the ion pump. The composition of gas phase was
monitored with a quadrupole residual gas analyzer.

3. Results

a. H_2/Fe(Polycrystalline)

Several different isothermal kinetic measurements are possible because H_2
adsorption is measurably reversible above about 250K. The measurements which
will be discussed are: (1) straight adsorption, where the coverage of hydrogen
adsorbed at constant H_2 pressure on an initially clean surface was measured
(by the resistance-change technique) up to an equilibrium coverage, and (2)
desorption, where the decay of an initial coverage of adsorbed hydrogen was
measured as hydrogen desorbs from the surface after rapidly evacuating the
gas phase H_2 to negligible pressure. Equilibrium measurements in the form of
coverage isotherms were, also, made at various temperatures.

A typical coverage isotherm for H_2 adsorption onto polycrystalline Fe is
shown in Fig. 1. At low pressures the coverage (resistance-change) is propor-
tioned to P, where p is the H_2 pressure. This behavior is characteristic of
dissociative adsorption. A value of 82.1 kJ/mole for the isosteric heat of

adsorption was obtained from a Clausius-Clapeyron analysis of coverage isotherms taken at tempera-
tures from 300K to 450K.
This value for the iso-
steric heat of adsorption
is in good agreement with
other measurements made
on polycrystalline Fe
films (5).

Analysis of a large
collection of kinetic data
shows that the approach of
the coverage to its final
equilibrium value follows
an exponential behavior
for both adsorption and
desorption. This char-
acteristic behavior is
shown in Fig. 2 for a particular set of adsorption
and desorption measure-
ments. The data have
been normalized so that
they fit on the same scale.

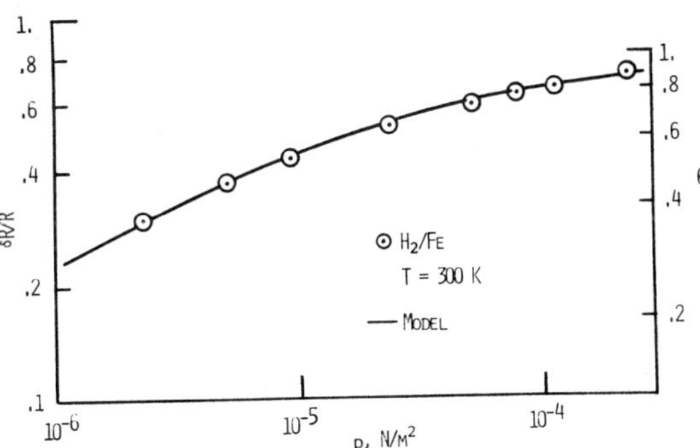

Fig. 1 - Isotherm obtained from resistance change
measurements for H_2 adsorbed onto a polycrystalline
Fe film at 300K. The measured fractional resistance change, $\delta R/R$, and the estimated hydrogen
coverage, θ, is indicated. The solid line is a
fit to the data using the precursor adsorption
model.

The rate constant for a particular kinetic process is taken to be the decay
constant of the exponential; the adsorption rate constant is denoted as τ_a^{-1}
and the desorption rate constant as τ_d^{-1}.

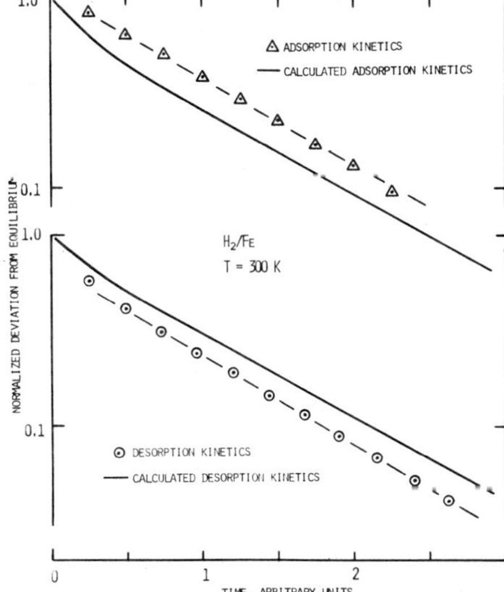

Fig. 2 - Examples of the approach to
equilibrium for H_2 adsorption and desorption on polycrystalline Fe at 300K.
The deviation from equilibrium and the
time-axis have been normalized to fit
both measurements on the same scale.
The solid lines are the calculated
kinetics based on the precursor adsorption model.

For desorption, τ_d^{-1} is temperature dependent with an activation energy of 28.9 kJ/mole (2).
Note that this energy is much smaller than the isosteric heat of adsorption (82.1 kJ/mole). This difference suggests that desorption is
not controlled by the recombination
of adsorbed hydrogen atoms, because
such a process would be expected to
have an activation energy much closer to the isosteric heat of adsorption.

For adsorption, τ_a^{-1} is observed to be both pressure and temperature dependent (2). The pressure
dependence of τ_a^{-1} at 300K is shown
in Fig. 3. At low pressures, τ_a^{-1}
is directly proportional to pressure
with a p = 0 intercept that equals,
within experimental error, τ_d^{-1}. At
higher pressures τ_a^{-1} attains a pressure-independent value as indicated
in Fig. 3. The magnitude of this
pressure-independent value is temperature-dependent, with the value
increasing with temperature (3).

A model where the dissociative chemisorption of H_2 occurs via a molecular precursor seems to provide the most consistent explanation of the experimental results. Schematically, the chemisorption process can be described by:

$$H_2(gas) \underset{k_d}{\overset{k_a p}{\rightleftarrows}} H_2{}^* \underset{k_2}{\overset{k_1}{\rightleftarrows}} 2(H-Fe) \tag{1}$$

where $k_a p$ is the rate of adsorption from the gas phase to the molecular precursor, $H_2{}^*$; p is the H_2 pressure; k_d is the rate constant for desorption from the precursor to the gas phase; k_1 is the rate constant for adsorption from the precursor to the dissociated chemisorbed state; and k_2 is the rate constant for desorption from the dissociated chemisorbed state to the molecular precursor. The quantity k_a is given by $k_a = s_o/\bar{n}\sqrt{2\pi m k_B T}$, where s_o is the sticking coefficient for adsorption onto a vacant precursor site, $\bar{n}$ is the average number of adsorption sites per unit area (the number of precursor sites is taken to be equal to the number of chemisorption sites), m is the mass of the gas molecule, k_B is Boltzman's constant, and T is the gas temperature. Note that the step between the precursor and the chemisorbed state, which is characterized by k_1, is a rate-limiting step for adsorption when $k_a p > k_1$. Adsorption into the precursor is determined by the flux of molecules hitting the surface, where this flux is proportional to $k_a p$. When $k_a p > k_1$, the step from the precursor to the chemisorbed state acts to limit the net adsorption rate compared with $k_a p$.

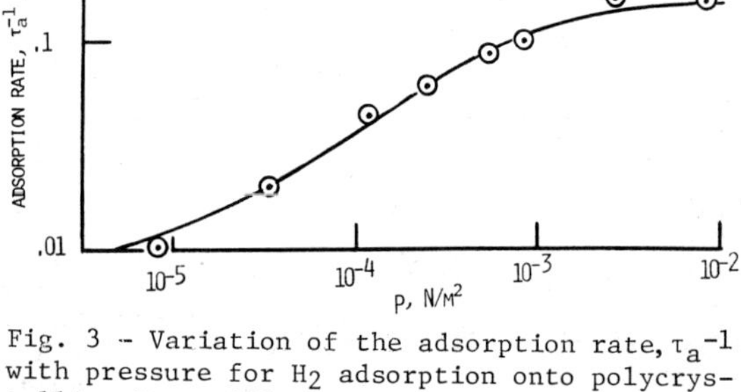

Fig. 3 -- Variation of the adsorption rate, τ_a^{-1} with pressure for H_2 adsorption onto polycrystalline Fe at 300K. The solid line is the adsorption rate calculated from the precursor adsorption model.

Analysis of the data within the framework of the precursor adsorption model leads to the following assignments for the rate parameters: k_a is the initial slope of τ_a^{-1} with p; k_d is approximately equal to τ_d^{-1}; k_1 is the pressure-independent value of τ_a^{-1} at high pressure; and knowing k_a, k_d, and k_1, k_2 is obtained from the equilibrium constant for the isotherm K, since $K=(k_a/k_d)(k_1/k_2)$.

Exact numerical calculations based on the precursor adsorption model are in good agreement with the data. The solid line in Fig. 1, for example, shows the agreement between the coverage (resistance) isotherms and the precursor adsorption model assuming that the resistance change per adatom and K are independent of coverage. The calculated approach to equilibrium for adsorption and for desorption is compared in Fig. 2 with the respective data. Finally, a comparison is presented in Fig. 3 between the calculated pressure dependence for τ_a^{-1} and the experimental results. The calculated adsorption–desorption kinetics are very sensitive to the rate constants employed in the calculations. The good agreement between the calculated kinetics and the data supports both the precursor adsorption model and the evaluation of the rate constants. The rate constants obtained from this study for H_2 adsorption onto polycrystalline Fe are presented in Table I.

TABLE I.   Rate Constants for H_2 and H_2S Adsorption

Rate Constant	H_2 [at 300K]	H_2S at [300K]
k_a, $(Nm^{-2}sec)^{-1}$	3×10^2	1.2×10^3
k_d, sec^{-1}	11T exp (-28.9/RT) [0.030]	[0.050]
k_1, sec^{-1}	604T exp (-34.7/RT) [0.16]	[0.22]
k_2, sec^{-1}	3.45×10^{13}T exp (-87.9/RT) [4.74]	[Zero]

R is in units of kJ/mole K and T is in degrees Kelvin.

b. H_2S/Fe(111)

It was possible to observe only straight adsorption kinetics with H_2S
since sulfur does not desorb from Fe at the temperature of these measurements
(near 300K). Although adsorption measurements were made on both polycrystalline as well as oriented Fe surfaces, only the more easily analyzed results
obtained from oriented surfaces is presented here (6). An initially clean
Fe surface was incrementally exposed to H_2S at a fixed pressure for varying
periods of time. After each exposure the sulfur Auger amplitude was recorded.
The H_2S adsorption measurements have been done at 300K and over a pressure
range from $1.33\times10^{-7}N/m^2$ to $1.33\times10^{-5}N/m^2$. A characteristic set of data for
H_2S adsorption to Fe(111) at 300K and p = $1.33\times10^{-7}N/m^2$ is shown in Fig. 4.
In this figure the normalized amplitude of the Auger measurement is plotted
as a function of the normalized exposure, X. The normalized exposure is
given by X = K*·L, where L is the total exposure to H_2S measured in Langmuirs
$(1.33\times10^{-4}Nsec/m^2)$ and K* is
the initial adsorption rate
determined near L=0. The
initial adsorption rate, K*,
is not constant, but has been
observed to vary with
pressure (6).

Low Energy Electron
Diffraction (LEED) obser-
vations were also made during the adsorption studies.
These observations show the
growth in intensity of a
$(\sqrt{3} \times \sqrt{3})R30^\circ$ pattern with
continued exposure to H_2S
up to saturation coverage.
This observation suggests
that at saturation cover-
age 1/3 of the available
adsorption sites are filled
with sulfur atoms. This
interpretation makes the
reasonable assumption that
the dominant electron scattering occurs from the sulfur adatoms.

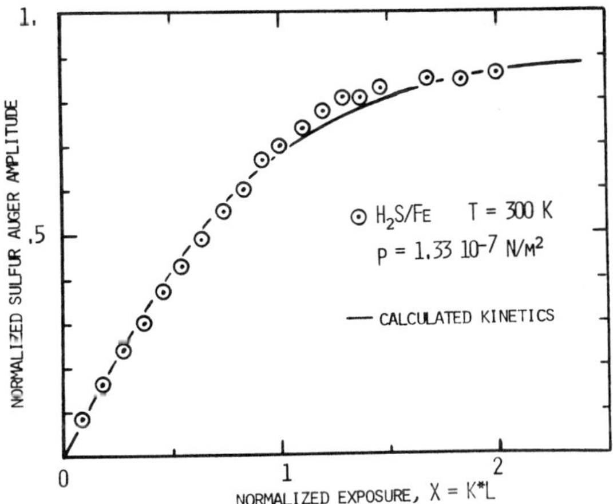

Fig. 4 - An example of the kinetics for H_2S adsorption onto (111) Fe at 300K. The sulfur
Auger amplitude has been normalized to its saturation value at high exposure. The solid line
is the result of exact numerical calculations
based on the precursor adsorption model.

Analysis of the data using a variety of simple adsorption models does not produce satisfactory agreement. Good agreement can be obtained, however, with the precursor adsorption model. Assuming that the dissociation products for H_2S are H-Fe and HS-Fe, then schematically, the adsorption on Fe of H_2S is as follows:

$$H_2S(g) \underset{k_d}{\overset{k_a p}{\rightleftarrows}} H_2S^* \overset{k_1}{\to} H\text{-}Fe + HS\text{-}Fe \tag{2}$$

The rate parameters have the same definitions as for H_2 adsorption. Note that unlike the precursor model for H_2 where k_2 was finite, the k_2 for H_2S is for all practical purposes zero since sulfur does not desorb to any measurable degree at 300K. The dissociation products for H_2S have not been clearly established, but there is evidence from preliminary thermal programmed desorption measurements and from permeation studies (7) for the presence of both H and HS on the surface.

An analytical solution to the equations describing the precursor model can be obtained from a steady state approximation ($d\Theta^*/dt=0$, where Θ^* is the precursor coverage)(8). The initial adsorption rate for this solution, K_{eff}, is pressure dependent and is given by

$$K_{eff}(L^{-1}) = 2\ k_a \times 10^{-6}/\Theta_{sat}[1 + (k_a p + k_d)/k_1] \tag{3}$$

where k_a is in units of $(Nsec/m^2)^{-1}$ and Θ_{sat} is the saturation coverage of the chemisorbed state. Assuming that the LEED structure is due primarily to the sulfur adatoms, then $\Theta_{sat} = 2 \times (1/3) = 2/3$.

Values for the rate constants can be obtained from least squares fitting the measurements of K* as a function of pressure to eq. (3) and also from fitting the general shape of the measured adsorption kinetics to the analytical solution for the model. A comparison between the results of this analysis and the data is shown in Fig. 4 where the solid line is the result of an exact numerical calculation to the model using the rate constants obtained from the analysis. The rate constants for H_2S adsorption onto (111) Fe surfaces at 300K are shown in Table I.

4.   Discussion

An important component in the hydrogen embrittlement process for ferrous alloys is postulated to be the concentration of protonic hydrogen in the lattice. The kinetics of hydrogen sorption from a gaseous hydrogen environment involve adsorption (chemisorption) and the subsequent absorption reaction. Absorption is the kinetic process which connects the adsorbed (chemisorbed) hydrogen with the sub-

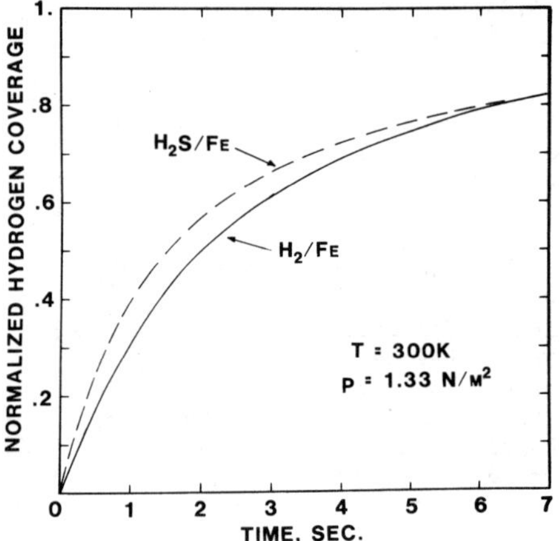

Fig. 5 – Comparison of the calculated adsorption kinetics on Fe of H_2 and H_2S at 300K. The rate constants were obtained from Table I. For H_2 the saturation coverage is essentially 1.0, whereas for H_2S it is 2/3 (of which only 1/3 is chemisorbed hydrogen).

surface hydrogen. The results of this study allow us to make a few comments about the role of adsorption in the sorption of hydrogen from gaseous H_2 and H_2S environments.

For both H_2 and H_2S, adsorption will be rate limited whenever $k_a p > k_1$. Significant differences between H_2 and H_2S in the rate-limited region near 300K are not expected since the k_1's are essentially equal with $k_1(H_2) \simeq$ .19sec.$^{-1}$ and $k_1(H_2S) \simeq$ .22sec.$^{-1}$. Small differences in the adsorption kinetics do exist as a result of the other rate parameters. These establish the overall manner in which equilibrium is achieved. This is demonstrated in Fig. 5 where calculated adsorption kinetics are shown for H_2 and H_2S at 300K and $p = 1.3N/m^2$. If the absorption kinetics from the chemisorbed state to the subsurface region are identical for hydrogen irrespective of its source, then significant differences (not larger than 2-3) in the overall sorption rate for hydrogen is not expected, nor should there be a difference in the embrittlement behavior if it is controlled by the adsorption kinetics.

The previous comments relate entirely to reactions which are occurring on essentially pure Fe surfaces. Clearly a ferrous alloy has a very complex surface. Recent studies have shown that certain surface impurities will selectively change k_1(9). Hence, certain impurities may enhance k_1 for H_2S and not for H_2 (and vice versa). Note, for example, that the HS species from adsorbed H_2S could act as such impurity to enhance the adsorption kinetics for H_2. Unless great care is taken H_2 will exist as an impurity in gaseous H_2S. Also, chemisorbed species often cause surface reconstruction. Such reconstruction could significantly change the absorption kinetics (as well as the adsorption kinetics) compared with unreconstructed surfaces. Clearly, additional work is required to establish the role of surface impurities on not only the adsorption kinetics, but also the absorption kinetics.

5. <u>References</u>

(1) Howard G. Nelson, Dell P. Williams, and Alan S. Tetelman, Metal. Trans. <u>2</u> (1971) 953.

(2) M. R. Shanabarger, Surface Sci. <u>52</u> (1975) 689.

(3) M. R. Shanabarger, in Proceedings 7th International Vacuum Congress and 3rd International Conference on Solid Surfaces, edited by R. Dobrozemsky (Vienna, 1977), p. 997.

(4) P. Wissmann, in "Springer Tracts in Modern Physics, Vol. 77," p. 57, Ed. G. Holder (Springer-Verlag, Berlin, 1975).

(5) G. Wedler and D. Borgmann, Ber. Bunseges, Phys. Chem. (Germany) <u>78</u> (1974) 67.

(6) M. R. Shanabarger, to be published in the Proceedings of the 8th International Vacuum Congress, 4th International Conference on Solid Surfaces and the 3rd European Conference on Surface Science (Cannes, 1980)

(7) H. G. Nelson, private communication.

(8) C. O. Steinbrüchel, Surface Sci. 51 (1975), 539.

(9) G. Ertl, M. Weiss, and S. B. Lee, Chem. Phys. Letters 60 (1979), 391.

A MODEL FOR DECREASING HYDROGEN ABSORPTION IN METALS APPLICABLE

TO ION IMPLANTED MEMBRANES[*]

H. W. Pickering and M. Zamanzadeh

Metallurgy Section
Department of Materials Science and Engineering
The Pennsylvania State University
University Park, PA   16802

A model is presented for decreasing hydrogen absorption by catalyzing the hydrogen evolution reaction (h.e.r.). Its salient features are the presence of a metal on the substrate surface with a higher exchange current density for the h.e.r. than that of the substrate, and a mechanism of hydrogen evolution for which the coverage of the surface is a function of the overpotential. Pt-implanted Fe membranes are evaluated with regard to their catalytic character for the h.e.r. using a Devanathan-Stachurski permeation cell. These data are in full accord with, and illustrate, the above model for decreasing H absorption.

Factors affecting the choice of implants for decreasing hydrogen absorption (via modification of the overpotential and the hydrogen coverage) include the nature of the implanted species for catalyzing the h.e.r.; the concentration profile of the implanted species, in particular its concentration in the surface atomic layer(s); and the tendency for enrichment of the surface in the implanted species, as occurs during open circuit corrosion. These factors are illustrated in the data.

[*]This work was supported by the Metallurgy Branch of the Office of Naval Research, Contract N000-14-75-C-0264.

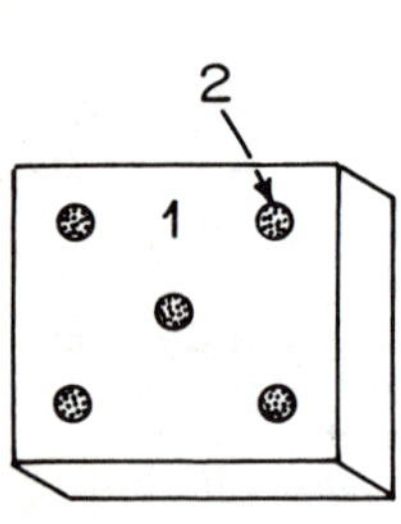

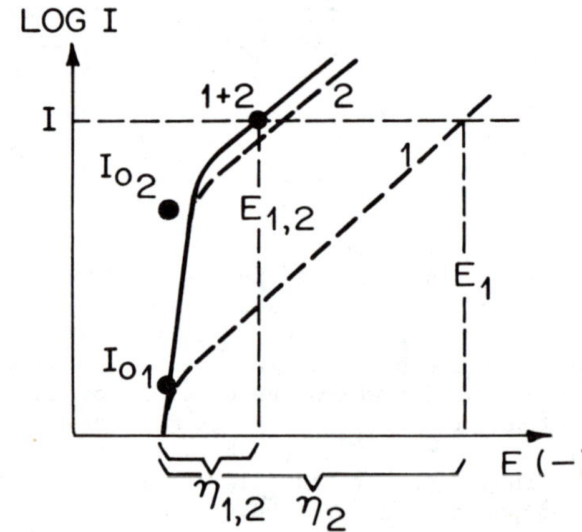

Fig. 1. Schematic Illustration of the polarization curves for hydrogen evolution on surface 1 alone and surface 2 alone (dashed curves) and on the composite surface 1 and 2 (solid curve) for the situation $(i_o)_2 \gg (i_o)_1$.

Introduction

Essentially two different approaches are available for decreasing hydrogen absorption depending on the conditions. One, catalytic in character, relies on a reduced coverage of adsorbed hydrogen for certain mechanisms of the hydrogen evolution reaction (h.e.r.). The other is the well known (and commercially used) barrier effect which becomes a factor as the thickness of a low permeability, dense, continuous coating increases. In this paper the catalytic mechanism is modeled and then evaluated using iron membranes sparingly covered with platinum on the input surface.

Catalytic Mechanism

The effectiveness of different substrates for promoting the h.e.r. may be described in terms of their exchange density, i_o, for the h.e.r. When the transfer coefficient of the h.e.r. does not change, the rate, i, of the h.e.r. at constant overpotential, η, on different substrates is simply in the ratio of their exchange current densities, i.e., $i_1:i_2:i_3\cdots=(i_o)_1:(i_o)_2:(i_o)_3\cdots$, where i_1,i_2,i_3 are the rates of the h.e.r. on substrates 1, 2, 3... This follows since the rate of the h.e.r. on substrate j at $\eta \gg RT/F$ is given as

$$i_j = (i_o)_j e^{-\vec{\alpha}_j \eta\ F/RT} \tag{1}$$

where $\vec{\alpha}_j$ is the transfer coefficient of the h.e.r. on substrate j, F is the Faraday Constant, R is the gas constant, and T is the temperature.

For a composite surface the total current I is given by

$$I = \sum_{j=1}^{n} I_j = \sum_{j=1}^{n} i_j A_j \tag{2}$$

where A_j is the area of substrate j. The total current and individual currents for a composite surface of two different substrates are shown as a function of the electrode potential, E, for the h.e.r. in Figure 1.

Hence, at a given charging current, I, which corresponds to the $E_{1,2}$ potential for the composite surface (Figure 1), most of the h.e.r. occurs on the substrate of higher exchange current density, 2, and the current on substrate 1 is $i_1=I-i_2$ whereas in the absence of substrate 2, I occurs wholly on substrate 1 at E_1. Addition of substrate 2, therefore, provides for a decrease in the overpotential ($\eta_{1,2}$ rather than η_1) for the charging current I.[†] This fact may be used to advantage for decreasing hydrogen absorption in substrate 1 as has been recently pointed out [1], since the coverage, and hence the absorption, of hydrogen is a function of the overpotential for certain mechanisms of hydrogen evolution [2]. This model presupposes that hydrogen absorption via substrate 2 is negligible as would be the case if the solubility or diffusion of hydrogen in substrate 2 is low.

The overall h.e.r. on iron in acid solution,

$$H_3O^+ + e^- \rightarrow 1/2\ H_2 + H_2O \tag{3}$$

can be written as discharge

$$H_3O^+ + e^- \xrightarrow{k_1} H_{ads} + H_2O \tag{4}$$

followed by chemical desorption

$$H_{ads} + H_{ads} \xrightarrow{k_2} H_2 \tag{5}$$

with both reactions (4) and (5) displaced from equilibrium and therefore both influencing the overall rate of reaction (3). Usually, only a small amount of adsorbed hydrogen, H_{ads}, enters the iron. To this extent reaction (4) is followed by the absorption step,

$$H_{ads} \underset{k_{-3}}{\overset{k_3}{\rightleftarrows}} H_{abs} \tag{6}$$

rather than by reaction (5).

Expressing the charging current density in terms of reactions (4) and (5) for hydrogen evolution on iron, one has

$$i = 2\ F\ k_1\ a_{H_3O+}(1-\theta)\ e^{-\beta\Delta\phi F/RT} \tag{7}$$

and

$$i = 2Fk_2\theta^2 \tag{8}$$

where k_1 and k_2 are the rate coefficients of reactions (4) and (5), a_{H_3O+} is the activity of H_3O^+, θ is the fraction of the iron surface covered with adsorbed hydrogen, β is the symmetry factor of the activation barrier for discharge, and usually has a value of about one-half, and $\Delta\phi$ is the potential gradient across the iron/solution interface. Rewriting equation (7) in terms of η, one has

---

†The change in area of substrate 1 is appropriately neglected when the area of substrate 2 is much smaller than that of substrate 1 and/or the exchange current density of the h.e.r. on substrate 2 is much greater than on substrate 1, $(i_o)_2 \gg (i_o)_1$.

$$i = (i_o)_{Fe} e^{-\beta\eta F/RT} \qquad (7a)$$

where $(i_o)_{Fe} = 2Fk_1 a_{H_3O^+}(1-\theta) e^{-\beta\Delta\phi_e F/RT}$

and $\Delta\phi_e$ is the potential gradient at equilibrium.

Combining equations (7a) and (8), and allowing $(1-\theta)$ to approach unity corresponding to low coverage by hydrogen, the relation

$$\theta = \left[\frac{(i_o)_{Fe}}{2Fk_2}\right]^{1/2} e^{-\beta\eta F/2RT} \qquad (9)$$

is obtained. Eq. (9) shows that by decreasing the overpotential η, e.g., by (sparingly) distributing over the Fe surface a metal with a higher exchange current density, the hydrogen coverage on the remaining uncovered iron surface decreases; for every 240 mV decrease in η, the coverage, θ, decreases by an order of magnitude. The solubility of hydrogen in equilibrium with the coverage also decreases according to reaction (6) and then so also does the permeation of hydrogen. This catalytic approach for lowering hydrogen uptake from electrolytic solutions is a function of the mechanism of the h.e.r. and, thus, may vary from metal to metal and even for the same metal under different electrolytic conditions. The explicit form of the dependency of θ on η will depend on the details of the h.e.r. just as equation (9) does in the case of iron.

There is usually little flexibility for decreasing the overpotential of a reaction occurring at a given rate. By utilizing a composite surface, however, one may accomplish this goal [1]. This model is supported by experimental results for Fe-Pt composite surfaces for which a sparse distribution of platinum over the iron surface causes a significant decrease in the overpotential at constant i and, hence, in the coverage, solubility and permeation of hydrogen, as described next.

<u>Experimental Verification of the Model</u>

<u>Pt Electrodeposits on Fe</u>

An earlier paper [1] evaluates the usefulness of metals more noble than iron, e.g., Pt, Cu and Ni, when present as particles or discontinuous layers on the surface, for decreasing the coverage and, hence, decreasing the entry of hydrogen into iron. The exchange current densities of the h.e.r. on these metals fall in two ranges: $i_o(Pt) >> i_o(Fe)$ and $i_o(Cu \text{ or } Ni) \approx i_o(Fe)$. The amount of hydrogen absorbed was determined from measurement of the permeation rate of hydrogen through the membranes using a Devanathan and Stachuroki cell [3]. This technique for obtaining a measure of hydrogen absorption requires that the diffusion of hydrogen controls the overall permeation process, in which case, in the limit, local equilibrium exists between adsorbed (coverage) and absorbed (solubility) states. Then a change in steady-state permeability is a direct measure of H absorption

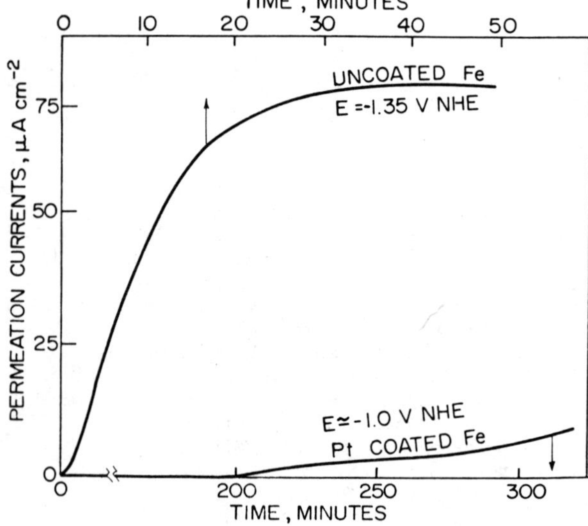

Fig. 2. Hydrogen permeation transients. Charging current, i, =2 mA cm^{-2} in 0.1N NaOH + 20 ppm As^{3+}. [1]

under constant exit conditions and a composition and position independent
H diffusivity.

A significant effect of Pt deposits for reducing hydrogen permeation
was observed, Figure 2. The upper curve shows a typical time-flux transient
for hydrogen permeation through iron in the absence of Pt. The lower curve
shows the effect of a Pt deposit on the iron. The amount of Pt deposited
was quite small so that a barrier effect due to the Pt could be neglected.
It is seen that the hydrogen permeability is much less in the presence of Pt.
The measured electrode potential of the composite Fe/Pt surface is considerably more noble than that of the iron surface, Figure 2. The potential
slowly drifted in the less noble direction, consistent with a decreasing
catalytic activity of the Pt sites (aging effect) which occurs especially
in the presence of cathodic poisons, e.g., As^{3+} ion. A shift of the electrode
potential in the less noble direction corresponds to a rising coverage of
hydrogen on the surface according to the model (Equation 9); this was confirmed by a rising permeation rate. Since all of the results are in accord
with the above model, the role of the Pt deposit is concluded to be one of
providing a more catalytic surface for the h.e.r., thereby reducing the
overpotential at both Fe and Pt sites (Figure 1). This, in turn reduces the
coverage of hydrogen on the Fe sites and its permeation through the iron
membrane.

Pt-Implanted Fe

Factors affecting the choice of implants for decreasing hydrogen
absorption by metals via modification of the overpotential and the hydrogen
coverage include the nature of the implanted species for promoting or catalyzing the h.e.r.; the concentration profile of the implanted species, in
particular its concentration in the surface atomic layer(s); and the tendency
for enrichment of the surface layer(s) in the implanted species during
anodic dissolution of the substrate, as occurs during open circuit corrosion.
The tendency for deactivation (aging) of the implant with regard to its
catalytic character is, in principle, also a factor. Recently, Zamanzadch
et. al., reported on the effect of implanted Pt on decreasing hydrogen
absorption in iron [4]. A summary of these results follows.

Iron membranes implanted with Fe, He or Pt on the charging side were
evaluated in the permeation cell in the as-implanted condition or following
controlled anodic (selective) dissolution of iron which increased the surface
Pt content, Figure 3. For the surface enriched, Pt-implanted Fe membranes
a significant decrease in permeation was observed, Figure 4 and Table 1.
The electrode potential for the charging current density, i, on the Pt-
implanted membrane is always noble to that on the unimplanted membrane. The
Tafel slope and the slope of a plot of the charging potential, E, and log
of the steady state permeation rate, i_∞, for both unimplanted and Pt-
implanted membranes are -122 ± 5mV and -225 ± 5mV, respectively. These
values are in rather good agreement with the model, e.g., the model predicts
$\partial E/\partial \log i$ = -120 mV and $\partial E/\partial \log i_\infty$ = 240 mV (extensions of Equations 1-9,
which are not shown). With increasing immersion time in the sulfuric acid
prior to the permeation measurement, i_∞ goes through a minimum and E goes
through a maximum, Figure 5. This behavior is expected as a consequence of
the fact that the surface concentration of Pt goes through a maximum as
shown for similar conditions in Figure 3. The steady-state permeation
current density is a linear function of the square root of the charging
current density with the curve passing through the origin, indicative of a
diffusion-controlled hydrogen permeation process [5]. These results are all
in accord with the above model in which surface Pt sites catalyze the h.e.r.,

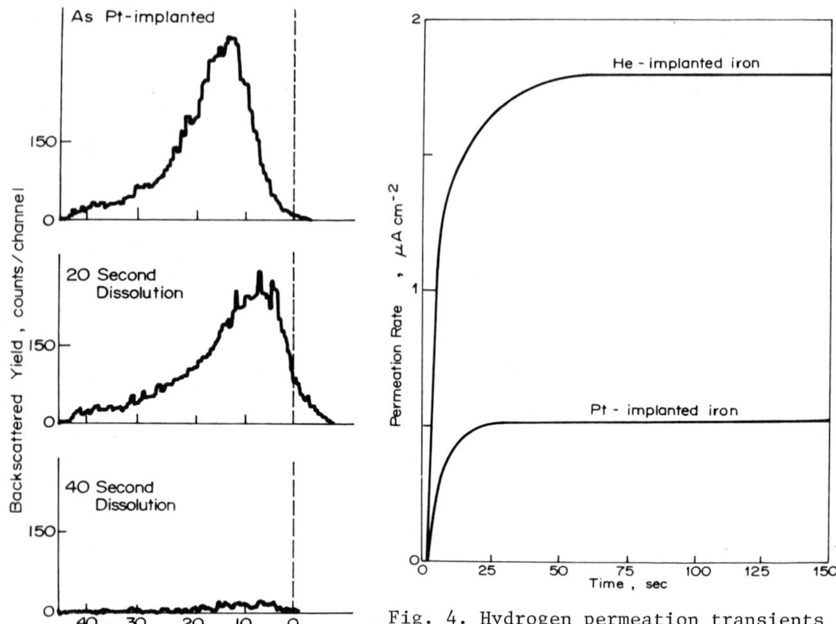

Fig. 3. Rutherford backscattering profiles of Pt-implanted Fe showing the surface concentration of Pt to go through a maximum with increasing time of immersion in 1M H_2SO_4. [4].

Fig. 4. Hydrogen permeation transients for condition b in Figure 3. i = 0.1 mA cm^{-2} in 0.1N NaOH. [4].

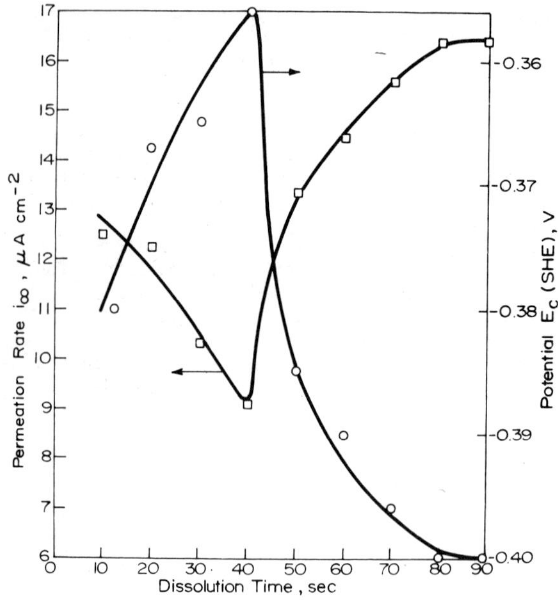

Fig. 5. Steady-state permeation rate (□) and charging potential (o) for a Pt-implanted Fe membrane as a function of additional immersion periods in 0.1N H_2SO_4 beyond the initial 20 sec immersion in 1M H_2SO_4. i = 0.6 mA cm^{-2}. [4]

Table 1

Variation of the steady state hydrogen permeation flux i_∞ and of the electrode potential E of the charging side with the charging current density i in 0.1N H_2SO_4.

$\dfrac{i}{mA/cm^2}$	unimplanted iron		Pt-implanted iron	
	i_∞ $\mu A/cm^2$	-E, V(SHE)	i_∞ $\mu A/cm^2$	-E, V(SHE)
1.0	22.0	0.425	16.2	0.407
0.8	20.0	0.415	14.5	0.395
0.6	17.5	0.400	12.5	0.380
0.4	14.2	0.380	10.5	0.360
0.2	10.2	0.345	7.5	0.325
0.1	7.3	0.310	5.5	0.290

thereby decreasing the overpotential and in turn the hydrogen coverage, absorption and permeation.

With implanted specimens a second possibility exists for decreasing hydrogen permeation through the membrane in the form of traps introduced by the implantation process. These may be (i) defects introduced by the implantation process and (ii) the implanted species. The trapping effect, has been observed for a Ti-implanted Fe membrane which shows a delayed permeation kinetics compared to Fe-implanted Fe membranes. So that the trapping effect was not obscured by the above described catalytic effect, a thin Pd coating was applied to the input surface and, as usual, to the exit surfaces of these membranes. In recent ion beam studies Myers et. al. [6], have observed trapping of deuterium-implanted iron. Ti, as a solute in iron, has been reported to bound with hydrogen with an interaction energy of 0.27eV [7].

Concluding Remarks

The above model predicts, and the experimental data, e.g., Figures 2, 4 and 5, show that a metal with a high exchange current density like Pt might be beneficially deposited on iron for purposes of decreasing hydrogen absorption. In this event, the bulk of the h.e.r. occurs on the Pt sites and the overvoltage decreases; thus, the coverage of hydrogen on the Fe surface is lower and, consequently, so also are the absorption and permeation rates. It follows that for implants to be effective for reducing hydrogen absorption, they should have a higher exchange current density than the substrate for the h.e.r., and be relatively inert in which case corrosion of the substrate can have the beneficial effect of increasing the surface concentration of the implanted species.

References

1. S. S. Chatterjee, B. G. Ateya and H. W. Pickering: Met Trans. A., 1978, vol 9A, pp. 389-393.
2. J. O'M. Bockris and P. K. Subramanyan: Electrochem Acta., 1971, vol. 16, pp. 2169-2179.

3.   M. A. Devanathan and Z. Stachurski: Proc. Roy. Soc., 1962, vol. A 270, p. 90.
4.   M. Zamanzadeh, A. Allam, H. W. Pickering and G. K. Hubler: J. Electrochem. Soc., 1980, vol. 127, pp. 1688-1693.
5.   J. O'M. Bockris, J. McBreen, and L. Nanis: J. Electrochem. Soc., 1965, vol. 112, pp. 1025-1031.
6.   S. M. Myers, S. T. Picraux and R. E. Stoltz: J. Appl. Phys., 1979, vol. 50, pp. 5710-5719; this Proceedings, pp.
7.   G. M. Pressouyre and I. M. Bernstein: Met. Trans. A., 1978, vol. 9A, pp. 1571-1580.

A Pulse Technique Employed for Studying Egress of

Hydrogen from Iron Polarized Cathodically in As^{3+} -Containing Solutions

P. Kedzierzawski, Z. Szklarska-Smialowska and M. Smialowski

Institute of Physical Chemistry, Polish Academy of Sciences

Warsaw, Poland

Studies of the diffusion of hydrogen in metals carried out by time lag electrochemical methods give S-shaped permeation vs. time curves. The detailed form of these curves depends on the boundary condition prevailing at the input side of the membrane. The influence of the boundary condition on the shape of the permeation curve has been discussed by Nanis and Namboodhiri (1) and Fullenwider (2). In these papers two types of limiting conditions at the input side have been considered.

In the first case, each variation of polarization conditions, for example of the cathodic current density, induces immediately an appropriate change in the concentration of hydrogen just under the surface of the metal. Such a condition takes place when the rates of ingress v_{in} and eggress v_{eg} of hydrogen at the input side of the membrane are relatively high: $v_{in} \approx v_{eg} \gg c^0 D/L$, where D – diffusion coefficient of hydrogen in metal, L – membrane thickness.

In the second case variation of polarization conditions produces relevant changes in the value of $(\partial c/\partial x)_{x=L}$ i.e. of the concentration gradient of hydrogen at the entry side of the membrane. Such a condition takes place when the rate of eggress of hydrogen at the polarized side is significantly lower, than the rate of diffusion: $v_{in} \approx c^0 D/L \gg v_{eg}$. Permeation curves predicted by the first condition have been observed on iron in the absence of promoters (1,3,4), and by the second - on palladium (2). The aim of this work is to elucidate which condition is fulfilled at the iron surface in the presence of arsenic in the electrolyte.

The principles of the method used.

In the presence of promoters in the solution, it is impossible to use the time lag methods because each change in the potential induces a change of the surface concentration of the promoter. Adsorption or desorption of the promoter is a slow step determining the shape of the transient. Therefore in this study a pulse technique has been applied. Experiments have been carried out by applying triangular pulses to the entry side of the iron membrane and by measuring for each applied pulse the changes in the rate of hydrogen ionization at the diffusive side of the membrane. Because the duration of the pulse was 0.01 to 0.03 s, in the calculations pulses could be approximated by Dirac delta functions $\delta(t)$. The system of equations to be solved is:

$$\frac{\partial^2 c}{\partial x^1} = \frac{1}{D} \frac{\partial c}{\partial t} \; ; \qquad c(0,t) = 0; \qquad c(x,0) = c^o x/L$$

The third boundary condition has the form: $c(L,t) = \lambda\delta(t)$ for the first case, and $(\partial c/\partial x)_{x=L} = \mu\delta(t)$ for the second case. Solution of these equations gives the dependence of the permeation P on the dimensionless parameter $k = Dt/L^2$:

$$P = FD(\partial c/\partial x)_{x=0} = -2\pi D^2 \lambda F K_1 /L^3 \qquad \text{for case 1 and}$$

$$P = -\pi D^2 \mu F K_2 /L^2 \qquad \text{for case 2}$$

Plots of K_1 and K_2 versus k are shown in Fig. 1. According to calculations, the half peak width is 1.4 in the first case and 4.3 in the second case.

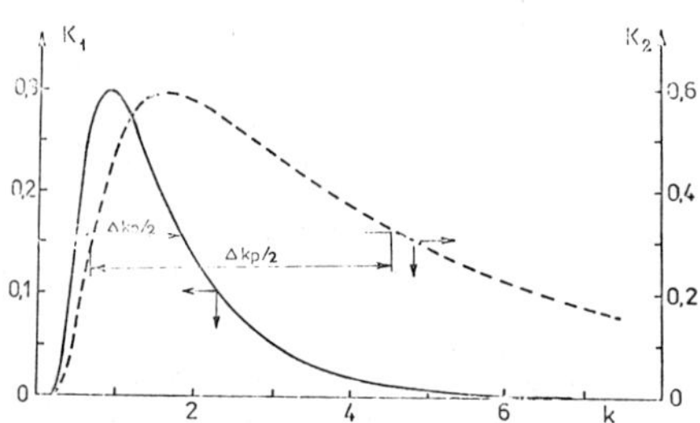

Fig. 1.   Variations in K_1 and K_2 with k.

Experimental.

The measuring system was similar to previously described (4) except that both sides of the membrane worked under potentiostatic conditions.

Initially the polarized side of the membrane was kept at a constant "starting potential" E_i (Fig. 3) of either -500 mV$_{NHE}$ (in 0.1 N H_2SO_4 + As^{3+}) or -1300 mV$_{NHE}$ (in 0.1 N NaOH + As^{3+}). Under these conditions the cathodic current densities and the hydrogen ionization rates were recorded. After the steady-state was established, a potential pulse, either cathodic or anodic, was applied to the entry side of the membrane. The triangular pulse lasted for 0.01 to 0.03 s. After attaining the peak at E_z (Fig. 2), the potential returned to its initial value of E_i. The change in the hydrogen ionization current occurring at the diffusive side in response to the potential pulse was recorded. Simultaneously with permeation measurements, changes in the cathodic current occurring after pulse application

were recorded. These measurements permitted detection of the promoter
desorption.

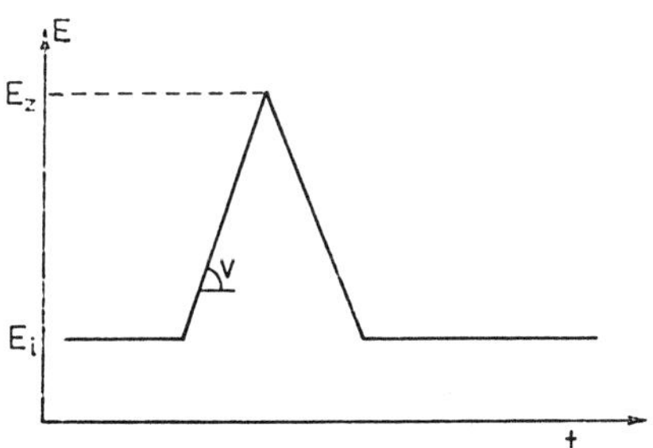

Fig. 2.   Shape of the potential pulse applied. E_i starting
potential; E_z -reversal potential; $v = (dE/dt)$
-sweep rate.

Results.

As shown in Figs. 3 and 4, in 0.1 N H_2SO_4 + As^{3+} at E_z potentials less
positive than 0.15 V_{NHE}, and also in 0.1 N NaOH + As^{3+} at E_z potentials
more negative than -0.5 V_{NHE} the time corresponding to the half-peak width
is close to the value calculated for the kinetically controlled removal of
hydrogen. At more anodic E_z potentials of the pulse wider peaks are observed. It can be seen that for E_z values more cathodic than 0.15 V_{NHE} in
H_2SO_4 and -0.5 V_{NHE} in NaOH there are no changes in the cathodic current
density (Figs. 5 and 6). At more anodic E_z values the current increases
after pulse application and then returns slowly to its initial values indicating respectively desorption and readsorption of the promoter.

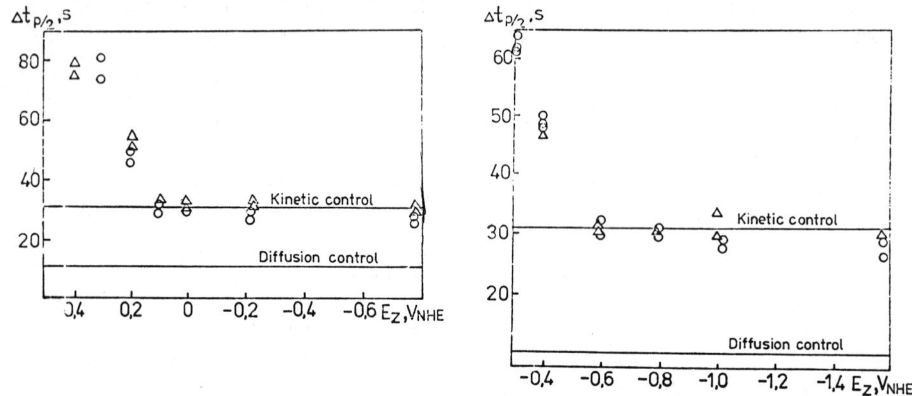

Figs. 3 and 4. Effect of reversal potential on width of half
peak of the P - t curve for membranes polarized
in 0.1N H_2SO_4 + As^{3+} (Fig. 3) and in 0.1N NaOH +
As^{3+} (Fig. 4).

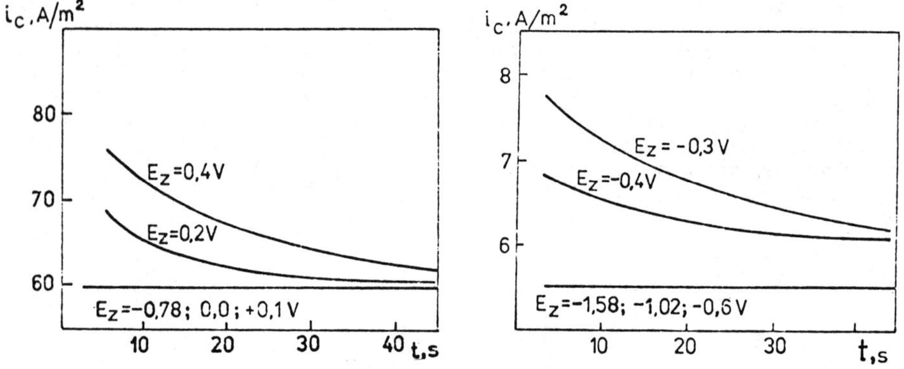

Figs. 5 and 6. Changes in cathodic current density with time
after application of potential pulse in
0.1N H_2SO_4 + As^{3+} (Fig. 5) and in 0.1N NaOH +
As^{3+} (Fig. 6).

Summary.

A pulse technique was used to study the removal of hydrogen from iron membranes whose one side was polarized cathodically in the presence of small amounts of As^{3+} in either 0.1 N H_2SO_4 or 0.1 N NaOH. In contrast with the results of Nanis and Namboodhiri (1), and others (2-4), who found that in pure (i.e. As^{3+}-free) solutions the egress of hydrogen from iron at the polarized side of the membrane was unhampered and almost as rapid as the ingress of hydrogen into iron, the results of the present study indicated that As^{3+} added to the catholyte at a concentration of the order of 10^{-5}M caused a nearly complete interception of the egress of hydrogen from iron. Under these conditions almost all the hydrogen which entered into iron at the polarized side of the membrane diffused out through its other side.

References

1. L. Nanis and T.K.G. Namboodhiri, J. Electrochem. Soc. 119, 693 (1972).
2. M.A. Fullenwider, J. Electrochem. Soc., 122, 648 (1975).
3. C.D. Kim and B.E. Wilde, J. Electrochem. Soc., 118, 202 (1971).
4. P. Kedzierzawski, K. Benczek and A. Sadkowski, Bull. Acad. Pol. Sci., Ser. Sci. Chim., 24, 595 (1976).

DEUTERIUM PROFILEMETRY AND DIFFUSION COEFFICIENT

IN ELECTROCHARGED STAINLESS STEEL*

M. B. Lewis and K. Farrell
Metals and Ceramics Division
Oak Ridge National Laboratory
Oak Ridge, Tennessee 37830

The method of nuclear microanalysis using the reaction $D(d,p)T$ is used to measure concentration-depth profiles of deuterium in austenitic type 310 stainless steel electrocharged in 1 N D_2SO_4 with arsenic poison at room temperature. The influence of aging after charging is examined. A near-surface density of 0.6 deuterium atoms per metal atom (i.e., 38 at. %) is found for the initially charged condition. Associated with this high gas concentration is considerable microstructural damage. An analysis of the deuterium depth profiles yields a room-temperature diffusion coefficient for deuterium of 1.4×10^{-16} m^2s^{-1} at high concentrations which is consistent with extrapolation of higher-temperature data available from other techniques for annealed material. At lower deuterium concentrations, $<10^{-2}$, in the aged material the extracted diffusion coefficient is a factor of three higher, implying a dependence on concentration. At concentrations $<10^{-3}$, diffusion may be hindered by dislocation trapping.

* Research sponsored by the Division of Materials Sciences, U.S. Department of Energy under contract W-7405-eng-26 with the Union Carbide Corporation.

Introduction

In hydrogen embrittlement and hydriding studies of metals, the hydrogen
is frequently introduced by cathodic charging in aqueous acid solution.
Usually a poison such as arsenic or sulfide ions is added to the electrolyte
to boost hydrogen uptake (1). Under these conditions it is expected that
the hydrogen concentration in the surface layers of the cathode will be
large in accordance with the expectedly very high fugacity of hydrogen at
the surface. Once absorbed, ingress of hydrogen proceeds by diffusion, the
rate of which may be different from that measured for the usual gas permeation conditions because of hydrogen-hydrogen interaction imposed by the high
concentration of hydrogen, and from trapping or pipe diffusion at microstructural damage (2—8) cointroduced during charging. Knowledge of the
hydrogen concentrations and diffusion associated with cathodic charging is
therefore needed to provide a better understanding of hydrogen behavior in
metals. We have measured these parameters for an austenitic stainless
steel cathodically charged in a poisoned electrolyte.

Procedure

Specimens were cathodically hydrogenated and their concentration profiles were determined by nuclear microanalysis using a probing beam of ions
from a particle accelerator. Details of the principles and practice of
this technique are available elsewhere (9—11). The particular conditions
used in the present experiments are described in Ref. 10. Briefly, a mono-
energetic beam of ions impinges at a known angle on a target surface where
it penetrates the surface and undergoes a nuclear reaction with the impurity
in the target. The reaction products are scattered and a fraction of them
are emitted from the target and captured in a detector which measures the
yield of reaction ions, $Y(E)$, and their energy, E. The yield is related to
the concentration, C, of the specific target element, and their emergent
energy is a function of the depth, x, at which the nuclear reaction occurred.
Thus, a concentration-depth, C_x, profile can be obtained. The depth of
measurement of the profile is limited to the penetration of the probing
ion (of the order of 1 μm in the present studies). If the target element
is mobile, the shape of the profile will change with time and temperature,
and small shifts of less than 0.1 μm are readily detected; analyses of these
shifts in terms of Fick's diffusion laws gives an effective diffusion
coefficient.

We used deuterium instead of hydrogen in order to utilize the well-
characterized $D(d,p)T$ reaction, where D denotes the target atom, d the
probe ion, and p and T the measured reaction products. The targets were
5-mm diam disks, 0.3 mm thick, of type 310 stainless steel (21.1 wt % Ni,
23.4 Cr, 2.1 Mn, 1.0 Si, 0.06 C) annealed at 1050°C in helium. Four disks
were mechanically polished on Syntron lapping wheels to a 0.1 μm diamond
finish and were spot welded on their edges to a cross of nickel wires. This
assembly was suspended in the electrolyte such that the polished face of
each disk was equidistant from a surrounding cylindrical platinum anode.
The electrolyte was 1 N D_2SO_4 poisoned with 2.5 g metallic arsenic per cubic
meter.* The bath was maintained at 24°C and was stirred gently with a
plastic-coated magnetic paddle. The DC current density was 1 kA m^{-2} at 3 V,
and the charging time was 6 ks.

*With no arsenic in the electrolyte, deuterium uptake in the cathode is
 significantly reduced.

Immediately after electrocharging, the specimens were washed in water and alcohol. One was quickly transferred to the ion scattering chamber which was then pumped to a pressure of less than 10^{-3} Pa to allow accurate energy and particle current measurements. The minimum time between cessation of charging and beginning of probing was usually 480 s. The time required for analysis was 60 to 300 s. The remaining specimens were allowed to age for periods up to several hours at room temperature and pressure. For specimens that were probed twice during the experiment, care was taken to insure that the beam spots never overlapped in order to avoid defects introduced by the ion beam from affecting deuterium egress (10). While defects are introduced during measurement, the measuring time is too short to affect the results.

Data Analysis and Results

The data from the nuclear reactions are pulses or counts (yield) per measurement channel (energy) and are stored in a computer. They must be converted to concentration-depth data which are then analyzed through suitable diffusion expressions to obtain diffusion coefficients.

Concentration-Depth Profiles

Simple concentration-depth data, C_x, are derived from the yield-energy data by the method of convolution integrals (12). A computer code estimates energy losses, straggling, multiple scattering, and other relevant kinematic information then calculates the convolution integrals

$$Y(E)/\Delta E = K \int_S P_{in} C_x b P_{out} dS \qquad (1)$$

where $Y(E)/\Delta E$ = the yield of reaction product,
         K = the normalizing constant including beam flux on the target,
         P = the Gaussian functions whose widths are determined by the overall resolution,
         b = the nuclear reaction cross section, and
         S = the space of energy and volume over which the reaction takes place.

The stopping powers are taken from Ref. 13, and Bragg's rule (9) is used to estimate stopping powers in the target. Cross sections are given by Ref. 14. To extract C_x, an analytic form for C_x, such as the error function, may be chosen and the appropriate error function parameter from Eq. (1) may be adjusted to fit the $Y(E)/\Delta E$ data. Alternatively, C_x may be factored out and solved for directly via a deconvolution procedure.

For the above analytic method, the gas distribution is assumed to be described by the usual diffusion laws such that the profiles should follow the aging pattern depicted schematically in Fig. 1. Here the boundary condition assumed during charging is

$$C = C_o \text{ for } x = 0, t_i > 0 , \qquad (2)$$

where C is the concentration (expressed throughout this work as gas atoms per metal atom) and t_i is the charging or ingress time. The solution for Fick's laws satisfying this condition would then be

$$C(x,t_i) = C_o \text{ erfc } [x/2(Dt_i)^{1/2}] , \qquad (3)$$

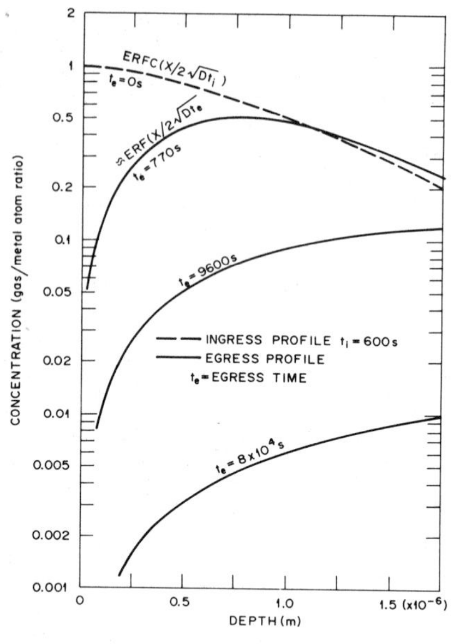

Fig. 1 - Schematic illustration of expected concentration profiles near the specimen surface at points in time during electrocharging and three points in time during release (decay) after a fixed electrocharging time.

during the electrocharging process, where D is the diffusion coefficient and x is the distance below the surface along the surface normal. As can be seen in Fig. 1, if the charging time is long compared to the time for release or egress, the region near the specimen surface can be assumed relatively uniform in concentration C_O when the charging ends (i.e., when the specimen is removed from the electrolytic bath). This point in time becomes $t_e = 0$ for the process of deuterium egress with boundary conditions

$$C = C_O \text{ for } t_e = 0; \qquad C = 0 \text{ for } x = 0, t_e > 0 , \qquad (4)$$

for which the solution satisfying Fick's laws is

$$C(x,t_e) = C_O \, erf[x/2(Dt_e)^{1/2}] . \qquad (5)$$

This gives the aging curves which show that on cessation of charging, the front of the profile continues spreading inwards but the deuterium in the immediate surface layers escapes to the atmosphere; the maximum concentration declines and moves inwards. Only for very long charging times compared with release times is the maximum concentration, approximately equal to the original immediate sub-surface concentration, C_O. Because of the delay between charging and measurement in the present work, deuterium release is significant. Nevertheless, C_O can be extracted by normalizing Eq. 5 to the measured profile.

An example of results for the analytic method is given in Fig. 2 for a specimen electrocharged for 6×10^3 s and probed 660 s later. Note that the specimen surface is at the right-hand side in the yield/channel plot and at the left-hand side of the concentration-depth plot. A value of $\sqrt{Dt} = 0.28 \, \mu m$ gives the closest fit to the data and provides the concentration-depth curve in the upper part of the figure. This curve is for an aging period of 660 s, and this is reflected in the curve at shallow depths, <0.5 μm, where much of the deuterium has escaped.

When the deconvolution method is used, the resulting C(first-order approximation) becomes the first trial function which is then renormalized to fit the yield data (more exactly, a second-order approximation). Concentration profiles obtained this way for the aged specimens are given in Fig. 3.

These ion beam measurements are quantitative in concentration and depth to an accuracy of 5–20%.

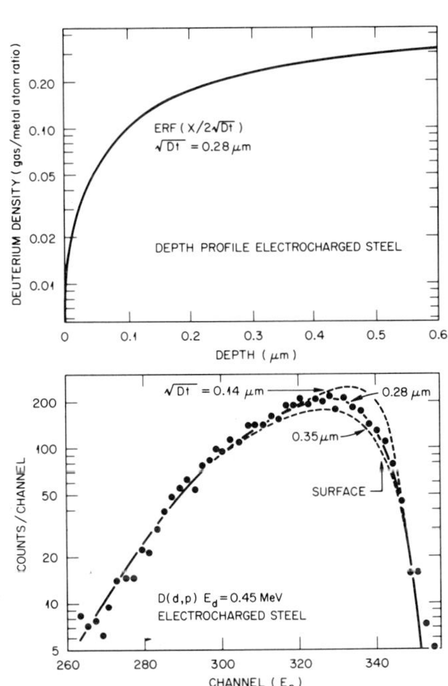

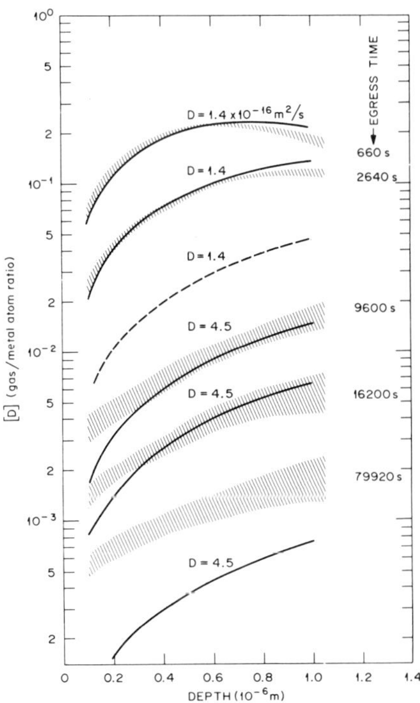

Fig. 2 - Data and calculated fits for a D(d,p)T reaction spectrum from a 0.45-MeV deuterium beam incident on 310 stainless steel electrocharged with deuterium for 100 min and aged 11 min before probing. The corresponding concentration profile is shown above the data. The right-hand edge of the spectrum corresponds to the surface of the specimen.

Fig. 3 - Deuterium depth profiles as a function of egress time following a 100 min electrocharge. Data shown hatched to indicate uncertainty.

Diffusion Coefficient

The best-fit value of $\sqrt{Dt} = 0.28$ μm for Fig. 2 gives a diffusion coefficient for deuterium in austenite at 25°C of 1.4×10^{-15} m²s⁻¹. This is based on the assumption that the initial deuterium concentration ($t_e = 0$) in Fig. 1 is uniform so that the analytic form $\text{erf}(x/2\sqrt{Dt})$ is an appropriate description of the concentration profile. While this may be satisfactory for short aging times, it will not be valid at longer times, $t_e \gtrsim 10^3$ s, and the diffusion equations must be solved more exactly to portray the true deuterium loss. To do this we substitute

$$C(x, t_e = 0) = C_o \, \text{erfc}[x/2(Dt_i)^{1/2}] \, . \tag{6}$$

With this condition, we have solved numerically the equation

$$\frac{\partial C}{\partial t} = D \frac{\partial^2 C}{\partial x^2} \ , \tag{7}$$

assuming the same diffusion constant for ingress and egress.

The solutions were obtained with the use of the differential equation solver GEARB (15), utilizing 150 grid points. The boundary condition $\partial c/\partial x = 0$ was applied at an internal point of 12 µm, sufficiently deep for the longest times to avoid interference in the 0–1 µm region from any "back reflected" deuterium. The corresponding diffusion coefficients are shown on the egress curves in Fig. 3 and are further considered in the discussion section.

The Initial Deuterium Concentration

When the solutions to the differential equation are adjusted to fit the data in Fig. 3, two parameters, C_O and D are determined. The overall normalization is C_O, the immediate subsurface concentration during electrocharging. This parameter is of special interest because it describes the local concentration of deuterium present in the steel during the charging. For the curves in Fig. 3, the appropriate C_O value is 0.50 ± 0.15 gas atoms/metal atom. For Fig. 2, C_O = 0.60 ± 0.15.

Discussion

This work illustrates the usefulness of nuclear microanalysis to gas/metal studies. In particular, the technique provides fast and quantitative information on near-surface concentrations and migration of hydrogen (deuterium) which is unavailable by other techniques. We find that the concentration of deuterium introduced into 310 austenitic stainless steel during cathodic electrolysis of sulphuric acid-D_2 containing small amounts of arsenic poison is high, about 0.6 deuterium/metal atoms or 37.5 at. %. This is considerably greater than the values of 1×10^{-2} (Ref. 16) and 2.5×10^{-2} (Ref. 8) hydrogen atoms/metal atoms quoted for bulk analyses of similarly charged materials which do not allow for a nonhomogeneous depth distribution of the hydrogen. However, Smialowski and his colleagues have long claimed high surface concentrations of hydrogen in α-iron (17) and nickel (18); in the latter surface concentrations as high as 0.7 were attained, and an unstable nickel hydride phase was identified (19,20). Other evidence supports high subsurface hydrogen concentrations in austenite charged under cathodic conditions like ours. For example, x-ray measurements show that the γ lattice is significantly expanded (2,6,7); interplanar spacings are reversibly increased by about 5% (Ref. 7). When this increase is compared with the linear increase in the host unit cell with hydrogen concentration in other single-phase fcc materials charged in high-pressure gaseous hydrogen (21), it corresponds to a hydrogen/metal ratio of 0.62. Also, the calculations and experiments of Atrens et al. (22) on rates of ingress and egress of cathodic hydrogen in 310 steel, indicate a subsurface hydrogen concentration of about unity. Our measurements confirm these high hydrogen levels. We have also done transmission electron microscopy studies of these specimens and find that there was dense microstructural damage created in the surface layers of our specimens. Details of this damage will be made available elsewhere. Briefly, the damage consisted of intense bands of dislocations and faults on slip planes, with cracks at a few of the slip bands and at some annealing twins, grain boundaries, and inclusions. The cracks were barely discernible by optical metallography but were easily revealed by electron microscopy. This damage commenced in very short charging times as dislocation loops emanating from the specimen surface and later from interfaces. With increasing charging time the damage moved slowly inwards as a front. In 24 h it penetrated only $\sim$5 $\times$ 10^{-6} m; serial sectioning and further deuterium profiling showed that the deuterium concentration also declined beyond this depth. We note, too, that if we use our extracted diffusion coefficient of 1.4 $\times$ 10^{-16} m^{2}s^{-1}, the penetration depth for a 24-h charge given by $x = \sqrt{Dt}$ is about 4 $\times$ 10^{-6} m. Charging in unpoisoned electrolyte, which induced lower deuterium levels, caused relatively little damage. The implication is that the near-surface damage structure resulted from the deuterium content or gradient exceeding some critical value in those regions. Such structures could develop from stress relief by dislocation generation and shear at steep deuterium gradients between a near-surface metal lattice expanded by deuterium and a relatively unexpanded substrate lattice.

In the light of these findings, it is reasonable to suppose that the plastic deformation, martensite formation and cracking observed by others (2–8) at the free surface of austenite after cathodic hydrogenation in poisoned electrolyte were also caused by very high hydrogen concentrations similar to those measured here. These "skins" of damage may contribute to the reported hydrogen embrittlement of 310 steel. In this alloy hydrogen embrittlement is noted only for conditions of cathodic hydrogenation in poisoned electrolyte (2,16,23,24) but not when charging is done in high-pressure hydrogen gas or in unpoisoned electrolyte at 280°C (28).

The diffusion coefficients determined from Figs. 2–3 are in agreement with the extrapolated values for a range of austenitic stainless steels (Table I) at 25°C. This suggests that most of the deuterium remains in atomic form even at concentrations of several atomic percent. A small iso-tope dependence is consistent throughout the table. Agreement with the literature despite the intense dislocation damage in our measurement area implies that the dislocations are not interfering with the migration of the deuterium at high concentrations. We take this to mean that there is suf-ficient deuterium present to saturate any trapping sites presented by the dislocations and to leave most deuterium free to migrate. The amount of cracking observed by TEM is not enough to seriously affect diffusion.

Table I. Comparison of Diffusion Coefficients of Hydrogen
and Its Isotopes in Austenitic Stainless Steels at 298 K

Type Steel	Isotope	Measurement Temperature Range (K)	D (298 K) ($\times 10^{-16}$ m^2/s)	Reference
316, 304	T	298–498	0.96	29
310, 309, 304	D	385–713	1.6	30
310	H/D	472–779	13.8/11.6	31
309	H/D	423–873	2.9/2.1	32
304	T	373–573	1.32	33
310	D	298	1.4 ± 0.2	this work $C_o \approx 0.10$
			4.2 ± 1.0	this work $C_o \lesssim 0.01$

Other sources of interference might arise from barriers at the free surface, and from deuterium interaction with impurities and with other deuterium atoms. We looked for evidence of deuterium blocked at the surface and found only very slight amounts. Interference by other impurities, if it occurred, is likely to be common to all the measurements in the table, so it would not show as a discernible difference in our value. We did find signs of a possible interaction between the deuterium atoms, as follows. When the diffusion coefficient of 1.4×10^{-16} m^2s^{-1} was applied to the set of curves in Fig. 3, good fits were obtained for the t_e = 660 s and 2640 s aging curves, but not for the curves for longer aging times. To fit the t_e = 9600 s and 16,200 s data, it was necessary to *increase* the diffusion coefficient by a factor of three. At these aging times the deuterium levels were also considerably lower than for the shorter times. Such an increase in diffusivity with decreasing concentration is observed in other hydrogen-metal systems, as recently summarized by Volkl and Alefeld (34). The explanation is that the effective diffusion coefficient has a concentration dependence through the activity coefficient, γ, thus (35)

$$D = D^* \ [1 + \partial(\ln\gamma)/\partial(\ln C)] \ . \tag{8}$$

With increasing concentration, the mutual interaction of atomic hydrogen in a metal will lower its activity coefficient thus lowering the hydrogen diffusivity D compared to the infinitely dilute case D* (sometimes referred to as the "tracer" or Einstein diffusion coefficient).

At the longest aging time (8×10^4 s) and corresponding low concentrations ($\lesssim 10^{-3}$), we find the amount of residual deuterium is somewhat higher than predicted. The most reasonable explanation of this is that at these low deuterium concentrations transitory trapping by the dislocation damage structure is being observed.

Conclusions

1. Nuclear microanalysis is a useful and fast technique for studying the distribution and migration of deuterium (hydrogen) in metals.

2. The near surface concentration of deuterium introduced into 310 austenitic stainless steel during cathodic charging in sulphuric acid-D_2 poisoned with 2.5 mg/l arsenic at 24°C is 0.6 deuterium atoms per metal atom.

3. Concurrent with this high deuterium concentration, intensive microstructural damage is developed in the form of dislocations and cracks.

4. Egress of deuterium after charging does not appear to be affected by the dislocation damage for egress times $\lesssim 10^4$ s (i.e., deuterium $\gtrsim 10^{-2}$) and the diffusion coefficient at 25°C is 1.4×10^{-16} m^2s^{-1}, for C $\approx$ 0.10.

5. At longer egress times (and lower deuterium content), the effective diffusion coefficient is increased threefold, compatible with a concentration dependence of the hydrogen activity coefficient.

6. At long egress time and low deuterium level ($\lesssim 10^{-3}$), the residual deuterium is observed to be less mobile as if trapped by dislocations or other defects.

7. Even at high concentrations most of the deuterium remains in atomic form.

Acknowledgment

We appreciate the assistance of J. T. Houston in these experiments, and the helpful discussions of hydrogen diffusion with L. K. Mansur.

References

1. F. Eisenkolb and G. Ehrlich, *Archiv. für das Eisenhüttenwesen,* 1954, 3–4, p. 187.

2. M. L. Holzworth and M. R. Louthan, Jr., *Corrosion*, 1968, vol. 24, p. 110.

3. M. L. Holzworth, *Corrosion*, 1969, vol. 25, p. 107.

4. K. Kamachi, S. Miyata, and M. Eto, *Japan Inst. Met.*, 1969, vol. 33, p. 824.

5. H. Okada, Y. Hosoi, and S. Abe, *Corrosion*, 1970, vol. 26., p. 183.

6. K. Kamachi, K. Akiyoshi, and H. Hayashi, paper 2C13 in *Deuxieme Congres International L'Hydrogene Dans Les Metaux,* Paris, 1977.

7. H. Mathias, Y, Katz, and S. Nadiv, *Met. Sci.*, 1978, p. 129.

8. P. Maulik and J. Burke, *Scripta Met.*, 1975, vol. 9, p. 17.

9. W. K. Chu, J. W. Mayer, and M. A. Nicolet, *Backscattering Spectroscopy,* Academic Press, New York, 1978.

10. M. B. Lewis and K. Farrell, in Proc. TMS-AIME Symp., *Advanced Techniques for Characterization of Microstructures,* J. A. Spitznagel and F. W. Wiffen, eds., Las Vegas, February 1980, in press.

11. M. B. Lewis and K. Farrell, *Appl. Phys. Lett.* 1980, vol. 36, p. 819.

12. M. B. Lewis, *Analysis of Nuclear Backscattering and Reaction Data by the Method of Convolution Integrals,* ORNL/TM-6697, Oak Ridge National Laboratory, 1979.

13. H. H. Anderson and J. F. Ziegler, *Hydrogen Stopping Powers and Ranges,* Pergamon Press, New York, 1977.

14. L. Ruby and R. B. Crawford, *Nucl. Instr. Methods* 1963, vol. 24, p. 413; R. B. Theus, W. J. McGarry, and L. A. Beach, *Nucl. Phys.*, 1966, vol. 8, p. 273.

15. A. C. Hindmarsh, *GEARB: Solution of Ordinary Differential Equations Having Banded Jacobians,* UCID-30059, Lawrence Livermore Laboratory, 1975.

16. J. A. Peterson, R. Gibala, and A. R. Troiano, *J. Iron Steel Inst.*, 1969, p. 86.

17. M. Smialowski, *Hydrogen in Steel,* Addison-Wesley, Reading, Mass., and Pergamon Press, New York, 1962.

18. B. Baronowski and M. Smialowski, *Intern. J. Phys. Chem. Solids,* 1959, vol. 12, p. 206.

19. A. Janko, *Bull. Acad. Polon, Sci., Ser. Sci. Chim.,* 1960, vol. 8, p. 131.

20. A. Janko and P. Michel, *Comptes Rendus,* 1960, vol. 251, p. 1001.

21. B. Baronowski, S. Majchrzak, and T. B. Blanagan, *J. Phys. F: Met. Phys.* 1971, vol. 1, p. 258.

22. A. Atrens, J. J. Bellina, N. F. Fiore, and R. J. Coyle, *The Metal Science of Stainless Steels*, p. 54, E. W. Collings and H. W. King, eds., AIME, 1978.

23. M. B. Whiteman and A. R. Troiano, *Corrosion*, 1965, vol. 21, p. 53.

24. M. L. Mehta and J. Burke, *Corrosion*, 1975, vol. 31, p. 108.

25. R. M. Venett and G. S. Ansell, *Trans. ASM*, 1967, vol. 60, p. 242.

26. R. J. Walter and W. T. Chandler, *Mater. Sci, Eng.*, 1971, vol. 8, p. 90.

27. M. R. Louthan, Jr., G. R. Caskey, Jr., J. A. Donovan, and D. E. Rawl,Jr., *Mater. Sci. Eng.*, 1972, vol. 10, p. 357.

28. R. Lagneborg, *J. Iron Steel Inst.*, 1969, p. 363.

29. J. H. Austin, T. S. Ellerman, and K. Verghase, *J. Nucl. Mater.*, 1973, vol. 48, p. 307.

30. M. R. Louthan, Jr., and R. G. Derrick, *Corros. Sci.*, 1975, vol. 15, p. 565.

31. N. R. Quick and H. H. Johnson, *Metall. Trans.*, 1979, vol. 10A, p. 67.

32. W. J. Kass and W. J. Andrzejewski, AEC Report No. SC-DR-72-0137, Washington, 1972, unpublished.

33. K. F. Chaney and G. W. Powell, *Metall. Trans.*, 1970, vol. 1, p. 2356.

34. J. Völkl and G. Alefeld, *Topics in Applied Physics*, vol. 28, p. 321, G. Alefeld and J. Völkl, eds., Springer-Verlag, Berlin and New York, 1978.

35. P. G. Shewman, *Diffusion in Solids*, McGraw-Hill, New York, 1963.

HYDROGEN PERMEATION OF STAINLESS STEEL UNDER STRESS*

P. Studebaker and C. Altstetter
University of Illinois at Urbana-Champaign
and
W. Conley
Illinois Tool Works
Elgin, Illinois

The hydrogen permeability, diffusivity, and solubility of Type 302 austenitic stainless steel are determined as a function of stress over the temperature range 200°C to 320°C using a gas-phase UHV technique. Permeability and diffusivity are found to be enhanced by stress, while solubility remains essentially unchanged.

* This work was supported by the National Science Foundation under grant number NSF-DMR 77-09808.

Introduction

In the study of stress-corrosion cracking and hydrogen embrittlement of austenitic stainless steels, it has become apparent that a more thorough knowledge of the effects of stress on the permeation behavior of hydrogen through these materials is necessary to further understand the mechanisms involved. To this end, an apparatus utilizing the gas-phase permeation technique has been designed with the capability of in-situ specimen deformation. A calibrated ion pump permits simultaneous measurement and exhaust of the entire permeating flux, offering high resolution and rapid response to changes in the permeation rate.

The apparatus was used to examine the permeability of Type 302 stainless steel over the temperature range of 200° C to 320°C in both unstressed and stressed conditions. The same specimen was used for all temperatures and stress levels.

Theory

Gaseous hydrogen permeates through metal by undergoing adsorption, dissociation and solution, transport, recombination, and desorption. To describe the permeation mathematically, several assumptions are made, namely, that the rate controlling step is transport through the metal, that the hydrogen-metal solution obeys Sieverts' Law, and that equilibrium exists between the gas phase and the solid solution. The most restrictive of these assumptions is that the hydrogen flux is rate limited by transport. To substantiate this assumption, it is common to use a coating such as palladium on the specimen surfaces. With these assumptions, Fick's Law can be applied to the experimental arrangement to obtain an expression relating the steady state of flux J_∞ to the partial pressures P_1 and P_2 on each side of a planar membrane of thickness h:

$$J_\infty = SD(P_1^{\frac{1}{2}} - P_2^{\frac{1}{2}})h^{-1} \tag{1}$$

where the product of the solubility constant S and diffusion constant D is defined as the permeation constant . Thus

$$\phi = Jh/(P_1^{\frac{1}{2}} - P_2^{\frac{1}{2}}) \tag{2}$$

and the permeability ϕ follows directly from the experimental measurement of the steady state flux through a membrane.

The diffusivity is determined from the lag time, L, which has been defined and related to the diffusion coefficient by Barrer (1), and developed by Devanathan and Stachurski (2) for use where permeating flux, rather than the total amount of permeating species, is the measured quantity. From Devanathan, the lag time, L, is found to be the time required for the transient flux to attain the value J_L, where:

$$J_L = .6299 \, J_\infty \tag{3}$$

and J_∞ is the steady state flux. Barrer has related the lag time to the diffusion coefficient D:

$$L = h^2/6D \tag{4}$$

By measuring the lag time, L, and the steady state flux, J_∞, over a range of temperatures, the temperature dependence of the coefficients of permeation, diffusion, and solubility is determined. These are usually expressed as Arrhenius equations, where:

$$\phi = \phi_0 \exp(-\Delta H_\phi/RT) \tag{5}$$

$$D = D_0 \exp(-\Delta H_D/RT) \tag{6}$$

$$S = S_0 \exp(-\Delta H_S/RT) \tag{7}$$

Arrhenius plots of the natural logarithms of ϕ, D, and S as functions of reciprocal temperature should yield straight lines with slopes equal to the respective enthalpies ΔH_ϕ, ΔH_D, and ΔH_S divided by the gas constant.

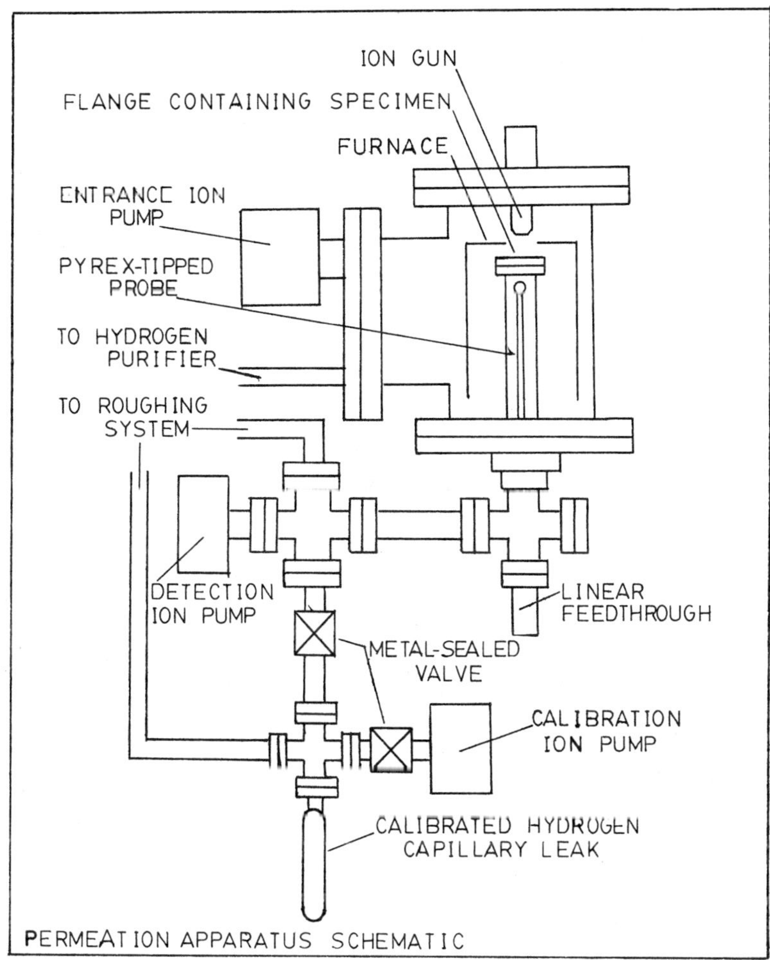

Figure 1. Permeation Apparatus Schematic.

Apparatus

The apparatus (Fig. 1) consists of three ultra-high vacuum chambers connected to a common roughing system but separately ion pumped. The entrance chamber is sealed from the detection chamber by the 3.65 cm diameter membrane specimen, which is clamped against a copper gasket in a modified UHV flange. The calibration chamber is connected to the detection chamber through a metal sealed valve. The entire high vacuum system is bakeable and pressures below 1 x 10^{-6} Pa (10^{-8} torr) are routinely achieved, assuring minimal contamination of the hydrogen, and maximum specimen cleanliness.

The entrance chamber includes a diaphragm pressure transducer for measurements in the KPa range, and manifolding is provided to admit purified hydrogen and any other gas of interest at pressures in the range of 10^{-6} Pa to 10 KPa. The specimen heating furnace is within the entrance chamber and is capable of maintaining the specimen at any temperature between ambient and 500°C within 0.2°C, as monitored by a thermocouple attached directly to the specimen. An ion gun is positioned above the entrance surface for studies involving surface film removal or addition, and injection of hydrogen or other gases into the specimen.

The detection chamber accommodates a micrometer linear motion feedthrough equipped with a strain gauge bridge which permits calibrated application of up to 50 kg load at the center of the specimen. The tip of the linear motion feedthrough is currently fitted with a Pyrex sphere 1.27 cm in diameter. Deflection of the membrane upward results in a decrease in thickness and an increase in area of the specimen. This is measured subsequently but results in a correction of less than 5%.

The calibration chamber consists of a calibrated capillary hydrogen leak which can be valved either into an ion pump (to prevent hydrogen build-up in the chamber) or into the detection chamber, where it provides a known flux for calibration of the detection ion pump.

Specimen Preparation

Specimen material was commercially available Type 302 stainless steel, received as sheet rolled to a thickness of 100 μm. After annealing (1000 C), the specimens were electropolished for 1.5 minutes at 30 volts in a chilled solution of perchloric and glacial acetic acids (1:4). Palladium was vapor-deposited on both entrance and exit sides of the specimen after argon gas plasma sputtering the surfaces for 2 hours at an average current density of 100 μA/cm^2, and a potential of 500 volts.

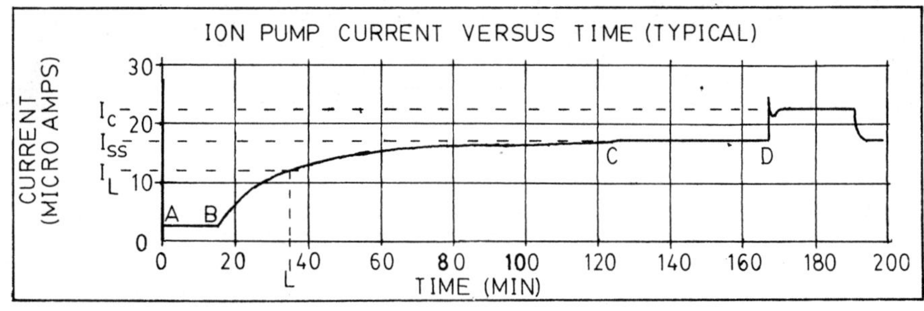

Figure 2. Trace of Ion Pump Current vs. Time. (typical)

Procedure

To determine the diffusion coefficient, D, the lag time, L, must be obtained. The coefficient of permeability, ϕ, depends on the steady state flux, J_∞. A typical plot of the current drawn by the detection ion pump as a function of time is shown in Fig. 2. At first, the ion pump current will be due to residual gas in the system, and constant with time (trace AB, Fig. 2). With the entrance chamber pumps turned off, the entrance side of the specimen is pressurized with purified hydrogen gas to a preselected pressure, usually between 6×10^2 Pa and 1×10^4 Pa. The time at pressurization (B, Fig. 2) serves as the zero point for lag time determination. After a short time interval, hydrogen appears at the exit side of the foil, as evidenced by the rising pump current (trace BC, Fig. 2). The steady state is achieved at point C, and the pump current becomes constant. This current, I_{ss}, is proportional to the steady state flux, J_∞, and is used to determine the lag time, L, as previously explained. The steady state flux is calibrated by superimposing the known flux from the capillary hydrogen leak onto the permeation flux (point D, Fig. 2). The pump current rise I_c is proportional to the known flux, and thus calibrates the permeating flux. The lag time L and steady state flux J_∞ are obtained, permitting calculation of the diffusion coefficient D, permeability coefficient ϕ, and solubility coefficient S for a given temperature.

Results

A test (3) of the assumption that bulk diffusion is rate limiting for this experimental arrangement was made, and the permeating flux found to be proportional to the square root of the input pressure over the range of experimental temperatures (4). Baseline determinations of the permeability, diffusivity, and solubility of the unstressed Type 302 stainless steel were made at four temperatures within the range 200°C to 320°C, and linear least-squares fit of the data yielded a correlation coefficient of at least .99 in all cases. For the unstressed specimen,

$$\phi = 1.24 \times 10^{-4}\ \exp(-65470/RT)\quad cm^3(STP)/cm\ s\ Pa^{\frac{1}{2}} \tag{8}$$

$$D = 1.65 \times 10^{3}\ \exp(-47850/RT)\quad cm^2/s \tag{9}$$

$$S = 7.52 \times 10^{-2}\ \exp(-17620/RT)\quad cm^3(STP)/cm^3\ Pa^{\frac{1}{2}} \tag{10}$$

where the ΔH's have the dimensions of Joules/mole. These values lie well within the scatter of values reported in the literature (4-6).

The specimen was then stressed in tension by means of a dead weight load applied to the probe. Loads of up to 33 kg were applied in increments, and the Arrhenius relations for ϕ_σ, D_σ, and S_σ were established under each load. Since the geometry of the loaded specimen results in nonuniform biaxial tensile stress, ϕ_σ, D_σ, and S_σ are determined as a function of probe load only.

The effects of stress are illustrated at a constant temperature of 240°C in Fig. 3, by plotting the normalized ratios of ϕ_σ/ϕ, D_σ/D, and S_σ/S as a function of probe load. The corresponding plots for other temperatures within the experimental range are similar. As can be seen in this figure, a rise in permation constant occurs upon reaching a critical load This rise is attributable to the change in diffusivity in the specimen, since the solubility constant is essentially unchanged by the applied load at this temperature. There is a tendency for the diffusivity to level off at higher loads. Upon removal of the load the diffusivity and permeation

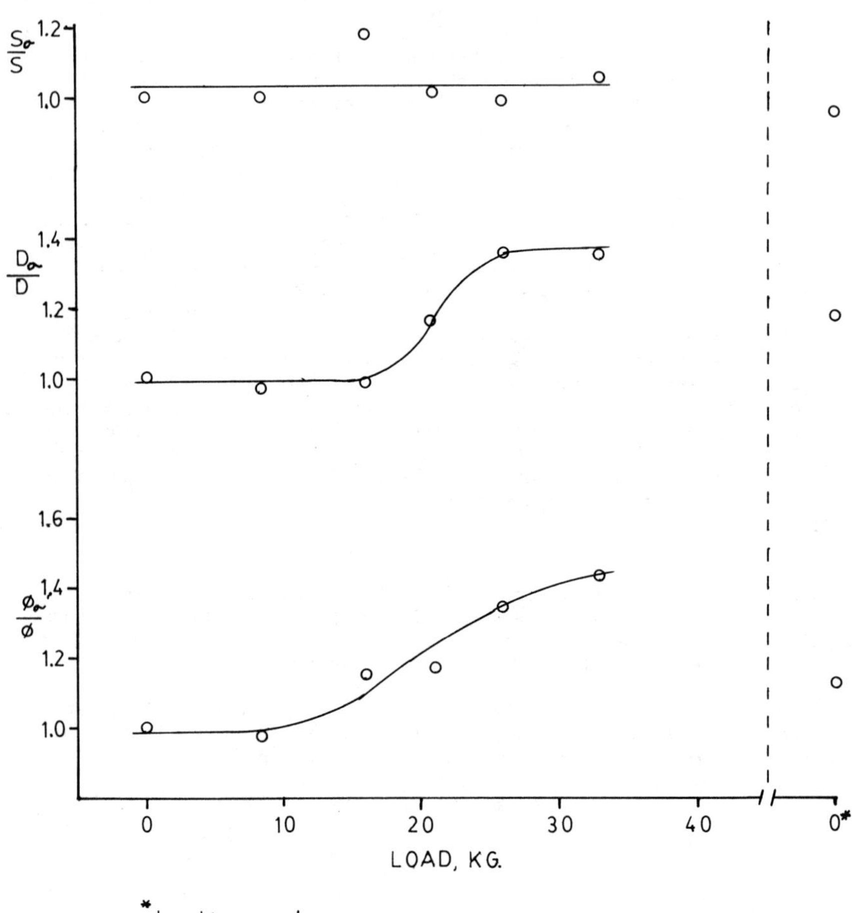

Figure 3.   Relative Permeability, Diffusivity, and Solubility vs. Load at
240°C.

constant are significantly reduced, whereas the solubility is essentially
unaffected. Preliminary results indicate that the no-load normalized diffusivity and solubility constants vary with temperature, whereas the permeability does not.

Also of interest is the transient behavior of the permeation flux upon
the application of load to the specimen. A typical plot of the ratio ϕ_σ/ϕ
and probe load against time is shown in Fig. 4. Rapid flux of hydrogen
through the specimen occurs with loading, followed by a rapid decrease in
the flux, and finally a slow rise to a higher permeation rate. Some plastic
deformation of the specimen occurs at all loading levels.

Similar effects have been observed using hydrogen in nickel by
Sidorenko et al. (7) and with tritium in a variety of materials by Donovan

(8). Both attributed the release of hydrogen to egressing dislocations.
Further work is under way to determine the generality of these effects.

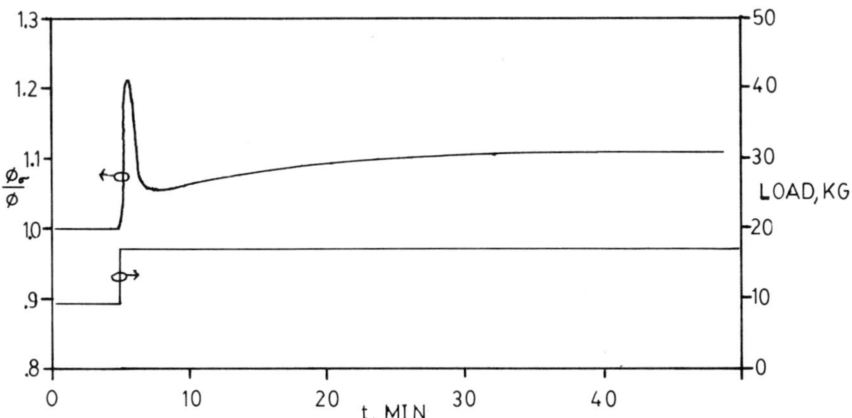

Figure 4. Relative Permeability and Load vs. Time. (typical)

References

1. R. M. Barrer, Diffusion In and Through Solids, pp. 18-19; Cambridge
   University Press, Cambridge, England, 1951.

2. M. A. V. Devanathan and Z. Stachurski, "The Adsorption and Diffusion of
   Electrolytic Hydrogen in Palladium," Proceedings of the Royal Society,
   A270 (1962) pp. 90-102.

3. M. R. Quick and H. H. Johnson, "Hydrogen and Deuterium in Iron,
   49-505 C," Acta Metallurgica, 26 (1978) pp. 903-907.

4. W. R. Conley, "Hydrogen Permeation in Austenitic Stainless Steels,"
   M.S. Thesis, University of Illinois at Urbana-Champaign, 1979.

5. M. R. Louthan, Jr., and R. G. Derrick, "Hydrogen Transport in
   Austenitic Stainless Steel," Corrosion Science, 15 (1975) pp. 565-577.

6. H. G. Nelson and J. E. Stein, Gas-phase Hydrogen Permeation Through
   Alpha Iron, 4130 Steel and 304 Stainless Steel from Less Than 100°C to
   Near 600°C; National Aeronautics and Space Administration Technical
   Report, NASA TND-7265, 1973.

7. V. M. Sidorenko, B. F. Kachmar and M. S. Borisova, "Effect of Deformation and Stress on the Diffusion Characteristics of Hydrogen in
   Metals," Fiziko-Khimicheskaya Mekhanika Materialov, 9 (5) (1973) pp.
   14-19.

8. J. A. Donovan, "Accelerated Evolution of Hydrogen from Metals During
   Plastic Deformation," Metallurgical Transactions, 7A (1976) pp.
   1677-1683.

STRESS-INDUCED HYDROGEN REDISTRIBUTION IN HIGH-PURITY

Ti-31V ALLOY

Robert L. Schulte, Philip N. Adler
Grumman Aerospace Corporation
Bethpage, NY   11714

&

Neil E. Paton
Rockwell International
Thousand Oaks, CA   91360

The partial molal volume of hydrogen, $\overline{V}_H$, in single phase Ti-31V was determined both by in situ measurements of hydrogen change in a stress gradient and by precision lattice parameter measurements. The hydrogen concentration change and the diffusion coefficients were measured using a nuclear microprobe technique in samples stressed under four-point bending. Samples contained 400 WPPM hydrogen, 750 WPPM oxygen, and 130 WPPM nitrogen. Lattice parameters were measured over a range from 0 - 4000 WPPM hydrogen. The good agreement between the values of the partial molal volume, i.e., $\overline{V}_H$ - 1.8 ± 0.2 cm^3/mol for the stress gradient measurements and 1.6 ± 0.2 cm^3/mol for the lattice parameter measurements, indicates that no significant trapping of available hydrogen occurs in this high-purity material. Hydrogen transport was consistent with diffusion in other β-phase Ti alloys.

Introduction

In recent investigations (1,2) of stress-induced hydrogen redistribution in β-phase, commercial titanium alloys the extent of hydrogen concentration change was 50 to 70% lower than predicted based on the formalism of Li, Oriani and Darken (3) and values of the partial molal volume of hydrogen, $\overline{V}_H$, in β-phase Ti. The measurement of $\overline{V}_H$, as deduced from the stress-induced change in hydrogen concentration, may be influenced by some form of trapping which limits the amount of hydrogen available for redistribution. If any portion of the hydrogen solute, for example that bound by trapping sites, is unable to migrate in the applied stress gradient, the value of $\overline{V}_H$ determined by a measure of the change in hydrogen concentration will be smaller than the partial molal volume. For this reason, a parameter, $\overline{V}_H^s$, is designated to distinguish the proportionality constant which describes the response of hydrogen to a stress gradient within a particular alloy from the actual partial molal volume, $\overline{V}_H$, of the alloy.

Since the effects of impurity trapping are minimized in a high purity material, the value of $\overline{V}_H^s$ would be expected to approach that of the partial molal volume. This paper reports on the investigation of the stress-induced hydrogen concentration changes for a high purity, β-phase, Ti-31 w% V alloy. The value of $\overline{V}_H$ deduced from these measurements is compared to a direct measure of the partial molal volume obtained from precision lattice parameter measurements made on a series of Ti-V alloys.

Experimental

Samples of the Ti-31V alloy used to measure hydrogen redistribution in a stress gradient were prepared from high purity bar stock containing 30.5 w% V. Samples were solution heat treated at 800°C for 1-1/2 h at 8 x 10^{-6} torr and He-gas quenched to room temperature; hydrogen charging was done in a Sievert's apparatus at 800°C, where the samples were rapidly cooled by removing the furnace from the sample chamber. Samples were charged to 400±10 WPPM hydrogen, and no significant change in oxygen or nitrogen concentration was introduced by charging. X-ray diffraction on the hyrogen charged samples indicated the presence of only the β-phase. Using optical microscopy, second phase particles, approximately 1.5 μm in size, were observed throughout the samples; grain size was of the order of 50 - 100 μm.

Hydrogen concentration measurements were made with a nondestructive nuclear probe which employs the $^1H(^7Li,\gamma)^8Be$ resonant nuclear reaction (4). Each specimen was depth profiled to establish that the hydrogen content was uniform throughout the first $1.2\,\mu$m and consistent with the hydrogen concentration determined by bulk analysis.

The specimen, 1.5 mm thick x 12.5 mm wide x 75 mm length, was mounted in a four point bending apparatus which provided an adjustable, uniaxial, linear stress gradient through the specimen thickness. A strain gauge, attached to the back surface of the specimen, was used to determine the strain (equal in magnitude, opposite in direction) at the front surface where the in situ measurement of the hydrogen concentration was made. Stress was determined from measured strain using an elastic modulus of 59 x 10^3 MPa; this value is based on ultrasonic measurements in a Ti-30V alloy as a function of temperature and hydrogen content (5). The surface measured was alternately stressed in uniaxial tension and compression, and a series of hydrogen measurements were made at a depth of $1.2\,\mu$m for each level of stress. The changing hydrogen concentration was monitored as a function of time following reversal of the applied load. After evaluation of hydrogen redistribution within the region of elastic strain, the specimens were plastically deformed at two levels of strain, and the hydrogen content was measured at each loading level.

For the lattice parameter measurements, copper $K\alpha$ radiation was used in conjunction with a precision diffractometer equipped with a crystal monochrometer. Alloys containing 20, 25, 30 and 40 wt. % V were examined. The (310) and (321) peak positions were measured as a function of hydrogen content over a range from 0 - 4000 WPPM hydrogen.

<u>Results</u>

Change in hydrogen concentration as a function of the applied strain was measured for two specimens of high-purity Ti-31V. The hydrogen measurements were made with the front surface of samples held at 440 MPa compression, 440 MPa tension and at no load. The hydrogen concentrations of the unstressed samples were found to be 390 ± 20 and 410 ± 20 WPPM using the nuclear probe. The data, shown in Figure 1, are the mean values of, typically, 10 measurements, and the standard deviation associated with each set of measurements is indicated. The data within the elastic region were fit to the uniaxial form of the relationship between hydrogen concentration change and the applied stress:

$$\ln\frac{C_s}{C_o} = \frac{\overline{V}_H^s\,\sigma_{xx}}{3RT} \tag{1}$$

where C_S and C_O are the hydrogen concentrations in the stressed and
unstressed region, respectively; σ_{xx} is the applied uniaxial stress; $\bar{V}_H^s$
is the effective partial molal volume as determined by the response of
hydrogen to a stress gradient; R is the gas constant and T is the absolute temperature. The least squares fit to the data is indicated by the
solid lines for each of the specimens, and the values for $\bar{V}_H^s$ obtained
from the slope are given in Table I. These results are for a measured
temperature of 360 K which was uniform to within 10 K through the specimen
thickness.

The response of hydrogen to increasing the applied strain into the
plastic region is indicated by the dashed line in Figure 1 for each specimen. At a strain of 2.2%, nearly 2/3 of the specimen thickness is plastically deformed with an elastic region comprising 1/3 of the thickness at
the center of the sample. Each data point within the plastic region is the
mean of four hydrogen concentration measurements. Within the statistical
fluctuations of the data there appears to be no further increase in hydrogen content when plastic strain levels are applied.

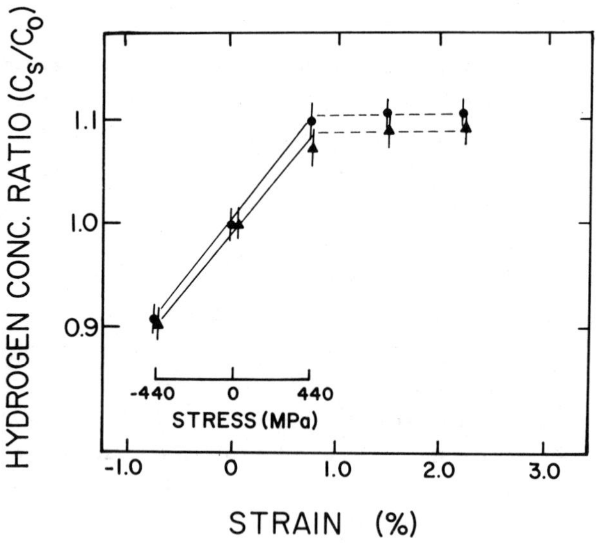

Fig. 1   Change in hydrogen concentration as a function of applied strain
for two samples of Ti-31V. The corresponding stress scale is
indicated within the elastic region of the alloy.

Table I. Summary of Results of Stress Induced Hydrogen Change for Ti-31V

Sample	Slope (MPa^{-1}) $\ln(C_s/C_o)$ vs σ_{xx}	$\overline{V}_H^{s*}$ (cm^3/mol)	$D*$ (cm^2/s)
A	$(2.09\pm.11) \times 10^{-4}$	$1.9\pm.1$	$(1.0\pm.5) \times 10^{-6}$
			$(1.5\pm.6) \times 10^{-6**}$
B	$(1.90\pm.32) \times 10^{-4}$	$1.7\pm.3$	$(0.9\pm.3) \times 10^{-6}$

* at 360 K
** measured during reversal of plastic strain

The time dependent concentration change associated with the hydrogen
redistribution following a reversal of the stress gradient was measured to
determine the diffusivity of hydrogen. The data shown in Figure 2 are the
average of three series of measurements following the reversal of the
applied stress from 440 MPa compression to 440 MPa tension. The solution
of the diffusion equation for constant D under the condition of a linear
distribution of hydrogen in a slab with impermeable surfaces (6) has been
fit to the data. Using a least squares criterion, the best fit to the data
yields a diffusion coefficient of 9×10^{-7} cm ^{2}/s. One of the samples
was also cycled between + 2.2% and − 2.2% applied strain at which point 2/3
of the thickness of the sample was plastically deformed. The fit to that
data resulted in a diffusion coefficient of 1.5×10^{-6} cm^2/s. The
results of all the diffusion measurements are compiled in Table I.

The lattice parameter for each alloy and hydrogen composition investigated was obtained using a least squares fit of the Nelson-Riley-
Taylor-Sinclair function (7) for the copper $K\alpha_1$ and $K\alpha_2$ peaks from
the (310) and (321) planes. The molar volume in each case was calculated, and these results are shown in Figure 3 for each alloy as a function of hydrogen content. A linear extrapolation of a least squares fit
was used to determine $\overline{V}_H$ for each alloy. As shown in Figure 3, the
values for $\overline{V}_H$ do not appear to vary significantly with vanadium content.

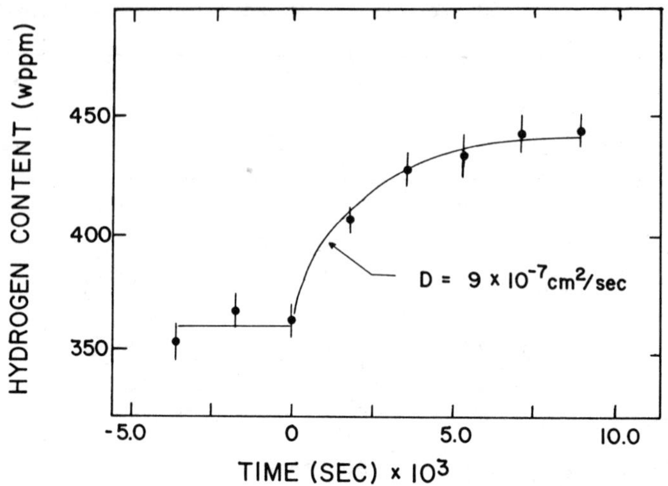

Fig. 2   Time dependence of the hydrogen content for Ti-31V following a reversal of the linear, uniaxial stress gradient from 440 MPa compression to 440 MPa tension.

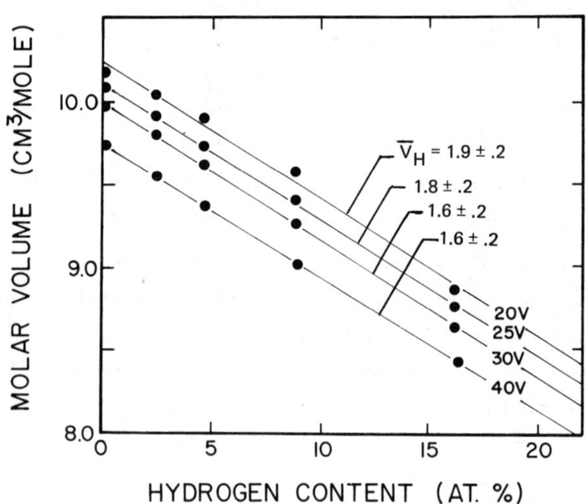

Fig. 3   Molar volume as a function of hydrogen content for high purity Ti-V alloys obtained from lattice parameter measurements.

Discussion

The average partial molal volume deduced from stress-induced hydrogen redistribution measurements, i.e. the value of 1.8 ± 0.2 cm^3/mol, is in excellent agreement with 1.6 ± 0.2 cm^3/mol obtained from precision lattice parameter measurements. The consistency between the measurement in a stress gradient and the measurement of the lattice dilation due to interstitial hydrogen in solid solution indicates that hydrogen responds in the expected stress-dependent manner for the high purity Ti-31V alloy. This contrasts with earlier work in three commercial β-phase titanium alloys where $\overline{V}_H^s$ values of 0.6 - 0.9 cm^3/mol were measured (see Table II). It is likely that the commercial alloys contain traps that reduce the mobility of a portion of the existing hydrogen, making that portion relatively unavailable for long range transport in a stress gradient. Using internal friction measurements to study hydrogen diffusion in bcc niobium, Baker and Birnbaum (8) found that hydrogen clusters exist which affect transport. These clusters, either hydrogen pairs in high-purity Nb or oxygen (nitrogen) hydrogen pairs in Nb containing interstitial solutes, reduce the diffusivity from that calculated based on Gorsky effect measurements (9,10,11). Since the Gorsky effect measurement involves long range hydrogen diffusion, single hydrogen atoms are probably involved. The presently reported stress gradient measurements are similar to the Gorsky effect measurement since long range transport is also required.

Table II. Comparison of High-Purity Ti-31V Alloy With Commercial Beta Phase Alloys

Alloy	$\overline{V}_H^s$ (cm^3/mol)	D (cm^2/s)	Interstitial Content (At.%)		
			H	O	N
Ti-31V	1.8±.2	(1±.5)x10^{-6}	1.90±.02	.24±.02	.05±.01
Ti-8M0-8V-2Fe-3Al	0.7±.1	(8±2)x10^{-7}	1.20±.10	.43±.01	.04±.01
Ti-3Al-8V-6Cr-4Mo-4Zr	0.6±.1	(7±2)x10^{-7}	1.30±.10	.34±.01	.02±.01
Ti-30Mo	0.9±.2	(1±.5)x10^{-6}	1.30±.10	.44±.02	.02±.01

As indicated in Table II, the hydrogen diffusivity at about 360 K in both the high purity Ti-31V alloy and the bcc phase commercial alloys is essentially the same, i.e., approximately 10^{-6} cm^2/s. This suggests

that the hydrogen that is mobile behaves in the same manner in both type
alloys. However, that portion of hydrogen that is trapped in the commercial alloys would have a reduced diffusivity by approximately two or
three orders of magnitude if comparison with Nb is valid (8). This would
result in an increase in diffusion time of a few orders of magnitude making long range transport of the trapped portion extend beyond the detection
limits of the present experiment. The interstitial solute content in the
Ti-31V alloy and the β-phase commercial alloys is listed in Table II, and
it appears that the oxygen content may be a factor contributing to the
inferred trapping phenomenon. Additional work, however, is required to
identify the source of trapping in these alloys.

No evidence of enhanced diffusion accompanying plastic deformation
has been observed in the present study or in previous measurements on the
commercial titanium alloys (1,2). Hydrogen transport by dislocations has
been proposed to explain enhanced diffusion effects (12). The fact that
the present measurement does not reflect an increased diffusivity that may
accompany plastic deformation can be explained by the loading configuration utilized, i.e., four-point bending to achieve a stress gradient.
When the strain at the outer surface is approximately three times the yield
strain, one-third of the sample thickness is still elastic. The overall
diffusivity, D_t, can be expressed by,

$$\frac{1}{D_t} = \frac{2}{3D_p} + \frac{1}{3D_e} \tag{2}$$

where

D_e = diffusivity under elastic loading

D_p = diffusivity with plastic deformation

The value of D_t is relatively insensitive to substantial changes in the
value of D_p, because diffusion through the specimen is dominated by the
diffusion through the thickness element that is under elastic strain. As a
result, the present approach cannot identify dislocation sweeping effects,
if they occur.

Conclusions

The extent of hydrogen redistribution in a uniaxial stress gradient
for high purity Ti-31V is in excellent agreement with the formalism of Li,
Oriani, Darken (3) wherein $\bar{V}_H^s$ in Equation (1) is equal to the partial
molal volume. This contrasts with the results of stress-induced hydrogen
changes measured in commercial β-phase titanium alloys. A decrease or
lack of impurity trapping of hydrogen in the high purity alloy may be the

reason that the predicted changes in hydrogen content are realized in this alloy, whereas, in the commercial alloys, trapping limits the hydrogen available for redistribution by lowering the diffusivity of the trapped portion by a few orders of magnitude.

Acknowledgements

This work was partially supported under AFOSR Contract No. F44620-76C-0071. The Ti-31V high-purity alloy samples that were used for the measurements in a stress gradient were kindly supplied by Professor Don Koss of Michigan Technological University. We also wish to thank Jonathan Montgomery for his assistance in computing the partial molal volumes from the lattice parameter data.

References

1. P.N. Adler and R.L. Schulte: Scr. Metall., 1978, vol. 13, p. 669.

2. P.N. Adler, R.L. Schulte, E.J. Schneid, E.A. Kamykowski, and F.J. Kuehne: Met. Trans., 1980, vol. 11A, p 1617.

3. J.C.M. Li, R.A. Oriani, and L.S. Darken: Z. Phys. Chem., Neue Folge, 1966, vol. 49, p.271.

4. E.A. Kamykowski, F.J. Kuehne, E.J. Schneid, and R. L. Schulte: Nucl. Instr. and Meth., 1979, vol. 165, p. 573.

5. O. Buck, D.O. Thompson, N.E. Paton and J.C. Williams: "Internal Friction and Ultrasonic Attenuation in Crystalline Solids", D. Lenz and K. Lucke, eds., vol. 1, p.451-458, Springer-Verlag, New York, 1975.

6. J. Crank: "The Mathematics of Diffusion", p. 58, Clarendon Press, Oxford, 1956.

7. C.S. Barrett and T.B. Massalski: "Structure of Metals", p. 145, McGraw-Hill, New York, 1965.

8. C. Baker and H.K. Birnbaum: Acta Metall., 1973, vol. 21, p. 865.

9. G. Schaumann, J. Volkl and G. Alefeld: Phys. Status Solidi, 1970, vol. 42, p. 401.

10. R. Cantelli, F.M. Mazzolai and M. Nuovo: Phys. Status Solidi, 1969, vol. 34, p. 397.

11. J. Volkl,, G. Schaumann, and C. Alefeld: J. Phys. Chem. Solids, 1970, vol. 31, p. 1805.

12. J.K. Tien, A.W. Thompson, I.M. Bernstein and R.J. Richards: Met. Trans., 1976, vol. 7A, p. 821.

HYDROGEN PERMEATION IN NIOBIUM*

R. Sherman and H. K. Birnbaum
Department of Metallurgy and Mining Engineering
University of Illinois
Urbana, IL 61801

The transport of hydrogen across Group Vb metals such as niobium is of
interest from a variety of scientific and technological reasons. Hydrogen
diffusion has been studied successfully below 300°C using Gorsky effect
techniques.(1) At higher temperatures this method does not work due to hydrogen loss from the specimens. Permeation studies have applied at temperatures above room temperature, but these have been characterized by large
errors due to surface barriers such as oxides.(2) We measured the permeability and diffusivity of hydrogen in niobium from 1000°C down to 450°C by
gaseous permeation methods in an ultra high vacuum system. The hydrogen flux
through the niobium sample was determined by measuring the pressure difference across a known conductance. The pressures were monitored as a function
of time by a LSI-11 minicomputer system and the steady state flux was related
to the permeability and the transient behavior was related to the diffusivity.

By use of the diffusivity measured by Alefeld and the measured solubility,(2) the expected permeability was calculated and a negative activation
energy of -6 kcal/mole H was found. This negative activation energy was due
to the high exothermic activation energy of solution compared to the endothermic activation energy of diffusion. Physically, this implies that as the
temperature increases, the permeating hydrogen flux will decrease. Our data
are in agreement with this expectation and with the calculated permeabilities.

Our measured permeabilities were highly sensitive to gas purity. At
$1\frac{1}{2}$ and 4 microns, our measured permeabilities were in reasonable agreement
with the expected values. The data was linear from 1000°C down to 600°C and
exhibited a negative deviation, i.e. a positive permeation enthalpy at lower
temperatures. At a pressure of 31 microns, our measured permeation was close
to the values of Rudd, Vose and Johnson with a positive activation energy
typical of an oxide coating on niobium.

Our diffusivity data agrees with the extrapolation of the low temperature diffusivity experiments of Alefeld to within 20% in the activation energy and pre-exponential values. Our diffusivity is linear from 1000°C to
800°C and then exhibits a negative deviation and higher activation energy
which is close to the activation energy of Albrecht's.(3) A growth at a surface barrier, such as an oxide barrier, may be the source of the decrease in
*This work was supported by DOE Contract No. DE-AC02-76ER01198.

the permeabilities and diffusivities.

To conclude, we performed high temperature hydrogen permeation experiments on niobium. We observed high temperature permeabilities which were in reasonable agreement with the low temperature extrapolation of the solubility and diffusivity data. We did observe the expected negative activation energy which has not previously been observed. Our diffusivities were in reasonable agreement with the low temperature surface independent Gorsky effect experiments. We believe that these are the first surface dependent experiments to measure diffusivities in niobium and agree with the low temperature surface independent experiments.

References

1. G. Alefeld and J. Volkl, <u>Diffusion in Solids: Recent Developments</u>, Eds. A. S. Nowick and J. J. Burten, Academic Press, NY, 1975.

2. R. E. Stickney, <u>The Chemistry of Fusion Technology</u>, Ed. D. M. Gruen, Plenum, NY, 1972, p. 241.

3. Albrecht, W. M., Goode, W. D. and Mollett, M. W., <u>J</u>. <u>Electrochem</u>. <u>Soc</u>., <u>106</u> (1959) p. 981.

EFFECTS ON DEFORMATION AND MECHANICAL PROPERTIES

SOFTENING AND HARDENING IN HIGH PURITY IRON AND ITS ALLOYS

CHARGED WITH HYDROGEN

H. Kimura and H. Matsui
Research Institute for Iron, Steel and Other Metals,
Tohoku University, Sendai, Japan

and

Akihiko Kimura, T. Kimura and K. Oguri
Graduate School of Engineering (Materials Science and Metallurgy)
Tohoku University, Sendai, Japan

Iron with a sufficient purity is softened by hydrogen charging at temperatures between 200K and 300K. This softening is caused by the increase in the
screw dislocation mobility by hydrogen. Hardening is observed below 190K.
This is caused by the interaction of hydrogen with edge dislocations and kinks
on screw dislocations. Another type of hydrogen induced hardening observed
in impure iron and mild steels is considered to be due to the interaction
among solute (impurity) atoms, hydrogen atoms and dislocations.

In order to obtain further information of the effect of hydrogen on the
plastic deformation of iron, high purity iron single crystals were tested
during hydrogen charging. At the initial stage of deformation, only the
primary screw dislocations are observed. This is contrary to the observation
that in specimens without hydrogen screw dislocations in more than two slip
systems exist at the macroscopic yielding as a result of pre-yielding motion
of edge dislocations in the secondary system. Under hydrogen charging, the
primary screws move before the secondary edges. This is a strong support to
the proposal that hydrogen enhances the mobility of screws. Investigation of
the orientation dependence of the slip plane and of the hydrogen softening was
performed. The results suggest that hydrogen reduces the Peierls stress of
a screw dislocation more effectively on {110} plane than on {112} plane.
Hydrogen softening is unobservable at a large strain where the motion of screw
dislocations no longer controls the deformation.

To confirm the hardening due to solute-hydrogen-dislocation interaction,
the hydrogen effect in high purity iron alloys were examined. When the
concentration of solutes are large enough, i.e., appreciably larger than the
concentration of solute hydrogen, hardening takes place by hydrogen charging.
A solute interacting strongly with hydrogen, e.g., Ti, causes large hydrogen
induced hardening, while molybdenum which hardly interacts with hydrogen
causes little hardening with hydrogen. Interstitial impurities, C and N, also
cause hydrogen induced hardening.

Introduction

The effect of hydrogen on the mechanical properties, especially the yield and flow stress, of iron and steels has been investigated quite extensively. Nevertheless, there have been many controversies on whether hydrogen causes hardening or softening. Many investigators have reported that the yield and flow stress of iron are increased by hydrogen (1-8), and considered the hardening to be solution hardening due to hydrogen. Softening has been reported occationally (9-11). But, one who considers hydrogen to cause hardening has argued that the softening is due to hydrogen induced damage; the specimens are charged with hydrogen electrolytically during tensile test to investigate the effect of hydrogen, so that the high effective hydrogen pressure induces damage to the specimens and produces dislocations.

Specimens used in those experiments mentioned above contained impurities of several hundreds ppm or more. The concentration of hydrogen in the tensile specimen can not be measured, but is estimated to be 100at.ppm or less (12) even under a very high effective pressure, more than 10^{10}Pa, produced by electrolytic charging (13). Thus, the concentration of hydrogen is much less than that of impurities. It is widely known that hydrogen in iron interacts with impurities. Hence, it is very likely that the hydrogen induced hardening mentioned above is due to a combined effect of impurities and hydrogen and not to the sole effect of hydrogen. In order to reveal the real nature of the hydrogen effect on the yield and flow stress of iron, one should use high purity iron in which the impurity concentration is smaller than the hydrogen concentration. Matsui, Moriya and Kimura (14) were the first to report a remarkable softening caused by hydrogen in a specially prepared high purity iron which contains less than about 10at.ppm impurities (15). The softening is most remarkable at around 200K. Cornet and Talbot-Besnard (16) also reported that a high purity iron is softened by hydrogen at above 273K.

Matsui, Kimura and their coworkers extensively investigated the effect of hydrogen (17-21). They extended the testing temperature below room temperature. In most of the previous research, the testing temperature was limited to near room temperature, because the electrolytic solution freezes at low temperatures. Matsui et al used a solution of $0.1N-CH_3OH-H_2O-H_2SO_4$ with $NaAsO_2$ so that they could investigate the hydrogen effect down to about 200K. Moreover, they quenched the hydrogen charged specimens to low temperatures, thereby lowered the testing temperature to 77K. The experiment in which hydrogen is charged to the specimens while they are strained will be referred to as the "during charging" experiment, and the experiment in which specimens are pre-strained, charged with hydrogen at room temperature and quenched to the test temperature will be referred to as the "after charging" experiment. They investigated the hydrogen effect for a wide range of temperature, purity, strain rate and charging current density, i.e., the hydrogen concentration. Their results of the "during charging" experiment are summarized as follows:——

(i) Hydrogen induces softening at above 200K, if the charging current density, i.e., the hydrogen concentration, exceeds a critical value.
(ii) The critical charging current density is smaller for the lower temperature, for the purer specimen and for the smaller strain rate.
(iii) Although an accurate determination of the hydrogen concentration in the specimen is not possible at present, the critical concentration is of the order of the concentration of impurities in the specimen.
(iv) The amount of hydrogen induced softening is larger for the purer specimens, and largest at around 200K. At and above 273K, the hydrogen induced softening is very small and observed only in very pure specimens, i.e., those with the residual resistivity ratio larger than 4000.
(v) The hydrogen induced softening is not due to damage which is often

claimed to be caused by hydrogen charging. No damage was detected in
the high purity specimen with optical microscopy, electronmicroscopy
or resistivity measurement in specimens to which hydrogen caused softening.
On the other hand, severe damage was observed in less pure specimens
under the same experimental conditions as for the pure specimen, and
still the specimen was <u>hardened</u> by hydrogen at room temperature. Thus,
the hydrogen induced softening is considered to be an intrinsic effect,
i.e., due to the interaction of hydrogen and dislocations. The hardening
observed in impure iron and steels is considered to be due to a hydrogen-
solute atom-dislocation interaction.
(vi) Below 190K, hydrogen causes hardening if the charging current
density exceeds a critical value. The critical current density, i.e.,
the critical hydrogen concentration, is smaller for the lower temperatures.

The results of the "after charging" experiment are essentially the same
to those of the "during charging" experiment. However, while in the "during
charging" experiment the hydrogen effects persist as long as hydrogen is
charged, the effects disappear soon after the initiation of tensile test in
the "after charging" experiment. In the "after charging" experiment hydrogen
in the iron lattice escapes out of the specimen before testing and hydrogen
deeply trapped at pre-existing dislocations is quenched and effective to the
motion of the pre-existing dislocations. The trapped hydrogen is not effective to the motion of dislocations generated by the tensile deformation after
quenching. The hydrogen effects also disappear if the charged specimens are
held at room temperature for more than 10min before quenching. This is also
evidence for the softening not to be caused by damage. Other hypotheses for
the softening, e.g., an electrochemical reaction at the surface, a gradient
in the hydrogen concentration and hydride formation, were also discussed and
ruled out. At below 200K, the microstrain prior to the macro yielding, which
is considered to be due to the motion of edge dislocation component, is
suppressed. This is evidence for the hardening below 190K to be due to
pinning or dragging of edge dislocation component by hydrogen.

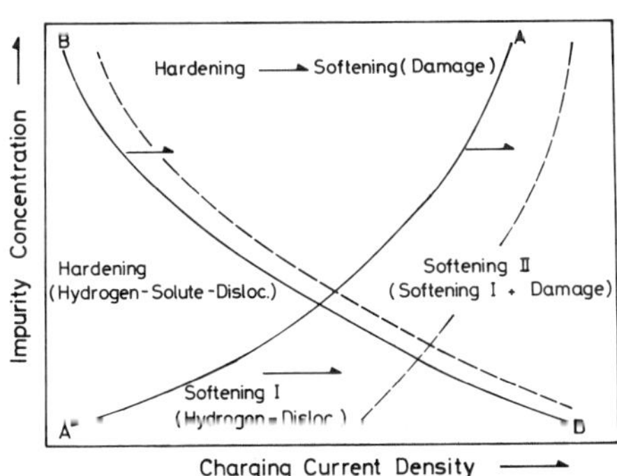

Fig. 1 - A schematic representation
of the effect of hydrogen on
the flow stress of iron, the
amount of impurity and the
charging current density
being taken as variables.
As the temperature increases,
curves A and B shift to the
right as shown by broken
lines.

These results of the "during
charging" experiment, except the
hardening below 190K, are schematically summarized in Fig. 1. The
curve A gives the boundary between
the intrinsic softening and the
hydrogen-solute hardening. The
curve B gives the critical purity
and the critical charging current
density for hydrogen damage to
occur. The positions of these
curves depend on the kind and state
of the solute. As the temperature
is raised, the curve A moves
markedly to the right, and the
curve B may slightly move to the
right because as the temperature
is raised the material softens and
becomes more endurable to damage.
In order to reveal the intrinsic
softening, the specimen purity and
the current density should be in
the proper region. The region
becomes smaller at higher temperatures. It is almost impossible to
observe unambiguously the intrinsic

softening in impure iron at near room temperature; with a small charging current density the hydrogen-solute hardening is observed and with a large current density the specimen is damaged and the damage induced softening may predominate.

The softening and hardening described above, except the hydrogen-solute hardening, are satisfactorily explained with the following mechanisms:——
(i) Above 200K, hydrogen is quite mobile and hydrogen trapped at screw dislocations moves along with them. It increases the mobility of screw dislocations by enhancing the rate of double kink formation.
(ii) Above 300K, edge dislocations contribute to the deformation, and the contribution of screw dislocations becomes small. Moreover, the motion of screw dislocations is largely aided by thermal energy. Hence, the hydrogen enhanced screw dislocation motion does not affect much the flow stress. Hydrogen does not appreciably affect the mobility of edge dislocations above about 200K.
(iii) Below 190K, hydrogen mobility decreases and hydrogen interacts with the edge dislocation component. Thus, hydrogen acts as pinning or dragging points to the edge dislocation. The double kink formation on screw dislocations is still enhanced by hydrogen below 190K, but the sideward motion of kinks is hindered by hydrogen and the hardening is resulted in.

In order to deepen our understanding of the effect of hydrogen, much research should be performed further. Especially the following three are important, (a) discussion of the mechanism of the enhancement of double kink formation rate by hydrogen, (b) detailed investigation of the hydrogen-solute hardening to discuss its mechanism, and (c) measurement of the interaction energy of hydrogen with dislocations and the state and the mobility of hydrogen at the dislocation core. The last problem will not be discussed in the present paper. The first problem may be approached with investigations of slip geometry, slip lines and dislocation configurations. Researches by single crystals are required, and being performed with the authors. The next chapter will review some of the results already published together with some new results. The second problem requires investigation with high purity iron alloys in which the kind, amount and state of a solute and/or solutes are well controlled. The third chapter will describe some results with iron alloys already published by the present authors together with some of their new results.

<u>Study of the Hydrogen Effect with High Purity Iron Single Crystals</u>

<u>Purpose of the Research</u>

The purpose of the research with single crystals is threefold. The first is to confirm further that the hydrogen induced softening in high purity iron is due to the enhanced mobility of screw dislocations and not to the hydrogen induced damage. If the damage causes softening by producing dislocations, a very high density of screw dislocations or an appreciable number of edge dislocations should be produced so as to meet the applied strain rate at the observed flow stress. Neither case was observed with polycrystalline specimens. The dislocation density in specimens deformed during hydrogen charging has been found to be somewhat larger than that in an uncharged specimen, but the difference is not enough to explain the observed softening. The character of dislocations has been found to be almost the same in both kinds of specimens. Thus, the hydrogen induced damage hypothesis for the softening seems to be quite unlikely. As already described, there is much other evidence which contradicts   the damage hypothesis. However, since the dislocation configuration in a deformed polycrystalline specimens is always complex, dislocations in single crystals should be investigated in order to confirm the results with polycrystals. Thus, transmission electron microscopy is performed with single crystals deformed to various tensile strains.

The second is to investigate the effect of hydrogen on the macroscopic characteristics of plastic deformation, such as the stress-strain curve, slip planes and their crystallographic orientation dependence. These properties cannot be investigated with polycrystals. The results of these investigations will further support the consideration that hydrogen enhances the mobility of screw dislocations, and may be useful to discuss the mechanism of the hydrogen enhanced mobility.

The third is to investigate the effects of hydrogen on the flow stress at a large strain and on the fracture. Polycrystalline specimens fracture along grain boundaries at a several percent tensile strain when tested during hydrogen charging, although uncharged high purity iron does not break along grain boundaries even at 4.2K (22). Hence, it is impossible to investigate the effect of hydrogen at large strains. Thus, single crystals are needed to avoid the grain boundary fracture. It is well known that precipitates and inclusions play important roles in the hydrogen embrittlement in steels. Previous investigations on hydrogen embrittlement, although important and fruitful, has not revealed the fracture response of iron crystal itself to hydrogen. Investigations with the present high purity single crystals may shed some light to this point. Most part of this chapter has been published elsewhere (23,24). Some unpublished results are included below.

Experimental Procedure

Single crystals were grown from the specially prepared high purity iron wire or ribbon with strain-anneal technique. The wire was drawn from rod of 5mm in diameter and recrystallized at 873K for 1h under flowing purified hydrogen gas. It was deformed in tension by 3% at a strain rate of $3.3\times10^{-5}\mathrm{s}^{-1}$, and then recrystallized to a single crystal by passing through a furnace at 1123K under purified hydrogen. The procedure for ribbon was similar to the above. The single crystals were hydrogen-treated to remove interstitial impurities and finally the surface was chemically polished. Crystals free from occluded small grains were used. The final size of specimens is 0.4mm diameter and 0.1mm×1.7mm for wire and sheet specimens, respectively. The gauge length is 10mm for both specimens. The residual resistivity ratio of the crystals was between 3600 and 5000.

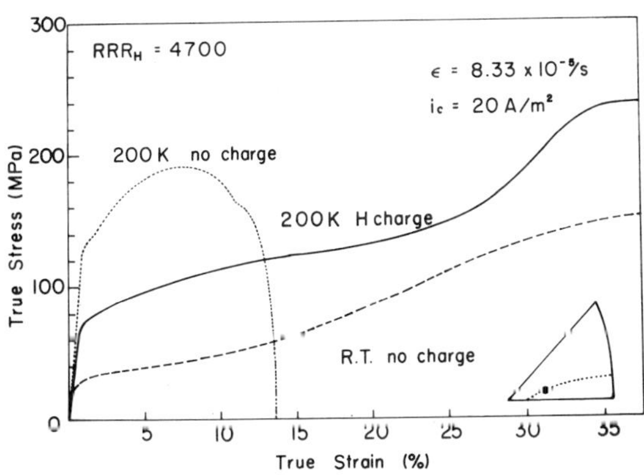

Fig. 2   Stress strain curves at 200K with and without hydrogen charging. The stress-strain curve at room temperature without hydrogen charging is also shown for comparison.

The method of hydrogen charging was the same as already described (14,19). The electrolytic solution was 0.1N CH₃OH H₂O H₂SO₄ with a small addition of NaAsO₂. The temperature of the bath was controlled within +0.5K. The strain rate was $8.3\times10^{-5}\mathrm{s}^{-1}$ unless otherwise stated.

Experimental Results

The Stress-Strain Curve.   Figure 2 shows the stress-strain curves of single crystals of which orientation is shown in the standard stereographic triangle. The specimen purity and the charging current density is shown in the figure. The full line is the stress-strain curve at 200K during hydrogen charging. Hydrogen charging was started before initiation of the tensile test and continued through the test. The dotted line is that at 200K but without hydrogen charging. The broken line is that at room temperature and without hydrogen. Hydrogen causes softening at small strains, and the stress-strain curve is of the three-stage hardening. The mechanism of the work-hardening under hydrogen charging is not known as yet and not necessarily the same as that of three-stage hardening in specimens deformed at room temperature without hydrogen. Hence, we shall denote the three stages of the stress-strain curve resulted in by hydrogen as stage I', II' and III', hereafter. The length of stage I' at 200K is longer for the lower current density and for the crystal orientation of larger range for the single glide.

Figure 3 shows the stress-strain curves at 170K with (full line) and without (dotted line) hydrogen charging. One specimen was pulled without hydrogen and charged with hydrogen at about 3% strain. It is obvious that hydrogen causes hardening at 170K. Note that the microstrain prior to the macroscopic yielding is reduced by hydrogen. The microstrain is considered to be due to the motion of edge dislocation components, and hence the reduced microstrain means that the motion of edge dislocation is hindered by hydrogen. The shape of the stress-strain curve is parabolic and similar to that without hydrogen.

Slip Lines and Slip Planes.   (a) Stage I' at 200K.   Slip lines in a hydrogen free specimen were so fine that they could be observed only near the necking point in a fractured specimen. In a hydrogen charged specimen, slip lines were relatively sharp and were usually observed all along the specimen at a few percent elongation. In both cases slip lines on the top surface (which is perpendicular to the Burgers vector) were wavy. Figure 4 shows the ψ-χ relation in stage I' of hydrogen charged specimens deformed at 200K (open circles), together with the results for uncharged specimens (closed circles). ψ is the angle between the observed slip plane and $(\bar{1}01)$ plane and χ is the angle between the maximum resolved shear stress plane (MRSSP) and $(\bar{1}01)$ plane. The result for uncharged specimens agrees with that due to Spitzig and Keh (25). Hydrogen charging makes the {110} slip easier than {112} slip.

(b) Stage II' at 200K. Both primary $(\bar{1}01)$ and conjugate (110) slip lines are observed at the end of stage II', 32% strain, in a hydrogen charged specimen. This result shows that at least at the later

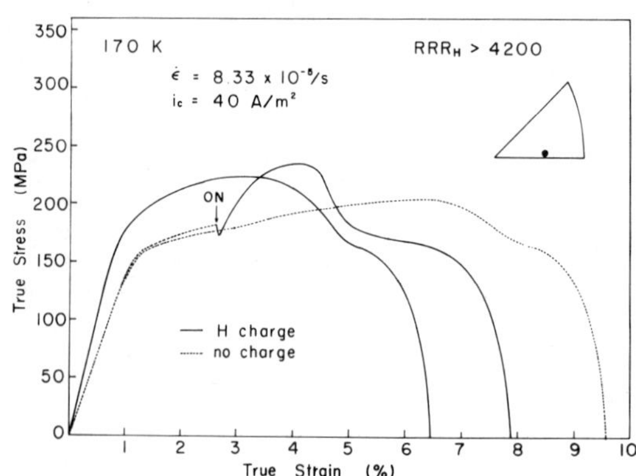

Fig. 3 – Stress-strain curves at 170K with and without hydrogen charging. One specimen is deformed without hydrogen charging and at about 2.5% strain hydrogen is charged.

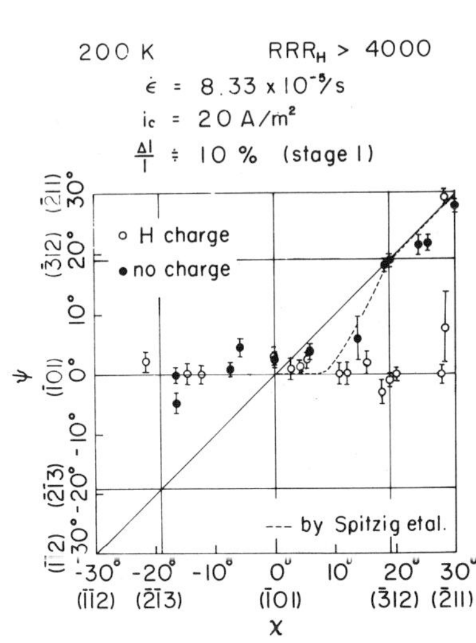

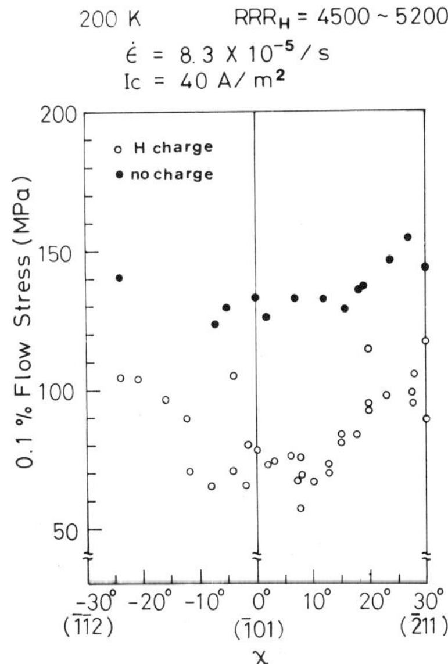

Fig. 4 - The ψ-χ relation in stage I' of hydrogen charged specimens deformed at 200K (open circles), and that of uncharged specimens (closed circles).

Fig. 5 - Orientation dependence of the yield stress at 200K. Closed circles are for uncharged specimens and open circles for charged specimens.

stage of stage II' the secondary slip operates.

(c) At 170K. In hydrogen charged specimens, the slip lines are straight and coarse, and the slip plane is of {110} type regardless the specimen orientation. In a specimen with $\chi=+30°$, both {110} and {112} type slip lines are observed.

<u>Orientation Dependence of Softening</u>. Figure 5 shows the yield stress of specimens with (open circles) and without (closed circles) hydrogen charging against the specimen orientation given by χ. Here, hydrogen charging was started before the tensile test and continued through the testing. Data scatter considerably, but the amount of softening seems to be larger for specimens of $\chi=0°$ than those near $\chi=30°$. In another series of experiment, tensile test was started without hydrogen charging, and at about 2% tensile strain the charging was started. In this experiment, the amount of softening was measured for each specimen and data scatter little. It is clearly revealed that specimens near $\chi=0°$ show more softening than those near $\chi=30°$ (23).

In the case of hardening below 190K, the amount of hardening does not depend on the specimen orientation; the critical shear stress resolved on the $(\bar{1}01)$-[111] slip system obeys the Schmid law.

<u>Transmission Electron Microscopy</u>. (a) Stage I' at 200K. Figure 6 shows dislocations in a specimen deformed to 7% tensile strain (stage I') at

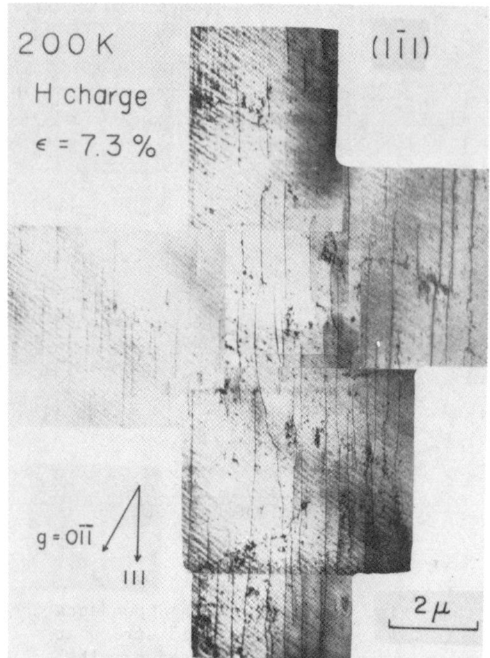

Fig. 6 – Transmission electron microscopy of a specimen deformed to 7% at 200K with hydrogen charging (stage I').

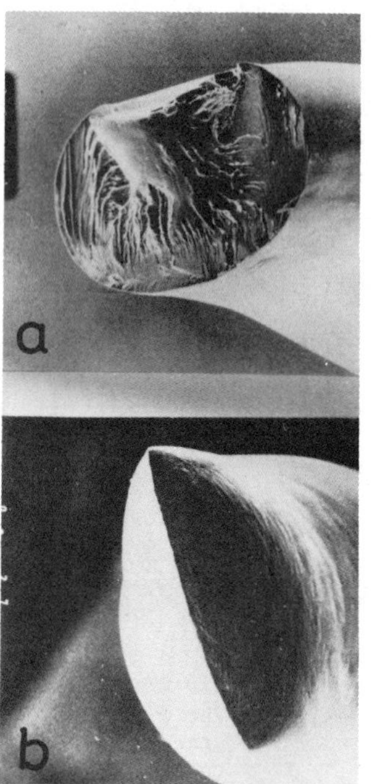

Fig. 7 – (a) Fracture end of a hydrogen charged specimen deformed at 200K with hydrogen charging. (b) Ductile fracture of a uncharged specimen deformed at 200K.

200K during hydrogen charging. The foil plane is (1Ī1). Dislocations are mostly primary screw dislocations, which are nearly straight. The secondary dislocations and dislocation tangles are very scarce. For comparison, dislocations were observed in a specimen deformed at 200K to 3% without hydrogen. Dislocations with several Burgers vectors are observed as usually so in iron deformed at low temperatures.

(b) Stage II' at 200K. A specimen was deformed to 22% strain at 200K during hydrogen charging. Screw dislocations of the primary slip system are predominant, but a considerable number of secondary screw dislocations are observed. They are also nearly straight and intersect with the primary screws. As deformation proceeds, dislocations become more tangled and finally they form cell structures.

<u>Effect of Hydrogen at Large Strains</u>. As already shown in Fig. 2, the flow stress of a hydrogen charged specimen exceeds that of an uncharged specimen at large strains. On-and-off experiment was performed with a specimen deformed to stage III' at 200K during charging; the charging current was put off at a strain in stage III'. The flow stress first decreases slightly and soon goes back to almost the same level as before. The variation in the flow

stress is very small compared to the variation at stage I'. More detailed
investigation is now in progress, and preliminary results show that the effect
of hydrogen becomes less and less as the deformation proceeds.

<u>Fracture Behavior</u>.   Figure 7 (a) is an example of micrographs of a
hydrogen charged and fractured specimen at 200K. Hydrogen was charged through
the whole course of deformation. The mode of fracture is not unique. It is
sometimes cleavage. Otherwise, the fracture surface appears to have resulted by the microvoid coalescence with a considerable loss in the reduction in
area, or mixture of cleavage and microvoid coalescence. The crystallographic
orientation of the fractured surface was difficult to determine because of
the severe deformation at the necking point. Without hydrogen, high purity
iron single crystals break in a completely ductile manner as shown in Fig. 7
(b). Hydrogen charging at 200K markedly reduces the ductility; the reduction
in area decreases from 100% for the hydrogen free specimens to about 50%.
At 170K, hydrogen causes hardening through the whole course of deformation,
and the specimen is completely ductile, being very similar to the uncharged
specimen.

<u>Discussion</u>

<u>Confirmation of the Proposed Mechanism for the Hydrogen Induced Softening</u>.
Transmission electron microscopy of the dislocations in stage I' shows
long and straight screw dislocations and almost no edge dislocations and the
slip lines are wavy at the top surface. These facts show that the motion of
screw dislocations controls the deformation of hydrogen charged specimens at
200K. The density of the screw dislocations is comparable to that in hydrogen
free specimens. Hence, hypotheses, such as hydrogen induced damage, which
assume edge dislocations or a large density of screw dislocations to be
produced and controll the deformation, are ruled out. Thus, the observation
of dislocations in single crystals confirms the mechanism proposed by the
authors that hydrogen increases the mobility of screw dislocations and causes
softening between 200K and 300K.

<u>Interpretation of the Deformation Characteristics of Hydrogen Charged
Crystals with the Proposed Mechanisms</u>.   (a) Stage I' and II' at 200K. At
low temperatures, the deformation of pure iron without hydrogen is characterized by a quasi-parabolic stress-strain curve, and at small strains long
screw dislocations belonging to two or three slip systems are observed. This is
interpreted as a result of different mobility of screw and edge dislocations.
The tensile stress to move edge and screw dislocations in primary and secondary
slip systems is schematically given in Fig. 8. Edge dislocations move under a
much smaller stress than screw dislocations in BCC metals.

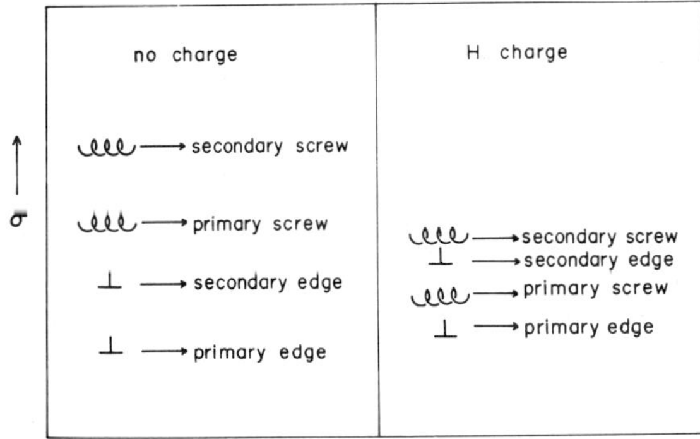

Fig. 8 - A schematic diagram of the tensile
stress to move various dislocations
with and without hydrogen.

The difference in the stress between primary and secondary dislocations is due to the difference in the Schmid factor. Edge dislocations in both systems moves before the primary screws and leave long screw dislocations of both types. At the yield stress the primary screw dislocations start to move, and their interaction with the secondary screw dislocations causes the rapid hardening. When hydrogen is charged the mobility of screw dislocations is increased and the primary screw dislocations move under a smaller tensile stress than the secondary edge dislocations. Hence, at a small strain in stage I' only primary screw dislocations are observed, and the work hardening is mainly due to elastic interaction among the primary screw dislocations, so that the work hardening rate is small. This consideration also agrees with the result that only the primary slip lines are observed.

As the deformation proceeds, cross-slip of the screw dislocations takes place as indicated by the wavy slip lines, and their mutual annihilation occurs. The cross-slip usually leave edge dislocations on non-primary plane. These edge dislocations move more slowly than those in the primary planes and may be pinned by hydrogen at 200K and below. These edge dislocations may act as additional and rather strong barriers to the primary screw dislocations. The edge dislocation pinning is more effective for the larger charging current density, i.e., hydrogen concentration, so that the length of stage I' becomes shorter as the current density increases. As the flow stress increases, the secondary edge dislocations start to move leaving long secondary screw dislocations. The secondary screw dislocations soon start to move and interaction of both screw dislocations causes rapid work hardening. This is stage II'. In the above consideration the hydrogen induced pinning of edge dislocations produced by cross-slip plays a role in determining the length of stage I'. Since the pinning should be temperature dependent, the length of stage I' should also be temperature dependent. More detailed investigation of the initiation and development of stage II' is in progress and the results will be published elsewhere.

(b) Stage III'at 200K. Dislocation configurations in stage III' at 200K are cell structures, and the flow stress is considered to be determined mainly by dislocation interactions. The screw dislocation mobility no longer determines the flow stress, so that hydrogen does not cause softening. Rather, hydrogen causes slight hardening because edge dislocations may predominantly controll the deformation at this stage.

(c) Hardening at 170K. Hydrogen induced hardening below 190K is due to pinning or dragging of edge dislocations, including kinks on screw dislocations, by hydrogen. Therefore, the hardening is observed through the whole course of deformation.

(d) Choice of the Slip Plane and Plastic Anistropy. The hydrogen induced softening is more marked in specimens with $\chi=0°$ than those with $\chi=30°$. This result implies that hydrogen enhances the rate of double kink formation more effectively in {110} plane than {112} plane. It has already been suggested that the hydrogen enhanced double kink formation would be due to the reduction in the Peierls stress and not to the elastic interaction of hydrogen with dislocations (19). Thus, it is further suggested that hydrogen reduces the Peierls stress more effectively in {110} plane than in {112} plane. In agreement with this consideration the active slip plane is $(\bar{1}01)$ regardless of the specimen orientation except for $\chi=30°$, where {112} type slip lines are also observed.

As the deformation proceeds, the screw dislocations play less and less role in controlling the deformation, so that the active slip plane becomes more and more close to MRSSP.

Conclusions

The experiments with single crystals give further and clear evidence to
support the mechanisms of hydrogen induced softening and hardening proposed
by the present authors. Other mechanisms so far proposed for the softening
due to hydrogen charging, e.g., the hydrogen induced damage or the electrochemical reaction at the surface, cannot explain the observed results so well
and comprehensively as the model by the present authors. It has been suggested that hydrogen reduces the Peierls stress of a screw dislocation more
effectively in {110} plane than {112} plane. It is found that the hydrogen
induced softening is sensitive to the dislocation configuration; it is marked
at small strains and becomes unobservable when dislocation cell structures
are formed and screw dislocations no longer controll the flow stress.

Study of the Hydrogen-Solute Effect

Purpose of the Research

As discussed in the proceeding section, there is little doubt in considering that the hydrogen induced softening in high purity iron above 200K is
due to the increased mobility of screw dislocations decorated with hydrogen.
Hydrogen causes hardening to the high purity iron at room temperature when
its concentration is below a critical value. The critical concentration is
larger for a less pure specimens. Thus, we are forced to consider that the
hydrogen induced hardening at room temperature in high purity iron charged
with a smaller current than the critical density and the hardening due to
hydrogen often observed in electrolytic iron and mild steels is due to some
combined effect of hydrogen and solute atoms to dislocations. Research should
be conducted with high purity iron doped with a solute or solutes of known
kind, amount and state to reveal the nature of the hydrogen induced hardening.

Some results have been reported by other investigators on substitutional
iron alloys (26) and iron-carbon system (27). Those investigations were only
at room temperature or above 273K. As seen in the research with high purity
iron, experiments under wide variaties of condition should be performed to
reveal the nature of the hydrogen effect. Especially, the test temperature
is the most important variable. The authors are investigating the hydrogen
effect with Fe-Ti, Fe-Mo (28) and interstitial impurity containing iron with
various purities (29). Results of these investigations will be reviewed in
this section.

Experimental Procedures

Preparation of Specimens. An Fe-Ti alloy was prepared by zone melting
a piece of high purity iron rod wound with high purity titanium wire under
high purity argon atmosphere. An Fe-Mo alloy was prepared similarly but under
ultra high vacuum instead of argon atmosphere. The alloy rods were drawn to
wire of 0.5mm diameter. The Fe-Ti wire was annealed at 1170K for 1h in dry
hydrogen to recrystallize fully. The grain size was 60 to 100µm. The Fe-Mo
wire was annealed at 1270K for 4h under wet and dry hydrogen to be decarburized
and recrystallized. The grain structure was of a bamboo type. Both alloys
are solid solutions, Ti being 0.6at.% and Mo 0.2at.% . The distribution of
molybdenum is fairly uniform along the specimen length but titanium distribution
is not so uniform. The titanium concentration varies from specimen to specimen
and the concentration 0.6at.% is the average. This variation, however, may
be disregarded for the present purpose of research because the yield stress
of specimens taken from various parts of the wire agrees with each other within
30% or less as shown in Fig. 9. In this figure are shown the temperature
dependence of the yield stress of the two alloys. The yield stress of alloys
are not much different from that of pure iron, i.e., the solid solution

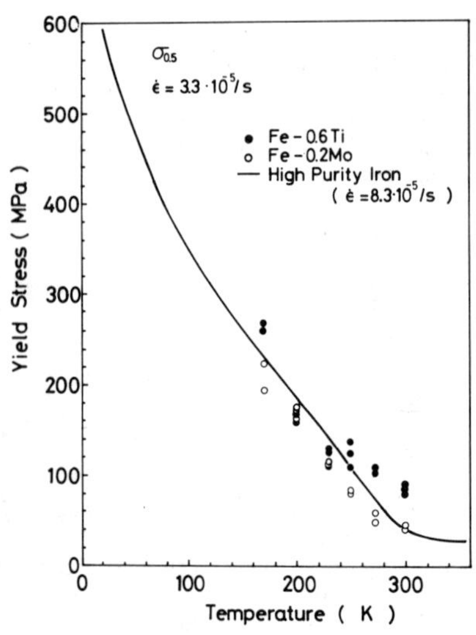

Fig. 9 – Temperature dependence of the yield stress of Fe-Ti and Fe-Mo alloy. The yield stress of high purity iron is also shown with a solid curve.

softening is not appreciable because of the small concentration of alloying elements.

In order to examine the effect of interstitial impurities on the hydrogen effect, three kinds of iron were used, an electrolytic iron (EI), a commercial high purity iron (Johnson Matthey iron, JMI) and the high purity iron same to those used by the present authors in their previous research (FZI).

EI contains 40, 10 and 100wt.ppm of C, N and O, respectively, and substitutional impurities probably more than 100wt.ppm. JMI contains 21, 224 and 47wt.ppm of C, N and O, respectively, as reported by the supplier, and substitutional impurities probably 10wt.ppm or more. Three kinds of heat treatments were applied (a) to remove interstitial impurities by thorough hydrogen treatment, (b) to precipitate interstitial impurities by slow cooling after a short anneal under dry hydrogen and (c) to dissolve them by quenching. These treatments will be referred to as H, A Q, respectively. FZI-H contains about 0.5wt.ppm or less carbon and has a resistivity ratio of about 4000. Carbon in FZI-A is less than 2wt.ppm.

Hydrogen charging and testing procedures were the same as in the case of high purity iron experiment. The strain rate was $8.3 \times 10^{-5} s^{-1}$ unless otherwise stated.

Results

Effects of Substitutional Solutes. Figure 10 shows the effect of hydrogen charging on the flow stress of the Fe-Ti alloy ($\dot{\varepsilon} = 3.3 \times 10^{-5} s^{-1}$). At all temperatures examined, the flow stress increases markedly by hydrogen charging (marked with ↓). The increase in the flow stress is about 100%. Upon putting off the current (marked with ↑), the flow stress recovers toward the uncharged level at a very slow rate. Figure 11 shows that it takes about 30min for the recovery in the flow stress to take place at room temperature. At the strain marked with ↑ the charging current was put off, the cross-head was stopped and the specimen was kept for 30min. The flow stress almost recovered to the uncharged level. The amount of hardening hardly depends on the temperature.

Figure 12 shows the effect of hydrogen in the Fe-Mo alloy ($\dot{\varepsilon} = 3.3 \times 10^{-5} s^{-1}$). The flow stress increases slightly (less than 10%) at room temperature but decreases at temperatures below 273K by hydrogen charging. Upon putting off the current the flow stress increases at below 273K. The recovery in the flow stress is as rapid as in high purity iron.

Effect of Interstitial Solutes. Figure 13 shows the effect of hydrogen charging (the current density of 10A. m⁻²) on the flow stress at 200K of irons with various purities and treatments. The yield stress is largest in the purest

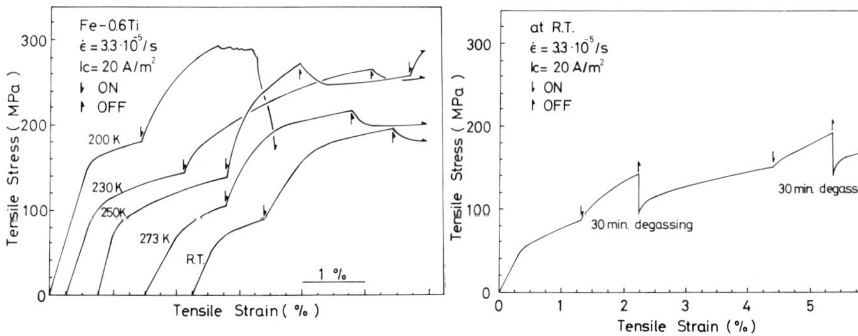

Fig. 10 – Effect of hydrogen
          charging on the flow
          stress of Fe-Ti alloy
          at various temperatures.

Fig. 11 – Effect of degassing at
          room temperature on the
          flow stress of Fe-Ti alloy.
          At the strain marked with
          ↑, the charging current
          is put off and the cross-
          head is stopped for 30min
          at room temperature.
          After 30min, the cross-
          head is moved again.

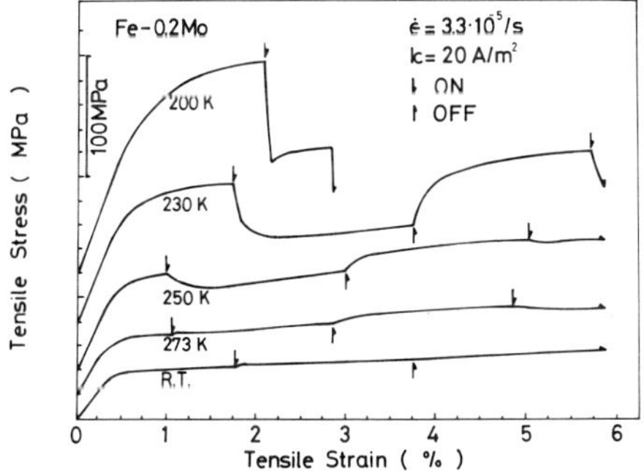

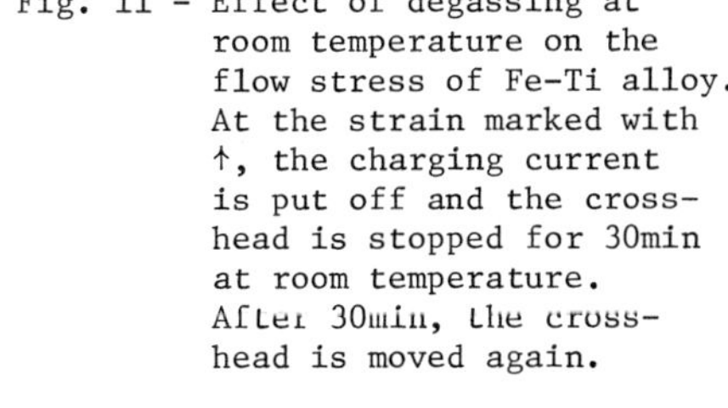

Fig. 12 – Effect of hydrogen
          charging on the flow
          stress of Fe-Mo alloy
          at various temperatures.

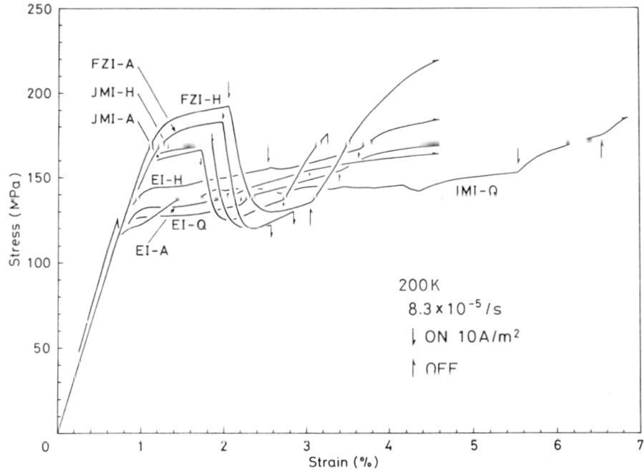

Fig. 13
    Effect of hydrogen
    charging on the flow
    stress of irons with
    various purity at
    200K. Note that
    impure irons have
    smaller yield stress
    than the purest iron
    (FZI-H); EI-Q, EI-A
    and JMI-Q show
    hardening by hydrogen
    charging.

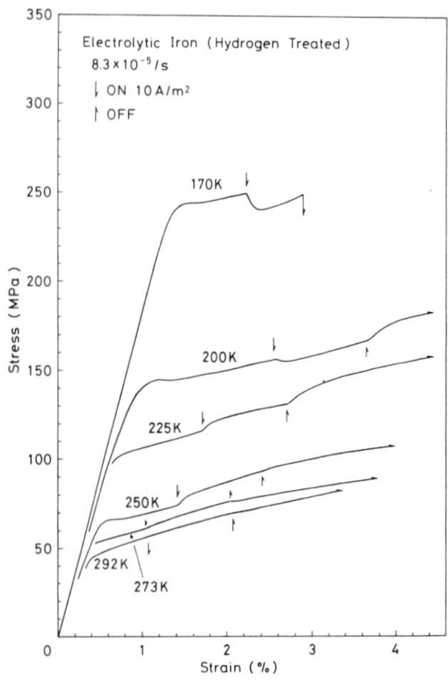

Fig. 14 – Effect of hydrogen charging on the flow stress of electrolytic iron (interstitial impurities having been removed) at various temperatures. Hydrogen causes softening at 200K and hardening at above 225K.

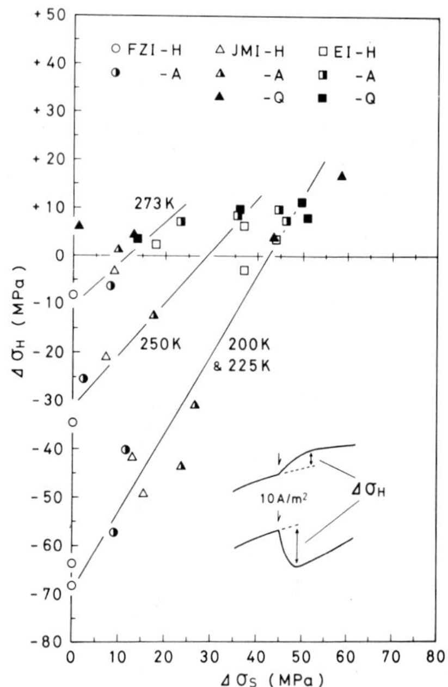

Fig. 15 – Relation between the amount of solution softening, $\Delta\sigma_S$, and the amount of hydrogen induced flow stress change ($\Delta\sigma_H$).

specimen (FZI-H), and smallest in the two least pure specimens (EI-Q and JMI-Q). The solution softening effect is evident. Hydrogen charging causes softening in pure specimens, e.g., FZI and JMI except JMI-Q. Impure specimens are hardened by hydrogen, e.g., EI-A, EI-Q and JMI-Q. As the temperature is raised, the hydrogen induced softening is replaced more and more by the hydrogen induced hardening. Figure 14 shows an example; EI-H specimens are tested at various temperatures. There is good correlation between the amount of solution softening and the hydrogen effect as shown in Fig. 15. Here $\Delta\sigma_H$ is the increase or decrease in the flow stress due to hydrogen as defined in the insert, and $\Delta\sigma_S$ is the amount of solution softening defined as the difference in the yield stress between FZI-H and other specimens. At below 170K, specimens fractured immediately after the start of charging and hence it could not be determined accurately whether hardening or softening occurs. It is expected from the results with high purity iron that they would show hardening if they had not fractured. Only JMI-Q specimens did not fracture and showed marked hardening, though not shown in a figure.

Discussion

It is now definitely shown that hydrogen causes hardening to iron containing solutes, and the hardening behavior depends on the amount and the kind of solutes as will be discussed below.

The present Fe-Ti alloy contains about 100at.ppm oxygen and 0.6at.% titanium. The oxygen is considered to be combined with titanium but most of titanium exists in a solid solution, so that the hydrogen induced hardening may be due to interaction of titanium, hydrogen and dislocations. Titanium is considered to interact with hydrogen in iron, since the titanium forms hydride and also interacts with other interstitials in iron. The amount of hardening is very large, about 100%, compared with the hardening found in other impure iron and mild steels, about 10% or less. There may be several reasons for this large hardening. The first is the large concentration of titanium, 0.6at.%. However, a mild steel containing about 0.2at.% carbon and a 17% Cr steel show only about 10% hardening under a similar charging conditions, i.e., in a solution of 1N H_2SO_4 with As_2O_3 at a current density of $10Am^{-2}$ (8). Hence, the rather large concentration of titanium is not likely the main cause of the large hardening. Secondly, the grain size should be discussed. It was found that the hydrogen induced hardening is larger for a smaller grain size specimen (10). However, the mild steel and the 17Cr steel mentioned above have grain size of 20µm and 10µm, respectively, and the Fe-Ti alloy has 60 to 100µm. Hence, the grain size is not the main reason. Thus, it is very likely that the large hydrogen induced hardening is due to the character of titanium, perhaps a strong interaction with hydrogen. The slow recovery in the flow stress after the current is cut off also shows a strong interaction between titanium and hydrogen, i.e., the reduced hydrogen diffusivity due to trapping, and its importance in hardening. The mechanism of hardening is not clear as yet, but the authors tentatively consider that titanium-hydrogen complex is formed, though its size or nature is not known, and acts as strong barriers to the dislocation motion. It should be noted here that the amount of hardening is almost independent of temperature.

Bernstein (10) has shown that vacuum-melted plastiron (45ppm C, 125ppm O, 10ppm N) with 0.15wt.% Ti is softened by hydrogen. In this case, titanium was added as a getter to the interstitial impurities. The titanium-interstitial impurity complex may cause some hardening, but the softening in the purified matrix may overcome the hardening.

The effect of molybdenum on the hydrogen effect is very small, or even null. The change in flow stress due to hydrogen in Fe-Mo alloy is somewhat similar to that in the high purity iron containing a small amount of carbon, i.e., hydrogen has almost no or very small effect at room temperature and causes softening at low temperatures (30). The observed hardening and softening in the Fe-Mo alloy could be due to residual carbon, although the alloy specimens were hydrogen treated. This point should be examined further. At any rate, the effect of molybdenum is very small. This is expected because molybdenum is considered not to interact with hydrogen; molybdenum element does not form hydride and its atomic size is not very different from that of iron.

Hydrogen treatment, i.e., removal of interstitial impurities, increases the tendency of hydrogen induced softening in each iron, FZI, JMI and EI; even EI shows softening at 200K when decarburized (Fig. 13), and FZI sometime shows hardening at room temperature if not decarburized. Thus, it is clear that interstitial impurities cause the hydrogen induced hardening. The hardening is larger when the interstitial solutes are in solution than in the form of precipitates, i.e., compare the annealed specimens to the quenched ones. The difference in the hydrogen effect among FZI, JMI and EI when they are decarburized is due to the difference in the substitutional impurity concentration.

The hydrogen induced hardening in iron with interstitial solutes has a good correlation with the solution softening. This is a remarkable difference from the hydrogen induced hardening in the Fe-Ti alloy. The Fe-Ti alloy shows no appreciable solution softening but a large hydrogen hardening, which is tentatively considered to be a sort of dispersion hardening. On the other hand, there are at least two possibilities in the case of iron with interstitial solutes; (a) hydrogen interacts with interstitial solutes to cancel their effect of solution softening, so that the hydrogen induced hardening is only apparent, and (b) hydrogen-interstitial solute complex is formed and acts as barriers to the dislocation motion. At present it is impossible to determine which effect is predominant. The specimens containing carbon sometimes show further hardening when the charging current is put off. This effect is not observed in the Fe-Ti alloy. This point may also be important to discuss the mechanism of hardening. More extensive research is now in progress with high purity iron containing controlled amounts of carbon to clarify the hardening mechanism.

Conclusions

It is confirmed that impure irons are hardened by hydrogen, and the hydrogen induced hardening so far reported on electrolytic iron and mild steels is not a sole effect of hydrogen but a combined effect of hydrogen with impurities.

The nature of the hydrogen hardening depends on the kind of solutes. A solute which interacts strongly with hydrogen seems to cause more hardening.

References

1. H. G. Vaughan and M. E. de Morton, "An Effect of Hydrogen on the Hardness of Mild Steel", J. Iron Steel Inst., 182 (1956) p.389.

2. A. M. Adair and R. E. Hook, "The Appearance and Return of a Hydrogen Yield Point in Iron", Acta Met., 10 (1962) pp.741-745.

3. S. Matsuyama, "Effects of Hydrogen on the Yielding and Strain-hardening of Iron Single Crystals", p.113 in Mechanism of Delayed Failure of Steels due to Hydrogen, Iron and Steel Institute of Japan, Tokyo, 1975.

4. S. Asano and R. Otsuka, "The Lattice Hardening due to Dissolved Hydrogen in Iron and Steel", Scripta Met., 10 (1976) pp.1015-1020.

5. Y. Tobe and W. R. Tyson, "Effect of Hydrogen on Yield of Iron", Scripta Met., 11 (1977) pp.849-852.

6. S. Asano and R. Otsuka, "Further Discussion on the Lattice Hardening due to Dissolved Hydrogen in Iron and Steel", Scripta Met., 12 (1978) pp.287-288.

7. H. Hagi, Y. Hayashi and N. Ohtani, "Effect of Dissolved Hydrogen and its Precipitation on Flow Stress of Electrolytic Iron", J. Japan Inst. Met., 42 (1978) pp.124-130.

8. S. Asano, Y. Nishino and R. Otsuka, "Lattice Hardening and Anomalous Softening of Iron and Steel Caused by Electrolytic Hydrogen Charging", J. Japan Inst. Met., 43 (1979) pp.241-248.

9. C. D. Beachem, "A New Model for Hydrogen-Assisted Cracking", Met. Trans., 3 (1972) pp.437-451.

10. I. M. Bernstein, "The Effect of Hydrogen on the Deformation of Iron", Scripta Met., 8 (1974) pp.343-350.

11. A. Gourmelon, "Influence de L'Hydrogène sur la Déformation Plastique et la Rupture der Fer", Mem. Sci. Rev. Met., 72 (1975) pp.475-489.

12. O. D. Gonzalez, "The Measurement of Hydrogen Permeation in α-Fe: An Analysis and Experiments", _Trans. Met. Soc. AIME_, 245 (1969) pp.607-612.

13. A. J. Kumnik and H. H. Johnson, "Steady State Hydrogen Transport through Zone Refined Irons", _Met. Trans._, 6A (1975) pp.1087-1091.

14. H. Matsui, S. Moriya and H. Kimura, "Effect of Hydrogen Charging on the Mechanical Properties of Very Pure Iron", Proc. 4th Int. Conf. Strength of Metals and Alloys, Nancy, France (1976) pp.291-295.

15. S. Takaki and H. Kimura, "Preparation of High-Purity Iron by Electron-Beam Floating-Zone Melting under Ultra-High Vacuum", _Scripta Met._, 10 (1976) pp.1095-1100.

16. M. Cornet and S. Talbot-Besnard, "Influence de L'Hydrogène sur le Glissement dans le Fer", _Compt. Rend. Acad. Paris_, 285B (1977) pp.223-226.

17. H. Kimura, H. Matsui and S. Moriya, "Comment on the Paper, "The Lattice Hardening due to Dissolved Hydrogen in Iron and Steel" by Asano and Otsuka", _Scripta Met._, 11 (1977) pp.473-474.

18. H. Kimura and H. Matsui, "Reply to "Further Discussion on the Lattice Hardening due to Dissolved Hydrogen in Iron and Steel" by Asano and Otsuka", _Scripta Met._, 13 (1979) pp.221-223.

19. H. Matsui, S. Moriya and H. Kimura, "The Effect of Hydrogen on the Mechanical Properties of High Purity Iron, I. Softening and Hardening of High Purity Iron by Hydrogen Charging during Tensile Deformation", _Mat. Sci. Eng._, 40 (1979) pp.207-216.

20. S. Moriya, H. Matsui and H. Kimura, "The Effect of Hydrogen on the Mechanical Properties of High Purity Iron, II. Effect of Quenched-in Hydrogen below Room Temperature", _Mat. Sci. Eng._, 40 (1979) pp.217-225.

21. H. Matsui, H. Kimura and Akihiko Kimura, "The Effect of Hydrogen on the Mechanical Properties of High Purity Iron, III. The Dependence of Softening on Specimen Size and Charging Current Density", _Mat. Sci. Eng._, 40 (1979) pp.227-234.

22. H. Matsui, S. Moriya, S. Takaki and H. Kimura, "Mechanical Properties of High Purity Iron at Low Temperatures", _Trans. Japan Inst. Metals_, 19 (1978) pp.163-170.

23. H. Matsui, Akihiko Kimura and H. Kimura, "The Orientation Dependence of the Yield and Flow Stress of High Purity Iron Single Crystals Doped with Hydrogen", Proc. 5th Int. Conf. Strength of Metals and Alloys, Pergamon Press (1979) pp.977-982.

24. Akihiko Kimura, H. Matsui and H. Kimura, "Effect of Hydrogen on the Plasticity of High Purity Iron Single Crystals at Low Temperatures", Proc. JIMIS-2, "Hydrogen in Metals", _Suppl. Trans. Japan Inst. Metals_, (1980) pp.541-544.

25. W. A. Spitzig and A. S. Keh, "Orientation and Temperature Dependence of Slip in Iron Single Crystals", _Met. Trans._, 1 (1970) pp.2751-2757.

26. S. Asano and K. Oguri, to be published.

27. M. Cornet and S. Talbot-Besnard, "Influence of Hydrogen on Mechanical Characteristics of Carbon-Doped Single Crystals of Pure Iron", Proc. JIMIS-2, "Hydrogen in Metals", _Suppl. Trans. Japan Inst. Metals_, (1980) pp.545-548.

28. H. Kimura, H. Matsui and T. Kimura, "Effect of Hydrogen on the Mechanical Properties of High Purity Iron", Proc. JIMIS-2, "Hydrogen in Metals", _Suppl. Trans. Japan Inst. Metals_, (1980) pp.533-540.

29. K. Oguri and H. Kimura, "The Effect of Hydrogen on the Flow Stress of
   Iron with Various Purities", Scripta Met., 14 (1980) No.9.

30. K. Oguri and H. Kimura, to be published.

Acknowledgment

Research performed by the present authors and reviewed in this article
was partly supported financially by Grant-in-Aid for Co-operative Research
(1987 and 1979) from the Ministry of Education, Science and Culture, Japan.

DISCUSSION

A. Atrens, Brown Boveri Research Center, Baden, Switzerland: Your H conditions were expressed as a cathodic current density. Have you done any experiments which would allow you to express this as a fugacity and if so,
what are the H fugacities.

H. Kimura: No, we have not done such an experiment. But, we measured the
concentration by extraction methods. he hydrogen concentration is estimated
to be several tens atomic ppm for the charging current densities we used.

D. L. Davidson, Southwest Research Institute: You showed hydrogen both increasing and decreasing the modulus of Fe in various temperature and purity
regimes. What do these modulus changes imply about the likely hydrogen-
dislocation interactions?

H. Kimura: Hydrogen increases the slope of stress-strain curves prior to
the macroscopic yielding below 190K. The slope is not the real Young's
modulus, but only apparent, and is determined by the motion of edge dislocations, i.e., micro-strain. Hydrogen is considered to hinder the edge
dislocation motion below 190K. We have not observed a decrease in the
apparent modulus. Of course, we have obtained no information of the effect
of hydrogen on the real elastic modulus.

M. Meshii, Northwestern University: The resolved shear stresses are not
much different between the primary and secondary slip system for your tensile
axis orientation. On the other hand, the stress to move screw dislocations
is 2~3 times that for edge dislocation. How can the primary screw dislocations become mobile before the secondary edge dislocations become mobile?

H. Kimura: We have other evidence to support our consideration that the
primary screw dislocation is as mobile as the primary edge dislocation when
the screw is associated with hydrogen, (see reference 20 in our paper). We
do not know as yet the exact mechanism of increasing the screw dislocation
mobility by hydrogen, but the experimental results clearly show that the
primary screw dislocations move before the secondary edge dislocation moves.

EFFECTS OF STRAIN RATE, PURITY AND THERMAL HISTORY ON

MECHANICAL BEHAVIOR OF CATHODICALLY CHARGED IRON*

K. S. Shin, C. G. Park, and M. Meshii
Department of Materials Science and Engineering
and
Materials Research Center
Northwestern University
Evanston, Illinois 60201

The effects of strain rate, specimen purity, carburization and cooling
rate were examined by a tensile test in pure iron being hydrogen charged
cathodically at 200 K. Intergranular fracture took place in polycrystalline
pure iron at low strain rates, while quasi-cleavage fracture dominated in
pure iron single crystals under the same condition. The critical stress
for intergranular fracture decreased rapidly with decreasing strain rate and
became considerably lower than the expected yield stress. The carburization
($\sim$185 at ppm) prevented this intergranular fracture. Both softening and
hardening effects of hydrogen charging were observed, depending on experimental conditions. The evidence indicates that two mechanisms due to hydrogen, can operate simultaneously and the observation of softening/hardening
depends on which mechanism dominates under a given condition.

*This work was sponsored by the U. S. Department of Energy and by the NSF-
MRL program through the Materials Research Center of Northwestern
University.

Introduction

The effect of hydrogen on mechanical properties of iron and steel is complex; it appears to possess multiple characters, one of which may manifest itself seemingly opposite to what one expects from another. In addition, various observations were carried out in different, often complex alloy specimens. Therefore it is generally difficult to comprehend various observations of hydrogen effects in a consistent and organized manner. In the present work, we took a step to minimize the complexity by choosing pure iron as the specimen material and by characterizing different manifestations of hydrogen effect with respect to conditions in a simple manner. Both softening and hardening effects of hydrogen were observed, depending on experimental conditions. The usual ductile fracture of pure iron was changed to intergranular fracture by the presence of hydrogen in polycrystalline iron specimens at low strain rates. This intergranular fracture, due to hydrogen, appears to be a property of pure iron. The carburization reduced the susceptibility to the hydrogen-induced intergranular fracture. On the other hand, hydrogen promoted quasi-cleavage fracture in pure iron single crystals at low strain rates. All the observations can be rationalized in a single consistent framework of hydrogen behavior in iron, although further substantiation of the proposed mechanism is clearly necessary.

Experimental Procedure

Single crystal specimens of MRC-VP iron and polycrystalline specimens of MRC-VP and Marz irons were used in this investigation. Shouldered sheet specimens were prepared by spark machining. The specimen dimensions in the guage section are 2.5 x 0.8 x 16.5 mm for polycrystalline specimens and 2.0 x 0.8 x 7.8 mm for single crystal specimens, respectively. The polycrystalline data reported here were taken from two groups of specimens with grain sizes of $\sim$90 and 250 μm, respectively. All the specimens were subsequently purified in a dynamic ZrH_2 atmosphere at 1133 K for 120 $\sim$ 144 hours and were furnace cooled. This treatment was expected to reduce the interstitial impurity contents to a level less than an atomic ppm (1). Some of the purified specimens were subsequently carburized at 833 K in a dry hydrogen saturated with the n-pentane atmosphere and were quenched into brine water. The carbon concentration was estimated to be $\sim$185 atomic ppm from the carburization temperature (2). Some specimens were quenched in the same manner without carburization. All the quenched specimens were stored in a liquid nitrogen bath until the tensile test, except for a few hours at room temperature for surface polishing and specimen handling.

All tensile tests were carried out in an electrolyte bath using an Instron machine, in the strain rate range from 2 x 10^{-6} to 5 x 10^{-4}/sec. Hydrogen was introduced into the specimens by cathodic charging with a current density of 60 A/M^2. The electrolyte was a H_2SO_4 - H_2O - CH_3OH solution containing a small amount of $NaAsO_2$. Most specimens were cathodically charged continuously from the onset of tensile tests, some were charged continuously a few hours prior to and during tensile tests, and some others were charged intermittently.

Results and Discussion

The hydrogen charging affects the mechanical behavior of iron in various ways, depending on test and specimen conditions. It can change failure mode drastically; both hydrogen-induced intergranular fracture and quasi-cleavage fracture were observed in specimens which were expected to fail in

a ductile manner without hydrogen charging. The hydrogen charging can result in softening or hardening, depending on experimental conditions. In
the following, the different manifestations of hydrogen effect are characterized with respect to experimental conditions. An attempt is made to rationalize these observations. It is hoped that the rationalization will enhance
the understanding of hydrogen damage in iron and will identify further works
to complete the understanding.

Strain Rate Dependence of Hydrogen Effect

In examining hydrogen effect by a tensile test, the strain rate has a
profound influence on the nature and the magnitude of the effect. Fig. 1
summarizes the strain rate dependence of yield stress and fracture strain of
specimens which were continuously hydrogen charged. The fracture mode observed at the highest strain rate used, 5.2×10^{-4}/sec., was ductile with
many dimples appearing on the fracture surface (Fig. 2). The failure was
always preceded by necking. This fracture mode was also observed in specimens deformed without hydrogen charging.

At lower strain rates, the fracture mode was quite different from that
observed at the high strain rate. No evidence of necking was recognized in
fractured specimens. The fracture mode was intergranular in polycrystalline
specimens (Fig. 3), while quasi-cleavage fracture was observed in single
crystals (Fig. 4). It is particularly noteworthy that the fracture strain
of polycrystalline specimens approached zero and the yield stress also was
reduced drastically with decreasing strain rate. The yield stress, therefore,
indicates the start of grain boundary decohesion instead of the onset of
massive dislocation motion in the bulk specimen. This is particularly true
in the low strain rate region where the yield stress of polycrystalline
specimens is significantly lower than that of single crystals.

Effect of Impurities on Fracture

The intergranular fracture was equally prominent in both VP and Marz
irons. It appears that nearly complete removal of interstitial impurities
by the purification treatment promoted this mode of fracture in pure iron.
The difference in residual impurities between VP and Marz (Table 1) apparently did not influence the occurence of the hydrogen-induced intergranular
fracture. In fact, the observation of intergranular fracture has been reported in irons of higher purities (3,4). Therefore it can be concluded that the
intergranular fracture is the direct result of the interaction of hydrogen
with grain boundaries in pure iron.

The susceptibility of pure iron to the hydrogen-induced intergranular
fracture can be eliminated by carburization. Fig. 5 illustrates the effect
of 185 atomic ppm carbon on the stress-strain curves of polycrystalline iron
specimens. The cathodic charging effect on the stress-strain curve of pure
iron is dramatic as can be seen from the comparison of curves (A) and (B).
The fracture mode for (B) is completely intergranular. The presence of 185
atomic ppm carbon changed the hydrogen effect drastically. The grain boundaries stopped being weak links even with hydrogen charging. Consequently,
the load bearing capability of the specimen returned and a significant
elongation was regained (Fig. 5D). A new type of transgranular fracture
dominates in this case as shown in Fig. 6. The structural change causing
the rapid work-hardening is also thought to be responsible for the
elongation being still smaller than that found in a specimen without cathodic
charging. The effect of hydrogen charging on work-hardening will be discussed
in the next section in conjunction with softening and hardening mechanisms.

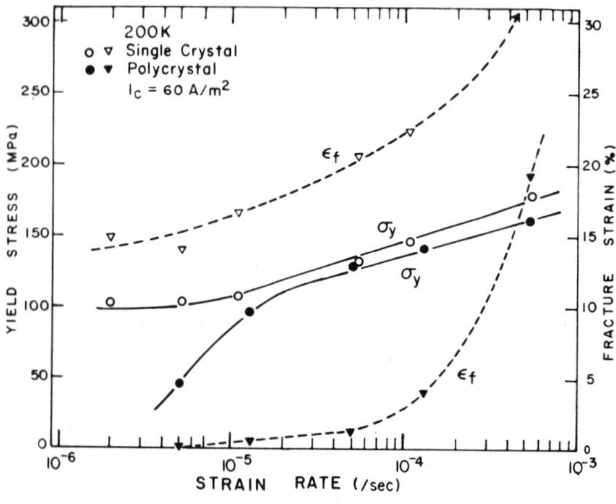

Fig. 1 – The yield stress and fracture strain of hydrogen charged iron are plotted against strain rate. The results obtained from single crystals and polycrystals are marked accordingly.

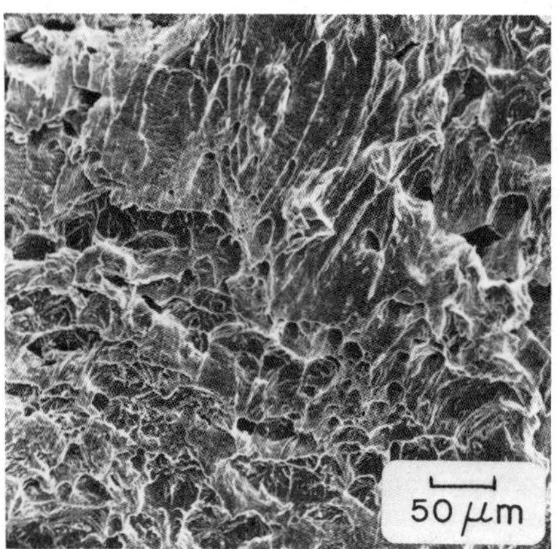

Fig. 2 – Micrograph of ductile fracture surface of an iron single crystal specimen which was cathodically charged, deformed to fracture with a strain rate of 1.1 x 10^{-4}/sec.

Fig. 3 – Micrograph of intergranular fracture surface of a polycrystalline specimen deformed to fracture with a strain rate of 5.1 x 10^{-5}/sec.

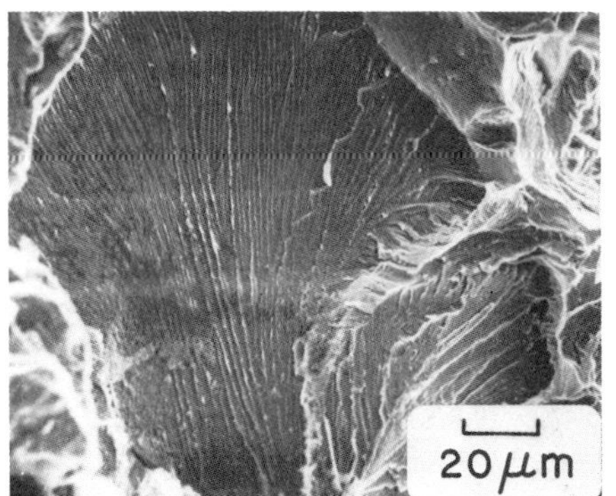

Fig. 4 - Micrograph of a quasi-cleavage fracture surface of an iron single
crystal which was cathodically
charged, deformed to fracture with a
strain rate of 5.1 x 10^{-6}/sec.

Table I. Impurity Content (ppm) in VP and Marz Irons

Elem	C	N	O	Mg	Si	Al	S	Ca
VP	18.0	<1.0	33.0	<10.0	50.0	60.0	40.0	<10.0
Marz	12.0	10.0	60.0	0.87	<0.1	<0.1	2.60	0.80

Elem	Ti	Cr	Ni	Cu	Mo	Ag	Sn	Pb
VP	<10.0	30.0	<10.0	30.0	30.0	<5.0	<30.0	<30.0
Marz	1.40	1.60	<0.1	0.60	<0.1	<0.1	<0.10	<0.10

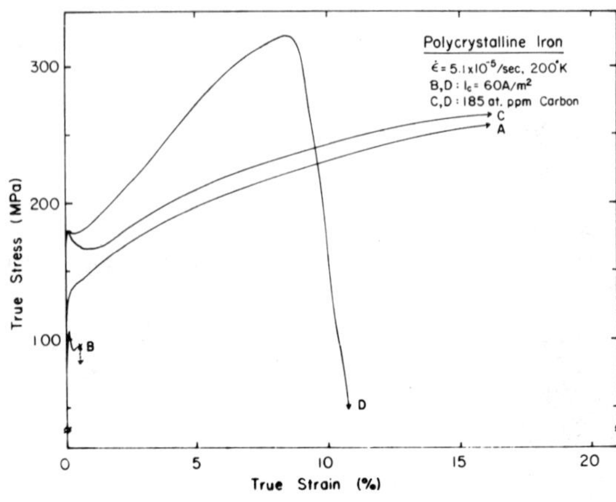

Fig. 5 – Stress-strain curves of polycrystalline iron specimens. (A) pure iron without cathodic charging, (B) pure iron being charged cathodically, (C) carburized iron without charging, and (D) carburized iron being charged cathodically (strain rate 5.1 x 10^{-5}/sec.)

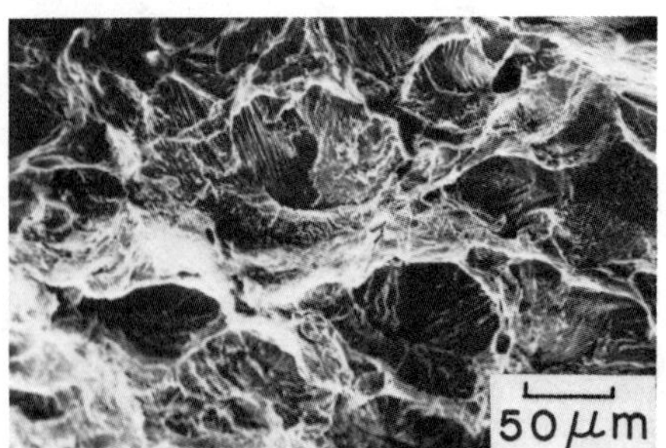

Fig. 6 – Micrograph of fracture surface of a carburized polycrystalline iron specimen deformed to fracture with a strain rate of 5 x 10^{-5}/sec. The surface resembles those of quasi-cleavage found in single crystal specimens.

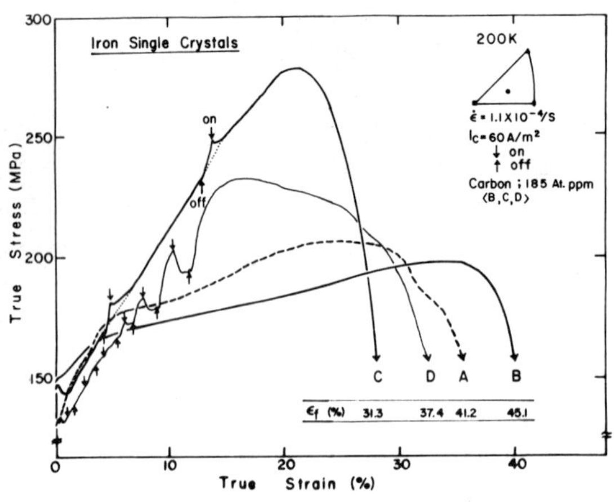

Fig. 7 – Stress-strain curves of single crystal iron specimens. (A) pure iron without cathodic charging (B) carburized iron w/o cathodic charging, (C) carburized iron being charged cathodically, (D) carburized iron with intermittent charging (strain rate 1.1 x 10^{-4}/sec.)

The effect of carburization on the fracture mode of single crystals was
not significant. The rapid work-hardening was prominent in single crystals
(Fig. 7) as in polycrystalline specimens (Fig. 5).

The effect of quenching was examined in polycrystalline pure iron specimens. The quenching from a temperature range between 833 and 993 K into brine
water did not alter the hydrogen-induced intergranular fracture mode of pure
iron. It should be pointed out, however, that an environment for quenching
may carburize iron or decarburize iron with a small carbon content,
resulting in a change in fracture morphology. We have encountered such
examples accidentally.

Softening and Hardening Effects of Hydrogen

The hydrogen-induced softening effect is now well established (5-7),
although there was a dispute over whether the real effect of hydrogen should
be softening or hardening (8,9). Experimentally, both softening (5-7) and
hardening (10,11) due to cathodic charging have been reported. In the
present work, both effects were observed, depending on experimental conditions.
Examining the hydrogen effect on flow stress, two methods can be employed.
In the first method, a set of two stress-strain curves, one from a hydrogen
charged specimen and another from a reference specimen, can be compared. In
the second approach, a stress-strain curve can be obtained from a specimen
which is hydrogen charged intermittently during the test. A softening effect
(or a hardening effect) is inferred from the observation of a decrease (or an
increase) in work-hardening rate after a restart of cathodic charging and of
an increase (or a decrease) of the rate after a suspension. If both methods
give the same result (sign and magnitude), the interpretation of the result
is straight forward. On the other hand, the results of these two methods
can be different as shown in Fig. 7. The comparison of (B) and (C) (in the
figure) indicates that the hydrogen charging appears to cause a hardening
effect. On the other hand, the intermittent charging appears to demonstrate
a softening effect (Fig. 7D). A similar conclusion can be drawn from the
effect of suspension of charging as shown in (C).

It is proposed, therefore, that two mechanisms involving hydrogen are
operating simultaneously. A mechanism is responsible for softening. This
is transient in nature as indicated by the rapid loss of softening upon
turning off the charging current. An increase in dislocation mobility due
to hydrogen in solution is often suggested for the softening mechanism.
Another mechanism involving hydrogen is responsible for hardening. This
appears to be more permanent in nature (12). The large work-hardening rate
observed in the carburized iron specimens being charged cathodically is
caused by this hardening mechanism. It is suggested that this hardening is
caused by an increase in dislocation density accompanying the void formation
due to hydrogen precipitation. The dislocation generation from the hydrogen-induced voids was directly observed in a TEM (13,14). Although a fraction
of these dislocations may contribute to an increase in mobile dislocation
density, the primary role of these dislocations is to become obstacles to
slip dislocations responsible for the observed macroscopic deformation.

The partition of hydrogen between the two mechanisms must depend sensitively on the impurity concentration. The presence of carbon promotes the
second mechanism (hardening) at the expense of the first mechanism (softening).
This can be seen clearly from the comparison of Fig. 7D with the corresponding
stress-strain curve of pure iron (MRC-VP) (7). In the extreme case, the
softening mechanism may become negligible. The failure of observing the
hydrogen softening in impure irons even by intermittent charging may be such

an extreme case.

By the same token, the second mechanism can become nearly negligible in high purity iron (5,6). Let us recognize two hydrogen-induced mechanisms operating simultaneously and the sensitivity of the hydrogen partition between the mechanisms to impurity content. It appears to explain a number of seemingly contradictory observations of hydrogen effects on the flow stress of irons of various purities.

Hydrogen-Induced Fracture

In polycrystalline iron, the strain rate dependence of yield stress does not change due to the hydrogen charging at higher strain rates; however, the apparent yield stress decreases rapidly with decreasing strain rate below 2×10^{-5}/sec. At 5×10^{-6}/sec., the intergranular fracture took place at one-half of the yield stress which is expected from the extraporation of the yield stresses at higher strain rates. The hydrogen-induced grain boundary separation can apparently take place at a stress significantly below the critical stress for the massive dislocation motion. Consequently, the specimens appeared completely brittle as the negligible fracture strain indicates. The fracture surface was smooth on each grain. On the other hand, a number of microvoids can be identified on the quasi-cleavage fracture surface.

The elimination of the intergranular fracture by carburization is an interesting observation. Further study should establish whether this is caused by the grain boundary strengthening or by the increased hydrogen traps in the matrix.

The quasi-cleavage fracture is the dominant fracture mode at low strain rates in single crystal specimens. At low magnification, the fracture surface resembles that of cleaved surfaces. At high magnifications, however, there are considerable differences between the two surfaces. The hydrogen-induced quasi-cleavage surface revealed more substructure. Some complicated internal small cracks were clearly identified on SEM pictures as reported earlier (15). The quasi-cleavage plane observed in pure iron single crystals was identified as being parallel to one of {110} planes which is the primary slip plane with respect to the tensile direction. The observation clearly indicates that the slip is responsible for this mode of fracture.

This is in contrast to the intergranular fracture which apparently can take place without plastic deformation. The observation of specimen surfaces indicates that a number of microvoids and small cracks form along active slip planes. The localization of heavy slip on a relatively limited number of slip planes may be encouraged due to hydrogen-induced softening and, thus, hydrogen voids are formed along these planes. The plane most populated with the voids will become the fracture plane. A similar surface morphology was also observed in carburized polycrystalline iron specimens deformed with cathodic charging. All the observations suggest that the quasi-cleavage fracture, similar to pure iron single crystals, took place in these specimens.

Concluding Remarks

The hydrogen effect manifests in a number of drastically different ways; hardening, softening, embrittlement due to intergranular fracture, and quasi-cleavage fracture. All the observations can be rationalized by considering hydrogen involving different mechanisms which can take place simultaneously and are extremely sensitive to impurity concentration and experimental conditions.

References

(1) D. F. Stein, J. R. Low, Jr., and A. U. Seybolt, "The Mechanical Properties of Iron Single Crystals Containing Less that 5 x 10^{-3} ppm Carbon," Acta Met., 11 (1963) pp. 1253-62.

(2) D. J. Quesnel, A. Sato, and M. Meshii, "Solution Softening and Hardening in the Iron-Carbon System," Mater. Sci. Eng., 18 (1975) pp. 199-208.

(3) S. Moriya, H. Matsui, and H. Kimura, "The Effect of Hydrogen on the Mechanical Properties of High Purity Iron, II Effect of Quenched-in Hydrogen Below Room Temperature," Mater. Sci. Eng., 40 (1979) pp. 217-25.

(4) M. Cornet and S. Talbot-Besnard, "Hydrogen Influence on Several Properties of High Purity Iron," in Environment-Sensitive Fracture of Engineering Materials, Z. A. Foroulis, ed.; TMS-AIME, New York, N. Y. (1978) pp. 411-430.

(5) H. Matsui, S. Moriya, and H. Kimura, "Effect of Hydrogen Charging on the Mechanical Properties of Very Pure Iron," Proc. 4th Int. Conf. on Strength of Metals and Alloys, E. N. S. M. I. M., Nancy, France (1976) pp. 291-95.

(6) M. Cornet and S. Talbot-Besnard, "Influence of Hydrogen on Characteristics of Carbon-Doped Single Crystals of Pure Iron," Proc. 2nd JIM Int. Symposium on Hydrogen in Metals, Minakami, Japan, Suppl. Trans. Japan Inst. Metals, 21 (1980) pp. 545-548.

(7) C. G. Park, K. S. Shin, J. Nagakawa, and M. Meshii, "Effect of Cathodic Charging on Creep and Tensile Deformation of Pure Iron," Scripta Met., 14 (1980) pp. 279-284.

(8) H. Kimura, H. Matsui, and S. Moriya, "Comment on the Paper, "The Lattice Hardening due to Dissolved Hydrogen in Iron and Steel," by Asano and Otsuka," Scripta Met., 11 (1977) pp. 473-74.

(9) S. Asano and R. Otsuka, "Further Discussion on the Lattice Hardening Due to Dissolved Hydrogen in Iron and Steel," Scripta Met. 12 (1978) pp. 287-8.

(10) S. Asano and R. Otsuka "The Lattice Hardening Due to Dissolved Hydrogen in Iron and Steel," Scripta Met., 10 (1976) pp. 1015-20.

(11) H. Hagi, Y. Hayashi, and N. Ohtani, "Effect of the Damage Due to the Hydrogen Precipitation on the Flow Stress of Electrolytic Iron," Proc. 2nd JIM Int. Symposium on Hydrogen in Metals, Minakami, Japan, Suppl. Trans. Japan Inst. Metals, 21 (1979) pp. 561-64.

(12) H. Wada and K. Sakamoto, "Size Dependence of the Effects of Hydrogen on the Mechanical Properties of Iron Single Crystals," Proc. 2nd JIM Int. Symposium on Hydrogen in Metals, Minakami, Japan, Suppl. Trans. Japan Inst. Metals, 21 (1980) pp. 553-56.

(13) H. Takahashi, T. Takeyama, and T. Hara, "Electronmicroscopic Study of Micro-crack Formed by Hydrogen Precipitation in Pure Iron," J. Japan Inst. Metals, 43 (1979) pp. 492-99.

(14) T. Takeyama and H. Takahashi, "Formation of Micro-crack Due to Hydrogen Aggregation in Iron and Steel," Proc. 2nd JIM Int. Symposium on Hydrogen in Metals, Minakami, Japan, Suppl. Trans. Japan Inst. Metals, 21 (1980) pp. 409-12.

(15) S. Hinotani, F. Terasaki, and F. Nakasato, "Fractographic Approach to Hydrogen-Embrittled Fracture in Iron and Steels," Proc. 2nd JIM Int. Symposium on Hydrogen in Metals, Minakami, Japan, Suppl. Trans. Japan Inst. Metals, 21 (1980) pp. 421-24.

DISCUSSION

<u>R. Gibala, Case Western Reserve University</u>: In the experiments that your group has done, as well as those by Kimura and co-workers, why should temperatures in the vicinity of $\sim$ 190-200K be a lower limit at which hydrogen softening is observed?

<u>M. Meshii</u>: I believe two factors are important in understanding this observation. The clustering tendency of hydrogen would increase with decreasing temperature. Hydrogen clusters act as hardeners. As the mobility of hydrogen decreases with decreasing temperature, screw dislocations cannot carry the hydrogen along as they move. Thus, the fraction of hydrogen participating in the hardening mechanism rapidly increases at the expense of that for the softening mechanism. 200K must be a critical temperature for this change-over.

EFFECTS OF HYDROGEN ON PLASTIC FLOW AND FRACTURE
IN IRON AND STEEL

C. J. McMahon, Jr.

Department of Materials Science and Engineering
University of Pennsylvania
Philadelphia, Pennsylvania

Summary

An attempt is made here to present a self-consistent synthesis of
apparently disparate observations on the effects of hydrogen on the
flow and fracture of iron and steel. The key features are: (1) that
hydrogen alters the binding energy of the crystal lattice in such a
way that the cohesive energy is reduced and screw dislocation mobility is increased, and (2) that hydrogen is trapped by particle/matrix
interfaces, thereby enhancing plastic rupture, and by dislocation
cores, thereby permitting dislocation sweep-in and glide-plane decohesion. It is argued that the main practical problem of hydrogen
embrittlement in quenched and tempered steels is the stress-controlled intergranular cracking due to impurity segregation, rather than
the plasticity-related, strain controlled cracking phenomena.

Introduction

With the large increase in the resources and effort recently devoted to the study of the effects of hydrogen on the deformation and fracture of iron and steel, the main features of this set of phenomena are becoming increasingly clear. Although in many respects detailed knowledge is lacking, it now seems possible to construct a self-consistent rationalization of practically the whole spectrum of experimental observations. The present paper will attempt to do this, drawing upon the observations of many workers and a number of the ideas they have expressed. The focus is entirely on the mechanisms, or the physical nature of the phenomena, rather than on their kinetics. The latter are complicated by the superposition of various surface reactions and transport and trapping phenomena and can vary greatly with the experimental conditions. The point of view taken here is that the physical nature of the influence of hydrogen on deformation and fracture will ultimately be understood in terms of the effects of hydrogen on bonding in the b.c.c. lattice and the interaction of hydrogen with lattice defects.

Effects on Deformation.

It is now generally accepted that dissolved hydrogen ($\underline{H}$) is strongly attracted to the cores of dislocations in iron with a binding energy large enough for the core concentration of $\underline{H}$ to be high at room temperature (1). There is much evidence, both circumstantial and direct, that $\underline{H}$ can be "swept-in" from the external surface by dislocations originating there (2) and can be "swept-out" by dislocations arriving at the surface (3) if the internal concentration is large enough (e.g., due to prior electrolytic charging). Since $\underline{H}$ is also strongly trapped by particle-matrix interfaces (1), this sweep-in can lead directly to a build-up of hydrogen in such interfaces, which can produce interface decohesion, void formation, and enhanced rupture (4).

From a number of careful studies (5-11) of single and polycrystalline irons of varying purity, it has been demonstrated that hydrogen (usually injected by simultaneous or prior cathodic charging) can either reduce the stress needed for plastic flow or increase it, depending on the purity, temperature, amount of prior strain, sample size, etc. These observations can be rationalized by postulating that the presence of sufficient $\underline{H}$ in dislocation cores can:

(a) increase the mobility of screw dislocations (5,6), and

(b) decrease the mobility of dislocations with some edge
     component (7).

It also appears that $\underline{H}$ tends to promote planar slip, presumably by inhibiting the cross-slip of screw dislocations (10). Non-screw dislocations are usually highly mobile in iron, except in special cases, such as Fe-3% Si (12). It is generally screw dislocations which control plastic flow at ambient temperatures and below, because of their lower mobility and because some cross-slip is needed for dislocation multiplication. Thus, it seems reasonable to treat the drag on non-screws by core-trapped $\underline{H}$ as peripheral to the overall effect of $\underline{H}$ on deformation in steel.

The enhancement of screw mobility of core-trapped $\underline{H}$ is postulated to be the cause of such diverse behavior as: reduction of the yield strength of high purity iron above ~190 K (5-8), enhanced creep-rate of iron at 200-300 K (9), the occurrence of local plastic instability at the notch root of a bend specimen of steel (13) and the analogous enhanced plastic shearing at the tips of pre-cracks of fracture toughness specimens, both on the surface (14) and internally (15). These latter effects ultimately lead to strain-controlled, Mode II cracking along surfaces of maximum shear strain (15-17).

<u>Effects on Fracture.</u>

In addition to the enhancement of void nucleation and rupture due to dislocation sweep-in (4), $\underline{H}$ has been found to produce brittle fracture by decohesion of grain boundaries already weakened by impurity segregation (18-23) and to cause glide-plane decohesion in iron and steel (24-31) a phenomenon apparently uniquely due to $\underline{H}$. The latter effect is characterized by apparently brittle cracking on $\{110\}$ or $\{112\}$ planes. It occurs in iron single and polycrystals (24-29) and in the plates or laths of tempered bainitic or martensitic steels (30,31).It seems fairly widely accepted that it is the result of the collection of $\underline{H}$ in heavily dislocated slip bands, presumably from sweep-in of core-trapped $\underline{H}$ as the slip bands form. This is, therefore, a strain controlled fracture mode that produces Mode II cracking.

The more commonly observed hydrogen-induced fracture mode in quenched and tempered steels is intergranular decohesion, usually along prior austenitic grain boundaries. The following characteristics of this phenomenon have been demonstrated (18-22):

1. The presence of segregated impurities promotes intergranular decohesion. At a fixed hydrogen fugacity the stress needed for such cracking decreases as the impurity concentration increases (Fig. 1). The causative impurities are those metalloid and metal-metalloid combinations which produce temper embrittlement and tempered martensite embrittlement (22).

2. For a given intergranular impurity concentration the stress (or stress intensity) needed for cracking decreases as the hydrogen concentration increases. As shown in Fig. 2, for a gaseous H_2 environment of pressure p, this is proportional to $(p)^{\frac{1}{2}}$ exp $\alpha Y\Omega/RT$ where Y is the yield stress of the steel, Ω is the relevant atomic volume of $\underline{H}$ in iron, and α is a constant ~ 3 which depends on details of the stress analysis.   This behavior is consistent with the model of Troiano and co-workers, (33,34) elaborated later by others (35-37,32).

A set of experiments just completed (15) has succeeded in putting the latter stress-controlled and the former strain-controlled fracture modes in perspective. It was found that when pre-cracked or notched specimens of a 5 pct Ni steel (HY 130) which did not

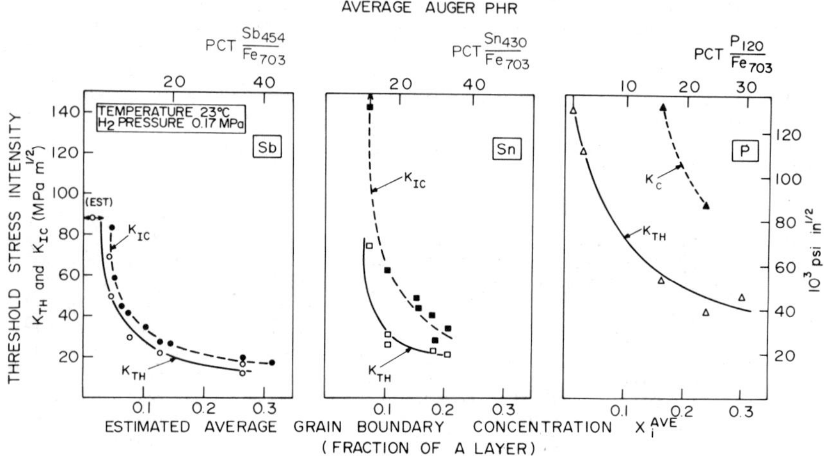

1. Effect of segregated Sb, Sn, and P in a Ni–Cr steel on
    reducing the fracture toughness, Kc, and the threshold
    stress intensity for cracking in H_2 gas, K_{TH}. For the
    pure steel Kc is 140 MPa m$^{1/2}$. After Kameda, Ref. 51.

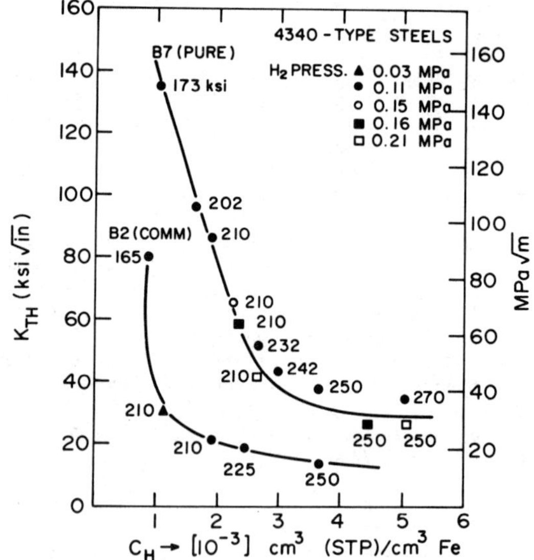

2. Dependence of K_{TH} on the calculated maximum H concentra-
    tion in 4340 – type steels of two levels of purity. Data
    points obtained by varing yield strength and hydrogen
    pressure independently. After Bandyopadhyay, Ref. 32.

contain segregated impurities were stressed in H_2 gas, plasticity-related hydrogen-induced cracking (PRHIC) occurred along surfaces of maximum (plane strain) shear stress. The fracture mode was a mixture of interface decohesion (apparently between martensitic laths and lath colonies) and transcrystalline cracking of laths, as shown in Fig. 3. The latter appeared to be dominant, and it had the same appearance as the glide-plane decohesion observed by others. In pre-cracked samples the PRHIC began at stress intensities well-below the load level corresponding to K_{Ic} in air, in which the fracture mode would have been rupture. However, the PRHIC produced bifurcation of the crack tip (Fig. 4) which increased in extent as the load was raised and continued to some extent when the loading was stopped and the displacement held constant. The crack tip bifurcation caused a significant decrease in the actual Mode I stress intensity K_I (38) and the occurrence of the PRHIC was generally not detected (by a change in specimen compliance) until it was well-developed.   The effect was to give anomalously high values of the nominal K_I at which crack extension was detected, designated K_{TH}.

When such a steel containing sufficient Mn or Si was aged to allow segregation of P and/or Ni+Si, a continuous, but rapid, transition occurred to the intergranular cracking mode: decohesion along prior austenitic grain boundaries. As the amount of this increased, the apparent K_{TH} value dropped from values as high as 200 MPa m$^{1/2}$ to as low as 20 MPa m$^{1/2}$; an example as shown in Fig. 5. The conclusion reached in this research was that the PRHIC represents the intrinsic behavior of steel stressed in hydrogen. The only remedy to this is to prevent the entry of hydrogen. On the other hand, the intergranular decohesion at low stress intensities is related to impurity segregation, and is in a sense a "disease" for which there is a "cure" other than exclusion of hydrogen. If this fracture mode could be eliminated, then the practical problem of hydrogen embrittlement of high strength steels would largely disappear.

<u>Binding Energy Effects</u>.

Certainly the most important effect of $\underline{H}$ in iron and steel is that it reduces the cohesive energy of the crystal lattice. The ideal work of fracture γ, which is the work done against the cohesive force, can control the overall fracture energy, which includes plastic work, γ_p, even though γ may be $\leqslant 0.1\,\gamma_p$. It has been shown that small reductions in γ can lead to much larger reductions in γ_p and thus large embrittlement effects (39-41). Such reductions in γ can be caused by segregation of impurities to grain boundaries or by a build up of hydrogen, as is evident from many experimental studies.

At present, our ideas about the physical basis of the effects of concentrated solutes on the cohesive energy of transition metals are very sketchy and for the most part speculative. There is as yet no known method of making a rigorous "first-principles" calculation for even a simple case, such as a single

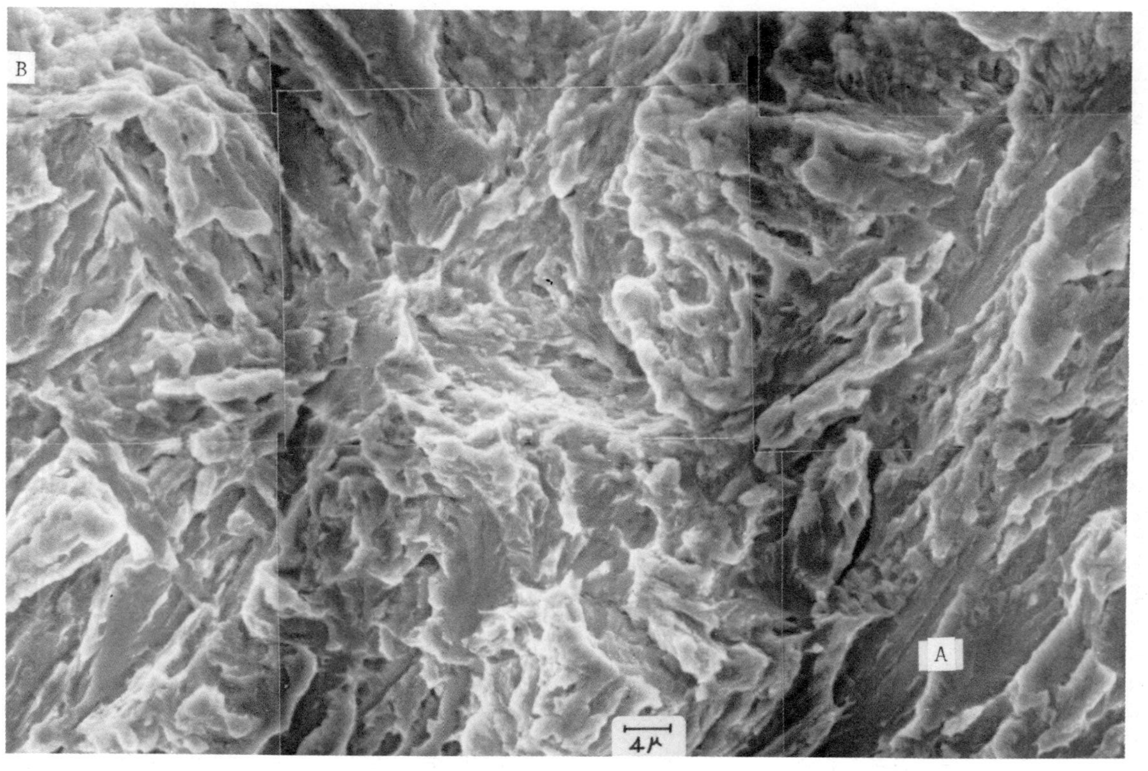

3.   Example of plasticity-related hydrogen-induced cracking (PRHIC) in a pure Hy 130-type steel with no embrittled prior austenitic grain boundaries, showing glide-plane decohesion (A) and interface decohesion (B).   After Takeda, Ref. 15.

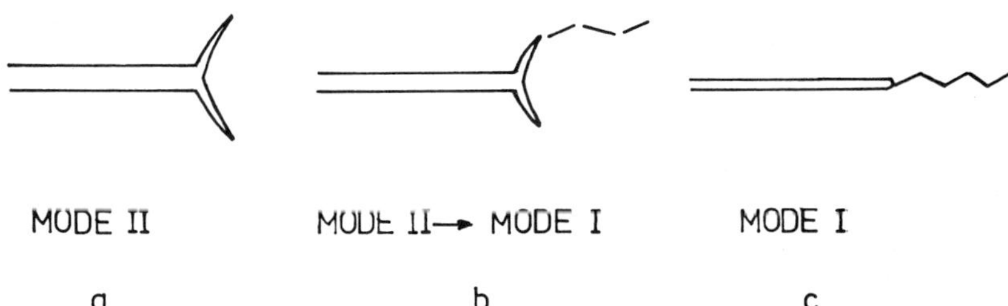

4. Schematic representation of observations of crack-tip
    phenomena in H_2 gas in HY 130-type steels showing (a) PRHIC
    in a pure steel with high apparent K_{TH}, (b) initial PRHIC,
    then intergranular cracking in a partly embrittled steel
    with intermediate K_{TH}, (c) intergranular cracking in a
    fully embrittled steel with low K_{TH}.

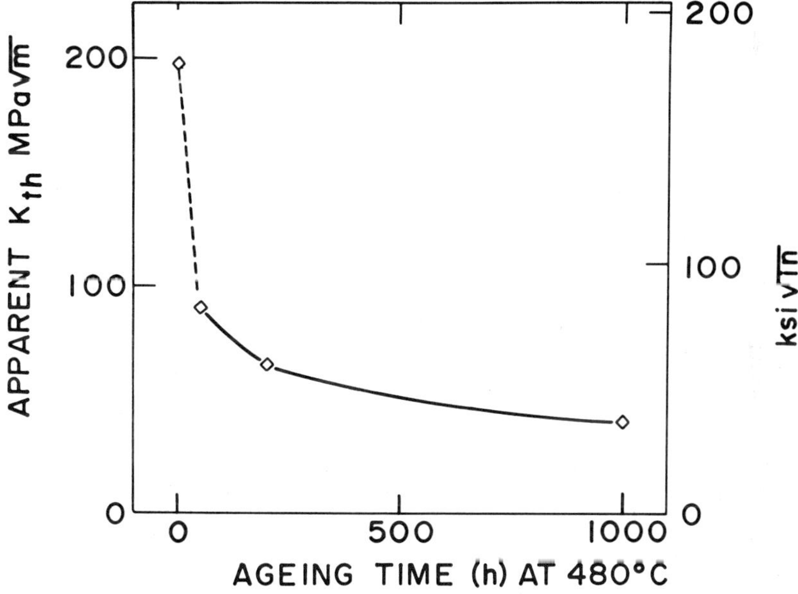

5. Examples of decrease in apparent K_{TH} with ageing time at
    480°C in Hy 130 steel. Transition, due to segregation of
    P and Ni+Si, corresponds to that shown in Fig. 4.
    After Takeda, Ref. 15.

solute concentrated in a pure tilt grain boundary in a bicrystal.
However, insights can be gained from some approximations, such as
the use of atom clusters for molecular orbital calculations (42),
and from observation of the electronic structure of atoms adsorbed
on crystal surfaces (43). These have already indicated a possible
explanation of the nature of embrittlement of Ni by S (42,43).

With regard to $\underline{H}$ in Fe, the critical issue is: where is the
1s electron of the $\underline{H}$? If it is delocalized in the valence band
of the Fe, as some recent surface observations by ultraviolet photo-
electron spectroscopy suggest (44), then the embrittlement may be
rationalized as an increase in the average local ion-core separation to maintain a constant volume per valence electron. The
observed lattice dilation due to $\underline{H}$ in Fe fits this picture. This
would reduce the bonding effect due to formation of the d-band and
valence band. However, the actual nature of the $\underline{H}$ effect could be
more complicated in ways not yet imagined, and this is obviously
an area requiring a great deal of attention, both theoretical and
experimental.

Although we cannot yet explain the effects of $\underline{H}$ on the binding
energy of Fe, we know they exist. Hence, it is to be expected that
$\underline{H}$ would have some effect on the mobility of screw dislocations.
This can be seen by considering the results of computer modeling
of dislocation cores using empirical pair-wise potentials. It has
been shown that the low mobility of screw dislocations in the b.c.c.
lattice is due to the non-planar configuration of the dislocation
core (45,46). The computer calculations have confirmed previous
conclusions from hard-sphere models (47) that the 1/2 <111> shear
is distributed on several {110} or {112} planes in sessile configurations. A large shear stress, or considerable thermal activation, is required for a sessile → glissile transition in the core.
The details of this transition, and thus the mobility, depend
sensitively on the potential chosen for the computer calculation (48).

We can use the results of Duesbery et al. (45) to illustrate
how a change in interatomic potential could lead to the kinds of
effects observed for $\underline{H}$ in Fe. To do this we select the potentials
labelled J_2 and J_3, shown in Fig. 6, and we imagine that the effect
of $\underline{H}$ is to alter the bonding energy from the "hard" potential J_3
to the "soft" potential J_2. In the unstressed state the displacements in the core of a screw dislocation in a b.c.c. crystal are
distributed along three intersecting {110} planes, rather than
along one {110} plane as for an edge or mixed dislocation. Since
a screw dislocation lies along a <111> zone axis, illustrated
for the case of the [111] zone in Fig. 7, one can represent the
core displacements in a (111) projection in the manner shown
in Figs. 8 and 9, which were calculated (45) for the J_3 and
J_2 potentials, respectively. Here, the lengths of the arrows
represent the magnitudes of the relative [111] displacements
between the atom pairs joined by the arrows; the points represent atoms in three adjacent (111) planes. The largest arrow,
i.e., one touching both atoms represents a displacement of
$b/3 = 1/6$ [111]; the others are drawn proportionately. The
small solid triangle, which

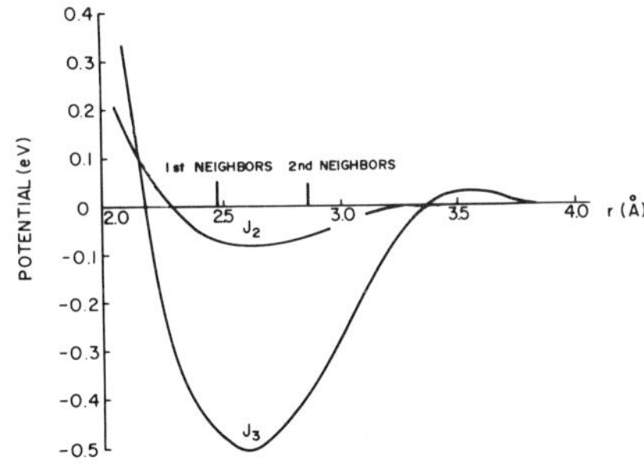

Fig. 6 Two of the interatomic potentials used to calculate screw dislocation core configurations in Ref. 45.

Fig. 7 [111] zone showing relevant planes.

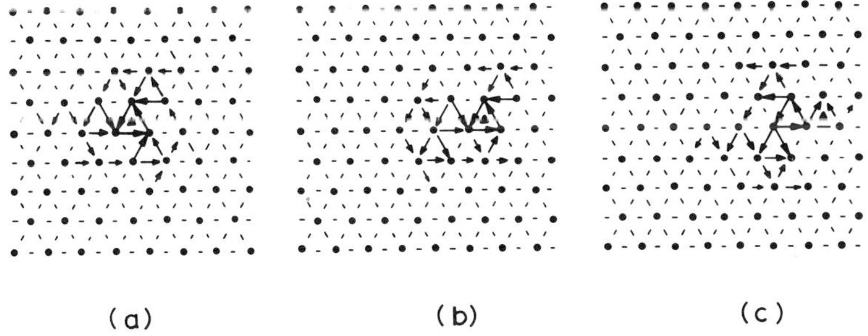

(a)　　　　　　(b)　　　　　　(c)

Fig. 8 Core structure changes under the influence of a shear stress on the (101) plane for the case of potential J_3: (a) zero stress, (b) 0.0235G, (c) 0.0265G (Ref. 45).

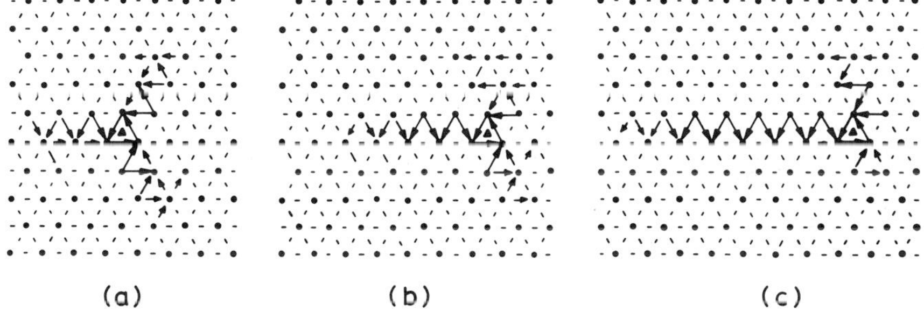

(a)　　　　　　(b)　　　　　　(c)

Fig. 9 As in Fig. 8 but for the case of potential J_2: (a) zero stress, (b) 0.0125G, (c) 0.0270G (Ref. 45).

remains in a fixed position, indicates the original center of the dislocation core.

Figures 8 and 9 indicate the responses of the two screw dislocations to a shear stress on the $(\bar{1}01)$ plane which tends to move the dislocations to the left. In the case of potential J_3 (Fig. 8) the core configuration begins to change at a stress of nearly 0.0200G (where G is the shear modulus), and at 0.0235G the trailing "fractional" dislocation on the $(\bar{1}10)$ plane has constricted almost into the original center (Fig. 8b). As the stress increases further the other trailing fractional on the $(0\bar{1}1)$ plane also constricts, and at 0.0265G we find two *leading* fractionals on the $(0\bar{1}1)$ and $(\bar{1}10)$ planes (Fig. 8c). Finally, at a somewhat higher stress (<0.027G) the dislocation moves freely through the lattice by a series of alternate unit translations on $(0\bar{1}1)$ and $(\bar{1}01)$ planes, which produces an average motion on the $(\bar{1}12)$ plane.

The case of a screw dislocation, similarly stressed, in a crystal with the J_2 potential is portrayed in Fig. 9. At zero stress the displacements are spread out farther on the $\{110\}$ planes (Fig. 9a). Contraction of the trailing fractionals is quite noticeable at 0.0115G. As the stress is increased further the trailing fractionals become rearranged (Fig. 9b) and the leading fractional spreads out on the $(\bar{1}01)$ plane, until at about 0.0275G the dislocation glides freely on that plane. Because the core rearrangement in the J_2 screw begins at a much lower stress (at OK) than for the J_3 screw, one would intuitively expect that the glide stress would drop more rapidly with increasing temperature for the J_2 screw than for the J_3. Hence at, say, 200K the mobility of the J_2 screw should be larger than that of the J_3; however, cross slip of the J_2 screw should be more difficult, since it is widely spread on the $(\bar{1}01)$ plane and because that is clearly such a preferred, well-defined glide plane. Thus, a change in potential from J_3 to J_2 would produce the kind of change in screw dislocation behavior such as deduced from the effect of H in Fe.

The point here is not to suggest that the particular potentials J_2 and J_3 actually represent Fe with and without H, but merely to show that a change in screw mobility follows from a change in interatomic potential and that it is feasible to view the effects of H in Fe in this way. Thus, a H-induced change in binding energy should also be reflected in a change in screw mobility. The direction of the change could not have been predicted; however, this knowledge has been provided by experiment. Of course, any solute which affects the binding energy would likewise affect screw mobility if it were mobile enough to be carried along by gliding dislocations. Hydrogen in iron at room temperature is unique in that respect. It should be noted that the argument works in both directions; the experimentally deduced effect on screw dislocations necessitates a binding energy effect, and experiment tells us the direction of that effect.

Glide-Plane Decohesion.

The existence of the $\underline{H}$-induced glide-plane decohesion has been recognized widely (24-31,15), but has been discussed only briefly in terms of mechanism (15,30). Recently (15), a model was proposed in which an analogy was made to the behavior of Zn at low temperatures. Gilman (40) demonstrated that, in an asymmetrical bicrystal of Zn at 77K, the blockage of edge dislocations at the grain boundary led to splitting along the glide plane, as shown in Fig. 10. This occurs because of:

(1) the effectiveness of the boundary in blocking dislocations
    (Glide occurs only on (0001) at low temperatures.)

(2) the low value of γ on the (0001) plane of Zn

(3) the build-up of stress normal to (0001) in a pile-up of
     edge dislocations.

The model proposed for $\underline{H}$ in Fe is similar, as shown in Fig. 11. The carbides formed in lath boundaries can provide effective dislocation barriers; the presence of $\underline{H}$ in the edge dislocation cores reduces γ. Thus, it was postulated (15) that the {110} or {112} "quasi-cleavage" (50) in iron and steel is a result of Zn-like behavior in Fe due to sweep-in of $\underline{H}$ along slip bands.

Conclusion

It now appears possible to construct a unified view of the effects of hydrogen on the mechanical behavior of iron and steel. The most important aspects of the hydrogen behavior are that it alters the binding energy of the iron lattice and that it is strongly trapped by dislocation cores and intercrystalline, or particle/matrix, interfaces. Experiments have shown that these lead to a decrease in lattice cohesion and an increase in the mobility of screw dislocations, but inhibited cross-slip, thereby promoting concentrated shear bands and ultimately glide-plane decohesion in the shear bands. The presence of impurity-weakened grain boundaries allows interface decohesion and brittle behavior at relatively low stress levels. The presence of distributed particles permits trapping of hydrogen in the particle/ matrix interfaces, and ultimately plastic rupture at reduced strains. It is argued that, from a practical standpoint, more attention should be directed toward the intergranular fracture problem in quenched and tempered steels, since this is how actual failures occur in such materials and since the problem has a solution other than the exclusion of hydrogen.

Acknowledgements

This paper was prepared under the auspices of the National Science Foundation MRL Program, grant no. DMR76-80994. Discussions with my co-workers, Drs. Jun Kameda, Y. Takeda, N. Bandyopadhyay, and K. A. Akhurst, and with Professor E. W. Plummer, and access to their experimental results are gratefully acknowledged.

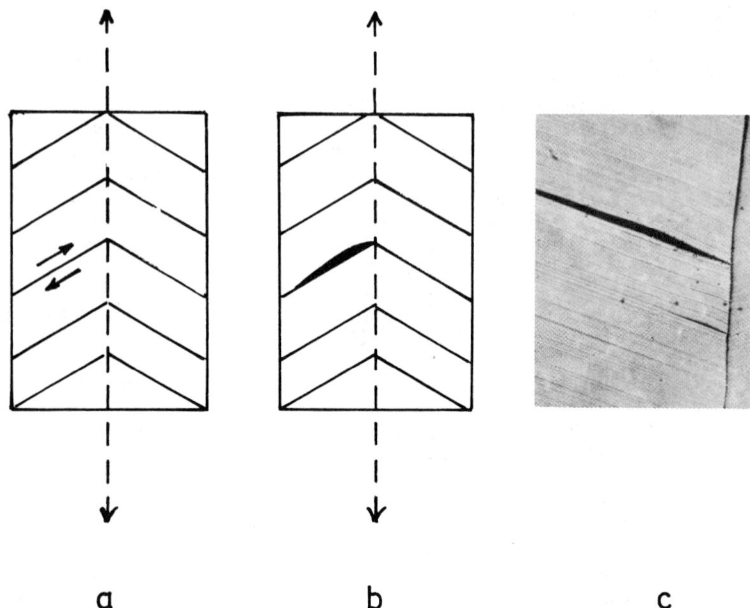

10. (a) Schematic diagram of bicrystal tested by Gilman
     (Ref. 49), exhibiting basal glide prior to (b) glide plane
     decohesion due to edge dislocation pile-up at grain boundary.
     (c) Examples of such cracking (Ref. 49).

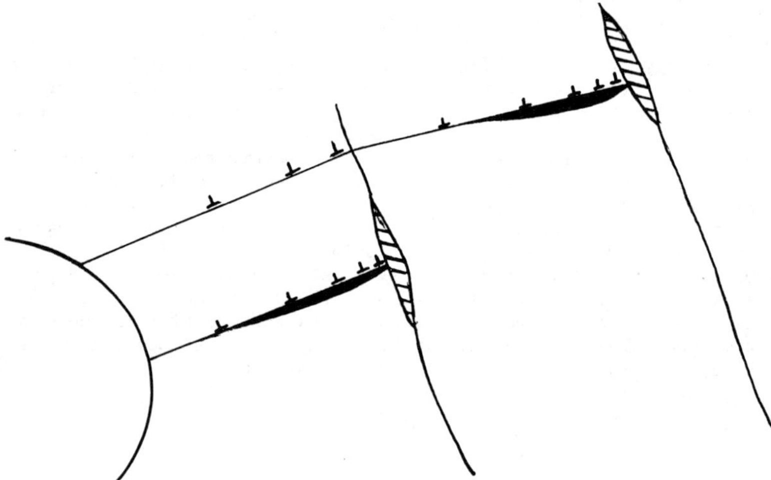

11. Schematic diagram of proposed model for formation of trans-
     crystalline cracks in martensitic or bainitic steels.
     Blockage of edge dislocations carrying $\underline{H}$ at interlath car-
     bides producing Zn-like glide plane decohesion.

References

1. J. P. Hirth: Met Trans A, 1980, vol. 11A, p. 861.

2. J. K. Tien, A. W. Thompson, I. M. Bernstein, and R. J. Richards: Met Trans A, 1976, vol. 7A, p. 821.

3. J. A. Donovan: Met Trans A, 1976, vol. 7A, p. 145.

4. J. K. Tien: in "Effect of Hydrogen and Behavior of Materials", A. W. Thompson and I. M. Bernstein, eds., TMS-AIME, New York, 1976.

5. H. Matsui, S. Moriya, and H. Kimura: Proc. 4th Int. Conf.; Strength of Metals and Alloys, 1976, vol. 1, p. 291.

6. H. Kimura, H. Matsui, and S. Moriya: Scripta Met, 1977, vol. 11, p. 473.

7. H. Mastsui, H. Kimura, and S. Moriya: Mat. Sci. Eng., 1979, vol. 40, p. 207, p. 217, p. 227.

8. A. Kimura, H. Matsui, and H. Kimura: Proc 2nd JIM Int. Symp., Hydrogen in Metals, suppl to Trans. Japan Inst. Met, 1980, vol. 21, p. 541.

9. J. Nagakawa, K. S. Shin, C. G. Park, and M. Meahii: Ibid., p. 549.

10. M. Cornet and S. Talbot-Besnard: Ibid., p. 545.

11. H. Wada and K. Sakamoto: Ibid., p. 553.

12. H. D. Solomon and C. J. McMahon, Jr.: Acta Met., 1971, vol. 19, p. 291.

13. T. D. Lee, T. Goldenberg, and J. P. Hirth: Fracture 1977, vol. 2, p. 243, Univ. of Waterloo Press, Waterloo, Canada, 1977.

14. Wu-Yang Chu, Chi-Mei Hsiao, and Shi-Qun Li: Scripta Met., 1979, vol. 13, p. 1063.

15. Y. Takeda and C. J. McMahon, Jr.: submitted to Met Trans.

16. T. D. Lee, T. Goldenberg, and J. P. Hirth: Met Trans A, 1979, vol 10A, p 199.

17. T. D. Lee, T. Goldenberg, and J. P. Hirth: Ibid., p. 439.

18. K. Yoshino and C. J. McMahon, Jr.: Met Trans, 1974, vol. 5, p. 363.

19. R. Viswanathan and S. J. Hudak, Jr: Met Trans A, 1977, vol. 8A, p. 1633.

20. C. L. Briant, H. C. Feng, and C. J. McMahon, Jr.: Met Trans A, 1978, vol. 9A, p. 625.

21. S. K. Banerji, C. J. McMahon, Jr., and H. C. Feng: Met Trans A, 1978, vol. 9A, p. 237.

22. Jun Kameda, N. Bandyopadhyay, and C. J. McMahon, Jr.: Proc. 2nd JIM Int. Sympos., Hydrogen in Metals, Suppl. to Trans Japan Inst. Metals, 1980, vol. 21, p. 437.

23. H. Mabuchi and C. J. McMahon Jr.: Ibid., p. 441.

24. I. M. Bernstein: Met Trans, 1970, vol. 1, p. 3143

25. I. M. Bernstein: Mater. Sci. Eng., 1970, vol. 6, p. 1.

26. F. Terasaki and F. Nakasoto: Proc. Conf. Mechanisms of Delayed Fracture Caused by Hydrogen, p. 165, Tokyo, 1975.

27. F. Nakasoto and I. M. Bernstein: Met Trans A, 1978, vol. 9A, p. 1317.

28. M. Nagumo, H. Morikawa, and K. Miyamoto: Proc 2nd. JIM Int. Sympos. Hydrogen in Metals, suppl. to Trans Japan Inst. Met, 1980, vol. 21, p. 405.

29. T. Takeyama and H. Takahashi: Ibid., p. 409.

30. T. Araki and Y. Kikuta: Ibid., p. 425.

31. S. Hinotani, F. Tarasaki, and F. Nakasoto: Ibid., p. 421.

32. N. Bandyopadhyay: Ph.D. Thesis, Univ. of Pennsylvania, 1980.

33. H. H. Johnson, J. G. Morlet, and A. R. Troiano: Trans TMS-AIME, 1958, vol. 212, p. 528.

34. A. R. Troiano: Trans, ASM, 1960, vol. 52, p. 54.

35. R. A. Oriani: Ber Bunsen Gesellsch. Phys. Chem., 1972, vol. 76, p. 705.

36. R. A. Oriani and P. H. Josephic: Acta Met, 1974, vol. 22, p. 1065.

37. W. W. Gerberich and Y. T. Chen: Met Trans, 1975, vol. 6A, p. 271.

38. V. Vitek: Int. J. Fracture, 1977, vol. 13, p. 481.

39. C. J. McMahon, Jr. and V. Vitek: Acta Met, 1979, vol. 27, 507.

40. M. L. Jokl, Jun Kameda, C. J. McMahon Jr., and V. Vitek: Met. Sci., 1980, vol. 14, in press.

41. M. L. Jokl, V. Vitek, and C. J. McMahon, Jr. Acta Met, in press.

42. C. L. Briant and R. Messmer: Phil. Mag. in press.

43. E. W. Plummer, B. P. Tonner, N. Holzworth, and A. Liebsch: Phys. Rev. B, 1980, in press.

44. E. W. Plummer: personal communication, University of Pennsylvania, 1980.

45. M. S. Duesbery, V. Vitek, and D. K. Bowen: Proc. Roy. Soc. 1973, vol. A332, p. 85.

46. V. Vitek: Proc. Roy. Soc., 1976, vol. A352, p. 109.

47. F. Kroupa and V. Vitek: Canadian J. Phys., 1967, vol. 45, p. 945.

48. V. Vitek: Crystal Lattice Defects, 1974, vol. 5, p. 1.

49. J. J. Gilman: Trans TMS-AIME, 1958, vol. 212, p. 783.

50. C. D. Beachem: Met Trans, 1972, vol. 3, p. 437.

51. Jun Kameda: Univ. of Pennsylvania, 1980, to be published.

DISCUSSION

R. P. Gangloff, Exxon Research and Engineering Company: It is reasonable
to expect that hydrogen can segregate to incoherent interfaces and grain
boundaries. What is the basis for precluding hydrogen effects on the bonding across pure grain boundaries, while postulating such a decohesion mechanism for glide planes?

C. J. McMahon: There is no intention to preclude hydrogen effects on pure
grain boundaries. It is recognized that grain boundaries provide strong
trapping sites for H, along with dislocation cores and particle matrix
interfaces. In fact, the interface decohesion shown in Fig. 3 may have
occurred along boundaries containing an insignificant amount of segregated
embrittling element. The phenomenon of H-induced glide-plane decohesion in
Fe is now rather well-documented (cf. Ref. 24-31 in my paper), so it is
probably time to elevate it above the rank of a mere postulate. Equally
well-documented is the phenomenon of impurity-related H-induced decohesion
along grain boundaries (cf. Ref. 18-23). However, very little is known
at present about possible H-induced decohesion of interfaces uncontaminated
by impurities.

John Tien, Columbis University: With respect to your model of glide plane
decohesion near a strong barrier like carbides, would you expect dislocation
sweep-in of hydrogen to play a role?

C. J. McMahon: Yes, the model depicted in Figure 11 specifically involves
the sweep-in of core-trapped H by the dislocations.

H. Birnbaum, University of Illinois: I am interested in the fact that in
the past few years there has been a complete shift in emphasis from a neglect
of plasticity effects in hydrogen embrittlement to an obsession with plasticity driven fracture. You also want to hold on to the concept of decohesion. Can you cite any direct evidence in support of H induced decohesion,
i.e. the reducation of the atomic bond energy due to hydrogen concentrations
of the order of those which occur during hydrogen embrittlement.

C. J. McMahon: That would depend on what you would accept as direct evidence.
For example, a routine method of increasing the rate of field evaporation of
atoms from a metal specimen tip in a field-ion-microscope is to admit a low
pressure of H_2 gas into the chamber. This could be considered such direct
evidence. However, someone whose tastes run counter to the idea of decohesion might prefer an alternative explanation. I cannot even conceive of
an experiment that would be convincing to everyone in a self-evident way.
This issue may finally have to be settled by some kind of first-principles
calculation. However, I don't think that we should hold our breath waiting
for it.

EFFECTS OF HYDROGEN ON THE PLASTIC RESPONSE OF STEELS

235

R. A. Oriani[*]

U. S. Steel Research Laboratory
Monroeville, Pennsylvania 15146

Our experiments on the effects of hydrogen injection on the straining
response, the stress relaxation, and the creep of medium-carbon steels are
reviewed and the internal consistency of these measurements is demonstrated. At input fugacities larger than a threshold value, hydrogen hardens
steels by increasing the impediment to cross slip. At input fugacities
larger than another threshold value, hydrogen enhances the microvoid
population produced at plastic incompatibilities, and thereby produces
indirectly a softening of the steel. Hydrogen of very large input
fugacity has the additional effect of supplying internal mechanical
stresses which produce an apparent softening of the steel. The implications of these results to the hydrogen embrittlement of steels are
discussed.

[*] Present address: Department of Chemical Engineering and Materials
Science, University of Minnesota, Minneapolis, MN 55455.

Introduction

The data in the literature on the effect of hydrogen on the plastic properties of iron and steels are contradictory and confusing. Even a cursory survey of the available information leads to the suspicion that important parameters exist which have not been controlled, the variation of which from one set of experiments to another has produced the observed diversity of results.

One of the poorly recognized important parameters is the thermo-dynamic fugacity of the hydrogen presented to the specimen. The work done at the U. S. Steel Research Laboratory, summarized in the present paper, is characterized by the use of controlled and measured (1) input fugacities of hydrogen, mostly in a range low enough so that the corresponding pressure of molecular hydrogen gas is lower than the flow stress of the steel. Another characteristic of our work is the use, whenever possible, of differential techniques. In this paper, observations on the effect of hydrogen on various plastic processes of medium-carbon steels are summarized, and the obvious phenomenological inferences are made. The phenomenology is interpreted in terms of dislocation dynamics, and the implications of the results to hydrogen embrittlement are discussed.

Effects of Hydrogen on Aspects of the Stress-Strain Curve

Tension specimens of AISI-1045 steel, either in the pearlitic or in the spheroidized condition, machined to 64-mm gage length, 13-mm width, and 0.25-mm thickness, were mounted in a table-model Instron testing machine. Hydrogen was charged in cathodically (1) from both sides of the specimen over the desired area of specimen before, or during deformation, or both, depending on the particular objective of the experiment. All deformation was done at room temperature and at a rate of 0.51 mm/min.

Luders Deformation (2)

Tension specimens marked with a photoengraved measuring grid on one side were mounted in the testing machine. Hydrogen at a measured input fugacity, f_i, was charged in for two hours prior to deformation and that charging was continued during deformation. However, hydrogen charging was restricted to one-half of the gage length. After straining the specimen into or somewhat beyond the Luders region, with typical results such as shown in Fig. 1, measurements of both halves of the photoengraved grid were made. These measurements, plus visual observations of the nucleation and propagation of Luders bands in both halves of the duplex specimen, permit the assertion that for this steel hydrogen raises the stress (see Table I) necessary both to nucleate and to propagate Luders bands.

Effect of Hydrogen After Interrupted Deformation (2)

A tension specimen of pearlitic 1045 steel with photoengraved grid was deformed in air to a plastic strain of 6.5% or less, after which the specimen was unloaded and the changes in grid dimensions were measured. After

Table I

Effect of Hydrogen on the Stress Required to Cause Luders Yielding in 1045 Steel

Microstructure	f_i, MN/m^2	Increase in Luders Yield, %, Caused by Hydrogen
Spheroidized	500	4.6
Pearlitic	200	4.0
Pearlitic	10^7	9.0
Pearlitic	10^3	6.0

Table II

Effect of Hydrogen on the Deformation of Pearlitic 1045 Steel After Interrupted Loading

Prior Plastic Strain, % (H-free)	f_i MN/m^2	Additional Overall Plastic Strain, %	$\Delta l/l'$, % H-free	$\Delta l/l'$, % With H	$\Delta w/w'$, % H-free	$\Delta w/w'$, % With H
6.5	200	6.4	8.5	3.5	−3.6	−1.4
6.5	200	1.8	1.8	1.4	−0.9	−0.6
4.6	10^7	0.8(fracture)	−	−	−	−
6.5*	200	1.8	1.8	1.4	−0.8	−0.6

* Entire specimen was strain aged at 155°C for two hours after the initial strain of 6.5%.

Note: Δl and Δw are the changes in length and width, respectively, of each half gage length referred to the length l' and width w measured after the prior strain.

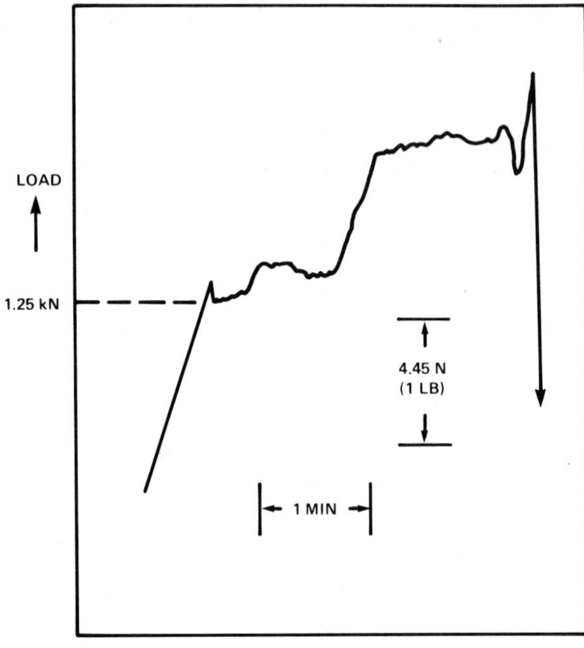

Fig. 1 – Tracing of section of Instron strip chart showing two-stage Luders deformation of pearlitic 1045 steel. Only one half of gage length received hydrogen at an input fugacity of 200 MPa.

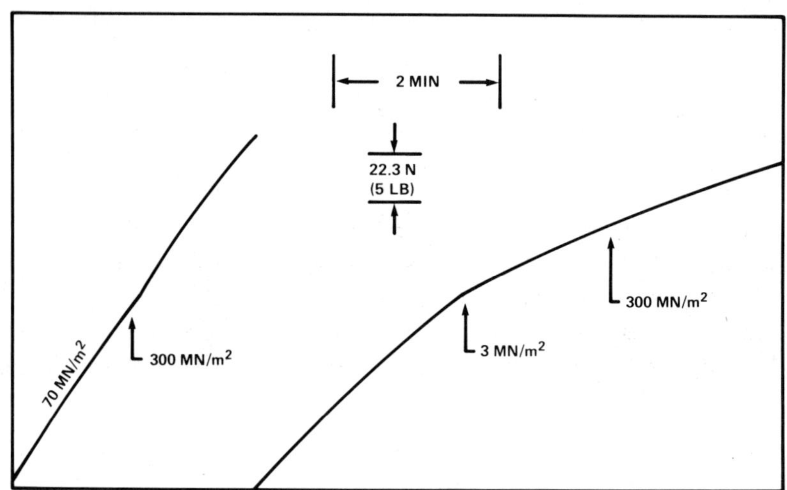

Fig. 2 –   Tracing of section of Instron strip chart, load vs. time, for pearlitic 1045 steel specimen being deformed at 0.51 mm/min. Vertical arrows indicate instants at which the input fugacity of the hydrogen was changed to the indicated value.

remounting in the testing machine, hydrogen at controlled f_i was charged
into one-half of the gage length of the specimen for two hours. This was
continued while the specimen was strained an additional amount. After
unloading, the dimensional changes were measured separately in the two
halves of the gage length.

The results shown in Table II demonstrate that the hydrogen-charged
section deforms much less, for virtually the same stress-time history, as
does the hydrogen-free section. The fourth experiment of Table II shows
that strain aging does not modify the hardening action of hydrogen.
Furthermore, the reloading curve was smooth, showing that hydrogen did not
introduce a yield point. This remark is equally applicable to the other
entries of Table II, including that for the largest input fugacity.

Effect of Hydrogen on the Slope of
the Stress-Strain curve (2)

Tension specimens of 1045-steel, both pearlitic and spheroidized,
were strained initially in the absence of hydrogen. During continuing
straining at a constant 0.51 mm/min, the cathodic charging current was
turned on, then decreased or increased to generate a variety of approximately known input fugacities of hydrogen over the entire gage length of
the specimens. Inspection of the load-time records of the Instron, an
example of which is shown in Fig. 2, revealed that an increase of f_i either
increased the slope of the load-time curve or had no perceptible effect.
Similarly, decreasing the input fugacity had either an imperceptible or a
negative effect on the slope. Thus, hydrogen increases by a small amount,
or leaves unaffected, the resistance of 1045 steel to plastic deformation.
This result is very different, both as to magnitude and sign, from the
results of Matsui et al (3,4) with high-fugacity hydrogen presented to
dynamically strained, high purity iron at room temperature. This difference is discussed in a subsequent section of this paper.

Residual Effect After Deformation With Hydrogen Charging

Fig. 3 illustrates the sequence of operations in an experiment with
pearlitic 1045 steel, tension specimen. After photoengraving a measuring
grid on one side of the specimen, one-half of the gage length was charged
with hydrogen at $f_i = 10^3$ MN/m^2, and with continuing charging into the same
half, the entire specimen was deformed by an overall amount of 6.0%. After
unloading, the grid dimensions of both halves of the duplex specimen were
determined, after which the specimen was allowed to lose hydrogen first by
exposure in air for 64 hours, then by anodic polarization in a suitable
electrolyte for 19 hours. The specimen was then strained an additional
overall 7.2%, after which the dimensions of both half-gage lengths were
measured.

The hardening produced by hydrogen both in the Luders yielding and in
the subsequent stage of the stress-strain curve were again observed, as
described above. The new results are that the removal of the hydrogen from
the previously charged half-specimen caused an immediate softening by about
4.9% (point B of Fig. 3) upon reloading, and that the section which had

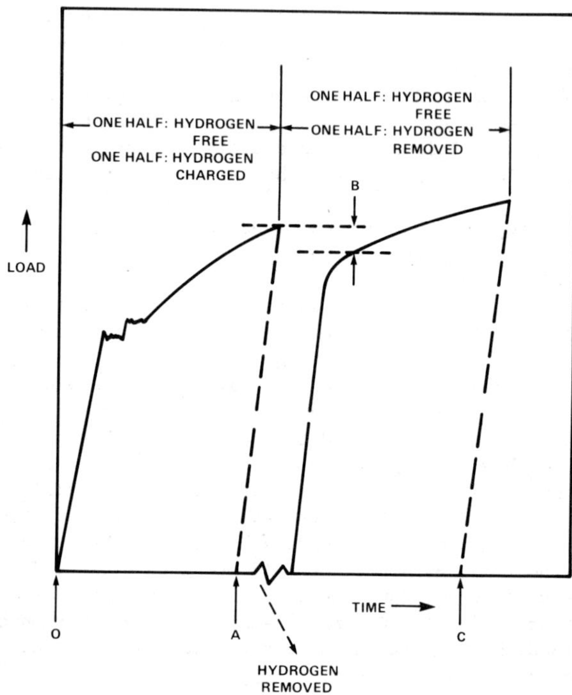

Fig. 3 - Protocol of experiment to show effect of the excess microvoid population produced by the decohering action of hydrogen. At point O, one half of the gage length was charged with hydrogen at $f_i = 10^3$ MPa, and charging was continued during deformation until specimen was unloaded at point A. After unloading, as much hydrogen as possible was removed. The specimen was then re-loaded and deformed further, then unloaded at point C.

Table III

Dimensional Changes in Pearlitic 1045 Steel Duplex
Specimen Put Through the Procedure Shown
in Fig. 3

	Hydrogen-Free Section	Section at First Hydrogenated	Section at First Hydrogenated Then Discharged
Changes at Stage A:	$\Delta\ell$: 0.1702 mm Δw: 0.3734	0.1346 mm -0.3073	
Changes at Stage C:	$\Delta\ell$: 0.2997 Δw: 0.6426		0.3658 mm -0.7620

Note: All Δ's are referred to the original length, ℓ, of the photoengraved
grid of 2.540 mm, and to the original width, w, of 1.273 cm of the
gage section of the specimen.

been initially deformed while charged with hydrogen and was then discharged
subsequently deformed much more than did the half-gage length which had
never been exposed to hydrogen. (See Table III). Thus, plastic deformation with hydrogen of moderate f_i leaves a marked softening influence that
is manifested when all mobile hydrogen is removed.

Effect of Hydrogen on 1020-Steel (2)

A thin tension specimen of pearlitic 1020-steel was photoengraved on
one side with a measuring grid. Half of its gage length was charged with
hydrogen at $f_i = 10^3$ MN/m^2 prior to straining, and that charging was
continued during deformation to an overall extension of about 4.0%. Upon
unloading, the hydrogen-charged section of the specimen was found to have
elongated by 2.3%, whereas the hydrogen-free section elongated by 3.1% for
the same load-time history, with corresponding differences in the changes
of the width dimension. The hydrogen-caused hardening contrasts strongly
with the marked softening reported by Beachem (5) who charged hydrogen at
very large, although unmeasured, f_i upon the outer surface of a torsion
specimen.

Effects of Hydrogen on Stress Relaxation (6)

Thin tension specimens similar to those described above were machined
from pearlitic AISI-1045 steel, with careful attention to avoiding imperfections along the edges. The specimens, fitted with electrolytic
charging cells and reference and auxilliary electrodes, were mounted in an
Instron testing machine totally enclosed in a constant-temperature environment. The specimen was strained in the absence of hydrogen at 0.51 mm/min
to the desired value of plastic strain, at which time the crosshead of the
testing machine was stopped and the relaxation of the applied load was
observed at room temperature. At the desired time, cathodic charging was
begun at a known f_i into both sides of the gage section of the specimen and
maintained for the desired duration. If no change was observed in the
character of the load-time curve, the input fugacity was increased and held
constant for another period of time.

Fig. 4 displays a typical result. The log of the load-relaxation rate
is linear in the log of time not only for the initial relaxation without
hydrogen subsequent to the straining without hydrogen, but also after the
presentation of hydrogen at input fugacities lower than some critical
value. However, at or above that value, a rapid increase in the relaxation
rate occurs. The threshold f_i is an increasing function of the initial
plastic strain.

It was found that the dynamics of the load relaxation at $f_i = 70$ MN/m^2
following the initial straining under the same input fugacity cannot be
distinguished from those of the relaxation without hydrogen which is subsequent to an initial straining also without hydrogen. In addition, if the
initial straining is restricted to 80 or 90% of the elastic limit the
dynamics of the subsequent load relaxation are about the same as for relaxation after initial straining past the elastic limit. The threshold input
fugacity for the sudden increase of relaxation rate is much larger when the

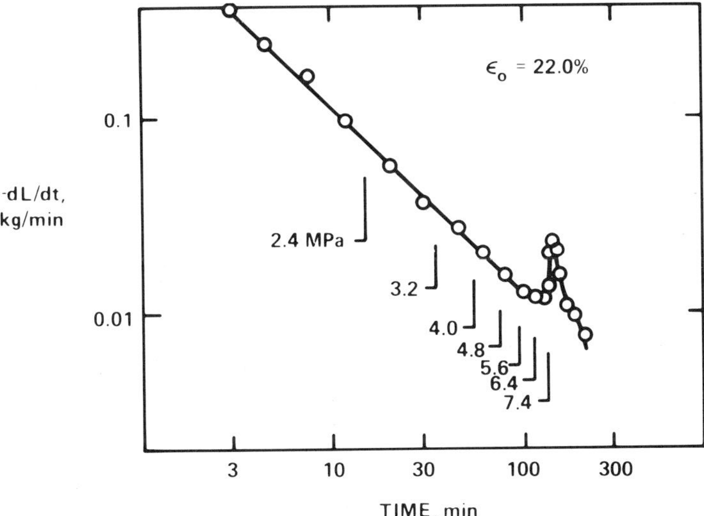

Fig. 4 – The rate of load relaxation at constant strain vs. time for pearlitic 1045 steel, initially deformed in the absence of hydrogen to 22.0% and allowed to relax initially in the absence of hydrogen. The vertical arrows indicate instants at which the input fugacity of hydrogen was changed to the indicated values, in MPa.

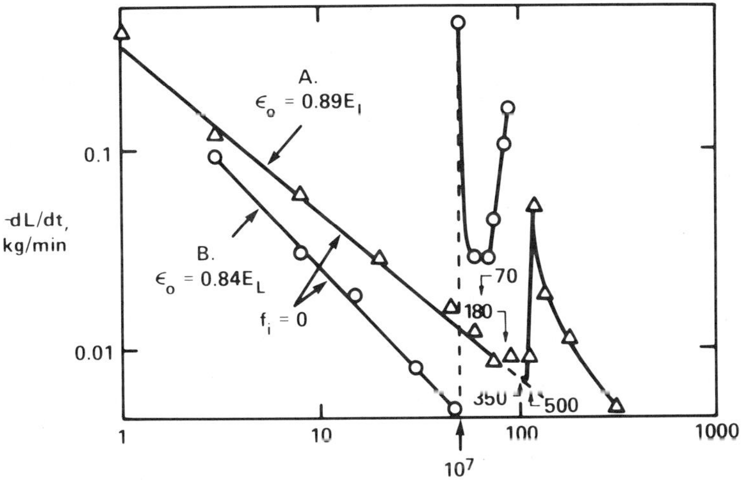

Fig. 5 – Stress relaxation of 1045 steel initially deformed to the indicated fraction of the elastic limit, E_L, in the absence of hydrogen, and allowed to relax at constant strain initially without hydrogen. The vertical arrows indicate values of applied f_i, in MPa. Curve A: Moderate values of input fugacity. Curve B: Very large value of f_i = 10^7 MPa.

initial strain is below than when it is above the elastic limit. Although
the decay of the peak of relaxation rate could not be studied in detail, it
was observed that if very large input fugacity is applied (Fig. 5), the
character of the post-peak relaxation rate-time curve is very different
from that with moderate values of f_i.

Effects of Hydrogen on Creep at Constant Stress (7)

Wire specimens, 0.12 mm in diameter, of spheroidized AISI-1040 steel
were mounted in a simple device by which the wire can be strained to a
desired value of plastic deformation with or without hydrogen charging.
The change in length of the wire can be followed at room temperature over
one or two days by an LVDT system, during which time the applied stress
changes by less than 0.7%. Hydrogen is presented to the wire at desired
times and at desired input fugacities by cathodic charging before and during
the initial straining, or during the constant-strain creep. Very often,
another plastic strain increment was applied after an extended creep
episode, and the subsequent creep was followed.

One characteristic result is illustrated in Fig. 6. The log of the
creep rate, $\dot{\varepsilon}$, is linear in the creep strain, and the slope is $-1/k$, where
k is defined by the logarithmic creep relation

$$\varepsilon - \varepsilon_o = k \ln(t - t_o) \tag{1}$$

in which $(t - t_o)$ is the elapsed time after the strain ε_o. The parameter
k for the creep episode directly following the initial prestrain to the
value ε_o increases with increasing ε_o, and is essentially the same whether
f_i during both prestraining and creeping is zero or has a finite value. As
in the stress-relaxation experiments, a larger value of f_i than some
threshold value is required to change the dynamics of the creep process.
Fig. 6 shows the threshold for affecting a sudden increase in the creep
rate; this is clearly analogous to the threshold effect discussed above in
stress relaxation.

The creep experiments clearly reveal a phenomenon that could not be
observed in the stress-relaxation work, that whenever increasing the f_i
causes a rapid jump in the creep rate there is also produced an immediate
increase in the slope $(-1/k)$ of the relation between $\ln\dot{\varepsilon}$ and $(\varepsilon - \varepsilon_o)$.
Considering that this means that a given $\Delta\varepsilon$ produces a greater decrease in
the creep rate after the jump in $\dot{\varepsilon}$ than before, one concludes that the
creep after the jump is characterized by a greater strain-hardening coefficient. This is quite analogous to the observation discussed in an
earlier section that an increase in f_i during dynamic straining sometimes
causes a small increase in the slope of the stress-strain curve, never a
decrease. Fig. 7 shows the variation of the logarithmic creep parameter k
with successive increments of input fugacity of hydrogen for six different
series of experiments.

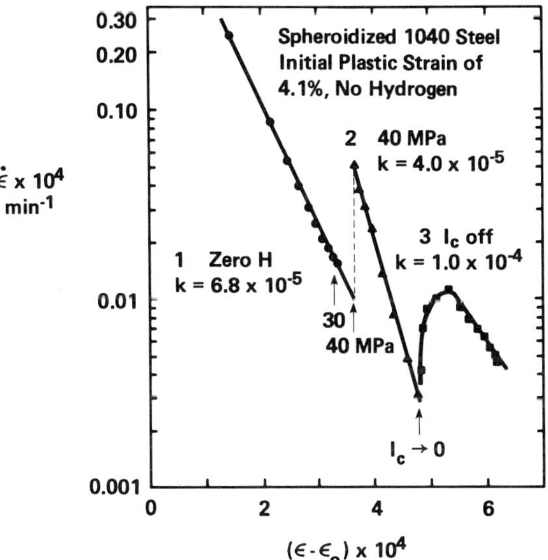

Fig. 6 – A typical creep rate vs. creep strain experiment
with spheroidized 1040 steel. The initial plastic
deformation of 4.1% was done in the absence of
hydrogen, as also was the first creep episode.
Vertical arrows indicate changes of the input
fugacity of hydrogen to the values given in MPa.
$I_c \to 0$ indicates that the cathodic charging
current was set to zero at the point indicated by
the arrow.

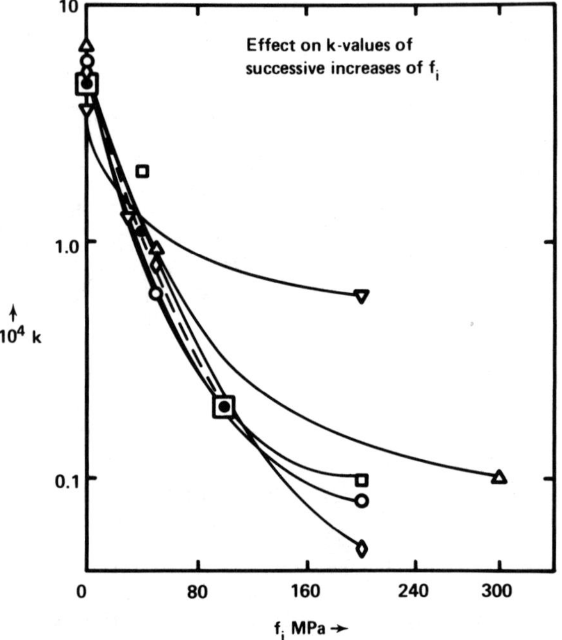

Fig. 7 – The variation of the logarithmic creep parameter
k with successive increments of input fugacity of
hydrogen for six different experiments.

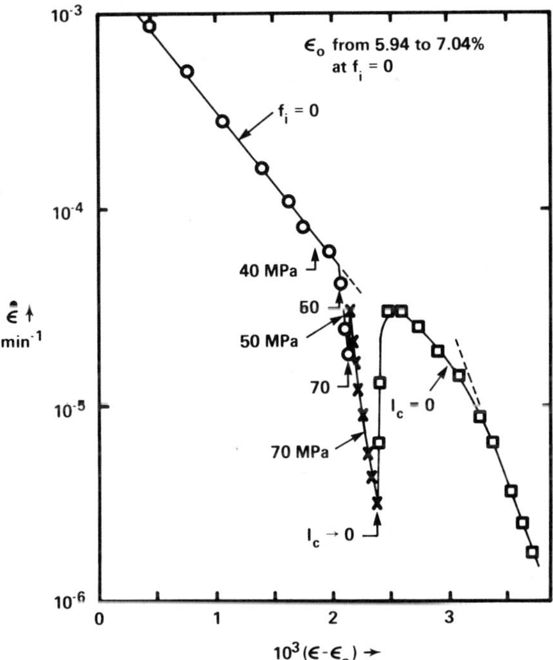

Fig. 8 – Creep of 1040 steel after an increment of loading to produce an additional plastic strain increment from 5.94 to 7.04 % in the absence of hydrogen. The vertical arrows mark the application of hydrogen at the indicated input fugacities.

Another kind of threshold is illustrated by Fig. 8, one which was not seen in the stress relaxation experiments. Whereas f_i = 40MPa did not change the creep dynamics from those of creep with f_i = 0, hydrogen input fugacity of 50 MPa caused a sharp increase in the strain hardening coefficient, but without the initial jump in the creep rate that is seen in Fig. 6, and which is the more frequently observed phenomenon.

Figs. 6 and 8 give examples of what happens when the cathodic charging current is turned of after a creep episode at f_i above the threshold value. Halting the further injection of hydrogen causes $\dot{\varepsilon}$ to increase (but more slowly than the $\Delta\dot{\varepsilon}$ that occurs upon increasing f_i), reach a maximum, and then decrease with increasing creep strain, and the dependence of $\ln\dot{\varepsilon}$ upon $(\varepsilon - \varepsilon_0)$ again approximates linearity. The maximum $\dot{\varepsilon}$ attained increases roughly linearly with increasing value of the input fugacity at the prior creep episode. Any amount of decrease of f_i after a creep episode at a finite input fugacity above the threshold value causes an increase in creep rate followed eventually by a linear decrease of $\ln\dot{\varepsilon}$ with creep strain. However, if the prior creep episode had taken place at f_i below the threshold value, then dropping f_i to zero does nothing to the dynamics of the creep.

Discussion

The increase in the slope of the dynamic stress-strain curve that was often seen (2) when hydrogen was presented is consistent with the observation of a greater coefficient of strain hardening caused by hydrogen in the creep experiments (7), and with the smaller amount of deformation in the hydrogen-charged half of the duplex specimens (2). The sudden increase in $\dot{\varepsilon}$ seen in creep experiments, caused by above-threshold fugacities, cannot be seen in the dynamic straining experiments because of the transient nature of that increase. The softening observed in dynamic straining when the hydrogen fugacity is reduced is in keeping with the large rise in creep rate shown in Figs. 6 and 8. In turn, these two effects are consistent with the marked softening observed in the half of the duplex specimen that had first been deformed with hydrogen and then discharged (2). The existence of a threshold f_i to produce peaks in the stress-relaxation curves (6) is clearly consistent with the threshold effect in the generation of sharp increases in creep rate (7). Agreement was also found for the independence of the creep dynamics and of the stress-relaxation dynamics from the magnitude of the f_i at which both the pre-strain and the first relaxation or creep episode was carried out.

The greater resistance to deformation and the greater rate of strain hardening has been attributed (2,7) to the increased difficulty of cross slip by screw dislocations owing to the widening of their cores. This is in turn a manifestation of the weakening by hydrogen of the attractive interactions between iron atoms. The hydrogen-loaded screw dislocations therefore produce a greater entanglement, and hence a greater internal stress σ_i, than when they move without carrying a load of hydrogen. Both our experimental results and the interpretation of increase of σ_i caused by hydrogen are consistent with those of Matsuyama (8) with decarburized carbon steel. Our finding that an above-threshold f_i is needed to produce an impediment to cross slip is not understood at present.

The observations of sudden increases in creep rate and in stress-relaxation rate caused by hydrogen have been interpreted (7,6) to be due to hydrogen-enhanced nucleation and decohesive growth of microvoids at regions where inhomogeneous plastic deformation has generated large stresses, which however are not sufficiently large to have nucleation or decohesive growth occur in the absence of hydrogen. Such processes relax the stresses that produced them so that σ_i is suddenly decreased, leading to a burst of dislocation activity. The existence of a threshold f_i is consistent with the basic tenet of the decohesion model (9) that bond breaking occurs only above a critical combination of local hydrogen concentration and local stress.

The microvoids generated with or without the aid of hydrogen not only decrease σ_i when they nucleate or grow decohesively but also serve to maintain the local stresses lower than they would be in their absence. This is because the presence of the void surfaces obviate the generation of plastic incompatibility stresses. As a result, the overall level of σ_i and its rate of increase with strain is lowered by the microvoid population. Hence, deformation under a finite f_i has two opposing effects, the softening effect of the hydrogen-enhanced, number and mean size of microvoids, and the hardening due to hydrogen-induced impediment to cross slip.

Our experimental results show that with medium-carbon steels the hardening effect tends to predominate, although by only a small amount, when the steel is dynamically strained under a sufficient magnitude of f_i. However, when the f_i is decreased to zero after an episode of creep at a finite f_i above the threshold value, the impeding of cross slip by hydrogen is gradually lifted since the newly formed dislocations move with thinner Cottrell atmospheres because the hydrogen already in the specimen must be shared by an ever increasing number of dislocations. In addition, the hydrogen-sparse dislocations are moving in a field characterized by a lower σ_i because of the enhancement of microvoid population that developed in the prior creep episode at finite f_i. For these two reasons, the creep rate increases when the charging current is shut off. Finally, as the condition of zero charging current is allowed to persist, only dislocations that are essentially free of hydrogen move, and another nearly linear dependence of log $\dot{\varepsilon}$ upon creep strain ensues. The fact that the maximum creep rate ratio attained, after shutting off the charging current, increases as the prior f_i increases is understandable since the larger f_i during the preceding creep episode has produced a greater residual softening effect.

Matsui et al (3,4) have clearly shown that the presentation of high-fugacity hydrogen during dynamic straining of iron of very high purity causes a marked decrease of the flow stress, and that the flow stress recovers reversibly when the hydrogen supply is terminated. This result is in apparent disagreement with ours on medium-carbon steels. However, different mechanisms are at work. In an iron of very high purity the

deformation rate is controlled by the mean velocity imparted to a dislocation by the effective stress, so that the height of the Peierls barrier is
important. Matsui et al argue that hydrogen in the Cottrell cloud about
the dislocation lowers the Peierls barrier because the dislocation core is
widened by the hydrogen. This is the same phenomenon that we have argued
increases the impediment to cross slip. In impure irons and in steels, the
difficulty of surmounting obstacles by cross slip controls the deformation
rate, not the mean velocity of a dislocation as it moves between obstacles.
Therefore, the identical action of hydrogen—that of decreasing the
attractive interactions between the iron atoms—leads to contrasting
phenomenologies depending on the degree of purity of the iron.

The use of very large input fugacities of hydrogen does not change
qualitatively the effects of hydrogen discussed up to this point. However,
if the mechanical pressure (10) that corresponds to the input fugacity is
large enough to overcome the flow stress of the metal about the microvoids,
then the molecular hydrogen gas that precipitates within such voids
produces plastic flow (11) that makes itself felt externally as an apparent
softening. This effect, which is artifactual from the point of view of
understanding the intrinsic action of hydrogen on steels, can be so large
as to submerge every other effect. It is probably responsible in large
measure for the variety of results on the effect of hydrogen on the plastic
properties of irons and steels since most work has been done with
uncontrolled and very large hydrogen fugacities. As an example, the
softening observed by Beachem (5) of 1020-steel charged at its outside
surface during the torsional deformation, which is in marked contrast with
the result (2) discussed above, is probably a result of the internal
pressurization effect. On the other hand, the data of Matsui et al (3,4)
are not vitiated by this effect despite the magnitude of the input fugacity
employed. The reversibility that they observed attest to that. Additional
support is provided by the observations of Kufukadis and Raczynski (12) and
Raczynski (13) on the difficulty of nucleating microvoids in iron as a
function of purity and amount of deformation.

<u>Implications for Hydrogen Embrittlement</u>

The primary mechanism by which hydrogen reduces the ductility of lower-
strength steels and increases the cracking of higher-strength steels is the
reduction of the cohesive strength of atom-atom bonds (6,14,15). Nevertheless, the modification of the plastic properties of steels by hydrogen in
the ways described above may be expected also to have an influence.

Since bond breaking and slip are alternative responses (16) to applied
forces, it is clear that the greater difficulty for cross slip induced by
hydrogen in steels must serve to aggravate the direct decohesive action of
hydrogen. As we have discussed, this is an effect that occurs immediately
upon presentation of hydrogen to a deforming steel. However, with
increasing time of deformation at $f_i \neq 0$ there develops an influence
towards easier slip due to the hydrogen-enhanced population of microvoids.
In particular, the generation of microvoids by the intersection of a slip
band with precipitate particles should facilitate more slip in that band.

This may be the explanation of some observations (17,18) that hydrogen promotes shear localization. This, however, opposes the tendency to delocalization of shear by whatever causes increased strain hardening, as hydrogen solute has been shown in the foregoing to do. Indeed, preliminary evidence has been obtained (19) that hydrogen at f_i = 10 MN/m^2 represses the tendency to localize shear at small deformations of a maraging steel of 1725 MN/m^2 strength at 0.2% offset. It is clear that a variety of results from experiments designed to elucidate the effect of hydrogen on shear localization is to be expected, depending on the f_i, the extent of deformation at f_i, and the nature of the steel. This remark is directed to the use of low-fugacity hydrogen. If higher values of f_i are used, such that the flow stress is overcome by the molecular gas pressure of hydrogen, additional effects will be observed that will not be easily relatable to primary mechanisms.

Conclusions

In medium-carbon steels, hydrogen has a direct hardening effect due to increasing the difficulty of cross slip, and an indirect softening effect from the increase of the number and mean size of microvoids caused by the decohesive action of hydrogen. Which of these effects predominates depends on the details of the experiment. Although the decohesive action of hydrogen is the primary mechanism by which the ductility of lower-strength steels and the cracking resistance of higher-strength steels are decreased, the modification of the plastic response of the steel also has an influence which can be mitigating or aggravating. Working out the details of the interrelationships remains a challenging task for any one steel.

References

1. R. A. Oriani and P. H. Josephic, Proc. Symp. Environment-Sensitive Fracture of Engineering Materials, AIME, Chicago 1977, pp. 440-50. Z. A. Foroulis, ed. The Metallurgical Society of the AIME, 1979.

2. R. A. Oriani, submitted to Met. Trans., Jan. 1980.

3. H. Matsui, S. Moriya and H. Kimura, Proc. 4th Int. Conf. Strength of Metals and Alloys, Nancy 1976, pp. 291-5.

4. H. Kimura, H. Matsui and S. Moriya, Scripta Met. 11, (1977) 437.

5. C. D. Beachem, Met. Trans. 3, (1972) 437.

6. R. A. Oriani and P. H. Josephic, Acta Met. 27, (1979) 997.

7. R. A. Oriani and P. H. Josephic, to be published Scripta Met.

8. S. Matsuyama, Proc. 2nd JIM Int. Symp. Hydrogen in Metals, Minakami 1979, Suppl. Trans. Jap. Inst. Metals 21, (1980) 557.

References (continued)

9. R. A. Oriani, Berichte Bunsen-Gesellschaft physik. Chem. <u>76</u>, (1972) 848.

10. B. Baranowski, Berichte Bunsen-Gesellschaft physik. Chem. <u>76</u>, (1972) 714.

11. A. S. Tetelman, C. N. J. Wagner and W. D. Robertson, Acta Met. <u>9</u>, (1961) 205.

12. W. Raczynski and A. Kufukadis, Czech. J. Phys. <u>B26</u>, (1976) 1360.

13. W. Raczynski, presentation at Symp. Hydrogen Embrittlement and the Corrosion of Metals, Jablonna, Poland, 1977.

14. R. A. Oriani, Ann. Rev. Mater. Sci. <u>8</u>, (1978) 327.

15. R. A. Oriani, "The Decohesion Theory of the Hydrogen Embrittlement of Steels, to be published in Proc. A. R. Troiano Honorary Symposium, Case Western Reserve University, June 1980.

16. J. R. Rice and R. Thomson, Phil. Mag. <u>29</u>, (1974) 73.

17. T. D. Lee, T. Goldenberg and J. P. Hirth, in Fracture 1977, Vol. 2, D. M. R. Taplin, ed., pp. 243–8, Pergamon, Oxford, 1977.

18. T. D. Lee, T. Goldenberg, and J. P. Hirth, Met. Trans <u>10A</u>, (1979) 439

19. R. A. Oriani, R. J. Sober, and W. A. Spitzig, to be published.

DISCUSSION

<u>H. K. Birnbaum, University of Illinois</u>: You suggest that H "broadens" the dislocation core by affecting the "cohesive energy" and leading to an increase in flow stress. A core broadening would be expected to decrease the Peierls-Nabano stress and to make dislocation motion easier and not more difficult. Can you explain how the hardening would occur.

<u>R. A. Oriani</u>: I suggest that a hydrogen-induced lowering of the cohesive force between iron atoms at the core of a screw dislocation is responsible both for a lowering of the Peierls barrier, as already suggested by Kimura, et al, and for increasing the difficulty of cross slip by broadening the core. The first effect increases the "free-flight" velocity of the dislocation, and is manifested in obstacle-free pure iron as a hydrogen-induced softening. The second is manifested in impure-iron and in steels, in which the deformation velocity is controlled by the difficulty of surmounting obstacles, not by the free-flight velocity, and hence results in a hardening by hydrogen.

<u>D. C. Langstaff, METCO</u>: You have emphasized the importance of adequately describing the fugacity of the hydrogen presented to a test specimen during charging. Would you comment on the significance of the chemical activity or chemical potential of the charged hydrogen located just beneath the surface of the specimen, i.e., the initial bulk chemical activity of the charged hydrogen?

<u>R. A. Oriani</u>: The activity, a_o, of the charged-in hydrogen just beneath the input surface is a significant and useful parameter. It sets the thermodynamic condition for the interactions of the dissolved hydrogen with structural and chemical singularities at that point within the metal, and elsewhere within the metal, when due account is taken of the diffusion with accompanying reactions under the appropriate boundary conditions. The a_o is the maximum activity that the hydrogen may attain anywhere in the metal, with the exception of transients associated, for example, with the annihilation of hydrogen-loaded dislocations.

<u>D. C. Langstaff, METCO</u>: Because hydrogen fugacity is an important parameter in hydrogen studies, a standardized method of determining hydrogen fugacity experimentally and calculating its value would be very useful. Would you recommend a standardized method for determining the absolute value of hydrogen fugacity?

<u>R. A. Oriani</u>: By input fugacity I mean the fugacity of hydrogen that at the given temperature would be in equilibrium with the thermodynamic activity of the hydrogen in the metal immediately below the input surface. That thermodynamic activity, a_o, is a property both of the lattice-dissolved hydrogen at concentration c_o and of whatever trapped hydrogen occurs in the sub-surface layer of atoms. It follows then that the input fugacity is related to c_o via Sievert's law, $c_o = s_t f^{1/2}$. The parameter s_t is independent of the number and type of traps within the metal and of surface contamination, whereas c_o (and a_o) is a strong function of the electrochemical parameters and surface composition and metallurgical state of the interface since c_o is determined by competition among various kinetic processes at the input surface.

At steady state, the permeation flux of hydrogen through a thin sheet of metal at whose output surface the hydrogen concentration is held at zero is given by $J_\infty = Dc_o A/\Delta X = \phi f^{1/2} A/\Delta X$, irrespective of the existance of traps. In these relations, D and ϕ are, respectively, the diffusivity and permeability of hydrogen undisturbed by traps, and A and ΔX are, respectively, the permeation area and thickness of the sheet. It follows that a measurement of J_∞ enables one to calculate c_o and f_i if D and ϕ, respectively, are known. Hence, to know these quantities for a mechanical test specimen under cathodic charging J_∞ may be measured at the temperature of interest for a thin sheet of the same material with identical surface treatment, using identical electrochemical charging conditions as for the mechanical test specimen, and palladizing the exit surface of the sheet to decrease the exit impedance to nearly zero. Clearly, it is necessary in general to measure D and ϕ for the specific alloy employed, usually at higher temperatures where the trapping phenomenon is negligible, followed by extrapolation to the temperature of interest. I caution that this approach cannot be used to obtain f_i within a crack at which cathodic charging is being done because in general the electrochemical conditions within such a crevice-like geometry are not simply related to those at a flat surface immersed in the same electrolyte.

In cases where the hydrogen is proced non-electrochemically, such as by reaction of water vapor with the metal, again, in general, kinetic factors control c_o and hence f_i, and a J_∞ measurement is necessary to establish the operating f_i. The maximum developable f_i may be calculated if the reaction product MO_y and its free energy of formation are known. Thus, if ΔG^o is the standard free energy change at the experimental temperature for the reaction $M + y\ H_2O(v) = MO_y + y\ H_2$, then the maximum input fugacity of hydrogen is given by $P_{H_2O(v)} \exp(-\Delta G^o/yRT)$.

<u>R. E. Stoltz, Sandia Laboratories</u>: Do you consider the samples as a composite: surface + bulk, especially in charging transient experiments, and have you looked at different thicknesses to separate the effect of hydrogen at the surface and in the bulk.

<u>R. A. Oriani</u>: I do not believe that my specimens should be regarded as composites insofar as the hydrogen concentration is concerned. Prior to deformation, pre-charging at the desired input fugacity was carried out long enough for the center of symmetry of the specimen to have reached at least 90% of the sub-surface concentration. When the creep rate or the stress relaxation rate changed in response to a change in the input fugacity of hydrogen, the change occurs so rapidly, in a fraction of a second, that my working hypothesis immediately became that dislocation drag was the transport mechanism during creep and during stress relaxation, and that, therefore, the change of input fugacity was felt by the entire specimen within a fraction of a second. We did not look into the effect of different thicknesses of specimens.

SOME NEW ASPECTS OF HYDROGEN DAMAGE

Chi-mei Hsiao and Wu-yang Chu

Beijing University of Iron and Steel Technology

Beijing, People's Republic of China

Abstract

Some of our salient experimental results regarding the
changes induced by hydrogen in high strength steels, austenitic
stainless steel, commercially pure titanium and silicon single
crystals will be presented. Based upon these results, the
related mechanisms of hydrogen damage will be discussed. The
hydrogen induced delayed plasticity and cracking of high strength
steel and its cause will be discussed in detail.

I Introduction

Hydrogen damage to the materials implies the loss of macroscopic ductility and/or inducing of cracking of the materials by
hydrogen. All these macroscopic phenomena are caused by the
changes of the materials induced by hydrogen, and these changes
include deformation-fracture, phase transformation and chemical
change.
Fundamental studies push and engineering experiences pull
the whole field of hydrogen damage, they all provide valuable
experimental and/or theoretical pieces which are interlocking
and make up this jigsaw puzzle game of hydrogen damage. Owing
to various reasons and motives, we and our colleagues in the
Department of Metal Physics, BIIST, have carried out some
research works on hydrogen damage in last few years. We will
present some of the salient experimental results and upon which
we will discuss the mechanisms of hydrogen damage.

II Hydrogen Induced Delayed Plasticity and Cracking (HIDP-C)

The occurrence of the HIDP in front of the crack of the
steel specimen requires three co-existing conditions to be
satisfied (1)-(3):
(a) the strength of the steel is higher than a critical

value,
(b) K_I is greater than K_{ISCC} or K_{IH},
(c) the supply of hydrogen is sufficient.

A typical example of the sequences of HIDP-C for ultra-high strength steel is shown in Fig. 1. The hydrogen had been charged electrolytically into the WOL type specimen (B=20mm) which was subsequently loaded in air. There was a visible plastic zone in front of the crack tip and it did not change if K_I is smaller than K_{IH}. However, if K_I is greater than K_{IH}, then the plastic zone enlarged with time (Fig. 1-1 and 1-2), i.e., HIDP was observed. After the closure of the delayed plastic zone, a HIDC was evident at its tip B (Fig. 1-3). These processes were repeated (Fig. 1-4 and 1-5), and the HIDC grew and joined one another (Fig. 1-5).

For lower strength steels, the progress of the HIDP-C was somewhat different. When the HIDP developed into a critical extent, the HIDC occured and propagated along the path near the border of the plastic zone(3).

These HIDP-C processes were also observed on the surfaces of the samples in which the hydrogen had been introduced during stress corrosion in flowing hydrogen gas, hydrogen sulfide gas, water, aqueous solution of saturated hydrogen sulfide or 0.1 N potasium dichromate.

There are many important questions to be aswered for this new phenomenon, particularly:
(a) Does the deformation at the surface is a consequence of crack growth from the interior region of the material?
(b) What is the reason for this unusual enlargement of the plastic zone in front of the crack?

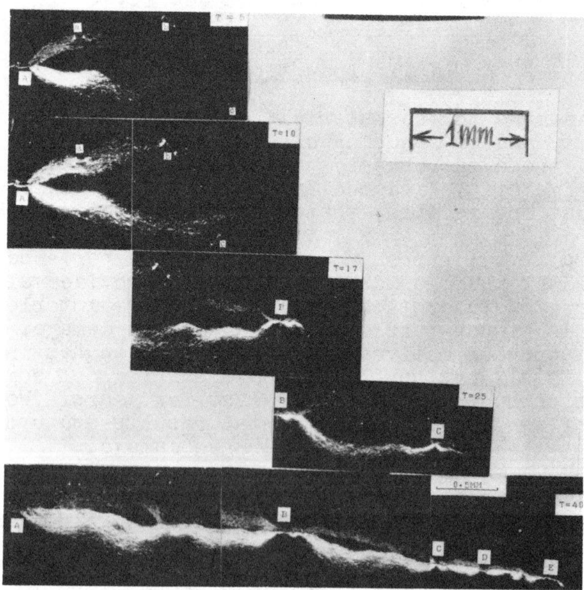

Fig. 1 The progress of HIDP-C of steel 30CrMnSiNi2 after charging at 20 mA/cm^2 for 6 hrs. (TS – 1580 Mpa.)

1   Mechanical Factor and Chemical Factor in Hydrogen Damage

From purely mechanical consideration, one might expect that
cracking leads in the mid-section region where the condition of
plain strain prevails. Thus the crack front is curved as usually
found in fatigued specimens. One may argue that the delayed
plastic flow of the external surface may reflect deformation
associated with the outward growth of the crack toward the
surface, and is therefore a consequence of crack growth rather
than its cause. However, while analyzing the hydrogen damage
problem, we are compelled to consider the important effect of the
chemical factor, i.e., the distribution of hydrogen within the
specimen. As shown in Fig. 2-1, we measured the P-V (crack open-
ning) and P-ε (strain at the crack tip region) curves of the WOL
specimens which had been charged electrolytically. The plastic
zone at point B (Fig. 2-2) is shown in Fig. 2-3 and it was enlarged by hydrogen. Once the loading had reached point B, the
specimen was immediately unloaded and then colored by heat-tinting
(350C-1.5hrs.), no HIDC was observed on either the free surface
or the fracture surface of the specimen (Fig. 2-4). However,
another duplicate specimen had been charged and loaded to point B
(Fig. 2-2), and maintained for 15 minutes, HIDC had been observed
on the free surface. The specimen was then unloaded and colored
by heat-tinting. On the fracture surface, the length of HIDC was
decreased from the surface layer toward the interior of the
specimen because of the difference in hydrogen concentration
from the surface to the interior during charging. These results
were reproducible and demonstrated that HIDP is the preceeding
step of HIDC, and HIDP is not associated with the outgrowth of
the crack from the interior region toward the surface.

As shown in Fig. 3, various shapes of crack fronts were
found for the specimens after stress corrosion in water or
saturated aqueous solution of hydrogen sulfide. The mid-thick-
ness region can either lead the cracking (A) or not (B, C and D).
This can be explained by the fact that the cathodic sites which
generate hydrogen can be anywhere on the crack front. It demonstrated once more that chemical factor is more important than
mechanical factor in hydrogen damage.

For ultra-high strength steels, the mode of fracture in all
the envirnments studied was intergranular, and micro-plasticity

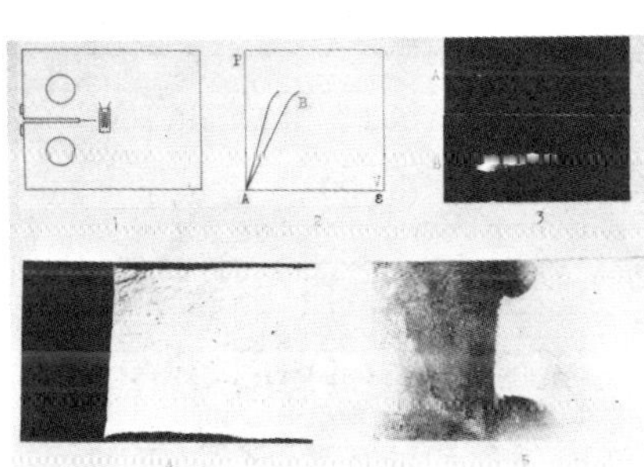

Fig.2 HIDP-C at surface and in
the interior region of steel
30CrMnSiNi2

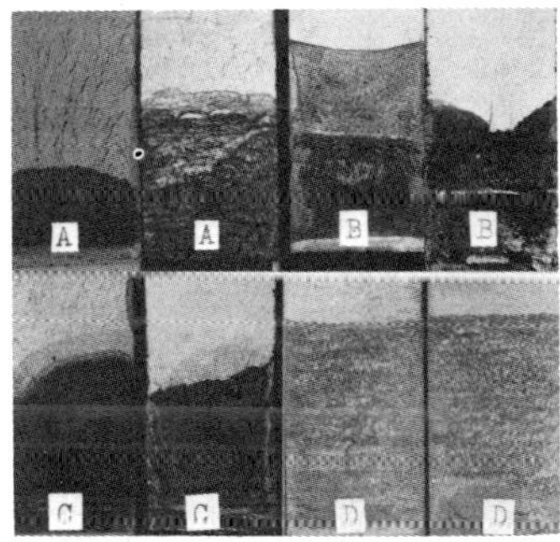

Fig.3 Macro-fractograghy of steel
30CrMnSiNi2 after SCC in aqueous
solutions

was evident in the intergranular facets as tear ridges or tiny dimples (Fig. 4).

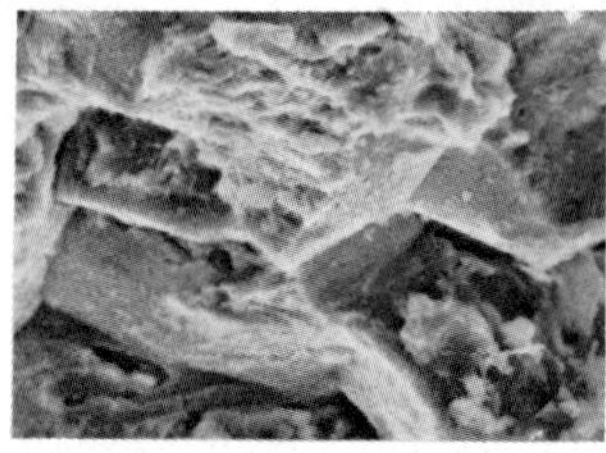 

(a) In H_2 gas (TS: 1580 MPa)   (b) In H_2S gas(TS: 1810 MPa)

Fig.4   SEM of steel 30CrMnSiNi2 after SCC

2 The Effect of Hydrogen on Local Yield or Shear Strength

The significant enlargement of the plastic zone or the enhancement of the plastic deformation by hydrogen implies a marked reduction of the local yield or shear strength of the steel by hydrogen.

All our experimental results are in agreement with the generally accepted idea that hydrogen can be enriched during loading by diffusion under stress gradient. The concentration of hydrogen at the sites where the hydrostatic   stress is σ_h can be expressed as:

$$C_\sigma = C_\sigma \exp (\sigma_h V_H / RT) \tag{1}$$

where C_o is the initially uniform concentration of hydrogen and V_H is the partial molar volume of hydrogen in iron. It is this enrichment of hydrogen which cause the reduction of local yield strength of the steel, the higher is the C_σ, the more will be the reduction. Since there is no stress gradient in smooth specimen under tensile test, as expected, the effect of hydrogen on the yield strength is small, i.e., less than 10% (Fig. 5). On the other hand, there does exist a stress gradient in either bending or WOL specimens, a reduction of local yield strength is expected and the higher is the C_o, the higher will be the C_σ, thus the more will be the reduction of the local yield strength (Fig.6,7). Within the plastic zone of the crack, the maximum σ_n is proportional to the yield strength of the steel (4), therefore C_o would be increased with the yield strength of the steel; this can be used to explain the dependence of the reduction of local yield strength with the strength of the steel (Fig. 6, 7). Since HIDP is the preceeding step of HIDC, if we accept the criterion of K_{IH} or K_{ISCC} is τ^*, the critical local shear strength of the steel, we can easily deduce that K_{IH} is proportional to τ^*, and would expect that the higher is the strength of the steel, the lower will be the K_{IH} as usually found and also in Fig. 8. Only the tensile stress component would contribute to σ_h, we would expect that enough K_I component is required to cause the reduction of torsional yield strength of the steel in mixed-mod (I + III) tests and the higher is the K_I, the more will be the reduction (Fig. 9).

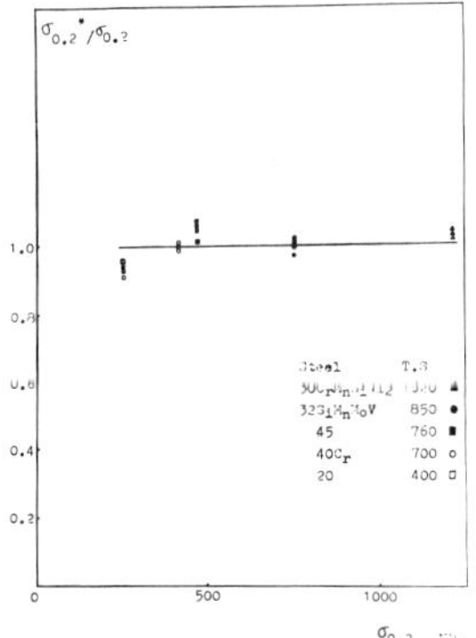

Fig.5 Variation of
ratio of the yield
strength of charged
to uncharged speci-
men with the yield
strength (i=40 to
80 mA, 6 to 8 hrs,
slowly loaded)
(Sooth tensile)

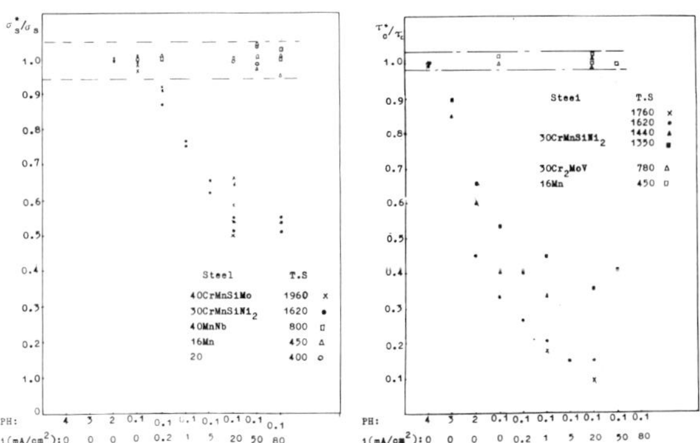

Fig. 6
Smooth bending

Fig. 7
Precracked WOL

Fig.6 to 7 Variation of ratio of the yield
or shear strength of charged to uncharged
specimens with the charging conditions

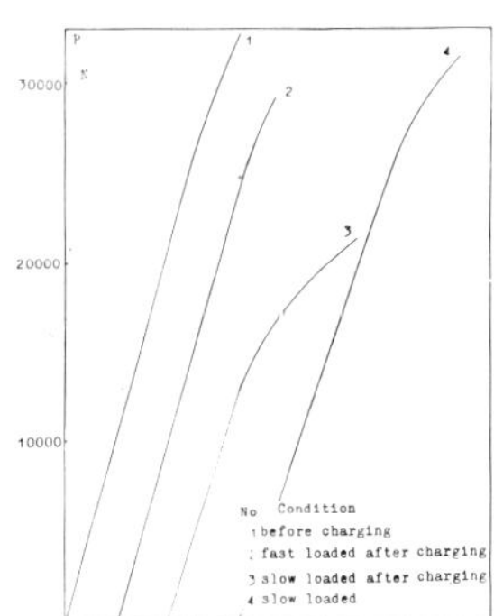

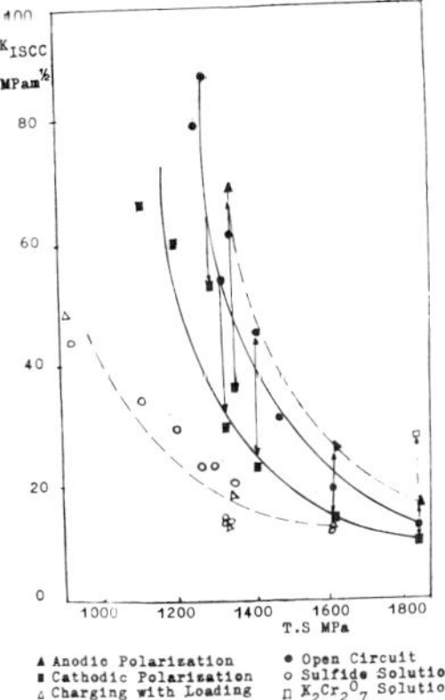

Fig.10 Load—strain curve of the
bending specimens in different
test conditions (30CrMnSiNi2,
T.S. — 1620MPa)

Fig.8 Variation of K_{ISCC} with
tensile strength (T.S.) of
steels in various environments

It is important to point out that the reduction of local
yield or shear strength is reversible and can be recovered by
heating at 240C for 48 hours to remove the hydrogen (Fig. 9).
This recovery effect can be further demonstrated in Fig. 10.
After severly charging, the P-ε curve for rapid loading (curve 2)
is about the same of that before charging (curve 1), but the P-ε
curve for slow loading (curve 3) manifests a significant reduction
of local yield strength. After heating the specimen at 280C for
30 hours, the P-ε curve for slow loading (curve 4) recovered to
that of before charging (curve 1). This experiment demonstrates

further that the reduction of the local yield or shear strength
is due to the <u>diffusible hydrogen atoms</u> rather than the possible
micro-cracks during charging. The effect of diffusible hydrogen
atoms should be influenced by the temperature and strain rate of
the tests. As shown in Fig. 11, the reduction of the local shear
strength could not be manifested if testing temperature is lower
than -90C or the crosshead rate is higher than 0.5 mm/min.

The recovery of the local yield strength to the original
value by heating (Fig. 9, 10) precludes the <u>pressure mechanism</u>
of molecular hydrogen gas to be operated. From the published
experimental evidences, we suggested(1) that the interaction
between hydrogen and iron reduces the local shear strength of the
iron as found in this paper, thus the plastic zone will be enlarged because of the easier nucleation and motion of dislocation.
Recently, Brockris-Flitt (5) reported V_H to be 2.67 cc/mole from
permeation studies, and suggested that there is much repulsion
between the cationic hydrogen and positively charged iron nuclei.
Also, Mr. Kao of our Department has found that hydrogen can
reduce the work function of iron in stainless steel by about 20%
from mass ion spectrography (Fig. 12), and it provides further
experimental support of the <u>decohesion mechanism</u> of hydrogen
damage in addition to the work of Lumarski et al. (6), who found
that the shear modulus of iron was reduced by hydrogen to 8% per
atomic % of hydrogen. However, such large reduction of the local
yield or shear strength as found in Fig. 6, 7, 9, 10 can't be
explained by decohesion mechanism alone, we believe that the
highly hydrogen-supersaturated region would form a "<u>platelet
hydrogen cluster or atmosphere</u>" similar to that proposed by
Fujita(7) and the shear component of the high pressure from this

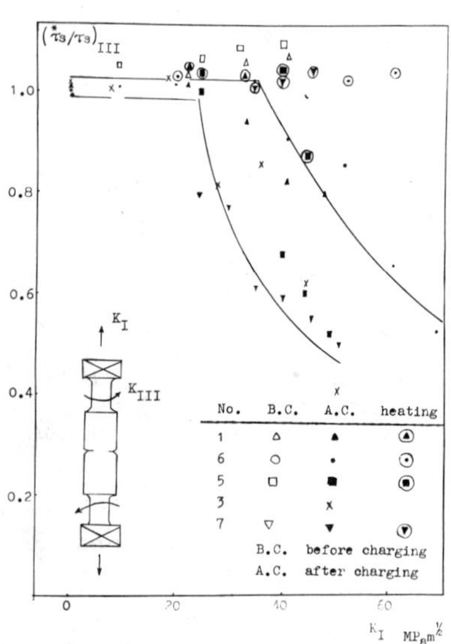

Fig.9 Variation of ratio of
shear strength of charged to
uncharged specimens with K_I
in mixed I + III loading
(Steel 30Cr2MoV, TS: 1820 MPa)

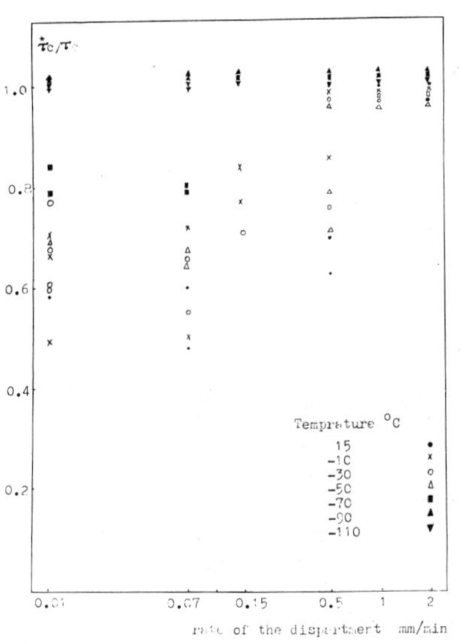

Fig.11 Variation of ratio of
shear strength of charged to
uncharged specimens with rate
of displacement & testing temp.
Steel 30CrMnSiNi2, TS: 1580MPa

cluster would aid to the external load for local yielding. The
measured load for local yielding will be greatly reduced, and we
denote the yield stress calculated from the external load as the
local yield strength. Such hydrogen atmosphere has also been
found in front of the crack tip of vanadium single crystal by
Suzuki(8) and of pure titanium polycrystal by us, the latter will
be described latter.

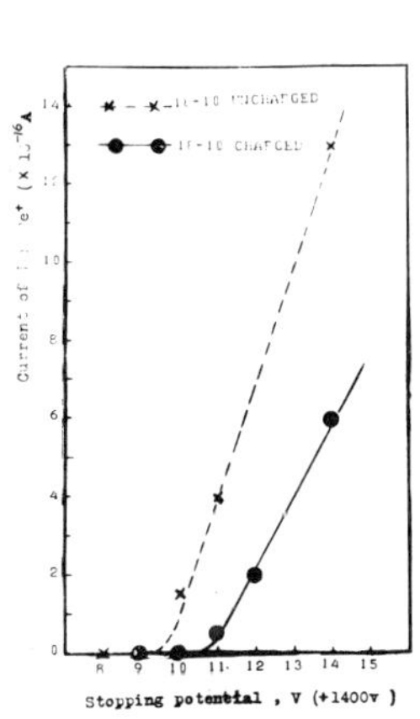

Fig.12 Effect of H on
the work function of
Fe in stainless steel

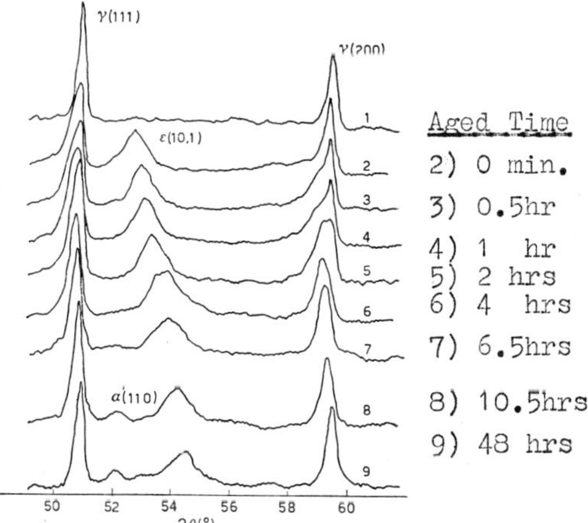

Fig.13 Structural changes during aging
after charging at CD of 0.2A/sq.cm.for
one hour (1-as annealed; 2-as charged)

III Hydrogen Induced Hydride Formation and Martensitic Transformation in Austenitic Stainless steel

The changes of the structures of stainless stell (similiar
to AISI 321) with aging at room temperature after electrolytical
charging in 1N sulfuric acid poisoned with 0.25 g As_2O_3/liter
for one hour with a current density of 0.2 A/cm² are shown in
Fig. 13. The asymmetry of (200)γ and (111)γ diffraction lines
is attributed to the appearence of a hydride phase MHx, most
probably NiHx (10). The lattice constants of the phases are
calculated and the results are listed in Table I. In calculating
a_0 for ε, $c_0/a_0 = 1.63$ is adopted (12).

During charging, the hydrogen content in the austenite is
increased, the lattice constant in the austenite is thereby increased. Hydrogen has a higher bonding energy with dislocation
and nickel, the distribution of hydrogen would be nonuniform
microscopically, the local stress field so generated, similiar
to cold working, would induce the martensitic transformation to ε.
If the amount of hydrogen dissolved exceeds the solid solubility,
hydride would be precipitated out; this hydride is also face
centered cubic, the lattice constant of which is larger than that
of austenite owing to higher content of hydrogen.

Table I   The Change of Lattice Constants

No.	Condition	$(200)_\gamma$	$(111)_\gamma$	$(200)_{MHx}$	$(10.1)_\varepsilon$	$(110)_{\alpha'}$
1	As-annealed	3.604	3.606	-----	-----	-----
2	As-charged	3.610	3.611	3.620	2.637	-----
3	2 +aged 30 min	3.610	3.611	3.627	2.626	-----
4	2 +aged 60 min	3.610	3.611	3.632	2.616	-----
5	2 +aged 120 min	3.610	3.611	3.636	2.607	-----
6	2 +aged 240 min	3.625	3.621	3.639	2.594	NC(*)
7	2 +aged 390 min	3.620	3.616	-----	2.584	NC(*)
8	2 +aged 630 min	3.620	3.616	-----	2.567	2.876
9	2 +aged 2880 min	3.612	3.613	-----	2.556	2.881

*NC -- Not Calculated

During aging, the lattice constant of ε is decreased with time, and that of MH_x is changed otherwise. The appearence of α' is accompanied by a loss of asymmetry of $(200)_\gamma$, implying the disappearence of MH_x. As shown in Fig. 14, ε is wavy and its surface relief is shallower than that of α'; while α' appears as thin plates in $\{111\}_\gamma$ with definite crystallographic orientation.

The formation of MH_x causes the expansion of the lattice in the surface layer during charging, it will impose a compressive stress in the surface layer. The decomposition of MH during aging causes the shrinkage of the surface layer and it will impose a tensile stress in the surface layer. It is this tensile stress which will bring about the martensitic transformation and the surface cracking. These cracks usually situated at α'/γ interface or appeared in the interior of the martensite (Fig.15)

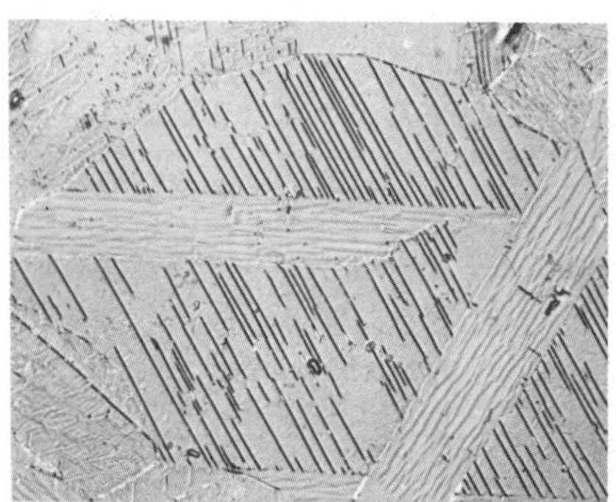

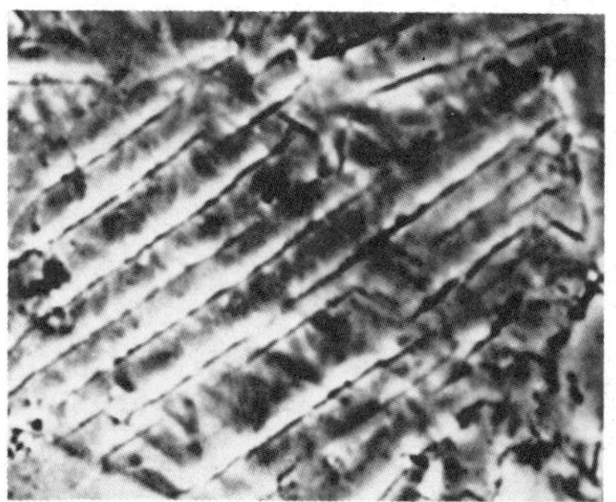

Fig.14 Photomicrograph – Aged   Fig.15 SEM – Aged at Room
    at RT for 24 hours (9)   Temerature for 2 weeks(9)
Fig.14-15   Appearence of the sample of austenitic stainless
    steel after charging at CD of 0.2 a/sq.cm.

IV Hydrogen Induced Precipitation in Commercially Pure Titanium

As reviewed by Paton-Williams(13), the equilibrium hydride phase γ is stable over the composition range from $TiH_{1.53}$ to

TiH$_{1.99}$, and the solubility of γ in αTi is about 20 to 200ppm hydrogen at room temperature. Paton(13) observed on foils prepared from the compact tension specimen after sustained load cracking tests that hydride had been formed ahead of the crack in Ti-4Al-100ppmH alloy. Koch-Bursle-Pugh(14) presented some direct evidence for the formation of fcc hydride phase at the fracture surfaces of the pre-cracked SEN specimens of Ti-8Al-1Mo 1V alloy during slow crack growing process in one atmosphere of hydrogen at room temperature, and their results provided strong support for Nelson's film-rupture model(15). In addition to the equilibrium fcc γ hydride, a strain induced hydride with bcc structure was observed by Boyd(16).

Recently Huang-Wang of Chinese Research Institute for Non-ferrous Metals in collaboration with us, have made in-situ HVEM (1000 KV electron microscope JEM-1000) observation of the kinetic process of the hydride formation ahead of the advancing crack during sustained tension of commercially pure titanium which had been pre-charged electrolytically in 10% HCl solution.

For specimens with about 100ppm hydrogen, "hydrogen cloud or atmosphere" was observed ahead of the crack, and it grew in size with time (Fig.16) and transformed into bcc (Fig.16) or fcc (Fig.17) hydride phase. Selective electron diffraction patterns were used for positive identification of the hydrogen cloud and the hydride phases which would exhibit hcp and bcc (or fcc) structures respectively. The cracks propagated along the hydride phase.

The formation of the hydrogen cloud in the plastic zone ahead the advancing crack is consistent with the previously discussed idea that hydrogen can be enriched by diffusion under the stress gradient, and similiar effect had been reported by Suzuki for vanadium single crystals(8).

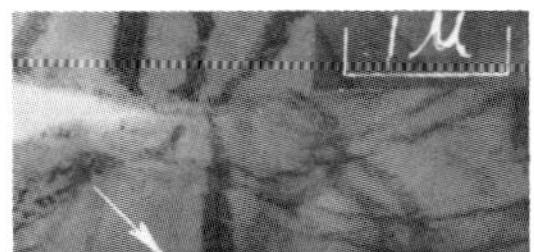

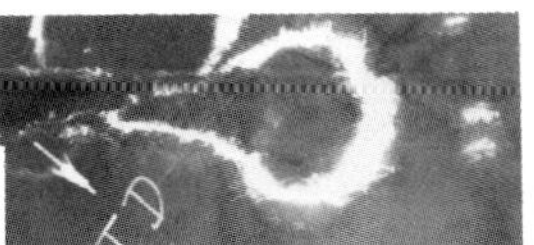

(a) 25 min.   (b) 73 min.   (c) 83 min.
Fig.16 TEM for Ti-H foil under sustained load for various times

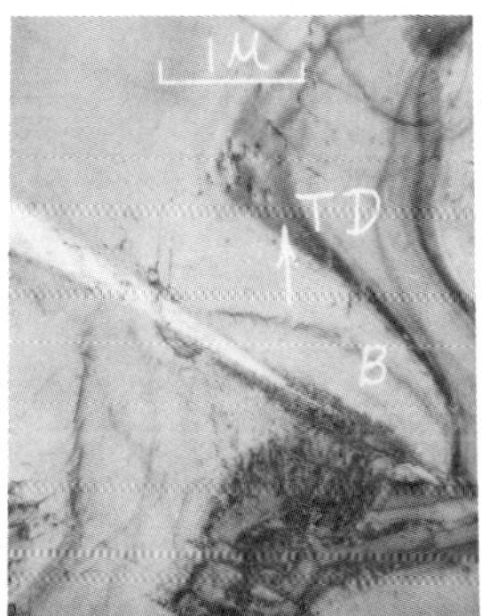

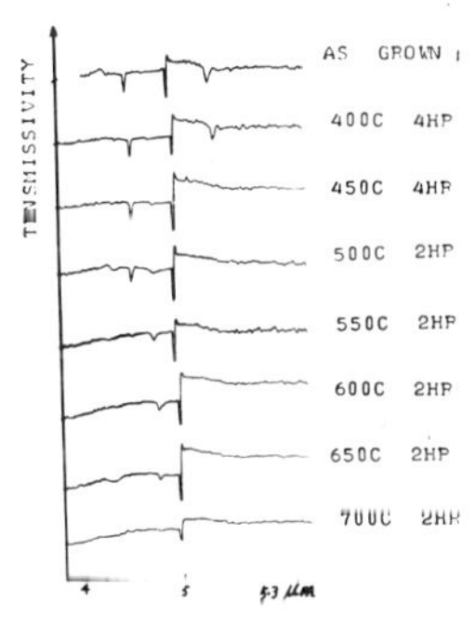

(a) bright field   (b) dark field   Fig.18 Infra-red
   Fig.17 TEM for Ti-H foil under sustained   absorption peaks for
            load   Si single crystal

<u>V Hydrogen Induced Precipitation
in Silicon Single Crystals</u>

The variation of solubility of hydrogen gas (C) and diffusion coefficient of hydrogen atom (D) with temperature in solid silicon crystal were determined by Wieringen-Wasmolts (17) as follows:

$$C = 2.4 \times 10^{21} \exp (-43000/RT) \ (molecules/cm^3) \qquad (2)$$

$$D = 9.4 \times 10^{-3} \exp (-11000/RT) \ (cm^2/sec) \qquad (3)$$

Because of the high heat of solution, C decreases rapidly with decreasing temperature, for instance, C are calculated to be 6.3×10^{15} and 1.7×10^9 molecules/cm^3 at melting point of silicon and 500C respectively. The solubility of hydrogen gas in molten silicon would be much higher than that in solid silicon crystal. Therefore the silicon crystal grown under pure hydrogen gas by floating zone method is a highly supersaturated solid solution of hydrogen in silicon, it would decompose under suitable conditions. The bond energy of Si-H bond is high, amounting to about 70 Kcal/mole, therefore the rate of precipitation of hydrogen will be affected by the decomposition of the Si-H bond and or diffusion of hydrogen.

The results of infrared studies (Fig. 18) indicated that between 4.5 to 5.2 microns, there were several absorption peaks which had been confirmed to be caused by the presence of Si-H bond. This bond is thermally unstable, and will be broken at high temperature. The infrared absorption peaks disappear completely at 700°C, i.e., the Si-H bond is completely broken. This characteristics can be used to explain the precipitation of hydrogen in silicon(18)(19).

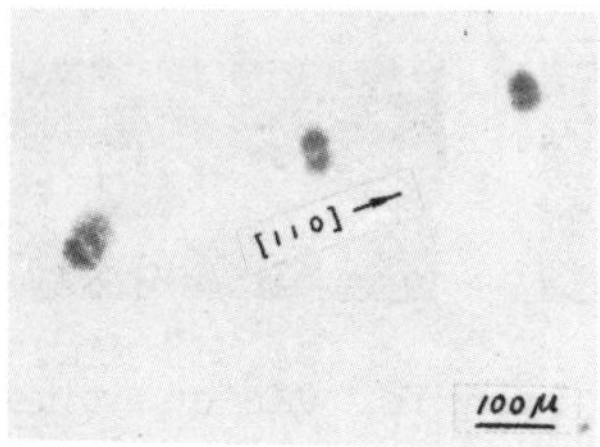

Fig. 19 550C-1hr.
X-ray topograph

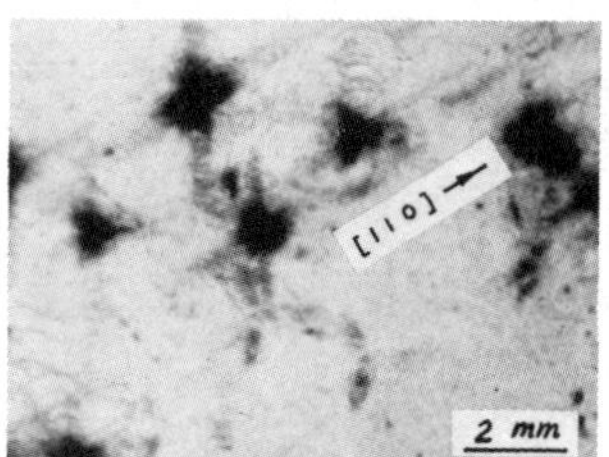

Fig. 20 1000C-1hr.
X-ray topograph

Fig. 21 700C-1hr.
Fractograph by SEM

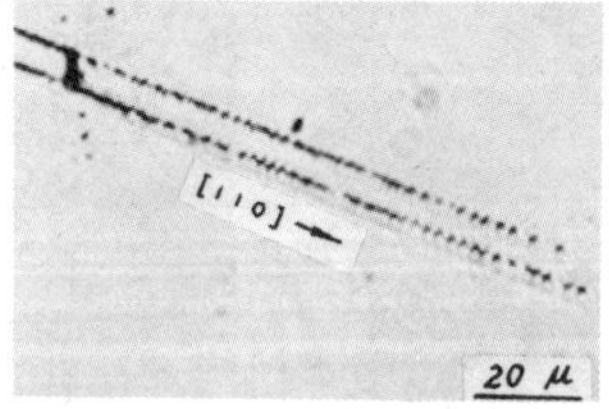

Fig. 22 700C-1hr.
Etch pits near crack

Fig. 19-22 Hydrogen induced precipitation and
defects in silicon single crystals

X-ray diffraction topography method developed by Lang and chemical etching method were used to investigate the defects and precipitates in silicon single crystals (19). The precipitation process at the temperature equal to and below 600C was studied by employing the size of strain region (d) (Fig. 19) as a parameter. The variation of d with time under isothermal condition indicated a constant rate of growth (V). A linear relation between lnV and 1/T was obtained, the activation energy of the precipitation Q was then calculated to be 56 ± 0.3 Kcal/mole (2.4ov). The activation energy of diffusion of hydrogen(17), interstitial silicon atom, vacancy and divacancy(19) were reported as 0.48, 0.51, 0.33, and 1.3 ev respectively, and all these values are much smaller than Q (2.4 ev). It can be concluded that the controlling step of the precipitation process is the breakage of the Si-H bonds.

The high stress field accompanying the large precipitates formed at temperatures equal to and above 700°C will generate prismatic dislocation loops (Fig. 20) and cracks (Fig. 21). From the number of the dislocation loops generated around the crack (Fig.22), Mr. Xiao in our department is attempting to calculate the pressure of the precipitate, which may provide some experimental data for pressure mechanism of hydrogen damage.

VI Concluding Remarks

From the general theory of decomposition of supersaturated solid solution, we may expect different decomposition products, for instance, GPI, GPII, θ' and θ in Al-Cu systems. The same should be true for mateials supersaturated with hydrogen. The enrichment of hydrogen during loading by diffusion under stress gradient in certain regions such as those in front of the crack or notch would greatly increase the supersaturation of the solid solution. Similar to GPI in Al-Cu system, the highly hydrogen supersaturated solid solution would facilitate the formation of platelet hydrogen cluster which pressure would aid to the external load for local yielding. This presumption can be used to explain the HIDP and the significant reduction of the local yield or shear strength; and the existance of the cluster was confirmed in single crystal of vanadium by Suzuki and in polycrystal of titanium by us. Our experimental results for high strength steel tend to support the presumption of Beachem (20), i.e., hydrogen aids plasticity. Other decomposition products such as hydrogen gas and hydrides would cause other hydrogen damage, such as flakes in alloy steel, blisters in aluminum alloys and cracking in titanium, vanadium etc. The decomposition of the unstable hydride NiH_x during aging will bring about the martenitic transformation of austenitic stainless steel from γ to α' and thus the surface cracking.

The strong bond of Si-H and Ti-H would suggest the good potentiality of using these elements in steel for trapping hydrogen. Counting the number of the prismatic dislocation loops generated from the pressure of the minute cracks in single crystals of silicon opens the possibility to calculate the pressure of the precipitate, most probably gas of molecular hydrogen within the crack from the experimental data.

We believe that the existing theories of hydrogen damage are not contradictory but complementary. The pressure from the

hydrogen cluster could be considered to be an alternative of the
pressure mechanism, althought its effect is reversible; the hydrogen cluster could be considered as one of the physical causes for
decoheson in addition to the electronic effect of hydrogen.

By all means we should investigate the <u>changes induced by
hydrogen</u> for particular system and decide which mechanism is
predominant and what practical measures should be taken to combat
the problem of hydrogen damage.

ACKNOWLEDGEMENTS

The authors are grateful to our colleagues, Gao Pei-yu,
Jiang Bai-lin, Xiao Zhi-gang, Tu Huan and also Mr. Huang Xian-
ya and Wang De-ming of Research Institute of Non-ferrous Metals
for discussion and preparation of this paper.

REFERENCES

1 Wu-yang Chu, Chi-mei Hsiao and Shi-qun Li, Scripta Met.,
   1979, Vol.13, pp 1057-62.
2 Wu-yang Chu, Chi-mei Hsiao and Shi-qun Li, Acta Met. Sinica
   1980, Vol.16, p.179.
3 Wu-yang Chu, Chi-mei Hsiao, Shi-qun Li and Cheng Wang,
   Accepted for publication in Acta Met. Sinica, 1980.
4 W. W. Gerberich, J. Garry and J. F. Lessar, in "Effect of
   Hydrogen on Behavior of Materials", Ed. by A. W. Thompson
   and I. M. Bernstein, AIME, 1976, p.70.
5 J. O'M Brockris and H. J. Flitt, in "Hydrogen in Metals",
   JIMIS-2, Japan, 1980, pp 9-16.
6 E. Lunarski etc., Acta Met., 1977, Vol.25, pp 305-8
7 F. E. Fujita, in "Hydrogen in Metals", Paris, 1978, 2B10.
8 T. Suzuki, same as (5), pp 17-24.
9 Peiyu Gao, Acta Met. Sinica, 1979, Vol.15, pp 323-8.
10 J. Saga and S. Miyata, Trans. ISIJ, 1978, Vol.18, p.206.
11 Kazuyoshi Kamachi, ibid, p.485.
12 H. M. Otte, Acta Met., 1962, Vol.10, P.614.
13 N. E. Paton and J. C. Williams, in "Hydrogen in Metals", Ed.
   by I. M. Bernstein and A. W. Thompson, ASM, 1974, pp 409-32.
14 C. H. Koch, A. J. Bursle and E. N. Pugh, same as (7), 3D4.
15 H. G. Nelson, Met. Trans. 1976, Vol. 7A, pp 621-7.
16 J. D. Boyd, Trans. ASM, 1969, Vol. 62, p.977.
17 A. Von Wieringen and N. Wasmolts, Physica, 1962, Vol. 22,
   p.1899.
18 H. J. Stein, J. Electro. Mater., 1975, Vol. 4, p.159.
19 Bai-lin Jiang, Shi-xiong Sheng, Zhi-gang Xiao and Jian-ying
   Bao, Accepted for publication in Acta Physica Sinica, 1980.
20 C. D. Beachem, Met. Trans., 1972, Vol. 3, p.437; also in
   "Hydrogen Damage" Ed. by C. D. Beachem, 1977, ASM, p.xii.

DISCUSSION

R. A. Oriani, University of Minnesota: I cannot conceive how a hydrogen-induced change in the electronic work function can be interpreted to yield information relative to a hydrogen-induced change in the cohesive force. Would Dr. Hsiao explain the reasoning involved, as well as the experimental detail?

S. B. Shanabarger, University of California: I believe you said that hydrogen chemisorbed on iron surface changes the clean surface work function by about 20%. You implied that this was a measure of the change in cohesion energy between iron surfaces due to hydrogen adsorbed along the grain boundary. I am not aware of any direct relation between changes in the work function and decohesion energy. Would you comment on the physical basis for this implied relationship.

M. R. Louthan, Jr., Virginia Polytechnic Institute and State University: I am continually surprised to find various investigations proposing hydrides in austenitic stainless steels. Now you are suggesting that the hydride is NiH_x. How do you propose to segregate enough nickel to cause hydride formation? What evidence do you have for NiH_x? Why do most investigators fail to find the hydride?

C-M. Hsiao: We would like to thank the comments and interests of our paper by Drs. Oriani, Shanabarger and Louthan. Their contributions will certainly clarify certain ideas mentioned in our paper.

(1) The term 'work function' used in our paper is not electronic work function, and is the work required to remove to infinity an iron ion from the surface. The experiments were carried out with Analyseur Ionioue (SMI-300). The specimen was bombarded with primary beam of argon with a spot size of about 200µm in diameter at 5.5kv, the secondary ion was deflected by magnetic prism so that only Fe^+ would strike the ion mirror. Plotting current of Fe^+ (I_R) versus the stopping potential (V_R) and at $I_R=0$, the corresponding V_R is the ionic work function. As shown in Figure 12 of our text, one volt higher is required for hydrogen charged 18-10 stainless steel than that for uncharged 18-10 stainless steel, implying that an iron ion is easier to remove in the presence of hydrogen in the steel. If this reduction of ionic work function of one volt is divided by the heat of sublimation H_S(4.13ev) of iron, a reduction of cohesion energy of iron with hydrogen is about 24%; if divided by the sum of H_S and the first ionization energy of iron (7.83ev), then the reduction is only 8.4%.

(2) Our experimental evidence for the presence of hydride is the asymmetry of (200), diffraction lines, which can be interpreted as the resultant superposition of two lines, one of which is austenite, and the other is hydride. Similar results were reported by Kazuyoshi Kamachi (Iron and Steel Institute, Japan, 18 (1978) 485) who used pseudo Kossel line patterns to identify the hydride phase. Since nickel has a stronger bond with hydrogen than iron, and the hydride is fcc in structure, we tentatively designated it as NiH_x. Certainly, additional work has to be done to differentiate between the so called NiH_x and simply highly hydrogen saturated austenite.

INFLUENCE OF HYDROGEN ON BETA PHASE TITANIUM ALLOYS

N. E. Paton, R. A. Spurling and C. G. Rhodes
Rockwell International Science Center
Thousand Oaks, CA   91360

Hydrogen is well known to be soluble in the beta titanium alloys to
levels in excess of 4000 ppm by weight (approximately 10 atom percent).
However, little information is available on the invluence of hydrogen on
aging kinetics, phase stability and mechanical properties of this class of
titanium alloys. In this study, the influence of up to 4000 ppm hydrogen
on aging kinetics and phase stability was examined by hardness measurements and transmission electron microscopy (TEM). Ti-20V-H alloys were
prepared with zero, 500, 1000, 2000 and 4000 ppm hydrogen and aged at
350°C for up to 28 hours. Hardness measurements indicated that hydrogen
retards the formation of omega phase; TEM showed that this occurs by
decreasing both the volume fraction and the number of particles formed at
a given aging time. The nucleation of omega phase was also retarded by
hydrogen, as evidenced by the omega phase having been observed in the "as
solution treated condition" in the sample with no hydrogen, but first
detected after 40 minutes at 350°C with 1000 ppm hydrogen, and not present
after 28 hours aging with 4000 ppm hydrogen. The delayed nucleation and
growth kinetics were accompanied by a change in particle morphology from
cubodial particles with no hydrogen to ellipsodial at the higher hydrogen
contents. This result indicates that hydrogen in the β lattice alters its
lattice parameter to reduce the misfit of the omega particles. The tolerance of beta titanium to large amounts of hydrogen in solution suggests
that beta titanium alloys may perform satisfactorily in a hydrogen atmosphere. However, tensile results of a beta titanium alloy in 5000 psi
hydrogen will be presented which indicate that though some degree of hydrogen tolerance is present, considerable embrittlement is still observed.

Introduction

The considerable solubility of hydrogen in beta phase Ti alloys makes these systems interesting materials in which to study the effect of H on physical, mechanical and electronic properties of bcc alloys. The relatively large effects of hydrogen on yield strength, elastic modulus, and lattice parameter of metastable beta phase Ti alloys have been the subject of previous investigations (1-4), and studies of these types are simplified because hydride formation does not interfere with the measurements; however, omega phase formation can and will influence the measurements. The primary purpose of the present investigation was to determine the extent to which properties of β titanium are affected by omega phase.

Previous results on a Ti-18 w/o Mo alloy indicated that both the proportional limit and 0.2% offset yield stress decrease linearly in proportion to the square root of hydrogen content (Fig. 1) (4). The modulus was shown to decrease in a similar manner (Fig. 2), and these results were interpreted as evidence for an electronic interaction of hydrogen with the host lattice (4). Although these previous experiments were conducted in a metastable beta titanium alloy in which discrete precipitation of omega phase does not form during quenching, it is well known that certain precursors to the omega phase can form in these alloys (5-6). If hydrogen has an influence on the relative stability of the beta and omega phases, then some of the observed effects of hydrogen on mechanical and physical properties of beta Ti alloys might be marked by this interaction.

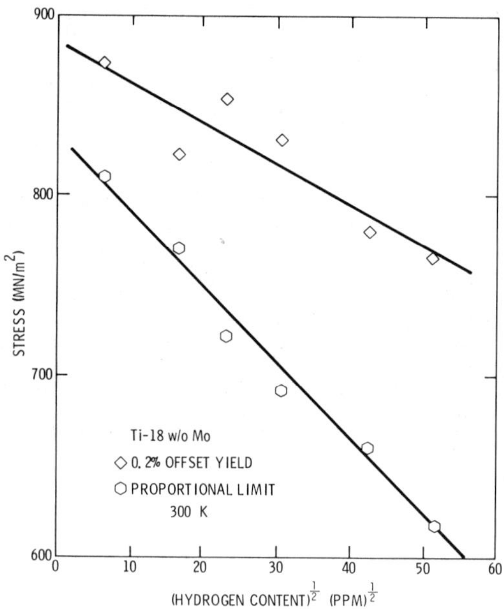

Fig. 1   Proportional limit and 0.2% offset yield stress of Ti-18 w/o Mo alloy at 300 K versus square root of H content.

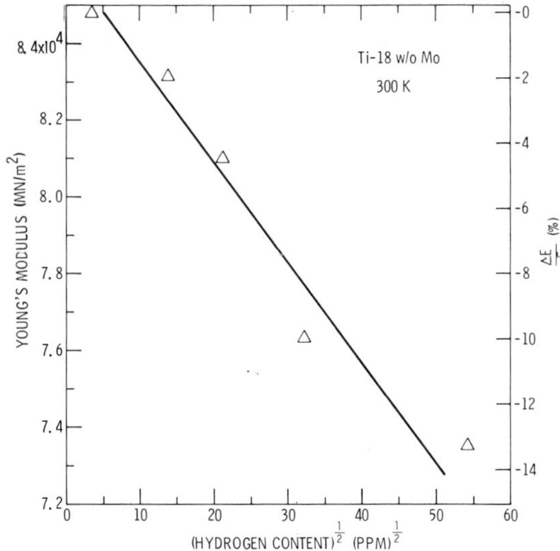

Fig. 2 Young's modulus at 300K versus square root of H content.

The present program sought to address these issues by studying the influence of hydrogen on the aging kinetics of Ti-20 w/o V H alloys. The Ti-20V system was chosen because the kinetics of omega phase formation in the anhydrogenated alloy have been well studied (7). Furthermore, data already exist indicating that hydrogen promotes the formation of omega phase in Ti-20V by raising the ω_s temperature (4), and that oxygen tends to retard the formation of omega phase in Ti-20V alloys (8). The former results were derived from measurements on the effect of temperature on modulus, and the latter were direct observations using electron diffraction methods.

The most unambiguous method for determining the influence of hydrogen on the relative stabilities of the beta and omega phase in these alloys is to study the aging kinetics. A delay in age hardening due to omega phase formation, together with transmission electron microscopy observations on size and volume fraction of omega phase, enables a definitive assessment of this important issue to be made. The most critical point at issue is the possibility that the precursor to the ω phase is giving rise to the high initial as-quenched hardness, and that subsequent decreases in both hardness and modulus could be a result of increasing beta phase stability with increasing H content rather than a result of a bond weakening due to an electronic interaction.

Finally, the abnormal tolerance for hydrogen of beta phase Ti alloys prompted the examination of the influence of high precursor hydrogen on tensile properties of a commercial beta Ti alloy; these results are also reported.

Experimental

The Ti-20V alloy was prepared by arc melting materials containing less than 1200 ppm O+N total. These buttons were swaged to rod solution treated in vacuum at 800°C and rapidly cooled to room temperature by removing the furnace from the vacuum tube and cooling with forced air. X-ray and TEM examination of the material in this condition revealed only the bcc beta phase with no evidence of omega phase.

Hydrogen was introduced into the base alloy in controlled amounts, by heating the specimens in a modified Sieverts apparatus (9) for 30 min at 1073K. The amount of H introduced into each specimen was determined by the pressure change in the system and subsequently verified on selected specimens by vacuum extraction.

Tensile properties of a commercial beta phase Ti alloy Beta III of composition Ti-11.5Mo-6Zr-4.5Sn were determined in both high pressure H and He at 5000 psi.

Hydrogen Effects

Omega Phase

The influence of hydrogen on omega phase nucleation and growth was studied in Ti-20 w/o. V charged with various levels of hydrogen up to 4000 ppm by weight. Following solution treatment the samples were aged at 350°C for times up to 28 hours and examined by hardness measurements, light microscopy and transmission electron microscopy. The hardness results are presented in Fig. 3 where it can be seen that the hardness is reduced by increasing hydrogen content. Furthermore, the rate at which the hardness increases with time also decreases with increasing H content. TEM reveals that the reduction in hardness corresponds to a reduction in the volume fraction of omega phase; for example, Fig. 4 shows a series of electron micrographs illustrating the distribution of omega phase in samples aged for 28 hours having four different levels of hydrogen. Not only does the volume fraction of omega phase decrease with increasing hydrogen content, but the number of particles is also significantly reduced with increasing hydrogen content. Nucleation of omega phase is retarded by the presence of hydrogen as evidenced by the observation that omega phase is observed in the "as solution treated" condition in the sample with no hydrogen, is first detected after 40 minutes of aging in the sample with 1000 ppm hydrogen, and is not present after 28 hours of aging in the sample with 2000 ppm hydrogen. Thus hydrogen clearly stabilizes the beta phase and retards formation of the omega phase.

Another interesting effect of hydrogen is its influence on omega particle morphology. In the alloy with no hydrogen, omega forms cuboidal particles, but when 500 ppm hydrogen is added, the particles are ellipsoidal, as illustrated in Fig. 5. It is well known that omega phase morphology is a function of the misfit between the parent β-phase lattice and the omega phase lattice. The result shown in Fig. 5 would indicate that the presence of hydrogen in β alters its lattice parameter toward that of omega phase. Precision lattice parameter measurements, which were beyond the scope of this program, would assist in resolving this point. Finally, it was also observed that the omega particles, which form initially as ellipsoidal particles, eventually grow into a more nearly cuboidal form, c.f. Figs. 5b and 4b. This effect is not the result of a loss of hydrogen.

In recognition of the fact that beta titanium alloys appear to have an exceptional tolerance for high levels of hydrogen in solution, it was

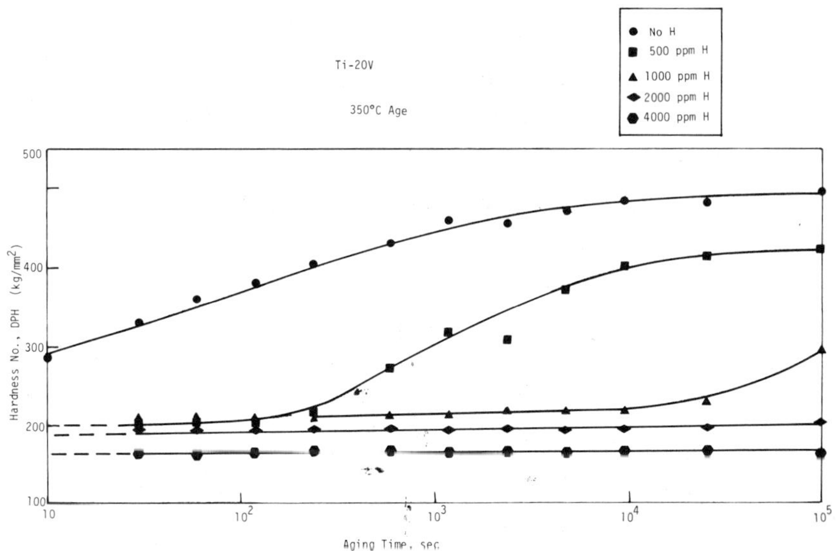

Fig. 3   Microhardness of Ti-20V, as a function of aging time at
350° for various levels of hydrogen.

decided that the tensile properties of a commercial titanium alloy tested
in high pressure hydrogen would be worthy of investigation. The tensile
properties of beta titanium in high pressure hydrogen were investigated in
a commercial alloy Beta III. Round tensile bars of a commercially
available Beta III titanium alloy, the composition of which was Ti-11.5Mo-
6Zr-4.5Sn, were tested in various heat treatment conditions in an atmosphere of 5,000 psi hydrogen and, for comparison, 5,000 psi helium. The
results of these tests are shown in Table I, and it can be seen that the
high pressure hydrogen environment caused a definite degradation in the
ductility of this titanium alloy. The degradation in ductility was
particularly evident in the results of the fractography performed on the
fracture specimens of the failed tensile samples. An example of this
fractographic evidence is shown in Fig. 6, where the classical ductile-
dimple rupture found in the high pressure helium environment is shown in
Fig. 6a, contrasted with the very brittle cleavage-like fracture found in
high pressure hydrogen environments shown in Fig. 6b. These results show
that, although beta titanium alloys can dissolve large amounts of hydrogen
without severely damaging tensile properties, high pressure hydrogen
environments apparently introduce sufficiently large quantities of
hydrogen into solution to cause severe degradation in ductility. The
possibility exists that residual properties in H may still be adequate for
certain applications; however, time dependent properties such as sustained
load cracking would also have to be evaluated prior to serious
consideration of these materials for such service.

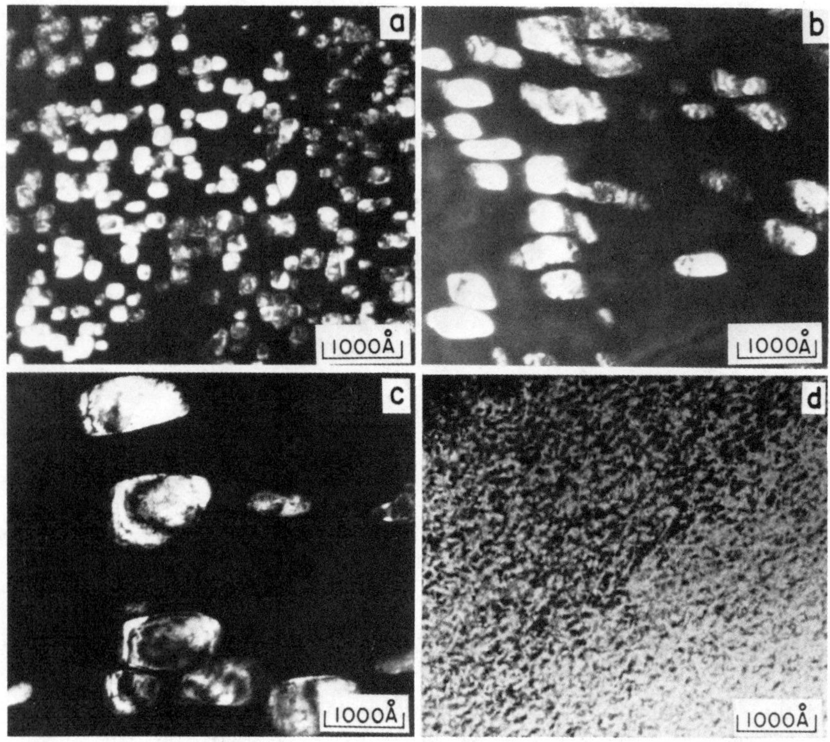

Fig. 4   Dark field transmission electron micrographs showing omega phase particles in Ti-20V with various levels of hydrogen aged at 350°C for 28 hours. (a) no H; (b) 500 ppm H; (c) 1000 ppm H; (d) 2000 ppm H.

Table I. Tensile Properties of Beta III in High Pressure Helium and Hydrogen

Heat Treatment	Gas	Test Pressure psia	Yield Strength ksi	Tensile Strength ki	R.A. %	Elongation %
ST 745°C/5 min/W.Q.	H₂	5000	78.8	97.5	13.1	4.0
	He	5000	79.8	113.7	63.4	22.0
ST 745°C/5 min/W.Q.	H₂	5000	143.4	154.0	6.7	4.0
+ Age 540°C/8 hrs	He	5000	143.1	152.3	16.6	7.2
ST 815°C/5 min/W.Q.	H₂	5000	80.5	96.8	6.3	3.8
	He	5000	85.4	110.7	72.0	31.6
ST 815°C/5 min/W.Q.	H₂	5000	143.4	150.4	7.1	6.6
+ Age 540°C/8 hrs	He	5000	137.6	147.2	13.8	8.8

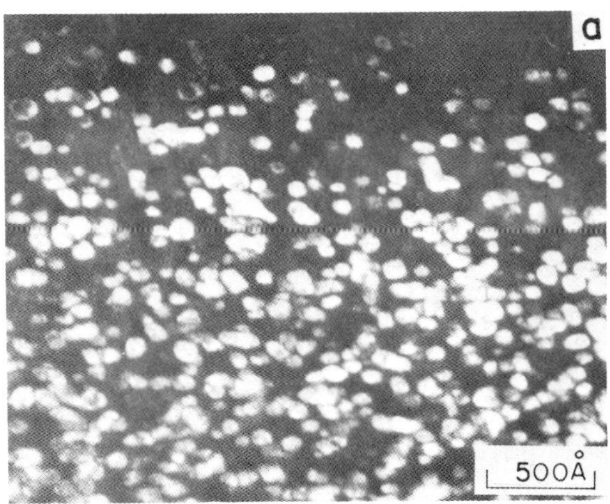

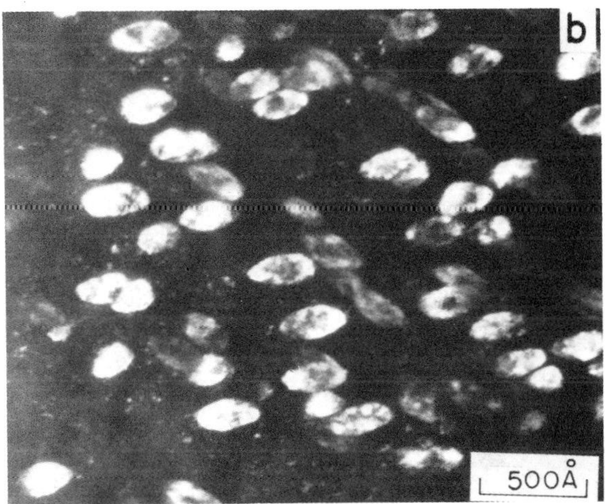

Fig. 5 Dark field transmission electron micrographs showing omega phase
particle morphologies in Ti-20V aged for 20 minutes at 350°C with
different levels of hydrogen. (a) no H; (b) 500 ppm H.

The finding that high pressure hydrogen seriously degrades ductility
of beta titanium, indicates that the tolerance of these alloys for
hydrogen is distinctly limited. Earlier results on Ti-Mo alloys also
showed that at very high hydrogen levels, ductility was severely limited
(13). This present result would indicate that high levels of hydrogen
are, in fact, reached in room temperature high pressure hydrogen tensile
tests, probably by dislocation transport of hydrogen into the interior of

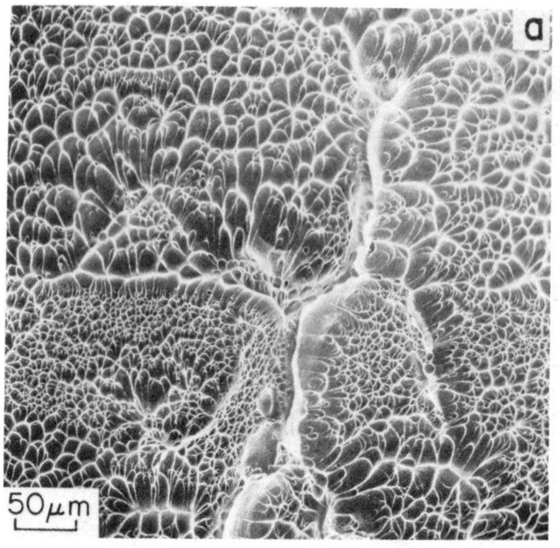

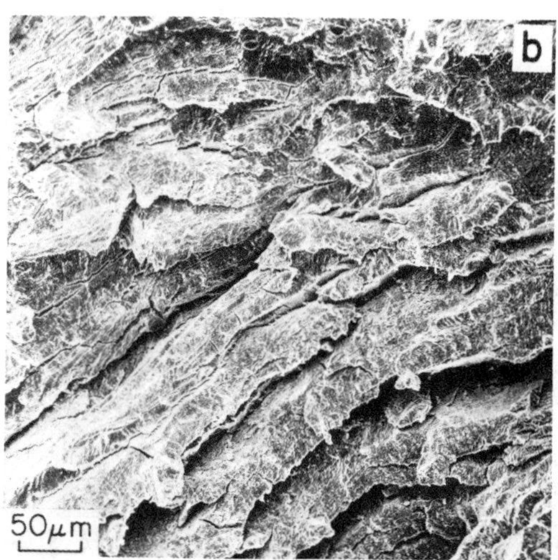

Fig. 6   Scanning electron micrographs of fracture surfaces of β-III
titanium alloy solution treated at 1375°F and tensile tested in
(a) 5000 psi He, (b) 5000 psi H_2.

the specimen as suggested by previous workers (14). These tests indicate
that beta titanium alloys are probably not useful as containment vessels
for high pressure hydrogen or for service in high pressure hydrogen
environment.

An attempt was made to establish the solubility limit of hydrogen in
a beta stabilized alloy. Ti-40 w/o. V was selected for this study and hydrogen was added by the Seiverts method at 800°C. Hydrogen levels of from

7000 to 19,782 ppm by weight were examined. The presence of hydrides was monitored by X-ray diffraction and by TEM.

No evidence of hydrides was seen in the X-ray diffraction patterns for all hydrogen levels. TEM revealed only beta phase in samples with hydrogen up to 12,884 ppm. The samples with 16,413 and 19,782 ppm H showed unusual linear features which dominated the microstructure and which gave a face centered cubic diffraction pattern with a lattice parameter of 4.5 Å, slightly higher than that of the hydride phase (4.3 Å) (10). The structure had the appearance of thinning artifacts, designated spontaneous relaxation (12), common to beta stabilized titanium alloys. However, since hydrogen has been shown to be a beta stabilizer, it is unlikely that this feature is the same as that previously identified as an artifact which generally occurs only in unstable alloys. The X-ray diffraction patterns from this alloy exhibited only a bcc structure, not fcc. Although precision lattice parameter measurements were not made, it could be estimated from the X-ray patterns that the lattice parameter increased from about 3.18 Å to about 3.42 Å as the hydrogen level increased from zero to 16,413 ppm.

Discussion

This study has confirmed that hydrogen suppresses omega phase formation. The hardness curves in Fig. 3 indicate that nucleation of omega phase is retarded by hydrogen, and the TEM results support this observation. In this respect, hydrogen acts in a manner similar to other beta stabilizers, even though hydrogen is an interstitial, rather than a substitutional, alloying addition.

The suppression of omega phase formation by H would tend to support the hypothesis that the decrease in strength and modulus of beta Ti alloys with H may at least in part be due to the reduced volume fraction of omega. The omega phase is well known to be a potent strengthening precipitate (7) and even small decreases in volume fraction would be expected to decrease both strength and modulus. Available data on the influence of increasing vanadium concentration in Ti-V alloys (15) indicated that solid solution softening occurs in the system; i.e., increasing V concentration increases the stability of the beta phase, reduces the volume fraction of omega phase (or incipient omega) and thus decreases strength. The hypothesis here is that H has the same effect.

The effect of hydrogen on omega phase particle morphology is similar to that observed for Zr additions in which it was shown that increasing the Zr content of a Ti-10V alloy altered the omega morphology from cuboidal to ellipsoidal (11). In that case, it was shown Zr expanded the beta phase lattice parameter, thereby reducing misfit between the beta and omega phases. Hydrogen is known to increase the beta lattice parameter (4).

Reversion of the omega morphology in the hydrogen bearing alloys from ellipsodial to cuboidal with aging time is somewhat curious. As noted earlier, this is not the result of a loss of hydrogen during the aging treatments. It is possible that 500 ppm hydrogen is sufficient to expand the beta lattice parameter so that near zero misfit exists between the omega and beta phases. However, as the volume fraction of omega phase increases with continuing aging (and rejects hydrogen), the beta matrix becomes further enriched in hydrogen and the lattice parameter expands such that the misfit is negative, using Hickman's designation (7).

Previous work (3) showed that hydrogen additions to the Ti-Mo system significantly increased the beta phase lattice parameter and extrapolation of that data to the current Ti-V system indicates the above speculation is reasonable.

Positive identification of hydrides in Ti-40V containing up to 19,789 ppm hydrogen was not obtained. However, features with the appropriate crystal structure but a slight discrepancy in lattice parameter from the hydride were observed at the highest hydrogen levels. The observed increase in lattice parameter of the beta phase from 3.18 Å to 3.42 Å represents a 7% increase as the hydrogen level increased from zero to ~ 16,000 ppm.

Conclusions

1. Hydrogen supresses omega phase formation in Ti-20V and retards the kinetics of omega precipitation on aging.

2. At least a part of the decrease in strength and modulus of beta titanium alloys containing H can be attributed to an increased stability of the beta phase and decreased volume fraction of omega.

3. Hydrogen is soluble in Ti-20V up to at least 12,884 ppm. Above 16,413 ppm, features which could have been hydrides were observed but were not positively identified as such.

4. The commercial beta titanium alloy, Beta III, suffers considerable but not total ductility loss in 5000 psi hydrogen as compared to 5000 psi helium.

Acknowledgement

The authors are grateful to Drs. O. Buck and B. MacDonald for participating in several stimulating discussions concerning interpretation of the results. Dr. R. Jewett assisted in conducting the tensile test in high pressure H and He.

This research was supported by the Office of Naval Research under Contract No. N00014-76-C09598.

References

1. A. W. Sommer, S. Motokura, K. Ono, and O. Buck, Acta Met. <u>21</u>, 489 (1973).

2. N. E. Paton, O. Buck, and J. C. Williams, Scripta Met. <u>9</u>, 687 (1975).

3. O. Buck, D. O. Thompson, N. E. Paton, and J. C. Williams, in <u>Internal Friction and Ultrasonic Attenuation in Crystalline Solids</u>, <u>1</u>, D. Lenz and K. Lucke, Eds., Springer Verlag, Berlin, P. 451 (1975).

4. N. E. Paton and O. Buck, in <u>Effects of Hydrogen on Behavior of Materials</u>, (A. W. Thompson and I. M. Bernstein, Eds.), AIME, New York, p. 83 (1976).

5. D. deFontaine, N. E. Paton, and J. C. Williams, Acta Met. <u>19</u>, 1153 (1971).

6. D. deFontaine and O. Buck, Phil. Mag. <u>27</u>, pp. 967-983, 1973.

7. B. S. Hickman, J. Mat. Sci. <u>4</u>, pp. 554-63, 1969.

8. N. E. Paton and J. C. Williams, Scripta Met. (1973) <u>7</u>, pp. 647-650, (D. Lenz and K. Luke, Eds., Springer-Verlag, Berlin pp. 451-58, 1974).

9. N. E. Paton, B. S. Hickman and D. H. Leslie, Met. Trans. <u>2</u>, pp. 2791-96, 1971.

10. T. S. Liu and M. A. Steinberg: Trans ASM <u>50</u>, p. 455, 1958.

11. J. C. Williams, B. S. Hickman and D. H. Leslie, Met. Trans. <u>2A</u>, pp. 477-84, 1971.

12. R. A. Spurling, C. G. Rhodes and J. C. Williams, Met. Trans. <u>5A</u> pp. 2597-600, 1975.

13. N. E. Paton and O. Buck, <u>Effect of Hydrogen on Behavior of Materials</u>, A. W. Thompson and I. M. Bernstein, Eds., AIME, New York, p. 83, 1976.

14. J. Tein, A. W. Thompson, I. M. Bernstein and R. J. Richards, Met. Trans., <u>7A</u>, pp. 821-29, 1976.

15. N. E. Paton and J. C. Williams, Proceedings of the Second International Conference on Strength of Metals and Alloys, Asilomar pp. 108, 1970.

DISCUSSION

<u>H. K. Birnbaum, University of Illinois</u>: Have you looked for hydride at the fracture surface of the β alloys tested in high pressure hydrogen?

<u>N. E. Paton</u>: We have not looked for hydrides on the fracture surface of the beta alloys tested in high pressure hydrpgen.

THE EFFECT OF HYDROGEN ON MAGNETIC AFTEREFFECT AND COLD-WORK RELAXATION IN

α-IRON

T.S. Kê[*], H. Kronmüller, A. Seeger, Z.Q. Sun[*]

Max-Planck-Institut für Metallforschung, Institut für Physik,
Heisenbergstr.1, 7000 Stuttgart 80, Postfach 800665, Fed.Rep.Germany

[*])Permanent address: Institute of Metal Research, Academia Sinica,
Shenyang, China

Abstract. The magnetic aftereffect and the internal friction of α-Fe after
cold-work and/or charging with hydrogen are reviewed and related to the in-
ternal-friction peaks introduced into Group-V and Group-VI transition metals
by cold-work. Kink-pair formation on non-screw dislocations (α-process) is
responsible for a relaxation process at ~20K in α-Fe and recently established low-temperature internal friction peaks in Nb and V. These peaks are
different from Chambers' α-peak, which has now been identified to be a hydro
-gen cold-work peak. The hydrogen cold-work relaxation in α-Fe consists of a
lower part (at~100K) due to kink-pair formation on non-screw dislocations in
the presence of mobile hydrogen atoms and a Debye-type upper part (~150K)of
not yet definitely established origin.

Introduction

Relaxation processes associated with hydrogen in α-Fe have been investi
-gated by many authors by means of internal friction (IF) and magnetic after
-effect measurements (MA).The analysis of the experimental results,however,
remained in an unsatisfactory state since the relaxation spectra were extremely sensitive to slight changes in the preparation and the impurity content
of the specimens.Only recently, after pure iron specimens with impurity contents <10 ppm became available,could significant progress in the understanding of the relaxation spectra be achieved.It has become evident that the
former difficulties in analyzing the spectra were due to the superposition
of intrinsic dislocation relaxations and of so-called cold-work peaks (CWP)
caused by hydrogen—dislocation interactions.

The problem of understanding the interaction of hydrogen with dislocations in α-iron should be seen in the wider context of the bcc transition
metals, in particular of V,Nb,Ta (Group-V) and of Mo,W (Group-VI). In the
Group-V transition metals hydrogen dissolves readily, and it is quite difficult to degas these metals to such an extent that even the hydrogen trapped
at dislocations is removed. In Mo and W the hydrogen solubility is low, but
little is known about the diffusivity of H at low temperatures, and well-
planned experiments on hydrogen—dislocation interactions are therefore diffi
-cult to perform. α-Fe combines low solubility (<1 atppm below room temperature) with a high and reasonably well studied diffusivity of hydrogen. From
the experimental view-point it offers the advantage that in addition to mechanical tests (internal friction, elastic modulus, and mechanical aftereffects) rather sensitive magnetic measuring techniques may be used, in

particular the magnetic aftereffect of the initial susceptibility. More about the latter will be said below.

In α-iron the relaxation processes relevant for the present topic occur in the temperature range up to about 400 K and will be classified by a mixture of the phenomenological classification of Chambers [1] and the interpretational scheme given by Seeger and Wüthrich [2]. We shall use the following notation (the temperature ranges refer to relaxation times between about 10^{-1}s and 10^{2}s i.e., to torsion pendulum and aftereffect experiments):

(i) 10-30 K: <u>α-peak</u>, formation of kink-pairs on non-screw(more specifically: 71°-)dislocations.

(ii) 70-150 K: <u>hydrogen cold-work peaks</u>, due to dislocation–hydrogen interactions.

(iii) 250-400 K: <u>γ-peaks</u>, formation of kink-pairs on $\langle 111 \rangle/2$ screw dislocations.

A key point for the understanding of the relaxation spectrum in α-Fe is the correct interpretation of the low-temperature relaxation (i). Only fairly recently have MA [3-5] and IF [6-8] measurements demonstrated it to be due to an intrinsic process involving dislocations, thus refuting earlier interpretations [9,10] in terms of a Snoek effect of hydrogen on interstitial sites of tetragonal symmetry.

The hydrogen cold-work peaks (ii) in the temperature range between 70K and 200 K have been studied by many authors, but not until recently could the preparation and measuring techniques be advanced to the extent that new features in the structure of this relaxation could be analyzed.

The extraordinary properties of the relaxation spectra of cold-worked hydrogen-charged α-iron are due to the extremely small activation energy of diffusion (50-100 meV) [11-13] as well as to the rather small solubility of hydrogen in α-iron at and below room temperatures (RT) (<1 ppm below RT).

<u>Measuring Techniques</u>

In ferromagnetic α-Fe CWP phenomena may be investigated by means of MA as well as IF. In both cases the observed relaxation is due to dislocation motion under the influence of external stresses (IF) or the magnetostrictive stresses of the domain walls (MA). The magnetic relaxations are due to changes in the domain-wall mobility caused by dislocation rearrangements within domain walls [14]. These changes can be measured with high accuracy by means of an LC-oscillator technique [15]as a time-dependent decrease of the initial susceptibility, χ. It is possible to study dislocation movements of as little as one Burgers vector.

From a theoretical point of view it is appropriate to consider the magnetic reluctivity

$$r(t;T) = 1/\chi(t;T) , \qquad (1)$$

since $r(t;T)$ is proportional to the curvature of the domain-wall potential. Experimentally one obtains <u>isothermal</u> relaxation curves for different temperatures T. From these <u>isochrones</u> of the relaxation amplitude Δr, measured in the time interval between t_1 and t_2, may be derived:

$$\Delta r(t_1,t_2;T) = r(t_2;T)-r(t_1;T) . \qquad (2)$$

If the relaxation process results from a thermally activated process characterized, say, by a relaxation time

$$\tau = \tau_\infty \exp(H/kT) \qquad (3)$$

with activation enthalpy H and frequency factor τ_∞^{-1}, the relaxation amplitude shows a maximum as a function of temperature. Shape and position
of the relaxation peak depend on the relaxation function r(t) and the parameters characterizing τ. Fig.1 shows schematically the derivation of relaxation isochrones from isothermal relaxation curves.

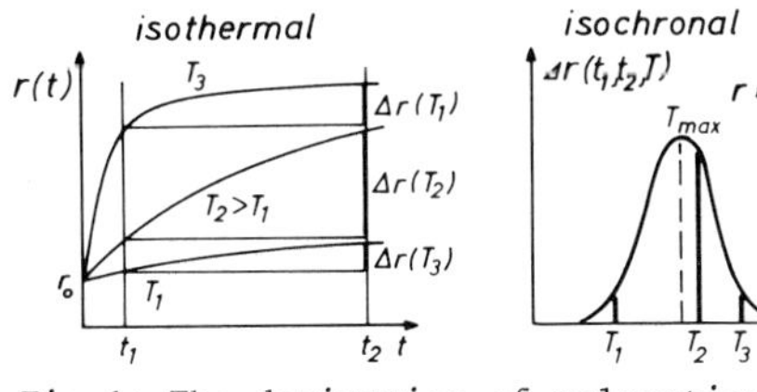

Fig.1: The derivation of relaxation
isochrones from isothermal relaxation
curves of the magnetic aftereffect.

With decreasing t_2 the maximum
shifts to larger temperatures and
the halfwidth of the maximum decreases. If the time dependence of the
relaxation process is known (e.g.,
from isothermal relaxation curves)an
analysis of the isochrones gives
numerical values for the activation
energy and the frequency factor. In
the case of MA a typical time interval lies between t_1=1s and t_2=180s.
This means that relaxation processes
with relaxation times of about 1 minute can be detected. MA peaks correspond
to a measuring frequency f≈0.01 Hz and are thus shifted to lower temperatures compared with IF measurements at a frequency f≳1 Hz. In principle MA
measurements give the same information as IF; in practice, however, both
their sensitivity and their resolving power exceed that of IF thanks to the
high sensitivity of the LC-oscillator technique and the smaller background
of MA measurements.

Magnetic Aftereffect Spectra

MA measurements on cold-worked and hydrogen-charged α-Fe crystals
[3-5,16] have revealed a series of relaxation peaks in the temperature range
between 4K and 200 K. Their interrelations depend on the hydrogen content
and the plastic deformation. Systematic variation of hydrogen charging and
plastic deformation is thus required.

Relationships between the α-peak and the hydrogen cold-work peak

The MA aftereffect spectra of an α-Fe single crystal after different
pre-treatments shown in Fig.2 reveal the following features:
a) After <u>tensile deformation at RT</u> a relaxation peak is observed at 18K.This
relaxation peak is rather broad and the extrapolation of the relaxation amplitude to 0 K does not vanish. The parameters of this relaxation peak are
H=0.048 eV and τ_∞=1.6x10⁻¹¹s [3]. Similar results were derived by Dufresne
et al [7] and Hivert et al [17]from IF measurements. We identify this peak
with the α-peak (i). b) After <u>electrolytic hydrogen charging</u> the peak at 18K
is completely suppressed and a hydrogen CWP appears at 130K. It shows a
steep low-temperature and an extended high-temperature flank. c) <u>Annealing
at RT</u> shifts the hydrogen CWP to lower temperatures. The α-peak reappears,
but shifted to lower temperatures. After annealing for 100 h at RT under a
pressure of 10⁻⁶ Torr the hydrogen CWP has completely disappeared and the
α-peak has shifted again to somewhat higher temperatures, thus approaching
its original position in the uncharged specimen.

Substructure of the hydrogen CWP and its dependence on hydrogen content

In the hydrogen CWP shown in Fig.2 no fine structure is detectable. The
extended high-temperature flank, however, suggests a possible contribution
of another relaxation process. In order to obtain more detailed information,

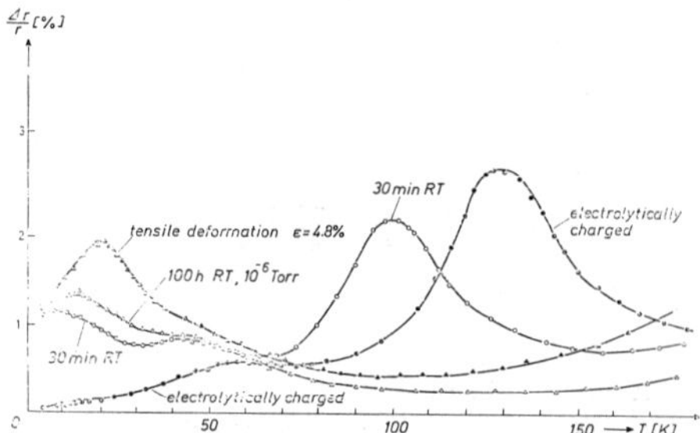

Fig.2: Magnetic aftereffect spectra of plastically deformed and hydrogen-charged α-Fe single crystals[3](▲ tensile deformation,ε=4.8% at RT;● electrolytically charged; O annealed 30 min at RT;△ annealed 100 h at 10^{-6} Torr).

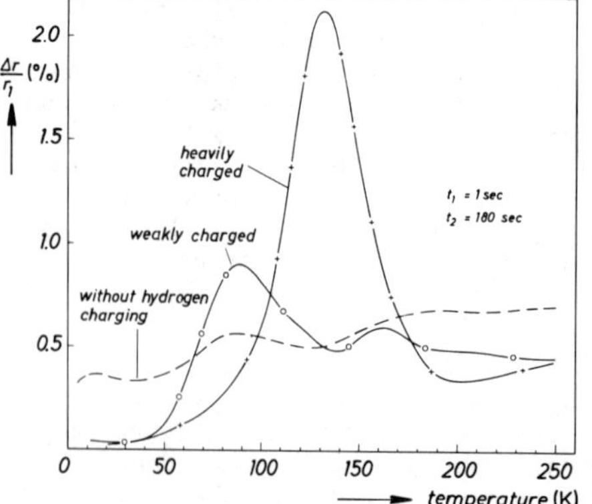

Fig.3: Magnetic aftereffect spectra of polycrystalline α-Fe after tensile deformation(ε=1.6%,RT;then ε=0.4%,77K) for weak and heavy hydrogen charging.

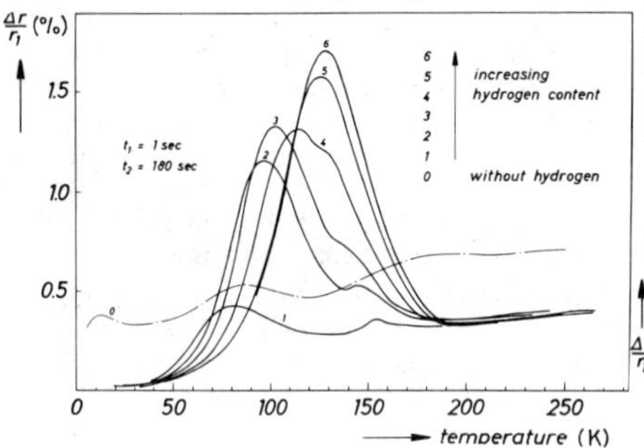

Fig.4: Magnetic aftereffect spectra of polycrystalline α-Fe after tensile deformation (ε=1.6%,RT)for increasing hydrogen concentration.

Fig.5(right hand):Magnetic aftereffect spectra of hydrogen-charged Fe-V alloys.

measurements were performed on α-Fe crystals with varying H concentrations [4,16].Fig.3 shows that the concentration of hydrogen does have a drastic effect on the structure of the hydrogen CWP. In weakly hydrogen-charged specimens two clearly separated relaxation peaks are observed at 80K and 155K, whereas in highly charged specimens only one broad peak appears at 130K.

Fig.4 shows the transition from a two-peak spectrum to the well-known hydrogen CWP for increasing hydrogen content. With increasing hydrogen concentration the lower peak shifts rapidly towards higher temperatures (from 80K to 130K), whereas the upper peak moves slowly towards lower temperatures (from 150K to 140K). Finally,in heavily charged specimens both peaks superimpose to one broad hydrogen CWP.As shown in Fig.5 for the example of Fe -V alloys, the 155K peak was also detected in dilute alloys of α-Fe with Co,Cr and V [16].

<u>Effect of deformation temperature [16]</u>

In Fig.6 the dependence of the amplitude of the hydrogen CWP on deformation temperature and degree of deformation is shown for α-Fe single crystals of the same orientation.The experimental sequence was as follows: Plastic deformation,electrolytic charging with hydrogen,MA measurements. Before the subsequent deformation steps the specimens were degassed at RT under vacuum (10^{-5}Torr)for at least 24 h.

For small tensile deformations, ε<2%, the relaxation amplitude of the

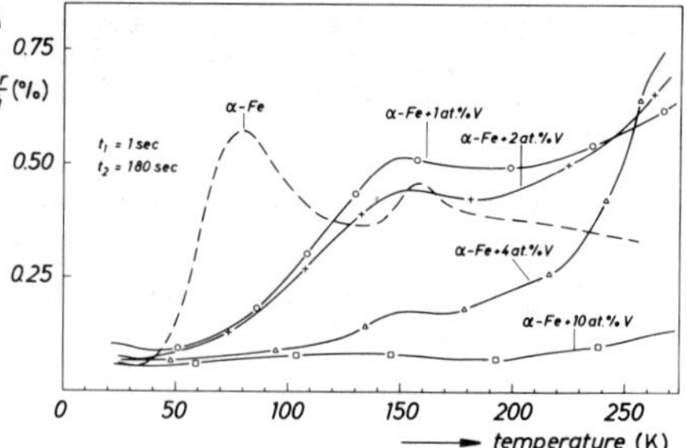

hydrogen CWP increases, up to a factor of two after low-temperature deformation, up to 30% after RT deformation. For deformations ε>2% the relaxation amplitude decreases and approaches the same value for both deformation temperatures.

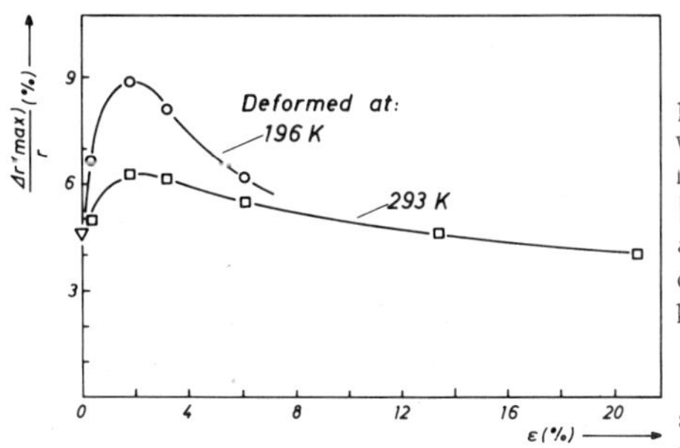

Fig.6: Dependence of the amplitude of the hydrogen CWP as a function of the degree of deformation for two different deformation temperatures.

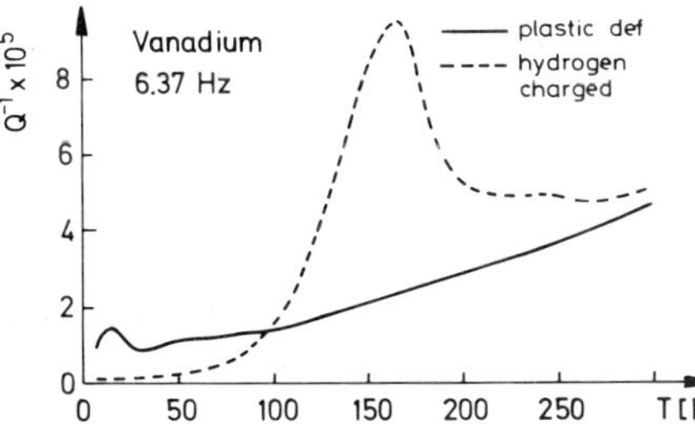

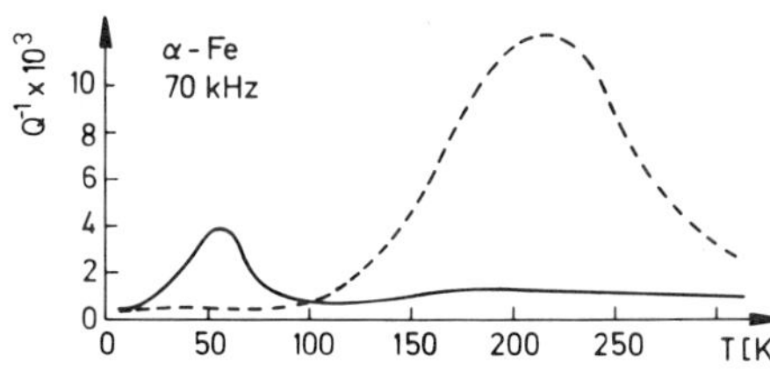

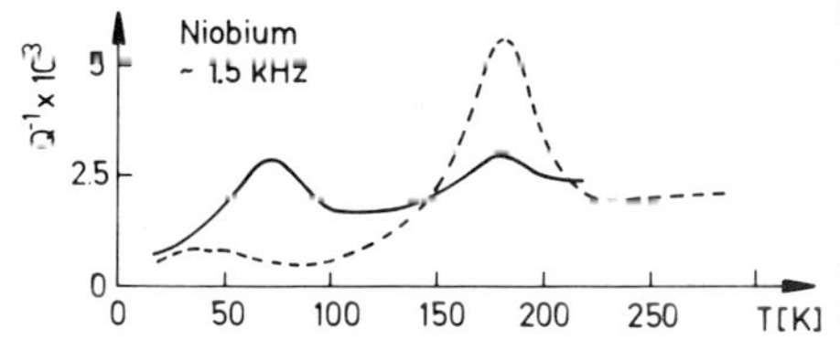

Fig.7: Comparison of IF-spectra of plastically deformed and deformed and hydrogen-charged transition metals.Vanadium[24](4.6% bent,RT,150ppm H);α-Fe [6](4.6% bent,RT,150ppm H);niobium single-crystal [25](~2%bent at 20K,1-5 ppm H; 1week UHV-degassed).

Internal-friction measurements

IF experiments on α-Fe after plastic deformation and/or charging with hydrogen have been performed by many authors. Gibala [10] and Heller [9] found an IF peak at 30K (f=1Hz) and 48K (f=80Hz),respectively. Later on this peak was also observed by Takita et al. [6,18], Dufresne et al. [7], Hivert et al.[17], Ritchie et al [8,19], and Astié et al.[20]. Fig.7 shows that hydrogen charging of a plastically deformed α-Fe specimen causes the low-temperature IF peak to disappear and that it creates the IF hydrogen CWP at higher temperatures [6]. These results are in good agreement with, though less detailed than, the MA measurements.

Chambers and collaborators[1,21] gave the name α-peak to an IF relaxation peak observed in plastically deformed Nb at about 130 K for f≈10Hz. Subsequently Mazzolai and collaborators observed an IF peak at low temperatures both in Nb [22] and V [23] which they called δ-peak and which they attributed to dislocation relaxation processes. Fig.7 shows that in analogy to α-Fe the low-temperature IF peak present in V and Nb after plastic deformation is suppressed by hydrogen charging and that a hydrogen cold-work peak appears at higher temperatures.(Note that the data of Fig. 7 were taken at widely different measurement frequencies so that in a comparison amongst the different IF measurements as well as between IF and MA data the appropriate shifts of the peak temperatures must be taken into account).

If we wish the common behaviour of α-Fe and of the Group-V transition metals to be reflected in a common terminology, we have to change either the assignments used in this and earlier papers [2-5] for α-Fe or the nomenclature proposed by Chambers for the Group-V metals[1].We choose to retain the α-Fe nomenclature since in the meantime it has become customary to attribute the names α- and γ-peak to relaxation phenomena associated with definite <u>dislocation mechanisms</u> as discussed in the Introduction.

Discussion

The interpretation of relaxation peaks in hydrogen-charged and/or cold-worked bcc metals has been controversal for many years.Recently it has become evident [26] that the confusing experimental results were largely due to the peculiar rôle of hydrogen in bcc metals.Nb,V, and Ta possess a large hydrogen solubility and are therefore rather difficult to degas,in particular if hydrogen atoms are trapped by dislocations.This has the consequence that in these materials the relaxation process associated with the formation of kink-pairs on non-screw dislocations is difficult to detect.According to Seeger [27] the formation of kink-pairs on dislocations lying along crystallographic directions with high Peierls energies is responsible for the Bordoni relaxation [28] observed at low temperatures in many fcc metals(for a recent review see [29].In bcc metals the analoguous process in <u>non-screw</u> dis -locations is expected to occur in approximately the same temperature range, i.e. under conditions under which hydrogen interacts strongly with dislocations. It is therefore not surprising that in the Group-V materials the intrinsic α-process could not be clearly identified for a long time and that the hydrogen CWP was erroneously attributed by many workers to the formation of kink-pairs on non-screw dislocations.

In α-iron the difficulties just mentioned are considerably reduced on account of the low solubility of hydrogen. It was in this material that a definitive assignment of a low-temperature peak to the α-process could first be given [3]. Whether the δ-peak observed in Nb and V [22,23] is due to the α-process is not easy to say since the material used contained a high concentration of oxygen and nitrogen, so that relaxation processes associated with defect complexes involving these impurities are feasible. On the other hand, the δ-peak temperatures are compatible with the α-peak and the high O and N concentrations may have helped in keeping the dislocation—hydrogen interactions small.

In Mo and V the hydrogen solubility is small, too. It appears that in Mo the α-peak has been observed at about 100K (f=19Hz)[26]. It has a complicated substructure which needs further elucidation.

Because of the particular structure of <111>/2 screw dislocations, in bcc metals the relaxation process due to kink-pair formation on screw dislocations (γ-process) is expected to occur at much higher temperatures than the α-process [2,30],so that dislocation—hydrogen interactions should be less important. The identification of this process [2,30] with the γ-peaks observed by Chambers and co-workers [1,21] is thought to be still valid and is in agreement with the current ideas on the flow stress and the mechanism of plastic deformation of bcc metals (for a review see[31].Recent work [20,32] showing that low-temperature deformation enhances the γ-peak and reduces the α-peak relative to RT deformation supports this model, since low-temperature deformation of bcc metals is known to favour the generation of extended dislocation loops consisting mainly of <111>/2 screw dislocations [31].

The Hydrogen Cold-Work Peak

Bruner [33] appears to have been the first to state that the IF peak in Nb subsequently named α-peak by Chambers was <u>not</u> a Bordoni-type peak and to propose that it was analoguous to the hydrogen CWP in α-iron observed by Weiner and Gensamer [34].Further evidence supporting this view was found on both Nb [22,25,35] and V [22,36].The close connection between the α-peak (in the present nomenclature) and the hydrogen CWP was detected first by magnetic aftereffect studies on α-iron single crystals after different kinds of pre-treatments as shown in Fig.2 [3-5]. Such measurements are also particularly suited for elucidating the fine structure of the hydrogen CWP.

According to Figs.3,4 the hydrogen CWP in α-Fe is split into two sub-peaks. This has recently been confirmed by low-frequency IF [36].Since the lower subpeak (CWP$_\ell$) shifts to higher temperatures with increasing hydrogen concentration,we suppose that it derives from the α-peak and that it is due to the interaction of non-screw dislocations with hydrogen.

The upper subpeak (CWP$_u$)has been attributed to the movement of geome-trical kinks on screw dislocations [4].The following experimental results support this proposal. (i) The relaxation strength of CWP$_u$ is enhanced by low-temperature deformation. (ii) With increasing hydrogen content CWP shifts to lower temperatures as a result of the reduction of the free-seg-ment lengths of the dislocations. (iii) In dilute Fe-V,-Cr, and -Co alloys CWP$_u$ remains nearly unaffected by the alloying component, whereas CWP$_\ell$ shifts to higher temperatures because of the stronger interactions of non-screw dislocations with foreign atoms (see Fig.5).

Activation Enthalpies and Comparison with Theory

In α-Fe the information on activation enthalpies and frequency factors is scarce. This is due to the fact that the frequency factor depends on the lenght of dislocation segments and that hence from measurements at one fre-quency or from measurements on different samples cannot deduce the activa-tion enthalpies.

Kronmüller et al.[3] showed that the $\underline{\alpha\text{-peak}}$ in α-iron is well described by (3) with H=48meV and $\tau_\infty=1.6\times10^{-11}$s. According to the present interpreta-tion H equals twice the enthalpy of kink formation, H_k, in 71° dislocations. The orders of magnitude of H_k and τ_∞ are as expected from theory.The rela-xation times deduced from IF^k[7,10,18] agree reasonably well with the extra-ipolation of the MA results.

The analysis of $\underline{CWP_\ell}$ is difficult, since its width indicates that it cannot be described well by (3).In the lower (T=80K) and in the upper (T=124K) temperature range activation enthalpies H=(0.2$\pm$0.08)eV and H=(0.29$\pm$0.09)eV, respectively, and an average value $\tau_\infty=3\times10^{-10\pm3.3}$s were obtained [37] by comparing the slopes of isothermal relaxation curves [38].

According to Seeger [39] the activation enthalpy of the relaxation pro-cess due to kink-pair formation in the presence of a high concentration of isolated mobile interstitial atoms is given by

$$H = 2 H_k + H^{M'} , \qquad (4)$$

where $H^{M'}$ is the enthalpy of interstitial migration close to the dislocation line. If we use $2H_k=0.05$ eV from the α-peak measurements and H=0.12 eV,the lowest value compatible with the MA data [37],we obtain $H^{M'}=0.07$ eV.This is compatible with what is known on hydrogen diffusion in α-Fe [12],so that it appears that in the low-temperature range of CWP$_\ell$ the assumptions of the theory [39] are satisfied. This is clearly not so at higher temperatures,and we have to assume that here we are dealing with less mobile $\underline{hydrogen}$ clusters on dislocations.

CWP$_u$ behaves like a Debye relaxation and is well described by (3),with H=(0.44$\pm$0.02)eV and $\tau_\infty=1.7\times10^{-13\pm0.2}$s in the case of weakly charged speci-mens.The small value of τ_∞ is difficult to understand in terms of the kink-migration mechanism discussed above and may mean that the process involves the reorientation of complexes consisting of hydrogen and intrinsic point defects or impurity atoms. Further work on the nature of CWP_u is clearly desirable.

Acknowledgements

The authors are grateful to Prof. H. Schultz for discussions and the communication of unpublished results.

References

[1] R.H.Chambers, Physical Acoustics, W.P.Mason, ed., vol.3A, p.123, Academic Press, New York and London, 1966.

[2] A.Seeger and C.Wüthrich, Il Nuovo Cimento 33B (1976) 38.

[3] H.Kronmüller, H.Steeb and N.König, Il Nuovo Cimento 33B (1976) 205.

[4] K.Vetter, H.Steeb and H.Kronmüller, in: Proc. 2nd Intern. Cong. on Hydrogen in Metals, nr. 2C5, Paris, June 1977.

[5] H.Kronmüller, "Magnetic Aftereffects of Hydrogen Isotopes in Ferromagnetic Metals and Alloys", pp.289-320 in Hydrogen in Metals I, J.Völkl and G.Alefeld, eds., Springer, Berlin-Heidelberg-New York, 1978.

[6] K.Takita and K.Sakamoto, Scripta Met. 10 (1976) 399.

[7] J.F.Dufresne, A.Seeger, P.Groh and P.Moser, phys.stat.sol.(a) 36 (1976) 579.

[8] I.G.Ritchie, J.F.Dufresne and P.Moser, phys.stat.sol.(a) 52 (1979) 331.

[9] W.R.Heller, Acta Met. 9 (1966) 600.

[10] R.Gibala, Trans.Met.Soc.AIME 239 (1967) 1574.

[11] H.K.Birnbaum and C.A.Wert, Ber. Bunsenges. Physik Chem. 76 (1972) 806.

[12] J.Völkl and G.Alefeld, "Diffusion of Hydrogen in Metals" pp.321-348 in Hydrogen in Metals I, J.Völkl and G.Alefeld, eds., Springer,Berlin-Heidelberg-New York, 1978.

[13] A.Seeger, "Positive Muons as Light Isotopes of Hydrogen",pp.349-398 in Hydrogen in Metals I, J.Völkl and G.Alefeld, eds., Springer,Berlin-Heidelberg-New York, 1978.

[14] H.Kronmüller, Appl.Phys. 4 (1974) 317.

[15] F.Walz, Appl.Phys. 3 (1974) 313.

[16] K.Vetter, Doctor rer. nat. thesis, Stuttgart University, 1976

[17] V.Hivert, P.Groh, P.Moser and W.Frank, phys.stat.sol.(a) 42 (1977) 511.

[18] K.Takita and K.Sakamoto, Scripta Met. 4 (1970) 403.

[19] I.G.Ritchie, J.F.Dufresne and P.Moser, Proc. VI Intern.Conf.on Internal Friction and Ultrasonic Attenuation in Solids, Tokyo, Japan, 1977, p.701.

[20] P.Astié, J.P.Peyrade and P.Groh, Proc.Third European Conf.on Internal Friction and Ultrasonic Attenuation in Solids, Manchester, July 1979.

[21] R.H.Chambers, W.D.Crie, T.E,Guthrie, H.T.Harper, H.H.Moellerand and T.A.Trozera, AEC Research and Development Report GA-7669 (1967).

[22] F.M.Mazzolai and M.Nuovo, Sol.State Comm. 7 (1969) 103.

[23] G.Cannelli and F.Mazzolai, J.Phys.Chem.Solids 31 (1970) 1913.

[24] H.Mizubayashi, M.Daikubara and S.Okuda, Proc. JIMIS-2, Hydrogen in Metals, Suppl. Trans.Jap.Inst.Metals 21 (1980) 217.

[25] M.Maul and H.Schultz, Max-Planck-Institut für Metallforschung, Stuttgart, personal communication July 1980.

[26] H.Schultz, M.Maul, U.Rodrian and R.Grau, Proc.5th Intern.Symp.on High-Purity Materials in Science and Technology, Vol.III, p.334, Dresden 1980.

[27] A.Seeger, Phil.Mag. 1 (1957) 651.

[28] P.G.Bordoni, J.Acoust.Soc.Am. 26 (1954) 495.

[29] G.Fantozzi, W.Benoit, C.Esnouf and J.Perez, Ann.Phys.Fr. 4 (1979) 7.

[30] A.Seeger and B.Šesták, Scripta Met. 5 (1971) 681.

[31] B.Šesták and A.Seeger, Z.Metallkde. 69 (1978) 195, 355, 425.

[32] N.Igata, H.Murakami and N.Masamura, Scripta Met. 14 (1980) 463.

[33] L.J.Bruner, Phys.Rev. 118 (1960) 399.

[34] L.C.Weiner and M.Gensamer, Acta Met. 5 (1957) 692.

[35] G.Ferron, M.Quintard and P.Mazot, Scripta Met. 12 (1978) 623.

[36] M.Shimada and K.Sakamoto, Proc.JIMIS-2, Hydrogen in Metals, Suppl.Trans.Jap.Inst. Metals 21 (1980) 569.

[37] R.Martinez, F.Walz and H.Kronmüller, Appl.Phys. 7 (1975) 107.

[38] A.Seeger, H.Kronmüller and H.Rieger, Zeitschrift Angew. Physik 18 (1965) 377.

[39] A.Seeger, phys.stat.sol.(a) 55 (1979) 457.

EFFECTS OF MICROSTRUCTURE

MICROSTRUCTURE AND HYDROGEN EMBRITTLEMENT

Anthony W. Thompson and I. M. Bernstein
Dept. of Metallurgy and Materials Science
Carnegie-Mellon University
Pittsburgh, PA 15213

Over the past several years, our work on hydrogen embrittlement has emphasized the role played in this phenomenon by metallurgical variables, particularly microstructure. We have reviewed the literature on this topic in some detail, for aluminum alloys and stainless steels (1,2), for ferritic and martensitic steels (2,3), and for titanium and nickel alloys (2), as well as considering the implications for alloy design (4) and for a variety of engineering applications (5-9). Consideration has also been given to questions of possible mechanisms of embrittlement (2,3,7,9-14), although in light of evidence that none of the proposed mechanisms can adequately account for the effects of metallurgical variables, detailed discussion of mechanisms is still likely to be premature.

An alternative to such discussion, and in fact a necessary preliminary to an informed discussion of mechanisms per se, is to identify and understand micromechanisms of fracture. By "micromechanisms" we mean the physical fracture processes which take place on the scale of the microstructure. In general, such processes can be divided into nucleation and propagation stages of the fracture event (15). For example, in the familiar case of ductile fracture, the micromechanisms are the nucleation, growth and coalescence of microvoids (hence the name, microvoid coalescence or MVC). For other fracture modes, however, the micromechanisms are rather less clearly known (2,7, 10, 16); moreover in these latter cases there is even less understanding of how hydrogen alters or assists the various micromechanisms. Nevertheless, the advantage to considering processes at this level is that it helps one gain insight into reasons for interactions with microstructure. It has been our view that a reasonably complete, albeit qualitative description of these processes, and their sensitivity to hydrogen, is an essential step in understanding hydrogen embrittlement.

We have been emphasizing for some time that strength level alone is not a particularly reliable indicator of susceptibility to hydrogen. That this is true even for low to medium strength levels in steels is illustrated in Fig. 1, with some 465 individual results (17). We presume that differences among the metallurgical characteristics of these steels, particularly microstructure, account for the broad scatter in Fig. 1. Microstructure has been explicitly treated as a variable for steels of similar strength levels by Cain and Troiano (3,14,18), who conducted delayed failure tests rather than

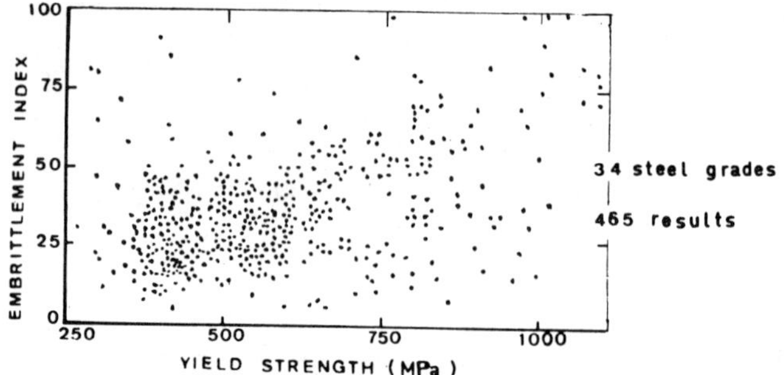

Fig. 1.   Embrittlement Index (per cent RA loss) as a function of yield
strength for 34 grades of low and medium strength steels, from
Zmudzinski, Bretin and Toitot (17).

the tensile tests of Fig. 1. Our own recent experimental work has been focused on obtaining more complete and explicit information on the relation
between microstructure and hydrogen embrittlement. Although detailed reports on this work have been or will be presented elsewhere, as we cite in
the following discussion, we summarize here the salient findings of several
experiments as a foundation for a discussion of our current understanding
of the role of microstructure in hydrogen embrittlement.

EXPERIMENTS

There are two broad types of experiments which we have been using to
study hydrogen embrittlement. One of these is the simple tensile test. Although care must be used in interpreting tensile data for some kinds of service applications (6,15,19,20), the widespread use of such tests, the large
body of existing test results, and the often complex and interesting phenomena which occur during fracture of tensile specimens (7,15,19) combine to
make tensile testing informative and useful.

The second test type is intended to examine more explicitly the processes of crack initiation and propagation. Crack initiation is being studied in notched bend bars, of the type shown in Fig. 2. Rice originally suggested (21,22) that this kind of experiment could best be conducted with
moderately notched specimens. Determinations of the local stress level as
a function of depth below the notch root are available (23), so that qualitative experiments on cracks below notches (24) can be replaced by more
quantitative ones. Such experiments are then in principal capable of identifying the cracking location through simple metallography and measurement.
As Rice pointed out (22), reduction in notch sharpness causes the location

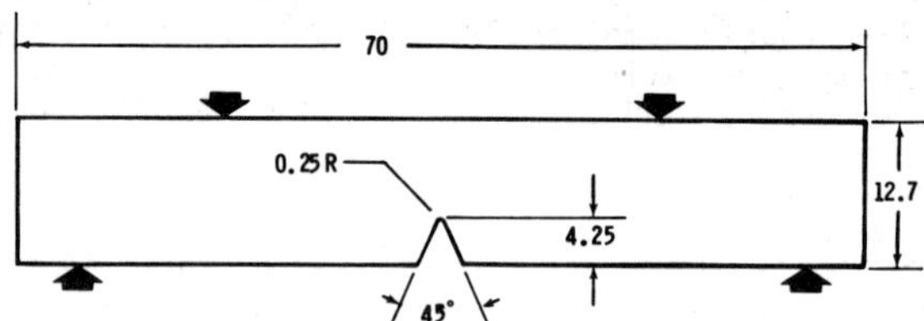

Fig. 2.   Specimen used for 4-point bend testing; dimensions in mm. Load
points indicated.

of maximum stress to lie farther below the notch root, thus facilitating crack examination and identification of the initiation location. In addition, as the distance from notch root to maximum stress location becomes large compared to the size of microstructural units, ease of identification of the cracking nucleation site increases. These expectations about the utility of moderately notched specimens have been borne out in several hydrogen cracking studies (25-28).

Crack propagation has been studied under conditions both of sustained load, and of cyclic load; the former condition has utilized both notched round and DCB (double cantilever beam) specimens, while the latter has employed conventional CT specimens conforming to ASTM E399 suggestions as to specimen proportions, except that H/W = 0.486 and B = 12.7 mm, in all cases.

Detailed fractographic study of all specimens tested to failure is an important part of the test results. Both low (ca. 200X) and high (ca. 2000X) magnification examination with the viewing direction normal to the macroscopic fracture plane has been performed in a scanning electron microscope (SEM). Sectioning studies of unfailed specimens have also been used to examine the conditions of fracture initiation and propagation.

SELECTED RESULTS

Tensile Tests

A simple kind of analysis which can be performed for tensile tests is to correlate the ductility (usually reduction of area, RA) with fractographic characteristics. For example, when the fracture mode is MVC both with and without hydrogen, the primary fractographic unit is the microvoid or dimple size, and the correlation of this parameter with RA has been explored at some length (29,30).

The potential difficulty in such an analysis, however, is that usually a substantial fraction of the microvoids are either nucleated or experience much of their growth in a strain interval which is relatively small and rather near the fracture strain. Thus the fracture surface data should be interpreted in light of information about the preceding range of strain. The circumstance being discussed is schematically depicted in Fig. 3, taken from the work of Garber, et al. (31). Examination of, for example, the effect of hydrogen on microvoid nucleation would have to be performed as a function of strain, rather than using fracture surface data alone; otherwise, results can be subject to uncertainties arising, for example, from dynamic effects occurring during final fracture separation.

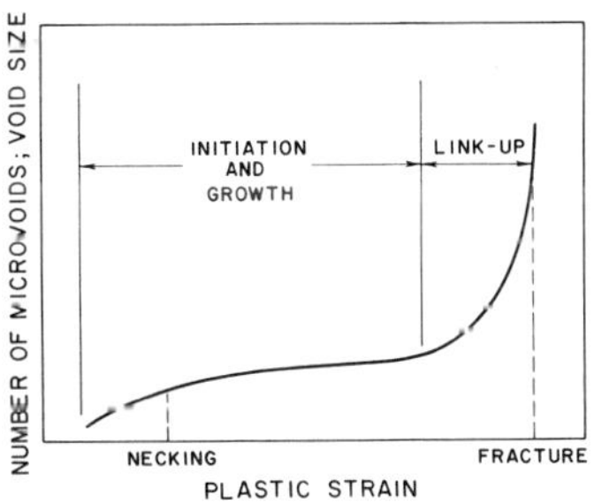

Fig. 3.   Schematic Illustration of the expected change in number and size of microvoids as a function of strain. The extent of the "link-up" stage is somewhat arbitrary, but can be recognized by a marked steepening of the curve in a plot of this type, usually at a large fraction of the fracture strain. From Garber, et al. (31).

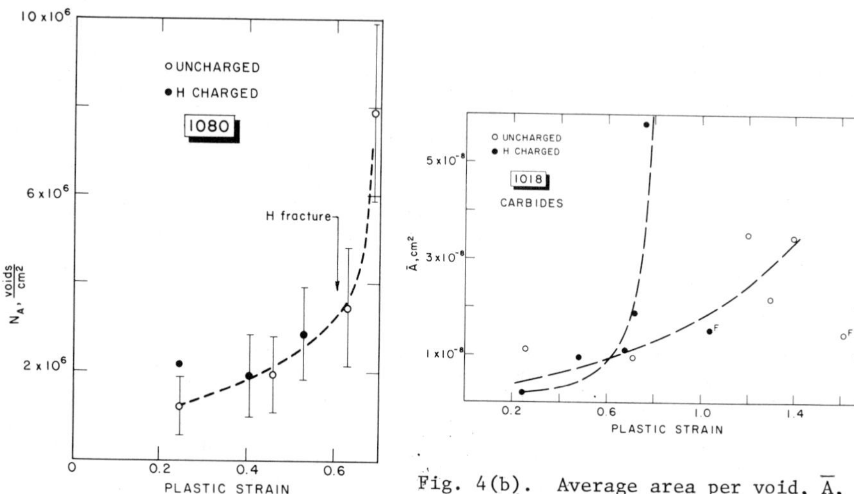

Fig. 4(a). Microvoid density, N_A, as a function of strain for uncharged and hydrogen-charged 1080 steel. Hydrogen fracture strain indicated by arrow. Data range for separate measurements shown. From Garber, et al. (31).

Fig. 4(b). Average area per void, $\overline{A}$, as a function of strain for uncharged and hydrogen-charged 1018 steel, for carbide-initiated voids. From Garber, et al. (31).

Sectioning studies to provide quantitative metallography as a function of strain are tedious, but provide helpful and necessary data of the kind described above. In Garber's work (31), both the nucleation of microvoids (expressed as the number of voids per unit area, N_A) and the growth of voids (expressed as average area per void, $\overline{A}$) were studied as a function of strain. Examples are given in Fig. 4 for spheroidized microstructures in two different steels, 1018 and 1080. There are certainly problems with scatter in data of this type, and these problems have been treated in a fairly complete statistical way (31) wherever appropriate. The trends, however, are fairly clear. In both examples shown, it will be noted that hydrogen has no very obvious effect --although in Fig. 4(b), the comparison to uncharged data must be made prior to "link-up" of the kind suggested in Fig. 3, i.e. below about $\varepsilon=0.7$. Even with that restriction, it is possible to conclude, from Fig. 4(b) and other data (31), that there is some acceleration of void growth by hydrogen, prior to link-up, in 1018. No such effect was evident for 1080.

For fracture modes other than MVC, the way in which to approach a microstructural analysis is less clear. But it would appear that a combination of detailed fractography and strain-dependent metallographic study will generally be of value, provided the nucleation and early propagation events (7,10,16) are on a scale amenable to such techniques.

<u>Crack Initiation and Propagation Tests</u>

A program of bend testing on the specimen type shown in Fig. 2 has been conducted on 1045 steel, which was chosen as a steel which is microstructurally versatile yet reasonably simple. Th ree distinct microstructures have been studied: quenched and tempered martensite; isothermally-transformed pearlite; and spheroidized. Details of heat treatment procedures and micro-structures have been presented elsewhere (28,32).

<u>Martensitic 1045 Bend Tests</u>

Several interesting results were obtained in an initial study by Costa (28) on the martensitic material. In uncharged bend specimens, cracking was found to initiate at the notch root, usually in a mode I or radial orientation. This observation differs from results of several earlier studies (25-27). When specimens were hydrogen charged, the cracking exhibited a mode II or shear orientation at notch roots, with a mode I orientation occurring below the notch; this is shown in Fig. 5. The observation is in agreement with previous work on martensitic 4340 steel (27). It appeared that the mode I crack occurred first, and evidently initiated near the location of maximum triaxial stress. Crack initiation loads were nearly independent of tempering temperature over the range studied, as illustrated in Fig. 6.

Hydrogen accentuated a tendency toward prior-austenite intergranular fracture in bend specimens, a fracture mode which was likely due to sulfur segregation to prior austenite boundaries (tempered martensite embrittlement); such segregation was observed by scanning Auger Spectroscopy. Tho hydrogen effects on fracture did not appear (28) to be additive to those of the sulfur segregation. Hydrogen also induced a transgranular, quasi-cleavage fracture mode in bend specimens near the maximum stress location; the tear ridges in this mode had a morphology and spacing which corresponded to the packet structure of the martensite. This reflection of the underlying microstructure on the fracture surface is illustrated in Fig. 7, and is believed (28) to result from ductile tearing at changes in crystallographic orientation between packets. This interpretation is consistent with recent views of the nature of quasi-cleavage fracture development (7,10,16).

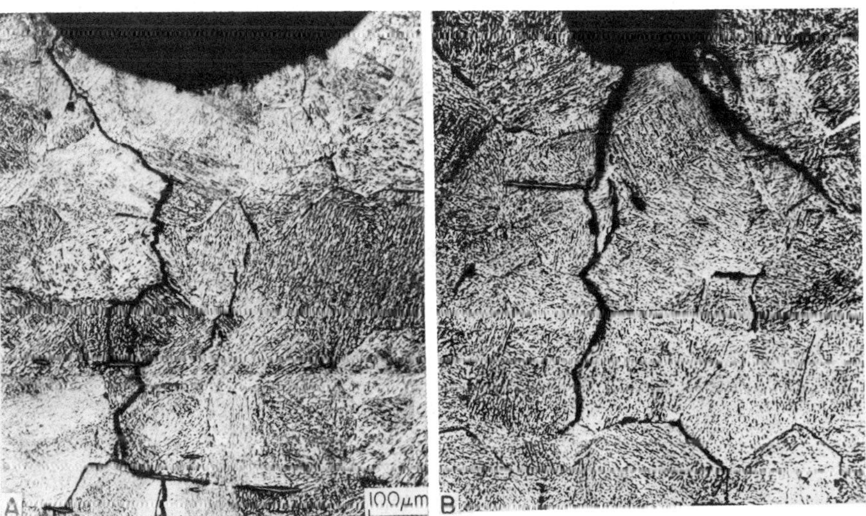

Fig. 5.   Fracture appearance in hydrogen-charged martensitic 1045 steel (notch root at top) at specimen mid-plane. Light microscopy. (a) 325°C temper; note mode I - mode II cracking. (b) As-quenched; note mode I crack. From Costa and Thompson (28).

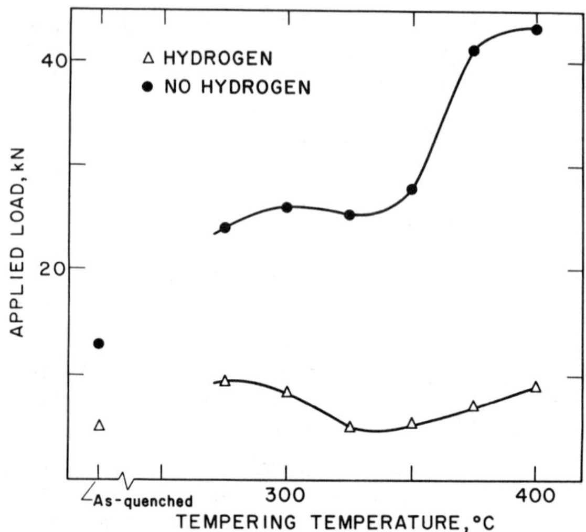

Fig. 6.   Applied load to cause crack initiation at room temperature in bend
specimens (see Fig. 2), as a function of tempering temperature for
both uncharged and hydrogen-charged specimens of martensitic 1045
steel.   From Costa and Thompson (28).

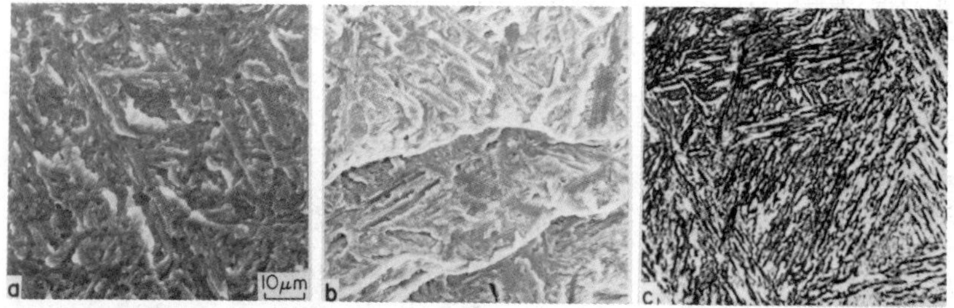

Fig. 7.   Example of quasi-cleavage fracture details in 1045 martensite
400°C temper is shown), all at same magnification.   (a) SEM of hydrogen-charged tensile specimen.   (b) SEM of hydrogen-charged bend
specimen.   (c) Light microscopy of microstructure at specimen mid-
plane; units visible were shown to be martensite packets in thin
foils.   From Costa and Thompson (28).

Pearlitic 1045 Bend Tests

A parallel investigation to the foregoing study on martensitic 1045 was
performed on fully pearlitic 1045, obtained by isothermal transformation in
molten salt at 805 K.   This microstructure exhibited strikingly different behavior from that of the martensitic material (33).   In tensile specimens,
hydrogen induced an increase in the extent of cleavage fracture.   In notched
specimens, crack initiation occurred in mode I, and loads were considerably
larger (as a fraction of the load for general yielding) than for tests on
martensite, both with and without hydrogen, yet fracture was almost exclusively by a cleavage mode (Fig. 8).   This emphasizes the profound role played
by carbide morphology in a steel of this type: despite constancy of composition and carbide volume fraction, the pattern of behavior in the pearlitic

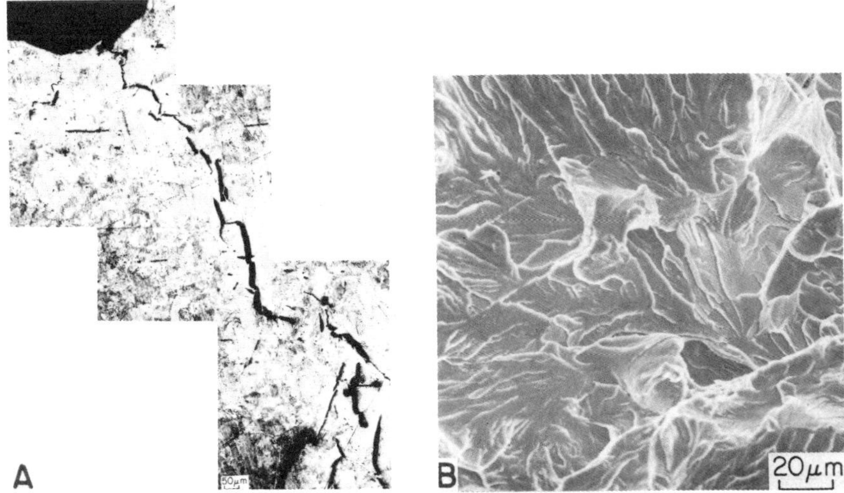

Fig. 8.   Fracture behavior of fully pearlitic 1045 steel.   (a) Light microscopy below notch, showing discontinuous crack appearance (notch
root at top).   (b) Fracture surface by SEM, showing typical cleavage
details.

microstructure is totally unlike that of either tempered martensite or of
spheroidized material.   It had been thought possible that a fractographic indication of the effect of hydrogen on propagation of cleavage could be obtained, using parameters such as facet size and density of river lines (34),
but none was found.   It is therefore inferred that hydrogen predominantly
affects cleavage nucleation, particularly in terms of the critical nucleus
size for cleavage (34).   This is consistent with the effect of hydrogen on
crack initiation loads in notched bend specimens.   Work on this topic is continuing.

<u>Spheroidized 1045</u>

A third study in this group was conducted on a spheroidized microstructure
of 1045.   In this material, large bend angles did not cause fracture, with or
without hydrogen, even when temperature was reduced to about 200 K.   However,
these notched bend tests showed that constrained deformation below the notch
root initiated at lower loads with hydrogen, consistent with earlier work on
notched specimens of spheroidized 1090 steel (35).   Work hardening in the
constrained plastic zone, however, was increased by hydrogen.   These two observations are shown in Figs. 9 and 10, as a function of testing temperature.
These observations appear consistent (36) with a variety of earlier studies
on deformation of low-purity iron (37).   It will also be noted in Figs. 9 and
10 that the maximum hydrogen effect occurs around -20°C (~250 K), a temperature which, possibly coincidentally, corresponds to the well-known temperature
of maximum hydrogen embrittlement in spheroidized microstructures (38).
It should be pointed out that in tensile tests on the spheroidized 1045,
little effect of hydrogen on strength properties was evident, although ductility (RA) was reduced by 72%.   Analysis of tensile fracture surfaces, which
were entirely composed of microvoid coalescence with and without hydrogen,
showed that hydrogen affected both nucleation and growth of microvoids but
only for strains near the fracture strain   (30).   This is consistent with

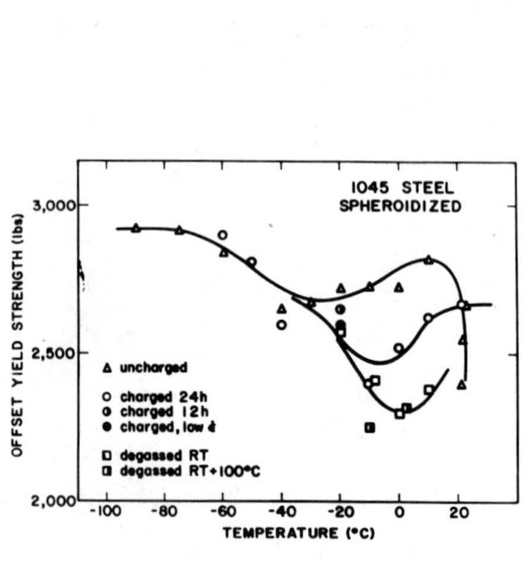

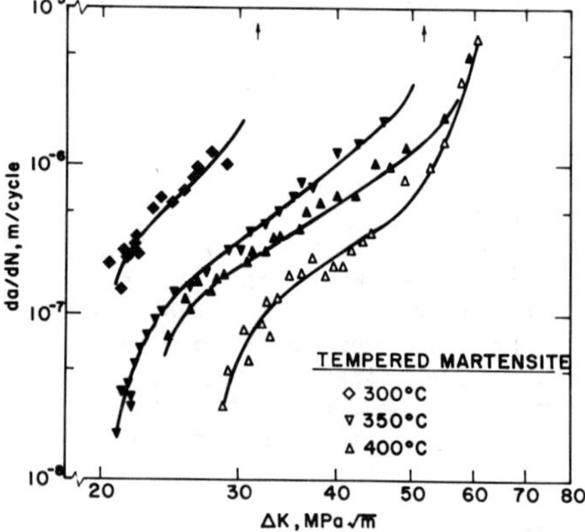

Fig. 9.   Constrained-plastic yield strength below notch root as a function of test temperature for spheroidized 1045. "Degassed" results indicate that hydrogen effects on yielding are not reversible. From Kosco and Thompson (36).

Fig. 10.   Work hardening in constrained plastic zone below notch as a function of test temperature for spheroidized 1045. "Degassed" results indicate that hydrogen effects on work hardening are reversible. From Kosco and Thompson (36).

Fig. 11.   Fatigue crack propagation rate, da/dN, in CT specimens as a function of stress intensity amplitude, ΔK, for martensitic 1045. Filled points are for tests in 1 atm. hydrogen gas. From McLaren and Thompson (41).

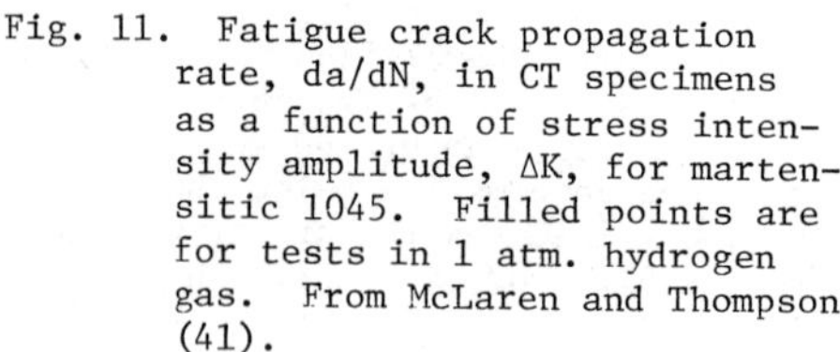

other work on notched specimens of 1045 steel (39), although other spheroidized steels have shown different effects (31).

Crack Propagation in 1045

The foregoing studies on crack initiation were supplemented by work on
crack propagation behavior in the presence of hydrogen. The CT specimens
described above were tested in ~0.1 MPa (1 atm) dry hydrogen gas, for all
three 1045 microstructures. For each microstructure, the fatigue crack propagation (FCP) rate was increased by hydrogen. The extent of increase was
about a factor of four in the mid-range of FCP rate, for both pearlitic and
spheroidized specimens. For martensitic material, the extent of FCP increase
was a function of tempering temperature. For 350°C and 400°C tempers, the
FCP rate increase was again about a factor of four in hydrogen, but the 300°C
temper exhibited a much larger increase, of a factor of twenty or more (Fig.
11). Moreover, the apparent "threshold" in FCP (growth rates were not sufficiently low to identify threshold ΔK values accurately) shifted from near
30 MPa $\sqrt{m}$ to below 20 MPa $\sqrt{m}$. These data therefore do not follow the trend
of Fig. 6 for initiation, in which the tempering temperature of minimum hydrogen resistance was 350°C.

Fractographically, the FCP specimens exhibited interesting patterns of
behavior. The pearlitic and spheroidized specimens showed essentially a
tearing topography or TTS fracture mode (40), both with and without hydrogen,
independent of ΔK. Note that this locally-plastic mode is consistent with
the test results described above for spheroidized material, but appears to
contradict the results for pearlite, in which cleavage would be expected.
However, the crack advance per cycle evidently was less than the critical
nucleus size for cleavage (34) until ΔK levels of about 50 MPa $\sqrt{m}$ were
reached, at which point cleavage facets began to be observed. In the martensitic specimens, substantial amounts of intergranular fracture were observed,
consistent with the notched bend results (28), both with and without hydrogen. As tempering temperature increased from 270°C to 400°C, however, increasing proportions of first TTS, and then MVC modes, were evident. The
fracture mode was also a function of ΔK, becoming more ductile as ΔK increased.

This work appears to show that crack growth, at least under fatigue conditions, can be assisted by hydrogen without major effects on crack path or
mode. Moreover, only in martensitic material was the apparent threshold ΔK
shifted downward by hydrogen. This contrasts with the results for the notched
bend specimens, in which behavior is expected to be dominated by initiation.
Although these results need to be checked by sustained-load tests on precharged specimens, they appear to indicate (41) that crack initiation phenomena can be assisted by hydrogen in more different ways than can crack propagation phenomena. Work on this topic is also continuing.

Bend Tests on 4140

In light of these results, it was felt that further work on an additional
microstructural type, bainite, would be of interest. There is relatively
little information available on the hydrogen performance of bainitic steels;
reviews have concluded (2-4) that bainites might perform as well as any other
steel microstructure (at a given strength level) in hydrogen. It therefore
seemed appropriate to conduct a small study on this microstructure to explore
whether performance would in fact compare well with that of tempered martensite.

As it did not prove possible to produce a fully bainitic microstructure in the 1045 steel (32), an alloy steel, AISI 4140, was chosen instead. Bend specimens for this study were standard Charpy specimens, according to ASTM E23; information on stress below this type of notch root is available (23, 42). Tensile and Charpy slow-bend specimens were austenitized at 1115 K (1545°F) for 1 hr. in argon, and then either quenched into molten salt at 670 K for isothermal transformation to bainite, or quenched into room temperature oil and promptly tempered in salt at 575 K for 1 hr. It was expected, on the basis of preliminary data, that these treatments would give comparable strengths of about 1250 MPa (180 ksi). In fact, however, strengths differed by about 25%, with the bainite being weaker.

Notched bend tests of hydrogen-charged and uncharged specimens were conducted as described in preceding sections. Behavior of the two materials was quite different (43), as Fig. 12 shows. While hydrogen scarcely reduced the crack initiation load in the bainite, it reduced it to about 10% of the uncharged value in the martensitic material. Fractographic observations showed a MVC fracture mode in uncharged specimens of both microstructures. This observation is consistent with results of a detailed study (44) on similarly austenitized and tempered 4140 specimens without hydrogen. Hydrogen, however, induced an intergranular mode near the notch root in both microstructures. Within the roughly 250 µm between the root and the location of the peak stress (23), this intergranular mode comprised about 25 areal percent in the bainite, and 75% in the martensite. Fig. 13 illustrates this observation. The areal percentages of brittle fracture induced by hydrogen in the two microstructures, together with crack initiation load data, Fig. 12, permitted us to model the fracture behavior as a composite behavior of brittle and ductile material, with the brittle material having negligible relative toughness. The ductile ligaments were then treated with the Dugdale model, as has been done elsewhere for PH 13-8 Mo steel (45). Good agreement with the load data, Fig. 12, was obtained.

This model, however, provides no <u>explanation</u> for the differences between the two microstructures, only a rationale for the connection between Figs. 12 and 13. The observed differences may have arisen from differing extents of metalloid element segregation at the bainite transformation temperature, compared to the martensite tempering temperature, although no Auger spectroscopy was performed. It might also reflect, of course, the indirect effects of

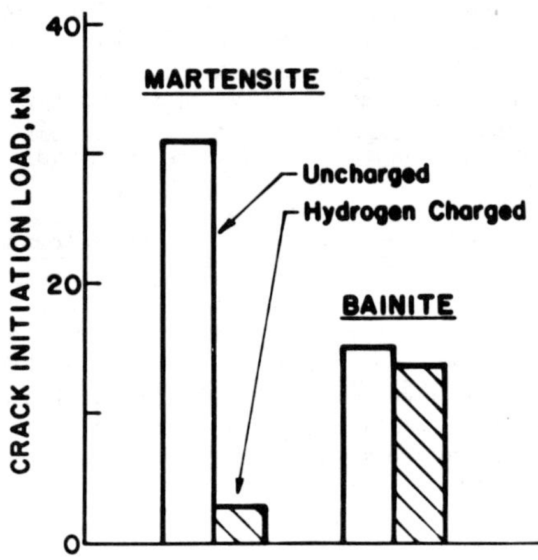

Fig. 12.   Load to cause crack initiation in slow-bend tests (see Fig. 2) of 4140 steel at room temperature. From Bevan and Thompson (43).

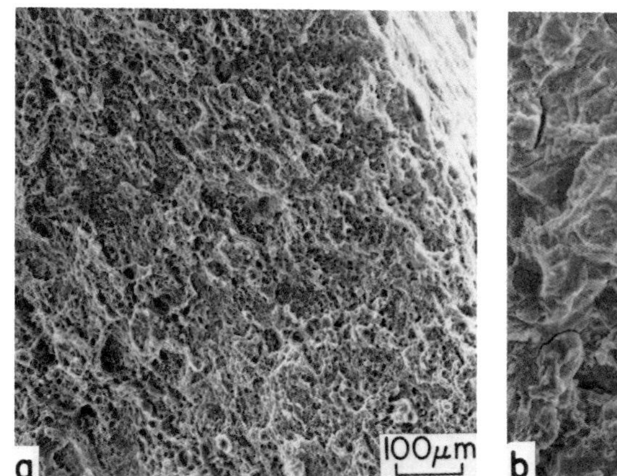
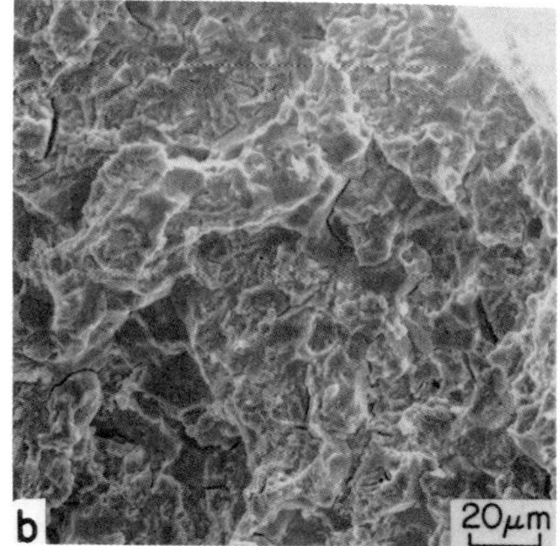

Fig. 13. Fractography (SEM) of bainitic 4140 steel. (a) Uncharged, notch
root at upper right. (b) Hydrogen-charged, notch root at upper
right. Fracture mode is about 25% intergranular. From Bevan and
Thompson (43).

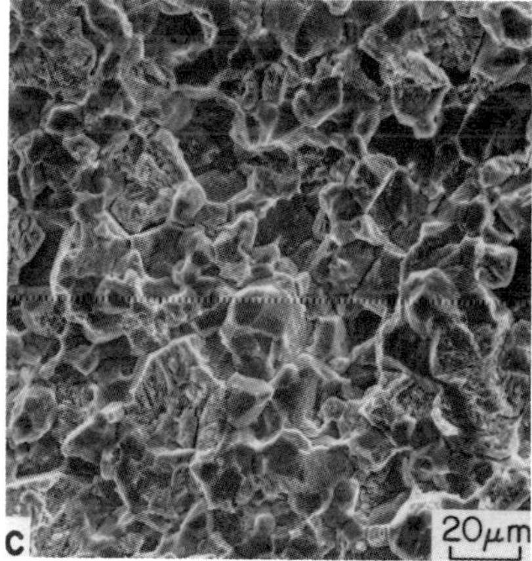

Fig. 13(c). Fractography (SEM) of martensitic 4140 steel, same magnification as (b); hydrogen-
charged. Fracture mode is
about 75% intergranular. From
Bevan and Thompson (43).

strength level on hydrogen embrittlement (2,3,46), although in this case the
origins of the effect are unclear. More work on bainitic microstructures seems
appropriate and is now underway in our laboratory using a 9% Ni-4% Co steel.
Extensive background information is available on this type of steel in both
microstructures (47-50), which should permit a focus on the hydrogen aspects
of the problem.

DISCUSSION

The most critical aspect of hydrogen effects in structural materials is
usually considered to be cracking which is either induced or assisted by hydrogen, the best known type being classical delayed failure or static fatigue,
in which a specimen, smooth or notched, fails suddenly after a delay period.
In such a process, a crack is first nucleated and then grows (15) until

failure occurs. Thus the problem of cracking caused by hydrogen has two intimately connected parts: hydrogen-induced crack initiation, and hydrogen-assisted crack growth. A study of either part alone would neglect an essential aspect of the problem, as reviewers have indicated (2-4,9,37,46,51,52). In studying them together, a concerted attack on the micromechanisms of hydrogen cracking can be made.

The work on crack initiation and propagation described above has been based upon the following considerations: there is evidence that cracks initiate below the surface of the material in the presence of internal hydrogen (24,27,28), and slow crack growth seems to occur by repeated re-initiation ahead of crack tips and joining up with the main crack (51,53). Thus, crack growth in delayed failure may be controlled by initiation events. Furthermore, there is evidence that in the presence of substantial pressures of external gaseous hydrogen, cracks can originate at the surface rather than in the sub-surface (54). This difference in initiation behavior needs to be confirmed by additional work, but probably arises from differences in relative hydrogen susceptibility (25-28); differences in hydrogen fugacity could also be partly responsible. If confirmed, the difference in initiation may point to a critical difference in the effects of internal and external hydrogen, since both delayed failure life and crack growth rate under sustained load may be affected by differences in crack initiation behavior. It is thus appropriate to study crack initiation in some detail.

When subsurface initiation does occur, there is evidence (24,27,28,51) that it occurs at locations of maximum triaxial stress. Whether hydrogen is present outside the specimen, or if hydrogen is initially present within the metal, it must be transported to zones of maximum stress (55). In either case, the transport can be by normal lattice diffusion or by dislocation transport (56).

Once the hydrogen is available at the maximum stress locations, a crack can potentially initiate there. This crack will occur in some particular location, which can proceed (using steels as an illustration) either by intergranular cracking (at ferrite boundaries or at prior austenite boundaries); by interphase cracking, e.g. in pearlite colonies; by an acceleration of normal MVC at carbide particles; or by transgranular cleavage. It is very important to know which of these locations (in each microstructure studied) is operative, since the fracture path and mode can vary in each instance. Such variations, in turn, correspond to differences in ease and rate of cracking. An additional contribution to differences in cracking rate arises from the differences in "trapping strength" for hydrogen of various microstructural features (7,11,14,32,46). Hydrogen can be trapped at several different types of defects in steel: dislocations; impurity atoms; inclusion particles; interfaces between the matrix and carbide particles; internal cracks and voids; and grain boundaries. Because of the wide range of heat treatments and mechanical treatments in any given steel, the extent of these internal defects can vary widely in a single steel and between different steels. The important aspect here is that these defects are porential fracture nuclei (2,9,15) and can play a pivotal role in a particular embrittlement process.

It should be emphasized that although there are a very large number of hydrogen embrittlement studies on record, the microstructure of the material(s) studied has usually been characterized superficially, and rarely altered (in order to serve as an experimental variable). Even the exceptions to this general description have rarely taken microstructural entities seriously enough to study their effect on the fracture process itself. Yet understanding hydrogen effects on fracture events which have the scale of the microstructure (which are called "micromechanisms" here) must form the core of any complete understanding of hydrogen embrittlement.

It should be pointed out that neither the previous research, nor the research program which we are currently pursuing, is oriented toward "mechanisms" of embrittlement, such as decohesion, enhanced plasticity, or others. There is currently a widespread viewpoint (2,3,9-11,37,46,57) that none of the extant mechanisms is yet complete enough to incorporate much in the way of microstructural variables, and furthermore, that most if not all of the five mechanisms usually listed (9,37,57) can each be dominant under suitably chosen experimental conditions. In this light, it would certainly seem premature to try to select the "best" mechanism. Instead, detailed data on microstructure-fracture interactions are needed, thus facilitating (a) an accurate, physical description of hydrogen effects on fracture, and (b) a rigorous test of the ability of any mechanism to describe the varying manifestations of hydrogen embrittlement. We therefore believe that work on micromechanisms is "mechanism-oriented" in the best sense of the term.

It was asserted in the Introduction that the only fracture mode for which one has much confidence in knowing the microstructural nuclei for fracture, as well as the propagation process, is MVC. However, there have been suggestions (7,10,11,16) of a general kind for some of the other fracture modes: cleavage, quasi-cleavage or QC, and intergranular. As a means of focusing a discussion of these modes, a schematic of several fracture modes is presented in Fig. 14, for a particular assumption about fracture nuclei, namely that they are particles or inclusions. This is not intended to exclude the possibility (or probability) that other microstructural nuclei exist, indeed, other reasonable assumptions can be and have been made (2,7,15,16,37).

Each individual sketch in Fig. 14 shows a possible nucleation event at the left, an indication in the center as to how propagation from the nucleus may begin, and at the right a rather schematic profile of the ensuing fracture surface. At the top of Fig. 14 is illustrated the MVC case (29). The next case below that is cleavage, here depicted as nucleating from cracking or fracture of an inclusion, a process for which particular stress-distance conditions may have to be met (58). Following the cleavage case is an illustration of QC fracture, which is here shown as possibly arising from either a cracked particle or from plastic microvoid growth on a more restricted scale than for MVC (7,16,59). The local plasticity depicted here, which gives rise to the typical small-scale tear ridges characteristic of QC (59,60), may be of a similar kind to that proposed as operative in TTS fractures (40). Finally, at the bottom of Fig. 14, intergranular fracture is depicted as orig-

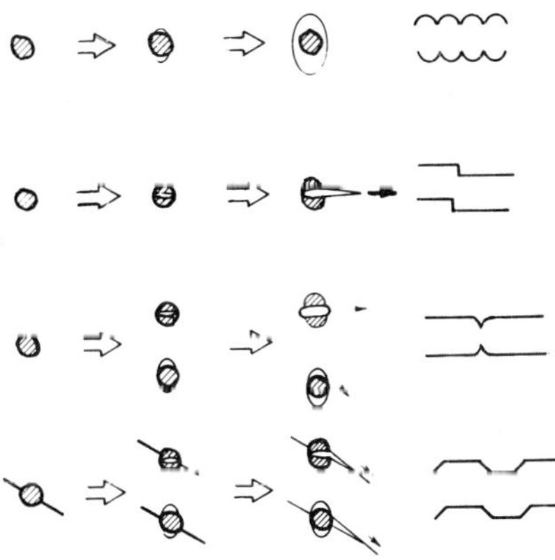

Fig. 14.   Schematic illustration of fracture mode development from a single type of fracture nucleus (particles or inclusions). Discussed in text.

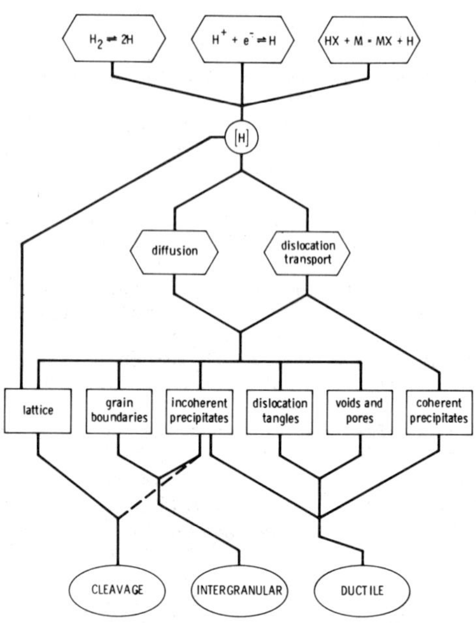

Fig. 15.   Summary of hydrogen processes: sources (gaseous, acqueous or chemical) leading to hydrogen in solution, transport by diffusion or
as dislocation atmospheres leading to accumulation at microstructural locations, and finally a microstructurally-controlled process of fracture. Dashed line refers to cleavage in hydrides.
From Thompson and Bernstein (2).

inating by either cracking or void growth at a grain boundary particle, the
latter case corresponding to what is usually called "ductile intergranular
fracture" (61).

Fig. 14 provides an example of the development of a viewpoint in which
the various fracture modes can be understood to develop from particular microstructural elements under particular degrees of local plasticity, interfacial
strength, and similar parameters. Fig. 14 itself represents, as mentioned, a
specific viewpoint concerned only with particle-nucleated fracture, and thus
cannot be general; but it does illustrate, for this particular assumption
about nucleation, understandable processes for the propagation of fracture.
It can, for example, be used to rationalize our experimental results on the
variety of steel microstructures reported above, even though it has little
predictive power without additional knowledge of other parameters such as
interfacial strength.

We devised some time ago (2) a "roadmap" to illustrate the broad patterns which occur in hydrogen embrittlement, as shown in Fig. 15.   We
have observed elsewhere (7) that Fig. 15 is deficient in omitting an explicit
step (the fracture process) preceding the fracture mode or type. The fracture process, in turn, should consist of nucleation and propagation steps,
in order to rationalize why particular accumulation sites for hydrogen (the

row of rectangular units across the center of the figure) are able to act as favored nuclei and/or propagation paths for hydrogen fracture, as well as indicating how they give rise to the different fracture modes at the bottom of Fig. 15. Although it is still not possible to present a detailed set of such fracture processes, Fig. 14 is a preliminary attempt to indicate ways in which to proceed toward such a presentation. It may well be, however, that numerous versions of Fig. 14 will prove necessary to a complete description of alloy systems (2), and in that case the apparent simplicity of Fig. 15 will be lost. At the risk of appearing pessimistic about understanding of hydrogen embrittlement, we think such a development is likely, if not probable. There now appears to be a considerable amount of experimental work needed before an adequate base for detailed modeling of hydrogen embrittlement is available. This prospect should be kept in mind in the design of hydrogen experiments as well as in the interpretation of experimental results.

ACKNOWLEDGEMENTS

We have benefited from discussions on the topic of this paper with our colleagues R. E. Stoltz, C. D. Beachem, J. P. Hirth, R. O. Ritchie, R. M. Pelloux, J. R. Rice and J. C. Williams.

We appreciate the provision of unpublished results by T. B. McLaren, M. B. Bevan and T. M. Kosco. The experimental work described herein was supported by the National Science Foundation under Grant Nos. DMR 75-10741 and DMR 78-00723, and by the Office of Naval Research under Grant No. N00014-75-C-0265; it has also benefited from the use of Central Facilities supported by the Center for the Joining of Materials at Carnegie-Mellon University.

REFERENCES

1. A. W. Thompson and I. M. Bernstein: Reviews on Coatings and Corrosion, 1975, vol. 2, pp. 3-44.

2. A. W. Thompson and I. M. Bernstein: Advances in Corrosion Science and Technology (M. G. Fontana and R. Staehle, eds.), vol. 7, pp. 53-175, Plenum, New York, 1980.

3. I. M. Bernstein and A. W. Thompson: Internat. Metals Reviews, 1976, vol. 21, pp. 269-87.

4. I. M. Bernstein and A. W. Thompson: Alloy and Microstructural Design (J. K. Tien and G. S. Ansell, eds.), pp. 303-47, Academic Press, New York, 1976.

5. A. W. Thompson and I. M. Bernstein: Int. J. Hydrogen Energy, 1977, vol. 2, pp. 163-73.

6. A. W. Thompson: Transmission and Storage, vol. 2 of Hydrogen: Its Technology and Applications (K. E. Cox and K. D. Williamson, eds.), pp. 85-124, CRC Press, Cleveland, 1977.

7.   A. W. Thompson: <u>Environmental Degradation of Engineering Materials</u> (M. R. Louthan and R. P. McNitt, eds.), pp. 3-17, V.P.I. Press, Blacksburg, VA, 1977.

8.   A. W. Thompson: <u>Plating and Surface Finishing</u>, 1978, vol. 65(No. 9), pp. 36-44.

9.   A. W. Thompson: <u>Environment-sensitive Fracture of Engineering Materials</u> (Z. A. Foroulis, ed.), pp. 379-410, TMS-AIME, Warrendale, PA, 1979.

10. A. W. Thompson and I. M. Bernstein: <u>Fracture 1977</u> (D. M. R. Taplin, ed.), vol. 2, pp. 249-54, Univ. of Waterloo Press, Waterloo, Ont., 1977.

11. I. M. Bernstein and A. W. Thompson: <u>Mechanisms of Environment-sensitive Cracking of Materials</u> (P. R. Swann, F. P. Ford and A. R. C. Westwood, eds.), pp. 412-24, Metals Society, London, 1978.

12. I. M. Bernstein and A. W. Thompson: <u>Hydrogen in Metals</u>, Suppl. to <u>Trans. Japan Inst. Metals</u>, 1980, vol. 21, pp. 429-32.

13. A. W. Thompson: <u>Mater. Sci. Eng.</u>, 1980, vol. 43, pp. 41-46.

14. I. M. Bernstein and A. W. Thompson: <u>Hydrogen Embrittlement and Stress Corrosion Cracking - A. R. Troiano Honorary Volume</u> (R. Gibala, ed.), Academic Press, New York, in press.

15. J. R. Low: <u>Prog. Mater. Sci.</u>, 1963, vol. 12, pp. 1-96.

16. A. W. Thompson and I. M. Bernstein: <u>Hydrogen in Metals (Paris)</u>, vol. 10, paper 3A-6, Pergamon Press, New York, 1977.

17. C. Zmudzinski, L. Bretin, and M. Toitot: <u>Hydrogen in Metals (Paris)</u>, vol. 3, paper 6A-2, Pergamon, New York, 1977.

18. W. M. Cain and A. R. Troiano: <u>Petroleum Engineer</u>, 1965, vol. 37, pp. 78-82.

19. R. E. Stoltz and A. J. West: this volume.

20. A. J. West and J. H. Holbrook: this volume.

21. J. R. Rice: <u>Corrosion</u>, 1976, vol. 32, pp. 22-26.

22. J. R. Rice: <u>Stress Corrosion Cracking and Hydrogen Embrittlement of Iron-base Alloys</u> (R. W. Staehle, et al., eds.), pp. 11-15, NACE, Houston, 1977.

23. J. R. Griffiths and D. R. J. Owen: <u>J. Mech. Phys. Solids</u>, 1971, vol. 19, pp. 419-31.

24. H. H. Johnson, J. G. Morlet and A. R. Troiano: <u>Trans. AIME</u>, 1958, vol. 212, pp. 528-36.

25. T. D. Lee, T. Goldenberg and J. P. Hirth: <u>Met. Trans. A</u>, 1978, vol. 9A, pp. 1663-71.

26. T. Goldenberg. T. D. Lee and J. P. Hirth: <u>Met. Trans. A</u>, 1979, vol. 10A, pp. 199-208.

27. T. D. Lee, T. Goldenberg and J. P. Hirth: Met. Trans. A, 1979, vol. 10A, pp. 439-48.

28. J. E. Costa and A. W. Thompson: Met. Trans. A, in press.

29. A. W. Thompson: Effect of Hydrogen on Behavior of Materials (A. W. Thompson and I. M. Bernstein, eds.), pp. 467-77, TMS-AIME, New York, 1976.

30. A. W. Thompson: Met. Trans. A, 1979, vol. 10A, pp. 727-31.

31. R. Garber, I. M. Bernstein and A. W. Thompson: Met. Trans. A, 1981, vol. 12A, in press.

32. W. M. Robertson and A. W. Thompson: Met. Trans. A, 1980, vol. 11A, pp. 553-57.

33. J. E. Costa and A. W. Thompson: Met. Trans. A, submitted.

34. Y. J. Park and I. M. Bernstein: Met. Trans. A, 1979, vol. 10A, pp. 1653-64.

35. T. D. Lee, T. Goldenberg and J. P. Hirth: Fracture 1977 (Proc. 4th Int. Conf. on Fracture), pp. 243-48, Univ. Waterloo Press, Waterloo, Ontario, 1977.

36. T. M. Kosco and A. W. Thompson: manuscript in preparation.

37. J. P. Hirth: Met. Trans. A, 1980, vol. 11A, pp. 861-90.

38. T. Toh and W. M. Baldwin: Stress Corrosion Cracking and Embrittlement (W. D. Robertson, ed.), pp. 176-86, Wiley, New York, 1956.

39. H. Cialone and R. J. Asaro: Met. Trans. A, 1979, vol. 10A, pp. 367-75.

40. A. W. Thompson and J. C. Chesnutt: Met. Trans. A, 1979, vol. 10A, pp. 1193-96.

41. T. B. McLaren and A. W. Thompson: manuscript in preparation.

42. W. L. Server: J. Eng. Mater. and Technol. (Trans. ASME, Series H), 1978, vol. 100, pp. 183-88.

43. M. B. Bevan and A. W. Thompson: unpublished research, Carnegie-Mellon University, 1980.

44. T. K. Odegaard: M.S. Thesis, Brown University, June 1979.

45. A. W. Thompson: Fracture 1977 (D.M.R. Taplin, ed.), vol. 2, pp. 237-42, Univ. of Waterloo Press, Waterloo, Ont., 1977.

46. J. P. Hirth and H. H. Johnson: Corrosion, 1976, vol. 32, pp. 3-26.

47. J. S. Pascover and S. J. Matas: Strength and Properties of Ultrahigh-Strength Steels (STP 370), pp. 30-46, A.S.T.M., Philadelphia, 1965.

48. J. S. Pascover, M. Hill and S. J. Mat as: Fracture Toughness Testing and Its Applications (STP 381), pp. 310-25, A.S.T.M., Philadelphia, 1965.

49. G. M. Waid and R. T. Ault: *Stress Corrosion - New Approaches* (STP 610), pp. 199-212, A.S.T.M., Philadelphia, 1976.

50. R. T. Ault: Republic Steel, personal communication, 1980.

51. A. R. Troiano: *Trans. ASM*, 1960, vol. 52, pp. 54-80.

52. H. H. Johnson, *Fundamental Aspects of Stress Corrosion Cracking* (R. W. Staehle, et al., eds.), pp. 439-45, N.A.C.E., Houston, 1969.

53. E. A. Steigerwald, F. W. Schaller and A. R. Troiano: *Trans. AIME*, 1959, vol. 215, pp. 1048-52.

54. R. J. Walter, R. P. Jewett and W. T. Chandler: *Mater. Sci. Eng.*, 1969/70, vol. 5, pp. 98-110.

55. J. C. M. Li, R. A. Oriani and L. S. Darken: *Zeits. f. Phys. Chem.*, 1966, vol. 49, pp. 271-90.

56. J. K. Tien, A. W. Thompson, I. M. Bernstein and R. J. Richards: *Met. Trans.*, 1976, vol. 7A, pp. 821-29.

57. H. K. Birnbaum: *Environment-sensitive Fracture of Engineering Materials* (Z. A. Foroulis, ed.), pp. 326-60, TMS-AIME, Warrendale, PA, 1979.

58. R. O. Ritchie, J. F. Knott and J. R. Rice: *J. Mech. Phys. Solids*, vol. 21, pp. 395-410.

59. C. D. Beachem and R. M. N. Pelloux: *Fracture Toughness Testing and Its Applications* (STP 381), pp. 210-44, A.S.T.M., Philadelphia, 1965.

60. C. D. Beachem: *J. Basic Eng.* (*Trans. ASME, Series D*), 1965, vol. 87, pp. 299-306.

61. A. W. Thompson: *Scripta Met.*, 1979, vol. 13, pp. 329-30.

MICROSCOPIC REDISTRIBUTION BEHAVIOURS OF HYDROGEN
AND FRACTURE MORPHOLOGY OF HYDROGEN-ASSISTED CRACKING
IN HIGH STRENGTH STEEL

Yoneo Kikuta and Takao Araki
Department of Welding Engineering
Faculty of Engineering
Osaka University
Suita, osaka, Japan

Present work has discussed a cooperative relation between microscopic
redistribution behaviour of hydrogen and fracture morphologies of HAZ cold
cracking in high strength steel. The authors carried out a tension test,
an internal friction experiment and an electro-chemical hydrogen flux
monitering test. From these experimental results, it may be concluded
that hydrogen accumulates at the neighborhood of the dislocation on slip
planes in grain boundaries. The microscopic redistribution behaviour
of hydrogen will hence affect the fracture morphologies and the progression
of fracture in steel

Hydrogen-assisted cracking result in formation of transgranular and
intergranular cracks. The transgranular cracks were often observed with
subcracks along the boundary of martensite lath and crystallographic,
forming on variants of {110} planes. The intergranular cracks (intergranular fracture of hydrogen embrittlement, named IG_{HE}) were observed a couple
of tongue like pattern and microcracks on a part of grain boundary surface.
The interfaces of the microcracks consisted of lath boundaries which was
same as the transgranular cracks.

Since, occurrence of quasicleavage fracture of hydrogen embrittlement
(named QC_{HE}) takes an average or long time to fracture, hydrogen first
diffuses toward dislocations on slip planes in lath at HAZ which is narrow
plastic deformed range, such as crack tips, by a external stress. Then,
a translath fracture (named C_{TL}) in QC_{HE} may be start from a region of the
highest hydrogen concentration which is generally located on the highest
dislocation density on slip planes ({110} in this study) in lath and the
identified striation like pattern also shows this occurrence. So, hydrogen
embrittlement is considered to be slide-plane decohesion.

If slip continues by lath and grain boundary source operation, these
fresh dislocation pick up hydrogen, because the rapid arrival of disloca
tions on the other side of lath and grain boundary supersaturates the
region locally with hydrogen. Then, it seems that a interlath fracture
(named C_{TL}) is caused by accumulated hydrogen at the lath boundary. With
increasing a progress time to fracture under lower applied stress, IG_{HE} is

dominant, because grain boundary dislocation densities have been estimated to be three order of magnitude higher than interior values at modest strain and hydrogen supersaturates in the region.

These morphologies will hence consist of serious of microcracks which are the embryos of hydrogen cracks with hydrogen gathering.

Introduction

It is generally said that steel can absorb hydrogen from production, processing, and use under natural environment, and thus have its properties impaired due to hydrogen-assisted cracking. Hydrogen-assisted cracking is recognized as hydrogen embrittlement, delayed cracking, weld cold crack- ing, lamellar tearing adjacent to inclusions elongated in the rolling direction, and environmental embrittlement occurring in use under natural environments or a hydrogen sulfide (H_2S) atmosphere.

Recently, electron fractography has been used to investigate the causes of various fracture morphologies. Although complex, the character- istic fracture morphology has been related to stress level and microstruc- ture. Therefore, causes and mechanisms of fracture can be suggested approximately by observations of fracture morphology. The morphology of hydrogen-assisted cracking includes dimple rupture (DR), quasicleavage fracture (QC), and intergranular fracture (IG). However, these fracture morphologies do not necessarily exhibit characteristics of fracture induced by only hydrogen but also are observed in the case of other cracking processes. It is difficult to isolate uniquely the characteristics of fracture morphology induced by hydrogen. However, it is important to establish the characteristics of the fracture morphology which are relevant to investigations of causes of premature failure.

Experimental Procedure

Present work was carried out for commercial high strength steels. These chemical compositions are given in Table 1.

To investigate the ordinary fracture morphology caused by hydrogen, V-notch specimens (depth: 2mm, angle: 45°) were used for the three-point bending test and delayed cracking test. The specimens used for the implant HAZ cold cracking test was prepared from sections in longitudinal direction of plate.

Specimens were hydrogenated by cathodically charging for the three- point bending test by a temperature hydrogenation treatment for the delayed cracking test and by manual bead-on-plate welding for the implant test. These hydrogenating conditions are given in Table 2.

Table 1. Chemical compositions of materials used, weight percent

Materials	C	Si	Mn	P	S	Al	Mo	Ni	Cr	Cu	V	B	Ti
HT80A	.17	.35	1.31	.021	.023	.03	.42	.32	.49	.26	.04	.003	.015
HT80B	.12	.29	.85	.014	.005	–	.42	.07	.81	.17	.05	.001	–
SM50	.14	.31	1.36	.021	.014	–	–	–	–	–	–	–	–

The three-point bending test was carried out with a bending rate of 1 mm/min by Instron type tensile machine. For the delayed cracking test and the implant weld cold cracking test, the applied nominal stress was maintained at a constant until complete failure occurred. Fracture surface

Table 2. Methods of hydrogenation and hydrogen contents

Methods of Hydrogenation	Conditions of Hydrogenation	Hydrogen Contents, ppm	Remark
Cathodic charging	current density: 80 mA/cm^2 for 3 h at 20°C electrolyte: 5% H_2SO_4 in Poison P (20 mg/litre)	8	three-point bending test
High temperature method	hydrogen atmosphere (1 atm) at 950°C for 2 h and water quenching.	4	three-point bending delayed cracking test
Shielded metal arc welding	used electrode: E11016 (4 mm) as received welding conditions: 25 V, 180 A and 150 mm/min speed	12	implant weld cold-cracking test

obtained by the tests were observed by scanning electron microscopy and two-stage replica method.   Crystallographic orientation on the fracture surface was measured using the etch pit method.

Results and Discussion

Relation between microstructure and fracture morphology due to hydrogen

The dominant fracture morphology of hydrogen-assisted cracking differs according to applied stress level in a delayed cracking test or strain rate in a tensile test.   QC and IG are dominant with a medium and lower constant applied stress within the range in which the delayed fracture occurred or with continuous slower bending and tension.   Fracture of hydrogen charged specimen of HT-80 steel with various microstructures tested by three point bending test is shown in Fig. 1.   In Fig. 1(a)~(e), equidistant subcracks oriented approximately vertical to the main crack surface are observed.

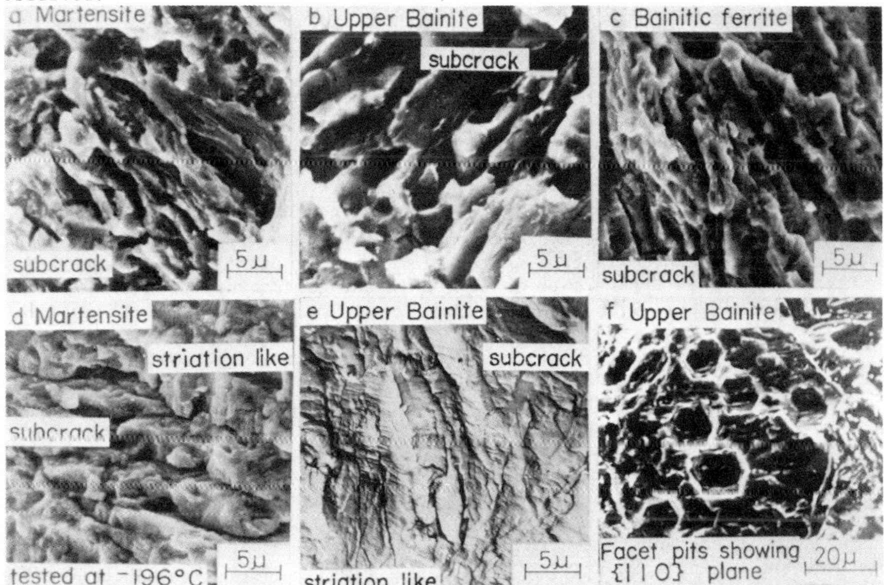

Fig. 1 Characteristic fracture modes of quasicleavage fracture due to hydrogen embrittlement, QC_{HE} tested at 25°C (HT80A)

The relation between fracture facet size and prior austenite grain size for martensitic structure is shown in Fig. 2.   It was clear that the distance between subcracks corresponded to martensite lath or ferrite lath

width. Facet size of the hydrogen embrittlement fracture was a third or
a half of that of brittle fracture [1]. It may be concluded that hydrogen
embrittlement fracture will hence consist of series of small fracture area.

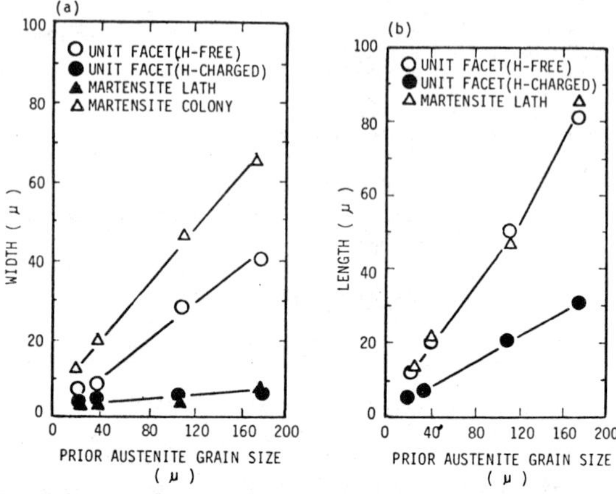

Fig. 2 Variation of unit facet size, martensite lath width,and
colony spacing with prior austenite grain for martensite
(HT80A)

Moreover, in Fig. 1(e), a striation-like or ripple pattern was
identified between subcracks. The formation of this pattern has not been
explained clearly, but it is considered to be due to collapse of slip
planes after considerable plastic deformation without the formation of
voids, such as dimples. The pattern is formed because materials deform
plastically not only on the slip plane but also on the cross-slip plane
within the individual laths. The crystallographic orientation of hydrogen
embrittlement fracture, for example, is shown in Fig. 1(f). Figure 1(f)
was a {110} plane on which the etch pits are hexagonal. It is well known
that the crystallographic orientation of cleavage is a {100} plane on
which etch pits are square. From the aforementioned results, it was found
that fracture morphology of hydrogen embrittlement was different from ductile fracture, brittle fracture and quasicleavage fracture [2]. The author-
s propose that this fracture be named quasicleavage fracture of hydrogen
embrittlement (QC_{HE}) [1].

Figure 3 exhibits the QC_{HE} in a weld cold-cracking which belongs to
the one of hydrogen-assisted cracking and fracture path was classified into
two types such as translath and interlath fracture (named C_{TL} and C_{IL}
respectively [3]). With an applied stress of 403 MPa, which was nearly a
critical lower stress level within the range of delayed fracture, IG were
dominantly observed as shown in Fig. 4. The intergranular fracture of
hydrogen-assisted cracking differed slightly from those of temperbrittle-
ness and quenched cracking. In Fig. 4(b) high magnification photographs
of (a), the intergranular fracture also included some parallel tongue like
patterns and microcracks between lath boundaries on a part of grain boundary
surface. Figure 5 exhibits the microcracks on a part of grain boundary and
schematic representation of fracture profile. The microcrack was located
between lath boundaries, like QC_{HE} and the interface of the crack will
hence consist of {110} plane. Then, the authors propose that the fracture
be named intergranular fracture of hydrogen embrittlement (IG_{HE}) and it
seems that the microcracks correspond to embryo of hydrogen embrittlement.
Takeyama [4] observed the microcrack on the grain boundary for hydrogenated

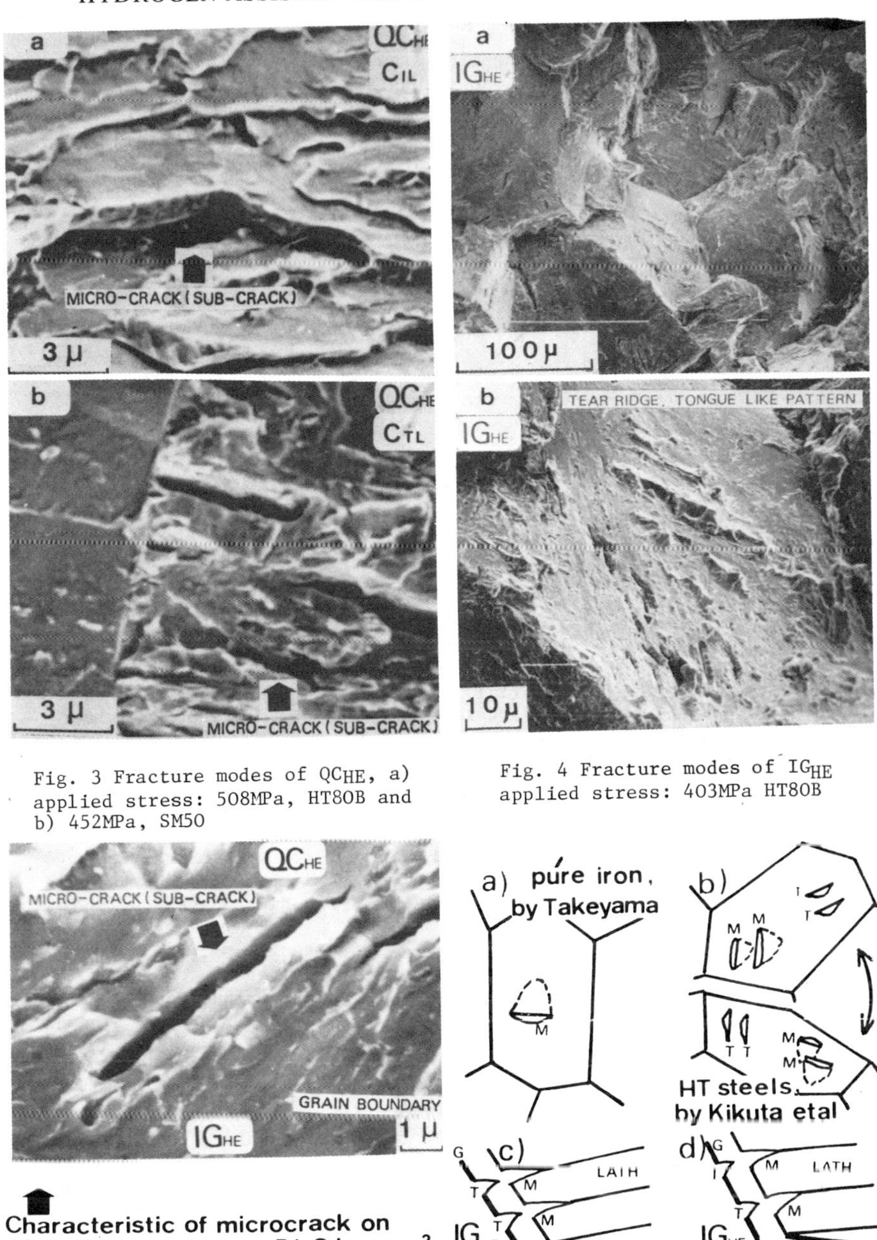

Fig. 3 Fracture modes of QC$_{HE}$, a) applied stress: 508MPa, HT80B and b) 452MPa, SM50

Fig. 4 Fracture modes of IG$_{HE}$ applied stress: 403MPa HT80B

Fig. 5 Fracture morphologies of microcracks in IG$_{HE}$

pure iron as shown in Fig. 6. The microcracks were generated at grain
boundaries and interface of the cracks consisted of zigzag steps on {100}
or {112} planes. Terasaki [3] and Bernstein and Thompson [5][6][7] had
also observed   the same fracture morphology for pure iron.

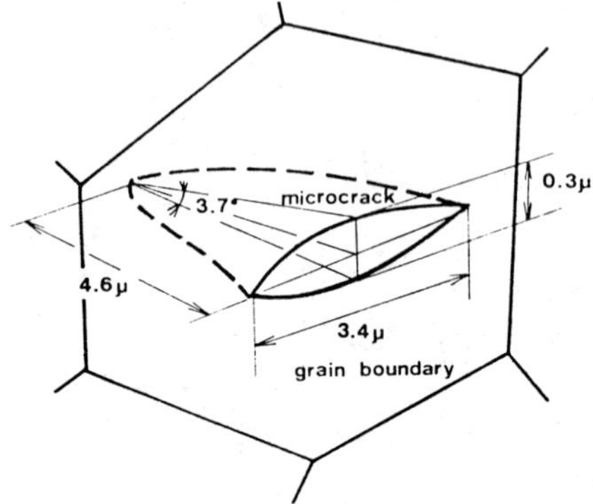

Fig. 6 Schematic representation of the microcrack (pure iron)

The role of hydrogen on fracture morphology

It has been proposed that hydrogen embrittlement is diffusion dependent. There is also much discussion in the literature concerning the exact
location of hydrogen in the material. Several workers have postulated
voids [8], second phases [9], vacancies [10], and dislocations [11] have
been suggested as possible traps. There is, however, no direct evidence
that interaction in a necessary part of the embrittlement process.

Recently, electron fractography has been used to investigate the causes
of various fracture morphologies. The characteristic fracture morphologies
have been related to stress level and microstructure. Therefore, causes
and mechanisms of fracture can be suggested approximately by observations
of fracture morphologies and aforementioned microscopic redistribution
behaviours of hydrogen.

The fracture morphologies of hydrogen-assisted cracking can be affected
by microscopic diffusion behaviour of hydrogen under various conditions,
such as applied stress, strain rate and progress time to fracture. On the
other hand, the authors carried out a tension test, an internal friction
experiment [12] and an electrochemical hydrogen flux monitering test [13].
For example, Fig. 7 shows a relationship between ductility and hydrogen
cold-work-peak height for hydrogenated HT-80 steels. When specimens of
HT-80A steel were hydrogen charged by the high temperature method, their
ductility decreased at an early stage of aging, steadied at 0°C, and then
went through a minimum. At a later stage of aging, it recovered the value
of the hydrogen-free specimen. Notch tensile strength (NTS) could be
regarded as a measure of ductility, and it was used as the most convenient
parameter for embrittlement. Hydrogen cold-work-peak height increased
with increasing aging time, went through its maximum, and decreased to a
vanishing small value. Parallelism was found between the drop in ductility
and the hydrogen cold-work-peak height.

Activation energy for hydrogen diffusion, Q is determined from temperature dependency of diffusion in the tensile test [14]. In the early stage
of aging, the activation energy for hydrogen diffusion was about 3200 cal/

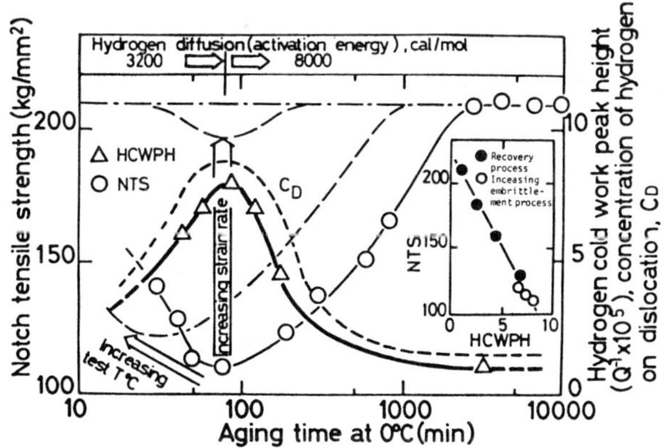

Fig. 7 Relationship between ductility and hydrogen cold-work-
peak height for hydrogenated steel (HT80)

mol. This value may express only the lattice diffusion of hydrogen. In
the latter stage of aging, it was about 8000 cal/mol. This value may result
from diffusion through microscopic defects such as dislocations and
vacancies.

From these experimental results, it may be concluded that hydrogen
cold-work-peak height can be proportional to hydrogen, C_D, in the neighborhood of the dislocation, as shown in Fig. 7 and that hydrogen first diffuse-
s toward dislocations and then away from dislocations.

It seems that the microscopic diffusion behaviour of hydrogen affects
fracture morphology of hydrogen embrittlement as observed in this study.

Also, Nagumo [15] defined clearily that the hydrogen accumulates at
slip planes in grain and grain boundaries for hydrogenated pure iron by
using the radioactive isotope of hydrogen as shown in Fig. 8. The whitish
areas indicate the site where tritium accumulates. Slip planes (central
area of Fig. 8) as well as grain boundaries and interface between non-
metallic inclusion and matrix (lower-right) are accumulation sites. This
observation suggests that defects which act as accumulation sites of
hydrogen are formed along slip planes with cooperation of hydrogen and
dislocation movement.

From the aforementioned results, the schematic representation of
fracture profile on hydrogen embrittlement is shown in Fig. 9. Since
occurrence of QC$_{HF}$ takes an average or long time, hydrogen first diffuses
toward dislocations on slip planes in lath which is narrow plastic deformed
region by a external stress   crack tip   Then, C_{TL} in QC$_{HE}$ may be start
from a region of the highest concentration which is generally located on the
highest dislocation density on slip planes ({110} plane in this study) in
lath and the identified striation like pattern, also, shows this occurrence.
So, hydrogen embrittlement is considered to be slide-plane decohesion. If
slip continues by lath and grain boundary source operation, these fresh
dislocations pick up hydrogen, because the rapid arrival of dislocations
on the other side of lath and grain boundary supersaturates the region
locally with hydrogen. Then, it seems that C_{TL} in QC$_{HE}$ is caused by
accumulated hydrogen at the lath boundary. With increasing fracture time

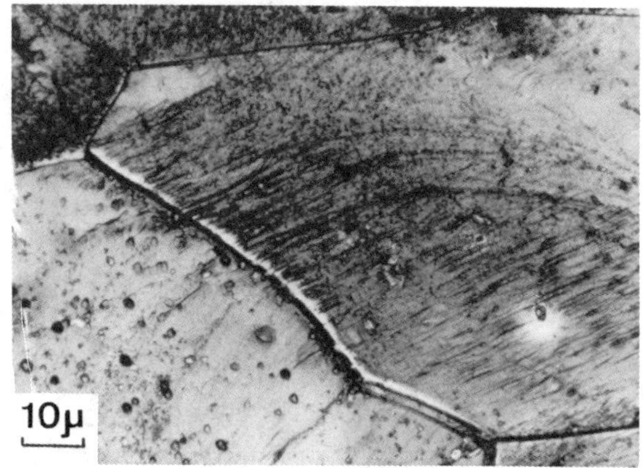

Fig. 8 Tritium color autoradiograph of pure iron strained under
       hydrogen charging by the optical microscopic method

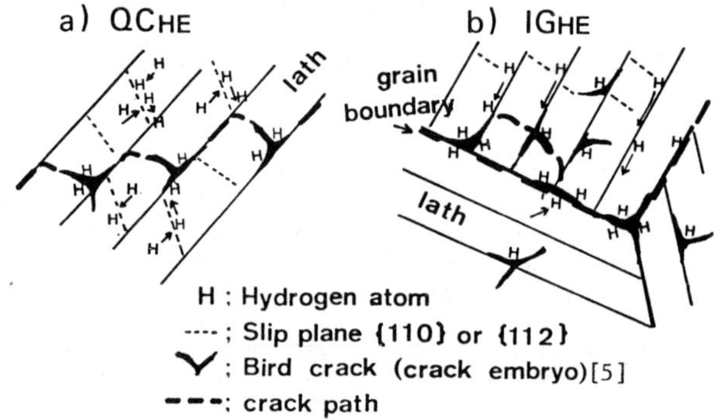

Fig. 9 Schematic representation of fracture profile on hydrogen
       embrittlement

under lower applied stress, IG$_{HE}$ is dominant because grain boundary
dislocation densities have been estimated to be three order of magnitude
higher than interior values at modest strain and hydrogen supersaturates in
the region. These morphologies will hence consist of series of microcracks
which are the embryos of hydrogen cracks.

Conclusions

Hydrogen-assisted cracking and HAZ cold cracking result in formation
of transgranular and intergranular cracks. The transgranular cracks were
often observed with subcracks along the boundary of martensite lath and were
crystallographic, forming on variants of {110} planes. The intergranular
cracks (named IG$_{HE}$) were observed a couple of tongue like pattern and
microcracks on a part of grain boundary surface. The interfaces of the
microcracks consisted of lath boundaries which was the same as the transgranular cracks.

Since, occurrence of quasicleavage fracture of hydrogen embrittlement
(named QC$_{HE}$) takes an average or long time to fracture, hydrogen first
diffuses toward dislocations on slip planes in lath at HAZ which is narrow
plastic deformed range, such as crack tips, by a external stress. Then,

a translath fracture (named C_{TL}) in QC_{HE} may be start from a region of the highest hydrogen concentration which is generally located on the highest dislocation densities on slip planes in lath and the identified striation like pattern, also, shows this occurrence. So, hydrogen embrittlement is considered to be slide-plane decohesion.

If, slip continues by lath and grain boundary source operation, these fresh dislocations pick up hydrogen, because the rapid arrival of dislocations on the other side of lath and grain boundary supersaturates the region locally with hydrogen. Then, it seems that a interlath fracture (named C_{IL}) is caused by accumulated hydrogen at the lath boundary. With increasing a progress time to fracture under lower applied stress, IG_{HE} is dominant, because grain boundary dislocation densities have been estimated to be three order of magnitude higher than interior values at modest strain and hydrogen supersaturates in the region.

These morphologies will hence consist of series of microcracks which are the embryos of hydrogen cracks with hydrogen gathering.

Reference

[1] Y.Kikuta, T.Araki, and T.Kuroda, ASTM,STP 645, 1978, p107

[2] Y. Kikuta, and T.Araki, Tech. Rep. of Osaka Univ., 26 1976, p53

[3] F.Terasaki, J. of ISI of Japan, 58, 1972 p1067

[4] T.Takeyama, J. of Japan Inst. of Metal, 43, 1979, p492

[5] I.M.Bernstein, Met. Trans., 1, 1970, p3143

[6] I.M.Bernstein and A.W.Thompson, Trans. of The Japan Inst. of Metals, 21, 1980, p429

[7] B.B.Rath and I.M.Bernstein, Met. Trans., 2, 1971, p2845

[8] H.H.Podgurski, Trans. Met. Soc. AIM, 221, 1961, p389

[9] F.R.Coe and J.Moreton, J of ISI, 207, 1966, p366

[10] W.R.Heller, in Stress Corrosion Crack Embrittlement, Wiley, New York, 1956, p163

[11] P.Bastien and P.Azou, Proc. First World Metal Cong. ASM for Metal, Cleveland, Ohio, 1951, p535

[12] Y.Kikuta, K.Sugimoto and S.Ochiai, Proc. First Int. Cong. Hydrogen in Metals, Paris, 1972, p144

[13] Y.Kikuta, T.Araki, and S.Ochiai, Proc. Second Int. Cong. Hydrogen in Metals, Paris, 1977, 3A3

[14] Y.Kikuta, T.Araki, and S.Ochiai, J. of Japan Welding Soc., 45, 1976 p1016

[15] M.Nagumo and I.Taguchi, Trans. of The Japan Inst. of Metals, 21, 1980, p225

DISCUSSION

<u>J. W. Morris, Jr., Lawrence Berkeley Lab.</u>:   It was not clear to me whether
your results and interpretations are consistent with the similar work presented by Professor McMahon.   (1) Could you elaborate on the <u>direct</u> microstructural evidence for interlath fracture?   (2) Are you and McMahon in
essential agreement on the details of the crack path?

<u>Takao Araki</u>:   (1) Yes, the direct microstructural evidence for interlath
fracture was reported in reference (1).   Moreover, recently, this was
directly observed for hydrogenated specimens, using an SEM equipped with a
delayed cracking tester, which was etched to reveal the crack path.
(2) Yes, we are in essential agreement on the crack path.

INTERACTIONS BETWEEN HYDROGEN AND AGEING PROCESS IN α-IRON AT 300°C (572°F).

M. Habashi[*], A. Elkholy[**], J. Galland[***], P. Azou[**]
[*]Centre National de la Recherche Scientifique, France
[***]Ecole Centrale des Arts et Manufactures,
Châtenay-Malabry, France
[****]Université Paris XI, France

After introducing hydrogen at 300°C in α-iron by the means of molten
salt bath electrolysis technique at -1.5 Volt/Ag, we have tensile tested the
metal at this temperature without interruption cathodic electrolysis. The
maximum strength σ_{max} and the reduction of area at rupture Σ_R are calculated
and compared with those of the metal tensile tested at the same temperature
and at its dissolution potential. The parameters studied are : holding time
and electrolysis (7.5 to 45 min) before tensile tests, and rate of hot strain-
ning at 300°C (1 to 3%) before electrolysis and tensile tests. The outgassing
quantities of desorbed hydrogen, under vacuum, are measured as function of electrolysis time, from 20° to 600°C, in the both cases, after electrolysis and
immediatly after tensile rupture. Comparative studies are carried out on α-iron
heated in furnace or electrolysed at 300°C during 45 min, cooled at room temperature, then tensile tested immediatly in furnace at 300°C.

The experimental results allow to give the following points :
1) At 300°C, for α-iron heated at its dissolution potential or cathodically
electrolysed σ_{max} decreases linearly with time. σ_{max} for hydrogenated metal
is always lower by about 5% than for the metal heated at this temperature.
The embrittlement factor is independent of electrolysis time and equal to 19%.
2) Outgassing, before and after tensile rupture, shows that the quantities of
interstitial and trapped hydrogen, measured just below 600°C, increase with
holding time. This result displays that trapped hydrogen does not lower the
embrittlement factor.
3) During cooling of the hydrogenated α-iron and reheating in furnace at 300°C ,
the most part of interstitials and trapped hydrogen is outgassed just below
300°C. The embrittlement factor is not affected, while σ_{max} is lowered by
about 26%. In this case, the ageing process observed on α-iron at 300°C is
cancelled in the presence of trapped hydrogen above 300°C.
4) When hot straining α-iron at 300°C is achieved before cathodic electrolysis
at this temperature, σ_{max} increases firstly with straining rate ($\leqslant$ 1%) and
then becomes constant. Outgassing at 600°C shows that the quantity of hydrogen
Q_{II} decreases, and becomes constant after 1% plastic deformation. Trapped
hydrogen by free dislocations increases σ_{max} and decreases Q_{II} when hot straining rate is achieved between 0 and 1%. Σ_R is not affected by straining rate.

Introduction

Mechanical stress σ_{max} variation measurements for plain steel in function of test temperature allow to put in evidence the ageing or hardening mechanisms. Indeed, such phenomenon becomes very active for temperature θ_v at which σ_{max} value is maximum. After the works of Grumbach and Sanz (1), ageing mechanisms levels and θ_v values depend on several parameters which are: cooling rate during heat treatment, tempering temperature, nitrogen and carbon contents, and of any element being able to pin up N and C, previous cold working, θ_v holding time and finally the plastic deformation rate. Let us take in account a steel having undergone a given heat treatment with C, N and Al. contents kept constant and having been tensile tested with a given plastic deformation rate $\dot{\varepsilon}$. Holding time at θ_v may first accelerate ageing mechanism inside a limit time; after that the hardening does not evolute, and even can lower slightly for an over passing the holding time limit. In a parallel direction the higher is the previous cold working at θ_v before tensile test at the same temperature, the lower is ageing process.

In this work, we want to study, using true stress-strain curve, $\sigma = f(\varepsilon)$, the mechanical properties of α-iron at 300°C (≈ 572°F) : σ_{max} and Σ_R (σ_{max} mechanical strength and Σ_R reduction of area at rupture) as function of heat holding time at 300°C and of the previous hot straining at the same temperature. Those properties will be studied with or not hydrogen induced cathodic charging. Indeed , in a previous work (2), we have studied the mechanical properties of α-iron at 300°C in a molten salts bath electrolysis (3), but after having cooling the previously hydrogen charged metal at room temperature before heating it again at 300°C tensile test temperature.

In such conditions, on one hand exists in the metal an hydrogen oversaturation and on other hand the surface concentration is none in relation with desorption in the atmosphere during the stay of some minutes at room temperature. After that, the hydrogen superficial concentration is always zero and hydrogen escapes during heating and tensile testing time. At the opposite, in this study, the tensile test is achieved during the 300°C cathodic charging, it means during the time when the gradient of hydrogen concentration is very low.

Experimental procedure

The chemical analysis of α-iron is as following (Table I)

C %	Si %	Mn %	S %	P %	Cu %	Ni %	Cr %	N %	O %	F %
0,012	0,01	0,07	0,06	0,008	0,025	0,02	0,015	0,042	0,072	balance

Tensile cylindrical specimens display a 4 mm diameter and 35 mm gauge length. All specimens undergo a normalized annealing at 950°C during one hour and then are slowly cooled in the furnace. Those heat treatments are achieved under vacuum (10^{-3} Pa).

The cathodic charging at elevated temperature, 300°C in a molten salts bath with injected water allowed us to introduce in the α-iron an hydrogen content of 15 ppm without noticeably any blistering.

With special device, figure 1, we have realized tensile test during the cathodic charging time in the molten salts bath. The cathodic polarization potential applied to the specimen during the charging is equal to -1.5 Volts with respect to silver reference electrode. Plastic deformation rate have been made $5\ 10^{-4}\ s^{-1}$. In this study, the influence of two parameters have been examined :

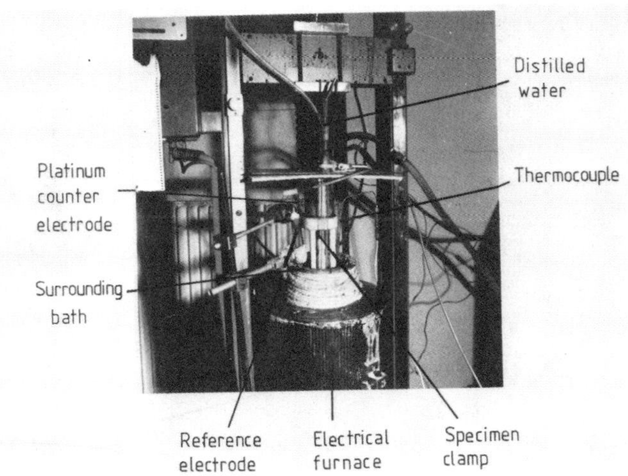

Fig. 1 – Special device for tensile testing during cathodic
polarization at 300°C.

1) Holding time at 300°C either at α-iron dissolution potential, or at
-1.5 Volts/Ag cathodic potential.
2) Different hot straining levels previously realized at 300°C before cathodic charging at the same temperature.

In both cases, we have followed the variation of σ_{max}, Σ_R and Q_H, which is
the hydrogen quantity desorbed immediately after rupture of cathodic charged
α-iron.

Experimental results and discussion

Holding time at 300°C.

We have choosen holding time values between 7.5 and 45 min before tensile test of α-iron, either at its dissolution potential, either at -1.5Volts/Ag
cathodic   charging. The cathodic charging is maintained during tensile
tests. Figure 2 shows, either hydrogenated α-iron or not, σ_{max} decreases
when holding time increases. Also, in presence of hydrogen σ_{max} evolves with
the same manner as the former case but its level is always   lower than that
α-iron without hydrogen. In the two cases Portevin-Le Chatelier phenomenon
is observed, figure 3. Unfortunately, one cannot distinguish any differences
between the absence or the presence of hydrogen effect.

Reduction of area at rupture Σ_R is dependent of holding time, figure 4.
Σ_R for hydrogenated metal is also lower than the α-iron tensile tested to
rupture at its dissolution potential. Embrittlement factor, in our study is
equal to about 19%.

In a previous work (?), we have shown that the ageing process is maximum
when α-iron is tensile tested at   300°C; figure 5, with deformation rate ε
equal to $2\,4\,10^{-3}\,s^{-1}$. Grumbach and Sanz (1) observed that the ageing process
is reduced when deformation rate decreases. In this study $\dot{\varepsilon} = 5\,10^{-4}\,s^{-1}$, and
ageing process is then attenuated. Nevertheless, the minimum holding time
used is 7.5 min which is greater than the limit time over which no hardening
evolution can occur .

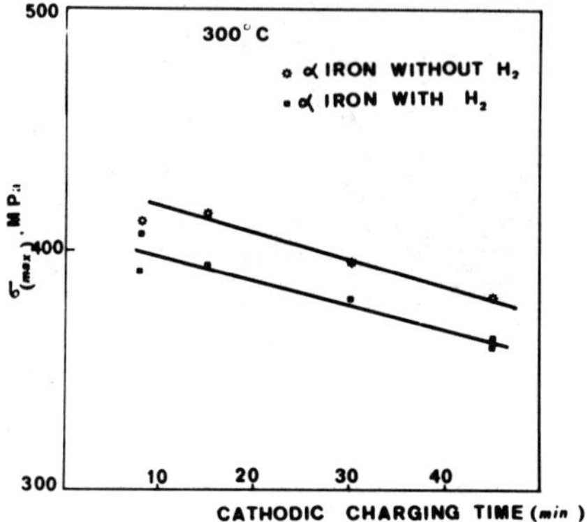

Fig. 2 - Mechanical strength σ_{max} versus holding time at 300°C for α-iron with or without cathodic hydrogen.

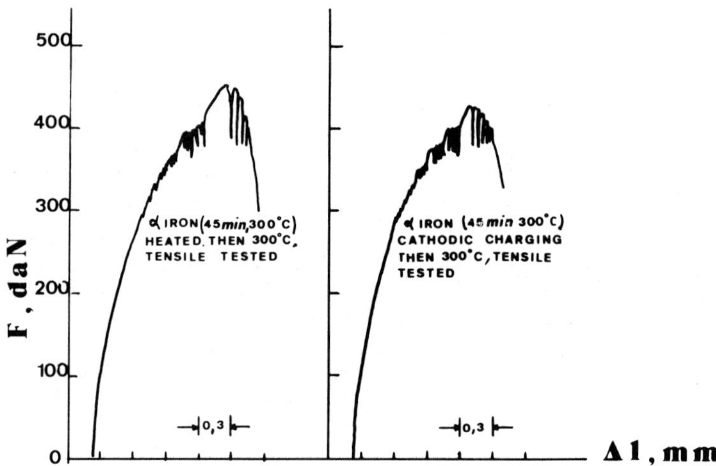

Fig. 3 - Tensile curves for α-iron at 300°C, at its dissolution potential and at -1.5 Volts/Ag.

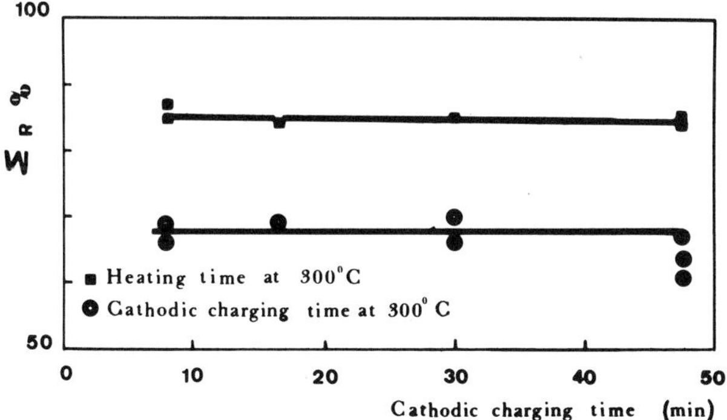

Fig. 4 - Reduction of area at rupture versus holding time at
300°C for α-iron with or without cathodic hydrogen.

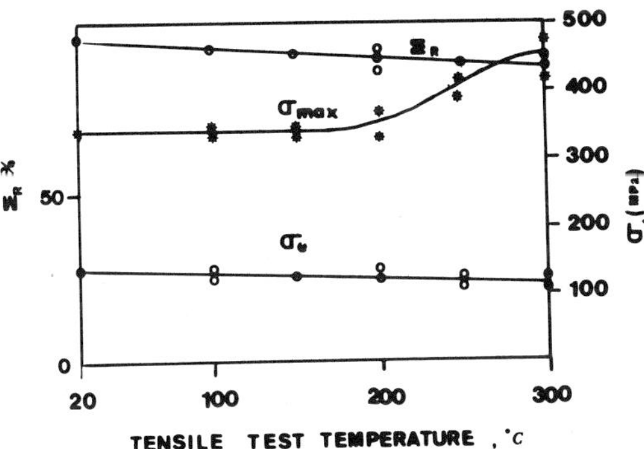

Fig. 5 - α-iron mechanical properties evolution in function
of tensile test temperature.

Without applied stress, hydrogen quantity desorbed at 300°C and 600°C
increases as the holding time at 300°C is increased, figure 6. On the opposite
side, after rupture of hydrogenated α-iron at 300°C, this quantity is decreased when holding time is raised, figure 6. Also, the difference gives an idea
about trapped hydrogen quantities at 600°C when applying high tensile stress
to α-iron during hydrogenation process at 300°C. One can conclude then that
changing trapped hydrogen quantities with holding time has no effect on the
evolution of σ_{max} and Σ_R when α-iron is tensile tested to rupture at 300°C,Opti-
cal micrographs 7, 8a and show that frequently at the grain boundaries, for

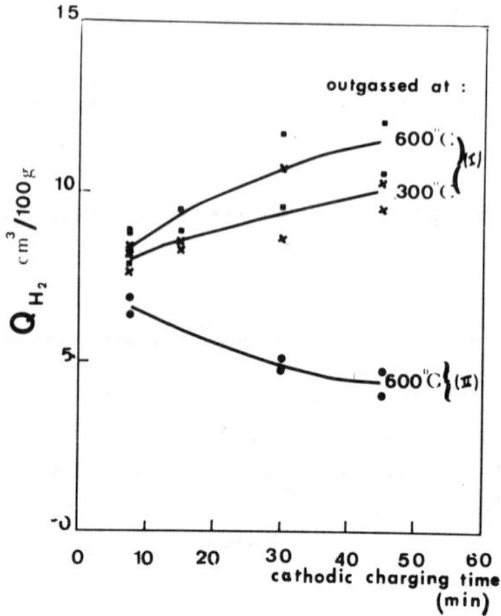

Fig. 6 - Hydrogen outgassed quantities versus holding time before and
after tensile tested α-iron hydrogenated cathodically at 300°C.

α-iron tensile tested at its dissolution potential, one can observe microcracks, figure 7, which are connected with cathodic component at this potential. Hydrogen quantity measured, under vacuum, in this case is negligeable.
When holding time is equal or superior at 7.5 min, the rupture surface of
α-iron hydrogenated at 300°C displays macro-cracks in the matrix and at the
grain boundaries, parallel to the deformation bands, figures 8a and 8b.

To explain the influence of interstitial and trapped hydrogen on the
mechanical properties of α-iron at 300°C, we have proceeded in the following
manner : after 45 min hydrogenation of α-iron at 300°C, and at -1.5 Volts/Ag,
the metal is quenched in water and immediately tensile tested to rupture
at 300°C in electrical furnace. During cooling in water, heating and tensile
testing, all interstitial hydrogen and trapped one, specially with decorated
dislocations (4) are desorbed. σ_{max} for hydrogenated metal is equal to
280 MPa, figure 9, and is lower by about 100 MPa. than that tensile tested
without hydrogen,Portevin-Le Chatellier phenomenon is also presented but
its frequency and amplitude are more attenuated than in the first case.
300°C cathodic charging time, and cooling at room temperature, reduces the
hardness of the metal after rupture (170 Hv for 7.5 min to 150 Hv for 45 min).
This is due to ageing process attenuation as it was in the former case. The
embrittlement factor in this condition is none ($\Sigma_R \approx 80\%$), and the rupture
surface contains few cracks, figure 10. Comparing these results with α-iron
hydrogenated before and during tensile test   in the molten bath without
cooling, σ_{max} = 360 MPa and Σ_R = 68% we could   show that the existence
of interstitial hydrogen and its interactions with free dislocations,
carbon and nitrogen increase σ_{max} and decrease Σ_R. The trapped hydrogen does
not alter the mechanical properties such as σ_{max} and Σ_R when one compares
the evolution of σ_{max}, Σ_R and Q_H with the holding time.

Different hot straining levels at 300°C before cathodic charging at the
same temperature.

Plusquellec and al. (5) showed that previous cold work ε_p on 0,08% C
steel before cathodic charging at 20°C increases σ_{max} firstly when ε_p value

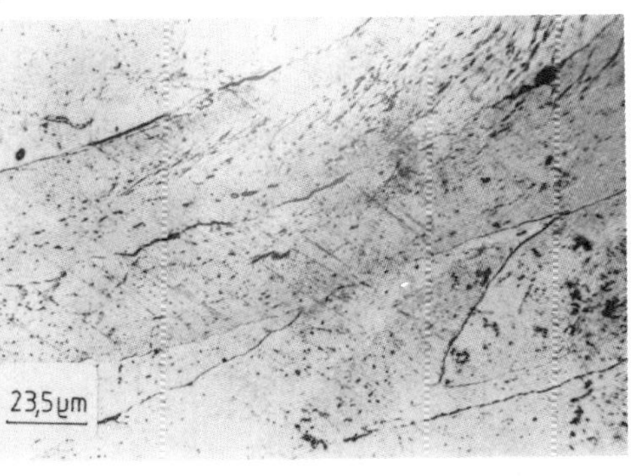

Fig. 7 – Rupture surface of α-iron tensile tested at its dissolution potential during 45 min at 300°C.

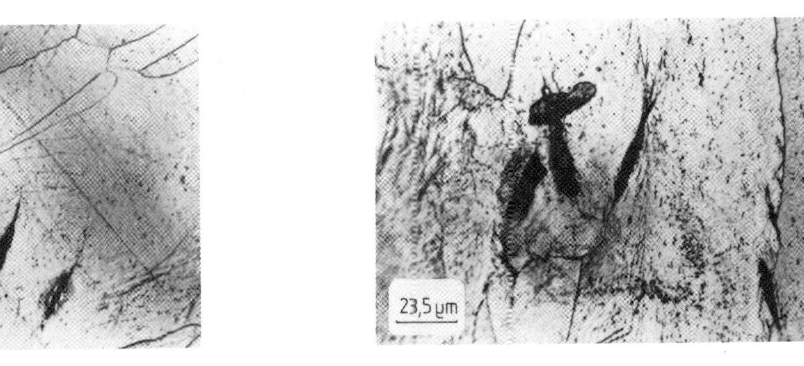

Fig. 8a – Rupture surface of α-iron tensile tested at –1.5 Volts/Ag during 7.5 min at 300°C.

Fig. 8b – Rupture surface of α-iron tensile tested at –1.5 Volts/Ag during 45 min at 300°C.

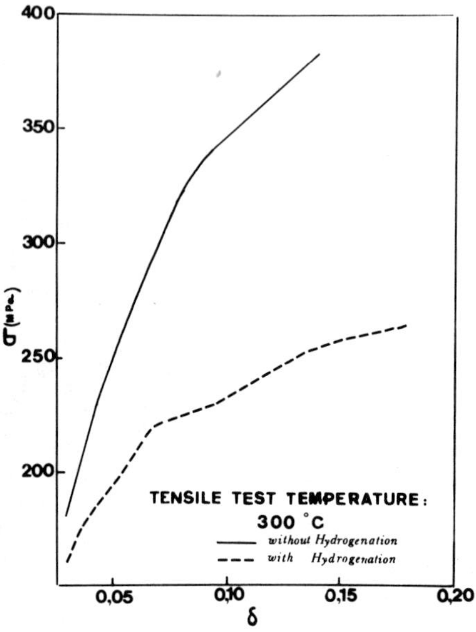

Fig. 9 - True stress-strain curves of α-iron hydrogenated at 300°C or heated at the same temperature, cooled at room temperature then tensile tested at 300°C in electrical furnace.

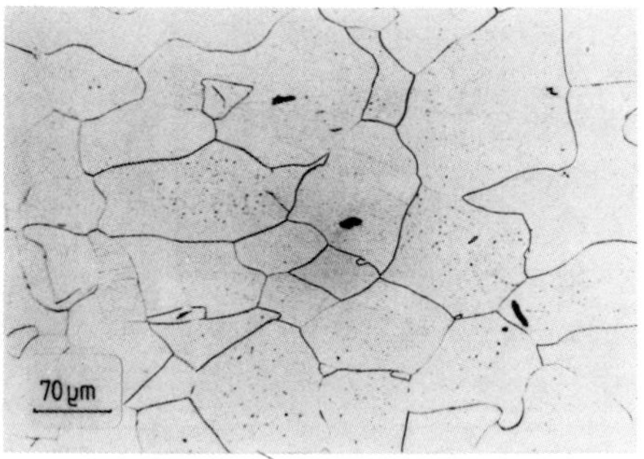

Fig. 10 - Rupture surface of α-iron hydrogenated at 300°C, cooled at 20°C then tensile tested at 300°C in electrical furnace.

increases up to about 6%, σ_{max} then decreased, and at $\varepsilon_p = 28\%$ σ_{max} is the same as for the metal without hydrogen. At the same time, Σ_R decreses in presence of hydrogen from 78% to 49%. Also Q_H remained constant up to $\varepsilon_p \approx 6\%$ and then linearly increased with increasing ε_p value. Those results show that up to a critical value of cold working ($\varepsilon_p \approx 6\%$) increasing interactions of interstitial hydrogen-free dislocations, carbon and nitrogen give rise to σ_{max} and reduce Σ_R. After this critical value, trapped hydrogen by free dislocations and other defects changes this tendancy.

In this work, we have pursued Plusquellec's point in studying the influence of hot straining ($\varepsilon_p \approx 0$ to 3%) on the mechanical properties of α-iron at 300°C before cathodic charging at this temperature, during 7.5 min

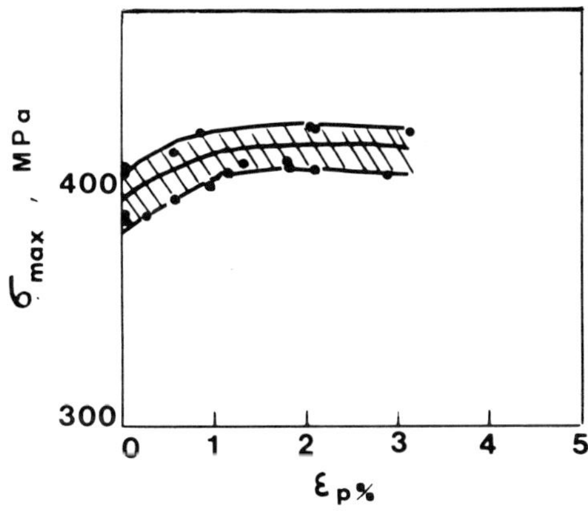

Fig. 11 – Mechanical strength σ_{max} versus previous hot straining levels for α-iron cathodically hydrogenated at 300°C.

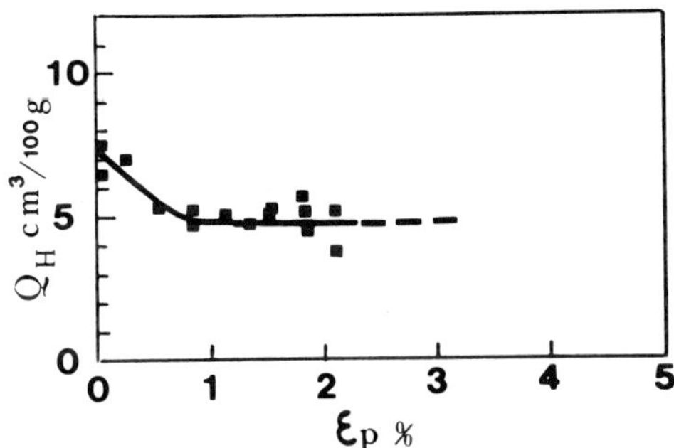

Fig. 12 – Hydrogen outgassed quantities versus previous hot straining levels for α-iron cathodically hydrogenated at 300°C.

Figure 11 shows that when increasing hot straining from $\varepsilon_p \approx 0$ to 3%, σ_{max} increases firstly, then becomes constant at $\varepsilon_p \approx 1\%$. Σ_R remains constant independantly of ε_p value, figure 12. The desorbed hydrogen quantity of tensile tested α-iron decreases to be constant when hot straining value is about 1% or more, figure 13. These results are the same as those found by Plusquellec (4) on 0,08% steel, with two differences; firstly, previous hot straining at 300°C may reduce the ageing process by attenuating σ_{max} evolution, secondly, interactions interstitial hydrogen-free dislocations increase σ_{max}. The result of these two actions is that σ_{max} firstly increase just to $\varepsilon_p \approx 1\%$.

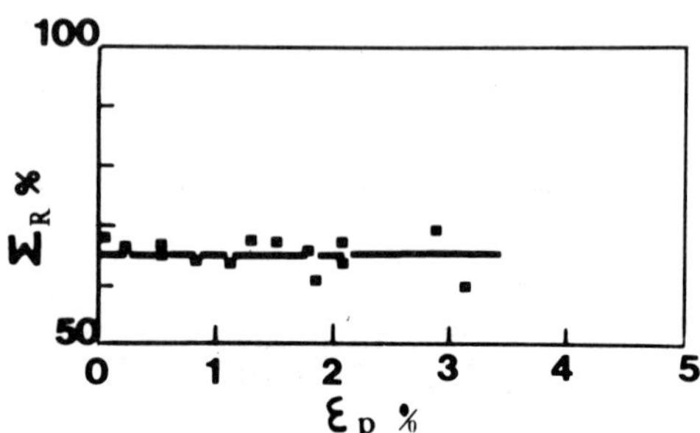

Fig. 13 - Reduction of area at rupture versus previous hot straining
levels for α-iron cathodically hydrogenated at 300°C.

as   trapped hydrogen by free dislocations and probably grain boundaries
(3) becomes active (see figure 11), but this trapped hydrogen does not alter
either σ_{max} nor Σ_R (see figures 11, 12 and 13).

Conclusions

Having cathodically introduced hydrogen at 300°C in α-iron by means of
molten salts bath electrolysis technique, and using true stress-strain curve,
$\sigma = f(\varepsilon)$, we have studied the evolution of the mechanical properties at
300°C : σ_{max} and Σ_R. Also, we have measured the hydrogen quantity desorbed
either before tensile or after tensile test of the hydrogenated metal. Two
parameters have been examined:holding time at 300°C either at α-iron dissolution potential or at−1.5 Volts/Ag; hot straining levels previously realized at 300°C before cathodic charging.

This work gives the following conclusions :

1) At 300°C, when holding time varies between 7.5 to 45 min either at α-iron
dissolution potential or at −1.5 Volts/Ag, σ_{max} decreases linearly when
holding time increases. σ_{max} with hydrogenation condition is lower than in
the absence of hydrogen.

2) Σ_R is not sensitive to holding time at 300°C, and the embrittlement
factor is about 19%.

3) When cooling hydrogenated α-iron (300°C) in water, then heating it at
300°C in electrical furnace, σ_{max} is   reduced by about 100 MPa with respect to the virgin α-iron. The hardness value decreases with holding time.
Embrittlement factor, in this case, is none. These results show that interstitial hydrogen has an important action, by mean of interactions with interstitial hydrogen-free dislocations, on the evolution σ_{max} and Σ_R: σ_{max} increased and Σ_R decreased.

4) On the opposite side, measuring Q_H at 600°C, under vacuum, before and
after tensile test at 300°C, shows that trapped hydrogen is increased with

holding time, i.e. the trapped hydrogen, in our conditions, has no effect on the mechanical properties of α-iron hydrogenated at 300°C.

5) Previous hot straining at 300°C followed by cathodic hydrogenation of α-iron at the same temperature attenuates the ageing process but in the same time interstitial hydrogen-free dislocations interactions increase σ_{max} slightly to $\varepsilon_p \approx 1\%$ and then becomes constant. The trapped hydrogen increased with ε_p but does not alter the reduction of area at rupture Σ_R.

This work allowed to show the actions of interstitial hydrogen and its interactions with free dislocations, and of trapped hydrogen during ageing process, on the mechanical properties at 300°C of α-iron hydrogenated at this temperature.

References

1. M. Grumbach, G. Sanz : Rapport IRSID, 1970, CIT, N° 5.

2. M. Habashi, A. Elkholy, J. Galland, P. Azou : Journées d'Automne, 1978 To be published in Mem. Sci. de la Revue de Métall., France.

3. A. Elkholy, J. Galland, P. Azou, P. Bastien : C. R. Acad. Sc., France, 1977, 284, Série c, p. 363.

4. M. Habashi, J. Galland, A. Elkholy, P. Azou : Second JIM Internat. Symposium (JIMIS 2), Nov 1979, Minakami, Japan.

5. J. Plusquellec, P. Azou, P. Bastien : Mém. Sci. de la Revue de Métall., France, 1963, LX, N° 2, 111.

DISCUSSION

Anton West, Sandia National Laboratories. In your outgassing experiments, you imply that hydrogen is associated with deeper traps when the specimen is strained prior to charging, compared to unstrained samples, and therefore, the amount of hydrogen outgassed at 600°C is less in strained samples. Have you vacuum extracted the deeply trapped hydrogen in the strained samples to show that the total amount of hydrogen charged into both strained and unstrained samples is the same? Is it possible that pre-straining affects the amount of hydrogen charged into the sample?

P. Azou: When deeply trapped hydrogen is involved, we have the experience that it is necessary to exceed 600°C to outgas the hydrogen. Moreover, when pre-straining has been carried out, the total amount outgassed above 600°C is higher than without pre-straining.

IMPLICATION OF STRENGTHENING MECHANISMS IN HYDROGEN

EMBRITTLEMENT SENSITIVITY OF HSLA STEELS

Michihiko Nagumo, Katsuyoshi Miyamoto and Takeshi Kubota

Fundamental Research Laboratories, Nippon Steel Corporation
1618 Ida, Nakaharaku, Kawasaki, JAPAN

Microstructures predominating in the sensitivity to hydrogen embrittlement and their roles in the fracture process have been examined with HSLA
steels of ferrite-pearlite as well as tempered martensite structures. In
the former steels, the pearlite fraction and the hardness of ferrite were
controlled by means of alloy compositions to cover a strength range. The
sensitivity to hydrogen induced cracking was measured by an immersion test
in synthetic sea water saturated with H_2S.

It was revealed that the ferrite hardness is the primary factor irrespective to the strength of the steel as a whole which is varied by pearlite
fraction and solution hardening of ferrite. Non-metallic inclusions act as
crack nucleation sites, but the crack length which is controlled by the
crack growth stage is mainly determined by the ferrite hardness. The
fracture mode was transgranular.

Observation by electron microscope was conducted on the microstructure
just below the crack surface. The area is high in dislocation density, and
dislocation cells which elongated almost parallel to the crack are developed.
Thus the crack growth is accompanied with heavy deformation and the cell
boundary is the probable crack path in accord with the previous findings in
pure iron. It suggests that the dislocation cell boundary decohesion
mechanism is operative in ferrite-pearlite HSLA steels.

Tempered martensite steels with tensile strength between 600 and 1200
MPa were examined by notched tensile test under cathodic hydrogen charging.
It was revealed quantitatively that the fraction of the intergranular
fracture is the main cause of the variation in the sensitivity to hydrogen
embrittlement among steels.

Introduction

Studies from many respects have been compiled previously about the effects of microstructures on hydrogen embrittlement of steels.[1] A view point which should be pursued is to reveal the prominent metallurgical variable for the embrittlement in order to make an optimised design of a required steel. Another point is to get insight about the significance of each variable in the fracture process under the existence of hydrogen.

Impurities such as P, N and B play an essential role in high strength steels with tempered martensite structure when intergranular fracture mode prevails.[2~5] In lower strength steels with ferrite and pearlite structures such as steels for pipes of API X-60 or X-70, the reduction and shape control of non-metallic inclusions have been major means for the suppression of hydrogen induced crackings.[6~8] The effects of solute alloying elements are more complicated.[1] In commercial steels, the microstructure is not homogeneous and the effects of metallurgical variables are subject to change by manufacturing procedures of steels. In this respect, not only the phenomenological knowledge but the insight on the mechanism of hydrogen embrittlement is requisite.

It has been recently well recognized that fracture path of iron at hydrogen embrittlement in transgranular mode is along slip planes although the fractographic features are different from dimple pattern.[9~11] The present authors recently revealed a microscopic process of void nucleation in the tensile deformation of iron under the existence of hydrogen.[12] Nucleation of voids along dislocation cell boundary during deformation process is likely to cause decohesion along cell boundary which coincides with slip planes, is consistent with fractographic features in hydrogen embrittlement. The present paper aims to distinguish the roles of microstructures on the hydrogen embrittlement and to examine the applicability of the fracture model to HSLA steels.

Ferrite-Pearlite Steels

Significance of Heterogeneity in Hydrogen Induced Cracking

Before stating about the respective structure, it will be demonstrated that a heterogeneity is the primary factor for the occurrence of hydrogen induced cracking in commercial steels. Fig. 1 is an example of cracks which were generated in a HSLA steel with the yield strength of 460 MPa. The chemical compositions were C:0.08, Si:0.25, Mn:1.50, Nb:0.03, V:0.05, P:0.016, S:0.005 (wt.%) and it was hot rolled to 20mm to give a fine grained ferrite and pearlite structure. The specimen of 50, 20 and 15mm in length, width and thickness, respectively, was immersed in a synthetic sea water saturated with H_2S for 96 hours. The crack within the specimen extends predominantly in transgranular mode.

The crack contains an elongated MnS inclusion which is the nucleation site of the crack. A feature to be noted is the fact that the crack extends along a zone of bainitic structure and the crack length is determined by that of the zone rather than MnS. The bainitic structure was produced by a segregation of C and Mn, and enrichment of P was also observed. Fig. 2 illustrates the distribution of P obtained with an Electron Probe Micro-Analyzer. The hardness within the zone is increased by about 150 in DPH from that at homogeneous area.

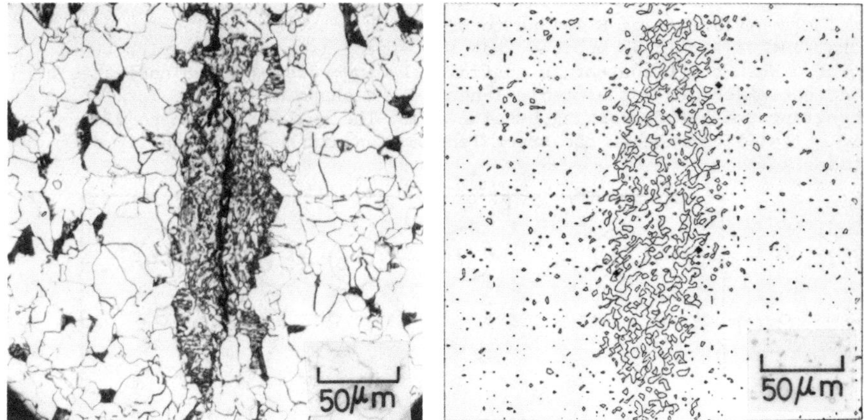

Fig.1 An Hydrogen Induced Crack De-   Fig.2 Distribution of Phospher in the
veloped Along a Segregation   Segregation Zone of Fig.1 by
Zone in a Commercial Ferrite-   EPMA. (The lines are equi-
Pearlite High Strength Steel.   concentration contours.)

It is a well established fact that the sensitivity of hydrogen embrittlement is strongly dependent on the strength of steels. So among three
factors pointed above, i.e. microstructure, alloying elements and local
hardness, which is predominating in the cracking sensitivity should be
clarified.

Hydrogen Induced Cracking Sensitivity

In order to separate the contributions of respective factors, a series
of steels tabulated below was prepared by vacuum induction heat of 50 kg to
minimize the segregation of alloying elements. Calcium was added to some of
steels to control the shape of sulfides. The steels were hot rolled at above

Table I   Compositional Ranges of Steels(wt %)

C	Si	Mn	S	P
0.13				0.001
0.30	0.25	1.6	0.001	ι
0.60	1.0	2.6	0.02	0.40

1073K to 20mm and air-cooled. The microstructure was the mixture of ferrite
and pearlite with almost uniform grain size of about 20μm in diameter.
Hydrogen induced cracking test was conducted in the same method as described
before. The frequency and the length of cracks were mainly determined by
the balance of P and S as shown in Fig. 3, where the crack length L defined
as the sum of length of each crack in unit area was taken as the measure of

the cracking sensitivity.

The contribution of S is due to MnS which acts as a nucleation site of cracks, but it should be noticed that the effect is concurrent with the P content. When the P content is sufficiently low, cracks which nucleated at MnS do not extend to a measurable length. The shape of sulfide is also a controlling factor of cracking sensitivity. The smaller length to width ratio of sulfides reduces the crack length even at the same quantity of sulfides.

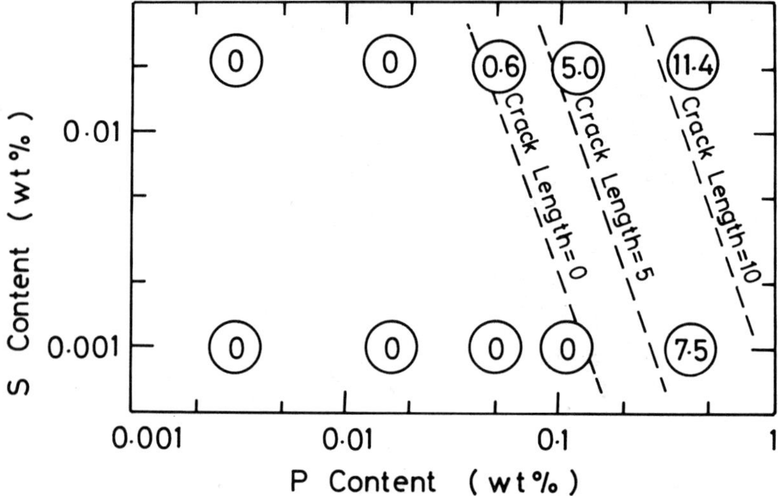

Fig.3   Effect of Balance of P and S on the Hydrogen Induced Crack Length Expressed by $L(mm/cm^2)$.(Numbers within the figure denote crack length at each composition, and lines show equi-crack-length contours.

Increase in C and Mn contents generally enhances the pearlite fraction. However, it hardly affects the cracking sensitivity so far as the structure is ferrite and pearlite although the strength of the steels is increased therewith. When the structure is changed to bainite or martensite by the addition of such elements, the cracking sensitivity is increased because of their high hardness. However, bainite or martensite is rather superior to ferrite and pearlite at a constant strength level below 750 MPa in yield strength. It may be due to grain size effect. This fact suggests that the hardness of ferrite constituent rather than the strength of the steel as a whole is the primary factor which controls the cracking sensitivity. It should be noted that the pearlite grains are not the preferred site of crack nucleation nor the crack path in spite of their high hardness.

In Fig. 4, the crack length is plotted against the hardness of ferrite of various steels. The hardness of ferrite was varied mainly by the quantities of P and Si, i.e. by means of solution hardening. In steels with high P concentration, significant crackings occur even at low sulfur content. The cracks nucleate at oxide inclusions or at grain boundaries in such situations. There remains a discussion concerning which of nucleation or growth stage of cracks is dominating in the hydrogen embrittlement of steels. The meaning of the separation of the two stages should be discussed, but the present findings indicate that the growth stage of cracks actually determines the hydrogen cracking sensitivity defined by the crack length in unit area.

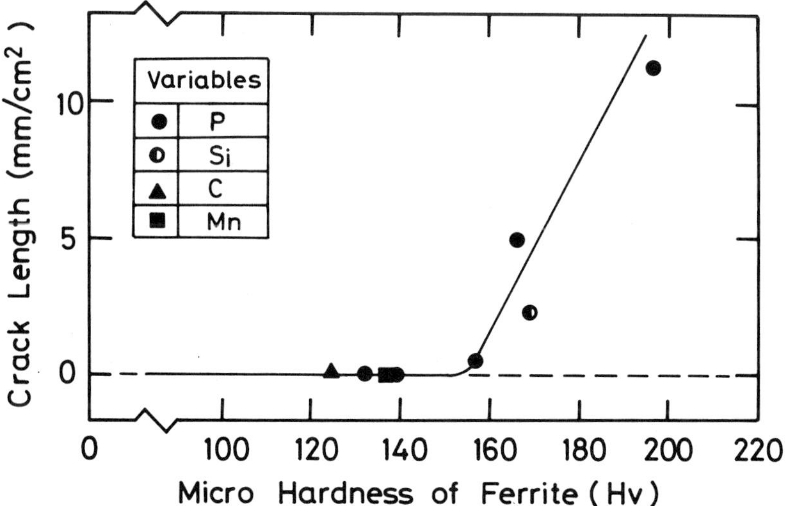

Fig.4   Hydrogen Induced Crack Length of Steels Listed in Table I Against
Micro-Hardness of Ferrite Constituent. (S = 0.02%)

<u>Microscopic Features of Hydrogen Induced Cracking</u>

The crack extension is almost parallel to the rolling plane. A specimen with cracks was fractured by impact at the liquid nitrogen temperature along the cracked plane. The hydrogen induced cracks can be easily
distinguished from the matrix by the fractographic features.

Observation by transmission electron microscopy was carried out with a
steel of 0.4% in P content to reveal the microscopic structure just below
the crack surface. There, dislocation density is very high and dislocation
cells elongated almost parallel to the crack surface were observed as shown
in Fig. 5. Dislocation density in an area fairly apart from the crack surface was much lower. This fact indicates that a hydrogen induced crack propagates associating with heavy deformation, although no external stress was
applied. The driving force for deformation in this case is presumably hydrogen gas pressure within the initial crack. This observation is in accord
with the previous one revealed in iron at tensile test under cathodic hydrogen charging.[10,12)

Dislocation cell boundaries which are parallel to the crack surface
suggests that the crack propagates along the cell boundaries in the similar
way as revealed in pure iron. In HSLA steels with ferrite-pearlite structure,
in which the transgranular crack growth through ferrite takes place, the
decohesion along the dislocation cell boundary is likely to be operative as
the mechanism of hydrogen embrittlement. The formation of precursing voids
along cell boundaries has not been definitely recognisable yet in HSLA
steels, while faint indications have been observed. Higher strength of
ferrite may not necessitate the stable growth of voids, but the details are
for further examination.

<u>Tempered Martensite Steels</u>

Tempered martensite is a standard structure which is employed for HSLA steels with tensile strength over 600 MPa. The alloy designs and heat treating conditions which are manifold according to specific usages result in various combinations of microstructural factors. In this type of steels, the fracture mode under the existence of hydrogen is a mixture of transgranular and intergranular ones, and the significance of each mode to the hydrogen embrittlement sensitivity has not been well established. Here, a preliminary result to separate factors which cause scattering of the sensitivity to the hydrogen embrittlement will be briefly presented.

<u>Contribution of Intergranular Fracture to Hydrogen Embrittlement Sensitivity</u>

A series of steels with combination of various alloying elements as shown below was prepared. Ice brine quenching from 1223 K followed by tempering at different temperatures was given to cover the yield strength levels from 550 to 1100 MPa. For some steels, the heating temperature was

Table II   Compositional Range of Steels Tested (wt %)

C	Si	Mn	P	S	Cr	Mo	V	Ti	N
	0.01				0	0	0	0	
0.15	0.25	1.50	0.002	0.004		0.30			0.004
	1.0				1.0	0.50	0.10	0.02	

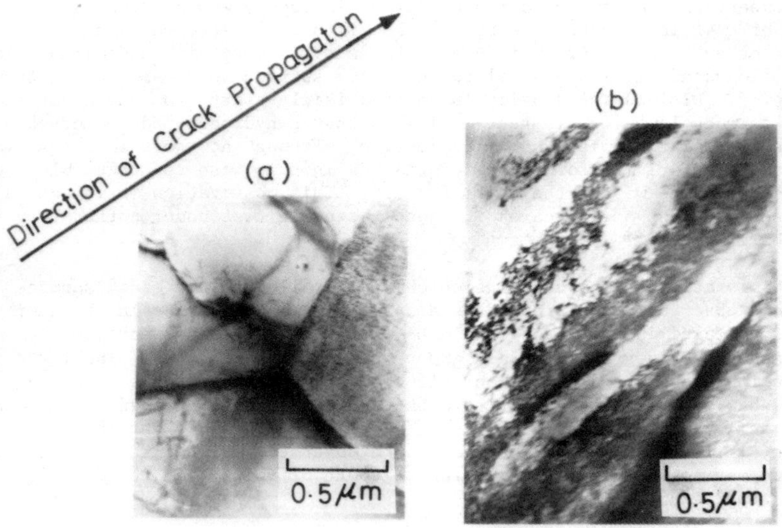

Fig.5 Electron Micrographs Showing (a) An Area Apart From the Crack and (b) Dislocation Cell Structures Developed Just Below the Crack Surface.

varied to see the effect of austenite grain size. The sensitivity to hydrogen embrittlement was evaluated in terms of the reduction of the notch tensile strength with a round specimens of 7 mm in diameter and with a circumferential notch of 1.5 mm in depth and the root radius of 0.25 mm. Hydrogen was cathodically charged in 0.1 N H_2SO_4 added with 5 mg/ℓ of AS_2O_3 at a current density of 100 A/m^2.

The sensitivities to hydrogen embrittlement of steels are summarized in Fig.6, where the parameter for the sensitivity λ is defined as $\lambda=(1-\sigma_{NH}/\sigma_N)$, where σ_{NH} and σ_N are notch tensile strengths with and without hydrogen, respectively. It was examined separately that the parameter λ shows good correspondence to the sensitivity to hydrogen cracking. The sensitivity varies by steels in a wide range at an intermediate strength range. The fraction of intergranular fracture area is the primary factor for the variation of the sensitivity, and the suppression of intergranular fracture is effective for the reduction of the sensitivity as shown in Fig. 7.

Thus the function of a metallurgical variable on the hydrogen embrittlement must be examined according to the fracture mode.

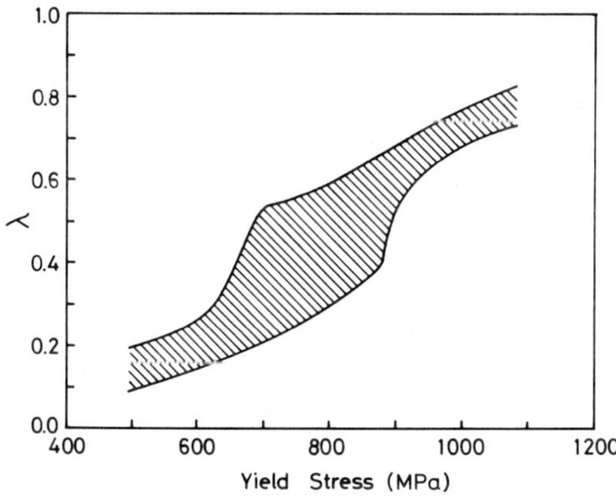

Fig.6   Hydrogen Embrittlement Sensitivity Parameter λ for Tempered Martensite Steels Listed in Table II and Controlled to Various Yield Strengths.

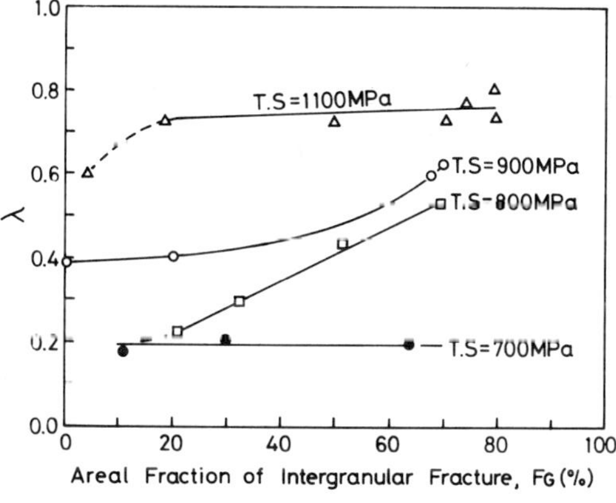

Fig.7   Contribution of Intergranular Fracture to the Hydrogen Embrittlement Sensitivity at Different Strength Levels.

In the absence of intergranular fracture, the sensitivity is also controlled by alloying elements and microstructures. The crack path in this case is mainly through interlath boundary. The compatibility of the fracture process observed in ferrite-pearlite steels with that in the martensite steels is for further examination.

Acknowledgements

The authors are indebted to Messrs. H. Morikawa and S. Sakata for the Electron Microscope observation. Their thanks are also due to Dr. H. Okada for his interests in this work.

References

1) I.M. Bernstein and A.W. Thompson: International Met. Rev. 1976, vol. 21, pp. 269

2) K. Yoshino and C.J. McMahon, Jr. : Met. Trans, 1974, vol. 5, pp. 363

3) C.L. Briant, H.C. Feng, and C.J. McMahon Jr. : Met. Trans. 1978, vol. 9A, pp. 625

4) T. Inoue, K. Yamamoto, M. Nagumo and K. Miyamoto : Proc. JIMIS-2, Hydrogen in Metals, Suppl. Trans. JIM, 1980, vol. 21, pp. 433

5) J. Kameda, N. Bandyopadhyay and C.J. McMahon Jr. : ibid. pp. 437

6) M. Iino, N. Nomura, H. Takezawa and H. Gondoh : Rev. Met. 1979, vol. 76, pp. 597

7) H. Nakasugi, T. Murata, H. Matsuda and S. Sugimura : NACE Middle East Conference, 1979

8) Y. Nakai, H. Kurahashi, T. Emi and O. Haida : Trans. Iron Steel Inst. Japan, 1979, vol. 19, pp. 401

9) F. Terasaki and F. Nakasato : Proc. Symp. Mechanism of Delayed Fracture Caused by Hydrogen, 1975, Iron Steel Inst. Japan, pp. 165

10) H. Takahashi, T. Takeyama and T. Hara : J. JIM, 1979, vol. 43, pp. 492

11) F. Nakasato and I.M. Bernstein : Met. Trans. 1978, vol. 9A, pp. 1317

12) M. Nagumo, H. Morikawa and K. Miyamoto : Proc. JIMIS-2, Hydrogen in Metals, Suppl. Trans. JIM, 1980, vol. 21, pp. 405

DISCUSSION

<u>M. Meshii, Northwestern University</u>:   I wonder if it is a matter of semantics
to define slip planes or cell boundaries as the planes on which microvoids
form (or quasi-cleavage occurs).   The important element is the high dislocation activity and probably annihilation.   Then, both can qualify, depending on the circumstance.   This generalization, I believe, is necessary,
for example, as quasi-cleavage takes place in iron around 200K, where cell
boundaries rarely form.

<u>M. Nagumo</u>:   The term 'slip plane decohesion' is sometimes ambiguous. A
model analogous to the hydride cracking in vanadium leads to a concept that a
cluster of hydrogen on a certain crystallographic plane, which happens to
be slip planes, causes weakening of atomic bond on that plane. Our concept,
however, is that the dynamical interaction of hydrogen with dislocation
cell formation is essential for void initiation and propagation to form a
crack.   So, the void and crack do not necessarily form along a fixed crystallographic plane, although the dislocation cell boundary which is a
favorite void nucleation site and crack propagation path is formed 'almost'
parallel to slip planes due to shear deformation.   In a sense that a plane
on which dislocation dynamics is active operates as the crack path, I agree
with you.   I have no idea about the behavior at low temperatures, since I
have made no observation.   It can be mentioned, however, that the loss of
local stress yield around the dislocations in the course of cell formation
is supposed to be the cause of hydrogen precipitation which leads to void
formation. So interactions between dislocations can result in the similar
effect as cell formation.

ANALYSIS OF TRAPPING EFFECTS ON HYDROGEN

EMBRITTLEMENT OF A HSLA STEEL

M. F. Stevens, I. M. Bernstein and W. A. McInteer
Carnegie-Mellon University, Pittsburgh, PA

The hydrogen trapping strength of various microstructural features in
ferrous alloys has been studied by varying the titanium carbide distribution
in a HSLA steel using quenching and subsequent tempering in the range 500° -
900°C. Resultant microstructures were characterized using optical and transmission electron microscopy, in particular the carbide size and shape. The
relative densities of reversible and irreversible trapping sites were determined using an electrolytic permeation method. Hydrogen embrittlement was
characterized both by reduction in area loss and changes in fracture mode.
The results showed that the peak aged (600°C) microstructure yielded the
greatest reduction in hydrogen susceptibility, corresponding to a fine distribution of semi-coherent TiC particles (20-40 Å). Also, it was found that
the trapping sites in this alloy could be saturated during cathodic charging,
whereupon the material was returned to its original susceptibility.

Introduction

Recent studies of the hydrogen embrittlement of ferritic and martensitic steels have investigated the role of microstructure and strength, as well as composition, on susceptibility(1,2). Although the embrittlement resistance is a complex function of many parameters such as grain size and crack path(3), identifiable beneficial features include refined grain size (4) and fine carbide sizes(5). The latter category is partially supported by results showing the enhanced resistance of bainitic and tempered martensitic microstructures(1), as well as the trapping analysis of Pressouyre (6,7). The latter studies showed that hydrogen trapping at TiC particles, as well as at Ti atoms in ferrite, could be accomplished innocuously if these traps were finely and homogeneously distributed. Similar results have been demonstrated in thoria dispersed nickel(8), and in a number of alloy steels(9). Although Pressouyre's findings are dependent on the initial state of the material(10), hydrogen transport mode, and precipitate size, the results suggest that through proper thermomechanical treatment, a trap density may be generated which can be exploited to modify the extent and kinetics of embrittlement.

This beneficial feature of trapping is likely related to the common requirement of the prevalent embrittlement mechanisms(11,12,13), that high, non-equilibrium concentrations of hydrogen must be developed at the potential failure site; traps somehow retard or prevent this accumulation step and embrittlement is reduced.

The present study has attempted to combine an analysis of embrittlement, as measured by tensile RA loss, with an electrochemical permeation study, in order to correlate mechanical response to hydrogen mobility and trapping density. A Ti-based high strength low alloy (HSLA) steel was chosen with the principal trap density consisting of TiC precipitates ranging from coherent to incoherent in nature. The carbide/matrix interface (14,15) has been shown to be an effective, "irreversible" trap for hydrogen, with a trapping energy as high as 94.6 KJ/mole H(6,10), thus capable of dramatically modifying H transport kinetics. The density and strength of this trap was controlled through quench-and-age heat treatments, thereby allowing both size and density dependency to be evaluated.

Experimental

The alloy used in this study has the following composition:

Table I   Alloy Composition

C	Mn	P	S	V	Ti
0.064	1.37	0.003	0.007	<002	0.22

In these steels, manganese is added for its beneficial effect on low temperature impact properties, for its modest solution strengthening, and also for its ability to depress the γ/α transformation. Silicon is added mainly as an oxygen getter prior to the addition of Ti, which itself is added as a carbide former and grain refiner. Nitrogen and sulfur levels are kept low in order to eliminate formation of TiN and Ti(C,S), both of which decrease the amount of Ti available for strengthening(16).

The alloy heat was received in hot-rolled form, with a microstructure

of 0.45%TiC in an acicular ferrite matrix. All specimens were then solutionized at 1200°C in argon for 1 hr, followed by an ice water quench. Selected
specimens were then tempered in salt baths at temperatures from 500°C to
900°C for 1 hr. Carbide precipitation was not observed below 600°C, suggesting that below this temperature any carbides present were coherent and less
than 30Å in diameter, as will be discussed. Transmission electron microscopic quantitative analysis showed that carbide densities in specimens aged
above 600°C corresponded to a constant volume fraction (0.4∿0.5%) of carbide,
indicating that carbide coarsening predominates above 600°C. Optical microscopy revealed that recrystallization occurs above 800°C, as observed by
other workers using electrical resistivity methods(17).

Uniaxial tension tests were conducted using specimens with a 25.4mm gage
length and 6.4mm diameter. Tests were carried out on an Instron testing
machine with a constant crosshead rate of 0.02mm/min. Duplicates of hydrogenated samples of each heat treatment were tested, after cathodic charging
in $1\underline{N}$ H_2SO_4 using a platinum anode. As_2O_3 and CS_2 were added to the bath as
recombination "poisons" to promote hydrogen entry. Although input fugacities
were not determined, two charging times were chosen, using a matrix diffusivity determined from the literature(18). To simulate low and high hydrogen
input situations, a 5 hr charge was used as a non-saturable condition, with
most trapping occurring at irreversible sites, while a 24 hr charge ensured
total saturation of all the trapping populations as well as the matrix. The
charging current in all cases was $10^{mA}/cm^2$. This value was chosen as the
maximum value for which no irreversible damage occurred, as determined by
outgassing and subsequent tensile testing; these latter tests showed that the
uncharged ductilities (as measured by RA charges) were obtained.

In the second part of this study, electrochemical permeation(19) experiments were carried out on identically heat treated coupons in order to obtain information on the active trapping densities in these microstructures.
The technique used followed that used by Pressouyre(5,6) and will thus not
be reviewed here. Diffusivities determined from these experiments were calculated using the time-lag method, an approach derivable from simple diffusion
theory(20).

Results

Tensile Tests

The carbide distribution in three of the final microstructures is shown
in Figure 1. Precipitation can be seen to occur preferentially along ferrite
lath boundaries, sub-grain boundaries, and dislocations. Previous field ion
microscopy studies(21) have determined that for carbide sizes between
30-150Å, the particles begin to lose their coherency with the formation of
interfacial dislocations; above 200Å, the particles became incoherent. The
coarsening of the carbide particles is evident from examination of these
micrographs.

Figure 2 summarizes the yield and ultimate strengths for each of the
microstructures. The gradual drop in strength beginning at the 700°C temper
is associated with the combined effects of grain growth and recrystallization,
with precipitate coarsening believed to play a secondary role. For reasons
which will become evident, subsequent analysis was centered on structures
with yield strength > 550 MPa.

Ductility measurements were made on hydrogen charged as well as uncharged tensile specimens, and are shown in Figure 3. Uncharged ductilities

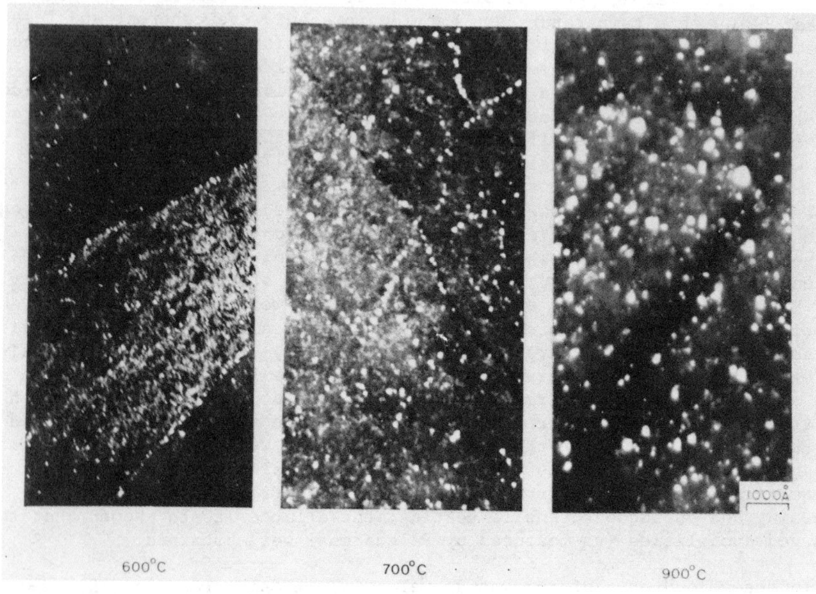

Fig. 1.   TEM dark-field micrographs of carbide distribution and sizes at aging temperatures of 600°C, 700°C and 900°C.

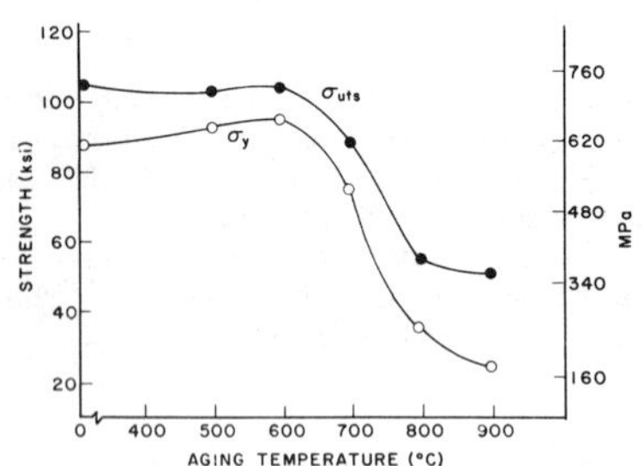

Fig. 2.   Tensile property response to aging temperature.

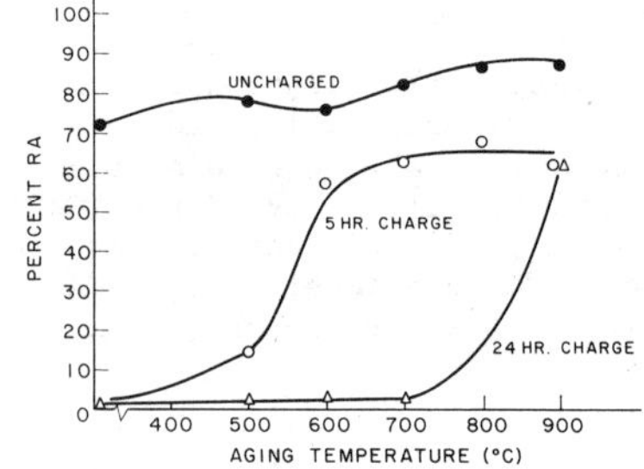

Fig. 3.   Effect of carbide morphology on tensile ductility at various bulk hydrogen concentrations.

were generally independent of annealing conditions, except for a slight dip
at 600°C, and a slight increase after recrystallization. After moderate
charging (5 hrs.), the microstructures free of visible precipitates were
virtually totally embrittled. The first increase in ductility accompanied
the formation of significant amounts of semi-coherent precipitates. After
total hydrogen saturation (24 hr charge), all microstructures were totally
embrittled, except for the recrystallized materials. This latter result is
probably due to the decreased strength levels and, therefore, the subsequent
inability of these samples to attain the critical stress levels needed for
cleavage(22).

Scanning electron micrographs of the as-quenched and 600°C fracture
surfaces are shown in Figures 4 and 5. These structures are compared since
they have equivalent strength levels. Both structures display a microvoid
coalescence failure mode in the uncharged condition. The as-quenched structure failed by a cleavage mode after both charging conditions, whereas the
600°C temper retained a predominantly MVC type fracture mode, gradually
developing a faceted intergranular failure along prior γ boundaries as the
hydrogen content increased.

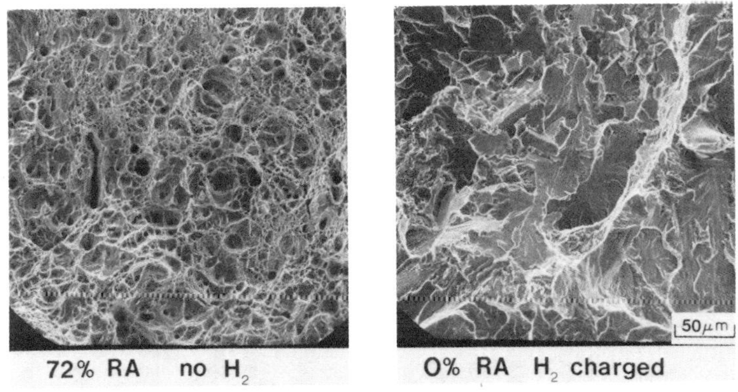

Fig. 4. SEM micrographs of as-quenched fracture surfaces for uncharged
and charged conditions.

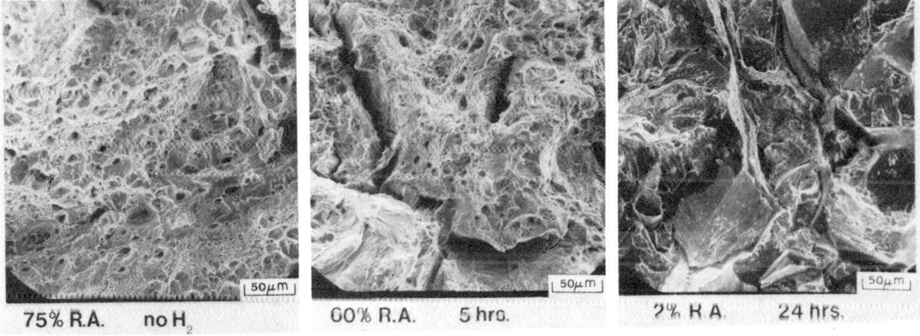

Fig. 5. SEM micrographs of 600°C aged fracture surfaces for uncharged and
charged conditions.

<u>Permeation Studies</u>

Time lag diffusivities are summarized in Figure 6. For each tempering condition, an effective diffusivity was determined using rise transients from both the first and second polarizations. The second polarization in each experiment was the subsequent permeation of hydrogen through the membrane after the charging current was removed following the first polarization and the coupon was allowed to outgas. The type of traps present in each structure can be deduced from a comparison of successive polarizations. The first polarization transient gives information on the concurrent behavior of irreversible and reversible traps; subsequent polarizations predominantly involve the resaturation of reversible traps. A substantial increase in the second apparent diffusivity relative to the first indicates the presence of significant irreversible trap density, as seen, for example, for the 600°C temper in Figure 6. The decrease in this relative difference at higher tempering conditions is consistent with the decrease in available carbide/matrix surface area that accompanies the coarsening process. These results are consistent with a strong influence of irreversible trapping in reducing the mobility and bulk transport of hydrogen in these alloys.

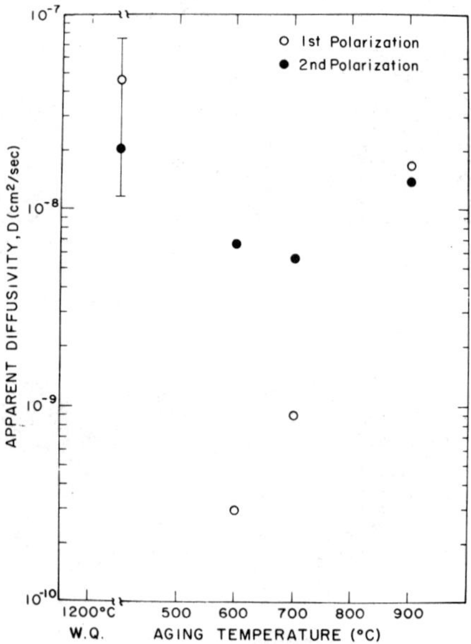

Fig. 6.   Diffusivity measurements from primary and secondary polarizations showing variation of trap type densities.

Discussion

Several features of the embrittlement susceptibility of this alloy can
be deduced from examination of the data. First, Figure 2 shows that the
strength levels obtainable in this alloy class are primarily a function of
grain (lath) size. Although metallographic examination indicates that the
as-quenched microstructure results from a shear assisted transformation, the
low carbon content suppresses the occurrence of the brittle behavior characteristic of untempered martensite. The high toughness and ductility values
found in these hydrogen-free alloys for all tempers has been attributed to
the high density of mobile dislocations found in the acicular ferrite matrix
as a result of the shear assisted transformation from austenite(23). Evidence
of secondary hardening associated with alloy (Ti) carbide precipitation can
be seen in the temperature range between 500°C and 600°C. As stated previously, the loss in strength at tempering temperatures $\geq$ 700°C is associated with recovery and recrystallization processes.

While it is possible that the observed hydrogen embrittlement in this
alloy may involve some effect of hydrogen on slip behavior, it is curious
that it manifests itself by different failure mode changes at equivalent
strength levels for different heat treatments. In the as-quenched structure, the observed cleavage may result from a surface or cohesive energy
suppression thus lowering the inherent toughness. The tendency for the
hydrogen saturated tempered microstructures to fail along prior γ boundaries may may reflect an effect of Si on intermetallic (S,P) segregation
or direct segregation of Mn to these boundaries during solutionizing treatments. It is significant, however, that the moderate charge condition in
the 600°C temper microstructure does not affect the explicit fracture mode
change in the same manner as it does in the as-quenched condition. Instead,
a ductile fracture mode continues to operate, presumably because of the continued strong influence of the carbide traps. Most of the dissolved hydrogen
is either preferentially trapped there prior to deformation or else the dislocations are able to deposit some of their hydrogen atmosphere at the interface when slip occurs, thereby preventing its later involvement in the embrittlement step.

The permeation studies when compared with the TEM results show clear
evidence that the strong irreversible trapping is attributable to the TiC
precipitation. The lack of ductility in all of the unrecrystallized structures after long term saturation indicates that these traps are saturable
and that excess dissolved hydrogen will be free to exploit the microstructural features open to it even in the absence of these traps.

Conclusions

It is clear that through the proper development of grain structure,
dislocation density and precipitate distribution, that the embrittlement
susceptibility of this alloy can be controlled and minimized. The refined,
lath-like ferrite formed during the quench-and-age treatment promotes increased strength, while the low carbon level precludes formation of brittle
martensite, thereby producing significant strength and ductility. The high
dislocation density insures both an ample mobile dislocation population for
ductility and also provides a large reversible hydrogen trapping population.
In addition, the TiC precipitates formed in this alloy were clearly associated with the strong irreversible trapping observed by hydrogen permeation
studies. Their presence in the microstructure was responsible for the in-

creased resistance to embrittlement. Through long-term charging, it was also shown that these latter traps were saturable, thereby returning the matrix to high concentrations of hydrogen and a greatly increased susceptibility to embrittlement.

Acknowledgments

This work was supported by the U.S. Office of Naval Research, Grant No. N00014-75-C-0265.

References

1. A. W. Thompson and I. M. Bernstein, "Advances in Corrosion Science and Technology, Vol 7" (Eds. Mars G. Fontana and Roger W. Staehle), Plenum Press, 1980.
2.° A. W. Thompson and I. M. Bernstein, "Proceedings 2nd Int. Conf. on Hydrogen in Metals," Paris, 1977, Vol. 10, Pergamon, Oxford 1978.
3. C. D. Beachem, Met. Trans., Vol. 3, 437 (1972).
4. B. B. Rath and I. M. Bernstein, Met. Trans., 2, 2845-2851 (1971).
5. G. M. Pressouyre, "The Role of Trapping on Hydrogen Transport and Embrittlement," Ph.D. Thesis, Carnegie-Mellon University, 1977.
6. G. M. Pressouyre and I. M. Bernstein, Met. Trans., 9A, 1571-1580 (1978).
7. G. M. Pressouyre and I. M. Bernstein, Corr. Science, 18, 819 (1978).
8. A. W. Thompson and B. A. Wilcox, Scripta Met., 6, 689-696 (1972).
9. Y. N. Archacov, et. al., Z. Metallov, 9, 228 (1973).
10. G. M. Pressouyre, Acta Met., 28, 895-912 (1980).
11. C. A. Zappfe and C. E. Sims, Metals and Alloys, 11, 145, (1940).
12. A. R. Troiano, Trans. AM Soc. Met., 52, 54 (1960).
13. R. A. Oriani, Ber. Bunsenges Ges., 76, 848 (1972).
14. J. P. Laurent, G. Lapasset, M. Aucouturier and P. Lacombe, in "Hydrogen in Metals," (Eds. I. M. Bernstein and A. W. Thompson), 559 (1974).
15. T. Asaoka, C. Dagbert, M. Aucouturier and J. Galland: Ser. Met., 1977, Vol. 11, 467-72.
16. E. E. Fletcher, A Review of the Status, Selection, and Physical Metallurgy of HSLA Steels, MCIC-79-37 (1979).
17. I. Takahashi, et. al., Hydrogen in Metals, Proc. JIMIS-2, Supp. to TRANS. JIM, Vol 21, 1980.
18. G. Alefeld and J. Volkl, Hydrogen in Metals, Topics in Applied Physics, Vol. 28, 1978.
19. M.A.V. Devanathan and Z.O.J. Stachursky, Proc. Roy. Soc., A270, 90 (1962).
20. J. Crank, The Mathematics of Diffusion, Oxford, Clarendon Press (1956).
21. A. Youle, B. Ralph, S. Freeman and R.W.K. Honeycombe, Metallography, Vol. 7, 1974, 333.
22. T. D. Lee, T. Goldenberg and J. P. Hirth, Met. Trans., 10A, 199-208 (1979).
23. M. R. Krishnadev and R. Ghosh, Met. Trans., 10A, 1941 (1979).

RELATIONSHIP BETWEEN MICROSTRUCTURE AND HYDROGEN SUSCEPTIBILITY

OF SOME LOW CARBON STEELS

L.N. Pussegoda and W.R. Tyson
Physical Metallurgy Research Laboratories
Canada Centre for Mineral and Energy Technology
Energy, Mines and Reşources, Canada.

INTRODUCTION

The susceptibility of steels to hydrogen embrittlement is generally enhanced as their strength increases. For example, it is well known that martensitic microstructures are embrittled by hydrogen[1], their susceptibility increasing with increasing carbon content and hardness. However, for a given hardness, different microstructures can have different susceptibilities to hydrogen embrittlement[2]. For example, most studies show that tempered martensite or bainite has superior resistance to hydrogen embrittlement compared with pearlite or pearlite-ferrite mixtures at equivalent strength levels[2].

In a series of recent investigations, we have examined the effects of hydrogen on steels of a variety of microstructures. Two classes of steels are involved: high-strength low-alloy (HSLA), and electroslag weld (ESW) metal. The prevention of cold cracking as a result of welding is of considerable practical importance in applications of these steels.

Strong, weldable, HSLA steels have recently been developed for structural applications. Such steels have low carbon contents to enhance weldability[3], and micro-alloy additions primarily for microstructural control. HSLA steels are normally air cooled following controlled rolling, but quenching with water sprays directly after rolling (i.e. direct quenching) or quenching and tempering improves strength and toughness[4]. We have studied the hydrogen susceptibility of such steels, including microstructures representative of the characteristic regions of the heat-affected zone (HAZ) of a manual field weld of a direct-quenched steel[5,6].

Recently, ESW metal has been observed to crack along the pro-eutectoid ferrite which delineates prior γ grain boundaries by a hydrogen-induced mechanism,[7,8] and referred to as grain boundary separations. Similar cracks along prior γ grain boundaries have been frequently observed in weld metal microstructures which simulate those found in arc welds[1]. We have studied the hydrogen susceptibility of ESW metal[9] as a contribution toward understanding this phenomenon.

In this work, samples were charged thermally[5] or electrolytically[6,9] with hydrogen and their susceptibility to embrittlement evaluated by slow tensile tests and metallographic examination of fractures[5,6,9]. In this paper a comparison is made of the major results from these investigations, in an attempt to identify embrittlement mechanisms.

EXPERIMENTAL DETAILS

The composition and mechanical properties of the steels studied are compared in Table 1. The DQ-CGHAZ microstructure was obtained by thermal cycle (Gleeble) simulation of a manual field weld.

Table 1 - Composition and Mechanical Properties of the Steels

Steel[a]	QT	DQ	DQ-CGHAZ	ESW
C	0.14	0.06	SAME AS	0.18
Mn	1.39	1.95		1.30
Si	0.37	0.11	IN	0.19
S	0.017	0.005		0.02
P	0.015	0.005	DQ	0.005
Mo	0.18	0.25		-
Cr	0.12	-		-
Nb	-	0.07		-
0.2% P.S.	730 MPa	800 MPa	820 MPa	380 MPa
D.P.H. No.	252	330	332	201 (b) 156 (b)

(a)   QT - Quenched and Tempered
   DQ - Direct Quenched
   DQ-CGHAZ - Coarse grain HAZ of DQ steel
   ESW - Electroslag weld

(b)   Micro-hardness
   D.P.H. = 201, for interior of prior γ grain (acicular ferrite)
   D.P.H. = 156, for boundary of prior γ grain (pro-eutectoid ferrite)

Miniature tensile test specimens with a gauge diameter of almost 3 mm were used. QT steel specimens were charged in an atmosphere of hydrogen gas, at 650°C for 3 hours, quenched into an ice/water bath, and stored in liquid nitrogen until testing. This treatment produced a uniformily distributed hydrogen concentration of ᴕ 1 ppm as analysed by hot extraction[5]. The gauge length of the other steel specimens were cathodically charged with hydrogen in a solution of 0.1N H_2SO_4 with 0.2 mg/l of As_2O_3. The solution was heated to 80°C to promote hydrogen diffusion and homogenization. The current density was varied to obtain a range of hydrogen levels from ᴕ 1 to to ᴕ 5 ppm[6,9]; a charging time of 2 hours was found to be sufficient to produce saturation (6). Specimens were stored in liquid nitrogen until testing.

Tensile tests of both charged and uncharged specimens were normally carried out at a strain rate of ᴕ 5 x 10^{-4}/s on an Instron testing machine at temperatures between -196 and 135°C. Charged specimens tested above ambient temperature were electroplated with a thin layer of cadmium to prevent hydrogen loss during testing. The reduction in area was calculated from measurements on the fracture surface with a binocular microscope.

Microstructures were examined by optical microscopy, TEM and SEM. Selected fractures were examined by SEM. Longitudinal sections of some broken tensile specimens were also examined.

MICROSTRUCTURES

The QT HSLA steel had a tempered martensite microstructure. TEM studies showed the lath morphology of quenched low carbon martensite with partially spheroidized carbides mainly at lath or other sub-grain boundaries (Fig. 1a). Some recovered fine polygonal ferrite sub-grains were also observed. The dislocation density of the lath ferrite was substantially lower than in a quenched structure.

The as-direct-quenched (DQ) HSLA steel had a fine-grained acicular ferrite microstructure. TEM studies show ferrite laths (i.e. elongated ferrite sub-grains) with high dislocation densities together with some small polygonal ferrite grains (Fig. 1b). Carbide plates at lath boundaries were occasionally seen. After thermal cycling to simulate the CGHAZ (DQ-CGHAZ), the DQ HSLA steel had a microstructure of martensite/bainite laths (Fig. 1c), with some regions of martensite/austenite. All carbides were in the form of plates, mainly between laths (upper bainite) although a few existed within laths (lower bainite) (Fig. 1c). Dislocations appeared to be more dense in ferrite laths than in the acicular ferrite structure.

The ESW metal had a duplex structure, with predominantly acicular ferrite and pearlite within the massive (columnar) prior γ grains and proeutectoid ferrite outlining prior γ grain boundaries (Fig. 2a). At high magnification, pearlite grains can be seen to contain fine lamellae (Fig. 2b). This indicates that the pearlite formed at low temperatures, in spite of the low cooling rate of the ESW weld. Pearlite nucleation was likely inhibited by the large grain size of the parent austenite.

RESULTS AND DISCUSSION

EFFECT OF HYDROGEN ON FLOW STRESS

Hydrogen increases the stress level in DQ specimens, but has little effect on flow in CGHAZ specimens[6]. The flow stress is raised by hydrogen in both QT steel[5] and ESW metal[9]. The effect of the temperature of testing on all the microstructures studied is shown in Fig. 3. Hydrogen increases yield stress in all cases, although only slightly on CGHAZ specimens (Fig. 3c), and the effect becomes negligible below -100°C. The increase in flow stress may be explained by a model based on hydrogen-dislocation interactions[10], i.e., solute atmospheres being dragged by mobile dislocations.

REDUCTION IN DUCTILITY

Reduction in the ductility of the tensile specimens caused by hydrogen was measured by changes in the true fracture strain $\varepsilon_f - \ln (A_o/A_f)$, where A_o and A_f are initial and final cross-sectional areas respectively. These results are presented in the form of an embrittlement index EI in Fig. 4 at a hydrogen concentration (C_H) of $\sim$ 1 ppm (wt %); EI = $(\varepsilon_{fu} - \varepsilon_{fc})/\varepsilon_{fu}$, where u and c designate uncharged and charged conditions respectively. The limited temperature range over which hydrogen embrittlement is operative has been observed in mild[11] and ultra-high-tensile steel[12], and by Graville et al[13] on notched tensile specimens. The effect has been ascribed by the present authors[6] to decreased hydrogen transport at lower temperatures and decreased hydrogen concentration at equilibrium in trap sites at higher temperatures.

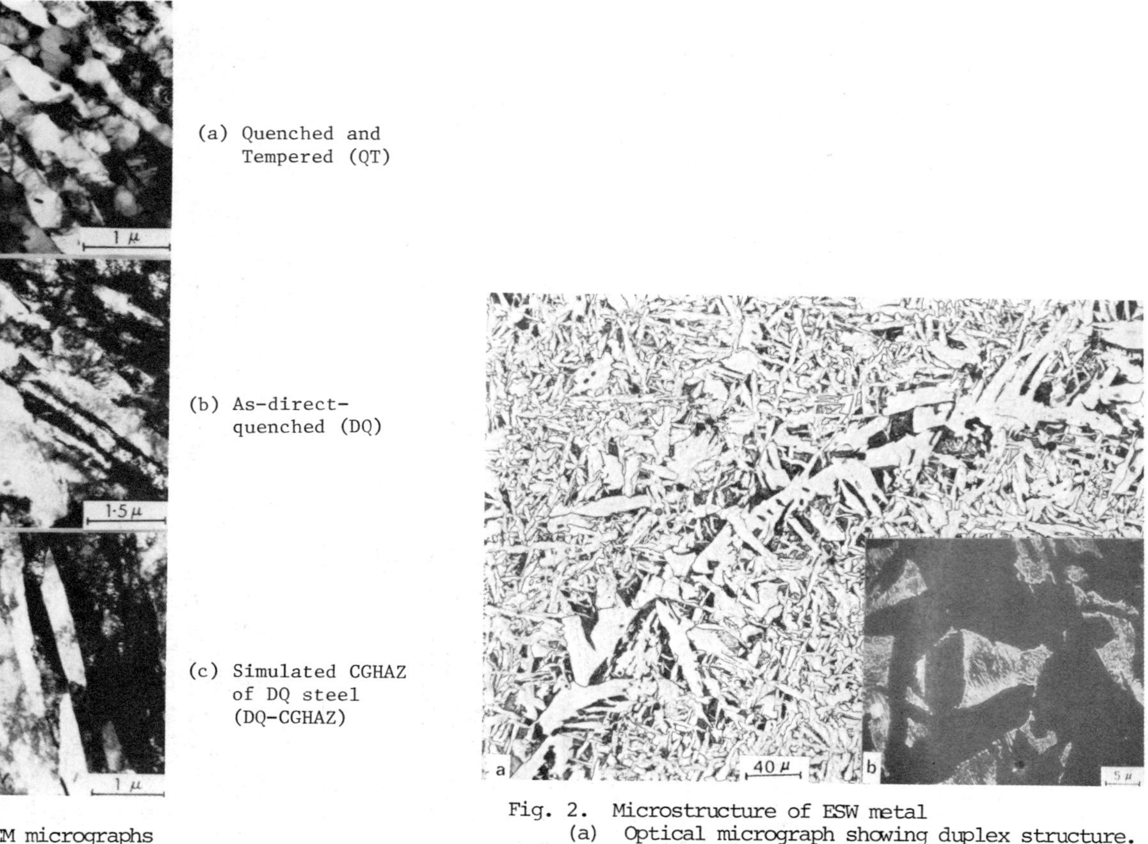

(a) Quenched and
    Tempered (QT)

(b) As-direct-
    quenched (DQ)

(c) Simulated CGHAZ
    of DQ steel
    (DQ–CGHAZ)

Fig. 1. Typical TEM micrographs
         of HSLA steels.

Fig. 2. Microstructure of ESW metal
    (a) Optical micrograph showing duplex structure.
    (b) SEM micrograph of a prior γ grain
         boundary area.

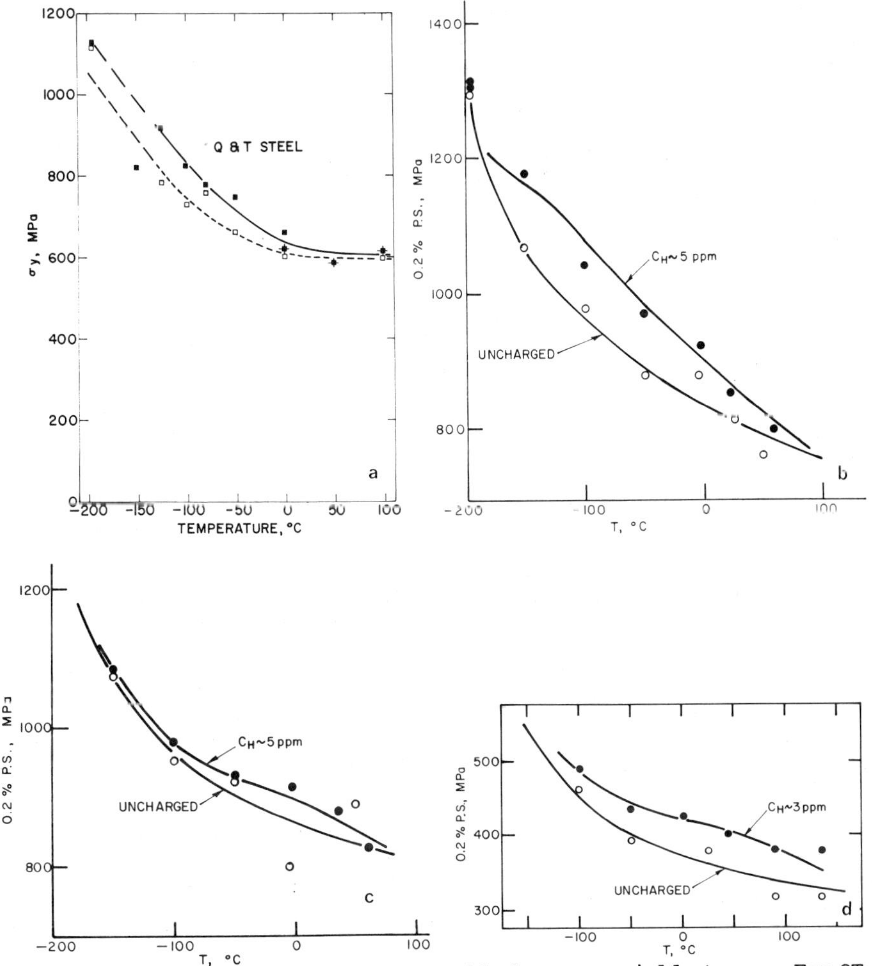

Fig. 3. Effect of test temperature T and hydrogen on yield stress. For QT steel σ_y is the upper yield stress; for the rest of the steels it is the 0.2% proof stress.
(a) QT steel, (b) DQ steel, (c) DQ-CGHAZ and (d) ESW metal

Figure 4a shows that hydrogen is less effective in promoting a loss of ductility in DQ microstructure than in either the martensite/bainite of the DQ-CGHAZ or the tempered martensite of the QT steel. The maximum embrittlement in the QT steel is less than in the DQ-CGHAZ but the EI valve is larger in the high temperature recovery range. Figure 4b clearly shows that ESW metal has significantly greater hydrogen susceptibility compared with the DQ steel. For comparative purposes, the results of Farrell and Quarrell[12] on an ultra-high-tensile steel (tensile strength 1700 MPa (250 ksi)) of very high hydrogen susceptibility is also shown in Fig. 4b.

With increasing C_H for a given steel, there is an increasing loss of ductility and a widening of the temperature range over which hydrogen embrittlement occurs[6,9].

FRACTURE MODES

The following hydrogen-assisted modes of failure were observed: cleavage, quasi-cleavage (i.e. predominantly cleavage accompanied by substantial plastic deformation evidenced by tear ridges) and microvoid coalescence. The operative mode depended on test temperature and hydrogen concentration. In general, as the hydrogen concentration increased, the operative mode changed from microvoid coalescence to cleavage or quasi-cleavage in the maximum embrittlement region[6,9]. In the higher temperature recovery range failure occurred by hydrogen-assisted microvoid coalescence.

At $C_H \sim$ 1 ppm, the fracture mode in the region of maximum embrittlement for the DQ steel, i.e., in the temperature range -100 to -50°C (see Fig. 4a), is primarily microvoid coalescence with some areas of quasi-cleavage (Fig. 5a). However, for the martensite/bainite of the CGHAZ, fracture facets were predominantly cleavage or quasi-cleavage (Fig. 5b). This observation is consistent with the greater susceptibility to hydrogen embrittlement of this microstructure as shown in Fig. 4a. Microcracks elongated in a direction normal to the tensile axis were observed near the fracture surface in longitudinal sections of DQ steel specimens ($C_H \sim$ 5 ppm) tested in the maximum embrittlement range[6] (Fig. 6a). No microcracks could be detected at the same C_H and temperature range in DQ-CGHAZ specimens[6]. Here the fracture profile was relatively flat (Fig. 6b) compared with the ragged profile on the DQ specimens (Fig. 6a).

Fractographic examination of the ESW metal specimens showed similarities with HSLA steels. Fractured surfaces had quasi-cleavage facets in the maximum embrittlement region and microvoid coalescence occurred in the higher temperature range. In the maximum embrittlement region specimens failed by quasi-cleavage primarily along prior γ grain boundaries (Fig. 7)[9].

MECHANISMS OF EMBRITTLEMENT

It is generally accepted that internal hydrogen embrittlement results from hydrogen transport to susceptible sites such as inclusions, interfaces, grain boundaries etc. However, it is becoming increasingly evident that a unique "hydrogen fracture mode" does not exist[2,14] and that failure can occur by different mechanisms that are dependent on composition, microstructure, density and type of trap sites, and test conditions.

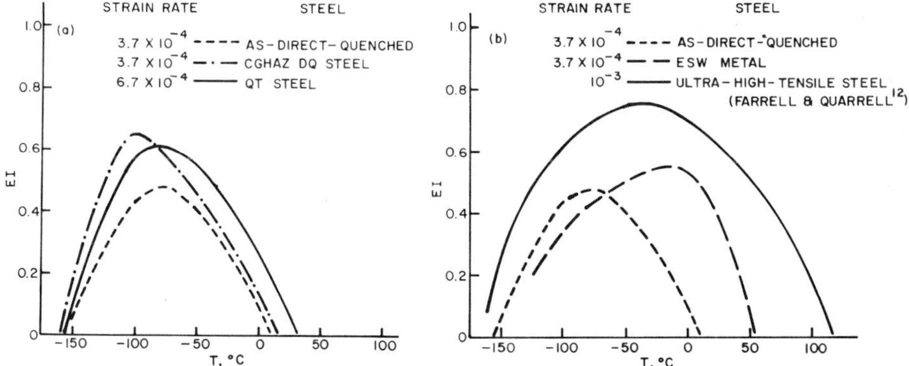

Fig. 4. Variation of embrittlement index EI (see text) with test temperature T at $C_H \sim$ 1 ppm.

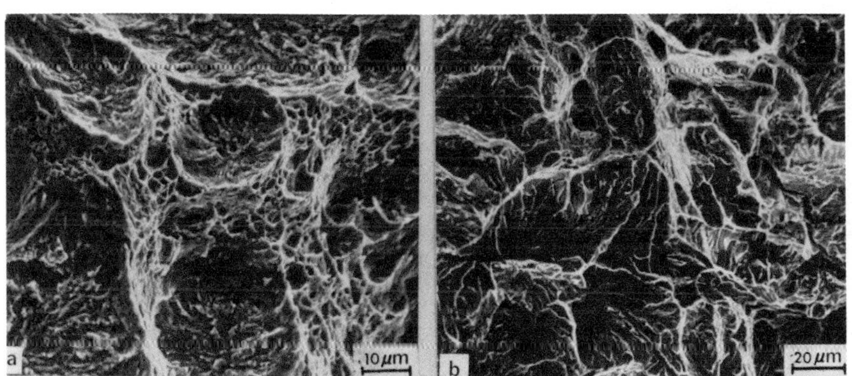

Fig. 5. SEM fractographs of hydrogen charged ($C_H \sim$ 1 ppm) specimens tested in the maximum embrittlement temperature range.
(a) DQ specimen   (b) DQ-CGHAZ specimen.

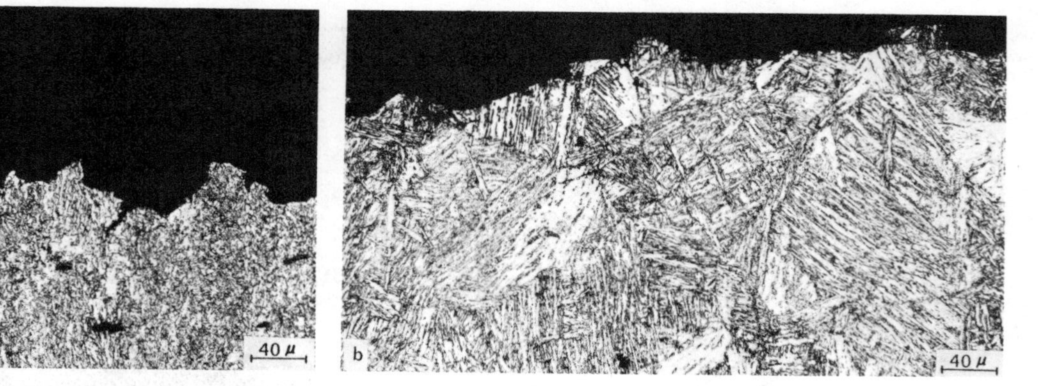

Fig. 6. Optical micrographs of longitudinal sections of fractured specimens ($C_H \sim 5$ ppm) tested in the maximum embrittlement temperature range.   (a) DQ specimen   (b) DQ-CGHAZ specimen.

Fig. 7. Fracture of a hydrogen charged ESW metal specimen tested in the maximum embrittlement temperature range.
A – prior γ grain boundary,
B – prior γ grain.

For the QT and DQ steels, the increase in yield and flow stresses[6,10] (Figs. 3a,b) is evidence of hydrogen-dislocation interaction. Such interaction can lead to transport of hydrogen by moving dislocations to sites of fracture nucleation as elaborated by Tien[15]. These sites are thought to be mainly non-coherent precipitates (carbides). Final fracture probably occurs by the joining of these cracks, causing a ragged fracture profile[5,6] (see Fig. 6a for DQ steel). For the DQ steel, it was seen that at two hydrogen concentrations, the temperature ranges for hydrogen-dislocation interaction (revealed by increase in yield stress) and hydrogen embrittlement coincide[6].

In the low-carbon martensite/bainite structures of the DQ-CGHAZ, hydrogen does not seem to produce significant drag on dislocations (Fig. 3c). This suggests that hydrogen is immobilized at traps other than mobile dislocations. These traps could be carbide/matrix or subgrain interfaces. Comparing the microstructures of Fig. 1, the DQ-CGHAZ structure (Fig. 1c) contains platelike carbides which have a larger area of interface per unit volume of carbide than the more equi-axed carbides of the other two structures, and a somewhat greater density of substructural boundaries (lath boundaries). Other workers have inferred that significant trapping exists in martensitic and bainitic structures[16]. If hydrogen is, indeed, strongly trapped at such interfaces, it will weaken interatomic bonds there and lead to easy crack nucleation. The crack could then propagate by cleavage which is a likely mode in this structure; only the DQ-CGHAZ structure failed by cleavage at -196°C in the uncharged condition. This is consistent with the relatively flat profile of fractures of charged DQ-CGHAZ specimens (Fig. 6b) at maximum embrittlement.

As shown in Fig. 4a, the DQ steel shows less hydrogen-assisted ductility loss than the QT steel of slightly lower strength level. This could be a result of the larger volume fraction of carbides in the QT steel which has a larger carbon content; the hydrogen trapped at these carbide interfaces could assist microvoid nucleation and/or growth, whereas the hydrogen trapped at substructural elements in the DQ steel may play a less direct rôle in the fracture process.

Summarizing our interpretation of the effects of hydrogen on the susceptibility to embrittlement of HSLA steels:
(1) For a given total concentration, the relative concentrations of hydrogen at different types of trap sites will depend on the microstructure, due to differences in density and type of trap sites;
(2) In the DQ-CGHAZ structure, most of the hydrogen is at trap sites where it is relatively immobile, and hence, does not increase the level of the flow stress; and
(3) The presence of hydrogen at interfaces in the DQ-CGHAZ structure, combined with the inherent low toughness of this structure, enhances the initiation and propogation of cleavage fracture.

In ESW metal specimens, in the maximum embrittlement range, fracture takes place primarily through pro-eutectoid ferrite at prior γ grain boundaries[9]. It has been suggested that this preferential fracture path may be due to preferential plastic deformation of the softer pro-eutectoid ferrite (see Table 1) and to a lower cleavage stress of this region compared to the interior of the prior γ grains[9]. This supports the view that hydrogen exploits pre-disposed fracture paths[14,17].

For all steels tested at temperatures above the region of maximum embrittlement fracture occurred by microvoid coalescence. At these temperatures, hydrogen enhances fracture by assisting either void nucleation (which leads to a decreased dimple size) or void growth (which leads to increased dimple size)[18]. No attempt was made in the present work to perform the extensive quantitative fractography required to identify which process pre-dominated.

CONCLUSIONS

This study on the effect of internal hydrogen in reducing the ductility of some low-carbon steels, as indicated by slow tensile testing, has shown the following:
(1) The low-carbon martensite/bainite structure of a simulated CGHAZ has a greater susceptibility to hydrogen embrittlement than the fine sub-grained acicular ferrite of the parent material, although both structures have sim- ilar hardness.
(2) For the HSLA steels, the toughest and the most refined microstructure (the DQ steel) has greatest resistance to hydrogen embrit- tlement, even though its hardness and strength are somewhat higher than the QT microstructure.
(3) Although the ESW metal has the lowest hardness it has the highest hy- drogen susceptibility at higher temperatures which correlates with a lower toughness of the operative crack path.
(4) Hydrogen can assist failure by cleavage, quasi-cleavage or microvoid coalescence, or cause a transition from microvoid coalescence to cleavage or quasi-cleavage, depending on test temperature, hydrogen concentration, and microstructure.

ACKNOWLEDGEMENTS

The authors wish to thank the staff at PMRL, EMR who have contributed to this work. Particular thanks are due to Mr. R. Desjardins for hydrogen analysis, Mr. M. Charest for making the thin foils, and to Drs. J.D. Boyd, M.J. Godden and J.T. McGrath for useful discussions. L.N.P. is grateful to NSERC and EMR of Canada for award of a Fellowship. The authors are grateful to Canadian Heat Treaters Ltd. and Dominion Bridge Co. Ltd. for the supply of materials.

REFERENCES

1. Boniszewski, T. and Watkinson, F.: Metals and Materials, 1973, vol. 7,
    pp. 90-96 and 145-51.

2. Bernstein, I.M. and Thompson, A.W.: Int. Met. Rev., 1976, vol. 21, pp.
    269-287.

3. Rothwell, A.B. and Bonomo, F.: Welding of Line Pipe Steels, (D.E.
    Osborne, ed.) pp 118-46, Weld. Res. Council, New York, 1977.

4. Boyd, J.D.: Materials Engineering in the Arctic, (M.B. Ives, ed.) pp.
    200-209, ASM, Metals Park, OH, 1977.

5. Tobe, Y. and Tyson, W.R.: 2nd Int. Congress Hydrogen in Metals, (P.
    Bastien and P. Azon, chairman), vol. 10, 9 pp, Paris, France, 1977.

6. Pussegoda, L.N. and Tyson, W.R.: PMRL Report MRP/PMRL 79-73 (J),
    CANMET, Energy, Mines and Resources, Canada, 1979.

7. Kunihiro, T. and Nakajima, H.: Proc. Japan - U.S. Seminar on
    Significance of Defects in Welded Structure, pp. 105-9, 1974.

8. Konkol, P.J. and Domis, W.F.: Welding Journal, 1979, vol. 58, pp. 161-S
    to 167-S.

9. Pussegoda, L.N. and Tyson, W.R.: PMRL Report MRP/PMRL 80-33 (J),
    CANMET, Energy, Mines and Resources, Canada, 1980.

10. Tobe, Y. and Tyson, W.R.: Scripta Met., 1977, vol. 11, pp. 849-52.

11. Toh, T. and Baldwin, W.M. Jr.: Stress Corrosion Cracking and
    Embrittlement, (W.D. Robertson, ed.) pp. 176-86, Wiley, New York, 1956.

12. Farrell, K. and Quarrell, A.G.: J. Iron Steel Inst., 1964, vol. 202,
    pp. 1002-11.

13. Graville, B.A., Baker, R.G. and Watkinson, F.: Brit. Weld. Journ.,
    1967, vol. 14, pp 337-42.

14. Bernstein, I.M. and Thompson, A.W.: Mechanisms of Environment Sensitive
    Cracking of Materials, (P.R. Swann, F.P. Ford and A.R.C. Westwood eds.)
    pp 412-25, Metals Society, London, 1977.

15. Tien, J.K.: Effect of Hydrogen on Behaviour of Materials, (A.W.
    Thompson and I.M. Bernstein eds.) pp. 309-26, TMS-AIME, New York, 1976.

16. Morris, J.G. and Roopchand, B.J.: Environmental Degradation of
    Engineering Materials, (M.R. Louthan Jr. and R.P. McNutt eds.) pp
    475-85, Proc of Conf. Blacksburg, Va., 10-12 Oct. 1977.

17. Bernstein, I.M. and Thompson, A.W.: Alloy and Microstructural Design,
    (J.K. Tien and G.S. Ansell eds.) pp. 303-45, Academic Press, New York,
    1976.

18. Thompson, A.W.: Met Trans, 1979, vol 10A, pp 727-31.

DISCUSSION

<u>J. Hewitt, Risco, Zimbabwe</u>: First an observation. I would congratulate the author in actually measuring hydrogen concentration, one of the rare papers in this conference to do so. I would further comment I find it unbelievable that research workers in this field can study the effect of hydrogen on material properties yet neglect to measure its concentration. Any correlation with models of embrittlement would seem to be based on an unacceptable degree of uncertainty.

I would ask the author if he has studied the embrittlement of these steel at temperatures up to about 130–150°C, particularly the electroslag HAZ material, where other workers have shown embrittlement is absent despite the presence of significant hydrogen concentrations?

<u>L. N. Pussegoda</u>: The accurate measurement of hydrogen by hot extraction, especially at low concentrations ($\sim$2 ppm), needs a fully stabilized measuring apparatus and, of course, substantial experience in using this specialized technique. With extremely carefully controlled charging conditions we achieved reproducible results ($\pm$ 0.5 ppm) after charging under the same conditions.

In the HSLA steel microstructures full recovery in embrittlement occurred well below 130°C, even at the highest concentration studied, i.e., $\sim$5 ppm. However, the ESW metal had a small reduction in ductility around 130°C at the higher hydrogen concentration, i.e., $\sim$5 ppm with full recovery taking place around 150°C. We have not studied the ESW HAZ material.

<u>B.R.W. Hinton, University of Manchester</u>: There is considerable evidence in the literature that the presence of hydrogen can either raise, lower or have no effect on the flow stress. The effect, you observe, of hydrogen on the flow stress seems very small. How sure are you that these effects are real? Could you indicate the approximate range of variation possible in the flow stress of uncharged specimens tested in air?

<u>L. N. Pussegoda</u>: We are aware of the different effects of hydrogen on flow stress from earlier literature and also from this conference. Factors such as the cleanliness of the steel, the variety of ways in which hydrogen can interact with dislocations, the state of the hydrogen, i.e., atomic or molecular, have been used to explain the above differences. See, for example, papers by H. Kimura and R. A. Oriani in this conference.

The effect we have reported of increased flow stress due to hydrogen we believe to be significant, although small. We have measured increases in the range of 50 to 100 MPa, with a typical reproducibility of $\pm$ 20 MPa.

EFFECTS OF CARBON CONTENT AND PRIOR AUSTENITE GRAIN SIZE

ON HYDROGEN EMBRITTLEMENT OF 1% Cr - 0.75% Mo - 0.025% Nb STEEL

IN AQUEOUS HYDROGEN SULFIDE ENVIRONMENT

R. Garber, P. J. Grobner, and D. L. Sponseller

Climax Molybdenum Company of Michigan, Ann Arbor, Michigan

Abstract

A series of 1% Cr - 0.75% Mo - 0.025% Nb steels with 0.15, 0.22, 0.30, 0.37, and 0.45% C was tested for sulfide stress cracking (SSC) resistance. Each steel was evaluated in the quenched and tempered condition at three yield strength levels in the 620 to 860 MPa range using the Shell bent-beam test method. In addition, a 0.30% C, calcium deoxidized heat was examined at three grain sizes, ASTM 1.4, 8.7, and 11.8. Carbon variation over the range of 0.15 to 0.37% did not markedly affect SSC resistance; the 0.45% C steel showed higher SSC resistance. Refining the prior austenite grain size increased SSC resistance.

Introduction

The presence of hydrogen sulfide in oil and natural gas wells imposes severe restrictions on the useful strength of casing and other oil field steel products because of sulfide stress cracking (SSC).

The mechanism of sulfide stress cracking has been recognized by Snape (1) as being closely related to hydrogen embrittlement. Hydrogen, produced by the corrosion reaction, enters the steel and diffuses inward, reducing the fracture toughness of the material. This effect generally increases with increasing strength of the steel. The strongest steels normally used to date for oil well casing in sour environments have been the L-80 and C-90 grades, with minimum yield strengths of 550 and 620 MPa, respectively. New SSC-resistant steels with a minimum yield strength of 690 or 725 MPa would do much to facilitate the economical production of oil and gas from deep sour wells.

Previous work at the Climax Laboratory has established an improvement in sulfide stress cracking (SSC) resistance of SAE 4130 and 4135 steels by modification with 0.75% Mo and 0.03% Nb (2). The effects of rolling temperature, cooling rate after austenitizing and after tempering on mechanical properties and SSC resistance were established in a subsequent study (3). Evaluation of samples of the Modified 4135 steel by twelve different laboratories using a wide variety of SSC test procedures confirmed the superiority of this composition over standard SAE 4100 series steels (4). Based on these results a detailed study of the effect of molybdenum content was conducted to determine the optimum composition for SSC resistance (5). The optimum molybdenum content was found to be 0.75% or 0.90%, depending on the type of SSC test employed. Similar optimum molybdenum contents were found by Waid and Ault (6) and Zikeev et al.(7) in higher chromium steels. The steel industry is producing oil country tubular goods of the standard and modified 4100 series steels in a broad range of carbon contents. Since carbon content significantly affects hardenability and temper resistance, an assessment of the effect of carbon content on the SSC resistance is therefore useful for determining the optimum composition.

It is also known that processing variables, such as deoxidation practice, forming technique, or heat treatment can affect the prior-austenite grain size; a deleterious effect of coarse grain size on stress corrosion resistance has been found by Proctor and Paxton (8), and a similar effect on SSC resistance was suspected.

The purpose of the present investigation was twofold: 1) to establish the effect of carbon on SSC resistance of steels with 1% Cr, 0.75% Mo, and 0.025% Nb in the range of 0.15-0.45% C, and 2) to assess the effect of prior austenite grain size on the SSC resistance of Mo-Nb modified SAE 4130 steel.

Laboratory Procedures

Materials

Six 57-kg heats with chemical compositions listed in Table 1 were air-melted in an induction furnace. The compositions are similar with the exception of carbon, which varies from 0.15% to 0.45%. Five of the heats, representing the carbon series, were deoxidized with aluminum, while one 0.30% C heat, designated for investigation of the effect of prior-austenite grain size, was deoxidized with calcium to avoid the grain refining effect of aluminum.

Rolled plates of all heats, 16 mm thick, were austenitized for 30 min

Table 1. Chemical Compositions of Steels

Steel	Heat Number	Element, %										
		C	Mn	Si	Cr	Mo	Nb	Al	P	S	N	O
0.15C	6144	0.15	0.66	0.19	1.05	0.78	0.026	0.047	0.023	0.020	0.015	0.010
0.22C	6139	0.22	0.58	0.16	1.05	0.75	0.024	0.034	0.024	0.019	0.014	0.012
0.30C	6140	0.30	0.65	0.19	1.02	0.76	0.024	0.043	0.023	0.019	0.014	0.009
0.30C-Ca	6141	0.31	0.60	0.15	1.02	0.75	0.022	0.003	0.026	0.021	0.014	0.018
0.37C	6142	0.37	0.67	0.20	1.02	0.75	0.024	0.050	0.023	0.021	0.020	0.007
0.45C	6143	0.45	0.60	0.15	1.00	0.76	0.023	0.042	0.022	0.021	0.015	0.010

at 900 C and water-quenched. Each steel was tempered at different temperatures with subsequent air-cooling to obtain a series of strength levels for
each carbon content (Table 2). The calcium-deoxidized heat was austenitized
in three different ways to obtain three levels of prior-austenite grain size.
One plate section was heat treated as described above for the other steels to
produce a medium grain size while another plate section was austenitized at
1260 C for two hours, furnace cooled to 900 C and held one hour before quenching to produce a coarse austenite grain size. A fine grain size was obtained
by four cycles consisting of austenitizing at 900 C for one minute and then
water quenching.

Table 2. Properties of Steels Investigated

Steel	Tempering Temp.,C	Time, h	0.2% Offset Y.S., MPa	Tensile Strength, MPa	Elonga- tion, %	Reduction of Area, %	ASTM[a] Grain Size	Critical Stress, MPa
0.15C	620	1	860	920	24.0	70.0		462
	655	1	745	825	23.0	69.5	8.3	772
	695	1	624	721	28.5	73.5		1386
0.22C	665	1	815	894	22.0	65.5		724
	705	3	665	721	24.5	70.5	8.0	1310
	705	6	646	715	25.5	71.5		>1930
0.30C	675	1	801	905	24.0	65.5		683
	705	1	698	812	26.5	68.0	8.4	979
	705	4	651	744	27.5	68.0		1055
0.37C	690	1	866	965	22.5	58.5		531
	705	2	727	827	26.0	62.0	8.2	910
	705	6	670	778	29.0	59.5		1100
0.45C	690	1	831	927	25.0	66.0	8.2	917
	705	1.5	738	836	25.0	65.5		1200
0.30C-Ca Fine Grain	670	1	909	988	22.5	64.0		462
	690	1	832	906	23.5	65.0	11.8	779
	705	9	671	727	28.0	67.5		1055
0.30C-Ca Medium Grain	670	1	881	987	21.5	65.0		434
	690	1	794	906	24.0	65.5	8.7	800
	705	9	642	748	29.5	68.0		1475
0.30C-Ca Coarse Grain	690	1	910	1021	15.0	31.0		365
	700	2	762	887	20.0	48.0	1.4	517
	705	9	638	766	24.0	61.0		986

[a]Measured on as-quenched specimens

SSC Testing

The SSC resistance of the steels was determined by the Shell bent-beam method described by Fraser, Eldredge, and Treseder (9). The stressed specimens were immersed in an aqueous 0.5% acetic acid solution. The solution was deaerated by purging with nitrogen for 30 min and then saturated with hydrogen sulfide for 20 min. In keeping with Shell's normal procedure, the solution was resaturated by charging with hydrogen sulfide for 30 min every Monday, Wednesday, and Friday. (These charging conditions are less severe than the continuous charging used in Reference 5.) The specimens were removed from the solution after 30 days and inspected for cracks.

The results were evaluated by the simplified statistical method recommended by Fraser, Eldredge, and Treseder (9) to obtain a critical stress (S_c) necessary to cause failure in 30 days with a probability of 50%. This method does not provide an estimate of the standard deviation of S_c. According to Fraser, Eldredge, and Treseder (9) the standard deviation of S_c is about 40 MPa.

RESULTS

Mechanical Properties

The tempering conditions and resultant tensile properties of the steels are shown in Table 2. A plot of the yield strength versus the tempering parameter $P_T = T(C+\log t)$ for all steels except 0.30% C-Ca is shown in Figure 1. The tempering constant C was assumed to depend on carbon content as $C = 21\frac{3}{16} - 5\frac{5}{8}$ (%C), as found by Hollomon and Jaffe (10). (This figure includes data from preliminary tensile tests not shown in Table 2.) Results for all but the 0.45% C steel can be described very well ($r^2 = 0.97$) by a quadratic least squares regression curve·

$$YS(MPa) = 23168 - 2198.0\ P_T + 53.519\ P_T^2$$

Results for the 0.45% C steel lie 40 to 100 MPa below the regression line as shown by the dashed line in Figure 1. This suggests a difference in carbide precipitation as compared with the other steels.

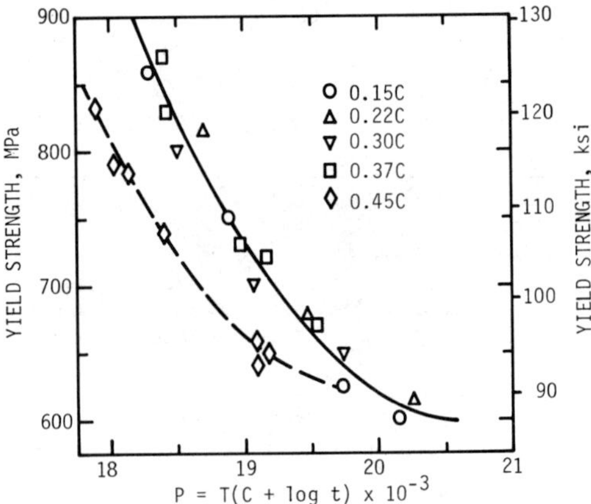

Fig. 1 – Effect of Tempering Parameter on the Yield Strength of Modified 4100 Steels. C varies with carbon content: C = 21.19-5.63x(%C).

Results for the 0.30% C-Ca steel used for the grain size study would lie
about 70 MPa above this regression line. The furnace cool and hold at 900 C
applied to the coarse grained material was intended to keep segregation at the
austenitizing temperature the same as in the medium grain size material, and
also to achieve a similar NbC morphology. Comparison of the yield strengths
of the 0.30% C-Ca steel tempered for one hour at 690 C shows that the coarse
grain material has a higher yield strength than the medium or fine grain
material. Therefore, it appears that the NbC precipitation makes a greater
contribution to strength in the coarse grained steel.

SSC Resistance

Results of the bent beam tests are given in Table 2. The effect of carbon content is shown in Figure 2 as a plot of S_c versus yield strength for
various carbon contents. By interpolating the data points in this figure, a
plot of S_c versus carbon content was constructed at constant yield strengths
of 700, 750 and 800 MPa. This plot is shown in Figure 3. The 0.22% C steel
seems to have some advantage over the 0.15% C as well as the 0.30% C steel,
but a really significant effect of carbon occurs when the carbon content is
increased from 0.37% to 0.45%, whereupon S_c increases by about 340 MPa.

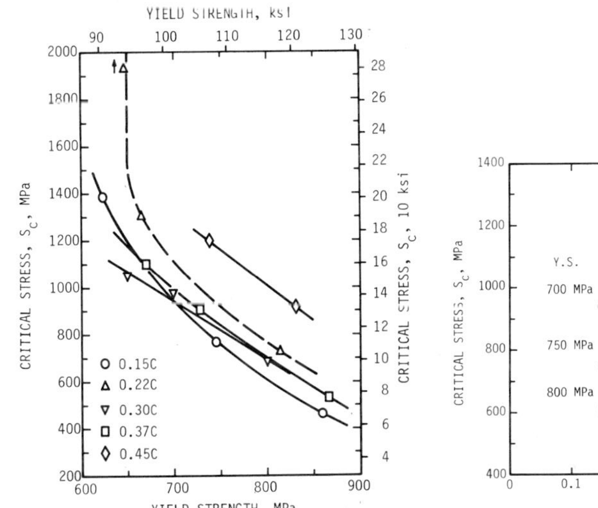

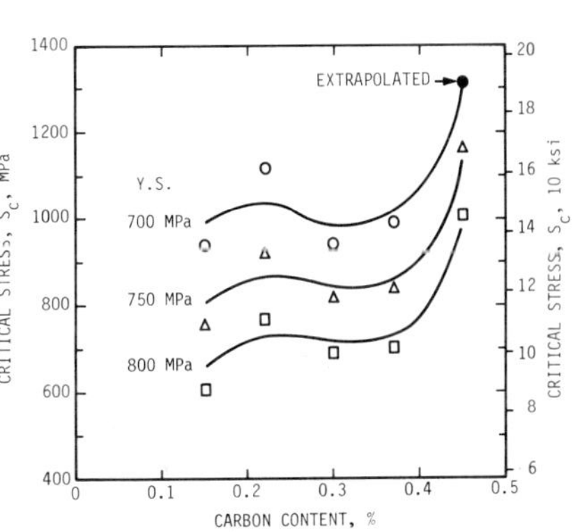

Fig. 2 – Effect of Yield Strength on
    Critical Stress, S_c, at Dif-
    ferent Carbon Concentrations

Fig. 3 – Effect of Carbon Content on
    Critical Stress, S_c, at Dif-
    ferent Yield Strength Levels

The effect of yield strength on SSC resistance of the 0.30% C-Ca steel
at three different prior austenite grain sizes is shown in Figure 4. Figure 4 also contains data for the 0.30% C steel. The medium grain size
0.30% C-Ca steel shows inconsistently high SSC resistance at the lowest
strength level. At the two higher strength levels the results are more consistent. By interpolating the data of Figure 4, plots of SSC resistance
versus ASTM grain size number can be constructed at various yield strength
levels. Such plots are shown in Figure 5. The critical stress, S_c, increases by about 60 MPa for an increase of one unit in the ASTM grain size
number.

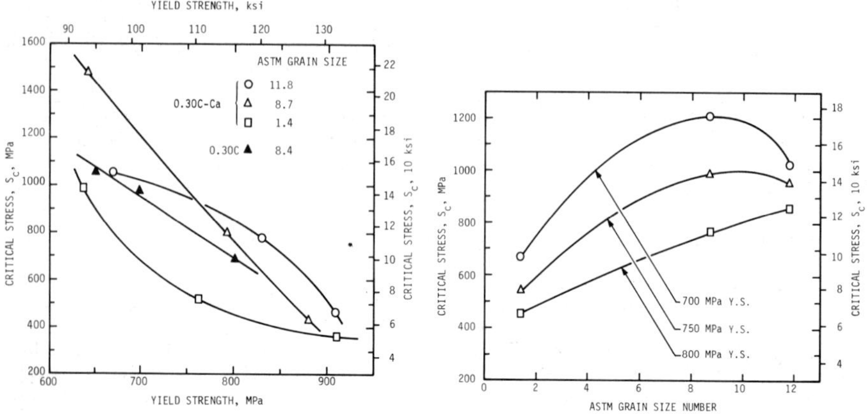

Fig. 4 – Effect of Yield Strength and
Prior Austenite Grain Size on
the Critical Stress, S_c, of
Modified 4130 Steels

Fig. 5 – Effect of Prior Austenite
Grain Size in Critical Stress,
S_c, of Modified 4130 Steel at
Various Yield Strengths

Metallography

The microstructure of the 0.45% C steel is shown in Figure 6.   X-ray analysis of extracted carbides confirmed that the fine rod-like carbide is M_2C. M_2C was also found in the 0.15% C steel.

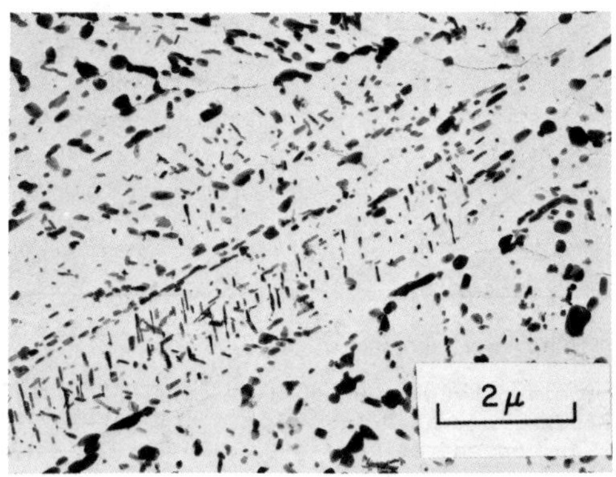

Fig. 6 – Extraction Replica Electron Micrograph of 0.45% C Steel, Tempered
at 705 C for 1.5 Hours

Discussion

The results of this investigation show that, over the carbon range from 0.15% to 0.37%, carbon has no marked effect on SSC resistance.   At 0.45% C however there is a marked increase in SSC resistance; at equal yield strength the critical stress of the 0.45% C steel is 340 MPa higher than the critical stress of the 0.37% C steel.   The improved SSC resistance is of some practical

interest for drilling equipment applications which require both high hardenability and high yield strength. The relative insensitivity of SSC resistance to carbon content over the range from 0.2% C to 0.3% C is of practical
importance since the lower carbon steel can be more easily tempered to the
C-90 strength range; indeed there may be some benefit in SSC resistance at
0.22% C versus 0.30% C. The improvement in SSC resistance obtained by refining the prior austenite grain size, as shown in Figure 5, may also be of
practical interest in helping to produce a satisfactory C-100 grade casing.

The mechanism for the increased SSC resistance of the 0.45% C steel is
not clear. One possibility is a change in the as-quenched martensite morphology occurring at about 0.45% C according to Maki _et al_. (11) A more
likely possibility is a difference in carbide precipitation behavior and
tempering response as is evident in Figure 1. At equilibrium, the carbide
sequence in the Fe-Mo-C system (12) is listed below, in order of increasing
molybdenum-to-carbon ratio:

$$1)\ Fe_3C,\ 2)\ M_{23}C_6(Fe_{21}Mo_2C),\ 3)\ Fe_2MoC,\ 4)\ M_2C,\ 5)\ M_6C$$

This is not the sequence observed at commercially relevant tempering times
in either Cr-Mo steels (5) or in Mn-Mo steels (13); the $M_{23}C_6$ and Fe_2MoC are
missing. The onset of M_2C coincides with the maximum in SSC resistance, so
that it is probable that the presence of excessive amounts of M_2C is responsible for the decrease in SSC resistance observed at molybdenum contents
above the optimum. In the present study, as the carbon content is increased
at constant molybdenum, the amount of M_2C relative to Fe_3C should decrease.

The data presented in Figure 5 show that S_c generally increases with decreasing prior-austenite grain size. The results obtained in this investigation may not apply to other alloys and thermal treatments. Comparison of the
coarse and medium grain sizes shows the effect of grain size when it is
changed by altering austenitizing temperature (and also the NbC precipitation). Comparison of the medium and fine grain sizes shows the effect of
thermal cycling. These effects are not equivalent. Lessar and Gerberich (14)
noted that the relationship between yield strength and prior-austenite grain
size was different when the grain size was changed by altering austenitizing
temperature and when changed by thermal cycling. Thermal cycling as used by
Proctor and Paxton can strengthen by a combination of grain size and transformation induced dislocations. Annihilation of these dislocations during
tempering could account for the change in S_c ranking of the medium and fine
grained steels shown in Figure 5.

<u>Conclusions</u>

The following conclusions may be drawn from this study of the effects
of carbon content and prior austenite grain size on the SSC resistance of
modified SAE 4100 steels:

1. Carbon does not markedly affect SSC resistance over a range from 0.15
    to 0.37%. A 0.22% C steel does show somewhat higher SSC resistance
    than the other steels.

2. A 0.45% C steel shows significantly higher SSC resistance than the
    lower carbon steels. For example, at a yield strength of 750 MPa it
    has a critical stress of 1170 MPa versus 830 MPa for the lower carbon
    steels.

3.   Refinement of the prior-austenite grain size increases the SSC resistance of Modified 4130 steel markedly. The critical stress, S_c, increases by about 60 MPa for an increase of one unit in the ASTM grain size number.

Acknowledgments

The authors are indebted to Dr. T. Wada for helpful discussion and to R. W. McConnell for careful SSC testing.

References

1.   E. Snape, Corrosion, 1967, vol. 23, pp. 154-172.

2.   P. J. Grobner, D. L. Sponseller, and W. W. Cias, Materials Performance, 1975, vol. 14, No. 6, pp. 35-43.

3.   P. J. Grobner, D. L. Sponseller, and D. E. Diesburg, Trans. ASME, J. of Engineering for Industry, 1976, vol. 98, No. 2, pp. 708-716.

4.   J. A. Straatmann, P. J. Grobner, and D. L. Sponseller, ASME Paper 77-Pet-48, (1977).

5.   P. J. Grobner, D. L. Sponseller, and D. E. Diesburg, Corrosion, 1979, vol. 35, No. 6, pp. 240-250.

6.   G. M. Waid and R. T. Ault, Corrosion '79, Paper No. 180, available from NACE, Publications Department.

7.   V. N. Zikeev, Yu. F. Myasnikov, and L. V. Popova, Metallovedenie i. Term. Obrabot. Metallov, No. 6, June 1979, pp. 59-60.

8.   R. P. M. Proctor and H. W. Paxton, Trans. ASM, 1969, vol. 62, pp. 989-999.

9.   J. P. Fraser, G. G. Eldredge and R. S. Treseder, Corrosion, 1958, vol. 14, No. 11, pp. 517t-523t.

10.   J. H. Hollomon and L. D. Jaffe, <u>Ferrous Metallurgical Design</u>, John Wiley New York, 1947, pp. 245-6.

11.   T. Maki, K. Tsuzaki and I. Tamura, Trans. ISIJ, 1980, vol. 20, pp. 207-214.

12.   B. Uhrenius in D. V. Doane and J. S. Kirkaldy, (editors) <u>Hardenability Concepts with Applications to Steel</u>, AIME, 1978, pp. 28-81.

13.   F. B. Fletcher and Y. E. Smith, ASME Paper No. 78-Pet-2, (1978).

14.   J. F. Lessar and W. W. Gerberich, Met. Trans. A, 1976, vol. 7A, pp. 953-960.

GRAIN BOUNDARY CHEMISTRY, FRACTURE MODE AND DUCTILITY COMPARISONS FOR

IRON AND NICKEL TESTED AT CATHODIC POTENTIALS*

R. H. Jones, S. M. Bruemmer, M. T. Thomas and D. R. Baer
Pacific Northwest Laboratory
Richland, Washington 99352

Introduction

Iron and nickel containing similar grain boundary sulfur concentrations
show similar changes in fracture mode and ductility at equivalent cathodic
potentials. Both iron and nickel show a change from microvoid coalescence
to intergranular fracture at grain boundary sulfur concentrations of about
0.1 monolayers accompanied by a decrease in the total elongation and reduction of area. It is not unexpected that sulfur at the grain boundaries of
iron and nickel promotes intergranular fracture in both materials at cathodic
potentials as others (1-6) have shown that grain boundary impurities enhance
the intergranular fracture of iron and nickel and their alloys. The similarity of the effect of sulfur in iron and nickel is unusual and it is this
aspect which this paper addresses.

Experimental Procedure

Vacuum melted iron and nickel were evaluated in this study. Glidden
A101 iron and carbonyl nickel were induction melted in a helium atmosphere.
The significant impurity concentrations are 50 appm S, 80 appm C, 120 appm
N and 1600 appm O in iron and 20 appm S, 110 appm C, 55 appm N and 180 appm
O in nickel. Grain boundary chemistries were determined by Auger Electron
Spectroscopy of specimens impact fractured at 170°K in an ultra-high vacuum
system. The nickel specimens were charged with hydrogen prior to fracture.
Straining electrode tests of iron and nickel were performed at cathodic
potentials with an inverted loading frame in 1N H_2SO_4 (pH = 0.3). Specimen
current densities ranged from 1 mA/cm^2 at 0.6 V to 130 mA/cm^2 at -2.0 V for
iron and 0.1 mA/cm^2 at -0.3 V to 80 mA/cm^2 at -1.7 V for nickel. The measured corrosion potentials for iron and nickel in 1N H_2SO_4 were -450 mV
(SCE) and -230 mV (SCE), respectively. More specific information concerning
experimental details can be obtained in References 7-9.

Grain Boundary Chemistry

Grain boundary chemistries observed for iron and nickel given heat
treatment A and B are listed in Table 1. Sulfur coverages observed after
heat treatments A and B are similar in iron and nickel; however, this is
fortuitous given the temperature differences for heat treatment A for iron
and nickel and the differences in bulk sulfur concentration, solubilities
and diffusivities. Because of the observed concentration variation between

* Supported by Office of Basic Energy Sciences, U.S. DOE, DE-AC06-76RLO-1830.

Table I. Average Grain Boundary Fracture Surface Coverage and Grain Size

| Material | Heat Treatment | Coverage Fractions of a Monolayer | | | | Grain Size |
		S	O	C	N	μm
Iron	A: 1/2 hr/800°C	0.09	0.26	0.19	--	90
	B: 1/2 hr/800°C + 240 hr/600°C	0.24	0.32	0.30	0.11	
Nickel	A: 1 hr/1100°C	0.11	0.27	0.18	0.07	120
	B: 1 hr/1100°C + 240 hr/600°C	0.22	0.16	0.47	0.02	

intergranular surfaces on a given sample, the average of at least ten analyses per fracture surface are given in Table I. The adsorption rates of oxygen and carbon on iron were too low to account for their observed concentrations; therefore, the concentrations given in Table I are due partially to segregation. The need to hydrogen charge nickel samples allowed less time for vacuum processing the AES system and, therefore, adsorption from the AES system vacuum accounts for the oxygen and some of the carbon concentrations reported in Table I. The large change in carbon concentration between heat treatment A and B suggests that carbon may have segregated in nickel.

Straining Electrode Behavior

Tensile Behavior in Air

The fracture mode of iron and nickel given heat treatment B and tested in air was a mixture of microvoid coalescence and shear as shown in Figure 1a, b. A small amount of intergranular fracture was found at the surface and the interior in nickel given heat treatment B and tested in air. No intergranular fractures were observed in iron tested in air. The strain to failure and reduction of area after heat treatment B were 35% and 85%, respectively, for iron and 42% and 40%, respectively, for nickel. Samples given heat treatment A were not tested in air but extrapolation of the cathodic potential results given in Figure 2 suggests that heat treatment B caused a decrease in the strain to failure of iron and nickel and the reduction of area of nickel relative to heat treatment A. The effect of heat treatment B on the ductility of nickel is possibly related to the intergranular failure.

Effect of Cathodic Potentials

An applied cathodic potential of -0.15 V resulted in 100% intergranular fracture in iron given heat treatment B, Figure 1c, while an applied potential of -0.3 V was needed to induce 90% intergranular fracture in nickel given heat treatment B, Figure 1d. At these potentials, the current densities were $\sim$1 mA/cm^2 for both iron and nickel; therefore, their fracture modes are consistent at equal current densities. The strain to failure and reduction of area of iron and nickel also showed a consistency with current density. A shift in the iron curves, Figure 2, by -0.15 V or the nickel curves by +0.15 V adjusts them to equal current densities and, therefore, closer in comparison. Nickel given heat treatment A showed a similar response with applied potential as that given heat treatment B except for a shift to higher ductilities at applied potentials of <-0.5 V (SCE). This increased ductility is similar in magnitude to that observed in air tests.

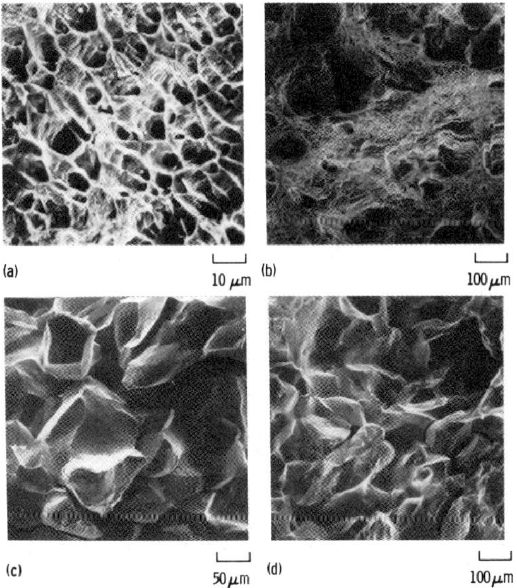

Fig. 1 - Scanning electron micrograph of iron and nickel heat
treated for 240 hr at 600°C and tested in air (a) iron,
(b) nickel in 1N H2SO4 at an applied potential of (c)
-0.15 V (SCE), iron and (d) -0.3 V (SCE) nickel.

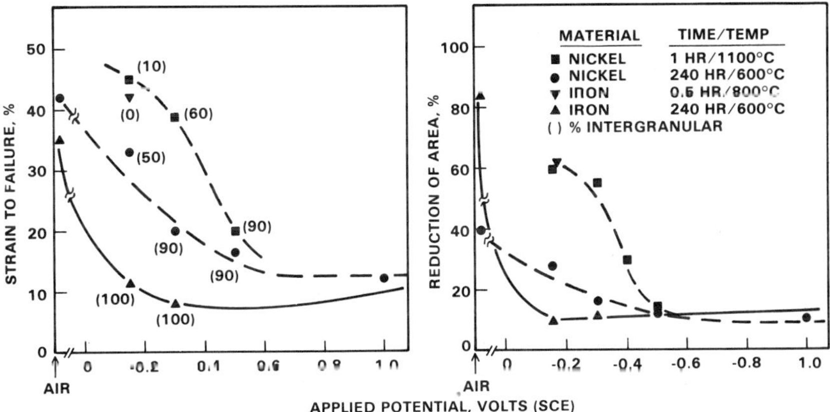

Fig. 2 - Strain to failure and reduction of area of iron and
nickel as a function of applied potential (E_{APP} =
E_{TEST} - E_c). Note that the % intergranular fracture
increases with applied cathodic potential.

<u>Effect of Grain Boundary Sulfur</u>

Iron and nickel show a similar transition in fracture mode and reduction of area with grain boundary sulfur concentration as shown by Figure 3.
Data for these curves was obtained from zone refined and vacuum melted iron
and nickel and has been reported elsewhere (8, 9). As seen in Figure 3a,

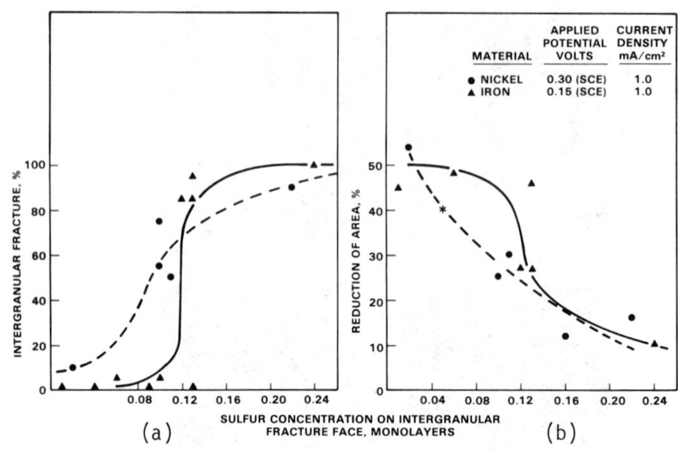

Fig. 3 - Percent intergranular fracture and reduction of area of
iron and nickel versus average grain boundary sulfur concentration. Grain boundary chemistries are average values
obtained from several ($\sim$10) intergranular surfaces.

the iron has a sharp fracture mode transition at about 0.12 monolayers of
sulfur while nickel has a more gradual increase in percent intergranular
fracture with increasing grain boundary sulfur concentration. Taking 50%
intergranular as the fracture mode transition, nickel shows a transition
at 0.09 monolayers of sulfur. The reduction of area results, Figure 3b,
coincide with the fracture mode results with the reduction of area decreasing with increasing grain boundary sulfur. The reduction of area of iron
shows a sharper change with sulfur than does nickel, which is consistent
with the fracture mode results.

Discussion

The similarity in the effect of grain boundary sulfur on the fracture
mode and ductility of iron and nickel tested at cathodic potentials suggests
that the same embrittlement mechanism may be operating. However, iron and
nickel have many differences in the physical parameters involved in these
mechanisms. A list of some of these parameters are given in Table II and
are the basis for the following discussion. It is important to note that
the straining electrode results are sensitive to hydrogen transport processes
and equilibrium hydrogen effects such as proposed by Troiano (10) and modified by Oriani (11) because deformation and cathodic potential are applied
simultaneously. Therefore, any conclusion regarding mechanisms involved in
the fracture mode transition must consider both effects.

The large difference in diffusivities of hydrogen in iron and nickel
is well known and representative values are given in Table II. It is difficult to explain the fracture mode transition in iron and nickel by a
decohesion model which requires hydrogen transport to the maximum stress at
the crack tip. This conclusion applies to both bulk and grain boundary
transport of hydrogen.

Dislocation transport can enhance hydrogen permeation in metals and it
is possible that this process enhanced the hydrogen permeation in nickel

TABLE II. COMPARISON OF HYDROGEN AND SULFUR ENHANCED FRACTURE PARAMETERS FOR IRON AND NICKEL

MODEL	PARAMETER	IRON		NICKEL	
		PROPERTY	REFERENCE	PROPERTY	REFERENCE
GENERAL COMPARISON	SULFUR/FRACTURE MODE TRANSITION, MONOLAYERS	0.12	---	0.09	---
	CRYSTAL STRUCTURE	bcc	---	fcc	---
	YIELD STRENGTH, MPa	82	---	22	---
	ULTIMATE TENSILE STRENGTH MPa	186	---	314	---
HYDROGEN TRANSPORT	HYDROGEN EXCHANGE CURRENT DENSITY IN H_2SO_4, 20°C	10^{-6}		10^{-6}	
	HYDROGEN DIFFUSIVITY				
	a. BULK, 20°C	2.3×10^{-5} cm²/s	13, 14	4.1×10^{-10} cm²/s	12
	b. GRAIN BOUNDARY	2.0×10^{-5} cm²/s	13	1.7×10^{-9} cm²/s	13
	c. GB/BULK	1		4	
	DISLOCATION TRANSPORT				
	X_C/SEC.	7 cm	15	1×10^{-2} cm	
	X_D/SEC.	3×10^{-3} cm		3×10^{-5} cm	
	SULFUR ENHANCED PERMEATION	30 mA/m²/mole/ℓ	16	?	
CRACK TIP PLASTICITY	a. SLIP PROCESSES	INCREASE σ	17	H DECREASES SFE	18
	b. WORK HARDENING	INCREASE OR DECREASE	19, 20	INCREASE	21
	c. SULFUR ENHANCED GB DISLOCATIONS	YES	22	PROBABLY	23
	SULFUR SOLUBILITY (600°C)		24	< 25 appm	25
	FRACTURE MODES				
	1. LOW TEMP. IMPACT* -196°C, LOWS	CLEAVAGE + IG	26 THIS WORK	DUCTILE	THIS WORK
	2. LOW TEMP. IMPACT* HIGH S	IG	THIS WORK	DUCTILE	
	3. HYDROGEN, RT LOW S	CLEAVAGE OR IG	THIS WORK	TG	THIS WORK
	4. HYDROGEN, RT HIGH S	IG	27	IG	THIS WORK
	5. RT, HIGH S	DUCTILE	THIS WORK	DUCTILE AND IG	THIS WORK
DECOHESION (SULFUR)	INTERGRANULAR COHESION	DECREASED BASED ON FATT	28	---	
	BONDING EFFECTS	S-Fe BONDS DECREASE Fe-Fe BONDS	BRIANT, MESSMER	S-Ni BONDS DECREASE Ni-Ni BONDS	29
INTERFACIAL FRACTURE (HYDROGEN)	$\Delta\gamma$ LATTICE	0	30	750 ergs/cm²	30
SURFACE ENERGY (SULFUR)	$\dfrac{d\gamma}{dX_c} = -\dfrac{RT}{X_c}\,\Gamma$	600°C 0.76 × 10⁶ cal/cm²	THIS WORK	600°C 1.83 × 10⁶ cal/cm²	THIS WORK
CHEMICAL STATE OF SULFUR	AES/ESCA	ELEMENTAL	THIS WORK	COMPOUND ELEMENTAL	32 33

*AUGER IMPACT SPECIMEN

more than iron. However, according to Tien (15), the velocity of a dislocation with a hydrogen cloud is proportional to the diffusivity. The parameters given in Table II compare Tiens (15) values for the penetration of
hydrogen by dislocation transport, X_C and lattice diffusion, X_D. As these
numbers indicate, dislocation transport does not make the hydrogen permeation
in iron and nickel equal. The kinetics of hydrogen transport and, therefore,
the embrittlement due to intergranular fracture should occur at lower current densities in iron than nickel. This is not supported by the present
data.

Enhanced permeation by impurities, whether by the hydrogen recombination poison process as proposed by Latanision and Opperhauser (5, 6); the
chemisorbed metalloid hydride process proposed by Zakroczymski, et al. (16);
or any cathode reaction, may be involved in the enhanced intergranular fracture of iron and nickel. Zakroczymski, et al. (16) have shown that metalloid
elements enhance the permeation of hydrogen through iron when added to the
aqueous solution. Sodium sulfite enhanced the hydrogen permeation by 30
$mA/cm^2/mole/\ell$ and it is reasonable to expect a similar enhancement in nickel;
however, data is not available to support this. This mechanism could explain
the similarity between the iron and the nickel results if the hydrogen
absorption reaction (HAR) and not transport is rate limiting. Equal amounts
of sulfur in the grain boundaries of iron and nickel could cause the HAR to
be equal in these two materials with hydrogen transport being sufficiently
fast in both materials at a given strain rate to result in equal intergranular crack growth rates.

Hydrogen has been shown to increase the thermally activated component
of the flow stress of iron (17) while it decreases the stacking fault energy
of nickel (18). Both of these effects may result in an increased work hardening rate and hence, smaller plastic zone size at the crack tip. Hydrogen
has been shown to cause a decreased work hardening rate in iron (20) under
certain conditions and, therefore, it is difficult to conclude that hydrogen
altered the crack tip plastic zone size in iron and nickel in an equal
manner.

Sulfur has been shown to cause an increased dislocation density near
the grain boundaries of iron when diffusion from the surface to the bulk
occurred via the grain boundaries (22). Krishtal (23) has generalized this
phenomenon to include the diffusion of low solubility surface active sub-
stances into metals. Sulfur in iron and nickel certainly satisfies this criteria as it is surface active on iron and nickel and equally insoluble in
both as shown in Table II. A high dislocation density adjacent to the grain
boundary could act to decrease the plastic zone size of an intergranular
crack and, therefore, favor intergranular fracture over transgranular fracture. A large dislocation density was not observed in preliminary TEM
examinations of iron and nickel with 0.2 monolayers of sulfur. More detailed
examination needs to be done before this mechanism can be seriously considered.

Results of the present study show similarities between low temperature
fracture modes and hydrogen induced fracture modes as a function of grain
boundary sulfur concentration, as shown in Table II. This comparison is
quite good for iron while the nickel required hydrogen charging to obtain
equivalent low temperature fractures. In any event, these similarities suggest that grain boundary decohesion is significant in both processes.
McMahon (28) has suggested that impurities contribute to hydrogen induced
grain boundary decohesion such that intergranular fracture can occur at a
lower critical stress when both are present. Sulfur is expected to have a
large effect on this critical stress in iron (28). From a more theoretical
standpoint, Briant and Messmer (29) concluded that sulfur contributes to

the intergranular fracture of nickel by the formation of strong Ni-S bonds in nearest neighbor atoms with a weakening of the next nearest Ni-Ni bonds. In the present results, it would be fortuitous that equal grain boundary sulfur concentrations in iron and nickel produced similar intergranular fracture tendencies in hydrogen. However, until this is disproven, it seems that the effect of sulfur on decohesion of iron and nickel in the presence of hydrogen must be considered as a viable process.

Another approach to interfacial cohesion has been presented by Rice (30) who evaluated this problem from the activation energy for dislocation nucleation at a crack tip. For $U_{act} > 0$, dislocation nucleation at an interfacial crack is not favored while an atomically sharp crack is stable. The result of this model suggests that hydrogen must reduce the $\Delta\gamma$lattice of nickel by 750 ergs/cm^2 for intergranular fracture to be favored while the grain boundaries of iron should be capable of separation without a change in $\Delta\gamma$lattice. Assuming that sulfur contributes to the intergranular decohesion of iron, it is difficult to account for the fracture mode transition with increasing grain boundary sulfur observed in the present work if intergranular fracture is always favored in iron as proposed by Rice.

An early model for the role of hydrogen on fracture stress was proposed by Petch (31) who suggested that hydrogen decreased the energy to produce new surfaces with a crack in the Griffith fracture stress expression. Both hydrogen and sulfur could contribute to this decrease. The change in surface energy of iron and nickel with hydrogen is very comparable (31) while the effect of sulfur on the surface energy of iron and nickel is given in Table II based on segregation measurements and the Gibbs isotherm. The change in surface energy per unit change in solute content is 2.4 times greater in nickel than in iron. This qualitatively supports the lower grain boundary sulfur concentration required for the fracture mode transition in nickel than in iron. The chemical state of sulfur at the grain boundaries of iron and nickel and the presence of a thin brittle phase could also be a factor in the intergranular fracture of iron and nickel. Auger electron spectroscopy of sulfur at the grain boundaries of iron gives every indication that it is elemental while in nickel there is some splitting of the nickel lines suggestive of compound formation. Thompson (32) concluded that a thin film of Ni$_3$S$_2$ was present in nickel following similar annealing conditions while Marcus, et al. (33) concluded from ESCA analysis of sulfur adsorption on the surface of nickel that sulfur is elemental up to about one monolayer and Ni$_3$S$_2$ at greater coverages. Since the grain boundary sulfur concentration in the nickel was less than one monolayer, it can be assumed that it was elemental and, therefore, the chemical state was equal in iron and nickel which supports the decohesion or the HAR explanation for the equal effect of sulfur on iron and nickel.

<u>Conclusions</u>

The fracture mode of iron and nickel tested at equal cathodic current densities was found to be a function of the grain boundary sulfur concentration with iron and nickel having transitions at 0.12 and 0.09 monolayers of sulfur, respectively. The reduction of area of iron and nickel tested at equal cathodic current densities also decreased similarly with increasing grain boundary sulfur concentrations.

The similarity in the effect of hydrogen and sulfur on iron and nickel can be explained by their role in intergranular decohesion. However, the different hydrogen transport rates in these materials suggests a rate controlling HAR which is altered similarly by sulfur in iron and nickel. This process requires that hydrogen transport by either dislocations or bulk diffusion is sufficiently fast that permeation is surface controlled.

References

1. I. M. Bernstein, Met. Trans., 1970, Vol. 5, p. 363.
2. K. Yoshino and C. J. McMahon, Jr., Met. Trans., 1974, Vol. 5, p. 363.
3. R. Viswanathan and S. J. Hudak, Jr., Met. Trans., 1977, Vol. 8A, p. 1633.
4. C. L. Briant, H. C. Feng and C. J. McMahon, Jr., Met. Trans., 1978, Vol. 9A, p. 625.
5. R. M. Latanision and H. Opperhauser, Jr., Met. Trans., 1974, Vol. 5, p. 483.
6. R. M. Latanision and H. Opperhauser, Jr., Met. Trans., 1975, Vol. 6A, p. 233.
7. R. H. Jones, S. M. Bruemmer, M. T. Thomas and D. R. Baer, submitted for publication in Met. Trans.
8. S. M. Bruemmer, R. H. Jones, M. T. Thomas and D. R. Baer, submitted to Scripta Met.
9. S. M. Bruemmer, R. H. Jones, M. T. Thomas and D. R. Baer, Scripta Met., Vol. 14, 1980, p. 137.
10. J. G. Morlet, H. H. Johnson and A. R. Troiano, J. Iron and Steel Inst., 189 (1958) p. 37.
11. R. A. Oriani, Bar Bunsenges, 76 (1972) p. 848.
12. W. M. Robertson, Z. Metallkde, 64 (1973) p. 436.
13. V. M. Sidoreules, I. I. Sidorak, Fiziko-Kinicheskaya Mekharika Materialov, Vol. 9, No. 4, 1973, pp. 12-16.
14. J. Volkel and G. Alefeld, "Diffusion in Solids - Recent Developments," pp. 272-302, A. S. Nowick and J. J. Burton, eds., Academic Press, New York, NY, 1975.
15. J. K. Tien, "Effect of Hydrogen on Behavior of Materials," 309, A. W. Thompson and I. M. Bernstein, eds., proceedings of an Intl. Conf., September 1975.
16. T. Zakroczymski, Z. Szklarska-Smiakowska and M. Smiakowski, Werkstoffe und Korrosion, 27 (1976) p. 625.
17. A. Gourmelon, Mém Sci. Rév. Metall., 6 (1975) p. 475.
18. A. H. Windle and G. C. Smith, Met. Sci. J., (1968) p. 187.
19. J. Clum, Scripta Met., 9 (1975) p. 51.
20. H. Matsui, S. Moriya and H. Kimura, 4th Intl. Conf. Strength of Metals and Alloys, Nancy, Vol. I, 1976, p. 291.
21. A. H. Windle and G. C. Smith, Metal Science Journal, 4 (1970) p. 136.
22. N. G. Ainslie, V. A. Phillips and D. Turnbull, Acta Met., 8 (1960) p. 528.
23. M. A. Krishtal, Fiziko-Khimicheskaya Mekanika Materialov, 5 (1969) p. 643.
24. N. G. Ainslie and A. V. Seybolt, J. Iron and Steel Institute, (London) 194 (1960) p. 341.
25. J. S. Kirkaldy, Scripta Met., Vol. 3 (1969) p. 947.
26. C. Pichard, J. Rieu and C. Goux, Mém. Sci. Rév. Metall., 70 (1973) 13.
27. M. Cornet, M. F. Trichet and S. Talbot-Besnard, Mém. Sci. Rév. Metall., 74 (1977) 307.
28. C. J. McMahon, Jr., Matl. Sci. and Engr., 42 (1980) 215.
29. C. L. Briant and R. P. Messmer, submitted to Phil. Mag.
30. J. R. Rice, "Effect of Hydrogen on Behavior of Materials," A. W. Thompson and I. M. Bernstein, eds., 1975, p. 455.
31. N. J. Petch, Phil. Mag., 1 (1950) 331.
32. A. W. Thompson, "Grain Boundaries in Engineering Materials," proceedings 4th Bolton Landing Conf., eds. J. L. Walter, J. H. Westbrook and D. A. Woodford, 1974, p. 607.
33. D. Marcus, J. Oudar and I. Oleford, Matl. Sci. and Engr., 42 (1980) 191.

DISCUSSION

<u>M. R. Shanabarger, University of California</u>: If sulfur is producing de-
cohesion at the grain boundary, why is the fracture mode transition so sharp
for Fe compared with Ni when the grain boundary sulfur concentration distributions for Fe and Ni suggest the opposite result?

<u>R. H. Jones</u>: I believe the sharp fracture mode transition in iron resulted
from the fact that iron exhibited both ductile and cleavage fracture modes
at low grain boundary sulfur concentrations, while nickel exhibited only
ductile fracture. Apparently, the balance between intergranular fracture
and cleavage or ductile fracture in iron is closer in iron than the balance
between intergranular and ductile fracture and a smaller change in grain
boundary sulfur concentration is needed to tip this balance to iron.

THE ROLE OF GRAIN BOUNDARY CHEMISTRY
AND THE ENVIRONMENT ON INTERGRANULAR FRACTURE

R.M. Latanision, M. Kurkela and F. Lee
Corrosion Laboratory
Massachusetts Institute of Technology
Cambridge, Massachusetts 02139

In this discussion the influence of grain boundary segregates and their interaction with the environment will be discussed with particular reference to hydrogen-induced intergranular embrittlement. Many of the elements which are found segregated at grain boundaries (i.e., the metalloid elements) are catalyst poisons for the hydrogen evolution reaction in electrolytes and, hence, stimulate hydrogen absorption. The use of glassy metal alloys (transition metal-metalloid type) as grain boundary analogs in studying hydrogen absorption and permeation is explored. Likewise, permeation experiments which distinguish between lattice, grain boundary and dislocation transport mechanisms will be discussed.

Introduction

The importance of impurity segregation at internal interfaces, particularly grain boundaries, in the phenomenology of various forms of embrittlement -- temper embrittlement (1), stress corrosion cracking (2), etc., -- had been proposed some time ago. More recently, surface analytical tools such as Auger electron spectroscopy have made it possible to substantiate such proposals. For example, temper embrittlement of steels appears related to the grain boundary accumulation of alloying elements such as Ni and Cr and impurity elements (metalloids) such as P, Sb, Sn, etc. (3-6). Intergranular embrittlement of W (7), Cu (8), and phosphor bronze (9) have been associated with segregated P,Bi, and P, respectively. In many cases embrittlement has been attributed to a reduction in (grain boundary) surface energy, a consequence of the segregation of solutes to the grain boundary, or to galvanic effects in electrolytes arising out of chemical inhomogeniety between grain boundaries and contiguous grains (2). Of course, depending upon the nature of the impurity element, increased or decreased reactivity may be expected to occur as a consequence of segregation. In the present discussion, attention will be focussed on the part played by segregated impurities in the intergranular hydrogen embrittlement of polycrystalline metals and alloys. In particular, we will describe in this paper recent work performed in the Corrosion Laboratory at M.I.T. on the influence of solutes in the absorption of electrolytic hydrogen. In addition, we will assess the effectiveness of grain boundary and lattice diffusion as well as of dislocation transport of hydrogen in the embrittlement process.

Impurity – Environment Interactions in Electrolytes

Impurities may play an important role in embrittlement induced by cathodically produced hydrogen, particularly if the segregated species happen to be metalloids such as those mentioned above which are known to be effective catalytic poisons for the hydrogen recombination reaction in electrolytes. The significance of recombination poisons was first pointed out by Latanision and Opperhauser (10-13) in a study of the intergranular embrittlement of nickel by cathodically produced hydrogen. Perhaps the most effective way to examine the influence expected of metalloid elements on cathodic kinetics is to consider Figure 1, taken from West (14), which shows the exchange current density for hydrogen evolution for some of the elements in the Periodic Chart. The exchange density may be considered a measure of the catalytic efficiency of a given element for the hydrogen evolution reaction at the reversible potential. Notice that the exchange current density for the hydrogen reaction is orders of magnitude higher on noble metals (typically ~10^{-3} A/cm^2) or other transition metal surfaces (~ 10^{-6} A/cm^2) than on metalloid surfaces (~ 10^{-13} A/cm^2). The overall hydrogen evolution reaction may be broken down into, for example, a proton reduction step, $H^+ +e = H_{ads}$, and a hydrogen adatom – adatom combination step. Metalloid elements effectively poison the reaction $H_{ads} + H_{ads} = H_2$ thereby increasing the population of adsorbed uncombined hydrogen on the electrode surface and, consequently, the probability that atomic hydrogen will be absorbed by the metal also increases. The influence of soluble metalloid salts in increasing the absorption of hydrogen from electrolytes at cathodic potentials is well documented (15-16). Hence, as suggested by Latanision and Opperhauser (10-13), one might expect metalloid-segregated grain boundaries to act as preferential sites for the absorption of cathodic hydrogen into polycrystalline metals, Figure 2. In essence, it was argued that the entry of hydrogen into nickel occurred preferentially in the proximity of the grain boundary intersections with the free surface due to the presence therein of Sn and Sb (and the absence of Cu) which act to poison the combination of hydrogen atoms formed by the discharge of protons from the electrolyte. At locations re-

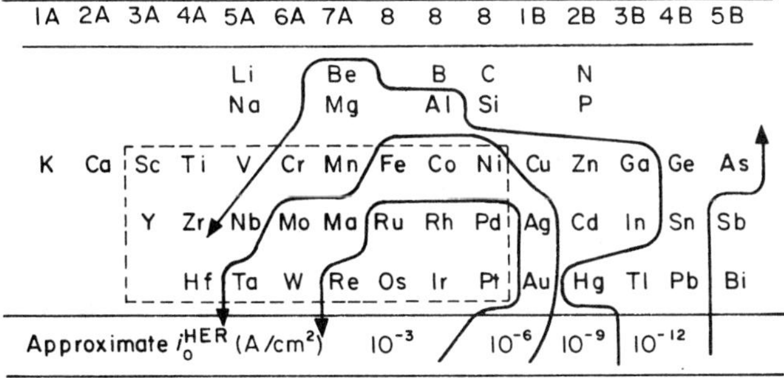

Fig. 1 – The exchange current density for the hydrogen evolution reaction, i_o^{HER}, for various elements in the Periodic Chart (after West (14)).

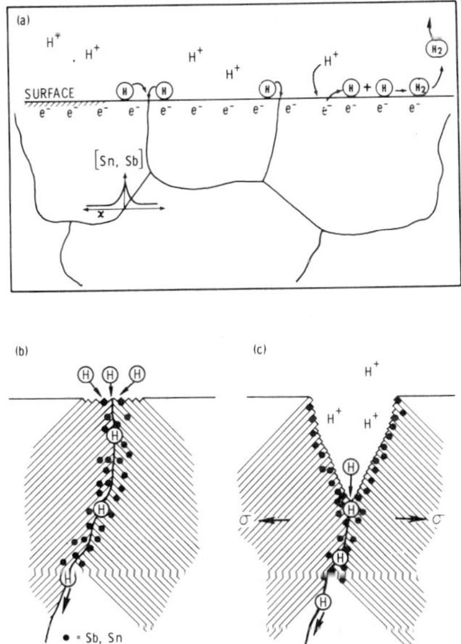

Fig. 2 – Schematic indicating the preferential absorption of atomic hydrogen along metalloid-segregated grain boundaries and subsequent intergranular embrittlement of polycrystalline metals and alloys (after Latanision and Opperhauser (12)).

mote from the grain boundary, protons are reduced forming hydrogen adatoms which in the absence of a poison have a high probability of combining to form molecular hydrogen and are, thus, subsequently evolved. In the vicinity of a grain boundary, in contrast, rather than evolving from the electrode surface in molecular form, uncombined hydrogen adatoms increase in number at the interface and the probability of their absorption into the metal lattice increases. It should be noted that the observed absence of copper, which would otherwise stimulate the combination reaction, further increases the probability of atomic hydrogen absorption. Hence one must consider the influence of species which are deficient as well as in excess (relative to the unembrittled condition) on the grain boundaries.

While we have to this point discussed the importance of impurities largely from the point of view of their influence on cathodic hydrogen embrittlement, it might be well to notice that those impurities which increase the rate of hydrogen absorption from electrolytes (i.e., metalloids such as Sn, Sb, S, P, As, etc.) typically have the concurrent effect of reducing the corrosion rate. Since such elements have typically low exchange current densities for the hydrogen evolution reaction, one expects that in acid or deaerated electrolytes the open circuit corrosion rate will be decreased in those regions where the segregation of such metalloids occurs. In effect, in alloys that are not sensitive to hydrogen, susceptibility to stress corrosion cracking, presuming that cracking occurs by a dissolution related mechanism, may be reduced. Preferential selective segregation may therefore provide a means of stress corrosion cracking control.

In a similar context, control of susceptibility in alloys sensitive to cracking during cathodic charging requires quite different control. One would in this case hope to decrease the concentration of metalloids at grain boundaries by, for example, reducing grain size, purification of the alloy, or by providing alternate and non-continuous interfaces as opposed to grain boundaries at which solutes might accumulate. Conversely, while metalloids would have the effect of increasing the rate of atomic hydrogen absorption, noble elements such as Ru, Pt, Pd, Rh, etc., would have the opposite influence with regard to cathodic charging. Hence, perferential segregation of noble metals to the grain boundaries of alloys typically embrittled intergranularly during cathodic charging may be beneficial. Indeed, the work reported by Pickering and Zamanzedeh (17) in the poster session at this conference demonstrates very nicely that Pt introduced into the surface by ion implantation decreases the entry and permeation of electrolytic hydrogen into iron. Somewhat different but related considerations apply if the source of hydrogen is molecular as has been discussed by Berkowitz et al. (18). In this case, noble metals (hydrogen dissociation catalysts) segregated to the grain boundaries would be expected to stimulate hydrogen absorption and subsequent cracking. On the other hand, catalytic poisons should retard cracking as shown in the interesting work of Liu et al. (19).

It should be appreciated that the embrittlement of nickel described above (10-13) is not the result of segregation of metalloids alone as indicated by the fact that identical tensile specimens deformed in the absence of hydrogen were not embrittled. Indeed embrittlement is associated with the interaction between segregated species and the surrounding environment. Such impurity-microchemistry interactions may apply to other metals and alloys as well. Grain boundary segregation of phosphorus, for example, has been observed in thermally treated nickel-base alloys such as Inconel 600 (20, 21) and Hastelloy C-276 (22), both of which are subject to intergranular hydrogen embrittlement. Likewise, recent extensive work by Bruemmer, et al. (23) has shown a similar fracture mode transition in hydrogen charged iron as a function of grain boundary sulfur segregation. In addition, of course,

it has been observed that the tempering temperature range producing shortest
life in steels exposed to cathodic hydrogen corresponds also to the typical
temper embrittlement range (5, 24). Moreover, it is known, as pointed out
earlier, that temper embrittlement is associated with grain boundary accumulation of Sb, P, Sn, etc., all of which are very effective hydrogen recombination poisons. Recognizing that water and water vapor are damaging to high
strength steels (25) -- presumably due to the presence of hydrogen produced
as the cathodic equivalent of the oxidation or corrosion of iron -- it has
been suggested (13) that temper embrittlement may be due not solely to the
accumulation of impurities at the grain boundaries but, as in the case of
nickel, in part as well to the local rate of absorption of hydrogen by the
matrix. It should be mentioned that Yoshino and McMahon (5) attribute the
increased hydrogen sensitivity of temper embrittled steels to the combined
effects of impurities and hydrogen in reducing the intergranular cohesive
strength. The point of the present discussion is to suggest that the accumulated impurities may be responsible as well for the presence of hydrogen in
the grain boundaries.

In short, identification and/or control of the partitioning of solutes--
alloying elements as well as metalloid impurities -- may well provide a clue
as to the nature of a particular embrittlement phenomenon (i.e., SCC vs.
hydrogen embrittlement) as well as suggest means of reducing susceptibility
by metallurgical treatment. Of course, we need at this stage to know much
more about the details of such segregation phenomena -- how might one treat
an alloy to induce selective segregation; what interactions might one expect
between segregated solutes; etc.? It is also clear that despite years of
activity we still do not understand on an atomic scale how and what embrittling species (hydrogen atoms, chloride ions, liquid metal atoms, etc.)
do to induce decohesion at a crack tip. This remains to be answered by
means other than traditional mechanical testing. Detailed studies of the
surface structure, electronic structure, and bonding of adsorbed hydrogen,
for example, on strained metal surfaces are just beginning to proceed (26)
but are crucial, we believe, to understanding such phenomena.

Given all of the above, some of the elements which may be involved
in the intergranular embrittlement of nickel are shown in the sequence in
Figure 3. Recent studies (see reference 27 for a review) suggest that
yielding begins in the surface grains of polycrystals through the action of
dislocation sources near the free surface. The result is that one expects
some hydrogen is dragged into the interior along with mobile dislocations
which may then interact with grain boundaries, Figures 3(a) and (b). Some
hydrogen is likely to enter the solid at other than poisoned grain boundaries,
evidence for which is the fact that serrated yielding has been observed in
large-grained polycrystals and in similar experiments with monocrystals (12,
28) cathodically charged and deformed simultaneously. The latter suggests
that dislocation-solute (hydrogen) interactions occur. Likewise, atomic
hydrogen which presumably enters the solid preferentially at grain boundary
intersections with the free surface may diffuse via the grain boundaries into
the solid. In the later stages of deformation, internal dislocation sources
become operational and the incidence of dislocation interactions with the
grain boundaries increases. It is conceivable, for example, that dislocations
generated by sources located at grain boundaries may sweep hydrogen into the
bulk. The attendant stress and the presence of hydrogen in the vicinity of
the grain boundaries may subsequently lead to embrittlement, Figure 3 (c),
perhaps as a result of the chemisorption induced reduction in the cohesive
strength of atomic bonds at regions of stress concentration.

In the following sections we will examine the above views with regard
to (a) the possibility of increased hydrogen absorption in the proximity of

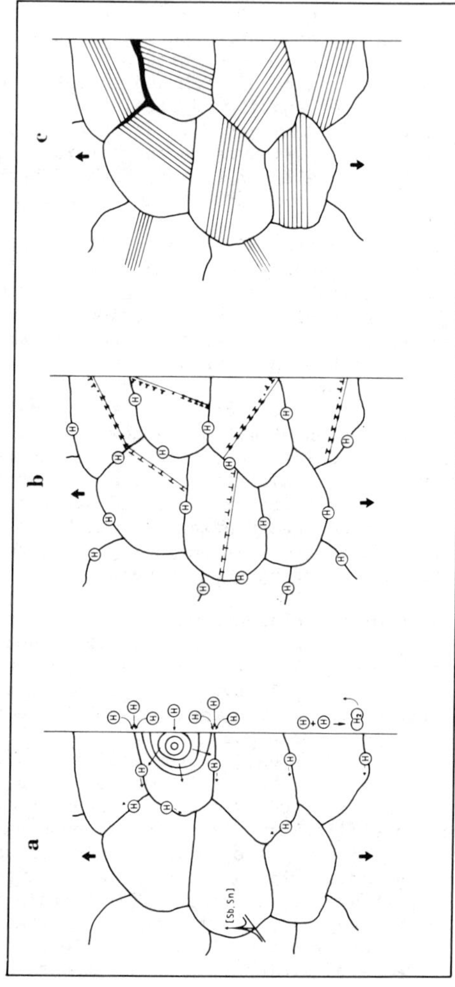

Fig. 3 – Schematic showing sequence perhaps involved in intergranular hydrogen embrittlement (after Latanision and Opperhauser (10)).

segregated grain boundaries; (b) the effectiveness of grain boundary diffusion; and (c) the likelihood of dislocation transport in nickel and nickel-base alloys.

Hydrogen Absorption at Metalloid-Segregated Grain Boundaries in Nickel-Base Alloys

Ideally, one would like to explicitly examine the behavior of the grain boundary - electrolyte interface with regard to the hydrogen evolution/absorption processes. Obviously, this is difficult to do _in situ_. On the other hand, we have recently explored the use of metallic glasses as a structural and chemical analog of segregated grain boundaries in electrolytic hydorgen permeation studies (29, 30). Structurally, Ashby et al. (31) have pointed out that grain boundaries may be described on an atomic scale as a packing of polyhedra, a model which has been used as well to effectively characterize the structure of metallic glasses (32). In effect, both grain boundaries and metallic glasses may be considered to possess a certain short range order on an atomic scale, but not the long range periodicity typical of perfect crystals. Chemically, the transition metal-metalloid type glasses (typically 80 atomic per cent transition metals, 20 atomic per cent metalloid compositions) are of compositions that are good approximations of the chemistry of solute segregated grain boundaries in polycrystalline metals and alloys. As mentioned earlier, grain boundary segregation of phosphorus has been observed in thermally treated nickel-base alloys such as Inconel 600 and Hastelloy C-276. Hence, Ni-P binary glasses may be considered to be good structural and chemical analog of grain boundaries in thermally treated nickel-base alloys. (For more detailed information on metallic glasses, see references 33 and 34.)

The studies of hydrogen absorption on metalloid containing glasses were pursued by means of hydrogen permeation measurements following the Devanathan technique (35). In this approach, a thin membrane is cathodically charged with hydrogen on one side while the other side is maintained at a constant potential sufficiently noble to oxidize all of the hydrogen diffusing out. The anodic current monitored between the membrane and a Pt counter electrode is a direct measure of the rate at which hydrogen permeates the membrane. Moreover, from permeation transients one may calculate both the diffusivity of hydrogen and the concentration of hydrogen just beneath the surface on the cathode side of the membrane. One expects, as explained earlier, that the presence of metalloids in some metallic glasses should increase the rate of absorption of atomic hydrogen into the membrane and, hence, affect the permeation flux. In effect, in a Fick's first law approximation, the flux of hydrogen at the anodic side of a membrane of thickness L expressed as a permeation current density is given by:

$$J = -nFD\left(\frac{\partial C}{\partial X}\right)$$

where C is the concentration of hydrogen, n the number of electrons involved in the hydrogen reaction, F, Faraday's constant and D the diffusion coefficient. Note that the measured current density is proportional to

$$-\left(\frac{\partial C}{\partial X}\right)_{X=L}$$

and this varies from zero to a steady state value of $\frac{C_o}{L}$. Hence, at steady state, the permeation flux is given by:

$$J_\infty = nFD\frac{C_o}{L}$$

Since Co is expected to increase in metalloid containing glasses, J_∞ is correspondingly likely to increase. Work on Fe-base glassy alloys (29, 30) has shown D_H to be of the order of 10^{-10} cm^2/sec and Co to be as high as 0.06 moles/cm^3. The latter is far higher than is typical of Ni or Pt (1.4 x 10^{-5} to 2.7 x 10^{-4} moles/cm^3, respectively) and is more typical of the hydride forming elements Ti, Zr or Ta (~0.1 moles/cm^3). Likewise, it was observed that if the cathodic charging current (or, correspondingly, hydrogen overpotential) exceeded a certain limit, abnormalities would occur in the permeation transients. Indeed, rather than achieving a steady state, the permeation flux began to decrease with time despite the fact that the charging current was maintained constant throughout. Similar observations have been reported by Bockris and Subramanyan (36) and have been associated with trapping of hydrogen within the solid. In most instances, the nature and identity of traps in solids is relatively uncertain. In the present case, the traps seem quite clearly to be associated with micron-order diameter voids introduced during fabrication (29). Ultimately, glasses exhibiting such analogous transients were found to fail catastrophically, despite the absence of any externally applied stresses.

The results of permeation experiments on polycrystalline high purity nickel, a $Ni_{92.7}Si_{4.5}B_{2.8}$ glass, and $Ni_{81}P_{19}$ glass are given in Figure 4. The work on $Ni_{81}P_{19}$ was intended to address the case of the metallic glass analog of phorphorus-segregated grain boundaries in nickel-base alloys. In effect, as described earlier with reference to Figure 2, it has been proposed that the segregated grain boundaries might act as preferential paths for the entry of hydrogen into the polycrystalline matrix. Notice that as the charging current increases the steady state permeation current also increases, but far more rapidly for the metalloid-concentrated glass as expected. Figure 5 shows the dependence of the permeation transient on charging current or, correspondingly, overpotential. As in the case of Fe-based glasses evidence of trapping is present above a critical overpotential.

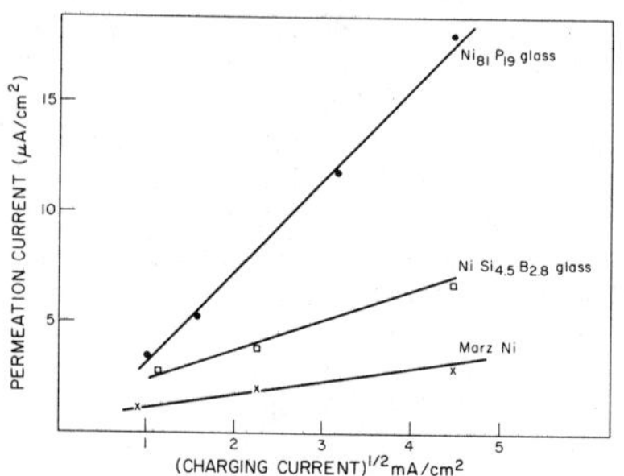

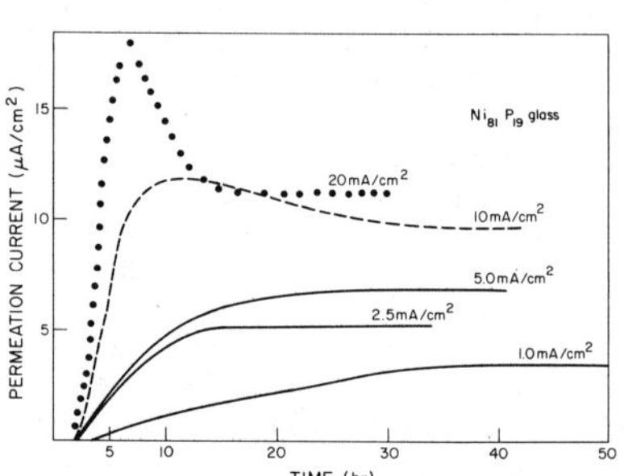

Fig. 4 - Steady state permeation flux as a function of charging current for various crystalline and glassy nickel-base materials (after Latanision, et al. (30)).

Fig. 5 - Permeation transients for $Ni_{81}P_{19}$ exhibiting trapping behavior at high charging current or, correspondingly, overpotential. (after Latanision, et. al. (30).)

Table I. summarizes the permeation information taken from this work.

Table I. Permeation Data

Material	I_c(mA/cm^2)	D_H(cm^2/s)	C_o(mol/cm^3)
$Ni_{81}P_{19}$	2.5	7.9×10^{-11}	2.6×10^{-3}
$Ni_{81}P_{19}$	5.0	8.6×10^{-11}	3.1×10^{-3}
$Ni_{81}P_{19}$	10.0	11.5×10^{-11}	4.0×10^{-3}
$Ni_{81}P_{19}$	20.0	13.8×10^{-11}	5.1×10^{-3}
$NiSi_{4.5}B_{2.8}$	1.25	7.0×10^{-11}	1.5×10^{-3}
$NiSi_{4.5}B_{2.8}$	5.0	8.4×10^{-11}	2.0×10^{-3}
$NiSi_{4.5}B_{2.8}$	20.0	10.0×10^{-11}	3.0×10^{-3}
Polycrystalline Marz Ni	0.8	1.9×10^{-10}	2.3×10^{-4}
" "	5.0	2.3×10^{-10}	3.1×10^{-4}
" "	20.0	3.1×10^{-10}	3.8×10^{-4}

In effect, the diffusivity of hydrogen in $Ni_{81}P_{19}$ is found to be $\sim 10^{-10}$ cm^2/sec., or roughly on the order of lattice diffusion in nickel. This is significant since it suggests, given the grain boundary-metallic glass comparison, that grain boundary diffusion of hydrogen may be unlikely in the segregated grain boundaries of nickel-base alloys. Indeed, parallel efforts in our laboratory to detect grain boundary diffusion by electrolytic permeation experiments on polycrystalline nickel and nickel-base alloys have been unsuccessful (37). Despite the free volume available in glasses, perhaps metalloid-related or other trapping phenomena are overwhelmingly more effective than transport phenomena. Work is in progress using means other than electrochemical to explore the possibility that grain boundary diffusion may occur.

In summary, it appears that metalloid containing glassy alloys do absorb considerable amounts of hydrogen, presumably catalytically stimulated as described above. In both Fe-and Ni-base glasses, the diffusivity of hydrogen at ambient temperatures is of the order of the lattice diffusion coefficient in fcc metals. Evidence of internal trapping, which leads in some cases to catastrophic embrittlement, is also observed. In short, it does indeed appear that metallic glasses may serve very well as grain boundary analogs in certain circumstances. Work is in progress now to further explore their utility in this regard.

<u>Dislocation Transport of Hydrogen in Nickel-Base Alloys</u>

As discussed with regard to Figure 3, the permeation of hydrogen into crystalline materials may occur in part by lattice diffusion, by grain boundary (or other short circuit) diffusion, or, in the case of specimens undergoing plastic deformation, by dislocation transport. In the latter instance, the transport of hydrogen occurs in the form of Cottrell atmospheres dragged by mobile dislocations. This was first suggested by Bastien and Azou (38, 39). Later, this concept was supported by observations of serrated yielding (28, 40-42), tritium release measurements during plastic deformation and autoradiographic techniques (43-45). Tien, et al. (46) and

Johnson and Hirth (47) have considered dislocation transport analytically. Recently, Kurkela and Latanision (48) have shown by more direct electrolytic permeation measurements that mobile dislocations transport hydrogen at rates much higher than lattice diffusion in nickel. The permeation of hydrogen was studied by the method developed by Devanathan and Stachurski (described earlier) modified to allow plastic deformation of the specimens during the introduction of hydrogen. Similar studies have been performed on Ni_2Cr by Berkowitz, et al. (49) and are reported in this Proceeding.

Typical permeation build-up and decay transients under plastic deformation are shown in Figures 6 and 7. Hydrogen permeates through the specimen in less than 10 seconds and a steady state is reached in less than a minute. On switching off the charging current, the anodic current decays to the value observed prior to charging. When deformation is stopped the anodic current decays back to the original background value. From Figures 6 and 7, it can be seen that increasing strain rate increases the observed permeation flux. Increasing charging current also gave higher permeation fluxes. The largest strain rate (4.2×10^{-4}/s) was below the critical strain rate, given by $\sim 10^{-11}\rho_H$ (ρ_H = density of hydrogen-carrying dislocations), above which hydrogen transport does not occur. Figure 7 shows decay transients for two different strain rates observed after switching off the charging current.   In the case of the higher strain rate, the decay is much faster.

The mathematics of hydrogen permeation has been dealt with elsewhere (35, 50). By applying Fick's second law with the appropriate initial and boundary conditions and using the half-rise time method, D_H can be evaluated. In plastically deformed nickel, the "effective" D_H was found to be of the order of $10^{-5}cm^2$/s in contrast to $10^{-10}cm^2$/s in unstrained nickel, the latter of which compares well with the findings of Robertson (51). Indeed, the effective diffusivity of hydrogen increases linearly with strain rate (37).

From these results, it is evident that in plastically deformed nickel dislocation transport of hydrogen is the predominant mechanism and that the transport rates are several orders of magnitude higher than in unstrained nickel where lattice diffusion is the predominant mechanism. It is interesting to note that according to the modeling of Tien, et al. (46), it should take about 5 seconds for hydrogen to travel through the specimens used in this work. Examination of Figure 6 shows that breakthrough times vary between 2.5 and 5.5 seconds and are functionally dependent on the strain rate. One should, in principle, be able to show that above a certain critical strain rate no dislocation transport occurs - i.e., dislocations outrun their atmospheres. In the case of nickel, the critical strain rate is about 10^{-1} sec^{-1}, far faster than is practical in our permeation experiments.

Dislocations cannot travel from one grain to another with their hydrogen atmospheres. When arriving at a grain boundary they are likely to "dump" the hydrogen there. Dislocation generated in the adjacent grain might then pick up the hydrogen and sweep it away. Based on this view, one might expect dislocation transport to be more difficult in a large grain size material where there are fewer obstacles to dislocation motion; i.e., dislocations have a larger mean free path. Figure 8 shows results for three different grain sizes. These results argee with the predictions - the larger the grain size the higher permeation flux.

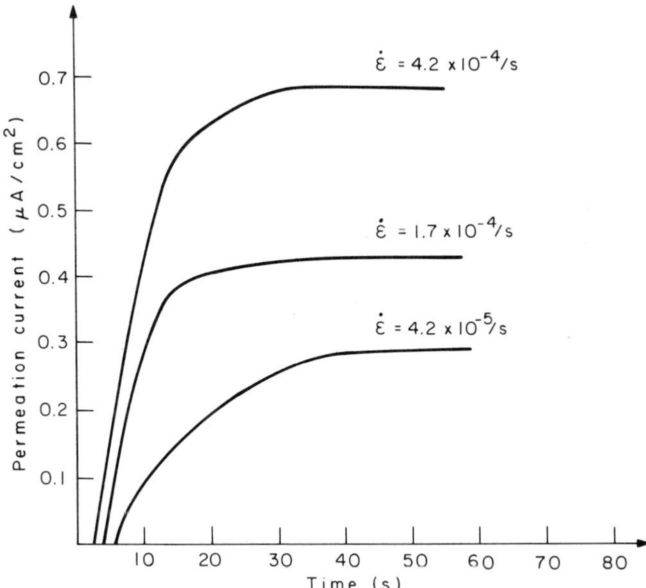

Fig. 6 – Effect of strain rate on permeation behavior (after
Kurkela and Latanision (48)).

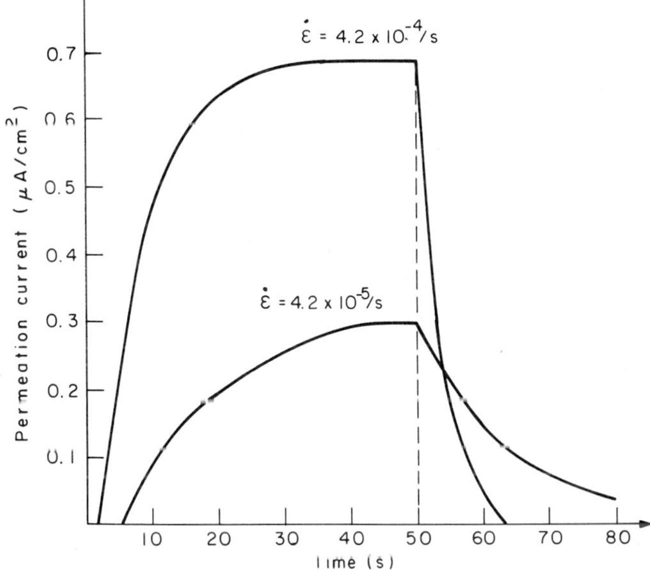

Fig. 7 – Effect of strain rate on the decay transient (after
Kurkela and Latanision (48)).

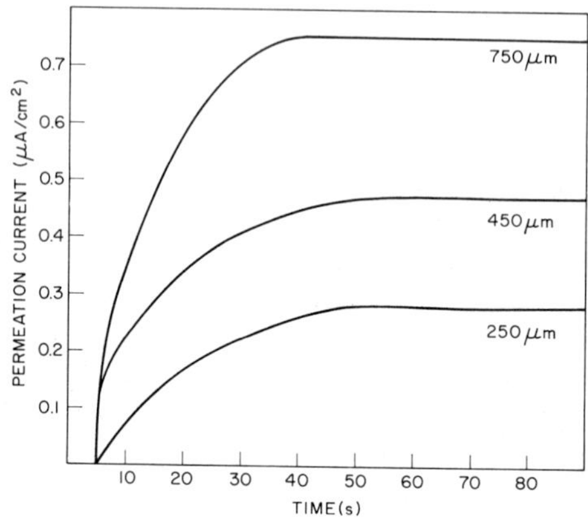

Fig. 8 - Effect of grain size on permeation behavior in
Ni 270 undergoing plastic deformation

Closing Remarks

We have learned quite a lot about the phenomenology of hydrogen embrittlement. It is now more than 100 years, however, since Reynolds (52)
and Hughes (53) and perhaps others first associated certain detrimental
effects on the ductility of iron with the presence of hydrogen. Not only is
hydrogen embrittlement still a major industrial problem, but it is safe to
say that in a mechanistic sense we still do not know what hydrogen (but not
nitrogen or oxygen, etc.) does on an atomic scale to induce this degradation.
The same applies to other examples of environmentally-induced fracture:
what is it about the ubiquitous chloride ion that induces premature catastrophic fracture (stress corrosion cracking) of ordinarily ductile austenitic stainless steels? Why, moreover, are chloride ions troublesome but
chemically similar iodides inhibitors for SCC in such stainless steels? In
short, despite all that we may know about the phenomenology of fracture on
a macroscopic scale, we know precious little about the atomistics of embrittlement phenomena such as those described above. In this regard one
interesting electronic model for hydrogen embrittlement (decohesion) was
proposed by Troiano in 1960 (54). More recent electronic structure calculations (55) involving the application of molecular orbital theory in modeling certain aspects of embrittlement phenomena seem promising.

On the other hand, it is interesting to note that physical chemists and
surface chemists also have interests in the same kinds of interactions that
occur on an atomic scale when metals such as nickel or platinum, for example,
are used as catalysts for chemical reactions. Such metals are very effective
catalysts for the dissociation of molecular hydrogen. Indeed, considerable
experimental study has been directed toward the surface chemistry of hydrogen adsorption, etc., on transition metal surfaces. But much of the same
uncertainty in terms of fundamental understanding pervades this area of
science. Why is it that nickel and platinum are effective hydrogenation

catalysts, but copper is not? And, why are other metals effective catalysts for other chemical reactions, but not for hydrogenation?

It seems likely that the development of fundamental understanding of the kind implicit in the above discussion of catalysis would impact, as well, understanding of the fundamentals of embrittlement, and vice versa. It is interesting, for example, that the dissociation of transition metal carbonyls such as cobalt carbonyl may be stimulated by the presence of hydrogen (56). Here is an example of a technology known to physical chemists, relatively unknown to materials scientists and solid state physicists, but of vital importance to both in the sense of atomic scale understanding of catalysis and embrittlement (the latter, in this case, the dissociation of atomic bonds between cobalt atoms in the cobalt carbonyl molecule). One wonders in a converse sense if the effectiveness of a catalyst might not be significantly increased if the catalyst were allowed to plastically deform while performing its chemical function.

At any rate, we believe that more direct and frequent communications must be stimulated between the kinds of disciplines described above. Indeed, it seems very clear now that much is known by physical chemists and quantum chemists that bears directly on problems of embrittlement and fracture of interest to materials scientists and those involved with the mechanics of solids. This information is often, however, not appreciated. Development of the atomic scale understanding which is sought is not only of academic significance but may well lead to technological advances in chemistry and materials science ranging from catalysts for use in energy conversion and storage, for example, to the improved lifetime and reliability of complex and expensive structures in aggressive environments. In this context, in fact, it seems to us that the technological significance of the interactions of hydrogen with materials will become increasingly more important in the decades to come. There is, for example, the issue of a hydrogen economy -- i.e., the use of hydrogen as an energy medium for this planet. This is not unlikely, particularly if effective (semiconductor) photoelectrodes can be developed to split water (57) since water and sunlight are both plentiful and, at the moment, free! In addition, of course, questions such as materials for storage and transmission or distribution of hydrogen (perhaps in this case materials that are already in place) will become increasingly important. We suspect that if the previous decades of research and engineering on hydrogen were considered exciting, the next decades will be even more so and the impact of this effort will be of even greater consequence.

Acknowledgments

The support of the Office of Naval Research Contract No. N00014-78 - C-0002, NR 036-127 is gratefully acknowledged. Special thanks are due to Dr. P. A. Clarkin of ONR for his continuing encouragement. Likewise, we are happy to acknowledge many exciting discussions over several years with Dr. T. E. Fischer of Exxon Research and Engineering Company regarding the atomistics of hydrogen embrittlement.

References

1. K. Balajiva, R.N. Cook and D.K. Worn: Nature, 1956, vol. 178, p. 433

2. R. Speiser and J.W. Spretnak: Stress Corrosion Cracking and Embrittlement, p. 92, Wiley, New York, 1956

3. H.L. Marcus and P.W. Palmberg: Trans. AIME, 1969, vol. 245, p. 1664

4. D.F. Stein, A. Joshi and R.P. Laforce: Trans. ASM, 1969, vol. 62, p. 776

5. K. Yoshino and C.J. McMahon, Jr.: Met. Trans., 1974, vol. 5, p. 363

6. R.O. Ritchie: Met. Trans., 1977, vol. 8A, p. 1131

7. A. Joshi and D.F. Stein: Met. Trans., 1970, vol. 1, p. 2543

8. A. Joshi and D.F. Stein: J. Inst. Metal, 1971, vol. 99, p. 178

9. J.M. Popplewell and J.A. Ford: Met. Trans., 1974, vol. 5, p. 2600

10. R.M. Latanision and H. Opperhauser, Jr.: Proc. 3rd Int'l Conf. on Strength of Metals and Alloys, vol. 1, p. 472, Institute of Metals and Iron and Steel Institute, 1973

11. R.M. Latanision and H. Opperhauser, Jr.: Hydrogen in Metals, p. 539, ASM, Metals Park, 1974

12. R. M. Latanision and H. Opperhauser, Jr.: Met. Trans., 1974, vol. 5, p. 483

13. R.M. Latanision and H. Opperhauser, Jr.: Met. Trans., 1975, vol. 6A, p. 233

14. J.M. West: Electrodeposition and Corrosion Processes, 2nd ed., pp. 57 and 139, Van Nostrand Reinhold, London, 1971

15. R.O. McCright and R.W. Staehle: J. Electrochem. Soc., 1974, vol. 121, p. 605

16. F. Zakroczymski, Z. Szklarska-Smialowska and M. Smialowski: Werkstoffe und Korrosion, 1976, vol. 27, p. 625

17. H.W. Pickering and M. Zamanzedeh: This Conference, Poster Session

18. B.J. Berkowitz, J.J. Burton, C.R. Helms and R.S. Polizzotti: Scripta Met., 1976, vol. 10, p. 871

19. H.W. Liu, Y.L. Hu and P.J. Ficalora: Engineering Fracture Mechanics, 1973, vol. 5, p. 281

20. N. Pessall, G.P. Airey and B.P. Lingenfelter: Corrosion, 1979, vol. 35, p. 100

21. G.S. Was: Sc. D. Thesis, M.I.T., June 1980

22. B.J. Berkowitz and R.D. Kane: Corrosion, 1980, vol. 36, p. 24

23. S.M.Bruemmer, R.H. Jones, M.T. Thomas and D.R. Baer: Scripta Met.,
     1980, vol. 14, p. 137

24. R. Perry: Metallurgical Developments in High Alloy Steels, Special
     Report 86, p. 227, British Iron and Steel Institute, 1964

25. G.G. Hancock and H.H. Johnson, Trans. AIME, 1966, vol. 236, p. 513

26. M.T. Thomas, Battelle Northwest Laboratories, Private Communication,
     October 1980

27. R.M. Latanision: Surface Effects in Crystal Plasticity, p. 3, Noordhoff
     International Publishing Co., Leyden, 1977

28. R.M. Latanision and R.W. Staehle: Scripta Met., 1968, vol. 2, p. 667

29. R.M. Latanision, J.C. Turn and R.C. Compeau: Proc. Third Int'l
     Conference on Mechanical Behavior of Metals, vol. 2, p. 475, Pergamon,
     Toronto, 1979

30. R.M. Latanision, C.R. Compeau and M. Kurkela: Proc. Alexander R.
     Troiano Symposium on Hydrogen Embrittlement and Stress Corrosion
     Cracking, in press

31. M.F. Ashby, F. Spaepen and S. Williams: Acta Met., 1978, vol. 26,
     p. 1647

32. D.E. Polk: Scripta Met., 1970, vol. 4, p. 117

33. Rapidly Quenched Metals, N.J. Grant and B.C. Giessen eds., M.I.T.
     Press, Cambridge, 1976

34. Metallic Glasses, J.J. Gilman and H.J. Leamy eds., ASM, Metals Park,
     1978

35. M.A. Devanathan and Z. Stachurski: Proc. Roy. Soc., 1962, vol. A 270,
     p. 90

36. J.O'M. Bockris and P.K. Subramanyan: J. Electrochem. Soc., 1971,
     vol. 118, p. 1114

37. M. Kurkela, T.S.F. Lee and R.M. Latanision: Fourth Technical Report
     to ONR, Contract No. N00014-78-C-0002, July 1980

38. P. Bastien and P. Azou: C. R. Acad. Sci. Paris, 1951, vol. 232, p. 1845

39. P. Bastien and P. Azou: Proc. 1st World Metallurgical Congress, p.
     535, ASM, Cleveland, 1951

40. B.A. Wilcox and G.C. Smith: Acta Met., 1965, vol. 13, p. 331

41. A.H. Windle and G.C. Smith: Metal Sci. J., 1970, vol. 4, p. 136

42. G.C. Smith: Hydrogen in Metals, I. M. Bernstein and A. W. Thompson
     eds., ASM, Metals Park, 1974

43. M.R. Louthan, G.R. Caskey, J.A. Donovan and D.E. Rawl: Mat. Sci.
     Eng., 1972, vol. 10, p. 357

44. J.A. Donovan: Met. Trans., 1976, vol. 7A, p. 145

45. L.M. Foster, T.H. Jack and W.W. Hill: Met. Trans, 1970, vol. 1, p. 3117

46. J.K. Tien, A.W. Thompson, I.M. Bernstein and R.J. Richards: Met. Trans., 1976, vol. 7A, p. 821

47. H.H. Johnson and J.P. Hirth: Met. Trans., 1976, vol. 7A, p. 1543

48. M. Kurkela and R.M. Latanision: Scripta Met., 1979, vol. 13, p. 927

49. B.J. Berkowitz, M. Kurkela and R. M. Latanision, This Conference, Poster Session

50. T.K. Govindan Namboodhiri and L. Nanis: Acta Met., 1973, vol. 21, p. 663

51. W.D. Robertson: Z. Metallkunde, 1973, vol. 64, p. 436

52. O. Reynolds: Manchester Lit. Phil. Soc., 1874, vol. 13, p. 93

53. D.E. Hughes: Scientific American Supplement, 237, July 17, 1880

54. A.R. Troiano: Trans. ASM, 1960, vol. 52, p. 54

55. C.L. Briant and R.P. Messmer, Private Communication, February 1980

56. D.M. Muetterties: Transition Metal Hydrides, Dekker, New York, 1971

57. L.A. Harris and R.H. Wilson: Ann. Rev. Mat. Sci., 1978, vol. 8, p. 88

DISCUSSION

<u>C. L. Briant, General Electric</u>: Hydrogen recombinations poisons also poison hydrogen dissociation. Therefore, if your argument is correct, Sn, Sb, etc. segregated to grain boundaries should improve the material with respect to <u>gaseous</u> hydrogen embrittlement. Is there any evidence for this?

<u>R. M. Latanision</u>: As far as I am aware, no one has demonstrated what you have suggested, but I agree with your expectation. The influence of dissociation catalysts(Pt, for example) should be opposite to the effect of recombination poisons in principle. In the latter instance, the work reported by Pickering in this conference shows very nicely that Pt implanted into iron decreases the rate of electrolytic hydrogen absorption. Conversely, elements like Pt and Pd are very good dissociation catalysts for molecular hydrogen and, as you are probably aware, Pd membranes are used to separate H_2 from gas mixtures based on the effective absorption of hydrogen from the gas phase.

<u>C. L. Briant,General Electric</u>: I have always thought of grain boundaries as fast diffusion paths, yet the amorphous metal results suggest this idea may be incorrect. Would you comment on this point.

<u>R. M. Latanision</u>: It seems to me that your perception is in general correct- i.e., grain boundaries are often fast diffusion paths. With regard to hydrogen diffusion along segregated grain boundaries in nickel or nickel-base alloys, our observations to this point indicate that such grain boundaries (or their metallic glass analogs) are not effective short circuits, perhaps because of some solute-hydrogen (or other) trapping interactions along the grain boundary.

<u>R. A. Oriani, University of Minnesota</u>: The larger permeability for hydrogen of the glassy metal may indeed be due in part to a higher input fugacity induced by the high metalloid content. However, I suggest that there may be another contributing factor to increase the solubility; that is the 30% smaller Young's modulus that characterizes metallic glasses, and to the extent that a grain boundary resembles a metallic glass may also be characteristic of grain boundaries. It may be possible to separate the Young's modulus effect from the fugacity effect by measuring the permeability through the metallic glass after it has been caused to crystallize by laser annealing.

<u>R. M. Latanision</u>: I agree with your comments. We have, in fact, attempted to examine the permeability of glasses after crystallization (not by laser annealing, but by other means). The problem here is that glasses of the kind we've examined became so brittle after crystallization that we have difficulty handling them. We plan to continue this effort.

HYDROGEN EFFECTS IN NICKEL -- EMBRITTLEMENT OR ENHANCED DUCTILITY?

J. Eastman, T. Matsumoto, N. Narita, F. Heubaum and
H.K. Birnbaum

Department of Metallurgy and Mining Engineering
University of Illinois at Urbana-Champaign
Urbana, Illinois 61801

In the present work we have attempted to study hydrogen-related fracture
of nickel using a variety of experimental methods with the principle aim of
establishing the mechanism of fracture. Fracture of nickel was studied in
low pressure hydrogen gas and for specimens containing solute hydrogen using
slow strain rate tension tests. Both intergranular and transgranular
fracture was observed on testing in hydrogen gas depending on the heattreat-
ment and on the presence of trace impurities such as S, Mn, and Mg. The
intergranular fracture mode was correlated with the segregation of S to the
grain boundaries as measured by Auger and SIMS techniques. Specimens
containing solute hydrogen fractured in an intergranular mode for all heat-
treatments. Detailed examination of both the intergranular and transgranular
fracture surfaces indicated that neither was a truly "brittle" fracture mode
but that both indicated a high degree of local ductility.

The effect of hydrogen on the plastic properties of nickel was examined
using stress strain techniques and in situ deformation in an environmental
cell in the HVEM. It was shown that hydrogen decreases the flow stress of
nickel. In situ environmental cell fracture studies in hydrogen gas in the
HVEM showed that the fracture process resulted from locally enhanced
deformation at the crack tip due to the presence of hydrogen. This was confirmed by in situ fracture studies in the SEM.

I.   Introduction

Embrittlement of nickel and its alloys by solute and gaseous hydrogen is well established and has been extensively studied. Polycrystalline nickel has been reported to fracture in a brittle intergranular (IG) mode when charged with hydrogen (1-3) while single crystals containing solute hydrogen fail in a completely ductile mode (3). Kamdar (4) has recently reported that single crystals of nickel fracture in a transgranular (TG) mode when tested in gaseous H_2. Both single and polycrystalline Ni specimens exhibit extensive plasticity prior to fracture so the process can hardly be considered a completely brittle one. Hydrogen effects have also been noted on the plastic response of nickel with solute strengthening and Portevin-Le Chatalier effects (5).

The mechanisms of the hydrogen-related fracture have not been established nor has it even been shown that the process is a brittle one. In the present paper we will summarize a series of experiments carried out at the University of Illinois which bear on the question of what determines the fracture path and what the fracture mechanism is. While a detailed discussion of mechanisms will be published elsewhere we will show in this paper that the fracture is largely a result of localized plastic processes which hydrogen assisted.

II.   Experimental Procedures

A variety of alloys of Ni were prepared to study the effects of trace elements on the fracture and deformation behavior. In the present paper we shall review the results obtained for the alloys specified in Table I. Since the fracture behavior was noted to be extremely sensitive to the S distribution , several classes of heat treatments will be discussed:

Series VR; anneal at 1573K in a vacuum of 1.3×10^{-3} Pa followed by a 200K/sec quench into silicone oil;

Series VS, anneal at 1573K in a vacuum of 1.3×10^{-3} Pa, quench into silicone oil, age at 1173K for 3.6Ksec, followed by cooling at a rate of 0.03K/sec;

Series HR; same as VR above except the annealing and quench took place in 10^2 KPa of H_2 gas which introduced solute H at the level of 700 at.ppm. (specimens were stored at 77K until used;

Series CR; same as VR above except the anneals and quenching took place in CO/CO_2 mixtures which were controlled to adjust the resulting C concentration;

Series CS; same as VS above except CO/CO_2 mixtures were used as annealing atmospheres.

Various specimen configurations including single crystals and relatively large grained polycrystals were used as required by the tests and these will be described in detail elsewhere. Tensile testing was carried out at strain rates of about 2×10^{-5} sec^{-1} (or as subsequently specified) in 10^2 KPa H_2 atmospheres, vacuum, helium or air as required.

Table I. Concentrations[*] of Significant Alloying
Elements in the Alloys Studied

Alloy/Heat Treatment	C	S	H	Mg	Mn
HP/VR	3200	44	0	0	0
HP/VS	3200	44	0	0	0
HP/CR	140	1.6	0	0	0
HP/CS	140	1.6	0	0	0
Cl/VR			0	2500	900
Cl/VS			0	2500	900
C2/HR	280	< 1	700	70	< 1
C3/CS					

[*] Atomic ppm; metallic elements by spark source mass spectroscopy, interstitials by vacuum fusion.

III. Results

A. Fracture Path

The results of fracture studies on polycrystalline specimens are shown in Table II where the values listed are the averages over many tests. After fracture detailed Auger analysis and depth profiling was carried out on the IG fracture facets and on the TG surfaces. The results of the Auger analysis clearly showed: (a) a high level of sulfur segregation on the IG fracture surfaces of specimens HP/VS and HP/CS; (b) no significant sulfur segregation on the IG fracture surfaces of specimens HP/VR, HP/CR or Cl/VR or Cl/VS; (c) no significant sulfur segregation at any of the TG fracture surfaces; (d) no significant sulfur segregation at the IG fracture surfaces at specimens C2/HR; (e) no elements other than sulfur exhibited any significant segregation at the fracture surfaces.

The extent of hydrogen embrittlement by gaseous H_2 or solute H_2 is clearly shown by the data of Table II. After a detailed analysis of the results (to be shortly published) we conclude the following:

(a) The fracture path in specimens containing solute hydrogen (C2/HR) is IG over the temperature range 300-123K and since these specimens do not have S segregation at grain boundaries the IG fracture does not depend on S segregation. This result is consistent with that reported for fatigue fracture of Ni (6). Nickel specimens containing solute hydrogen exhibit IG fracture even when tested in 10^2 KPa of H_2. The IG fracture is observed even at temperatures (123K) where H diffusion is very slow and hydrogen diffusion to the fracture tip can not take place.

(b) The fracture mode of Ni tested in gaseous H_2 atmospheres depends on the distribution of S in contrast to the behavior of specimens containing solute hydrogen. Specimens treated to segregate S at grain boundaries exhibit predominantly IG fracture (HP/VS, HP/CS) while in the absence of S segregation TG fracture is observed (HP/VS, HP/CR). The strain to failure is significantly less when the fracture mode is IG.

(c) The presence of Mg and Mn in solid solution serve to trap the S and prevent segregation at grain boundaries. In these alloys (Cl/VR, Cl/VS) predominantly TG fracture is observed even in specimens (Cl/VS) given an anneal which would cause S segregation and IG failure in the absence of Mg and Mn.

(d) At the level of C concentrations used in the present experiments no significant effects are noted on the ductility of Ni specimens in the absence of H. When tested in gaseous H_2 the high C alloys (HP/VR, HP/VS) exhibited less ductility loss than the low C alloys (HP/CR, HP/CS).

(e) The difference in energy between the IG and TG fracture paths must be relatively small as the fracture surface contains elements of both paths in all cases.

(f) In all specimens testing in air or inert gases resulted in a completely ductile behavior with failure by microvoid coalescence. Sulfur segregation in the absence of solute H or gaseous H_2 had no effect on the fracture at the concentration levels of the present experiments.

(g) The TG fracture surfaces had surface intersections which were parallel to the $\{111\}$ slip plane traces. By carrying out two surface analyses of the fracture planes of single crystals tested in gaseous hydrogen the fracture plane was determined to be the $\{111\}$. This contrasts to the previous report of fracture along the $\{110\}$ (7).

<u>Table II. Fracture Characteristics</u>

Alloy/ Heat Treatment	Test Atmosphere	% Reduction in Area	Fracture Mode
HP/VR	H_2	75	20% IG 80% TG
	Air	95	Ductile Microvoid
HP/VS	H_2	35	70% IG 30% TG
	Air	90	Ductile Microvoid
HP/CR	H_2	55	20% IG 80% TG
	Air	95	Ductile Microvoid
HP/CS	H_2	14	80% IG 20% TG
	Air	95	Ductile Microvoid
Cl/VR	H_2	55	25% IG 75% TG
	Air	90	Ductile Microvoid
Cl/VS	H_2	60	20% IG 80% TG
	Air	90	Ductile Microvoid
C2/HR	Vac.	26	$\sim$ 100% IG
	H_2		$\sim$ 100% IG

B.   Fractography

A large number of interesting features were noted on the fracture surfaces using high resolution SEM methods and these will be discussed in subsequent publications. At the present time we will focus on the generic features. Both IG and TG fracture surfaces were not truly "brittle" but showed many features characteristic of a high degree of local plasticity. The IG fracture surfaces obtained in specimens containing solute H or fractured in gaseous H_2 were qualitatively the same. There were no evident features which could be ascribed to the segregation of S at the grain boundaries.

Some features of the {111} TG fracture surfaces are shown in Fig. 1. Traces of intersecting {111} slip lines were observed (Figs. 1a and 1c) as were small hillocks and depressions, the sides of which appeared to correspond to the intersecting {111} (Figs. 1c and 1d). No evidence for microvoids were observed in contrast to a previous report (7) (Fig. 1b). Secondary cracking was often observed (Figs. 1e and 1f). TG cracking in polycrystalline specimens showed many of the same features (Figs. 1f, 1g and 1h) but since the {111} fracture plane was less well aligned with the tensile axis, the TG surfaces were often more highly stepped. The surfaces shown in Fig. 1 are clearly not the result of a true cleavage process (they are of the type often designated as "quasi-cleavage") and give the impression of a high degree of local plasticity. Identifiable "river lines" are observed but are significantly less sharply defined than in true cleavage.

Similar conclusions may be drawn from the IG fracture surfaces as shown in Fig. 2. Slip line markings (Fig. 2a) are observed as are patches of highly deformed regions (Fig. 2b). Rarely is the grain boundary surface revealed by IG fracture smooth and planar as would be expected if the fracture were due to "decohesion" along the boundary. The IG fractographs shown in Fig. 2 are typical of those produced by solute and by gaseous hydrogen.

C.   In-situ Fracture Studies

The course of fracture was studied using in-situ SEM and HVEM fracture studies. SEM photographs of the tips of propagating cracks for Ni-H alloys are shown in Fig. 3. Studies of the video tapes of IG fractures clearly show the occurrence of a large amount of slip in both grains which is manifested in the large COD's shown in Figs. 3a-3c. Appreciable shear across the grain boundary often occurred in front of the crack tip (Fig. 3c). The crack would often proceed rapidly along a grain boundary until it intersected a triple point at which it would stop until sufficient grain boundary shear occurred in one of the boundaries in front of the crack tip. In many cases a new crack would open along the grain boundary in front of the main crack and then the two would grow together (Fig. 3a).

Similar SEM experiments were carried out in H_2 gas to produce TG fracture as shown in Fig. 3d. Large amounts of slip can be seen at the surface, some of which is due to the deformation obtained before the fracture began. Again there clearly is a good deal of plasticity accompanying the crack propagation as evidenced by the COD. No evidence for microvoid formation was seen.

A more detailed study of the TG crack tip processes was carried out using a tensile stage in an atmosphere of about 10^4 Pa H_2 gas in the HVEM and will be published in detail shortly. Some of the results are shown in Fig. 4. The tip of a fracture occurring in vacuum is shown in Figs. 4a and 4b which show a noncrystallographic fracture with a large

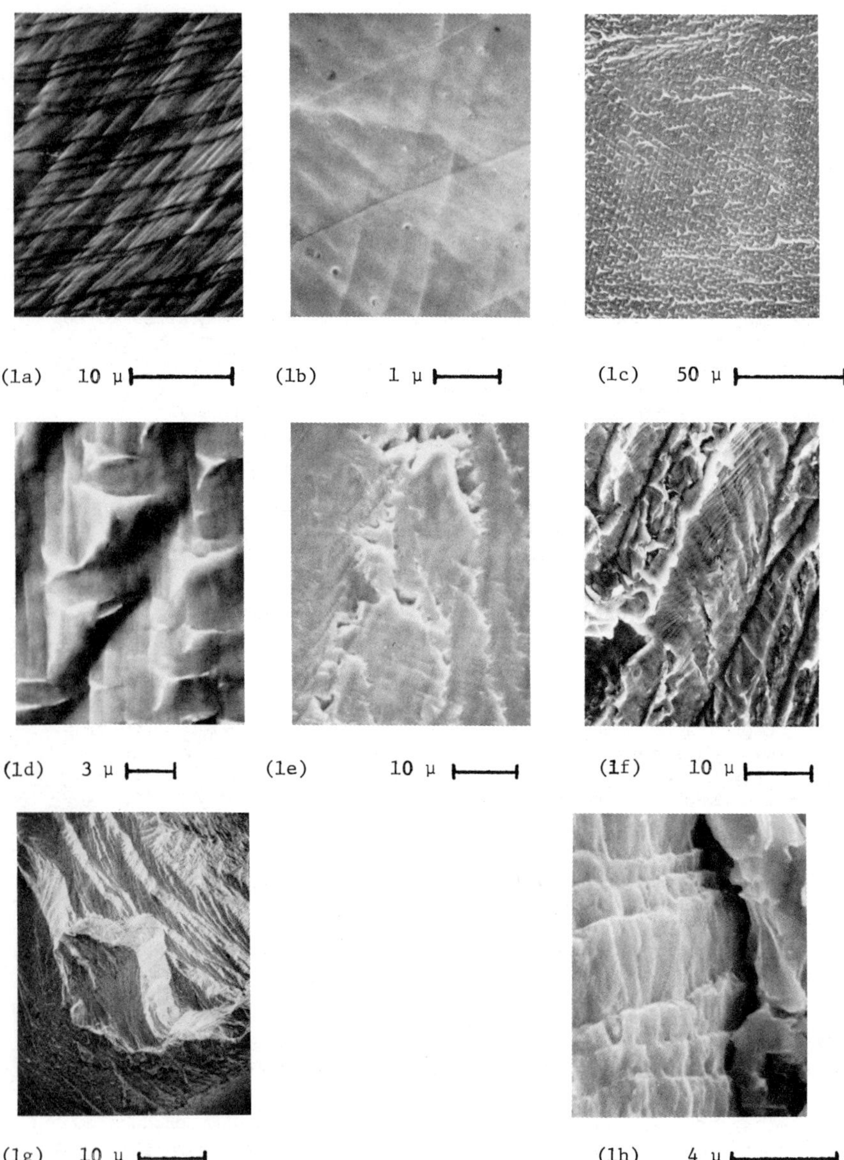

(1a)   10 μ ⊢————⊣   (1b)   1 μ ⊢——⊣   (1c)   50 μ ⊢————⊣

(1d)   3 μ ⊢——⊣   (1e)   10 μ ⊢——⊣   (1f)   10 μ ⊢——⊣

(1g)   10 μ ⊢————⊣   (1h)   4 μ ⊢————⊣

Fig. 1 – Characteristic transgranular fracture surface features.
    (1a) – {111} fracture surface showing intersecting {111} slip
        plane traces. Fracture in H_2 gas; single crystal.
    (1b) – Same as (a) at high magnification. No evidence is found for
        microvoid formation.
    (1c) – Hillocks formed by intersecting {111} slip planes.
    (1d) – Same as (c) at high magnification.
    (1e) – Secondary cracking on transgranular fracture surface formed
        by stressing in H_2 gas.

(1f) – Stepped transgranular fracture surface formed in polycrystalline specimens stressed in H_2 gas. The surface is of
the "quasicleavage" type showing "cleavage" steps and
apparent crack arrest markings.

(1g) – "Quasicleavage" fracture surface formed in polycrystalline
specimens stressed in H_2 gas.

(1h) – Same as (g) at higher magnification.

(2a)   20 μ ⊢————⊣   (2b)   20 μ ⊢————⊣

Fig. 2 – Characteristic intergranular fracture surface features. Similar
features are observed for specimens fractured in H_2 gas or
containing solute hydrogen.
(2a) – Slip line traces.
(2b) – Localized regions of high ductility.

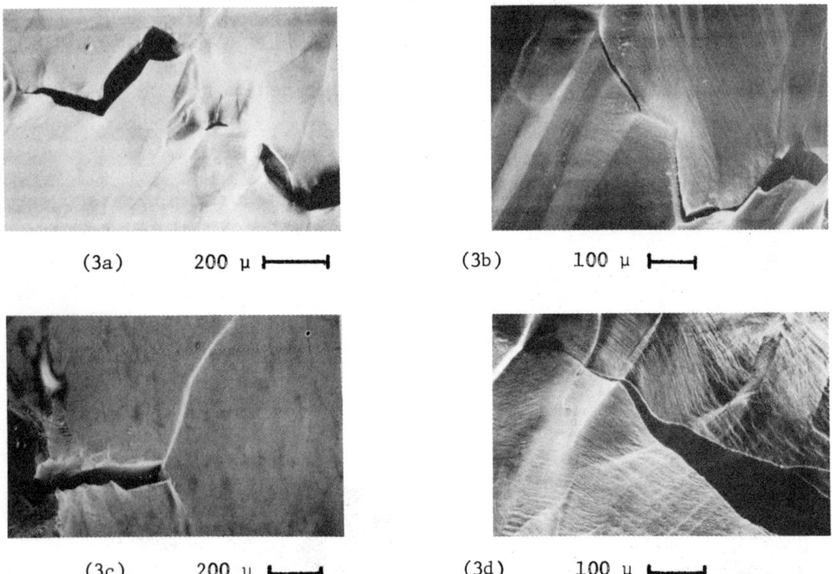

(3a)   200 μ ├────┤   (3b)   100 μ ├────┤

(3c)   200 μ ├────┤   (3d)   100 μ ├────┤

Fig. 3 - In-situ SEM micrographs of propogating cracks.
  (3a) - Intergranular fracture in specimen containing solute
         hydrogen. Micrograph shows crack nucleation in front of the
         main crack; large COD.
  (3b) - Intergranular fracture showing crack nucleation in front of
         the main crack and the large slip zone in front of the crack.
  (3c) - Intergranular fracture showing large COD and large shear
         displacements along adjacent grain boundaries.
  (3d) - Transgranular fracture showing large COD and the slip zone
         in front of the crack.

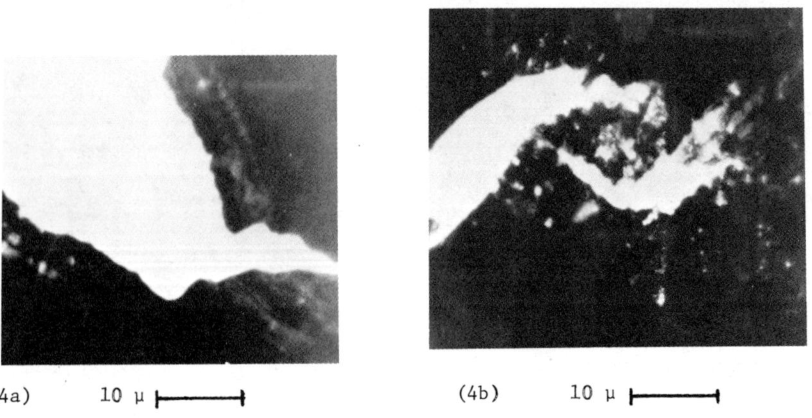

(4a)   10 μ ├────┤   (4b)   10 μ ├────┤

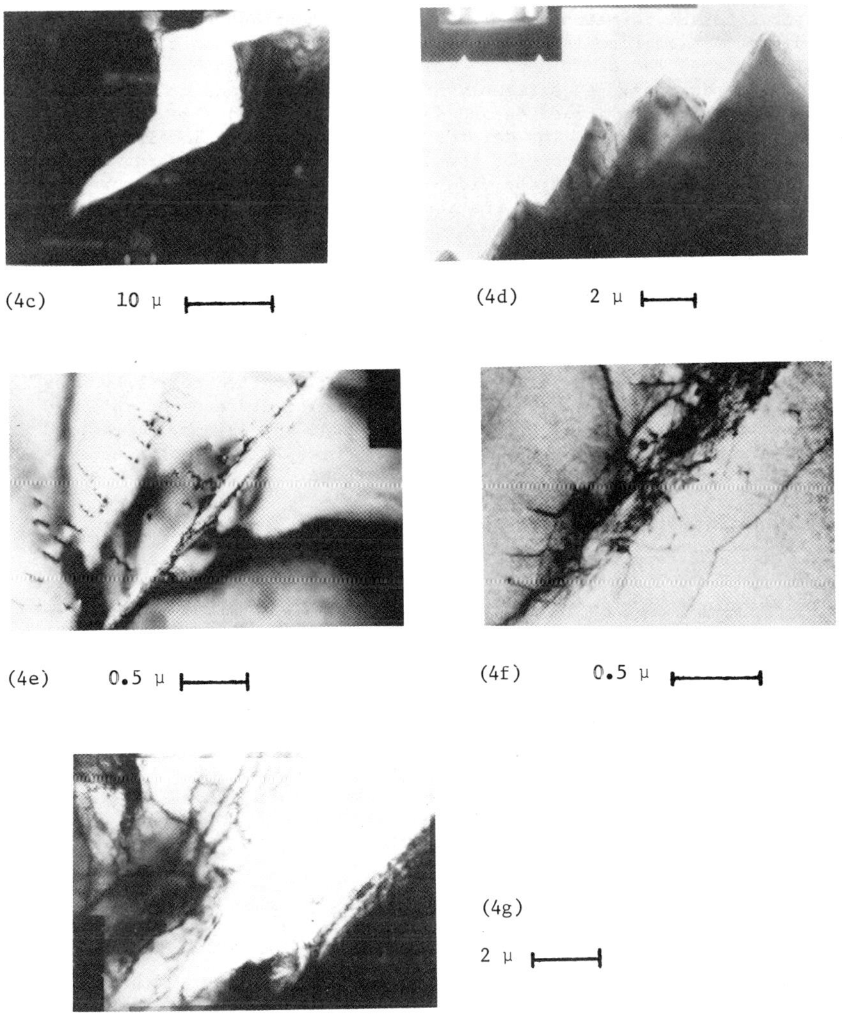

Fig. 4 - HVEM micrographs of propogating cracks.
  (4a) - Noncrystallographic fracture for specimens stressed in
         Vacuum.
  (4b) - Same as (a). Crack tip region showing "hole" formation in
         front of the main crack.
  (4c) - Crystallographic fracture for specimens stressed in H_2 gas.
         Fracture surfaces are the {111}.
  (4d) - Same as (c). The thinned regions along the fracture surfaces
         result from slip along the {111} causing the sawtooth
         appearance.
  (4e) - Dislocation structure in front of a crack tip formed in H_2
         gas showing intense slip in front of the crack.
  (4f) - Same as (e). Region further along in front of the crack.
  (4g) - High dislocation densities localized along the fracture
         surfaces of specimens fractured in H_2 gas.

deformation zone and "holes" opening up in front of the main crack. This corresponds to the formation of microvoids. In contrast to this the crack tip and fracture surface in H_2 gas is highly crystallographic (Figs. 4c and 4d) with the fracture planes parallel to the {111} as observed in macroscopic specimens. An extremely high dislocation density forms in front of the crack tip (Figs. 4e and 4f) and is observed all along the crack surfaces (Fig. 4g). The deformation is very highly localized along the {111} slip planes in front of the crack tip rather than being broadly distributed. Examination of the video tapes shows that the fracture results from this localized deformation by a shearing of one-half of the crystal relative to the other. This fracture by slip on {111} results in a thinned region along the fracture surface which is bounded by {111} plane intersections as shown in Fig. 4d.

While the detailed relationships between the HVEM results and the fracture of macroscopic specimens cannot be discussed in detail, it is clear that both are the result of highly localized plastic processes which occur in the presence of H solutes or H_2 gas. A detailed correspondence between the fracture details is not expected due to differences in stress states but the sawtooth {111} fractures (Figs. 4d and 4h) may correspond to the features shown in Figs. 1c and 1d. The results strongly suggest that slip is enhanced by the presence of hydrogen.

D. Hydrogen Effects on Plastic Deformation

Since the fracture process is closely related to a localized plasticity induced by hydrogen we examined this question using HVEM and tensile testing techniques the results of which will be reviewed. Plastic deformation in the HVEM in vacuum resulted in dislocation motion and generation as expected (Fig. 5a). The specimen displacement was held constant and H_2 gas was admitted to the environmental chamber. At about 10^3 Pa of H_2 dislocations were observed to begin to move and at about 10^4 Pa of H_2 rapid dislocation generation and motion occurred resulting in the structure shown in Fig. 5b. Similar experiments using He gas had no effect on the dislocation structure and since the specimen which was held at a constant displacement experienced a decreasing stress due to the dislocation motion, we conclude that the dislocation generation and motion was due to H introduced as a solute from the gas phase.

As a further demonstration of solute softening due to H, low strain rate tensile tests of 25 μm sheet specimens were carried out in 10^2 KPa of H_2 and He gas atmospheres with the typical results shown in Fig. 6. At high strain rates, $\dot{\varepsilon} = 7\times10^{-5}$ sec^{-1}, the stress strain curves are identical for He and H_2 atmospheres. In contrast, slower strain rates, such as $\dot{\varepsilon} = 10^{-7}$ sec^{-1}, showed lower flow stresses for specimens tested in H_2 as compared to those tested in He gas. This softening is observed in the microstrain range and in the post yield region. At higher strains the softening effect due to H_2 is reduced.

The cause of this decrease in flow stress due to H_2 is not yet established but it has the general characteristics of that observed on electrolytically charging Fe with H during deformation (8). We have measured the activation volume, $V^* = (RT)^{-1} \partial \ln \sigma / \partial \dot{\varepsilon}$, and found no difference between the two atmospheres. One significant difference is the "internal stress"

<table>
<tr><td>(5a)</td><td>0.5 μ ⊢———⊣</td><td>(5b)</td><td>0.5 μ ⊢———⊣</td><td>(5c)</td><td>0.5 μ ⊢———⊣</td></tr>
</table>

Fig. 5 — HVEM micrographs of dislocation structures formed during insitu
deformation.
    (5a) – Specimen stressed in vacuum.
    (5b) – Same area as (5a) after the introduction of H_2 gas while the
        specimen is held under stress at constant displacement.
    (5c) – Dislocation structure resulting from the introduction of H_2
        gas after the specimen was stressed in vacuum. Prior to the
        introduction of H_2 gas no dislocations were visible in this
        region.

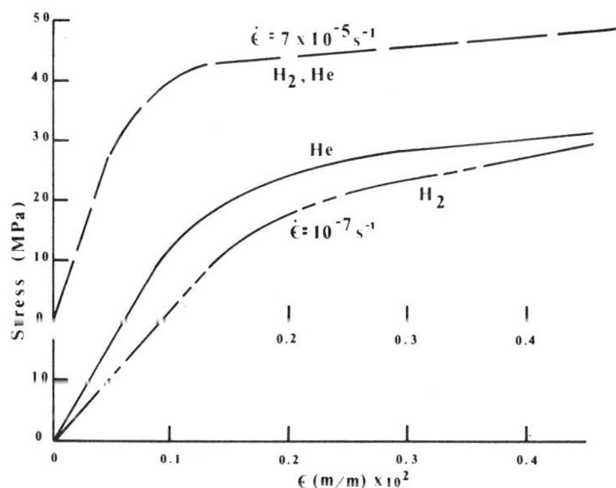

Fig. 6 — Initial portions of the stress strain curves for specimens tested
in He or H_2 gas environments at "high" and "low" strain rates.

as measured by the "stress dip test". Specimens deformed in He exhibited $\sigma_{int}/\sigma_{flow} \approx 0.85$ independent of the ε as expected for fcc metals. In contrast, deformation in H_2 gas resulted in $\sigma_{int}/\sigma_{flow} \approx 0.65$ at $\varepsilon = 10^{-3}$ which increased to 0.85 at about $\varepsilon \simeq 3.5 \times 10^{-3}$. The lower σ_{int} for H_2 atmospheres is consistent with a marked difference in the dislocation structure as observed in the HVEM experiments and the increase in σ_{int} is also consistent with H enhanced plasticity. Thus, while the mechanisms are not well understood, it appears that small concentrations of H in solid solution have a major effect on enhancing dislocation motion.

IV. Conclusions

While only a brief summary of our results have been presented we can draw the following conclusions concerning hydrogen related fracture of Ni:

(a) IG fracture of Ni in H_2 gas occurs when S segregation exists at grain boundaries. In the absence of S segregation a TG mode is observed.

(b) IG fracture of Ni containing solute hydrogen occurs whether or not S segregation at grain boundaries occurs.

(c) Mg and Mn can act as traps for S to prevent segregation at the grain boundaries.

(d) The IG and TG fracture processes result from a localized deformation process at the crack tip. The importance of H enhanced plastic processes at crack tips was first suggested for steels by Beachem (9).

(e) Hydrogen in solution at low concentration levels results in enhanced dislocation generation and motion.

(f) The TG fracture mechanism is related to hydrogen enhanced plasticity along 111 planes in front of the crack tip.

V. Acknowledgements

This work was supported by the Office of Naval Research Contract N00014-75-C-1012 and by the National Science Foundation Grant DMR 77-09808. We would like to acknowledge the support of the HVEM facility at the Argonne National Laboratory and to particularly thank Mr. A. Philippides and Mr. E. Ryan for their assistance with the HVEM experiments.

References

1. G. C. Smith, pp. 485-511 in *Hydrogen in Metals*, I. M. Bernstein and A. W. Thompson, eds.; ASM, Metals Park, 1974.
2. T. Matsumoto and H. K. Birnbaum, *Hydrogen in Metals*, Proceedings of the Second J.I.M. International Symposium, *Transactions of the Japan Institute of Metals Supplement*, 21 (1980) 493.
3. A. M. Windle and A. C. Smith, Journal of Material Science, 4 (1970) 136.

4. M. H. Kamdar, Paper 3D10 in _Proceedings of the Second International Congress on Hydrogen in Metals_; Pergamon, Oxford, 1977.
5. B. A. Wilcox and G. C. Smith, _Acta Metallurgica_, _12_ (1964) 371.
6. R. H. Stulen, _Hydrogen in Metals_, Proceedings of the Second J.I.M. International Symposium, _Transactions of the Japan Institute of Metals Supplement_, _21_ (1980) 501.
7. S. P. Lynch, _Scripta Metallurgica_, _13_ (1979) 1051.
8. H. Matsui, H. Kimura and S. Moriya, _Materials Science and Engineering_, 40 (1979) 207; 217.
9. C. D. Beachem, _Metallurgical Transactions_, _3_ (1972) 437.

EFFECT OF ORDERING ON THE HYDROGEN
PERMEATION AND EMBRITTLEMENT OF Ni$_2$Cr

B. J. Berkowitz
Exxon Research and Engineering Company
Corporate Research Science Laboratories
Linden, NJ 07036

and

M. Kurkela and R. M. Latanision
Corrosion Laboratory
Department of Materials Science and Engineering
Massachusetts Institute of Technology
Cambridge, Massachusetts 02139

Introduction

It has been found that highly cold worked HASTELLOY C-276 is susceptible to hydrogen embrittlement after aging at temperatures in the range
of 200 to 500°C.(1)   It was further shown that the grain boundary segregation of phosphorus is a key factor for the increased susceptibility to
hydrogen induced intergranular failure.(2)   In contrast, it has been
suggested by Asphahani(3) that an ordering reaction, as opposed to
segregation, is responsible for the increased susceptibility.   It was
hypothesized that A$_2$B type order in the form of Ni$_2$ (Cr, Mo) nucleated
during aging of HASTELLOY C-276 and that it was the growth of this
ordered phase which caused an increase in the hydrogen embrittlement
susceptibility.(3)   In order to study this phenomenon as a possible
mechanism for increased embrittlement, Berkowitz and Miller(4) investigated the effect of ordering on the hydrogen embrittlement
susceptibility of Ni$_2$Cr, a model ordering alloy.

That study showed that there is very little effect of ordering on
ductility when tested in air.   However, as shown in Figure 1, hydrogen
charging produces a dramatic effect.   The maximum embrittlement
susceptibility exists in the disordered and highly ordered samples.
They each display a loss in reduction in area (RA) of approximately 70%.
Upon ordering, there is a sharp decrease in embrittlement susceptibility
with increasing order to about 40-50% order followed by an increase in
embrittlement thereafter.   There is a minimum in susceptibility at
40-50% order.

Scanning electron microscopy (SEM) revealed much about the mode of
failure.   The air-tested samples typically exhibited transgranular dimpled
ductile failure as shown in Figure 2, indicating substantial ductility.
All the air tested samples displayed the same type of ductile fracture
independent of the amount of ordering.   The hydrogen charged specimens,
on the other hand, varied considerably in failure mode.   The disordered

<u>Table I</u>

The Effect of Ordering on the Hydrogen
<u>Permeation of Ni$_2$Cr During Plastic Deformation</u>

THICKNESS: 0.2 mm
Current Density: 200 mA/cm^2

TREATMENT*	Strain Rate (s-1)	J_∞ (µA/cm^2)	D_{eff} (cm^2/s)
as quenched	8.3 x 10^{-5}	0.06	7.0 x 10^{-6}
(unordered)	4.2 x 10^{-4}	0.22	1.1 x 10^{-5}
2 hr/500°C	8.3 x 10^{-5}	0.10	7.0 x 10^{-6}
(6% order)	1.7 x 10^{-4}	0.16	9.2 x 10^{-6}
25 hr/500°C	8.3 x 10^{-5}	0.16	7.0 x 10^{-6}
(44% order)	1.1 x 10^{-4}	0.28	9.2 x 10^{-6}
	4.2 x 10^{-4}	0.61	1.1 x 10^{-5}
100 hr/500°C	8.3 x 10^{-5}	0.06	7.0 x 10^{-6}
(70% order)	1.7 x 10^{-4}	0.09	9.2 x 10^{-6}
	4.2 x 10^{-4}	0.24	1.1 x 10^{-5}

* All samples were first heated at 900°C for 2 hrs. followed by
  water quench.

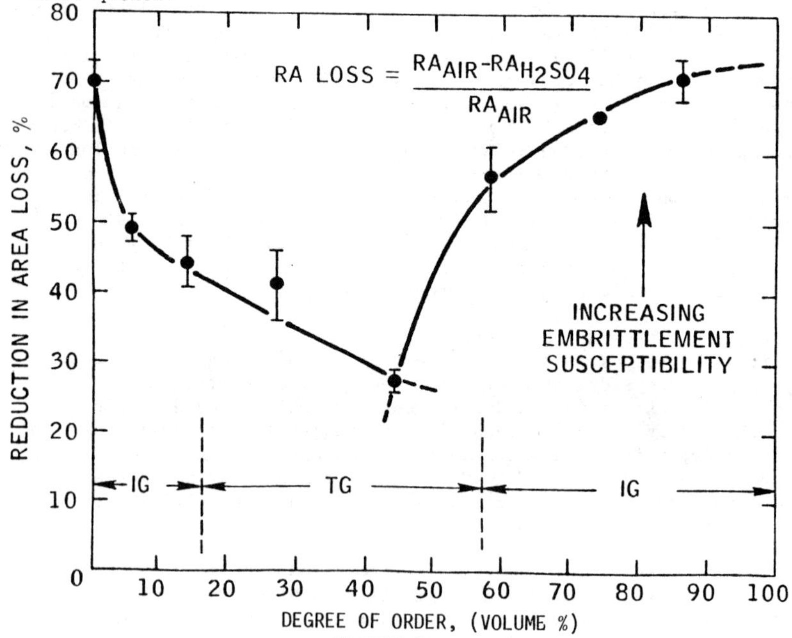

<u>FIGURE 1</u>
The Effect of the Degree of Order on the Embrittlement
Susceptibility of Ni$_2$Cr. Regions of Intergranular (IG)
and Ductile Transgranular Failure (TG) are Shown.

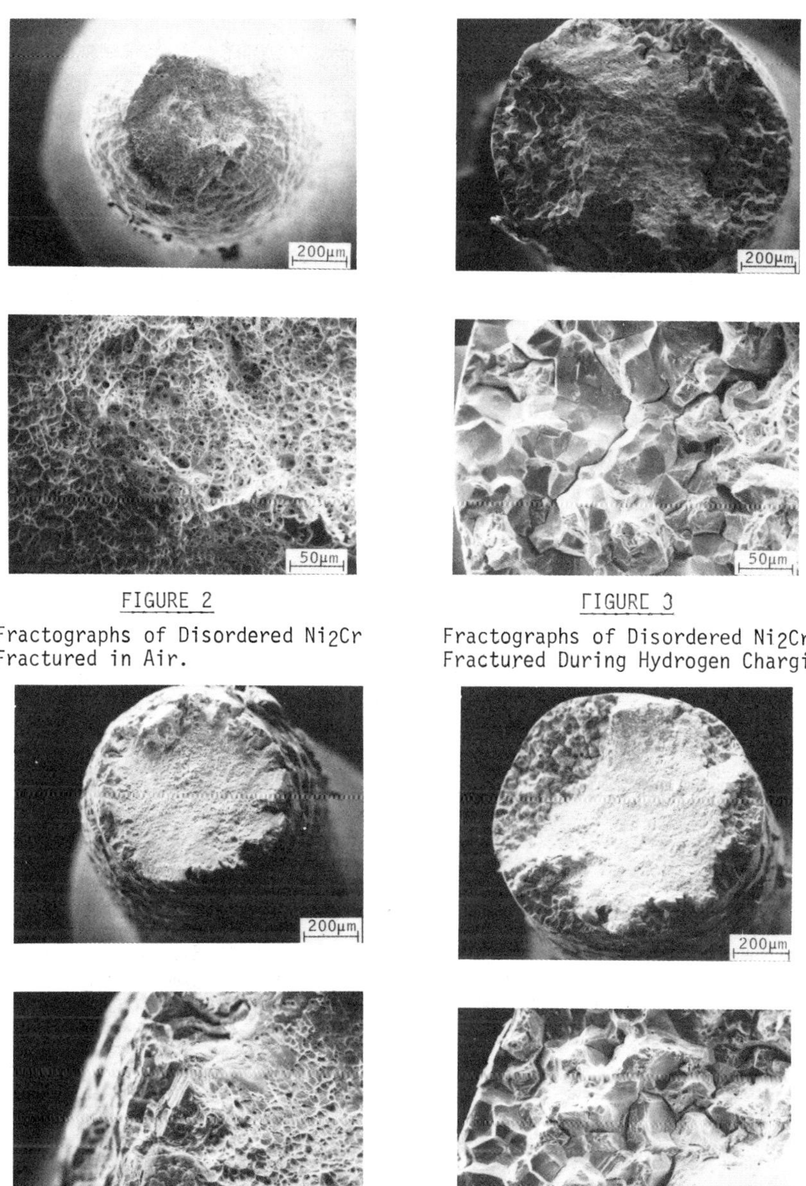

FIGURE 2

Fractographs of Disordered Ni2Cr
Fractured in Air.

FIGURE 3

Fractographs of Disordered Ni2Cr
Fractured During Hydrogen Charging.

FIGURE 4

Fractographs of Ni2Cr Ordered 44 pct,
Fractured During Hydrogen Charging.

FIGURE 5

Fractographs of Ni2Cr Ordered 74 pct,
Fractured During Hydrogen Charging.

specimen showed ductile transgranular failure in the center and intergranular failure along the perimeter as exhibited in Figure 3. The percentages of intergranular and ductile failure varied in amount and related directly with the degree of embrittlement resulting from hydrogen
charging. Figures 4 and 5 show the fracture surfaces of hydrogen charged
specimens which were ordered 44% and 74%, respectively. The depth of
penetration of brittle failure is highest for disordered and highly
ordered specimens. Samples which showed less effect of hydrogen (e.g.,
14-44% order) displayed much less intergranular failure. It appears
that quasicleavage may be the mode of failure at the extreme perimeter
of these samples. Above 44% order, the brittle regions are intergranular and increase in area and penetration with increasing order.

A number of possible mechanisms for these results were hypothesized.(4) One proposal was that the nucleation and growth of ordered
domains may affect dislocation motion, and, if dislocation motion is
responsible for hydrogen transport(5,6), changes in size and distribution
of ordered domains could control the delivery of hydrogen to the grain
boundaries. Thus, the purpose of this study was to learn more about the
role of dislocations on hydrogen transport in Ni_2Cr as a function of
ordering.

Experimental Procedure

A review of the ordering characteristics of Ni_2Cr are described in
the previous paper.(4) Simply stated, Ni_2Cr is disordered after heating
at 900°C for 2 hours followed by a water quench. Subsequent isothermal
aging at 500°C for various periods of time results in the volume fraction
of ordered phase as shown in Figure 6.(7)

The permeation of hydrogen was studied by the method developed by
Devanathan and Stachurski(8), modified to allow plastic deformation of
the specimen during the introduction of hydrogen. Figure 7 shows a
schematic of the apparatus.(6) The specimen was mounted between the two
compartments giving a 0.3 cm^2 area of exposure. Strips of Ni_2Cr were
cold rolled to a thickness of 0.20 mm prior to heat treatment. Tensile
bars with a gage length of 5 cm and reduced section width of 1.875 cm
were machined from these strips. They were polished with emery papers
and one side (anodic) was coated with a thin layer of palladium by
electroless deposition from "Pallamerse" solution. The specimen-cell
assembly was positioned between the crossheads of an Instron machine.
The anodic side was potentiostated at a potential above the reversible
potential for the hydrogen electrode. Cathodic charging was performed
galvanostatically. Specimens were deformed at a constant extension
rate. An immediate increase in the background anodic current is observed at the onset of macroscopic plastic deformation, perhaps due to
accelerated anodic dissolution in the presence of plastic deformation or
to disruption of surface films. After some time, the current reaches a
steady state value and at this point hydrogen charging is begun. Permeation currents and stress-strain curves were recorded. All experiments
were run at room temperature.

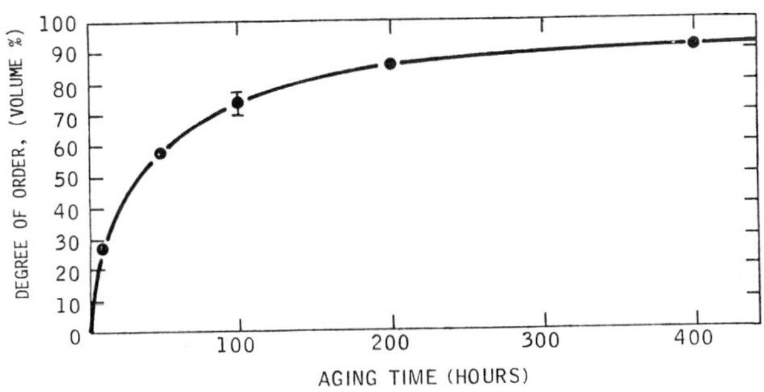

FIGURE 6

The Relationship Between the Degree of Long Range
Order and Time of Isothermal Aging at 500°C for Ni₂Cr.(7)

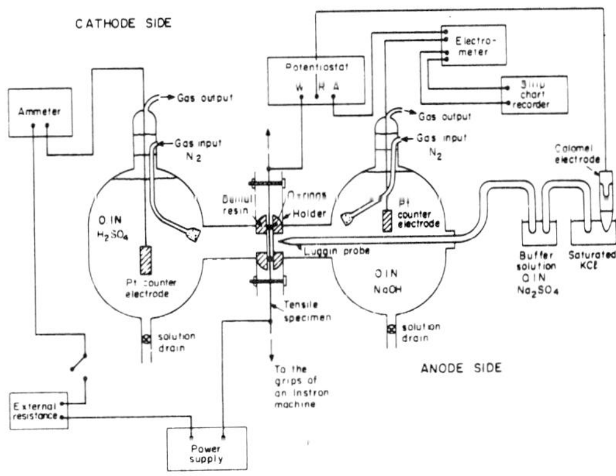

FIGURE 7

Hydrogen Permeation Cell Modified to Allow
Plastic Deformation of the Specimen.(6)

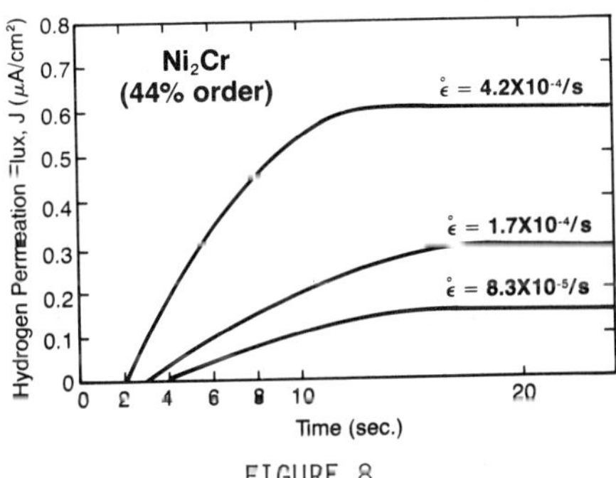

FIGURE 8

Hydrogen Permeation Transients for Various Strain Rates

Results and Discussion

The results of the hydrogen permeation experimentation are tabulated
for various degrees of order and strain rates in Table I.   There are a
number of trends which will be shown individually in the following
figures.   Figure 8 shows typical hydrogen permeation transients as a
function of strain rate for the samples aged for 25 hrs. (44% order).
From this set of transients, as an example, one may calculate effective
hydrogen diffusion coefficients using the half rise time approximation.(8)
These are shown in Table I.   Note that both the steady state permeation
flux, J_∞, and effective hydrogen diffusion coefficients, D_{eff} increase with
strain rate.   In fact, J_∞, is linearly related to the strain rate, as
shown in Figure 9.   This data suggests that dislocation transport of
hydrogen occurs during plastic deformation.   There is no other explanation
for this behavior.

Figure 10 shows that there is a maximum in J_∞ as a function of the
degree of order.   This is shown at a strain rate of 8.3 x 10^{-5}/s but is
true for all strain rates.   The most striking observation is that the
maximum in permeability shown in Figure 10 (44% order), corresponds to
the minimum in hydrogen embrittlement susceptibility shown in Figure 1.
That is, in Ni_2Cr, when the material is most permeable to hydrogen, during
dynamic straining, it is most resistant to embrittlement.   It should be
noted that this observation is contrary to those made in the case, for
example, of high strength steels(9), and high strength aluminum alloys(10)
in which maximum susceptibility corresponds to maximum permeability.   The
latter emphasizes the point that high hydrogen permeability does not
necessarily lead to enhanced embrittlement.   Indeed, what is required
is some degree of permeability in a susceptible microstructure.

This behavior suggests a hypothesis for a possible mechanism for the
embrittlement of Ni_2Cr as a function of ordering.   One might expect that
the distribution of slip (spacing of active slip planes) should be affected
by ordering.   In the disordered material, these are likely to be many,
closely spaced slip planes.   Although, the ordered alloy is somewhat
harder(4), fine slip is again expected.   On the other hand, in the case
where an intermediate volume fraction of the alloy is ordered, plastic
deformation is likely to be localized to the softer disordered regions,
resulting effectively in coarse slip.   In effect, in order to accommodate
plastic strain, relatively few dislocations move on many closely spaced
slip planes in the disordered and highly ordered alloys, while relatively
many dislocations move on few widely spaced slip planes in the partially
ordered alloy.   Given the evidence that dislocation transport of hydrogen
occurs (Figures 8 and 9), it is not unlikely that more hydrogen will
permeate at a given strain when the deformation is localized.   In the
disordered and highly ordered alloys, the dislocations will distribute
the hydrogen homogeneously over a wide area.   As the dislocations reach
the grain boundaries, the boundaries will become saturated and will
fracture uniformly under the applied loads resulting in intergranular
failure.   Conversely, if hydrogen is transported along narrow channels
as a result of coarse slip, only a small segment of the grain boundary
will be occupied by hydrogen and most of the hydrogen will be transported
from one grain to the next, resulting in principally ductile rupture.

While we believe this is a possible mechanism for this behavior,
studies are in progress to examine distribution of slip and dislocations as
a function of ordering.   This will be reported in a subsequent publication.

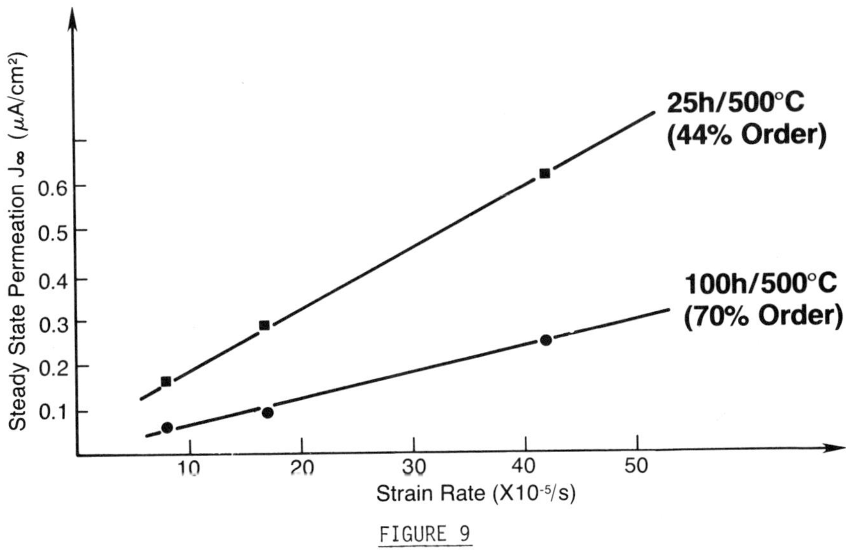

<u>FIGURE 9</u>

Effect of Strain Rate on Steady State Hydrogen Permeation

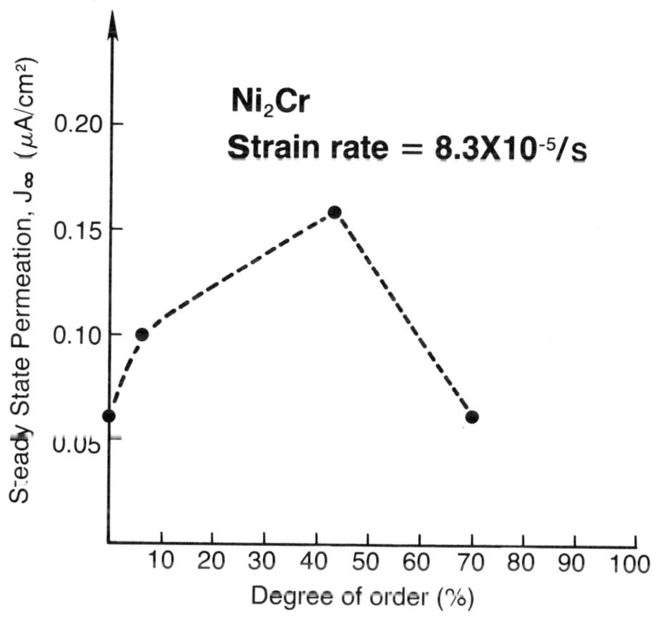

<u>FIGURE 10</u>

Effect of Order on the Steady State Permeation

Acknowledgements

It is a pleasure to acknowledge the support and encouragement provided by the Office of Naval Research and the Center for Materials Science and Engineering at M.I.T. Likewise, the M.I.T. authors wish to acknowledge the partial support provided by an Exxon Education Foundation grant.

References

1.　R. D. Kane, M. Watkins, D. F. Jacobs and G. L. Hancock, Corrosion, 1977, Vol. 30, No. 9, pp. 309-.

2.　B. J. Berkowitz and R. D. Kane, Corrosion, 1980, Vol. 36, No. 1, pp. 24-29.

3.　A. I. Asphahani, Hydrogen in Metals, Proceedings of the 2nd Int. Congress, Vol. 4, sect. 3C, paper #2, pp. 1-10, Pergamon Press, Paris, June 1977.

4.　B. J. Berkowitz and C. Miller, Met. Trans., in press.

5.　J. K. Tien, A. W. Thompson, I. M. Bernstein and R. J. Richards, Met. Trans. A, 1976, Vol. 7A, pp. 821-.

6.　M. Kurkela and R. M. Latanision, Scripta Met., 1979, Vol. 13, pp. 927-932.

7.　G. I. Nosova and N. A. Polyakova, Fiz. Metal. Metalloved., 1971, Vol. 32, pp. 825-830.

8.　M. A. Devanathan and Z. Stachurski, Proc. Roy. Soc., 1962, Vol. A270, pp. 90-.

9.　C. F. Barth, E. A. Steigerwald and A. R. Troriano, Corrosion, 1969, Vol. 25, pp. 353-.

10.　M. O. Speidel, in Hydrogen in Metals, p. 249, ASM, Metals Park, 1974.

11.　G. I. Nosova, N. A. Polyakova and E. A. Medvedev, Fiz. Metal. Metalloved., 1974, Vol. 38, pp. 176-182.

HYDROGEN EMBRITTLEMENT OF Ni-Al-Mo EUTECTIC COMPOSITES*

L. J. Klein and N. S. Stoloff
Rensselaer Polytechnic Institute
Troy, New York   12181

Room temperature tensile and delayed failure tests to detect hydrogen embrittlement were conducted on two directionally solidified $\gamma/\gamma'-\alpha$ (Ni-Al-Mo) eutectic alloys. Delayed failure experiments conducted by charging while testing notched specimens revealed much more severe embrittlement than that resulting from precharging followed by tensile testing.

Transverse secondary cracking was extensive in both tensile and delayed failure samples; however, no recurring cracking patterns were found. A flat, featureless region of the fracture surface resulted from hydrogen charging the delayed failure specimens. Fracture of the latter appears to be a result of embrittlement and cracking of the Mo fibers, followed by crack propagation through the embrittled nickel-rich matrix.

* This work was supported by the Office of Naval Research, under Contract No. N00014-79-0583.

Introduction

Eutectic composites, with their aligned, semi-coherent interfaces, offer convenient microstructures in which to establish the importance of interfaces as sites for hydrogen trapping or hydrogen assisted cracking. Previous work (1) has shown that lamellar Ni-Nb-Al (γ/γ'-δ) eutectic alloys are susceptible to reversible embrittlement by precharging with hydrogen. Hydrogen appeared to enhance cracking along the interface between the reinforcing γ phase and the γ/γ' matrix, but this was believed to be of only secondary importance in embrittlement, since most cracks were associated with slip bands. Garmong (2) has reported severe embrittlement in three nickel-base aligned eutectics in hydrogen gas which was attributed to facilitated cracking along slip planes as well as weakened fiber-matrix interfaces.

The present investigation had as its principal aim clarification of the relative importance of slip bands and other microstructural features as sites for hydrogen embrittlement. The alloy system chosen for study was fibrous Ni-Al-Mo which consists of a γ/γ' (Ni/Ni$_3$Al) matrix, reinforced with ductile α-Mo fibers. Two alloys were chosen to study the effect of microstructural variation (different volume fractions of the three phases) on hydrogen susceptibility.

Experimental Procedure

Materials and Directional Solidification

Master alloy ingots of two Ni-Al-Mo alloys, AG-15 and AG-34, approximately 1.8 cm in diameter and 16 cm long, were purchased from the General Electric Company. The composition of AG-15 is 65.5w%Ni, 8.1%Al, and 26.4%Mo and that of AG-34 is 62.5w%Ni, 6.3%Al and 31.2%Mo. The alloys were induction melted under argon in alumina crucibles and directionally solidified in a water-cooled Bridgeman apparatus. AG-15 was solidified at a rate of 1.9 cm/hr, while AG-34 was grown at 0.76 cm/hr; similar interfiber spacings resulted. The principal differences in the two alloys are a higher volume fraction of Mo fibers and a higher residual γ content in AG-34.

Sample Preparation

Cylindrical tensile bars were electro-discharge machined from the directionally solidified ingots; the final specimen gage section was $\sim$ 0.30 cm dia by 1.7 cm long. The gage section was mechanically polished and then electro-polished prior to testing. Flat samples for delayed failure experiments also were machined from directionally solidified ingots. A gage section 1.7 cm long and 0.46 cm wide with a 0.25 cm deep edge notch was introduced into 0.27 cm thick slabs by grinding. Samples were then mechanically polished.

Hydrogen Charging

Either precharging or simultaneous charging and testing was conducted in 1N H$_2$SO$_4$. In early experiments 0.004 gm As$_2$O$_3$ per liter in solution was added as a cathodic recombination poison. In later experiments, 100 ppm of a more soluble poison, NaAsO$_2$, was used in place of the As$_2$O$_3$. All charging was done at room temperature at a fixed current density of 50 mA/cm^2

Testing

Tensile tests were conducted at room temperature in air at a strain rate

of 6.67 x 10^{-4} sec^{-1}. Delayed failure tests were carried out with a plexiglass chamber surrounding the specimen containing the charging solution and
the platinum screen cathode. A wire was soldered directly to the specimen
to achieve electrical contact. A microswitch and digital timer were employed
to record time to failure. Stresses were applied parallel to the fiber axis.

<u>Experimental Results</u>

<u>Precharging and Tensile Testing</u>

Preliminary testing to establish a hydrogen embrittling effect in AG-15
and AG-34 was conducted by cathodically precharging and then tensile testing
cylindrical samples. Table I shows the results of these tests. (These samples actually had a cellular microstructure, but this did not affect the
validity of the tests in establishing a hydrogen effect.) Precharging at 25°C
lowered the reduction in area of AG-15 from 19.5% in air to 15.0% in hydrogen
and from 21.1% in air to 13.6% in hydrogen for AG-34. The ultimate tensile
strength of AG-15 was reduced by 8% and that of AG-34 by 24%. The yield
strengths were not changed by hydrogen.

The charging conditions for one sample of AG-15 were altered to 75°C,
100 mA/cm^2 in an effort to increase the hydrogen input to the specimen by
increasing the charging temperature and current. Table I shows no effect upon
the tensile characteristics compared to room temperature precharging at 50
mA/cm^2.

Baking a sample of AG-15 at 500°C in air for 3 days following an 8 day
room temperature precharge restored ductility and UTS to the levels observed
in uncharged samples.

Hydrogen caused transverse secondary cracking to occur during tensile
testing in both AG-15 and AG-34; cracks were more frequent in AG-34. No
cracks were observed in uncharged samples. Fig. 1 (a) shows the general

Table I. Tensile Data

Condition	Y.S. (MPa)	UTS (MPa)	%RA
AG-15			
uncharged	619	1517	19.5
charged 8 days, 25°C, 50 mA/cm^2	---	1400	15.0
charged 8 days, 75°C, 100 mA/cm^2	---	1537	16.1
charged 8 days, 25°C, 50 mA/cm^2, baked 3 days @ 500°C	---	1614	20.3
AG-34			
uncharged	697	1927	21.1
charged 10 days, 25°C, 50 mA/cm^2	---	1741	13.6

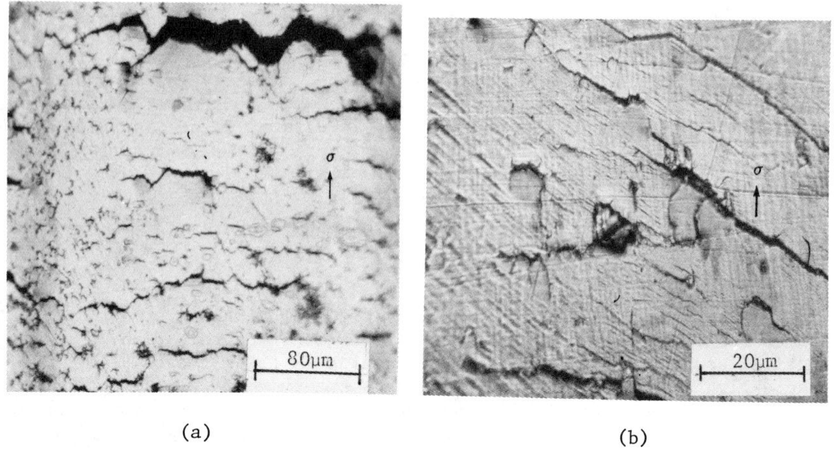

(a)   (b)

Fig. 1.   Transverse secondary cracks on surface of polished tensile specimens.
(a) AG-15, precharged 8 days
(b) AG-34, precharged 10 days.

nature of secondary cracking on the surface of the polished gage sections of
AG-15 after testing. The sample of AG-15 that was baked at 500°C in air for
3 days exhibited a small amount of secondary cracking; however, the crack
density was much less than in the sample which was precharged but not baked.
This indicated that most, but not all, of the hydrogen was removed by baking.
Although much of the secondary cracking found in both alloys did not follow
any specific microstructural paths, some evidence of crystallographic cracks
were found in an AG-34 sample. Fig. 1 (b) shows an example of cracks running
parallel to slip bands, evidence that the cracks are crystallographic; no
evidence of interphase cracking was found in this sample.

Fig. 2 shows secondary cracks in a precharged AG-15 tensile sample.
Cracks initiate at the surface and have a tendency to be diverted longitudinally into the fiber growth direction.

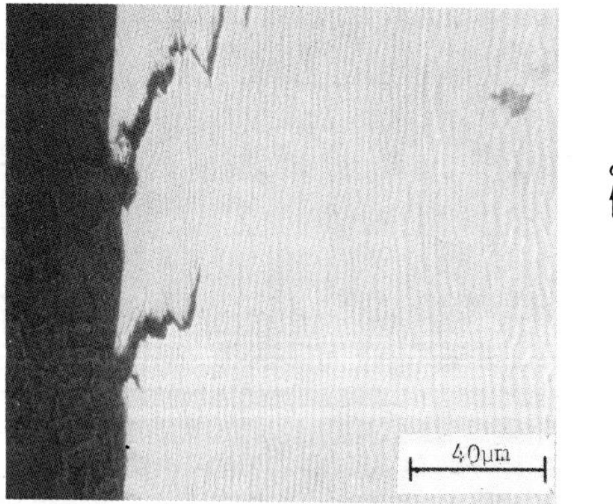

Fig. 2.   Secondary cracks in precharged AG-15 tensile sample.

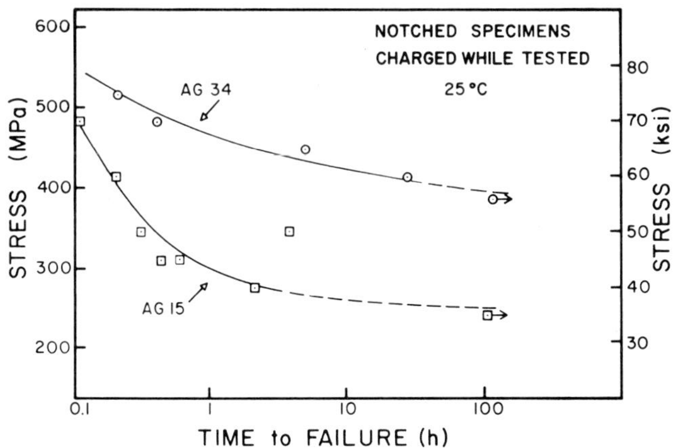

Fig. 3.   Delayed failure of γ/γ'-Mo.

Delayed Failure Testing

Fig. 3 shows the results of delayed failure experiments conducted while cathodically charging notched samples of AG-15 and AG-34. Both curves fall entirely below the unnotched 0.2% offset yield strengths of 619 MPa and 697 MPa for AG-15 and AG-34, respectively. The notches magnify the stress at the notch root by a factor of about three, thus causing the material at the tip of the notch to undergo plastic deformation upon loading.

The reason for the superior properties of AG-34 is unknown. AG-34 has an ultimate tensile strength about 8% higher than AG-15, not enough to account for the difference. A possible explanation is that whereas the fracture surfaces of samples of AG-15 were generally flat and perpendicular to the stress axis, those of AG-34 tended to be more jagged, with the final fracture occasionally occurring away from one of the notches.

The jagged fracture mode in AG-34 is a result of the crack diverting longitudinally into the fiber growth direction, see Fig. 4. This, in effect, blunts the crack tip and delays the eventual failure of the sample. To a lesser degree longitudinal cracks occur also in AG-15. For that matter, some longitudinal cracks were noted even in uncharged Ni-Al-Mo alloys when tested in tension. The reason why more longitudinal cracking and therefore more blunting occurs in AG-34 than in AG-15 is not understood.

Fig. 5 shows an SEM fractograph of the flat fracture region (near the notch) of an AG-34 sample. As with AG-15, the only features visible in the flat region at the bottom of the figure are river patterns and some longitudinal cracks which intersect the fracture surface. The fracture surface becomes more irregular as the overload region is approached, as seen in the top of the figure.

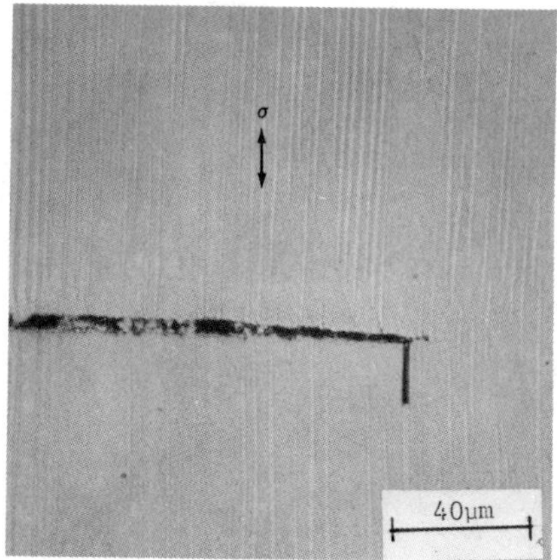

Fig. 4.   Crack blunting along fibers,   Fig. 5.   SEM fractograph of AG-34
          delayed failure in AG-34.   delayed failure; flat region
                                                      below notch.

Fig. 6 shows the hydrogen-induced fracture regions of an AG-34 specimen. The fracture surface in Fig. 6 is much smoother than in the overload zone. The broken fibers in this figure appear to have pulled away from the matrix, but since this fiber-matrix separation also occurs in the overload zone, it is not believed to be a direct result of the presence of hydrogen. Some of the fibers in Fig. 6, e.g. at position A, appear to have fractured in a chisel-type, ductile manner, while others, e.g. at B, appear to have a flat, brittle-cleavage appearance. This dual fiber fracture morphology was confirmed by replicating the fracture surface; often noted, also, were fibers that appeared to have split longitudinally. No such split fibers have been seen in uncharged γ/γ'-α alloys.

Transverse secondary cracking was prevalent in the simultaneously charged and tested delayed failure samples as it was in the precharged tensile tests; the secondary cracks generally occurred in the gage section beneath the tensile

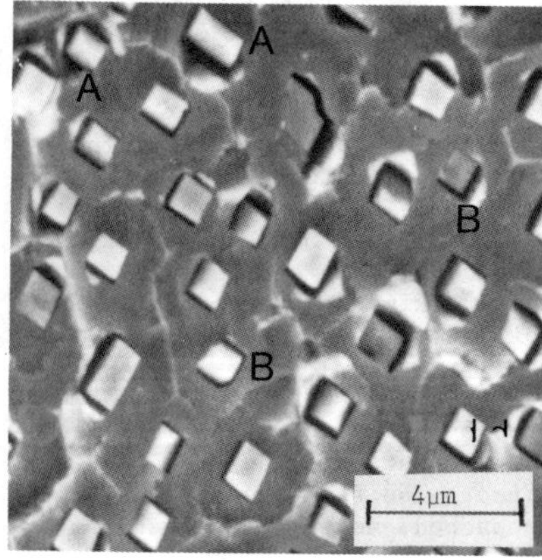

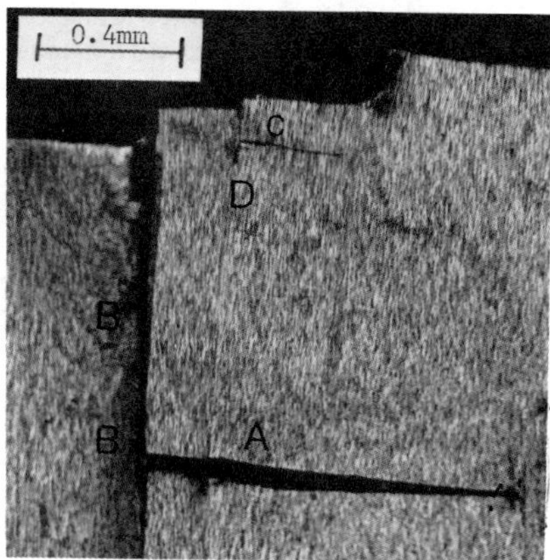

Fig. 6.   SEM fractograph of AG-34   Fig. 7.   Crack paths in AG-15.
          delayed failure.

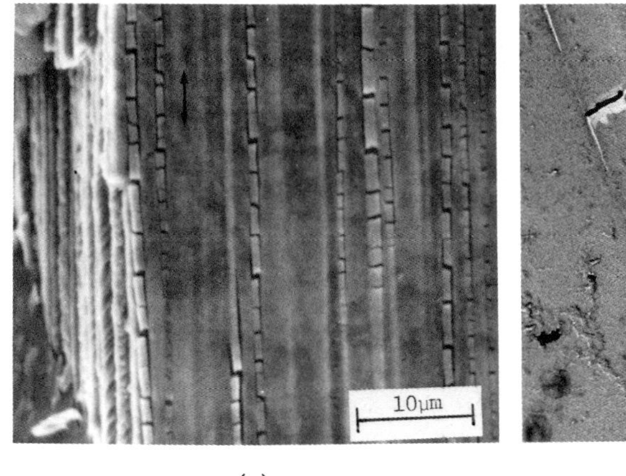

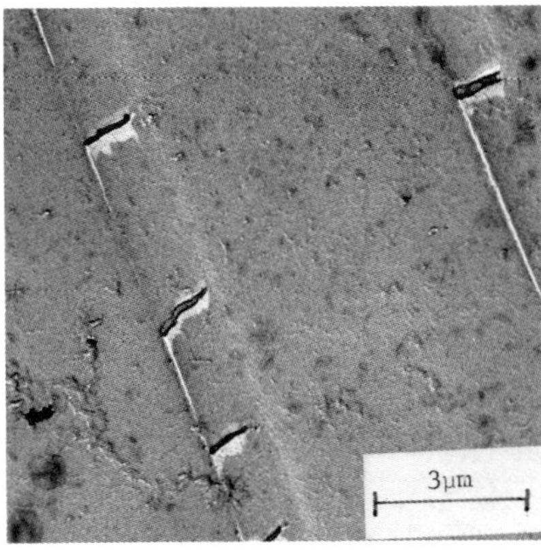

(a)   (b)

Fig. 8.   Cracked Mo fibers in AG-15.
          (a) SEM, longitudinal surface
          (b) TEM of replica showing partially cracked fibers.

overload region of the delayed failure specimen rather than below the hydrogen
cracked region.

Transverse and longitudinal cracks were examined in an effort to find
some relationship between either grain boundaries, interphase boundaries, or
deformation bands and the hydrogen-induced failure. Secondary cracks in AG-34
often blunted at grain boundaries. In Fig. 7, a transverse crack in AG-15,
labeled A, originating at a longitudinal crack, labeled B, has propagated
practically unimpeded through a grain boundary and into the next grain. At
the top of Fig. 7 is another secondary crack, labeled C, which has initiated
at a grain boundary and propagated into the adjacent grain. Note, also, that
in this figure the longitudinal crack at the top of the figure, labeled D,
followed a grain boundary. In summary, no specific relationships between
cracking and grain or interphase boundaries were found in either alloy, nor
were cracks associated with slip bands.

Cracked fibers were detected in the hydrogenated delayed failure samples
of both alloys, e.g. Fig. 8 (a). The normally ductile molybdenum fibers have
cracked in a brittle manner. This type of fiber fracture has never been
observed in uncharged specimens. Cracking of fibers was verified in another
AG-15 sample, as shown in Fig. 8 (b). Note that some of the fibers are not
completely cracked, indicating that fiber cracking is an initiation and growth
process rather than simply an elastic breakdown into critical lengths on loading. Hydrogen charging itself did not cause the fibers to crack, but the combination of applied stress (or strain) and charging was necessary.

Discussion

Previous investigations (1,2) of several aligned eutectics tested either
after precharging or in hydrogen gas had demonstrated their susceptibility to
hydrogen embrittlement under tensile loading. The present work points up the
critical importance of selecting proper test conditions to achieve embrittlement. When tested in tension, precharged γ/γ'-α alloys reveal modest losses

in ductility and tensile strength. Both properties can be restored easily by baking to remove hydrogen. Much more dramatic embrittlement is achieved by simultaneous charging and static loading on notched samples. The critical feature in the reproductibility and magnitude of the effect of the latter experiments seems to be the presence of a notch. Without the notch, excessive scatter in results was observed, although some embrittlement was noted.

The loss in ductility for Ni-Al-Mo alloys tested in tension was less severe than had been found previously for $\gamma/\gamma'-\delta$, for which an 8 day precharge had reduced the %RA from 15% to zero (1). Garmong (2) also reported nearly zero %RA for $\gamma/\gamma'-\delta$ tested in hydrogen gas, while a Ni-45.5%W eutectic exhibited a drop in %RA from 73.3% to 30.7%. The varying susceptibilities of the different alloys to hydrogen undoubtedly depends on several factors: ductility of the alloy in air, method of hydrogen ingress, strain rate and perhaps hydrogen diffusivity in the lattice and along interfaces.

The lack of more than a few well-defined cracks along microstructural features such as slip bands in $\gamma/\gamma'-\alpha$ [Fig. 1 (b)] contrasts with the previous results on $\gamma/\gamma'-\delta$, for which slip band cracking in the matrix seemed to predominate. However, when the latter was tested in hydrogen gas, severe degradation of reinforcement-matrix interfaces was noted (2); thus it may be concluded that the path as well as the severity of cracking depends upon the means by which hydrogen is introduced into a sample.

It is clear that hydrogen must be present simultaneously with plastic deformation in order for embrittlement to occur. The more severe embrittlement noted in our delayed failure tests, in which charging was carried out concurrently with load application is at least partly due to the interaction between hydrogen diffusion and plastic deformation. The presence of a notch increased the localized plastic deformation and further added to embrittlement.

It appears that hydrogen embrittlement in $\gamma/\gamma'-\alpha$ occurs by crack initiation in the fibers with subsequent propagation through the embrittled Ni-rich matrix, at least during simultaneous charging and delayed failure testing. Birnbaum and Wadley (3) have induced embrittlement in bulk molybdenum by the same technique; under ordinary conditions of precharging followed by tensile testing molybdenum is not embrittled. Hydrogen also seems to have an effect on the strength of grain and interphase boundaries in $\gamma/\gamma'-\alpha$ alloys, but this is only of secondary importance to the ultimate failure of the material. The alignment of the grains and phases, nearly parallel to the applied stress, does not permit failure to occur solely by embrittlement of such boundaries; rather cracks are deflected down these boundaries, thereby increasing the time to failure.

Conclusions

Hydrogen embrittlement of two directionally solidified $\gamma/\gamma'-\alpha$ alloys has been demonstrated. Embrittlement is most severe when notched samples are tested under static loading as hydrogen charging occurs. Embrittlement in delayed failure tests is initiated by the hydrogen-induced brittle fracture of molybdenum fibers.

References

1. N. S. Stoloff, L. Klein, H. Marcus, and J. E. Grossman, *Scripta Met.*, V. 10, 1976, p. 889.

2. G. Garmong, *Met. Trans.*, V. 8A, 1977, p. 535.

3. H. K. Birnbaum and H. Wadley, *Scripta Met.*, V. 7, 1973, p. 409.

THE INFLUENCE OF MICROSTRUCTURE ON THE HYDROGEN

EMBRITTLEMENT OF PURE AND COMMERCIALLY-PURE TITANIUM

K.J. Puttlitz
IBM Data Systems Division
Hopewell Junction, New York 12533

A.J. Smith
Professor Emeritus of Metallurgy
Michigan State University
East Lansing, Michigan

The hydrogen embrittlement of pure and commercially-pure (CP) titanium under dynamic load conditions is well documented in the literature.

By cross-correlating microstructural features, fracture morphology, and fracture toughness it was possible to determine the role of micro-structure during the ductile-to-brittle transition (DTBT). The accurate fracture energy measurement system utilized in this study was capable of discriminating small changes in fracture toughness; thus the effects of microstructure beyond the transition were determined as well.

Hydrogen dissolved within the α phase of the pure metal had no noticeable effect on absorbed energy during impact. However, there was a marked reduction in absorbed energy corresponding to the initial precipitation of hydrides within the α matrix. The progress of pure titanium's transition (degree of brittleness) was determined to be related to the hydride's growth and development.

A hydrogen concentration of about 50 ppm, this study's maximum limit, was insufficient to cause embrittlement of CP titanium if the retained β phase was disbursed within the matrix. There was a progressive embrittlement as the β phase agglomerated within the α grain boundaries forming a continuous network.

Lamellar α-β structures failed in a manner similar to that noted for the pure metal's alpha matrix containing precipitated hydride platelets. However, in this case the crack followed retained beta platelets across the grain.

It will be shown that further reductions in energy beyond the transition for both pure and CP titanium were related to platelet development.

Introduction

The manner in which hydrogen embrittlement exhibits itself in a metal depends on whether it is either an endothermic or exothermic type occluder. The former, which forms true solid solutions which hydrogen, embrittles under static load conditions. This sensitivity decreases, however, with a quickened loading pace (1).

Titanium, like all group IV-A transition metals, forms nonstoichiometric hydrides with a maximum hydrogen-to-metal atomic ratio of 2. Based on the fluorite structure, exothermic occluders dissolve initial additions of hydrogen as a solid solution (2,3). More important is the formation of the hydrides as the hydrogen content is increased beyond a solid solubility limit, as in the case of alpha titanium (4,5). The hydrogen solubility between 44 and 300°C has been determined to be c(ppm) = 8.60 x 10^4 exp (-2.52.X 10^3/T) for high purity α-Ti (6).

That pure α-titanium's tensile properties are relatively unaffected by modest hydrogen additions {up to 1.0 a % (200 ppm)} was demonstrated by Jaffee and Campbell (7).

However, Craighead, et al (8), and the early work of Pitler, cited by Jaffee and Campbell (7), demonstrated that a 1.0 a % H addition to α-titanium reduced the impact strength appreciably. Slowly cooled α-titanium containing 2.0 a % H is completely embrittled, whereas a quenched 1.0 a % alloy initially possesses an impact fracture toughness similar to hydrogen-free pure titanium. The enhanced toughness is lost upon aging at room temperature (9). This temporary behavior is due to the fine dispersion of the hydride during quenching which then agglomerates upon aging (10).

Purity of the α-titanium has also been found to be important for impact toughness. Commercially-pure α-titanium requires only about one half as much hydrogen (1.0 a %) as does the pure metal to undergo the ductile-to-brittle transition (9). Rylski (11) noted that titanium's embrittlement could occur in the absence of a hydride if the structure contained retained beta phase.

The present study was undertaken to determine the microstructure's role of pure and CP titanium in bringing about DTBT. Its contribution to embrittlement beyond the transition was of interest as well.

Materials

High Purity Titanium - Electrorefined (HP-BM)

An ingot made by remelting electrorefined cyrstals, supplied by the Bureau of Mines (U.S. Dept. of the Interior) at Boulder City, Nevada, was forged at and rolled by Reactive Metals, Inc. into a 13.8 mm thick plate designated as HP-BM.

Commercially-Pure Titanium (CP), ASTM Grade II

A plate (25 mm thick) of ASTM Grade II (designation B265) commercially-pure titanium was procured from Titanium Product Corp., Detroit. The chemistry of the alloy is given below.

Chemistry - percent by weight of commercially-pure titanium, ASTM Grade II.

Fe	C	N	O	H
0.11	0.026	0.012	0.10	0.004

Hydrogen Charging

Apparatus

The hydrogen treatments were carried out in a glow-bar tube furnace system similar to those commonly described (12, 13) for gas treatment where achieving and maintaining a high degree of purity is important. Care was taken to exclude oxygen and nitrogen. They interstitially solution harden titanium, reducing its static ductility properties (14) and resistance to impact failure (15). Nitrogen's influence on inducing brittleness is far greater than oxygen's (16).

Treatment Procedure and Schedules

Various (HP-BM) specimens were hydrogenated at 825, 890, and 945°C for two-hour treatment periods. Some specimens were subjected to a 975°C treatment for two, four, and six hours.

The ASTM Grade II alloy test bars were treated for two-hour periods at: 710, 770, 825, 895, 955, 975, and 1010°C within the hydrogen-treatment furnace.

All test bars were furnace cooled.

Testing

Instrumented Impact

The standard Charpy test was instrumented as described in the literature (17). The fracture energy (initiation plus propagation energy) was computer calculated from an analog-to-digital conversion of the data stored on high speed FM tapes. This technique allowed small changes in the fracture energy to be correlated to microfeatures of the fracture morphology. The Charpy bar lateral expansion, determined according to ASTM Standard A370, served as a much coarser measurement of impact ductility.

Results and Discussion

The Transition of Hydrogenated Pure Titanium (HP-BM)

The dynamic ductility of the HP-BM titanium was noted to markedly decrease between the 945 and 975°C, two-hour treatments. The lateral expansion was decreased from 1.73 to 1.14 mm with an attendant 40-percent reduction in the energy absorbed during failure (Table I). These changes indicated the initial stage of the transition. The four-hour 975°C treatment sample (48 ppm hydrogen) exhibited a complete ductile-to-brittle transition (DTBT) as evidenced by the drastic reduction in absorbed energy. The transition is easily identified from a sequence of instrumented impact load vs. time oscilloscope traces (Fig. 1). The beginning is marked by the shortened plastic segment of the 975°C two-hour treatment trace, whereas the four-hour treatment's decreased load maximum (pulse height) and greatly abbreviated fracture time (pulse length) is characteristic of a brittle material.

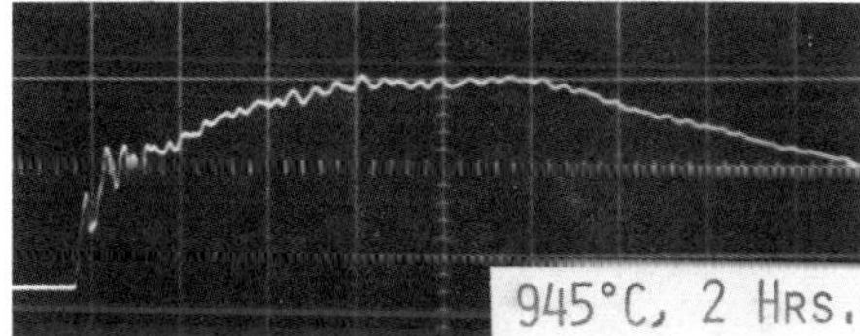

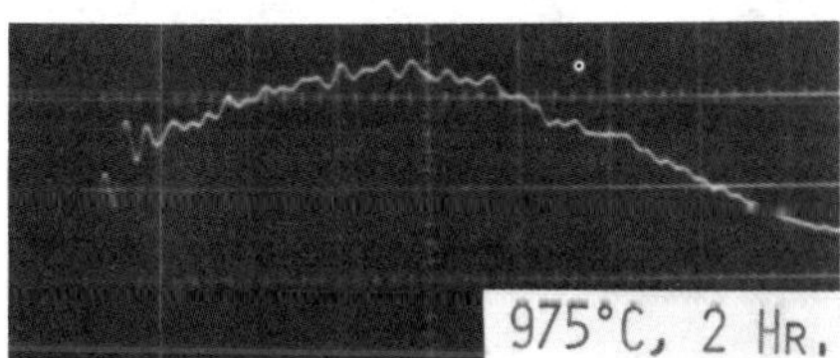

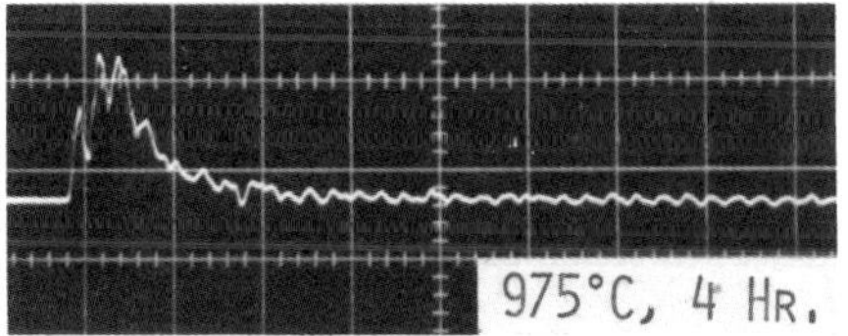

Fig. 1   Instrumented Charpy Impact Load vs. Time Oscilloscope Traces of Hydrogenated High Purity α-Titanium (HP-BM).

Horizontal = 200 μ-Sec/Major division.
Vertical   = 465 Kg (f).

TABLE I: Pure Titanium (HP-BM) Fracture Energy Summary

Treatment* Temp, °C	Hydrogen Content, ppm	Lateral Expansion, mm	Propagation Energy, Joules	Total Fracture Energy, Joules
825	39	1.65	26.44	93.01
890	36	1.65	32.94	86.00
945	26	1.73	41.38	98.16
975	38	1.14	18.82	56.18
975-4 hrs.	48	0.20	3.29	6.14
975-6 hrs.	46	0.20	1.90	4.62

(*All treatments for 2 hr. period unless otherwise indicated.)

The conditions which gave rise to the transition were determined by correlating the phase and fracture morphology features with absorbed fracture energy. The test bars through the 945°C treatment failed in a ductile manner by massive twinning. Their fractures were fibrous in appearance consisting of a fine deformation structure. The first signs of embrittlement coincided with the initial precipitation of hydrides (975°C two-hour treatment). Also, there was no evidence of twinning. Although still ductile and fibrous in appearance (Fig. 2a), the deformation structure was coarser than noted for the test bars treated at lower temperatures.

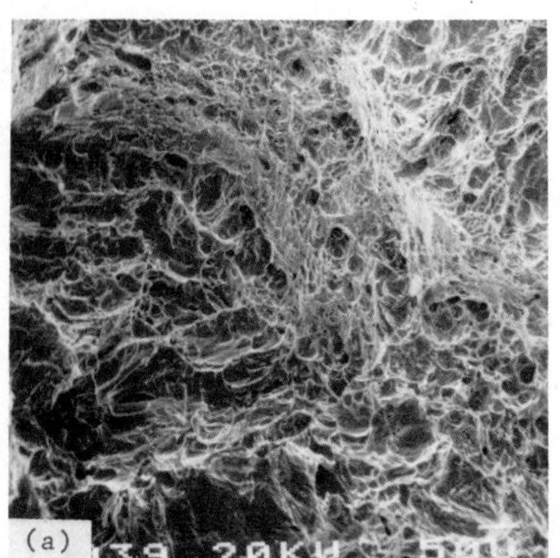
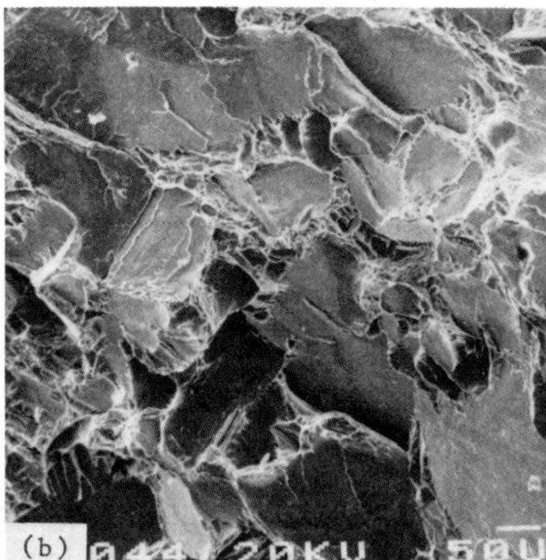

Fig. 2   SEM Photographs of the Fracture Morphology of α-Titanium (HP-BM). Test Bars Hydrogenated at 975°C for (a) 2 hrs., and (b) 4 hrs.

A microstructure favorable to embrittlement was formed by increasing the treatment time to four hours (975°C). Grain growth increased as did the hydride platelets which extended across most grains. Most of the fracture surface consisted of low energy cleavage areas formed by separations along the intragranular hydrides (Fig. 2b). Beyond the DTBT, further decreases in fracture energy were achieved by increasing the hydride size. Thus, the cleavage area increased at the expense of the ductile failure component located between the cleavage areas.

A minimum hydrogen concentration between 38 and 46 ppm was found necessary to undergo the DTBT. However, increasing the hydride size results in further embrittlement with no additional hydrogen required.

The Failure of Hydrogenated Commercially-pure Titanium (ASTM Grade II)

Commercially-pure titanium is the designation for a series of very dilute interstitially strengthened alloys available on the market. These alloys often contain a high temperature iron-stabilized titanium polymorph, beta phase. The interstitials, oxygen, carbon, and nitrogen, stabilize the alpha phase. Due to the small iron content and the fact that iron is a sluggish eutectoid former, the ASTM Grade II alloy does not contain intermetallics. Retained beta phase is present in the microstructure as a second phase.

The manner in which the CP alloy embrittles and the resulting fracture morphology depends upon microstructure and thus upon the phase field to which the test bar was heated during hydrogen charging.

Alpha-Field Soak

The microstructure remained fine grained for all treatments within the alpha field. Retained beta phase was distributed throughout the alpha grains with some agglomeration in the grain boundaries noted after the 770°C treatment (Fig. 3a), which also coincided with the initial embrittlement of the alloy (Table II). The transition was complete after the 825°C treatment because the beta phase within the grain boundary network was continuous (Fig. 3b).

TABLE II: Commercially-pure Titanium (ASTM, Grade II) Fracture Energy Summary

Treatment Field	Treatment* Temp., °C	Hydrogen Content, ppm	Fracture Toughness, mJ/mm^2
	710	32	131.7
ALPHA	770	42	103.5
	825	48	20.8
ALPHA + BETA	895	40	19.9
	955	41	28.5
BETA	975	39	23.4
	1010	39	17.0

(*All treatments for 2 hr. period.)

Unlike pure titanium, the commercially-pure alloy's transition was not hydride related. The microstructure was not found to contain hydrides. Rylski (11) postulated that such a transition is due to the embrittlement of the beta phase since its hydrogen solubility is greater than alpha on a comparative basis. Craighead et al suggested that the embrittlement of α-β alloys is promoted by the presence of an interface. They observed that all beta alloys, some containing up to 2000 ppm H, were not markedly embrittled compared to α-β alloys, with no hydrides present in either (18). Concluding a strain aging effect, they proposed that hydrogen diffused, probably

as protons, to the α-β interface, segregated, and ultimately precipitated
(19). Although the fracture in the present study has been observed to occur
at the interface between the two phases, no hydride precipitates were found
at these locations. Also, in this regard, the failure of α-β alloys is complicated by the fact that beta's own hydrogen tolerance depends upon the
β-stabilizing addition and even the presence of a third alloying element
(18). The tolerance of the iron-stabilized beta in commercially-pure α-β
alloys is probably significantly less than the all-β and α-β alloys investigated by Craighead, et al.

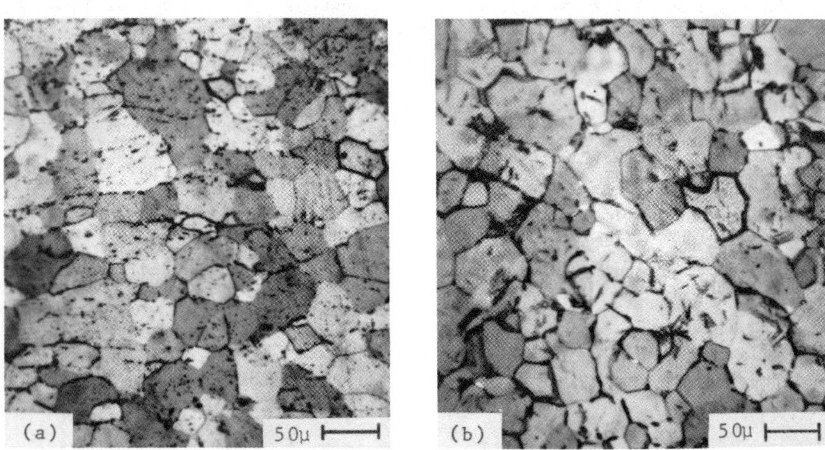

Fig. 3   Darkfield Photomicrographs of Hydrogenated CP Titanium
(ASTM Grade II) Treated Within the Alpha Field.
(a)   770°C Treatment, Beta Phase Collecting Within the GB Marks the
Start of DTBT.
(b)   825°C Treatment, Beta Phase Mostly in GB, Transition Complete.

The fractures of test bars soaked within the alpha field increased in
granularity upon progressing through the transition, since the fracture
followed the beta within the grain boundaries.

(α + β) Field Soak

The fracture appearances of embrittled test bars treated within the
α and (α + β) regions were similar as were their fracture energies. The
increased grain size due to the higher temperature (890°C) soak caused the
fracture's granular appearance to be coarsened. Also, a few beta platelets
formed within and stretched across the alpha grains. The introduction of
flat cleavage areas on the fracture surface (Fig. 4b) stems from the fracture locally following the intragranular beta platelets.

These platelets (Fig. 4a) are very similar in appearance to the well-
formed hydrides which caused pure titanium to embrittle.

β-Field Soak

Test bars treated within the beta field (945°C and higher in this
study) exhibited a distinctly different fracture morphology, having an

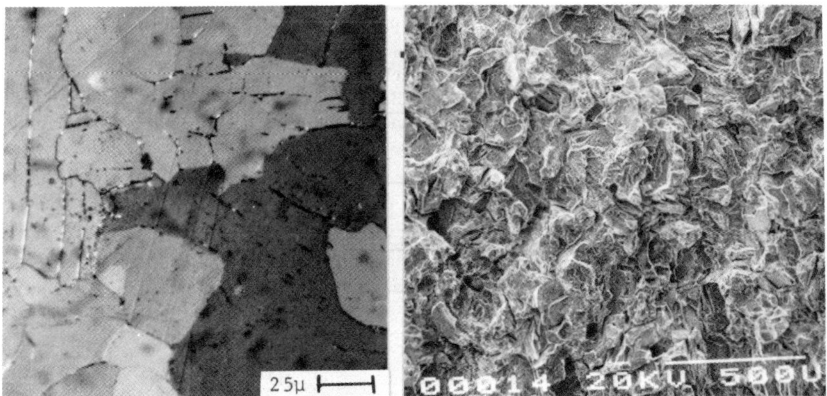

Fig. 4   Dark Field Photomicrograph and SEM Fractograph of CP Titanium (ASTM Grade II) Treated within the α & β Field (890°C, 2 hrs.).

appearance likened to broken shale. The fracture consists mostly of large flat areas with some evidence of intergranular failure. The completely lamellar sturcture provides many opportunities for the fracture to locally advance unimpeded along the beta lamellae. Many parallel secondary cracks are formed by separations along beta lamellae aligned at an angle to the main fracture path (Fig 5a).

Similar to the pure titanium case cited earlier, further reductions in fracture toughness beyond the DTBT were achieved by increasing the grain size, so too the length of the beta lamellae. The cleavage fracture area is thereby increased at the expense of ductile tearing required as the fracture advances from one beta lamella to the next.

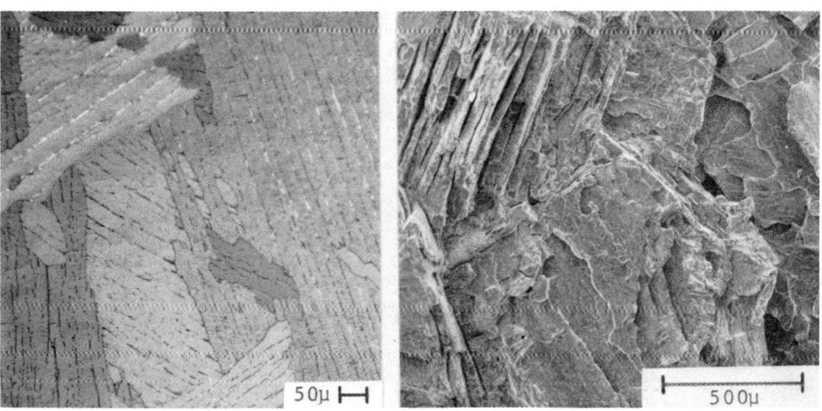

Fig. 5   Photomicrograph and SEM Fractograph of CP Titanium (ASTM Grade II) Treated within the Beta Field (945°C, 2 hrs.)

Note that the hydrogen content was about constant, varying between 39 and 41 ppm. Although fracture toughness reductions are small beyond the DTBT, it does, however, indicate the role of platelet development in failures of this type.

Pure titanium's DTBT was easily determined on the basis of fracture appearance alone; not so, however, CP titanium. A visual fracture-appearance determination can only be made in the case of an embrittled beta-soak microstructure (completely lamellar). However, the pre- and post-transition fracture appearances resulting from alpha and (α + β) soak microstructures are very similar. It is difficult to determine the transition on the basis of fracture appearance in these cases.

Acknowledgments

The authors are grateful to the IBM Data Systems Division and the Division of Engineering Research (MSU) for their financial support of the research. Appreciation is also expressed to Mr. Blue (Bureau of Mines, Department of the Interior, Boulder City, Colorado) for supplying the pure electrolytic titanium; and to Dr. H. Baumberger and his staff (Reactive Metals, Inc., Niles, Ohio) for their cooperation in remelting and working the pure titanium.

Cited Literature

1. A.H. Cottrell, <u>Dislocations</u> and <u>Plastic</u> <u>Flow</u> in <u>Metals</u>, Clarendon Press, Oxford, 1953, pp. 133-150.

2. G.G. Libowitz, J. Nuclear Mater., 1960, Vol. 2, pp. 1-22.

3. P.E. Irving and C.J. Beevers, Met. Trans., 1971, Vol. 2 (2), pp. 613-615.

4. E.A. Gulbransen and K.F. Andrew, J. Metals, 1949, Vol. 1, pp. 741-748.

5. A.D. McQuillan, Proc. Roy. Soc. (London), 1950, Ser. A204, pp. 309-323.

6. R.S. Vitt and K. Ono, Met. Trans., 1971, Vol. 2 (2), pp. 608-609.

7. R.I. Jaffee and I.E. Campbell, Trans. AIME, 1949, Vol. 185, pp. 646-655.

8. C.M. Craighead, G.A. Lenning and R.I. Jaffee, Trans. AIME, 1952, Vol. 194, pp. 1317-

9. G.A. Lenning, C.M. Craighead and R.I. Jaffee, Trans. AIME, 1954, Vol. 200, pp. 367-376.

10. R.I. Jaffee, "The Physical Metallurgy of Titanium Alloys," <u>Progress</u> in <u>Metal</u> <u>Physics</u>, Vol. 7, Pergamon Press, New York, 1958, pp. 65-163.

11. O.Z. Rylski, Ottawa, Canada, Dept. of Mines and Technical Surveys, Mines Branch, Phys. Met. Div., 1956, PM203, 77 pp.

12. J.L. Wyatt and N.J. Grant, Trans. A.S.M., 1954, Vol. 46, pp. 540-565.

13. R.W. Hanzel, Metal Prog., 1954, Vol. 65 (3), pp. 89-96.

14. W.L. Finlay and J.A. Snyder, Trans. AIME, 1950, Vol. 188, pp. 277-286.

15. E.H. Rennhack, Trans. TMS-AIME, 1958, Vol. 212, pp. 487-489.

16. A.E. Jenkins and H.W. Worner, J. Inst. Metl., 1951-1952, Vol. 80, pp. 157-166.

17. H.P. Tardif and H. Marquis, Can. Met. Quart., 1963, Vol. 2 (4), pp. 373-398.

18. C.M. Craighead, G.A. Lenning and R.I. Jaffee, Trans. AIME, 1956, Vol. 206, pp. 923-928.

19. R.I. Jaffee, G.A. Lenning and C.M. Craighead, Trans. AIME, 1956, Vol. 206, pp. 907-913.

STUDIES ON ALUMINUM ALLOYS

STUDIES OF HYDROGEN EMBRITTLEMENT AND STRESS-CORROSION
CRACKING IN AN ALUMINUM-ZINC-MAGNESIUM ALLOY*

S. W. Ciaraldi, J. L. Nelson, R. A. Yeske, and E. N. Pugh**
Department of Metallurgy and Mining Engineering
University of Illinois
Urbana, Illinois 61801

Tensile tests have been carried out on a high-purity Al-5.6 Zn-2.6 Mg
alloy hydrogenated by exposure to moist air. The results indicate that
internal hydrogen embrittlement occurs by the formation and rupture of a
stress-induced hydride at the grain boundaries. The hydride, identified
by electron diffraction as AlH_3, is shown to be unstable in laboratory
air, reverting to aluminum. The hydride phase was not detected in specimens failed by SCC, despite evidence that hydrogen is transported ahead of
advancing stress-corrosion cracks, and this leads to the possibility that
a basically different mechanism may be responsible for SCC in this alloy.

*This work was supported in part by the U.S. Department of Energy, Con-
tract DOE EY-76-C-02-1198, and by the Air Force Office of Scientific
Research under Grant AFOSR 77-3175.

**The authors are now at Amoco Research Center, Naperville, IL (SWC);
INCO Research and Development Center, Suffern, NY (JLN); Westinghouse
Research and Development Center, Pittsburgh, PA (RAY); and National
Bureau of Standards, Washington, DC (ENP).

Introduction

Following the work of Gruhl and Brungs (1), it has become well established that hydrogen introduced into aluminum-zinc-magnesium alloys leads to significant embrittlement (2-12). Hydrogen has been introduced electrolytically (2,3,7,10), by exposure to moist air at room temperature (the pre-exposure effect) (4,6,8) or at 70 °C (5,9,12), by hydrogen-ion bombardment (11), and by polishing with aqueous slurries (12), and the path of fracture is generally intergranular. The mechanism of embrittlement has not been established, and is a primary objective of the present work. In particular, the present studies were aimed at determining the nature and role of a fragmented layer which had been observed at the fracture surfaces of hydrogen-embrittled specimens (6,8,11).

The fact that hydrogen can induce brittle intergranular cracking makes it attractive to speculate that intergranular SCC, a major practical problem in these high-strength alloys, is also due to hydrogen embrittlement. However, there is no direct evidence for a causal relationship at present, and this question is also addressed in this paper.

Experimental

The experiments were carried out on a single high-purity ternary alloy, kindly provided by Mr. D. O. Sprowls, Alcoa. The alloy contained 5.63 wt pct zinc and 2.59 wt pct magnesium, and the concentration of each of the elements silicon, iron, copper, manganese, chromium, nickel, and titanium was < 0.01 wt pct. The cast ingot was rolled to sheet of thicknesses 0.305, 0.076, and 0.038 cm, respectively, with intermediate and final anneals at 470 °C in air. The air-formed film produced at 470 °C is known to play an important role in the embrittlement phenomenon (5,6). Two types of specimens were cut from the sheet:

(i) Tensile specimens of thickness 0.076 and 0.038 cm with a gauge length 0.635 cm long and 0.318 cm wide. No machining was carried out on the rolled surfaces.

(ii) Bend specimens of cross section 0.25 cm x 0.25 cm.

The specimens were solution treated for 1 h at 470 °C in dry argon and water quenched to room temperature. The 0.076 cm-thick tensile specimens were then aged for 60 d at room temperature in either dry or moist air. Based on previous work (4,6,8), the latter treatment was expected to introduce significant amounts of hydrogen into the specimens. After natural aging, the specimens were tested to failure at room temperature in dry argon at a strain rate of 1.3×10^{-4} s^{-1}.

After quenching, the 0.038 cm-thick tensile specimens and the bend testpieces were aged to peak hardness (24 h) at 130 °C in a silicone-oil bath. The tensile specimens were cleaned with acetone and, using the method described by Scamans and Tuck (9), they were then hydrogenated by exposure to water vapor saturated air (WVSA) at 70 °C for various times up to 7 d. The hydrogen content of certain specimens was measured by means of a LECO Hydrogen Determinator. Other tensile specimens were tested to failure as in the case of the 0.076 cm-thick specimens. The bend specimens were electropolished and stressed under constant deflection in a four-point bend jig in either aqueous 3 pct sodium chloride solution at room temperature, or in 70 °C WVSA.

Electron-diffraction studies were carried out to identify the fragmented layer found on certain fracture surfaces. These were performed using a JEOL 200 TEM operated at 200 kV. Diffraction patterns were obtained either by transmission through pieces of the layer extracted by replica techniques, or directly from the fracture surface, either by glancing-incidence diffraction or by transmission through fragments partially detached from the surface. Diffraction was carried out at the eucentric position of the microscope to minimize error in the use of the camera constant. A densitometer was used to scan the diffraction patterns. One series of diffraction experiments was carried out on the naturally aged tensile specimens, 0.076 cm thick, fractured at a strain rate of 1.3×10^{-4} s^{-1} in argon and subsequently stored in laboratory air before being placed in the TEM. A second series was carried out on the peak-aged tensile specimens, 0.038 cm-thick, which were exposed to 70 °C WVSA for 7 d. These specimens were etched with Tuckers etch to remove any surface films, rinsed with acetone, and dried. They were then placed in a small tensile machine enclosed in a glove bag which was also attached to the port of the TEM. The glove bag was purged of air by means of flowing argon and the specimens fractured at a strain rate of 1.3×10^{-4} s^{-1}. One half of the fracture surface was transferred to the TEM within 5 min of fracture without exposure to air.

Results and Discussion

Tensile Tests

Data from tensile tests on 0.076 cm-thick specimens naturally aged for 60 d in dry or moist air are given in Table I.

Table I

Results of tensile tests on specimens naturally aged for 60 d in dry or moist air. Strain rate 1.3×10^{-4} s^{-1}

Aging Environment (dry or moist air)	Ultimate Tensile Strength (MN/m^2)	Total Strain to Failure (pct)
Dry	285	42.4
	274	39.4
	261	40.5
	287	41.4
Moist	238	22.1
	205	15.0
	222	19.0

Exposure to moist air can be seen to have caused significant reductions in tensile properties compared to those for specimens aged in dry air. The latter underwent ductile transgranular fracture. Specimens aged in moist air also failed predominantly by the ductile mode, but, in addition, the fracture surfaces exhibited up to 30 pct intergranular fracture, Figure 1(a). At higher magnifications, several of these intergranular

facets were found to exhibit the fragmented layer, reported earlier (6,8),
Figures 1(b)-(d). Examination of the opposing fracture faces indicated that
part of the layer was frequently attached to one side and the remainder to
the other, Figures 1(c) and (d), indicating clearly that the layer was not
produced after fracture. The maximum thickness of the layer was ∿ 1 μm.
Electron-diffraction studies on extracted pieces yielded ring patterns which
were indexed as FCC aluminum, see below. In no cases were patterns corresponding to aluminum hydroxides or oxides observed. It should be pointed
out that the fragmented layers had been exposed to laboratory air for at
least 2 to 3 d before the diffraction experiments were carried out.

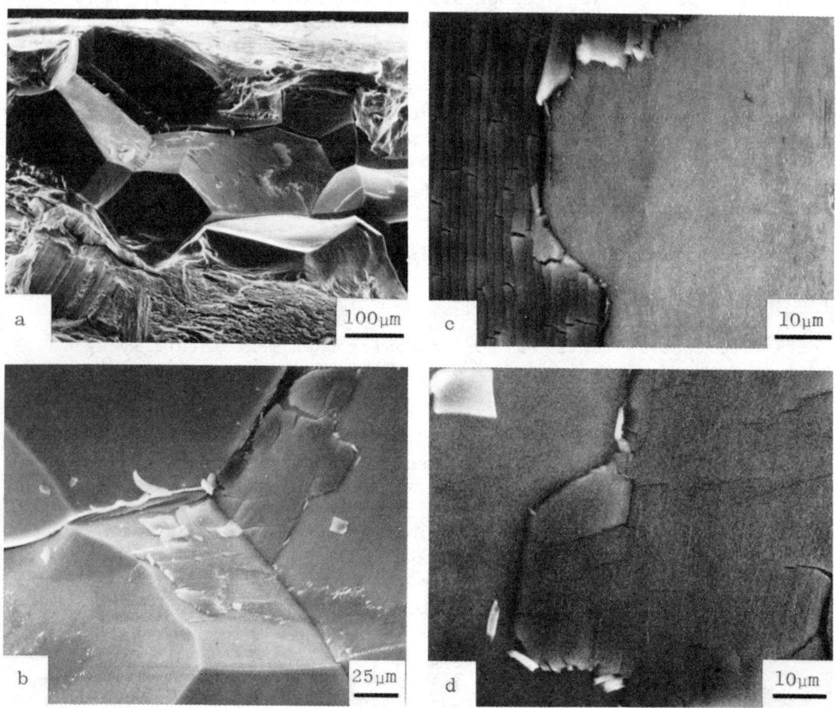

Figure 1.   Scanning electron micrographs of fracture surfaces of a tensile
specimen naturally aged for 60 d in moist air. The intergranular facets
in (a) are shown at high magnifications in (b)-(d), revealing the fragmented layer. The opposing fracture faces are illustrated in (c) and (d),
indicating the matching and interlocking nature of the layer.

As heat-treated specimens naturally aged in dry and moist air were
also tested at a higher strain rate, viz. 10^{-2} s^{-1}. The results, presented in detail elsewhere (6), demonstrated that pre-exposure embrittlement
was completely suppressed at this strain rate, no significant differences
in mechanical properties being observed, and ductile fracture occurring in
all cases with no evidence of the fragmented layer. These results indicated that the layer observed in tests at 1.3 x 10^{-4} s^{-1} is not formed
by penetration of grain boundaries by a corrosion process during aging,

but that it is produced during straining. The necessity for slow strain
rates also suggests that formation of the layer requires diffusion in the
lattice, presumably of hydrogen.

To investigate the nature of the film, further tests were carried out
on 0.038 cm-thick tensile specimens, aged to peak hardness at 130 °C and
hydrogenated by exposure to WVSA at 70 °C. The hydrogen content of these
specimens was found to increase progressively with exposure time, attaining
$\sim$ 300 ppm ($\sim$ 1 at pct) after 7 d (12). The presence of the hydrogen led
to increasing embrittlement, manifested by reductions in ultimate tensile
strength and plastic strain, ε_{pl}, defined as

$$\varepsilon_{pl} = \frac{\text{crosshead displacement after yielding}}{\text{gauge length}} \times 100 \text{ pct.}$$

It can be seen in Figure 2 that embrittlement for exposure times up to
approximately 4 d was characterized by large reductions in ε_{pl} but

little decrease in UTS, whereas there were large reductions in UTS for
longer exposure times. These two regimes are subsequently referred to as
Stage I and Stage II. Increasing exposure times were associated with
increasing amounts of brittle intergranular fracture, and the intergranular facets exhibited the fragmented layer of the type illustrated in
Figure 1. The amount and thickness of the layer increased with increasing
exposure time. The maximum thickness of the layer was $\sim$ 2.5 μm.

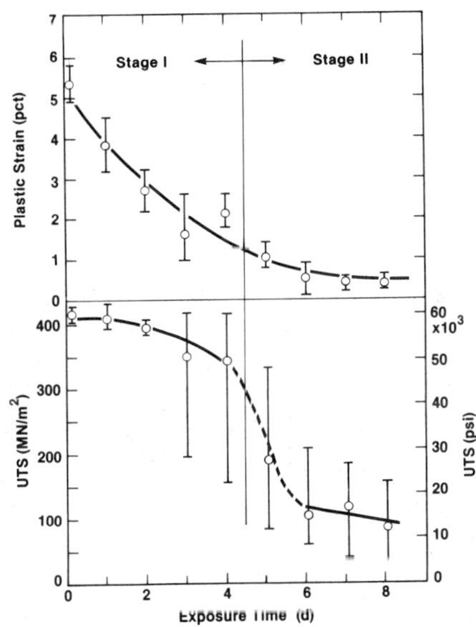

Figure 2. Effect of exposure to 70 °C WVSA on the mechanical properties
of 0.038-cm thick tensile specimens tested in dry argon at a strain rate
of 1.3×1^{-4} s^{-1}.

Studies were made of the effects of strain rate and thermal treatments
designed to outgas the hydrogen. These results, described in detail else
where (12), indicated that Stage I embrittlement could be completely suppressed by testing at a strain rate of 1.3×10^{-2} s^{-1}, whereas testing

at this increased rate did not influence the degree of embrittlement in Stage II.   In fact, brittle intergranular facets and the fragmented layer were produced in Stage II specimens fractured under impact conditions.   It should be recalled that embrittlement was suppressed in specimens hydrogenated by exposure to moist air at room temperature, suggesting that the hydrogen contents of those specimens corresponded to Stage I of those exposed to 70 °C WVSA.   Outgasing hydrogen by a complete re-heat treatment, i.e., solution treatment at 470 °C in argon and aging at 130 °C, completely recovered the mechanical properties of both Stage I and II specimens, and demonstrated the reversibility of the phenomenon.   TEM studies of samples produced by thinning specimens exposed for 7 d at 70 °C WVSA revealed no evidence for the presence of any layer or other unusual structures at the grain boundaries, suggesting that the fragmented layer is not formed prior to stressing the specimens.

Electron-diffraction patterns were taken directly from the fragmented layer on intergranular facets in specimens which had been exposed for 7 d to 70 °C WVSA.   As described above, these specimens were fractured in dry argon and transferred to the TEM quickly and without exposure to laboratory air.   This procedure was adopted because all forms of aluminum hydride have been reported to be unstable, particularly in moist environments (13).   Under these conditions, the patterns obtained were quite different from those obtained for the specimens embrittled by room-temperature exposure, in which the fracture surfaces had been exposed to laboratory air for long periods before the diffraction patterns were obtained.   The patterns, Figure 3(a), were characterized by continuous rings, indicating a very fine grain size, and the plane spacings obtained from these rings were found to be consistent with those for a hexagonal aluminum hydride, AlH_3, with a = 2.90 Å and c = 4.55 Å (14).   Table II lists the predicted plane spacings for this structure and the spacings obtained experimentally for several specimens.   It can be seen that good correlation exists, although several lines are missing.   The detailed structure of AlH_3 is not known, so that it is not possible to calculate structure factors and to determine which reflections would have low intensities.   The spacings were also compared to those of various forms of alumina and aluminum hydroxide but no correspondence was found.

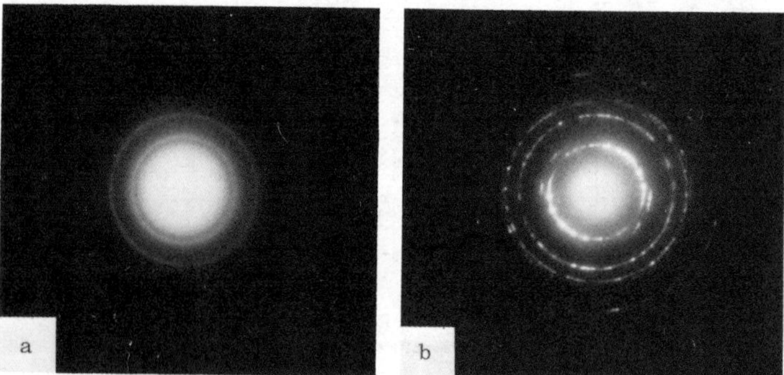

Figure 3.   Electron-diffraction patterns obtained from a fragmented layer on the fracture surface of a peak-aged specimen exposed to 70 °C WVSA for 7 d and tested in argon at a strain rate of 1.3×10^{-4} s^{-1}.   Pattern (a) was obtained approximately 0.5 h after fracture without exposure to air and (b) was obtained from the same fragment after subsequent exposure to laboratory air for 150 h.

Table II

Comparison of plane spacings for AlH_3 with typical values
obtained from electron-diffraction patterns such as that
shown in Figure 3(a).

Reflection	Plane Spacing for AlH_3[a] (A^o)	Plane Spacings Obtained from Typical Diffraction Patterns		
		1	2	3
0001	4.55			
$01\bar{1}0$	2.51	2.53	2.54	2.53
0002	2.28			
$01\bar{1}1$	2.20	2.19	2.16	2.21
$01\bar{1}2$	1.69	1.70		
0003	1.52	1.49	1.52	1.51
$11\bar{2}0$	1.45			
$11\bar{2}1$	1.38	1.41	1.42	
$01\bar{1}3$	1.30	1.33		1.33
$02\bar{2}0$	1.26			
$11\bar{2}2$	1.22			
$02\bar{2}1$	1.21	1.20		1.21

[a] Calculated for hexagonal structure with a = 2.90 Å and c = 4.55 Å (14).

Several fractured specimens for which the AlH_3 structure had been
observed were subsequently exposed to laboratory air for approximately
150 h, and further electron diffraction patterns were obtained. In no
cases were patterns consistent with AlH_3 observed, all the patterns being
indexed as polycrystalline aluminum. For example, Figure 3(b) shows a
pattern obtained from the fragmented layer from which the pattern shown
in Figure 3(a) had been obtained previously. The polycrystalline nature
of the aluminum pattern is of particular interest, since patterns taken
from the fracture surfaces of non-embrittled specimens corresponded to
single-crystal aluminum. Such single-crystal patterns would be expected
since the grain size of the alloy was large ($\sim$ 300 μm) and since the beam
diameter was small (200 to 300 Å).

Tests on Bend Specimens

Stress-corrosion tests were carried out on peak-aged bend specimens
tested under constant deflection in 3 pct aqueous sodium chloride (6).
Crack-growth rates in these tests varied between 4.2 x 10^{6} cm s^{-1} and
2.0 x 10^{5} cm s^{-1}, depending on deflection. The resulting fracture surfaces
consisted entirely of brittle intergranular facets except for several

grains on the compression side which exhibited ductile transgranular fracture. SEM examination of the intergranular SCC facets revealed no evidence of the fragmented layer.

In one test, a specimen was unloaded prior to complete fracture, dried with acetone, and then immediately strained to failure in tension at a strain rate of $1.1 \times 10^{-5} \text{ s}^{-1}$ in dry air. Fractographic examination of the transition region from I-SCC to ductile overload indicated an intervening intergranular region, 200 to 300 μm wide, which exhibited the fragmented layer, Figure 4(a). No electron-diffraction studies were carried out on this layer, but its characteristics were the same as the layer observed in the preceding section. It is considered that this region was also produced by internal hydrogen embrittlement, the hydrogen in this case being introduced from the aqueous solution at the tip of the advancing stress-corrosion crack.

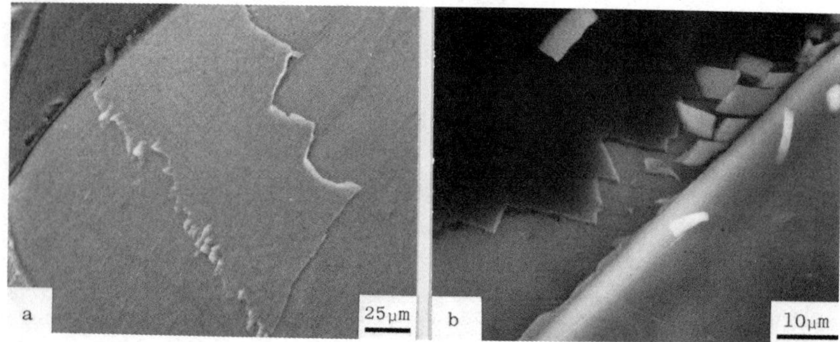

Figure 4. Scanning electron micrographs of fracture surfaces of peak-aged bend specimens. (a) A region in the transition zone between I-SCC and ductile overload for a specimen partially fractured in aqueous sodium chloride and then strained to failure in tension in dry air. (b) Taken from a specimen tested under constant deflection in 70 °C WVSA.

Constant-deflection bend tests were also conducted in 70 °C WVSA (12). Slow crack growth again occurred, with complete failure taking place within 24 h; the crack velocity was not determined in these tests. The primary fracture mode was again brittle intergranular but, in contrast to the tests in aqueous sodium chloride, the fragmented layer was detected on many of the intergranular facets, Figure 4(b). No attempts were made to carry out electron-diffraction studies on these layers.

General Discussion and Conclusions

The major finding of these studies is the occurrence of a layer of AlH_3 at the intergranular fracture surfaces produced in tensile tests on hydrogenated specimens. The results established that the hydride is not produced by a corrosion process prior to testing, and strongly suggest that it is formed during stressing. Thus it is concluded that the hydride is stress-induced, and that its formation is responsible for internal hydrogen embrittlement (IHE) in this alloy, by a mechanism similar to that proposed for zirconium (15) and niobium (16).

The occurrence of two stages of embrittlement for specimens hydrogenated by exposure to 70 °C WVSA, Figure 2, can be rationalized in terms of the several different roles of stress in the embrittlement process.

When the hydrogen concentration is low, Stage I, it is considered that diffusion of hydrogen is necessary into the region ahead of the advancing crack tip to provide a sufficient concentration to produce AlH_3. Thus the large fracture stresses in Stage I are considered to correspond to the stresses necessary to concentrate sufficient hydrogen by the Troiano-Oriani mechanism. This view is consistent with the observed strain-rate dependence of embrittlement in Stage I.

In Stage II, it is considered that sufficient hydrogen is already present at the grain boundaries to form AlH_3, so that diffusion of hydrogen over long distances is not required during stressing—hence the absence of a strain-rate dependence. Given the measured bulk hydrogen concentrations, this view requires that virtually all the hydrogen in the alloy is at grain boundaries for stoichiometric AlH_3 to be formed. Under these conditions, the formation of AlH_3 requires only the stress necessary to induce the phase transformation. It is assumed here that the transformation is martensitic in nature, requiring only local rearrangement of atoms, and that it can be stress induced. It should be emphasized that the details of the phase transformation are not known, but, by analogy with other hydrides, one can expect ordering of hydrogen and correlated shears of the aluminum lattice to be involved; stress might also be important in accommodating the volume change associated with the transformation (16). The third stress to be considered is that to fracture the hydride, or, more probably, the hydride-matrix interface. Thus the low fracture stresses in Stage II correspond to either the stress required to induce the transformation or the fracture stress, whichever is larger.

Detection of the fragmented layer on the fracture surfaces of all specimens which underwent IHE suggests that the phenomenon can be attributed to a hydride-rupture mechanism (15,16), and that there is no necessity to seek other embrittling mechanisms in this system. Conversely, the absence of the layer on the fracture surfaces of specimens which underwent SCC in aqueous sodium chloride questions whether hydrogen plays a role in the stress-corrosion process. The experiment in which a stress-corrosion crack was halted and the specimen then fractured in dry air, Figure 4(a), clearly demonstrates that sufficient hydrogen is transported ahead of the advancing stress-corrosion crack to produce IHE; a similar effect has been observed by other workers (17,18). Further, the occurrence of slow crack growth in specimens stressed in 70 °C WVSA and the presence of the fragmented layer on the resulting fracture surfaces indicates that the hydride-rupture mechanism can operate when the source of hydrogen is external. Thus the possibility must be recognized that stress-corrosion cracks in this alloy may propagate by a competing and basically different mechanism. For example, it has been argued that the film-rupture mechanism can operate in this system--that is, localized plastic deformation within the soft PFZ at grain boundaries can lead to rupture of the passive film and thus to crack propagation by preferential anodic dissolution (19).

Pursuing the possibility that a second mechanism of slow crack growth can operate in this alloy in aqueous solutions in addition to hydride rupture, it is interesting to note that double plateaus have been reported in the V-K curves for aluminum-zinc-magnesium alloys in some cases (20-22). It may be speculated that this results from the overlapping of two different V-K curves, one corresponding to HE and the other to the second mechanism. In the majority of cases, a single plateau is observed for these alloys, suggesting that one of the V-K curves becomes completely submerged by the other, so that the corresponding mechanism plays no role in the fracture process. The present observations could be interpreted to indicate that HE is the "submerged" mechanism in this specific case. Further work is clearly necessary to pursue these possibilities. Moreover, it should be recognized that any conclusions for this high-purity alloy cannot necessarily be extended to other high-strength aluminum alloys, particularly to the commercial alloys.

It is concluded that IHE of the high-purity ternary alloy investigated in the present work occurs by the formation and rupture of a stress-induced hydride, AlH_3. Experiments in 70 °C WVSA indicate that the hydride-rupture mechanism can also be responsible for slow crack growth in this alloy. However, failure to detect the hydride phase in specimens failed by SCC in aqueous sodium chloride questions that hydrogen plays a role in the stress-corrosion process.

References

1. W. Gruhl and D. Brungs: <u>Metall.</u>, 1969, vol. 23, p. 1020.

2. R. Gest and A. R. Troiano: <u>Corrosion</u>, 1974, vol. 30, p. 274.

3. R. M. Pelloux and J. A. Van den Avyle: <u>Hydrogen in Metals</u>, p. 275, ASM, Metals Park, OH, 1974.

4. L. Montgrain and P. R. Swann: <u>ibid.</u>, p. 575.

5. C. M. Scamans, R. Alani, and P. R. Swann: <u>Corros. Sci.</u>, 1976, vol. 16, p. 443.

6. J. L. Nelson: Ph.D. Thesis, University of Illinois at Champaign-Urbana, 1976.

7. J. Albrecht, B. J. McTiernan, I. M. Bernstein, and A. W. Thompson: <u>Scripta Met.</u>, 1977, vol. 11, p. 893.

8. J. L. Nelson (reported by E. N. Pugh): <u>Mechanisms of Environment Sensitive Cracking of Materials</u>, p. 493, The Metals Society, London, 1977.

9. G. M. Scamans and C. D. S. Tuck: <u>Environment-Sensitive Fracture of Engineering Materials</u>, p. 464, TMS-AIME, Warrendale, PA, 1979.

10. J. Albrecht, A. W. Thompson, and I. M. Bernstein: <u>Met. Trans. A</u>, 1979, vol. 10A, p. 1759.

11. G. H. Koch: <u>Corrosion</u>, 1979, vol. 35, p. 73.

12. S. W. Ciaraldi: Ph.D. Thesis, University of Illinois at Champaign-Urbana, 1980.

13. <u>Metal Hydrides</u>, p. 573, Academic Press, New York, NY, 1968.

14. M. Appel and J. P. Frankel: <u>J. Chem. Phys.</u>, 1965, vol. 42, p. 3984.

15. D. G. Westlake: <u>Trans ASM</u>, 1969, vol. 62, p. 1000.

16. H. K. Birnbaum: <u>Environment-Sensitive Fracture of Engineering Materials</u>, p. 326, TMS-AIME, Warrendale, PA, 1979.

17. B. F. Brown: Private communication, The American University, Washington, DC, 1980.

18. J. E. Morral: University of Connecticut, Storrs, CT, 1980.

19. A. J. Sedriks, P. W. Slattery, and E. N. Pugh: Trans. ASM, 1969, vol. 62, p. 238.

20. M. O. Speidel: The Theory of Stress Corrosion Cracking in Alloys, p. 289, NATO, Brussels, Belgium, 1971.

21. P. K. Poulouse, J. E. Morral, and A. J. McEvily: Met. Trans., 1974, vol. 5, p. 1393.

22. R. Haag, J. E. Morral, and A. J. McEvily: Environmental Degradation of Engineering Materials, p. 19, VPI Press, Blacksburg, VA, 1977.

DISCUSSION

M. P. Shanabarger, University of California: My question concerns the amount of hydrogen required to form your AlH_3 flakes. The "flake" of AlH_3 you showed appears to be of a reasonably large size, and to be composed entirely of AlH_3 implies that a very large amount of hydrogen must be present in the lattice during fracture. This does not seem to be consistent with the low solubility of hydrogen in Al. Would you comment on the source of this large source of lattice hydrogen.

E. N. Pugh: We show in the paper that the hydride layer, $\sim 1\mu m$ thick, requires all the measured H content to be at the grain boundaries if the hydride is stoichiometric. The reason that no precipitation of hydride occurs before stressing may be due to the large volume change associated with the phase transformation; this is also discussed in the paper.

ENVIRONMENT-SENSITIVE EMBRITTLEMENT OF

HIGH STRENGTH ALUMINIUM ALLOYS*

N.J.H. Holroyd and D. Hardie
Department of Metallurgy and Engineering Materials
The University
Newcastle upon Tyne NE1 7RU
England

The results of extensive slow strain rate testing of 7000 series alloys
with and without prior exposure to hot tap water or cold sea water are assessed in terms of the loss in ductility promoted. The more important factors
affecting the embrittlement include the strain rate (which must be very slow),
the immersion time (because of time-dependent embrittlement), the innate
aggressiveness of the test environment (because of the possibility of time-
dependent recovery), the condition of the alloy and the potential. As long
as all of these variables are borne in mind, the strain rate at which detectable embrittlement occurs in laboratory air may be employed for comparing
the susceptibility of various alloys to stress corrosion failure.

Testing at controlled potentials has allowed some assessment of susceptibility to hydrogen embrittlement and to anodic dissolution, as well as some
indication of the relevance of these processes to stress corrosion failure at
the free corrosion potential.

Published work on similar alloys has been examined, and in certain cases
re-interpreted, on a basis of the conclusions reached.

* This work was supported by the Science Research Council and Alcan
Research Laboratories at Banbury.

Introduction

The practical problems encountered because of the stress corrosion failure of high-strength aluminium alloys have caused a resurgence of interest in the mechanisms by which failure is instigated. Despite the observation by Gruhl[1] that prior exposure to a solution of sodium chloride led to a reduction in the stress corrosion life and ductility of a high-purity Al-5Zn-3Mg alloy, hydrogen embrittlement was not felt to pose any particular problem until quite recently. However, since Gest and Troiano[2,3] found that they could introduce a temperature-dependent embrittlement into commercial 7075 by cathodic charging and Scamans et al[4] observed a reversible embrittlement of high-purity Al-6Zn-3Mg containing 0.14% chromium and of 7075 on pre-exposing to water-vapour-saturated air, later work[5,6] has confirmed the important contribution that hydrogen may make to the stress corrosion process.

Extensive testing of a wide range of alloys at very slow strain rates in carefully controlled environments, both with and without prior exposure, has allowed a greater understanding of the critical processes involved in embrittlement and a rationalisation of many results that were previously believed to be inconsistent. It has become abundantly clear that the lack of reproducibility and apparent contradictions inherent in much published work resided in a lack of appreciation of the importance of certain variables such as strain rate and the aggressiveness (due to moisture content) of laboratory air. The main aspects of these interesting phenomena that will be discussed here involve a study of the importance of strain rate when examining the effects of pre-charging or when testing in an aggressive environment, the observation of time-dependent recovery from the effects of pre-charging and the variation in behaviour introduced by potentiostatic control in aqueous environments.

Experimental

The technique developed at Newcastle involves slow strain rate testing of standard tensile specimens, normally machined with their principal axis parallel to the short transverse direction of plate material, in carefully controlled environments. Those environments that have proved most useful, if not essential, to a deeper understanding of the processes involved are vacuum ($<10^{-5}$torr), dry air (using a packing of anhydrous magnesium perchlorate around the specimen), laboratory air (relative humidity 58-60%), tap water (pH = 7.2 and Cl ion content 22 mg/1) and natural sea water (pH between 8.1 and 8.3 and retained constant to ±0.1 by inherent buffering). Certain tests have also involved prior exposure either in warm tap water or sea water at ambient temperature. All specimens were mechanically polished to 000 finish, using emery papers soaked in paraffin, and then degreased before introduction to the environment for pre-soaking or testing.

Although many tests have been conducted without potentiostatic control, the latter has been fruitfully employed as a means of assessing the relative susceptibilities of different materials to hydrogen embrittlement and/or anodic dissolution stress corrosion. Extraneous galvanic potentials between the short specimens and the grips that were inevitably also immersed in the environment were avoided by manufacturing grips from similar materials to the specimens, as well as lacquering all but the gauge lengths of the test pieces.

Plastic reduction in area at fracture proved to be the most dependable and meaningful parameter for assessing the degree of embrittlement. The alternatives of plastic elongation and the related fracture energy both suffer from the effects of secondary cracking, which introduces a spurious contribution to the measured elongation. The cross sectional area at fracture was measured perpendicular to the tensile axis by a combination of techniques - both weighing a projected image (x20) outlined on tracing paper and direct measurement by a calibrated measuring eyepiece. In general, a "reduction in area ratio" (RA ratio), dividing the value obtained in the particular environment by that measured for a test in vacuum or dry air at the same strain rate, has been employed for representing the degree of embrittlement.

Discussion

Effect of Strain Rate

The importance of the strain rate used in conducting pre-exposure and stress corrosion tests on high strength aluminium alloys has not been generally appreciated. Scamans et al[4] found no difference in the strain rate sensitivity of Al-Zn-Mg specimens stressed in vacuum and in air, after 3 days pre-exposure to water-vapour-saturated-air at 90°C, at strain rates from 4.2×10^{-4} to 4.2×10^{-2}/s. It is therefore understandable when later workers have assumed that this would apply over a greater range of testing speeds. However, both increases[7] and decreases[8,9] in stress corrosion susceptibility have been reported for 7075-T651 tested at decreasing strain rates below about 10^{-6}/s.

Early attempts to induce pre-exposure embrittlement in a relatively susceptible commercial alloy (7179) also met with little success until testing was conducted at strain rates below 10^{-4}/s. This not only established that the rate of straining was important but indicated that, even when tested parallel to the short transverse direction (acknowledged to be the most susceptible), commercial alloys were less susceptible to embrittlement effects than their high-purity analogues. Subsequent investigations have confirmed this trend and emphasised the importance of comparing the more resistant commercial alloys at particularly slow strain rates. In fact, the susceptibilities of these materials to hydrogen embrittlement and stress corrosion may be compared in terms of the strain rate at which the ductility of specimens that have been subjected to pre-exposure or are tested in an aggressive environment differs significantly from that observed in an inert environment.

Such a comparison needs to make no assumptions concerning the source of embrittlement in stress corrosion tests, although pre-exposure effects are known to be attributable to hydrogen assimilated into the alloys as a result of interaction with the environment[4-6,9]. Where hydrogen is indeed involved, the effect of strain rate may readily be interpreted in terms of the time required for hydrogen to diffuse to those sites where its concentration becomes sufficient to induce brittle decohesion of the material. This seems to give rise most frequently to an intergranular mode of failure although transgranular failures have been observed[3,6,16]. Most mechanisms of stress corrosion failure by anodic dissolution would predict an increase in ductility when testing is conducted at even slower strain rates but any such effects reported for high-strength aluminium alloys can usually be attributed to other phenomena.

One such anomaly arises because of the aggressive nature of undried air and the use of tests in laboratory air as a reference has bedevilled certain investigations of these alloys. Not only did Speidel[10] stress that water vapour was the only gas that had been shown to cause initiation and/or propagation of stress corrosion cracks in aluminium alloys but Scamans et al[3] carried out a detailed assessment of the effect of relative humidity (at 50°C) on the embrittlement of an Al-6Zn-3Mg alloy and also emphasised that air-embrittlement occurs even at 20°C. Despite this, results have subsequently been published that have not taken such factors into account. Where Scamans et al employed a ductility ratio to express their results, they used a value in vacuum as the reference but Buhl[11] compared the fracture energy for 7075-T6 in aqueous sodium chloride with that measured in laboratory air. As a result, he obtained maximum embrittlement at a strain rate between 10^{-6} and 10^{-5}/s and an apparent recovery at slower testing speeds. Testing of a similar alloy (7049) at Newcastle produces the same trend when the RA ratio employed to assess ductility involves a result in laboratory air as the reference instead of the RA under vacuum[6], but this is purely due to the onset of embrittlement in laboratory air between 10^{-5} and 10^{-6}/s (Figure 1). The variation in sus-

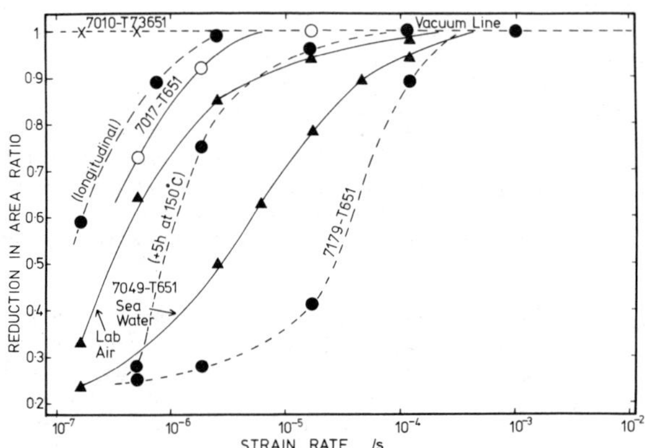

Figure 1. Variation of reduction in area ratio (using a result in vacuum/dry air as the standard) with strain rate in laboratory air for a range of commercial high strength aluminium alloys: 7179–T651 ●, 7049–T651 ▲, 7017–T651 O and 7010–T73651 X.

ceptibility with orientation of specimens removed from commercial bar or plate, due no doubt to the incomplete recrystallisation and the occurrence of a "pancake" grain structure, also becomes readily apparent. Those specimens having their principal axis parallel to the longitudinal (rolling) direction of plate do not exhibit brittle failure until much slower strain rates e.g. 7179. Whether or not the effect may be attributed entirely to simultaneous occurrence of a recrystallised grain structure is debatable, but results presented in the literature for high purity analogues of commercial alloys invariably appear more susceptible to embrittlement than their commercial counterparts and therefore suffer embrittlement at faster strain rates. Some support for the idea that the recrystallised structure is at least partially responsible for this difference is provided by Lynch's report[13] of embrittlement of commercial 7075 in laboratory air at a strain rate of 1.05×10^{-3}/s, because his alloy showed a recrystallised microstructure. The strain rate at which embrittlement in laboratory air becomes noticeable does, of course, vary with the inherent susceptibility of the material and may be employed to provide a comparison between a wide variety of materials (Figure 1). In their testing of a series of high-purity Al–Zn–Mg alloys, Miller and Scott[12] compared the mechanical behaviour of materials tested in synthetic sea water with that in laboratory air at a strain rate of 10^{-5}/s, at which they would certainly be picking up extensive embrittlement because the air was not dried. An assessment of the effect of strain rate on susceptibility may also provide the basis of a comparison of the aggressiveness of different environments, e.g. the line on Figure 1 representing the testing of 7049 in sea water may be compared with the results for the same material in laboratory air – embrittlement occurs at faster strain rates in sea water than in laboratory air (as would be expected).

Examples may also be selected from reported results in the literature
to demonstrate the way in which the embrittlement of materials by cathodic
charging has been misunderstood because charged material was tested in laboratory air. Close inspection of the early results of Gest and Troiano[2] for
7075-T651 suggests that the reduction in area at fracture for the "reference"
material at a deformation rate of 5×10^{-3} cm/min (a strain rate of about
3×10^{-5}/s if a 1 in. gauge length is assumed) is lower than would be expected
from extrapolation of the three points obtained at faster strain rates. Work
at Newcastle on 7049 (a very similar alloy) suggests that the material would
suffer from slight embrittlement if tested in laboratory air at this strain
rate. Since they used a faster strain rate (8.3×10^{-4}/s), Albrecht et al[14]
avoided this complication in their reproduction of the earlier charging
experiments. However, when Taheri et al[15] later investigated the effect of
heat treatment of 7075 on its response to hydrogen charging, the problem was
reintroduced by employing 3.3×10^{-5}/s as their slowest strain rate. Again
the reduction in area observed in uncharged specimens tested at this strain
rate is less than would be expected from results at faster rates of strain
and the authors indicate this by a dip in their curve through the points (as
did Gest and Troiano earlier). The way in which this invalidates certain
interpretations of the results of cathodic charging experiments will be discussed later, because recovery is also involved.

<u>Time of Immersion</u>

Since the strain rates required to detect embrittlement resulting from
exposure or testing in an aggressive environment involve long times to
failure, even in slow strain rate testing, the period of immersion in an
aggressive environment during testing can itself make a significant contribution to the embrittlement. The effects of pre-exposure and the test
environment tend to be additive and this has been clearly demonstrated in
tests on 7049-T651 (Figure 2). Although testing of specimens of this alloy

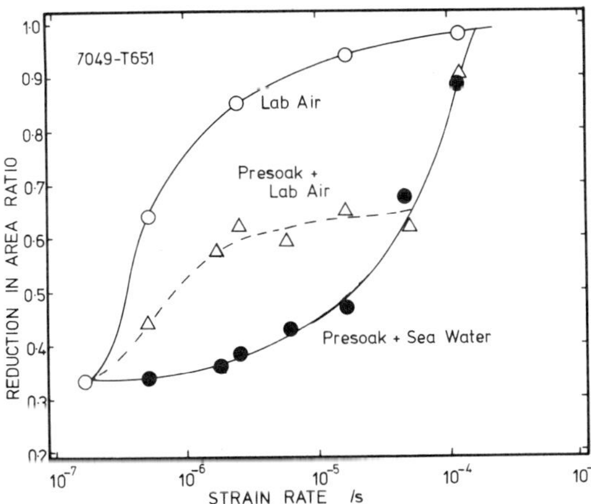

Figure 2. Effect of strain rate on the reduction in area in laboratory air or
sea water compared with that in vacuum for specimens of 7049 T651 with and
without prior immersion in sea water for 72h.

after prior immersion in sea water gives the same ductility at faster strain
rates ($>5 \times 10^{-5}$/s) whether the test environment is laboratory air or sea water,
RA ratios diverge at slower strain rates because of the more aggressive nature
of sea water. In fact, the embrittlement of the pre-exposed specimens tested

in laboratory air changes very little between 5×10^{-5}/s and 2.5×10^{-6}/s because the test environment makes little additional contribution, whereas testing in sea water causes increasing embrittlement. However, below 2.5×10^{-6}/s the contribution made by laboratory air to the embrittlement becomes appreciable and the results for pre-exposed specimens tested in both environments converge. Testing a specimen in laboratory air with or without prior exposure produces the same degree of embrittlement when the strain rate is 1.6×10^{-7}/s.

Recovery

If, as now seems firmly established, the embrittlement induced in susceptible alloys by prior exposure to warm tap water or cold sea water may be attributed to hydrogen taken up during pre-exposure, then re-distribution of this hydrogen might be expected to give rise to the recovery of ductility that is frequently observed. Of course, such re-distribution could involve loss of hydrogen to the environment (which would generally be restricted to the surface layers of thick specimens but would be more complete in thin sheet) or diffusion of the hydrogen to innocuous traps where it is prevented from further participation in embrittlement.

A detailed investigation of recovery effects in two commercial alloys has revealed pronounced differences in the time-dependent recovery of ductility. Samples of 7179 appeared to undergo a two-stage process after exposure to hot tap water[5], which involved a gradual improvement in ductility (without applied stress) during the first 50h in laboratory air followed by a more pronounced recovery between 50 and 100h, even though testing this material in laboratory air does itself promote some degree of embrittlement. By contrast, 7049 did not appear to undergo any recovery in laboratory air after exposure to sea water at ambient temperature (Figure 2) but did show recovery effects when strained in dry air[6]. Since both recovery and embrittlement are time-dependent, a wide variety of behaviour may be induced by varying the period of immersion in the environment before or during testing.

The recovery of ductility in 7049 that occurs when pre-exposed specimens are tested in an inert environment has been examined in an attempt to ascertain whether the applied stress or slow straining play an essential role in the recovery process. Specimens that had been pre-soaked in sea water were tested in dry air at a strain rate of 1.7×10^{-5}/s after allowing various periods for recovery in dry air, either unstressed or stressed to just above the yield (to 525 MN/m^2). None of these showed any sign of recovery of ductility within 250h, whereas a specimen tested at 5.2×10^{-7}/s (with a total test time of 100h) exhibited a ductility almost equal to that achieved in vacuum or dry air without pre-exposure, and specimens tested at 1.9×10^{-6}/s and 2.6×10^{-6}/s had intermediate degrees of embrittlement. It seems therefore that continuous straining, or at least a continually increasing stress, is essential to promote recovery in this material. If testing of these pre-exposed specimens is carried out in laboratory air that has not been dried, no recovery is then apparent and accurate prediction of the extent of embrittlement in pre-exposed specimens so tested is possible by simple addition of the loss in RA ratio resulting from pre-exposure and that occurring during the actual straining in laboratory air alone[6].

Earlier results quoted[3] for a similar alloy (7075) indicate that similar recovery from the embrittlement induced by cathodic charging techniques occurs, even though Taheri et al[15] considered that such recovery effects were absent. Unfortunately for them, their conclusion was based upon the results of two interrupted tests that took 3h to impose 6% strain (equivalent to an average strain rate of 5.6×10^{-6}/s). At such a slow strain rate, the embrittlement induced by laboratory air (already detectable in their tests at 3×10^{-4}/s) is

considerable and compensates for the loss of the charged hydrogen. This is
why they observed essentially the same ductility, or rather embrittlement,
in such an interrupted test on a charged and an uncharged sample. If their
results are handled rather differently, and the ductility expressed as an
RA ratio (using the extrapolated line of best fit through their points at the
three higher strain rates as a normalising parameter) a strain rate (time-)
dependent recovery similar to that observed in pre-exposed specimens can be
seen quite clearly (Figure 3). Thus the occurrence of a maximum embrittle-

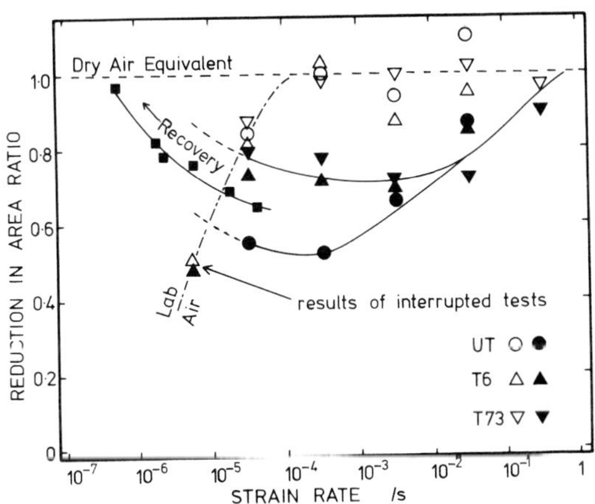

Figure 3. Reassessment of the results obtained by Taheri et al[15] for 7075, in
various heat-treated conditions, cathodically charged with hydrogen. Results
obtained[6] for 7049-T651 tested in dry air after pre-soaking in sea water (■)
are added for comparison of recovery. Open symbols represent uncharged and
closed cathodically charged specimens.

ment at an intermediate strain rate, which varied with the heat-treated condition of the alloy, is merely a fortuitous reflection of a strain-rate-
dependent embrittlement and a time-dependent recovery.

   The revised plot of the results for 7075 (Figure 3) suggests that recovery
from the embrittlement induced by cathodic charging occurs at a similar rate
in the T6 and T73 conditions but rather more slowly in the underaged (UT) condition i.e. after only 24h at 100°C. This could very well be associated with
the very different precipitate dispersion, which may affect the provision of
innocuous traps or the permeation rate for hydrogen. However, a recent
examination of the embrittlement of an Al-6Zn-3Mg alloy[16] suggests that some
recovery occurred because of the nucleation of hydrogen bubbles at grain boundary precipitate particles, but no bubbles were observed below 120°C. It is
still, therefore, not clear whether trapping at innocuous sites or loss of
hydrogen from the surface is the major contributor to the recovery of ductility
observed in commercial alloys. Nevertheless, the similarity between the rates
of recovery of 7075 from cathodic charging and 7049 from pre-exposure to sea
water (Figure 3) does seem to indicate similar mechanisms.

Effect of Potential

   Stress corrosion tests covering a range of commercial high strength aluminium alloys have been carried out at various strain rates and controlled
potentials to assess the contribution made by hydrogen to the loss of ductility.
Slow strain rates are again essential if embrittlement effects are to be

maximised[5,6], probably because of the time-dependent processes involved i.e. hydrogen absorption at strongly cathodic potentials and anodic dissolution at anodic potentials. The measured free corrosion potential is generally in the range −600 to −1000 mV (sce), but always below −800 mV (sce) for bare metal, and results from many alloy systems suggest that both hydrogen embrittlement and anodic dissolution may be involved here, particularly since the evolution of hydrogen from water at pH = 7.2 may be expected on thermodynamic grounds at potentials more cathodic than −664 mV (sce). Because it is much less aggressive than sea water, tap water has proved a most useful environment for studying potential effects.

Following the success of potentiostatically controlled tests on 7179−T651 in tap water[5] and the discovery that similar trends are discernible in 7049−T651, tests at −4V (sce), the free corrosion potential and −250 mV (sce) have been chosen to assess the susceptibility of a variety of alloys (in terms of an RA ratio) to hydrogen embrittlement and anodic dissolution. The results can be tabulated in terms of the potential conditions imparting embrittlement (Table 1).

Table 1. RA ratio for alloys tested at various potentials at 2×10^{-6}/s

Alloy	−4V	FCP (Sea Water)	−250 mV
Al−Cu 2024−T351	0.91	0.73	0.98
Al−Mg−Mn N8	0.99	0.92	0.07
Al−Mg−Si HE30	1.00	1.00	0.41
Al−Zn−Mg 7010−T73651	0.52	0.98	0.28
7017−T651	0.38	0.42	0.24
7049−T651	0.50	0.10	0.08

The utility of such tests at controlled potentials has been further demonstrated by an examination of the effect of overaging 7049−T651. The material investigated was given an additional treatment of 6h at 150°C before testing and this had a pronounced effect on its suceptibility to stress corrosion at different potentials (Figure 4). It is particularly notable that

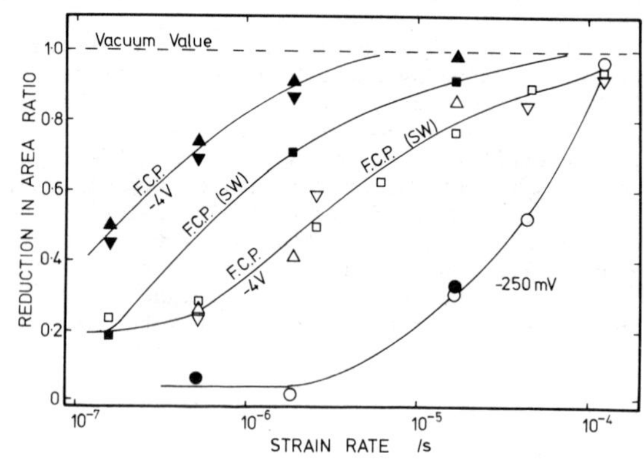

Figure 4. Comparison of the RA ratio obtained in testing 7049 in the T651 (open symbols) and overaged (closed symbols) conditions, in tap water at −4V sce (▽), the free corrosion potential (△) and −250 mV sce (○) or in sea water at the free corrosion potential (◻).

overaging has no effect upon the behaviour of the material in tests conducted at -250 mV (sce), where hydrogen is unlikely to be involved and embrittlement is due solely to anodic dissolution, but reduces the degree of embrittlement if tests are carried out at -4V (sce), where anodic dissolution is not involved but hydrogen most definitely is. These observations prompt the suggestion that the reduced embrittlement also observed at the free corrosion potential (in both tap water and sea water) is also attributable to a reduction in the susceptibility to hydrogen embrittlement. It could well be that the permeation rate for hydrogen is reduced for the overaged condition, as has been suggested by certain results for high purity alloys, and the time required for equivalent embrittlement thereby increased.

Summary of Conclusions

1.   Extensive investigation of the behaviour of commercial high strength aluminium alloys of the 7000 series confirms that hydrogen assimilated from the environment induces pre-exposure embrittlement and may also make a significant contribution to stress corrosion failures at the free corrosion potential.

2.   Extremely slow rates of strain are essential to the detection of embrittlement in the less susceptible alloys, particularly in the more resistant orientations of commercial alloys or in the less aggressive environments.

3.   Because of the embrittling role of moisture, the only safe comparisons of embrittlement employ ductility measurements in dry air or vacuum as a standard. This has, in the past, led to some confusion in the interpretation of experimental data because of the moisture content of laboratory air.

4.   A further complication is introduced by the time-dependent recovery that occurs under strain from both pre-exposure embrittlement and that introduced by cathodic charging. As a result, unless the correct environment is employed to prevent it, significant recovery may occur at the slow strain rates necessary to detect embrittlement and this causes the embrittlement-strain rate curve to pass through a spurious maximum.

5.   Slow strain rate tests at selected controlled potentials may be employed to provide a guide to the susceptibility of alloys to hydrogen embrittlement and anodic dissolution, and perhaps to indicate the important process at the free corrosion potential.

References

1.   W. Gruhl: Z.Metallkunde, 1963, vol.54, pp86-91.
2.   R.J. Gest and A.R. Troiano: L'Hydrogene dans les Metaux, pp427-432, Editions Science et Industrie, Paris, 1972.
3.   R.J. Gest and A.R. Troiano: Corrosion, 1974, vol.30, pp274-279.
4.   G.M. Scamans, R. Alani and P.R. Swann: Corros.Sci, 1976, vol.16, pp443-459.
5.   D. Hardie, N.J.H. Holroyd and R.N. Parkins: Met.Sci, 1979, vol.13, pp603-610.
6.   N.J.H. Holroyd and D. Hardie: in press, Corros.Sci.
7.   G.M. Kuln: Aluminium, 1977, vol.53, pp301 307.
8.   H. Buhl: Werkstoffe u Korrosion, 1978, vol.29, pp725-731.
9.   L. Montgrain and P.R. Swann: Hydrogen in Metals, pp575-584, ASM, 1974.
10.   M.O. Speidel: Met.Trans, 1975, vol.6A, pp631-651.
11.   H. Buhl: Stress Corrosion Cracking - the Slow Strain Rate Technique, pp333-346, ASTM-STP665, 1976.
12.   W.S. Miller and V.D. Scott: Met.Sci, 1978, vol. 12, pp95-101.
13.   S.P. Lynch: Met.Sci.J, 1973, vol.7, pp93-99.
14.   J. Albrecht, B.J. McTiernan, I.M. Bernstein and A.W. Thompson: Scripta Met, 1977, vol.11, pp893-897.
15.   M. Taheri, J. Albrecht, I.M. Bernstein and A.W. Thompson: Scripta Met, 1979, vol.13, pp871-875.
16.   L. Christodoulou and H.M. Flower: Acta Met, 1980, vol.28, pp481-487.

EFFECT OF STRESS STATE ON THE STRESS CORROSION CRACKING

OF 7075 ALUMINUM

R. E. Swanson, A. W. Thompson, I. M. Bernstein and J. L. Maloney, III
Department of Metallurgy and Materials Science
Carnegie-Mellon University
Pittsburgh, PA 15213

Introduction

The phenomenon of stress corrosion cracking (SCC) is usually regarded
as an expression of several types of events. These events can be classified into two broad classes: anodic dissolution and hydrogen embrittlement.
A variety of criteria for distinguishing between the contributions of these
two classes of events have been proposed over the years (1-3); one of the
more recent proposals is the loading mode technique (3,4). Certain alloys
which are considered prone to hydrogen-induced embrittlement have been shown
to be more susceptible to cracking in aqueous environments when loaded in
tension (Mode I) than when loaded in torsion (Mode III).

Green, et al. (4) suggested that the decreased susceptibility under
torsional loading may be understood in terms of a hydrogen embrittlement
mechanism since, for torsional loading, the hydrostatic component of stress
is greatly reduced. Since that component is believed responsible for driving hydrogen into the region ahead of the crack tip (5,6), cracking under
torsion should be assisted by hydrogen to a much smaller extent than could
occur in Mode I. If, on the other hand, a material tends to fail by anodic
processes such as film rupture or stress-assisted dissolution, then failure
should occur at similar rates in both tension and torsion, since stressed
material is exposed to the environment in both cases (4). This technique
has been applied to steels, which (particularly at higher strengths) are
generally accepted to be largely or entirely embrittled in aqueous solutions by a hydrogen mechanism. Results showed a large difference in cracking threshold between Mode I and Mode III loadings (3,7). In particular,
there was virtually no cracking sensitivity in Mode III, indicating an ab
sence of any measurable contribution to cracking from anodic processes (3,4).

This testing technique, and particularly its interpretation as a dis-
criminator between hydrogen embrittlement and anodic dissolution, has a-
roused some controversy. The usual concern is that a pure Mode III test
would be difficult to conduct. Specimen gripping and loading must be very
accurately performed to avoid a bending component in the load; moreover, the
grain boundaries, inclusions and precipitates in a real material would per-
turb the stress locally, and (in general) introduce non-Mode III components.
Even beyond those two problems, it appears (8) that a hydrostatic component
can arise under pure torsional loading for materials with particular combinations of elastic constants, i.e. anisotropy. There is, however, no dispute that a nominal Mode III loading should give rise to a greatly reduced
hydrostatic component of stress, compared to Mode I. Thus the

interpretation that hydrogen embrittlement is favored by Mode I loading relative to "Mode III" remains sound. If it did not, it would be difficult to explain the large Mode I—Mode III cracking differences for steel (3,7).

Tests of this type have also been performed on 7075 aluminum, and it was shown (4,9) that 7075 is more susceptible to cracking in aqueous solution when loaded in Mode I than in Mode III. Cracking was not absent in Mode III, however, so while these results indicate that a hydrogen embrittlement mechanism may be predominant in the SCC of 7075 aluminum, it appears that anodic processes are also involved (2,4,9,10). The alternative interpretation would be that hydrogen is also responsible for the "Mode III" cracking, for one of the reasons listed above, although we prefer at present the anodic view.

Additional testing has now been conducted to investigate the effects of microstructure found in the previous study (9). Also, straining electrode tests (SET) have been conducted to supplement the previous work, since such tests are believed to maximize hydrogen uptake in aluminum (11). Thus, the purpose of this investigation was to use the loading mode technique to compare in more detail the roles of hydrogen embrittlement and anodic dissolution in the SCC of a commercial high-strength aluminum alloy.

Experimental

Commercial, 63.5 mm thick, 7075 aluminum plate material, having a typical disk-shaped grain structure, was used in this investigation. Two specimen types were machined from the plate: a smooth tensile specimen with 5.39 mm gage diameter and 25 mm gage length, and a round specimen with 60° vee-shaped notch of ligament diameter 2.69 mm and notch root radius 0.15 mm. Tensile axes of the notched specimens were parallel to the longitudinal (rolling) direction of the plate, while smooth specimens were oriented parallel to the short transverse (thickness) direction of the plate.

All specimens were solution heat treated at $465 \pm 5^{\circ}C$ for 20 minutes in a salt bath, and then quenched into ice water. The smooth specimens were solution heat treated after mechanical polishing to eliminate the effects of any deformation produced during machining (12), while the notched specimens were solution heat treated before notching to minimize bending.

After solution treating, all specimens were ultrasonically cleaned in acetone and pre-aged at room temperature for at least 50 hours to optimize mechanical properties (13). After pre-aging, one of three artificial aging treatments was performed in an air furnace: T6, or peak strength temper, 24 hours at $120^{\circ}C$; T73, or overaged temper, 24 hours at $160^{\circ}C$; and UT, or underaged temper (12), 24 hours at $100^{\circ}C$. The microstructures produced by these heat treatments have been presented elsewhere (12).

Testing was divided into two types: constant load tests, and straining electrode tests (SET). Constant load tests included both Mode I and Mode III loading. Mode I testing was conducted on a modified constant-load, lever-arm stress rupture testing machine. Mode III testing was conducted in an Instron machine having a torsional load cell and load cycling capability. The torque force was kept within 1% of the original torque by electronically comparing the load cell signal with the original value, and automatically correcting for differences by twisting the specimen until the signals matched. This procedure is virtually the same as dead loading. All constant load tests were conducted in 3.5% NaCl solutions at $50^{\circ}C$, and time to failure was recorded. Failure was defined as fracture resulting in complete separation.

The SET tests were conducted (11) with notched longitudinal specimens in Mode I loading in pH 1 HCl solutions at room temperature. A cathodic potential of -150 mV (relative to open circuit) was applied to produce nascent hydrogen, which was available to be swept into the specimen at the notch tip. Hydrogen entry should be facilitated by the clean metal surface produced by the plastic deformation; subsequent transport of hydrogen is

expected to occur in the form of atmospheres associated with mobile dislocations (10-12,14) generated near the surface. Crosshead speed was about
10^{-4} cm/sec.

Failed specimens were ultrasonically cleaned, then examined with a
scanning electron microscope (SEM). Viewing direction was always parallel
to the tensile axis.

<u>Results</u>

Time-to-failure results for the constant load tests are presented in
Figures 1 and 2. The ordinate of each figure shows the initial applied
stress level of the specimen, in normalized form; for short transverse
specimens (Fig. 1), it is given as the ratio of the applied stress to the
0.2% yield stress for a similar specimen tested in silicone oil at 50°C.
For notched longitudinal specimens (Fig. 2), the ordinate shows the ratio of
the applied stress intensity to the failure stress intensity for a similar
specimen tested in silicone oil, again at 50°C. The stress intensity values
in Mode I were calculated from a published equation (15) for notched specimens, while a graphical method (16) was used for determining K_{III}.

Figure 1 shows that for all three tempers, significant differences
exist between Mode I and Mode III loading for short transverse specimens.
These differences were found to be statistically significant. The UT temper
appears to be superior in cracking resistance to T6 in Mode III and inferior
to T6 in Mode I loading. As discussed below, this difference in ranking of
UT material as a function of loading mode may result from hydrogen having a
greater effect in the SCC of the UT temper material.

Figure 2 shows that the stress intensity ratios for notched longitudinal specimens exhibit the same trend as the short transverse specimens for
increasing resistance to SCC (UT to T6 to T73).

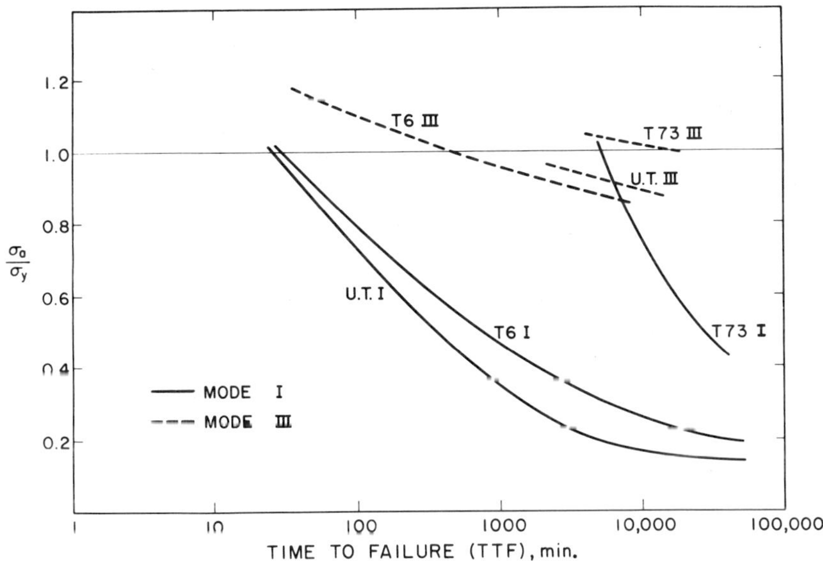

Figure 1. Time-to-Failure Results for Notched Longitudinal Specimens

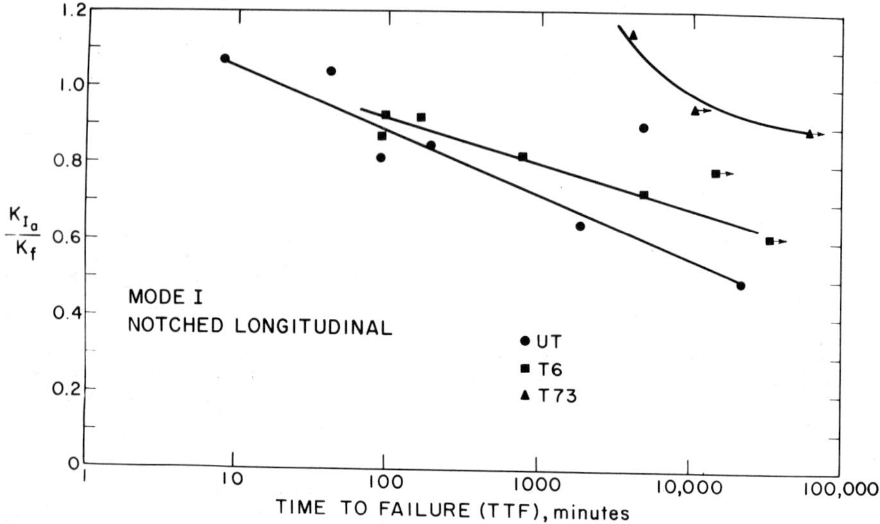

Figure 2.   Time-to-Failure Results for Short Transverse Specimens

Figures 3 and 4 show fractography of both short transverse and longitudinal specimens tested in Mode I. Each specimen had a flat, featureless region near the circumference of the fracture surface. The rest of the surface had an appearance which suggested a ductile overload region comprising dimples or microvoids. The surfaces were similar for short transverse and longitudinal specimens, and all showed only slight amounts of secondary cracking. Specimens tested in torsion had similar fracture surfaces, except that the dimples in the torsion specimens were typically elongated, i.e. were shear dimples.

The SET Mode I tests for notched longitudinal specimens produced fracture surfaces similar to those of the constant load tests, except that the SET specimens had somewhat wider (300 µm vs. 200 µm) flat regions and more secondary cracking. SET failure stress intensity values were similar to those for specimens tested in air for both the T6 and T73 tempers, as shown in Table I. The failure stress intensity values for the UT temper, however, were

Table I.   Failure Stress Intensity Values (MPa$\sqrt{m}$) for Longitudinal

Specimens

	Constant Load (air)	SET (pH 1 HCl)
UT	24.3	16.6
T6	25.2	25.3
T73	23.1	23.1

reduced about 32% in the SET test. This suggests more susceptibility to hydrogen embrittlement in the UT temper than in either the T6 or T73 tempers, as has been noted before (11,12). Since such susceptibility is sensitive to strain rate (17), this result indicates that the effective notch root strain rate was less than $1 \times 10^{-2}\mathrm{s}^{-1}$. Moreover, because the UT temper was exposed in the SET process to a similar amount of available, total hydrogen, it can be inferred that a lower amount of hydrogen was sufficient to cause embrittlement of the UT temper material.

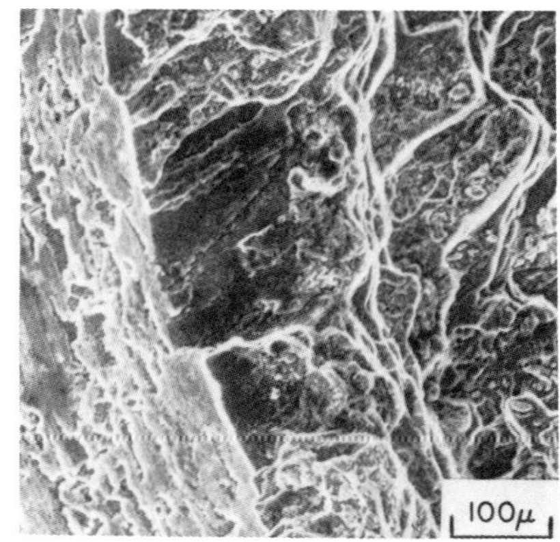

Underaged Temper

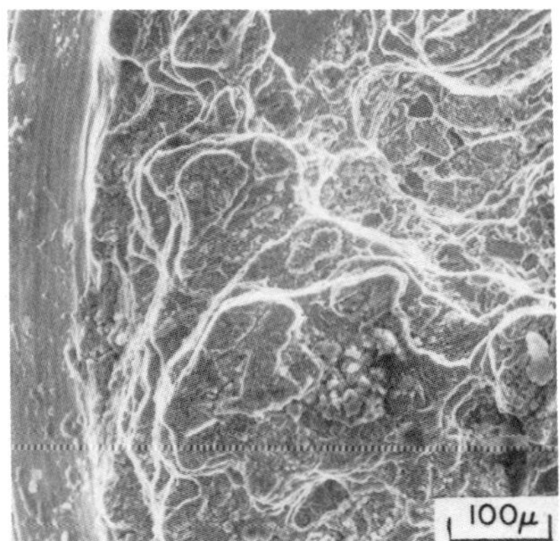

T6 Temper

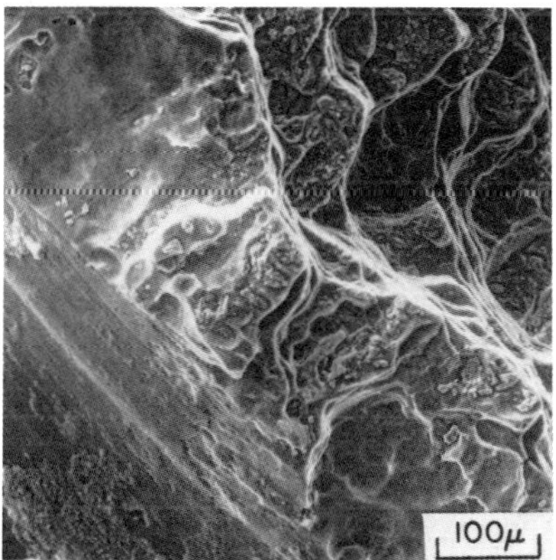

T73 Temper

SHORT TRANSVERSE TENSILE SPECIMENS

3.5% NaCl Solution, 50°C

Figure 3.

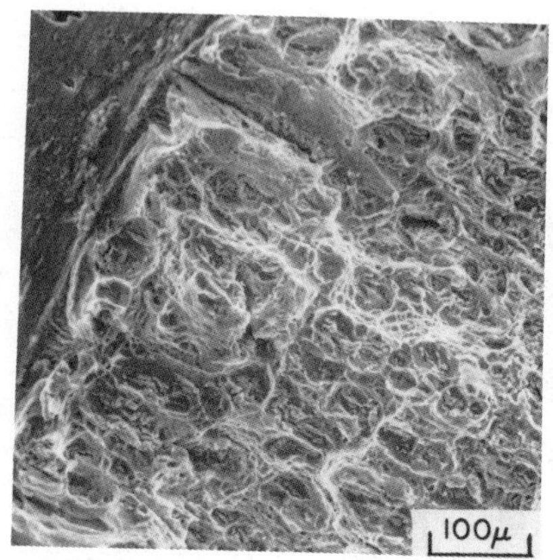

Underaged Temper

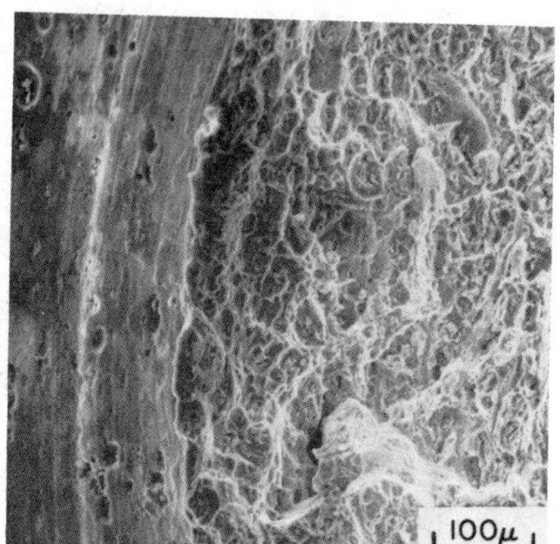

T6 Temper

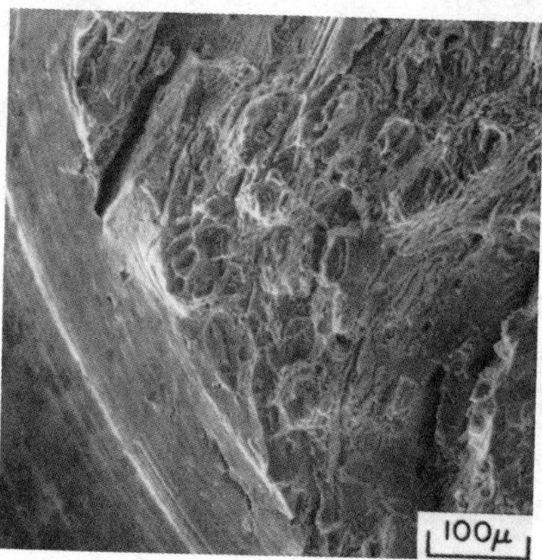

T73 Temper

LONGITUDINAL TENSILE SPECIMENS
3.5% NaCl Solution, 50°C

Figure 4.

Discussion and Conclusions

The SCC behavior of 7075 aluminum in commercial hot rolled plate form has been shown to be affected by the mode of loading. Time-to-failure curves showed that specimens loaded to comparable stress levels failed at times which were as much as several orders of magnitude sooner in Mode I than in Mode III. This effect is present for all three tempers, including the commercially important, "environmentally-resistant" T73 temper.

A second important point is that the extent of the loading mode effect depends on microstructure. Increased tempering, in the sequence from UT to T6 to T73, produced increasing resistance to cracking in Mode I (and probably in Mode III, although the results are less clear-cut). Interpreting this behavior as an effect of hydrogen, i.e. as a manifestation of a predominant hydrogen contribution to SCC in the environment used, would be consistent with earlier evidence that hydrogen embrittlement susceptibility alone, without any "SCC process", is also decreased by the added tempering from UT to T6 to T73 (9,11,12,17). The precise extent of the SCC sensitivity should not, however, be correlated with the hydrogen sensitivity alone (for each microstructure) unless the anodic contribution to SCC is either independent of microstructure, or accurately parallels the microstructure dependence of hydrogen embrittlement. Neither of these two possibilities is particularly likely, because both the size and composition of precipitates in 7075, and also the local matrix composition near grain boundaries, vary widely between UT, T6 and T73; anodic processes should be sensitive to these parameters, and almost certainly in ways which differ from the sensitivity to hydrogen embrittlement. The UT temper ranking relative to the T6 temper in Fig. 1 probably reflects this point, comparing the two tempers in Modes I and III.

The fact that a nominal Mode III loading of a real material is likely to produce a small hydrostatic component of stress near the crack tip was mentioned in the Introduction. This fact, however, should not obscure the main issue: that the Mode III hydrostatic component is greatly reduced (whether or not it is zero) relative to Mode I. Consequently, there appears to be no error in evaluating a difference in SCC sensitivity between Modes I and III as reflecting a _relative_ contribution of hydrogen embrittlement to SCC. It should be noted that the extent of this relative contribution is minimized, because the small hydrostatic stress in the real Mode III test presumably assists a correspondingly small hydrogen contribution to SCC. If a pure Mode III test could be conducted, it follows that the difference between Modes I and III would increase. Fig. 1 of the present work thus indicates a minimum or conservative extent of the contribution of hydrogen embrittlement to SCC of 7075 aluminum in the UT, T6 and T73 tempers.

The SET tests produced slightly more severe cracking, as evidenced by fractography, than the constant load tests, although fracture stress intensity values for the T6 and T73 tempers were virtually unchanged from the all test results. The UT temper, however, exhibited a reduction in fracture stress intensity. This coincides with the constant load test result, that the UT temper is more susceptible to a hydrogen effect than either the T6 or T73 temper. Additional tests will be conducted at lower strain rates to verify this point.

Acknowledgments

We appreciate the support of this work by the U. S. Air Force Office of Scientific Research under Grant No. F49620-77-C-0003, and the encouragement of A. H. Rosenstein.

References

1. R. M. Latanision, O. H. Gastine, and C. R. Compeau, *Environment-Sensitive Fracture of Engineering Materials* (Z. A. Foroulis, ed.), pp 48-70, TMS-AIME, Warrendale, PA, 1979.

2. A. W. Thompson, *Mater. Sci. Eng.*, 1980, vol. 43, pp 41-46.

3. H. W. Hayden and S. Floreen, *Corrosion*, 1971, vol. 27, pp 429-434.

4. J. A. S. Green, H. W. Hayden, and W. S. Montague, *Effect of Hydrogen on Behavior of Materials* (A. W. Thompson and I. M. Bernstein, eds.), pp 200-217, TMS-AIME, New York, NY, 1976.

5. A. R. Troiano, *Trans. ASM*, 1960, vol. 52, pp 54-80.

6. J. C. M. Li, R. A. Oriani and L. S. Darken, *Z.f. Phys. Chem.*, 1966, vol. 49, pp 271-90.

7. C. St. John and W. W. Gerberich, *Met. Trans.*, 1973, vol. 4, pp 589-594.

8. Y. T. Chou and R. P. Wei, Lehigh University, Personal Communication, 1980.

9. J. L. Maloney, "The Effect of Microstructure and Loading Mode on the Stress Corrosion Cracking of 7075 Aluminum", M. S. Report, Carnegie-Mellon University, 1979.

10. A. W. Thompson and I. M. Bernstein, *Corrosion Science and Technology* (M. G. Fontana and R. W. Staehle, eds.), vol. 7, pp 53-175, Plenum, New York, 1980.

11. J. Albrecht, I. M. Bernstein and A. W. Thompson, in preparation.

12. J. Albrecht, A. W. Thompson and I. M. Bernstein, *Met. Trans. A*, 1979, vol. 10A, pp 1759-66.

13. H. Y. Hunsicker, *Aluminum*, vol. 1, p 154, ASM, Metals Park, OH, 1967.

14. J. K. Tien, A. W. Thompson, I. M. Bernstein and R. J. Richards, *Met. Trans. A*, 1976, vol. 7A, pp 821-29.

15. W. F. Brown and J. E. Srawley, *Plane Strain Crack Toughness Testing of High Strength Materials* (STP 410), pp 15-16, A.S.T.M., Philadelphia, 1966.

16. H. Tada, P. Paris and G. Irwin, *The Stress Analysis of Cracks Handbook*, Del Research Corp., Hellertown, PA, 1973.

17. M. Taheri, J. Albrecht, I. M. Bernstein and A. W. Thompson, *Scripta Met.* 1979, vol. 13, pp 871-75.

HYDROGEN INDUCED FRACTURE OF ALUMINIUM ALLOYS

G.M. Scamans

Alcan Laboratories Limited
Southam Road, Banbury, Oxon, England, OX16 7SP

Introduction

Certain important groups of aluminium alloys, notably those based on
AlZnMg, AlZnMgCu, AlMg, AlCuMg and AlCu are susceptible to failure by stress-
corrosion cracking.   Of these groups of alloys AlZnMg, AlZnMgCu and AlMg
suffer a reversible embrittlement by hydrogen (1-3) whereas AlCu and AlCuMg
are resistant to hydrogen introduced either by cathodic charging or by
unstressed exposure to a water-containing environment (pre-exposure
embrittlement) (4).   This may indicate a critical difference in failure
mechanism for the two groups of alloys under conditions likely to be
experienced in service.

Significantly, recent high resolution fractography of stress-corrosion
fracture surfaces of AlZnMg alloys has demonstrated that failure occurs
discontinuously and may be interpreted as repeated pre-exposure embrittlement (5).   Precisely matched striations which are contours of crack arrest
are observed on both fracture surfaces and are formed from the corrosion
product of the hydrogen generation reaction (6).   Such matched arrest
markings are observed both under environmental conditions of total immersion
in chloride containing electrolytes and in water vapour and are apparent
both on test specimens and on service failures (7).   A typical example of a
striated stress-corrosion fracture surface is shown in Figure 1.   Fracture
essentially occurs by hydrogen-induced grain boundary decohesion once all
available sites for hydrogen trapping are saturated.   Trapping of the
embrittling atomic species may be achieved either by discharge to form
bubbles of molecular hydrogen (8)(9)(10) or as has been recently proposed (11)
by hydride formation at specific sites which have been identified, in the
case of bubbles, as grain boundary precipitates, intermetallic dispersoids
or dislocation nodes.   Increased resistance to stress-corrosion cracking
results in an increased propensity for hydrogen trapping and a decrease in
the permeation rate of hydrogen through unstressed membranes (12).   The
beneficial effects of slow quenching, overageing and low level additions of
copper may be understood by this mechanism.   It has also been established
that hydrogen entry at grain boundary/surface intersection (pre-exposure
embrittlement and crack initiation) and along grain boundaries at crack tips
is controlled by the level of grain boundary magnesium segregation (13)(14).
High temperature solution heat treatment, slow quenching and overageing
may reduce the level of segregated magnesium and hence may reduce the hydrogen

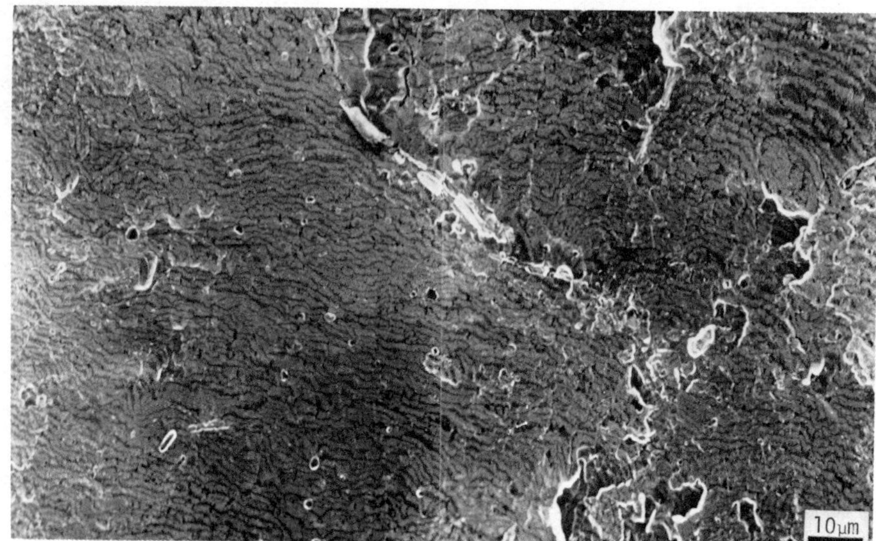

Fig. 1 - Striated stress-corrosion fracture surface of a DCB specimen of a
CWQ 7017 T651 alloy exposed to 95% RH at 40°C.   (K = 7.9 MPa m$^{\frac{1}{2}}$,
v = 2.0 x 10^{-2} mm/h crack direction (←).)
Note the extreme irregularity of the crack front.

entry rate. Alternatively, preferential hydrogen discharge may be achieved
away from the critical grain boundary region by additions of chromium (10).

Since both hydrogen entry and trapping phenomena are now well documented
it is the purpose of the present paper to examine their influence directly
on the propagation of stress-corrosion cracks by a detailed examination of
striated fracture surfaces using the AlZnMg alloy family as a model system
for the alloy groups susceptible to hydrogen embrittlement.

<u>Experimental</u>

Standard bolt-loaded double cantilever beam (DCB) (15) specimens were
exposed to water vapour (95% Relative Humidity RH) at 40°C. Test pieces
were stressed in the short transverse direction (S) with crack growth in the
longitudinal direction (L), i.e. parallel to the rolling plane and in the
rolling direction.   Initial crack lengths were of the order of 30 mm after
pop-in and tests were continued until crack velocities had dropped to below
5 x 10^{-3} mm/h. After testing the stress-corrosion fracture surfaces were
separated by mechanical overload and then carefully sectioned to fit the
specimen stage of the JEOL 100C Temscan electron microscope.   Regions of
known stress intensity and crack velocity could then be examined directly
with a resolution of better than 50Å in the secondary electron imaging (SEI)
mode. The composition limits of the alloys used are detailed in Table I.

<u>Table I.   Composition Limits of 7017 and 7179 Alloy Plate</u>

Alloy	Si	Fe	Cu	Mn	Mg	Cr	Zn	Ni	Zr	Ti	Others Each	Total
7017	0.25	0.45	0.20	0.05-0.50	2.0-3.0	0.35	4.0-5.2	0.10	0.10-0.25	0.15	0.05	0.15
7179	0.15	0.20	0.40-0.8	0.10-0.30	2.9-3.7	0.10-0.25	3.8-4.8	-	-	0.10	0.05	0.15

Results

Effect of Stress Intensity on Striation Spacing

Crack velocity versus stress intensity curves (v-K curves) for the DCB specimens are shown in Figure 2(a) and (b).   Cracking in all cases followed the familiar pattern observed with many alloy systems (16) of an initial period of rapid crack growth at high stress intensities (although a Region II plateau could not be strictly identified) followed by a rounding-off of the curve at intermediate stress intensities and finally a further deceleration at low stress intensities until a threshold stress intensity (K_{ISCC}) could be determined by extrapolation.

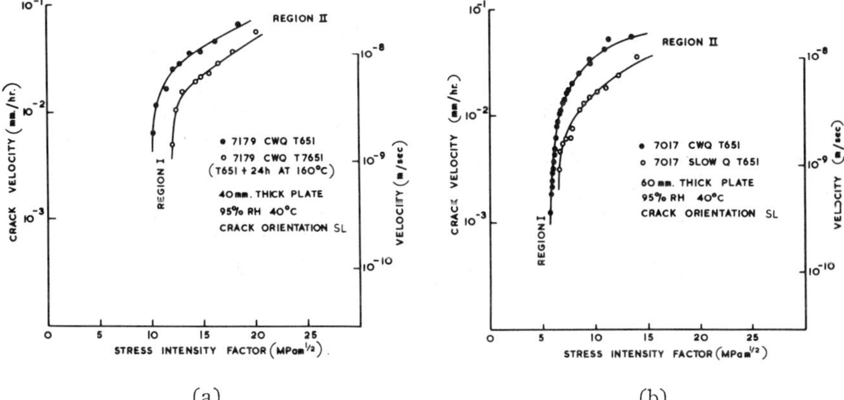

(a)   (b)

Fig. 2 - Stress-corrosion crack velocity versus crack-tip stress intensity curves for (a) 7179 and (b) 7017 alloys.

Regions of the fracture surface corresponding to various positions on the v-K curve for the 7179 T651 alloy are shown in the series of micrographs in Figure 3 (a)-(d).   Figure 3(a) shows the transition from the ductile "pop-in" crack to the striated stress-corrosion failure.   Stress-corrosion cracks nucleate at many points along the ductile failure front and then rapidly radiate and coalesce to form a broad crack front within the first few microns of crack growth.   Since the very first crack jumps can be observed it should therefore be possible to study the effect of different starting cracks, e.g. fatigue or machined notch etc., on crack initiation. Even at the highest stress intensity the crack is essentially brittle. The particles on the fracture surface in each of the micrographs in Figure 3 are grain boundary $MgZn_2$ precipitates formed during the cold water quenching of the 60 mm plate.   Clearly discontinuous cracking often at an angle to the main direction of propagation occurs over the whole range of applied stress intensity factors. Figures 3(b) and (c) show that at similar stress intensities cracks can propagate both independently of microstructure (3(b)) or with distinct interaction and microstructural pinning (3(c)). Striations were observed on all areas of the fracture surface although in some cases they were partially obscured by post-fracture corrosion.   In certain areas striations were continuous and parallel across several grains indicating that a single crack advance had occurred along a wide front of up to several hundred microns.   In other areas cracks had obviously only advanced very locally particularly where the grain boundary microstructure had a strong influence.

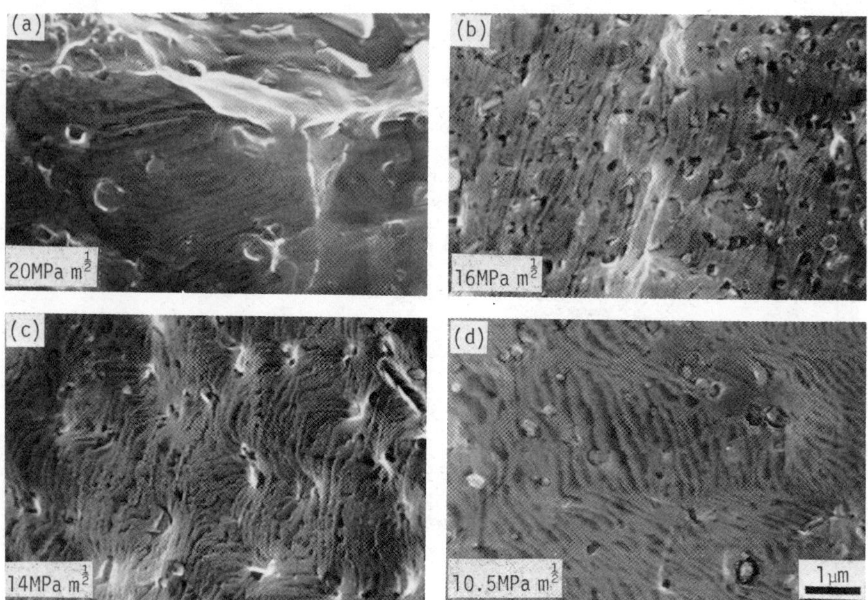

Fig. 3 - Effect of stress intensity on striation morphology for alloy 7179 T651 (a) $v = 8 \times 10^{-2}$ mm/h (b) $v = 4 \times 10^{-2}$ mm/h (c) $v = 3.5 \times 10^{-2}$ mm/h (d) $v = 1.1 \times 10^{-2}$ mm/h; crack direction (↓).

A detailed example of this interaction is shown in the stereo-pair of micrographs in Figure 4 of the 7017 T651 CWQ alloy in this case from a region of low stress intensity.   The crack front has degenerated from propagation on a broad front to short local cusps of advance.

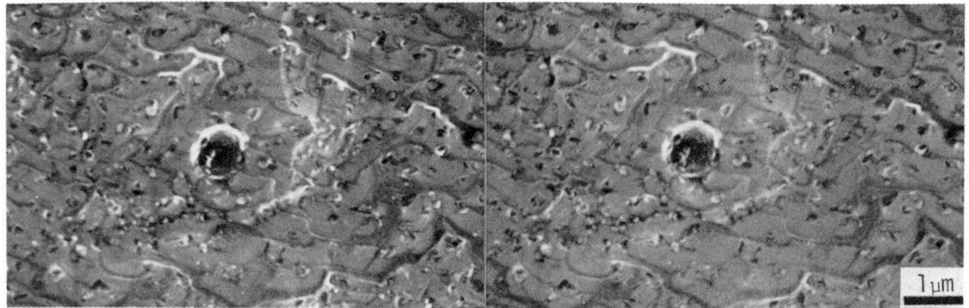

Fig. 4 - Stereo-pair of striation pinning in alloy 7017 T651 CWQ (K = 5.7 MPa m$^{\frac{1}{2}}$, $v = 1.3 \times 10^{-3}$ mm/h, crack direction (↓)).

The grain boundary surface shown in Figure 4 is extremely irregular and often cracks have been arrested at grain boundary steps or ridges. Stereo-imaging is clearly essential to interpret the convolutions of the crack front in such regions.   There is no evidence of ductile dimple formation in the region of fracture between striations.   The fine precipitates which are just visible (250Å) are $MgZn_2$ developed during T651 ageing.

The results of measurements of average striation spacings on micrographs taken from different regions of the stress-corrosion failures are shown in Figure 5. For both the alloys in the CWQ T651 condition the average striation spacing as measured from striation centre to striation centre is not a strong function of the applied stress intensity factor and average values of crack jump of 1350Å (7179) and 3700Å (7017) could be determined although the data for the 7017 alloy showed consider-able scatter. The v-K curves shown in Figure 2 can therefore be considered as plots of crack jump rate versus stress intensity. For example in Figure 2(a) the crack jump rate falls from an initial value of 500 jumps/hour down to 50 jumps/hour, i.e. crack arrests increase from 7.2 seconds up to 72 seconds.

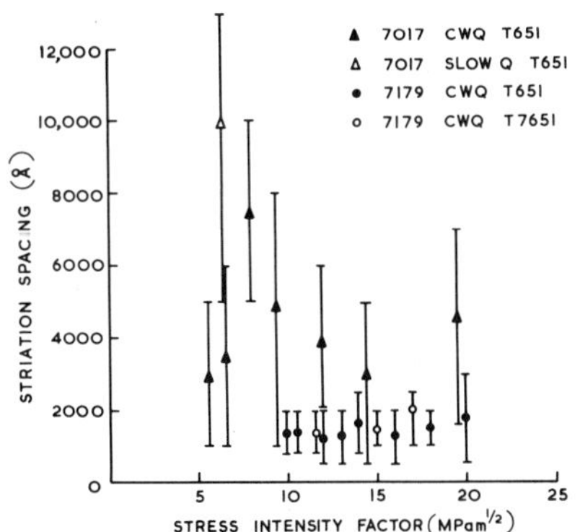

Fig. 5 - Effect of stress intensity on average striation spacing (average values are 1350Å for 7179 and 3700Å for 7017).

Effect of Overageing

Figure 6 shows the effect of 24 hours at 160°C (T7651) on the striation morphology of the 7179 T651 alloy. Note that the only difference between Figure 6(a) and 6(b) is an increase in the fine grain boundary precipitate ($MgZn_2$) size. This has increased from an average value of 250Å up to 500Å. The particle in the top right hand corner of Figure 6(b) has the characteristic shape of a Cr-rich dispersoid particle. The average striation spacing is not affected (see Figure 5) over the entire v-K curve. Figure 2(a) can hence be interpreted as showing that the 24-h treatment at 160°C has effectively halved the crack jump rate for any particular value of K at high and intermediate stress intensities.

Effect of Quench Rate

A comparison of cold water quenched and slow furnace cooled 7017 T651 plate is shown in Figure 7(a)(b). Note that the slow cooling has increased the size range of the grain boundary $MgZn_2$ precipitate from 500-1500Å (CWQ) to 1000-5000Å (slow Q). Figure 5 shows that the average striation spacing may in fact be increased by this extreme practice although further measure-ments are required due to the wide range of scatter associated with striation measurements in this alloy. Quench rate reduction has its major effect at high stress intensities rather than on K_{ISCC} as is shown in Figure 2(b) where again in this region the crack jump rate has been approximately halved. During the periods of crack arrest delineated by the striations in Figure 6(b) significant oxide film growth has occurred and there is some evidence of water vapour penetration and subsequent hydroxide precipitation ahead of the crack front. The hydroxide has the characteristic morphology of pseudoboehmite (14).

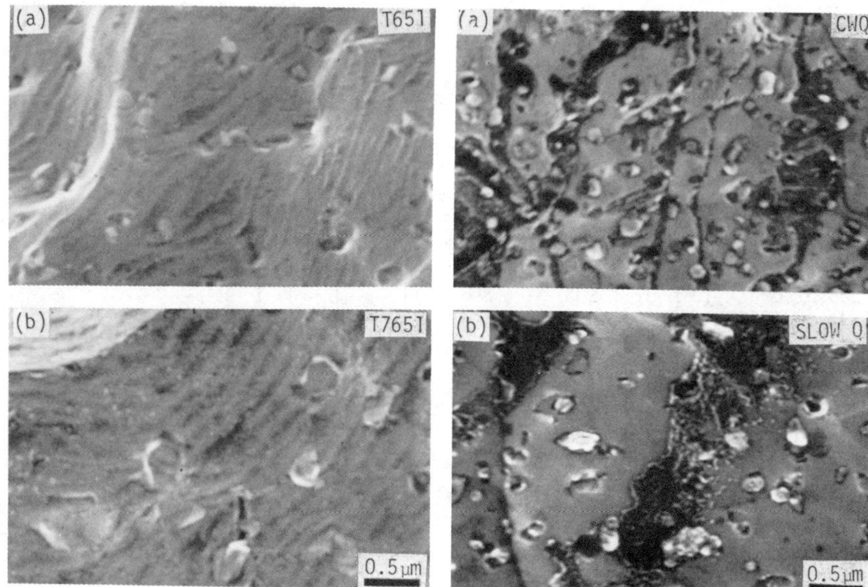

Fig. 6 - Effect of overageing on striation morphology (a) 7179 T651, K = 20 MPa m², v = 8 x 10⁻² mm/h (b) 7179 T7651, K = 12 MPa m², v = 5.6 x 10⁻³ mm/h, crack direction (↓).

Fig. 7 - Effect of quench rate on the striation morphology (a) 7017 CWQ, K = 7.9 MPa m², v = 2.0 x 10⁻² mm/h (b) 7017 slow Q, K = 6.3 MPa m², v = 3.3 x 10⁻³ mm/h, crack direction (↓).

A high magnification examination of a striation characteristic of slow crack growth for the slowly cooled alloy is shown in the stereo-pair of micrographs in Figure 8. In this case the oxide film has adhered to the opposite fracture face leaving the crystallographic etching which has occurred during crack arrest clearly visible. Note that at several points along the crack front dissolution has apparently penetrated ahead of the crack. Such features were only observed in regions characteristic of very slow crack growth. Note the very low density of grain boundary ageing precipitate caused by excessive solute depletion of zinc by growth of coarse $MgZn_2$ particles during the slow quench.

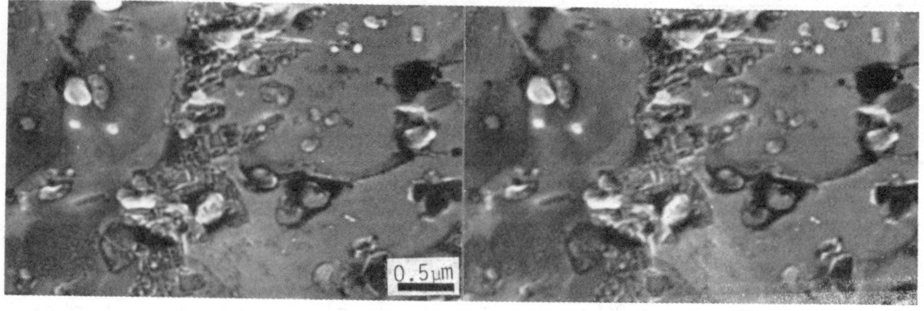

Fig. 8 - Stereo-pair of striation detail in alloy 7017 T651 slow Q (K = 6.3 MPa m², v = 3.3 x 10⁻³ mm/h, crack direction (↓)).

Discussion

The effect of stress intensity on stress-corrosion cracking of AlZnMg alloys by the hydrogen embrittlement mechanism is clearly to increase the rate of crack jumping rather than to alter the magnitude of the crack advance. If the number of available trap sites or local getters for the diffusing embrittling species is fixed by the alloy composition and its thermal history then the operating stress across the grain boundary must affect the grain boundary diffusion rate such that all available trap sites are saturated more rapidly and brittle grain boundary failure can be induced by any further accumulation of atomic hydrogen. This idea is supported by the results of Gruhl et al (17) which demonstrate that the kinetics of pre-exposure embrittlement are accelerated by applied stress.   In later work however the effect of stress is interpreted as progressively increasing the grain boundary hydrogen solubility enabling only boundaries with a perpendicular tensile stress component to absorb sufficient hydrogen to become embrittled (18). The precise mechanism of this decohesion is not understood although an interaction between planes of segregated magnesium atoms and the atomic hydrogen seems the most likely explanation. Whether this can be described as a chemisorbed monolayer of hydrogen on magnesium or as magnesium hydride will require further careful experimentation to establish.   The results also suggest that once a specific concentration of atomic hydrogen occurs on the grain boundary plane ahead of the crack then the grain boundary fracture strength is suddenly and drastically reduced probably to the level of the threshold stress intensity.

Overageing and slow quenching presumably effectively reduce the level of segregated magnesium at grain boundaries by growth of large $MgZn_2$ precipitates.   This may result both in a reduced grain boundary diffusion rate for atomic hydrogen and/or an increase in the number of trap sites either of which could explain the observed increase the time interval between bursts of crack growth.

For crack growth rates of the order of 10^{-1} mm/h crack jumps must occur every six seconds if each jump is of the order of 1350Å (i.e. for 7179 T651). This requires a grain boundary hydrogen diffusivity of 3×10^{-9} cm^2/s. Furthermore since cracking in this alloy can be a factor of 10^3 faster than this in aqueous chloride, bromide or iodide solutions (16) assuming that the jump distance is constant then a grain boundary diffusivity of 3×10^{-6} cm^2/s is required.   Hardie et al (2) have estimated a diffusion coefficient of 10^{-7} cm^2/s for 7179 T651 from their stressed recovery experiments in vacuum after charging with hydrogen by pre-exposure to various aqueous environments. Measured values from permeation experiments with unstressed membranes have been 2×10^{-9} cm^2/s (19) and 10^{-10} cm^2/s (12).   This again indicates that stress can have a significant effect on hydrogen diffusivity and that values as high as 10^{-6} cm^2/s are by no means unreasonable.

It would now be very informative to monitor acoustic emission variation during similar stress-corrosion cracking tests particularly where cracking is occurring on a wide front.   Where such data already exists cracking has been reported to be discontinuous and to increase in frequency with increasing stress intensity in agreement with the present results (20). At very high stress intensities and high growth rates cracks were reported to occur at intervals of 0.1 to 0.5 seconds with a calculated jump distance of 3300Å (7075 T6); however these were interpreted as ductile rupture rather than brittle fracture events (20).

It is important that this work should now be continued to include a range of AlZnMg, AlZnMgCu and AlMg alloy compositions in different conditions of stress-corrosion susceptibility.   Particularly it is necessary to include a wider range of environments and to study the effect of anodic and cathodic polarisation.   The role of anodic potential in accelerating cracking may perhaps be understood by a progressive increase in penetration of corrosion ahead of the crack tip with increased anodic potential until this equals the rate of hydrogen penetration and cracking eventually becomes wholly by dissolution.   The reduction in cracking rates on application of cathodic polarisation suggests that hydrogen discharge from the alloy surface is encouraged rather than the stimulation of its entry at crack tips.

Conclusions

Stress-corrosion cracking in AlZnMg alloys and by inference in AlZnMgCu and AlMg alloys occurs discontinuously by hydrogen embrittlement. For AlZnMg alloys the magnitude of each crack advance depends on alloy composition but is insensitive to the applied stress intensity and to heat treatment variables which modify susceptibility, e.g. overageing and slow quenching, which rather have their effect on the time interval between each burst of wholly brittle intergranular failure.

References

1. G.M. Scamans, R. Alani and P.R. Swann, Corros. Sci., 1976, Vol.16, pp.443-459.

2. D. Hardie and N.J. Holroyd, Metal Sci., 1979, Vol.13, pp.603-610.

3. T. Ohnishi, K. Higashi, N. Inoue and Y. Nakatani, J. Jap. Inst., 1980, Vol.30, pp.263-270.

4. N.J. Holroyd, Unpublished research, Newcastle University, England, 1980.

5. G.M. Scamans, Scr. Met., 1979, Vol.13, pp.245-50.

6. G.M. Scamans, Met. Trans. A., 1980, Vol.11, pp.846-850.

7. G.M. Scamans, Unpublished research, Alcan Laboratories Limited, Banbury, England, 1979.

8. R. Alani and P.R. Swann, Br. Corros. J., 1977, Vol.12, pp.80-85.

9. G.M. Scamans, J. Mat. Sci., 1978, Vol.13, pp.27-36.

10. L. Christodoulou and H.M. Flower, Acta Met., 1980, Vol.28, pp.481-487.

11. C.D.S. Tuck, Unpublished research, Alcan Laboratories Limited, Banbury, England, 1979.

12. G.M. Scamans and C.D.S. Tuck, "Environment-Sensitive Fracture of Engineering Materials," paper presented at TMS-AIME Fall Meeting, Chicago, Illinois, 1977, pp.464-483.

13. J.M. Chen, T.S. Sun, R.K. Viswanadham and J.A.S. Green, Met. Trans. A., 1977, Vol.8, pp.1935-1940.

14. G.M. Scamans and A.S. Rehal, J. Mat. Sci., 1979, Vol.14, pp.2459-2470.

15. M.O. Spiedel, "The theory of stress corrosion cracking in alloys" ed. J.C. Scully, NATO Brussels, 1971, pp.289-344.

16. M.O. Spiedel, Hydrogen in Metals, ASM, Metals Park, Ohio, 1974, p.249.

17. W. Gruhl and D. Brungs, Metall., 1969, Vol.23, pp.1020-1026.

18. L. Ratke and W. Gruhl, Metall., 1979, Vol.33, pp.644-648.

19. R. Gest and A.R. Troiano, Corrosion, 1974, Vol.30, pp.274-279.

20. W.E. Wood and W.W. Gerberich, Met. Trans., 1974, Vol.5, pp.1285-1304.

<u>DISCUSSION</u>

J. A. Moskovitz, Kaiser Aluminum: If the striations are due to oxide formation, why is their thickness not affected by crack opening displacement, or
by the time between jumps?

G. M. Scamans: If a constant amount of hydrogen is assumed to be responsible
for each crack advance, i.e. when all available trap sites have been saturated and atomic hydrogen is then available to interact with segregated Mg
to embrittle the boundary, this would be supplied by a constant amount of
crack tip reaction which is visualized as the striations on the fracture
surface. This assumes that the ratio of hydrogen absorbed/hydrogen evolved
is not affected by the COD. A wider range of stress intensities must now be
studied. If the highest crack velocities are to be explained by stress
assisted grain boundary hydrogen diffusion it is possible that grain boundary
dislocation transport may be required.

P. N. Adler, Grumman Aerospace: I recall that others mechanisms have reported
discontinuous SCC in Al Zn Mg alloys and have proposed models to explain this
behavior, specifically, a film-rupture model.

G. M. Scamans: None of the models would explain the wholey brittle (in appearance) crack jumps observed in the present work. The fracture after rupture
of a brittle film would, I believe, proceed by ductile IG failure with evidence
of dimple formation. This would be a very localized plastic failure but would
in no way provide perfectly matching fracture faces.

E. N. Pugh, National Bureau of Standards: I have a comment rather than a
question. (i) We have detected discontinuous acoustic emission during intergranular SCC in our Al-5.6Zn-2.6Mg alloy (J. L. Nelson, Ph.D. Thesis, University
of Illinois, 1976), but found no evidence for crack-arrest markings of the type
shown in this paper. (ii) It is surprising that you observed no dependence of
the crack-advance distance on K in Stage I. In other cases where the crack advance distance has been determined, it increases with increasing K in Stage I
and attains a constant value in Stage II (D. Beggs, M. Hahn and E. N. Pugh,
A. R. Troiano Conference, June 1980.)

THE ROLE OF HYDROGEN ON ENVIRONMENTAL FATIGUE OF HIGH

STRENGTH ALUMINUM ALLOYS*

R. J. Jacko and D. J. Duquette
Rensselaer Polytechnic Institute
Troy, New York 12181

An extensive series of experiments have been performed on a 7075
aluminum alloy and its high purity analogue, Al5Zn2.5Mg, in aqueous environments, to determine the role of cathodically evolved hydrogen on
fatigue crack initiation and early propagation. Variables in the program
include the effects of applied potential, hydrogen damage reversibility,
solution pH and mechanical loading modes. Results obtained in the program refute the concept that anionic species need be specific to cause
enhanced susceptibility to fatigue cracking, but that hydrogen, both in
the presence or absence of significant corrosion, is indeed a potent
species for the degradation of fatigue properties. Evidence will be
presented that hydrogen, presumed to be associated with the plastic zone
at crack tips causes increased crack propagation rates and also results
in significant changes in fracture surface morphologies. Evidence will
also be described which strongly suggests that the ingress of hydrogen
may be related to the structure and properties of the oxide film(s)
which are formed under given environmental circumstances.

* This work was supported by the Office of Naval Research under
   Contract #N00014-75-C-0466.

Introduction

The role of hydrogen in embrittlement of high strength Aluminum alloys, particularly those which are hardened by the $MgZn_2$ precipitate, has been a subject of controversy for some time. For monotonic loading cases the evidence of hydrogen embrittlement is still being debated, particularly since it is often difficult to unambiquously resolve the differences between what might be dissolution controlled stress corrosion cracking, and true hydrogen embrittlement. Under cyclic loading conditions however, many of the effects of environmental fracture may be accentuated and, assuming that a strong link can be forged between monotonic and cyclic loading, a considerable amount of information may be gained from an analysis of fatigue results in the Al alloys. The results discussed herein are a brief summary of a large number of experimental variables which have been examined to attempt to understand environmental degradation of the fatigue properties of high strength aluminum alloys. This environmental degradation is similar to that generally reported for monotonic loading, however there are several significant differences which are generally observed. Principally these differences include the fact that, for commercial alloys such as 7075 and 7050, in the peak hardened condition, fatigue cracks are transgranular. In some cases stage I cracking is observed, and in other cases only stage II cracking is observed. Also, because of the transgranular nature of cracking, coupled with sufficiently low crack growth rates when stage I cracking occurs, the crystallographic crack path has been identified as following either {100} or {112} planes in contrast to the {111} crack propagation plane observed in dry air. These observations will be discussed in more detail in this presentation.

Experimental

The majority of the results discussed herein were obtained on a commercial extruded 7075 Al alloy, heat treated to the T6 condition and stressed in the longitudinal direction. In cases where it is relevant, the results of some experiments on a high purity Al5Zn2Mg alloy will be discussed. All experiments were performed on electropolished specimens under load control at 30 Hz. Environments included dry and moist air, aerated and deaerated 0.5 $\underline{N}$ NaCl, 0.5$\underline{N}$ Na_2SO_4 and 0.5 $\underline{N}$ H_2SO_4. Anodic and cathodic polarization experiments were also performed. Additionally some specimens were pre-exposed to either NaCl or H_2SO_4 and subsequently tested for fatigue damage reversibility.

Results and Discussion

Sensitivity to Water Vapor

Under monotonic loading conditions moist air has been suspected to embrittling high purity Al alloys, although there is little evidence of such embrittlement in commercial alloys. Limited ductility of high purity ternary alloys which exhibit intergranular failure has been reported (1). The effect of water vapor is particularly important under cyclic deformation conditions, and as early as 1960, the decreased fatigue resistance of dur-**aluminum in moist air was associated with hydrogen evolved from the hydrolysis** of water in growing cracks (2). Several workers have since shown that water vapor, in concentrations less than could produce capillary condensation in growing cracks, is a potent enhancer of fatigue crack propagation rates (3-5).

Wei and co-workers, in particular, have demonstrated not only that water vapor is a damaging species but that surface reactions producing hydrogen may be the rate determining step for crack growth although the observed embrittlement is a bulk phenomenon (as contrasted to surface energy reduction due to adsorption of water vapor). There is no model for dissolution controlled stress corrosion cracking which can explain the results of water vapor on fatigue crack initiation or propagation.

Crack Path and Crystallography

Most studies of environmentally assisted crack propagation in high strength Al alloys have been performed in the crack propagation regime where stage II cracking dominates, and little can be said about the specific crystallographic morphology of crack propagation. Interestingly, however, some studies of very early crack propagation in water vapor in single crystals of AlZnMg alloys have shown that the surface topographies of stage I cracks are significantly different (6,7). Additionally these studies have shown that a shift in crack propagation path from stage I to stage II with decreasing frequency occurs only in water vapor. These results suggest that lower frequencies allow increased bulk diffusion to regions ahead of the crack tip where hydrogen presumably exerts its damaging effect. Under more severe conditions such as exposure to NaCl solutions or to other solutions under cathodic changing conditions, initial crack propagation in commercial alloys occurs in an essentially "featureless" zone which is approximately normal to the applied stress direction and appears to be noncrystallographic (Fig. 1). The extent of this region appears to be relatively independent of charging conditions and cracking appears to occur discontinuously after a considerable number of cycles (8). The featureless region is converted to a quasi-cleavage region and finally to stage II crack propagation. In high purity analogues with equiaxed grain structures only the quasi-cleavage region is observed, although at low applied stresses

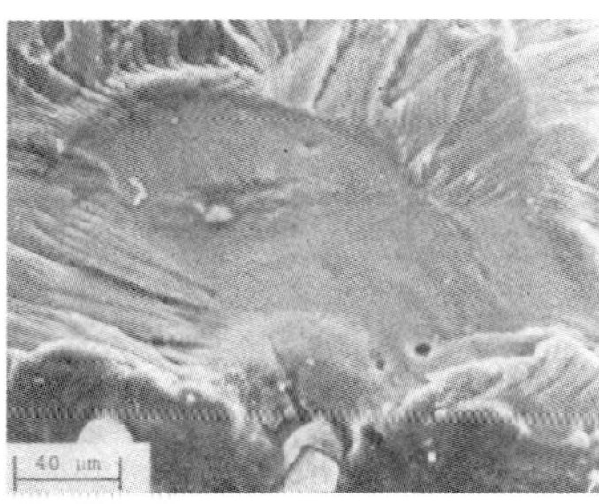

Fig. 1. "Featureless" early crack propagation zone associated with environmental fatigue of 7075-T6 Al, initiated at an inclusion. Similar zones can be identified for free corrosion or cathodic charging conditions in virtually all electrolytic solutions.

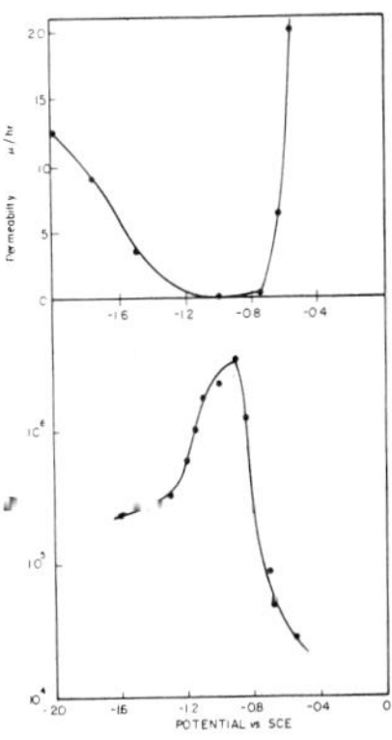

Fig. 2. Correlation of hydrogen permeability experiments (11) with fatigue life in 0.5 N NaCl as a function of applied potentials. $\phi_{corr} \cong -0.7V$ vs SCC.

and/or severe charging conditions intergranular crack initiation precedes transgranular propagation (9). The results of these experiments cannot be associated with a dissolution process unless it is assumed that dissolution is the rate determining step for hydrogen evolution. Rather these observations are consistent with a mechanism where a dissolved solute in the metal (presumably hydrogen) reaches a critical concentration leading to cracking, and is only indirectly associated with deformation. For example, at moderate to high stress levels, or when stress concentrations are sufficiently large, extensive slip is believed to result in dislocation transport of hydrogen into the lattice, while at lower stresses the preferred transport path is along grain boundaires. This latter hypothesis may in fact be the link between "SCC" and "corrosion fatigue" of these alloys since cyclic deformation could be expected to be a much more effective transport mechanism of hydrogen into the lattice for sufficiently high stresses or hydrogen fugacity. It may also be significant that the reported fracture path in single crystals of these alloys appears to be either along {112} or {100} planes. These planes have been reported to be the interfacial planes between the $MgZn_2$ precipitates and the matrix and there is a temptation to consider that these interfaces would be potential sinks for hydrogen. This process would be analogous to that reported by Swann and co-workers for monotonic loading, where the grain boundaries become the preferred enhanced diffusion path and separation is observed along non-coherent precipitate matrix interfaces, with no appearent corrosion attack of either phase at electron optical resolutions.

Hydrogen Damage Reversibility

Precharging of 7075 T6 Al with hydrogen by cathodic charging in either 0.5 $\underline{N}$ NaCl or in 0.5 $\underline{N}$ H_2SO_4 results in a significant reduction in fatigue resistance. In the former case significant corrosion occurs at non-metallic inclusions while for the latter case, little or no corrosion of the alloy occurs. For both cases, however, post charging heat treatments, consisting of resolutionizing of the alloy at 460°C followed by re-aging at 120°C, results in siginficant recovery of fatigue properties. Increasing the resolutionizing time at temperature results in still larger degrees of fatigue resistance recovery, although for the NaCl case the damage associated with the inclusions precludes total recovery (Table I).

Table I.   The Effect of Re-Heat Treatment Duraction on Subsequent Fatigue
           Properties in Air

Pre-Corrosion: 24 hours in aerated 0.5 NaCl

Re-heat treat: Solutionize at 470°C for x hours
                Age at 121°C for 24 hours

Fatigue in air: Mean stress 276 MPa, Cyclic Stress $\pm$ 96 MPa

Time at 470°C	N_f
0 hours	30,000 cycles
3 hours	85,000 cycles
6 hours	101,000 cycles
24 hours	> 13,000,000 cycles*

*Equivalent to tests in dry air without pre-exposure.

It is important to note that control specimens, fatigued in dry laboratory
air after similar "baking" experiments show no effect of cyclic heat treatments. Thus, it may be concluded that the hydrogen effect in these alloys
is essentially reversible, presumably by removing dissolved hydrogen from
the alloy lattice. No classical mechanism of stress corrosion cracking, or
of corrosion assisted fatigue can explain these results.

Electrochemical Parameters

     It has been shown that environmentally affected fatigue of these alloys,
under cathodic charging conditions is virtually independent of the type of
anion, eg. Cl^- or $SO_4^=$ or of the anionic strength of the solution eg. pH
0.6 to 10 (10). Additionally it has been shown that a maximum in fatigue
resistance is observed at a potential which is virtually identical with that
where hydrogen permeation has been shown to be a minimum in these alloys
(Fig. 2). Thus, the presence of hydrogen is sufficient to cause significant
decreases in fatigue resistance independent of chemical attack of the protective passive films. It cannot be concluded however, that chemical attack
of passive films is not important to fatigue resistance, _per se_, since it
has also been shown that, in the absence of externally produced hydrogen
(charging), chloride ion is certainly more damaging than sulfate ion (Fig.
3a), which is in turn more damaging than de-ionized water. However, acidic

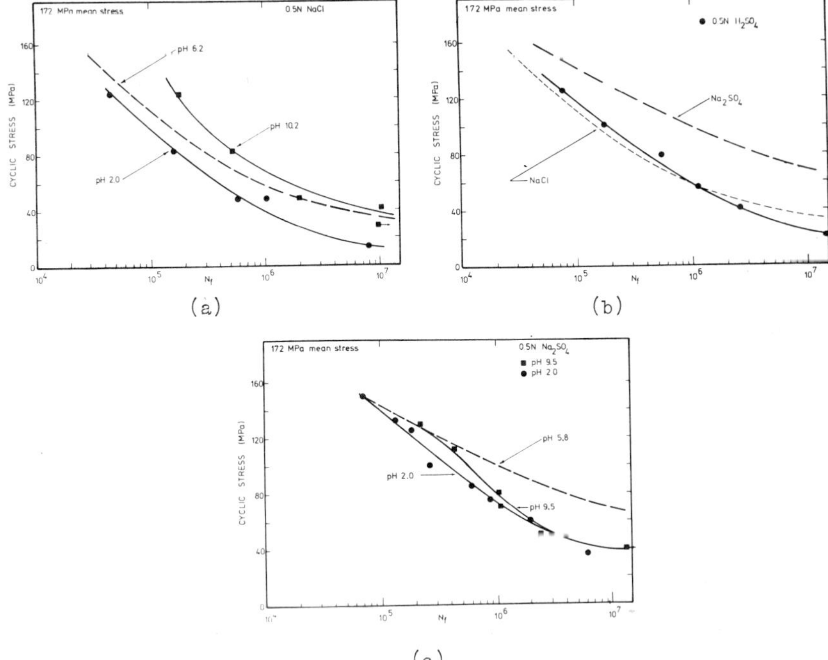

(a)   (b)

(c)

Fig. 3. The effects of anion species and pH on the fatigue resistance of
         7075 Al alloy. a) effect of pH in NaCl solution, b) comparative
         effects of neutral Na_2SO_4, neutral NaCl and 0.5 N H_2SO_4 (pH 0.6)
         c) effects of pH in Na_2SO_4 solutions. In general, corrosion rates
         are least at neutral pH and are equal for pH 2 and ~pH 10 in Cl^-
         and $SO_4^=$ solutions.

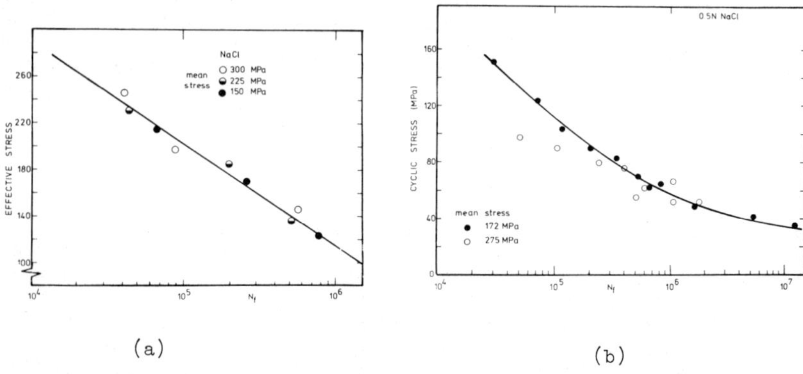

(a)   (b)

Fig. 4.   a) Correlation of effective stress, normalized for different mean
stress on fatigue resistance of 7075 Al alloy in 0.5 $\underline{N}$ NaCl solution b) Fatigue resistance of 7075 Al in 0.5 $\underline{N}$ NaCl solution as
f (σ_{cyclic}) for two mean stresses. The divergence at high cyclic
stresses can be eliminated by correlation with "effective" stress
range (12).

solutions are not necessarily more damaging than neutral solutions, under
free corrosion conditions (Fig. 3b). Neutral sulfate solutions of pH2 and
pH 9.5 at low stresses are an exception in that both environments lead to
equivalent decreases in fatigue resistance (Fig. 3c).

Mechanical Parameters

Normalization of mean stress effects on environmental fatigue of 7075 Al
in distilled water, NaCl and Na_2SO_4 solutions under freely corroding conditions, and under cathodic changing conditions, indicate that neither mean
stress or maximum stress in a given cycle control fatigue damage (Fig. 4a).
Rather, these results indicate that the cyclic stress range is the critical
mechanical parameter which controls fatigue resistance (Fig. 4b).

Additionally, experiments performed under nominal Mode III (torsional)
cyclic loading indicate that there is little effect of corrosion or of
cathodic charging on fatigue resistance for this loading condition (Fig. 5).
These results are to be expected if it is assumed that significant hydrogen
transport or accumulation in the alloy necessitates a hydrostatic stress
component in the vicinity of the alloy surface or near a crack tip. However, if metal dissolution is the critical process in environmental fatigue,
no appreciable difference in Mode I vs Mode III loading should be expected
(neglecting transport in the liquid inside the crack).

A Model for Hydrogen Assisted "Corrosion Fatigue" of Al Alloys

The results of these and other investigations strongly suggest that dissolved hydrogen in the alloy lattice is a critical species in determining
the fatigue resistance of high strength aluminum alloys. Under corrosion
conditions hydrogen may be effectively blocked from entry to the alloy by

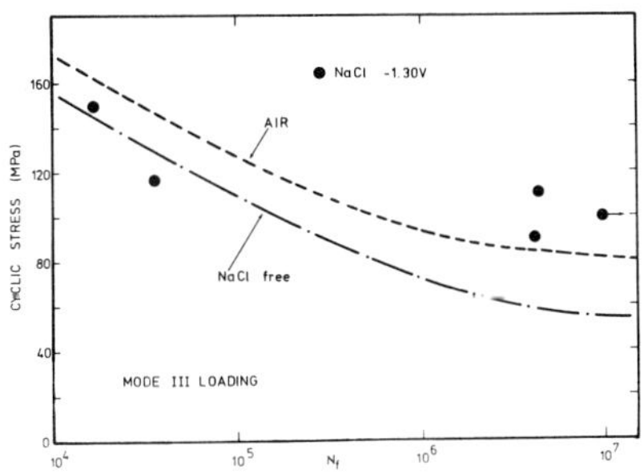

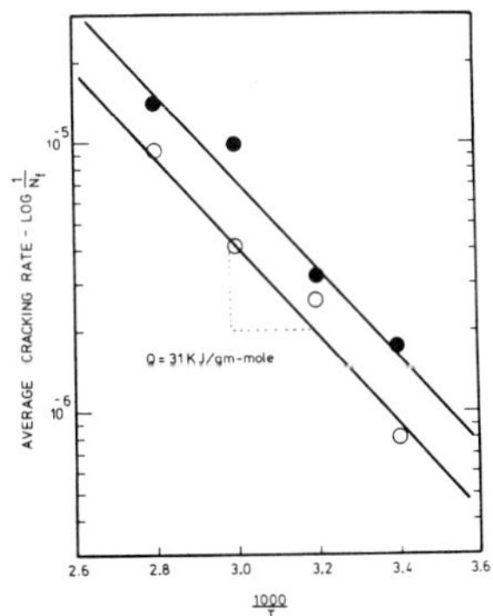

Fig. 5. Effect of Mode III loading in fatigue resistance of Al alloy in air, NaCl solution under free corrosion and hydrogen charged in NaCl solution showing little or no effect of environment.

Fig. 6. Activation energies for apparent crack propagation in Al alloys pre-exposed in 0.5 N H_2SO_4 for one hour (open circles) and for 24 hours (closed circles) and subsequently fatigued in distilled water.

the presence of the adherent passive film which is naturally formed on the alloy. Exposure to NaCl, or to solutions of high or low pH effectively damages this film allowing ready access of bare metal for hydrogen entry. Thus fatigue lives are improved in less agressive solutions such as distilled water or neutral Na_2SO_4. In these cases hydrogen entry is apparently blocked during the initiation and early crack growth stages, however as the crack propagates, hydrogen entry is facilitated by mobile dislocations in the vicinity of the crack tip. It is significant to note that experiments conducted either by pre-exposure to H_2SO_4 solutions for 4 hours and for 16 hours followed by fatigue in distilled water as a function of temperature show essentially the same apparent activation energy for average crack propagation rate ($1/N_f$) (Fig. 6). The apparent cracking rate is $\sim$ 2x indicating a strong correlation with a diffusion process obeying a parabolic relation with diffusional distance.

To summarize, it appears that the rate determining process in environmental fatigue of high strength Al alloys is the absorption of hydrogen into either the alloy surface or into the plastic zone ahead of a growing crack. Hydrogen ingress during initiation may be facilitated by chemical means such as Cl^- environment, high or low pH, cathodic charging, etc., or by mechanical means such as slip band rupture of protective films. During the early propagation stages hydrogen ingress is further facilitated by crack tip chemistry and the large amounts of deformation associated with crack tip stress and strain concentrations. In every case, it is believed that a critical rate determining step is important to the process and that the rate determining step may be different for different environmental conditions. Some of the steps which might be involved could be hydrogen dissociation, absorption diffusion, concentration etc. and these steps may be further determined by hydrogen fugacity, chemical or mechanical rupture of oxide films, or even a sufficient local corrosion rate to support sufficient amounts of hydrogen generation and absorption. However, while the rate determining step may differ for different chemico-mechanical conditions, the fact remains that the evidence presented here strongly supports a hydrogen embrittlement phenomenon as the cause of environmental fatigue of these high strength Al alloys. The vexing problem remains, however, that while dissolved hydrogen is an embrittling agent, there is still no direct evidence for the specific mechanism(s) associated with this phenomenon.

References

1. G. Scamans, R. Alani and P. Swann, Corr. Sci. (1976), vol. 16, pp. 443-459.

2. T. Broom and A. Nicholson, J. Inst. Metals, 1960, vol. 89, p. 183.

3. R. P. Wei, Eng. Fract. Mech., 1971, vol. 1, p. 633.

4. M. Nageswararao, R. Meyer, M. Wilhelm and V. Gerold in "Mechanisms of Environment Sensitive Cracking of Materials" ed. P. R. Swann, F. P. Ford and a.r.c.Westwood,pp. 383-389, The Metals Society, London, 1977.

5. R. P. Wei in "A. R. Troiano Honorary Symposium-Hydrogen Embrittlement and Stress Corrosion Cracking" ed. R. Gibala and R. Hehemann, Case Western Reserve University, Cleveland, 1980.

6. M. Nageswararao and V. Gerold, Met. Trans., 1976, vol. 7A, p. 1847.

7. M. Nageswararao and V. Gerold, Met. Sci., 1977, vol. 11, p. 31.

8. L. V. Corsetti and D. J. Duquette, Met. Trans. 1974, vol. 5, p. 1087.

9. E. F. Smith, III, R. Jacko and D. J. Duquette in "Effect of Hydrogen on Behavior of Materials", A. W. Thompson and I. M. Bernstein eds. p. 218, AIME, NY, 1976.

10. R. J. Jacko, Ph.D. Dissertation, Rensselaer Polytechnic Institute, Troy, NY 1978.

11. R. J. Gest and A. R. Troiano, Corrosion, 1974, vol. <u>30</u>, p. 274.

12. K. Walker, in "Effects of the Environment and Complex Load History on Fatigue Life" ASTM STP 462, Phila, 1970.

MECHANISMS OF CORROSION FATIGUE CRACK PROPAGATION

OF 7XXX ALUMINUM ALLOYS IN AQUEOUS ENVIRONMENTS*

Fu-Shiong Lin and E. A. Starke, Jr.
Fracture and Fatigue Research Laboratory
Georgia Institute of Technology
Atlanta, Georgia   30332

The fatigue crack propagation (FCP) behavior of four Al-6Zn-2Mg-xCu alloys has been studied in dry air, distilled water and a 3.5% NaCl solution. In addition, extensive studies were made in the alloy containing 1% Cu in oxygen-saturated and nitrogen deaerated solutions, and solutions containing poisonous inhibitors in order to elucidate the role of hydrogen in the corrosion FCP process of these materials. The curves for the ratio of $(da/dN)_{H_2O}/(da/dN)_{dry\ air}$ vs stress intensity range (ΔK) shows that the peak aged condition is most susceptible to environmental effects and that the environmental sensitivity and the ΔK value for maximum sensitivity increase with decreasing copper content. However, a comparison of the curves of the FCP rate for both aged conditions in distilled water indicates that the over-aged condition does not improve the resistance of corrosion FCP for $\Delta K < 7$ MPam$^{1/2}$. These effects are controlled by the slip mode which is controlled by copper content and heat treatment. The mechanism studies indicate that hydrogen plays a dominate role in corrosion FCP of high strength aluminum alloys. The extent of the effect depends on the rate of formation of hydrogen at the crack tip and the repassivation rate of the particular alloy. The association of environmental sensitivity with slip mode supports a mobile dislocation transport mechanism.

*This research was sponsored by the Air Force Office of Scientific Research, Air Force Systems Command, USAF, Under Grant No. AROSR-78-3471, Dr. Alan H. Rosenstein, Program Manager.

1. Introduction

High strength Al–Zn–Mg alloys show enhanced fatigue crack propagation in aggressive environments. Numerous experiments have been performed to investigate the mechanisms of corrosion fatigue in aluminum alloys, and those parameters which control the environmental effects (1-10). The results of these investigations suggest a number of possible mechanisms for corrosion fatigue in aqueous environments: [a] fracture due to a surface energy reduction at the crack tip associated with preferential adsorption of environmental species (1,4,6), [b] fracture associated with dissolution of active slip bands at the crack tip (5,7), and [c] fracture associated with hydrogen embrittlement phemonena (2,8-10).

Corrosion fatigue of metals in aqueous environments involves electro-chemical processes which include anodic and cathodic reactions. The former is related to an anodic dissolution mechanism and the latter is associated with hydrogen embrittlement. Since both reactions occur simultaneously with equal intensity, it is impossible to separate one from another. However, it is possible to add other types of cathodic reactions which do not affect hydrogen fugacity by saturating oxygen in the test solution. This will result in an increase in the amount of anodic dissolution. The research described in this paper was designed to investigate the effect of the corrosion rate and the passivation behavior of the fatigue crack growth rate (FCGR). In addition, the influence of deformation mode on the corrosion fatigue crack propagation (FCP) behavior was studied. We hope that the results obtained will aid in clearfying the mechanisms of corrosion FCP of aluminum alloys in aqueous solutions.

2. Experimental Procedures

The composition of the alloys used in this study are shown in Table I. Their processing history and microstructures have been described elsewhere (11,12). They were tested in the peak strength, T651, and overaged, T7351, conditions. The monotonic properties are given in Table II.

Table I. Chemical Composition (Weight Percent)

Alloy	Cu	Zn	Mg	Zr	Ti	Fe	Si	Al
0.01 Cu	0.01	6.41	2.08	0.11	0.02	0.05	0.05	bal.
1.0 Cu	0.98	6.10	2.20	0.12	0.02	0.05	0.05	bal.
1.6 Cu	1.56	6.07	2.24	0.12	0.02	0.05	0.06	bal.
2.1 Cu	2.11	5.97	2.11	0.12	0.02	0.07	0.06	bal.

FCP measurements were made in dry air (H_2O content < 3 ppm), laboratory air with 45% relative humidity (R.H.), 100% R.H. air, distilled water, a 3.5% NaCl solution deaerated with N_2, a similar solution saturated with O_2, a salt solution with 50 ppm As, and a 2.5 pH solution (adjusted by HCl) deaerated with N_2 and a similar solution saturated with O_2. The test cell was a plexiglass tube with a 2.2 cm inner diameter and 16.0 cm long. The test solution was not circulated.

Table II. Monotonic Mechanical Properties

Alloy		YS (MPa)	TS (MPa)	Reduction in area (%)	K_Q (MPam$^{1/2}$)
0.0 Cu	T651	482.2	551.7	31.2	49.8
	T7351	403.8	440.0	45.8	75.5
1.0 Cu	T651	522.3	588.0	32.5	57.6
	T7351	432.2	483.2	40.7	72.8
1.6 Cu	T651	529.2	613.5	32.1	58.0
	T7351	462.2	510.6	39.3	59.0
2.1 Cu	T651	545.9	613.5	36.4	46.8
	T7351	430.2	484.1	39.2	55.0

Duplicate 3.5 x 2.0 x 0.72 cm samples of the 1.0% Cu alloy were cut from
the plate for corrosion tests. They were mechanically polished through 600
grit emery paper, degreased, weighed, and immersed in the solution which had
been purged with N_2 or O_2 for 20 minutes. The gases were passed into the
solutions continuously during the tests. After 50 hours the samples were
removed, washed with acetone and weighed.

Compact-tension samples for FCP tests were cut from the aged plate with
starter notches parallel to the rolling direction and a T-L crack plane
orientation (long transverse-longitudinal). The samples were 7.2 mm thick
and had a H/W of 0.6 (H and W are height and width of the samples). The tests
were performed on a closed-loop servohydraulic MTS Machine using tension-
tension loading, a minimum-maximum load ratio of 0.1 and a frequency of 10 Hz.
The crack length was measured with a travelling microscope to an accuracy of
$\pm$ 0.01 mm. The Secant method was used to calculate the growth rate. The
fracture surfaces of selected FCP specimens were examined by SEM.

3. Experimental Results and Discussion

3.1 Corrosion Tests

The results of corrosion tests for the 1.0% Cu alloy in various environments are presented in Table III. The corrosion rate for the oxygen-saturated
solution is higher than that for the N_2 deaerated solution. For the 2.5 pH
solution, the difference in corrosion rate increases with increasing oxygen
content, i.e., O_2 saturated > mechanically stirred > N_2 deaerated. The
difference is about a factor of three following this order. No significant
corrosion was found for the salt solution with 50 ppm As.

The increase in the corrosion rate is a result of saturation of the
solution with O_2 is due to the higher anodic polarization of the dissolution
reaction caused by an increase in the cathodic reaction rate. The additional
cathodic reaction is $O_2 + 2H_2O + 4e \rightarrow 4(OH)^-$ (13). Because of the highly
negative values of the corrosion potentials the hydrogen reduction reaction

still takes place, but its rate is reduced because its overpotential is reduced. The poisonous inhibitor (As) retards the rate of formation of hydrogen molecules, and therefore decreases the corrosion rate. When the 3.5% NaCl solution was saturated with O_2 a dark hydroxide film almost completely covered the surface. In addition, a grease-like corrosion product appeared on the dark film. This product is believed associated with the dissolution of aluminum hydroxide caused by a higher local pH value on the metal surface due to oxygen reduction (14,15).

Table III.   Corrosion Rate for the 1.0% Cu Alloy (T651)
Immersed in Various Environments for 50 Hours

Environment	Corrosion Rate (cm/yr)
pH = 2.5 (O_2 saturated)	0.098
pH = 2.5 (N_2 deaerated)	0.018
3.5% NaCl (O_2 saturated)	0.065
3.5% NaCl (N_2 deaerated)	0.011
3.5% NaCl (Mechanically Stirred)	0.033
3.5% NaCl + 50 ppm As (N_2 deaerated)	Nil

3.2   Anodic Dissolution and Corrosion Fatigue Crack Propagation

In discussing these results, it is necessary to assume that qualitatively the corrosion behavior within the crack of the FCP samples in the various environments is analogous to that which occurs in the corrosion tests described previously. However, the electrochemical reactions between the newly created surface and the environment at the crack tip are more intense (16).

The fatigue crack growth rates (FCGR's), da/dN, plotted against stress intensity range (ΔK), as affected by environmental conditions, are shown in Fig. 1 along with the FCGR in dry air as a reference. The FCGR's for the 2.5 pH solution O_2 saturated and N_2 deaerated, respectively, are the same, although the corrosion rate differed by a factor of six. Fig. 1(b) also indicates the same trend for the 3.5% NaCl solution with and without deaeration. However, the FCGR is maximum in the salt solution with 50 ppm As and minimum in the O_2 saturated solution. The corrosion rates in these environments varied in the opposite direction to the FCGR. These results show that the dissolution rate can not be directly related to the corrosion FCGR.

Although As decreases the corrosion rate in the salt solution, it accelerates the FCGR since it retards the rate of formation of molecular hydrogen and thus increases the accumulation of adsorbed hydrogen atoms at the crack tip. Consequently, the diffusion rate of hydrogen atoms into the plastic zone ahead of the crack tip increases. Jacko and DuQuette (2) observed a similar effect on the high cycle fatigue behavior of aluminum alloys tested in the same environment. As discussed previously, the salt solution with O_2 has the greatest effect on increasing the corrosion rate and reducing the

FCGR. The hydroxide film which forms in this environment probably acts as a
diffusion barrier and inhibits hydrogen entry into the metal lattice. These
results support the hydrogen embrittlement concept of corrosion FCP in high
strength aluminum alloys.

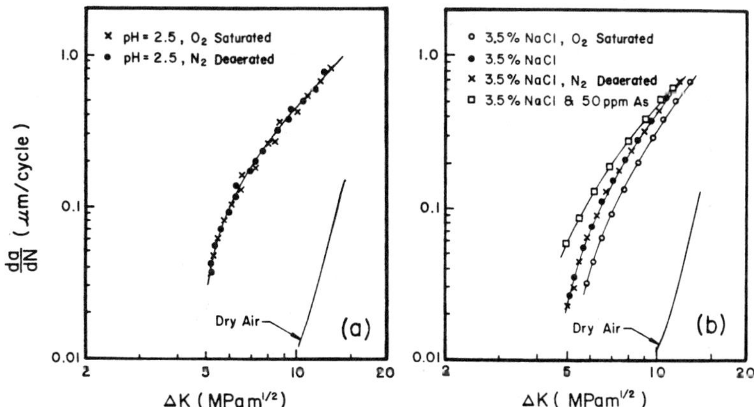

Figure 1 - Effect of environmental conditions on the FCGR
of the 1% Cu alloy in the T651 condition.

3.3 Effect of Chloride Ion and Hydrogen Concentration on the Growth Rate

The effect of the chloride ion on the corrosion FCGR is presented in Fig.
2(a). The growth rate in a 3.5% NaCl solution is approximately the same as
that in distilled water. This supports our earlier contention (11,12) that
adsorption of a damaging species, e.g. the chloride ions, at the crack tip
(1,4,6) or the acceleration of anodic dissolution by the presence of chloride
ions (5,7), is not of critical importance in the corrosion FCP of high strength
aluminum alloys. The growth rates for the 45 and 100% R.H. air are very close
to those of distilled water and the salt solution. The water vapor in the
45 and 100% R.H. air reacts rapidly with the bare surface at the crack tip
resulting in the formation of aluminum hydroxide and the evolution of hydrogen. The time needed for condensation probably accounts for the slight
difference in FCGR. These results also support the hydrogen embrittlement
theory and suggest that the main source of hydrogen is the reaction of water
vapor or water with the fresh metal surface at the crack tip.

Our results do not contradict the observation that high pressure hydrogen gas doen not embrittle aluminum alloys (17,18). Hydrogen gas must
dissociate into hydrogen atoms to diffuse into the metal. Hydrogen molecules
are very stable and dissociation is difficult and the process too slow for
embrittlement to occur. The water-aluminum surface interaction at the crack
tip produces hydrogen atoms. Although recombination of hydrogen atoms occurs,
we believe some hydrogen atoms diffuse into the metal lattice before recombination is complete. As mentioned previously, we believe the role of the
poisonous inhibitor is to retard the rate of formation of molecular hydrogen,
thus enhancing its entry into the metal and resulting in an increase in the
FCGR.

It is not surprising that a salt solution has a dramatic effect on the
stress corrosion crack velocity (Region II) as compared with a distilled
water environment (19), but no significant effect on corrosion fatigue. Even
though the fracture features for SCC and corrosion FCP are different (inter-

granular vs transgranular), the repassivation rate is important for both cases. For static stress corrosion cracking tests, film rupture and repassivation occurs repeatedly. The chloride ions delay repassivation, and increase the stress corrosion crack velocity (20). However, for the fatigue test conditions employed, a fresh metal surface is created faster during cyclic loading than repassivation can occur. As a result, the function of chloride ions is masked during cyclic loading and no significant difference in the corrosion FCGR occurs between distilled water and salt water.

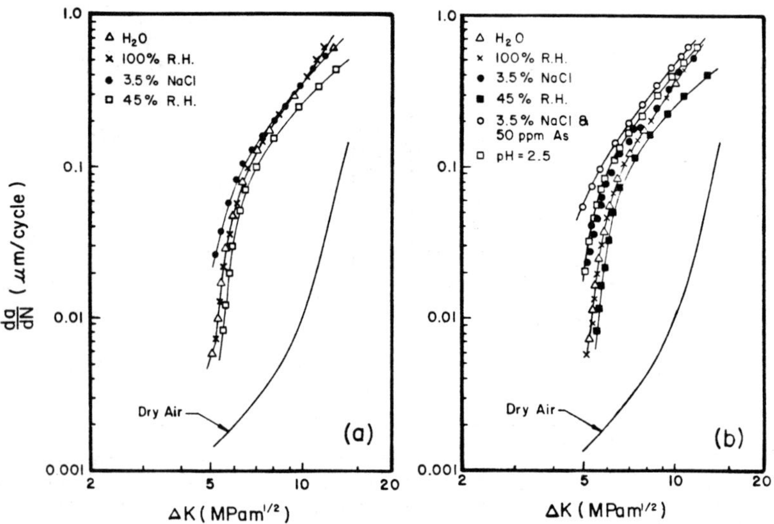

Figure 2 – Effect of (a) chloride ion and (b) the evolved hydrogen concentration of the FCGR of the 1% Cu alloy in the T651 condition.

Generally, if hydrogen embrittlement is the major mechanism in corrosion fatigue the FCGR would be expected to depend on the pH value and could be limited by the amount of water at the crack tip. The results shown in Fig. 2(b) appear generally to follow this trend, but the differences are not significant. This may be attributed to the water content of all environments being above the critical value, and the pH at the crack tip being stabilized by a balance between the hydrogen ion producing and consuming reactions.

3.4 Effect of Deformation Mode on the Corrosion Fatigue Properties

Al–6Zn–2Mg–xCu alloys are age hardenable and their deformation mode depends on the degree of coherency of the strengthening precipitates. The precipitates are either sheared or looped and bypassed by dislocations during plastic deformation. Coherent precipitates can be sheared by glissile dislocations leading to the formation of concentrated slip bands; whereas the incoherent precipitates are looped and bypassed by the dislocations resulting in more homogeneous deformation. Our previous studies (11,12) have shown that for the T651 condition the degree of coherency of the precipitates in these alloys decreases with increasing copper content from 0% to 2.1% Cu. In other words, the homogeneity of plastic deformation continuously increases with copper content. For the T7351 condition the precipitates become incoherent for all four alloys, resulting in uniform plastic deformation. The effect of deformation behavior, as influenced by both copper additions and aging treatments, on the corrosion FCP has been evaluated in this study.

In order to discuss the mechanism of corrosion FCP, it is convienent to express the FCGR in terms of the difference in the growth rate between the corrosive and inert environments and/or the ratio of the growth rate between these two environments. The ratio of the FCGR in distilled water to that in dry air, as a function of ΔK is shown in Fig. 3. This figure shows that the environmental sensitivity decreases when the copper content increases or when the alloys are overaged. Thus the principal conclusion of these curves is that the environmental sensitivity is a function of deformation mode. This effect is largest for the highest degree of planar slip, such as the 0% Cu alloy in the T651 condition, and decreased with increasing copper content. The smallest effect is for the over-aged samples regardless of the copper content since they all exhibit uniform deformation. These finding are consistent with a hydrogen embrittlement mechanism for corrosion fatigue, which relates the degree of susceptibility to slip planarity (21-24).

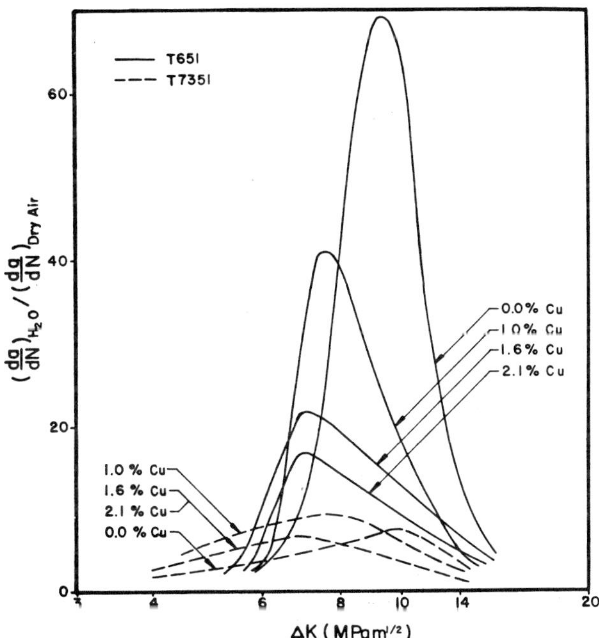

Figure 3. The environment sensitivity of the four alloys in the T651 and T7351 conditions.

The correlation between deformation mode and hydrogen embrittlement susceptibility suggests that hydrogen is transported by mobile dislocations in aluminum alloys (22-24). For the planar slip materials, the density of mobile dislocations on the localized slip plane within the plastic zone is considerably greater than in other areas and more hydrogen atoms can be transported along these localized regions. Consequently, the combination of highly localized slip and highly localized hydrogen fugacity leads to a higher susceptibility to hydrogen embrittlement. On the other hand, in the alloys undergoing homogeneous deformation the hydrogen atoms are transported uniformly within the plastic zone. This reduces the susceptibility to hydrogen embrittlement.

Acknowledgements

The authors would like to thank Dr. M. Marek for reviewing the manuscript and for his helpful suggestions.

References

1. R. E. Stoltz and R. M. Pelloux: Corrosion, NACE, 1973, Vol. 29, No.1, p. 13.
2. R. J. Jacko and D. J. DuQuette: Met. Trans., 1977, Vol. 8A, p. 1821.
3. M. O. Speidel, NATO Advanced Study Institute of SCC, 1975, Copenhagen, Denmark.
4. R. M. Pelloux: Fracture, Proc. of the Second Int. Conf. of Fracture, 1969, Brighton, p. 731.
5. F. P. Ford: Corrosion, NACE, 1979, Vol. 35, No. 7, p. 281.
6. R. J. Selines and R. M. Pelloux: Met. Trans., 1972, Vol. 3A, p. 2525.
7. T. Pyle, V. Rollins and D. Howard: Proc. of the Int. Conf. on Corrosion Fatigue, 1971, Storrs, Conn., p. 312.
8. T. Broom and A. Nicholson: J. Inst. Metals, 1960, Vol. 89, p. 183.
9. M. O. Speidel: Hydrogen in Metals, Proc. of the Int. Conf. on the Effects of Hydrogen on Materials Properties, 1973, p. 249.
10. R. P. Wei: Int. J. Frac. Mech., 1969, Vol 5, p. 71.
11. Fu-Shiong Lin and E. A. Starke, Jr.: Mater. Sci. and Eng., 1979, Vol. 39, p. 27.
12. Fu-Shiong Lin and E. A. Starke, Jr.: Mater. Sci. and Eng., 1980, Vol 43, p. 65.
13. H. H. Uhlig: Corrosion and Corrosion Control, McGraw-Hill, New York, 1963.
14. J. W. Diggle and A. K. Vijn: in "Oxide and Oxide Film" by R. S. Alwitt, Marcel Dekker, Inc., New York, 1976, Vol. 4.
15. W. Vedder and D. A. Vermilyer: Trans. of the Faraday Soc., 1969, Vol. 65, Part 1, p. 561.
16. F. P. Ford: Metal Science, 1978, Vol. 12, p. 326.
17. M. O. Speidel and M. V. Hyatt: Advances in Corrosion Science and Technology, 1972, Vol. 2, Plenum Press, p. 115.
18. F. J. Bradshaw and C. Wheeler: Applied Materials Research, April 1966, p. 112
19. M. O. Speidel: The Theory of Stress Corrosion Cracking in Alloys, Nato Scientific Affairs Division, Brussels, 1971, p. 289.
20. J. C. Scully: Metal Science, 1978, Vol. 12, p. 290.
21. Fu-Shiong Lin and E. A. Starke, Jr.: Mater. Sci. and Eng., 1980, in press.
22. J. A. Donovan: Met. Trans., 1976, Vol. 7A, P. 1677.
23. M. R. Louthan, Jr., F. R. Caskey, Jr., J. A. Donavan and R. E. Rawl, Jr.: Mater. Sci. Eng., 1972, Vol. 10, p. 357.
24. M. R. Louthan: Hydrogen in Metals, 1974, ASM, Metal Park, p. 53.

IN-SITU H.V.E.M. OBSERVATIONS OF HYDROGEN EMBRITTLEMENT

IN Al-Zn-Mg ALLOYS

L. Christodoulou[*] and H. M. Flower
Imperial College of Science and Technology
London SW7 2BP

[*]Now at Carnegie-Mellon University
Pittsburgh, PA 15213

Synopsis

An investigation into the effect of exposure of Al-Zn-Mg alloys to water vapour has shown that hydrogen, produced by the reaction is absorbed by the alloy. The amount of absorbed hydrogen depends on the local chemistry at the metal/oxide interface and is influenced by alloying additions. Attainment of a critical concentration of dissolved hydrogen in the boundaries leads to brittle intergranular fracture on the subsequent application of stress. Attainment of a critical hydrogen concentration in the matrix leads to transgranular cleavage fracture.

Depending upon the alloy composition and heat treatment bubbles of gaseous hydrogen can be formed at suitable sites within the alloys such as grain boundary particles. The formation of grain boundary bubbles reduces embrittlement by reducing local concentration of dissolved hydrogen.

It is proposed that the interaction of stress and environmental variables with the above criteria governs the environmental failure of Al-Zn-Mg alloys in aqueous and water vapour containing environments. It is suggested that such failures are hydrogen embrittlement dominated.

Introduction

It has been shown previously that the exposure of Al-Zn-Mg alloys to moist environments results in severe embrittlement of the grain boundaries(1, 2, 3, 4, 5) and under certain conditions of the alloy matrix (3,4). The characteristics of the phenomenon-strain rate sensitivity and reversibility of the embrittlement, formation of hydrogen bubbles at grain boundaries and the evolution of hydrogen during fracture - strongly suggest that it is controlled by hydrogen penetration and embrittlement of the alloy (3, 4, 5).

It was the purpose of the work reported here to examine the conditions of hydrogen generation and its entry into the metal. Although some information had been obtained in the previously reported studies using scanning and transmission electron microscopy direct observation of the phenomenon is much more informative and in the present work conditions were established whereby this could be achieved.

Experimental Technique

The experimental twchnique employed involves the use of a differentially pumped, aperture type environmental cell placed in the objective pole piece of an AEI EM7 High voltage electron microscopy (6, 7). The experimental arrangement allows the environment of interest, in this case water vapour saturated helium, to be introduced directly into the specimen chamber of the microscope (Fig. 1). The specimen chamber is separated from a second, differentially pumped chamber by two 100μm apertures: the second chamber being separated from the high vacuum in the microscope column by two 200μm apertures. The residual leak rate into the column is thus kept to a value which can be adequately dealt with by the column pumping system. The high pressure region, therefore, is maintained at a state of dynamic equilibrium by balancing the inlet gas flow to the leak rate through the apertures. Typical pressure conditions across the cell are shown in Figure 1.

The thin electropolished samples were introduced into the specimen chamber on a liquid nitrogen cooled stage so that the temperature could be directly monitored and controlled between room temperature and 213K. The chemical reaction between the thin foil and the water vapour was observed using accelerating voltages in the range of 100 - 500 KV.

Companion studies were carried out on both thin foil and bulk material exposed to water vapour outside the microscope and subsequently examined using standard SEM/STEM techniques. The effect of this exposure on the properties of the alloys was monitored by mechanical testing of tensile samples similarly treated.

Three alloys were employed in this work. A high purity Al-6%Zn-3%Mg

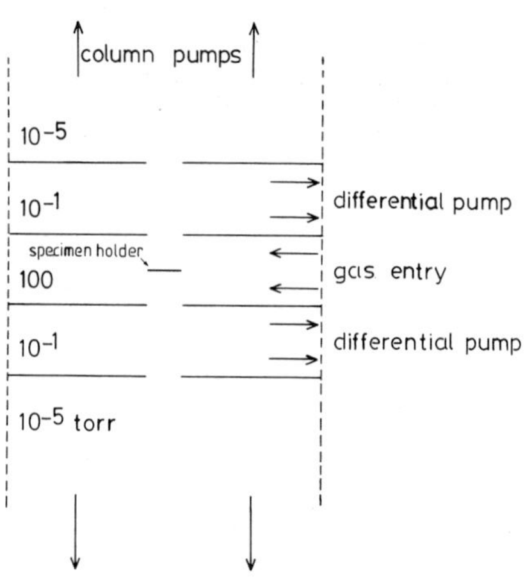

Fig. 1.   Schematic representation of the environmental cell.

alloy (by weight) and two quaternary alloys based on the above but containing 1.7%Cu and 0.14%Cr, respectively.

Results

The "in-situ" reaction between all the alloys and water vapour saturated helium was found to occur preferentially at sites under electron irradiation (Fig. 2). The reaction, in the first instance, took the form of surface blistering whereby the oxide film on the foil partially decohered from the metal substrate forming hemispherical blisters.

The reaction occurred at all accelerating potentials employed and there was no measurable difference in the rate of reaction over the range of potentials of 100 - 500KV. The reaction was highly sensitive to the electron flux density which was maintained at approximately $10^{-3} Amm^{-2}$ to provide adequate image brightness together with managable reaction rates.

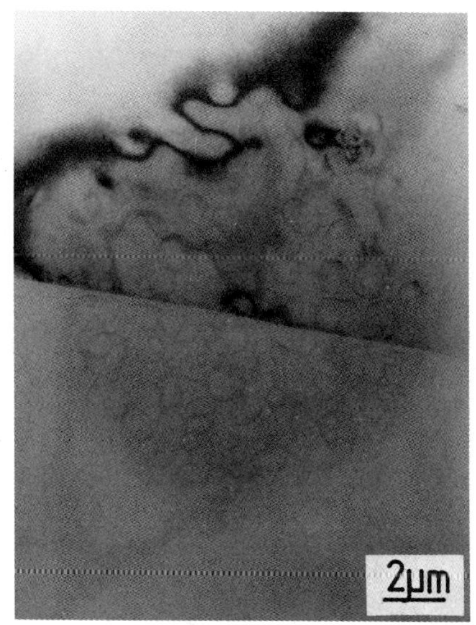

Fig. 2. Surface blisters formed in the environmental cell under the influence of the electron beam and water vapour. Electron flux density = $5x10^{-3} Amm^{-2}$, Temperature = 253K.

The rate of the reaction was also dependent on the temperature. The reaction at 293K was very fast and thus most experiments were carried out in the temperature range of 233 - 273K.

The response of the different alloys and heat treatment conditions may be summarized as follows:

1. The reaction between all alloys and water vapour occurred preferentially and precisely at sites under the electron beam.

2. There existed a degree of preferential reaction at the grain boundaries (Fig. 3). This selective attack, however, was limited to the initial stages of the reaction. The high purity base and Cu containing alloys showed similar blistering characteristics and kinetics. In contrast, the Cu quaternary exhibited blistering behaviour of increased severity.

3. Following the formation of surface blisters internal hydrogen bubbles were found to occur at the grain boundaries of the alloys (Fig. 4). The bubbles are contained entirely within the thickness of the foil and are imaged via their strain fields.

In the case of the high purity alloy the formation of such hydrogen bubbles was a strong function of microstructure. Bubbles were only formed in microstructural states which exhibited grain boundary precipitates of size greater than 200A. Cu containing alloys on the other hand exhibited grain boundary bubbles in all microstructural states including the as-quenched condition. In contrast to this behaviour, the Cr quaternary showed

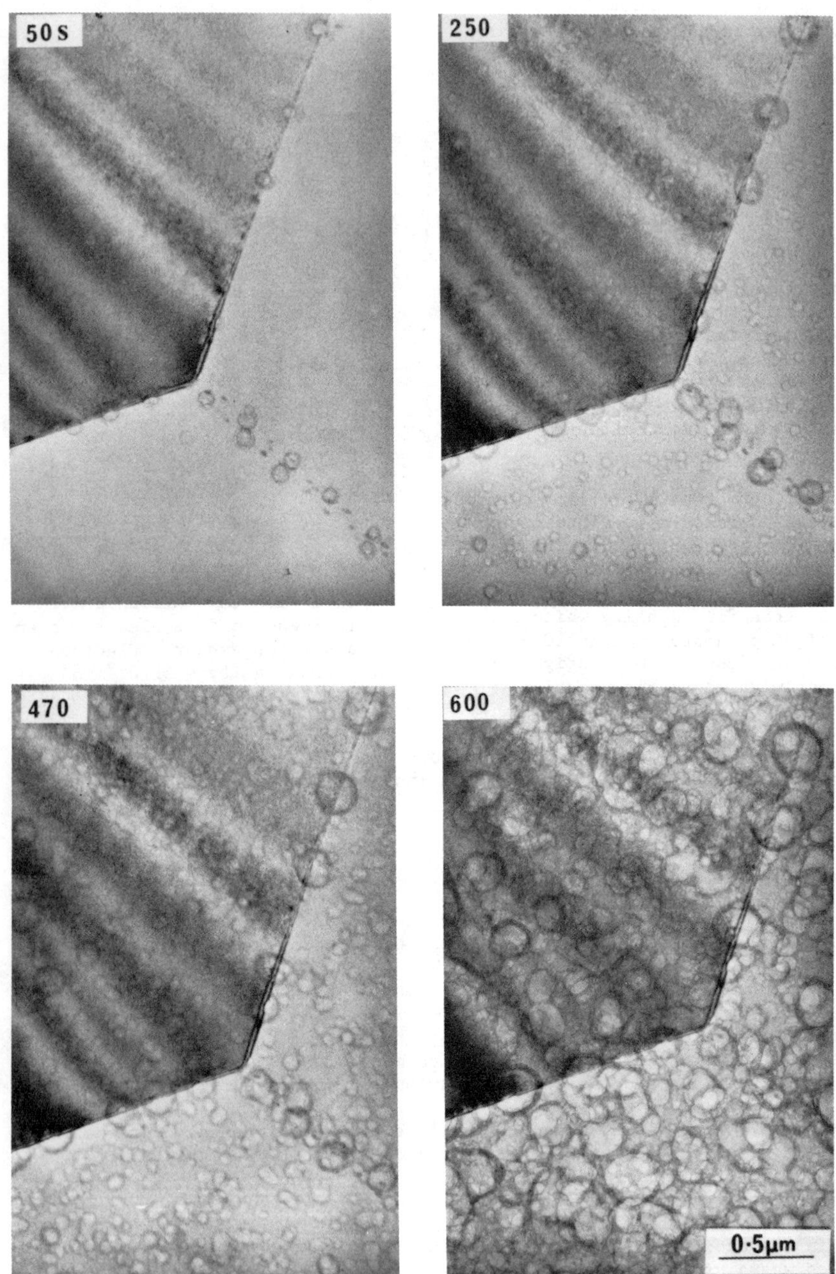

Fig. 3.   Preferential blistering reaction at the grain boundaries of a Al-6%Zn–3%Mg–1.7%Cu alloy. Blisters form on both sides of the boundaries. Electron flux density = 10^{-3} Amm^{-2}, Temperature = 233K.

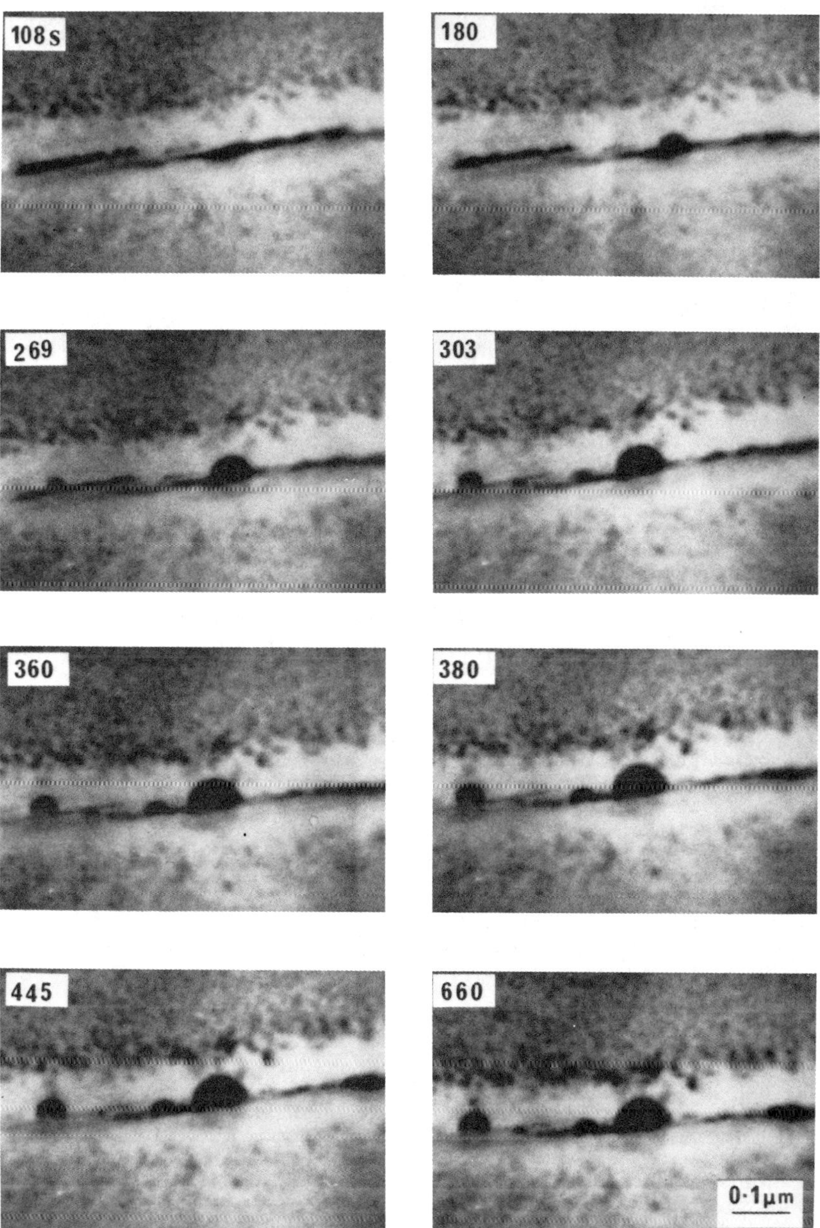

Fig. 4. Stills from a video sequence showing the development of internal
hydrogen bubbles in the Al-6%Zn-3%Mg-1.7%Cu alloy (Time in seconds).
Electron flux density = 10⁻3 Amm⁻², Temperature = 255K.

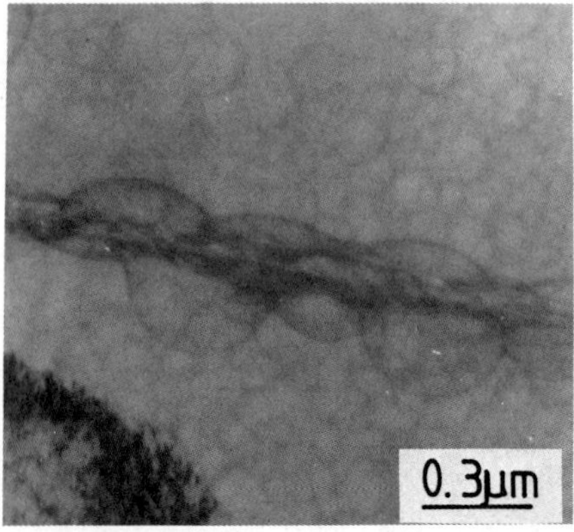

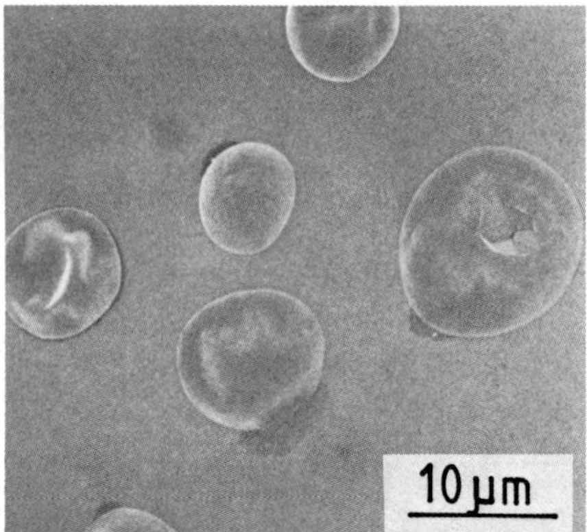

Fig. 5. Preferential surface attack at the grain boundary of a stressed sample in the environmental cell. Microcracks are evident in the boundary.

Fig. 6. Surface blistering in a <u>bulk</u> sample of the 6/3 base alloy exposed to water vapour in an autoclave for 2 minutes.

minimal amounts of bubble formation after considerable irradiation and blistering. When bubbles did form these were confined to grain boundary precipitation or other intermetallics.

In some experiments a stress was applied across the foil during its exposure to irradiation and water vapour in the high voltage microscope at room temperature. The result of such action was to induce the formation of small cracks at the grain boundaries after the surface blistering reaction had taken place for some time. The newly exposed surfaces thus produced rapidly blistered and generally exhibited enhanced reactivity (Fig. 5). The cracks formed prior to the observation of grain boundary bubbles in the absence of grain boundary precipitation in the base alloy.

In all the in-situ experiments the reaction between water vapour and the metal was clearly stimulated by the electron irradiation and took place very much more slowly outside the electron beam exposed areas. Since the reactivity was equally great at 100KV as at 500KV it cannot be due to atomic displacement damage in aluminum and alumina since the threshold voltages for both lie above 100KV. The reaction must, therefore, be stimulated by an ionization damage mechanism (4, 7).

Although the in-situ experiments are clearly influenced by electron irradiation the results of bulk studies show that the types of reaction and the sequences observed are the same as those of bulk unirradiated material. Surface blistering (Fig. 6) is observed and internal hydrogen bubbles are formed (Fig. 7). Most interestingly bubble formation in the high purity alloy is only detected when grain boundary precipitates over 200A in size are present: a result identical with that of the in-situ experiments. The effects of copper and chromium additions were also consistent with the in-situ results. It is, therefore, believed that the role of electron irradiation is simply to stimulate the series of events which normally occur rather than introducing different and, therefore, spurious reactions. Thus the in-situ gas reaction studies can be used to interpret the results of bulk studies. For example, it was found that embrittlement of solution treated material could be retarded, but not eliminated entirely, either by the addition of copper or by aging to produce particles at grain boundaries

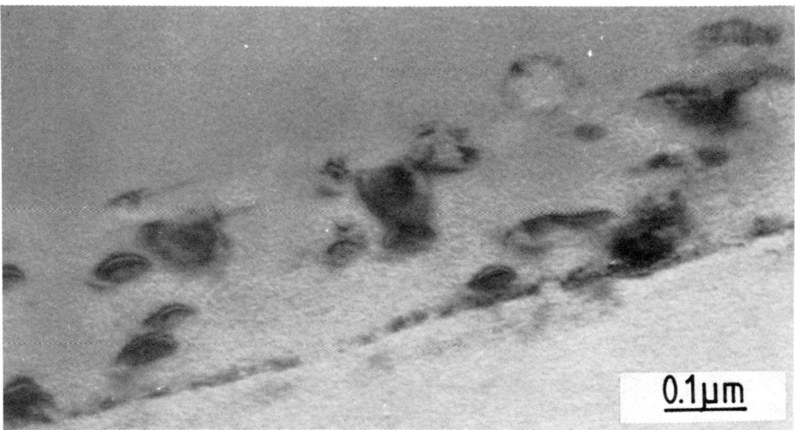

Fig. 7. Hydrogen bubbles at grain boundary precipitates of an Al-6%Zn-3%Mg
alloy tensile sample dry aged at 120°C for 24 hours and subsequently
exposed to water vapour saturated air for 72 hours.

greater than 200A in size (3). The in-situ studies showed that in both cases
hydrogen bubble nucleation was enhanced and occurred at an earlier stage in
the reaction. Such removal of hydrogen from solution is clearly beneficial.
However, the in-situ studies of bubble growth also showed that bubbles ceased
to grow beyond a certain size (as a result of matrix constraints) and, therefore, trap site saturation must eventually occur. Hence, beyond this point
the concentration of hydrogen in solution will rise with increased reaction
time and embrittlement will ensue, as observed in practice.

Discussion

The in-situ gas reaction studies in the high voltage electron microscope have clearly shown that the reaction of Al-Zn-Mg alloys with water
vapour leads to the formation of hydrogen at the metal/oxide interface and
its subsequent penetration into the alloy substrate.

High purity material solution treated and aged so that the grain boundary precipitate size was less than 200A showed no internal bubbles but exhibited severe embrittlement. Material with grain boundary particles greater than 200A showed formation of bubbles and, at least initially, no embrittlement (3, 4,). Since no differences were evident in the surface reaction as a function of aging treatment (i.e. the hydrogen generation step),
it is concluded that embrittlement is not caused by bubbles themselves but
by hydrogen in solution. It is the dissolved hydrogen which induces brittle
decohesion and suitably sized grain boundary precipitates are beneficial by
acting as traps for the absorbed hydrogen. Formation of bubbles leads to a
reduction in the amount of dissolved hydrogen and, therefore, increases the
overall tolerance of the alloy to the presence of hydrogen. Similarly, the
beneficial effect of copper addition on pre-exposure embrittlement and S.C.C.
can be explained by the ready formation of hydrogen bubbles in all heat
treated conditions, including the solution treated state. The role of cop
per cannot, therefore, be simply modification of the grain boundary precipitates to produce more effective nucleation sites (although this may also
occur): possibly copper segregation is present at the grain boundaries,
abinitio and catalyses hydrogen recombination to molecular and ultimately
gaseous form.

The influence of chromium addition, on the other hand, appears to be surface related. The extreme blistering tendency observed, together with the absence of internal hydrogen bubbles suggests that while hydrogen is readily generated at the metal/oxide interface little is absorbed. The reduction of hydrogen entry may be due to chromium catalyzing the formation of molecular, rather than atomic hydrogen at the metal surface: the molecular form would not then be expected to penetrate into the metal but would remain to give rise to the blistering observed.

Conclusions

It has been demonstrated in this work that exposure of Al–Zn–Mg alloys to water vapour leads to the formation of hydrogen at the metal/oxide interface and its subsequent introduction in the alloys.

The amount of hydrogen introduced into the samples is affected by minor alloying additions. Specifically Cr additions reduce the amount of absorbed hydrogen by catalyzing the formation of gaseous hydrogen and its subsequent removal into the surface blisters and the atmosphere.

The internal tolerance of the alloys to the presence of hydrogen is determined by the availability of trap sites for the formation of gaseous hydrogen bubbles. In the high purity alloy grain boundary precipitates of size 200A or greater constitute such sites. In the case of Cu bearing alloys nucleation and growth of bubbles occurs even in the absence of grain boundary precipitates. It is to this effect that these alloys owe their superior embrittlement resistance.

In all cases embrittlement is caused by dissolved hydrogen. Internal hydrogen bubble formation leads to lowering of the concentration of dissolved hydrogen and increases the resistance to embrittlement.

Acknowledgments

This work was partially supported by funds provided by the United States Army through the European Research Office.

References

1. L. Montgrain and P. R. Swann, Int. Confer. on Hydrogen in Metals (ed. A. W. Thompson and I. M. Bernstein, p 375, ASM, Metals Park, OH, 1974.

2. G. M. Scamans, R. Alani, and P. R. Swann, Corrosion Science, vol 16, (1976) p 443.

3. L. Christodoulou and H. M. Flower, Acta Metall, vol. 28, (1980) p 481.

4. L. Christodoulou, Ph.D. Thesis, London University, March 1980.

5. G. M. Scamans, J. Mat. Science, vol. 13 (1978) p 27.

6. H. M. Flower, N. J. Tighe, and P. R. Swann, Proc. 3rd Int. Conf. on H.V.E.M. (ed. P. R. Swann et al), p 383, Academic Press, London, 1974.

7. H. M. Flower, Radiation Effects, vol. 33 (1977) p 173.

DISCUSSION

<u>H. K. Birnbaum, University of Illinois</u>: In the past, workers in the field of hydrogen embrittlement have obtained similar results and then have spent their time arguing about the mechanisms of fracture. Now we have, in this paper and the several preceding ones, similar experiments, on similar alloys which give different results; aluminum hydride, bubbles, or neither, and we have arguments about the results as well as the mechanisms. Is there something, perhaps alloy chemistry or charging conditions which can be identified as giving rise to these divergent observations?

<u>L. Christodoulou</u>: As indeed you have indicated, there exists considerable disagreement as to the exact mechanism of environment sensitive failure of most materials including high strength aluminum alloys. The results presented during this session all agree on a hydrogen embrittlement dominated event and differ only as to the precise role of hydrogen. This is not exclusive to aluminum alloys.

Our work has shown that the overall hydrogen effect depends mainly on two factors. Firstly, on the generation and entry of hydrogen into the the material and secondly, on the tolerance of any one microstructure (through its ability to form bubbles etc) to the hydrogen presence. Both of these processes have been shown to be sensitive to the presence of minor quaternary alloying additions. Thus, it is not surprising that nominally similar materials behave in different ways.

<u>C.D.S. Tuck, Alcan Labs</u>: It is well known that the addition of copper to magnesium catalyzes the formation of magnesium hydride when the metal is reacted with pressurized hydrogen. Do you see this process suggesting an explanation of why copper-containing Al-Zn-Mg alloys previously reacted with water vapor will readily form internal hydrogen bubbles and are able to recover their ductility under ambient temperature and pressure conditions?

<u>L. Christodoulou</u>: Our results show that the addition of copper to Al-Zn-Mg alloys greatly enhances the formation of hydrogen bubbles at the grain boundaries. We found no evidence of hydrides (magnesium or aluminum) and thus I cannot give any support to your proposal.

<u>J. A. Moskovitz, Kaiser Aluminum</u>: My comment is in response to Howard Birnbaum's question as to why many researchers get different results on the same system: Some work, which will be reported by Joe Pickens of Martin Marietta, seems to indicate that the oxide film formed during solution heat treating and aging can affect H entry into the metal, as the composition of the oxide varies significantly with solution heat treat temperature in these alloys This may explain in part the concern expressed by Dr. Birnbaum.

<u>L. Christodoulou</u>: Similar suggestions as to the entry step being influenced by solution heat treatment temperature and oxide composition have been made previously by Scamans, Alani and Swann (Ref. 2).

EVIDENCE FOR THE FORMATION OF MAGNESIUM HYDRIDE
ON THE GRAIN BOUNDARIES OF Al-Mg AND Al-Zn-Mg ALLOYS
DURING THEIR EXPOSURE TO WATER VAPOUR

C.D.S. Tuck

Alcan Laboratories Limited
Banbury Research and Development Centre
Banbury, Oxon., OX16 7SP, England

Differential Scanning Calorimetry (DSC) and Gas Chromatography were used to investigate and quantify reactions occurring when Al-5%Mg and Al-4½%Zn-1½%Mg alloys, previously exposed to water-vapour-saturated air (WVSA), were heated from ambient temperature to 600°C. Hydrogen evolution was found to occur, mainly between the temperatures 400°C and 500°C, and this was paralleled by decomposition peaks in the same temperature range on the DSC traces. A similar investigation carried out on magnesium hydride showed that decomposition occurred in the same temperature region.

A parallel investigation involving transmission electron microscopy of magnesium hydride showed that decomposition to magnesium and hydrogen occurred readily after short exposures to the electron beam. When thin foils prepared from Al-5%Mg and Al-4½%Zn-1½%Mg previously exposed to WVSA were placed in an electron beam, bubbles of hydrogen developed on the grain boundaries, showing a localisation of the decomposition reaction to the boundary region. Using the DSC data and quantifying the volume of hydrogen produced by electron irradiation of the exposed alloys, the magnesium hydride was calculated to occur on the grain-boundary surfaces to a depth of $\sim$ 25 Å in alloys reacted with water-vapour-saturated air at 70°C for about 50 days. It is suggested that its presence plays a prominent role in the pre-exposure embrittlement and stress-corrosion cracking of Al-Mg and Al-Zn-Mg alloys.

Introduction

Extensive investigations have shown with little doubt that Al-Zn-Mg (7000 series) and Al-Mg (5000 series) alloys suffer hydrogen embrittlement when exposed to environments which react with the metal and generate hydrogen (1-13). This effect is not dependent upon there being a stress present on the specimen during exposure and for this reason it has been termed pre-exposure embrittlement. Hydrogen has been observed to enter the alloys when they are undergoing corrosion reactions in salt solution with or without impressed currents (1,5,7-10), in water vapour (2,3,4,11) and during bombardment with ionised hydrogen gas (6), resulting in the total brittle fracture of grain boundaries when tensile specimens of charged material are subsequently fractured in air. When the pre-exposed samples are fractured under vacuum (2,5), hydrogen is evolved and a completely ductile fracture is obtained. This return to ductile fracture behaviour is also observed when the pre-exposed specimens are re-solution-heat treated (1,2). With Al-8%Mg (8), Al-6%Zn-3%Mg-0.14%Cr and 7075 alloy (2), this recovery of pre-exposed material is reported to occur at room temperature with the concurrent loss of hydrogen by the specimens (8,9).

There have been some suggestions that hydrides could be formed on the grain boundaries of Al-Zn-Mg during pre-exposure embrittlement (14). If this were so, the hydride formed could be one or more of three possibilities: aluminium hydride, magnesium hydride or zinc hydride. All three hydrides are quite stable at room temperature, decomposing, according to the available literature (15), at $\sim 200^{\circ}$C, 287°C (16) and 90°C respectively. Magnesium hydride is the only one which can be prepared directly from its constituent elements, the reaction occurring at 380°-450°C, with hydrogen at 100-200 atmospheres, continuous grinding of the magnesium, and iodine added as a catalyst (17). Thus, if hydride compounds were to be formed during pre-exposure embrittlement of Al-Mg and Al-Zn-Mg they would be more likely to be those involving magnesium, especially as it has been shown that alloys of magnesium with copper or nickel react with hydrogen gas quite readily (350°C at 20 atmospheres pressure) to yield magnesium hydride (18,19) [cf the stress-corrosion resistance of Al-Zn-Mg-Cu alloys (20)].

There has hitherto been no systematic work on what form the embrittling hydrogen species is on the grain boundaries of pre-exposed Al-Mg and Al-Zn-Mg. The work described here gives an indication of both the form of the embrittling hydrogen on the grain boundaries and the mechanism of embrittlement, by investigating the evolution of hydrogen from exposed specimens when they are heated.

Experimental

The alloys used were Al-5%Mg (5.11%Mg, 0.11%Fe, 0.055%Si, 0.01%Ti, 0.001%Zn, 8ppm B) and Al-4½%Zn-1½%Mg (4.30%Zn, 1.5%Mg, Fe<0.01%, Si<0.01%, Ti<0.010%) which were cold-rolled to 0.5 mm. The sheet was solution-heat treated either at 450°C for 30 minutes (Al-5%Mg) or 465°C for 30 minutes (Al-4½%Zn-1½%Mg), cold-water quenched, and given a light chemical polish in Phosbrite (chromic/phosphoric acid).

The sheets were either exposed to water-vapour-saturated air (WVSA) for a number of days at 70°C or treated for the same length of time at 70°C in laboratory air. After exposure to WVSA the surface of the exposed metal was covered with a white film of hydroxides which is known to contain excess water molecules which would react with the aluminium to give hydrogen during any subsequent heating of the specimen. Thus it was removed by dissolving in chromic/phosphoric acid and subsequently abrading the surface. Coupons measuring 60 mm x 5 mm were cut from the pre-exposed and non-exposed sheet samples and placed at the closed end of a narrow silica tube of volume 20 ml. A long glass tube extended through a rubber bung at the mouth of this tube to the specimen and a rubber septum was placed on the external end of this tube. The septum enabled syringe samples of the gaseous contents of the silica tube to be taken for analysis in a gas chromatograph. The silica tube was placed in a furnace and whilst the specimens were heated from ambient temperature to 600°C samples of the atmosphere of the silica tubes were taken after each 25 deg.C temperature rise and analysed for hydrogen, oxygen and nitrogen. The hydrogen-evolution characteristics observed during heating the pre-exposed alloys were compared to those exhibited when non-pre-exposed mateial was heated.

A comparison of these results was made with magnesium hydride (obtained in a powder form >85% pure from Lancaster Synthesis Limited, Lancaster, England) heated over the same temperature range.

Differential Scanning Calorimetry (DSC) was carried out by means of a Dupont 910 Differential Scanning Calorimeter, comparing pre-exposed Al-Mg and Al-Zn-Mg with non-exposed material. The samples were heated at a fixed heating rate (50 deg.C/min) from ambient temperature to 570°C under a flow of high-purity argon, and the resulting temperature differential curves plotted. Again the results were compared with those obtained by heating magnesium hydride (in an enclosed aluminium container under argon).

Some magnesium hydride powder was placed on a copper grid and viewed in a JEOL 120C electron microscope under electron transmission. Diffraction patterns were taken of the same area after exposure to the beam for short times, in order to detect any change in composition during electron irradiation.

Results

Figure 1 shows the hydrogen evolution characteristics during the heating of Al-5%Mg sheet which had been exposed on both sides to WVSA at 70°C for 40 days and Figure 2 shows the evolution of hydrogen with temperature from Al-4%Zn-1½Mg sheet, one side of which had been exposed to WVSA at 70°C for 40 days. A comparison was made with alloy samples

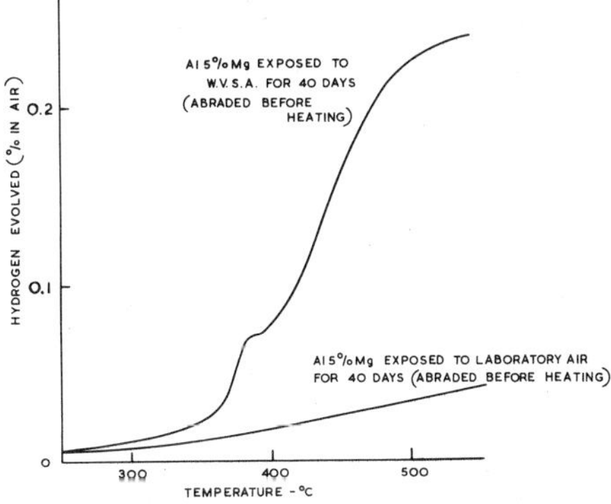

Figure 1. The hydrogen-evolution characteristics with temperature on heating Al-5%Mg exposed to water-vapour-saturated air at 70°C for 40 days and the same material exposed to laboratory air at 70°C for 40 days. The temperature was raised at a rate of 8 deg.C per minute.

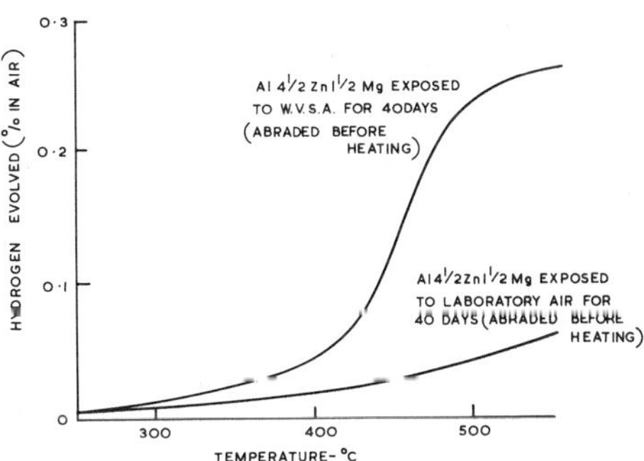

Figure 2. The hydrogen-evolution characteristics on heating Al-4½Zn-1½Mg which had been exposed to water-vapour-saturated air at 70°C for 40 days, and the same material exposed to laboratory air at 70°C for 40 days. (Heating rate 8 deg.C/min.)

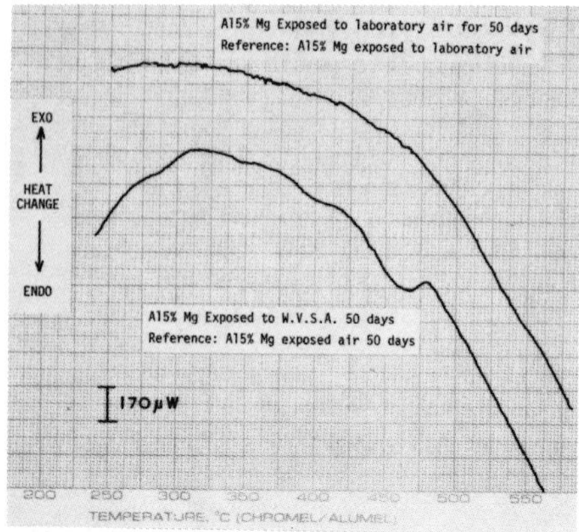

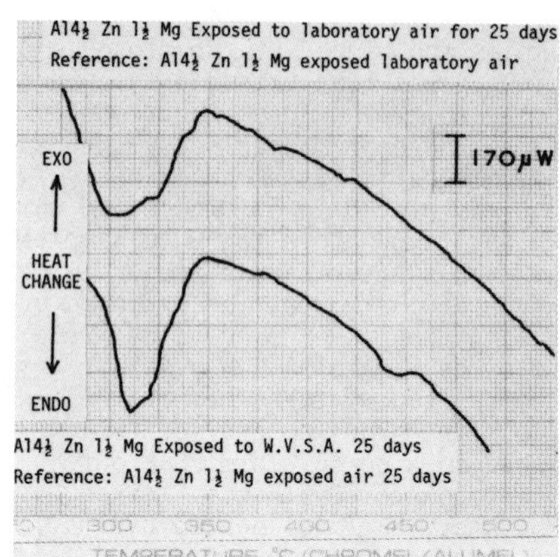

Figure 3. DSC traces obtained by heating Al-5%Mg (6mm x 0.5mm discs), both sides of which had been exposed at 70°C for 50 days either in WVSA or laboratory air using the latter samples as references. All samples had the surface hydroxides removed before heating.(Heating rate 50 deg. C per min.

Figure 4. DSC traces obtained by heating Al-4½Zn-1½Mg (6 mm x 0.5 mm discs), one side of which had been exposed at 70°C for 25 days either in WVSA or laboratory air using the latter samples as references. All samples had the surface hydroxide removed before heating. (Heating rate 50 deg.C/min.

exposed to laboratory air at 70°C for 40 days. Hydrogen evolution from the pre-exposed alloys was seen to occur mainly in the temperature range 400°-500°C, the maximum evolution rate occurring at around 450°C. The non-exposed samples also displayed hydrogen evolution characteristics, but on a smaller scale. This is probably due to hydrogen contained in the samples as a result of their casting and the reaction of the metal with water-vapour contained in the atmosphere. The volume of hydrogen evolved by the pre-exposed specimens in the temperature range 400°-500°C was about 0.04 ml ($\pm$ 0.01 ml)

Figure 3 and Figure 4 show DSC traces of Al-5%Mg exposed to WVSA at 70°C for 50 days and Al-4½Zn-1½%Mg exposed similarly for 25 days. In both cases the reference samples were exposed to laboratory air for the same time periods. Control experiments were also conducted on the samples exposed to laboratory air to determine the shape of the base line, which was found to be highly curved towards the high-temperature end, caused by slight differences in heat emissivity between the sample and reference. The Al-5%Mg exposed to WVSA showed an endothermic peak in the temperature range 430°-480°C and possibly two other endothermic peaks with minima at 345°C and 400°C. Pre-exposed Al-4½Zn-1½Mg exhibited two endothermic peaks at temperature ranges 275°C-350°C and 435°C-455°C. The first of these is due to dissolution of the equilibrium ageing precipitate [η-phase (21)].

The results of similar experiments carried out on magnesium hydride are shown in Figure 5 and Figure 6. Figure 5 shows the hydrogen evolution with temperature obtained by heating a sample under argon (in order to minimise hydrolysis by water vapour and reaction with oxygen) at 2 deg.C/min in the temperature range 200°-500°C. The hydrogen is evolved between the temperatures 400°C and 470°C with the maximum reaction rate occurring at 440°C. The DSC curve of magnesium hydride sealed in an aluminium container under argon and heated at 20 deg.C/min is shown in Figure 6. It shows the presence of two small peaks at 225°C and 340°C and a large endothermic peak in the range 490°-525°C. The latter appears to be composed of two subsidiary peaks.

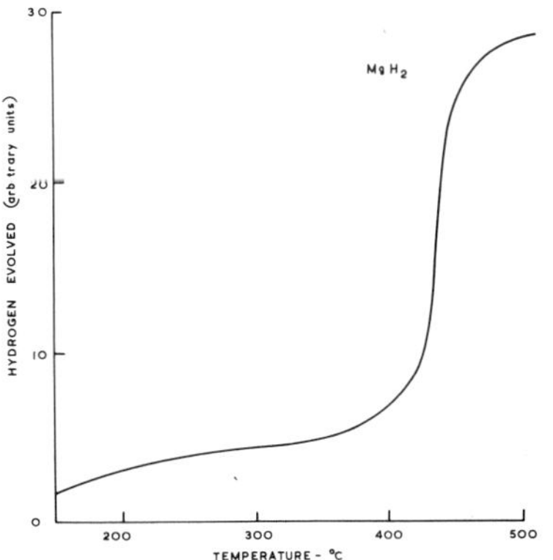

Figure 5. The hydrogen evolution characteristics with temperature on heating magnesium hydride under argon.(Heating rate 2 deg.C per minute.)

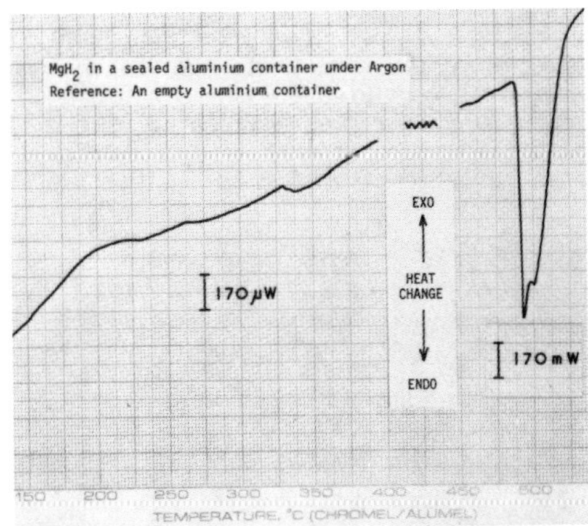

Figure 6. A DSC trace obtained by heating a sealed aluminium container filled with magnesium hydride under argon, with an empty aluminium container as reference. (Heating rate 20 deg.C per min.)

Fractography of pre-exposed Al-4½Zn-1½Mg subsequently heat-treated was undertaken using a scanning electron microscope. Specimens which had been exposed to WVSA at 70°C for 40 days were heated to 350°C, 400°C and 450°C, quickly cooled and bent. Fracture occurred in the surface regions of the specimen and Figure 7 shows the types of fracture observed. In all cases the fracture is seen to be intergranular, but the specimen heated to 350°C shows a much more brittle fracture mode.

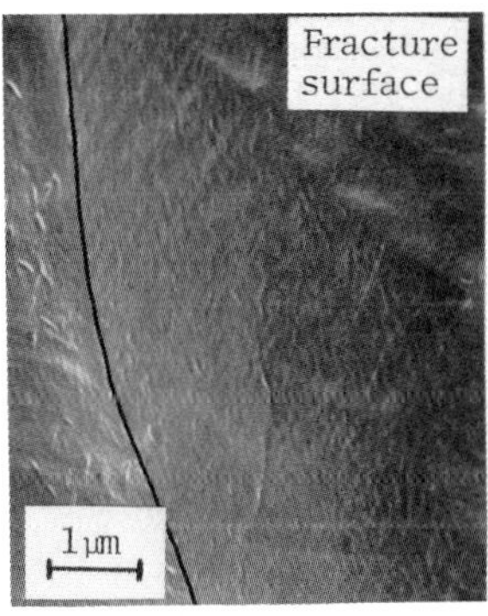

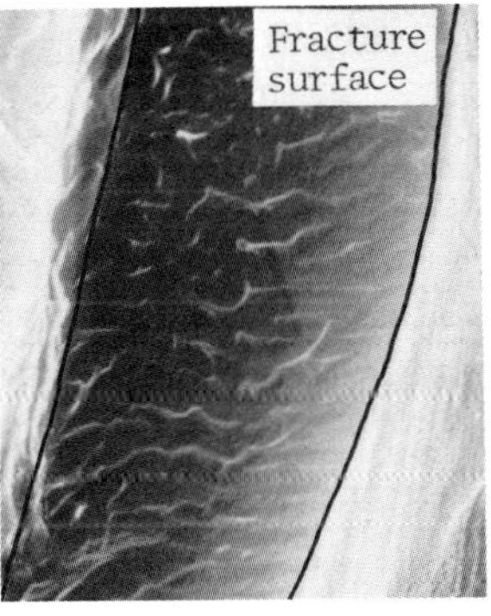

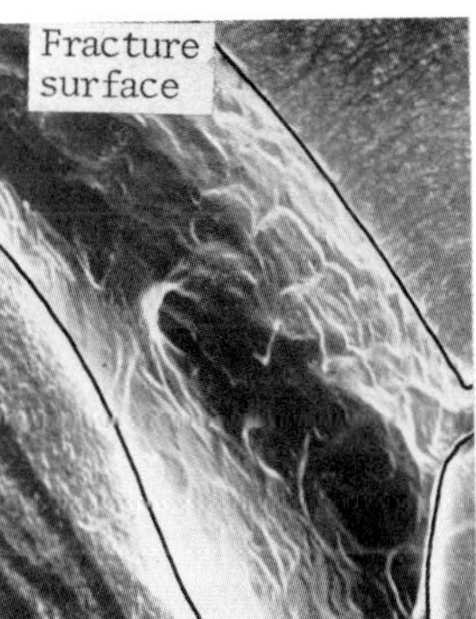

(a)   (b)   (c)

Figure 7. Fractography of Al-4½Zn-1½Mg alloy previously exposed to WVSA at 70°C for 40 days, heated to the temperature shown, quickly cooled and then bent to induce fracture in the surface region (a) 350°C, (b) 400°C, (c) 450°C

Magnesium hydride is not very stable under electron-beam irradiation, as shown in Figure 8. The selected area diffraction pattern (SADP) in

Figure 8(a) was obtained from a particle of magnesium hydride located under the beam, and shows strong reflections from 110 planes (3.2 Å).   After 30 seconds beam exposure during which time there were vigorous internal disturbances in the particle, the SADP shown in Figure 8(b) was taken. The rings observed are solely from magnesium metal.   Thus, the decomposition $MgH_2 \rightarrow Mg + H_2$ occurs by electron beam irradiation.   This behaviour can be compared to the formation of hydrogen bubbles on grain boundaries of pre-exposed Al-Mg.   Figure 9 shows a grain boundary in Al-5%Mg exposed to WVSA for 50 days.   It was irradiated by the electron beam for one minute, during which time hydrogen bubbles were nucleated and grew, generating their characteristic strain fields in the lattice around the boundary.

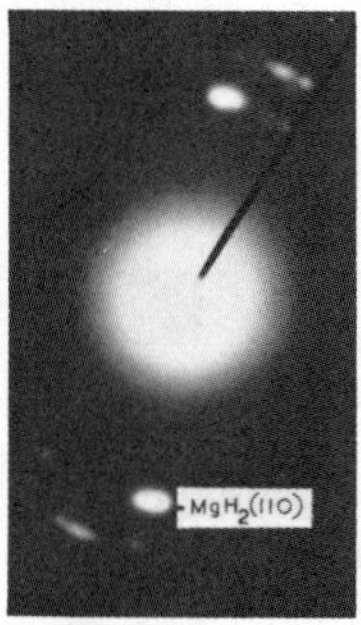

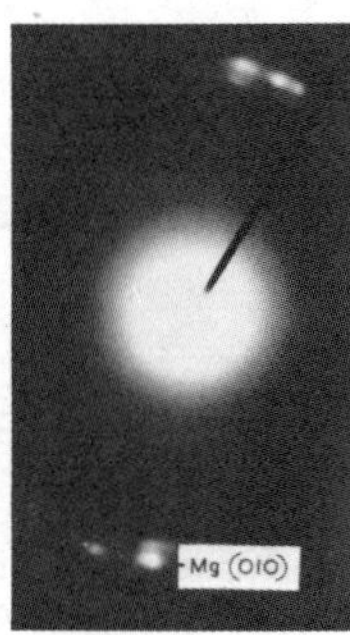

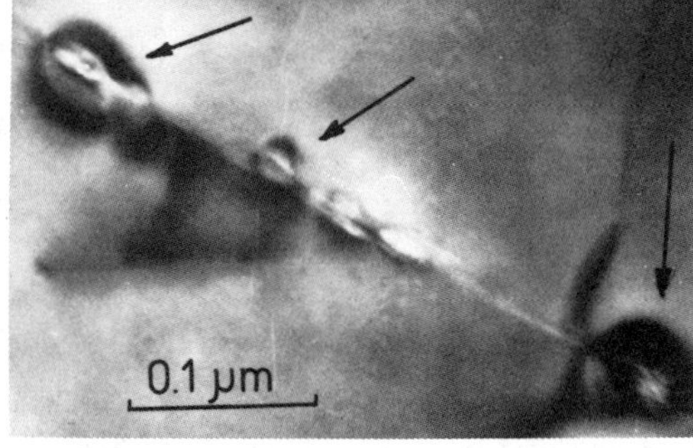

Figure 8.   Selected area diffraction patterns of magnesium hydride powder supported on a copper grid after different times of exposure to the electron beam. (a)   after 5 seconds exposure (b)   after 30 seconds exposure

Figure 9. Hydrogen bubbles with associated strain fields (arrowed) formed after 60 seconds electron beam exposure of a grain boundary in a sheet of Al-5%Mg which had been exposed to WVSA at 70°C for 50 days.

Discussion

There is good correlation between the position of peaks in the DSC traces and the temperature range in which hydrogen is evolved from pre-exposed Al-5Mg and Al-4½Zn-1½Mg.   The correlation between the decomposition temperature range of magnesium hydride and the endothermic peak position on the DSC trace is less good, which is due primarily to two factors.   The first is that the hydride was sealed in an aluminium container when the DSC experiment was carried out and the second is that the heating rate in the DSC cell was ten times that of the furnace used in the hydrogen-evolution experiment.   Both these factors cause the temperature of decomposition observed in the DSC to increase by tens of degrees (22).   The available literature (15,16) suggests that the decomposition temperature of magnesium hydride at atmospheric pressure should be 300°-330°C, a much lower temperature than that observed here.   Perhaps the hydride becomes more stable on prolonged storage as there is no reported decomposition behaviour under conditions similar to those used in the present work;   all are careful to use freshly prepared hydride.

The enthalpy of decomposition of magnesium hydride was calculated using the area of the large peak in Figure 6 and the appropriate calibration factors for the DSC instrument, and was found to be + 8.2 kcal/mole (± 0.5 kcal/mole).   The values quoted in the literature (15) are around -18 kcal/mole for the heat of formation of MgH_2 at 298°K.   Using values of heat capacities for Mg, H_2 and MgH_2 from tabulated data, the enthalpy of decomposition of the hydride at 500°C was found to be +11 kcal/mole. There is still a significant discrepancy between this figure and the experimental result obtained from the DSC experiment and this is probably due to the use of thermodynamic data for MgH_2 which have only been found applicable below 350°C.

Using data from Figure 3 the mass of magnesium hydride responsible for the peak at 470°C can be calculated from the enthalpy value obtained above. The result is that in a 0.5-mm sheet of Al-5%Mg, area 0.29 sq.cm, exposed both sides to WVSA for 50 days, there is 5.0 µg (- 0.3 µg) of magnesium hydride. This should produce 0.005 cc of hydrogen when decomposed. Three sq. cm. of Al-5%Mg exposed to WVSA at 70°C for 40 days produces 0.04 cc (- 0.01 cc) of hydrogen between the temperatures 400°C and 500°C giving some quantitative evidence that the hydrogen produced in this temperature range is due to magnesium hydride decomposition. The DSC data for Al-4½Zn-1½Mg shows that a sheet sample (area 0.28 cm^2) exposed on one side to WVSA for 25 days contains 1.0 µg (-0.3 µg) of magnesium hydride. Using the value of 5 µg of magnesium hydride in the Al-5%Mg sheet sample (0.29 sq.cm x 0.5 mm), an estimate of its distribution on the grain-boundary surfaces was obtained. The grain size of this material is 50 µm and if individual grains are approximated to flat hexagonally-faced pillars the total grain-boundary surface area represented by the grains is 12.8 sq.cm. The density of MgH_2 is 1.5 g/cm^3, thus 5 µg must occupy an average depth of 25 Å on each grain surface. In practice the magnesium hydride would probably be unevenly distributed, there being generally more towards the surface and an uneven distribution amongst the grain boundaries, depending on their different chemical properties.

Data from the hydrogen bubbles generated when an exposed boundary is subjected to electron irradiation (Figure 9) leads to an estimate of the hydrogen volume which would be produced by the decomposition reaction of the boundary MgH_2. It has been difficult in the past to judge the hydrogen pressure manifest in these bubbles, but occasionally they have been observed to punch dislocations into the grains. The yield stress of Al-5%Mg is 100 MPa which is equivalent to 1000 atmospheres pressure. This behaviour of the bubbles does not often take place, thus the pressure is probably more of the order of $\sim$ 100 atmospheres. The volume of hydrogen giving the bubbles in Figure 9 is then $\sim10^{-14}$ cc. Using the equation $MgH_2 = Mg + H_2$, this volume of hydrogen must have been produced by $\sim10^{-17}$ g of MgH_2. The area of boundary represented by Figure 9 is $\sim10^{-10}$ sq.cm which would have this quantity (i.e. $\sim10^{-17}$ g) on it to a depth of about 10 Å which agrees fairly well with the 25 Å thickness calculated from the DSC data.

Taking the free energy of formation of MgH_2 at 298°K ΔG°_{298} as -8.7 kcal /mole, the fugacity of hydrogen required for its formation at 70°C is 10^5 atmospheres. In the reaction of Al-5%Mg with WVSA, the equation $2Al + 6H_2O \rightarrow 2Al(OH)_3 + 3H_2$ can be used to approximately describe the main chemical reaction occurring. If equilibrium conditions were to prevail, and the fugacity of water were 1 atmosphere, the fugacity of hydrogen would be 10^{182} atmospheres. In the case of Al-5%Mg the reaction with water is very localised at grain boundary/surface intersections (13) and, because reaction product builds up in this intersection, the hydrogen is produced in a limited space. It is thus quite probable that high fugacities of hydrogen are produced in practice which would lead to the formation of MgH_2 with excess magnesium in the grain-boundary region (23). This process occurs with or without the application of stress to the material. When stress is applied the boundaries containing the hydride are placed in tension and it is suggested that if enough hydride is present it causes the boundary surfaces to separate in a brittle manner. Deeper in the material the concentration

of hydride on the boundaries is less and any decomposition could be accommodated by the formation of hydrogen bubbles exhibiting very high fugacities.
This could explain why the fracture reverts to the normal ductile intergranular mode in this region of the specimen. If pre-exposed Al-Zn-Mg specimens
are fractured in vacuum, hydrogen is evolved and the fracture is ductile (2,
5). This could be explained by a fast decomposition of the magnesium hydride
on the grain boundaries near the surface (caused by the low external pressure)
before fracture occurs.

References

1. W. Gruhl and D. Brungs. Metall, 23 (1969) pp.1020-1026.

2. G. M. Scamans, R. Alani and P. R. Swann. Corros. Sci., 16 (1976) 443-459

3. G. M. Scamans and C.D.S. Tuck. TMS-AIME Meeting on Environment-Sensitive
    Fracture of Engineering Materials. Chicago (1977) pp.464-483.

4. R. Alani and P.R.Swann. Br.Corros.J., 12 (2) (1977) pp.80-85.

5. D. Hardie, N.J.H.Holroyd and R.N.Parkins. Metal Science (1979) pp 603-610

6. G. H. Koch. Corrosion, 35 (2) (1979) pp.73-77

7. T. Ohnishi and Y.Nakatani. J.Jap.Inst.Light Metals, 27 (1977) pp.224-231

8. T. Ohnishi, T.Hara and Y.Nakatani. J.Jap.Inst.Light Metals, 27 (1977)
    pp.367-373

9. T. Ohnishi, T. Hara, Y. Mohri and Y. Nakatani. J.Jap. Inst. Light Metals,
    27 (1977) pp.473-479.

10. T. Ohnishi and Y. Nakatani. J. Jap.Inst.Light Metals, 27 (1978) 123-129

11. Leo Christodolou and Harvey M.Flower. Acta Met., 28 (1980) pp.481-487

12. G. M. Scamans. J. Mat.Sci., 13, (1978) pp.27-36

13. G. M. Scamans and A.S. Rehal. J. Mat.Sci., 14 (1979) pp.2459-2470.

14. G. M. Scamans and C.D.S.Tuck. International Conference on Mechanisms of
    Environment-Sensitive Cracking of Materials. London (Metals Society)
    (1977) (Discussion) pp.493-495

15. K. M. Mackay. Comprehensive Inorganic Chemistry. Ch.2. Pergamon 1975.

16. R.H. Wiswall and J.J.Reilly. 7th Int. Energy Conversion Engineering
    Conference, San Diego (1972) pp.1342-1348.

17. T.N. Dymova, Z.K. Zterlyadkina and V.G.Safronov. Russ. J. Inorg. Chem.,
    6 (1961) pp.389-392.

18. J. J. Reilly and R. H. Wiswall. J.Inorg. Chem., 6 (1967) pp.2220-2223

19. J. J. Reilly and R. H. Wiswall. J.Inorg. Chem., 7 (1968) pp.2254-2256

20. C. Peel and P. Poole. International Conference on Mechanisms of environ-
    ment-sensitive cracking of materials. London (Metals Society) (1977)
    pp.147-157

21. A. Zahra, C. Y. Zahra, M. Laffitte, W. Lacom and H.P. Degischer.
    Z. Metallkde 70 (1979) pp.172-179.

22. W. W. Wendlandt. Thermal Methods of Analysis. Ch.5 John Wiley and Sons
    Inc., New York, 1974

23. J. M. Chen, T.S. Sun, R.K. Viswanadham, J.A.S. Green, Met. Trans. 8A
    (1977) pp.1935-1940

DISCUSSION

<u>P. N. Adler, Grumman Aerospace Corporation</u>: For your model of hydride
formation to be consistent with the stress corrosion behavior of Al-Mg-Zn
alloys, aging would have to deplete Mg at the grain boundary. Is this the
case?

<u>C.D.S. Tuck</u>: Aging AlZnMg alloys results in the formation and growth of
$MgZn_2$ phase precipitates on the grain boundaries. The growth of these
particles decreases the local grain boundary magnesium concentration which
would tend to decrease the cracking susceptibility of the boundaries. I
also believe that the $MgZn_2$ particles can act as traps for incoming hydrogen which would otherwise cause grain boundary embrittlement.

<u>T. Schober, KFA , Jülich, Germany</u>: In my opinion the DSC peaks are due to
the reaction $Mg(OH)_2 \rightarrow MgO$ rather than the decomposition of the hydride
MgH_2. $Mg(OH)_2$ is always found on such samples. As to your mention of the
instability of MgH_2 in the electron beam I would like to add that MgH_2 is
stable in the TEM at temperatures below $-100°C$.

<u>C.D.S. Tuck</u>: I was aware of the possibility that magnesium and aluminum
hydroxides could interfere with any DSC peaks for hydride decomposition,
so I was very careful to remove the hydroxide film on preexposed specimens,
first by dissolution in chromic/phosphoric acid and then by abrading the
surface with silicon carbide. The latter was necessary to remove any
hydroxides formed internally in grain boundaries near the metal surface.
Thus I am confident that the peaks observed in this work are due to hydride
decomposition. Your work showing the stability of MgH_2 below $-100°C$ would
suggest that further TEM studies, directed towards direct observation of
the hydride on grain boundaries of embrittled alloys, should be carried out
using a TEM cold stage, imparting stability to the hydrides under electron
irradiation.

IMPROVED SCC RESISTANCE OF Al–Zn–Mg ALLOYS BY CONTROL

OF Mg CONTENT IN THE BULK METAL AND IN THE OXIDE FILM

J.R. Pickens
D. Venables
J.A.S. Green
Martin Marietta Laboratories
Baltimore, Maryland

Abstract

Previous work has shown that Al–Zn–Mg alloys undergo intergranular
stress-corrosion cracking (SCC) by a mechanism that involves hydrogen
embrittlement, and that enrichment of Mg in the oxide film occurs during
heat treatment. It has been proposed that the Mg plays a role in the
SCC mechanism, specifically, by facilitating entry of hydrogen into
the alloy, possibly by formation of a Mg–H complex. In an attempt to
substantiate the adverse effect of Mg on SCC behavior, high-purity
Al–Zn–Mg alloys were fabricated having constant alloying content (wt%
Zn + wt% Mg), but various Mg to Zn wt% ratios (Mg/Zn). To compare the
effects of bulk composition on SCC susceptibility in aqueous chloride
environments, the thermal (Mg-rich) oxide formed during solution heat
treatment was removed. At a constant solute level, alloys having
higher Mg/Zn and aged to equivalent hardness levels were found to be
more susceptible to SCC.

To investigate the importance of Mg in the oxide, the thermal
oxide was removed from one alloy and a Mg-free oxide was formed on it
by anodization in tartaric acid. Compared to specimens of the same
alloy having thermal oxide films, these specimens with Mg-free oxides
were less susceptible to SCC. This suggests that removal of the Mg-
rich film may decrease the SCC susceptibility of commercial 7xxx
alloys.

Examination with the scanning transmission electron microscope
revealed that the thermal oxide films were significantly more porous
than the Mg-free films. No differences in mechanical cracking were
detected between the thermal and Mg-free oxide films after they were
loaded to the yield stress in air.

The results of these experiments show the harmful effect of Mg
on SCC behavior and are consistent with the view that SCC in Al–Zn–Mg
alloys is facilitated by a Mg–H interaction. However, no direct
evidence of a Mg–H interaction is presented.

Introduction

In earlier work, Green et al. (1) investigated the effect of loading mode on the intergranular stress-corrosion cracking (SCC) of a commerical Al-Zn-Mg alloy and concluded that both dissolution and hydrogen embrittlement (HE) are involved, but that HE is more dominant in the mechanism of cracking. Others have also found evidence of HE in aluminum-base alloys (2-3).

Evidence exists that Mg and H have a mutual affinity (4, 5). For example, the temperature for solution treatment that results in the highest Mg concentration in the oxide film (5) coincides with the temperature for maximum pre-exposure embrittlement (3). Viswanadham, Green, and co-workers (6, 7) found that the grain boundary regions in Al-Zn-Mg alloys are enriched in magnesium, in both the as-quenched and the quenched and aged conditions. They also found that the oxide film formed during solution heat treatment is enriched in magnesium (8). These studies showed large Mg concentrations in the region where SCC occurs, which led Viswanadham et al. (6) to postulate that a magnesium-hydrogen interaction is involved in the SCC mechanism.

The research reported here seeks to elucidate the effect on SCC behavior of magnesium content in the bulk and in the oxide film. It is hoped that control of magnesium content in the oxide film can be used to reduce the susceptibility of Al-Zn-Mg alloys to SCC in aqueous chloride-containing environments.* This work demonstrates the adverse effects of magnesium in the bulk and in the oxide film.

Materials

Ingots of each of four high-purity Al-Zn-Mg alloys (alloys A, B, C, and D), purchased from Metals Specialties Company, Fairfield, Connecticut, were rolled to 0.0635 cm. The alloys contain two different solute levels (wt% Zn + wt% Mg), each level having one alloy at a high magnesium-to-zinc weight ratio (wt% Mg ÷ wt% Zn; Mg/Zn hereafter) and one at a low Mg/Zn. It was later found that these four alloys had some porosity, so an additional high-purity Al-Zn-Mg alloy (alloy G) was purchased as sheet from Reynolds Metals Company, Richmond, Virginia. The compositions of these five alloys (all used in this work) are given in Table I.

Alloy B has a measured grain size of 9 x 17 μm, and alloy G has a coarser and more equiaxed grain size of 170 μm. Qualitative microscopic inspection shows that alloys A, C, and D have grain sizes similar to that of alloy B.

Sheet tensile specimens (7.6 x 1.3 x 0.10-cm) were machined from the five alloys with the longitudinal axis of the specimens parallel to the rolling direction. Specimens were ground to a 600-grit finish before heat treating.

---

* In this paper, the term susceptibility will only be used to describe vulnerability to cracking in chloride-containing aqueous environments.

Table I. Alloy Composition

Element*	Alloy A	Alloy B	Alloy C	Alloy D	Alloy G
Mg (wt%)	2.5	3.7	2.6	4.0	3.7
Zn (wt%)	5.3	4.4	7.7	6.4	4.4
Si (wt%)	< 0.01	< 0.01	< 0.01	< 0.01	0.01
Solute content	7.8	8.1	10.3	10.4	8.1
Mg/Zn	0.47	0.84	0.34	0.63	0.84

* In all five alloys, the wt% of Cr, Cu, Ni, Mn, Fe, and Ti are all
< 0.01 wt%.

Experimental Procedures

Effect of Bulk Composition

To permit investigation of the effect of bulk Mg and Zn composition
on susceptibility, the oxide films on specimens of alloys A, B, C, and D
were chemically removed by the procedure of Pistulka and Lang (9) ("pickling" hereafter). The pickling was performed after the specimens of
the four alloys were slightly underaged to equivalent hardnesses.

The composition of the surfaces and that of the thermal oxide were
characterized by Auger depth profiling. The specimens were then tested
in tension at room temperature in a solution of 30 g NaCl + 80 ml 0.5 N
CH_3COONa + 20 ml 0.5 N CH_3COOH (acetic acid-brine), which has a pH of
~ 5.1 (9). Stress vs time-to-failure (ttf) curves were generated.

Mg-Free vs Mg-Rich (Thermal) Oxide Comparison

An experiment was performed to establish whether Mg enrichment in
the oxide film, which occurs during solution heat treatment, increases
susceptibility. The possibility of decreasing susceptibility by removing the Mg-rich film was also explored.

The thermal film was removed from several specimens of alloy G by
mechanical polishing. On some of these specimens, an alumina film, of
the same thickness as the thermal film, was formed by tartaric acid
anodizing (TAA). The oxide created by TAA has been shown to closely
resemble the natural oxide (10). These specimens were stressed in
tension in acetic acid-brine, and stress vs tff curves were developed.
Again, Auger depth profiling was used to measure the composition and
thickness of the TAA film.

Oxide Characterization

Experiments were performed to elucidate the effect of film porosity and its mechanical cracking behavior on susceptibility. The porosities of various films were qualitatively characterized by observation using the scanning mode of the STEM (scanning transmission electron microscope). To determine whether differences in mechanical cracking behavior of the thermal and Mg-free films exist, two specimens of alloy G, one with a thermal film and one with a TAA film, were loaded to the yield stress in air. The oxides were then observed in the STEM after unloading to note differences in cracking behavior.

Results

Effect of Bulk Composition

Variations in film composition were successfully eliminated by pickling, which removed the Mg-rich oxides and replaced them with an alumina film. (Compare Auger depth profiles in Figs. 1 and 2, and note that since the Auger energy of Na and Zn are the same, their curves coincide.) The profiles, which are shown in Figs. 1 and 2 for alloy B, are also typical of alloys A, C, and D. Pickling replaces the thermal film (Mg-rich) with a thinner "natural" film.

The stress vs tff curves for the bulk composition experiments (Fig. 3) show that susceptibility increased with increasing solute content. In addition, at a similar solute content, susceptibility increased with increasing Mg/Zn (see Table I and Fig. 3). For example, the curve given by the linear regression for alloy A (Mg/Zn = 0.47) is displaced to the right (greater ttf at a given stress) by more than a factor of two with respect to that of alloy B (Mg/Zn = 0.84). The curves in Fig. 3 show considerable scatter, which is attributed to porosity in the starting materials, but it is clear that at each solute content, the materials with higher Mg/Zn display greater susceptibility.

Mg-Free vs Mg-Rich (Thermal) Oxide Comparison

A comparison of the relative susceptibility of specimens having a Mg-rich, thermal vs a Mg-free film that forms naturally by pickling is not an ideal comparison because great differences in film thickness exist. This factor was compensated for by anodizing TAA films to the same thickness as the thermal films ($\sim$ 2000 Å) on specimens of alloy G that had the thermal film removed by polishing.

The TAA film obtained was essentially Mg-free, and was verified to be $\sim$ 2000-Å thick by 1) examining Auger depth profiles and 2) observing, with the scanning mode of the STEM, a metallographic section containing the oxide.

Figure 4 shows linear regressions of stress vs ttf data for alloy G with three different films: formed after polishing (i.e., mechanically polished) and TAA (both Mg-free), and thermal (Mg-rich). The mechanically

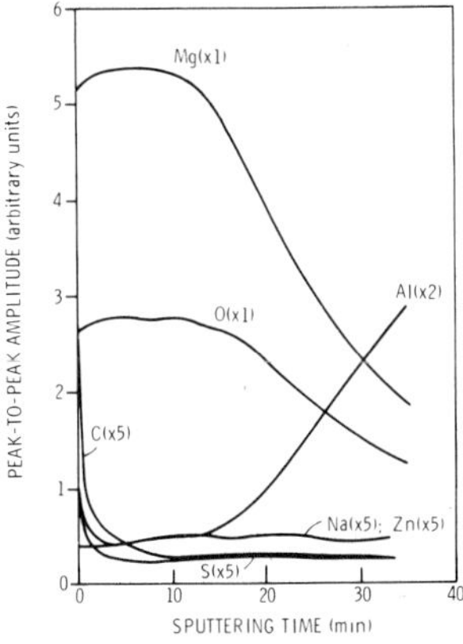

Fig. 1. Auger depth profile of thermal oxide in alloy B, sputtered at ~ 30 Å/min.

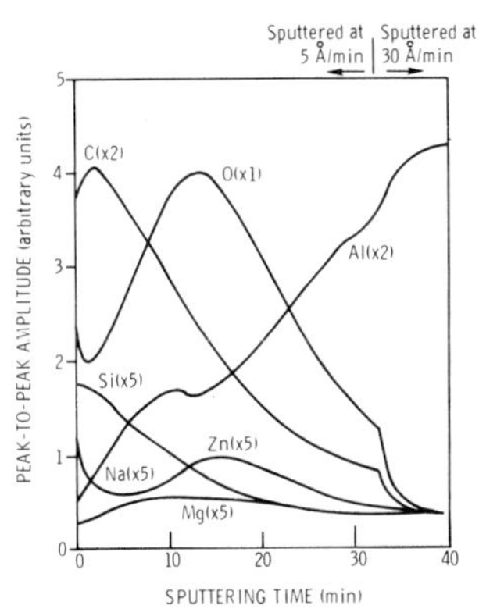

Fig. 2. Auger depth profile of room temperature-formed oxide on alloy B. The thermal oxide was stripped by the procedure of Pistulka and Lang (9).

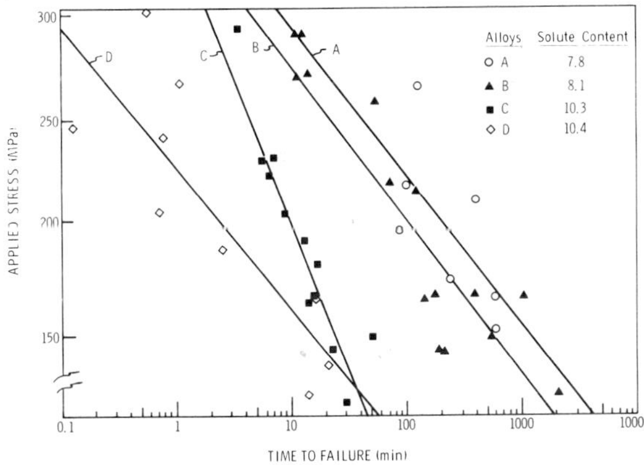

Fig. 3. Comparison of susceptibility of alloys A, B, C, and D in acetic acid-brine.

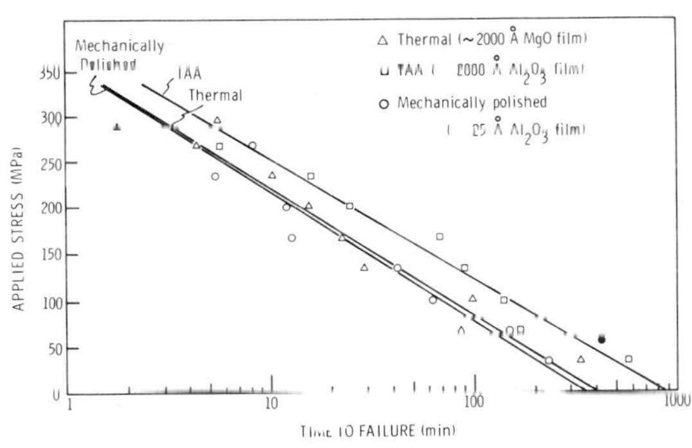

Fig. 4. Comparison of susceptibility of alloy G specimens having three different oxide films.

polished specimens have Mg-free films that are only 25-Å thick but
have approximately the same susceptibility as specimens with 2000-Å
thermal films. When the Mg-free film is grown to ~ 2000 Å by TAA, ttf at
a given stress increases by a factor of two.

Oxide Characterization

Film porosity as well as its Mg content might contribute to susceptibility. A thermal and a Mg-free (TAA) film of alloy G were observed
with the scanning mode of the STEM. The thermal film was quite porous,
having pores on the order of 2 μm in diameter spaced about 10 μm apart.
The TAA film, however, was essentially pore free. These observations
suggest that differences in film porosity may partially explain the differences in susceptibility between the Mg-free and thermal films.

Because we felt that differences in the tendency for films to
crack under purely mechanical influence also might explain differences
in susceptibility, we loaded two specimens of alloy G (one with a
thermal film and one with a TAA film) to the yield stress in air, and
then examined them in the STEM. No signs of mechanical cracking were
observed, indicating that the differences in susceptibility were likely
caused by factors other than differences in the mechanical properties
of the film.

Discussion

Effect of Bulk Composition

An increase in susceptibility to SCC with increasing solute content
is expected. However, the finding that susceptibility increases with
increasing Mg/Zn at a constant solute content disagrees with some
earlier work and may clarify an existing controversy in the literature.
Some investigators have claimed that low Mg/Zn in commercial Al-Zn-Mg
alloys produces greater strength and increased SCC susceptibility than
found in alloys with high Mg/Zn (11). In addition, Sakena et al. (12)
claim that low Mg/Zn improves strength but impairs SCC resistance, and
Swann (13) discussed preliminary evidence that HE in 7xxx alloys may be
related to Zn content and that aluminum alloys not containing Zn may be
immune to cracking. These views are inconsistent with the results of our
bulk composition experiments.

On the other hand, Staley (14) conducted experiments indicating
that low Mg/Zn causes an increase in SCC resistance in commercial 7xxx
alloys, which is consistent with our findings for high-purity Al-Zn-Mg
alloys. Also, the general trend in the development of newer 7xxx alloys
with enhanced SCC resistance (e.g., 7050, 7475, 7049), has been towards
lower Mg/Zn.

The earlier work may have suggested that low Mg/Zn causes increased
susceptibility because the effects of film composition were not controlled. Viswanadham et al. (5) showed that slight changes in solution

treatment temperature can greatly alter the Mg content of the oxide film. The results of this paper demonstrate that Mg content in the film greatly affects susceptibility. Thus, variations in the Mg content of the oxide films might account for the conflicting information in the literature. In addition, Staley claims that Cu content, which he found enhances SCC resistance in commercial 7xxx alloys, often obscures the effect of Mg/Zn (15). We plotted Staley's measure of SCC resistance, the mean critical yield strength (MCYS; Ref. 14) vs Mg/Zn for alloys of similar composition. Scatter was surprisingly low and revealed a definite trend toward decreased MCYS with increasing Mg/Zn.

Our findings are consistent with the suggestion of Viswanadham et al. (6) that a Mg-H interaction is involved in the SCC of Al-Zn-Mg alloys. This result does substantiate the adverse effect of Mg on susceptibility but, unfortunately, does not provide any direct evidence of a Mg-H interaction. However, it does have interesting practical applications. It suggests that Mg/Zn could be decreased, which still permits heat treatment to high strength levels, but reduces susceptibility.

Mg-Free vs Mg-Rich (Thermal) Oxide Comparison

Mechanical polishing and polishing plus TAA were both successful in removing the Mg-rich, thermal oxide and replacing it with a Mg-free film formed at room temperature. Other results (to be published) show that chemical techniques, as opposed to mechanical polishing, can also remove the thermal film, suggesting that there is commercial potential for decreasing susceptibility by control of Mg in the oxide film. We are conducting tests on commercial 7075-T6 to see if our technique works on an alloy of commercial purity.

The finding that specimens with essentially Mg-free oxide films have decreased susceptibility is again consistent with the model of Viswanadham et al. (5, 6); but, no direct evidence of a Mg-H interaction in the cracking process has been found.

Oxide Characterization

Since the TAA film is less porous than the thermal film, the oxide morphology may also be responsible for the decreased susceptibility of specimens having a Mg-free film. Unfortunately, Mg-free and thermal films of equivalent porosity and thickness could not be made.

Since the TAA and the thermal film both showed no signs of purely mechanical cracking when loaded to the yield stress, it is likely that differences in susceptibility are not due to differences in purely mechanical behavior of the films. This is consistent with the work of Bubar and Vermilyea (16) who demonstrated that alumina films have substantial ductility. (It is recognized that the oxide films were not examined at high magnification under load.)

The observed differences between the thermal and Mg-free films --
in Mg content and porosity -- are the two factors that most likely explain
the differences in susceptibility.

Conclusions

Magnesium has been shown to have an adverse effect on the SCC behavior of high-purity Al-Zn-Mg alloys. The effect is particularly damaging when a small fraction of the bulk Mg content finds its way to the
oxide film. This information ultimately may be useful in decreasing
the SCC susceptibility of commercial 7xxx alloys. Based on this work,
our conclusions are as follows:

1. When the effects of oxide film composition are eliminated
    for alloys of constant solute content (wt% Mg + wt% Zn),
    susceptibility to SCC increases with increased Mg/Zn.

2. When the Mg-rich film formed during solution heat treatment
    is properly removed and a Mg-free film is formed by anodiz-
    ing in tartaric acid, the susceptibility to SCC decreases.

3. The Mg-free films formed by TAA had significantly less
    porosity than did the thermally generated films.

4. The above observations, and the failure to observe oxide
    cracking of thermal or Mg-free films stressed to the
    yield strength in air, indicate that Mg in the oxide
    film, and perhaps film porosity, play a role in the SCC
    mechanism.

5. These results suggest that the susceptibility of commercial
    Al-Zn-Mg alloys may possibly be reduced by control of the
    composition of the oxide film.

Acknowledgements

The bulk-composition experiments were designed by R.K. Viswanadham,
and some of the data presented were generated by C. Eichleman, working
under his auspices. R. Butler performed the Auger experiments, which
were interpreted by T.S. Sun.

The support of the Office of Naval Research and especially of
Dr. P. Clarkin are greatly appreciated.

References

1. J.A.S. Green, H.W. Hayden, and W.G. Montague: in *Effect of Hydrogen
    on Behavior of Materials*, A.W. Thompson and I.M. Bernstein, eds.,
    pp. 200-218, AIME, Philadelphia, PA, 1976.

2. M.O. Speidel: in *Hydrogen in Metals*, I.M. Bernstein and A.W. Thompson,
    eds., pp. 249-276, American Society for Metals, Champion, PA, 1973.

3. G.M. Scamans, R. Alani, and P.R. Swann: Corros. Sci., 1976, Vol. 16, No. 7, pp. 443–459.

4. A. Csanady and D. Martin: J. Mater. Sci., 1979, Vol. 14, pp. 2289–2295.

5. R.K. Viswanadham, T.S. Sun, and J.A.S. Green: Corrosion., 1980, Vol. 36, No. 6, pp. 275–278.

6. R.K. Viswanadham, T.S. Sun, and J.A.S. Green: Metall. Trans. A, 1980, Vol. 11A, pp. 85–89.

7. J.M. Chen, T.S. Sun, R.K. Viswanadham, and J.A.S. Green: Metall. Trans. A, 1977, Vol. 8A, pp. 1935–1940.

8. T.S. Sun, J.M. Chen, R.K. Viswanadham, and J.A.S. Green: J. Vac. Sci. Technol., 1979, Vol. 16, No. 2, pp. 668–671.

9. W. Pistulka and G. Lang: Aluminium (Dusseldorf), 1977, Vol. 53, No. 6., pp. 366–371.

10. J.C. Grosskreutz and G.G. Shaw: J. Appl. Phys., 1964, Vol. 35, No. 7, pp. 2195–2197.

11. L.F. Mondolfo: Aluminum Alloys Structure and Properties, Buffersworth, London, pp. 844–ff, 1976.

12. B.K. Sakena, K. Lal, C.S. Sivaramakrichman, and R. Kumar: Light Met. Age, 1979, Vol. 37, pp. 24–29.

13. P.R. Swann: Hydrogen in Metals, Ibid., p. 274 (discussion in Ref. 2).

14. J.T. Staley: Met. Eng. Q., 1973, Vol. 13, No. 4, pp. 52–57.

15. J.T. Staley, Alcoa Technical Center, private communication, 14 Aug 1980.

16. S.F. Bubar and D.A. Vermilyea: "Deformation of Passive Films," J. Electrochem. Soc., 1967, Vol. 114, No. 9, pp. 882–885.

DISCUSSION

<u>J. A. Moskovitz, Kaiser Aluminum</u>: Was the tartaric acid anodic film sealed and if not, what was the thickness of the barrier layer portion of the anodic film:

<u>J. R. Pickens</u>: No, the sealing step was not necessary because the film formed by TAA does not have appreciable porosity, Work by Grosskreutz(10) showed that when anodizing was performed in a tartaric acid-ammonia mixture buffered to pH = 5.5, no pores were found to the limit of his resolution (10–15Å). Our STEM examination of the TAA surfaces revealed no large pores and only very fine porosity.

Thus, the barrier layer of the anodic film is, for practical purposes, the entire 2000Å reported in the text because of the essentially pore-free nature of the TAA film.

<u>E. A. Starke,Georgia Institute of Technology</u>: My comment has to do with the statement concerning the Zn-Mg ratio -- and the SCC resistance of 7xxx alloys. The work referred to (Staley, Pickens, Ref. 14) really concerns the effect of copper on the SCC resistance of these alloys -- as the Zn-Mg ratio is increased, the amount of Cu that can be put in solid solution at the solutionizing temperature increases. (Therefore, more Cu can participate in the precipitation process during low temperature aging).

Early work at Alcoa -- e.g., by E. H. Hollingsworth -- had shown that the resistance to SCC increases as the Cu concentration increases in 7xxx alloys.

<u>J. R. Pickens</u>: Professor Starke is correct that Staley's work also concerned the effect of Cu on SCC resistance and that the amount of Cu that can be put into solid solution is a function of Mg/Zn. Staley believes, however, that there is an independent benefit of a low Mg/Zn on SCC resistance(15). He found that the SCC resistance of his 7xxx alloys decreases with increasing Mg/Zn for constant Cu content(15).

Hollingsworth's finding that copper has a beneficial effect on SCC resistance of 7xxx is consistent with our model(1,6) concerning hydrogen embrittlement in Al-Zn-Mg alloys. It is worth noting that Cu-containing precipitates are in general noble with respect to the matrix and, therefore, may promote hydrogen recombination and subsequent removal of hydrogen gas from the alloy. Thus, the beneficial influence of Cu in improving resistance to SCC may be accounted for in an electrochemical sense.

<u>A. W. Thompson, Carnegie-Mellon University</u>: Your comments about the role of the surface film in SCC susceptibility imply that the SCC is initiation controlled, since once the crack is present it no longer interacts with the film. Do you have any direct indications that SCC is initiation controlled in your Al-Zn-Mg alloys, and if so, would you expect the same to occur in Al-Zn-Mg-Cu alloys?

<u>J. R. Pickens</u>: I believe that SCC susceptibility is both initiation and propagation dependent. The initiation step is clearly important based on our stress-ttf data of films of varying Mg content. After the film is penetrated, the alloy loses one defense against SCC; however, early work at MML showed that the rate of crack propagation along the grain boundaries

is controlled by the precipitate size and spacing,[*] and probably Mg content(6). Thus, the time for the stress-corrosion crack that has initiated in the film to grow sufficiently long to cause failure at a given load, is determined by the microstructure and composition of the grain boundary region, which is a propagation dependent.

We currently have only limited results of our SCC experiments on an Al-Zn-Mg-Cu alloy (commercial 7075). We hope to decrease the susceptibility of commercial 7075 by controlling the composition of the oxide film as we have for the Al-Zn-Mg aloy.

---

[*]A. J. Sedriks, J. A. S. Green, D. L. Noval, Proc. U. R. Evans Int. Conf. on Localized Corrosion, p. 569, etd. R. W. Staehle, Williamsburg, VA, NACE 1971.

STUDIES ON STAINLESS STEELS

HYDROGEN ASSISTED CRACKING OF AUSTENITIC STAINLESS STEELS

527

C. L. Briant
General Electric Company
Corporate Research and Development Center
P. O. Box 8
Schenectady, New York 12301

In this paper results are given which show that the presence of deformation induced martensite is essential for the hydrogen cracking of low yield strength austenitic stainless steels. The ease of forming martensite also explains the difference between sensitized and unsensitized materials. For alloys that do not transform to martensite the susceptibility correlates well with the yield strength.

Introduction

The determining factor which controls the hydrogen cracking susceptibility of austenitic stainless steels has remained unclear despite many previous studies. Early work suggested that the formation of deformation induced martensite was important (1-4). However, this mechanism cannot explain all results, since hydrogen cracking has been reported for steels which have stable austenite phases (5-10).

Hydrogen embrittlement in austenitic stainless steels has also been explained on the basis that reduced stacking fault energy and the accompanying prevalence of planar slip increase the susceptibility to hydrogen cracking (7,9,11). The idea in this model is that planar slip concentrates the hydrogen along slip planes, whereas cross slip dilutes the hydrogen throughout the matrix. This model assumes that hydrogen will be bound to the dislocations and move with them during slip.

Recently, additional evidence has shown that the formation of deformation induced martensite may be extremely important in the hydrogen cracking of austenitic stainless steels (12-16). This paper demonstrates the importance of martensite formation in determining the susceptibility of these alloys to hydrogen cracking. In particular, it is suggested that this mechanism is crucial for the lower yield strength alloys. As the yield strength increases, martensite formation is less important.

Experimental

In this work a cathodic charging experiment was used to measure the susceptibility of materials to hydrogen cracking. The samples were placed in a plastic container and surrounded by a solution of $0.1N$ H_2SO_4 which contained 500 mg/l of sodium arsenite. A current density of 40 ma/cm^2 was applied and produced hydrogen charging. A fixed mechanical load was immediately applied and the time to failure measured. If a sample did not fail within 48 hours, the test was stopped.

The samples used in these experiments were 0.3175 cm diameter rod. A gauge section of 0.1524 cm was electropolished into the rod. All samples were initially annealed for one hour at $1100^\circ C$ in argon and water quenched. Samples from each alloy were tested in this solution annealed condition. Other samples were given an additional aging heat treatment of $650^\circ C$ for 24 hours and then tested.

The chemical compositions of all alloys used in this work are given in Table I. Alloy A is a high purity laboratory heat of 304 stainless steel and alloy B is a laboratory heat of 304 stainless steel doped with nitrogen. Alloys C and D are laboratory heats of 304L stainless steel. Alloy E is a commercial 304 stainless steel and alloy F is a commercial 316 stainless steel. Alloy G is a laboratory steel with a base composition similar to 304 except for the high Ni content.

Table I. Chemical Composition of Alloys Used

Alloy	Ni	Cr	C	P	S	Si	N	Mn	Mo
A	9.4	18.6	0.069	0.003	0.009	0.01	0.002	---	---
B	9.6	18.6	0.058	0.003	0.003	0.01	0.033	---	---
C	9.2	18.5	0.022	0.004	0.006	0.01	0.01	---	---
D	9.5	18.5	0.028	0.003	0.003	---	---	---	---
E	8.49	18.1	0.078	0.027	---	0.41	0.05	1.12	---
F	12.77	17.14	0.057	0.035	0.025	0.54	0.03	1.67	2.21
G	13.8	18.6	0.063	0.003	---	---	---	---	---

<u>Results</u>

Table II lists the minimum stress required for failure during cathodic charging for a series of austenitic stainless steels given the single solution anneal treatment of $1100^{\circ}C$ for 1 hour. The point to be noted about these results is that only alloys C and D, the low carbon alloys, failed within 48 hours at a stress below 413 MN/m^2, the maximum stress used for these tests.

Table II. Minimum Stress Which Produced Failure Within 48 Hours
For Samples Heat Treated $1100^{\circ}C$ for 1 Hour

Alloy	Stress (MN/m^2)
A	>413
B	>413
C	275-310
D	372-379
E	>413
F	>413
G	>413

The minimum stresses required for failure of samples given an additional heat treatment of $650^{\circ}C$ for 24 hours are listed in Table III. This heat treatment considerably altered the relative susceptibilities. The 316 commercial steel (alloy F) still did not fail the test during 48 hours with an applied stress of 413 MN/m^2. In contrast the 304 type steels were the most susceptible to hydrogen cracking. The high nickel 304 stainless steel was intermediate between the 304 and 316 alloys. The susceptibilities of the low carbon alloys were only slightly affected.

Table III. Minimum Stress Required to Cause Failure in 48 Hours
For a Sample Aged at $650^{\circ}C$ for 24 Hours

Alloy	Stress (MN/m^2)
A	206-241
B	206-241
C	275-310
D	289-324
E	241-286
F	>413
G	324-344

The solution annealed samples that fractured during the cathodic charging test failed primarily along grain boundaries. An example is given in Figure 1. The intergranular cracking occurred near the edge of the sample. Many of the exposed boundaries had a striated appearance as can be seen in Figure 1a, although some regions of smooth boundary fracture could also be seen, Figure 1b. The center portion of the sample failed by ductile tearing, the air fracture mode of these alloys. Those samples which did not fracture were not immune to cracking. Examination of the unfailed gauge sections after the test always showed cracks similar to those in Figure 2. Both intergranular and transgranular cracks can be seen. However, cross sections of these samples showed that the penetration of the surface cracks was not deep in samples of alloys that did not fracture during the test at the maximum applied stress.

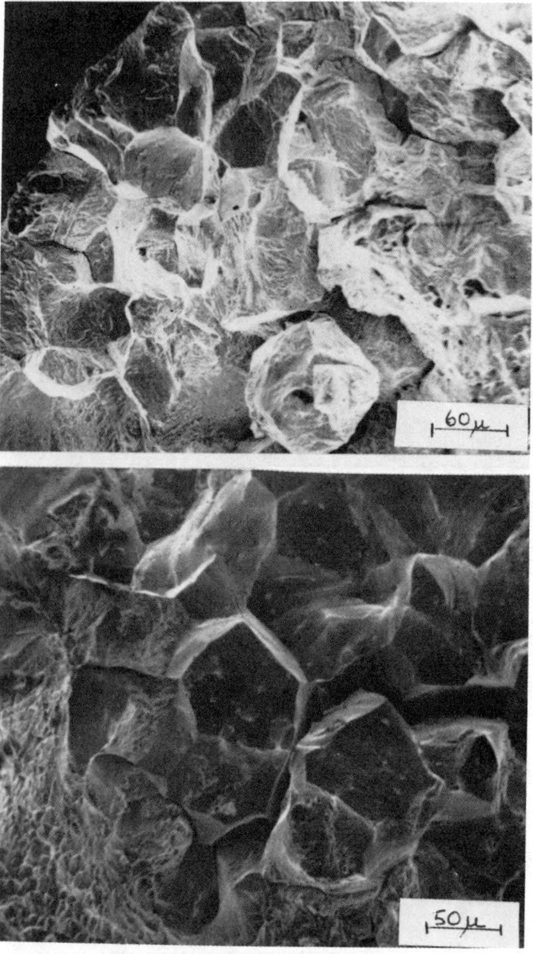

Fig. 1 –   An example of the intergranular fracture observed in solution annealed, low carbon steels.   Many boundaries appeared striated and had plastic tearing on them, Figure 1a, although patches of very smooth intergranular fracture could also be found, Figure 1b.

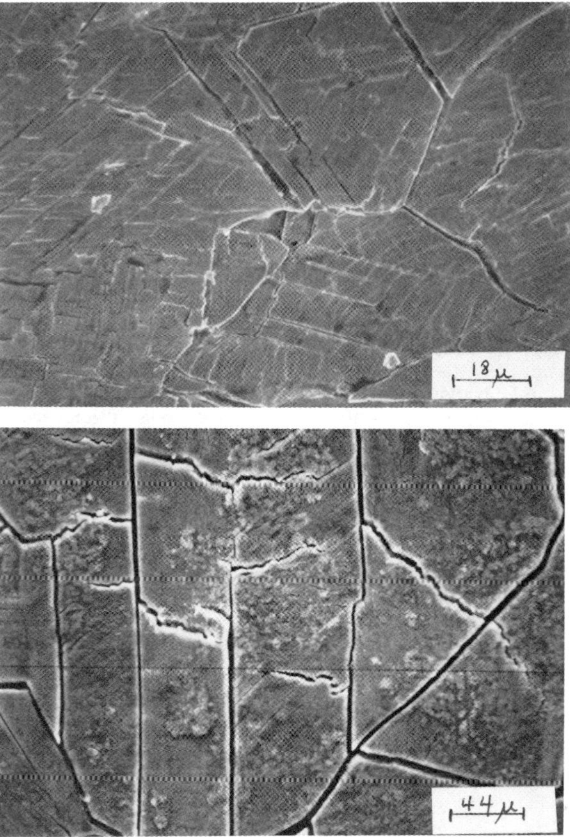

Fig. 2 -   An example of cracks observed on the sides of solution annealed samples that did not fracture during the cathodic charging test with an applied stress of 413 MN/m^2.

The additional heat treatment of 24 hours at 650°C caused many more of the alloys to fracture during the test.   In all cases the fracture was predominantly intergranular.   The gauge section of the 316 stainless steel, which did not fracture at the maximum applied stress of 413 MN/m^2 looked similar to that shown in Figure 2.

Discussion

Many papers (2,3,14-16) in the past years, including those by this author, have strongly advocated that the presence of deformation induced martensite greatly increases the susceptibility of austenitic stainless steels to hydrogen cracking.   I will now show that this idea explains the results presented above.   Afterwards, I will rationalize these results with many others reported in the literature, and also discuss steels which are clearly affected by hydrogen but are stable with respect to martensite formation.

To help interpret these results I will use a formula developed by Angel (17).   This formula was derived from a multiple regression analysis of many heats of stainless steel and gives the temperature at which fifty per cent of the austenite has transformed to martensite during a tensile test at a true strain of 0.3. The formula is given by

$$M_D^A(^oC) = 413 - 462[C+N] - 9.2[Si] - 8.1[Mn] - 13.7[Cr] - 9.5[Ni] - 18.5[Mo] \quad (1)$$

where the concentrations of elements are given in weight per cent. It must be stressed that given differences in testing conditions and the fact that this formula is derived from a multiple regression analysis, the actual temperatures it gives cannot be taken literally. However, it does provide a convenient scale for ranking the various alloys with respect to austenite stability. The higher the temperature calculated from the formula, the more unstable the austenite should be.

First consider the results of the solution annealed alloys. Two of the seven alloys fractured during the test. As can be seen in Table IV these are the alloys with the highest M_D^A temperatures. The ferrofluid technique for delineating martensite (18) in a polished cross section showed that deformation had produced extensive martensite formation in these alloys. As discussed previously (16) these grain boundaries were largely transformed to martensite. This transformation allowed the intergranular fracture to occur, and also explains the striated appearance of many of the grain boundaries (16).

Table IV. M_D^A Temperature for Solution Annealed Samples
and the Minimum Stress Required to Cause Failure in 48 Hours

Alloy	M_D^A (^oC)	Stress (MN/m^2)
A	36	>413
B	25	>413
C	57	275–310
D	56	372–379
E	12	>413
F	-42	>413
G	-2	>413

Table V. M_D^A Temperature for Volume of Material Near the Grain Boundary
After Aging at 650°C for 24 Hours and the
Minimum Stress Required for Failure Within 48 Hours
A chromium depletion of 5 wt% is assumed.

Alloy	M_D^A (^oC)	Stress (MN/m^2)
A	104.8	206–241
B	93.5	206–241
E	80.5	241–286
F	26	>413
G	66	324–344

The increased susceptibility of the sensitized materials can also be explained by martinsite formation through chromium depletion. The additional aging treatment of 650°C for 24 hours causes chromium carbides to precipitate along the grain boundaries of the higher carbon steels, and deplete the nearby regions of chromium (19). The grain boundary regions will then transform readily to martensite because of their low chromium content. Table V lists the M_D^A temperatures for the grain boundary region assuming a decrease of 5 wt% chromium for this region. As can be seen this depletion considerably raises M_D^A and the susceptibility to martinsite formation. The highest temperatures now are those for the high purity and nitrogen doped laboratory heats of 304 stainless steel. The next highest temperature is for the commercial 304 steel, followed by the high nickel 304 steel, and finally the commercial 316 stainless steel. This ranking agrees with the mechanical test results. Note that the alloy content of the 316 stainless steel is high enough that even with the chromium depletion the M_D^A temperature is below that of the solution annealed lower carbon alloys. Therefore, we would not expect the steel to be susceptible to hydrogen cracking. The presence of martensite at the grain boundaries of the 304 stainless steel after aging was verified both by the ferrofluid technique and by transmission electron microscopy, Figure 3 and 4. No martensite could be found in the 316 stainless steel.

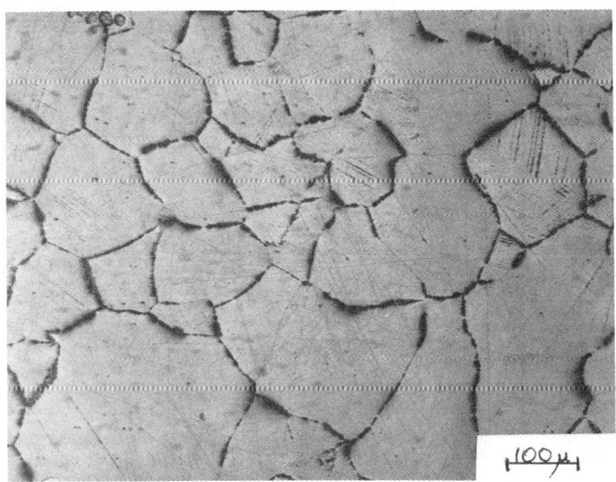

Fig. 3 -   A micrograph of an aged and deformed sample of 304 stainless steel, covered with ferrofluid and placed in a magnetic field. Note that the magnetic particles which identify the location of martensite are concentrated at the grain boundaries.

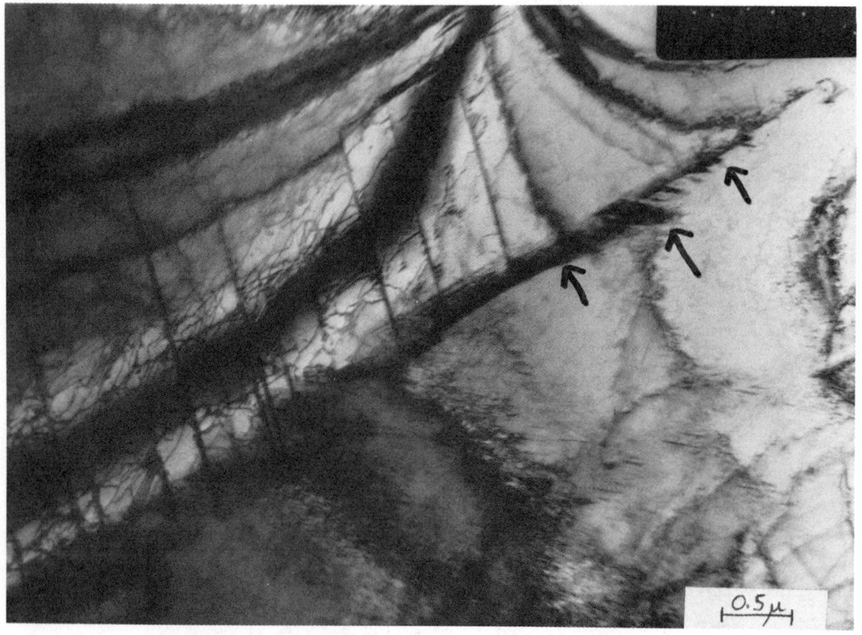

Fig. 4 -   A transmission electron micrograph showing the presence of martensite at
the grain boundaries of aged and deformed 304 stainless steel. The arrows
denote the α-martensite.

Figure 5 summarizes all of the data for these steels, both in the annealed
and aged conditions. In this figure the minimum stress required to cause failure within
48 hours during cathodic charging is plotted as a function of M_D^A. It is quite clear
that below a certain value of M_D^A we did not observe failures within 48 hours at the
maximum stress used in this study.   For steels with higher susceptibilities to
martensite formation and higher values of M_D^A, the fracture stress required to cause
failure within 48 hours dropped dramatically. It should also be noted that the grain
boundary carbide density was always equivalent in both the aged 304 and 316 alloys.
Therefore, a high carbide density at grain boundaries is of much less importance than
has been proposed by other workers (20).

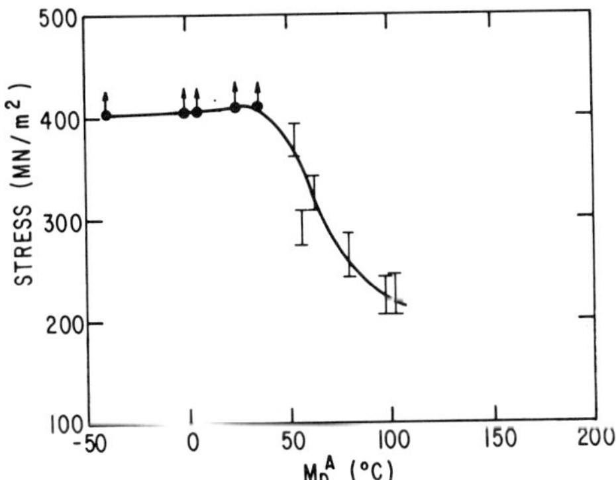

Fig. 5 -   A plot of M_D^A as a function of the minimum stress required to cause fracture during the cathodic charging test.

There are many reports in the literature of hydrogen cracking of austenitic stainless steels. Unfortunately, a variety of parameters have been used to measure the effects of hydrogen on materials. It is thus quite difficult to compare results from various workers. However, one of the most common tests used has been to pull smooth tensile bars to failure in 69MPa of hydrogen and measure the reduction in area. One can then discuss the loss in reduction in area caused by testing in hydrogen. Data obtained in this way has been used below for discussion.

Many of the studies have compared the behavior of type 304L stainless steel, which is known to transform to martensite upon pulling in tension, with steels which did not transform. It was generally found that the 304L stainless steels were more susceptible than were the stable austenitics. If martensite is an important factor then these results on 304L should depend on M_D^A. Only two studies are available which give accurate compositions and measure the percent reduction in area in high pressure hydrogen gas (7,9). One of these steels had a loss in reduction in area of 41% and an M_D^A of 27°C. The other had a loss in reduction in area of 45% and an M_D^A of 33°C. Although the data is certainly limited, these two results do stack up in the correct order. (The loss in reduction in area is the reduction of area in air or helium minus the reduction of area in H_2.)

The lower yield strength steels which do not transform to martensite upon deformation have been found to be only slightly susceptible to hydrogen cracking. These steels include types 316, 316L, 309S and 310. Also, because of the high alloy content of 309S, sensitization does not cause α-martensite to form at the boundaries and hence does not greatly affect the mechanical properties of the steel in hydrogen (8). Hydrogen induced brittle failures have been reported for 316 (10) and 310 stainless steel (5,6) but only when the test samples were extremely thin. These results are consistent with the fact that I and others (14,21,22) have observed surface cracking of the more stable steels but did not record failures or slow crack growth. This result further suggests that one role of martensite is to quicken hydrogen diffusion in the material.

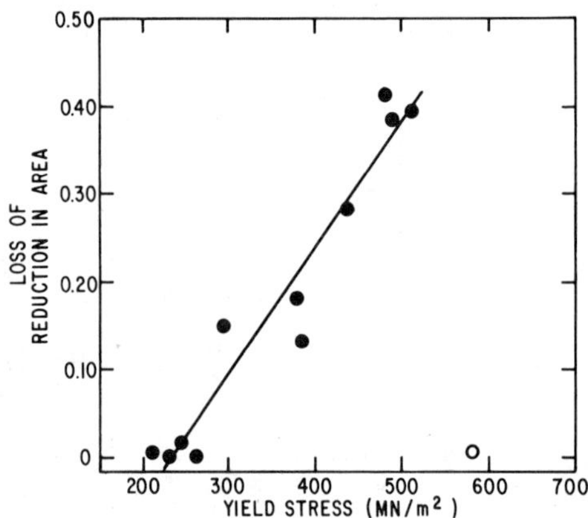

Fig. 6 -   A plot of the loss of reduction in area at 69MPa hydrogen as a functon of yield strength. Note that with the exception of one point there is a correlation between the two variables.

Experiments have clearly shown that some very stable austenitic stainless steels such as 21-6-9 are quite susceptible to hydrogen cracking (9,11). However, these steels have considerably higher yield strengths than do the 300 series. It is well known from work on body-centered-cubic alloys that increasing the yield strength increases the susceptibility of the steels to hydrogen cracking. Therefore, one might well expect such a relationship to be found for the stable austenitic steels as well. Figure 6 shows that such a relationship generally holds, although one data point is clearly out of line. The data in this figure were taken from references 2, 8, 9 and 23. As the yield strength is raised, the fracture mode tends to change from ductile tearing to intergranular fracture. It has been suggested that the hydrogen cracking in these alloys arises from the fact that the dislocation motion is in a planar mode rather than a cross slip mode and this in turn helps concentrate the hydrogen at the interface. However, the fracture mode in these high strength alloys tended to be intergranular and in the low strength alloys microvoid coalescence. Of all the dislocations piled up at the grain boundary or particle, only the leading one could deposit its hydrogen at the interface.   Therefore, if planar slip is important, it is probably because the dislocation pile-ups concentrate stress at the interfaces and help nucleate a crack there.   A similar idea has been suggested by West and Louthan (24). They further showed that raising the yield strength of the 300 series steels increased the susceptibility of these steels to hydrogen cracking (24). The very susceptible 304 samples used in this study which had been aged at 650°C for 24 hours showed cross slip.   This fact shows that in the low yield strength steels planar slip is not as important as the formation of martensite.

In correlating data from the literature it is difficult to assess the effects of grain size and grain boundary composition on hydrogen cracking susceptibility. Presumably, decreasing the grain size would decrease the susceptibility to hydrogen cracking as in other systems (25).   Grain boundary embrittlers may also affect susceptibility to hydrogen cracking. I have previously shown that 0.06 P has only a slight effect on the susceptibility of 304 stainless steel (16). This paper shows that 0.03 N also does not have an observable affect, and other results in our laboratory show that the same is true for an addition of 0.03 S. However, in the higher strength steels these elements could be much more deleterious (26,27).   Also, arrays of precipitates at grain boundaries such as are found in overheated alloys could be much more deleterious in higher strength steels.

Another unresolved question is why the martensite is so important in the lower strength alloys. Certainly, one reason could be that it provides a high diffusivity path for hydrogen permeation, and it may affect the localized stress at the crack tip. More probably, however, the answer lies in the way hydrogen interacts with the two different crystal lattices. This idea is supported by the fact that hydrogen appears in many cases to first induce the transformation to martensite and then to crack it (21,22). Also unresolved is the relative importance of α and ε martensite. In this work, only α martensite was detected, and Thompson has provided evidence that the presence of ε martensite does not affect hydrogen cracking susceptibility (8). In contrast, Inoue, et.al., (13) have suggested that the presence of ε may be quite important.

Conclusions

1. In low yield strength austenitic stainless steels the formation of deformation induced martensite greatly increases the susceptibility to hydrogen cracking. This fact appears to be independent of the mode of dislocation motion.

2. Chromium depletion at the grain boundaries can further lower the resistance to hydrogen cracking because the decrease in chromium enhances martensite formation near the grain boundaries.

3. If the steel is sufficiently alloyed, such as in 316, sensitization does not destroy the resistance to hydrogen cracking.

4. In stable austenitic stainless steels the yield strength greatly affects susceptibility. One would also predict that in higher yield strength steels the presence of other grain boundary embrittlers could be important.

Acknowledgments

The author would like to thank Ms. Ann Ritter for performing the transmission electron microscopy and Mrs. Jacqueline Briant for reading the manuscript critically.

References

1.   J. D. Hobson and J. Hewitt, J.I.S.I., 1953, vol. 173, p. 131.

2.   R. B. Benson, Jr., R. K. Dann, and L. W. Roberts, Jr., Trans TMS-AIME, 1968, vol. 242, p. 2199.

3.   R. M. Vennett and G. S. Ansell, Trans ASM, 1967, vol. 60, p. 242.

4.   M. R. Louthan, Jr., J. A. Donovan, and D. E. Rawl, Jr., Corrosion, 1973, vol. 29, p. 108.

5.   M. L. Holzworth, Corrosion 1969, vol. 25, p. 107.

6.   M. B. Whiteman and A. R. Troiano, Corrosion, 1965, vol. 21, p. 53.

7.   Anthony W. Thompson, Met. Trans. 1973, vol. 4, p. 2819.

8.   Anthony W. Thompson, Mat.Sci.Eng., 1974, vol. 14, p. 253.

9.   B. C. Odegard, J. A. Brooks, and A. J. West, Effect of Hydrogen on Behavior of Materials, A. W. Thompson and I. M. Bernstein, eds., p. 116, AIME, New York, 1976.

10.   H. Hanninen and T. Hakkarainen, Met. Trans. A, 1979, vol. 10A, p. 1196.

11.   M. R. Louthan, G. R. Caskey, J. A. Donovan, and D. E. Rawl, Mat.Sci.Eng., 1972, vol. 10, p. 357.

12.   H. Hanninen and T. Hokarainen, Corrosion, 1980, vol. 36, p. 47.

13.   A. Inoue, Y. Hosoya, and T. Masumoto, Trans. ISIJ, 1979, vol. 19, p. 170.

14.   D. Eliezer, D. G. Chakrapani, C. J. Altstetter, and E. N. Pugh, Met. Trans. A, 1979, vol. 10A, p. 935.

15.   C. L. Briant, Scripta Met., 1978, vol. 12, p. 541.

16.   C. L. Briant, Met. Trans. A, 1979, vol. 10A, p. 181.

17.   T. Angel, J.I.S.I., 1954, vol. 177, p.165.

18.   R. J. Gray, Proc. Fourth Ann. Tech. Meeting. Inst. Microstructural Anal. Soc., J. L. McCall, ed., p. 141, Int. Mic. Anal. Soc., Denver, 1971.

19.   E. C. Bain, R. H. Aborn, and J. J. B. Rutherford, Trans. Amer. Soc. Steel Treating, 1933, vol. 21, p. 481.

20.   J. P. Fidelle, L. R. Allemand, C. Roux, and M. Rapin, L'hydrogene dans les Metaux, J. P. Fidelle and M. Rapin, eds., p. 137, Centre d'Etudes CEA, Bruyeres-le-Chatel, 1967.

21.   M. L. Holzworth and M. R. Louthan, Jr., Corrosion, 1968, vol. 24, p. 110.

22.   H. Okada, Y. Hosoi, and S. Abe, Corrosoin, 1970, vol. 26, p. 183.

23.   A. W. Thompson and I. M. Bernstein, Rev. Coatings Corrosion, 1975, vol. 2, p. 5.

24.   Anton J. West and MacIntyre R. Louthan, Jr., Met. Trans. A, 1979, vol. 10A, p. 1675.

25.   C. L. Briant and S. K. Banerji, Int. Met. Reviews 1978, vol. 23, p. 164.

26.   S. K. Banerji, H. C. Feng, and C. J. McMahon, Jr., Met. Trans. A, 1978, vol. 9A, p. 625.

27.   R. A. Mulford, C. J. McMahon, Jr., D. P. Pope, and H. C. Feng., Met Trans A, vol. 7A, p. 1183.

D.C. Langstaff, METCO, Kennewick, WA: You mention the lack of tests available
to determine susceptibility of (unstable) austenitic stainless steels to hydrogen embrittlement. Soliciting input on suitable tests with the objective
of establishing a standardized test would be a commendable effort toward the
resolution of this difficult question. Would you suggest a test or investi-
gative method to use to determine the susceptibility of austenitic stainless
steels to hydrogen embrittlement?

C.L. Briant: I feel that a notched sample in which crack growth could be
measured would be best. However, only after extensive testing of various
types of samples could one definitely answer this question.

C.A. Grove, Knolls Atomic Power Lab., Schenectady, NY: What is your feeling
on hydrogen induced martensite formation? If it occurs, does hydrogen itself
induce martensite or does hydrogen enhance localized plastic deformation resulting in martensite formation?

C.L. Briant: I do not know the correct answer to this question, nor has anyone done experiments which would answer it clearly. My gut feeling is that
it occurs by localized plastic deformation but I have no real proof of this.

A.W. Thompson, Carnegie-Mellon University, Pittsburgh, PA: It's been known
for a long time, of course, that hydrogen susceptibility of austenitic stainless steels is roughly correlated with stability of the austenite. The degree
of stability, however, is not a reliable indicator, since extensive ductility
losses occur in some steels which form little or no martensite, while some
other steels which form profuse martensite show only moderate embrittlement.
There is no controversy about the observation that the presence of martensite
can exacerbate embrittlement, for example when martensite occurs preferentially near grain boundaries. However, the lack of exact correlation makes
one wonder whether martensite is really causative, and whether some other
characteristic such as slip mode, which is also correlated with stability,
might be a better basis for explaining the behavior. Would you comment on
this problem?

C.L. Briant: I agree that there are many cases where completely stable
steels have been found susceptible to hydrogen cracking. This point is discussed in the paper. However, I believe that most steels that form α' martensite are quite susceptible to hydrogen cracking. The idea that cross
slip improves the response to hydrogen does not hold in all cases because
the most susceptible 304 steels tested in this study cross slipped during
deformation. I do feel that planar slip may be important in hydrogen cracking because it will allow very high stress concentrations at interfaces and
help nucleate the cracks there.

H.K. Birnbaum, University of Illinois, Champaign, IL: I appreciate your emphasis on the relation between austenite stability and hydrogen embrittlement
but would like to suggest that the effect of hydrogen in solid solution is
to increase the M_d temperature. We have shown that the γ phase in 304 and
310 type steels transforms to martensite phases in the presence of high hydrogen concentrates and stress while the γ phase is stable (at room temperature) when they contain hydrogen alone. The presence of the martensite product during deformation in high hydrogen fugacities results from this increase
in M_d temperature. One must look for the martensite at the fracture surface.
In the stable steels the reversion back to the γ is fast at room temperature
(and much slower at lower temperatures) and results from the loss of hydrogen

through the surface. Therefore one must look for the martensite product
(ϵ or α) at the fracture surface before it reverts back to the γ phase.
(bear in mind it is hard to trap a Maxwell Demon.)

C.L. Briant: I fully agree that hydrogen raises the M_d temperature and that
one must look carefully for martensite on the fracture surface. However,
there appears to be good evidence of hydrogen cracking of stainless steels
when no martensite has been formed. My conclusion is that if martensite
forms, the hydrogen cracking susceptibility is greatly increased. However,
it is not clear that martensite formation is a necessary requirement to observe any hydrogen cracking.

J.P. Fidelle, Commissariat à l'Energie Atomique, Bruyères-le-Châtel, France:
A comment on Howard Birnbaum's statement that HE of even so called stable
austenitic SS such as 310 can be contributed by a transient ϵM formation –
Yet A 286 is fairly stable and does not transform when strained at 4°K.
However it can be affected by internal or H_2 gas embrittlement, because of
the hardening phases precipitated during aging before HE testing. Has some-
body carried out tests on unaged A 286? It must be rather small, unless
testing conditions are severe.

C.L. Briant: I do not know if anyone has done a study of unaged A 286.
(Ed. note: see Hydrogen in Metals, I.M. Bernstein and A.W. Thompson, eds.,
p. 91, ASM, Metals Park, 1974).

M.F. Stevens, Carnegie-Mellon University, Pittsburgh, PA: You concluded
that in austenitic alloys where the austenite was stable, that the HE was a
strong function of yield strength. In the range of yield strengths studied,
did you see any changes in slip mode (SFE)?

C.L. Briant: All of the data I presented on the correlation of hydrogen
embrittlement and yield strength was taken from the literature. Different
slip modes were reported. Slip mode was particularly well characterized in
reference 9.

B.M. Gordon, General Electric Company, San Jose, CA: Since in your investigations type 304L stainless steel is less resistant to hydrogen assisted
cracking than type 304, and the reverse is true for stress corrosion cracking in high temperature oxygenated water environments, would you consider
it correct to assume that the latter phenomenon is not due to a hydrogen
mechanism?

C.L. Briant: Type 304L is less resistant than solution annealed 304 but more
resistant than sensitized 304. At this point I do not believe we can state
whether or not hydrogen is playing a role in stress corrosion cracking in
high temperature oxygenated water.

B.M. Gordon, General Electric Company, San Jose, CA: Do you have any plans
to study alternate heat treatments in your test program? It may be considered prudent to investigate thermally disturbed material by welding (no
filler metal to maintain the purity of your laboratory heats) or lower temperature sensitized material.

C.L. Briant: These experiments would be interesting. I have done a few
experiments looking at other sensitizing heat treatments for 304 stainless
steel and the results fall in line with those reported here, i.e. increased
sensitization leads to increased susceptibility to hydrogen cracking. However, at this point I am not planning to look at welds or lower sensitizing
temperatures.

HYDROGEN ASSISTED FRACTURE IN FCC METALS AND ALLOYS

R. E. Stoltz
and
A. J. West
Sandia National Laboratories
Livermore, CA 94550

The fracture behavior of a series of FCC metals and alloys in hydrogen
at pressures up to 207 MPa has been studied. Transgranular slip band cracking
was observed in tensile tests of low strength nickel and Nitronic 40 while
intergranular fracture occurs in a higher strength precipitation-strengthened
alloy, JBK-75. Hydrogen induced flow localization on the slip plane leads
to slip band cracking in lower strength alloys. Intergranular cracking occurs
in the higher strength alloy. Sustained load cracking in 207 MPa hydrogen was
observed in iron-based alloys with yield strengths above 700 MPa. Alloys with
clean grain boundaries all exhibited intergranular fracture and all fall on a
common plot of threshold versus yield strength regardless of strengthening
mechanism. No crack growth was observed in 304L (380 and 590 MPa yield
strength) nor in 2124 or 7475 aluminum. The results suggest that hydrogen
induced decohesion at austenite grain boundaries is the controlling step in
sustained load cracking at these hydrogen pressures.

Previous investigations of tensile behavior in 69 MPa hydrogen gas (1-4)
showed that among the FCC materials studied, nickel and nickel base superalloys are severely affected by hydrogen. Unstable austenites are moderately
embrittled while stable austenites and aluminum alloys are only slightly
embrittled. Crack growth measurements under sustained load in 35-69 MPa
hydrogen have been previously made on a more limited number of alloys (2,5-8).
Sustained load cracking has been observed in IN718 and A286 but not in 2219
aluminum (2,5,8). Additionally, Loginow and Phelps (6) report no crack
growth in 304 at 69 MPa while Eliezer and co-workers (7) report crack growth
304 in 0.10 MPa (15 psi) hydrogen.

In this paper, we extend these observations to 207 MPa (30 ksi) hydrogen
pressure. Tensile tests have been performed on pure nickel, Fe-Ni-Cr-Mn stainless steels, and γ' strengthened Fe-Ni-Cr and steels. In a second section, a
summary of the sustained load cracking (SLC) behavior of selected alloys is given.
Excellent discussions of the effect of strength, chemistry, dislocation density,
and precipitate distribution on hydrogen degradation of stainless steels have
been previously presented (9-12). We will make use of these discussions in
interpreting the current work.

Experimental Details

All tensile tests were performed in purified hydrogen with measured oxygen and water vapor impurity concentrations of less than 1 ppm, at crosshead speeds of 1.27 mm/min (0.05 in./min). Sustained load cracking experiments were performed with bolt-loaded compact tension samples 2.2 cm (0.875 in.) thick. Decreasing load values were measured continuously throughout the test and converted to crack lengths using the known compliance of the specimens. Crack growth threshold stress intensities were defined either at crack arrest, or at propagation rates of 10^{-11} m/s.

Results

Tensile Tests

The composition and mechanical properties of all alloys described in this section on tensile testing in high pressure hydrogen are listed in Table I. The alloys represent a range of ductilities from ~90% reduction in area for pure nickel to ~50% for JBK-75. As a rule, the severity of hydrogen embrittlement was better reflected in the nature and extent of fracture mode change, rather than in the loss in ductility. Areal percentages of fracture mode change, either transgranular or intergranular, varied from 100% for nickel to less than 5% for JBK-75.

Table I.   Tensile Alloy Composition and Mechanical Properties

Nickel 270:   Ni-130 ppm C-68 ppm O-<10 ppm S (weight)
Nitronic 40:   Fe-19.2Cr-7.0Ni-0.24Si-0.27N-0.017C (wt%)
JBK-75 (Mod A286): Fe-14.0Cr-29.6Ni-1.3Mo-2.10Ti-0.16Al-0.35V-0.019C (wt%)

	Yield MPa (ksi)	Ultimate, MPa (ksi)	Total Elong,%	RA, %
Ni 270:	33(5)	359(52)	55	89
Nitronic 40:	462(67)	800(116)	56	78
JBK-75, (Aged 993K/16 hr):	717(104)	1131(164)	28	51

<u>Nickel 270</u>. The results of high-pressure hydrogen tests on Ni 270 are given in Table II. Increasing the test pressure from 3.5 MPa (5 ksi) to 172 MPa (25 ksi) had no effect on the ductility loss, which equaled ~50% in all cases. Prestraining 15% or 30% in air followed by testing in 3.5 MPa hydrogen also had little effect on the final ductility. Finally, prestraining in 3.5 MPa hydrogen followed by fracture in air did not change the final ductility. Additionally, hydrogen surface crack initiation was observed to occur at between 15 and 30% strain.

Fracture surface morphology varied from over 95% transgranular to a mixed 80% transgranular-20% intergranular mode. Intergranular fracture was most often as secondary cracks. Transgranular fracture extended entirely across the grain facets and exhibited a variety of features including triangular tongue-like regions eminating from crack arrest lines, Figure 1a, and well-defined striation features perpendicular to the crack growth direction, Figure 1b. Matching photos of both halves of the fracture surface, Figure 1c, show that the striations match one-to-one, but additional striations exist on one side of the fracture surface and not the other.

Table II. Nickel 270 Tensile Results

(Ni 270: Annealed 1273K/4 hr, furnace cooled, 850 μm grain size)

Test Condition	Ductility RA, %	Test Condition	Ductility RA, %
Air	89	Prestrain 15% Air - Test 35 MPa H_2	45
35 MPa H_2	47		
69 MPa H_2	48	Prestrain 30% Air - Test 35 MPa H_2	53
103 MPa H_2	48		
138 MPa H_2	46	Prestrain 15% 35 MPa H_2 - Test Air	91
172 MPa H_2	47		
		Prestrain 30% 35 MPa H_2 - Test Air	89

(a)

(b)

(c)

Fig. 1 - Fracture Features of Ni 270 in 3.5 MPa Hydrogen

(a) Crack arrest marks; (b) Striations perpendicular to crack growth direction; (c) Matching halves of fracture surface. (Arrows indicate crack growth direction.)

Nitronic 40 (Fe-21Cr-6NI-9Mn). Samples from a cross-rolled and annealed plate of Nitronic 40 were used in the high pressure tests described below. The nominal fracture ductility was 65-78% RA, depending on orientation within the plate. These values are lower than Ni 270 due to the higher inclusion content of the alloy. Though the stacking fault energy of the alloy is relatively high, (32 mJ/m^2) (13), extensive deformation twinning occurs at strains above 20%. Figure 2 is an example of the twinned substructure developed at ~40% strain. The presence of hydrogen had little measurable effect on twin nucleation or twin density.

Figure 3 shows the variation in RA with hydrogen test pressure for samples taken in the short transverse direction. All samples were precharged to the profile used in Ref. 16. The hydrogen concentration varied from 2700 appm at the surface to 500 appm at the centerline. The RA changes progressively from 69 to 120 MPa with a small drop to 172 MPa. The areal percentage of fracture mode change is also indicated. While the percentage of hydrogen affected fracture increased with test pressure, the fracture path did not change. Fracture surface features of Nitronic 40 are very similar to those seen in Ni 270. Secondary intergranular cracks are mixed with a "woody" transgranular mode, Figure 4a. The latter is composed of short, flat segments and intervening tear ridges. In some regions, the transgranular facets are entirely flat and stepped, Figure 4b. The height of the steps corresponds to the spacing of the deformation twin bands in Figure 2, suggesting hydrogen assisted fracture occurs along the twinned regions.

Previous work (14) indicated that increasing the nitrogen content of these steels increased the slip planarity and increased the hydrogen sensitivity. Recent work measuring the stacking fault energy (SFE) versus nitrogen level in these steels (13) supports the idea of increased planarity (lower SFE) affecting hydrogen embrittlement, as shown in Figure 5. One point not emphasized previously is that increased strength also accompanies the increase in nitrogen content. A shift from 0.31 to 0.47 wt % nitrogen increases the strength from 434 to 510 MPa without changing the SFE (13, 14). However, the fracture mode does shift from transgranular twin band fracture shown in Figure 4 to intergranular at the higher strength level (14). Similarily, increasing the strength by high energy rate forging results in a greater susceptibility and a greater tendency toward intergranular fracture. A more complete description of the effect of increased strength on changes in fracture mode is given in (15). The observation is, however, that not only increasing the slip planarity but also increasing the strength will lead to increased intergranular fracture.

Fig. 2 - Deformation Twins in Nitronic 40
After 40% Tensile Strain

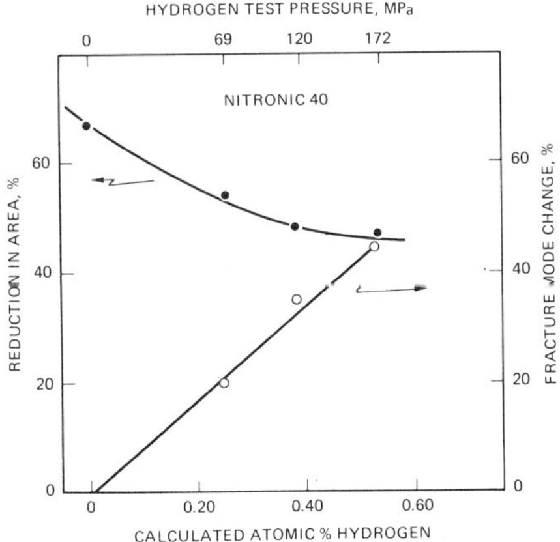

Fig. 3 - Reduction in Area and Percent Fracture Mode Change
vs Calculated Hydrogen Concentration for Nitronic 40

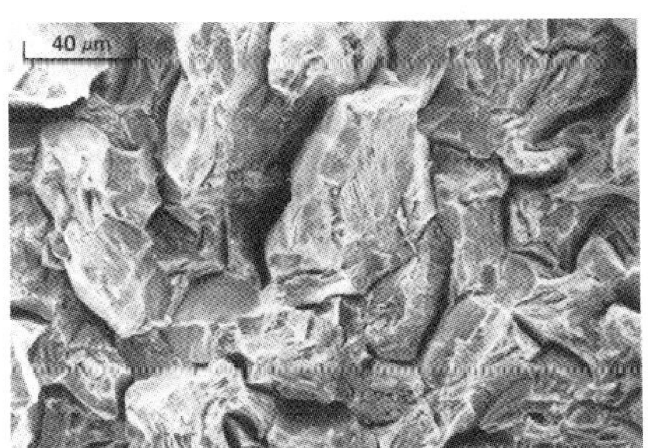

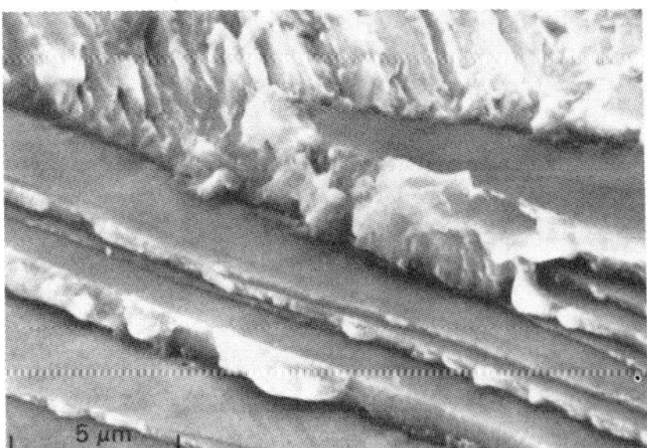

Fig. 4 - Fracture Surface of Nitronic 40 Tested in 172 MPa Hydrogen

(a) Mixed transgranular-intergranular mode;
(b) Stepped transgranular fracture along deformation twins

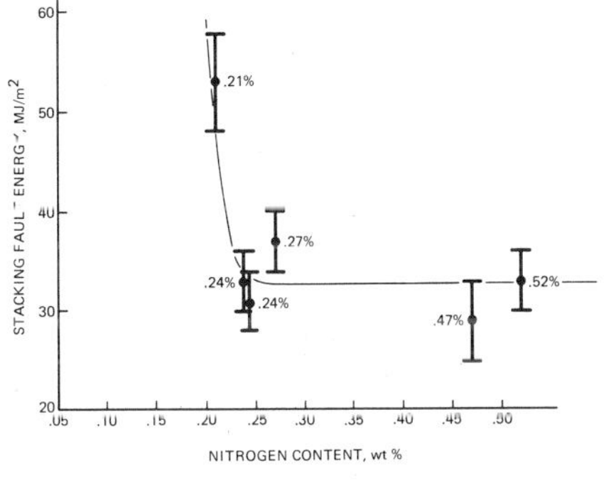

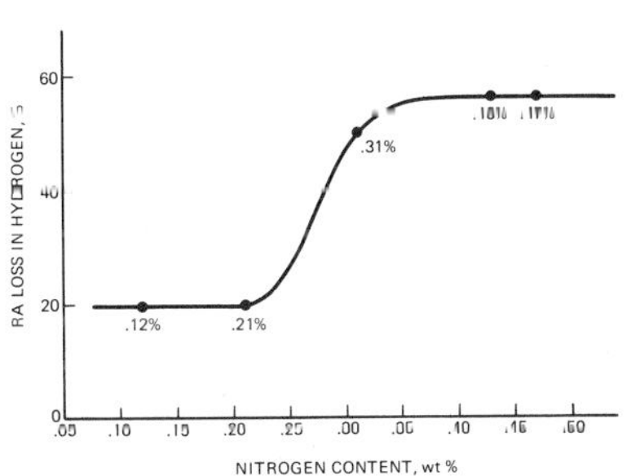

Fig. 5 - Correlation of Stacking Fault Energy and RA Loss In Hydrogen
With Nitrogen Content in Nitronic 40

(a) From Ref. 13; (b) Adapted from Ref. 14

JBK-75. Previous experiments on an alloy similar to JBK-75 indicated that no loss in ductility resulted from testing in 69 MPa (10 ksi) hydrogen (16). In the solution-treated and aged condition (993K/16 hrs), JBK-75 exhibits a fracture ductility of 51% RA in air. Testing in 172 MPa hydrogen resulted in a 47% fracture ductility or an 8% ductility loss. The fracture surface in hydrogen was entirely void growth and coalesence, except for isolated regions near the surface which exhibited an intergranular mode. In contrast, in a sample precracked in rotating bending fatigue to introduce a sharp circumferential flaw, tensile testing in 172 MPa hydrogen resulted in a 100% intergranular fracture mode.

Nucleation Strain. Smooth tensile bars of Ni 270, Nitronic 40, and JBK-75 were pulled in a number of high pressure hydrogen tests to strains short of failure and examined both optically and by SEM for surface crack nucleation. The surface finish was 16 RMS (machined finish) in all cases. Results are given in Table III along with the percentage fracture mode change, RA loss in the environment, and the nominal air ductility.

Table III. Surface Crack Nucleation Strains

Alloy	Environment	Nucleation Strain	Fracture Mode Change	RA Loss	Air Ductility
Ni 270	35 MPa H_2	22%	100%	50%	89%
Nitronic 40	69 MPa H_2	60%	10%	15%	78%
	172 MPa H_2	54%	40%	25%	78%
JBK-75	172 MPa H_2	45%	<5%	8%	51%

Sustained Load Cracking

Table IV summarizes the alloy compositions, heat treatments, and tensile properties of all the materials used in the sustained load cracking tests. Experiments were performed on samples initially loaded in air and tested in 207 MPa (30 ksi) hydrogen. Table V gives the initial stress intensity values K_O, and the threshold stress intensities K_{TH} for alloys which exhibited crack growth and the K_O and test times for alloys in which no growth was observed. Alloy 316 has not reached a threshold, but a value less than 125 MPa $\sqrt{m}$ is indicated. The threshold data for all the steels tested is given in Figure 6 where K_{TH} is plotted against yield strength. For clarity, the aluminum data was not included in the plot.

The fracture path in all alloys which showed crack growth was intergranular. At higher K levels, some ductile tearing occurred in regions adjacent to grain boundaries. Near K_{TH} the crack path was entirely along the grain boundary interface, with little evidence of plastic deformation accompanying the fracture process.

Optical and transmission electron microscopy was used to determine the microstructure of the grain boundary regions. The JBK-75 heat treatment C exhibited a high density of grain boundary precipitates, and these are responsible for the anomalously low K_{TH} value compared to the other steels.

Details of the behavior of JBK-75 in sustained cracking is given in Ref. 17. However, all of the alloys which do fall on the trend line of K_{TH} versus σ_y in Figure 6 exhibit clean grain boundaries free of any major concentration of inclusions or precipitates. Auger analysis of the JBK-75 and Nitronic 40 samples showed no segregation of alloying elements or metalloid impurities.

Table IV. Chemistry (Wt%), Treatment, and Properties of Sustained Load Cracking Alloys

304L:	Fe-18.3Cr-11.9Ni-1.77Mn-0.57Si-0.027C
JBK-75:	Fe-15.5Cr-30.7Ni-1.2Mo-0.26V-0.2Al-2.1Ti-0.017C
Nitronic 40:	Fe-19.9Cr-6.7Ni-9.2Mn-0.37Si-0.22N-0.03C
IN903:	Fe-37.8Ni-15.2Co-1.33Ti-0.9Al-3.07Nb-0.04C
316:	Fe-17.4Cr-13.8Ni-2.25Mo-1.5Mn-0.06C
2124:	Al-4.24Cu-1.38Mg-0.46Mn
7475:	Al-5.69Zn-2.36Mg-1.60Cu

	Yield MPa (ksi)	Ult. MPa (ksi)	Elong. %	RA %
304L				
A: Forged 1113K/WQ*	593(86)	752(109)	20	66
B: Forged 1253K/WQ	386(56)	656(95)	39	70
JBK-75				
A: Forged 1243K/WQ + Aged 948K/8 hr + 873K/8 hr	855(124)	1179(171)	18	37
B: Forged 1243K/WQ + Aged 948K/32 hr	924(134)	1221(177)	13	38
C: Soln 1253K/WQ + Aged 993 K/16 hr	717(104)	1131(164)	28	51
Nitronic 40				
Forged 1113K/WQ	828(120)	986(143)	15	36
IN903				
A: Soln 1213K/WQ Aged 993K/8 hr + 893K/8 hr	1055(153)	1317(191)	10	40
B: Soln 1213K/WQ Aged 993K/16 hr	917(133)	1179(171)	12	45
316				
Forged	903(131)	966(141)	15	70
2124-T851				
LT Orientation	430(62)	475(68)	9	20
ST Orientation	411(59)	456(66)	5	18
7475-T7351				
LT Orientation	445(64)	510(74)	14	38
ST Orientation	432(62)	500(72)	12	26

*WQ: Water quenched

Table V.   Crack Growth Data

[207 MPa (30 ksi) Hydrogen]

	Initial Ko, MPa $\sqrt{m}$	Final K_{TH}, MPa $\sqrt{m}$	Test Time, hrs
At Threshold			
JBK-75 A	160	110	
JBK-75 B	122	65	
JBK-75 C	125	45	
Nitronic 40	132	100	
IN903 A	125	30	
IN903 B	132	80	
Cracking			
316	132	<125	4200
No Crack Growth			
304L A	110	–	5000
304L B	50	–	5000
2124 LT	23	–	3000
2124 ST	17	–	3000
7475 LT	40	–	3000
7475 ST	30	–	3000

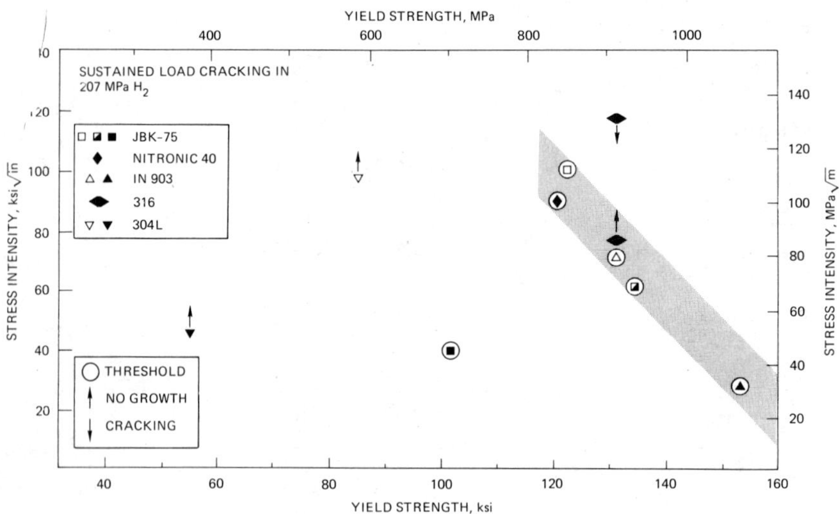

Fig. 6 - Crack Growth Threshold Stress Intensity (indicated by circles)
or Initial Stress Intensity (arrows) Versus Yield Strength For
All Steels Tested in 207 MPa Hydrogen

Discussion

Tensile Fracture Behavior

The fracture behavior of Ni 270 in high pressure hydrogen supports the observations of others in 0.10 MPa hydrogen (18,19). Lynch (18) reports that in bend tests the fracture plane is (100) while Clark and co-workers report a (111) fracture plane in tension. Transgranular fracture is explained either by hydrogen enhanced dislocation nucleation at the crack tip (18) or by localized work hardening on the slip plane (19). The two explanations are not mutually exclusive and the differences in crack plane may be due to differences in loading mode (20).

Though the fracture plane orientation was not determined in this work, tensile deformation was used, and the general fracture features in Figure 1 support the observations of Clark that (111) is the fracture plane. The sketch in Figure 7 is a proposed model which can be used to explain the details of the fracture features in Figure 1: (a) the primary slip plane (111) is intersected by a secondary system (1$\bar{1}$1). This intersection produces offsets on the primary plane and an increase in the local work hardening (b). As fracture proceeds on the primary slip plane, the crack can deviate and fracture along the secondary system, leaving the striation step (c). Alternatively, fracture may proceed first on the primary plane and, subsequently, secondary slip steps are produced after the crack has passed. This is especially likely if the crack plane is inclined to the tensile axis, and, if secondary slip occurs on one half and not the other, the striations will not correspond one-to-one (d). Stereographic analysis

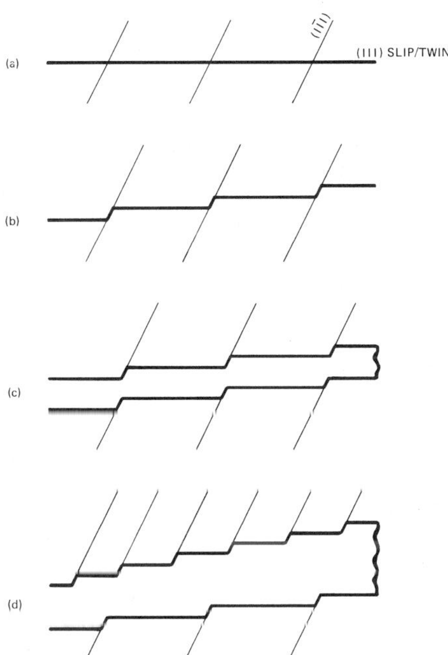

Fig. 7 - Schematic of Slip Band/Twin Band Cracking Process in Low Strength FCC Alloys

of matching surfaces as in Figure 1c shows the striations to match peak-to-trough as shown in Figure 7c. This model shows how the crack may stop and re-initiate at the striations, Figure 1a, and additionally predicts that the striations are parallel to a [110] direction. The latter observation is common to the work of Lynch (18) and Clark (19).

In order for slip band cracking to occur in preference to the normal rupture process, hydrogen gas must accelerate the nucleation of dislocations at a limited number of sites on the fracture surface, or hydrogen in solution must increase the local slip plane dislocation density by adjusting the internal arrangement of dislocations. As stated, both effects have been reported for nickel (18,19). Additional support for the latter observation is given in Figure 8. Shown are two adjacent regions of a sample tested to 30% strain in 35 MPa hydrogen and subsequently failed in air. The slip bands are straighter in the region in contact with hydrogen during deformation, suggesting a more planar dislocation arrangement.

Transgranular fracture similar to that in Ni 270 was observed in Nitronic 40. Hydrogen-induced transgranular fracture is associated either with slip or deformation twin bands both on (111) planes. Since the stacking fault energy for Nitronic 40 is much lower, 32 mJ/m^2 (13) versus 125 mJ/m^2 for nickel (21), the alloy deforms in a more planar manner independent of hydrogen. For this reason the fracture surfaces exhibit fewer features associated with secondary slip, Figure 4. It is important to note that this transgranular fracture mode is associated with deformation twins and not pre-existing annealing twins as shown by Caskey (22). Plastic deformation within the twinned regions combined with hydrogen in the environment leads to the transgranular fracture mode.

Ductility Losses and Fracture Mode in High Pressure Hydrogen

As indicated in Table III, the ductility loss in the three alloys studied varies from 50% for Ni 270 to 15-25% for Nitronic 40 to 8% for JBK-75. This variation can be explained as follows: Failure in a tensile test in air is due to nucleation of voids forming an internal crack and the growth of the crack from the centerline of the bar outward. The inward growth of hydrogen-induced cracks from the surface competes with this ductile process. In Ni 270, few inclusions exist at which voids can nucleate, as evidenced by the 90% RA in air. Though over 20% strain is required to nucleate a surface crack in

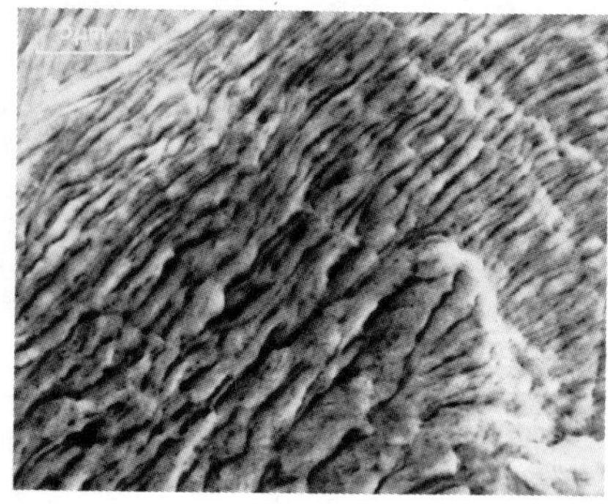

Air3.5 MPa Hydrogen

Fig. 8 - Comparison of Surface Slip Markings in Air and 3.5 MPa Hydrogen From Directly Adjacent Regions in Nickel 270. (Hydrogen promotes a more planar slip arrangement than observed in air.)

hydrogen, no void process competes with this external crack growth, leading to a 100% fracture mode change and a 50% loss in ductility. This large ductility loss in nickel is a reflection not only of its hydrogen sensitivity, but also of the lack of a competing fracture mode.

In Nitronic 40, strains of from 54-60% are required for surface crack nucleation. This is well beyond the 40% strain for the onset of necking, and thus even before a hydrogen surface crack has formed, the internal void processes are well underway. The final ductility is a result of competition between the two processes. The data of Table III also show that the larger RA loss and greater percent fracture mode change at 172 compared to 69 MPa (Figure 3) is due to earlier surface crack nucleation at the higher pressure. A critical combination of surface strain and hydrogen appears necessary for transgranular crack nucleation.

The RA loss of JBK-75 in 172 MPa hydrogen is less than that for Nitronic 40, suggesting a more resistant alloy. On the other hand, the critical strain for crack nucleation is lower. The air ductility of JBK-75 is lower than for Nitronic 40 (51 vs 78%) primarily due to its higher strength and as a result the crack nucleation strain in hydrogen almost equals the fracture strain in air. Thus any hydrogen-induced cracking is completely dominated by the ductile fracture process, and little RA loss or mode change is observed. Precracking overcomes the nucleation stage, and crack growth in hydrogen proceeds in a completely intergranular manner. The conclusion is that in order for hydrogen sensitivity to be manifested in the tensile ductility, either the ductile fracture process must be suppressed--for instance, by torsion testing--or a specimen geometry should be used which is more sensitive to surface cracking. Plane strain samples have been proposed as an alternative to cylindrical tensile bars (23) for the latter reason.

Finally, the presence of intergranular fracture in JBK-75 and absence of transgranular cracking as seen in Ni 270 and Nitronic 40 suggest a competition between hydrogen-induced fracture modes. As the yield strength increases, the resolved stress on the grain boundaries increase proportionately. Hydrogen-induced grain boundary decohesion can thus occur at a lower strain than transgranular slip band cracking, and an intergranular fracture path predominates.

<u>Sustained Load Cracking</u>

The alloys which fall along the common band of K_{TH} vs σ_y in Figure 6 derive their strength from a variety of mechanisms: warm work (316), hot work plus precipitation (JBK-75), hot work plus solid solution strengthening (Nitronic 40), and precipitation alone (IN903). The fact that their resistance to crack growth is not a function of their overall microstructure suggests that the local microstructure along the crack path is important. The common feature of all the alloys is that, at threshold, the crack path is directly along the grain boundary interface, and, additionally, no segregation or significant precipitation occurs at these boundaries. Hydrogen thus interacts with the clean austenite grain boundaries to cause failure much in the same way hydrogen can cause crack growth in non-temper embrittled 4340 (24,25). Since no transgranular cracking was observed in the sustained load cracking studies of austenitic steels, hydrogen-induced decohesion at the grain boundaries is the likely mechanism for crack growth in this pressure range.

If one extends the trend line of K_{IH} vs σ_y to lower strengths, then the initial stress intensity for the two 304L alloys is below the predicted K_{TH} and no crack growth is expected. This lack of crack growth is in contrast to Eliezer and co-workers (7) who report crack growth in only 0.10 MPa hydrogen.

The differences may be in the test method: in the current work, crack growth occurs in a fixed displacement, decreasing K test. In the earlier work (7), each increment of crack growth is followed by an increase in load to restart the crack growth process. As such, the test is similar to a fatigue test with high mean load, and the actual K at the crack tip may be as high as that indicated to be necessary in Figure 6.

Finally, no crack growth was observed in 2124 or 7475 aluminum alloys. This supports the review of Speidel (26) where he indicated that water vapor is necessary for crack growth in aluminum alloys so as to assist in the dissociation and absorption of hydrogen gas at the crack tip.

Conclusions

1. Transgranular slip band fracture occurs in Ni 270 and Nitronic 40 in the presence of hydrogen gas at pressures from 35 to 172 MPa. Intergranular fracture occurs in JBK-75 at similar hydrogen pressures. Hydrogen induced flow localization along the slip plane accelerates slip band cracking in the lower strength alloys, while grain boundary decohesion is the fracture mechanism in the higher strength JBK-75.

2. Degradation in tensile ductility in high pressure hydrogen does not always correlate to hydrogen cracking sensitivity. This is due to the large strains required to nucleate a surface crack and the competing ductile fracture process which masks the effect of hydrogen.

3. Sustained load cracking in 207 MPa hydrogen was observed in austenitic alloys with yield strengths greater than 700 MPa. All alloys with clean austenite grain boundaries fall on a common band of crack growth threshold vs yield strength, regardless of strengthening mechanisms, suggesting decohesion at grain boundaries is the controlling step in crack growth. No sustained load cracking was seen in 304L (380 and 590 MPa yield strength) nor in 2124 or 7475 aluminum alloys.

Acknowledgments

The authors would like to thank A. J. Salmi and J. C. O'Connor for technical assistance, and acknowledge helpful discussions on hydrogen mechanisms with M. R. Louthan, A. W. Thompson, H. K. Birnbaum, and W. W. Gerberich. This work was supported by DOE Contract No. DE-AC04-76DP00789.

References

1. R. J. Walter and W. T. Chandler, ASM Tech. Report No. D 8-14.2, ASM, Metals Park, OH, 1968.
2. R. J. Walter and W. T. Chandler, NASA Report No. CR-124410, Rocketdyne Div., Rockwell International, Canoga Park, CA, Oct. 1973.
3. W. T. Chandler and R. J. Walter, Hydrogen Embrittlement Testing, ASTM STP 543, ASTM, 1974, pp 170-197.
4. R. P. Jewett, R. J. Walter, W. T. Chandler, and R. P. Frohmberg, NASA Report CR-2163, Rocketdyne Div., Rockwell International, Canoga Park, CA, 1973. (Available NTIS: N73-21444).
5. P. M. Lorenz, Boeing Report No. D2-114417-1, The Boeing Co., Seattle, WA, Feb. 1969. (Available NTIS: N69-19152)
6. A. W. Loginow and E. H. Phelps, Corrosion, 31, 1975, pp 404-412.
7. D. Eliezer, D. G. Chakrapani, C. J. Altstetter, and E. N. Pugh, Met. Trans., 10A, 1979, pp 935-941.
8. R. J. Walter and W. T. Chandler, Environmental Degradation of Engineering Materials, M. R. Louthan and R. P. McNitt, Eds., V.P.I., Blacksburg, VA, 1977, pp 513-522.
9. A. W. Thompson and I. M. Bernstein, Advances in Corrosion Science and Technology, Vol. 7, Plenum Press, NY, 1976.
10. I. M. Bernstein and A. W. Thompson, Alloy and Microstructural Design, G. S. Ansell and J. K. Tien, Eds., Academic Press, NY, 1976, p. 303.
11. M. R. Louthan, Hydrogen in Metals, A. W. Thompson and I. M. Bernstein, Eds., ASM, Metals Park, OH, 1974, pp 53-78.
12. A. J. West and M. R. Louthan, Met. Trans., 10 A, 1979, pp 1675-1682.
13. R. E. Stoltz and J. B. VanderSande, Met. Trans., 11 A, 1980, pp 1033-1037.
14. B. C. Odegard, A. J. West, and J. A. Brooks, Effect of Hydrogen on Behavior of Materials, A. W. Thompson and I. M. Bernstein, Eds., TMS-AIME, NY, 1976, pp 116-125.
15. B. C. Odegard and A. J. West, this conference.
16. A. W. Thompson and J. A. Brooks, Met. Trans, 6A, 1975, pp 1431-1442.
17. M. W. Perra and R. E. Stoltz, this conference.
18. S. P. Lynch, Scripta Met., 13, 1979, pp 1051-1056.
19. E. Clark, J. Eastman, T. Matsumoto, N. Narita, and H. K. Birnbaum, this conference.
20. S. P. Lynch, Aeronautical Research Laboratories, Melbourne, Australia, private communication.
21. C. B. Carter and S. M. Holmes, Phil. Mag., 35, 1977, pp 1161-1172.
22. G. Caskey, Environmental Degradation of Engineering Materials, M. R. Louthan and R. P. McNitt, Eds., V.P.I., Blacksburg, VA, 1977 pp 437-450.
23. R. E. Stoltz, "Plane Strain Tensile Testing for Measuring Environment Sensitive Fracture," accepted for publication, Met. Trans., 1980.
24. D. M. Moon, Scripta Met., 8, 1974, pp 1435-1438.
25. R. Viswanathan and S. J. Hudak, Met. Trans., 8A, 1977, pp 1633-1637.
26. M. O. Speidel, presented at Troiano Festschrif, Case Western Reserve University, Cleveland, OH, June 1980.

HYDROGEN INFLUENCE ON MECHANICAL CHARACTERISTICS OF MARTENSITIC, STABLE

AND UNSTABLE AUSTENITIC PURE STAINLESS STEELS.

by D. Nejem[*], M. Habashi[++], J. Galland[+], S. Talbot-Besnard[++] and P. Azou[*].

[*]Ecole Centrale des Arts et Manufactures, Chatenay-Malabry, France.
[+]Université Paris XI, France.
[++]Centre National de la Recherche Scientifique, France.

Although hydrogen embrittlement of austenite has often been studied, the mechanism involved is not well understood. After Coriou and Grall (1) stable austenite embrittlement was said to come from an hydride formation. Benson(2) attributes to martensite, the embrittlement of strained unstable austenite. Mehta and Burke (3) think that embrittlement is an intrinsic effect and not the consequence of a phase change.

These differences appear because studies were made on industrial alloys with varied compositions and with different methods of hydrogen introduction resulting in different states of hydrogen in metals.

We have chosen to simplify the study, as we usually do, by studying pure metals. Thus, we avoid hydrogen recombination into molecular hydrogen which occurs on metal inclusions and defects. This permits a study of reversible embrittlement due to solid solution hydrogen.

First, we have studied the behavior of γ austenite (face centered cubic structure) then that of α' martensite (body centered cubic structure). Comparisons were made with an unstable austenite possibly transformed into martensite by cold working.

Fig.1 shows the curve of the resulting structure versus the Ni composition for 18% Cr. The two curves are related to two cooling methods.

We have manufactured three alloys by vacuum melting and casting from pure commercial metals (Table 1). We have chosen 14% of Ni to obtain a stable austenite with 18% Cr, 6% Ni for a martensite and 10% Ni which gives raise to an austenite which is able to give a martensite by cold working. Homogeneity was controlled by electronic microprobe (Table II). After cutting the ingot and cold rolling, tensile samples are prepared (16 mm gage length, 2 mm width, 1 mm thickness) followed by vacuum annealing for 1 h at 1000°C under vacuum.

Starting metals	C	Cu	Ti	Al	S	Si	Na	Pb
Fe	60	3	250	3	35	0	0	0
Cr	10	0.1	0.1	0	0.3	0.1	0.1	0.1
Ni	37	10	10	30	0	25	30	30

Table I : Chemical composition of starting metals used to manufacture the alloys in 10^{-6} (ppm by weight)

Alliages	Cr %	Ni %
Austénite "18-14"	17 $\pm$ 0.87	14 $\pm$ 0.55
Martensite "18-6"	18.1 $\pm$ 0.533	5.5. $\pm$ 0.2
Austénite "18-10"	18. $\pm$ 0.2	10 + 0.1

Table II : Local concentration variation of the alloys.

Analysis shows that the quantity of C is 120 ppm and O 148 ppm.

Specimens were then electropolished in a 90% monobutylic ester of ethyleneglycol, 10% perchloric acid bath.

We have checked the austenitic stainless steel with 14% Ni to see if it remained wholly austenitic when strained. An austenitic structure was observed by optical microscopy, X-rays and magnetism with "ferritest" apparatus. The structure was also checked after failure in a tensile test and after cathodic charging. In all cases no martensite structure was detected.

The austenite, which has a very low hydrogen diffusion coefficient was charged in molten salts at 200°C. Water was injected in acid and neutral sodium and potassium sulfate mixed at it was pointed out previously at the Ecole Centrale (4).

In the chosen experimental conditions we put approximately 47 ppm H_2 dissolved into the metal and measured by degassing at 600°C.

Degassing was done by the usual method at the Laboratory (5) under vacuum with a temperature increase rate of 200°C/H. Tensile tests results are plotted in Fig.2.,3.,4. We see hydrogen does not decrease strain, thus there is no hydrogen embrittlement of pure and stable austenite if this is chosen as an embrittlement criterium. The consolidation rate curve versus strain shows that there is one slide mechanism and this is not modified by hydrogen.

Strain rate has no influence on stress. Fig. 5 shows a typical ductile fracture surface with empty dimples found both in the presence or absence of hydrogen. This is further indication that embrittlement has not occurred.

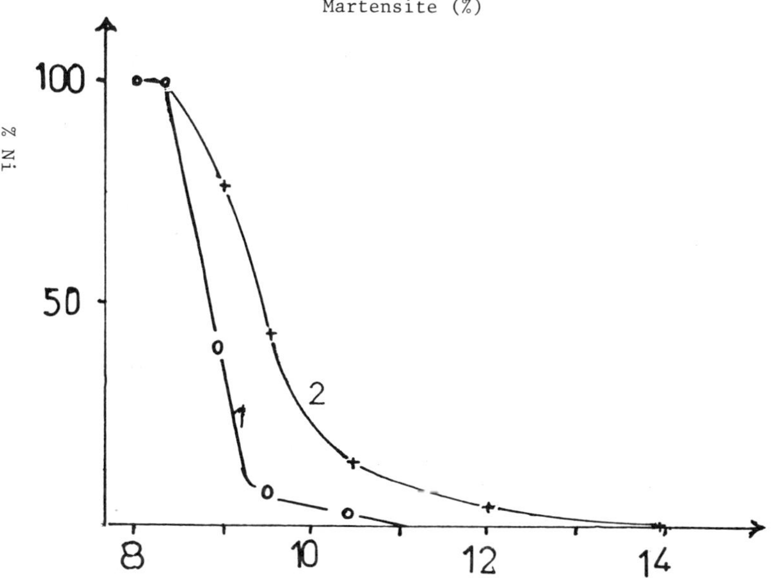

Fig. 1. Quantity of martensite in alloys with 18% Cr and various Ni curve 1 quenched samples.
curve 2 quenched samples cooled in liquid nitrogen.

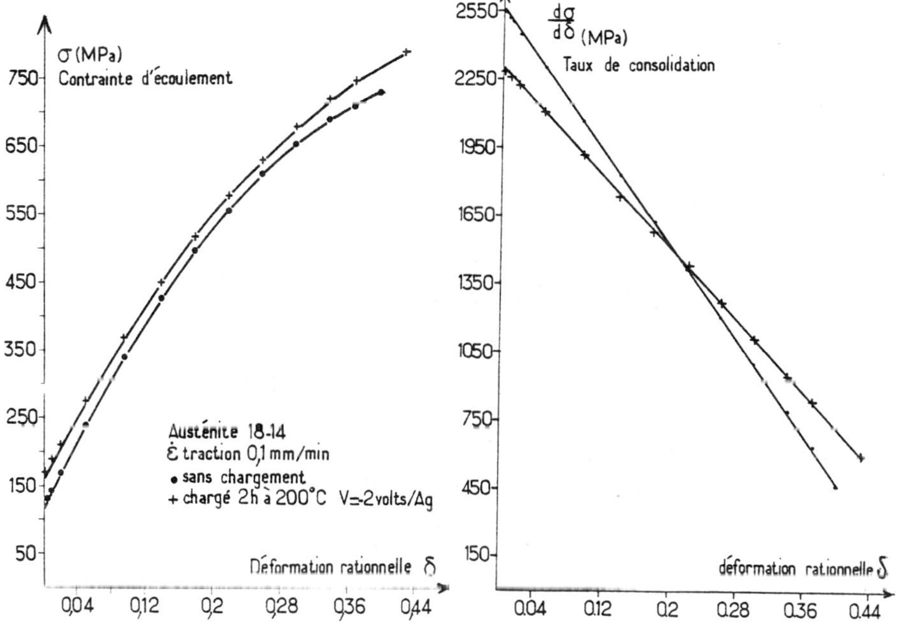

Fig. 2 - Flow stress of 18-14 austenite versus true strain.
Without hydrogen

Fig.3 - Work hardening rate of 18-14 austenite versus true strain.
+ With hydrogen.

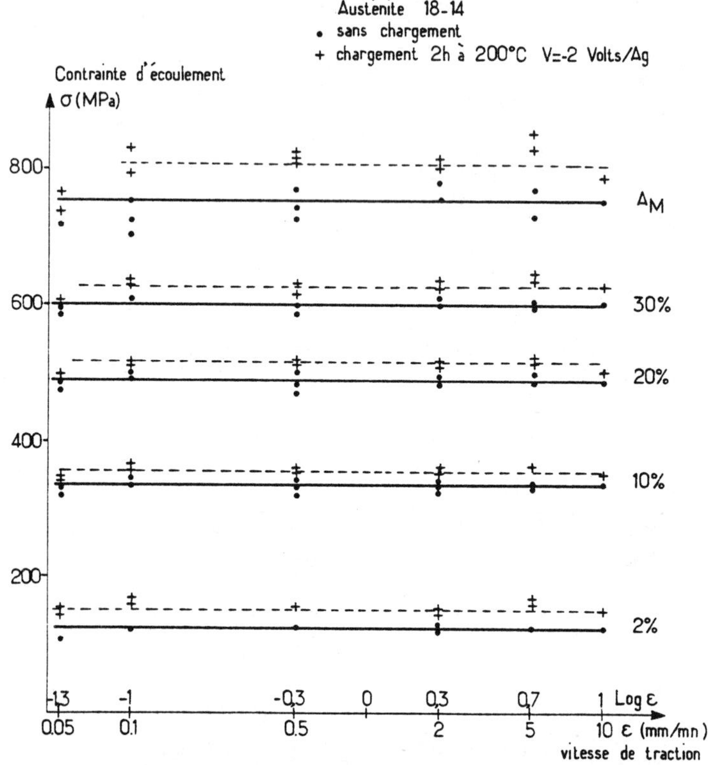

Fig. 4 – Flow stress versus strain rate.

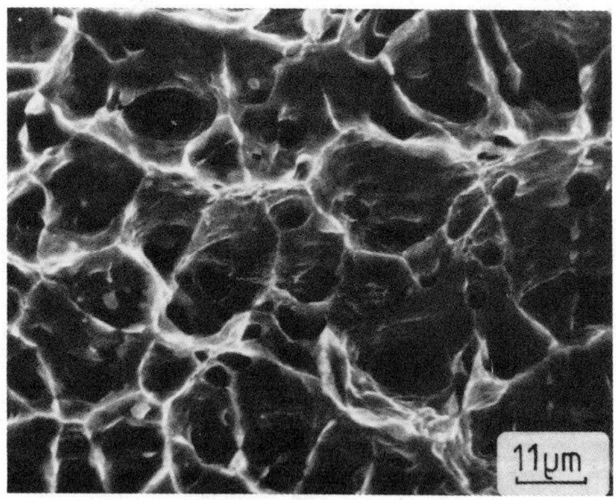

Fig. 5 – Fracture surface of 18-14 austenite with or without hydrogen.

Necking lightly decreases with hydrogen :

$$F \% = \frac{\Sigma_R - \Sigma_{RH}}{\Sigma_R} = 10\%$$

The second alloy tested had a martensitic structure. As before the structure was characterized by optical micrography, and X-rays. In addition to dilatometry since X-rays cannot distinguish between ferrite and martensite having the same cell parameters. The procedure involves heating the sample to more than 950°C at a rate of 150°C/h to obtain an austenitic structure. Then it is cooled at two different rates : a slow rate of 150°C/h and a quench by moving away the furnace keeping the sample under a static argon atmosphere. In both cases one phase transformation at 200°C corresponding to M_S was observed. So we are sure that we have only a martensitic structure without ferrite. Cathodic charging was performed in solution at ambient temperature because diffusion is easy in this structure. 9 ppm hydrogen were introduced.

Mechanical properties are presented in Fig. 6. The data indicates hydrogen increases yield stress and flow stress and greatly decreases strain in the martensitic samples. Strain rate variation does not influence the results. Failure wich was ductile without hydrogen becomes brittle in the presence of hydrogen : intergranular and transgranular. Embrittlement degree F % rises to 86 %. Martensitic structure is highly embrittled by hydrogen.

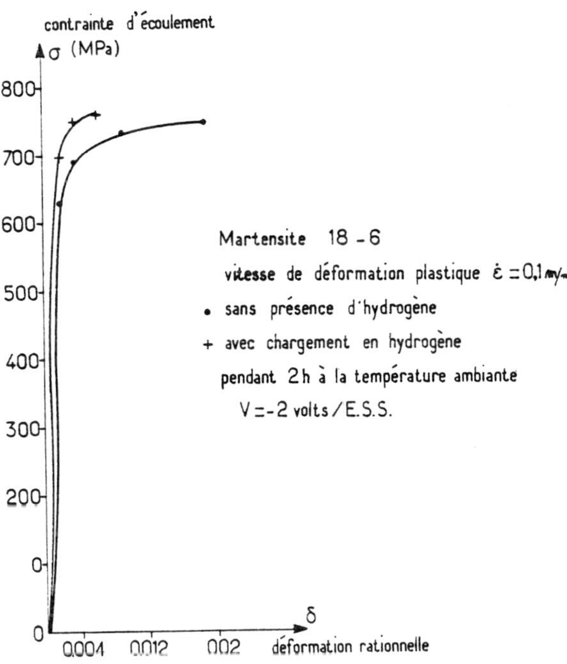

Fig. 6. Flow stress of 18-6 martensite versus true strain.

The third alloy studied was an unstable austenitic stainless steel with 10% Ni. X-rays show only austenitic peaks. This structure was charged in molten salts at 200°C · 45 ppm hydrogen were introduced reversibly.

Austenite destabilisation was studied by X-rays (Table III).

18-10	γ	α'	ε
without H_2 with H_2	+		
With cold working A = 48 %		+	+
With cold working A = 17,8 % With or without hydrogen	+		+

Table III   + Structure seen by X-rays.

In the case of the 17,8 % cold worked alloy it remains a part of austenitic structure. α' is not seen by X-rays but hindered perhaps by cold working. α' is present and seen by magnetism.

Fig. 7 shows α' martensite formation measured by magnetism versus strain. It appears that martensite formation is easier when the strain rate is low. However hydrogen does not influence the quantity of martensite produced.

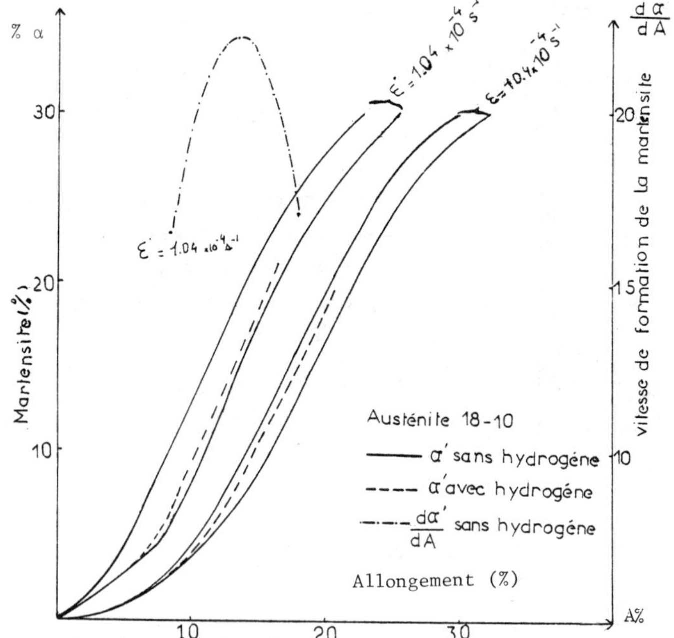

Fig. 7 –   martensite quantity formed by cold working of austenite 18-10 versus strain.

Hydrogen increases both yield and flow stress and decreases strain :
A : 20% with hydrogen 49% without hydrogen.

Mechanical properties are presented in Fig. 8-9-10. We see on Fig.
8 the curve slope increases from approximately δ = 0,1. This can be attributed to the highest rate of α' martensite formation as we have seen it
on the previous curve. (δ = 0,1 $\sim$ A = 11 %)

Fig. 9 shows that different stages of deformation occur in 18-10 as they
do in single crystals. This was not the case for 18-14 austenite which presents one slide mechanism as shown in fig. 3. Usually unstable austenite
gives raise to a strain higher than in stable austenite. Here it is the
contrary. This must be attributed to the purity of the metal. The introduction of hydrogen does not modify the plastic deformation mechanism but
facilitates easy glide.

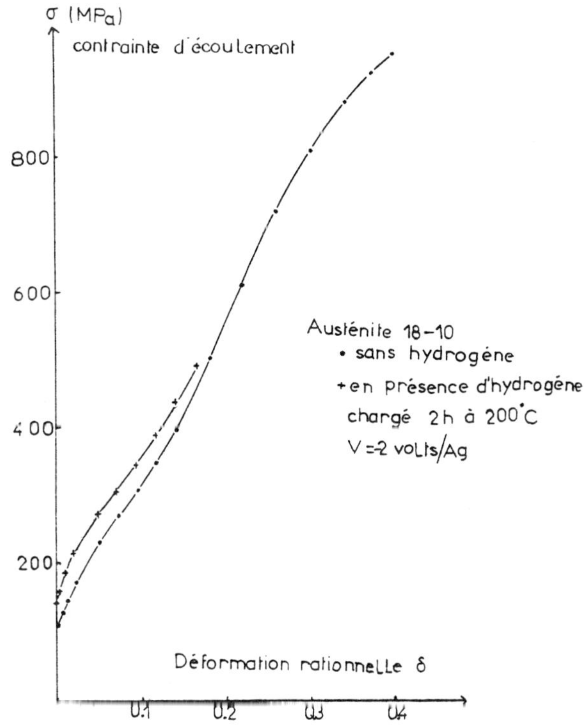

Fig. 8. - Flow stress of 18 10 austenite versus
true strain.

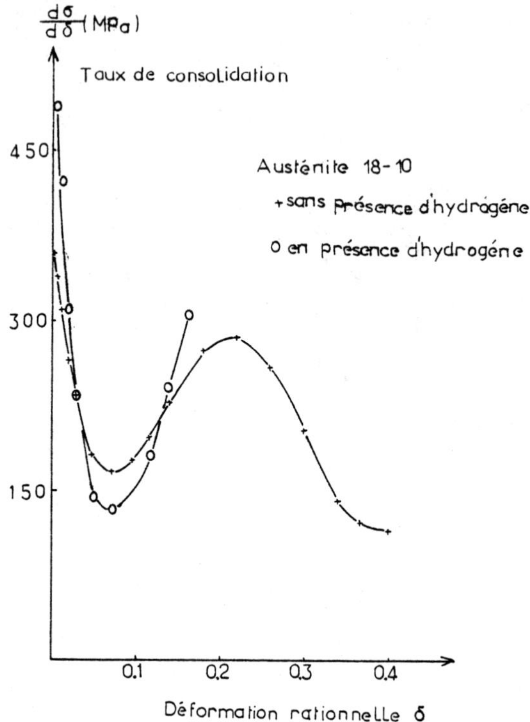

Fig. 9 — Work hardening rate of 18-10 austenite versus true strain.

Fig. 10 shows that strain rate influences the results. The yield stress is not changed when the strain rate varies, but the flow stress at 12,5 and 30 % strain, without hydrogen increases when strain rate decreases because the transformation of austenite to martensite is easier when rate is low (6 - Fig. 7).

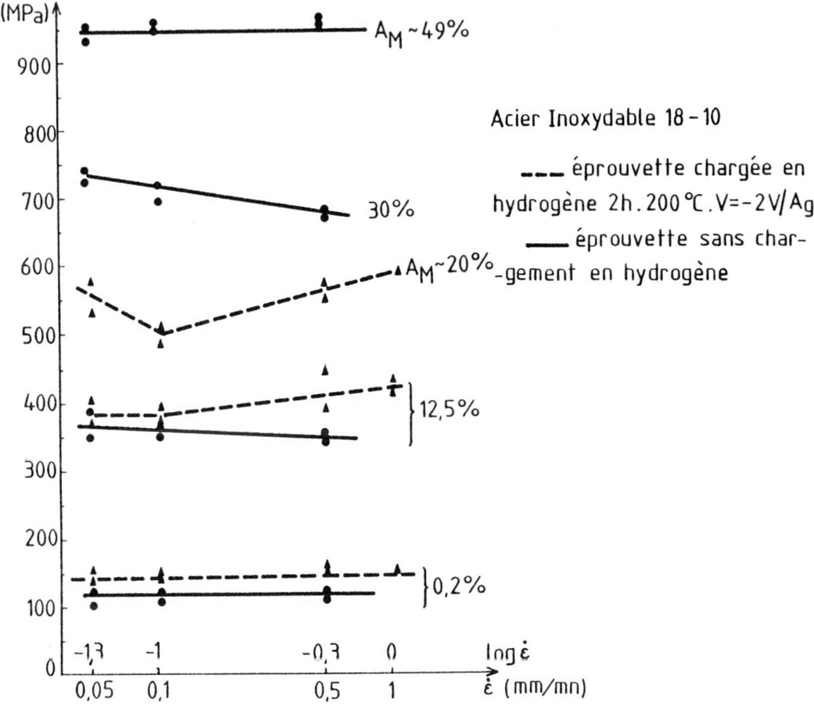

Fig. 10 - Flow stress of 18-10 austenite versus strain rate.

When cold working is high (45 - 50 %) the highest quantity of martensite is formed whatever the rate is. When the tensile test is done with hydrogen, steel behavior is influenced by austenite destabilisation and hydrogen dislocation interactions boundel with relative rates of strain and hydrogen diffusion. At low rates, destabilisation is the main phenomenon and as it is the case without hydrogen, the flow stress decreases when strain rate increases. Beyond $\dot{\varepsilon}$ - 0,1 mm/mn (1,04 10⁻⁴ ε⁻¹) dislocation hydrogen interactions give their own variation and the flow stress increases when strain rate increases. 18-10 unstable austenite without hydrogen fails with dimples. With hydrogen failure is brittle intergranular as we have often observed in pure metals (7).

The reversible hydrogen embrittlement of unstable austenite is well explained by the austenite martensite transformation occuring by cold working. However stable austenite is not embrittled.

<u>Bibliography.</u>

1. M.L. Mehta, J. Burke , Corrosion Nace, <u>31</u>, (3),108-110, (1975)·

2. R.B. Benson, R.K. Dann, L.W. Roberts, Trans. Asm, <u>242</u>, pp. 2199 (1968).

3   H. Coriou, L. Grall, A. Besnard, G. Pivard, Congrès Hydrogène dans les Métaux,p. 241, Paris(1972).

4. A. Elkholy, Thèse de 3ème cycle Univ. P. et M. Curie (1976).

5. J. Plusquellec, P. Azou, P. Bastien : Mem. Sci. Rev. Met., <u>60</u>, p. 11 (1963).

6. J. Galland, M. Habashi, H. Andriamiharisoa, Congrès International sur la rupture, Cannes (1981).

7. A. Gourmelon, Thèse Docteur-Ingénieur, Univ. Paris VI, (1974).

HYDROGEN ASSISTED CRACKING

IN TYPE 304L AND 316L STAINLESS STEEL

D. Eliezer
Materials Engineering Department
Ben-Gurion University of the Negev
Beer Sheva, Israel.

Hydrogen embrittlement of a type 304L and type 316L austenitic stainless steel has been studied by charging thin tensile specimens with hydrogen through cathodic polarization. Hydrogen induced slow crack growth was observed at room temperature when type 304L was stressed while undergoing cathodic charging. Room temperature cathodic charging of unstressed specimens for extended periods resulted in intergranular attack in both 304L and 316L types.

Throughout this study we have compared type 316L solution annealed samples with samples given the additional sensitizing treatment. Sensitization of 316L type lowers the mechanical properties of this material in hydrogen and the sensitized steel is more susceptible to hydrogen assisted cracking. The martensite formation was detected in connection with the hydrogen embrittlement.

Introduction

The occurrence of hydrogen embrittlement in austenitic stainless steels has been manifested by a reduction of ductility in dynamic tensile tests, and by the appearance of nonductile fracture surfaces. It has been shown that the ductility of austenitic stainless steels is reduced by charging with hydrogen through cathodic polarization or in a hydrogen atmosphere (1-6). From a practical point of view, it is more relevant to establish whether delayed failure occurs under static stresses. One of the objectives of the present work was to investigate this possibility.

A further objective was to study the effect of sensitization on the hydrogen cracking susceptibility of type 304L and 316L austenitic stainless steels. Recently, Briant (7) reported that the occurance of intergranular hydrogen assisted fracture in type 304 stainless steel was dependent upon the heat treatment which the material had received. Thompson (8) examined sensitization of type 309 stainless steel and found that it did not affect the mechanical properties of this material in hydrogen. This work reports a further study of hydrogen cracking in type 304L and type 316L austenitic stainless steel. It will be shown that a 316L type austenitic stainless steel can be severely embrittled by hydrogen.

Experimental

The studies were carried out on a type 304L and a type 316L austenitic stainless steel. These steels were of commercial grade, of compositions shown in Table 1, and were received in the form of sheets 0.2 mm thick.

Tensile specimens were prepared according to ASTM standard E8. All of the samples used in these experiments were first solution annealed for one hour at 1100°C and then water quenched. Some samples were given a further heat treatment at 650°C for 24 hours. This latter heat treatment caused the precipitation of chromium carbides along the grain boundaries. We have compared solution annealed samples with samples given the additional sensitizing treatment of 650°C for 24 hours. The cathodic charging of the specimens was carried out at room temperature in a 1N H_2SO_4 solution containing 0.25 g of $NaAsO_2$ per liter. A platinum counter electrode and a current density of 50 mA/Cm^2 were used.

Table I.   Chemical Composition of Alloys, Wt. Pct.

Element	304L	316L
Cr	18.68	17.65
Ni	9.02	11.10
Mn	1.80	1.80
C	.025	.026
Si	.64	.50
Cu	.06	.21
Co	.16	.35
P	.02	.03
N	.035	.061
S	.004	.009
Mo	.07	2.08

The following series of tests were conducted on each steel:

(i) Specimens were tested at room temperature under constant load in air after cathodic charging in the absence of external stress.

(ii) Specimens were tested under constant load at room temperature while undergoing cathodic polarization.

(iii) Tensile specimens were tested at an extension rate   of 0.005 cm/min., after cathodic charging in the absence of external stress.

(iv) Specimens were tensile tested at an extension rate of 0.005 cm/min, while undergoing cathodic polarization.

(v) For comparison purposes, specimens were tensile tested in air at room temperature.

After failure the fracture surfaces were examined with a scanning electron microscope. A measure of the amount of $\alpha`$  martensite present at the fracture surfaces was obtained using a commercial ferrite detector with a sensitivity of 0.1 vol pct of $\alpha`$ martensite. Hydrogen effects on the martensitic transformation of the austenite phase to $\alpha`$ or ϵ phases have been studied by using transmission electron microscopy (TEM).

Results and Discussion

Room temperature cathodic charging of unstressed specimens for extended periods (1 to 15 days) resulted in intergranular attack in both types 304L and 316L. Figs. 1(a) and (b) depict intergranular cracks and secondary transgranular cracks along crystallographic planes. These secondary cracks were more pronounced in the type 304L steel, which all-together was attacked at a faster rate than the 316L type steel.

Stressing at room temperature after charging did not produce slow crack growth (SCG) in either type of steel. One probable explanation for the absence of SCG in specimens tested after charging is that the hydrogen is located near the free surfaces and not in the interior.

Specimens were tested while undergoing cathodic charging. In the case

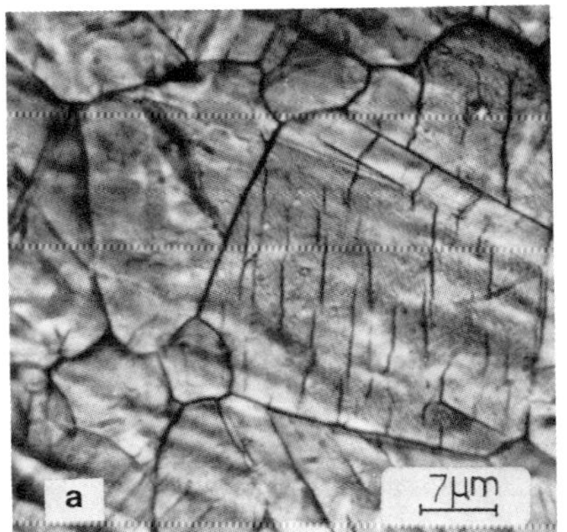

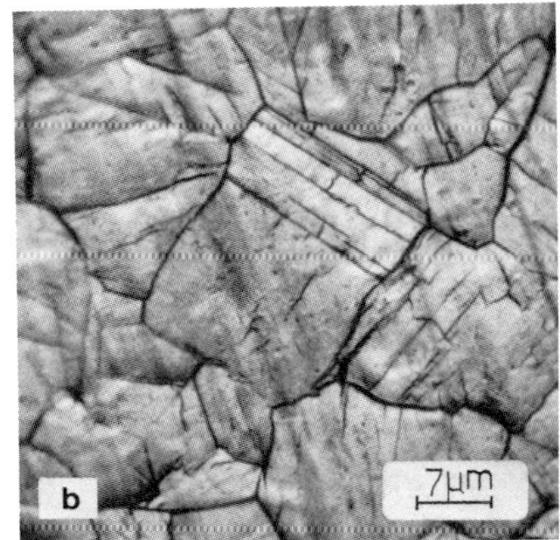

Fig. 1 - Surface attack caused by cathodic charging of
          unstressed specimens: (a) type 304L:cathodic
          charging for 24 hr. (b) type 316L:cathodic
          charging for 48 hr.

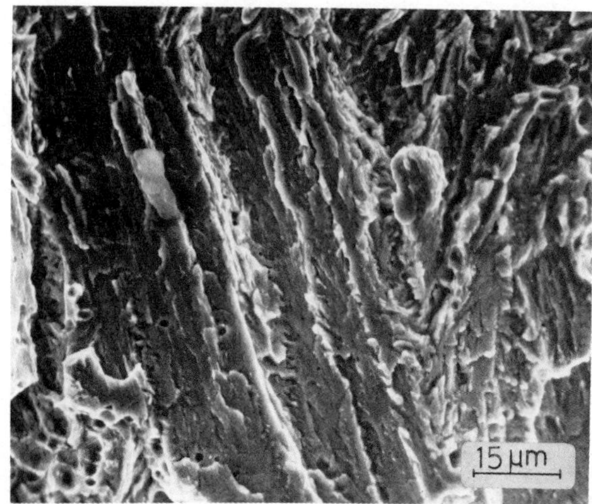

Fig. 2 - Scanning electron micrograph of fracture surfaces
of type 304L specimen tested while undergoing
cathodic charging.

Table 2. Percent Reduction in Ductility Compared to Uncharged
Specimens as a Function of Heat Treatment and Method
of Testing.

Heat Treatment \ Method of Testing	Cathodic charging for 15 days and tensile testing in air	Tensile testing while cathodically charging
1100°C for 1hr.	8	21
1100°C for 1hr. 650 for 24hr	25	49

of type 304L slow crack growth was observed and α˅ martensite was detected
on the fracture surfaces. Examination with the scanning electron microscope (SEM) indicated that these surfaces were transgranular and cleavage-
like (Fig. 2). No slow crack growth and no formation of α˅ martensite
were observed in type 316L specimens which were stressed, while being
cathodically charged. This topic will be dealt with in future work.

Type 316L solution annealed samples with samples given the additional
sensitizing treatment were compared. The results of the tensile tests
show that the room temperature tensile yield and ultimate strengths were
not affected considerably by sensitization (Table II). This presents
significant changes in the ductility of the steel depending upon the heat
treatment which the material had received and the method of charging.
Examination of the fracture surfaces of the annealed 316L specimens that
were cathodically charged for 15 days and tensile tested in air, shows
massive regions of microvoid coalescence producing ductile rupture (Fig.
3(a)). Several workers have reported that the fracture of specimens
which showed hydrogen induced ductility losses was completely ductile and
that the presence of hydrogen reduced the size of the dimples characteristic of ductile fracture (8,9). Possible mechanisms of this effect have
been discussed by Thompson (10). Typical fracture surfaces of the sensitized type 316L specimens after cathodic charging for 15 days are shown
in Fig 3(b). The initial hydrogen fracture was predominantly intergranular
and some regions of microvoid coalescence were observed.   The 316L type
specimens that were tested while cathodically charged show considerable
differences in the fracture surfaces between the annealed and the sensitized specimens. The sensitized specimens were predominantly intergranular
(Fig. 4) while the annealed fracture specimen shows microvoid coalescence
producing ductile rupture.

No evidence of α˅ martensite was found in the annealed or sensitized
316L specimens. The transmission electron microscopy (TEM) shows clearly
that no α˅ martensite was formed in connection with the hydrogen embrittlement process. TEM studies of the fracture surfaces have revealed dislocation structures which suggest the existence of a metastable phase,
probably ε martensite during the cathodic polarization.

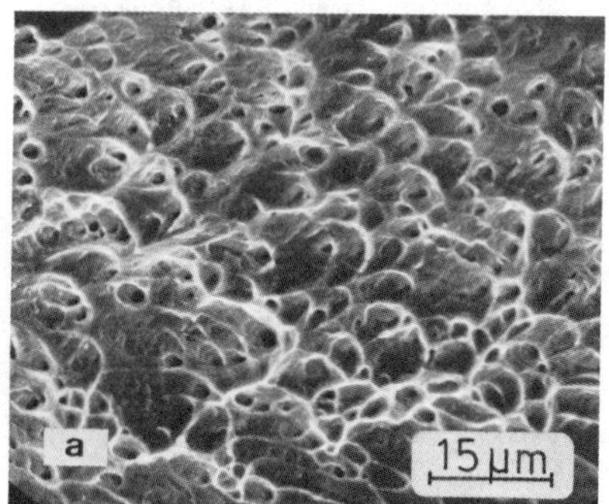 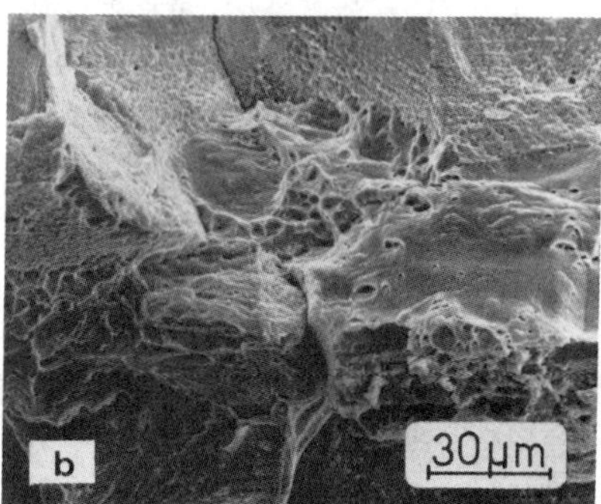

Fig. 3 - Fracture surfaces of type 316L specimens stressed
in air after 15 days of cathodic charging.
(a) solution annealed specimen (b) sensitized
specimen.

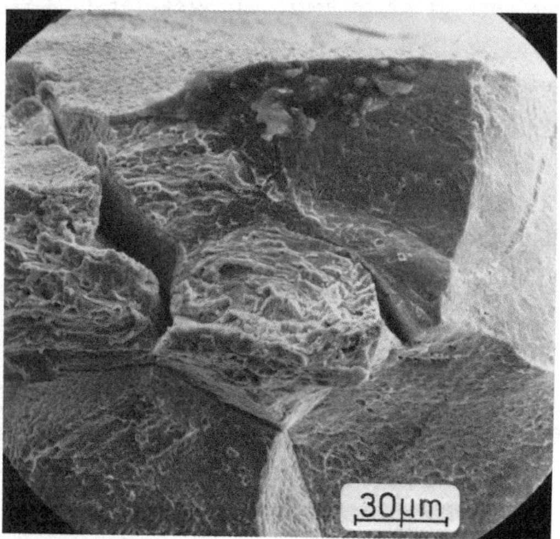

Fig. 4 - Fracture surface of type 316L sensitized specimen
tested while undergoing cathodic charging.

Conclusions

This paper reports a study of hydrogen assisted cracking in type 304L and 316L stainless steels. The conclusions from this study are the following:

(1)   Room temperature cathodic charging of unstressed specimen produced intergranular attack in both types 316L and 304L, the latter was attacked at a faster rate than the 316L type.

(2)   Hydrogen induced slow crack growth was observed at room temperature when type 304L was stressed while undergoing cathodic charging.

(3)   The results of the tensile tests of type 316L show that at room temperature tensile yield and ultimate strengths were not affected considerably by sensitization. Significant changes in the ductility were found due to the heat treatment and the method of charging.

(4)   No α' martensite could be detected in the type 316L steel after hydrogen charging.

(5)   Examination of the fracture surfaces of type 316L specimens tested while cathodically charged shows considerable differences in the fracture surfaces between the annealed and the sensitized specimens. The sensitized specimens were predominantly intergranular, while the annealed specimens fracture shows massive regions of microvoid coalescence producing ductile rupture.

References

1.   M.L. Holzworth: Corrosion, 1969, Vol. 25, pp. 107-15.

2.   R. Lagnebrog: J.I.S.I, 1969, Vol. 207, pp. 363-66.

3.   M.B. Whiteman and A.R. Troiano: Corrosion, 1965, Vol. 21, pp. 53-56.

4.   J. Kolts: Stress Corrosion - New Approaches, 1976, ASTM STP 610, ASTM, Pennsylvania, pp. 336-380.

5.   A.W. Thompson: Hydrogen in Metals, pp. 91-102, ASTM, Cleveland, OH, 1974.

6.   H. Hanninen and T. Hakkarainen: Corrosion, 1980, Vol. 36, pp. 47-51.

7.   C.L. Briant: Met. Trans. A, 1979, Vol. 10A, pp. 181-189.

8.   A.W. Thompson: Mater. Sci. Eng., 1974. Vol. 14, pp. 253-64.

9.   M.R. Louthan, G.R. Caskey, J.A. Donovan and D.E. Rawl: Mater. Sci. Eng. 1972, Vol. 10, pp. 357-368.

10.   A.W. Thompson: Met. Trans. 1979, Vol. 10A, pp. 727-731.

DISCUSSION

P. Azou, Laboratoire de Corrosion, Ecole Centrale des Arts et Manufactures, Chatenay-Malbray: Do you think that you have introduced hydrogen in the grain with your charging at room temperature? I think the diffusion coeffic- ient at room temperature is so low in 304L or 316 steels that I fear that no significant hydrogen may enter except at the grain boundaries.

No response by author.

EFFECT OF AGEING ON EMBRITTLEMENT AND MICROSTRUCTURES IN HYDROGEN CHARGED
THIN SPECIMENS OF AUSTENITIC STAINLESS STEEL

Hannu Hänninen, Tero Hakkarainen* and Pertti Nenonen*

Metals Laboratory
Technical Research Centre of Finland
Espoo, Finland

*Laboratory of Metallurgy and Mineral Engineering
Technical Research Centre of Finland
Espoo, Finland

Introduction

Austenite stability in hydrogen charged austenitic stainless steels
has been a controversial issue when hydrogen embrittlement (HE) of these
materials has been studied. The question is, whether the homogeneous
austenite (FCC) can be embrittled by hydrogen or is a transformation to
either a martensitic (ε- or α'-martensite) or some hydride phase necessary. Because the loss of ductility of hydrogen charged stainless steels is
reversible, the possible structural changes during charging and during ageing may be of crucial importance. After charging and some ageing, extensive surface cracking has usually been observed. This cracking is mainly
transgranular and crystallographic and does not require external stresses.

A number of authors (1-4) have studied hydrogen charged austenitic
stainless steels by X-ray diffraction. Phase transformations are observed
to occur at the surfaces of the specimens depending on the stability of the
austenitic stainless steels. The hydrogen-induced phase transformations
generally observed in AISI 304 and AISI 316 steels occur in the sequence:
$\gamma \rightarrow \varepsilon \rightarrow \alpha'$, AISI 316 steel being the more stable. AISI 310 steel shows no
α'-martensite but expanded austenite (γ^*) and expanded ε-martensite (ε^*)
according to Mathias et al. (3). The transformation $\gamma \rightarrow \gamma^*$ is reported to
be reversible at room temperature. In addition to those transformations,
Kamachi (4) has reported a metastable FCC-hydride and an intermediate hexagonal ε-phase. The above mentioned hydrogen-induced phase transformations
have been detected by X-ray diffraction near the outer surfaces of these
steels and therefore their signifigance in hydrogen-induced crack propagation is still unknown. No detailed TEM observations of these phases are
available at present.

Experimental

The materials used in this study were commercial AISI 304, AISI 316 and AISI 310 stainless steels (Table I). Thin tensile specimens (Fig. 1) were prepared from cold rolled foils (about 0.3 mm thick). The foils were annealed in vacuum for 1 hour at 1373 K and quenched into room temperature water. The final thinning was performed by grinding with a 600 grit paper.

Table I. Chemical Compositions of the Steels, Wt Pct

Material	C	Cr	Ni	Mn	Si	S	P	Cu	Mo
AISI 304	0.07	18.0	8.5	1.44	0.36	0.030	0.016	0.24	0.20
AISI 316	0.035	17.0	11.0	1.35	0.55	0.020	0.037	0.24	2.40
AISI 310	0.05	24.8	19.1	1.23	0.53	0.017	0.033	0.19	0.38

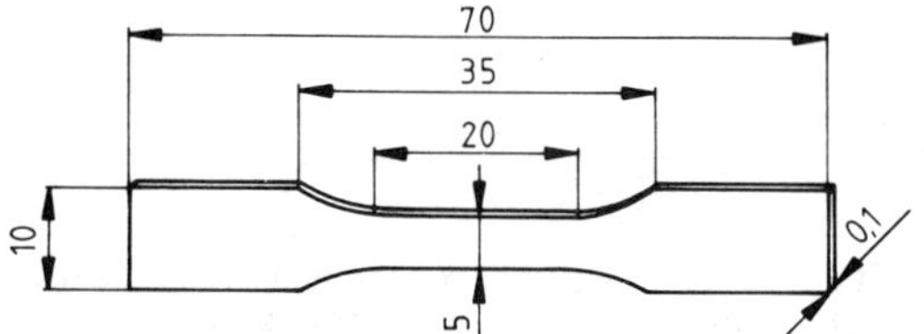

Fig. 1 – The thin tensile specimen for cathodic charging tests.

The thin tensile specimens were cathodically charged at 295 K in a 1 N H_2SO_4 solution containing 0.25 g/l of $NaAsO_2$. The current density was 50 mA/cm^2. A platinum counter electrode was used. The hydrogen charging time was 6 hours. After charging the specimens were aged for different times from 5 min to several months at room temperature or at elevated temperatures. Then they were tensile tested at a cross-head speed of about 5 cm/min.

The microstructures of the tensile specimens were studied with transmission electron microscopy (TEM). The TEM specimens were cut after tensile testing as near the fracture surfaces as possible. Because this revealed only the microstructures in the middle of the specimens, also ready made TEM specimens were hydrogen charged in order to examine the microstructures corresponding to the surface layers. The same charging parameters were employed as before. The charging times were 1 hour or less.

No attempts were made to measure the hydrogen contents of any of the charged specimens.

Results

The effect of ageing time after cathodic hydrogen charging, i.e. the hydrogen content, to room temperature tensile test results of thin AISI 316 steel specimens is presented in Table II. In this study only single specimens for each test were used, except for 5 min ageing. The reproducibility of tensile test results seems in this case to be poor. From Table II it can be observed that the ductility recovers relatively slowly at room tempera-

ture; in this study a clear elongation to fracture was discerned first
after ageing of one week. Elevated temperature ageing enhances the
recovery of mechanical properties.

Table II. Tensile Test Results of Hydrogen Charged AISI
316 Steel Specimens after Various Ageing Times

Ageing time after charging	$R_{p0.2}$ N/mm^2	R_m N/mm^2	A %	Ageing time after charging	$R_{p0.2}$ N/mm^2	R_m N/mm^2	A %
Reference (no charging)	300	480	30				
1 min	–	255	0	168 h	400	450	3
5 min	–	290	0	236 h	420	470	2
5 min	–	80	0	1416 h	370	500	20
5 min	–	400	0	7632 h	280	400	15
35 min	–	465	0	88 h at 373 K	330	430	15
25 h	–	360	0	1 h at 773 K	340	480	15

The fracture surfaces of hydrogen charged specimens were mainly
brittle. Details of hydrogen-induced fracture surfaces of AISI 316
steel have been presented by Hänninen and Hakkarainen (5). In Fig. 2
fracture surfaces and corresponding microstructures are shown after
various ageing times for hydrogen charged (6 h) AISI 316 steel specimens. The microstructures of hydrogen charged specimens after tensile
testing show a low dislocation density when the ageing time before tensile testing has been short. Some thin bundles, possibly ε-martensite,
can be observed as well as single stacking faults. When the ageing
time was one week, an elongation to fracture was observed. This led
also to an increased dislocation density (Fig. 2f). Extended dislocations as well as stacking faults can be seen. When the ageing time
still increased, almost fully ductile dimpled fracture surface was
formed and a cellular dislocation structure and deformation twins were
produced. Surface cracks formed during charging and subsequent outgassing of hydrogen were present in all specimens studied. These
cracks are formed in a narrow surface layer which was not seen in the
TEM study where only the structure in the middle of the specimen was
examined.

Microstructures of hydrogen charged TEM specimens of AISI 316 and
AISI 310 steels are shown in Fig. 3. Both steels showed after charging
buckling of thin areas. Stacking faults and ε-martensite in AISI 316
steel are shown in Fig. 3(a). In thicker sections the dislocation density increased and in some parts also the ε-martensite content was
higher compared to Fig. 3(a). In the AISI 310 steel specimen
ε-martensite was also formed, but single stacking faults were present
to a lesser extent (Fig. 3b). The habit plane of ε-martensite was confirmed to be $\{111\}_\gamma$: the analysis is presented in Figs. 3(c) and (d).
The orientation relationship of ε-martensite was analyzed to be:

$$\{111\}_\gamma \; // \; \{0001\}_\varepsilon$$

$$\langle 110 \rangle_\gamma \; // \; \langle 11\bar{2}0 \rangle_\varepsilon$$

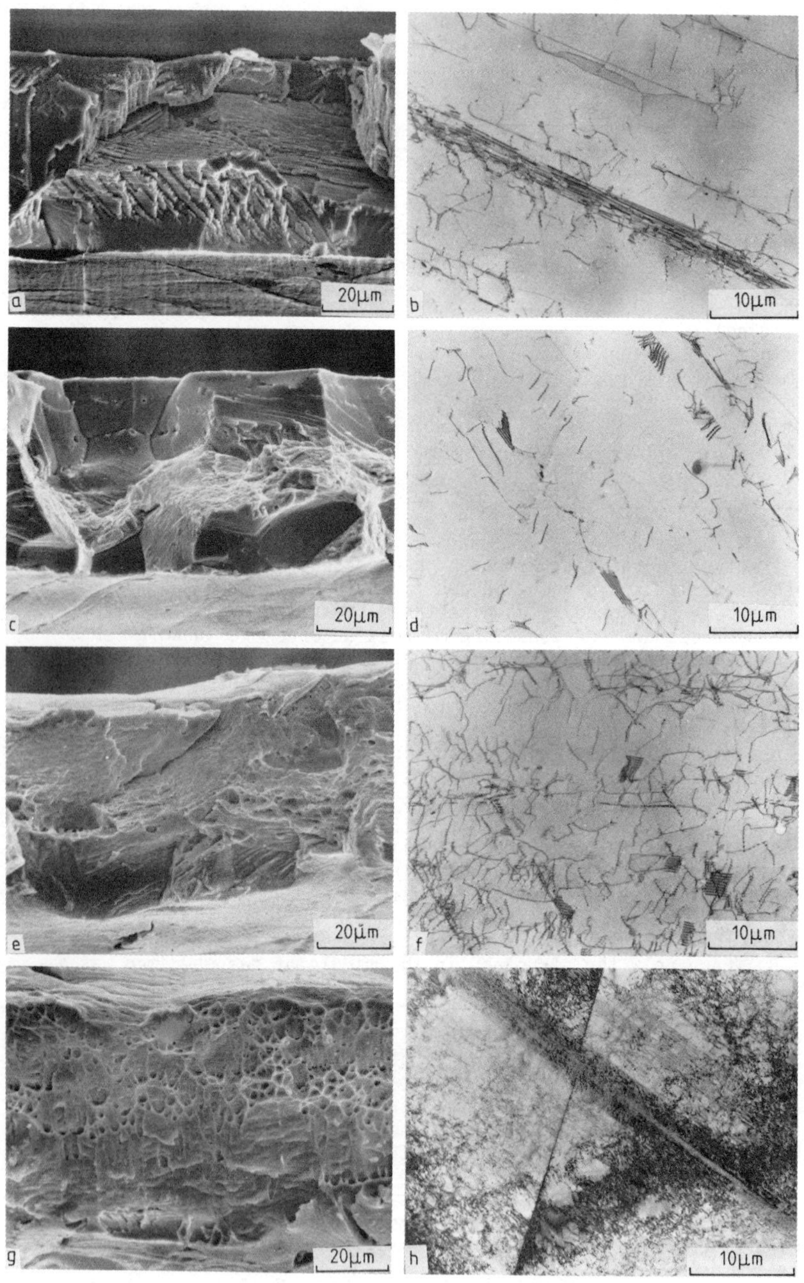

Fig. 2 – Fracture surfaces and corresponding microstructures in AISI 316 steel when aged for various times after hydrogen charging (6 h). (a) and (b) Ageing 1 min. (c) and (d) Ageing 48 h. (e) and (f) Ageing 168 h. (g) and (h) Ageing 7632 h.

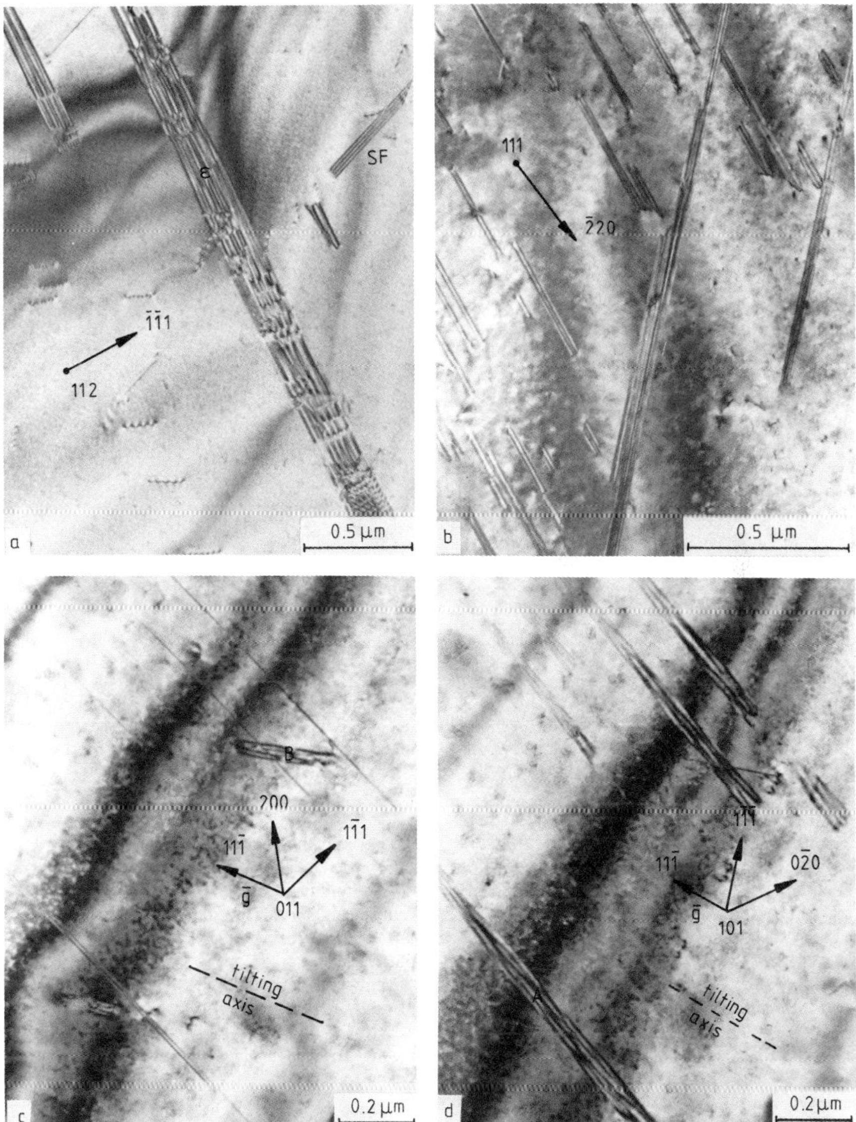

Fig. 3 – (a) Microstructure of AISI 316 steel after half an hour hydrogen charging: stacking faults (SF) and ε-martensite (ε). (b) AISI 310 steel after 1 hour hydrogen charging: ε-martensite and stain formed on the surface during charging. (c) and (d) ε-martensite on {111}$_γ$ planes in AISI 310 steel. The beam direction in (c) is $[011]_γ$ and in (d) $[1\bar{1}01]_γ$. The wide ε-bundles (A) in (d) are according to Fig. 3 (c) on {111}$_γ$ planes and ε-bundles in the other direction (B) are according to (d) on {1$\bar{1}$1}$_γ$ planes.

After charging of 1 hour the AISI 304 steel specimen was very severely buckled. Both ε-martensite and α'-martensite were present in abundance. These phases were also observed in specimens charged for 15 min. In thin sections of the specimens ε-martensite was present on several $\{111\}_\gamma$ planes, Fig. 4(a). In thicker areas the ε-martensite bundles were denser and generally on only one $\{111\}_\gamma$ plane inside an austenite grain. In many places the amount of ε-martensite was so high that no separate spots of austenite could be observed by electron diffraction. α'-martensite was observed only in thick areas of the specimens, where it was formed from ε-martensite, Fig. 4(b). The thickness of these areas was usually more than 3000 Å. A more detailed analysis of the hydrogen-induced ε- and α'-martensite transformations is given elsewhere (6).

In the thin edges of TEM specimens examined there were straight cracks sometimes changing their directions. These cracks were parallel to the ε-martensite and usually ε-martensite could be observed to be associated with them, Fig. 4(c). In the area shown in Fig. 4(c), the ε-martensite content is high. In AISI 304 steel cracks were also observed in thicker α'-martensite containing sections, Fig. 4(d). These cracks are parallel and because they cross the α'-martensite lenses only at few places, they are probably formed along the previous ε-martensite.

First observations on the microstructures of hydrogen charged TEM specimens were made within one hour after charging. The microstructures remained unchanged at least for several days.

Discussion

After hydrogen charging, the recovery of ductility occurs during long time ageing (outgassing of hydrogen) even at room temperature confirming the earlier results (7, 8). This indicates that the dissolved hydrogen in the steels causes the embrittlement in tensile tests.

The TEM observations do not indicate any hydrogen-induced phase transformations of the austenite in tensile specimens, whereas the observations from hydrogen charged TEM specimens definitely do. The usually almost completely brittle fractures of the thin tensile specimens show that the effects of hydrogen after charging extend throughout the specimens. The TEM samples of tensile specimens do not represent the microstructures next to the brittle fracture surfaces, but were taken as near as possible of the main cracks. Therefore, the eventual phase transformations at the crack tip region were not revealed. The hydrogen charged TEM specimens represent the surface layers of the specimens and confirm that there phase transformations really occur: in AISI 304 steel $\gamma \rightarrow \varepsilon + \alpha'$ and in AISI 316 and AISI 310 steels $\gamma \rightarrow \varepsilon$. No hydride formation was observed inside the matrix by TEM. Extensive surface cracking observed after charging even without external stresses seem to be dependent on the stability of the steel: AISI 310 steel showed only minor cracking. Therefore, the phase transformations in surface layers seem to be the probable cause of surface cracking.

Various reasons for austenite decomposition have been proposed in the literature: (i) hydrogen may affect the transformation temperatures (M_s and M_d); (ii) some indirect observations suggest that hydrogen can affect the stacking fault energy (SFE) of austenite (1); (iii) internal stresses and strains due to hydrogen distribution provide a driving force for phase transformations. The amount of ε-martensite in hydrogen charged TEM specimens was markedly higher than usually observed in cold worked materials. This may be due to different strain conditions (hydrostatic), effect of

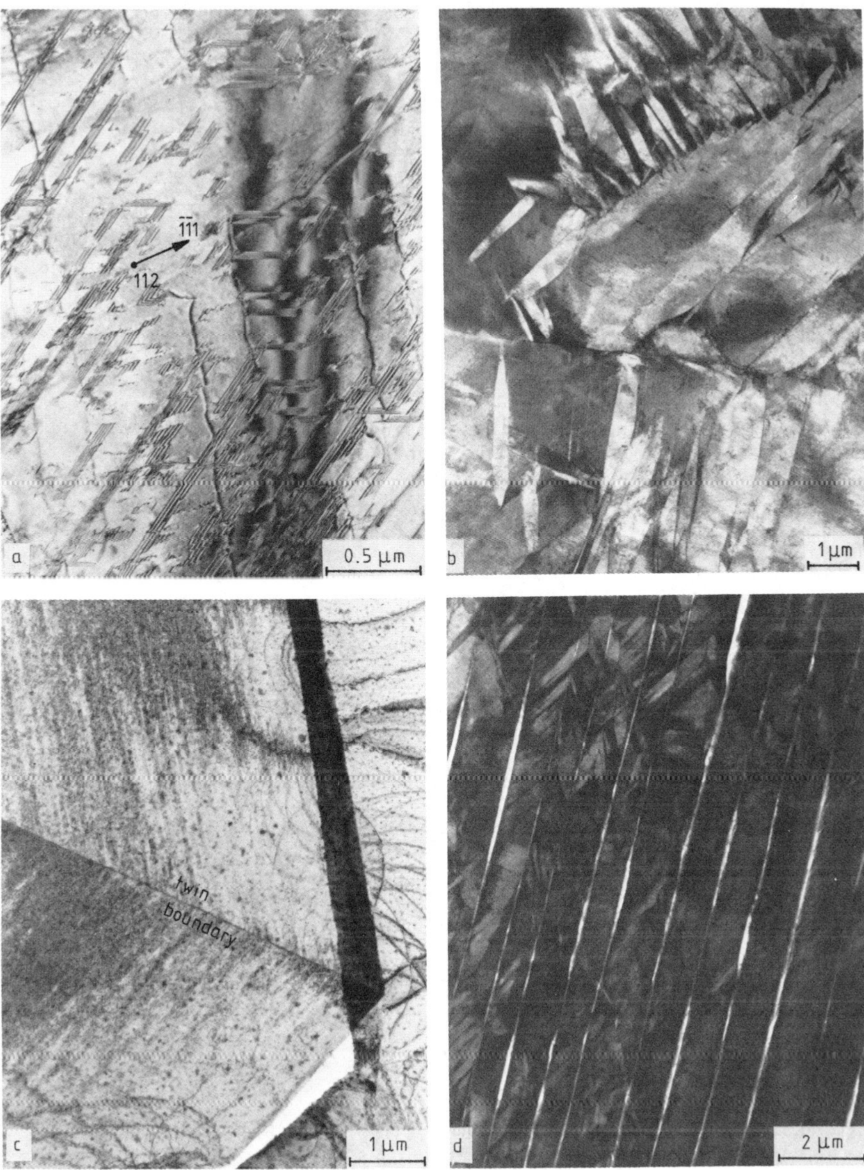

Fig. 4 – (a) AISI 304 steel after 1 hour hydrogen charging: ε-martensite in
a thin section. (b) α'-martensite in the same specimen in a
thicker section. (c) A crack parallel to ε-martensite near the
edge of the AISI 304 steel specimen (15 min hydrogen charging).
(d) Parallel cracks in α'-martensite in a thicker section of an
AISI 304 steel specimen (1 hour hydrogen charging).

hydrogen in SFE or thermodynamic reasons. We believe, that hydrogen favours the ε-martensite formation by lowering the SFE. The severe buckling of the foil edges indicates that considerable stresses are also associated with the penetration of hydrogen into the austenite in TEM specimens. The thin edges are free to deform due to small matrix constraint and thus only minor stresses are needed for buckling. In the thicker areas of the TEM speci- mens and at the surfaces of tensile specimens plastic deformation is induced, resulting in the formation of ε- and α'-martensite.

Various crack paths have been suggested for transgranular hydrogen- induced cracking: (i) α'-martensite or α'-martensite/austenite phase bound- aries (8); (ii) ε-martensite/austenite phase boundaries (9); and (iii) re- versible hydrogen-induced hydrides (4). The shrinkage of hydrogen expanded lattice near the outer surface during hydrogen outgassing produces strains between the surface and interior which can produce cracking because dif- fusion of hydrogen in austenite is slow. The results of the present study suggest that the cracks in hydrogen charged TEM specimens and the surface cracks in bulk specimens follow the phase boundary between austenite and hydrogen-induced ε-martensite. The reason that the cracks follow the γ/ε phase boundary may be due to hydrogen trapping or large internal shear stress induced by volume change which can be relatively large if hydrogen is unevenly distributed between the two phases. The TEM observations of ten- sile specimens, however, do not confirm this observation, because only main- ly unfaulted austenite was found. Therefore also the austenite phase may be prone to brittle cleavage-like cracking.

Conclusions

This study has shown that hydrogen-induced embrittlement in austenitic stainless steels is reversible. Hydrogen induces phase transformations (ε- and α'-martensites) in TEM specimens. These transformations were not ob- served in the middle of tensile specimens. In thin TEM specimens the cracks seemed to follow the γ/ε-interphase. Because hydrogen-induced brittle cracks penetrated through the tensile specimens, it seems possible that any of the phases present (γ, ε- and/or α'-martensite) can be embrittled.

Acknowledgements

This paper is based on work supported by the Ministry of Trade and Industry in Finland under Reactor Materials Research Project. Helpful comments of Prof. J. Forstén are gratefully appreciated.

References

1.   M.L. Holzworth and M.R. Louthan, Jr.: Corrosion, 1968, vol. 24, pp. 110-124.
2.   P. Maulik and J. Burke: Scr. Metall., 1975, vol. 9, pp. 17-22.
3.   H. Mathias, Y. Katz and S. Nadiv: Metal Sci., 1978, vol. 12, pp. 129-137.
4.   K. Kamachi: Trans. ISIJ, 1978, vol. 18, pp. 485-491.
5.   H. Hänninen and T. Hakkarainen: Met. Trans., 1979, vol. 10A, pp. 1196-1199.
6.   P. Nenonen, H.E. Hänninen and T.J. Hakkarainen, 1980, To be published.
7.   M.B. Whiteman and A.R. Troiano: Corrosion, 1965, vol. 21, pp. 53-56.
8.   M.L. Holzworth: Corrosion, 1969, vol. 25, pp. 107-115.
9.   A. Inoue, Y. Hosoya and T. Masumoto: Trans. ISIJ, 1979, vol. 19, pp. 170-178.

HYDROGEN TRANSPORT BY DEFORMATION AND

HYDROGEN EMBRITTLEMENT IN SELECTED STAINLESS STEELS

J. Chêne, M. Aucouturier

Laboratoire de Métallurgie,associé au CNRS N° 177

Université de Paris-Sud,91405 Orsay Cedex (France)

R. Arnould-Laurent, P. Tison, J.-P. Fidelle

Service Métallurgie,CEA, BP. 561,92542 Montrouge (France)

Abstract

Typical S.S. are investigated : stable 316 L, metastable 304 and
302 L, γ-M cw. 304 L, α-γ 326 (IN 744). The influence of δ ferrite and
carbides are considered also. High Resolution Tritium Autoradiography is
performed on disks strained by pressurizing T_2. It is correlated with
External and Internal Hydrogen Embrittlement (EHE,IHE) tests carried out at
various ε and temperatures on disks or cathodically charged in salt baths,
tensile specimens. Results are discussed in terms of trapping and strain -
induced H transport, related to deformation products and especially moving
dislocations.

Introduction

The low HE sensitivity often noticed for γ SS is not only due to a
lower sensitivity of the γ matrix but also to the low permeability under
normal conditions, near room temperature (RT).However the entry of large
enough H quantities can be achieved in significant volumes of material by
plastic straining in a H environment or by cathodic charging at elevated
temperatures; non austenitic, more H-sensitive phases can be generated.
Then one gets significant HE, an interpretation of its mechanism and the
role of metallurgical variables.

Materials

Three typical kinds of steels are selected :
- fully γ 316 L quenched from 1100°C , grain size G $\simeq$ 100 µm,
- γ and austenomartensitic (γ+ M) grades of 304, 304 L and 302 L,
- austenoferritic (α+γ) 326 (IN 744) typically cold rolled and annealed 4 mn
  at 870°C. G $\simeq$ 2 µm.

Characteristics appear on Tab. I,II. We emphasize 304 L and 316 L for
which the plastic deformation mechanisms are well known (1, 2) in order to
illustrate the influence of γ stability and the role of deformation
products in HE.

Steel		Cr%	Ni%	Mn%	Mo%	Si%	C	S (ppm)	P	N	Use
	(A)	17.2	10.3	1.7			200	250	250	370	H^+, T^+
304 L	(B)	18	10		nominal		300				DPT H_2, T_2
302 L	(C)	17.52	8.19	1.75	0.015	0.293	270	190			DPT H_2
304		18.51	9.6	1.62			700				DPT H_2
	(A)	17.16	13.73	1.60	2.48	0.37	100	30	170	90	$H^+ T^+$ DPT H_2
316 L	(B)	16.75	13	1.54	2.68	0.34	200	50	180		DPT $H_2 T_2$
	(C)	17.12	12	1.54	2.2	0.44	190	190	360		DPT H_2
326	*	26.5	6.6	0.42		0.50	600	700	30	0,037	DPT $H_2 T_2$

Table I. Chemistry and use of investigated SS.

* also includes 0.2% Ti.
H^+, T^+ = cathodic charging with H^+ or tritiated molten salts.
DPT H_2, T_2 = Disk Pressure Test with H_2 or T_2 gas.

Steel	(approx. S.F.E. 20 mJ/m²)		304 L (A)			304 L (B)		302 L	304	
Treatment	Quenched from 1050°	Quenched + c.w. 5%,15%,40%	Quenched + sensitization 700° 1h	800° 1h	650° 260h	Q. from 1050 + c.w	Q. from 1050 + weld	Q. from 1050	Q.+ sensitization 700°C 3h	870° 2×5mn
Structure	γ(+1% ferrit.- δ)	γ+ε γ+α' γ+ε+α'	γ+ fine Cr23C6	γ+coarse Cr23Cε	γ+Cr23C6 +α'at G.B	γ, γ+ε. γ+ε+α', γ+α'	γ+δ	γ	γ+fine continuous Cr23C6	γ+coarse discontinuous Cr23C6
Average Grain size(μm)	100					50	50	20	20 − 50	

<u>Table II</u> .   Treatments and structural characteristics of 302 - 304 SS.

Main Techniques

<u>The Disk Pressure Test (DPT)</u> used for embrittlement (3-5), permeation (6, 7) and T segregation on strained materials (8). We consider : π_0 ,pHe ,pH$_2$, pH$_{2c}$ the pressure at which one <u>locally</u> exceeds the onset of plastic flow, the rupture pressures under He-reference gas - and H$_2$ gas   for uncharged and H-charged metal, the ratios pHe/pH$_2$, pH$_2$/pH$_{2c}$.

For growing strain-induced H entry and distribution studies, T isotope is pressurized by cryogenics (9). The pressure increase rates $\Delta p/\Delta t$ used (Tab. 3) favor strain-induced absorption over regular diffusion. To avoid spurious surface contamination by adsorbed T, each face of samples taken out from tested disks is polished before HRTA.

Steel	304 L (B)		316 L (A)				326	
Structure	40% c.w.: γ+α'		γ				54% α + 46% γ	
Thickness (mm)	0,75		0,75		0,5		0,5	
π_0 (bars)	14,5		1,4		≃ 1		13	
max T pressure (bars)	3	40	3	40	0,7	34	3	40
$\frac{\Delta p}{\Delta t}$ (bar.mn^{-1})	7,5	35,3	7,5	35,3	14	0,8	335	7,5	35,3
t_{tot} (mn)	30	3	30	3	30	46	2	30	3

<u>Table III</u>. Disk Pressure Test conditions with pure T.

<u>HRTA (High Resolution Tritium Autoradiography</u> (10,11) evidences the micro-structure-related H distribution at the electron microscope scale. We also use it after the following charging technique.

High-Temperature Cathodic Hydrogenation (HTCH) at 130-250°C in an eutectic mixture of acid molten salts (12) (Tab. 4). The effects of temperature and high fugacities cooperate. CHENE estimates the flux in 304 L at 150°C to be 3000 times the one at RT. After 1 h theoretical penetration reaches 1 µm at RT, 25 µm at 150°C. Very thin specimen are no longer necessary, parasit phenomena are avoided. A quantitative exploitation of the rational tensile curves of minispecimens (1 mm thick, 2 mm large, useful length : 16 mm) and rupture appearances afford to characterize HE . Specimens are tested in air at RT at a $\dot{\varepsilon} = 10^{-3}s^{-1}$, unless specified otherwise.

Charging conditions	Trapping observations		Embrittlement tests	
Temperature	RT	150°	RT	150°
Specimen shape	Massive-Prethin foil	Massive-Prethin foil	Tensile specimen	Tensile specimen
Cathodic bath	$HTSO_4$ 1 N specific activity 1 Ci.cm^{-3}	Tritiated Molten salts NaHSO$_4$-HTO 57% KHSO$_4$ 43% specific activity 0,15 Ci.cm^{-3}	H_2SO_4 1N+ As$_2$O$_3$	Molten salts NaHSO$_4$-H$_2$O 57% KHSO$_4$ 43%
Electrochemical conditions	$d = -20$ mA.cm^{-2} 20h (without As$_2$O$_3$)	$V/_{Ag/Ag^+}$: -850mV 1^h	$V/_{ECS}$: -850 mV 20^h	$V/_{Ag/Ag^+}$: -850mV 5^h

Table IV. Cathodic charging conditions.

H   Transport and Trapping

Dislocations are most common trapping sites in a solution treated and rapidly cooled structure (13, 14). To better understand this inter-action and its role in HE we investigate H transport and segregation due to a growing plastic strain.

T Localisation after straining in the presence of T_2 gas

β^- counting and HRTA are used to examine tested disks,downstream. HRTA is carried out with SEM on massive specimens; Ag grains tracing T localization appear white on microstructure.

- Stable 316 L Lot (B). Strain does not transform it to martensite at RT. A large T amount is detected downstream as π_0 is exceeded (Fig. 1-a). The quasi uniform distribution agrees with the trapping by dislocations noted by (15). The T amount downstream appears closely related to the gross $\dot{\varepsilon}$: it is the lower, the higher this rate (Fig.1-a,b,c). Besides, low ε seem to favor preferential localization at such interfaces as grain boundaries (GB) (Fig.1-a). With no plastic strain, during a pressure hold up of 30 mn too short to allow regular permeation, no T is detected downstream or upstream after elimination of the adsorbed layers. Considering the low RT T diffusi-vity, the results demonstrate the existence of a very important phenomenon of H-dragging by dislocations.

- Cold rolled 304 L Lot (B). This is metastable and with a 40% cw has a γ+M structure, where BCC α'_M is the main martensitic constituent. Below π_0, little T is observed downstream or even upstream. H trapping by preexisting defor-mation products can hardly be noticed. Conversely when disks are strained above π_0, a large T amount is detected downstream (Fig.1-e) related to deformation products (ε_M , α'_M), especially in areas in tension : downstream pole (Fig. 1-f).

For the present test conditions a short circuit effect by the initially present deformation products is unlikely. Again the role of a concomittant plastic strain on enhanced T uptake is clear; it can be ascribed to dragging by ingressing dislocations and to transport by freshly strain-induced

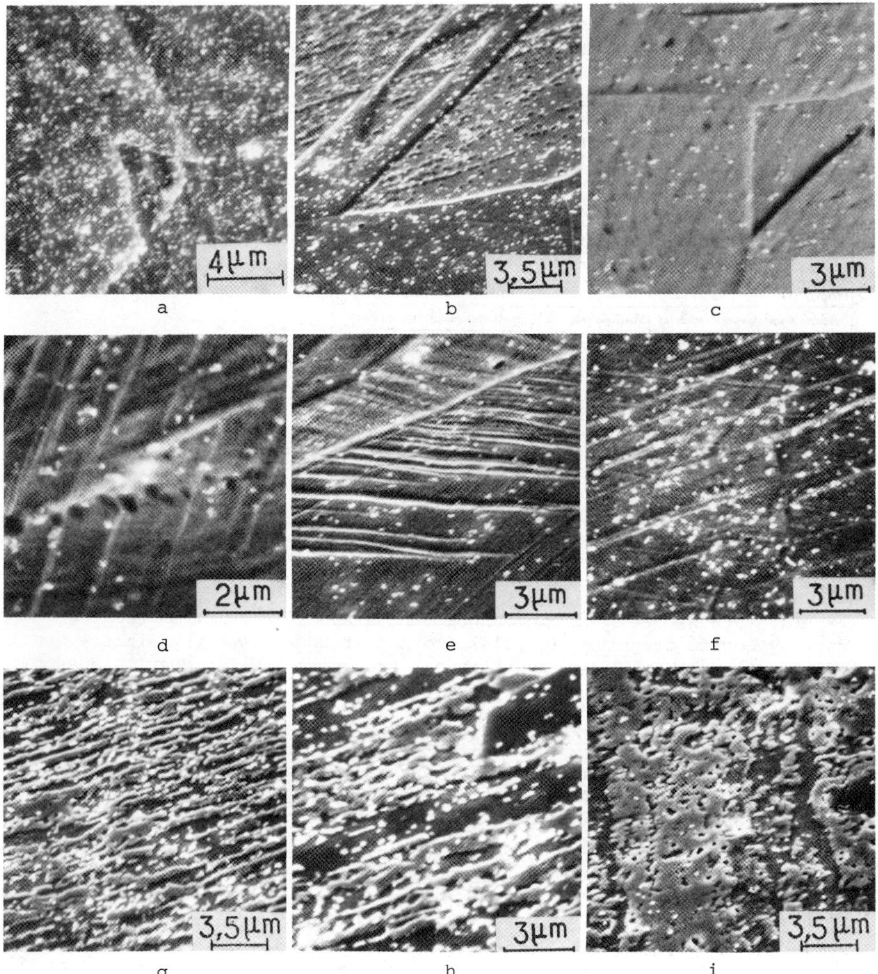

Fig. 1   S.E.M. autoradiographs showing T localization in different SS.
downstream : D ,   upstream : U

			max.pressure (bars)	test time (mn)	$\Delta p/\Delta t$ (bars/mn)
316 L (γ)	a	D	35	46	0.77
	b	D	3	30	7.5
	c	D	33.5	2	335
304 L (γ+M)	d	U	40	3	35.3
	e	D	–	–	–
	f	D	18	30	18
326 (γ+α)	g	D	40	3	35.3
	h	D	–	–	–
	i	D	3	30	7.5

martensites. As the strain modes of such a steel always associate an important dislocations movement to martensite formation by strain (1), H-dislocations interactions are at any rate causative reasons for transport processes and account for preferential trapping on this martensite up and downstream (Fig. 1-d,f).

- 326 . When $pT_2 < \pi_0$ a small T amount is trapped selectively at γ/α interfaces,the only irreversible traps, up- and down stream (Fig.1-i). T goes downstream through α; the absence of T in α is due to outgassing before autoradiographs are laid, P and D coefficients being 2 orders of magnitude higher than in a fully γ steel (16). After plastic straining a large T amount is noted downstream, preferentially localized in γ phase or at γ/α interfaces (Fig. 1-g,h) thus plastic strain-enhanced H absorption is also confirmed here.

Discussion

The demonstration of H transport between the in- and outgoing faces of a x thick disk, during a t_d long deformation, allows to compute an apparent D coeff.$D^* \geq x^2/ 2t_d$.

At the highest $\dot{\varepsilon}$ used for 316 L, the low T amount at the downstream pole indicates that this rate is close to the critical value beyond which H transport by dislocations ceases; one can estimate the maximum value of $D^*_{RT} = x^{*2}/ 2t_d \simeq 2.10^{-4}$ cm^2.s^{-1} in a x^* thick disk, and the mean limiting rate value of dislocations dragging their H atmosphere;$\overline{V}_c$ exp $= x^*/t_d \simeq 8.3 \times 10^{-3}$ cm.s^{-1} . Taking $[D_{H-\gamma}]_{RT} = 4.10^{-12}$cm^2.s^{-1} (16), to which is associated a maximum penetration distance $x_{H-\gamma}$, ratios $D^*/D_{H-\gamma} = 5.10^7$ and $x^*/x_{H-\gamma} \simeq 7.10^3$ for $t_d = 6$ s. This emphasizes the transport acceleration due to dislocations. Moreover a mean density of mobile dislocations ρm of 10^{-8} cm^{-2}(17) and using Taylor's equation , it is possible to figure out the critical $\dot{\varepsilon}_{c\ exp} = \rho m\ b\ \overline{V}_{c\ exp} \simeq 2.10^{-2}$.s^{-1} with Burger's vector b $= 2.5 \times 10^{-8}$ cm . Although sensibly superior $\dot{\varepsilon}_{c\ exp}$ does not strikingly conflicts with what derived from the study of the $\dot{\varepsilon}$ influence in 304 L (Fig. 7) (13). Assuming that the HE mecanism is contributed by H transport and local enrichment, conceivably transport can still be active although accumulation is no longer sufficient to cause HE. Therefore the critical rate derived from HE tests will be necessarily lower than the limit value supplied by the transport mechanism. However, $x^*/x_{H-\gamma}$ is several orders of magnitude higher than the value computed for a γ structure admitting a simplified Cottrell's model (18). This discrepancy questions the validity of the static model generally used, which, in the transport mechanism does not separate the dislocations speed role from the influence of an interaction between mobile or immobile dislocations.

Investigation of the deformation of previously charged specimens

HRTA is applied to (JT + Q) 304 L mini tensile specimens,cathodically charged in tritiated salt baths, then plastically strained to various levels. Before straining, T distribution is homogeneous (Fig.2-a) in agreement with trapping on dislocations already noted. The solution isotopic ratio T/H is $\simeq 10^{-4}$, yet the amount of T absorption is similar to which got during the DPT with pure T gas at RT, and the (H + T) surface concentration is even much higher; this confirms the high hydrogenation capacity of the H.T.C.H. technique.

After straining, one notes a significant preferential T segregation at slip lines and an activity increase at the metal surface,due to T dragging by egressing dislocations; it corresponds to a coplanar motion in agreement with the low SFE of this 304 L (Fig. 2-b). Because of T outgassing by dislocations transport both the surface activity (at high strains, Fig. 2-e) and the inner activity (Fig.2-c) decrease. In the specimen volume T is redistributed and trapped on deformation products (α'_M; ε_M, Fig. 2-c) whose γ/α' and mainly γ/ε interfaces have a high dislocation density (1). As shown by Fig. 2-c and f, this dynamic segregation effect compares with the trapping found on specimens plastically strained BEFORE H charging. The necked zones of specimens strained to rupture do not show the important accumulation , which one expects to explain crack nucleation (Fig. 2-d,e); but in such a metastable alloy these zones are almost completely martensitic and thus a T enrichment can have largely disappeared by outgassing during the test end and the few hours delay before autoradiographs are laid.

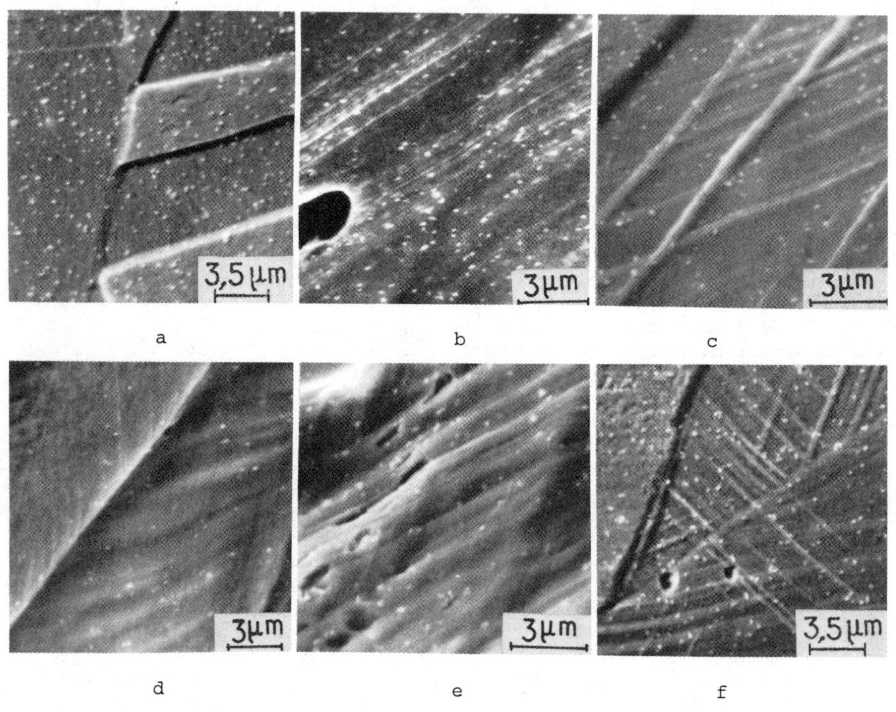

Fig. 2

S.E.M. autoradiographs showing T localization in 304 L before (f) or after cw in tension at various levels (b to e) (T charging at 150°C, 1 h).
    a) ST+Q uniform distribution (cw = 0% reference).
    b) 9% cw, not polished after straining. T associated with slip lines.
    c) 17% cw, polished and etched after straining. Trapping on cold work martensite.
d and e) Necked zone of a strained to rupture specimen not polished after straining. Segregation on G.B. and slip lines.
    f) 17% cw before H charging. Trapping on cw martensite.

Main results confirm the role of dislocations in H transport and trapping on freshly generated deformation products envisionned for our experiments where H enters DURING deformation. These are no basic differences on H transport and localization whether "internal " or initially " external ".

Hydrogen Embrittlement of Selected Stainless Steels

To what said on HE of γ SS in the introduction, we add the next general features :

After severe cathodic charging, HE of (ST + Q) γ SS displays surface microcracking, whatever structure stability (19). Ductility loss depends on structure stability and H pick up (13). Rupture mode changes progressively from dimples to cleavage as H charging grows (13, 20). A detailed analysis of stress strain curves indicates that H decreases correlatively metals plasticity and the consolidation coefficients at the various deformation stages; it is noted, whatever structure stability, and is similar to what observed for ductile BCC structure such as α Fe, and may be due to a parallel mechanism (19, 21 ,22).

Influence of Hydrogenation and deformation conditions

DPT under increasing He or H_2 pressures

- 316 L RT fairly stable grades are not or only little sensitive to Hydrogen Gas Embrittlement "HGE" (Fig. 3 on Lot B) and IHE by 50 ppm. Lot C is not affected but disks welded with a filler metal, giving $\simeq$ 8.5% δ in the weld exhibit HGE and at faster $\Delta p/\Delta t$ than lot B (Fig. 4).

As test temperature decreases unwelded (uw) lot C disks start to get embrittled and 2-3% α'_M - not to mention ε_M - is detected on disks strained to rupture by H_2 . HE of welded disks is even larger.

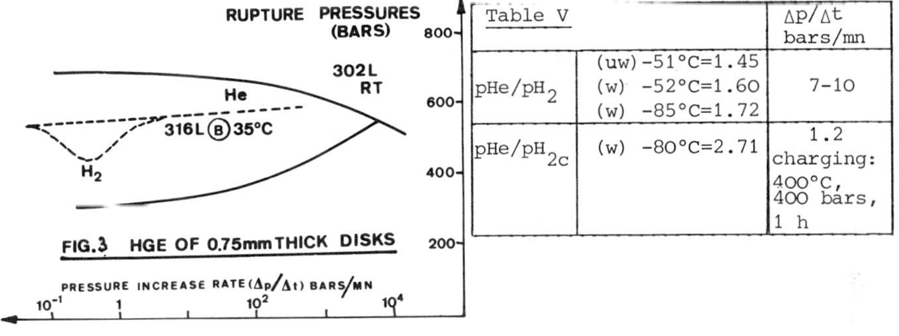

Table V		$\Delta p/\Delta t$ bars/mn
pHe/pH_2	(uw) $-51°C=1.45$ (w) $-52°C=1.60$ (w) $-85°C=1.72$	7-10
pHe/pH_{2c}	(w) $-80°C=2.71$	1.2 charging: 400°C, 400 bars, 1 h

These ratios may not represent maximum HE as a specimen shortage prevented looking for the influence of $\Delta p/\Delta t$. However one sees that the effect of low temperatures has to be considered for rating the HE sensitivity of γ SS. Not only the α'_M amount increases as temperature decreases but also its H permeability (23).

- 304 L Typical HE of (ST+Q) 304 L does not exist! Depending on chemistry and resulting stability, sensitivity varies from fairly high to small. pHe/pH_2 ratios ranging from 1.10 to 1.15 are currently noted at 5 bars/mn for rather stable grades purposed for H_2 service (lot B); it can go up to 1.25 with purified gas at 1.5 bar/mn (24).

However cold work (cw) before testing can generate martensite and later severe HGE even at relatively high $\Delta p/\Delta t$ (3) ; machining can give : pHe/pH_2 = 1.8 to 2 and rolling : pHe/pH_2 = 4.62, a value ordinarily found for high strength low alloy steels!

Fig. 5 shows the influence of cw% on pHe, pH_2 and magnetic phases in Lot B; although not clearly explained the plateau is similar to what one finds when Poore's (25) curve on " relative variation of magnetic permeability of a 316 L as a function of cw " is drawn, passing by the middle of the confidence intervals. Like for 316 L, δ ferrite can promote HGE (pHe/pH_2 = 1.30 tc 1.38) under conditions of no embrittlement for the unwelded metal (3).

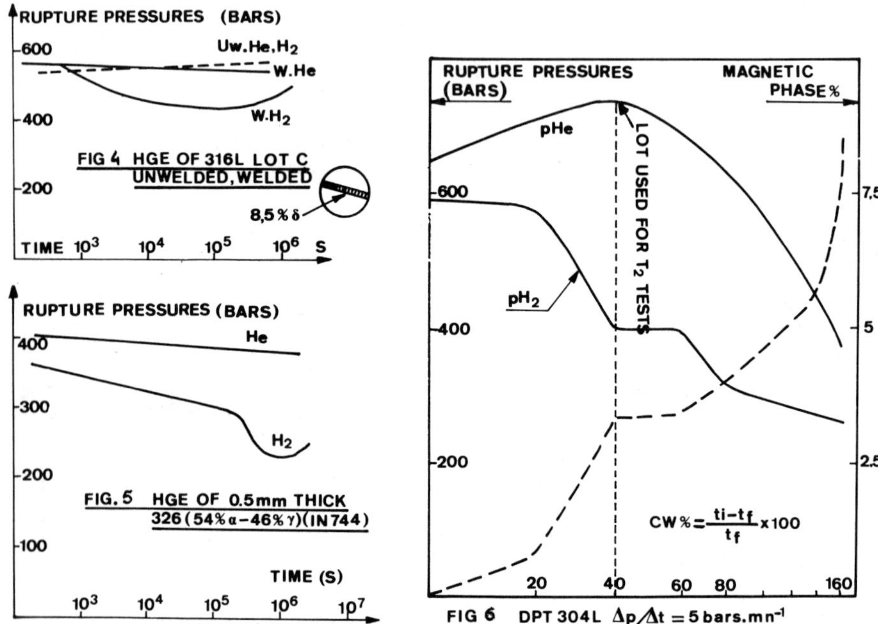

- 302 L   As 304 L stability decreases by changing its composition to that of a true " 18-8", it becomes increasingly sensitive to martensite formation (and HE) during testing. Fig. 5 compares the HGE ranges for 316 L (B) and 302 L, which is much more and faster affected (26). Like for 8.5 % δ-including 316 L weld, HGE increases as temperature decreases. Tab. VI shows that pHe/pH_2 at 10 bars/mn is larger at – 52°C than at RT and attains a value reached at RT, only at a slow rate of 0.3 bars/mn. Such a combined effect of $\dot{\epsilon}$ and temperature is due to the generation of more fresh martensite during testing as lower temperatures.

Table VI	10	0.3	10
Temp.	RT	RT	– 52°C
pHe/pH_2	2	2.3	2.3

- 326   This ($\alpha+\gamma$) structure is definitely sensitive to HGE: $(pHe/pH_2)_{max}$ = 1.63 at $\simeq$ 0.012 bars/mn. It appears at relatively high $\Delta p/\Delta t$ (Fig.6) confirming the effect of ferrite noted earlier. Yet this HGE is not very high, considering the relatively high $\sigma y_{0.2} \simeq$ 800 MPa and is smaller than HGE found in many ($\gamma+M$) structures with much lower martensite contents. However the considered steel has a very fine grain size $\simeq$ 2 μm, which might help to give a smaller HE than in an $\alpha-\gamma$ SS with a coarser grain.

Tensile Tests on Cathodically Charged Specimens (Tab. VII,VIII)

HTCH charging of (ST+Q) 304 L and 316 L induces significant ductility losses of tensile specimens even for 316 L (A).   In the other hand, in spite of the greater severity of stressing conditions DPT have evidenced no HGE of this 316 lot whereas a parent of the moderately HGE sensitive lot (B) was not found embrittled by 54 ppm at 5 bars/mn. This can be ascribed to the differences of H quantities absorbed or present in the material and illustrates the high charging efficiency of the salt bath technique, which appears to be able to supply an H charging with a surface concentration larger than 50 ppm.

Materials / Mechanical Charact.	σ_e MN/m²	σ_v MN/m²	E_H %	E_T %	Σ %	E %	$\dfrac{S_1}{S_2}$	Strain hardening exponent n_1	n_2	n_3
304L. 1050°C. 1h U.	$180^{\pm4}$	$1050^{\pm10}$	$72^{\pm1}$	$75^{\pm1}$	$78^{\pm2}$	—	1	0.3	0.5	0.7
304L. 1050°C. 1h C.	$180^{\pm5}$	$750^{\pm20}$	$40^{\pm2}$	$43^{\pm2}$	$41^{\pm2}$	$43^{\pm1}$	2.1	0.15	0.3	0.55
316L. 1100°C. 2h U.	$170^{\pm3}$	$840^{\pm10}$	$63^{\pm2}$	$70^{\pm2}$	$74^{\pm2}$	—	1	0.2	0.4	0.5
316L. 1100°C. 2h C.	$175^{\pm5}$	$860^{\pm10}$	$58^{\pm1}$	$62^{\pm1}$	$64^{\pm1}$	14	1.12	0.1	0.3	0.5

Table VII. Influence of cathodic charging in molten salts on 304 L (A) and 316 L (A) : U = uncharged, C = charged, E_H = homogeneous elongation, E_T = total elongation. S_1 : surface obtained by integrating the $\sigma_v = \mathcal{J}(\varepsilon_v)$ curves of the non hydrogenated specimens in the homogeneous deformation field. This surface is equivalent to the homogeneous deformation work of a non hydrogenated specimen. S_2 : surface obtained by integrating the $\sigma_v = \mathcal{J}(\varepsilon_v)$ curves of the hydrogenated specimens in the homogeneous deformation field. $E\% = \Sigma_0 - \Sigma_C / \Sigma_0$ x 100 : embrittlement index . Σ_0 and Σ_C : reduction of area of non hydrogenated and charged specimens.

Materials / Mechanicals aracteristics	σ_e MN/m²	σ_v MN/m²	E_H %	E_T %	Σ %	F %	$\dfrac{S_1}{S_2}$
304L U.	$180^{\pm4}$	$1050^{\pm10}$	$72^{\pm1}$	$75^{\pm1}$	$78^{\pm2}$	—	1
304L C.RT	$180^{\pm3}$	$980^{\pm10}$	$70^{\pm2}$	$74^{\pm2}$	$74^{\pm2}$	$4^{\pm2}$	1.1
304L C.MS	$180^{\pm5}$	$750^{\pm10}$	$40^{\pm2}$	$43^{\pm2}$	$41^{\pm2}$	$43^{\pm2}$	2.1
304L C.MS + O. (150°-5h)	$180^{\pm5}$	$1060^{\pm10}$	$72^{\pm1}$	$76^{\pm2}$	$79^{\pm2}$	—	1
304L U. tens.test. -196°	$300^{\pm5}$	$2060^{\pm10}$	$45^{\pm1}$	$49^{\pm1}$	$63^{\pm2}$	—	1
304L C.MS 5.tens.test. -196°	$340^{\pm2}$	$2020^{\pm20}$	$45^{\pm1}$	$48^{\pm2}$	$61^{\pm1}$	1.5	~1

Table VIII. Influence of hydrogenation conditions: U, C_{RT}, C_{HTCH}, O(outgassed) on (ST+Q) 304 L tested at RT and -196°C.

Structure		Mechanical Characterist.	σ_e MN/m²	σ_v MN/m²	E_H %	E_T %	Σ %	E %	$\dfrac{S_1}{S_2}$
AS QUENCHED STRUCTURE	304L 1050°-1h	U.	$180^{\pm4}$	$1050^{\pm10}$	$72^{\pm1}$	$75^{\pm1}$	$78^{\pm2}$	—	1
		C.MS	$180^{\pm5}$	$750^{\pm20}$	$40^{\pm2}$	$43^{\pm2}$	$41^{\pm2}$	$43^{\pm1}$	2.1
SENSITIZED STRUCTURE 304L 1050°~-1hour +	700°C 1h	U.	$180^{\pm5}$	$1050^{\pm10}$	$69^{\pm1}$	$72^{\pm1}$	70	—	1
		C.MS	$182^{\pm5}$	$740^{\pm10}$	$47^{\pm6}$	$49^{\pm3}$	$43^{\pm3}$	$42^{\pm2}$	1.75
	800°C 1h	U.	$181^{\pm5}$	$1100^{\pm20}$	$77^{\pm1}$	$82^{\pm2}$	$72^{\pm1}$	—	1
		C.MS	$182^{\pm4}$	$840^{\pm10}$	$50^{\pm4}$	$51^{\pm4}$	$44^{\pm1}$	$35^{\pm2}$	19
	650°C 260h	U.	$174^{\pm1}$	$1020^{\pm10}$	$69^{\pm1}$	$74^{\pm1}$	$69^{\pm1}$	—	1
		C.MS	$160^{\pm5}$	$630^{\pm15}$	$37^{\pm4}$	$39^{\pm4}$	$41^{\pm2}$	$41^{\pm3}$	2.15

Table IX. Influence of sensitization on H.E. of 304 L.

Next, macroscopic characteristics of RT cathodically charged 304 L are not influenced in spite of surface microcracking. This outlines the H pick up and distribution influences on this little notch sensitive material.

Fig. 7 shows the influence of ε on 150°C H^+ charged 304 L. Although phase transformation of this metastable structure is ε dependent (1,19 and pHe of 302 L on Fig.6) the HE decrease at high ε is large enough to be primarily associated with an efficiency decrease of H dragging and local accumulation, in line with HGE of 302 L and materials which are HE sensitive without transforming under strain. If specimens are outgassed at the same temperature, for the same time as during charging, a full recovery of mechanical properties takes place.

Influence of Different Metallurgical Factors

Whatever test conditions, <u>Stacking Fault Energy</u> has a paramount influence direct or indirect on HE susceptibility (27). Low SFE result in <u>Structure</u>

m̍etastability. At RT ductility loss of metastable 304 L (SFE $\simeq$ 20 mJ/cm^2) is much larger than that of 316 L (SFE $\simeq$ 45 mJ/cm^2) (Tab. VII); it is signifi-cantly contributed by strain-induced martensite. Such a transformation can also appear on our 316 L below RT and has been noted for disks; but the mo-derate ductility loss of tensile specimens at RT rather is due to a H-dislo-cation interaction in this H charged austenite. SFE has a primary role on the displacement mode of dislocations(27); it affects the local strain - induced H enrichment. 316 L grades most frequently appear as less susceptible than 304 grades, because of the beneficial Ni and Mo-induced SFE raise,but they are also less susceptible than 304 N (28); the latter with a coplanar motion of dislocations favoring H accumulation, can be severely affected in spite of its N-induced stability.

Cold work generated neither deformation products, nor HE of 316 L (A) disks although its effect was obvious for machined or cold rolled 304 L disks. Various microstructure of 304 L (A) :γ, $\gamma+\varepsilon$, $\gamma+\varepsilon+\alpha´$ and $\gamma+\alpha´$ are hydrogena-ted in molten salts before tensile testing. Their sensitivity appearing on Fig. 8 can be correlated to the phases nature and quantity present in the structure (29). The sensitivity peak found at low ε levels corresponds to the maximum amount of ε_M in the structure, which illustrates the deleterious ε_M behavior,evidenced by microcracking studies (19).

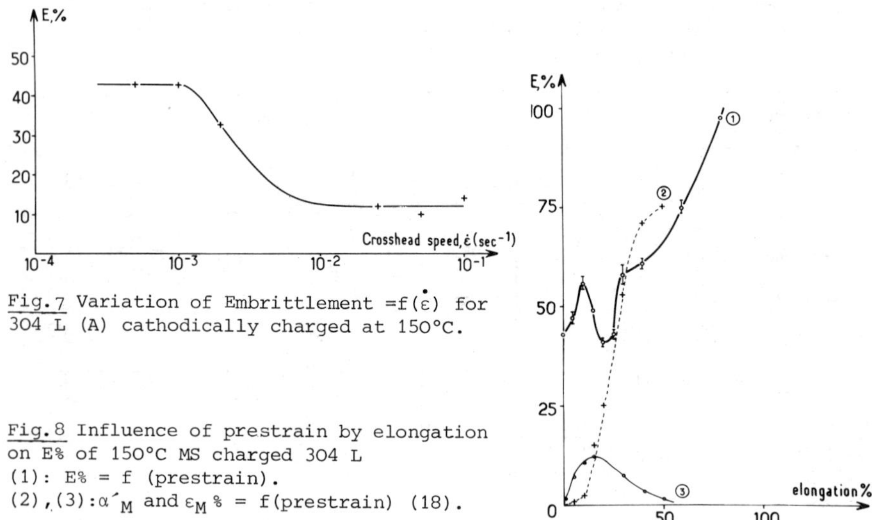

Fig.7 Variation of Embrittlement =f($\dot{\varepsilon}$) for 304 L (A) cathodically charged at 150°C.

Fig.8 Influence of prestrain by elongation on E% of 150°C MS charged 304 L (1): E% = f (prestrain). (2),(3):$\alpha´_M$ and ε_M % = f(prestrain) (18).

Sensitization of 304 disks (Tab.I,II) for 3 h at 700°C generate a continuous $M_{23}C_6$ network at GB and a remarkable HGE : pHe/pH$_2$ = 1.84 (30, 31),whereas 2x5 mn at 870°C give coarse scattered carbides and little HGE : pHe/pH$_2$ = 1.10). On 304 L the effects of sensitization treatments are closely related to the associated microstructural modifications. Short treatments giving a fine discontinuous Cr carbides precipitation at GB appear to be favorable (Tab. IX). In the other hand, HE increases and rupture mode changes (Fig. 9) after long treatments conducing to C and Cr depletions which favor a martensite border at GB (19, 32, 33).

Discussion

As a whole, reported results support the idea that H-dislocations interactions play a basic role in HE of Stainless Steels. DPT clearly

confirms,albeit qualitatively, H trapping and accelerated transport by dis-
locations.Such a mechanism is generally well accepted now,but the related
local enrichment process has been questionned (18,34) and this points to the
need of making transport tests quantitative.

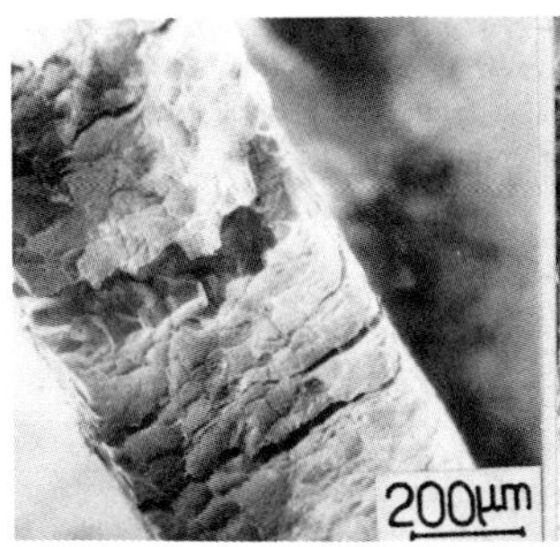
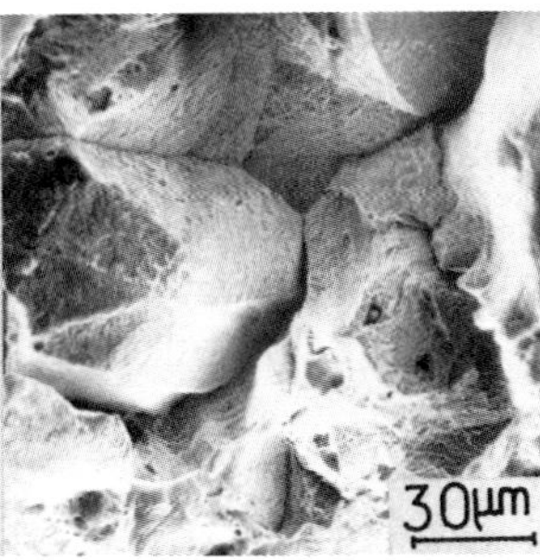

Fig.9 Intergranular
rupture of 260 h
sensitized at 650°C
304 L, H charged
5 h at 150°C. :

a) Specimen edge,

b) Fracture.

From a phenomenological stand point it is possible to propose a coherent interpretation of the whole of our HE results according to a mechanism
based on H-dislocations interactions including a transport and an accumulation step; the last one, beyond a critical concentration,conducing to microcracking initiation, whatever microcracking mechanism envisionned.
The role of transport stage is shown by the influences of temperature
(Tab.VIII) and especially ε in the presence of both internal (Fig.7)and external H
(Fig.3,4,6);on these the minimum of rupture pressures pH_2 is attributed to a
maximum embrittling effect of dislocations (4,5).Moreover, after a plastic
strain in a proper ε range,one directly observes a preferential H segregation on dislocations rich interfaces: γ/α' and especially γ/ε (Fig.1,2).The
local accumulation resulting from transport thus appears more pronounced
than that which might be ascribed to a stress - and strain - concentration
induced solubility increase (4,5).
The accumulation stage and the role of local enrichment are further
illustrated by the results on the influence of metallurgical parameters,
among which the HE increase with the possibility of coplanar dislocations
motion and H dragging, as SFE decreases.Likewise, the specific role of deformation products in 304 L can be interpreted in terms of local accumulation
if one correlates H enrichment associated with trapping on dislocations and
H ability to be distributed in the structure. So one can interpret the especially noxious role of HCP ε_M needles (Fig.8),knowing the existence of high
shear stresses and large dislocation densities at γ/ε interfaces, and assuming that H diffusivities in the FCC and HCP structures are close : See f.i.
Ref. 6 comparing γ(FCC)+ε(HCP)Phynox Co alloy to other FCC alloys. Due to
the effects of trapping and transport (Fig.1, 2) the large H accumulations on
preexisting or recently formed γ/ε interfaces combined with the limited diffusion possibilities in the present phases, can lead to the preferential
initiation of microcracking at these boundaries.
The "beneficial" effect of α'_M dispersed in small quantities in the γ
matrix (Fig. 8) is interpreted assuming that the higher H diffusivity in the
BCC phase favors H disribution trhoughout α' volume and a lowering of the
chemical potential in the structure, which restrains cracking. Using the
language of Pressouyre's theory of trapping (35), the role of homogeneously
distributed α' phase rather appears as a " good short diffusion path "
than a " good trap ".
Likewise, it seems that the sensible HE decrease due to a fine and
discontinuous Cr carbide precipitation stands as an experimental illustration

of the effect of precipitates on H transport by dislocations previously advo cated (35,36). Conversely,a continuous $M_{23}C_6$ carbides network in a sensitized 304 would favor accumulation and enhanced HGE.

Experimental and Scientific Conclusions

Charging in molten salts appears as quite an effective H charging technique,also able to introduce significant T quantities under good safety conditions. The use of the DPT is especially interesting as it involves a biaxial homogeneous stress state and observations are easily made on both in- and outgoing sides.

The investigation has clearly shown how much H transport by dislocations during plastic strain increases H mobility in a range of SS structures including γ phase alone or present along other phases of varied natures. HE of these structures can be large and closely depends on :
- the amount of H absorption,
- microstructure : main phases and precipitates,
- structure stability with respect to strain and temperature,
- the stressing conditions : rate, temperature ... Noteworthy is the HE increase whenever decreasing temperature triggers phase transformations.

Experimental data, among which the paramount influence of microstructure condition support the view that dynamic HE is based on a mechanism of H dragging by dislocations, driving to local enrichments responsible for crack initiation.

In several instances, the effects of H, whether internal or external have been related.

References

1) D. Rousseau, G. Blanc, R. Tricot, A. Gueussier , Mem. Sci. Rev. Mét., 67 (1970) ,pp.315-334.
2) B. Thomas, J. Microsc. Spectrosc. Electr., 1 (1976), pp.623-636.
3) J.P. Fidelle et al., ASTM STP 543 " HE Testing ",(1974) pp. 31-47,221-253,267-272.
4) " Disk Pressure Tests ", Ed. J.P. Fidelle CEA (1975) 112 pages.
5) J.P. Fidelle et al.," Effect of H on behavior of materials " Jackson Lake,1975,
   Ed. A.W. Thompson, I.M. Bernstein, AIME (1976), pp. 91-97, 507-515.
6) P. Tison, R. Broudeur, J.P. Fidelle, B. Hocheid , Paper presented at Internat.
   Congr. " H in Metals ", Chatenay-Malabry, Fr. , 1977, 1 A 4.
7) P. Tison, J.P. Fidelle, " Permeability an investigation technique of H behavior
   in Metals ". Paper presented at Congr. : Progr. in Metals Investigation Techniques,
   Saint-Etienne, Fr., 1978.
8) R. Clermont, J. Ovejero-Garcia, J. Chêne, M. Aucouturier, C. Pirrovani, P. Tison,
   J.P. Fidelle, Proc. JIMIS - 2, Minakami, Jap. Inst. Metals, 1980, pp. 233-236.
9) R. Broudeur, J.P. Fidelle, M. Rapin, High Temp. ,High Press., 6 (1974),173-176.
10) T. Asaoka, G. Lapasset, M. Aucouturier, P. Lacombe, Corrosion,34 (1978),39-47.
11) M. Aucouturier, G. Lapasset, T. Asaoka, Metallography, 11 (1978), 5-21
12) J. Chêne , Métaux Corr. Ind. , 623-624 (1977), 262-279.
13) J. Chêne, Y. Shehu, M. Aucouturier, paper presented at 19th Special Steels Days,
   Saint-Etienne, 1980, 16-1, 16-12.
14) J. Ovejero-Garcia, J. Chêne, M. Aucouturier, Proc. JIMIS -2, Minakami, Jap. Inst.
   Metals, 1980, pp. 525-529.
15) J. Chêne, J. Ovejero-Garcia, C. Paes de Oliveira, M. Aucouturier, P. Lacombe.
   J. Microsc. Spectrosc. Electr., 4 (1979) pp. 37-50
16) P. Tison, J.P. Fidelle, This Conf.
17) J.J. Gilman , Micromechanisms of Flow in Solids, McGraw Hill, N.Y. 1969.
18) A.J. West, M.R. Louthan, Met. Trans., 10 A (1979), pp. 1675-1682.
19) Y. Shehu, J. Chêne, M. Aucouturier , Paper presented at ICF-5 Cannes, Fr.,1981.
20) H. Hänninen, Int. Met. Rev., 24 n°3 (1979).
21) A. Kimura, M. Matsui, H. Kimura, Proc. JIMIS - 2, Minakami, Jap. Inst.Metals (1980),541-544.
22) M. Cornet, S. Talbot-Besnard, Ibid., 545-548
23) J.P. Bricout, R. Missiaen, C. Moriamez, Paper presented at 19th Special Steels Days,
   Saint-Etienne, Fr., 1980, 15-1 , 15-5.
24) Private Communication, C. Zmudzinski, L. Bretin, Creusot-Loire Steelmaking Co.
25) Private Communication, M.W. Poore, UG-25-Y-1871,Oak Ridge, 1973.
26) Private Communication, Pascale Azou, CEA Bruyères-le-Châtel, Fr, 1980.
27) B.C. Odegard, J.A. Brooks, A.J. West, " Effect of H on behavior of Metals ",
   A.W. Thompson, I.M. Bernstein, AIME (1976),pp. 116-125.
28) T.L. Capeletti, M.R. Louthan, J. Eng. Mtls & Technol. (Trans ASME),99 (1977),pp.153-158
29) P. Lacombe, M. Aucouturier, J. Chêne ,"A.R.Troiano Honorary Symposium",Cleveland,Ohio, June 1980.
30) J.P. Fidelle, L.R. Allemand, C. Roux, M. Rapin, Symp. " H in Metals ", Valduc, Fr,
   Ed. J.P. Fidelle, M.Rapin, CEA Bruyères-le-Châtel, 1969, pp. 131-172.
31) J.P. Fidelle, J. Legrand, C. Couderc " A Fractographic of HGE in Steels ", Paper
   F 71-3 presented at AIME Fall Meeting, Detroit, USA, 1971.
32) C.L. Briant , Met. Trans. 10 A (1979) pp. 181-189.
33) G.M. Gordon , H in Metals, Seven Springs, Fd. I.M. Bernstein, A.W. Thompson,ASM (1974).
34) H.H. Johnson, J.P. Hirth , Met. Trans. 7 A (1979), pp. 1675-1682.
35) G.M. Pressouyre, Met. Trans. , 10 A (1979) pp. 1571-1573.
36) J.K. Tien, A.W. Thompson, I.M. Bernstein, R.J. Richards, Met. Trans. 7 A (1976),pp.821-829.

<u>C. Alstetter, University of Illinois, Urbana, IL</u>:   What is the minimum concentration level of tritium that you can detect?   Can you determine the enhancement ratio of tritium at dislocations or other traps to the untrapped tritium?

<u>J. Chene</u>:   It is impossible to answer completely in a few lines to these two very interesting questions, which have been discussed in specialized publications (A,B,C).   The following indications can be given.

1)   The detection limit of autoradiography is theoretically a few atoms of radioisotope; but the exposure time cannot be indefinitely prolonged, and all photographic emulsions have an intrinsic background level. Thus, if one takes into account the statistical aspect, both in time and in space, of the detection process, <u>localized</u> concentrations of the order of one part per million of hydrogen <u>+ tritium</u> ($\simeq 10^{-4}$ppm tritium) can be considered to be easily detectable.

2)   High resolution autoradiography is a "binary" detection method: a radioactive atom cluster <u>is</u> or <u>is not</u> detected as a silver grain of the photograph. Moreover, the interpretation of the images is the statistical result of a great number of observations on different specimens and/or for different exposure times. If these precautions are fulfilled it has been proved possible to detect segregations with an enrichment factor of 3 to 5, defined as the <u>local</u> concentration ratio between the defect and the matrix.

In the present results, the enrichment factors are much larger. They have not yet been calculated here. Previous experiments, on ferritic materials (D), showed that the ratio of trapped hydrogen to lattice hydrogen can be as large as 10^{10}.

A.   A.W. Rogers, Techniques of Autoradiography, Elsevier, 1967.
B.   G.W. Prabhu-Gaukar, A.M. Huntz, P. Lacombe, Int. J. Appl. Rad. Isot., <u>30</u> (1979), 601 and 761.
C.   J.P. Laurent, G. Lapasset, Ing. J. Appl. Rad. Isot., <u>24</u> (1973), 213.
D.   T. Asaoka, G. Lapasset, M. Aucouturier, P. Lacombe, Corrosion, <u>34</u> (1978), 39.

THE EFFECT OF η PHASE ON THE HYDROGEN COMPATIBILITY OF A MODIFIED A-286

SUPERALLOY: MICROSTRUCTURAL AND MECHANICAL PROPERTIES OBSERVATIONS*

B. C. Odegard, Jr. and A. J. West

Sandia National Laboratories
Livermore, California 94550

The influence of η phase on the hydrogen compatibility of a modified
A-286 stainless steel was investigated using tensile tests conducted in air
and hydrogen. Smooth bar tensile specimens with microstructures exhibiting
two morphologies of η were thermally charged to 5500 appm hydrogen and
tested in 172 MPa hydrogen. The solutionized and aged specimens containing
grain boundary η showed significant decreases in uniform elongation and RA.
The fracture mode was intergranular and directly relatable to the η phase.
The cold worked and aged specimens containing cellular, intragranular η
exhibited no change in uniform elongation and losses in RA similar to the
solutionized and aged samples. The fracture mode was transgranular with a
mixed mode containing regions of boundary separation and tear ridges. A
rationale for the observed behavior in hydrogen is presented based on the
hydrogen sensitivity of the η phase and the interaction between two competing
fracture processes. The rationale is consistent with fractographic observations.

Introduction

The demand for structural materials suitable for use in hydrogen
environments has increased with the increased interest in hydrogen as an
energy source. The 300-series austenitic stainless steels have proven only
mildly affected by hydrogen and consequently are candidates for several long
term applications.(1-8) Unfortunately, they are limited in their application
by a low weld strength(9). A possible improvement is a precipitation-
hardened austenitic matrix which might offer both high strength and acceptable hydrogen compatibility and therefore have a wider application for
hydrogen service. One candidate is A-286. However, A-286 is highly susceptible to weld hot cracking(10-12). In an effort to improve weldability and
maintain hydrogen compatibility, a new alloy, JBK-75, has evolved from
selected chemistry modifications of A-286(13-15). JBK-75 achieves its
strength (700 MPa yield strength) by precipitation of an FCC γ' precipitate
[Ni₃(Ti,Al)]. The precipitation characteristics of JBK-75 have been well
characterized by Thompson and Brooks(15). These studies show that con
comitant with the γ' precipitation, many other phases can form with only

*This work supported by U.S. Dept. of Energy, DOE, under Contract Number
DE-AC04-76DP00789.

minor changes in either the heat treatment or chemistry. One of these phases, eta (η), has a similar chemistry to γ' but has an HCP (Ni_3Ti) structure. This phase has been associated with increased sensitivity to hydrogen in JBK-75(13-14). η is a hard, brittle phase which, if allowed to form as a continuous phase, could reduce the inherent ductility of the alloy. Similar to γ', η precipitates from the austenite matrix during the normal aging treatment in one of two forms. The aging characteristics result in either a grain boundary precipitate, or a precipitate having a cellular intragranular morphology. The morphology and quantity are directly related to the chemistry, mechanical processing and aging temperature.

To achieve additional strength over the solutionized and aged (SA) condition, thermomechanical processing (TMP) can be employed. However, working of the microstructure affects the aging kinetics and, in particular, the η phase morphology and concentration. Heat treating a worked structure at the standard aging temperature (720°C) used for solutionized material can result in an overaged structure containing increased concentrations of cellular η. Aging at a lower temperature (i.e., 675°C) slows the aging kinetics and reduces the level of η in the microstructure. TMP can produce yield strengths ranging from 800-1100 MPa and weld strengths of 700 MPa.

Slow crack growth (SCG) tests are of interest in studying the hydrogen degradation of materials because of their ability to measure crack growth in a controlled environment under static loading. This test measures the stress intensity, K_{TH}, below which hydrogen-induced slow crack growth does not occur. Although the influence of η has not been studied using SCG tests in any detail, some observations of interest have been made. Tests have been conducted on both SA and TMP JBK-75(16) and the results suggest that η affects K_{TH}. Solutionized and aged material gave low K_{TH} values in 172 MPa H_2 while TMP processing where a lower aging temperature was employed resulted in higher K_{TH} values.

Another form of deformation which can influence the aging characteristics and possibly the compatibility of JBK-75 is the localized cold work produced by machining. Recent studies(17) have shown that machining damage can accelerate the η precipitation at the 675°C aging temperature. The result is a high volume fraction of η within a narrow surface region which, if sensitive to hydrogen as suggested earlier, could accelerate the initiation of SCG in the parent material.

The following study is concerned with the influence of high concentrations of grain boundary and cellular η morphology on the tensile properties and fracture mode of JBK-75 after thermal charging with hydrogen.

<u>Experimental Procedure</u>

Material for this study was obtained in the form of 152 mm bar stock produced by a vacuum arc remelt process. The bar stock was solutionized at 980°C for one hour and water quenched. The chemistry of the material used is shown in Table I.

TABLE I

<u>Bar Composition (w/o)</u>

Cr	Ni	N	C	Al	V	Mn	Mo	Ti	Si	S	Fe	B	Fe
15.3	29.8	.004	.012	.3	.42	.011	1.2	2.1	.075	40 ppm	100 ppm	11 ppm	balance

Following the solutionizing treatment, a section of the bar was subjected to varying aging times at 720°C (standard aging temperature for A-286). The microstructure of solutionized and aged specimens exhibited no η phase using optical microscopy. Transmission electron microscopy, however, showed a fine dispersion of η at the grain boundaries which increased in contiguity as the aging time increased from 4 hrs. to 16 hrs. Portions of bar were cold swaged at room temperature to diameter reductions of 8, 20 and 36%. Subsequently, the swaged bars were aged at times of 1, 4 and 8 hours at 675°C to produce varying concentrations of cellular η phase. Depending on the amount of deformation and aging time, the areal fraction η could be varied between 8 and 76% (Note Fig. 1a and 1b). The areal fraction η was determined by the average linear intercept method.

The tensile bars were thermally charged to produce a hydrogen concentration gradient. The charging conditions of 200°C for 10 days at 69 MPa hydrogen produced a hydrogen surface concentration of approximately 5500 appm and a center axis concentration of ∿100 appm in a 5 mm gage diameter tensile bar. Tensile tests were conducted in 172 MPa hydrogen at room temperature at a strain rate of 8.3 x 10^{-4} sec^{-1}. The mechanical properties of charged specimens were contrasted to uncharged samples tested in air.

Microstructural examination was conducted using both optical and electron microscopy and the fracture mode was characterized with the scanning electron microscope.

Experimental Results

Tensile Properties

The results from the tensile tests for both the solutionized and cold worked specimens are shown in Table II. The uncharged, solutionized and aged specimens showed increasing strength and decreasing ductility with aging time consistent with the aging response of precipitation-hardened alloys. The yield strength as a function of aging time is shown in Figure 2. The yield strength varied from 240 MPa (solutionized condition) to 716 MPa (solutionized and aged 16 hrs. at 720°C) in the uncharged samples. Hydrogen charging produced no increase in either the yield strength or the ultimate tensile strength of these specimens. However, a significant decrease in ductility was observed in the charged compared to uncharged specimens. Losses in uniform elongation of ∿30% and in RA of ∿60% were observed. The decrease in uniform elongation is particularly significant because losses this large are not typical in hydrogen charged austenitic stainless steels except for alloys like 304L which exhibit a strain-induced phase transformation(18) or in heavily sensitized alloys. Interestingly, the RA loss in the SA alloys was 55-60% independent of the yield strength and concentration of grain boundary η.

The cold worked and aged specimens generally exhibited a decrease in yield strength with aging time for the uncharged samples. The magnitude of this change was a function of the amount of deformation. Both the 20% and 36% cold worked specimens reached peak strength in less than 1 hour at 675°C and longer aging times produced strength reductions indicative of overaging. The 8% cold worked sample reached peak strength in approximately four hours. These test results are shown graphically in Figure 2. The uniform elongation of the uncharged specimens remained constant for the 8% cold worked condition and increased slightly for both the 20% and 36% cold worked condition. Hydrogen charging and testing at 172 MPa hydrogen produced no change in the uniform elongation for the 8% and 20% cold worked conditions compared to uncharged samples. The uniform elongation for the

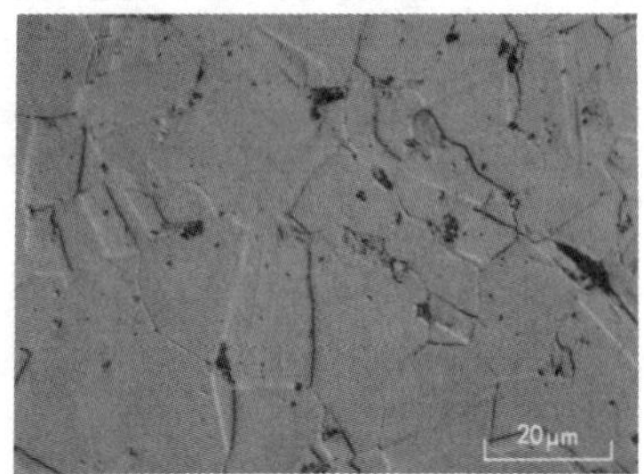

Figure 1a - Photomicrograph of microstructure following an 8% deformation and aged at 675°C for 1 hr. (~8% η)

Figure 1b - Photomicrograph of microstructure following a 36% deformation and aged at 675°C for 8 hrs. (~76% η)

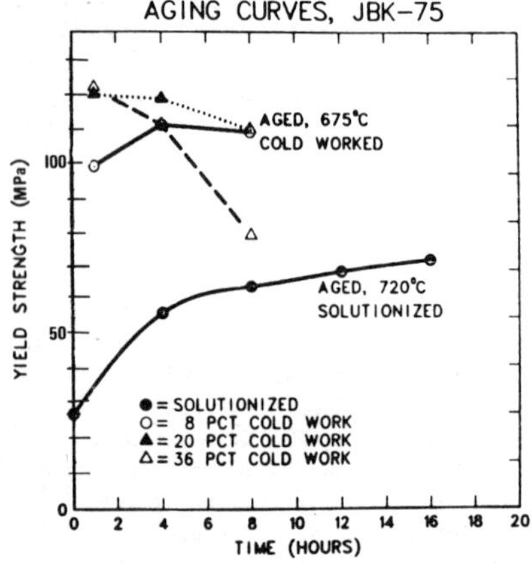

Figure 2 - Plot of yield strength as a function of aging time for the solutionized and cold worked material.

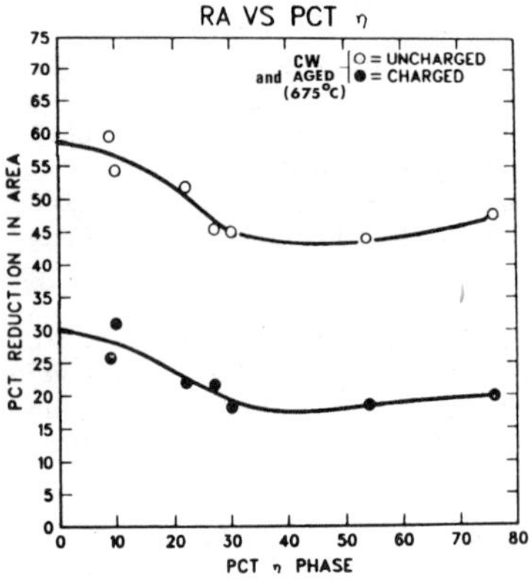

Figure 3 - Plot of RA as a function of %η for the cold worked and aged material.

charged 36% cold work condition tended to decrease. The RA decreased as the
aging time increased for both the air and hydrogen tests. There was approximately 55% RA loss when thermally charged with hydrogen as was observed in
the solutionized and aged specimens. The RA data for both the solutionized
and cold worked conditions is shown as a function of aging time on Figure 3.
As in the SA case, the RA loss due to hydrogen charging was constant ($\sim$55%)
independent of yield strength and η concentration.

<u>Microstructure</u>

Light microscopy revealed a clean, single phase austenitic structure
for the solutionized and aged specimens, independent of aging time at 720°C.
Occasionally, bands of intragranular carbides retained from the billet
casting process were observed which is typical of γ' precipitation-hardened
steels. The grain boundaries were clean with no evidence of η. However,
when examined using transmission electron microscopy (TEM), a fine grain
boundary precipitate identified as η (note Figure 4) was revealed. The η
phase was observed to increase in concentration as the aging time increased.
No grain boundary phases were observed in the solutionized condition.

Large concentrations of the cellular η were observed in the cold worked
and aged specimens. The microstructure of the 8% cold worked specimen aged
one hour at 675°C is shown in Figure 1a. The η appears as isolated cellular
colonies growing from the austenitic grain boundaries. The 36% cold worked,
eight-hour age microstructure is shown in Figure 1b. Here the cellular η
is seen as nearly continuous and consumes approximately 76% of the grain
area.

<u>Fractography</u>

All the uncharged SA specimens failed by dimple rupture similar to the
charged, solutionized sample in Figure 5a. The SA, charged specimens
failed by a predominantly intergranular mode as noted in Figures 5b-5e.
With increasing aging time, the intergranular facets of the fracture exhibited smaller areal fractions of ductile fracture. The fracture of the
specimen aged 16 hrs. was entirely intergranular with minimal evidence of
ductility on the grain facets.

Similar to the SA, uncharged samples, the uncharged cold worked and
aged samples exhibited a dimple rupture. However, the dimples were less
equiaxed (Figure 6a). The charged, cold worked and aged samples did not
show a dramatic difference in the fracture topography as the η concentration increased. The features shown in Figure 6b are typical of all
charged samples. The fracture is a mixed mode consisting of regions of
boundary separation and tear ridges.

<u>Discussion</u>

Hydrogen degradation has been observed to manifest itself in one of
several forms in austenitic stainless steels. The fracture process in
mildly embrittled alloys such as 310 and 316 is typically by void growth
and coalescence. Hydrogen in these alloys assists both the nucleation and
growth stages but does not change the fracture process. In less stable
austenites or austenitic alloys that are sensitized or laden with inclusions, hydrogen manifests itself by assisting interfacial fracture (twin
boundary, α'-γ boundaries, grain boundaries, etc). Type 304L, 21-6-9, and
JBK-75 fit into this category. In 304L, the presence of α' martensite has
been associated with the large losses in RA and fracture mode changes

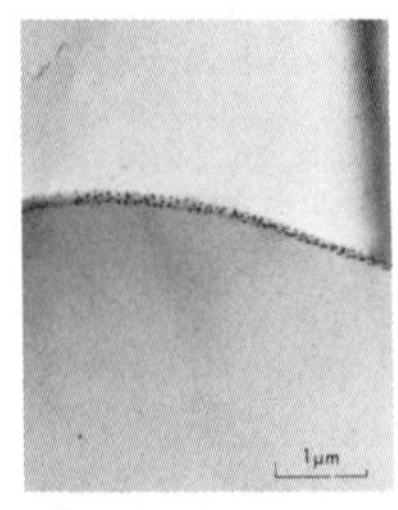

Figure 4 - Micrograph showing grain boundary
η following 16 hrs. at 720°C.

Figure 5 - Fractographs of charged solutionized and aged material tested
in 172 MPa hydrogen.

a. Solutionized   b. 4 hrs. at 720°C   c. 8 hrs. at 720°C

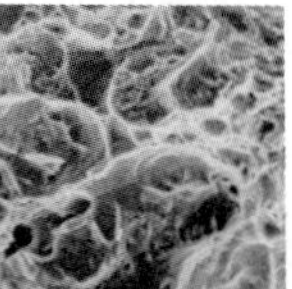   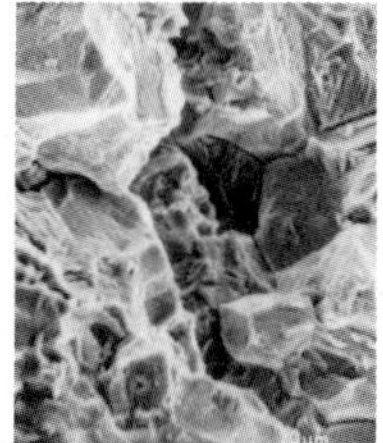   

d. 12 hrs. at 720°C   e. 16 hrs. at 720°C

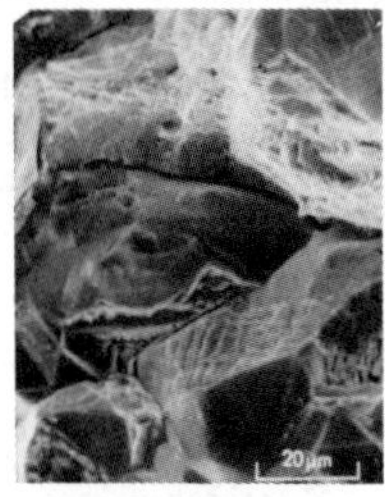   

Figure 6 - Fractographs of 36% cold worked material aged in 8 hrs. at 675°C.

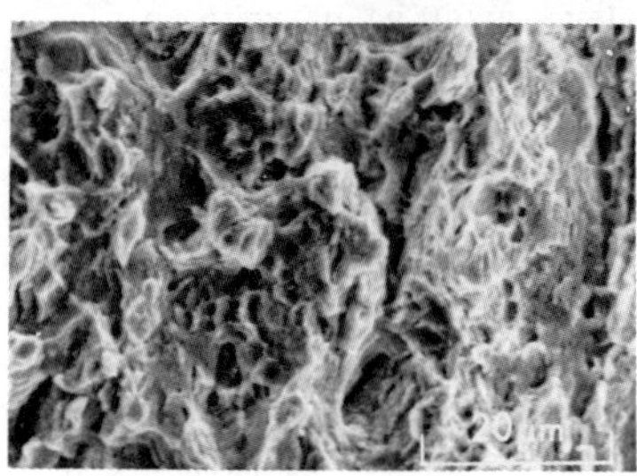   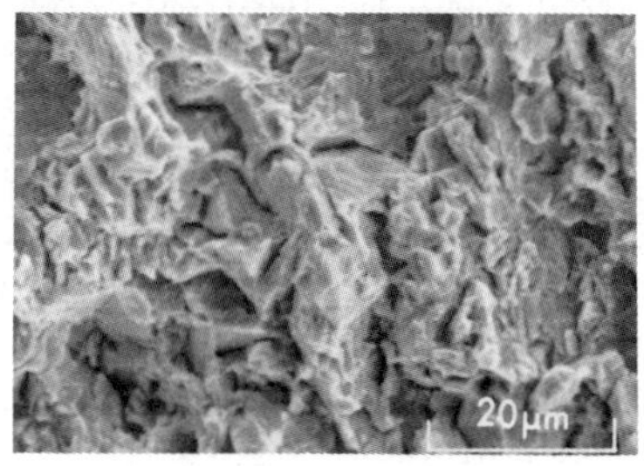

6a - uncharged,   6b - charged, tested
tested in air.   in 172 MPa hydrogen.

observed(17). The α' itself is sensitive to hydrogen and behaves in a brittle manner as do other BCC ferrites. Therefore, larger quantities of α' in 304L are expected to produce greater hydrogen sensitivity. Similar arguments would appear to hold for the effect of increasing concentrations of η on the compatibility of JBK-75.

This rationale appears to hold for the SA material test results presented in Table II. The tensile behavior of the SA specimens illustrates both the nature of the competitive fracture processes in JBK-75, and the importance of interfaces or second phases in hydrogen degradation. In the charged samples void growth and coalescence competes with grain boundary fracture which in turn appears to be dependent on the amount and continuity of η in the grain boundaries. As the fractography and TEM results show, increased concentrations of disc-shaped η in the grain boundaries produces increasingly "more brittle" intergranular fracture. Using replication techniques, Stoltz(19), has shown a one-to-one correlation of disc-shaped η particles with markings on intergranular fracture facets. The observations of ductile tearing along or near the grain boundaries in the shorter aged specimens is consistent with reduced η concentrations, and with "brittle" intergranular fracture in the regions of the grain boundary where sufficient η has precipitated. This supposition is further supported by the fact that the solutionized and unaged specimen, which contains no η in the grain boundary, failed by ductile void growth and coalescence with virtually no loss of RA. Thus, fracture of hydrogen charged SA specimens appears to be controlled by the presence and amount of disc-like η phase in the grain boundary. In these samples, failure by decohesion of the η-γ interface in the grain boundary intervenes before void growth and coalescence can occur. The decrease in uniform elongation with aging time, which is not typical of austenitic stainless steels, is consistent with the intervention of a fracture process other than void growth and coalescence.

The cold worked and aged specimens provide an interesting contradiction. If one assumes that increasing the concentration of a hydrogen sensitive phase like η is detrimental and that increasing the yield strength of an alloy increases its sensitivity to hydrogen, one would expect the presence of cellular η in JBK-75 to increase the ductility losses observed in hydrogen. However, the opposite appears to be true. The RA loss of charged, cold worked and aged specimens is approximately 55% independent of either yield strength or η concentration. Additionally, the RA losses are of similar magnitude to that observed in the SA samples which contain no cellular η. Therefore, the ductility losses in hydrogen are controlled by a process which is not totally dependent upon the presence of η.

One explanation of this process is based upon crack initiation phenomenon. Within experimental error, the presence of hydrogen has little effect on the uniform elongation of the cold worked and aged specimens. In a tensile test, reaching the uniform strain level typically marks the onset of plastic instability and void initiation and ductile fracture in uncharged specimens. In charged specimens, η associated fracture clearly competes with void growth and coalescence and thus the presence of ductile, as well as, interface associated (Figure 6b) fracture characteristics would be expected. The insensitivity of RA loss to the concentration of η could suggest that the initiation, rather than propagation, stage of the fracture process is controlling. Stoltz(20) points out that uncharged SA JBK-75 exhibits little RA loss in 172 MPa hydrogen. He attributes this relative insensitivity to hydrogen to the fact that initiation of a surface crack requires a strain level that is a large percentage of the total strain to fracture and therefore an initiating crack has little opportunity to grow

before void coalescence and final fracture. Further, Perra and Stoltz (16)
have shown that when the initiation stage is preempted by precracking of a
WOL specimen, intergranular fracture proceeds in a relatively short time.
Therefore, it is possible that the constant RA loss observed in this study
reflects an ability of JBK-75 to resist crack initiation until late in the
tensile test. If fracture of JBK-75 in tensile specimens is controlled by
surface crack initiation, then no RA dependence on the concentration of
intragranular η would be expected. Further, since the cold-worked specimens
were aged at a lower temperature than the SA specimens little or no disc-
shaped grain boundary η is present to interact with the hydrogen and cause
premature intergranular fracture.

Conclusions

The following conclusions are drawn from this study:

1. The presence of η phase in the grain boundaries strongly influences
    the fracture mode of hydrogen charged JBK-75. The presence of grain
    boundary η is strongly dependent on the aging temperature;

2. In the absence of grain boundary η, the concentration of η in the grain
    interior has little effect on the ductility losses in hydrogen charged
    JBK-75;

3. The ductility losses in JBK-75 appear to be controlled by surface crack
    initiation rather than by crack propagation. This causes RA losses
    that are independent of yield strength and η concentrations.

TABLE II
Mechanical Test Results

Solutionized (980°C) and Aged (720°C)

Aging Time	Text Condition	YS (MPa)	UTS (MPa)	Uniform Elong. (%)	RA (%)	RAL*
Solutionized	uncharged	241	620	35	70	
4 hrs	uncharged	565	1058	26.0	60.7	
8 hrs	uncharged	632	1091	24.0	56.7	
12 hrs	uncharged	672	1131	21.5	51.4	
16 hrs	uncharged	716	1130	22.0	51.0	
Solutionized	charged	245	618	34	67	4
4 hrs	charged	560	1012	18.4	23.7	61
8 hrs	charged	640	1063	16.5	23.4	59
12 hrs	charged	683	1092	16.5	21.1	59
16 hrs	charged	723	1137	16.5	23.5	54

* RA loss when charged and tested in 172 MPa hydrogen.

TABLE II
Mechanical Test Results
(continued)

Cold Worked and Aged (675°C)

%CW/ Aging Time	Test Condition	YS (MPa)	UTS (MPa)	Uniform Elong. (%)	RA (%)	% η-phase	RAL*
8/1 hr	uncharged	987	1169	12.8	59.5	9	
8/4 hr	uncharged	1100	1288	10.8	51.8	22	
8/8 hr	uncharged	1083	1302	11.3	45.0	30	
8/1 hr	charged	1054	1216	12.2	25.1	9	58
8/4 hr	charged	1136	1282	10.1	23.0	22	56
8/8 hr	charged	1089	1295	12.5	18.3	30	59
20/1 hr	uncharged	1196	1306	6.9	54.3	10	
20/4 hr	uncharged	1178	1340	8.5	45.4	27	
20/8 hr	uncharged	1085	1304	9.5	39.7	54	
20/1 hr	charged	1226	1325	7.8	31.0	10	43
20/4 hr	charged	1192	1326	9.2	21.6	27	52
20/8 hr	charged	1123	1295	9.6	18.6	54	53
36/1 hr	uncharged	1212	1337	5.8	50.0		
36/4 hr	uncharged	1029	1269	10.5	44.1		
36/8 hr	uncharged	785	1169	15.0	47.7		
36/1 hr	charged	1240	1350	7.2	21.2		58
36/4 hr	charged	1075	1268	9.5	19.2		56
36/8 hr	charged	878	1152	12.0	20.0	76	58

* RA Loss when charged and tested in 172 MPa hydrogen.

REFERENCES

1. R. M. Vennett and G. S. Ansell: Trans. ASM, 1967, vol. 60, pp. 242-51; ibid., 1969, vol. 62, pp. 1007-13.

2. R. J. Walter and W. T. Chandler: Effects of High-Pressure Hydrogen on Metals at Ambient Temperatures, Report R-7780-1 (Access No. N70-18637), Rocketdyne Division, Rockwell International, Canoga Park, Calif., Feb. 1969.

3. R. J. Walter, R. P. Jewett, and W. T. Chandler: Mater. Sci. Eng., 1969-70, vol. 5, pp. 99-110.

4. M. R. Louthan, G. R. Caskey, J. A. Donovan, and D. E. Rawl: Mater. Sci. Eng., 1972, vol. 10, pp. 357-68.

5. A. W. Thompson: Met. Trans., 1973, vol. 4, pp. 2819-25.

6. A. W. Thompson: Mater. Sci. Eng., 1974, vol. 14, pp. 253-64.

7. A. W. Thompson: Hydrogen in Metals, pp. 99-101, ASM, Metals Park, 1974.

8.   M. R. Louthan: <u>Hydrogen in Metals</u>, (I. M. Bernstein and A. W. Thompson, eds.) p. 53-75, ASM, Metals Park, 1974.

9.   J. A. Brooks and A. J. West: "Hydrogen Induced Ductility Losses in Austenitic Stainless Steel Welds", <u>Met. Trans.</u> (paper accepted).

10. J. J. Vagi and D. C. Martin: <u>Welding J.</u>, Research Suppl., 1956, vol. 35, p. 137-s.

11. B. S. Blum and R. H. Wit: <u>Welding J.</u>, Research Suppl., 1963, vol. 42, p. 365-s.

12. J. F. Radavich: <u>Advances in X-ray Analysis</u>, Plenum Press, New York, 1973, vol. 3, p. 365.

13. J. A. Brooks and R. Krenzer: <u>Welding J.</u>, Research Suppl., 1974, vol. 53, p. 242s-245s.

14. J. A. Brooks: <u>Welding J.</u>, 1974, vol. 53, p. 517s-523s.

15. A. W. Thompson and J. A. Brooks: <u>Met. Trans. A.</u>, 1975, vol. 6A, p. 1431-42.

16. M. W. Perra and R. E. Stoltz: this volume.

17. S. L. Robinson: Private Communication, Sandia National Laboratories, Livermore, CA, May 1980.

18. A. J. West and J. H. Holbrook: this volume.

19. R. E. Stoltz: Private Communication, Sandia National Laboratories, Livermore, CA.

20. R. E. Stoltz and A. J. West: this volume.

HYDROGEN IN AUSTENITIC STAINLESS STEELS:

EFFECTS OF PHASE TRANSFORMATIONS AND STRESS STATE*

A. J. West and J. H. Holbrook[†]

Sandia National Laboratories
Livermore, California 94550

[†]Battelle Columbus Laboratory
Columbus, Ohio 43201

The effect of hydrogen on the deformation properties of 304L containing 1400 appm H_2 and 21-6-9 containing 6000 appm H_2 were investigated. Smooth bar specimens of each alloy exhibited ductility losses due to hydrogen saturation. 304L had an RA loss compared to air of 60% while 21-6-9 lost only 25% in RA. Both alloys exhibited small increases (12%) in yield strength with no change in strain hardening exponent. Extensive fracture mode changes from dimpled rupture to a transgranular, brittle appearing faceted fracture associated with the formation of strain induced α' martensite was observed in the 304L specimens. Little tendency for fracture mode changes were noted in the smooth 21-6-9 samples. In notched samples, however, a dramatic reversal occurs. Notched 304L tensile bars exhibit less RA loss than smooth specimens while notched 21-6-9 showed an increased RA loss over the smooth bars. Also, a dramatic increase in notch flow stress was observed in both alloys. The fracture mode in the 304L notched specimens was entirely dimple rupture with no indication of hydrogen influencing the fracture process except for a slight decrease in dimple size. In 21-6-9 notched specimens an increased tendency for fracture mode changes ahead of the notch was observed. X-ray analysis of charged 304L specimens suggest that hydrogen assists the α' martensite transformation, while the constraint due to a notch retards α' transformation.

Introduction

The effects of dissolved hydrogen on the deformation, flow, and fracture properties of materials has been the subject of an extremely large number of investigations over the past decade.(1-19) Several studies have culminated in models for hydrogen degradation which are based on dislocation - hydrogen interactions(1) and the subsequent effect on plastic flow.(2) In particular, the dislocation transport model originally proposed by Bastien and Azou(3,4) and later modified by Tien et al(1,5) predicts a local concentration buildup of hydrogen at embrittled microstructural sites through dislocation transport of hydrogen "clouds." Clearly, the effectiveness of dislocations in assisting (deposition of hydrogen at twin or grain boundaries) or retarding (trapping at small particles) hydrogen induced degradation is dependent on

*This work supported by U.S. Dept. of Energy, DOE, under Contract Number DE-AC04-76DP00789.

607

how hydrogen affects the ease of dislocation motion, the slip character, or the local flow stress. If hydrogen effectively pinned dislocations thereby impeding their motion, the contribution of dislocation transport to the total degradation is reduced. On the other hand, if the presence of hydrogen in a material assists microplasticity ahead of a crack tip as suggested by Beachem(6) , transport of hydrogen by dislocations can be a major contributor to degradation. Several investigations have clearly established dislocation transport of hydrogen. Donovan's(7) outgassing experiments show a significant increase in outgassing rates in several alloys at the onset of yielding, during serrated yielding and at fracture. Birnbaum(8) has observed extensive dislocation rearrangement in thin stressed nickel foils after exposure to less than one atmosphere of hydrogen gas. This observation could indeed suggest a reduction in the lattice resistance to dislocation motion.

The confusion associated with hydrogen hardening or softening a material is matched by the lack of agreement among experimental data available in the literature. In 304L stainless steel for instance, Vennett and Ansell (9) have reported lower yield strengths due to testing in high pressure whereas Odegard, Brooks, and West(10) show an increase in yield strength and a decrease in UTS. The latter study also reported no increase in yield strength in 21Cr-6Ni-9Mn due to hydrogen. Similar conclusions were reported for another single phase austenitic stainless steel, 22Cr-13Ni-5Mn (11). In high purity nickel, both increases (Holbrook)(12), Boniszenski(13) and decreases (Birnbaum)(14) in flow stress with the addition of hydrogen have been observed.

Phase transformations during straining can also influence a materials behavior in hydrogen. This is particularly true in 304L where the α' transformation has been associated with the faceted fracture mode change observed in hydrogen charged samples(15). In this case, hydrogen may influence the α' transformation by affecting the slip mode since α' originates from ε martensite which is formed in slip bands. This transformation has been described as a primary contributor to the high ductility losses in 304L compared with other more stable 300 series stainless steels such as 310 and 316. As a result of ductility losses in short term tensile tests of smooth specimens, 304L is often considered less hydrogen resistant than the more stable austenitic alloys. However, in slow crack growth tests conducted by Perra(16) in 210 MPa hydrogen, 304L has exhibited a greater resistance to hydrogen than more stable austenites. These results suggest a reexamination of the effects of phase transformations and stress state may elucidate these apparent contradictions. In this paper we discuss the tensile behavior of annealed austenitic 304L and 21-6-9 stainless steels that have been thermally charged to saturation with hydrogen. The tensile behavior is examined in both the smooth and notched bar condition. The purpose of the investigation is to determine the effect of hydrogen on the mechanical behavior of austenitic stainless steels in terms of phase transformations, the role of stress state and the flow and fracture properties.

Experimental Procedure

Tensile specimens of 304L and Type 21-6-9 stainless steel were machined from the short transverse direction of the 4.5 cm thick cross-rolled plate having an average grain size of 50 microns. The smooth tensile specimens had a 25.4 mm gage length and 5.0 mm diameter. Notched specimens had similar dimensions with a 0.13 mm maximum root radius and 3.95 mm root diameter. Prior to thermal charging, the tensile bars were polished with 600μ silicon carbide paper to remove the machine damaged surface layer. Because

of difficulties in the uniformly polishing the notch region, notch specimens
were not polished. The specimens were ultrasonically degreased in acetone,
rinsed with ethyl alcohol, and thermally charged in a high pressure
autoclave. Saturation charging of the specimens was achieved by exposing
the specimens to 69 MPa H_2 at 300°C for 340 hours. These charging conditions produce a hydrogen concentration of 6200 appm as determined from
Sieverts law and Louthan's(17) relation for hydrogen solubility in 304L
stainless steel. Hydrogen concentration measurements were made on each
alloy by vacuum extraction at 750°C. An average concentration of approximately 6000 appm was measured in 21-6-9 specimens. In 304L, the hydrogen
level was 1400 appm, less than expected from solubility calculations.
Similar concentration levels were expected in both alloys because both
materials exhibit similar permeation rates(17) and were thermally charged
simultaneously. The observed concentration differences are likely due to
different surface oxide barriers in the two alloys as has been observed by
Louthan, et al(18). Tensile tests were conducted in air at room temperature. Smooth specimens were strained at a crosshead speed of 0.02"/min.
($\dot{\varepsilon}$ = 4.3 x 10^{-4} sec^{-1}). Notched specimens were strained at a slower rate of
0.005"/min. crosshead speed to achieve a strain rate in the notched region
similar to that of the smooth specimens.

Selected specimens were examined using x-ray and reflection diffraction
microscopy to characterize the amount of α' martensite as a function of
distance from the fracture surface. The specimens were sectioned perpendicular to the tensile axis, polished to remove the cutting damage, and
examined using x-ray diffraction. The structure of the fracture surface
was made using an electron reflection diffraction technique(19) which
analyses the local surface structure by diffraction of electrons through
asperities on the fracture surface.

<u>Experimental Results</u>

<u>Tensile Properties</u>

A summary of the tensile properties for both smooth and notched bar
specimens are presented in Table I. In the annealed uncharged condition,
both 304L and 21-6-9 exhibit extensive ductility, especially after necking
occurs. Both alloys elongate approximately 25% after necking commences.
Typical smooth bar tensile curves for each alloy are shown in Figure 1.
Hydrogen charging caused a 10% increase in yield strength in both alloys.
However, within experimental error, little change in ultimate tensile
strength (UTS) was observed. The largest effects of internal hydrogen were
manifested in the ductility parameters; in Type 304L, a 44% loss in uniform
elongation, ε_u, was observed associated with an RA loss of 59%. In Type
21-6-9, a 26% loss in RA with no loss in ε_u was noted. The shape of the
tensile curve in the post-necking region for charged 21-6-9 was the same as
the uncharged curve suggesting the RA loss was due to hydrogen assisted
void growth. In 304L, on the other hand, the fact that uniform elongation
is reduced by 44% and that no post neck ductility is observed suggests the
intervention of a fracture process other than void coalescence. Further,
the steep stress drop after the UTS is reached in 304L is consistant with
the intervention of another fracture process. There appears to be little
hydrogen induced change in strain hardening rate in either alloy and thus
hydrogen only causes a moderate lattice hardening.

Typical load-displacement curves for notched specimens are shown in
Figure 2. The displacements have been corrected for elastic strains in
the unnotched portion under the clip gage and to a first order the curves
shown represent the uniaxial σ-ε behavior of the notch. Two dramatic
differences were observed in notched specimens compared to smooth specimens.

TABLE I
Tensile Properties of Hydrogen Saturated Stainless Steels

304L*		σ_y MPa	UTS, MPa	ε_u	ε_t	RA	RA Loss	Fracture Mode
Smooth	Unc	255	690	70	93	84		MVC
	C	285	672	39	40	34	59	Faceted Transgranular
Notched	Unc		687			29		MVC
	C		748			17.5	40	MVC
21-6-9*								
Smooth	Unc	360	805	45	71	65		MVC
	C	395	847	45	62	48	26	MVC
Notched	Unc		929			38		MVC
	C		955			24.5	36	MVC

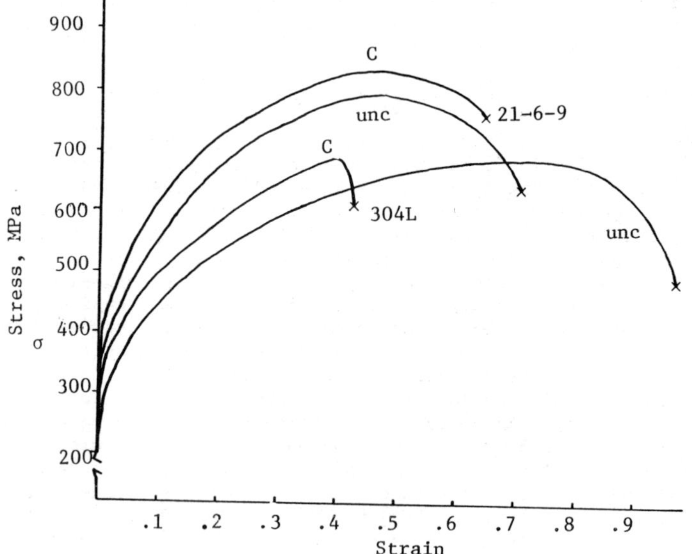

Figure 1.   Stress-strain behavior of smooth specimens of 304L and 21-6-9 in the uncharged condition and after thermal charging with hydrogen.

First, both alloys showed larger increases in flow stress compared to uncharged specimens than was observed in smooth specimens. In particular,
apparent flow stress increases as high as 50% were observed in notched 304L
specimens while increases as large as 30% were noted in notched 21-6-9.
Secondly, the dramatic post-instability stress drop which was observed in
charged 304L smooth specimens was not seen in the notched specimens. Both
alloys exhibited similar losses in notch RA (40%) due to hydrogen. In 304L
this represented a decreased RA loss compared to smooth specimens while in
21-6-9 an increased RA loss compared to smooth specimens was observed. The
notched tensile strength in 304L increased by approximately 10% while no
change was observed in 21-6-9. In 304L this decreased RA loss in notched
compared to smooth specimens was unexpected because the presence of a notch
and its associated stress concentration was expected to magnify, as in
21-6-9, rather than reduce any hydrogen assisted loss in ductility.

Fracture Characteristics

     Fractured specimens from each set of experimental conditions were ex-
amined to characterize the effect of internal hydrogen and stress state on
the fracture mode. In the uncharged condition both alloys fracture by
ductile void growth and coalescence independent of the presence of a notch.
Hydrogen charging in smooth 304L specimens produced substantial changes in
fracture characteristics as shown in Figure 3. Examination at low magnification revealed approximately 60% of the fracture surface was perpendicular to the tensile axis and contained numerous secondary cracks compared
with the cup-cone type fracture in uncharged specimens. The predominant
fracture mode in this region is transgranular and heavily faceted with
little indication of ductility in the form of void growth and coalescence
(Figure 3a). Secondary cracking appeared to be intergranular with some
tendency to crack along twin boundaries. This fracture mode is similar to
that in hydrogen saturated 304L tested at -77°C where the primary fracture
mode has been observed by Caskey(20) and Holbrook and West(21) to be
associated with α' martensite. The surface crack formation is similar to
that reported by West and Louthan(22). The fracture appears to initiate
from a surface crack and to propagate inward until overload and shear
fracture of the remaining ligament take place.

     The fracture characteristics of saturated smooth bars of 21-6-9 are
also included in Figure 3. Surface cracking was noted along the gage
section away from the neck region indicating cracking occurred at stresses
below the tensile strength as was observed in 304L. However, only a small
portion of the fracture surface exhibited a fracture mode other than void
growth and coalescence. These regions were all associated with surface
cracks and seldom penetrated more than 250μ. The small regions that exhibited mode changes (Figure 3b) are typically transgranular with a fibrous
fracture mode.

     Notched and hydrogen saturated 304L tensile specimens exhibited re-
markable and unexpected fracture characteristics. The fracture surface of
charged notched specimens was completely ductile even at the tip of the V
notch (Figure 3c). Only very subtle changes in fracture mode near the
tip of the V notch could be discerned. These changes consisted primarily
of planar regions of very fine dimples that may be associated with either a
twin or grain boundary. The only significant change in the fracture morphology was an increase in dimple size of approximately 70% compared to uncharged notch bars.

     Notched 21-6-9 specimens revealed different fracture characteristics
than 304L. In the presence of a notch, 21-6-9 exhibited an increased

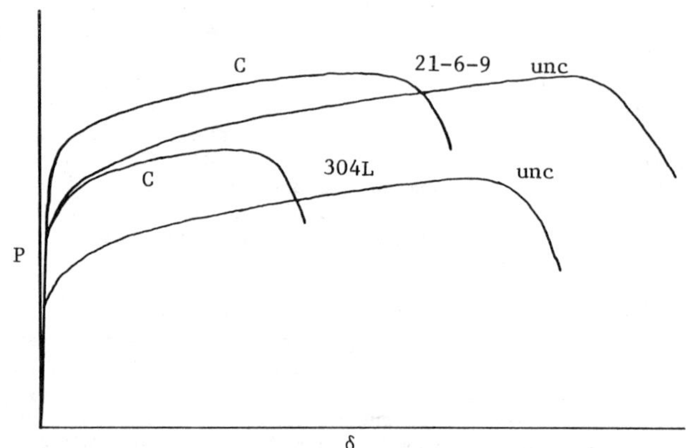

Figure 2.   Load-displacement curves for notched specimens
of 304L and 21-6-9 in the uncharged condition
and after thermal charging with hydrogen.

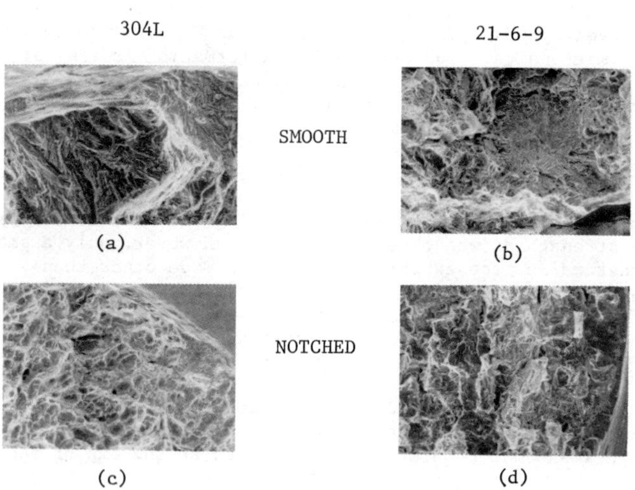

Figure 3.   Fracture modes of smooth and notched specimens of 304L and
21-6-9 after hydrogen charging.

tendency for surface cracking near the notch and hydrogen induced fracture
mode changes (Figure 3d). Below the notch, the fracture was transgranular
and faceted containing few dimples. Cracking along twin boundaries was
observed in addition to a small amount of secondary cracking. The hydrogen
affected region was approximately 800μ deep and did not typically contain
microvoids. Thus, the fracture characteristics in notched specimens are
consistent with the trends exhibited by RA loss. A reduced tendency for RA
loss in notched and saturated 304L was accompanied by an increased tendency
towards ductile fracture compared with smooth specimens. In 21-6-9 increased RA loss in notched specimens was accompanied by a larger fraction of
faceted, less dimpled fracture.

Diffraction Results

The x-ray diffraction results are presented in Figure 4 where the
intensity ratio of $I[\gamma(111)]/I[\alpha'(110)]$ is plotted as a function of distance
from the fracture surface. Both uncharged and charged specimens exhibited
only a α' adjacent to the fracture surface. No intensity above background
was observed for the $(111)\gamma$ peak indicating the near fracture region is
almost entirely α'. Unfortunately, the curves for the charged and uncharged
specimens are not directly comparable at a given distance from the fracture
surface because at any distance the local strain is significantly higher in
the uncharged specimen which necked and the charged specimen which did not
neck. However, a reasonable estimate of the local strain in the charged
specimen near the fracture surface (Figure 4, point 1) is approximately
0.4. The strain in the uniform strain region of the uncharged specimen
(point 2) is at least .7. Comparing these two points, the x-ray data
suggest a significantly higher concentration of α' martensite in the charged
304L specimen compared to the uncharged specimen which was strained to a
higher level. Similar x-ray diffraction scans on both charged and uncharged
304L notch specimens failed to detect any α' peaks at distances within 0.36
mm (.014") of the fracture surface.

The inability to detect a α' in notched specimens in the x-ray
diffraction tests suggests the possibility that α' formation is confined
to the near fracture surface region and is not easily analyzable by the
x-ray diffraction technique. The local structure was examined using a reflection diffraction technique(19) which is sensitive to phases located in
the first few microns of the fracture surface. Both smooth and notched
fractures were examined. The smooth specimens were used as references to
check the reflection diffraction technique since x-ray measurements clearly
showed the presence of α' near the fracture surface. Interestingly, all
four fracture surfaces exhibited the presence of α' indicated by BCC rings
in the reflection diffraction pattern (Figure 5). The diffraction patterns
contained both FCC and BCC rings. Since the diffraction measurements are
not sensitive to differences between BCC α iron and BCC α' martensite, it
was not clear which phase was responsible for the BCC rings. Consequently,
thin foils were removed from the unstrained grip section and found to be
fully austenitic. Therefore, no α iron was present in the as-received
plate and thus the BCC rings are due to the presence of α' martensite. It
appears that in both charged and uncharged notched 304L specimens, the α'
martensite formed during straining is confined to the near fracture surface
region.

Discussion

The observation of α' formation in the notched samples in the near
fracture surface areas accompanied by dimpled rupture even in the presence
of high levels of internal hydrogen and high triaxial stresses, allow a

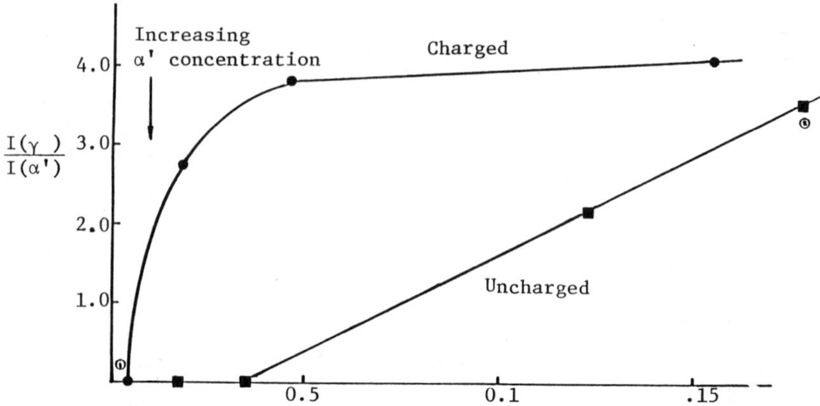

Figure 4.   X-ray diffraction ratio, $I\gamma(111)/I\alpha'(110)$ from charged and uncharged 304L smooth specimens as a fraction of distance from the fracture surface.

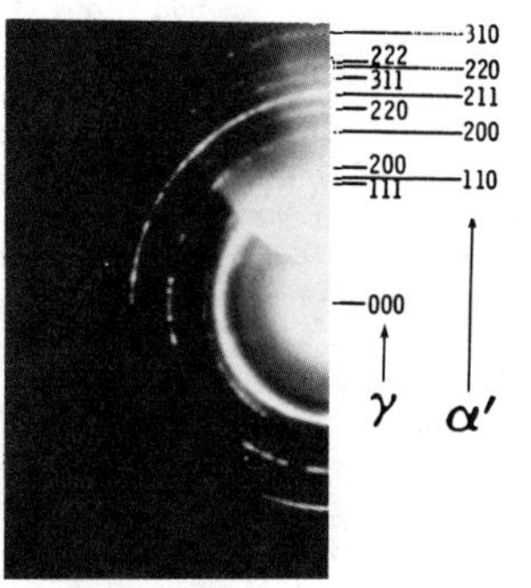

Figure 5.   Reflection diffraction pattern typical of that observed in both smooth and notched, hydrogen charged 304L specimens showing the presence of α' martensite.

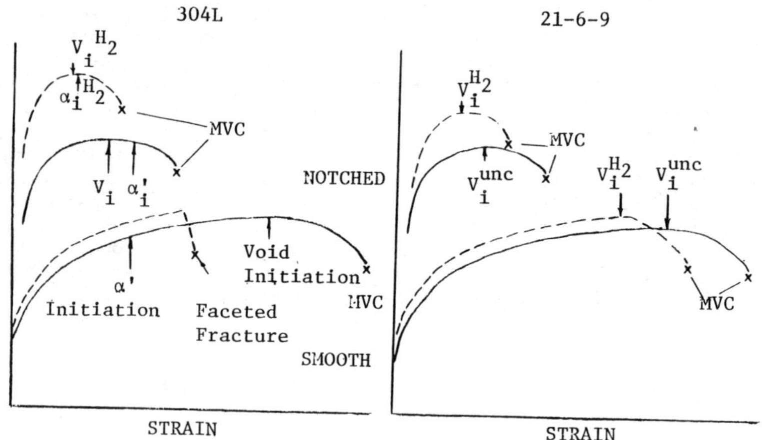

Figure 6.   Schematic stress-strain curves which describe the competitive fracture process in hydrogen stainless steel.

new interpretation of the role of α' martensite in the degradation of 304L
stainless steel. These results suggest that α' martensite only contributes
significantly to degradation when it forms well before fracture occurs, as
in a smooth bar test. In smooth specimens where the material flow is un-
constrained, α' formation occurs early in the plastic deformation history
providing new surface to trap hydrogen. Further, since BCC martensite is
sensitive to hydrogen(23), its presence early in the test allows hydrogen
to be transported to and trapped at this sensitive boundary. In the case
of notched samples, however, the constraint produced by the presence of
the notch reduces the local shear stresses and plastic flow such that α'
formation is retarded until void nucleation and growth intervenes. Strain
induced martensite formation in 304L is dependent on the deformation mode.
Specifically, shear deformation is required since α' formation appears to
proceed by the formation of epsilon martensite. This suggests that the
martensite transformation is strain, not stress, dependent and that the required strain level is not obtained in 304L notched specimens before microvoid initiation occurs. The level of uniform tensile strain required to
initiate α' in smooth specimens was approximately .3. Similar strain
levels were observed by Caskey(20) in smooth tensile specimens of 304L and
by Swearengen(24) in thin 304 torsion specimens. It follows then, that
the α' observed on the fracture surfaces of notched 304L specimens forms
after void initiation occurs because the stress to nucleate voids is
achieved before a strain of .3 is obtained. The free surface formed after
void initiation can then deform by shear processes and α' transformation
will follow. However, failure due to void coalescence occurs fairly
rapidly in the notch hydrostatic stress field before sufficient α' can form
and interact with the internal hydrogen to produce the faceted fracture
observed in charged smooth specimens. Therefore, hydrogen induces
ductility losses in notched 304L specimens by assisting void nucleation
and growth rather than by embrittling the α' martensite as in smooth specimens. This idea appears consistent with slow crack growth data presented
in these proceedings by Stoltz et. al.(11) and Perra(25). In particular,
no indication of slow crack growth was observed at stress intensities of
50 and 70 MPa$\sqrt{m}$ after approximately 10 months of exposure to 210 MPa
hydrogen.

The question of whether or not hydrogen suppresses the formation of α'
martensite has been examined in several previous investigations. Caskey(19)
observed no significant effect of hydrogen on α' formation at 248° and 298°K
and measured a significant retardation at 198°K. Caskey's samples were calculated to contain 200 mol/m^2 (3000 appm) of hydrogen. Thompson(26) also
reported no effect at room temperature and low hydrogen concentrations
($\sim$30 mol/m^2 or 450 appm). Further, Burke(27) found no change in the martensite start temperature (M_s) in several metastable austenitic steels.
These results suggest that only at low temperature does hydrogen affect α'
formation and there a significant retardation is observed. However, a
closer examination of Caskey's results suggests an alternate conclusion.
At both 248 and 298°K, his curves show consistently increasing concentrations of α' with strain level in charged compared to uncharged specimens.
At both temperatures, the magnetic readings are not only higher in charged
specimens but exhibit an increasing difference over uncharged specimens
with increasing strain. This increase with strain suggests hydrogen assists
the α' formation near room temperature. Additional support for this conclusion is available in a stress corrosion study by Asaro(28) et al. In
chloride stress corrosion tests of 304L at 154°C, α' martensite was
observed on the fracture surface using reflection diffraction techniques
while specimens pulled to fracture at 154°C in air exhibited no α'
associated with the fracture. Since 154°C is above the 304L M_d temperature,

the formation of α' in chloride solutions is likely due to hydrogen produced at the tip of the stress corrosion crack. The x-ray diffraction results reported in this investigation offer a similar conclusion.

At this time, it is unclear why hydrogen causes larger flow stress increases in notched specimens compared to smooth ones. Only a 10% increase would be expected from the increase in yield strength observed in charged smooth specimens. Further, we are unable to explain the larger increase in 304L notched specimens compared to 21-6-9 notched specimens although both alloys exhibit similar flow stress increases in charged smooth specimens. An explanation associated with the slip characteristics of each alloy is possible but requires further support. The higher stacking fault energy of 21-6-9 may reflect an ability to more easily accomodate a particular plastic stress state than 304L. Slip or deformation accomodation of the applied loads is more difficult in 304L which could be viewed as a resistance to strain accommodation or alternatively an increase in effective flow stress.

Fracture in hydrogen charged austenitic stainless steels can be viewed as a competition between several fracture processes. In austenitic stainless steels, uncharged specimens typically fail by ductile microvoid coalescence while other processes such as enhanced void growth, fracture of α' martensite, and decohesion of internal boundaries (grain boundaries, twin boundaries $\alpha'-\delta$ interfaces, inclusion matrix interfaces) occur in hydrogen charged specimens. The specific process that controls fracture in hydrogen depends on many factors including hydrogen concentration, stress state, stacking fault energy, strain rate, test temperature, composition, and microstructure among other variables. Competitive failure process in notched and smooth tensile specimens can be illustrated by noting where a specific mode initiates during the test (Figure 6). In 304L, the competition between microvoid coalescence and α' associated failure is shown for both charged and uncharged conditions. In smooth specimens, microvoid coalescence is favored in the absence of hydrogen. However in the presence of hydrogen, failure associated with the α' martensite occurs before void initiation and controls the fracture of 304L. In 21-6-9, no other failure processes intervene, and microvoid coalescence, assisted by hydrogen, controls the fracture.

The effect of stress state can influence the fracture process by assisting or retarding a specific failure process. Clearly, hydrostatic stress fields assist void growth and the presence of a notch should accelerate sensitivity to hydrogen in materials where microvoid coalescence controls fracture in hydrogen as in notched 21-6-9 specimens. However, since the transformation of γ austenite to ϵ and α' martensite requires shear deformation, the formation of α' is effectively retarded by the high hydrostatic stress field in the notch. This retardation, therefore, reduces the hydrogen $-\alpha'$ failure process which occurs in smooth specimens and allows fracture by microvoid coalescence to proceed as in uncharged notch specimens (Figure 4c). Only after void initiation does sufficient shear deformation occur to permit extensive strain-induced α' transformation. However, at this stage in the test void growth predominates and the effect of the formation of α' near the fracture region is minimal.

The results of the notch study suggest that reconsideration of the validity of short term tests to qualify materials for hydrogen or tritium service is in order. Certainly the observation of ductile failure processes in saturated 304L notched specimens and that of no slow crack growth in 210 MPa in WOL specimens suggest that 304L is a compatible alloy. In fact, in the absence of α' martensite 304L appears to be hydrogen

compatible. Further, since hydrogen effects tend to manifest themselves
in a reduction of post necking ductility in short term tests, the use of
forced overload (tensile) tests to assess compatibility is somewhat questionable. Therefore, hydrogen embrittlement theories which use short term
data for validation should only be applied to long term engineering
applications with extreme caution. Clearly, the use of more "static"
long term tests in future hydrogen degradation studies is warranted.

Conclusions

Examination of two hydrogen saturated austenitic stainless steels has
resulted in the following conclusions:

1. In sufficient concentrations, hydrogen increases the flow stress
in both smooth and notched specimens of 304L and 21-6-9.
2. Hydrogen assists the strain-induced transformation of α' martensite.
3. The α' transformation is strongly retarded by a hydrostatic stress
field.
4. 304L deforms and fractures in a ductile manner in the presence of
internal hydrogen when the α' transformation is retarded.

References

1. J. K. Tien: Effect of Hydrogen on Behavior of Materials, p. 309,
The Metallurgical Society of AIME, NY, 1976.
2. R. C. Frank: Internal Stresses and Fatigue in Metals, p. 411,
Elsevier Press, NY, 1959.
3. P. Bastien and P. Azon: C. R. Acad. Sci. Parris, 1951, vol. 232,
p. 1845.
4. P. Bastien and P. Azon: Proc. First World Metallurgical Congress,
p. 535, ASM, Metals Park, OH, 1951.
5. J. K. Tien, A. W. Thompson, I. M. Bernstein, and R. J. Richards,
Met. Trans. A., 1976, vol 7A, p. 821.
6. C. D. Beachem: Met. Trans. 1972, vol. 3, p. 437.
7. J. A. Donovan: Met. Trans. A., 1976, vol. 7A, p. 145.
8. H. K. Birnbaum, Private Communication, Sandia National Laboratories,
Livermore, CA, 1979.
9. R. F. Vennett and G. S. Ansell: Trans ASM, 1967, vol. 60, p. 242.
10. B. C. Odegard, J. A. Brooks, and A. J. West: Effect of Hydrogen
on the Behavior of Materials, p. 116, The Metallurgical Society of AIME, NY,
1976.
11. B. C. Odegard and A. J. West: Mater. Sci. Eng., 1975, vol. 19,
p. 261.
12. J. H. Holbrook: Private Communication, Sandia National Laboratories, Livermore, CA, 1979.
13. T. Doniszenski and G. C. Smith. Acta Met., 1963, vol. 11, p. 165.
14. see H. K. Birnbaum: this volume.
15. C. R. Caskey: Scripta Met., 1977, vol. 11, p. 1077.
16. R. E. Stoltz and A. J. West: this volume.
17. M. R. Louthan, Jr., and R. G. Dinick: Corrosion Science, 1975,
vol. 15, p. 565.
18. M. R. Louthan, Jr., J. A. Donovan, and G. R. Caskey, Jr.: Nuclear
Technology, 1975, vol. 26, p. 192.
19. J. S. Holliday and R. C. Newman: Brit. J. Appl. Physics, 1960,
vol. 11, p. 158.
20. G. R. Caskey, Jr.: Environmental Degradation of Engineering
Materials, p. 437, Virginia Polytechnic Institute, Blacksburg, VA., 1977.

21.   see J. H. Holbrook and A. J. West: this volume.

22.   A. J. West and M. R. Louthan, Jr., <u>Met. Trans A</u>, 1979, vol. 10A.

23.   R. A. McCoy and W. W. Gerberich: <u>Met. Trans</u>, 1973, vol. 4, p. 539.

24.   J. C. Swearengen: Private Communication, Sandia National Laboratories, Livermore, CA, 1980.

25.   see M. Perra: this volume.

26.   A. W. Thompson and O Buck: <u>Met. Trans. A</u>, 1976, vol. 7A, p. 329.

27.   J. Burke, A. Jickels, P. Maulik, and M. L. Mehta: <u>Effect of Hydrogen on Behavior of Materials</u>, p. 102, The Metallurgical Society of AIME, NY, 1976.

28.   R. Asaro, A. J. West and W. A. Tiller: <u>Stress Corrosion Cracking and Hydrogen Embrittlement of Iron-Base Alloys</u>, p. 1115, NACE, Houston, TX, 1977.

DISCUSSION

<u>Not Identified</u>: The difference in energy between γ and α' is likely to be small (see W.D. Wilson); this is presumably why the shear modulii c_{ij} are found experimentally to be considerably lowered near a phase transformation (for cubic, i=j=4). Hydrogen may enhance the phase transformation (thus c_{ij} would be expected to be lowered, making the shearing process to get α' easier). Is there evidence that C_{44} for γ 304 SS is lowered by hydrogen?

<u>A.J. West</u>: No.

EFFECTS OF INTERNAL AND ENVIRONMENTAL HYDROGEN

ON FATIGUE CRACK GROWTH RATE IN METASTABLE AUSTENITIC STRUCTURE.

H. Andriamiharisoa [+], M.Habashi [++], S. Talbot-Besnard [++],
J. Galland [+++], P. Azou [+].

+   Ecole Centrale des Arts et Manufactures, Chatenay-Malabry, France.
++ Centre National de la Recherche Scientifique, France.
+++ Université Paris XI, France.

Introduction of hydrogen before or during fatigue tests into austenitic Fe-23% Ni-0.47%C alloy and 304L austenitic stainless steel, does not
produce the same crack growth rate da/dN at the same stress intensity
factor ΔK. The introduction of hydrogen before fatigue test (internal hydrogen) is achieved by means of molten salt bath electrolysis techniques at
300°C (572°F). The quantity of desorbed hydrogen are about 40 ppm for 304L
and Fe-23% Ni-0.47%C alloy. The cathodic electrolysis during fatigue test
(environmental hydrogen) is produced at room temperature with a current
density of 5mA/cm^2, in an aqueous solution of HCL (N) + N_2H_4(4g/1). The
quantity of desorbed hydrogen is too low to be measured, while the gradient
of hydrogen concentration within the crack tip is high. In this study,
single-edge notched specimen 10 x 25 x 200mm and symmetrical center-cracked
specimen 5 x 50 x 400mm are tested.

The results show that, environmental hydrogen increases fatigue crack
growth rate in the both metals and specially at small values of ΔK. Internal
hydrogen decreases this rate in the case of 304L austenitic stainless steel
and does not alter crack growth rate in Fe-Ni-C austenitic alloy, with
respect to that of the metal without hydrogen.

Measurement of the volume percentage of martensitic phase α', by X-rays
techniques, on the surface of fatigue crack propagation, shows that environmental hydrogen reduces the martensitic transformation independently of ΔK
values. On the other hand, internal hydrogen does not change this transformation with respect to that measured in the virgin metals.

The reduction of α' percentage is more pronounced when cathodic
polarization potential at crack tip is hight (-0.7to 3 volts/SCE).

These results suggest that it is important to distinguish between
environmental and internal hydrogen because their influences on fatigue
crack propagation are different.

Introduction

Fatigue crack growth rate versus cyclic stress intensity factor ΔK had been studied on many steels in function of such parameters as : cyclic stress, mean stress, stress ratio, frequency, waveform. ... Several crack growth dependent rate laws had been established without taking care or controlling the environmental parameters. Recent works have shown that fatigue crack growth rate is very sensitive to those parameters and the rate evolution with respect to ΔK depends on metal-environment system. Corrosion fatigue studies through the use of fracture mechanics have led to an improved understanding of interactions (1, 2), superposition (3) or competition (4) between mechanical and environmental variables which control corrosion fatigue damage. The threshold K_{1scc} is found to be an important point to separate between two mechanisms : when applied $\Delta K/(1-R)$ is below K_{1scc} (R, stress ratio), crack growth rate is enhanced by a synergistic action of corrosion and cyclic loading — true corrosion fatigue (4) ; at $\Delta K/(1-R)$ values greater than K_{1scc}, crack growth rate evolution describes static load stress corrosion under fatigue conditions (stress corrosion fatigue).

In an inert environment, Paris law $da/_{dN} = C (\Delta K)^m$ describes perfectly fatigue crack growth per cycle evolution in function of ΔK. In true fatigue corrosion, the response curves $da/_{dN} = f(\Delta K)$ in inert and corrosion environment do not cross excepting at ΔK high value. When the material does not exhibit environmental enhancement until K_{1scc} is reached, the mechanism is stress corrosion fatigue and is characterised by a "bump" (4) with a plateau rate as observed in stress corrosion crack growth. Almost all materials exhibit, at the same time, true fatigue corrosion and stress corrosion fatigue.

It is now established that hydrogen embrittlement may occur in metastable austenitic structures (5, 6) in presence of sufficiently high hydrogen concentration. It has been reported (5) that no hydrogen embrittlement occurs in Fe - Ni - Cr alloys with chromium content of about 18% and nickel content between 15 to 35%. Below or above these contents the material is embrittled. The embrittlement at low nickel content is connected to $\gamma \to \varepsilon \to \alpha'$ transformation of the unstable austenite during plastic deformation (7, 8). In austenitic Fe - Ni alloys (9) cathodically hydrogenation arises partial transformation of austenite to martensite (ε and α'). The mechanism of this transformation and the susceptibility to hydrogen embrittlement is influenced also by nickel content : at nickel content lower than 31%, the martensitic transformation is induced by local lattice distorsion due to interstitial hydrogen atoms. Also, hydrogen in solid solution eliminates the martensitic transformation and lowers M_s when hydrogenated metal is quenched at low temperatures (9).

In this work, we have studied crack growth enhancement with respect to ΔK in cathodically hydrogenated (environmental hydrogen) metastable austenitic structure. This structure has low yield strength σ_y ; with plastic deformation at crack tip, α' transformation occurs, and σ_y increases. Generally, a reaction occurs between an aqueous environment and a freshly exposed crack tip surface leading to the adsorption of atomic hydrogen which diffuses within crack tip plastic zone, i. e. in martensitic phase. When hydrogen is previously introduced by energetic way (by example at high temperature) in the austenitic structure before fatigue test (internal hydrogen), the crack tip surface reaction is not active and then only hydrogen within crack tip zone will be responsable of crack growth rate variation. Another factor which may change corrosion fatigue damage rate is the polarization potential in relation with ΔK value at constant electrochemical conditions, frequency and waveform.

Experimental procedure

The present work has been carried out using two metastable austenitic alloys, Fe – Ni – C and 304 L stainless steel. The chemical analysis are given in table 1.

Table 1 : a) chemical analysis of Fe – Ni – C alloy.
b)   "   "   "   304L stainless steel.

a)

% C	Mn	Si	P	S	Ni	Fe
0.47	0.43	0.41	0.016	0.017	22.8	balance

b)

% C	Mn	Si	Ni	Cr	Mo	Ti	Fe
0.023	1.74	0.38	9.36	18.30	0.05	–	balance

Specimens of single-edge notched plate (tension) of 10 x 25 x 200 mm are used in the case of internal and environmental hydrogen, while symmetrical center-cracked plate (tension) 5 x 50 x 400 mm are machined to study the influence of polarization potential variation on crack growth rate evolution with respect to ΔK. Cyclic stress intensity factors are calculated using the relation $K = \Delta \sigma_y \sqrt{a}$, with $\Delta \sigma$ cyclic stress (MNm^{-2}) , Y compliance function, W specimen width (mm) and a total crack length (mm).

For single edge notched plate :

$$Y = 1.99 - 0.41 \, (a/w)^{1/2} + 18.70 \, (a/w)^{3/2} - 38.48 \, (a/w)^{5/2} + 53.85 \, (a/w)^{7/2}$$

For centre cracked plate :

$$Y = \left(\frac{w}{\Pi} \, \tan \, \frac{\Pi a}{w} \right)^{1/2}$$

All specimens undergone heating at 1050°C during two hours in an inert atmosphere (Argon) and then water quenched. Before hydrogen cathodic charging and fatigue tests, the involved surface is carefully polished and the notch length is measured up to $\pm$ 5 m.

Environmental hydrogen is achieved during fatigue tests, at room temperature (20°C) by applying constant current density equal to 5mA/cm^2 and in aqueous solution $HCL(N)+N_2H_4 (4g/1)$ .Internal hydrogen is preliminary introduced by means of molten sults bath electrolysis technique at 300°C (10). A constant polarization potential equal to −2volts with respect to silver reference electrode (10) is acheived during two hours. Hydrogen desorbed quantities, measured under vacuum, are about 40 ppm for 304L stainless steel and Fe-23% Ni-0.47%C alloy.

Fatigue tests has been carried out using a 200KN electro-hydraulic machine, under constant cyclic charge, with 1 Hz frequency and sine wave-form signal. Stress ratio R values are 0.05 for 10mm specimen width (B), and 0.11 for B = 5mm.

Crack length is measured using optical method with $\pm$ 30μm. Crack growth rate and ΔK are calculated by means of polynomial and deriving function program with correlation coefficient greater than 0.990.

Results and discussion.

Environmental hydrogen enhances crack growth rate in both 304L aus-tenitic stainless steel and Fe-23% Ni-0.47%Caustenitic alloy, figures 1 and 2 . In all cases, environmental hydrogen is active at low ΔK levels, special-ly in stage I, either the specimen thickness is 5 or 10mm. The threshold value in presence of environmental hydrogen $K_{I.S.H.}$ (K_{ISCC}) is very high for austenite structure. These results show no crossing between inert and envi-ronmental curves within $\Delta K \leqslant$ 40 MNm$^{-3/2}$. Crack growth rate is enhanced by synergistic action:hydrogen, cyclic loading. Fatigue damage is achieved by true corrosion fatigue (4). On the other hand, enviromnental hydrogen has no effect on crack growth rate, at frequencies 0.5Hz and 1Hz in an 316 stable austenitic steel in which no martensitic transformation occurs. On the op-posite direction, internal hydrogen may reduce crack growth rate in the case of 304L austenite, figure 1, and does not alter this rate for Fe-23% Ni-0.47%C

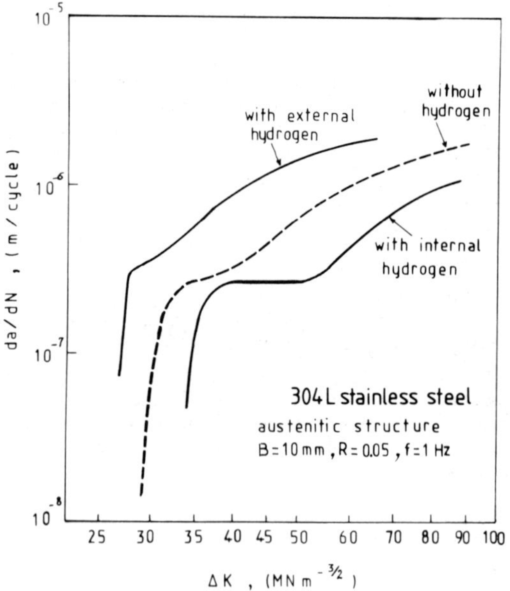

Fig.1 : Internal and environmental hydrogen effect on crack growth rate evolution w.r.t ΔK. Austenitic stainless steel 304L.

alloy, figure 2. One can conclude that internal hydrogen in metastable austenite and environmental hydrogen in stable austenite do not increase crack growth rate and may reduce it with respect to virgin metal. Introducing

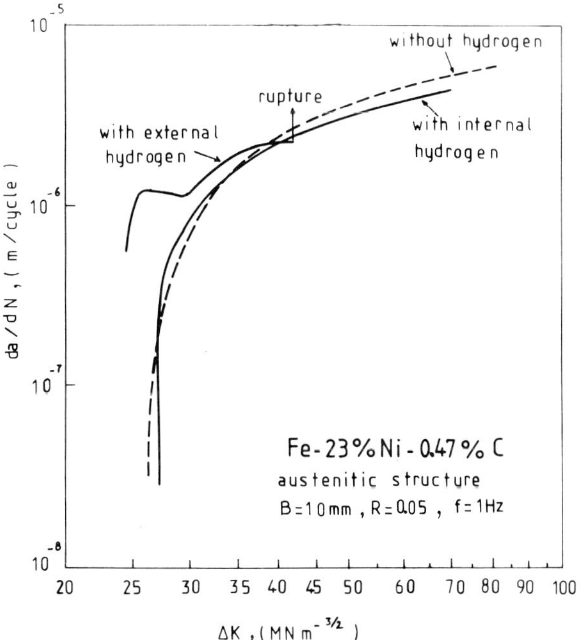

Fig. 2 : Internal and environmental hydrogen effect on crack growth rate evolution w.r.t ΔK. Fe-23% Ni-0.47%C, austenitic alloy.

hydrogen in the austenite at high temperature (300°C) induces energetic traps and, during fatigue test, interstitial hydrogen atoms, which are responsable of α' suppression (9), are desorbed and trapped hydrogen has no action on crack growth rate evolution. In the case of 304L, it reduces this rate : this reduction is due to high local plasticity promoted by those traps at crack tip and on the other hand to the stabilisation effect of hydrogen on austenite structure.

Environmental hydrogen introduces high interstitial gradient concentration at the crack tip during applying cyclic stress, consequently while $\gamma \rightarrow \alpha'$ transformation occurs. For austenitic stainless steel, Inoue and al. (11) has found that the plastic strain induced around crack and the work required for fracture along interface $\gamma \longrightarrow \varepsilon$ martensite decrease drastically as the hydrogen content increases ; the $\gamma \longrightarrow \alpha'$ transformation is restrained while ε formation is promoted by lowering stacking fault energy of austenitic structure.

To verify such new data we have measured quantitatively α' percentage transformation by means of X-ray technique : $(111)_\gamma$ and $(110)_{\alpha'}$ peak intensities are measured on the surface fracture, after fatigue test, at different crack lengths using $\lambda_{K\alpha}$ ray of cobalt. (12). Figure 3a and 3b show that α'

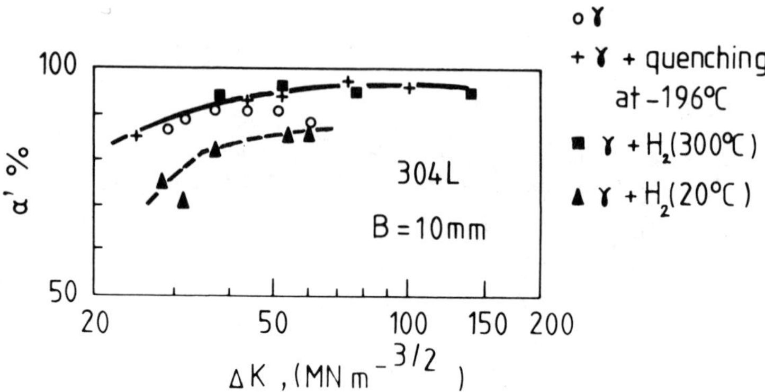

Fig. 3a : α' percentage evolution w.r.t ΔK in the cases : internal and environmental hydrogen – Austenitic stainless steel 304L.

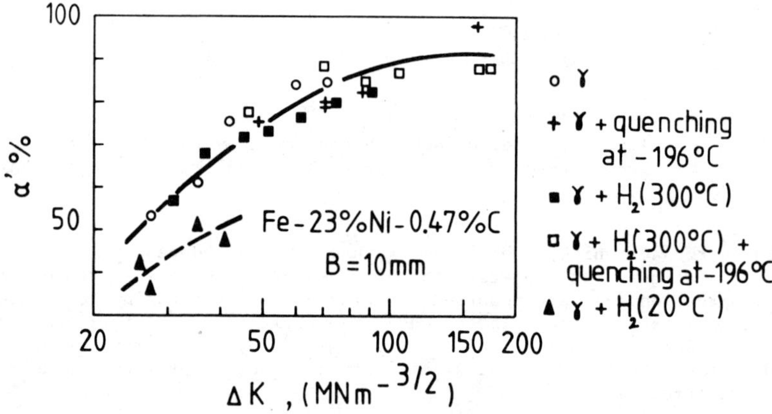

Fig. 3b : α' percentage evolution w.r.t ΔK in the cases : internal and environmental hydrogen. Fe-23% Ni-0.47%C austenitic alloy.

percentage increases with ΔK level and does not change with quenching at -196°C or internal hydrogen charging in the austenitic structure. On the opposite side, either 304L or Fe-23% NiO.47%C, environmental hydrogen decreases α' transformation.

Without hydrogen or with internal hydrogen condition, $\gamma \rightarrow \alpha'$ transformation within plastic zones is shown in figure 4a. Environmental hydrogen induced crack branching and $\gamma \rightarrow \alpha'$ transformation is observed mainly along these cracks, figure 4b.

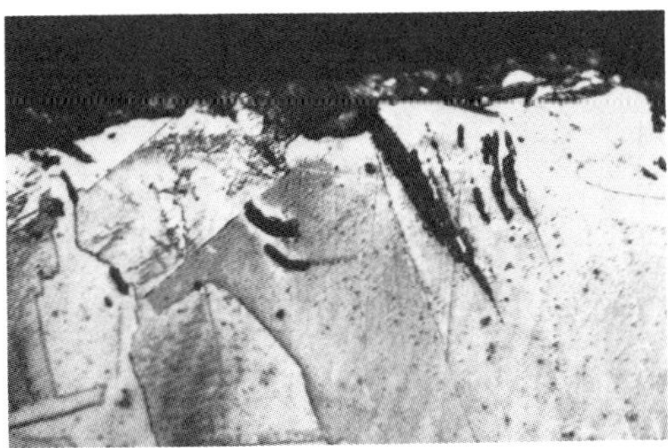

Fig. 4a : $\gamma \rightarrow \alpha'$ transformation within plastic zones in the case of internal hydrogen.

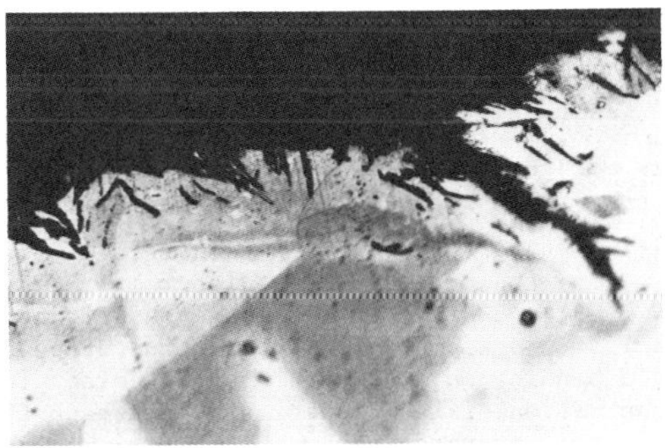

Fig. 4b : $\gamma \rightarrow \alpha'$ transformation within plastic zones and crack branching in the case of environmental hydrogen.

Generally it has been proved difficult to distinguish between slip dissolution and hydrogen embrittlement mechanisms in aqueous environments because they have many common features ; it has been suggested that the mechanism operating crack growth in a given material may depend on the applied potential (13)

With this background, we have studied the influence of cathodic potential on crack growth rate for 304L austenitic stainless steel in an aqueous solution $HCL(N) + N_2H_4(4g/1)$, with pH ≈ 1. The values of cathodic polarization potential with respect to saturated calomel electrode are 0.7, 1.4, 2.2 and 3.0 volts. The first value lies in region where slip dissolution and hydrogen embrittlement mechanisms may occur (14). Figure 5, shows that when cathodic polarization potential is applied, crack growth rate is enhanced for ΔK values varying between 17 and 25 $MNm^{3/2}$ This enhancement depends on polarization potential,(V). It increases firstly when(V) = 0.7V/SCE, i.e when the two mechanisms are operative. From 1.4 V/SCE to higher cathodic poten-

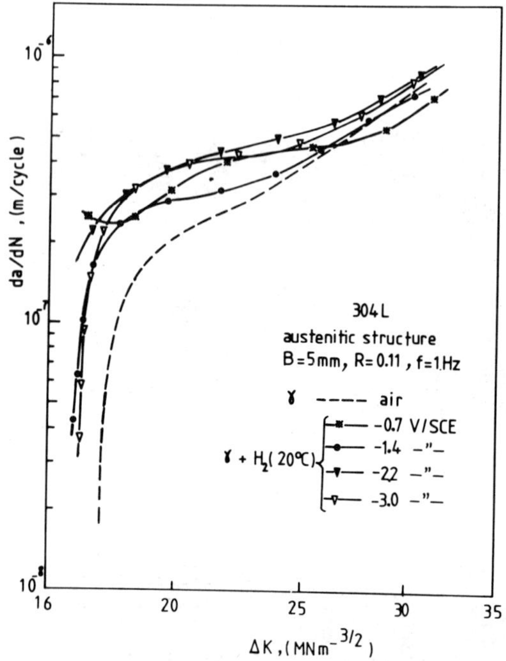

Fig. 5 : Surface cathodic polarization potential effect on crack growth rate evolution. Austenitic stainless steel 304L.

tial, hydrogen embrittlement increases crack growth rate ; this is due to increasing hydrogen fugacity and consequently higher hydrogen concentration. ATEYA with PICKERING (15) and MELVILLE (16) have shown that when applying polarization potential at crack mouth, some potential drop in noble direction occurs at crack tip, depending on crack tip opening, crack length, current density, pH and concentration of ionic species H^+ at crack tip. At larger applied cathodic potential, hydrogen gas bubbles are initiated at crack tip at which the effective cathodic potential is lower than the surface one. ATEYA and PICKERING (15) have measured the variation of potential in slot in iron and nickel under very hard hydrogen charging conditions in 1M $HClO_4$. The slot has 0.5mm wide and 12mm depth with a cathodic current density 5.2mA/cm². The potential drop measured at 1mm deeper to slot mouth was 0.1 volt, while it was about 0.4 volt when measurement has been achieved at 12mm deeper. In our case, potential drop must be more pronounced at larger surface applied potential (V≥2.2. volts/SCE) and at higher ΔK value, and one can expect that crack growth rate is no more increased at high applied cathodic potential values.

γ ⟶ α' transformation, in function of cathodic polarization potential and ΔK value, figure 6, is reduced when crack tip cathodic polarization potential increases showing that higher absorbed hydrogen quantity in austenitic structure restrains γ ⟶ α' transformation (11).

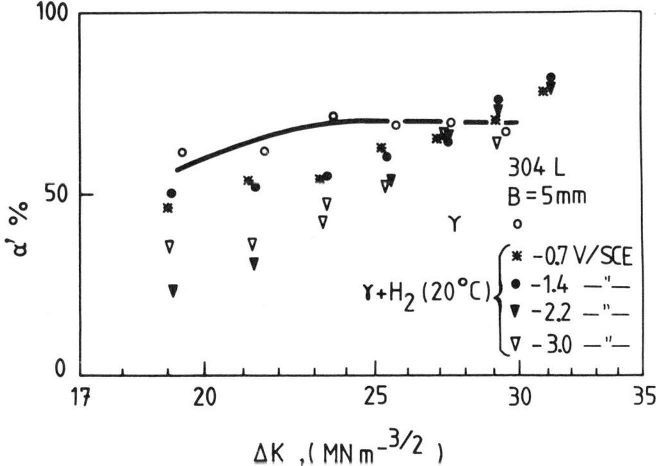

Fig. 6 : γ ⟶ α' transformation within plastic zones and evolution with surface cathodic polarization potential.

Conclusions.

Introduction of hydrogen before or during fatigue tests into aus-tenitic Fe-23% Ni-0.47%C alloy and 304L stainless steel, do not produce the same crack growth rate at the same ΔK value. The introduction of hydrogen before fatigue test (internal hydrogen) is achieved by means of molten salts bath electrolysis technique at 300°C (572°F). Desorbed hydrogen quantity is about 40 ppm for both metals. The cathodic electrolysis during fatigue test (environmental hydrogen) is performed at room temperature with a current density of 5mA/cm2, or with cathodic polarization potential varying between 0.7 to 3 volts/SCE in an aqueous solution of HCl(N) + N_2H_4 (4g/1). Desorbed hydrogen quantity is too low to be measured, while hydrogen gradient con-centration at crack tip is high. In this study, single-edge notched 10 x 25 x 200mm and symmetrical centered cracked 5 x 50 x 400mm specimens are used.

From the results obtained, we can present the following conclusions:
1) Environmental hydrogen increases fatigue crack growth rate in both Fe-23% Ni-0.47%C and 304L austenitic structure and specially at small ΔK values. Crack growth process is similar to true fatigue corrosion one.

2) Internal hydrogen decreases this rate in the case of 304L auste-nitic stainless steel and does not alter fatigue crack growth rate in Fe-Ni-C austenitic alloy, with respect to metal without hydrogen.

3) In the case of environmental hydrogen and on the external lateral surface within crack tip, we observe crack branching, which was absent in internal hydrogen case.

4) At dissolution potential or when cathodic polarization potential at crack tip increases, crack growth rate is enhanced within ΔK values between 17 and 26 MNm⁻³/² . High potential drop at crack tip, due to hydrogen bubbles formation may reduce this rate.

5) Measurement of volume percentage of martensitic phase α', by x-ray

technique on the fracture surface, shows that environmental hydrogen restrains $\gamma \longrightarrow \alpha'$ transformation. This reduction is more pronounced when increasing crack tip cathodic polarization potential. On the other hand, internal hydrogen does not change this transformation with respect to that measured in the virgin metal. In the first case, interstitial hydrogen promotes ε martensite formation and reduces rupture energy at crack tip, consequently α' is reduced. In the second one, trapped hydrogen does not change $\gamma \longrightarrow \alpha'$ formation mechanism.

Références

1. J.M. BARSON - Eng. Fract. Mech., (1971),vol.3, p.15.

2. J.D. ATKINSON and T.C. LINDLEY - I. Mech. Conf., LONDON, (1977).

3. R.P. WEI and J.D. LANDES - Mat. Res. and Standards, (1969), vol.9,n°7.

4. I.M. AUSTEN and E.F. WALKER - I. Mech. Conf., LONDON, (1977).

5. R. LAGNEBORG - Journal of the iron and steel institut, (1969), March, p.363.

6. P.A. BLANCHARD and A.R. TROIANO - Mém. Sci. Rev. Métal (1960),57, p.409.

7. H. ANDRIAMIHARISOA, M. HABASHI, J. GALLAND and P. AZOU - 5th international conference on fracture, CANNES, FRANCE (1981).

8. K. KAMACHI and S. MIYATA - J. Soc. Mat. Sci, JAPAN, (1969), vol.18, p. 1140.

9. M.L. HOLZWORTH and M.R. LOUTHAN - Corrosion, (1968), vol.24, p. 110.

10 A. ELKHOLY, J. GALLAND, P. AZOU and P. BASTIEN - C.R. Acad. Sci, FRANCE, (1977) Vol.284, Série C, p. 363.

11 A. INOUE, Y. HOSOYA and T. MASUMOTO - Trans. ISIJ, (1979),vol.19, p. 170.

12 M. HABASHI - Thesis Doctorat ès Sciences, (1977), PARIS VI, FRANCE.

13 W.A. VAN DER SLUYS - Eng. Fract. Mech., (1969), vol.1, p. 447.

14 R.N. PARKINS - 5th symp. on line pipe research, (1974), Amer. GAZ ASSOC., cat. No. L 30174.

15 B.G. ATEYA and H.W. PICKERING - J. Electrochem. Soc., (1975), vol.122, n°8.

16 P.H. MELVILLE - Br. Corros., J., (1979), vol. 14, N°1, p.15.

THE EFFECT OF AUSTENITE STABILITY ON SUSTAINED LOAD CRACKING AND

FRACTURE MORPHOLOGY OF STAINLESS STEELS IN 1 ATMOSPHERE HYDROGEN

John J. Lewandowski and Anthony W. Thompson
Department of Metallurgy and Materials Science
Carnegie-Mellon University
Pittsburgh, PA 15213

Introduction

Austenitic stainless steels normally contain 15-25% Cr, 7-25% Ni and minor alloying elements such as C, Mn and Si. Although these steels exhibit predominantly austenitic microstructures at room temperature, the degree of austenite stability is highly dependent on composition (1). Metastable austenite may transform to ε(hcp) and/or α'(bcc) martensite during cold work, depending on the M_d temperature (2-5). Martensitic phases have also been detected after either thermal or electrochemical hydrogen charging (6-8).

Combining plastic deformation with exposure to a hydrogen environment has been shown to be detrimental to a number of austenitic stainless steels (9-15). However, the role of the martensitic transformation products is not clear. Some have concluded that α' is causative in the hydrogen embrittlement (HE) of austenitic stainless steels (10,11,16,17). Others have argued and attempted to show that α' is neither necessary nor sufficient to explain embrittlement in these steels (13,14,18,19), although α' may _enhance_ embrittlement (19,20).

Although studies of this phenomenon have continued, the exact role of the martensitic phase transformation has not been resolved. For example, sustained load cracking has been reported in various hydrogen environments, including 0.1 MPa gas, for thin (1.6 mm) edge-notched sheet tensile specimens of AISI 304 (21). Fracture surfaces were transgranular, brittle, and reportedly associated with the presence of α'. Similarly, sustained load cracking was observed in notched (ligament diameter 1.9 mm), sensitized AISI 304 in ~ 0.2 MPa hydrogen (17,22); the mode of fracture, however, was intergranular.

Similar testing of thicker specimens of AISI 304L at hydrogen pressures in excess of 172 MPa failed to show any sustained load cracking (13,23).

The present work was intended to investigate the effect of austenite stability on the behavior of stainless steels in 0.1 MPa hydrogen gas. In particular, the phenomenon of sustained load cracking (21,22) for an unstable stainless steel (AISI 304L) was investigated.

<u>Experimental Procedure</u>

Table I lists the compositions of the AISI 310 and the two heats (called "A" and "B") of AISI 304L which were tested. The true volume grain size (24) of each steel, which had been cold worked and annealed, was ~20 μm. After machining, 304L-B samples were also given sensitizing heat treatments of 1100°C/3 hrs/AC followed by 650°C/15 hrs/AC. This heat treatment causes severe sensitization as measured by standardized ASTM tests (25,26). The high temperature solution anneal (1100°C/3 hrs) increased the grain size in the 304L-B to ~180 μm.

<u>Table I.   Composition of Steels Tested</u>
(wt. %)

	Cr	Ni	Mn	C	N
304L-A	18.77	10.25	1.77	0.027	–
304L-B	18.50	11.70	1.72	0.020	0.054
310	24.39	20.82	1.35	0.040	–

	Si	S	P	Mo	Co
304L-A	0.52	0.008	0.012	–	–
304L-B	0.45	0.016	< 0.04	< 0.25	< 0.20
310	0.67	0.030	0.025	0.27	0.37

The tensile specimens were cylindrical, either 4.95 or 6.4 mm in diameter. Notched samples contained a 60° vee-notch with ligament diameters of 2.54 or 4.27 mm; the ratio of these diameters to the gross diameter was ~1/3 to 1/2. Notch root radii were about 0.10 mm. Control tests on each steel were conducted in air and vacuum (0.13 mPa or 1×10^{-6} torr) at room temperature. Environmental tensile testing was performed in 0.1 MPa (1 psig) high purity hydrogen (Linde Ultra-High Purity Grade – 99.999% pure, min.) at room temperature. The testing procedure involved evacuating a bakeable, stainless steel chamber to 0.13 mPa before backfilling the chamber with hydrogen gas. Two kinds of tests were performed in hydrogen. The first involved straining notched specimens to failure at a crosshead speed of 8.5×10^{-5} cm/s. The second type was sustained load cracking, which was conducted by straining notched samples in hydrogen at the same speed to a load corresponding to 80-90% of the notched UTS. The samples were then held at this load for 168 hrs (1 week) before continuing straining to failure.

Fracture surfaces of all samples were examined in the scanning electron microscope (SEM). X-ray diffractometry of fracture surfaces was performed to detect the presence of γ, ε and α' (4) since this technique provides a qualitative surface phase analysis. A Siemens diffractometer with either Co or Cu radiation, and graphite monochromator, was used for identifying phases on fracture surfaces.

<u>Results</u>

Table II summarizes the tensile data obtained. For notched samples a true strain of 0.0005 was used for calculating notched yield strengths. Table II indicates that the mechanical properties of each steel were independent of test environment. Further, no sustained load cracking was obtained.

Table II. Mechanical Properties of Steels Tested

304L-A	d/D	σ_y (MPa)*	UTS(MPa)	RA(%)	True (MPa) fracture σ
Air	1	442	669	81	1787
H_2(0.1 MPa)	1	442	683	80	1828
Air	.54	862	987	60	1573
Vacuum (0.13 mPa)	.53	904	1021	60	1546
H_2(0.1 MPa)	.52	821	959	59	1380
H_2(SLC)**	.53	–	987	60	1587
304L-B					
Air	1	276	600	75	1532
Vacuum (0.13 mPa)	.51	518	800	51	1208
H_2(0.1 MPa)	.53	573	814	54	1228
Vacuum (0.13 mPa)	.67	386	766	42	1028
H_2(0.1 MPa)	.69	366	780	42	1001
H_2(SLC)**	.68	–	787	45	1035
Sensitized 304L-B					
Air	.53	173	538	80	1615
Vacuum (0.13 mPa)	.53	373	697	61	1187
H_2(0.1 MPa)	.54	386	697	58	1132
310					
Air	1	531	690	70	1553
Air	.53	1139	1270	33	1380
Vacuum (0.13 mPa)	.52	1208	1318	29	1387
H_2(0.1 MPa)	.54	1159	1249	31	1346

*Based on 0.002 true strain (smooth: d/D=1) or 0.0005 true strain (notched).

**In sustained load tests, if no cracking was detected after 168 hrs (1 wk.) in hydrogen, specimens were strained to failure at the same crosshead speed used throughout this investigation.

Similarly, SEM fractography revealed no effect of test environment on fracture morphology. Figure 1 shows the fracture surface of notched 304L-A tested in vacuum. There appeared to be two fracture zones on the surface of this specimen, as well as on all other notched specimens. Figure 2 shows that the outer zone consisted of elongated shear dimples (27), typical of shear fracture. Matching surface fractography confirmed this as the fracture mode in the outer zone. The central zone exhibited equiaxed dimples as shown in Figure 3. These two-zone fracture surfaces are typical of "cup-cone" failures in notch-strengthened materials such as austenitic stainless steels.

Examination of notched 304L-A samples tested in hydrogen revealed features very similar to those in Figures 1-3. The relative proportion of each fracture zone, and the size and appearance of individual features (such as dimples) within each zone, were unchanged for samples tested in hydrogen.

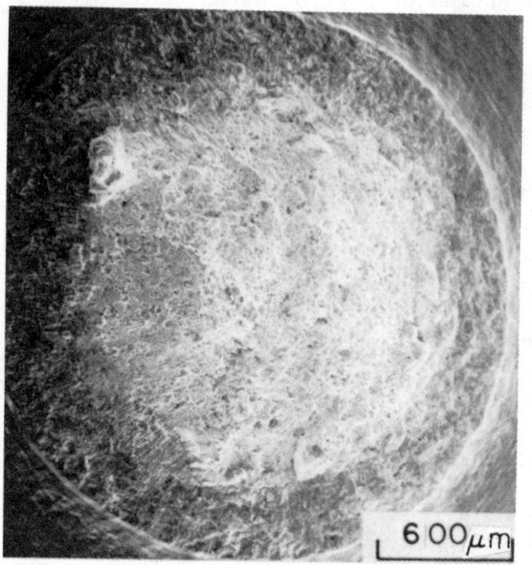

Fig. 1-Complete fracture surface of notched 304L-A tested in vacuum. Note outer "rim" of different fracture appearance.

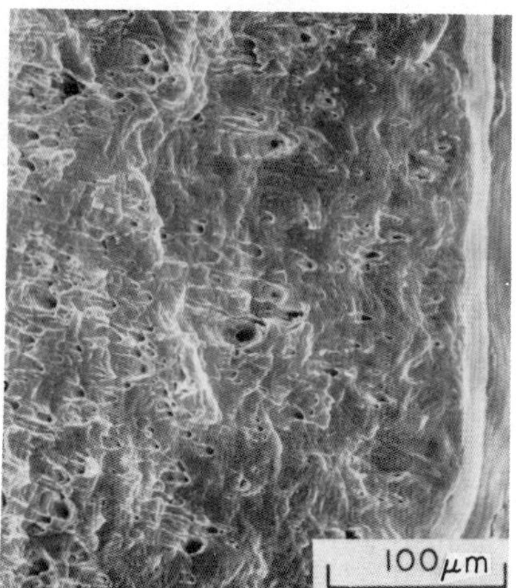

Fig. 2- Fractograph of notched vacuum 304L-A showing outer fracture zone (cf. Fig. 1) consists of elongated shear dimples.

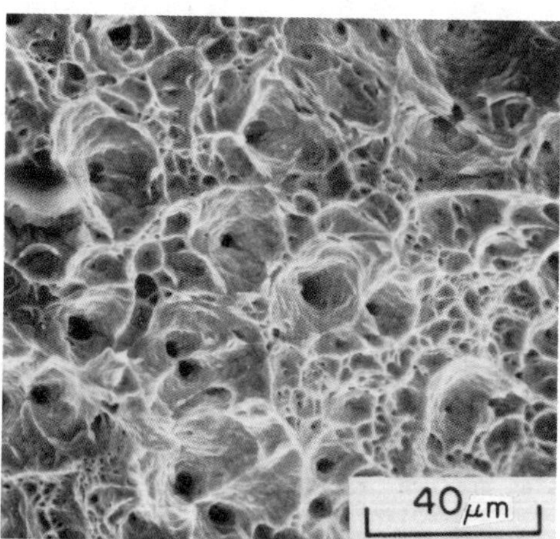

Fig. 3- Central fracture zone of Fig. 1 at higher magnification, exhibiting equiaxed dimples.

The 304L-B and 310 fracture surfaces similarly showed no effect of test environment. Both exhibited the two-zone fracture noted above, although the outer zone was somewhat smaller in extent. Typical outer zone widths in the 304L-A steel were ~350 μm compared to ~100 μm for the 304L-B and 310.

Results of fracture surface X-ray diffractometry were generally consistent with earlier work (2-6). Figure 4 is an X-ray trace for the notched 304L-A vacuum sample. Although considerable background exists, peaks for γ, ε, and α' phases can be discerned, as expected (2-6). Figure 5 presents the results for notched 304L-A in hydrogen; these data are similar to Figure 4. Further, peak heights and areas for each phase were compared between hydrogen and vacuum tests; there was no change with environment, which suggests that the relative amount of each phase was similar in vacuum and hydrogen. However, this measurement is only semi-quantitative due to the rough topography of the fracture surfaces.

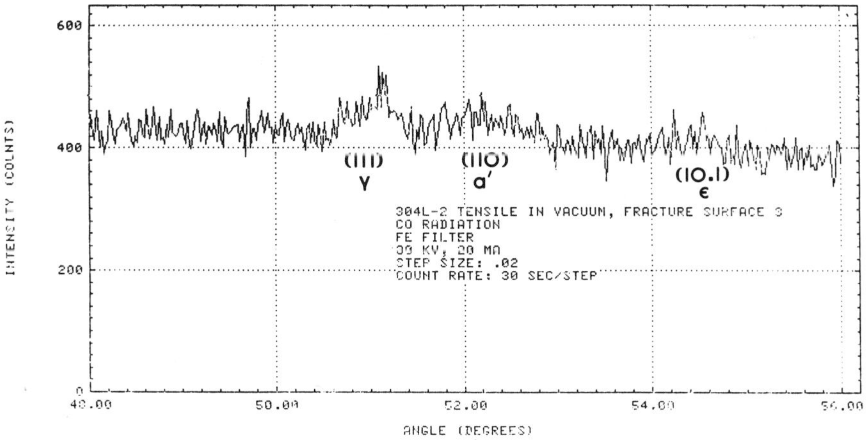

Fig. 4 X-ray diffractometer trace of notched 304L-A tested in vacuum showing γ,ε, and α' phases present on fracture surface.

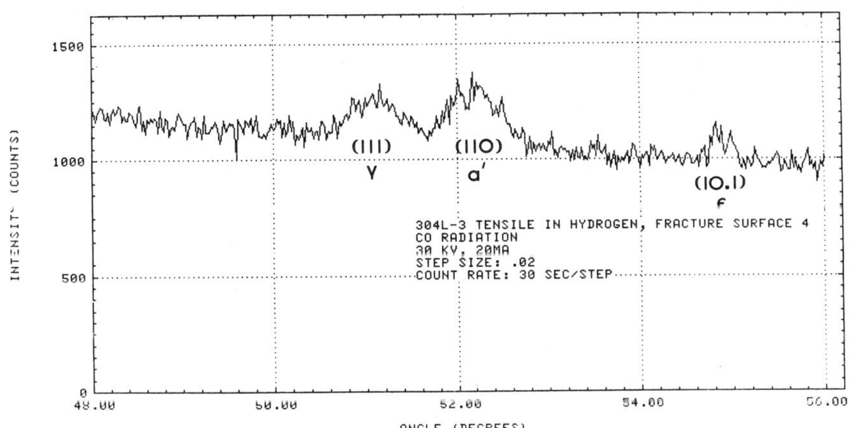

Fig. 5 X-ray diffractometer trace of notched 304L-A tested in 0.1 MPa hydrogen.

Diffractometer traces of 304L-B fracture surfaces indicated a wholly austenitic structure regardless of test environment (Fig. 6). This is somewhat surprising as 304L is normally considered unstable. However, examination of the composition of 304L-B in Table I shows a high Ni content (relative to 304L-A) and the presence of N, both of which evidently stabilize the austenite enough to prevent detectable decomposition to ε or α' during plastic deformation. It is of course possible that such decomposition occurred on a very small scale "at" the fracture surface (28); the present X-ray technique would not detect that. Diffractometry of 310 fracture surfaces also showed a wholly austenitic surface, in this case as expected.

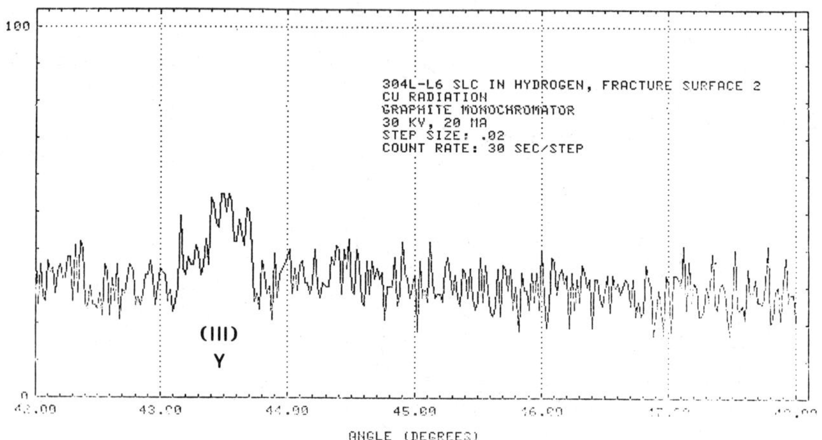

Fig. 6   X-ray diffractometer trace of notched 304L-B tested in 0.1 MPa hydrogen showed (within resolution of technique) a wholly austenitic fracture surface.

Sensitized 304L-B (1100°C/3 hrs + 650°C/15 hrs) similarly showed no effect of hydrogen on mechanical properties, fractography or diffractometry of fracture surfaces. The fracture surfaces of notched specimens were similar to those reported for as-received specimens, and were completely austenitic, as revealed by diffractometry.

Discussion

The fracture surfaces of specimens in this work exhibited a two-zone appearance. Others have observed the presence of two fracture zones when testing stainless steels in hydrogen environments (11,18). However, these investigators observed flat fractures, not "cup-cone" fractures, and identified the outer zone as resembling quasi-cleavage, evidently due to hydrogen embrittlement. This is clearly not the case in the present work. The outer fracture zone here is due to shear, typical in "cup-cone" type fractures.

Fracture morphology was independent of test environment for the steels tested, despite the presence of ε or α' in the 304L-A. This indicates that the formation of ε or α' does not necessarily cause or enhance embrittlement in 0.1 MPa hydrogen. However, this may not be true at higher hydrogen pressures.

The failure to obtain sustained load cracking in the unstable 304L-A
may seem puzzling since others have observed this in both solution annealed
and sensitized AISI 304 (16,21,22). 304L-A should be more susceptible than
304 because the removal of C additionally destabilizes the austenite. Yet
in this work, sustained loads up to 90% of the notched UTS failed to produce
cracking after 168 hrs (1 week) in 0.1 MPa hydrogen. Some previous work involved "load cycling" to compensate for testing system relaxation under load
(28). But in the present work, deliberate cycling of loads between 80 and
90% of the notched UTS before holding at 85% UTS was attempted, and similarly produced no cracking.

One clue to the behavioral differences is the specimen size effect.
Authors reporting sustained load cracking in 0.1-0.2 MPa hydrogen have
utilized either 1.6 mm sheet or notched round (ligament diam 1.9 mm) tensile
specimens (16,21,22) while the present work and earlier work with thicker
specimens in higher pressure hydrogen (13,23) revealed no sustained load
cracking.

The role of thickness in these tests may involve the ability of the
notched tensile specimen to deform and subsequently fracture by a shear process. As thinner samples are used, the constraint of plasticity by the triaxial stress field of the notch is reduced, and the relative amount of shear
deformation and fracture volume increases. Presumably the α' martensitic
transformation, being shear in character, can only take place above the M_s

in the presence of imposed shear stresses (2,3). As thickness is increased,
the converse holds, and less transformation to the martensitic phase is expected. Thus, use of thick or severely notched specimens may reduce the extent of α' formation and any accompanying exacerbation of hydrogen embrittlement. West and Holbrook (29) have made a similar suggestion.

One further consideration is the difference in the austenite stability
for each investigation. The earlier reports of sustained load cracking were
made for stainless steels containing 8.5-9.2 wt. % Ni (16,21,22), a more
unstable alloy than used in this investigation. The combined effects of
thinner specimens and less stable austenite appear to explain the differences
in results. This emphasizes not only the profound role played by microstructure (19), but also the fact that a variety of testing variables can affect
sensitivity to hydrogen.

It should be noted that the 304L-A results support the contention that
the presence of martensite in an austenitic stainless steel is not, of itself,
sufficient for a manifestation of HE. Since it appears that no martensite
is formed in some reported cases of HE in these steels (13-15), including
examples of sustained load cracking in 316 stainless (30), it continues to
seem correct to assert, as in earlier cases, that martensite may assist or
exacerbate HE, but of itself is neither necessary nor sufficient.

<u>Summary</u>

The role of martensitic phase transformations in the hydrogen embrittlement (HE) susceptibility of austenitic stainless steels has been a source of
controversy. Tests were conducted in 0.1 MPa hydrogen to determine the influence of austenite stability on fracture and mechanical behavior. Slow
strain rate notched tensile and sustained load tests in 0.1 MPa hydrogen
were performed on AISI 310 (stable) and two heats of AISI 304L (unstable).
Sensitizing heat treatments were performed on the 304L to further destabilize
the austenite before testing. The results of mechanical testing, fractography, and X-ray diffractometry of fracture surfaces failed to show any effect of 1 atmosphere hydrogen on any of the steels tested.

References

1. Metals Handbook, ASM, Cleveland, OH, 1948, p. 1261.

2. P. L. Mangonon Jr. and G. Thomas: Met. Trans., 1970, vol. 1, p. 1577–86.

3. P. L. Mangonon Jr. and G. Thomas: Met. Trans., 1970, vol. 1, p. 1587–94.

4. R. P. Reed and C. J. Guntner: Trans TMS-AIME, 1964, vol. 230, p. 1713–20.

5. J. F. Breddis and W. Robertson: Acta Met., 1963, vol. 11, p. 547–59.

6. M. L. Holzworth and M. R. Louthan: Corrosion, 1968, vol. 24, p. 110–24.

7. P. Maulik and J. Burke: Scr. Metall., 1975, vol. 9, p. 17–22.

8. H. Mathias, Y. Katz and S. Nadiv: Metal Science, 1978, p. 129–37.

9. A. R. Troiano: Trans. ASM, 1960, vol. 52, p. 54–80.

10. R. M. Vennett and G. S. Ansell: Trans. ASM, 1967, vol. 60, p. 242–51.

11. R. B. Benson, R. K. Dann, and L. W. Roberts: Trans. TMS-AIME, vol. 242, p. 2199–2205.

12. R. J. Walter, R. P. Jewett and W. T. Chandler: Mat. Sci. Eng., 1969/70, vol. 5, p. 99–110.

13. M. R. Louthan, G. R. Caskey, J. A. Donovan and D. E. Rawl: Mat. Sci. Eng., 1972, vol. 10, p. 357–68.

14. A. W. Thompson: Hydrogen in Metals (I. M. Bernstein and A. W. Thompson, eds.), ASM Metals Park, OH, 1974, p. 91–106.

15. M. R. Louthan: Hydrogen in Metals, (I. M. Bernstein and A. W. Thompson, eds.), ASM Metals Park, OH, 1974, p. 53–78.

16. C. L. Briant: Met. Trans. A, 1979, vol. 10A, p. 181–89.

17. C. L. Briant: this volume.

18. A. W. Thompson: Mat. Sci. Eng., 1974, vol. 14, p. 253–64.

19. A. W. Thompson and I. M. Bernstein: Advances in Corrosion Science and Technology (M. G. Fontana and R. G. Staehle, eds.), vol. 7, Plenum, New York, 1980, pp. 53–175.

20. B. C. Odegard, J. A. Brooks, and A. J. West: Effect of Hydrogen on Behavior of Materials, (A. W. Thompson and I. M. Bernstein, eds.), TMS-AIME, New York, 1976, p. 116–25.

21. D. Eliezer, D. G. Chakrapani, C. J. Alstetter and E. N. Pugh: Met. Trans. A, 1979, vol. 10A, p. 935–41.

22. C. L. Briant: Met. Trans. A, 1978, vol. 9A, p. 731–33.

23. R. E. Stoltz and A. J. West: this volume.

24. A. W. Thompson: Metallography, 1972, vol. 5, p. 366–69.

25. ASTM Standard 262, Practice E.

26. J. J. Lewandowski: Unpublished research, Carnegie-Mellon University and Standard Oil of California, 1979.

27. C. Beachem: J. Basic Eng. (Trans. ASME, Series D), 1965, vol. 87, p. 299–306.

28. C. Alstetter: University of Illinois, personal communication, 1980.

29. A. J. West and J. Holbrook: this volume.

30. D. Eliezer: this volume.

J-CONTROLLED CRACK GROWTH AS AN INDICATOR
OF HYDROGEN-STAINLESS STEEL COMPATIBILITY*

M. R. Dietrich,** G. R. Caskey, Jr., and J. A. Donovan[†]

E. I. du Pont de Nemours & Co.
Savannah River Laboratory
Aiken, South Carolina 29808

The J-integral was evaluated as a parameter to characterize fracture
of stainless steels and as a measure of hydrogen damage. C-shaped specimens
of Type 304L, 316, and 21-6-9 stainless steels were tested in high pressure
helium and hydrogen. The critical force for crack initiation (Jm), and
tearing resistance (dJ/da) were decreased by hydrogen in all three alloys.
The J-integral appears useful as a measure of hydrogen compatibility because
it is sensitive to both test environment and microstructure.

 * The information contained in this article was developed during the
   course of work under Contract No. DE AC09 76SR00001 with the U. S.
   Department of Energy.

** Babcock & Wilcox Research Center,
   Lynchburg, Va.

 † Mechanical Engineering Department
   University of Massachusetts
   Amherst, Ma. 01003

Introduction

A wide variety of mechanical tests have been applied to the study of hydrogen compatibility of metals. Crack velocity, threshold stress intensity, and fracture toughness are valid measures of hydrogen compatibility with high strength and low ductility alloys. Concepts of linear elastic fracture mechanics are inadequate to characterize fracture properties of stainless steel in either inert or hydrogen environments, because stainless steels are ductile and have relatively low strength. The J-integral (1, 2) was developed for fracture accompanied by large scale plasticity. Thus the J-integral provides a basis for characterization of crack initiation, stable crack growth, and plastic instability of stainless steels.

This report describes an investigation at the Savannah River Laboratory into utilization of the J-integral as a parameter to describe initiation of ductile cracks in Type 304L, 316, and 21-6-9 stainless steels. The tearing resistance (dJ/da), describing stable crack growth, also was estimated from the load-deflection data.

Measurement of J-Integral

"C"-shaped fracture specimens (3) approximately 3.8 mm thick and 25.4 mm outer diameter were machined from high energy rate forged (HERF) bar stock. Type 316, 304L, and 21-6-9 stainless steels were studied. Orientations of the specimens and the V-notch with respect to the forging flow line patterns are shown in Figure 1. Fatigue cracks approximately 2.5 mm long were generated at a maximum stress intensity of K = 44 MPa $\sqrt{m}$. The specimens were tested in high pressure (69 MPa) helium or hydrogen in the as-received HERF condition or in high pressure hydrogen after prior exposure to deuterium at 69 MPa pressure for seven days at 520 K.

The J-integral was evaluated by determining the area under the load-deflection curve out to the maximum load. The usual methods of detecting crack propagation were not applicable to the high pressure test apparatus. Therefore, the critical force at the maximum load (Jm) was determined rather than the critical force to initiate a ductile crack (Jc). Correction to Jm due to the combined bending and tensile loading on the 'C' shape specimens could increase J by about ten percent but was not made because of the larger uncertainty in the load and deflection at crack initiation (4).

Tearing resistance (dJ/da) has been suggested to characterize stable crack growth and as a parameter for predicting plastic instability (5). In the present study dJ/da was measured at the maximum in the load-deflection curve. The data are internally consistent but may not be directly comparable to dJ/da evaluated in other studies.

Hydrogen Effects on Fracture

The test environment and deformation pattern caused by the HERF processing affected the J-integral at maximum load and the tearing resistance of the three austenitic stainless steels. Generally, Jm was less for tests in hydrogen than for tests in helium. The tearing resistance was lower for tests in hydrogen than in helium in Type 304L and 21-6-9 stainless steels in all orientations (Figures 3b and 4b). The results for Type 316 stainless steel differed from the above in several ways: 1) Jm was the same in hydrogen and helium when the notch was parallel to the flow lines, 2) Jm was less in helium than hydrogen when the notch was at 45° to the flow lines, and 3) the tearing resistance was the same in both environments (Figure 2b).

A. Cross Section of Bar
Showing Forging Flow Lines

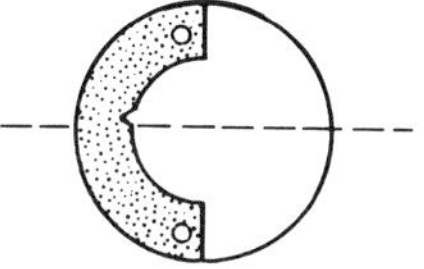

B. Parallel Orientation of Notch

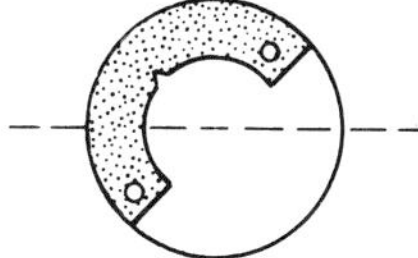

C. 45° Orientation of Notch

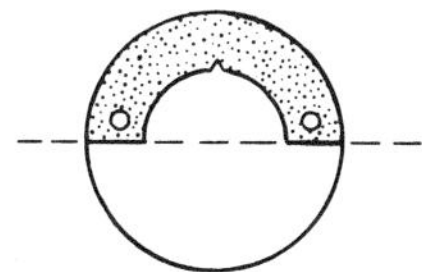

D. 90° Orientation of Notch

Fig. 1. Forging Flow Lines of HERF Stainless Steel
and Notch Orientations Relative to Flow Lines

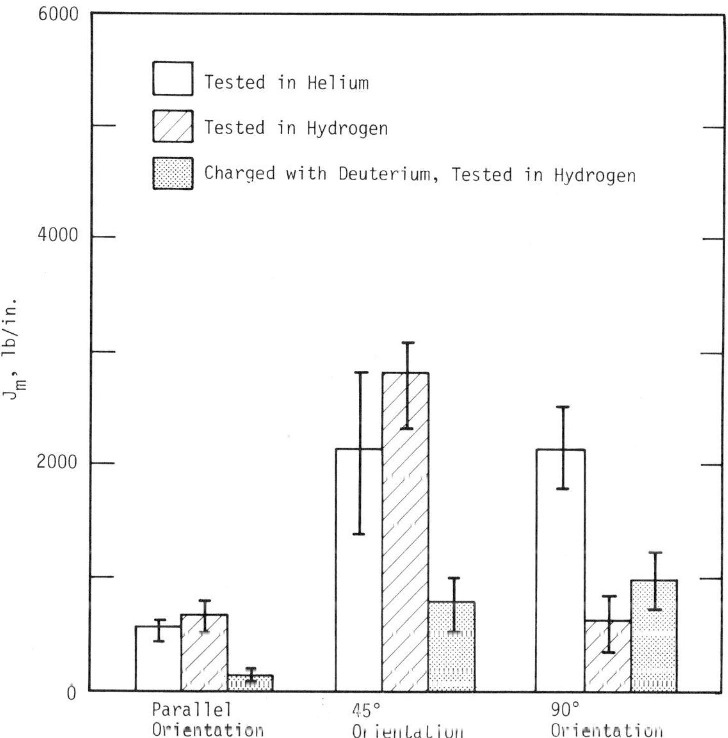

Fig. 2a. Fracture Parameters for Type 316 Stainless Steel

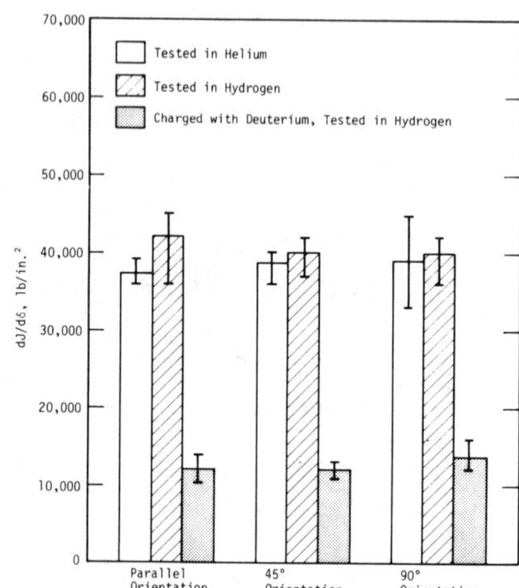

Fig. 2b. Fracture Parameters for Type 316 Stainless Steel

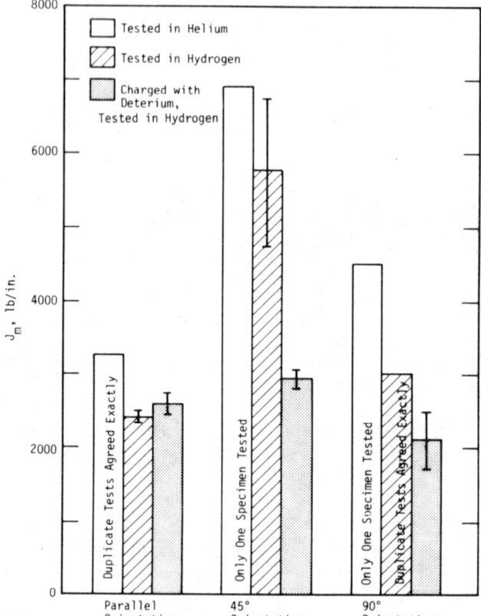

Fig. 3a. Fracture Parameters for Type 304L Stainless Steel

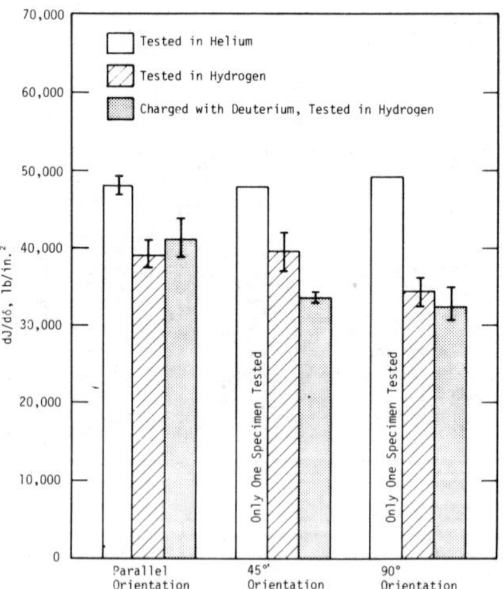

Fig. 3b. Fracture Parameters for Type 304L Stainless Steel

Regardless of the test environment, notch orientation relative to the HERF flow pattern affected Jm. The lowest values of Jm were for notches parallel to the flow lines, an effect that was especially pronounced in Type 316 stainless steel where inclusions were aligned along the forging pattern. Values of Jm were about the same for crack orientations 90 or 45° to the flow lines for Types 316 and 21-6-9 stainless steels (Figures 2a and 4a). Jm for Type 304L stainless steel was greatest when the crack was at 45° to the flow lines (Figure 3a).

Tearing resistance was unaffected by orientation of the crack relative to the flow lines for all tests in helium (Figures 2b, 3b, and 4b). Only small changes in tearing resistance with orientation were observed for tests in hydrogen.

Charging the specimens with hydrogen prior to testing in a hydrogen environment caused changes in both Jm and dJ/da which were alloy and orientation dependent. Hydrogen exposure of Type 316 stainless steel decreased Jm and dJ/da (Figure 2). For Type 304L stainless steel Jm was decreased for orientations of 90 and 45° but dJ/da was not altered significantly (Figure 3). In contrast, hydrogen exposure increased Jm for Type 21-6-9 stainless steel although dJ/da was unaffected (Figure 4). The increase in Jm is possibly due to microstructural changes that occurred during hydrogen charging.

Fracture modes were altered by the test environment but were independent of the crack orientation relative to the flow lines. All three alloys failed by microvoid coalescence when tested in helium as shown in Figure 5a. When tested in hydrogen, the fracture mode of Type 21-6-9 stainless steel changed to intergranular separation (Figure 5b), whereas the failure mode of Type 304L stainless steel changed to a brittle tearing mode (Figure 5c). The failure mode of Type 21-6-9 stainless steel charged with hydrogen and tested in hydrogen was a mixture of tearing and intergranular separation. Some intergranular separation was observed also in Type 316 stainless steel charged with hydrogen and tested in hydrogen.

Measures of Hydrogen-Stainless Steel Compatibility

Susceptibility to hydrogen damage may be evaluated by comparing Jm values for tests in helium with tests in hydrogen for the three types of steel. Type 304L stainless steel ranks better than Types 316 and 21-6-9 steel on this basis, regardless of which specimen orientation is chosen for the comparison. On the other hand, the relative ranking of Types 316 and 21-6-9 stainless steels depends upon specimen orientation. In addition, the data for Type 316 stainless steel is biased by the high inclusion content of these specimens. The Type 304L and 21-6-9 stainless steel specimens are much cleaner.

Tensile ductility has been used extensively as a measure of hydrogen damage (7). Based on plastic strain to failure (reduction in area), Type 304L stainless steel is more susceptible to hydrogen damage than Type 21-6-9 stainless steel in direct opposition to the ranking based on Jm. These measures of hydrogen damage are biased differently, however, with respect to deformation and fracture. Jm depends on flow stress and plastic strain up to the point where crack advance is assumed to begin. Plastic strain to failure, on the other hand, is affected strongly by the ability of the alloy to neck down without breaking and is effectively sampling a region of deformation not included in Jm. Furthermore, compact tensile specimens such as the C-shaped specimens of this study are not susceptible to unstable plastic tearing, whereas tensile specimens can be driven to tearing instability by internal microcracks in the necked region (5).

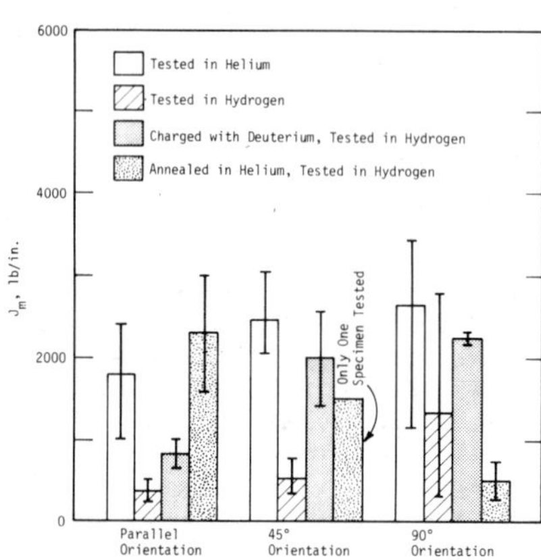

Fig. 4a. Fracture Parameters for Type 21-6-9 Stainless Steel

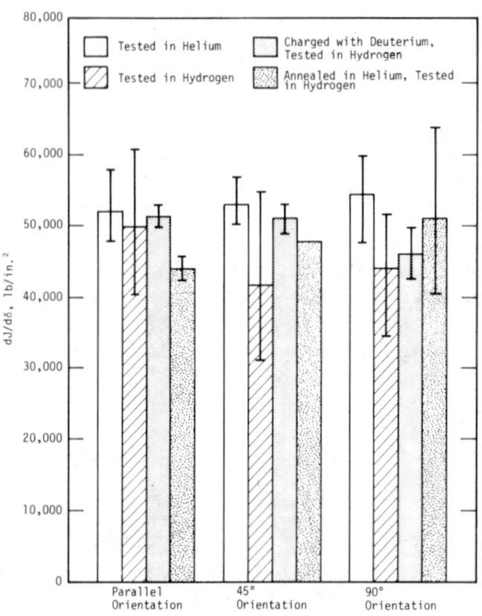

Fig. 4b. Fracture Parameters for Type 21-6-9 Stainless Steel

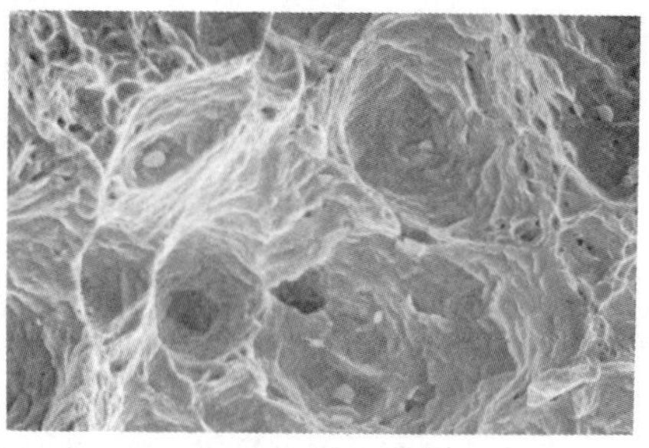

Fig. 5a.   HERF 21-6-9 Orientation 1 Dimple Fracture in Helium

Fig. 5b.   HERF 21-6-9 Orientation 1 Intergranular Fracture in Hydrogen

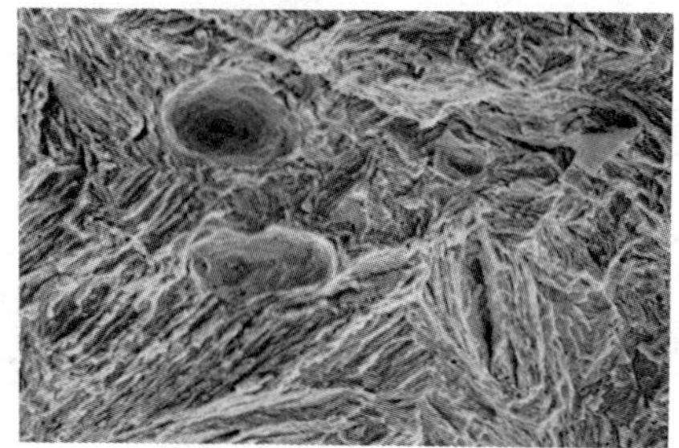

Fig. 5c.   HERF 304L Orientation 1 Brittle Fracture in Hydrogen

Fig. 5.   Typical Fractures

There is agreement in ranking of Type 304L and 21-6-9 stainless steels between tensile ductility and tearing resistance, dJ/da. In both cases, Type 21-6-9 steel is less susceptible to hydrogen damage. In this comparison both parameters are influenced by the deformation and crack growth processes that succeed the onset of crack growth on load maximum.

The method for determining Jm and dJ/da employed in this study was a variation of the procedure suggested by Paris (5). The results are in good agreement with results on Type 304L and 316 stainless steels reported elsewhere (8). Our results are an initial evaluation of J-integral methods for developing valid ductile fracture parameters for studies of hydrogen degradation of tough alloys such as the austenitic stainless steels. Further work is needed on experimental and analytical procedures as well as continued investigation of microstructural effects introduced by mechanical and thermal processing.

References

1. J. R. Rice: Jnl of Applied Mechanics, pp. 379-386, 1968.

2. J. A. Begley and J. D. Landes: ASTM STP 514, pp. 1-23, 1972.

3. E399, Standard Method of Test for Plane Strain Fracture Toughness of Metallic Materials, 1979 Annual Book of ASTM Standards, ASTM, 1979.

4. C. H. Laforce and J. Morrison: Jnl Eng. Matls and Techn., 100, pp. 248-252, Trans ASME, 1978.

5. P. C. Paris, H. Tador, A. Zahoor, and H. Ernst: ASTM STP 668, pp. 5-36, 1979.

6. J. W. Hutchinson and P. C. Paris: ASTM STP 668, pp. 37-64, 1979.

7. M. R. Louthan, Jr., G. R. Caskey, J. A. Donovan, and D. E. Rawl: Mater. Sci. & Eng., 10, pp. 357, 1972.

8. W. H. Bamford and A. J. Bush: ASTM STP 668, pp. 553-577, 1979.

SUSTAINED-LOAD CRACKING OF A PRECIPITATION-STRENGTHENED

AUSTENITIC STEEL IN HIGH-PRESSURE HYDROGEN[*]

M. W. Perra and R. E. Stoltz
Sandia National Laboratory, Livermore, CA

Sustained-load crack velocities and threshold stress intensities were
measured for a γ'-strengthened alloy similar to A-286 in gaseous hydrogen
at pressures of 100 and 200 MPa. The measured threshold values
corresponding to three different thermomechanical treatments ranged from 45
MPa$\sqrt{m}$ to 100 MPa$\sqrt{m}$. The fracture surfaces exhibited both intergranular
separation and ductile tearing, with intergranular separation predominating
at the lower K levels. Variations in crack growth velocity and threshold
stress intensity were related to the distribution of grain boundary
precipitates of the h.c.p. Ni_3Ti phase, η. As the number of grain
boundaries containing dense η precipitation increased, the hydrogen
fracture path became more intergranular, the crack velocity for $K > K_{th}$
increased, and the K_{th} value decreased. Yield strength, grain size, and
dislocation densities appear to have secondary effects on the crack growth
response. Calculations for the hydrogen diffusion distance ahead of a crack
growing in an austenitic material indicate that, for the highest measured
velocities of 10^{-5}m/sec, hydrogen penetration by diffusion cannot be
anticipated over more than a few atomic spacings ahead of the crack tip.
These results suggest either that the actual site of embrittlement for
$K > K_{th}$ can be far closer to the crack tip than it is to the site of
peak stress, or that bulk diffusion is not the rate-limiting transport step
in the embrittlement process.

Introduction

Among the austenitic alloys of interest for high-pressure hydrogen
applications is A-286, a γ'-strengthened iron-nickel-chromium superalloy,
which can be heat treated to strengths over 690 MPa. Early work by Walter
and co-workers (1) indicated that the tensile ductility of A-286 is only
slightly affected by 69 MPa hydrogen. Further testing of preflawed samples
in 34.5 MPa hydrogen (2) showed that A-286 exhibits sustained-load cracking
at stress intensities above 110 MPa$\sqrt{m}$.

[*]This work supported by U. S. Department of Energy under contract number
DE-AC04-76DP00789.

More recent alloy modifications to A-286 have been proposed that reduce the effect of thermally precharged hydrogen on tensile ductility (3). A patent has been awarded for this modified A-286, hereafter referred to as JBK-75, which specifically notes its improved weldability and hydrogen compatibility (4). The fundamental microstructural change to which the improved compatibility is attributed is a greater mismatch between the matrix and the γ'-strengthening phase. This larger mismatch causes the γ' to act as a more effective hydrogen trap, thus preventing large ductility losses due to intergranular fracture. Our work focuses on the sustained-load cracking susceptibility of JBK-75. Bolt-loaded wedge-opening load (WOL) samples were exposed to hydrogen pressures of 100 and 200 MPa. Three thermomechanical treatments were used to vary the yield strength from 717 to 924 MPa. Crack growth rates and threshold stress intensities were measured and related to the alloy microstructure.

Materials and Experimental Procedure

Composition, Heat Treatments, and Mechanical Properties

The chemical composition of the JBK-75 used in this study is given in Table I. Segments of bar were forward extruded in a single forging operation at 970°C to form rectangular test forgings. Specimen blanks were cut from these forgings with the crack growth direction parallel to the forging direction.

The samples with heat treatments A and B (Table II) retained strength from the forging operation and were additionally strengthened by precipitation hardening. The samples with heat treatment C were solution treated and then aged. (This is a standard aging treatment for A-286-type steels.) The tensile properties of the three sample groups are also given in Table II. Light microscopy showed macroscopically single-phase austenite with a few nonmetallic inclusions and second-phase particles. Heat treatments A and B both exhibited a duplex grain structure and a strong alignment of the grains in the forging direction. Heat treatment C exhibited large, equiaxed recrystallized grains with no discernable texture remaining from the forging operation.

Specimens and Test Environment

Sustained-load crack growth studies of three sample groups were performed with precracked, bolt-loaded WOL specimens that were 22.2 mm thick. A compliance method similar to that employed by Loginow and Phelps (5) was used to remotely monitor the onset and progress of crack propagation during testing. The smallest increment of crack growth that could be reliably detected was approximately 0.1 mm. The test duration was typically 5000 h (7 months). Average crack velocities as small as 10^{-11} m/sec could be resolved over this test period.

Table I. Chemical Composition of JBK-75

Chemical Composition, Wt. Pct

Cr	Ni	Mn	Ti	C	S	N
15.5	30.7	.053	2.1	.017	13ppm	.004

O	Mo	V	Al	Si	P	B	Fe
<10ppm	1.2	.26	.20	.032	<.002	<5ppm	bal

Table II. Heat Treatments Studied and Associated
Room Temperature Tensile Properties

Sample group	Heat treatment	0.2% σ_y, MPa (ksi)	σ_{uts}, MPa (ksi)	ε_u, %	ε_t, %	RA, %
A	Aged 8h/675°C+ 8h/600°C	855 (124)	1180 (171)	18	21	37
B	Aged 32h/675°C	924 (134)	1220 (177)	13	17	38
C	ST 1h/980°C. Aged 16h/720°C	717 (104)	1130 (164)	22	28	51

All test were performed at room temperature (295±4 K) and at hydrogen pressures of either 100 or 200 MPa. Hydrogen gas purification was accomplished with a 0.5 nm molecular sieve bed and liquid nitrogen cold trap on the high pressure side of the compressor. Analysis of the gas from within a test chamber showed that it contained approximately 4 ppm water vapor and < 0.5 ppm oxygen.

Results

Crack Growth Kinetics and Threshold Data

A total of 18 tests were conducted to characterize the influence of stress intensity and hydrogen pressure on crack growth in the three heat treatments of JBK-75. The log v vs. K data (Fig. 1) show that at a K of 120 MPa$\sqrt{m}$, approximately four orders of magnitude in crack velocity separate the most susceptible material (heat treatment C) from the least susceptible material (heat treatment A). Heat treatment C was the only sample group to show two clearly distinguishable stages of crack growth; namely Stage I, where the rate is strongly dependent on K and Stage II, where the rate is relatively independent of K.

Threshold stress intensities could be unambiguously defined for heat treatments B and C within the 5000 h test period. Measurable crack growth for these sample groups typically stopped after several hundred to several thousand hours. The samples with heat treatment A, however, usually continued to exhibit slow, steady crack growth even after considerably longer periods of time; a distinct threshold value was not apparent for this material within the 5000 h test period. For all materials, the stress intensity at which the crack velocity decreased below 10^{-11} m/sec was designated as the threshold stress intensity. These results indicate that K_{th} is approximately 100 MPa$\sqrt{m}$ for heat treatment A, 65 MPa$\sqrt{m}$ for heat treatment B, and 45 MPa$\sqrt{m}$ for heat treatment C.

The data in Fig. 1 show that hydrogen pressure was not a significant variable for any of the sample groups. No pressure dependence either of the threshold values or of the crack velocities for K > K_{th} was observed. The variations in heat treatment (microstructure) led to much larger changes in the crack growth response than did the variation in pressure from 100 to 200 MPa.

Microstructure and Fractography

Scanning electron micrographs of the specimens with heat treatments A and C are shown in Figs 2a and 2b, respectively. (Micrographs for heat treatment B generally exhibited an intermediate appearance and will be

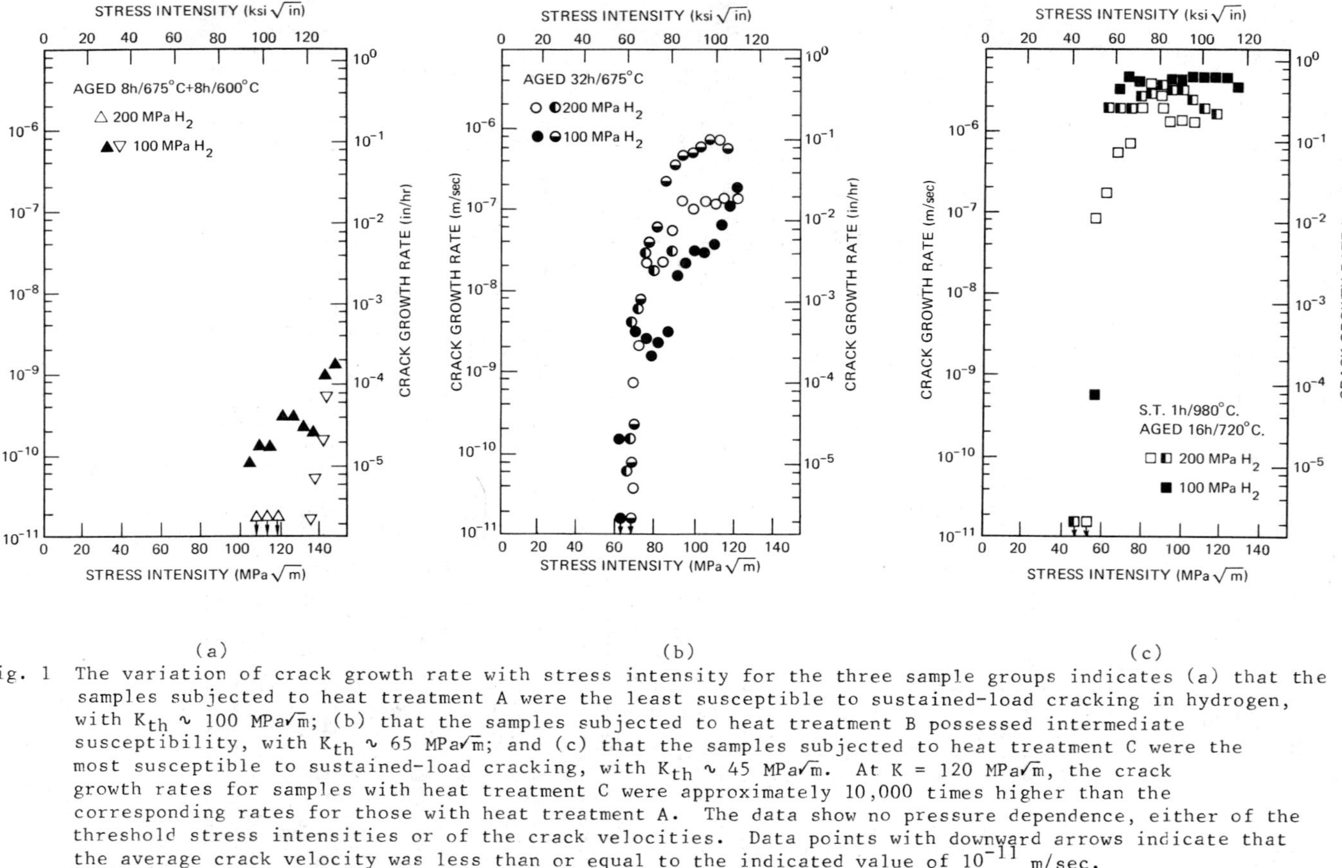

Fig. 1   The variation of crack growth rate with stress intensity for the three sample groups indicates (a) that the samples subjected to heat treatment A were the least susceptible to sustained-load cracking in hydrogen, with $K_{th} \sim 100$ MPa√m; (b) that the samples subjected to heat treatment B possessed intermediate susceptibility, with $K_{th} \sim 65$ MPa√m; and (c) that the samples subjected to heat treatment C were the most susceptible to sustained-load cracking, with $K_{th} \sim 45$ MPa√m. At $K = 120$ MPa√m, the crack growth rates for samples with heat treatment C were approximately 10,000 times higher than the corresponding rates for those with heat treatment A. The data show no pressure dependence, either of the threshold stress intensities or of the crack velocities. Data points with downward arrows indicate that the average crack velocity was less than or equal to the indicated value of 10^{-11} m/sec.

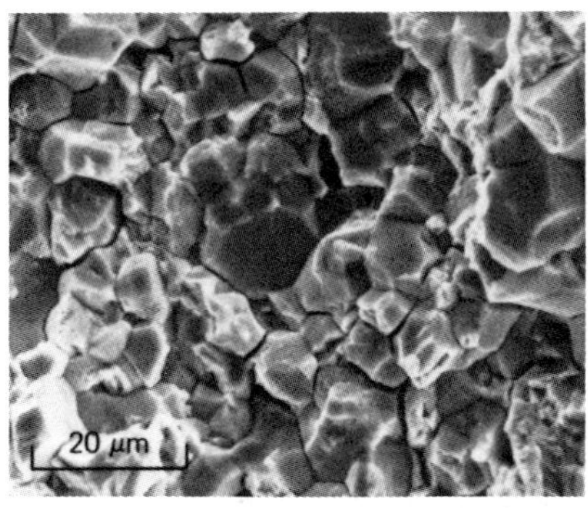

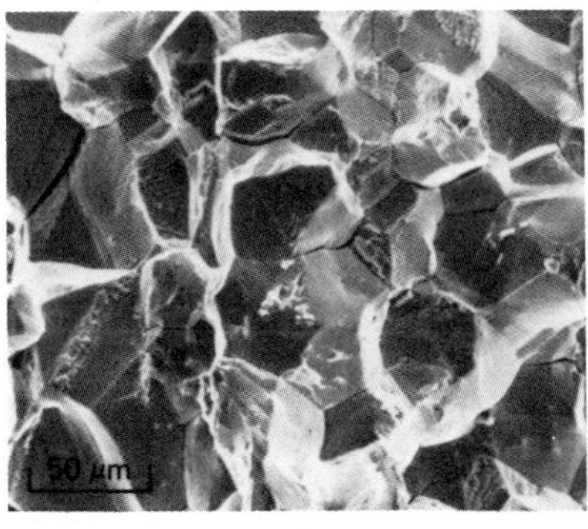

(a)   (b)

Fig. 2   Scanning electron microscopy of the fracture surfaces revealed both
         intergranular separation and ductile tearing for all sample groups.
         Heat treatment A samples (a) exhibited a larger fraction of ductile
         tearing than did heat treatment C samples (b).

omitted for brevity.) The fracture surfaces for the three sample groups all
exhibited both intergranular separation and ductile tearing, but in different proportions; the material with the lowest threshold value showed the
highest fraction of intergranular separation (Fig.2b). Secondary cracking
was a characteristic feature of all hydrogen fracture surfaces. The three
sample groups all produced dimpled rupture fracture surfaces in air.

The rapid crack growth regions (at the higher K levels) exhibited a
larger component of ductile tearing than the threshold regions (at the lower K
levels) for all sample groups. At very high K levels hydrogen caused rapid
cracking of the weakest grain boundaries, and the intervening ligaments--which
were largely unaffected by hydrogen--were then torn apart. At lower K levels,
however, fewer of these intervening ligaments were torn because of the lower
crack driving force. The resulting crack retardation allowed more time for
hydrogen to diffuse, and permitted a larger fraction of grain boundaries to
reach a critical hydrogen concentration for intergranular fracture. This has
been reported previously for several other materials, including HY130 (9).

Transmission microscopy observations revealed several characteristic
phase morphologies. By selected area diffraction and STEM non-dispersive
x-ray analysis we identified the predominant grain boundary precipitate to be
the hexagonal Ni_3Ti phase, η. The TiC or Ti(C,N) phase occurred as 0.1 to
1μm cuboidal particles, but this phase was far less abundant than the η
and did not vary in abundance among the sample groups. Two phase morphologies
for η were found. In one form, η occurred as 0.1 to 0.5 μm diameter
circular discs lying in the grain boundary plane. The thickness of the discs
ranged from approximately 10 to 20 nm. The second form consisted of colonies
of cellular plates emanating from the grain boundaries. The three heat
treatments all contained various proportions of both η morphologies. Group
A samples contained relatively low concentrations of both disc-like and
cellular η. This material possessed a heavily worked dislocation
substructure retained from the forging operation (Fig. 3a). Group B samples
contained an intermediate concentration of disc-like η and the highest
concentration of cellular η. Group C samples contained the highest concentration of disc-like η and a relatively low concentration of cellular η.

Fracture surface replica results were consistent with the above
microstructural observations. A dense array of second-phase precipitates

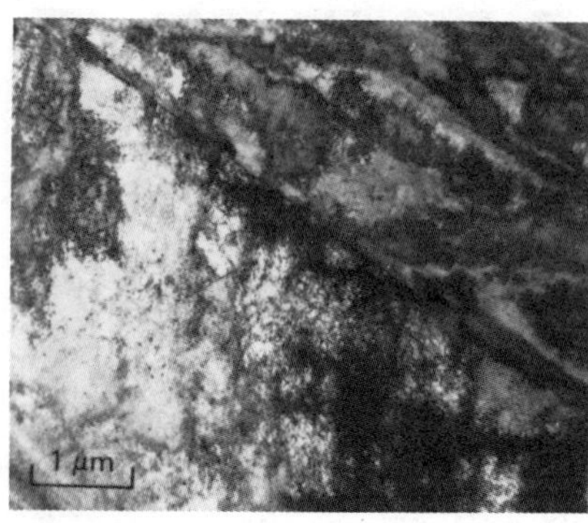

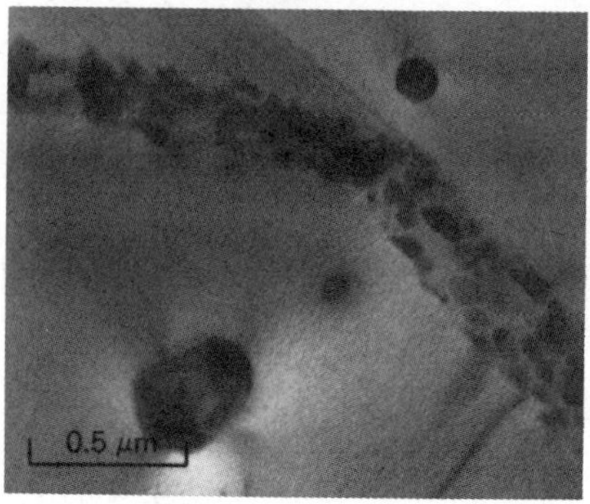

(a)   (b)

Fig. 3   TEM examinations showed that group A samples (a) possessed a heavily
worked dislocation substructure and relatively few grain boundary
precipitates. Most grain boundaries in group C samples (b) contained closely spaced discs of η lying in the grain boundary plane.

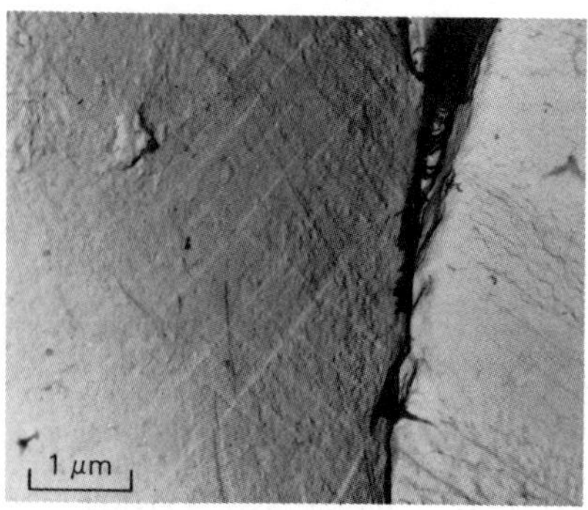

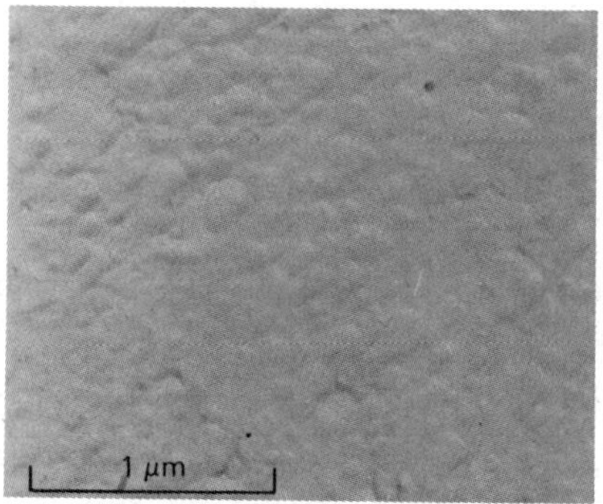

(a)   (b)

Fig. 4   Replication of heat treatment A fracture surfaces (a) revealed
mainly slip traces and deformation markings on the grain boundary
facets. Heat treatment C samples (b) showed evidence of dense
precipitation on more than 50% of the exposed grain boundaries.

was observed on more than 50% of the grain-boundary facets in heat treatment
C (Fig. 4b). The size and shape of these precipitates corresponded directly
with the size and shape of the grain boundary η discs seen in the TEM
observations (Fig. 3b). Heat treatment A samples, however, exhibited mainly
slip traces and deformation markings on grain-boundary facets (Fig. 4a).
Fewer than 20% of the facets in this sample group showed indications of
dense second-phase precipitation.

When the number of grain boundaries containing dense η precipitation
was low, sustained-load cracking in hydrogen occurred at high K levels with
considerable local plasticity and ductile tearing. As the number of grain
boundaries containing dense η precipitation increased, the crack path
became more closely associated with grain-boundary interfaces, the crack
velocity for for a given K level increased, and the threshold stress
intensity decreased.

Discussion

Crack Growth Resistance and Grain Boundary Precipitation

In view of the observed differences in the precipitate morphology and fracture surface appearance, it seems very likely that the substantial differences in hydrogen cracking resistance among the three groups are due to differences in the distribution of grain boundary η precipitates. The fracture surface replica results suggest that the γ-η phase boundaries are the preferred path for cracking; the γ-η phase boundary is apparently much more readily weakened by the presence of hydrogen than the γ-γ phase boundary. Thompson and Brooks (3) showed correlations between cellular η and intergranular fracture in tensile tests. Smugeresky (6) found that the decrease in ductility of iron-base superalloys in the presence of internal hydrogen is dictated by the morphology of η. We found that the size or the volume fraction of η does not by itself determine susceptibility: group C samples were more susceptible to sustained-load cracking than group B samples even though the latter contained a larger volume fraction of optically observable η. The disc-like η particles in group C samples were distributed over a large percentage of the grain boundary area, thereby permitting relatively easy crack propagation. The main factor affecting susceptibility appeared to be the distribution of grain boundary precipitates. Yield strength, grain size, and dislocation density appear to have secondary effects.

Location of the Fracture Process Zone

The fact that the local stresses reach a maximum ahead of the crack tip (7) has led several investigators to suggest that the site of embrittlement lies some distance ahead of the crack tip (8,9). If this is the case, an increment of crack growth will occur when a critical combination of local stress and hydrogen concentration is reached over a sufficiently large area of material on a susceptible interface ahead of the crack tip.

If crack growth in gaseous hydrogen is not strictly a surface phenomenon, hydrogen must penetrate some distance ahead of a moving crack tip either by diffusion or by dislocation transport. Johnson (10) has addressed the question of how far hydrogen can diffuse ahead of a growing crack tip in an analysis that permits hydrogen to enter the material only at the tip of a blunted crack. For a bulk hydrogen diffusivity of 10^{-15} m^2/sec (a representative value for hydrogen in austenitic steels) and for a steady-state crack velocity of 10^{-5} m/sec (approximately the highest velocity observed for JBK-75), Johnson's analysis indicates that hydrogen penetration by diffusion cannot exceed 0.5 nm (about two atomic spacings) ahead of the crack tip. By way of comparison, the peak stress corresponding to K = 110 MPa$\sqrt{m}$ and σ_y = 690 MPa occurs at a distance of approximately 50 μm, or 5 orders of magnitude farther from the crack tip.

These results suggest either that the actual site of embrittlement can be considerably closer to the crack tip than the site of peak stress, or that bulk diffusion is not the rate-limiting transport step in the embrittlement process. The above calculations, of course, ignore any possible influence of the crack-tip stress field on the rate of hydrogen diffusion. Wilson and co-workers (11), however, have found theoretically that the enhancement of hydrogen diffusivity due to crack-tip stress fields is small in f.c.c. materials. Additionally, hydrogen diffusion along grain boundaries in f.c.c. nickel is not appreciably faster than bulk diffusion (12). Hydrogen transport ahead of moving crack tips by dislocations could conceivably cause faster hydrogen penetration than diffusion, but the role of dislocation transport in embrittlement processes remains an open question.

Pressure Effects and Trap Saturation

Measurements of crack growth often indicate that the pressure dependence of K_{th} is weak at high pressures. Walter and Chandler (2), for example, found K_{th} to be virtually independent of hydrogen pressure from 0.034 to 34.5 MPa for AISI 4340, and from 21 to 69 MPa for Inconel 718. Our tests of JBK-75 at 100 and 200 MPa suggest the pressure independence of K_{th} in this pressure range.

The solution and diffusion of hydrogen in the crack-tip region is almost certainly influenced by trapping effects at grain boundaries, dislocations and second-phase interfaces; trap saturation at high pressures may be responsible for the observed pressure independence of K_{th}. At crack arrest, equilibrium exists between hydrogen in the gas phase and hydrogen in the near-tip region. If crack growth in hydrogen is governed by the properties of specific microstructural features or interfaces in the immediate vicinity of the crack tip (e.g., by the properties of the γ-η interfaces in JBK-75) the saturation of these traps above a critical pressure could lead to the pressure independence of K_{th}.

Conclusions

The principal factor affecting resistance to sustained-load cracking in JBK-75 appears to be the distribution of grain-boundary η precipitates. As the number of grain boundaries showing dense η precipitation increased, the fracture path became more intergranular, crack velocities for $K > K_{th}$ increased, and K_{th} decreased. The crack velocities attained in the most susceptible material (group C) were high enough to preclude bulk hydrogen diffusion over more than a few atomic spacings ahead of the crack tip. Trap saturation in the crack-tip region may be responsible for the pressure independence of K_{th} in the range of 100 to 200 MPa.

Acknowledgments

We gratefully acknowledge the skillful assistance of B. D. Schoeneman with the experimental work and F. Greulich and C. Karfs with electron microscopy. Discussions with A. T. Jones and A. J. West were most valuable.

References

1. R. J. Walter, R. P. Jewett, and W. T. Chandler: Mater. Sci. Eng., 1969, vol. 5, pp. 98-110.
2. R. J. Walter and W. T. Chandler: NASA Report CR-124410, Rocketdyne Division, Rockwell International, Canoga Park, CA, 1973.
3. A. W. Thompson and J. A. Brooks: Met. Trans., 1975, vol. 6A, pp. 1431-42.
4. J. A. Brooks and R. W. Krenzer: U. S. Patent No. 3,895,939.
5. A. W. Loginow and E. H. Phelps: Corrosion-NACE, 1975, vol. 31, no. 11, pp. 404-412.
6. J. E. Smugeresky: Met. Trans, 1977, vol. 8A, pp. 1283-1289.
7. J. R. Rice and M. A. Johnson: in Inelastic Behavior of Solids, M. F. Kanninen et al., Eds., p. 641, McGraw-Hill, New York, 1970.
8. A. R. Troiano: Trans. ASM, 1960, vol. 52, pp. 54-80.
9. C. L. Briant, H. C. Feng and C. J. McMahon: Met. Trans., 1978, vol. 9A, pp. 625-633.
10. H. H. Johnson: in Hydrogen in Metals, I. M. Bernstein and A. W. Thomspon, Eds., p. 34, ASM, Metals Park, Ohio, 1974.
11. M. I. Baskes and C. F. Melius: Z. Phys. Chem. Neue Folge, 1979, vol. 116, pp. 289-299.
12. W. M. Robertson: Jül-Bericht Jül-Conf-6, vol. 2, p. 449, KFA-Jülich, 1972.

J.P. Hirth, Dept. of Metallurgical Engineering, Ohio State Univ., Columbus:
Of importance for crack-tip plastic mechanisms is the sign of the shear
offsets at matching sides of an arrested crack tip. Your schematic cross-
section indicated that both surfaces had offsets corresponding to shears of
the same sense. Do you have experimental evidence for this view?

R. Stoltz: Yes - stereographic observation of both halves of the fracture
surface indicated that the matching is interlocking on peak-to-trough.

A.W. Thompson: I'm intrigued by your observation of apparent slip plane
cracking in 21-6-9 (Nitronic 40). Two questions: first, have you shown that
the flat facets are in fact {111}? second, how do you think this cracking
proceeds? is it restricted slip throughout the test, or an unstable shear
band process which is stimulated by hydrogen, or some other phenomenon?

R. Stoltz: No, we have not identified the facets as (111) planes, although
their spacing corresponds to the twin spacing. Also, fracture face etching
shows a correspondence between the underlying slip/twins revealed by etching
and the fracture facets. As far as the cracking process itself, I believe
hydrogen accelerates the accumulation of flow in the twinned region, much as
it appears to localize flow along slip bands in Ni 270. The question still
open is does hydrogen act only to localize flow, so that the material fractures like Zn at low temperatures, or does it participate directly in the
slip band decohesion process. The latter is attractive in that it would provide a link between slip band cracking or decohesion and grain boundary
cracking or decohesion which occurs in higher strength alloys.

B. Morris, Univ. of California, Berkeley, CA: You show two distinct morphologies of the η phase - disc and cellular. Could you comment on the relative
formation criteria and consequence of these?

R. Stoltz: The grain boundary (disc) η phase is much more prevalent in the
720°C aged material. After 16 hrs. at 720°C little or no cellular phase was
detected. On the other hand, in the forged material, heat treating 32 hours
at 675°C lead to ∿5 vol% cellular η in the structure. It does not appear that
the cellular phase nucleates on the grain boundary (disc) morphology, although
this cannot be ruled out. All of the cellular phase did grow from a grain
boundary, and no intergranular η phase was observed.

THE EFFECT OF TEMPERATURE AND STRAIN RATE ON THE TENSILE PROPERTIES

OF HYDROGEN CHARGED 304L, 21-6-9, AND JBK 75

J. H. Holbrook

Battelle's Columbus Laboratories
Columbus, Ohio 43201

and

A. J. West

Sandia National Laboratories
Livermore, California 94550

INTRODUCTION

Numerous studies have concentrated on the effects of temperature and
strain rate on the sensitivity of metals to hydrogen embrittlement. These
effects have been investigated for both internal and external hydrogen, for
smooth tensile specimens and notched or cracked geometries, and for various
alloy systems including steels and alloy steels,(1-6) stainless steels,(7-10)
titanium alloys,(11) nickel alloys,(12-14) and aluminum alloys.(15) A surprising similarity of observed temperature and strain rate effects on the severity of hydrogen embrittlement exists for all the alloys mentioned. In all
the materials a temperature, or range of temperatures, exists where the ductility (or toughness in the case of fracture studies) loss due to hydrogen is
maximum. Above and below this maximum embrittlement temperature (MET), the
influence of hydrogen is less severe. Moreover, all the materials exhibit
the same strain-rate dependence on ductility loss: monotonically increasing
loss with decreasing strain rate.

This similarity of temperature and strain rate effects on hydrogen-
induced ductility loss for the variety of alloys cited suggests that possibly
a single mechanism of hydrogen embrittlement is operative (although hydride
formation may be operating in the titanium alloys). Several mechanisms have
been suggested as rationales for the observed effects including the void
pressure theory of Zapffe and Sims,(16) the surface energy theory of Petch
and Stables,(17) and the dislocation transport model as suggested by Louthan
(0) and formalized by Tien(19). The first two theories, however, do not adequately predict the observed temperature dependence, that is, an intermediate temperature where embrittling effects are greatest. This is because the
embrittling mechanism of these theories is lattice diffusion - controlled.
Thus both theories predict that with less time (high strain rates) or slower
diffusion (low temperatures), embrittlement is less severe. This is consistent with observed behavior below the MET but not above.

The dislocation transport model of embrittlement has greater potential
for explaining the experimental results. This model predicts decreasing embrittlement with temperature below the MET due to the inability of hydrogen
atmospheres to "keep up" with moving dislocations as temperature is lowered.
Above the MET, hydrogen concentrations around dislocations are reduced with
temperature, allowing less hydrogen to be delivered to potential fracture
sites, therefore reducing embrittlement. There are, however, some experimental results associated with temperature and strain rate dependence that

655

the dislocation transport model has not been able to predict. The first of
these is remarkable observation that the MET of all the alloy systems mentioned fall within a relatively narrow temperature range of 200K to 300K
when one considers that hydrogen diffusivities, and therefore, mobilities
vary by as much as 7-10 orders of magnitude in these alloys, depending on the
temperature considered. A second observation is that the transport model,
in its original form, does not adequately account for the experimentally
observed strain rate dependence above the MET. Critical strain rates for
"dumping" atmospheres are very large above the MET,(19) thus ductility loss
should be strain rate insensitive below these critical rates. Experiments
show that less hydrogen is transported per unit strain at higher strain
rates(9) (well below critical strain rates), and increasing strain rate
produces less ductility loss.(1,2)

The experiments reported here were designed to examine if an MET exists
for three quite different stainless steel alloys in the temperature range
200K - 300K and to study the strain rate dependence of the hydrogen induced
ductility loss above the MET.

EXPERIMENTAL DETAILS

Alloys

Three stainless steels were selected for this study: 304L, a metastable
austenitic stainless that exhibits deformation-induced martensite particularly at low temperatures; 21-6-9, a stable, high-manganese, high-nitrogen
austenitic; and JBK 75, a modified A286 iron-base superalloy strengthened by
γ' precipitates. Compositions of these alloys are given in Table 1.

Cylinderical tensile specimens with diameters of 5 mm and nominal gage
lengths of 25 mm were machined from cross-rolled plates of each material.
The 304L and 21-6-9 specimens were then annealed at 1323K for two hours.
The JBK 75 specimens were solutionized at 1473K, water-quenched, and aged at
1033K for 16 hours. Hydrogen was introduced into each material by gas-phase
charging at 573C for 14 days with a hydrogen pressure of 69MPa. These conditions were sufficient to produce a constant hydrogen concentration profile
in all three alloys. Due to different hydrogen solubilities and surface conditions the resulting hydrogen concentrations were different for each alloy:
1400 appm for 304L, 6000 appm for 21-6-9, and 3000 appm for JBK 75. Charged
specimens were maintained in liquid nitrogen until testing.

Tensile Property Studies

Tensile tests were performed on a screw-driven Instron TTCL test machine. Load measurement resolution was better than 10 pounds (44.5N) corresponding to a stress resolution of 2.3 MPa. Tests were conducted at 77K,
196K, and 298K for all three alloys. Additional tests for JBK 75 were performed at 373K. For tests at temperatures other than 298K, the temperature
was maintained by immersion in equilibrium baths (liquid nitrogen, dry ice/
ethanol, boiling water). Crosshead speeds of 8.5 X 10^{-4} mm/sec and the three
decade values greater resulted in nominal strain rates of 4 X 10^{-5} sec^{-1} to
4 X 10^{-2} sec^{-1}. Tests were conducted at different strain rates at 298K only.
For the other temperatures, a fixed crosshead speed of 8.5 X 10^{-3} mm/sec was
used. After tensile testing, specimens were held at 77K until hydrogen content determinations using vacuum extraction techniques were performed.

RESULTS

The stress-strain behavior of the charged and uncharged 304L is shown
in Figure 1. The associated fractographs of charged specimens are shown in
Figure 2. The uncharged curves in Figure 1 illustrate a slightly decreasing
fracture elongation with decreasing temperature and the appearance of an in-
flection point in the stress-strain response that is associated with the onset of strain-induced martensite(10). The elongation of charged specimens
is not monotonic but exhibits a minimum at 196K (see Figure 7). An inflec-
tion point in the stress-strain response of charged 304L is not observed at
196K but is observed at 77K. The fractographs in Figure 2 indicate different
fracture morphologies for charged 304L at each temperature: ductile rupture
at 77K, a faceted structure at 196K, and quasi-cleavage at 298K. The faceted
fracture at 196K is believed associated with α' martensite platelets. (10)
Fracture morphologies for uncharged material (not shown) were ductile at all
temperatures.

The stress-strain behavior for 21-6-9 is shown in Figure 3. This alloy
shows a monotonic reduction in tensile ductility with temperature in both
charged and uncharged conditions. This is also shown in Figure 7. The fractographs for 21-6-9, Figure 4, indicate similar fracture morphologies at all
three temperatures with a increasing tendency toward intergranular fracture
with decreasing temperature. Uncharged 21-6-9 fracture surfaces were ductile
rupture at all temperatures.

The stress-strain behavior of JBK 75 is shown in Figure 5. Both charged
and uncharged specimens show a decreasing fracture elongation with increasing
temperature (see also Figure 7). This behavior is entirely different than
the other two alloys. The fractographs for charged JBK 75 are given in Figure 6. The morphologies are primarily intergranular with an increasing a-
mount of ductile rupture with decreasing temperature. Uncharged specimens
of JBK 75 exhibited a fracture morphology consisting of intergranular facets
with a substantial amount of ductile rupture.

Figure 7 summarizes the results of temperature dependence of fracture
elongation. Clearly, all three of the alloys exhibit widely different behavior. The relative ductility loss versus temperature is shown in Figure 8.
We have found it useful to devise a new ductility parameter, RA', for making
ductility comparisons between charged and uncharged tensile tests results.
RA' is defined as the reduction in area beyond that associated with uniform
elongation. That is, $RA' = (A_u - A_f)/A_u$, where A_u is the uniform area in
the gage length at the onset of necking and A_f is the minimum area in the
neck at fracture. It is easy to show that

$$RA' = 1-(1 + e_u) (1 - RA)$$

where e_u is the uniform elongation and RA has its usual meaning of $(A_o - A_f)/$
A_o and A_o is the undeformed specimen area. RA' has the convenient property
of equalling zero for a specimen that fails with negligible localized deformation. In Figure 8, RA' loss as a result of hydrogen charging is plotted
versus temperature. It is seen that each alloy exhibits an MET, about 200K
for 304L and 21-6-9 and about 300K for JBK 75. Since RA' represents the extent of necking, it can be seen that all three alloys lose practically all
capacity for necking at their respective METs.

The strain rate sensitivity of RA' loss at 298K is shown in Figure 9
for all three alloys. Ductility is regained at higher strain rates for 304L
and 21-6-9. The ductility loss of charged JBK 75, however, is practically
independent of strain rate over the experimental range of strain rates.

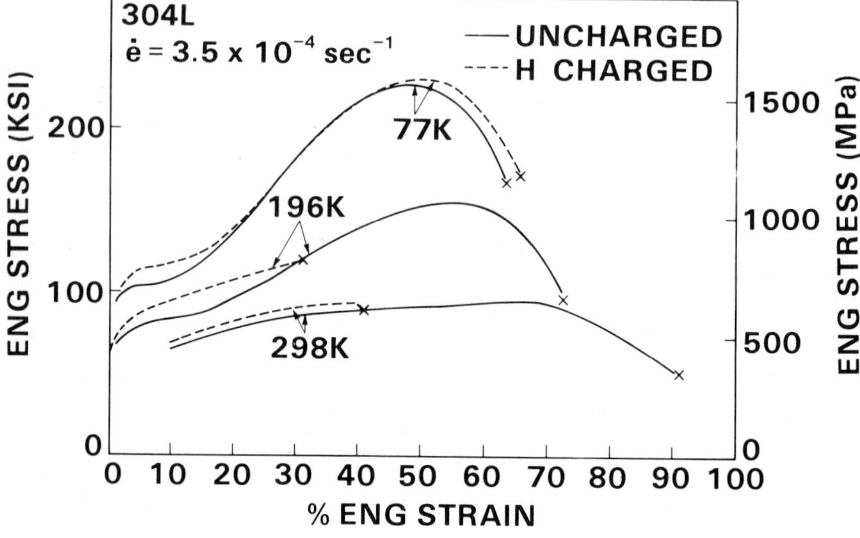

FIGURE 1.   STRESS-STRAIN BEHAVIOR OF HYDROGEN-CHARGED AND UNCHARGED 304L

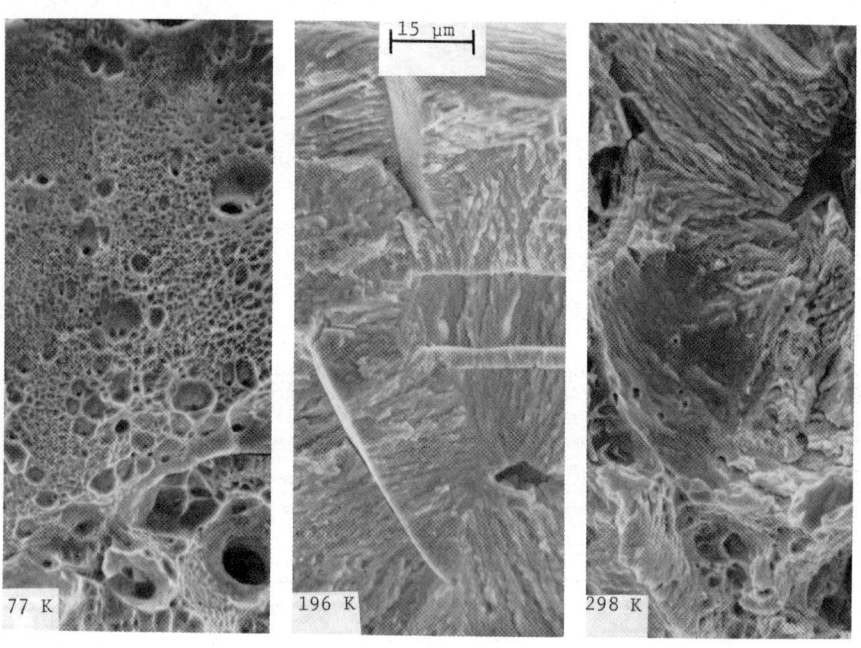

FIGURE 2.   SCANNING ELECTRON MICROGRAPHS OF CHARGED 304L
SPECIMEN FRACTURE SURFACES

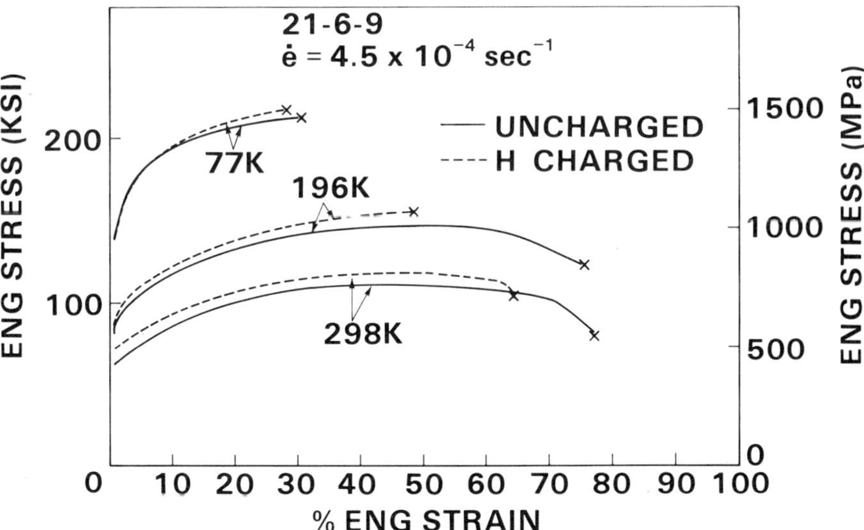

FIGURE 3.   STRESS-STRAIN BEHAVIOR OF HYDROGEN-CHARGED AND UNCHARGED 21-6-9

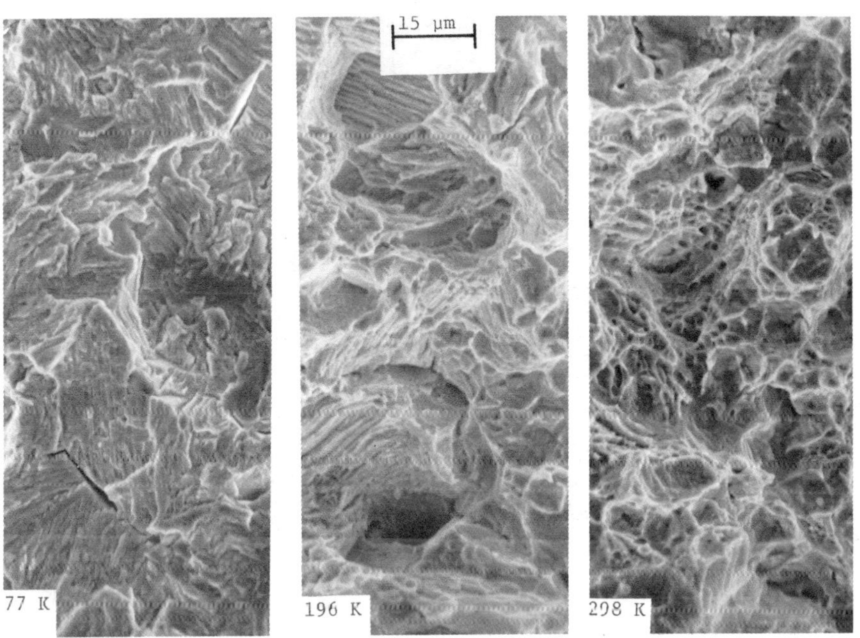

FIGURE 4.   SCANNING ELECTRON MICROGRAPHS OF CHARGED 21-6-9
SPECIMEN FRACTURE SURFACES

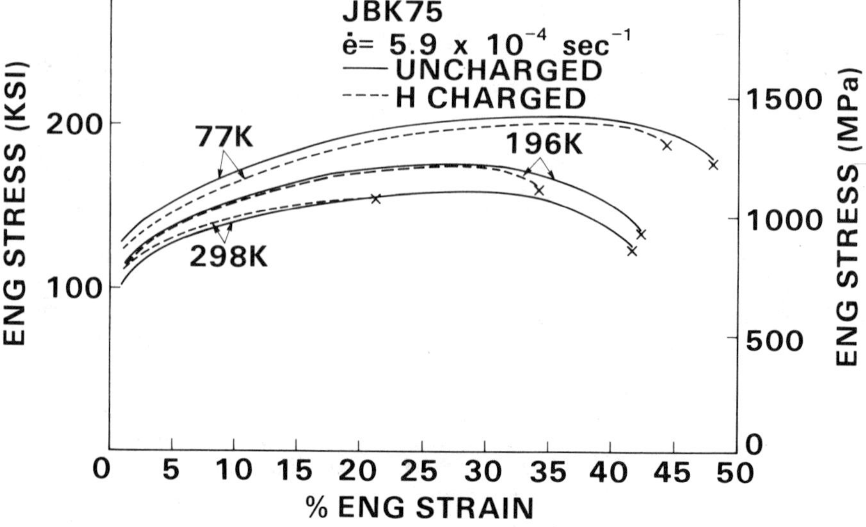

FIGURE 5.   STRESS–STRAIN BEHAVIOR OF HYDROGEN–CHARGED AND
UNCHARGED JBK 75.   NOTE DECREASING ELONGATION
TO FAILURE WITH INCREASING TEMPERATURE IN BOTH
CHARGED AND UNCHARGED MATERIAL

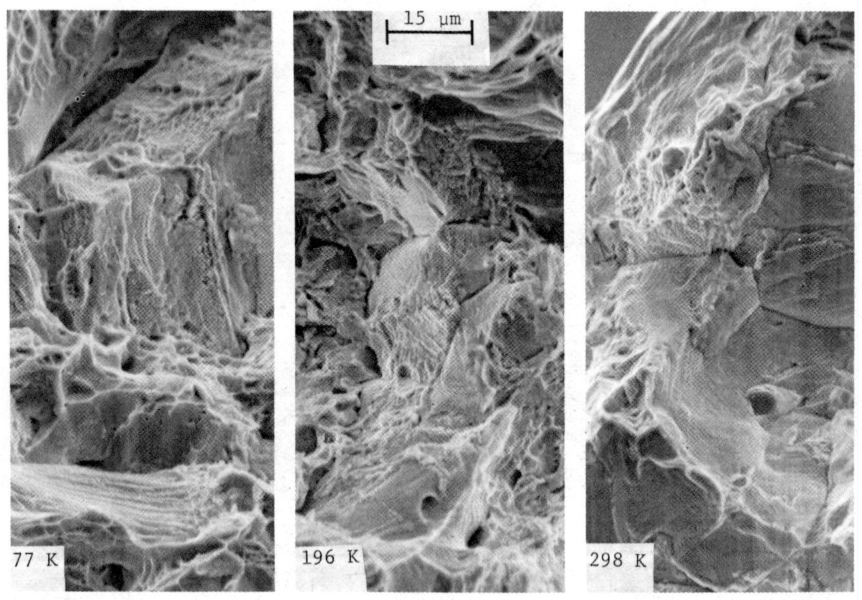

FIGURE 6.   SCANNING ELECTRON MICROGRAPHS OF CHARGED JBK 75
SPECIMEN FRACTURE SURFACES

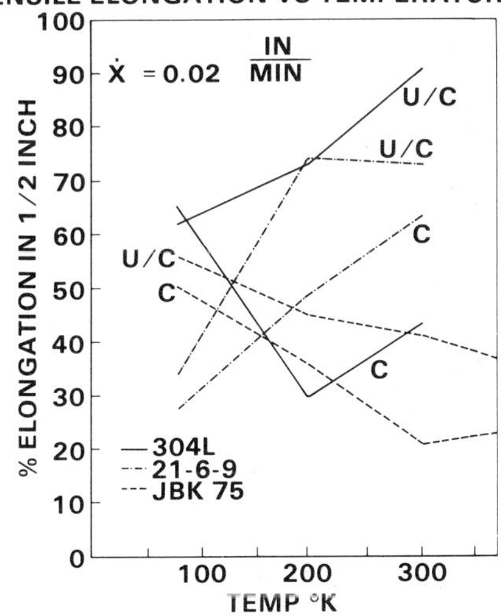

FIGURE 7.   TEMPERATURE DEPENDENCE OF ELONGATION TO FAILURE OF
CHARGED AND UNCHARGED SPECIMENS AT A FIXED CROSS-
HEAD SPEED.   STRAIGHT LINES HAVE BEEN USED TO
CONNECT DATA POINTS AT TEST TEMPERATURES OF 77,
196, 298, AND 373 K

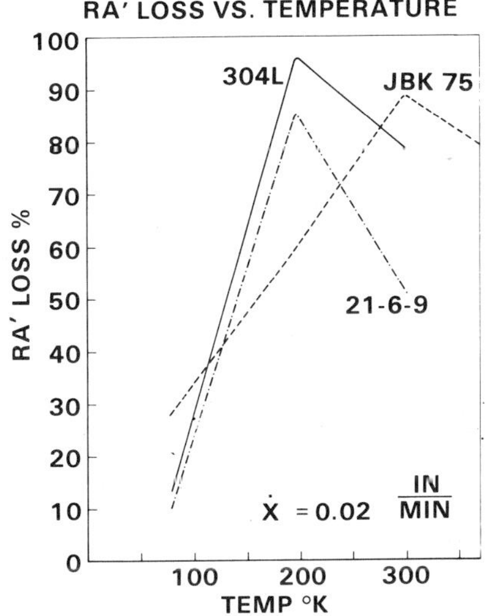

FIGURE 8.   TEMPERATURE DEPENDENCE OF
RA' LOSS AT A FIXED CROSS
HEAD SPEED.   RA' IS DEFINED
IN THE TEXT.   STRAIGHT LINES
HAVE BEEN USED TO CONVERT
DATA PONTS AT 77, 196, 298,
AND 373 K

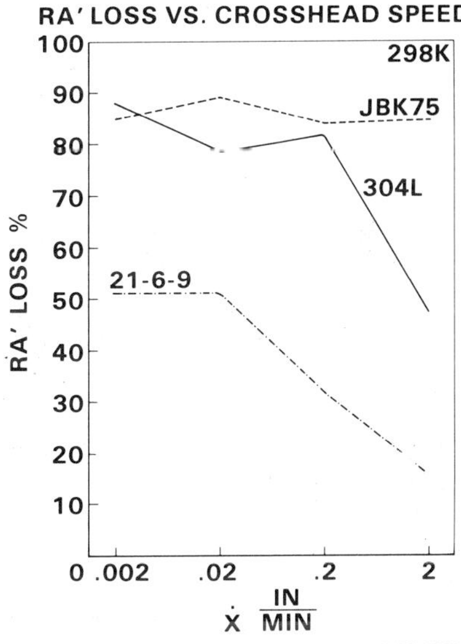

FIGURE 9.   CROSSHEAD SPEED DEPEND-
ENCE OF RA' LOSS AT 298K.
FOR A CROSSHEAD SPEED OF
0.02 IN/MIN, STRAIN RATES
WERE 3.9, 4.5, AND 5.9 X
10^{-4} SEC^{-1} FOR 304L, 21-
6-9, AND JBK 75

DISCUSSION

The results shown in Figure 8 indicate that the three alloys are exhibit MET values in a temperature range comparable to a variety of other alloys. The 100K difference in MET for JBK 75 compared to the other two alloys is not surprising in light of similar differences in METs for other stainless steels reported in the literature(7,8,10). The lack of ductility recovery of charged JBK 75 at high strain rates is unusual and the reason for this is unclear. It is entirely possible that ductility would be recovered at strain rates larger than those imposed in the present study. Alternatively, the mechanistic details of hydrogen embrittlement of JBK 75 could be different than in 304L and 21-6-9 by virtue of the strengthening precipitates. Other differences exist between the effect of hydrogen on properties of JBK 75 compared to its effects on 304L and 21-6-9 in terms of flow stress and strain rate sensitivity. These will be discussed in a separate publication but appear to suggest that hydrogen does not strongly interact with dislocations in JBK 75. In this event, the dislocation transport mechanism is not applicable for describing hydrogen embrittlement of this alloy.

REFERENCES

1. J. T. Brown and W. M. Baldwin, Jr., Trans AIME, 200, 298 (1954).

2. T. Toh and W. M. Baldwin, Jr., in Stress Corrosion Cracking and Embrittlement (W. D. Robertson, editor) J. Wiley and Sons, New York, p 176 (1956).

3. B. A. Graville, Brit. Welding Journal, 15, 191 (1968).

4. D. P. Williams and H. G. Nelson, Met Trans, 1, 63 (1970).

5. H. G. Nelson, D. P. Williams, and A. S. Tetelman, Met Trans, 2, 953 (1971).

6. W. Hoffman and W. Rauls, Welding J., 44, 225S (1965).

7. M. B. Whiteman and A. R. Troiano, Corrosion, 21, 53 (1965).

8. M. R. Louthan, Jr., in Hydrogen in Metals (I. M. Bernstein and A. W. Thompson, editors) ASM, Metals Park, Ohio, p 53 (1974).

9. J. A. Donovan, Met Trans, 7A, 1677 (1976).

10. G. R. Caskey, Jr., in Environmental Degradation of Engineering Materials (M. R. Louthan, Jr., and R. P. McNitt, editors) VPI, Blacksburg, Va., p 437 (1977).

11. D. N. Williams, J. Inst. Metals, 91, 147 (1962-63).

12. A. H. Windle and G. C. Smith, Met. Sci. J., 2, p 187 (1968).

13. G. C. Smith, in Hydrogen in Metals (I. M. Bernstein and A. W. Thompson, editors) ASM, Metals Park, Ohio, p 485 (1974).

14. P. Blanchard and A. R, Troiano, Mem. Sci. Rev. Met., 57, 409 (1960).

15. R. J. Gest and A. R. Troiano, in L'Hydrogène dans les Métaux, Editions Science et Industrie, Paris, p 427 (1972).

16. C. A. Zapffe and C. E. Sims, Trans AIME, 145, 225 (1941).

17. N. J. Petch and P. Stables, Nature, 169, 842 (1952).

18. J. K. Tien, in Effect of Hydrogen on Behavior of Materials, (A. W. Thompson and T. M. Bernstein, editors) AIME, New York, p 309 (1976).

TABLE 1. ALLOY COMPOSITIONS, WEIGHT PERCENT

	Cr	Ni	Mn	C	Si	P	S	N
304L	19.0	11.0	1.8	0.02	0.5	0.04	0.015	0.05
21-6-9	20.2	6.2	9.0	0.03	0.5	0.02	0.015	0.25
JBK 75	14.7	30.0	0.2	0.02	0.1	0.01	0.006	0.01

JBK 75 also contains 1.2 Mo, 0.3V, 2.2 Ti, and 0.25 Al.

INTERGRANULAR STRESS CORROSION CRACKING OF SENSITISED AUSTENITIC STAINLESS

STEEL - A MICRO COMPOSITIONAL AND STRUCTURAL INVESTIGATION

M.G. Lackey and F.J. Humphreys

Department of Metallurgy and Materials Science,
Imperial College,
London, S.W.7.

INTRODUCTION.

Most viable theories concerning the mechanism(s) of stress corrosion cracking in austenitic stainless steels fall into one of two categories. These are active path mechanisms, of which slip dissolution is the most popular, and the hydrogen embrittlement models. Both have been invoked to explain the SCC behaviour of austenitic stainless steels, but, as of yet, there has been no concensus of opinion as to which mechanism(s) is controlling.

It is well known that heat treatment of austenitic stainless steel in the temperature range $\sim500^{\circ}C$ - $\sim800^{\circ}C$ results in the precipitation of chromium-rich carbides ($M_{23}C_6$) type) along grain boundaries with concommitant depletion of chromium in the grain boundary region. The susceptibility of this material to intergranular attack and stress corrosion cracking has frequently been attributed to the existence of a zone adjacent to the grain boundary where the chromium content is less than the 12 wt.% required for passive film formation (1). Active path mechanisms rely on the compositional gradient at the sensitised grain boundaries, the importance of film formation, and the role of emergent slip steps to explain inter-granular SCC in these materials.

It has been proposed that the presence of martensite in metastable austenitic stainless steels is related to the stress corrosion and hydrogen embrittlement behaviour of the material (2,3,4). Electrochemical conditions are important in a hydrogen embrittlement mechanism of SCC in determining the rate of hydrogen evolution on the alloy surface. It is very likely that the hydrogen produced at the crack tip could be absorbed, and produce a mechanically weak path which would result in crack propagation.

It was the purpose of this investigation : 1) to characterise the chromium depletion which develops during the sensitisation process, and 2) to evaluate the SCC behaviour of sensitised 304 in dilute H_2SO_4 + NaCl solutions, and investigate the effect of Cl$^-$ and anodic potential on fracture.

EXPERIMENTAL.

Miniature tensile specimens (6.0mm width x 0.5mm thickness x 20mm gauge length) and 3.0mm diameter discs were punched from 0.5mm thick cold-rolled Type 304 sheet (0.07C, 1.55Mn, 19.0Cr, 9.0Ni, 0.025P, 0.03S, 0.25Mo). All specimens were annealed in vacuo for $\frac{1}{2}$ hour at 1120°C (grain size $\sim$50μm diameter), water quenched, and subsequently sensitised at 675°C for varying periods of time.

After sensitisation, TEM thin-foil specimens were prepared by electro-polishing in 95% Acetic Acid - 5% Perchloric Acid solution at 15°C using a commercial jet polishing unit. Stress corrosion specimens were abraded with 600 grit SiC papers, then electropolished over the gauge length.

Electron microscopy and quantitiative microanalysis of thin foils were conducted on a JEOL 120 CX Temscan equipped with Link Energy Dispersive X-ray Microanalysis facilities. The operating voltage for all micro-analytical experiments was 100kV, and the probe diameter employed was approximately 200 Å. The count time for each analysis was 100 seconds. All specimens were tilted $\sim$40° towards the detector for optimised X-ray collection. Only boundaries parallel to the incident beam were analysed.

The stress corrosion behaviour of sensitised Type 304 steel was evaluated through slow strain rate testing in dilute H_2SO_4 + NaCl solutions 0.01M H_2SO_4 + 0.002 to 0.02M Na Cl pH=2 at 20°C. All tests were performed on an Instron Testing machine at a strain rate of 2×10^{-6} sec.$^{-1}$. Experiments were conducted at the free corrosion potential (-420mV SCE) and under an anodic potential (-140mV SCE)

RESULTS AND DISCUSSION.

1. MICROSTRUCTURE OF THE SENSITISED TYPE 304 STEEL.

    1.1. Stem Quantitative Microanalysis.

        The results of the spot analyses at sensitised grain boundaries are presented in Figure 1. Each experimental point represents the average composition over a cylinder $\sim$500 Å in diameter and $\sim$1000 Å in height. Error bars on the individual analyses indicate one standard deviation.

        It was found that no unique solute profile could adequately characterise a particular sensitisation treatment; instead a range of solute profiles were obtained for each heat treatment examined. The nature of the chromium concentration profile which develops during sentisation is related to the type of grain boundary (and grain boundary plane). The chromium profiles contained in Figure 1 represent the most severe depletions detected for each sensitisation treatment.

        The development of the chromium depleted zone as a function of sentisation time is consistent with predictions by Stawstrom and Hillert (5), and Tedmon et. al.(6). The experimentally determined widths of the depleted zone for sentisation times < 24 hours are, in fact, larger than the predicted values. For longer sentisation treatments, however, the size of the zone was found to be narrower than indicated by these diffusion calculations. These

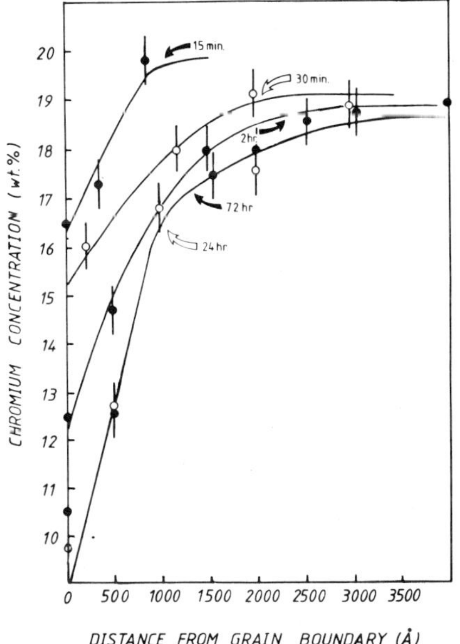

Figure 1.   Chromium concentration profiles for sensitisation treatments at 675 °C.

deviations imply that the grain boundary diffusion of chromium becomes more important with time in governing the supply of solute to the growing carbides, a factor neglected in the Stawström and Hillert calculations.

1.2.   Preferential Formation of Martensite.

Type 304 austenitic stainless steel is metastable with an M_s temperature of $\sim -200^{\circ}C$.   Examination of heavily sensitised specimens in the transmission electron microscope revealed the existence of a narrow band of martensite adjacent to certain grain boundaries (Figure 2).   The structure exhibited a Nishiyama-Wasserman relationship with the austenite :
$$(011)_{bcc} \; / \! / \; (111)_{fcc} \quad and \quad [100]_{bcc} \; / \! / \; [110]_{fcc}.$$

Quantitative microanalysis confirmed the existence of a compositional gradient within the martensite.   An example of such a profile is presented in Figure 3.   The martensite formed where the chromium content was less than 16 wt.% (7).   Reference to Figure 1 shows that martensite is to be expected for sensitisation times greater than 2 hours at $675^{\circ}C$.   The depletion of chromium effectively raises the M_s temperature in the grain boundary region so that the material can transform on cooling (8).   Each wt.% decrease in chromium content results in an increase of approximately 30 C° in the M_s temperature   (9).

Grain boundary martensite formation was enhanced by plastic deformation.

2.   INTERGRANULAR STRESS CORROSION CRACKING.

Material sensitised for 72 hours at $675^{\circ}C$ failed prematurely when tested in 0.01 M H_2SO_4 + 0.002 M NaCl (pH=2) solution at room temperature.   The concentrated H_2SO_4 + NaCl solutions used by previous investigators (10,11) led to significant corrosive attack of the fracture surface.   The more dilute solutions employed in the present investigation showed minimal post-fracture attack, making a detailed study of the fracture surfaces possible.   The fracture mode was primarily intergranular with small regions of 'feathery' or cleavage-like transgranular fracture (Figure 4).   Tests with an increased chloride ion concentration (0.02 M NaCl) yielded similar results.

Active path mechanisms do not adequately explain the observed fractography.   Although the material should be susceptible to intergranular dissolution on basis of the experimentally determined chromium concentration profiles, the transition of the fractography from intergranular to cleavage-like is inconsistent with dissolution models for stress corrosion cracking of sensitised austenitic stainless steels.

The preferential formation of martensite in the chromium depleted zone can provide a very susceptible path for a crack propagating intergranularly by an unspecified mechanism of hydrogen embrittlement (12). It is proposed that the regions of intergranular failure (Figure 4a) are due to such a mechanism.   As discussed in section 1.2., the martensite is not formed at all grain boundaries, and it is thought that the cleavage-like fracture (Figure 4b) corresponds to failure in the austenite.   Such fracture morphologies have been shown to be characteristic of hydrogen failure (13, 14).

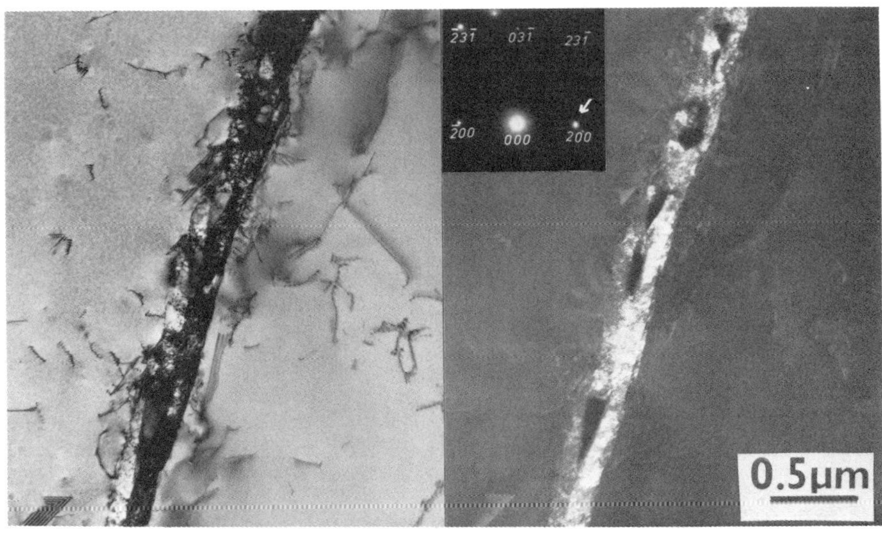

Figure 2.   Electron micrographs of martensite formed in the chromium
depleted zone of a specimen sensitised at 675 °C for 72 hours.
A) Bright field micrograph   B) Centred Dark Field   (200)

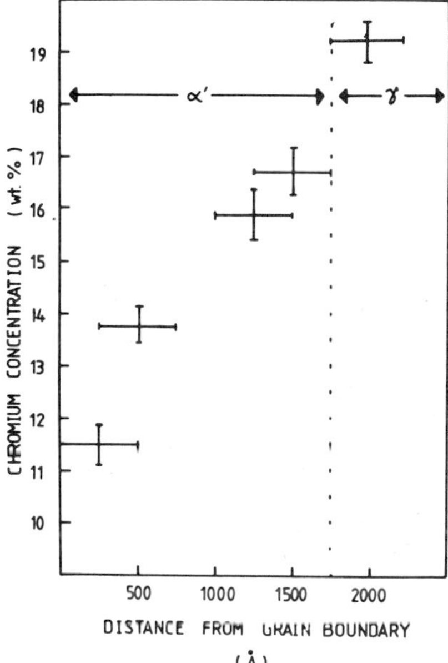

Figure 3.   Chromium concentration profile across a band of martensite
adjacent to a sensitised grain boundary.

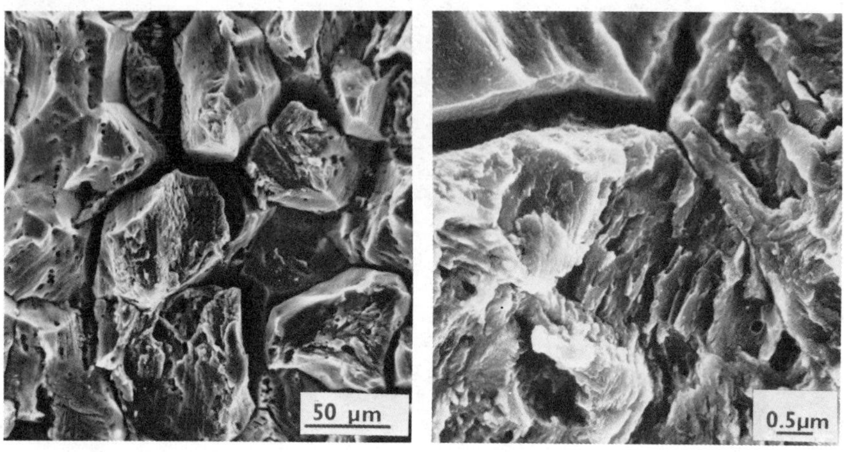

A.   B.

Figure 4.   Representative micrographs of free corrosion SCC tests.
A) Intergranular fracture   B) Cleavage-like Fracture

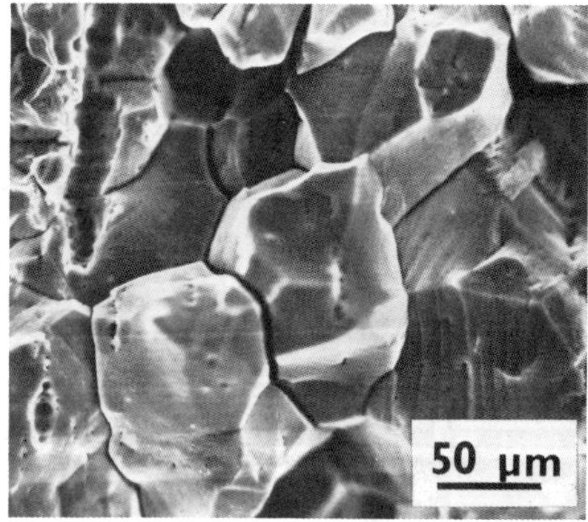

Figure 5.   Intergranular failure in 0.01M H_2SO_4 + 0.002M NaCl   (pH=2)
tested under anodic conditions (≃140 mV SCE)

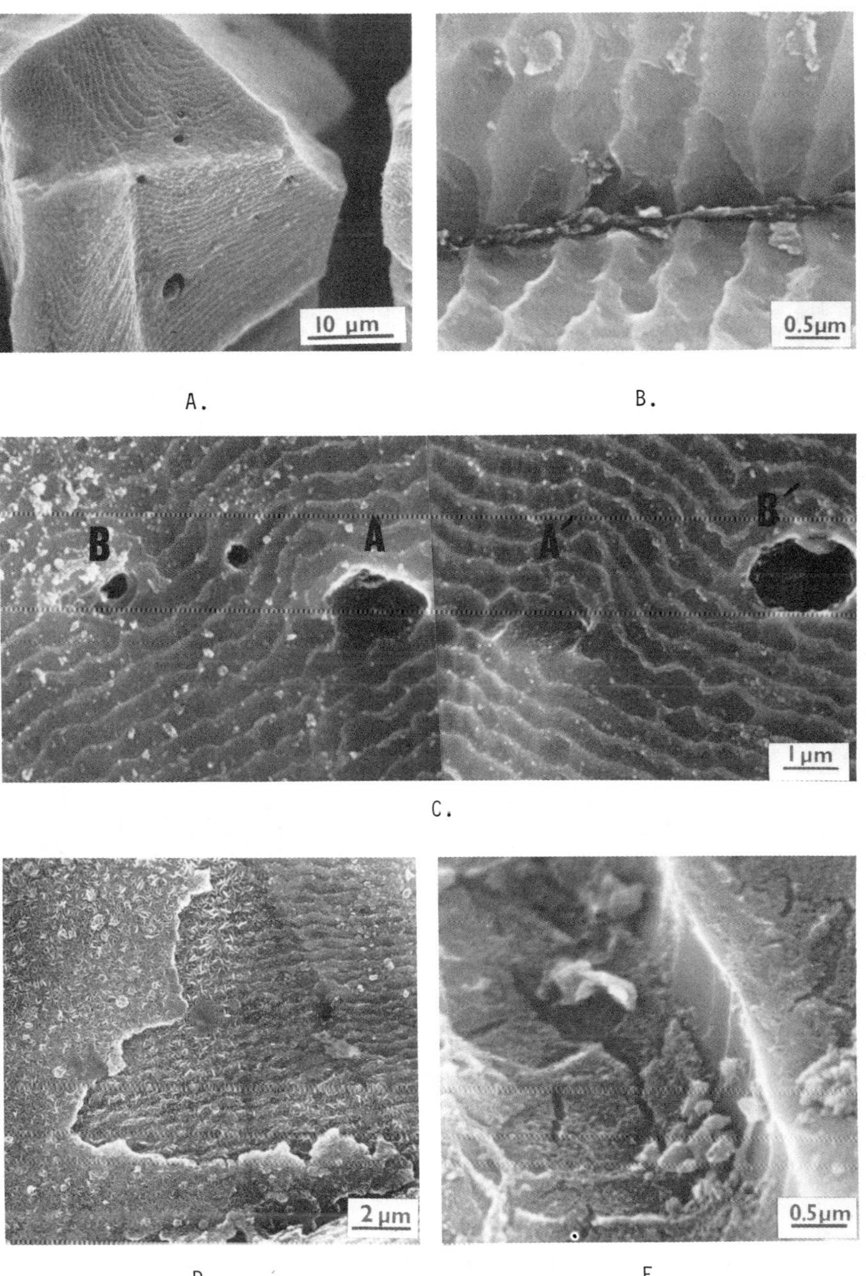

Figure 6.   Intergranular fracture surfaces observed in anodically polarised specimens tested in the 0.02M NaCl solution. A) & B) Intergranular striations; C) Matching fracture surfaces; D) & E) Corrosion product on striations.

The effect of applied anodic potential (-140 mV SCE) on the stress corrosion behaviour was evaluated for the 0.002M and 0.02M chloride ion concentrations. The ductility of the specimens was slightly reduced by anodic polarisation. In the 0.002M NaCl pH=2 solution, SEM examination a predominantly intergranular fracture morphology (Figure 5).

Increasing the chloride ion concentration from 0.002M to 0.02M NaCl (pH=2) did not significantly alter the ductility. However, the increased chloride ion concentration did cause a change in the fractography. Striations were observed in the 0.02M NaCl solution. These appeared as parallel grooves extending continuously across the fracture surface. The striation spacing varied from ∼1.5μm to∼0.5μm. It is thought that the striations are indicative of discontinuous crack propagation. Figure 6 contains representative micrographs of these features.

Discontinuous mechanisms of SCC include active path and hydrogen embrittlement models. Vermilyea (15,16) proposed that film rupture initiated active path dissolution which was subsequently halted by repassivation. Alternatively, in the Decohesion theory (17,18), the cohesive strength of the metal atoms ahead of the crack tip (in the plastic zone) is reduced as a result of the preferential accumulation of hydrogen in the plastic zone (region of maximum triaxial stress). Fracture then occurs when the local stress exceeds the interatomic cohesive strength. The intervals between cracking correspond to the periods during which hydrogen accumulates in the region of high triaxial stress. When the critical concentration of hydrogen is reached, the crack advances but is stopped as it enters the unembrittled metal. It has been demonstrated that hydrogen can be generated and absorbed under anodic potentials (19), and that martensitic steels are susceptible to hydrogen induced failure (20). The crack arrest markings may be accentuated by subsequent dissolution of the clean metal.

The observation of crack arrest markings in the 0.02M NaCl solution and their absence in specimens tested in the 0.002M NaCl solution suggests that the striation formation is a post-crack advance dissolution process which is favoured by high chloride ion concentrations. All fractographic observations are consistent with a hydrogen embrittlement mechanism of crack propagation.

<u>ACKNOWLEDGEMENTS</u>.

The authors are indebted to K. Richter for technical assistance and guidance with the STEM microanalysis. Invaluable discussions with R.A.H. Edwards and E.P. Butler are gratefully acknowledged. This research was supported by a grant from the U. S. Army.

<u>REFERENCES</u>.

(1)   E.C. Bain, R.H. Aborn, and J.J.B. Rutherford, <u>Trans. American Soc. Steel Treating</u>, <u>21</u> (1933) p. 481

(2)   D.A. Vaughan, D.I. Phalen, C.L. Peterson, and W.K. Boyd, <u>Corrosion</u>, <u>19</u> (1963) p. 315

(3)   S.S. Birley and D. Tromans, <u>Corrosion,</u> <u>27</u> (1971) p. 63

(4)   C.L. Briant, <u>Met. Trans.</u>, <u>9A</u> (1978) p. 731

(5) C. Stawstrom and M. Hillert, JISI, 217 (1969) p. 77

(6) C.S. Tedmon, D.A. Vermilyea, and J.H. Rosolowski, J. Electrochem. Soc., 118 (1971) p. 192

(7) S. Floreen, Trans. Met. Soc. AIME, 236 (1966) p. 1429

(8) E.P. Butler, M.G. Lackey, and K.B. Guy, Proc. 6th Int'l Conf. on HVEM- Antwerp 1980 (in press)

(9) G.M. Gordon in SCC and Hydrogen Embrittlement of Iron Base Alloys (NACE, 1977) p. 893

(10) J.D. Hartson and J.C. Scully, Corrosion, 25 (1969) p. 493

(11) S.J. Acello and N.D. Green, Corrosion, 18 (1962) p. 286

(12) C.L. Briant, Met. Trans., 10A (1979)

(13) A.J. Bursle and E.N. Pugh, in Environment-Sensitive Failure of Engineering Materials (AIME,1979) p. 18

(14) D. Eliezer, D.G. Chakrapani, C.J. Altstetter, and E.N. Pugh, Met. Trans., 10A (1979) p. 935

(15) D.A. Vermilyea, J. Electrochem. Soc., 119 (1972) p. 405

(16) D.A. Vermilyea, Corrosion, 29 (1973) p. 442

(17) A.R. Troiano, Trans. ASM., 52 (1960) p. 54

(18) R.A. Oriani, Ber. Bun. Gschft. Physcalishe Chem., 76 (1972) p. 848

(19) C.F. Barth, E.A. Steigerwald, and A.R. Troiano, Corrosion, 25 (1969) p. 353

(20) H.H. Johnson, J.G. Morlet, and A.R. Troiano, Trans. Met. Soc. AIME, 212 (1958) p. 528

CRACK PROPAGATION

RATE CONTROLLING PROCESSES AND CRACK GROWTH RESPONSE

R. P. Wei
Department of Mechanical Engineering and Mechanics
LEHIGH UNIVERSITY
Bethlehem, PA 18015 USA

Based on a series of parallel fracture mechanics and surface chemistry
studies on an AISI 4340 steel, exposed to hydrogen containing gases (hydrogen, hydrogen sulfide and water vapor) and to water, a number of processes
that control the rate of crack growth under sustained load has been identified. Crack growth response is shown to be determined principally by the
pressure and temperature dependence of the underlying rate controlling process (*viz.*, gas transport, surface reaction and diffusion), and can change to
reflect the transfer of control from one process to another as environmental
conditions are altered. Crack growth response and conditions for the transfer of control are discussed by means of illustrative examples. Crack
growth in binary mixtures, in which one of the components acts as an inhibitor, is considered on the basis of its competition with the aggressive component for surface adsorption sites. Extension of the concepts and understanding developed thus far to fatigue crack growth is briefly discussed.
The overall concept of rate controlling processes in providing a framework
for the understanding of environment assisted crack growth in high-strength
alloys and for the utilization of data in design is discussed.

Introduction

Environment assisted subcritical crack growth in high-strength steels and other high-strength alloys (particularly in hydrogen and hydrogenous environments) is an important technological problem of long standing. It relates directly to the issues of structural integrity, durability and reliability. Quantitative understanding of the overall characteristics of and mechanisms for environment assisted crack growth is essential for the development of alloys and inhibitors to provide improved cracking resistance, and of procedures for the utilization of materials characterization data in design. A series of parallel fracture mechanics and surface chemistry studies on an AISI 4340 steel, exposed to hydrogen containing gases (such as, hydrogen, hydrogen sulfide and water vapor) and to water, has provided a fairly comprehensive set of experimental data and a reasonably coherent framework for the understanding of environment assisted crack growth [1-15]. Principal findings from these studies have been summarized in a recent review paper [16]. In this paper, a more detailed examination of the role of rate controlling processes for crack growth in determining changes in crack growth response with environmental conditions is made. Extension of the understanding developed thus far to the consideration of crack growth in binary gas mixtures is described. To provide background for these considerations, the rate controlling processes for crack growth and some important findings from the previous studies are reviewed briefly.

Rate Controlling Processes and Relevant Findings

The overall crack growth response is governed by one or more of a number of processes operating in sequence in conjunction with the mechanical driving force for crack growth (the mechanical driving force being characterized by either the applied stress or the crack tip stress intensity factor K). The various processes that might be involved in the embrittlement of ferrous alloys by hydrogen and hydrogenous gases (such as H_2O and H_2S) are illustrated in Figure 1 and are as follows [17, 18]:

o   Transport of the gas or gases to the crack tip.

o   Sequential processes involved in the reactions of the gas or gases with newly created crack surfaces to evolve hydrogen (*viz.*, physical and dissociative chemical adsorption).

o   Hydrogen entry (or, absorption).

o   Diffusion of hydrogen to the fracture (or embrittlement) site.

o   Hydrogen-metal interactions leading to embrittlement (*i.e.*, the embrittlement reaction).

Only embrittlement at low temperatures is considered here. As such, high temperature processes (such as, internal oxidation, sulfidation and hydrogen attack) are not included in the aforementioned embrittlement sequence.

A comprehensive series of experiments on high-strength steels and supporting analytical studies were carried out to identify the rate controlling processes for crack growth and to provide insight into the potential mechanisms for such growth [1-15]. It is now clear [16] that environment assisted crack growth results from embrittlement by hydrogen that is produced by the reactions of the hydrogenous environments (H_2, H_2O and H_2S) with the freshly created crack surfaces. The extent of surface reaction is limited, and

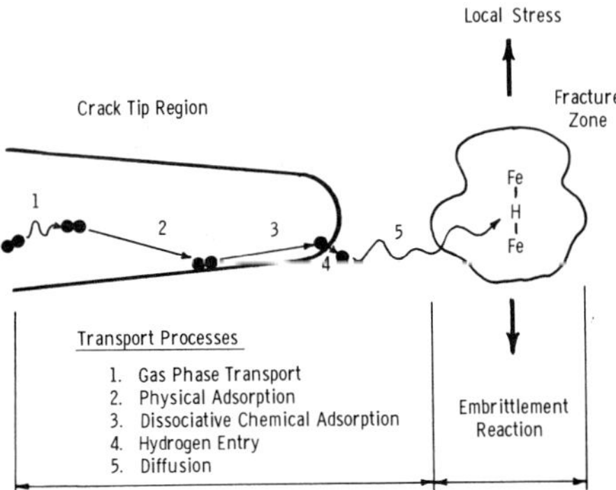

Fig. 1 – Schematic illustration of various sequential
processes involved in embrittlement of ferrous
alloys by external environments. (Embrittlement reaction is depicted schematically by
the Fe–H–Fe bond.) [17, 18]

amounts to no more than one or two atom layers in the case of H_2O and H_2S.
In other words, embrittlement results from hydrogen that is produced by the
reactions of the order of 10^{15} molecules of gas per square centimeter of crack
surface. More rapid surface reactions tend to produce greater embrittlement
or higher growth rates. For example, the three gases used in these studies
[16] rank, in decreasing order of severity, as $H_2S > H_2 > H_2O$, with the reaction rates associated with hydrogen production differing by about 9 orders
of magnitude at room temperature. The actual crack growth rates are determined, however, by the rate controlling process and the embrittlement mechanism (which is believed to be sensitive to microstructure). The crack growth
response, in turn, principally reflects the response of the relevant rate controlling processes [16]. Some of these processes have been identified and
confirmed, and are illustrated in Figure 2.

For the less reactive gases (such as water vapor and hydrogen) at low
pressures, crack growth is controlled by the rate of surface reactions. For
highly reactive gases (such as hydrogen sulfide), crack growth may be controlled by the rate of transport of the environment to the crack tip at low
pressures. At sufficiently high pressures, crack growth may become controlled
by the rate of diffusion of hydrogen into a region of material ahead of the
crack tip in both cases. The actual pressure and temperature dependence for
crack growth, therefore, reflects the dependence of the rate controlling processes on these parameters, and reflects also the transfer of control from one
process to another. A more detailed consideration of this transfer of control
is given in the following section.

Embrittlement apparently involves a region of material ahead of the crack
tip; that is, it occurs by the so-called "volume embrittlement" mechanism.
Although the actual mechanisms of embrittlement were not considered in these
studies, the cracking data indicate that they must cause fracture to occur
rapidly and with a very low activation energy ($i.e.$, less than about 5 kJ/mol).
As such, none of the observed pressure and temperature dependence has been related to the embrittlement mechanism per se.

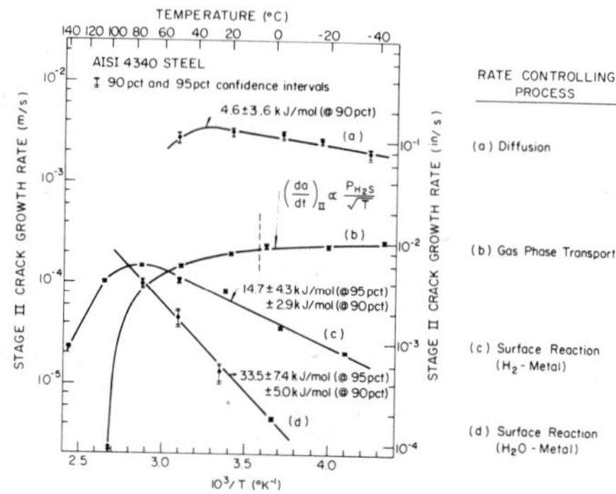

Fig. 2 - The temperature dependence and the corresponding rate controlling processes for Stage II crack growth in various hydrogenous environments: (a) H_2S at 2.66 kPa (20 torr), (b) H_2S at 0.13 kPa (1 torr), (c) H_2 at 133 kPa (1000 torr), and (d) H_2O (liquid). [1, 8, 9]

Changes in Crack Growth Response

From Figure 2, it can be seen that environment assisted crack growth is controlled by a number of processes in the embrittlement sequence, with a particular process controlling over a given range of environmental conditions of pressure and temperature. The possible transfer of control from one process to another and the resultant change in crack growth response, as one changes environmental conditions, must be clearly recognized both for the understanding of embrittlement mechanisms and in extrapolating and interpreting experimental data for design and analysis. To emphasize this point, a number of examples are considered here, using results from the recent studies [1, 8, 9].

These studies have shown that crack growth may be controlled by the rate of transport of an aggressive gas to the crack tip, the rate of surface reaction of the gas with the newly created surfaces at the crack tip, or the rate of diffusion of hydrogen to the embrittlement region ahead of the crack tip [1, 8, 9, 16]. Modeling of crack growth for each of the controlling processes has been made [9, 11, 16]. The results are as follows:

$$\text{Transport Control:} \quad (da/dt)_{II} = C_t P_o/\sqrt{T}; \quad \text{(for Knudsen flow [19])} \tag{1}$$

$$\text{Diffusion Control:} \quad (da/dt)_{II} = C_d\sqrt{P_o}\,\exp(-E_d/2RT) \tag{2}$$

$$\text{Surface Reaction Control:} \quad (da/dt)_{II} = C_s P_o\,\exp(-E_s/RT) \tag{3}$$

The constants C_i contain chemical and physical quantities that relate to gas transport, surface reaction, etc., and reflect the susceptibility of specific

alloys to embrittlement by specific environments. For the present discussion,
they will be regarded simply as empirical constants that are not sensibly
altered by the environmental variables, *i.e.*, by external pressure (p_o) and
temperature (T). E_s and E_d are the activation energies for surface reaction
and hydrogen diffusion, respectively, and R is the gas constant. If one assumes that eqns. (1) to (3) remain valid over the ranges of temperatures and
pressures of interest, then transfer of control from one process to another
may be considered.

Crack growth response for a hypothetical system, which reflects transport
and diffusion control, is illustrated schematically in Figure 3. Line a-a
that separates the region of transport control from that of diffusion control
is obtained by equating eqns. (1) and (2), and is expressed by eqn. (4)

$$p_o = (C_d/C_t)^2 \, T \, \exp(-E_d/RT) \tag{4}$$

It is seen that over the temperature range T_A to T_B, crack growth response
is governed by gas transport at the lowest pressure (p_1) and by hydrogen
diffusion at the highest pressure (p_3). At the intermediate pressure (p_2), it
is governed by gas transport at the higher temperatures and by diffusion at
the lower temperatures. This transfer of control is consistent with crack
growth data for an AISI 4340 steel in hydrogen sulfide at 0.133 kPa and 2.66
kPa (see Figure 2) [9]. Transfer of control from gas transport to diffusion
occurred at about 310 K for H_2S pressure of 2.66 kPa, and is estimated to
occur at about 180 K for 0.133 kPa. At higher temperatures, for both pressures, yet another process intervenes which leads to an apparent loss of hydrogen embrittlement susceptibility. This drop-off in crack growth rate is
believed to be associated with gas-adsorbate equilibrium [7, 20], and still
needs to be studied in detail.

Similarly, crack growth response involving transport and surface reaction control, and diffusion and surface reaction control are illustrated
schematically in Figures 4 and 5, respectively. In the former case, transfer
from transport to surface reaction control is defined by eqn. (5), based on
the assumed rate controlling processes that govern eqns. (1) and (3).

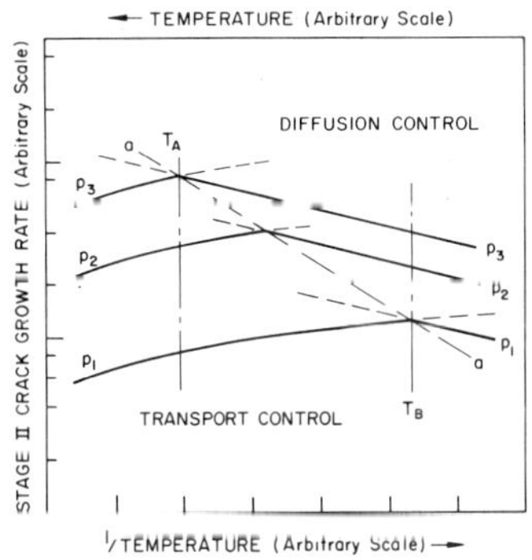

Fig. 3 – Schematic illustration of crack
growth response under transport
or diffusion control.

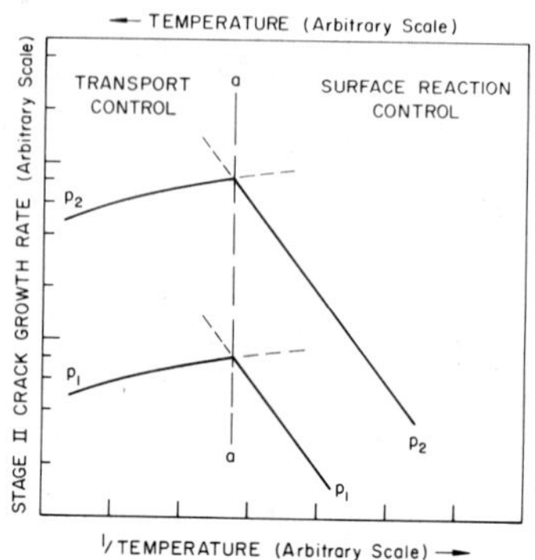

Fig. 4 – Schematic illustration of crack growth response under transport or surface reaction control.

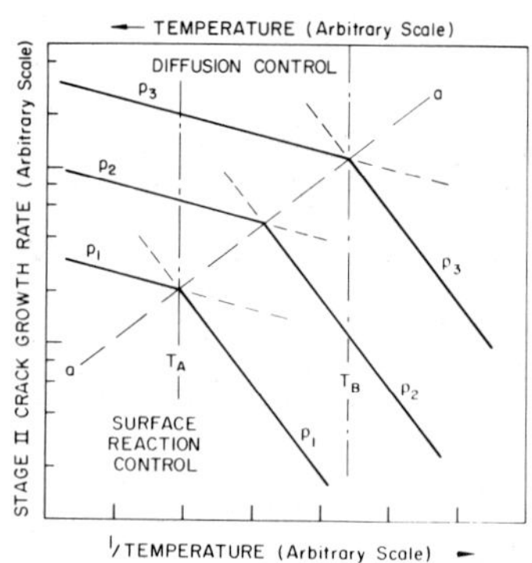

Fig. 5 – Schematic illustration of crack growth response under diffusion and surface reaction control.

$$C_t/C_s = \sqrt{T} \, \exp(-E_s/RT) \qquad (5)$$

Equation (5) indicates that the transfer would occur at a prescribed temperature for all pressures. The data reported by Johnson and Willner [21] for crack growth in an H-11 steel exposed to water vapor may be interpreted in terms of this transfer of control. For transfer from diffusion to surface reaction control, the line a–a in Fig. 5 may be obtained from eqns. (2) and (3) and is defined by eqn. (6).

$$P_o = (C_d/C_s)^2 \, \exp[(2E_s - E_d)/2RT] \qquad (6)$$

Equation (6), and hence line a-a, would have a different form if the pressure dependence for the surface reaction is different from the first power dependence assumed here. Again, it can be seen that in the temperature range from T_A to T_B, depending on pressure, crack growth response may be either controlled entirely by diffusion or by surface reaction, or by a combination of the two processes. Under certain circumstances, three different processes may control crack growth over a given range of temperatures at a prescribed pressure. A hypothetical case is illustrated schematically in Figure 6.

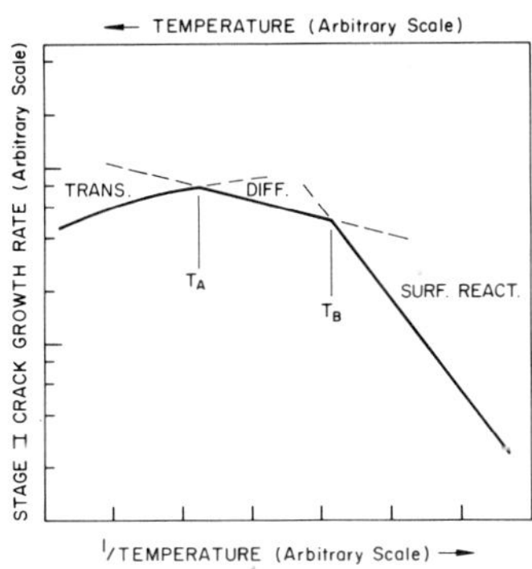

Fig. 6 – Schematic illustration of crack
growth response under transport,
diffusion or surface reaction
control.

From these illustrative examples, it is clear that crack growth response directly reflects the response of the underlying control processes to changes in environmental variables. Understanding of cracking response and extrapolation of experimental data, therefore, must be based on a clear understanding and identification of the processes that control crack growth. Meaningful information can be obtained only from fundamental and quantitative studies of the relevant chemical and physical processes that affect crack growth.

Crack Growth in Binary Gas Mixtures

For most practical cases, more than one gas is encountered in a given environment. It is useful therefore to extend the understanding to the case of environment assisted crack growth in gas mixtures. Experimental evidence obtained to date [16] clearly indicates that the rate of environment assisted crack growth is dependent on the rate of hydrogen production at the crack surfaces. This fundamental evidence provides the basis for considering crack growth in gas mixtures, more specifically, for examining the effect of an inhibitor gas (such as, oxygen or carbon monoxide). For simplicity, the case of a binary mixture at low temperatures is considered; that is, the temperatures are to be sufficiently low to preclude significant interactions between the two component gases. One of the component gases is taken to be an inhibitor (that is, a gas which will react with the clean metal surface but will not enhance crack growth). As such, the results may be used, for example, to examine the influence of oxygen (an inhibitor) on crack growth in humid air, where water vapor is the aggressive component. Crack growth under sustained-load is considered here. Consideration of fatigue crack growth in gas mixtures is given elsewhere [22].

It is assumed that (i) both gases are strongly adsorbed on the clean
metal surface, (ii) chemical adsorption of either gas at a given surface site
precludes further adsorption at that site, (iii) the ratio of partial pressures
of the gases at the crack tip is essentially the same as that of the surrounding environment, and (iv) no capillary condensation of either gas occurs at
the crack tip. The rate of crack growth in the gas mixture, $(da/dt)_m$, is assumed to be proportional to the extent of surface reaction with the aggressive
gas during a prescribed time interval, or to θ_a. If one now assumes, again
for simplicity and as an illustration, that the kinetics of surface reaction
of both gases are first order with respect to pressure and to available surface sites, then the rate equation for the surface reactions may be written
as follows [22]:

$$\frac{d\theta_a}{dt} = k_a p_a (1 - \theta) \tag{7}$$

$$\frac{d\theta_i}{dt} = k_i p_i (1 - \theta) \tag{8}$$

where the subscripts a and i denote the aggressive and inhibitor gases respectively. The quantities k_a, p_a, k_i and p_i are the reaction rate constants
and partial pressures of the gases at the crack tip, respectively. The coverages θ_a and θ_i denote the fraction of surface that has reacted with the aggressive and inhibitor gases, respectively. Coverage for both gases is denoted by θ, where $\theta = \theta_a + \theta_i$ and $0 \leq \theta \leq 1$.

Equations (7) and (8) can be solved straightforwardly to obtain the extent of reaction of a fresh surface with each gas after being exposed to the
gas mixture for a time t [22].

$$\theta_a = \frac{k_a p_a}{k_a p_a + k_i p_i} \{1 - \exp[-(k_a p_a + k_i p_i)t\} \tag{9}$$

$$\theta_i = \frac{k_i p_i}{k_a p_a + k_i p_i} \{1 - \exp[-(k_a p_a + k_i p_i)t]\} \tag{10}$$

It follows that the fraction of surface that has reacted with the aggressive
gas (θ_a) in relation to the total surface coverage (θ) after any time t is
given by eqn. (11).

$$\frac{\theta_a}{\theta} = \frac{\theta_a}{\theta_a + \theta_i} = \left(1 + \frac{k_i p_i}{k_a p_a}\right)^{-1} \tag{11}$$

Regardless of the rate controlling process, the total surface coverage (θ)
will tend to unity (*i.e.*, $\theta = 1$) for the case of sustained-load crack growth.
Thus, the fraction of crack surface that reacts with the aggressive gas to
contribute to crack growth is given by eqn. (12).

$$\theta_a = \left(1 + \frac{k_i p_i}{k_a p_a}\right)^{-1} ; \quad (\theta = 1) \tag{12}$$

Since the rate of environment assisted crack growth is proportional to the extent of surface reaction (or to hydrogen production), the crack growth rate in the mixed gases, $(da/dt)_m$, can be obtained from the corresponding rate in a pure gas, $(da/dt)_p$, and eqn. (12).

$$\frac{(da/dt)_m}{(da/dt)_p} = \left(1 + \frac{k_i p_i}{k_a p_a}\right)^{-1} = \left(1 + \frac{k_i p_{io}}{k_a p_{ao}}\right)^{-1} \tag{13}$$

The subscript o is used to denote pressures in the external environment, and reflects the assumption that the partial pressure ratio at the crack tip is the same as that in the external environment. It should be re-emphasized that only the competition between two gases has been modeled here, and eqn. (13) is expected to apply equally well to transport, surface reaction and diffusion controlled crack growth.

Based on the foregoing considerations and on recent studies of the influence of water vapor on fatigue crack growth in a 2219-T851 aluminum alloy [23], a re-examination of Speidel's data for the influence of relative humidity on crack growth in a 7075-T651 aluminum alloy has been made [24]. It is assumed that the rate of crack growth in <u>pure</u> water vapor would be independent of pressure over the range of water vapor partial pressures or relative humidities [23], and would be equal to the rate at 100 pct relative humidity (which presumes capillary condensation) or for tests in distilled water (or $\dot{a}_s$). With this assumption, it is implicitly assumed that crack growth is controlled by either hydrogen diffusion or the embrittlement process itself. By expressing the water vapor pressure (p_d) in terms of the saturation vapor pressure (p_{as}) and relative humidity (ϕ), the rate of crack growth in humid air ($\dot{a}$) can be rewritten from eqn. (14) as follows:

$$\frac{\dot{a}}{\dot{a}_s} = \left(1 + \frac{k_i p_i}{k_a p_{as} \phi}\right)^{-1} \tag{14}$$

Taking oxygen partial pressure in air to be 20.2 kPa, and saturation vapor pressure at 23°C to be 2.83 kPa, and assuming that a ratio of k_i/k_a of 0.14 (which approximates the ratio between the reaction rate constants for oxygen and water vapor over a 2219-T851 aluminum alloy [23, 25]) may be used, eqn. (14) is used to compare with Speidel's data on a 7075-T651 aluminum alloy (see Figure 7). Excellent agreement is seen, with capillary condensation indicated at relative humidity values above about 60 pct.

The inhibiting effect of oxygen on crack growth in high-strength steels exposed to water vapor or hydrogen may be readily understood in relation to eqns. (12) to (14) [26]. For the case of water vapor and oxygen, the ratio of k_i/k_a is of the order 10^9 [1-3]. Thus, the proficiency of oxygen in inhibiting crack growth in the absence of capillary condensation, can be easily demonstrated. Although eqn. (13) cannot be used directly to estimate the influence of oxygen in inhibiting crack growth in hydrogen (because the assumed pressure dependence differs from that of hydrogen reactions), the proficiency

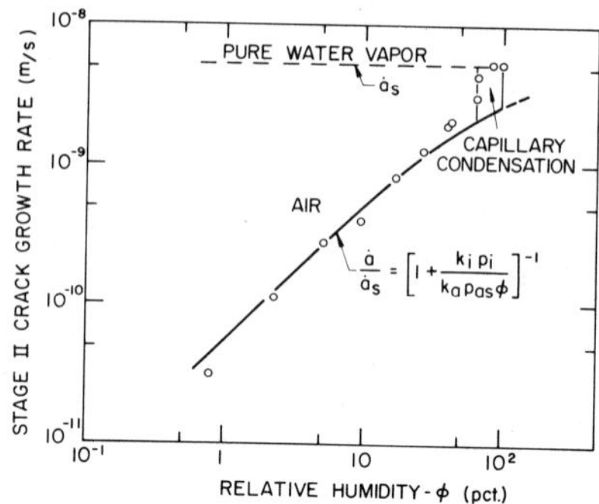

Fig. 7 – Influence of relative humidity on
sustained-load crack growth in a
7075-T651 aluminum alloy (SL orientation) in humid air at 23°C.
K = 19 MPa√m̄ [24].

of oxygen nevertheless may be qualitatively understood in terms of this
equation and the known reaction rates of oxygen and hydrogen with clean steel
surfaces [2, 3, 7].

Similar considerations of fatigue crack growth in binary gas mixtures
[22] show that crack growth response can be described by a relationship that
is comparable to eqn. (13). Comparison with available experimental data on
aluminum alloys [27, 28] indicates that the influence of relative humidity on
fatigue crack growth can be described reasonably well also by the competition
between atmospheric moisture (the aggressive component) and oxygen (the inhibitor) for adsorption sites on the new crack surfaces [22]. Preliminary
data on fatigue crack growth in binary mixtures of hydrogen sulfide and carbon monoxide [30, 31] suggest that the concept may be applied equally well to
the more complex situations.

Discussion

It is seen that the crack growth response is determined by a number of
processes, operating in sequence, that are responsible for the production and
transport of atomic (or protonic) hydrogen to the embrittlement region. These
processes include transport of the aggressive gas to the crack tip, surface
reactions of the gas with the newly created crack surfaces at the crack tip
to evolve hydrogen, and diffusion of hydrogen to the embrittlement region
ahead of the crack tip. All of them precede the embrittlement reaction, and
appear to be slower than the embrittlement reaction itself. The operation
of a given process as the rate controlling process depends on the particular
combination of pressure and temperature. Crack growth response reflects directly the pressure and temperature dependence of the operating rate controlling process, and can change as the rate controlling processes change
with environmental conditions. The functional dependence used in the illustrative examples applies to specific material-environment combinations and
processes, and can be expected to have different forms for different combinations of material, environment and environmental conditions. The specific
forms and rate controlling processes must be established through quantitative

measurements of all the relevant processes and quantitative analyses. Although a sharp boundary is used to delineate the transfer of control from one process to another in the examples, in reality such transfers are expected to be less abrupt.

The concepts and understanding that have been developed from the work on high-strength steels, under sustained loading [1-16], have now been extended successfully to the understanding and quantitative modeling of environment assisted fatigue crack growth [22, 23, 32, 33], and of crack growth in binary gas mixtures under both sustained loading and in fatigue [22, 30, 31]. These concepts and models appear to be broadly applicable over a range of materials and environments. Thus, it appears that a consistent framework for the understanding of environment assisted crack growth, and for the interpretation and extrapolation of data for design, has been firmly established. Meaningful progress in understanding can be made, and can be achieved only through coordinate and cooperative efforts by researchers in chemistry, mechanics and materials science (including solid state physics), vis-à-vis, the traditional approach involving individual investigators. To be meaningful, these efforts must be made on a quantitative basis.

For complex environments and for conditions of high pressure and temperature, similar types of quantitative, systematic investigations are needed. Internal reactions of the environment with the material and interferences caused by the corrosion products must now be considered. A concerted effort must be made to begin to unravel the many complex interactions if significant advances in the resolution of materials problems are to be accomplished in the coming decades.

Acknowledgement

Partial support by the Office of Naval Research, under Contract N00014-75-C-0543, NR 036-097, in the preparation of this paper is gratefully acknowledged. Helpful discussions by his colleague, Dr. Gary W. Simmons, are very much appreciated.

References

1. G. W. Simmons, P. S. Pao and R. P. Wei: Met. Trans. A, 1978, Vol. 9A, p. 1147.

2. G. W. Simmons and D. J. Dwyer: Surface Sci., 1975, Vol. 48, p. 373.

3. D. J. Dwyer, G. W. Simmons and R. P. Wei: Surface Sci., 1977, Vol. 64, p. 617.

4. D. J. Dwyer: Ph.D. Dissertation, Lehigh University, 1977.

5. R. P. Wei and G. W. Simmons: in Stress Corrosion Cracking and Hydrogen Embrittlement of Iron Based Alloys, ed. J. Hochmann, J. Slater and R. W. Staehle, NACE, 1978, p. 751.

6. P. S. Pao, W. Wei and R. P. Wei: Environment Sensitive Fracture of Engineering Materials, ed. Z. A. Foroulis, TMS-AIME, New York, 1979, p. 565.

7. N. H. Chan, K. Klier and R. P. Wei: in Hydrogen in Metals, Suppl. to Trans. Japan Inst. Metals, 1980, Vol. 21, p. 305.

8.   M. Lu, P. S. Pao, N. H. Chan, K. Klier and R. P. Wei: in Hydrogen in Metals, Suppl. to Trans. Japan Inst. Metals, 1980, Vol. 21, p. 449.

9.   M. Lu, P. S. Pao, T. W. Weir, G. W. Simmons and R. P. Wei: "Rate Controlling Processes for Crack Growth in Hydrogen Sulfide for an AISI 4340 Steel", submitted for publication to Metallurgical Transactions A, 1980.

10. R. P. Wei and G. W. Simmons: Scripta Met., 1976, Vol. 10, p. 153.

11. M. Lu and R. P. Wei: "A Model for Diffusion Controlled Crack Growth in Gaseous Environments", submitted for publication to Scripta Metallurgica, 1980.

12. Y. T. Chou and R. P. Wei: Acta Met., 1975, Vol. 23, p. 279.

13. Y. T. Chou, K. Y. Tsao and R. P. Wei: Mat. Sci. Eng., 1976, Vol. 24, p. 101.

14. Y. T. Chou, R. S. Wu and R. P. Wei: Scripta Met., 1978, Vol. 12, p. 249.

15. M. C. Lu, Y. T. Chou and R. P. Wei: "Tetragonal Distortion and Drift Flow of Solute Atoms Near a Crack Tip", submitted for publication to Materials Science and Engineering, 1980.

16. R. P. Wei, K. Klier, G. W. Simmons and Y. T. Chou: "Fracture Mechanics and Surface Chemistry Investigations of Environment-Assisted Crack Growth", to be published as Proceedings of Alexander R. Troiano Honorary Symposium on Hydrogen Embrittlement and Stress Corrosion Cracking, 1980.

17. D. P. Williams, III, P. S. Pao and R. P. Wei: in Environment-Sensitive Fracture of Engineering Materials, ed. Z. A. Foroulis, TMS-AIME, New York, 1979, p. 3.

18. R. P. Gangloff and R. P. Wei: Met. Trans. A, 1977, Vol. 8A, p. 1043.

19. S. Dushman: Scientific Foundations of Vacuum Techniques, 2nd ed., ed. J. M. Lafferty, Wiley, 1962, p. 88.

20. N. H. Chan, K. Klier and R. P. Wei: Scripta Met., 1978, Vol. 12, p. 1043.

21. H. H. Johnson and A. M. Willner: Appl. Mater. Res., 1965, p. 34.

22. R. P. Wei and G. W. Simmons: "Surface Reactions and Fatigue Crack Growth", 27th Sagamore Army Materials Research Conference on FATIGUE-ENVIRONMENT AND TEMPERATURE EFFECTS, Bolton Landing, New York, 14-18 July 1980.

23. R. P. Wei, P. S. Pao, R. G. Hart, T. W. Weir and G. W. Simmons: Met. Trans. A, 1980, Vol. 11A, p. 151.

24. M. O. Speidel: in The Theory of Stress Corrosion Cracking in Alloys, ed. J. C. Scully, NATO, Brussels, 1971, p. 289.

25. R. G. Hart, G. W. Simmons and R. P. Wei: "Studies of the Reaction of Oxygen and Water Vapor With 2219-T851 Aluminum Alloy", submitted for publication to J. Electrochem., 1980.

26. G. G. Hancock and H. H. Johnson: Trans. Met. Soc. AIME, April 1966.

27. F. J. Bradshaw and C. Wheeler: Intl. J. Fract. Mech., 1969, Vol. 5,
     p. 255.

28. A. Hartman, F. J. Jacobs, A. Nederveen and R. DeRijk: NLR Tech. Note
     No. M2181, National Aerospace Laboratory, Amsterdam, The Netherlands,
     1967.

29. J. A. Feeney, J. C. McMillan and R. P. Wei: Met. Trans., 1970, Vol. 1,
     p. 1741.

30. R. P. Wei and G. W. Simmons: "Fracture Mechanics and Surface Chemistry
     Studies of Steels for Coal Gasification Systems", Final Tech. Rept.,
     Contract No. EX-76-S-01-2527, DOE-Fossil Energy Research, Lehigh
     University, Bethlehem, PA, 1980.

31. R. L. Brazill, G. W. Simmons and R. P. Wei: unpublished results,
     Lehigh University, Bethlehem, PA, 1980.

32. T. W. Weir, G. W. Simmons, R. G. Hart and R. P. Wei: Scripta Met., 1980,
     Vol. 14, p. 357.

33. R. L. Brazill, G. W. Simmons and R. P. Wei: J. Eng. Matls. & Tech.,
     ASME, 1979, Vol. 101, p. 199.

DISCUSSION

<u>R.H. Jones, Battelle Northwest, Richland, WA</u>:   Does the fracture mode depend
on the rate controlling process such as transport, diffusion or surface
reaction rate as does the stage II crack growth rate?

<u>R.P. Wei</u>:   There should not be a change in the mode of fracture with the
transfer of rate controlling process; for example, from gas transport to surface reaction.   Available fractographic evidence tends to support this viewpoint.   If one considers a sufficiently broad range of environmental conditions (i.e., temperatures and pressures), over which the amount of available
hydrogen changes drastically, different fracture modes can be expected for
the same or different rate controlling processes.

<u>C. Alstetter, University of Illinois, Urbana, IL</u>:   In the case of change of
control of crack growth rate how do you determine the active surface area
from experiment?

<u>R.P. Wei</u>:   The amount of active surface area does not enter explicitly into
the consideration of transfer of control from one rate determining process
to another.   In the case of gas-transport-controlled crack growth, for example,
the rate of creation of new (active) surfaces is determined by the rate of
supply of the gas and the rate of consumption (reaction) by the active crack
surface, and is given by the crack growth rate, plate thickness, and a surface
roughness factor.

<u>A. Atrens, Brown Boveri Research Center, Baden Switzerland</u>:   Your work shows
that when gas transport is controlling, the crack velocity decreases with
increasing temperature.   What is the physical basis of this?   I would have
thought that gas phase diffusion would be higher at higher temperature. Could
you comment on how narrow the crack must be and what the hydrogen pressure
must be for these capillary effects to dominate over normal gas phase diffusion?

<u>R.P. Wei</u>:   The model used for illustration assumed that gas transport occurred
by molecular (Knudsen) flow through a narrow capillary (the crack).   The volumetric flow rate, which is determined by the mean gas velocity, is proportional to $T^{\frac{1}{2}}$ and does indeed increase with temperature but, because the
number of gas molecules per unit volume decreases with temperature (i.e.,
being proportional to T^{-1} according to the ideal gas law), the molecular flow
rate is proportional to $T^{-\frac{1}{2}}$.   Since the crack growth rate is based on a consideration of mass balance, the latter dependence on temperature was obtained.
The validity for using the molecular flow model is limited to those cases
where the crack opening is small in relation to the mean free path for the gas
molecules.   The maximum crack opening, therefore, must be determined for each
case.

<u>J.P. Fidelle, Commissariat à l'Energie Atomique Bruyères-le-Châtel</u>: Not only
your lecture on the various embrittlement stages will help a better understanding of the phenomenon, but this should also usefully help the design of
countermeasures.

AN EVALUATION OF HYDROGEN EMBRITTLEMENT

IN Cr-Mo PRESSURE VESSEL STEELS[*]

B. J. Shaw and E. W. Johnson
Westinghouse Research and Development Center
Pittsburgh, Pennsylvania 15235

Abstract

Commercial $2\tfrac{1}{4}$ Cr-1 Mo low strength steel specimens have been tested to
measure their susceptibility to hydrogen embrittlement in an environment of
H_2S at 50 psig. It was found that two factors, viz. i) the plane stress
zones on the crack front in compact tension specimens and ii) incubation
time effects, seriously confounded measurements on these steels when tested
by conventional rising load experiments. Because of the incubation time
effect, K_{or} (the stress intensity at which cracking starts in a rising load
test) is a loading rate dependent variable and is usually significantly
greater than the arrest stress intensity, K_{arr} in a bolt loaded test. K_{arr}
must therefore be used as a measure of hydrogen resistance. The incubation
time has been significantly reduced by cyclicly loading in the environment
to initiate the crack and K_{arr} has been measured by holding the specimen in
constant displacement immediately after crack initiation. The plane stress
problem has been eliminated by deeply side grooving the compact tension (CT)
specimens.

As an example of the importance of these effects a 3T CT smooth sided
specimen was compared with a side grooved 2T CT specimen of the same steel.
Whereas the K_{or} value for the smooth 3T was approximately 150 ksi $in^{\frac{1}{2}}$ the
K_{arr} value for the side notched 2T was approximately 20 ksi $in^{\frac{1}{2}}$. A study
of the effect of strength level is included.

[*]This work was supported by the Department of Energy Division of Coal
Conversion, DOE Contract DE AC05-780R13513.

Introduction

Hydrogen embrittlement of low alloy steels can be divided into three broad categories: high strength steels, with yield strengths (σ_y) of order 175 ksi or 1200 MPa which are prone to cracking in pure hydrogen gas; intermediate strength steels (σ_y ~120 ksi or 850 MPa) which are less susceptible to fracture in hydrogen gas but crack readily in H_2S gas; and low strength steels (σ_y ~70 ksi or 500 MPa) which will only crack in hydrogen charged aqueous solutions. These three categories have been delineated in earlier papers (1,2).

In this paper we report the methods developed to assess the hydrogen embrittlement susceptibility of low strength 2¼ Cr-1 Mo steels in commercially pure H_2S gas. The area explored lies in the range in which the propensity to cracking is limited and consequently the experimental technique had to be carefully refined. The experiments were designed to measure the threshold stress intensity (K_{th}) which is also commonly designated K_{Iscc} where scc is sub-critical cracking. Since our experiments measured the stress intensity at which the crack apparently arrested, we use both the terms K_{arr} and K_{Iscc} for our results.

The most important factors recognized in obtaining K_{arr} were the influence of incubation time and the role of the plane stress zones present on either side of the compact tension (CT) specimens used in the tests. Since the incubation time for crack initiation is usually large in low strength steels, it was not possible to estimate K_{Iscc} from a simple rising load test. The rising load method yields a satisfactory value of K_{Iscc} if the incubation time is very short and is sometimes referred to as the "apparent K_{Iscc}" (3). A detailed discussion of the influence of incubation time in the rising load test is presented elsewhere (4).

The effect of the plane stress zones is essentially that of pinning the crack on either side of the CT specimen. The conventional bolt loaded CT specimen, for example, would have a "tunneled" crack front, in which the center of the crack in plane strain has advanced much further than the sides. This effect is discussed by Gerberich (5). The evaluation of K_{Iscc} from a specimen with a tunneled crack configuration is extremely difficult.

The experiments presented here incorporate various methods for overcoming these difficulties and may be regarded as a hybrid of the more standard techniques. The measured K_{arr} values in many instances are surprisingly small. However, from the point of view of the industrial designer, it should be pointed out that the mechanical history of the specimen tested in the laboratory is unlikely to be found in normal commercial applications.

Discussion of the Experimental Technique for Testing
Low Strength Steels in H_2S

One of the objectives of this study was to evaluate K_{Iscc} as rapidly as possible. Whereas this is a relatively easy task in high strength steels, which have small incubation times and very small plane stress zones, it is increasingly difficult as the yield stress is decreased. The constant displacement or bolt load test described by Novak and Rolfe (6) has been demonstrated to yield a K_{arr} which is identical to K_{th} measured in constant load tests. After an incubation time t_i (in this test), the crack grows at decreasing K values until arrest occurs at K_{arr}. There are difficulties in using this test for low strength steels. First, the incubation time is usually very long and hence one is tempted to start at an applied K very much greater than K_{arr}. However, if K is too great compared with K_{arr}, the crack approaches the back face of the specimen on arrest. The experiment cannot then yield a valid K_{arr}, since the plastic zone size may be comparable with the remaining ligament. A second problem lies in the excessively large plane stress zone r_y produced in low strength steels. An approximate value for r_y is given by (7) $r_y = \dfrac{1}{2\pi}\dfrac{K^2}{\sigma_y^2}$, where K is the applied stress intensity and σ_y is the yield stress. It will be shown below that plane stress zones effectively pin the crack and lead to considerable overestimates of K_{arr}.

The two primary problems, plane stress zone pinning and incubation time, were reduced as follows.

Plane Stress Zone Elimination

In the first test described below on a 3T CT specimen, it was found that the plane stress zone was 0.5 in or 12 mm maximum (Fig. 1). It was decided therefore to incorporate side grooves on specimens with a depth approximately half of the maximum plane stress pinning zone. The result of this specimen modification can be seen in Fig. 2; the crack front is reasonably straight, thus implying to a first approximation that the entire crack front is in plane strain.

Estimates of the stress intensity K in a side grooved specimen are not as precise as those for a conventional smooth sided specimen. The simplest

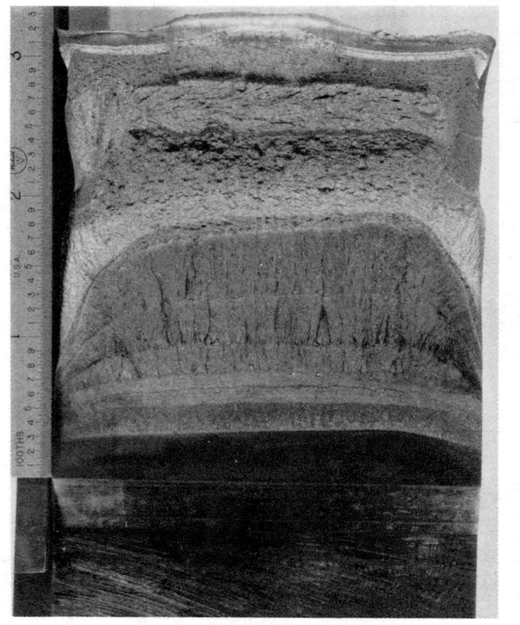

Fig. 1 – Fracture surface of 3T compact tension specimen 54H tested in 50 psig H_2S. This was a smooth sided specimen which exhibited the effect of the plane-stress zones on either side. The crack front pinning in the plane-stress zones is very striking.

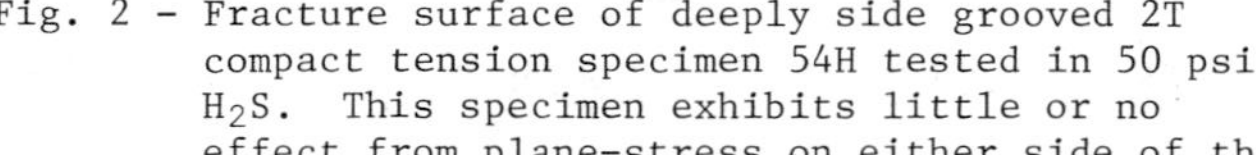

Fig. 2 – Fracture surface of deeply side grooved 2T compact tension specimen 54H tested in 50 psig H_2S. This specimen exhibits little or no effect from plane-stress on either side of the crack front.

approach for an estimate of K (8) is to assign an effective thickness to the specimen, B_{eff}, which is the geometric mean of the overall thickness, B, and the crack plane thickness, B_N, or $B_{eff} = (B \times B_N)^{\frac{1}{2}}$. This estimate for B_{eff} was used by Novak and Rolfe (6). Pollock (9) in later work concluded that the effect of side groove depth for B/B_N in the range 1.1 to 2.9 on the crack growth rate in stress corrosion tests was not significant.

Throughout the text values for K at various stages of crack growth will be given. These values are <u>estimates</u> based upon the following methods: (a) If the compliance of the specimen is measured at any stage an estimate of the crack length is obtained from methods given by Saxena and Hudak (10). [This applies for both grooved and ungrooved specimens.] (b) For a given crack length, either measured or estimated in an ungrooved specimen, K can be evaluated directly from the load P (10). (c) For a given crack length in a side grooved specimen K is estimated by replacing the specimen thickness B by an effective specimen thickness B_{eff}.

Wherever possible the crack length at any given stage is taken directly from the specimen fracture surface.

Reduction of the Incubation Time (t_i)

Once the plane stress problem is removed, the incubation time is the primary obstacle to evaluating K_{arr}, since we require to evaluate K_{arr} as rapidly as possible. There are probably a large number of factors, with varying degrees of importance contributing to t_i. These include (a) hydrogen sulfide reaction with a surface oxide film near the crack tip, (b) H_2S reaction with the fresh metal surface in the crack tip vicinity, (c) hydrogen entry into the metal and diffusion to the region of maximum triaxial tensile stress, (d) hydrogen accumulation in the triaxially stressed region until a concentration sufficient for internal crack nucleation is attained, (e) nucleation of the internal crack and growth at a rate determined by the rate of supply of hydrogen in the previously described processes, and finally (f) breakthrough of the internal crack to join the existing crack.

Of these the surface oxide film at the crack tip after air pre-fatigue cracking is probably one of the most important factors.

In order to minimize incubation time effects we therefore fatigue pre-cracked specimens in the H_2S environment. Since we are no longer concerned with measurements of K_{or} the maximum fatigue stress level is <u>not</u> a consideration as it would specifically be for measurements of K_{Ic} (i.e. K_{or} measured in air). The procedure developed for crack initiation test was to cyclicly load the specimen in the environment at 0.5 Hz, by increasing K_{max} values[*] in steps until crack growth was clearly discernible by a change in the specimen compliance. Each step was conducted at a constant maximum displacement so that crack growth was exhibited by a decreasing maximum load. A decreasing K_{max} with crack advance is automatically brought about by this method.

Immediately after the environmental cyclic loading crack growth started, unoxidized metal was exposed and the specimen could be tested to evaluate K_{arr} without an excessive incubation time. The method used to arrive at K_{arr} was simply to hold the specimen at a constant displacement v starting at a displacement slightly higher than that associated with the last K_{max}. This is essentially a simulated bolt loaded test in which the load drop at constant displacement can be monitored. Typically in $2\frac{1}{4}$ Cr-1 Mo specimens, tempered to an ultimate strength level of order 105 ksi (724 MPa), the time for the crack to grow to K_{arr} is 2 to 3 days.

Experimental Results

Preliminary Tests: Sample No54H

In this section we present the data generated from two CT specimens (2T and 3T) taken from one steel sample. The sample is $2\frac{1}{4}$ Cr-1 Mo plate donated to the program by the American Petroleum Institute and is designated as No54. It was re-austenitized and cooled to produce a bainitic structure and subsequently tempered to a Rockwell C hardness of order 19.5. This is equivalent to a U.T.S. level of order 110 ksi (758 MPa). The heat treatments and sample analytical chemistry are given in Table I. The hardness of this sample is the same order of magnitude as the peak hardness found in the heat affected zones (HAZ) around welds in $2\frac{1}{4}$ Cr-1 Mo plate (11).

Sample No54H, 3T CT Smooth Sided Specimen. The 3T specimen was machined according to the ASTM E399 specifications. It was machined without side notches since it was large enough to have at least 60% of the crack front in a plane strain condition.

The specimen was cyclicly loaded at 0.5 Hz in 50 psig H_2S (446 kPa) environment at a maximum of 77 ksi $in^{\frac{1}{2}}$ to produce approximately .17 in of crack growth. It was then loaded and held at constant displacement in four stages of increasing applied K = 102, 124, 130 and 133 ksi $in^{\frac{1}{2}}$. At

---

[*]K_{max} is the upper K associated with the cyclic loading.

Table I. Heat Treatment and Analytical Chemistry of Sample 54

Heat Treatment of 6-in thick plate:

    As received   1750°F -   8 hr - water quench
                    1290°F -   8 hr - water quench
    Second Heat Treatment
      pre-heat   900°F - 15 hr +
               1700°F - 3 hr
    Repeated water dip quench*
        to   900°F - 1 hr - air cool
        temper   1100°F - 9 hr - air cool

*Not less than 700°F
Rockwell Hardness R_C 19.5; Equivalent UTS 108 ksi

Analytical Chemistry

Element	C	Cr	Mo	Mn	Si	Ni	Cu
Weight Percent	.125	2.19	.93	.47	.47	.13	.09

Element	P	S	Sn	Sb	O	N	As
Parts per million	143	42	106	21	51	142	249

Conversions: 1 hr = 3.6 x 10^3 s
                t°C = (t°F - 32)/1.8
               1 ksi = 6.895 MPa

each stage the compliance check indicated no crack growth had taken place. In the final stage of this set of experiments the specimen was held at a K of 133 ksi $in^{\frac{1}{2}}$ for 175 hr or just over one week.

In order to be sure that the crack tip had not blunted in these experiments, the specimen was cyclicly loaded again at a maximum K of 130 ksi $in^{\frac{1}{2}}$. It was then reloaded at a constant displacement rate until crack growth started at a K of approximately 148 ksi $in^{\frac{1}{2}}$. A series of tests at constant displacement were then carried out. In the final test the crack advanced from an applied K of 143 to 125 ksi $in^{\frac{1}{2}}$ in 71 hr, representing a constant crack growth rate of 5.5 x 10^{-3} in/hr.

At this point the tests were concluded, even though the crack was still growing extremely slowly, and the specimen was pulled apart in air. Figure 1 shows the fracture surface upon which the various test stages are reasonably clearly delineated. This one test showed that there was a significant plane-stress component of approximately 0.5 in on each side, and that crack growth rate was very slow.

Throughout this section please convert:
    1 hr = 3.6 x 10^3 s
    1 in = 2.54 x 10^2 m
    1 ksi $in^{\frac{1}{2}}$ = 1.099 MPa $m^{\frac{1}{2}}$

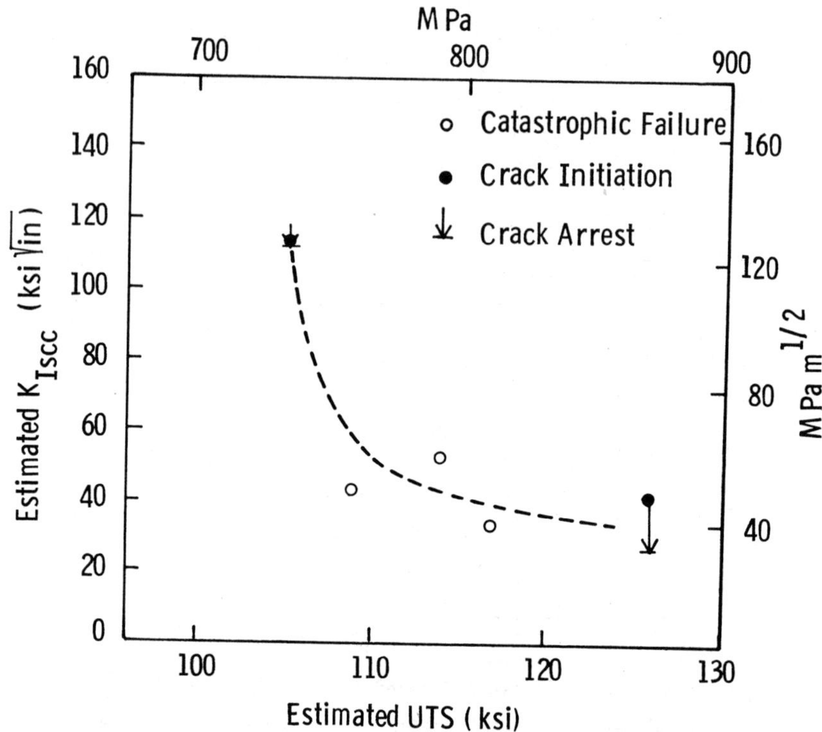

Fig. 3 - K$_{Iscc}$ as a function of the ultimate tensile strength for sample No. 74. The K$_{Iscc}$ was estimated from tests in 50 psig H$_2$S on 2T CT side grooved specimens and the UTS from the Rockwell Hardness measurements.

Table II. Results of H$_2$S Environment Tests on Sample No. 74

Specimen No.	Approximate UTS (ksi)	Estimated K$_{Iscc}$ (ksi√in)	Comments
74-3	126	28	Slow crack growth from 42 ksi√in in 54 hr.
74-1	117	34	Catastrophic fracture; specimen completely failed.
74-5	114	52	No failure after 17 hr at 50 ksi√in. Catastrophic at 52.
74-4	109	42	Catastrophic fracture.
74-2	106	113	Very small crack growth (.05 in) in 28 hr.

Conversions:
$1 \text{ in} = 2.54 \times 10^{-2} \text{ m}$
$1 \text{ ksi} = 6.895 \text{ MPa}$
$1 \text{ ksi}\sqrt{\text{in}} = 1.099 \text{ MN/m}^{3/2}$
$1 \text{ hr} = 3.6 \times 10^{3} \text{ s}$

<u>Sample No54H, 2T CT 0.25 Side Groove Test</u>. The standard 2T CT specimen was modified by the introduction of a 0.25 in side groove on the specimen side. The specimen was cyclicly loaded at 0.5 Hz to a final K_{max} of 48 ksi in$^{\frac{1}{2}}$. It was then loaded to a K of 105 ksi in$^{\frac{1}{2}}$ and held at constant displacement. After 70 hr the crack had extended to 0.35 in from the back of the specimen and had a K estimated between 20 and 25 ksi in$^{\frac{1}{2}}$. A plot of the crack length as a function of time was almost linear giving $\dot{a}$ = 22$\times$10^{-3} in/hr. The photograph of the fracture surface shows that the crack has a straight front indicating that the entire test was under the plane-strain conditions. It should be noted however that the fatigue crack front was not straight but had a "reverse-tunnel" shape (Fig. 2).

<u>Sample No74, Effect of Strength Level</u>. The effect of strength level, using the same test technique, was carried out on sample No74. The heat treatments and analytical chemistry were similar to that given in Table I, except that tempering times at 1100°F were varied to produce a range of strength levels. The details will be presented elsewhere (12).

The results of the tests are given in Table II. The highest strength specimen, which was untempered, cracked slowly from a maximum applied stress intensity K_{or} of 42 ksi in$^{\frac{1}{2}}$ to crack arrest at K_{arr} = 28 ksi in$^{\frac{1}{2}}$. The three specimens with UTS in the intermediate range of this series cracked catastro-phically. One of the three, 74-5, was held at constant displacement for 17 hr at an applied stress intensity of 50 ksi in$^{\frac{1}{2}}$ without any indication of cracking. A slight increase in K to 52 ksi in$^{\frac{1}{2}}$ resulted in an immediate catastrophic failure. In comparison, the lowest strength specimen of the set exhibited a very small amount of crack growth at K   = 113 ksi in$^{\frac{1}{2}}$.

These results indicate that, for this one sample of steel, there is a fairly sharp transition from a susceptibility to hydrogen embrittlement to relatively high resistance. This transition corresponds to the UTS range 106-109 ksi. (Fig. 3)

Conclusion

It has been shown that for low strength 2¼ Cr-1 Mo steels the plane stress zones in a conventional compact tension specimen serve to pin the crack front and thereby make the measurement of the threshold stress inten-sity in H_2S very difficult. The introduction of side grooves on the specimen substantially reduces this effect. The incubation time for crack initiation can be reduced by starting the crack with cyclic loading in the environment.

The effect of strength level on K_{Iscc} has been determined, using the modified test approach. Based upon the samples evaluated in this study it appears that hydrogen embrittlement in H_2S is a sensitive function of the strength level and that the steel is relatively immune if the UTS is below 100 ksi (690 MPa).

Acknowledgments

This work was performed on the DOE Contract DE-AC05-780R13513. The steels used in the study were provided by the API. The environmental tests were conducted by R. J. Smykal under the supervision of L. Ceschini at the Westinghouse Research and Development Center.

References

1. H. Okada, NACE-5 International Corrosion Conference Series, p. 124 (1973).

2. M. O. Speidel and P. M. Fount, ibid., p. 57.

3. W. G. Clark, Jr. and J. D. Landes, ASTM STP 610, p. 108 (1976).

4. E. W. Johnson and B. J. Shaw, submitted for publication to Met. Trans. Also Final Report U.S. Navy Contract N00014-77-C-0372 (1979).

5. W. W. Gerberich and Y. T. Chen, Met. Trans. A, p. 271 (1975).

6. S. R. Novak and S. T. Rolfe, Corrosion 26, p. 121 (1970).

7. D. Broek, Elementary Engineering Fracture Mechanics, Noordhoff (1974).

8. C. N. Freed and J. M. Krafft, Mater. 1, p. 770 (1966).

9. W. J. Pollock, Materials Note 121, Australian Dept. of Defense, Aero. Res. Labs (1979).

10. A. Saxena and S. J. Hudak, Jr., Int. Jnl. Fract. 14, p. 453 (1978).

11. R. K. Nanstad and D. A. Canonico, ORNL-5238, Dec. (1976).

12. B. J. Shaw and E. W. Johnson, Topical Report for DOE Contract DE-AC05-780R13513, Oct. (1980).

DISCUSSION

<u>A.D. Wilson, Lukens Steel Company, Coatesville, PA</u>: The crack front curvature that you showed in the fractures of your smooth CT specimens appears very similar to those we've seen in the fatigue crack propagation fractures of quenched and tempered steels. Is there a possibility that residual stresses from the quenching operation are influencing your results? Since most Cr-Mo steels are eventually used in the stress-relieved condition, it would not be a concern for service applications, but it may be of concern in your experimental program.

<u>B.J. Shaw</u>: Large steel samples, typically 6 in. thick plate, were quenched from 1700°F to 900°F and transferred to a furnace at 900°F to effect a transformation to bainite. After tempering treatments at 1100°F specimens were machined from the center of the samples. I believe sufficient stress relief would take place during the tempering treatments at 1100°F, even though postweld heat treatments are normally at 1250–1275°F.

INFLUENCE OF MICROSTRUCTURE ON

THE FATIGUE CRACK GROWTH OF A516 IN HYDROGEN

Harry F. Wachob

Failure Analysis Associates
Palo Alto, CA 94304
&
NASA-Ames Research Center

Howard G. Nelson

NASA-Ames Research Center
Moffett Field, CA 94035

Introduction

Some day hydrogen may be used as a viable energy storage and transport
medium within the United States. Hydrogen gas may be used to dilute and extend our present methane supply as a blend or may even be used in its pure
elemental form as a primary fuel (1). Independent of the methods of production, storage, and distribution, the interaction of hydrogen with its containment material will play an integral role in the success of a hydrogen energy
program. Presently, the selection of hydrogen containment materials can be
made such that the material will remain reasonably free from environmental
degradation; however, costly alloying additions are required. Unfortunately,
high alloy steels are economically prohibitive when large-scale hydrogen
energy storage, transmission, and conversion systems are desired. Therefore,
in order to implement such hydrogen energy systems in the future, existing
low-cost materials must be improved via mechanical, thermal, or thermo-mechanical processing methods or new low-cost materials which are compatible with
hydrogen must be developed.

Originally, low strength, low alloy steels at room temperature were
thought to be immune to hydrogen gas embrittlement, since no sustained load
crack growth is observed (2). However, results of Clark (3) in HY80 and Nelson (2) in SAE 1020 have shown that the fatigue crack growth rate can be
greatly accelerated in the presence of hydrogen gas. In recent results reported by Louthan (4) and Mucci (5), the smooth bar fatigue life of an A106B
pipeline steel was reduced up to a factor of ten when the tests were performed in a 13.8 MPa hydrogen environment. These results suggest that the selection of material for structures designed to operate in hydrogen under cyclic
loads must include consideration of hydrogen/metal fatigue interaction.

Although the hydrogen/metal fatigue interaction can be severe in low
strength low alloy steels, the degree of degradation may be altered by the
underlying ferrous microstructure. At present, no correlation between microstructure and degree of hydrogen susceptibility exists for low strength
steels. However, in high strength steels, susceptibility to hydrogen embrittlement has been shown to be strongly sensitive to the metallurgical microstructure (6, 7). In addition, compositional effects and grain size can

modify any specific susceptibility (7). In general, the best environmental resistance seems to be provided by a well tempered martensitic or bainitic microstructure, which has been ausformed to produce refined plate sizes and finely dispersed array of fine carbides (8, 9). A spheroidized structure of uniformly dispersed carbides in second best (9, 10) with a normalized structure being the poorest (9). Unfortunately, a similar ranking is not available for medium and low strength low alloy steels.

The objective of the current research program was to determine the role of ferrous microstructure on the near threshold fatigue crack growth rate for a low to medium strength low alloy steel. From these results, one can then identify the ferrous microstructure which is least susceptible to hydrogen degradation.

Experimental Procedure

Materials

Alloy A516-G70 plate chosen as a representative steel for this research program, was heat treated to obtain a variety of mechanical properties and microstructures. The plate, 1.25cm thick, was austenitized at either $900^{\circ}C$ or $1200^{\circ}C$ for 45 minutes, then icewater quenched, isothermally quenched, or furnace fan cooled. Some austenizing treatments were followed by tempering at $450^{\circ}C$ for 1.5 hours. The ambient mechanical properties, microstructures, and respective processing schedules are listed in Table I.

TABLE I.   PROPERTIES OF HEAT TREATED A516

CONDITION	YS, MPa	γ-CS, μM	MICROSTRUCTURE
NORMALIZED (AS RECEIVED) 900 C	330	35	PEARLITE-FERRITE
MARTENSITIC γ-900 C IWQ TEMPERED 1.5 hr @ 450 C	905	30	TEMPERED MARTENSITE AND BAINITE
NORMALIZED 1200 C	305	180	PEARLITE-FERRITE
BAINITIC γ-1200 C ISQ - TEMPERED 1.5 hr @ 450 C	415	200	BAINITE WITH CONTINUOUS α-FERRITE AT GRAIN BOUNDARIES (10%)
MARTENSITIC γ-1200 C IWQ TEMPERED 1.5 hr @ 450 C	820	200	TEMPERED MARTENSITE AND BAINITE

Testing Procedure

Tensile specimens and WOL - compact tension specimens (1.25cm thick) were cut from the as-received plate; the tensile axis was parallel to the long transverse rolling direction and the WOL specimens were in the L-T orientation.

Tests in high pressure hydrogen were performed using a stainless steel high pressure chamber. After the chamber was sealed and evacuated, it was purged and backfilled slowly with hydrogen. As part of each charging procedure, the high purity hydrogen was passed through a coil submerged in liquid nitrogen to lessen water and other active gas contaminants. The chamber valve was then closed upon reaching the desired pressure, and the specimen and crack opening displacement gage were allowed to come to equilibrium with the hydrogen environment for 8×10^5 seconds prior to initiating the test.

The fatigue crack growth tests were performed according to the guidelines of ASTM 647-78 (1978) (11) with the exception that the Newmann equation for stress intensity (12) was used as an alternative to the ASTM calculation. Fatigue tests were performed under constant crack opening displacement; this resulted in an ever decreasing load increment ΔP, $(P_{max}-P_{min})$, or ΔK as the crack front extended. This load decay, as a function of cycles, was monitored and later used in the data reduction program. The fatigue crack growth tests were performed using a 1Hz haversine waveform and a constant R ratio (P_{min}/P_{max}) of $\sim$0.15. Variations in cyclic loading rate from 0.1 to 10Hz,

did not have a significant effect on the observed fatigue crack growth rate
in air or in hydrogen.

Fractographic and metallographic analyses were performed after testing
was completed. If etching of the fracture surface or metallographic section
was desired, a solution containing 4 grams of picric acid in 100 ml of meth-
anol was used. This etching procedure permitted the delineation of several
microstructural features on the fracture surface.

Results

Fatigue Crack Growth Results

The fatigue crack growth rate (FCGR) of this A516 steel was influenced by
a high pressure hydrogen environment and by variations in ferrous microstructure. The FCGR increased and the fatigue threshold value (ΔK_o) decreased in
high pressure hydrogen compared with those values obtained in air. In addition, the near-threshold fatigue behavior was strongly influenced by the final
austenitizing temperature or austenitic grain size. Details for a given environment and microstructure are present below.

Baseline fatigue tests were performed in laboratory air (23°C and 45% RH).
Fatigue crack growth rates were obtained between 10^{-9} m/cycle (lower limit of
experimental resolution) and 10^{-6} m/cycle. At $\Delta K \simeq 16$ MPa.$\mathrm{m}^{\frac{1}{2}}$, the FCGR appeared to be independent of ferrous microstructure or austenitizing tempera-
ture, as can be seen in Figure 1. The slope of this linear regime in air is
3.8. Below $\Delta K = 16$ MPa.$\mathrm{m}^{\frac{1}{2}}$, however, the effect of final austenitizing temperature and strength level on the crack growth behavior becomes apparent. As ΔK
decreased, the FCGR decreases rapidly for those specimens austenitized at
1200°C. Threshold values (± 1 MPa.$\mathrm{m}^{\frac{1}{2}}$) in air were 14 MPa.$\mathrm{m}^{\frac{1}{2}}$ for the martensitic structure, 13 MPa.$\mathrm{m}^{\frac{1}{2}}$ for the
bainitic structure, and 12 MPa.$\mathrm{m}^{\frac{1}{2}}$ for
the normalized ferritic-pearlitic
structure. One important point to
note is that the martensitic structure
had the highest threshold value yet
had a 50% higher yield strength than
the normalized or bainitic structures.

Specimens austenitized at 900°C
exhibit a linear crack growth behavior
with ΔK above $\simeq 11$ MPa.$\mathrm{m}^{\frac{1}{2}}$ (Fig. 1).
Below this limit, the FCGR drops a-
bruptly and approaches threshold val-
ues of 9.5 MPa.$\mathrm{m}^{\frac{1}{2}}$ for the martensitic
structure and 9 MPa.$\mathrm{m}^{\frac{1}{2}}$ for the ferritic-pearlitic structure. Again, a
slightly higher threshold is obtained
in the higher strength martensitic
structure.

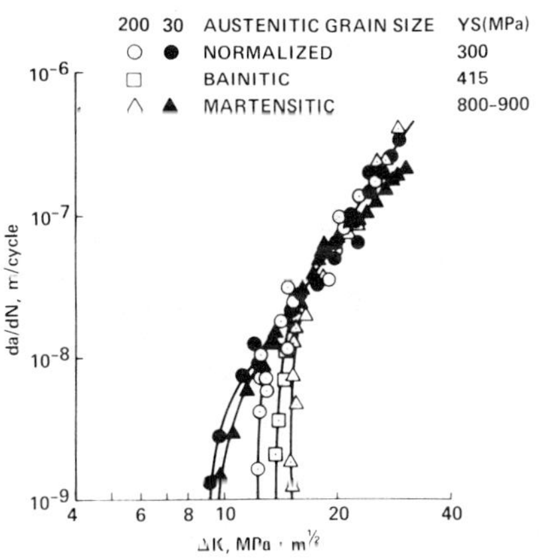

Fig. 1. The influence of austenitic
grain size and yield strength on the
FCGR of A516 at R = 0.15 and ν = 1 Hz
in air.

High pressure hydrogen is found
to alter the fatigue behavior of the
A516 steel as shown in Fig. 2. For a
given ΔK, the FCGR increases above
that observed in air when specimens
are tested in hydrogen. Also, the fatigue threshold values are lower in hydrogen than in air for all microstructures tested. The general features and
shape of the FCGR curves remain similar to the air data as is indicated in
Fig. 2.

Above $\Delta K \approx 13$ MPa·m$^{\frac{1}{2}}$ the FCGR in hydrogen appears to be independent of ferrous microstructure and prior austenitic grain size (Fig. 2). The slope of this linear ΔK curve is about 5.5 which is approximately 50% greater than the slope of 3.8 observed in air. As observed in air, the fatigue thresholds are higher for specimens austenitized at 1200°C than for those austenitized at 900°C. The threshold values for the 1200°C austenitized microstructure are 13 MPa·m$^{\frac{1}{2}}$ for the martensitic structure, 11.5 MPa·m$^{\frac{1}{2}}$ for the bainitic structure, and 10.5 MPa·m$^{\frac{1}{2}}$ for the ferritic-pearlitic microstructure. The threshold values for the 900°C austenitizing microstructures are 9 MPa·m$^{\frac{1}{2}}$ for the martensitic microstructure and 8 MPa·m$^{\frac{1}{2}}$ for the ferritic-pearlitic microstructure. Similar to that observed in air, a slight increase in threshold occurs with increased yield strength.

<u>Fractographic Results</u>

The effect of hydrogen, as seen fractographically, was to reduce the features typically associated with plasticity. The failure mode in air was transgranular ductile tearing with evidence of fatigue striations. In hydrogen, the failure mode is either transgranular cleavage or intergranular separation. Two microstructures (ferritic-pearlitic normalized at 900°C and the martensitic austenitized at 1200°C) exhibit all of the various fractographic features and will be used as examples.

The greatest change in failure mode occurred for the 900°C normalized microstructure. The primary fractographic features after fatigue in air were ductile tearing and fatigue striation formation; secondary cracking along α-ferrite-pearlite interfaces; and out-of-plane crack initiation at MnS rolling inclusions. These features were developed in air over the entire ΔK range investigated (Fig. 1). However, as ΔK decreased, less overall gross plasticity was associated with the fracture surface.

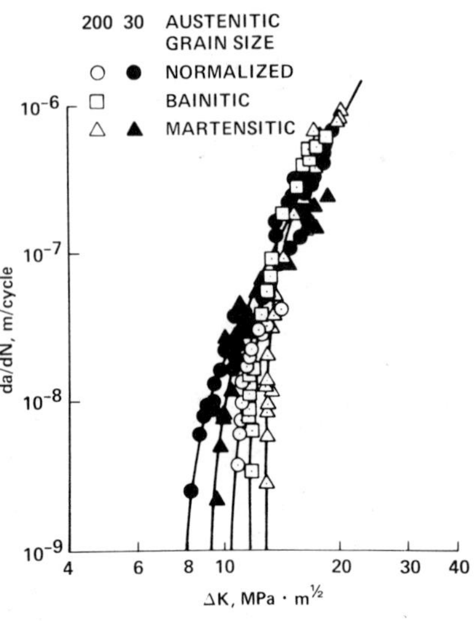

Fig. 2. The influence of austenitic grain size and microstructure on the FCGR of A516 at R = 0.15 and ν = 1 Hz in 6.9 MPa hydrogen.

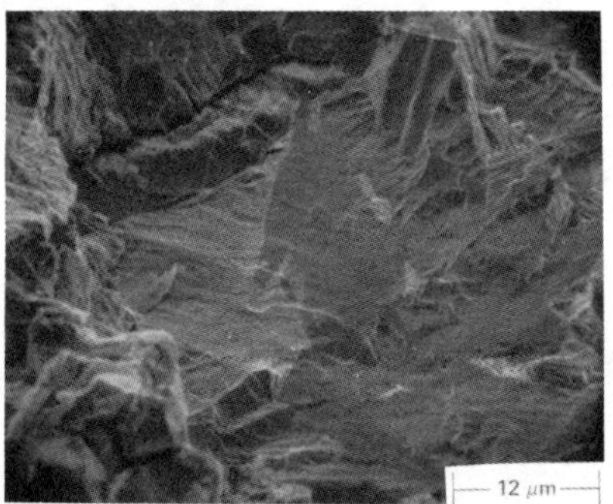

Fig. 3. Fractograph of the transgranular fracture of normalized A516 produced during fatigue at $\Delta K \approx 24$ MPa·m$^{\frac{1}{2}}$ in 6.9 MPa·H$_2$.

In hydrogen, the fracture mode of the normalized specimens gradually changed from a transgranular cleavage (Fig. 3) failure at $\Delta K \approx 24$ MPa·m$^{\frac{1}{2}}$ to an intergranular failure (Fig. 4) $\Delta K \approx 9$ MPa·m$^{\frac{1}{2}}$.

The majority of the intergranular
failures occurred at ferrite-pearlite
boundaries. (The fracture surfaces
have been etched to discern ferrite
and pearlite regions). Evidence of
gross plastic deformation, secondary
cracking, and MnS inclusion effects
was absent or substantially reduced
in hydrogen.

In the 1200°C martensitic microstructure, the failure mode was primarily transgranular in both air and
hydrogen. The fatigue surfaces produced in air show evidence of transgranular martensitic packet failures,
minor amounts of intergranular failure,
and local areas of plastic deformation
and striation formation. A typical
fracture surface produced by fatigue
in air is illustrated in Fig. 5. In a
hydrogen environment, the transgranular failures exhibited less plasticity
as seen in Fig. 6. Additionally, in
hydrogen the amount of intergranular
separation increased slightly to 10%
of the total failure surface.

No microstructural component was
consistently associated with intergranular failures in the martensitic
specimens. In the bainitic microstructure, however, intergranular
separation was found to be associated
with the pearlite-ferrite boundary,
similar to that seen in the 900°C
normalized material.

<u>Discussion</u>

The influence of microstructure
of A516 steel in air parallels previous
observation for a wide variety of plain
carbon steels. The FCGR in air at high
ΔK values is in good agreement with the literature results on SAE 1020 (2),
A516-G60 (13) A516-G70 (14), and several specially prepared Fe-C alloys (15-
16). The threshold stress intensity value at 10^{-9} m/cycle for the 900°C
normalized microstructure is very similar to that obtained for a variety of
ferritic-pearlitic steels (15-16). No threshold data is presently available
for the quenched and tempered martensitic structures or for the larger prior
austenitic grain sizes.

However, Minakawa and McEvily (17) reported recently on a duplex for-
ritic-martensitic microstructure which had threshold values of $\Delta K_{o} \simeq 14$ MPa.m$^{\frac{1}{2}}$.
This is in good agreement with the threshold values obtained for the marten-
sitic and bainitic microstructures in the present study that were austenitized at 1200°C.

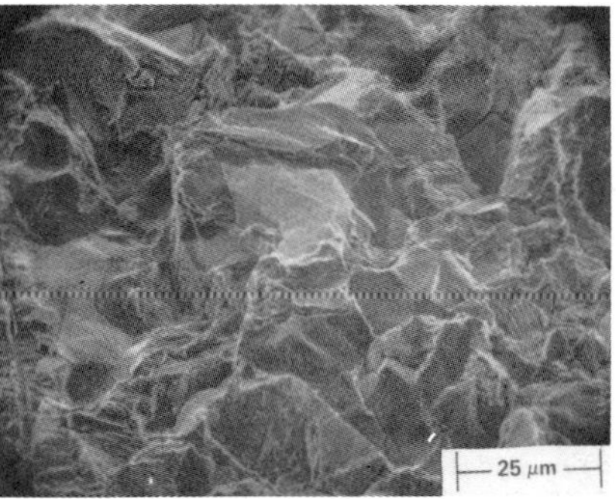

Fig. 4. Fractograph of intergranular separation of normalized A516
produced during fatigue at $\Delta K \simeq 10$
MPa.m$^{\frac{1}{2}}$ in 6.9 MPa H_2.

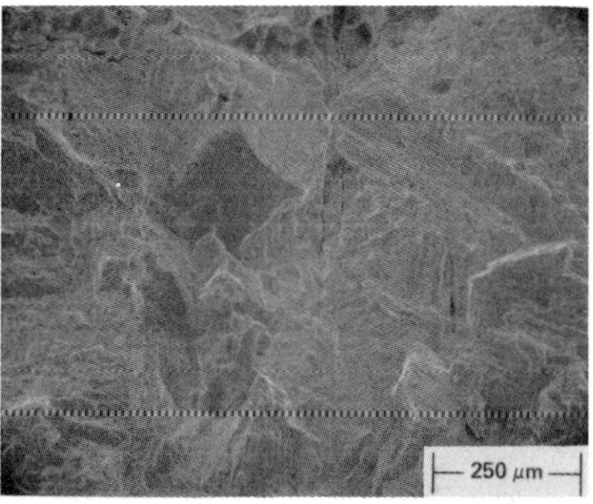

Fig. 5. Fractograph of the trans-
granular fracture of martensitic
A516 produced during fatigue at
$\Delta K \simeq 24$ MPa.m$^{\frac{1}{2}}$ in air.

The fatigue crack growth behavior of A516 steel is greatly influenced by the presence of a high-pressure hydrogen environment. The FCGR is increased by as much as a factor of ten above that observed in air (Figs. 1 and 2). This is in good agreement with earlier results of Nelson (2) on a ferritic-pearlitic SAE 1020 steel. The threshold stress intensities for fatigue crack growth are also lowered in hydrogen compared with air for all microstructures investigated (Fig. 1 and 2). Ritchie has measured the fatigue threshold in 0.1 MPa (1 atm) hydrogen for an X-60 linepipe steel and reported a value of 5.9 $MPa.m^{\frac{1}{2}}$ at 10^{-13} m/cycle (18). Additionally, he has also studied $2\frac{1}{4}$ Cr.-1Mo steel (19) and found $\Delta K \approx 9$ $MPa.m^{\frac{1}{2}}$ at a crack growth rate of 10^{-9}m/cycle. Both materials were austenitized in the 900°C range. These results are in good agreement with the values measured in the present study for a ferritic-pearlitic microstructure austenitized at 900°C.

Fig. 6. Fractograph of the transgranular fracture of martensitic A516 produced during fatigue at $\Delta K \approx 15$ $MPa.m^{\frac{1}{2}}$ in 6.9 MPa H_2.

The decrease in threshold observed in hydrogen was more severe in the ferritic-pearlitic microstructures (12% decrease) than in the martensitic microstructures (6% decrease). The fact that martensitic microstructures exhibit higher thresholds and therefore have a higher resistance to early crack growth may be a partial explanation for the observed fact that in high strength steels, the best resistance to hydrogen in an aqueous environment is found in the well tempered martensitic and bainitic microstructures (8, 9). However, comparison of these results may not be valid, because the mechanisms of hydrogen degradation in the high strength steels under a static load may not be the same as for this low strength steel under dynamic loads. Further experimentation is required to fully understand the additional hydrogen resistance observed in these cyclic crack growth tests.

The single most important microstructural variable found in this study is the prior austenitic grain size. The fatigue threshold stress intensity value is increased with increasing austenitic grain size. This has been observed by others and is the general behavior for high strength steels (7), precipitation hardened ferrites (20), and plain carbon steels (16). Present results indicate the same trend occurs for the hydrogen fatigue threshold as well. In both air and hydrogen, the fatigue threshold is observed to increase a minimum of 25% for a factor of ten increase in austenitic grain size. Finally, these results suggest that in welded structures, material experiencing the lowest austenitizing temperatures, such as in the base plate or at the parent metal – HAZ interface, may initiate flaws earlier and at lower stress levels, than in that material which has experienced higher austenitizing temperatures such as in the HAZ and fusion zone.

Another parameter found to influence near threshold fatigue behavior is the strength level. In the present study as the strength level increases for a given austenitizing temperature, the threshold fatigue value also increases. This increase, although small, is found to exist in both air and hydrogen. This shift in threshold is contrary to what has been seen by others in both low-strength (16, 19, 20) and ultra-high strength steels (7). However, Minakawa and McEvily (17) have reported an increase in the threshold with in-

creasing strength of duplex ferritic-martensitic microstructures. They
suggested that a large crack closure occurs due to back stresses and con-
finement of plastic deformation associated with the hard martensite. Their
analysis indicates that ΔK_{eff} is approximately the same for the variety of
duplex ferrous microstructures tested. A similar situation may exist in the
present microstructure whereby the quenched and tempered microstructures con-
strain the amount of plastic deformation that can occur at the crack tip.
This constraint is also suggested by evidence of reduced plastic deformation
processes on the fracture surface of the quenched and tempered specimens
(Figs. 5 and 6). However, no measurements of crack closure or effective ΔK
were made in this study to verify their proposal.

Conclusions

Hydrogen has been shown to degrade the near-threshold fatigue properties
of a low strength, low alloy steel and to be independent of ferrous microstructure. Hydrogen accelerates the fatigue crack growth rate and decreases the
threshold fatigue value as compared to that observed in air. An austenitic
grain size dependence and a strength level dependence of the near-threshold
fatigue behavior, has also been observed. The fatigue threshold increases
with increased grain size and increased yield strength.

From these results, martensitic microstructures produced at higher austenitizing temperatures provide the best static strength and near-threshold
fatigue behavior of A516 low alloy steel in both high pressure hydrogen and
air environments. On the other hand, the normalized microstructure produced
at a low austenitizing temperature results in both the lowest static strength
and the lowest fatigue threshold in both high pressure hydrogen and air environments.

Acknowledgements

This program was partially funded by the U. S. Department of Energy,
Office of Advanced Conservation Technologies, Thermal and Chemical Storage
Branch.

References

1. W. W. Scherenbach, Proc. of the DOE Chemical/Hydrogen Energy Systems
    Contractor Review, pp. 67-85, U.S. Department of Energy, Washington,
    D.C., Nov. 27-30, 1978.

2. W. G. Clark, Jr., Hydrogen in Metals, A. W. Thompson and I. M. Berstein,
    eds., p. 149, TMS-AIME, New York, NY, 1974.

3. H. G. Nelson, Proc. of 2nd Int. Conference on Mechanical Behavior of
    Materials, pp. 690-699, Boston, August 1976.

4. M. R. Louthan, Jr., Annual Report on "The Roles of Rate and State of
    Stress in Controlling Susceptibility to Hydrogen Embrittlement," VPI&SU,
    Blacksburg, VA, October 1979.

5. J. Mucci, Final Report on "Evaluation of Laser Welding Techniques for
    Hydrogen Transmission," Pratt and Whitney Aircraft Group, West Palm
    Beach, Florida, May 1980.

6. I. M. Bernstein and A. W. Thompson, International Metals Reviews, Vol. 21,
    1976, pp. 267-86.

7. R. O. Ritchie, Metal Science, 1977, Vol. 11, pp. 36881.

8. J. D. Hobson and C. Sykes, JISI, 1951, Vol. 169, p. 209.

9. E. Snape, Corrosion, 1968, Vol. 24, pp. 251-261.

10. G. Sandoz, in Stress Corrosion Cracking of High Strength Steels and in Aluminum and Titanium Alloys, B. F. Brown, ed., p. 79; Naval Research Lab., Washington, DC, 1974.

11. 1978 Annual Book of ASTM Standards, Vol. 10.

12. J. C. Newman, Jr., ASTM STP 560, p. 105, ASTM, Philadelphia, 1974.

13. A. M. Sullivan and T. W. Crooker, NRL Report No. 8004, Naval Research Laboratory, Washington, DC, August 1976.

14. A. D. Wilson, J. Eng. Mat. & Tech., 1979, Vol. 101, pp. 265-74.

15. C. R. Aita and J. Weertman, Met. Trans. A, 1979, Vol. 10A, pp. 535-44.

16. J. Masounave and J. P. Bailon, Scr. Met., 1976, Vol. 10, pp. 165-70.

17. K. Minakawa and A. J. McEvily, Proc. 5th Int. Conf. on the Strength of Metals and Alloys, P. Haasen, ed., Pergamon Press, New York, 1979.

18. R. O. Ritchie, MIT, Boston, MA, private communication, 1980.

19. R. O. Ritchie, MIT Fatigue and Plasticity Lab Report No. FPL/R/80/1032, MIT, Boston, MA, April 1980.

20. J. P. Benson, Metal Science, 1979, Vol. 11, pp. 535-39.

DISCUSSION

<u>B.R.W. Hinton, University of Manchester, UK.</u>: Why does the fracture mode, for tests with the normalised structure in hydrogen, change from intergranular to transgranular cleavage with increasing ΔK?

<u>H.F. Wachob</u>: We observed a gradual increase of the IG mode of failure as ΔK decreased from 24 MPa.m$^{\frac{1}{2}}$. Finally, at the threshold fatigue stress intensity, the failure mode was almost entirely IG. The pearlite-ferrite interface appears to be an inherently weak interface, since we observed IG separation at this interface in both the normalized and bainitic specimens. Thus, the lowest energy form of crack propagation appears to change from transgranular to intergranular when one approaches the threshold.

<u>A. Atrens, Brown Boveri Research Center, Baden, Switzerland</u>: How certain are you in the very existence of a fatigue threshold stress intensity in fatigue crack growth measurements?

<u>H.F. Wachob</u>: The existence of a true threshold stress intensity is unimportant from an experimental and engineering standpoint. The word "threshold" is probably a poor word choice. Threshold, as defined in the present work, is the alternating stress intensity that exists at a da/dN of 10^{-9} m/cycle. Other researchers define "thresholds" at a given crack growth rate or after a given number of cycles where no perceptible crack extension has been observed. In all cases, time or an engineering design limit have been the overriding factor for the determination of the threshold.

<u>J.P. Fidelle, Commissariat à l'Energie Atomique, Bruyères-le-Châtel</u>: You find different fatigue thresholds for H$_2$ and air. Fatigue being a cumulative process of microplastic damage, as H essentially decreases ductility, a simple reasoning is that although H enhances crack growth rates, the thresholds should be the same unless oxygen has some positive effect. Experiments under high vacuum would be necessary to supply a baseline and show whether H can affect threshold by a mechanism not envisioned above.

<u>H.F. Wachob</u>: We agree with your statement. There are indications from our unpublished work and from that of Ritchie and his co-workers (19), that the fatigue thresholds obtained in hydrogen and inert environments (either argon or helium) are the same. Thresholds obtained under typical laboratory conditions are raised above those obtained in hydrogen. When cycling near the threshold under lab conditions (i.e., where water, water vapor, and oxygen are present), an oxide debris forms near the crack tip which reduces the effective crack tip opening displacement, and therefore reduces the effective alternating stress intensity. Therefore, at the fatigue threshold in air, the applied or observed ΔK threshold is above that value obtained in hydrogen or helium. However, the effective ΔK at the crack tip is most probably equal to that threshold value obtained in hydrogen. A more comprehensive model which includes these variables as well as R ratio effects has recently been proposed by Ritchie (19). To your second comment, we strongly support the necessity for obtaining baseline fatigue threshold data. However, one typically obtains the highest fatigue threshold under conditions of a high vacuum. We feel that the complicated events such as rewelding of the crack tip during fatigue under vacuum conditions increases the difficulty of interpretation. Therefore, we suggest that baseline threshold fatigue tests be performed in high purity, well-characterized helium or other inert environments; thereby eliminating the possibility of crack tip rewelding.

Automated Real-time Analysis of Crack-Growth Rates of

Steels Tested in a Severe H_2S Environment

C. G. Interrante and S. R. Low

National Bureau of Standards
Washington, D.C.

Electrical resistance measurements of crack size are used in a multi-point data-acquisition system that uses direct current, checks for drift using a standard specimen, computes the stress-intensity factor (K), and the crack-growth rate (à), and records the results. An improved environmental chamber for fully instrumented tests of double-cantilever-beam (DCB) specimens has been developed. Using this improved design and the data-acquisition system, 26 DCB specimens were tested in a severe H_2S solution at room temperature. Stage II crack-growth-rate data are analyzed for the following testing variables: wax coating of the specimen, hardness in the range 22 to 38 HRC, and orientation of the specimen with respect to the principal rolling direction. Well documented computer codes, written for the data-acquisition system using the HP-Basic language are available from the authors.

Introduction

Electric potential techniques have been used effectively to monitor subcritical crack growth for over twenty years. Using notched tensile specimens, Barnett and Troiano (1) were first to use this method in laboratory studies of hydrogen-induced crack initiation and propagation behaviors. Techniques by which the method is applied to fracture problems have expanded through the years to applications that include fatigue, fast fracture, creep, and slow-crack-growth in sustained-load tests of hydrogen-embrittlement and stress-corrosion behaviors. The basic technique involves passing a known level of current through a test specimen, which contains a notch or crack, so that the resistance of the test specimen can be monitored directly by measurement of a potential. Crack extension is monitored from a change in potential.

Earlier techniques used high levels of direct current (≥ 10 amperes) to furnish sufficient changes in potential for the measurements of small crack extensions. Today, even small changes in potential are accurately measured with dc microvoltmeters; these high currents are not needed except in tests of crack growth in specimens of large cross section, such as the compact tension specimen. Even with this type of specimen, crack growth can be monitored using less than one ampere of current, provided that an ac measurement system is used (2).

Various analytical studies have been made for facilitating the accurate measurement of crack length for specimens of the many types to which the electric potential method is applied. Citing just a few, the method of conjugate functions has been used in calibrations of the center cracked panel (3), conformal mapping techniques have enabled accurate expressions to be produced for the crack length ahead of semi-elliptical or V notches (4), and finite element analyses have been applied to the problem for the single-edge-notched and the compact-tension specimens (5). Further, for the compact tension specimen, methods have been developed to optimize the accuracy, sensitivity, and reproducibility of measurements and to maximize voltages (6). With the various calibration curves developed by these methods, the measured voltage values can be used directly to compute crack length, with the computed length of the crack being independent of material properties, test-piece thickness, and magnitude of the current. To make this possible, the calibration curves use dimensionless ratios of voltages measured at some starting crack length and again at some increment of crack extension. Crack length is computed from the ratio of these voltages and the ratio of the crack length to the width of the test specimen.

A detailed method for hydrogen-embrittlement studies conducted in sulfide environments using the DCB specimen was first given by Heady (7). He found that the DCB specimen was particularly suitable for these studies. It is small enough to be cut from common tubular products used in the oil industry. In addition, it is rugged, easy to use, inexpensive, and yields quantitative design data, with the principal design parameter being K_{It}, the critical plane-strain mode-one fracture toughness for a given test environment. As a practical matter, Heady used this parameter to compare the relative resistance of two materials to environmental cracking. The measured parameter is compared as a ratio for materials A and B. This ratio equals the ratio of the loads that can be carried safely by the two materials. In other applications, the parameter K_{It} is also used for computation of allowable flaw sizes.

There is increasing recognition of the importance of the kinetics of crack growth in environmental testing. This is evidenced by the work of a joint task group that is developing a test method for the determination of stress-corrosion-cracking rate as a function of applied stress intensity using a DCB specimen. In the proposed method (8), crack length measurements are to be made as a function of exposure time by means of a visual or equivalent technique capable of resolving the position of the crack front to ± 0.1 mm. For studies conducted in sulfide environments, the use of optical methods is not practical, not only because of the dangers associated with removal of the test specimen from the sulfide environment, but also because the interaction of the environment with the test specimen totally blackens the test specimen and makes the aqueous test solution opaque.

The object of the present work was to develop an automated method for sequentially monitoring the crack-growth rate in a multiplicity of DCB test specimens in an aggressive environment, and to apply the method to tests of DCB specimens of a $2\frac{1}{4}$ Cr - 1 Mo steel of various orientations with respect to the principal rolling directions and of various hardness levels.

<u>Experimental Procedures</u>

The test methods and crack-growth monitoring procedures used were patterned after those described by Heady (7). Notable modifications are described below. These include the environmental test chamber; equipment for automation of the crack length measurements; methods for loading the test specimen, attaching the lead wires, minimization of thermal noise, and computation of crack length and crack-growth. Taken from Heady's method are procedures for computation of K for the mode-one loaded DCB specimen, and the apparatus for removal of H₂S from effluent gases. Due to limitations on the length of this publication, most of the adopted procedures are not described here.

Crack length was computed from the results of measured resistance of the test specimen, using a calibration relationship developed from four DCB specimens that were fatigue cracked to appropriate lengths so as to encompass the usable crack-length range of the test specimen. This relationship is a linear, least-squares fit of the dc resistance and an optical measurement of crack length taken at the side grooves of the four specimens. No attempt was made to account for any effects of non-linearity of the crack front.

The specimens were machined to the following dimensions: B = 0.250 in (6.35 mm), B_N = 0.170 in (4.32 mm), 2h = 1.00 in (25.4 mm), W = 5.25 in (133 mm), side grooves are 45° with a root radius = 0.010 in (0.25 mm). In preliminary studies (9), it was found that the resistance of the test specimen increases linearly with crack length. This is in agreement with the findings of Burns and Bilek (10), who conducted dynamic fracture tests using double-cantilevered specimens that were wedge loaded with a 50 lb Charpy hammer fitted with a nonconducting wedge.

It was found (9) that the slope of the linear relationship is independent of the position of the lead wires, for three sets of lead-wire positions: the set shown in figure 1, and two other sets near the chevron. The spacings between the lead wires for current and the probe wires for voltage were held reasonably constant. These wires were offset from the wedge by about 3 mm for the current leads and 6 mm for the voltage probes. Because the slope was found to be independent of lead-wire position, no attempt was made to determine whether this spacing significantly affects the measured voltage for the DCB specimen. This spacing is reported (11) to be significant for the compact tension specimen. These preliminary studies also disclosed that lead wires attached to the specimen by resistance spot welding have much lower resistance when compared with those attached by soldering. In the present work, the resistance welding method was used.

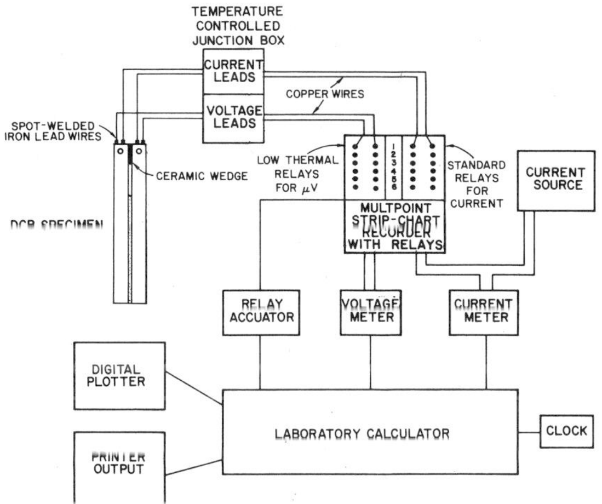

Fig. 1. Real-time multipoint crack-growth monitoring system.

After fatigue precracking, load was applied to the specimen using a nonconducting wedge of silicon nitride, Si_3N_4. The wedges were ground to a double-tapered (10/1 taper) shape from a sheet of this ceramic in a hot-pressed (at 1700 °C) and doped (with MgO) condition. The room-temperature properties are as follows: Knoop hardness = 2300, K_{IC} = 5 MPa(m)$^{\frac{1}{2}}$, and flexure strength (bending) = 700 MPa. This ceramic was found to be suitably strong, unyielding, and tough for this application. This wedge can be inserted with a bench vice without cracking, whereas single-tapered wedges of the same material tend to crack and require a tensile test machine to insert them--this could lead to undesirably high loading of the crack-tip region. With a conducting metal wedge, crack-growth rate can still be computed from formulas that account for the two parallel resistances, but error in the measured values of crack length could result either from positioning and corrosion of the wedge or from fouling from nonconducting corrosion products.

Crack-growth rates were measured using the real-time multipoint crack-length monitoring system shown as figure 1. This system uses a programmable laboratory calculator, coupled with two multimeters that monitor direct current and voltage, a digital plotter that displays crack-growth rate as a function of K, and a printer that displays various parameters measured and computed during the test. In this system, an iron-copper junction occurs unavoidably where iron wires from the specimen are attached to copper wiring of the measurement system. This junction was found to give rise to sudden and large thermal errors in measurements of crack length. A thermally insulated junction box decreases these errors considerably.

This system sequentially measures the crack length of up to five test specimens, checks for drift using a reference DCB specimen of constant resistance, and computes the values of K and crack-growth rate ($\dot{a}$). For multiple specimen tests, the rate at which measurements were taken was arbitrarily chosen to be 100 s per test specimen. The maximum sampling rate that could be obtained in single specimen tests is the rate set by the refreshing rates of the multimeters, which are 320 and 500 ms for the meters used to measure voltage and current, respectively, in this system. Data that are stored on tape include the initial parameters and the measured and computed values for a preselected increment of crack growth. For each specimen, when the computed crack length equals or exceeds the value needed for a computed crack extension of 0.3 mm, the data are stored and this computed increment of crack extension is noted both on a plot of the type shown as figure 2 and on a printer output that is not shown here. The printer output includes all measured quantities, as well as various computed values, such as resistance, crack length, crack-growth rate, and information taken on a reference DCB specimen that is used to indicate electronic and thermal drift (e.g., resistance and the standard error of the resistance).

The method for sequential monitoring of the crack-growth rate of a multiplicity of DCB test specimens was applied in tests conducted in the standard NACE solution for evaluations of sulfide stress cracking at room temperature (12), except that salt was not added to the test solution. The pH of this acetic acid solution was monitored and it varied from 2.9 to 3.9 during a seven-day exposure period of the test. Room temperature was maintained within the range of $22\frac{1}{2} \pm 1$ °C. This was done to minimize the effects of temperature on the crack-growth behavior and on the thermal noise of the crack-length measurements. After the test solution was purged with nitrogen gas (7), H_2S gas was bubbled through the solution at one atmosphere for a 15-minute period at the start of the test and for five minutes every 24 hours during a seven-day total exposure. Crack-growth rate measurements were made for about two days in this environment before fouling by sulfide buildup along the crack plane of the DCB specimen precluded further meaningful measurements.

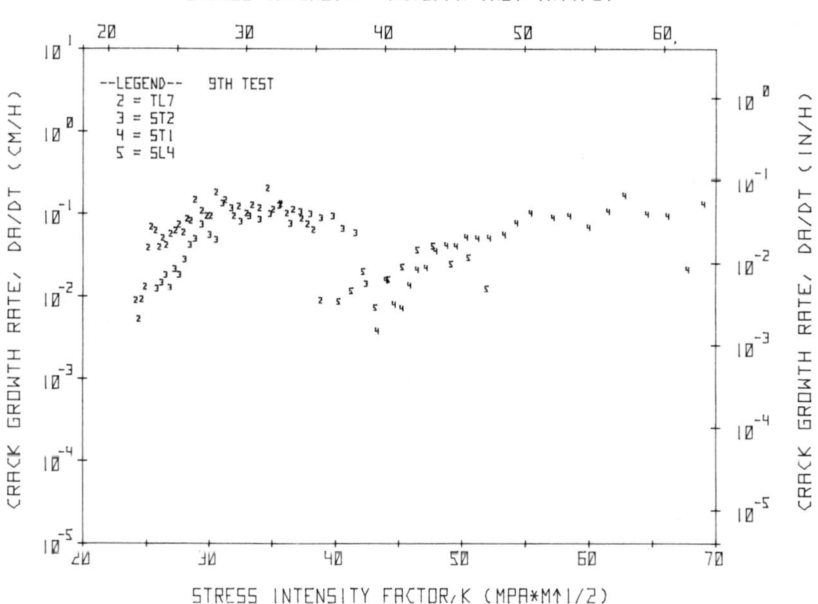

Fig. 2.   Real-time plot of crack-growth rate (å) versus stress-intensity
          factor (K). The numbers represent different specimens as noted
          in the legend.

     The material tested is a 2¼Cr - 1 Mo steel that was austenitized at
940 °C (1725 F), oil quenched to 38 HRC and then tempered to various hardness
levels. Four crack-plane orientations were tested; these are designated LT,
TL, ST, and SL, with the first letter representing the direction normal to
the crack plane and the second letter designating the expected direction of
crack propagation. In later text, the second letter is sometimes omitted,
for brevity and for combining ST with SL.

     In an attempt to prevent branch crack formation, selected test specimens
were coated with wax, using the following procedure. A plastic coated wire
(0.9 mm OD) is used to prevent wax from covering the region of the chevron
notch and the side grooves. Melted wax is brushed onto all surfaces to a
thickness of 1 to 2 mm. After the wax hardens, the wire is slowly removed.
The groove is scraped clean of any wax that might be present in the root of
the side grooves.

     All tests were conducted in a glass desiccator that was fitted, as shown
in figure 3, with a Plexiglas lid, which contains ports for sampling the
test solution, for entry and exit of gasses, and for insertion of test
specimens. Each of up to five test specimens is inserted through a 6 cm
circular cut-out, after being fitted into a rubber stopper that contains a
4.5 cm OD glass tube. The specimen is contained within the glass tube by
another stopper that is slotted for this purpose. This slotted region is
covered with wax to a level that covers the lead wires. The wax prevents
escape of gas from the environmental chamber, protects the lead wires from

attack by effluent gasses, and stabilizes the temperature at the contact points on the end of the test specimen.

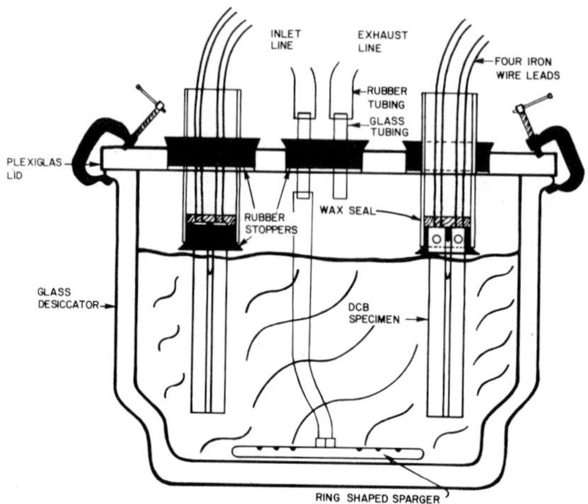

Fig. 3.   Environmental test chamber showing 2 DCB specimens mounted in lead-wire protection tubes.

Results

To assess the value of crack extension (Δa) that is considered significant in DCB specimens using this measurement system, the standard error for measured resistance of the reference DCB specimen, S_r, was used to determine the short- and long-term operating resolution of the system, using a result from the calibration curve, e.g., resistance of the DCB specimen is 7.72 $\mu\Omega$ per mm of Δa. Precision of the resistance measurements of crack length was governed by the sensitivity of the microvoltmeter, $\pm$ 0.1 μV, or 0.025 mm of crack length. Over a short-term interval of several minutes, S_r was 0.2 $\mu\Omega$ (or 0.026 mm) and this is a measure of the short-term operating resolution, or noise. Over a long-term interval of several days, S_r equals 1.0 $\mu\Omega$ (or 0.130 mm) and this is a measure of errors due to noise, thermal effects, and electronic drift in the system. A significant Δa was chosen to be one equal to or exceeding 2.3 times the long-term S_r, i.e., $\geq$ 0.30 mm. This increment of Δa was adequately small for the purposes of this study. It is concluded that smaller increments of 0.05 or 0.08 mm could be used with this system, provided that crack-length measurements are corrected by measurements of the instantaneous resistance of the reference DCB specimen. It is noted that some commercially available relays are rated at 2.0 amperes, as compared with 0.5 ampere for the relays used in this system. With the higher capacity relays, the increment of Δa could be decreased further, by about a factor of four.

Many of the specimens tested formed branch cracks just beyond the fatigue precracked region. Specimens containing branch cracks are not included in these results, as they are considered unsuitable for analyses of crack-growth rates. All unwaxed specimens of the LT orientation branched and numerous tests were conducted on waxed LT specimens. The S and T orientations were much less prone towards branching and waxing was not needed to prevent branch crack formation. Only two waxed specimens of these orientations could be tested due to a limited availability of test specimens of these

orientations. This circumstance precludes an adequate test matrix for
unequivocal determinations of the effects of waxing and test specimen orientations, without further testing. Nevertheless, the results of tests of 26
DCB specimens (unbranched) are instructive and they are used here to furnish
a preliminary assessment of the significance of four test variables on the
crack-growth-rate behavior. These variables are the stress-intensity
factor (K) at the start of the test, specimen hardness, surface waxing, and
specimen orientation with respect to the plate rolling direction. In the
text to follow, the subscript w represents the waxed condition.

From data plots of the type shown as figure 2, the range of K values
for the stage II period was established. This was done by inspection,
eliminating the first crack-growth-rate ($\dot{a}$) measurement and all measurements
contained within a sharply decreasing $\dot{a}$ leg of the data set for a test
specimen. The remaining data were averaged to furnish $\bar{\dot{a}}$, a mean value that
is used to represent the stage II behavior of the specimen. The $\bar{\dot{a}}$ results,
for the 26 DCB specimens tested, indicate that no crack-growth-rate dependence on K at the start of the test exists for the broad range of initial K
values of 28 to 72 MPa(m)$^{\frac{1}{2}}$ (26 to 66 ksi (in)$^{\frac{1}{2}}$). The effects of the other
three test variables were assessed from computed values of $\bar{\dot{a}}$.

The effect of hardness level of the test specimen on $\bar{\dot{a}}$ is shown in
figure 4, which gives the results for all 26 test specimens plotted as a
function of hardness (HRC), with hardness varying from 23 to 38 HRC. This
plot and other plots of various subsets of the data indicate that specimen
hardness does not significantly affect the mean stage II crack-growth rate.
The plot does illustrate a possible effect of waxing the test specimen.

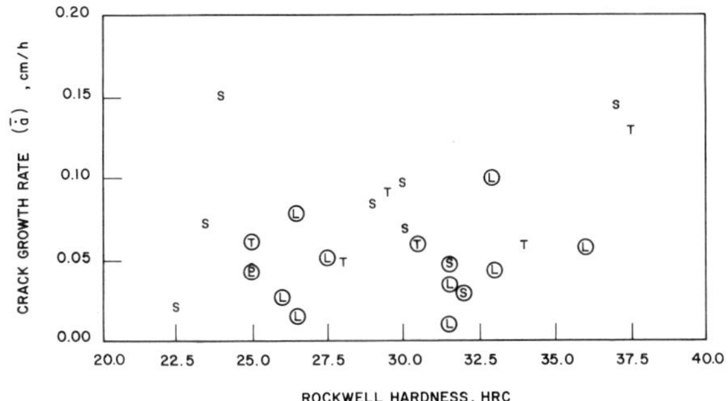

Fig. 4.   Plot of mean stage II crack growth rate ($\bar{\dot{a}}$) versus hardness (HRC)
          for specimen orientations L, T and S. Circled letters denote
          specimens having surface waxing.

The effects of waxing and of orientation of the test specimen were
evaluated by the "student" t test, which was used to determine whether
significant differences exist between subsets of the data. The null hypothesis (H_0) is that no significant difference exists between the two means, Y
and μ_0. Using 95 percent confidence limits, the value of t (of a random
sample of size N, selected from a normal population) falls within these
limits (with a probability equal to 0.95) when the hypothesis is accepted.
For an outlying value, H_0 is rejected indicating a rare event for this
confidence interval.

It is concluded that for the available data there is a significant difference between waxed and unwaxed specimens. Inspection of figure 4 indicates that waxed specimens tend to have lower crack-growth rates when compared with unwaxed specimens. In addition, test results on the significance of the effect of waxing are given in table Ia,

Table Ia - Significance of waxing.

Hypothesis H_o: $Y = \mu_o$ $t = \dfrac{Y-\mu_o}{S_Y/\sqrt{N_Y}}$

Y	μ_o	N_Y	N_{μ_o}	Accept H_o or reject H_o
S_w	S	2	8	ACCEPT
T_w	T	2	4	REJECT
S_w+T_w	$S+T$	4	12	REJECT
$S+T$	$S_w+T_w+L_w$	12	14	REJECT

Table Ib - Significance of orientation.

Hypothesis H_o: $Y = \mu_o$ $t = \dfrac{Y-\mu_o}{S_Y/\sqrt{N_Y}}$

Y	μ_o	N_Y	N_{μ_o}	Accept H_o or reject H_o
ST	SL	5	5	ACCEPT
SL	ST	5	5	ACCEPT
S_w	L_w	2	10	ACCEPT
T	S	4	8	ACCEPT
$T+T_w$	$S+S_w$	6	10	ACCEPT
$T+T_w$	L_w	6	10	ACCEPT
$S+S_w$	$T+T_w+L_w$	10	16	ACCEPT
T_w	L_w	2	10	REJECT

Table I.   Results of statistical "student" t tests of the hypothesis that the means of selected data subsets are not significantly different.

for various subsets of the data (Y for the smaller subset and μ_o for the larger subset). These results indicate that the null hypothesis of no significant difference is rejected for three of the four tests conducted, and it is accepted only for the case of $Y = S_w$ and $\mu_o = S$. As there are only two test results for test specimens in the S_w condition, this single acceptance of the null hypothesis is questionable.

It is concluded that for the available data there is no significant effect of orientation. The test results on the significance of the effect of orientation are given in table Ib. These results indicate that H_o is accepted in seven tests of significance and rejected in only one test. In the case for which H_o is rejected, T_w specimens are compared with L_w specimens; it is noted that there are only two test results in the T_w condition and both have about the same value of $\bar{a}$ (see figure 4). Thus, the value of their standard error (S_Y) is very small. In itself, this is an unusual event, and it leads to a large value of t and the rejection of H_o.

Conclusions

A multipoint measurement system developed for measurement of crack extension using dc resistance and a maximum current of 0.5 amperes is adequate for increments of 0.30 mm crack extension in DCB specimens. It is estimated that this increment could be decreased to less that 0.03 mm by using a reference DCB test specimen, and by increasing the current (dc) to 2.0 amperes.

From the results of 26 DCB specimens tested using this system, it is tentatively concluded that crack-growth rate in stage II is (1) independent of K at the start of the test, specimen hardness, and specimen orientation; and (2) significantly affected by the presence of wax on the surfaces of the specimen.

<u>References</u>

1. W.J. Barnett and A.R. Troiano: Trans AIME vol. 209, 1975, pp. 486-494.
2. R.P. Wei and R.L. Brazil: Fatigue Crack Growth Meas. and Data Anal. (in press) ASTM, Phila., PA.
3. H.H. Johnson: Matls. Res. and Stds, 1965, pp. 442-445.
4. G. Clark and J.F. Knott: J. Mech. Phys. Solids, 1975, vol. 23, 265-276.
5. R.O. Ritchie and K.J. Bathe: Intnl. J. Fract., vol. 15, 1979, pp. 47-55.
6. G.H. Aronson and R.O. Ritchie: J. Testing and Eval., vol. 7, 1979, pp. 208-215.
7. R.B. Heady: Corrosion, vol. 33, 1977, pp. 98-107.
8. ASTM Joint TG E24.04.02/ G01.06.04: 1980 draft, ASTM, Phila., PA.
9. C.G. Interrante, and G.E. Hicho: unpublished work presented at ASTM Sym. on Computer Automation of Matls. Testing, ASTM, Phila., PA., 1978.
10. S.J. Burns, and Z.J. Bilek: Met. Trans. 1973, pp. 975-984.
11. R.O. Ritchie, G.G. Garrett, and J.F. Knott: Intnl. J. of Fract. Mech., 1971, pp. 462-467.
12. J.B. Greer: see TG T-1F-9, Matls. Performance, vol. 16, 1977, pp. 9-15.

<u>Appendix</u>

Certain commercial equipment, instruments, or materials are identified in this paper in order to adequately specify the experiment procedure. In no case does such identification imply recommendation or endorsement by the National Bureau of Standards, nor does it imply that the materials or equipment identified is necessarily the best available for the purpose.

Laboratory Calculator - Hewlett-Packard (HP) 9830A calculator, with the following ROM's: matrix operations (HP-11270B), plotter control (HP-11271B), extended input/output (IO) (HP-11272B), string variables (HP-11274B), advanced programming I (HP-11279B), printer control (HP-11283B).

Memory - 30,144 total bytes of read/write memory (2 bytes per word).

Interfaces: HP-59045A HP-1B calculator interface, HP-11202A TTL IO interface, HP-9862A option 20/30 calculator plotter IO, HP-11336A printer interface, HP-11284A data-comm interface, HP-11203A BCD input interface.

Peripherals - HP9871-A printer, HP9862A calculator plotter.

Clock - A digital oscilloscope with "handshake" capability and a calculator interface, or a digital clock.

Digital Voltmeter - Keithley 174 digital multimeter (4 1/2 digit) with a Keithley 1723 ISB interface to the HP-IB.

Digital Ammeter - Keithley 160 digital multimeter (3 1/2 digit).

Relay Switches - HP 59306A relay actuator, and an L&N Speedomax W multipoint recorder with Flexelect "B" pushbutton point selector modified to be switched by the relay actuator.

An uninterruptable power system--Unit is used only under laboratory conditions that may lead to temporary loss of power such as in thunderstorms.

Software (HP Basic) - "Hydrogen-Induced Crack-Growth Testing", a testing routine that uses 23,500 bytes memory, 625 program lines.

COMPUTER MODEL FOR HYDROGEN EFFECTS ON

CRACK PROPAGATION IN B.C.C. IRON*

P. J. White

Polyatomics Research, Inc.
Mountain View, CA 94043

L. R. Kahn

Battelle Columbus Laboratories
Columbus, OH 43201

ABSTRACT

Atomistic calculations of the effects of hydrogen atoms on crack
propagation in body-centered-cubic iron are presented. The model used consists of a discrete region surrounded by an elastic continuum. The Fe-Fe
interaction is described by the Johnson interatomic potential and the Fe-H
interaction is described by a potential obtained using data from <u>ab initio</u>
calculations on Fe_4H clusters. Comparison is made with the results of calculations using other Fe-H potentials.

* This work was supported by National Aeronautics and Space Administration
contract #NAS2-10612 and the National Research Council.

I.   INTRODUCTION

Environmental enhancement of crack propagation in metals is a matter of considerable technological consequence which must be understood quantitatively before design, operating procedures, and safety factors can be optimized. The need for a theoretically sound description of the atomic processes involved in failure mechanisms such as hydrogen embrittlement of iron has prompted investigations using computer simulation techniques. Previous attempts to model the effects of hydrogen atoms on bonding in the vicinity of a crack tip in alpha iron, however, have caused some authors(1) to call for more realistic Fe-H potential functions. This paper reports some results observed using a new two-body potential energy function which was developed by doing a spline fit to energy vs. separation data gleaned from ab initio calculations on Fe_4H clusters.

The model used in these calculations is a slightly modified version of one developed at the Battelle Columbus Laboratories by P. C. Gehlen and others. It is well documented (2 - 6) and will be described only briefly in section II. Results obtained with the Battelle model using different Fe-H potential functions are mentioned and some results from this work are given in section IV. Section III outlines the manner in which the present potential function was obtained, and a summary is presented in section V.

II.   MODEL

The model employed in this work is a modified version of a comprehensive mathematical model developed at the Battelle Columbus Laboratories specifically for the purpose of analyzing crack growth on an atomistic scale. As shown in Figure 1, it is assumed that a single crystal contains a crack which has a length much greater than the radius of a cylindrical region which is to be modeled. A stress, σ, is applied so as to give mode I loading. Plane strain conditions are assumed since they suppress deformation at the crack, thereby enhancing susceptibility to environmental effects.

The cylindrical region of the model is subdivided into a discrete region, a boundary region, and a continuum region. The atoms in the vicinity of the crack tip (discrete region) have no restriction on their optimal positions except for that dictated by the interatomic forces. The Johnson-I potential (7) is used to simulate the Fe-Fe interactions, a Morse potential to simulate the H-H interactions, and a new potential, described in the next section, to simulate the Fe-H interactions. The outermost region of the model is treated as a continuum with iron atoms embedded. The positions of the embedded atoms are obtained using the elasticity equations given in reference 2. These equations provide the $[1\,0\,0]$ and $[0\,1\,0]$ displacements respectively from the perfect lattice positions when the crystal is under a stress described by a stress intensity factor, k:

$$\delta_{100} = k\sqrt{r/2}\left\{ (1/\alpha)(ba_{11} - a_{12}) \left[(1 - \alpha\sin 2\theta)^{\frac{1}{4}}\cos\varphi_2 - (1 + \alpha\sin 2\theta)^{\frac{1}{4}} \right.\right.$$

$$\left.\left. \cdot\cos\varphi_1 \right] + 2a_{11}\beta\left[(1 - \alpha\sin 2\theta)^{\frac{1}{4}}\sin\varphi_2 + (1 + \alpha\sin 2\theta)^{\frac{1}{4}}\sin\varphi_1 \right] \right\}$$

$$+ \ldots \tag{1}$$

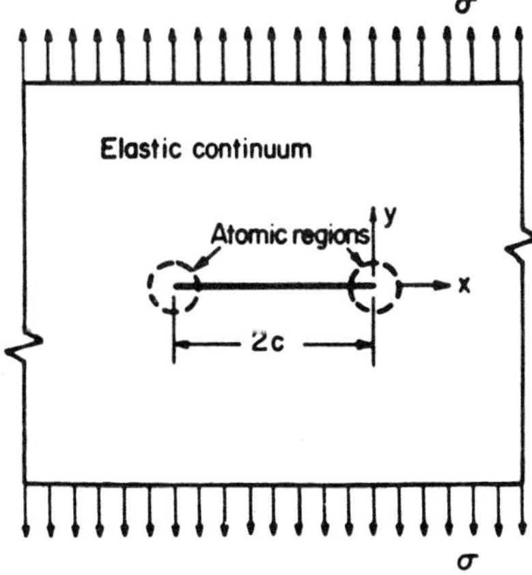

Figure 1. The configuration of the crack in the loaded crystallite.
The crack plane is a (100) plane. (After Ref. 5)

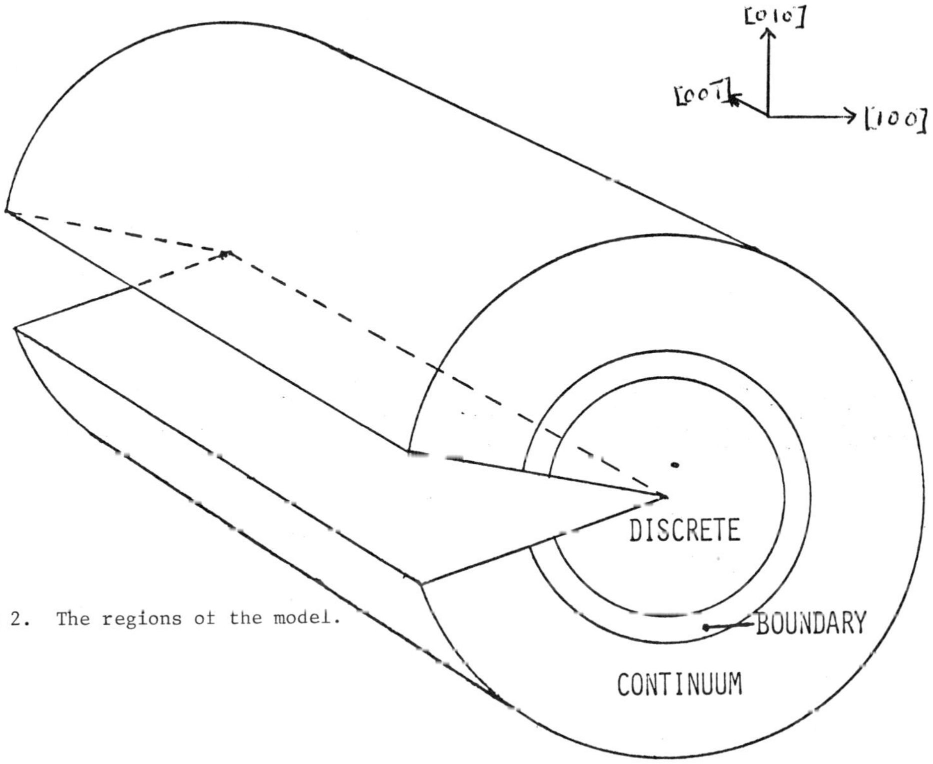

Figure 2.   The regions of the model.

$$\delta_{010} = k\sqrt{r/2}\left\{(a_{11} + a_{12})\left[(1 - \alpha\sin 2\theta)^{1/4}\cos\varphi_2 + (1 + \alpha\sin 2\theta)^{1/4}\cos\varphi_1\right]\right.$$

$$+ (a_{11} - a_{12})(\beta/\alpha)\left[(1 - \alpha\sin 2\theta)^{1/4}\sin\varphi_2\right.$$

$$\left.\left. - (1 + \alpha\sin 2\theta)^{1/4}\sin\varphi_1\right]\right\} + \dots \tag{2}$$

where the a_{ij} are the elastic compliances appropriate for the Johnson potential, and (r,θ) are the polar coordinates for the continuum region atoms relative to an origin at the crack tip. The other parameters are given by

$$b = (2a_{12}+a_{44})/2a_{11}, \quad \alpha = \sqrt{(1-b)/2}, \quad \beta = \sqrt{(1+b)/2},$$

$$\varphi_1 = \tfrac{1}{2}\tan^{-1}\left[\sin\theta/(\cos\theta + \alpha\sin\theta)\right], \text{ and } \varphi_2 = \tfrac{1}{2}\tan^{-1}\left[\sin\theta/\right.$$

$$\left.(\cos\theta - \alpha\sin\theta)\right].$$

Calculation of the atomic configuration around a crack is initiated by applying equations (1) and (2) throughout the model to obtain a starting configuration, R^o, which also serves as a reference configuration. The energy of the model is written as the sum of the atomic interaction energies for the atoms in the discrete region plus the difference in the (divergent) energies of the continuum configuration, R, and the reference continuum configuration. The equilibrium configuration is then obtained using the Sinclair-Fletcher conjugate gradient algorithm(8). The assumption is made that the discrete region can be made large enough that the positions of the continuum atoms are accurately described by a linear elastic displacement field (such as the one given above). If the parameters for that field are $t = \{t_1, t_2, \dots, t_n\}$, then the energy of the crystal can be minimized by taking to zero the forces on the individual atoms of the discrete region and the generalized forces given by

$$-\frac{\partial E}{\partial t_\alpha} = -\sum_k \nabla_k E \cdot \frac{\partial R_k}{\partial t_\alpha}. \tag{3}$$

This approach gives the model a flexible boundary between the discrete and continuum regions, thereby providing a medium for transfer of information from region to region about the state of each.

III.   POTENTIAL

The Fe-H two-body potential function used in this study was constructed using data obtained from _ab initio_ calculations on Fe_4H clusters. A hydrogen atom was located at various sites within a square array of iron atoms and the energy of each configuration was calculated using the unrestricted Hartree-Fock method. Because of the high CPU time requirements of these calculations only a few points were determined in this manner. The cluster energies calculated were subsequently decomposed into Fe-Fe and Fe-H contributions. Next, the Fe-H portion was represented by a sum of Fe-H pair

INTERVAL $(\times 10^{-10} \text{ m})$	a	R	b	c	d
$r \leq 0.96$	3.21×10^{-19}	0.96×10^{-10}	-4.63×10^{-10}	-4.04×10^{20}	
$0.96 \leq r \leq 1.47$	-1.07×10^{12}	0.96×10^{-10}	2.15×10^{2}	-1.49×10^{-8}	3.21×10^{-19}
$1.47 < r \leq 1.55$	-8.22×10^{11}	1.47×10^{-10}	5.21×10^{1}	-1.33×10^{-9}	-2.02×10^{-20}
$1.65 < r \leq 2.18$	-2.84×10^{10}	1.65×10^{-10}	6.91	-2.54×10^{-10}	-3.23×10^{-20}
$2.18 < r \leq 2.43$	1.97×10^{11}	2.18×10^{-10}	2.45	2.37×10^{-10}	-3.06×10^{-20}
$2.43 < r \leq 2.56$	-9.19×10^{11}	2.43×10^{-10}	1.72×10^{1}	7.26×10^{-10}	-2.02×10^{-20}
$2.56 < r \leq 3.17$	2.26×10^{-30}	2.56×10^{-10}	-1.83×10^{20}	1.76×10^{10}	

TABLE 1. Parameters for spline fit equations given as (4), (5), and (6) in the text. r is the Fe-H interatomic separation. Energy calculated with these parameters is in Joules.

contributions chosen so that the resultant discrete set of effective pair
energies were smoothly varying and possessed a single minimum. The effective
pair potential function was then obtained by doing a cubic spline fit to that
energy vs. separation data.

$$E = a(r - R)^3 + b(r - R)^2 + c(r - R) + d \tag{4}$$

E is the energy of the two-body interaction and r is the atomic separation.
The values of the parameters in the above equation are given in Table 1.
This function was knotted to an exponential function representing the inter-
action at short separations,

$$E = a \exp \left\{ b(r - R) - c(r - R)^2 \right\} \tag{5}$$

and also knotted to a higher order polynomial near the cutoff separation,
facilitating cutting off the interaction with zero slope and zero curvature
at 3.17×10^{-10} m. (6 Bohr radii).

$$E = a(r - R)^5 + b(r - R)^4 + c(r - R)^3. \tag{6}$$

IV. DISCUSSION

Previous calculations using the Fe-H potentials of Olander (9) and of
Walker et al. (10) have been done using a model very similar to that used in
the present work. With the Walker et al. potential, bond rupture was observed
to occur when hydrogen was present at a stress intensity factor about 10% less
than that giving bond rupture when no hydrogen was present (5). This potential
was formulated to describe the FeH molecule, however, and its validity in
describing the interatomic interactions in the lattice has been questioned by
some (see ref. 1). The Olander potential was formulated to describe the Fe-H
interaction rather than the FeH molecule, but proved insufficient for modelling
purposes because it is so strongly repulsive at the separations of interest that
the iron lattice rejected the hydrogen atom (1). Subsequently constraining the
hydrogen atom near the crack tip exhibited no effect on the stress intensity
factor necessary to cause bond rupture.

The potential function used in this work falls between the two potentials
mentioned above. Unlike with the Olander potential, a hydrogen atom is readily
accepted by the iron lattice in the present calculations. The results observed
here are not as dramatic as those obtained with the Walker et al. potential, but
they may be more reasonable. A single hydrogen atom in the vicinity of a crack
tip in α-iron is apparently insufficient to cause bond rupture at a lower stress
intensity level than for the pure iron. However, two hydrogen atoms can act in
a cooperative manner so as to reduce the stress intensity needed to rupture the
first bond by as much as 7%. No greater reduction in the necessary stress intensity has yet been observed using up to six hydrogen atoms in the vicinity of the
crack tip.

The results of the present calculations are believed to be more reasonable
than those of the earlier calculations using the Walker et al. and Olander
potentials. There are, as yet, no concrete indications that this Fe-H potential
provides an accurate representation of the interaction of hydrogen with an iron
lattice. However, the absence of unreasonable results is encouraging and enables
proceeding with further development of the model. Ultimately, the model should
account for temperature effects and the interaction of hydrogen with defects.

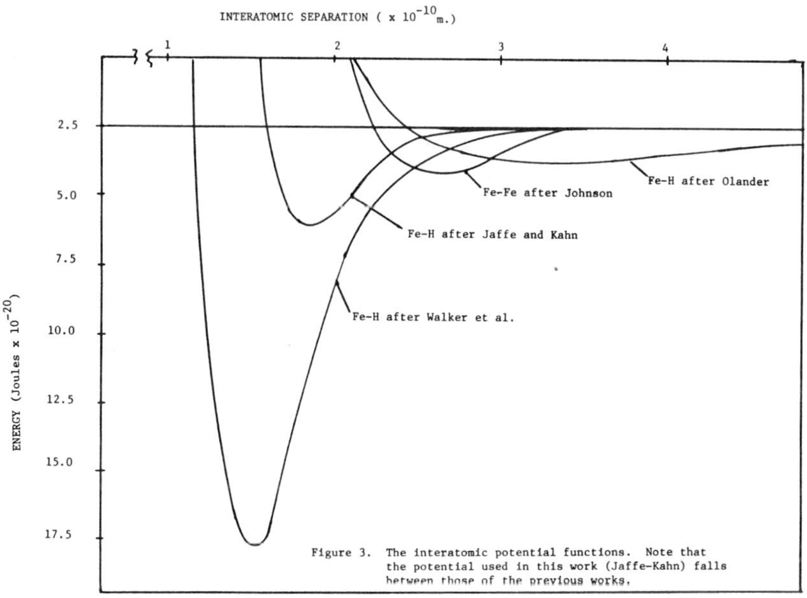

Figure 3.   The interatomic potential functions. Note that the potential used in this work (Jaffe-Kahn) falls between those of the previous works.

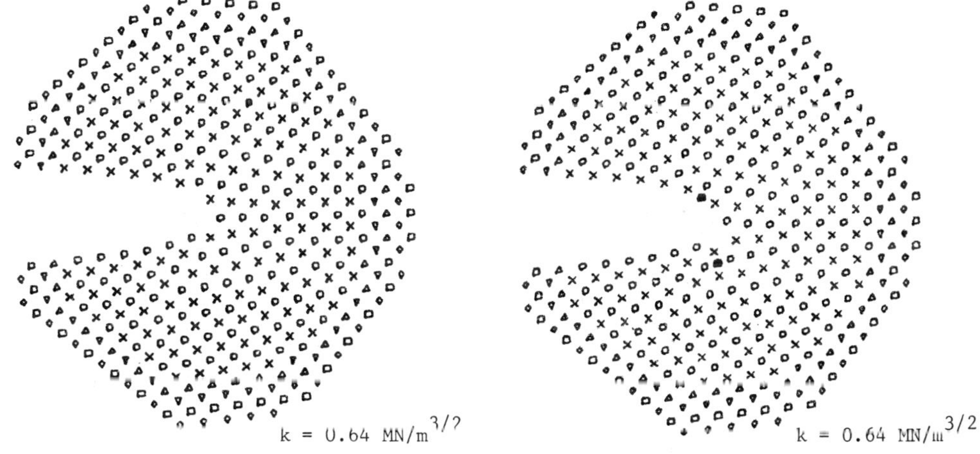

Figure 4.   Equilibrium crack configurations for pure iron (a) and iron with two hydrogen atoms near the crack tip (b). The plane of the page is the (001) plane. The x's and o's represent atoms in adjacent planes of the discrete region. The triangles represent boundary atoms and the squares represent continuum atoms. The two darkened squares in (b) represent the hydrogen atoms, which lie in the plane of the o's.

V.   SUMMARY

Calculations have been done using a new Fe-H potential function in an attempt to accurately describe the environmental effect of hydrogen on crack propagation in iron. Although the results presented are believed more realistic than those of previous calculations, they do not reveal much about the embrittling mechanism: (1) A single hydrogen atom in the iron lattice seems to have no effect on crack propagation. (2) Two hydrogens can cause up to a 7% reduction in the stress intensity required to rupture the next bond. (3) Addition of more hydrogen atoms to the lattice has not yet yielded a lower stress intensity required for breaking the next bond than is observed for two hydrogen atoms. Additional calculations with highly localized concentrations of hydrogen are in progress. (4) An accurate representation of the embrittling mechanism is not likely possible without a model which accounts for temperature effects and interaction of the hydrogen atoms with crystal defects.

ACKNOWLEDGMENT

The authors wish to acknowledge Dr. R. L. Jaffe of the N.A.S.A. Ames Research Center for providing the Fe-H potential used in these calculations and for his many helpful comments and suggestions.

REFERENCES

1. P. C. Gehlen, A. J. Markworth, and L. R. Kahn:Computer Simulation for Materials Applications, R. J. Arsenault, J. R. Beeler, and J. A. Simmons, eds.,1976.
2. M. F. Kanninen and P. C. Gehlen:Interatomic Potentials and Simulation of Lattice Defects, P. C. Gehlen, J. R. Beeler, and R. I. Jaffee, eds., Plenum Publishing Co., 1972.
3. P. C. Gehlen, G. T. Hahn, and M. F. Kanninen:Scripta Met.,1972, p. 1087.
4. P. C. Gehlen:Scripta Met.,1973, p. 1115.
5. A. J. Markworth, M. F. Kanninen, and P. C. Gehlen:Proc. Intern. Conf. on Stress Corrosion Cracking and Hydrogen Embrittlement of Iron Base Alloys, 1973, p. 447.
6. P. C. Gehlen, A. R. Rosenfield, and G. T. Hahn:J. Appl. Phys., 1968,vol. 39, p. 5246.
7. R. A. Johnson:Phys. Rev., 1964, vol. 134, p. A1329.
8. J. E. Sinclair and R. Fletcher:J. Phys. C., 1974, vol. 7, p. 864.
9. D. R. Olander:J. Phys. Chem. Solids, 1971, vol. 32, p. 2499.
10. J. H. Walker, T. E. H. Walker, and H. P. Kelly, J. Chem. Phys., 1972, vol. 57, p. 2094.

HYDROGEN IN α/β AND ALL β TITANIUM SYSTEMS:

ANALYSIS OF MICROSTRUCTURE AND TEMPERATURE INTERACTIONS ON CRACKING

W. W. Gerberich, N. R. Moody, C. L. Jensen,
C. Hayman and K. Jatavallabhula

University of Minnesota
Minneapolis, Minnesota

In an attempt to understand the rate-controlling mechanism for sustained load and fatigue crack growth, as well as those temperature and solubility factors that affect threshold, a four-year program has been in progress. This has involved mechanical property evaluation of Ti-6AL-6V-2Sn as a mill-annealed microstructure, Ti-5AL-4Mo as β annealed Widmanstatten α/β microstructures and Ti-30Mo as a medium grained all-β microstructure. In addition, resistivity, X-ray and transmission microscopy studies have been partially completed to quantify possible phase transformation effects.

Extensive flow, brittle fracture, threshold, and crack growth rate studies have been performed at temperatures from 180 to 340 K with hydrogen concentrations ranging from 30 ppm (wt) to 1500 ppm (wt). Continuum models incorporating microstructural effects associated with grain-orientation control and hydrogen effects on α/β interface cracking were then developed. Analytical relationships for predicting such effects on fatigue thresholds are described and reasons for inconsistencies are suggested.

Introduction

Hydrogen effects over the entire stress range of environmentally-enhanced sustained-load crack or fatigue crack growth have been quantified in numerous studies, but there are still enormous uncertainties as to how microstructure is partially responsible for the sigmoidal shape of the crack growth rate versus stress intensity curve(1). Microstructure usually exerts a greater effect near threshold and then, again, near failure. Significant effects in these two regimes have been reported by Beevers(2-4), Yoder, et al.(5,6), and Chesnutt, et al.(7). Recent studies (2,7-9) have shown this to be a complex interaction of metallurgical variables with the most important aspect being crack path tortuosity. Nevertheless, it is not clear how either test temperature or internal hydrogen level affects the process. In addition, it is still not clear what phase or phases are most responsible for detrimental hydrogen effects in Widmanstatten α/β alloys or near β alloys. To approach some of the unanswered questions, an extensive program involving three α/β and all β titanium alloys was initiated. This involved X-ray, resistivity, SEM and TEM studies for phase and slip-mode identification while tensile, compact tension and slow-notch bend testing characterized mechanical behavior.

Material and Test Procedures

The materials used in this study were the experimental all β Ti-30Mo and α/β Ti-5AL-4Mo alloys. In addition some previous data on a commercial α/β Ti-6AL-6V-2Sn alloy is cited. As received alloys were annealed for four hours at 950^{0}C at 1×10^{-5} torr or less. Samples were then either cathodically charged in 10% H_2SO_4 or gas phase charged in a Sieverts apparatus at 800^{0}C. After equilibration the samples were force-air cooled to RT in ½hr. or heat-treated above the β transus. A deviation in this procedure was to cool the α/β alloy more slowly so that both fine and coarse microstructures could be obtained. From both optical and transmission microscopy, the average Widmanstatten colony sizes were 30 μm for the fine and 55 μm for the coarse. The fine had nearly twice the volume fraction of β phase, 8.7 percent as opposed to 4.8 percent. Hydrogen analysis was performed on each sample using the hot vacuum extraction technique with a nominal precision of two percent. X-ray diffraction analysis was performed on the high resolution (.002^{0}2θ) Siemens D-500 diffractometer equipped with a graphite monochromator in the output stage.

Mechanical testing consisted of slow-notch bend testing, as discussed in detail elsewhere (10), tensile testing and fatigue crack propagation evaluation. Internal stresses and tensile properties were determined using standard flat tensile specimens fabricated so that the load is applied in the same TL orientation as the compact tension (CTS) samples. The internal stresses were determined at various strains using the decremental unloading technique (11). Fatigue testing was conducted under tension-tension sinusoidal loading in air at a frequency of 25 Hz   with a load ratio of 0.05.

Results and Discussion

All β Ti-30 Mo

Mechanical Properties. The yield strengths of the all β alloy were measured in compression at temperatures from 123 to 363 K with hydrogen contents ranging from 22 to 1400 ppm (wt). With hydrogen additions, yield strengths tended to go through a maximum at values of C_0 ranging from 100 ppm to 600 ppm, depending upon test temperature. These data are presented in detail in a forthcoming paper (10). With them, slow-notch bend ductile-brittle transition data could be converted to cleavage fracture stresses as shown in Figure 1. Although not shown clearly here, the values of cleavage fracture stress, σ_f^*, decreased four times as rapidly when cathodic charging was used instead of gas-phase charging.

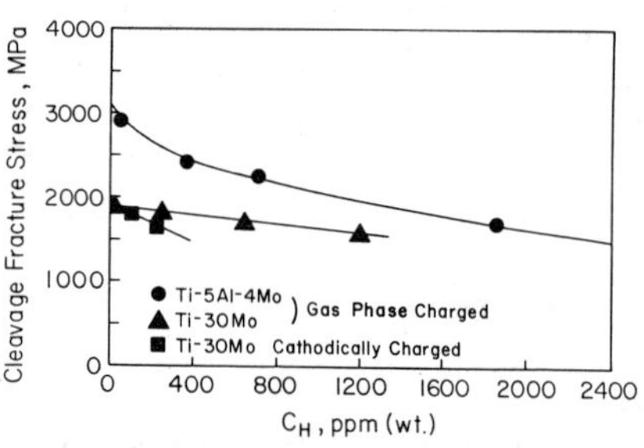

Fig. 1: The effect of hydrogen concentration on cleavage fracture stress.

Partial Molar Volume. Lattice parameters as a function of hydrogen were estimated from d(110) measurements, the results being shown in Figure 2. It is significant that the slopes are different for electrolytic versus gas-phase charging. This indicates that some phase may have either precipitated or that there is at least a pre-precipitation rearrangement of hydrogen and/or molybdenum atoms. The difference is consistent with the different cleavage fracture stress dependencies. With the data in Figure 2 and one additional set of data, the misfit parameter $\bar{\varepsilon}_{H-\beta}$, was determined from $(1/a_0)(da/dC_H)$ to be 0.043 ± 0.003. This was then converted to a partial molar volume from

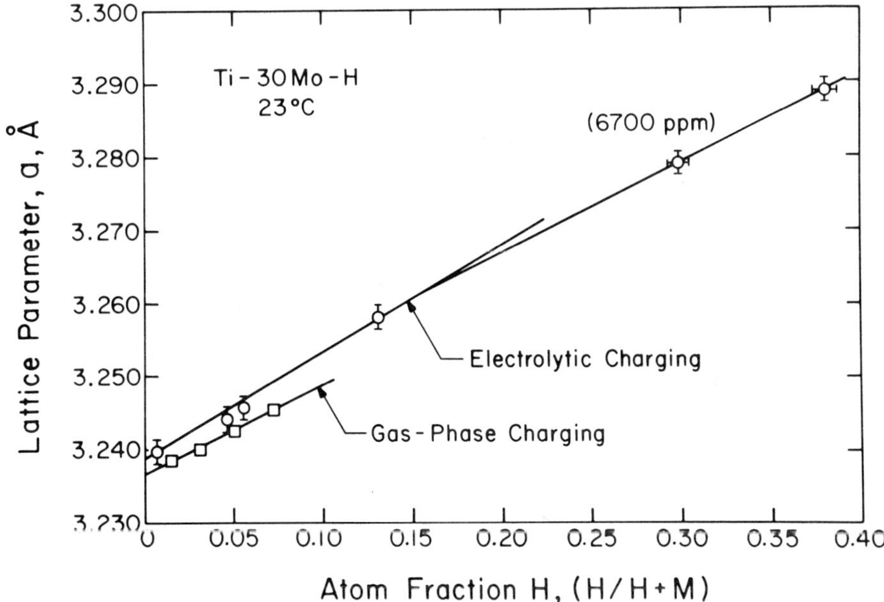

Fig. 2: The effect of hydrogen on β phase lattice parameter in Ti-30Mo for two hydrogen charging techniques.

$$\overline{V}_{H-\beta} = 4\pi\overline{\varepsilon}_{H-\beta}r_1^{\;3}(R/k) \tag{1}$$

using an interatomic radius of 1.622 Å and noting that R/k is the ratio of the appropriate gas constants. The value determined was 1.39 cm^3/mol which is reasonably close to those reported by others (12,13).

<u>Phase Identification</u>. The first inference that hydrogen was inducing a phase change was from low temperature resistivity studies where an annealed sample charged to 1350 ppm (wt) was observed to undergo a resistivity slope change and some hysteresis during thermal cycling near -20°C. This was followed up by an extensive X-ray diffraction analysis. Martensitic α''(orthorhombic) phase was found in both gas-phase and electrolytically charged samples at high hydrogen concentrations greater than 3000 ppm (wt). Detection of α'' in non-hydrogenated Ti alloys is relatively easy because of the appearance of five lines in the 30-45°2Θ region as opposed to three lines for α', hydride or ω(omega). Retention of the broadened β^{110} line in hydrogen containing alloys makes this analysis more difficult since it overlaps two of the five α'' peaks. Nevertheless, the present results are identical to the diffraction pattern found for α'' in α/β Ti-6AL-4V by Y. Mahajan, et al.(14).

At lower hydrogen levels, cooling to -20°C produces a different diffraction effect as represented by a Ti-30Mo sample containing 1260 ppm (wt). As seen in Figure 3, the initial dotted X-ray curve shows a very broad peak at 39.11° with a $\Delta 2\Theta = 0.258$. Upon cooling, the β^{110} peak appears to split into three lines and shift up in 2Θ indicating a rejection of hydrogen. The splitting of this peak is about the same as observed in Ti-18Mo (15) upon formation of ω phase (3 peaks: $(110)^{\beta}$, $(1011)^{\omega}$, $(1120)^{\omega}$). There is also a severe reduction in line broadening. This is consistent with an observed resistivity change near -20°C. While these data do not necessarily indicate the formation of athermal ω, they are not inconsistent either. For if hydrogen elevates the omega start temperature as Buck, <u>et al</u>. (16) suggests, then ~1,200 ppm may be sufficient to elevate the ω_s temperature to -20°C, from -120°C at nominal hydrogen as reported by Ono, <u>et al</u>. (17) for Ti-29Mo.

α/β Systems

   <u>Partial Molar Volume</u>. Chemical compositions and hydrogen extraction analysis for samples used in the fatigue crack propagation study are shown in Table 1.   Also shown are the α and β phase lattice parameters and their corresponding hydrogen concentrations, based upon the partitioning method (18). Of course, the hydrogen partitioning would have to be severely modi-fied if one considers the interface phase to be hydrogen enriched. Never-theless, this may not drastically reduce the hydrogen in the α phase and using the X-ray and hydrogen analyses, a partial molar volume calculation similar to that in Eq. (1) gives $\overline{V}_{H-\alpha}$ as 1.68 ± 0.14 cm^3/mol.

   <u>Tensile Properties</u>.   Tensile data for the fine Widmanstatten micro-structure are shown in Table 2 and reflect the normal temperature depen-dence with yield and ultimate strengths increasing as temperature decreases. The effect of large hydrogen additions was to decrease the elastic moduli and increase the ultimate strengths and work-hardening exponents, n, at all test temperatures. The most interesting aspect of the data is a crossover effect exhibited by the yield strengths. At 340K, the high hydrogen yield strengths are considerably less than the low hydrogen values while at lower temperatures the reverse is true. Coupled with the fact that hydrogen in-creased ultimate strength at 340K, the result was an approximate doubling of the work-hardening exponents. This is also clearly seen in the short-time internal stress measurements, when determined as function of prior plastic strain. In Figure 4, it is seen that the high hydrogen internal stress value was lower than the low hydrogen value at 340K for small

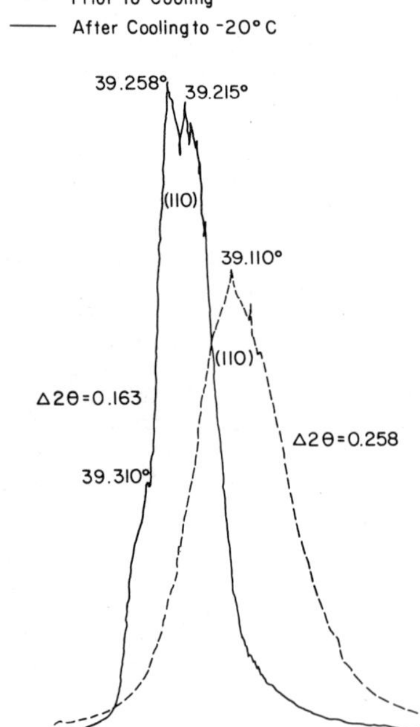

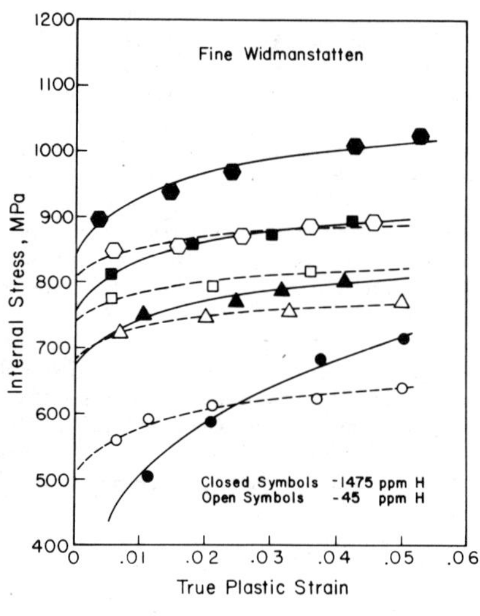

Fig. 4: Effect of temperature (340, 300, 260, 220K) and hydrogen on in-ternal stress.

Fig. 3: Shift in diffraction peak and line width on cooling Ti-30Mo-0.13H.

Table 1:   Chemical Composition of Ti-5Al-4Mo, (wt.%), Lattice Parameters and Hydrogen Concentrations.

Al	Mo	Fe	O	C	Balance
4.9	3.7	0.04	0.083	0.03	Ti

Fine Widmanstatten

Sample H concentration ppm (wt.)	α-H concentration ppm (wt.)	β-H concentration ppm (wt.)	a-α(Å)	a-β(Å)
65	43	250	2.927	3.242
167	118	575	2.927	3.246
1400	1030	4490	2.936	3.273

Coarse Widmanstatten

25	17	150	2.926	3.242
150	107	850	2.928	3.251
1555	1207	7250	2.940	3.300

strains. At larger strains, there is a crossover in the internal stress behavior. The result is that $d\sigma_i/d\varepsilon_p$ is more than doubled by the addition of hydrogen.

Slow-notch bend transitions and σ_f^*. Slow-notch bend testing produced brittle fracture at temperatures near 80K for the low hydrogen samples but at temperatures near 300K for those containing high hydrogen. The cleavage fracture stresses were determined using the plastic constraint approach and the von Mises yield criterion (19). As seen in Figure 1, 1850 ppm hydrogen lowers σ_f^* from about 3000MPa to 1700MPa.

Crack growth studies. The major emphasis in the present study was associated with crack propagation and threshold stress intensities under fatigue conditions. For three gas-phase charged conditions, four decades of crack growth rate were determined as a function of ΔK from 180 to 340 K. Typical examples of data at four test temperatures are shown for samples containing 45 ppm (wt) hydrogen in Figure 5 and 1475 ppm (wt) H in Figure 6. Fatigue crack growth rates for the low hydrogen samples were well behaved with the highest temperature (lowest yield strength) material producing the fastest growth rates over all stress intensity ranges. With high hydrogen contents, however, the growth rates exhibited substantially different behavior at low stress intensities with the two intermediate test temperatures producing faster growth rates than either temperature extreme.

Table 2:   Ti-5Al-4Mo Mechanical Properties (Fine Widmanstatten)

H ppm (wt.)	T,K	E,GPa	σ_{ys},0.2%,MPa	σ_{ute},MPa	n
45	340	115	708	792	0.045
	300	118	784	850	0.049
	260	120	823	923	0.060
	220	122	869	947	0.055
1475	340	107	672	843	0.097
	300	108	793	921	0.080
	260	109	854	1002	0.089
	220	110	922	1117	0.088

With regards to microstructural effects, there was a definite improvement in fatigue crack propagation resistance upon increasing the Widmanstatten colony size from 30 μm for the fine to 55 μm for the coarse. As is seen in Figure 7, this results in about a 20 percent improvement in ΔK at the same crack growth rate. Such findings are consistent with extensive studies by Yoder, <u>et al.</u>(6,20). This microstructural superiority carried over to the hydrogenated samples where threshold stress intensities for the coarse Widmanstatten charged to high hydrogen levels and tested at 220K had a lower ΔK_{TH} than its fine Widmanstatten counterpart. The reason lies in the greater susceptibility of the coarse microstructure which is probably related to the smaller volume fraction of β phase (4.8% coarse and 8.7% fine) and the correspondingly larger amount of hydrogen partitioned to the β as was indicated in Table 1. In fact, it is seen in Figure 8 that hydrogen concentrations near 160 ppm (wt) barely affect the fine Widmanstatten thresholds whereas the same hydrogen levels substantially reduce thresholds in the coarse Widmanstatten at 260 and 300K.

<u>Electron fractography and transmission microscopy</u>. For the present study, fractography was concentrated on the most microstructurally-sensitive regime near threshold. Threshold in the low hydrogen coarse Widmanstatten samples is characterized by faceted fracture approaching 100%. This is evident at all temperatures as typified by Figure 9(a) and 9(b) and is seen to result in severe out-of-plane deviations. On these cystallographic facets one notes evidence of parallel surface markings in Figure 9(b). These "striation-like markings" are not fatigue striations but intersections of intense slip bands with the fracture surface. At this point the crack was growing at 0.001 μm/cycle whereas these slip band intersections are spaced at about 1.5 μm intervals. The fracture appearance in the low hydrogen fine microstructure differs considerably in that there is a mixture of intergranular fracture along prior β boundaries and extensive α/β interface

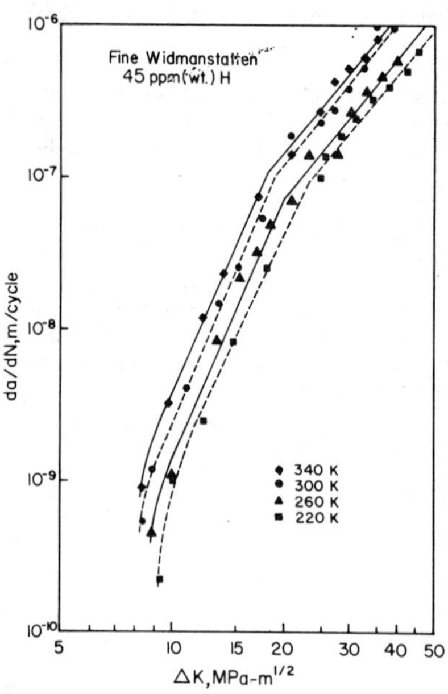

Fig. 5: Fatigue crack growth rate for fine Widmanstatten Ti-5Al-4Mo containing 45 ppm (wt) H.

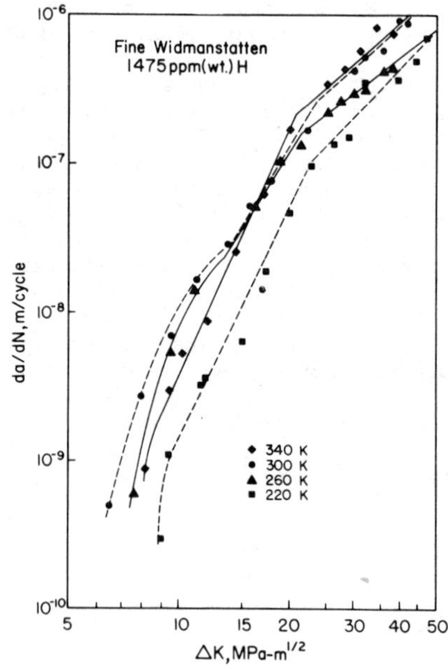

Fig. 6: Fatigue crack growth rate for fine Widmanstatten Ti-5Al-4Mo containing 1475 ppm (wt) H.

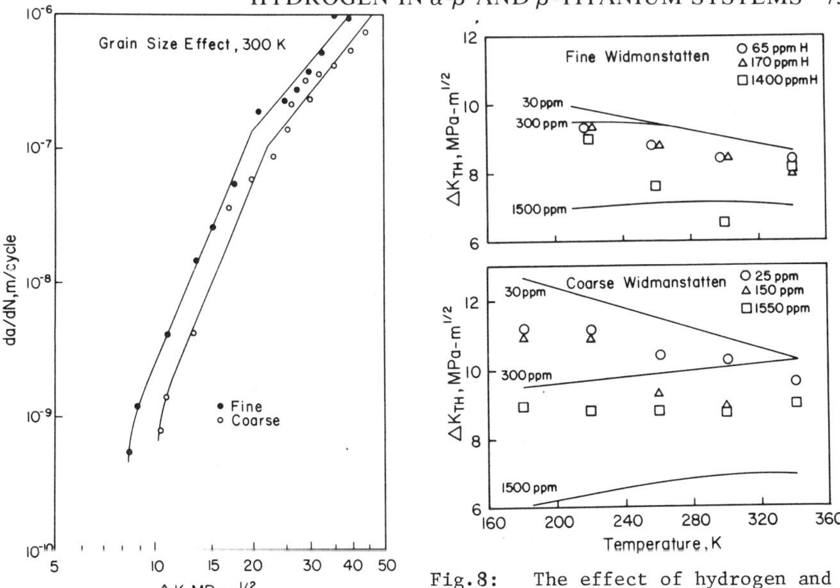

Fig. 7: Grain size effect on da/dN.

Fig.8:   The effect of hydrogen and temperature on fatigue thresholds. Solid curves are theoretical model.

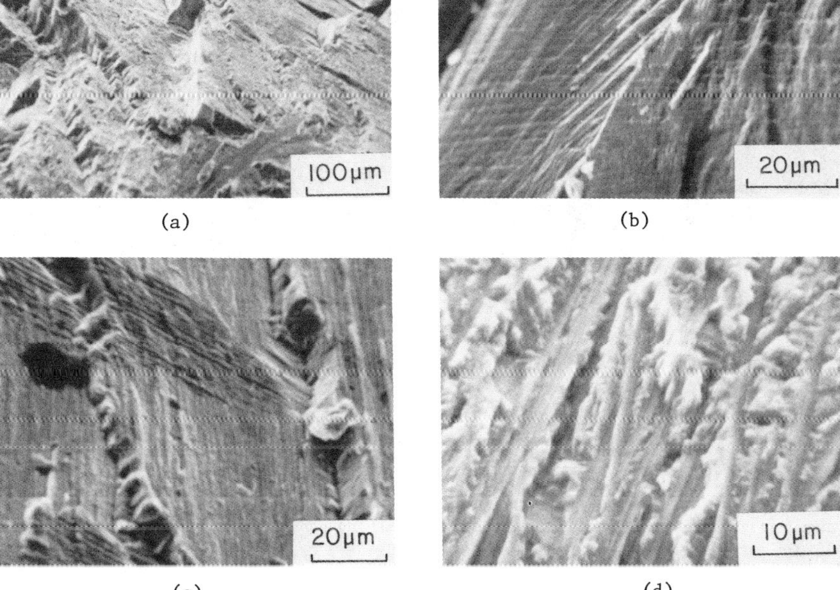

Fig. 9: Fracture surfaces at threshold: (a) 40 ppm (wt) at 300K, coarse; (b) 45 ppm (wt) at 220K, coarse; (c)1475 ppm (wt) at 300K, coarse; (d) 1475 ppm (wt) at 260K,fine.

fracture. With high hydrogen additions, the fracture appearance is still crystallographic but is now strongly associated with interface cracking in both microstructures. The cracking along α/β interfaces in the Widmanstatten colonies is seen clearly in Figures 9(c) and 9(d). When a crack encountered a colony in an orientation where the interface planes were nearly parallel to the advancing crack planes, slightly out of plane cracking was observed. This is clearly seen at the top of Figure 9(c) for the coarse microstructure. It should be noted that these hydrogen-induced out-of-plane deviations are small, however, compared to those obtained in Figure 9(a) due to slip-band orientation cracking. The later deviations are much greater compared to the smoother facets in Figures 9(c) and 9(d) and give some support to the crack path tortuosity theories.

Crystallographic slip studies via transmission microscopy showed that basal slip occurred at 340K but only prism and pyramidal slip occurred at lower temperatures. Slip band spacings determined from a number of micrographs for each condition decreased with increasing temperature, similar to previous findings (21). There were also indications that cross slip was more prevalent at higher temperature. In Figure 10(a), both the β and interface phase are imaged in dark field. Indexing showed that slip occurred on the (110) plane, which corresponds to the primary slip mode. A recent study of slip in Ti-6Al-4V showed that {101} slip in the bcc β is parallel to pyramidal slip in the adjacent α phase (22). It is also interesting that prismatic slip has difficulty traversing the β phase. This results in dislocation pileups and the associated stresses could enhance cleavage in the interface or β phase as suggested by Ruppen, et al. (22). In Figure 10(b) a bright field shot of a high hydrogen sample indicates the much larger interface phase. This is consistent with Hammond and Hall's suggestion (23) that this phase is hydrogen stabilized.

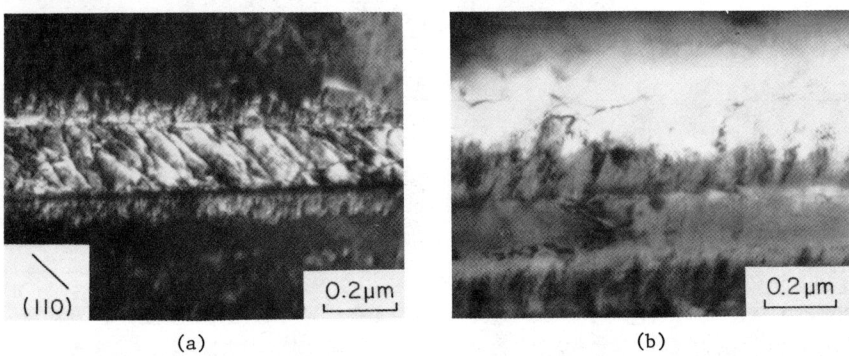

(a)   (b)

Fig. 10: Slip in β phase is shown by dark field in (a) along with the interface phase for fine Widmanstatten containing 45 ppm (wt.). (b) The effect of 1555 ppm (wt.) hydrogen content on the interface phase for coarse Widmanstatten at 260K.

Mechanistic Interpretations

There are many possible hydrogen mechanisms which might be invoked, particularly with regards to acceleration of crack growth under fatigue conditions. Several of these might be loosely grouped together as effects on dislocation substructure. Other mechanisms might be considered as having the common effect of requiring a reduced local stress for fracture of an inherently brittle or an embrittled primary phase, precipitated phase or interface.

Deformation substructure effects

With regards to this mechanism, it does seem possible that hydrogen enhancement of persistent slip bands at low temperatures could accelerate interface decohesion while enhanced dislocation mobility might provide easier interfacial shear at higher temperatures. Sliding of α/β interfaces has been recently observed by Baeslack and Mahajan (24) in a 4.5AL-5Mo-1.5 Cr α/β Widmanstatten microstructure containing 36 ppm (wt) hydrogen. There is some support for a change in slip character in Figure 4 where the internal stress seems to signal a difference in dislocation mechanism at 340K as opposed to lower temperatures. That is, for small strains there is a hydrogen-induced softening at high temperatures and hardening at low temperatures. This internal stress observation is most likely due to an inherent dislocation mobility or substructure effect. In addition, we did find some evidence that hydrogen affected slip band spacings. This was particularly evident in the coarse Widmanstatten microstructure where basal slip combined with prism and pyramidal slip to reduce the slip band spacing by 40 percent. Thus the initial reduction in internal stress by hydrogen at higher temperature may be attributed to either an activation of an additional slip system or an increased ease of deformation on the slip planes.

Our tentative conclusion was that hydrogen induced slip changes could affect the propensity towards decohesion or interface sliding but by itself could not explain the several effects observed on crack growth and thresholds in these two microstructures.

Proposed Microfracture Model

For ease of incorporation of microstructural features into a continuum analysis, we propose a modified Dugdale-Barenblatt model. All that is done is to extend previous plasticity concepts to include a semi-cohesive microfracture zone. Basically, this is nothing more than an analytical treatment of grain orientation control features that others had observed previously (3,25). The basic concept is that there is a fictitious elastic crack† of 2a being pulled apart by a series of forces. If the crack is static, these elastic forces must be in equilibrium with the internal forces pulling back on the "crack" faces. With only a homogeneous deformation zone at the crack tip, the internal forces are represented by the yield strength and the Dugdale solution for the plastic zone size results. If on the other hand, one has a semi-cohesive microfracture zone between the crack and the plastic zone, then there are additional internal forces or traction forces pulling back on the crack faces. At equilibrium, one simply sets the summation of forces in the direction of pull equal to zero (26-30).

Application to static thresholds. Extensive static threshold studies have been conducted on Ti-6Al-6V-2Sn in the mill annealed condition as reported elsewhere (26). It was established that with only 38 ppm (wt) hydrogen, that the alloy was very susceptible to hydride-induced slow crack growth. In fact, static thresholds in the range of 30-40 MPa-m$^{\frac{1}{2}}$ were calculated and predicted using a thermodynamic model which follows the development of Birnbaum (31) and Paton (32) but is modified by the fact that some favorably oriented grains or grain boundaries undergo slow crack growth while others do not. The resulting void fraction in the semi-cohesive zone controls the zone strength and hence partially controls the threshold.

Application to fatigue thresholds--no hydrogen. The first application of this model (27) to microstructural effects on fatigue thresholds was to both α/β and near α alloys under the assumption that the semi-cohesive zone

---

†The fictitious crack includes the real crack, 2c, as well as all deformation and microcrack zones.

size, ℓ^*, was a grain or colony size. This leads to

$$\pi\sigma - 2\sigma_{sc}\cos^{-1}\left(\frac{c}{a}\right) - 2(\sigma_{ys_\pm} - \sigma_{sc})\cos^{-1}\left(\frac{b}{a}\right) = 0 \qquad (2)$$

where σ is the applied stress, σ_{ys} is taken as double the monotonic strength, c is the half-crack, b-c is the semi-cohesive zone, ℓ^*, and a contains all three zones, $c + \ell^* + R_{p_\pm}$. The semi-cohesive zone strength is

$$\sigma_{sc} \simeq [1-f]\sigma_{ys_\pm} \qquad (3)$$

where f is the void fraction in the semi-cohesive zone. The procedure is to calculate the Dugdale plastic zone at the fatigue crack tip using

$$R_{p_\pm} = \frac{\pi}{8}\left\{\frac{\Delta K_{\ell^*\to 0}}{\sigma_{ys_\pm}}\right\}^2 \qquad (4)$$

where $\Delta K_{\ell^*\to 0}$ represents ΔK_{TH} at equilibrium for the inherent resistance of the alloy to fatigue crack propagation. Note that for $\ell^*\to 0$ that Eq. (2) reduces to the original Dugdale analysis (33) with f equal to zero and $\sigma_{sc} = \sigma_{ys}$. One other assumption that was made was that the semi-cohesive or microfracture zone contained 50 percent voids. Thus, with a single choice of $\Delta K_{\ell^*\to 0}$ and identification of the appropriate grain or colony size, the value of σ may be calculated from Eq. (2) knowing c, b and a. This is then converted to an applied stress intensity threshold, ΔK_{TH}, using the infinite plate solution with c, as above, taken to be 10 mm. (The choice of c is inconsequential as long as
c $>>R_{p_+} + \ell^*$). Since the original ap-
plication of this model, a considerable amount of new evidence has been
obtained where careful attention was
paid to identifying the microstructural size parameters using scanning
electron microscopy (7,20,26). In
addition, we include the present investigation of Ti-5Al-4Mo with average
colony sizes of 30 and 55 μm for the
fine and coarse microstructures.
Using an average value of the monotonic tensile yield strength to be 890
MPa and a value of $\Delta K_{\ell^*\to 0}$ to be 4
MPa-m$^{\frac{1}{2}}$, the theoretical curve fits the
data very well in Figure 11.

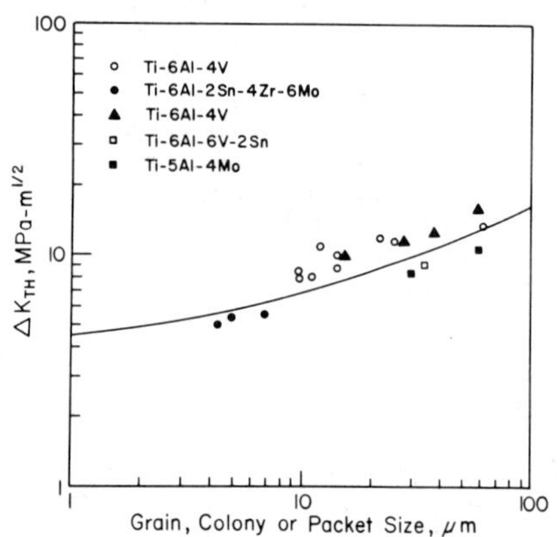

Fig. 11: Thresholds as a function of grain size along with the prediction.

<u>Application to fatigue thresholds --with hydrogen</u>. The concept for hydrogen-affected thresholds is interpreted in terms of either the β phase or the interface phase becoming embrittled. As discussed above, this could be due to either an α" or ω phase in the β or hydrogen absorption into the interface phase. Development of the concept is in three stages:

(i) Hydrogen affects static threshold in Ti-5Al-4Mo as is modelled in terms of the ability of hydrogen to collect at the stress field of a crack tip(3) and lower the fracture stress, σ_f^*;

(ii) Fatigue thresholds in the absence of hydrogen are modelled in terms of yield strength, as affected by test temperature, and Widmanstatten colony size;

(iii) Fatigue thresholds in the presence of hydrogen are affected by hydrogen inducing microfracture which reduces the semi-cohesive zone strength and thus ΔK_{TH}.

Stage (i) has partially been accomplished by determination of σ_f^* from the plastic constraint approach. The Ti-5Al-4Mo data can be quantitatively described by

$$\sigma_f^* = \sigma_{f_o}^* - \alpha_o C_H^{1/2} \tag{5}$$

which is the solid curve in Fig. 1 with $\sigma_{f_o}^*$ being 3100 MPa and α_o being 33 MPa-ppm$^{-1/2}$ for the fine Widmanstatten microstructure. In the absence of any test data for coarse microstructures, we simply scaled α_o to represent a larger partitioning of hydrogen to the lower volume fraction of β which gave α_o to be 44 MPa-ppm$^{-1/2}$, assuming $\sigma_{f_o}^*$ to be the same. By taking into account this effect of hydrogen on $\sigma_{f_o}^*$ and the fact that the triaxiality at the crack tip collects hydrogen, Eq. (5) finally leads to (10)

$$K_{TH} = \frac{1}{\alpha_1} \left[\sigma_{f_o}^* - \sigma_{ys} - \alpha_o c_o^{1/2} \exp\left\{ \frac{(-\frac{\sigma_{ys}}{2} + \alpha_1 K_{TH})\overline{V}_{H-\beta}}{2RT} \right\} \right] \tag{6}$$

Here, α_1 is 12.55 m$^{-1/2}$ from the plastic constraint approach and $\overline{V}_{H-\beta}$ is 1.39 cm^3/mol.

Stage (ii) has already been presented in the above section and thus the remaining description is stage (iii). The first assumption is that there is a fatigue crack nucleation threshold, K_n, below which hydrogen has no effect and that this scales to the static threshold, giving

$$K_n = K_{TH}/\alpha_2 \quad ; \quad \alpha_2 > 1 \tag{7}$$

In general, α_2 is thought to be much greater than one since hydrogen effects under slow frequency repeated loading are much more severe than static conditions. Using K_n to describe the onset of interface fracture, it is possible to interpret the orientation† effect as has been done previously for misoriented cleavage plane fractures (28,35). This gives the void fraction of interface fracture to be

$$f_H = \frac{1}{\pi} \sec^{-1} \left(\frac{\Delta K}{K_n} \right) \tag{8}$$

The choice of $\frac{1}{\pi}$ is consistent with the assumption that hydrogen-induced fractures are additive to the existing fatigue-induced void fraction. That is, at $\Delta K = K_n$, f_H is zero as it should be and as $\Delta K/K_n \rightarrow \infty$, f_H approaches a maximum value of 0.5. With the original assumption of a fatigue induced void fraction of 0.5, this later limit would reduce the semi-cohesive zone strength to zero. Thus, the appropriate void fraction in the presence of hydrogen is now

$$f_V = f_{V_o} + f_H \tag{9}$$

---

†It is quite clear that if the interface, for example, were oriented 0° to the crack plane, that it would more easily fracture than if it were 90°.

It is next possible to apply the concept to the data since the only adjustable parameter is α_2. As an oversimplification, α_2 was taken to be constant and equal to 10. Iteration on Eqs. (2-9) then gives ΔK_{TH} as a function of yield strength, temperature and hydrogen content. In the absence of hydrogen, Eq. (2) agrees very well with the experimental data as might be expected from the agreement already seen in Fig. 11. Thus, the trend in the low hydrogen predictions is very good in Figure 8. With intermediate hydrogen additions, relatively little effect is seen in the fine microstructures as predicted in Fig. 8. On the other hand, for the coarse microstructures a large effect is predicted as observed. However, the qualitative nature of the prediction is incorrect in that large effects are predicted at very low temperatures with smaller effects at higher temperatures, just the opposite of the observation. At the highest concentrations, the trend is reasonably predicted for the coarse microstructure but not for the fine. Such inconsistencies are probably due to the fact that hydrogen does have an effect on dislocation substructure which has an influence on the propensity to nucleate fracture near the α/β boundary. The decrease in susceptibility at 340K is probably due to a decrease in planar slip character which effectively makes it more difficult to have an internal stress concentration nucleate cracking. This is the same as saying that α_2 in Eq. (7) is not a constant and decreases above 300K.

Summary and Conclusions

Internal hydrogen causes substantial effects on the cleavage fracture stresses of all β and α/β titanium alloys. It also promotes second phases such as α'' martensite and ω in the Ti-30Mo all β alloy and interface phase in the α/β alloy. In addition, it affects the ease of dislocation motion and cross slip at temperatures greater than 300K

Such substructural deformation and morphology effects may be qualitatively incorporated into a more quantitative continuum model which can be utilized to understand threshold stress intensity. In this way some rationalization can be made of why high hydrogen but not low hydrogen affects a fine Widmanstatten microstructure while a wider hydrogen concentration range affects a coarse Widmanstatten structure. It also gives a rational framework for understanding microstructural size effects on fatigue thresholds as a function of both temperature and hydrogen concentration. Observations and theoretical predictions demonstrate that even 1500 ppm (wt) hydrogen only decreases fatigue thresholds by about 30 percent.

Acknowledgements

Three of us, W. W. G., N. R. M. and K. J. would especially like to thank the long term support of this project by the Air Force Office of Scientific Research, under contract AFOSR-77-3133.

References

1. R. O. Ritchie: ASME J. Eng. Mater. Technol., 1977, vol. 99, p. 195
2. C. J. Beevers: Metal Science, 1977, vol. 11, p. 362.
3. P. E. Irving and C. J. Beevers, Mat. Sci. Eng., 1974, vol. 14, p. 229.
4. F. L. Robinson and C. J. Beevers: Metal Science, 1973, vol. 7, p.153.
5. G. R. Yoder, L. A. Cooley and T. W. Crooker: Met. Trans. A., 1977, vol. 8A, p. 1737.
6. G. R. Yoder, L. A. Cooley and T. W. Crooker: Trans. ASME, 1979, vol. 101, p. 86.
7. J. C. Chesnutt, A. W. Thompson and J. C. Williams: Final Tech. Report AFML-TR-78-68, Rockwell Science Center, Thousand Oaks, CA, 1978.
8. A. J. McEvily: Metal Science, 1977, vol. 11, p. 274.
9. R. O. Ritchie: Int. Metals Rev., 1979, vol. 24, p. 205.
10. W. W. Gerberich, K. Jatavallabhula, K. A. Peterson and C. L. Jensen in Proc. Fifth Inter. Conf. on Fracture, ICF 5, Cannes, France, 1981, to be published.
11. H. Conrad and K. Okazaki: Scripta Met., 1970, vol. 4, p. 250.
12. H. Margolin: Met. Trans. A., 1976, vol. 7A, p. 1233.
13. J. L. Waisman, G. Sines, and L. B. Robinson: Met. Trans., 1973, vol. 4, p. 291.
14. Y. Mahajan: Scripta Met., 1979, vol. 13, p. 695.
15. R. M. Wood, Acta Met., 1963, vol. 11, p. 907.
16. O. Buck, D. O. Thompson, N. E. Paton and J. C. Williams: Fifth Inter. Conf. on Internal Friction and Ultrasonic Attenuation in Crystalline Solids, 1973, p. 451.
17. K. Ono, L. A. Rosales, S. Motokura and A. W. Sommer: Mechanical Behavior of Materials. I, Soc. of Materials Science, Japan, Kyoto, 1972, p. 66.
18. J. D. Boyd: Trans. ASM, 1969, vol. 62, p. 977.
19. H. J. Rack: Second Int. Conf. on Mech. Beh. of Matls., ICM II, Am. Soc. for Metals, 1976, p. 1196.
20. G. R. Yoder, L. A. Cooley and T. W. Crooker: Met. Trans. A., 1978, vol. 9A, p. 1413.
21. J. C. Williams, A. W. Sommer and P. P. Tung: Met. Trans., 1972, vol. 3, p. 2979.
22. J. Ruppen, P. Bhowal, D. Eylon and A. J. McEvily: Fatigue Mechanisms, ASTM STP 675, 1979, p. 47.
23. I. W. Hall and C. Hammond: Metal Science, vol. 12, p. 339.
24. W. A. Baeslack III and Y. Mahajan: Met. Trans., 1980 vol.11A, p.1234.
25. J. D. Boyd: Met. Trans., 1973, vol. 4, p. 1037.
26. N. R. Moody and W. W. Gerberich: Met. Trans. A., 1980, vol.11A, p.973.
27. N. R. Moody and W. W. Gerberich: Metal Science, 1980, vol.14, p. 95.
28. W. W. Gerberich and N. R. Moody: Fatigue Mechanisms, ASTM STP 675, 1979, p. 292.
29. N. R. Moody and W. W. Gerberich, ICM 3, vol. 2, 1979, p. 513.
30. W. W. Gerberich: Int. J. Fracture, 1977, vol. 13, p. 535.
31. H. K. Birnbaum, M. L. Grossbeck and M. Amano: AFSOR Tech. Report, Contract NOOO 14-75-C-1012, Univ. of Illinois, Urbana, IL, 1976.
32. N. E. Paton: AFSOR Tech. Report, Contract F44620-76-C-0025, Rockwell Science Center, Thousand Oaks, CA, 1977.
33. D. S. Dugdale: J. Mech. Phys. Solids, 1960, vol. 8, p. 100.
34. R. A. Oriani: Ber. der Bunsen-Gesellsch. f. phys. chem., 1972, vol. 76, p. 848.
35. W. W. Gerberich and N. R. Moody: Fracture, 1977, vol. 2 ICF4, Waterloo, Canada, p. 829.

DISCUSSION

<u>P. Adler, Grumman Aerospace Corporation, Bethpage, NY</u>: In the determination
of partial molar volume, your lattice parameters appear to be based exclusively on the (110) peak, i.e. $2\theta \cong 39°$. If this is the case, I would question the accuracy of your results and it's appropriate to use planes of much
higher 2θ, i.e. at least (310) or (321) in β-Ti. This is especially important to your calculations because the partial molal volume enters in the
exponential term.

<u>W.W. Gerberich</u>: The normal extrapolation of lattice parameter as determined
from each d_{hkl} was not possible since the line broadening of the higher 2θ
reflections made peak position determination impossible. For the gas phase
charged samples at ~1700 ppm wt.H there is a discontinuous drop in line width
(probably associated with ω phase formation) which allows position determination and an error analysis. By plotting the diffractometer error function,
$\cos^2\theta/\sin\theta$ for the (110), (200), (211), (220), (222) and (310) for the 1750
ppm wt.H and extrapolating to 90°, an error of 0.003Å is calculated. If it
is now assumed that this same error persists at lower hydrogen concentrations,
then a new value of $\bar{V}^\beta_H$ can be calculated. This was accomplished for both gas
phase charged and electrolytically charged samples with the following results:

	V^β_H (original determination)	V^β_H (corrected to $\theta = 90°$)
gas phase	1.70 cm^3/mol	1.41 cm^3/mol
electrolytic	1.30 cm^3/mol	1.18 cm^3/mol

The value of 1.41 cm^3/mol is close to the average value reported in the paper
(1.39 cm^3/mol) and in agreement with Lunch <u>et al</u>.† determination of $\bar{V}^\beta_H$ =
1.47 cm^3/mol for gas-phase charged T_i - Mo, independent of Mo content.

†J.F. Lunch and J. Tanaka, <u>Scripta Met</u>. 13 (1979), pp. 599-604.

CRACK PROPAGATION ALONG CRYSTALLOGRAPHIC SLIP BANDS

AND HYDROGEN EMBRITTLEMENT

J. R. Wilcox[†] and D. A. Koss
Department of Metallurgical Engineering
Michigan Technological University
Houghton, MI 49931

The propagation of an inclined crack along a coplanar slip band is examined with regard to the conditions usually associated with hydrogen embrittlement. The analysis shows that large normal stresses are located very near to the crack-tip, despite the mixed mode nature of such cracks. This state of stress, the presence of the planar slip at the crack-tip, and a small but probably significant amount of non-coplanar slip all contribute to a set of conditions which are very favorable to hydrogen embrittlement. In view of the analysis, the effect of gaseous environment on the fatigue crack propagation behavior of an age hardenable metastable β-phase Ti-30V alloy has been examined. Tests on both single crystal and polycrystalline specimens show that, consistent with the analysis, gaseous hydrogen accelerates the fatigue crack growth rates <u>only</u> when Stage I fatigue occurs along a crystallographic {112} slip plane.

†Presently at: General Electric Co., Research and Development Center, Schenectady, NY 12345

Introduction

The propagation of a crack along a crystallographic slip band occurs frequently in alloys susceptible to planar, non-uniform slip. Such crystallographic cracking occurs in large grain or single crystal material and on planes inclined to the stress axis and is most commonly observed under cyclic loading conditions in high strength Ni-, Al-, and Ti alloys. In all of these materials, cracks propagating along slip bands usually have a substantially brittle or cleavage-like fracture appearance, which can be sensitive to the environment.[1-4] This suggests that normal stresses and environmental effects are important in this fracture process.

Certain conditions for the propagation of a crack along a coplanar slip band (i.e., the slip plane and the slip vector are parallel to the plane of the crack, although not necessarily in the crack plane) have been analyzed by Koss and Chan.[5,6] In their model, a relaxed "elastic-plastic" state of stress exists ahead of a crack propagating along a slip band. Calculations show that this stress state is not conducive for activating secondary slip with a Burger's vector inclined to the crack plane.[5] As a result, large normal stresses can be generated near the tip of an inclined, mixed-mode crack propagating along a crystallographic slip band, particularly for basal plane cracking in an hcp material.[6] The purpose of this paper is to examine crack propagation along coplanar slip bands with regard to the conditions favoring hydrogen embrittlement. We will also show that, consistent with the model, an alloy (bcc Ti-30V, in this case) can be susceptible to hydrogen-assisted cracking if it is heat-treated to promote crack propagation along crystallographic slip bands.

Crack Propagation Along Planar Slip Bands and the Resulting Stress Stage

The basis for understanding crack propagation along a coplanar slip band is contained in Fig. 1 and is discussed in more detail by Koss and Chan.[5,6] If the crack-tip plasticity is dominated by coplanar slip whose Burger's vector is also coplanar with the crack, then the planar slip plastic zone extends to distance r_p ahead of the crack-tip. The singularity in the elastic stress field near the tip of a mixed mode crack dictates that if r is the radial distance from the crack-tip, then $\sigma_{ij} \propto 1/\sqrt{r}$, where σ_{ij} is a stress component. A key feature shown in Fig. 1 is the recognition that coplanar slip relaxes the shear stresses σ_{xy} and σ_{yz} to values (σ_{xy}^* and σ_{yz}^*) roughly determined by critical resolved yield stress τ_y of the material since $\tau_y = \sigma_{xy}^* \sin\phi + \sigma_{yz}^* \cos\phi$ at $r \leq r_p$.[†] However, the coplanar slip and the resulting relaxation of the shear stresses (the associated displacements are akin to sliding a deck of cards) does not produce displacements which are dilational. Thus, the normal stress components σ_{ii} continue to increase until very close to the crack-tip when at $r = r_p^s$, non-coplanar slip is activated. This is shown in Fig. 1. It is important to realize that this relaxed,

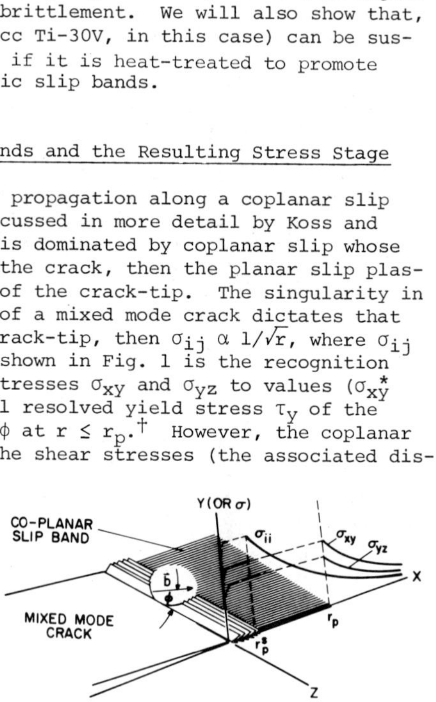

------

†Note that in the plane of a crack, $\sigma_{xz} = 0$ and that ϕ is the angle between the crack front (z axis) and the slip vector of the coplanar slip band.

Fig. 1. A schematic illustration of the stress state ahead of a mixed mode crack with a coplanar slip band characterized by a slip vector $\bar{b}$ at angle ϕ to the crack front. The coplanar slip extends to r_p which non-coplanar slip is activated to r_p^s (after Koss and Chan).

elastic-plastic stress state is not conducive for activation of non-coplanar slip (i.e., in plane strain, the normal stresses are nearly balanced triaxial; $\sigma_{xx} = \sigma_{yy} = 3/2\ \sigma_{zz}$). However, this stress state _is_ well-suited for the development of large normal stresses near the crack-tip.

Ignoring plastic relaxation, crack-tip blunting, work hardening effects, and assuming the normal stresses to obey the same functional dependence on r and elastic stress intensity factor regardless of any relaxation of σ_{xy} and σ_{yz}, Koss and Chan have estimated the magnitude of the maximum normal stresses for three different crack orientations and for fcc, hcp, and bcc slip crystallography.[6] Those calculations show that for 45° inclined, mixed-mode cracks, the maximum value of the hydrostatic stress σ_H^m is much greater for crack propagation along a slip band than if multiple slip occurred ahead of the crack.

The effect of the crystallography of slip in fcc, hcp, and bcc alloys on σ_H^m has also been examined.[6] As might be expected, very large hydrostatic stresses are predicted to be present near the tip of a crack propagating along a basal plane in an hcp crystal. This is due to the difficulty of activating slip or twinning which causes an extension (or relaxation of the normal stresses) in the C direction, normal to the (0001) plane. Extension along the C axis by twinning on the $\{10\bar{1}2\} <\bar{1}101>$ system, which is very common to most hcp alloys, is probably the most likely means that hcp metals have of relaxing the large normal stresses near the crack-tip. Observations of Stage I fatigue along the basal plane in hcp Ti alloys show that $\{10\bar{1}2\}$ twins are indeed found near the tip of crystallographic cracks.[7,8] Another interesting result is that, despite many available secondary slip systems, relatively large hydrostatic stresses can also be generated by crystallographic slip band cracking in both fcc and bcc material.[6] An implication of these calculations is that a cubic alloy might be sensitive to hydrogen embrittlement effects if it is heat-treated to promote planar slip and crack propagation occurs along a coplanar slip band (assuming that the kinetics of hydrogen transport are sufficiently rapid).

The Maximum Stress Location x_m

The position x_m of the maximum hydrostatic stress ahead of a crack-tip is important in hydrogen embrittlement because it will influence the time scale that H must diffuse (or be swept in). The methodology used in the above calculations assumes that the maximum normal stresses occur near the elastic-plastic interface for non-coplanar slip (i.e., at $r = r_p^{\ S}$). While crude, this method yields a value for a mode I crack which agrees well with continuum plasticity estimates for a von Mises material.[9] However, the continuum plasticity estimates based on multiple slip and a mode I crack place the largest value of the maximum principal stress at a distance x_m being two times the crack-tip opening displacement between the upper and lower crack faces.[9] In the absence of detailed continuum plasticity calculations in the present case, it is reasonable to conclude that the maximum stress position x_m for this form of crystallographic cracking is:

$$2\delta \lesssim x_m \lesssim r_p^{\ S} \tag{1}$$

where δ is the normal component of the crack-tip opening displacement for a mixed-mode crack and therefore is related to the extent of non-coplanar slip. Physically Eq. (1) implies that as one approaches the crack-tip in a non-hardening material, the normal stresses σ_{ii} in Fig. 1 are constant within the coplanar plastic zone from $x = r_p^S$ to $x = 2\delta$ at which point the stresses begin to relax due to the surfaces of the crack.

The important feature to recognize in this analysis is that Eq. (1) places the maximum normal stresses very near the crack-tip and thus results in relatively short H diffusion/sweep in distances. The values of r_p, r_p^S, and therefore x_m for a given crack geometry and crystal system can be calculated using the method outlined in the Appendix of Ref. 6. For example, in the case of a mixed mode I/II crack inclined 45° across the width of a plate in plane strain, $x_m \leq .07\ r_p$ for an fcc crystal cracking along a {111} plane, and $x_m \leq .02\ r_p$ for an hcp crystal with a basal crack. The crack-tip opening displacement criteria would result in an even smaller value of x_m. A rigorous calculation of δ is beyond the scope of this paper; it would involve knowing the detailed shape of the secondary slip plastic zone which in turn requires an accurate knowledge of <u>radial</u> distribution of stresses for the stress state shown schematically in the plane of the crack at $\theta = 0°$ in Fig. 1. An estimate of δ can be made by assuming $r_p^S \neq f(\theta)$, in which case $\delta \cong 2(\tau_y/\mu)r_p^S \int_0^\pi \sin\theta d\theta = 4(\tau_y/\mu)r_p^S$, where μ is the shear modulus. Given that $\tau_y/\mu \approx 10^{-2}$ for a high strength alloy, Eq. (1) becomes: $10^{-1}\ r_p^S \leq x_m \leq r_p^S$. In view of the magnitude of r_p^S, this places x_m very close to the crack-tip, minimizing the diffusion/sweep in distance of an external hydrogen source.

In summary, the large normal stresses, the presence of planar slip which enhances any film rupture and hydrogen sweep in process, and the location of the maximum stress being near the crack-tip are all conditions which favor hydrogen embrittlement phenomena, especially due to external hydrogen. It is also likely that the small amount of non-coplanar slip present near the crack-tip is important in providing sufficient crack opening separation of the upper and lower crack surfaces to allow an environmental interaction and to hinder slip reversibility within the planar slip band. All of the above suggests that an alloy which might be otherwise immune to hydrogen embrittlement may become susceptible to hydrogen-assisted cracking if it is heat treated so that cracking along a crystallographic slip band occurs.

<u>Hydrogen-Assisted Fatigue Crack Propagation in a Beta Titanium Alloy</u>

In view of the above analysis, the effect of a gaseous environment on the fatigue crack propagation behavior of an age hardenable, metastable β-phase (bcc) Ti-30V alloy has been examined. Beta Ti alloys are usually considered to be immune to hydrogen embrittlement effects.[10] However, the Ti-30V alloy can be heat-treated in the temperature range of 300°C to precipitate very fine, coherent ω-phase particles. Such precipitation is known to result in a transition from fine, wavy slip in the solution-treated condition to planar, non-uniform slip after age hardening by the ω-phase particles.[11] In fatigue, the precipitation of the ω-phase can also cause a transition in the plane of fatigue crack propagation. Stage II fatigue (i.e., propagation involving multiple slip and on a plane normal to the stress axis) dominates Ti-27V solution-treated material while Stage I fatigue (crack propagation along a crystallographic slip plane inclined to the stress axis) frequently occurs in the ω-phase hardened condition.[12] This study utilizes the age hardening behavior of the Ti-30V alloy and the use of single crystal specimens to examine the influence of gaseous hydrogen on both Stage I <u>and</u> Stage II fatigue behavior.

Polycrystalline samples with an average grain size of 0.057 mm and seeded single crystals in the form of identical strips $\sim$1.5 mm thick were tested. The configuration of the test samples consisted of a gauge length

25.4 mm long x 6.35 mm wide with a
centrally located hole which acted as
a crack starter. The orientation of
the single crystal samples is shown
in Fig. 2 and was chosen so that
crack propagation along the most
highly stressed {211} slip plane
should result in a mixed mode I/II
crack inclined across the width of
the sample. This crystal orientation
makes through-thickness slip very
difficult and assures predominantly
plane strain at small stress intensity levels even with the relatively
thin samples. Plane strain also dominates the polycrystalline samples at
least to stress intensity ranges up

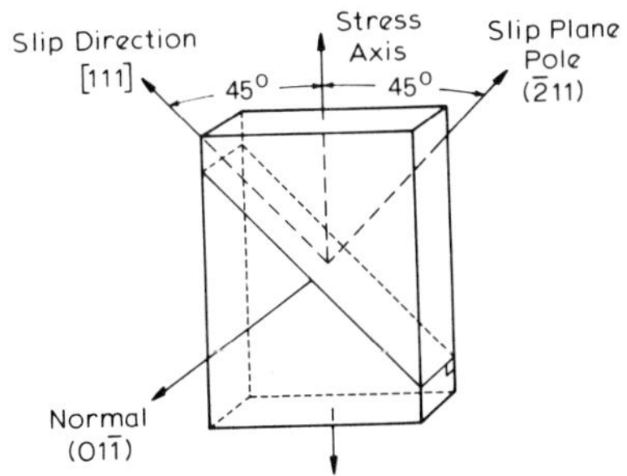

Fig. 2. The orientation of the single crystal specimens examined in this study.

to $\Delta K = 20$ MPa$\sqrt{m}$; this is based on cyclic plastic zone size calculations
(r_p < .04 x thickness) and visual observations. Prior to testing, all of
the samples were He gas quenched from 800°C, and the single crystals were
age hardened at 300°C for 53 hours, which increased the hardness from 230 DPH
to 350 300 DPH. The Ti-30 at.% V (31.5 wt.%) single crystal samples contained (in wt. ppm): 950 O, 41 H, 260 N, and 240 C while the polycrystalline
specimens contained: 720 O, 31 H, 320 N, and 190 C.

Fatigue testing was performed at 30 Hz in tension-tension at R = 0.1
in an environmental chamber. The environmental chamber permitted evacuation
before back filling with ultra high purity H or He gases which were also
passed through a liquid nitrogen trap. Crack lengths were measured using
macrophotography of the specimen at selected time intervals. For a given
crack, the stress intensity range ΔK was calculated utilizing a finite width
correction factor for an eccentric mode I crack.[13] The inclined cracks all
involved a mixture of modes I and II in which case summation of strain energy
densities gives an "effective" stress intensity of $\Delta K = (\Delta K_I^2 + \Delta K_{II}^2)^{\frac{1}{2}}$ from
the mode I stress intensity ΔK_I and that of mode II, ΔK_{II}.

The fatigue crack propagation behavior of the solution-treated polycrystalline specimens (DPH $\simeq$ 215) and the age hardened Ti-30V alloy single
crystals (DPH $\simeq$ 365) is shown in Figs. 3 and 4. As can be seen, a definite
acceleration of the fatigue crack growth rate da/dN occurs in gaseous hydrogen only when crack propagation takes place in a crystallographic Stage I
manner. In this case, propagation occurs in an extended manner on a plane
within 2° of the most highly stressed {211} slip plane, the (2̄11) slip plane
in Fig. 2. It is also possible to obtain crystallographic Stage I fatigue
in dry He in these age hardened crystals. However, Fig. 4 shows that da/dN
for Stage I fatigue in He is identical in the Paris Law regime with that
obtained by testing in laboratory air.*

As in most other cases involving crystallographic Stage I fatigue on an
extended scale, the fracture surfaces appear cleavage-like at low magnifications when Stage I cracking occurs.[1-3,12] At higher magnifications, there
is evidence of ductility on the surfaces of the Stage I cracks, especially
in the samples tested in He, but there is no evidence of striations. The

---

*Data of Salyat and Russ[12] is for Ti 27V single crystals of the same orienta-
tion and heat treated also at 300°C but for 18 hours which results in
slightly lower hardness (DPH ∿ 340) but still predominantly Stage I fatigue
crack growth. That data is for samples tested in air at R.H. < 30%.

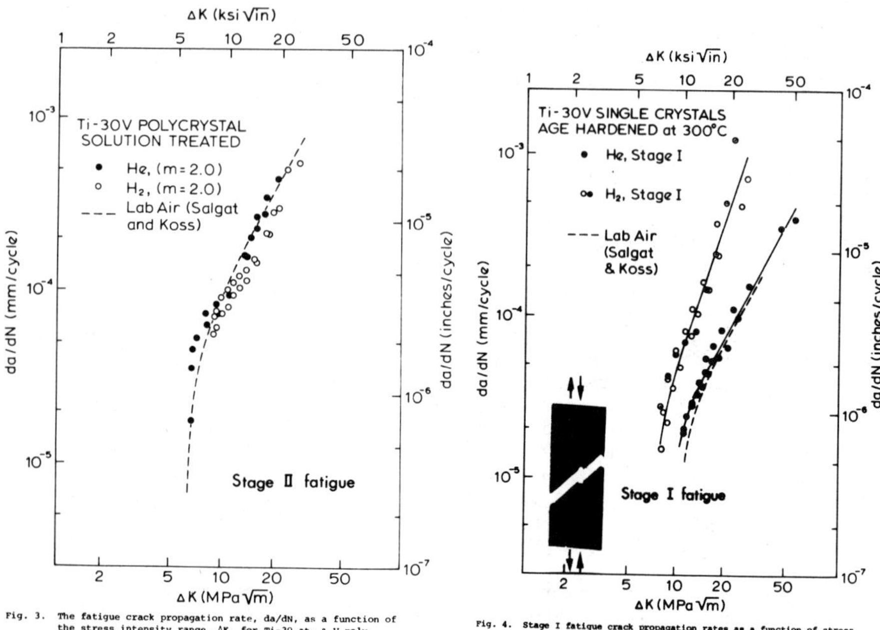

Fig. 3. The fatigue crack propagation rate, da/dN, as a function of the stress intensity range, ΔK, for Ti-30 at. % V polycrystalline samples helium gas quenched from 800°C. Hardness ≈ 215 HN.

Fig. 4. Stage I fatigue crack propagation rates as a function of stress intensity range for Ti-30 at. % V single crystal specimens quenched from 800°C and aged 53 hours at 300°C. Hardness ≈ 365 HN.

samples tested in H_2 show microcracking on the bottom face of an upward sloping crack, probably due to hydrogen embrittlement and the stress state behind the tip of a mixed mode crack.[14] The Stage II cracks exhibit striations on the fracture surface, the striations being aligned near a <110> direction and reflecting propagation by intersecting pairs of {211} slip planes.[15]

Correlation of the Fatigue Behavior and the Model

A principal result of this study is that the embrittling effects of gaseous hydrogen occur only when cracking occurs in the age hardened condition along a crystallographic slip band (Stage I fatigue). When the cracking is non-crystallographic (Stage II fatigue), the effect of hydrogen on crack propagation is negligible for the β Ti system studied. We believe that this behavior is a consequence of crack propagation along crystallographic slip bands as described in the first part of this paper. Concerning the hydrogen embrittlement by gaseous hydrogen in the present case, we believe that the key ingredients of this form of cracking are: the relatively large hydrostatic stresses, the location of the maximum in normal stresses being very near the crack-tip, and the presence of planar slip accompanied by a small but significant crack-tip opening displacement normal to the crack faces.

Figure 5 depicts approximately the situation described in the present experiments. It should be noted that Fig. 5 is based on previous calculations[5,6] which are modified for the case of cyclic loading by doubling the yield stress.[16] As is shown in Fig. 5, a substantial hydrostatic stress should exist near the crack-tip, $\sigma_H^m \approx 6.3\ \tau_y^C \approx 2.0$ GPa, where τ_y^C is the cyclic shear stress for yielding which is ≈ 330 MPa for age-hardened crystals, assuming $\tau_y^{\,c} \simeq \tau_y$. The location of the maximum stress is also very close to the crack tip ($x_m \simeq$ .4μm @ ΔK = 10 MP√m) so that diffusion or sweep-in distances are short. It is noteworthy that these values of $\sigma_H^{\,m}$ and x_m for the Stage I crack in this bcc alloy do not differ greatly from the magnitude of $\sigma_H^{\,m}$ (= 5.3$\sigma_y^{\,c} \simeq$ 2.6GPa, assuming $\sigma_y^{\,c} \simeq \sigma_y$ = 500MPa[12]) and x_m = 2δ ($\simeq$ 1μm at ΔK = 10 MP√m) for the Stage II fatigue which shows no hydrogen embrittlement effects. This may be interpreted that either: (a) as suggested previously, planar slip when combined with a small out-of-plane crack opening displacement is critical to hydrogen embrittlement either by causing film rupture or by enhancing dislocation sweep-in of hydrogen onto the coplanar slip bands[17] (which may result in easier slip band decohesion); or (b) that we have underestimated $\sigma_H^{\,m}$ and therefore overestimated x_m for the Stage I cracks in the age-hardened β Ti alloy. Such errors in $\sigma_H^{\,m}$ and x_m will occur if a significant amount of cyclic softening occurs causing $\tau_y^{\,c} \ll \tau_y$; experimentally cyclic softening does occur in β Ti-V alloys exhibiting planar slip.[18] Thus conditions which are favorable to hydrogen embrittlement exist when crack growth occurs along a crystallographic slip band. These experiments and this analysis do not, however, identify the mechanism by which such embrittlement occurs.

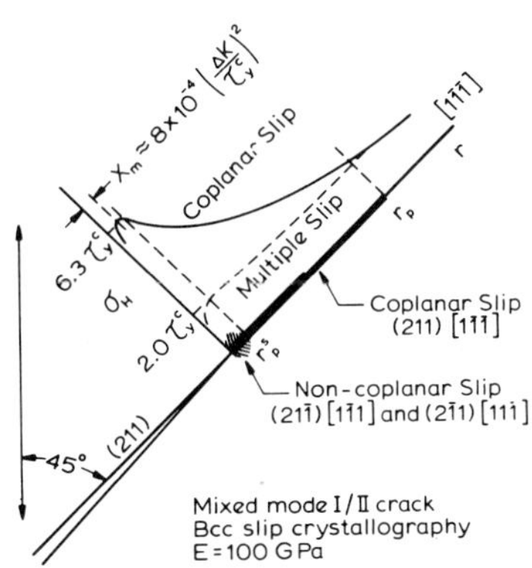

Fig. 5.   The approximate distribution of the hydrostatic stress component σ_H as a function of the distance r ahead of the crack-tip in the plane of the crack. Both coplanar and multiple slip situations are depicted for a 45° mixed mode I/II crack in a material with Bcc slip crystallography.

Acknowledgements

The authors wish to thank Kwai Chan for helpful comments. We would also like to acknowledge the support of the Air Force Office of Scientific Research through Grant No. 74-2596 and of the Office of Naval Research through Contract No. N00014-76-C-0037.

References

1. D. J. Duquette and M. Gell, Met. Trans., 1971, vol. 2, p. 1325.

2. P. E. Irvin and C. J. Beevers, Met. Trans., 1973, vol. 5, p. 391.

3. M. Nageswararao and V. Gerold, Met. Trans. A, 1976, vol. 7A, p. 1847.

4. P. Neumann, H. Vehoff and H. Fuhlrott: in Fracture 1977, vol. 3, p. 1313, University of Waterloo Press, Waterloo, Canada, 1977.

5. D. A. Koss and K. S. Chan, Acta Met. (in print).

6. D. A. Koss and K. S. Chan: in Acta/Scripta Met. Int. Conf. on Dislocation Modelling of Physical Systems, Gainesville, FL, 1980 (in print).

7. K. S. Chan and D. A. Koss, Mat'l. Sci. and Eng., 1980, vol. 32, p. 177.

8. G. P. VanderVelde, M. S. Thesis, Michigan Technological Univ., 1979.

9.   J. R. Rice in Stress Corrosion Cracking and Hydrogen Embrittlement, p. 11, NACE, Houston, 1977.

10. N. E. Paton and J. C. Williams in Hydrogen in Metals, p. 409, ASM, Metals Park, 1974.

11. A. Gysler, G. Lutjering and V. Gerold, Acta Met., 1974, vol. 22, p. 901.

12. G. Salgat and D. A. Koss, Mat'l. Sci. and Eng., 1978, vol. 35, p. 263.

13. G. C. Sih in Handbook of Stress Intensity Factors, p. 1.4.3-3, Lehigh Univ. Press, Bethlehem, PA, 1973.

14. M. G. Stout and D. A. Koss, Met. Trans. A., 1978, vol. 9A, p. 835.

15. J. A. Carlson and D. A. Koss, Acta Met., 1978, vol. 26, p. 123.

16. J. Rice in Fatigue Crack Propagation, ASTM STP 415, p. 247, ASTM, Philadelphia, 1967.

17. J. K. Tien, S. V. Nair, and R. R. Jensen, in the Proc. of the Third Int. Conf. on Effect of Hydrogen on Behavior of Materials, Jackson Lake Lodge, Wyoming, Aug. 26-31, 1980.

18. S. B. Chakrabotty, T. K. Mukhopadhyay and E. A. Starke, Acta Met., 1978, vol. 26, p. 909.

DISCUSSION

H.K. Birnbaum, University of Illinois, Champaign, IL: You showed that the yield stress in the crack plane is limited by the planar slip band in front of the crack. How are the shear stress components at other positions around the crack affected by this slip band?

J. Wilcox: The "sliding of a deck of cards" analogy still holds. Slip on a plane which is parallel to the crack plane but not in the crack plane should also be relaxed according to: $\tau_y = \sigma^*_{xy} \sin\phi + \overline{\sigma^*_{yz}} \cos\phi$ at $r \leq r_p$. However, the relative values of σ^*_{xy} and σ^*_{yz} (as well as r_p) will depend on the position around the crack. The magnitudes of these shear stress components could be crudely estimated as a function of position by utilizing linear elastic fracture mechanics, but one should recognize that the relaxation due to the initial coplanar slip band will tend to decrease the out-of-plane values of σ_{xy} and σ_{yz} at a given r and θ and thus act to hinder additional planar slip.

FRACTURE

A NEW CONCEPT OF RELATION BETWEEN REDISTRIBUTION
BEHAVIOUR OF HYDROGEN AND EMBRITTLEMENT INDUCED
BY HYDROGEN IN STEEL

Yoneo   Kikuta
Department of Welding Engineering
Faculty of Engineering
Osaka University
Suita, Osaka, Japan

Introduction

Many different embrittlement mechanisms have already been proposed by
various researchers. Most of the mechanisms propose hydrogen diffusion to
defective regions inside material, its high concentration there, and the
subsequent embrittlement. Especially the way hydrogen gathers at defective
regions is a matter of discussions. Some explain the process by just using
the word "Stress Induced Diffusion"[1] and others are more challenging by
assuming a mechanism where hydrogen is transported by dislocation motions
[2]. It is very plausible that there exists stress-field interaction
between hydrogen and dislocation, although it might be very small as
compared with the case in carbon and nitrogen.

Generally hydrogen behaviours in steels, especially its diffusional
behaviours, have been treated by the use of Fick's second law. The analysis
deals with hydrogen flow due to volumetric concentration gradient and gives
the macroscopically correct picture of the hydrogen behaviours.

However hydrogen embrittlement of steels is induced by hydrogen
diffusion and concentration at microscopic singularities. And we are
required to clarify the microscopic diffusion to such singularities (present
author considers them dislocation sites) as well as overall and macroscopic
diffusion.

According to the above mentioned standpoint, present author clarified
experimentally as well as theoretically the hydrogen-dislocation interaction
and the hydrogen redistribution behaviours according to the stress(strain)
fields.

Hydrogen-Dislocation Interaction

Specimen materials and tensile testing

The chemical compositions of the present pure iron and high strength
steel are given in Table 1. The charging conditions were chosen so that
hydrogen is uniformly saturated within the specimens at 950°C. The amount

of hydrogen occlusion thus achived was found to be 4.4 ppm. Then specimens
were aged for various time intervals before testing and hydrogen was allowed
to diffuse to newly multiplied or newly depinned (by mainly martensitic
transformation) dislocation sites. For tensile testing, specimens whose
shapes are shown in Fig. 1 were prepared, and tested at roughly the same
and slow enough loading speed such that the degree of embrittlement is
revealed at its maximum. And the fracture stress (notch tensile strength :
NTS) was recorded.

Internal friction experiments

Internal friction was measured by bending vibration method (electrostatic drive and detection). The specimens of 10mm width, 100mm length,
and 2 or 4mm thickness were hydrogen charged, 0°C aged for various time
intervals, mounted on the sample holder, and internal friction Q^{-1} was
measured during cooling at the rate of 1°C/min while the specimens were
kept in resonance with a friquency of 1-2kc. To observe hydrogen cold-
work-peak (here after, H.C.W.P.) Q^{-1}, has to be measured at various
temperatures in the neighborhood of peak temperature, and one run for the
measurement requires a considerable amount of time. H.C.W.P. of pure iron
and steels is smaller in magnitude when compared with its counterpart by
carbon and nitrogen (the order of 10^{-5} when compared with 10^{-3} of carbon
and nitrogen C.W.P.). According to Schoeck's model on cold-work-peak[3],
the peak height (in large displacement case) is proportional to $\Lambda \cdot l_0^2$ where
Λ is dislocation density and l_0 is mean dislocation loop length. Thus
higher dislocation density increases the peak height.

N.T.S. change during aging time at 0°C

When specimens of pure iron and HT-80 steel are hydrogen charged by
high-temperature method, their notch tensile strength decreases at the
early stage of aging, goes through the minimum, and at the later stage of
aging it increases gradually and recovers the value of the specimens of
same thermal treatment under 1 atom of argon (Fig. 2). The both processes
are very sensitive to aging temperature and we expect that these processes
are controlled by diffusion rate. In fact, we carried out an experiment
where measurements were done on diffusion controlled hydrogen evolution
from as-hydrogen charged specimens(Fig. 3), and we see a perfect
simultaneity of NTS recovery curves in Fig. 2 and hydrogen evolution curves
in Fig. 3. In the increasing embrittlement processes, hydrogen is considered to be redistributing itself in the neighborhood of defects. These facts
are compatible with our microscopic interpretation of two processes, where
hydrogen diffuses to dislocation sites at first during the increasing
embrittlement stage and, when large concentration gradient appears by long
enough aging, hydrogen diffuses out at the recovery stage.

Internal friction changes during aging time

In the case of pure iron, H.C.W.P. changes its height during aging at
0°C and its height increases till it reaches the maximum at around 90min,
but further aging reduces it such that the peak disappears at around 10^3min
(see Figs. 4(a) and (b)). HT-80 steel exhibited the similar manner in its
H.C.W.P., whose maximum height appeared at around 100min (Figs. 4(c) and (d)
). These results correspond well to the tensile data obtained in the
preceding paragraph.

Further we obtained hydrogen-dislocation interaction energy by already
mentioned peak-shift method. Our value for activation energy (7500 cal/

Table 1.   Chemical compositions of present pure iron and HT-80 steel

Materials	C	Si	Mn	P	S	Cu	Ni	Cr	Mo	V	B
Pure iron	0.003	0.01	0.01	0.002	0.006	—	—	—	—	—	—
HT-80 steel	0.108	0.26	0.95	0.01	0.01	0.30	0.82	0.55	0.49	0.004	0.0048

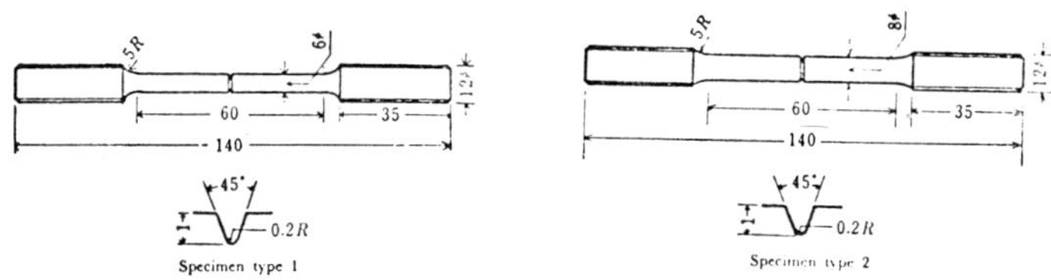

Fig. 1.   Specimen dimensions of tensile test pieces

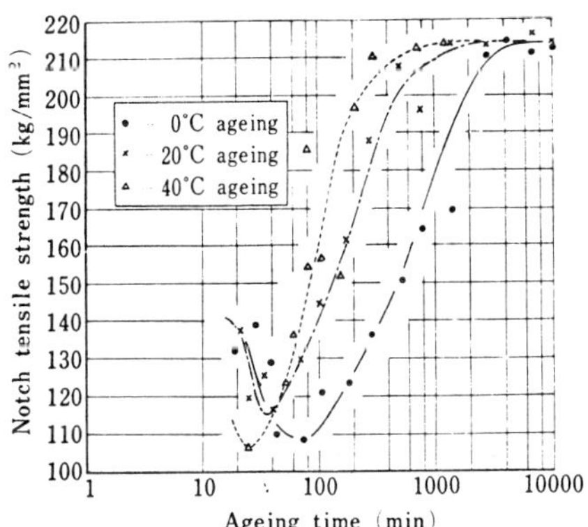

Fig. 2   N.T.S. of thermally hydrogen-charged HT-80 steel decreases with increasing aging time at first, goes through the minimum, and finally recovers the value of the hydrogen-free material

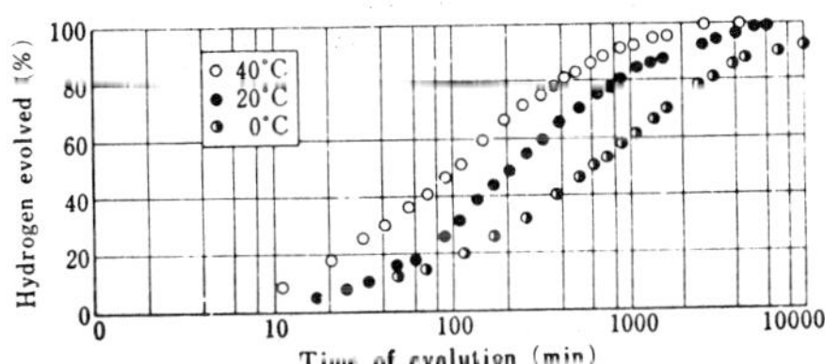

Fig. 3   Hydrogen evolution curves for thermally hydrogen-charged HT-80 steel

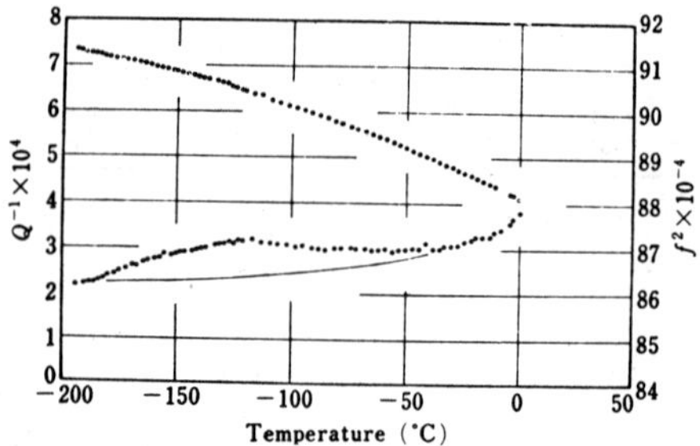

Fig. 4(a)   Internal friction ($Q^{-1} \times 10^4$) and modulus ($f^2 \times 10^{-4}$) of the pure iron, which was thermally hydrogen-charged and was 10% cold-rolled the right after

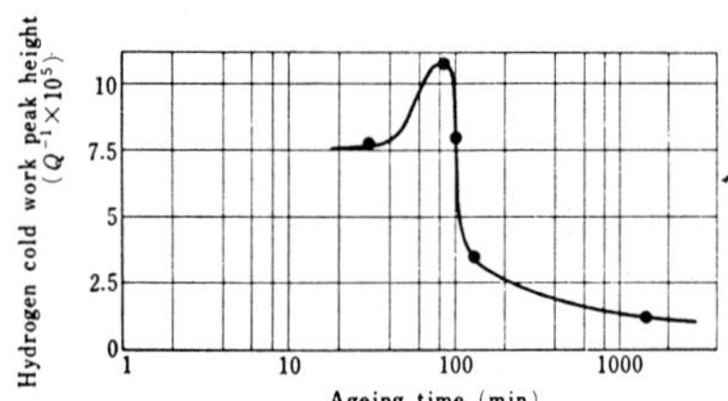

Fig. 4(b)   Hydrogen cold-work-peak height *vs.* effective 0°C aging time for thermally hydrogen-charged and subsequently cold-rolled pure iron

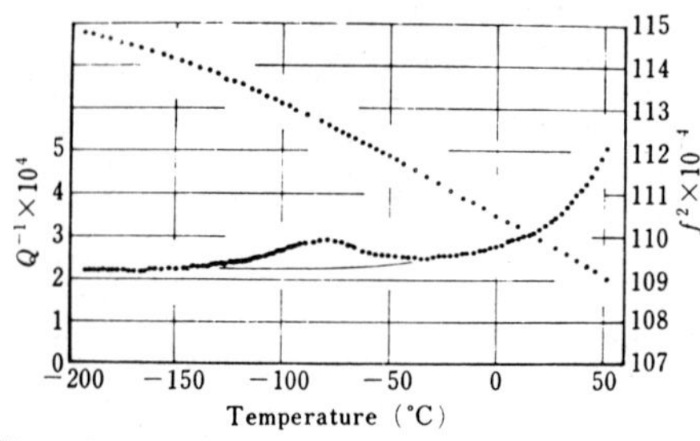

Fig. 4(c)   Internal friction ($Q^{-1} \times 10^4$) and modulus ($f^2 \times 10^{-4}$) of thermally hydrogen-charged HT-80 steel

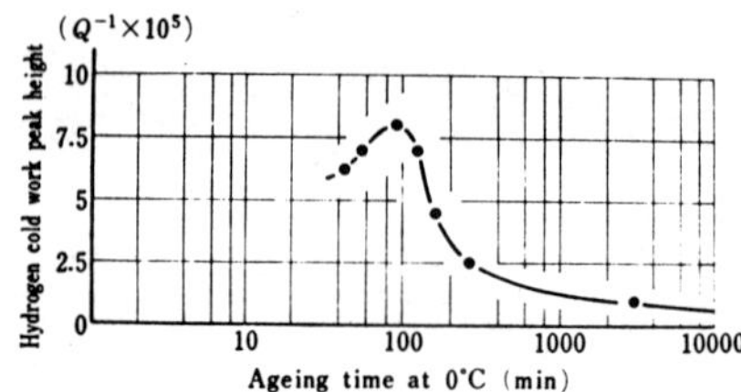

Fig. 4(d)   Hydrogen cold-work-peak height *vs.* effective 0°C aging time for thermally hydrogen-charged HT-80 steel

Table 2.   Activation energy for the recovery process of hydrogen embrittlement for HT-80 steel

Obtained from tensile test data	8 600 cal/g·atom
Obtained from hydrogen evolution data	7 200 cal/g·atom
Activation energy from lattice diffusion plus hydrogen-dislocation binding energy	7 500 cal/g·atom*

* Assumed to be same with the one for pure iron

g·atom) obtained for the relaxation process (dislocation-hydrogen binding
energy for hydrogen lattice diffusion). The activation energy is the
requirement for hydrogen to jump from one dislocation site to the other,
and should be compatible with activation energies of hydrogen diffusion
process and N.T.S. recovery process. Table 2 is such comparison of these
energies of thermally hydrogenated HT-80 steel. And we see that they are
compatible to each other.

<u>Parallelism between N.T.S. drop and the increase in H.C.W.P. height</u>

We have observed so far that the N.T.S. behaviour of hydrogen charged
specimen is very much related to that of the H.C.W.P.. Fig. 5 is such
correlation between H.C.W.P. heights at various aging times and the N.T.S.
values at the corresponding times. We see one-to-one correspondence
between the two experiments both at increasing embrittlement process and
the recovery process. Further the drop in N.T.S. is linearly related to
H.C.W.P. height.

Hydrogen Transport by Dislocation Motion

Hydrogen embrittlement of steels is induced by hydrogen-dislocation
interaction in steels as mentioned in the preceding paragraphs. Thus it is
very important to know how hydrogen, trapped at dislocation sites, diffuses
or alternatively is transported to strain(stress) concentrated region.

Fig. 6 is the tensile test results where the specimens thermally
hydrogen charged are aged for various time intervals and tested at
different cross head speeds. Note hydrogen embrittlement is most severely
revealed when both approximately 80min aging (maximum hydrogen trapping
at dislocation site) and slow strain rate are applied. While no appreciable
embrittlement is detected when tested at high strain rate. This shows the
importance of hydrogen transport by way of dislocation motion during tensile
testing (at high strain rate, hydrogen can not follow dislocation motion
and its binding to dislocation is broken).

<u>Hydrogen effusion phenomena</u>

Fig. 6(b) shows the specimen geometry which is used for hydrogen
effusion measurement by electro-chemical method using capillary tubes
positioned at various locations of the surface of cathodically hydrogen
charged specimens (Fig. 7 and Fig. 8). The measurements are done both
during the periods of aging and tensile testing. As is show in the example
(Fig. 8), hydrogen effusion from notch root is more distinct when measured
during tensile testing.

And Fig. 9 shows the effects of aging time as well as tensile test
strain rate on such measurements of hydrogen effusion from various location-
s. A combination of approximately 80min aging and slow strain rate tensile
testing gives the maximum hydrogen effusion and this is due to hydrogen
transportation by dislocation motion. While high strain rate tensile
testing does not induce any change in hydrogen evolution curves irrespective
of aging time differences. And this indicates ordinary hydrogen evolution
due to hydrogen broken away from dislocation trapping. These phenomena
could be more clearly seen in Fig. 10 and Fig. 11. Specimens which are aged
for 70min after charging and are tested by slow strain rate tension are
remarkably embrittled because of hydrogen transportation by dislocation
motion. But during short aging time or at high strain rate tensile test,
specimens shows ordinally delayed failure phenomena, because only a little

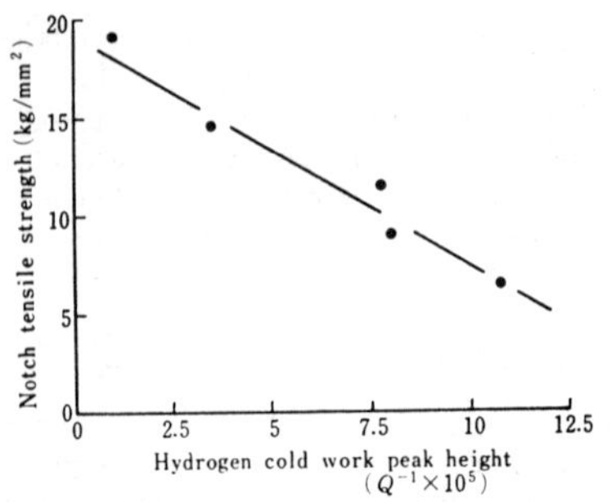

Fig. 5(a) Parallelism between notch tensile strength and hydrogen cold-work-peak height for thermally hydrogen-charged and subsequently cold-rolled pure iron. Notch tensile strength is directly influenced by hydrogen population at dislocation sites (assumingly proportional to hydrogen cold-work-peak height).

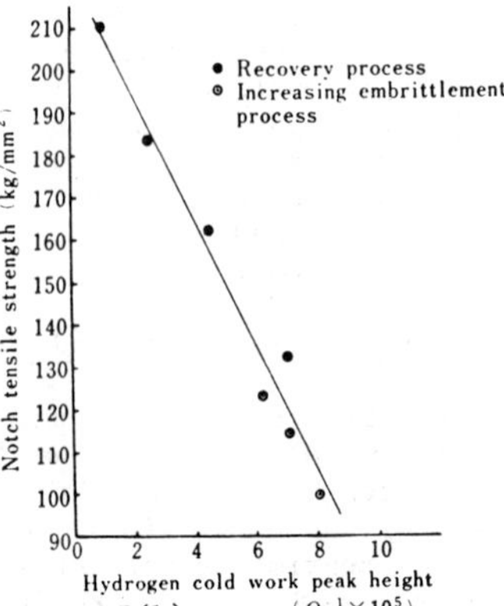

Fig. 5(b) Parallelism between notch tensile strength and hydrogen cold-work-peak height for thermally hydrogen-charged HT-80 steel

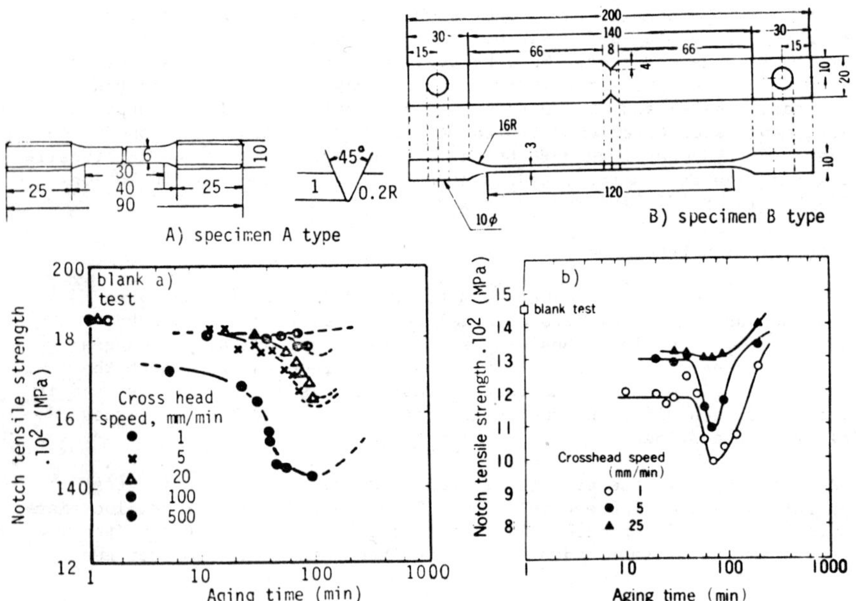

Fig. 6 Effect of aging time and cross head speed on notch tensile strength. a):specimen A type (steel A) and b):specimen B type (steel B)

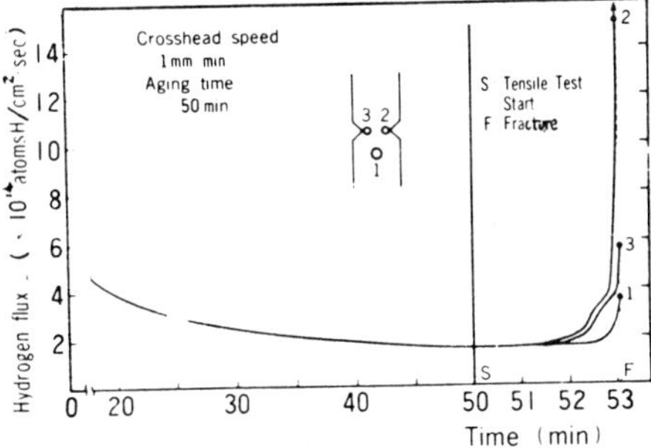

Fig. 7 Evolution transient during tensile test

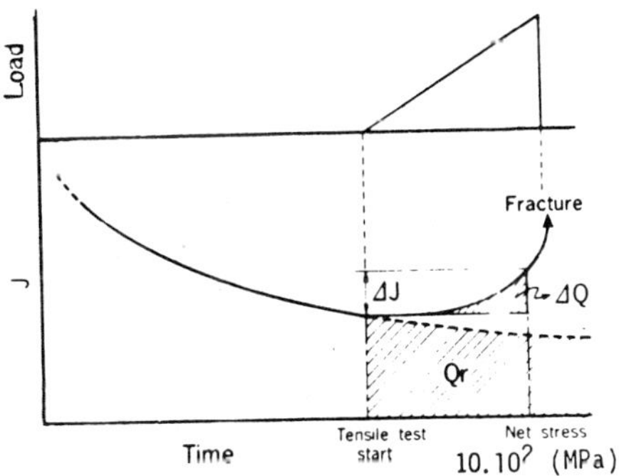

Fig. 8 Schematic diagram of evolution transient during tensile test

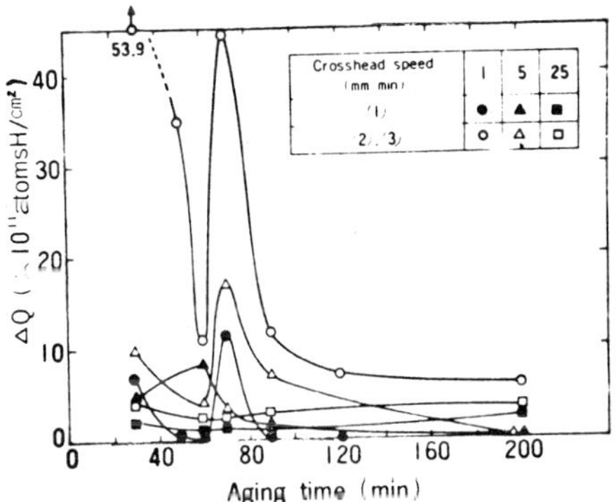

Fig. 9 Effect of aging time and cross head speed on ΔQ (specimen D)

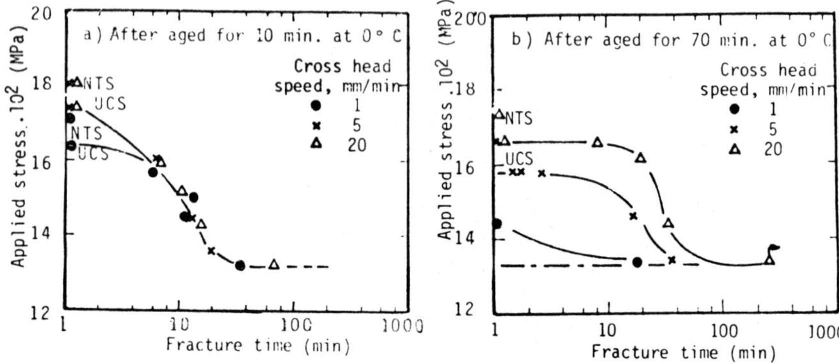

Fig. 10 Effect of cross head speed on delayed failure curve (specimen A)

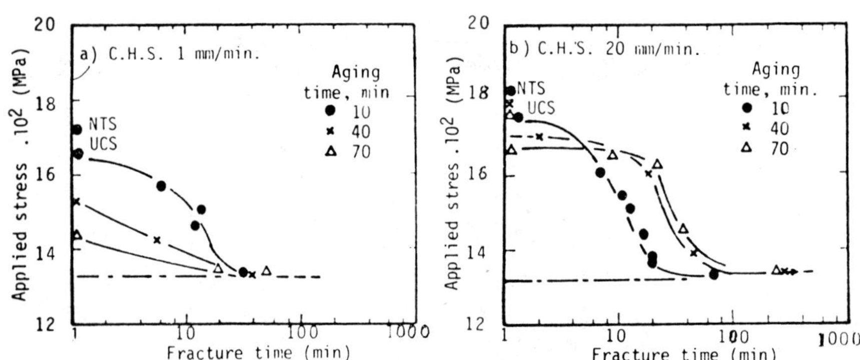

Fig. 11 Effect of aging time before loding on delayed failure curve (specimen A)

volume of hydrogen are trapped in dislocation sites or hydrogen are broken
away from dislocation trapping.

Analysis of Hydrogen Accumulation and Fracture Morphologies

Hydrogen diffusion equation for its accumulation

In practical frame, we use the engineering term of equivalent plastic
strain $\bar{\varepsilon}_p(x,y,z)$ to represent such strain. Augumenting the effect of this
plastic strain distribution in diffusion equation, the following hydrogen
stress(strain)-induced diffusion equation is formulated by analogy with
heat flow equation[4].

$$\frac{\partial C^*(x,y,z,t)}{\partial t} = \frac{\gamma D}{\Lambda(x,y,z)} \nabla^2 C^*(x,y,z,t)$$

where C^* is effective hydrogen concentration which is defined as $C^* = N/L$
with N and L being nembers of hydrogen atoms and dislocation length
respectively. Ordinary hydrogen volume concentration C can be readily
computed from C^* using the following relation.

$$C^* = \frac{N}{\Lambda V} = \frac{C}{\Lambda}$$

where V and Λ are volume and dislocation density respectively. Λ is assumed
in the first order approximation to be linearly related with the foregoing
equivalent plastic strain $\bar{\varepsilon}_p^*$.

$$\Lambda = \alpha\bar{\varepsilon}_p(x,y,z) + \beta$$

We can deduce from the above equation the ordinary diffusion equation when
the materials or structures are elastically or homogeneously plastically
deformed.

Hydrogen delayed cracking and its accumulation behaviours

Typical hydrogen delayed cracking are observed when hydrogen charged
specimens (either thermally or cathodically) are statically loaded and kept
for their fractures. During the incubation periods hydrogen redistributes
itself and concentrates near notch root.

Strain(stress)-induced diffusion of hydrogen is computer simulated by
use of Finite Element Method where hydrogen effusion from the surface and
strain(stress)-induced diffusion toward plastically deformed notch root are
taken into account at the same time. Typical simulation results is shown
in Fig. 12 where hydrogen volume concentration increases up to several times
of the initial value at the moment of fracture during the static fatigue
with circularly notched bar specimens. The location is somewhat inside
(a few tenths of a millimeter) from notch root. It also clarifies hydrogen
depleted regions just around the hydrogen peak and along specimen surface
due to diffusion. Further computer graphic technique is employed to make
easy access to displays of time-to-time hydrogen behaviours.

This fact coincides with experimental observations of crack initiation
locations as shown in Fig. 13 and Fig. 14. Fig. 13 shows a typical
initiation micro-crack at notch tip detected by the first initiation
Acoustic Emission signal, when specimen was delayed cracked through
incubation time. And Fig. 14 shows a fracture morphology at notch tips.

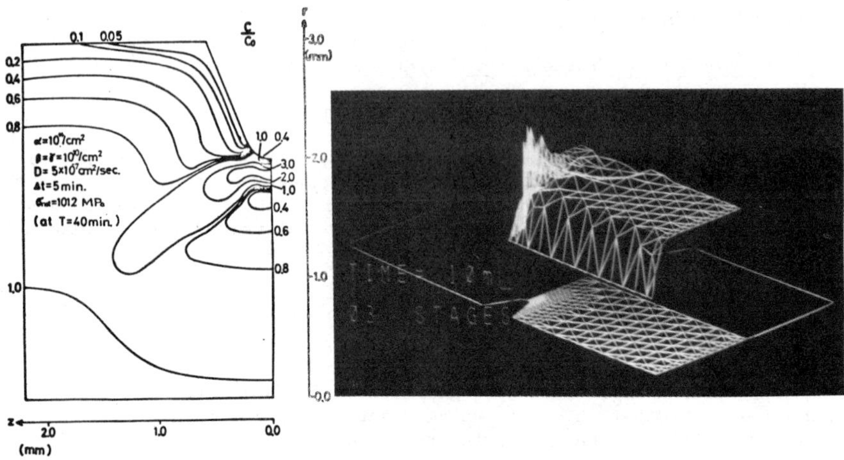

Fig. 12   Computer simulated hydrogen behaviours around notch root of circularly notched bar specimen

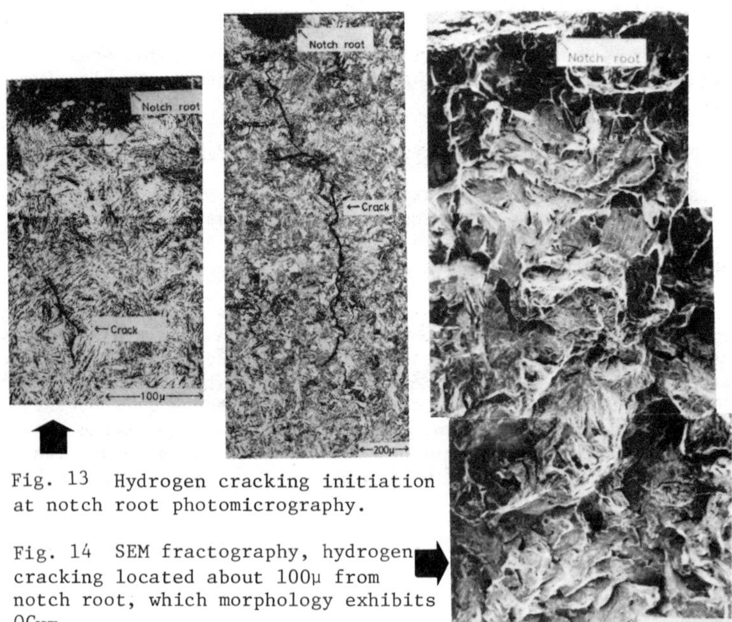

Fig. 13   Hydrogen cracking initiation at notch root photomicrography.

Fig. 14   SEM fractography, hydrogen cracking located about 100μ from notch root, which morphology exhibits QC_{HE}.

The initiation crack located inside about 100 ~ 200m from notch root and its fracture morphology exhibits the quasicleavage crack of hydrogen embrittlement, which can be distinguished from the cleavage fracture at -196°C.

Conclusion

The authors carried out a tension test, an internal friction experiment and an electro-chemical hydrogen flux monitering test. From these experimental results, it may be concluded that hydrogen accumulates in the neighborhood of the dislocation on slip planes near grain boundaries. The microscopic redistribution behaviour of hydrogen will hence affects the fracture morphologies of hydrogen and then incubation time to fracture in steel.

And we have newly formulated the stress(strain)-induced diffusion equation in the augmented form of Fick's second law, which equation takes into account such dislocation density gradient in terms of plastic strain distribution. And now our proposed equation for the non-steady state hydrogen gathering was solved by the use of finite element method for circularly-notched bar specimens with equally distributed initial hydrogen and under various applied stresses for hydrogen delayed cracking initiation.

Reference

[1] A.R.Troiano. <u>Am</u>. <u>Soc</u>. <u>Metals</u>, <u>Trans</u>. <u>Quart</u>., <u>52</u> (1960), 54

[2] P.Bastien and P.Azou, <u>Rev</u>. <u>Met</u>., <u>49</u> (1952), 837

[3] G.Schoek, <u>Acta</u> <u>Met</u>., <u>11</u> (1963), 617

[4] Y.Kikuta, S.I.Ochiai, and T.Kangawa, Proc. of Int. Conf. on Computer Simulation for Materials Applications, N.B.S. at Gaithersburg (1976), 709

HYDROGEN EFFECTS IN THE

FRACTURE OF SPHEROIDIZED PLAIN CARBON STEELS

H. Cialone and R. J. Asaro
Brown University
Providence, Rhode Island

Although hydrogen assisted fracture of metals can take many forms, the
way in which the fracture process of a metal is affected by hydrogen falls
into one of two categories: 1) the fracture mode which would normally be in
effect is somehow accelerated by the presence of hydrogen, with no change in
the microscopic features of the fracture process, or 2) hydrogen causes a
transition from a normally ductile fracture mode to a more brittle fracture
mode. An example of the latter effect is commonly found when high strength
steels are exposed to gaseous hydrogen, with the result that the normally
occurring fracture mode involving microvoid formation and growth is super-
seded by cleavage fracture, ostensibly due to the relaxation of local fracture
stresses brought on by hydrogen segregation and accumulation at internal interfaces [1]. If the same alloy were to normally (that is, in the absence
of hydrogen) fracture by intergranular cleavage due to prior embrittling effects such as temper embrittlement, for example, exposure to hydrogen would
further relax the local fracture stresses but would leave the fracture mode
unchanged [2]. Similarly, when low and medium strength plain carbon steels
are exposed to sufficiently large amounts of hydrogen, for instance by electrochemical charging [3], large losses in ductility can be attained. Yet,
while these fractures appear to be macroscopically brittle, microscopic examination reveals them to have occurred in a ductile manner [4]. In tensile
tests, hydrogen causes fracture with less non-uniform reduction in area (RA),
i.e. with a less severe neck, than in the uncharged condition, thus causing
a loss in macroscopic ductility.

This paper summarizes the results of experiments on commercial heats of
1017 and 1045 steels as well as a nominally clean laboratory heat of 1015
steel.* Both initially smooth and circumferentially notched tensile specimens were electrochemically charged with hydrogen at various cathodic current
densities and then uniaxially strained various amounts. Several important
microscopic features were quantified and compared to those occurring in uncharged specimens. The results of these comparisons showed the effect of
hydrogen on the events leading up to fracture in tension. In order to better
understand these effects, it is necessary to first briefly review the important microscopic and macroscopic events involved in ductile tensile fracture
in the absence of hydrogen.

* This paper is in fact a synopsis of a more complete paper to be published.

Ductile Tensile Fracture in Spheroidized Plain Carbon Steels

It is generally accepted that ductile tensile fracture begins with the formation of voids within the specimen; in spheroidized plain carbon steels voids initiate at second phase (i.e. cementite) particles [5]. Void initiation occurs either by particle cracking, particle-particle separation, or particle-matrix separation, which was in fact the primary mechanism observed. Voids then begin to grow into the matrix by plastic deformation which, along with void initiation to an even greater extent, is greatly augmented within the triaxial stress field imposed by the neck [6]. Although the earlier stages of void growth occur by plasticity controlled growth, the latter stages have been observed to occur preferentially along grain, and possibly subgrain boundaries [7,8]. An example of this type of void growth in a 1017 steel is given in Figure 1. That voids grow along boundaries does not necessarily indicate that grain boundaries are weaker than the grain matrix. Rather, in both cases cited, the void initiating sites happened to be situated on grain boundaries. Furthermore, at the strains at which void growth along boundaries occurs, most of the grain boundaries within the neck are aligned with the tensile axis, which is the direction in which voids experience the most growth [6].

Voids growing along grain boundaries may eventually coalesce by direct impingement. In addition, void coalescence may be further accelerated by the formation of void sheets by localized deformation over extended regions in the matrix [8]. One other way in which voids may coalesce is by internal necking between nearly contiguous voids [9]. Void coalescence leads to the formation of a large central crack within what is at this stage in the deformation a rather deep neck [10]. This crack then propagates in a ductile manner toward the lateral surface of the specimen, resulting in final separation with a cup-cone type fracture. All of the steps just described may be affected by hydrogen, but only the steps leading up to the formation of a central crack are felt to be most significantly affected [11].

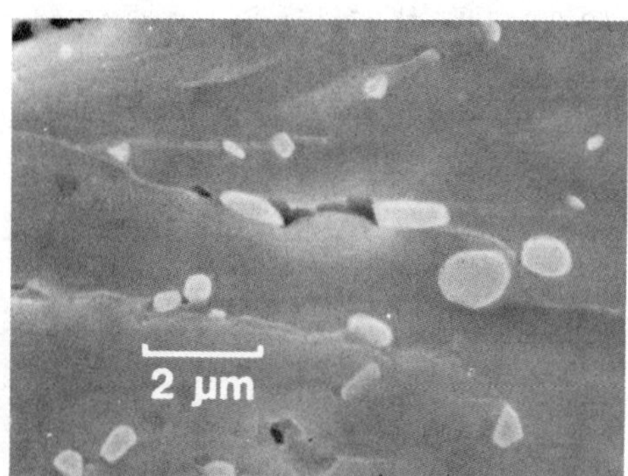

Figure 1.   Example of voids growing and coalescing along a grain boundary. The tensile axis orientation is horizontal.

Experimental

Experiments were carried out on a laboratory heat of 1015 steel and commercial heats of 1017 and 1045 steels. The steels were heat treated to have microstructures as described in Table 1. With the exception of steel D, all steels were heat treated by simple quench and temper methods, which resulted in microstructures with larger carbides on ferrite-grain boundaries and smaller carbides in the grain matrices, with a network of subgrain boundaries interlinking the particles. Steel D was thermally cycled after quenching in order to eliminate most of the subgrain boundary formation [12].

Table I. Microstructural Dimensions

Steel	Ferrite Grain Size	Particle Diameter
A (1015)	14 μm	0.9 μm
B (1017)	9.0 μm	1.0 μm
C (1017)	4.2 μm	0.4 μm
D (1017)	4.4 μm	0.4 μm
E (1017)	11 μm	0.4 μm
F (1045)	4.7 μm	1.0 μm

Axisymmetric tensile specimens were machined after heat treatment. Hydrogen was introduced by an electrochemical charging technique using cathodic current densities varying from 0 mA/cm^2 (i.e. uncharged) to 6 mA/cm^2. The charging time was 24 h and all tensile tests were conducted at room temperature immediately after charging. Charging at 4 mA/cm^2 was 100 percent reversible when the hydrogen was baked out for 400 h at 423 K (150°C). However, charging at 6 mA/cm^2 caused surface blisters in some specimens. Polished and etched sections of these blisters showed them to be associated with inclusions. After each test, the RA was measured and scanning electron microscope (SEM) fractographs were taken. Longitudinal sections along the center axis were cut out, polished, and then etched first with 4 percent Picral followed by a 3 percent Nital rinse. Ultrasonic cleaning in ethanol was employed after each step in the polish and etch procedure to remove flowed metal from the voids. The surfaces were then examined in the SEM, and several important microscopic quantities were determined and plotted vs. the distance from the plane containing the minimum section of the neck (or from the fracture surface for most of the specimens).

Results and Discussion

Hydrogen charging over the range of cathodic current densities described caused a range of ductility losses, shown in Figure 2 for steels A through E. The loss in RA at fracture for steel F charged at 4 mA/cm^2 was 22 percent, considerably less than for steels A and B. This is contradictory to the generally accepted feeling that hydrogen induced ductility losses become more severe with increasing strength level, since steel F (1045 steel) is stronger than steels A or B. The reasons for this unexpected behavior will be discussed later in this paper.

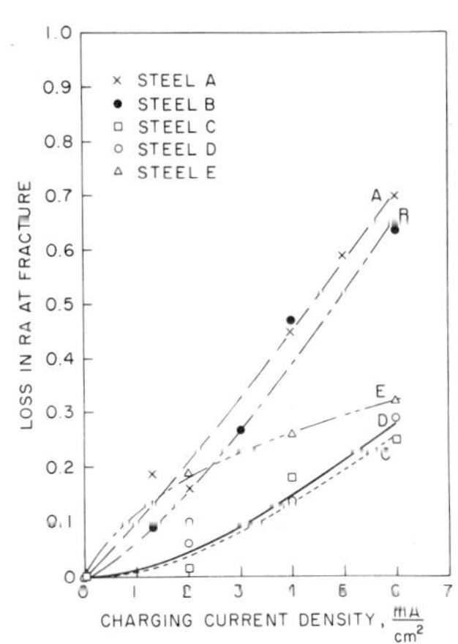

Figure 2. Loss in RA (nonuniform reduction in area) at fracture for steels A-E, vs. charging current density, as compared to the uncharged condition.

Void Initiation

Voids usually initiated at the particle-matrix interfaces; Figure 3 shows how the void density varied with axial plastic strain for steels B and F. The figure shows that after equivalent amounts of deformation, the hydrogen charged specimens have greater void densities than the uncharged specimens, indicating that hydrogen facilitates void initiation. An attempt was made to quantify this effect by straining charged and uncharged circumferentially notched specimens various amounts and determining the precise point along the stress and strain gradient at which void initiation just begins to occur. This is done for steels B and F in Figure 4. As has been noted in an earlier study [7], there exists a "base line count" of voids which apparently initiate from the start of the deformation process due to separation of particle clusters or cracking of extremely elongated particles. By de-

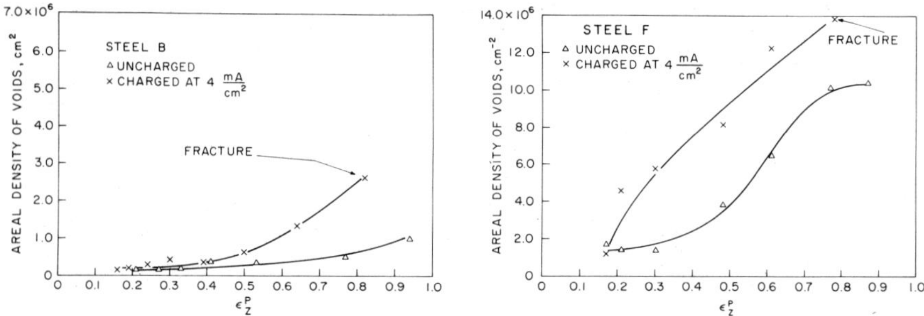

Figure 3.   Areal density of voids vs. axial plastic strain for steels B and F.

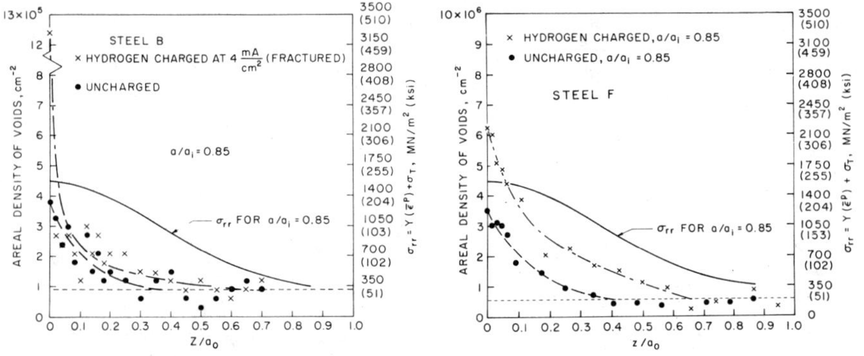

Figure 4.   Areal density of voids vs. vertical distance from minimum section of neck normalized by initial radius away from neck (Z/a_o) for steels B and F. Interfacial stress σ_{rr}, computed using the model of Argon et al. [13], is also shown.

termining the interfacial stress σ_{rr}, at the point where the void density profile intersects this base line count, a critical interfacial stress for void initiation can be found. Table II shows how hydrogen affects the critical interfacial stress for void initiation and significantly accelerates the initiation of voids. Hydrogen may enhance void initiation by adsorbing onto particle-matrix interfaces, causing losses in cohesion [14].

Table II. Void Initiation

Steel	Critical Interfacial Stress
B – uncharged	1055 MN/m^2
charged at 4 mA/cm^2	560 MN/m^2
F – uncharged	965 MN/m^2
charged at 4 mA/cm^2	593 MN/m^2

<u>Development of Voids</u>

Voids initially grew into grain matrices by plastic deformation and then appeared to coalesce along grain and subgrain boundaries. Figure 5 shows how the void area fraction (or volume fraction) varied with axial plastic strain for steels B and F. The figure indicates that after equivalent amounts of deformation, a charged specimen will have a greater void volume fraction than an uncharged specimen. The increase in void volume fraction cannot be accounted for on the basis of plasticity controlled growth [7,15]. That is, although the typical void in a charged specimen has had more time (actually strain) to grow than the typical void in an uncharged specimen (since hydrogen facilitates void initiation), the extra growth it experiences is in excess of the growth predicted by plasticity controlled growth laws. Therefore hydrogen in some way accelerates void growth as well as void initiation.

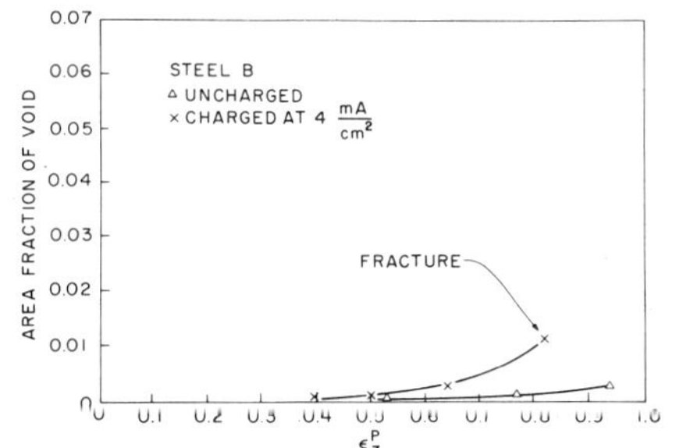

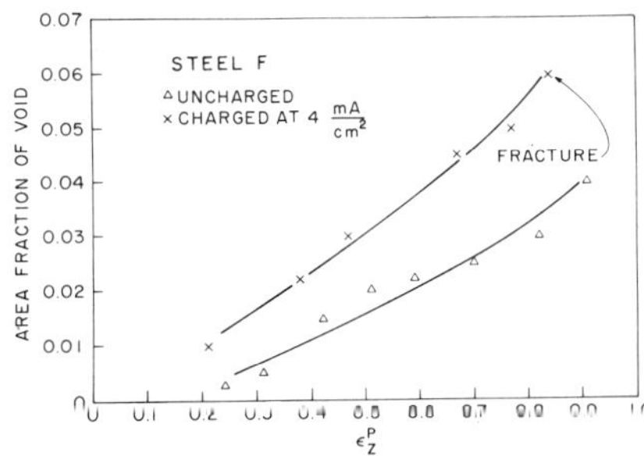

Figure 5. Area fraction of voids vs. axial plastic
strain for steels B and F.

The mechanism by which hydrogen accelerates void growth may be understood once the manner in which voids grow is clearly established. Voids appear to grow along grain and subgrain boundaries after an initial stage of plasticity controlled growth into the matrix. If a significant amount of void growth occurs along boundaries, then the fraction of boundary separated (the length of separated boundary over the total boundary length in a two-

dimensional section) should exceed the void area fraction after a significant amount of void growth has occurred [16]. Grain boundaries in a two-
dimensional section (viz. a photomicrograph) can be considered to be preferentially oriented test lines in the determination of the lineal fraction of
void (which is roughly equal to the area fraction of void [17]). Thus, even
though the early stages of void growth occur by plastic deformation of the
matrix, and voids which grow along boundaries have a finite thickness
(i.e. test lines <u>not</u> intersecting boundaries may intersect these voids), if
void growth along boundaries does play a significant role in the fracture
process, the fraction of boundary separated must be greater than or at the
very least comparable to the area fraction of void near the center of the
neck. Figure 6 contains these two quantities plotted vs. axial plastic strain
for steel B. The figure indicates that a significant amount of void growth
does in fact occur along grain boundaries and that hydrogen facilitates the
separation of boundaries. This too is consistent with a mechanism of hydrogen induced losses in cohesion by adsorption on interfaces (i.e. grain boundaries).

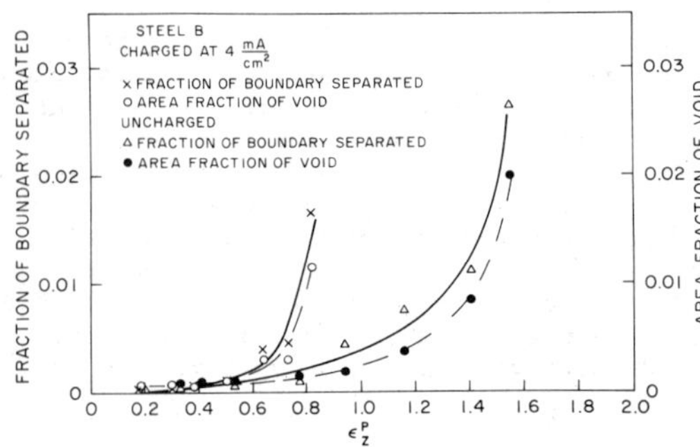

Figure 6. Fraction of boundary separated vs. axial plastic strain and area fraction of void vs. axial plastic strain for steel B.

Fractography

In most cases, hydrogen charging had little effect on the fracture sur-
faces--they consisted of dimples of approximately the same size as those of
the uncharged specimens. However, when steels A and B were charged at cathodic current densities in excess of 2 mA/cm², the fracture surfaces were often
found to contain regions of quasi-cleavage fracture. In most instances these
regions were found to be centered on a large inclusion, particularly MnS,
although specimens charged at 6 mA/cm² were found to exhibit quasi-cleavage
regions which did not appear to be associated with inclusions. These "fish-
eye" regions gave rise to large losses in ductility, as can be seen by comparing Figure 2 with Figure 7. Steel F exhibited no quasi-cleavage when
charged at current densities up to 4 mA/cm².

Table III gives the amount of interfacial area per cm³ of specimen for
steels A through F. When compared with Figure 7 the table indicates that
the amount of quasi-cleavage fracture occurring decreases with increasing
interfacial area, particularly particle-matrix interfacial area. This suggests that these interfaces act as preferential sinks for hydrogen in much
the same way as the dispersoids in a Ni-2 ThO$_2$ alloy [18]. While the presence of hydrogen causes an acceleration of the normal fracture mode of microvoid formation and growth, a sufficiently large number of sinks may hinder

the transition to a much more brittle mode of failure. The way in which
this failure occurs is not clear but it would seem reasonable that whatever
hydrogen is not trapped by interfaces and is in excess of the lattice solubility must accumulate near inclusions, notably MnS. Although no data were
available, the solubility for hydrogen in MnS should be larger than in ferrite since the tetrahedral interstices are some 35 percent larger by volume.

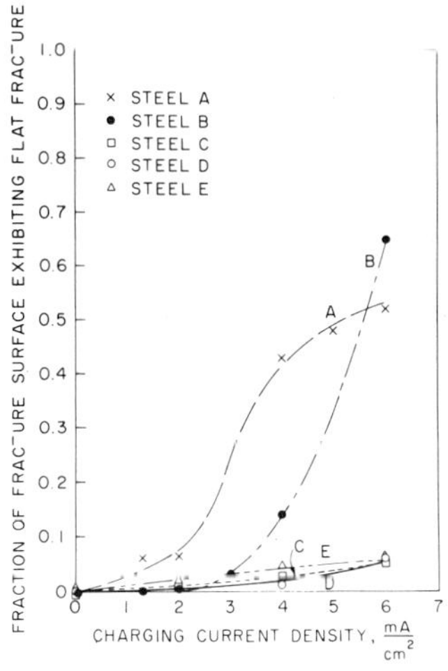

Figure 7. Fraction of fracture surface exhibiting flat quasi-cleavage
fracture features vs. charging current densities for steels A-E.

Thus MnS inclusions could also act as hydrogen sinks during charging. When
the specimen is strained, and the lattice dilated due to the hydrostatic
stresses imposed by the neck, hydrogen could diffuse from the inclusions
into the matrix in sufficient amounts to cause quasi-cleavage fracture to
occur. The effect is similar to one of dynamic charging, which has been
shown to cause quasi-cleavage fracture to occur in 1090 steel which would
ordinarily fail by a process of ductile hole growth when uncharged or when
precharged [19].

Table III. Interfacial Area

Steel	Particle-Matrix Inter-facial Area*	Grain Boundary Inter-facial Area*	Total Interfacial Area*
A	1.25×10^{11}	4.20×10^{11}	5.45×10^{11}
B	1.31×10^{11}	6.53×10^{11}	7.84×10^{11}
C	3.27×10^{11}	1.40×10^{12}	1.73×10^{12}
D	3.27×10^{11}	1.34×10^{12}	1.67×10^{12}
E	3.27×10^{11}	5.35×10^{11}	8.62×10^{11}
F	3.75×10^{11}	1.20×10^{12}	1.58×10^{12}

* $\mu m^2/cm^3$ of specimen

Conclusions and Summary

Hydrogen charging at low cathodic current densities facilitates void initiation at cementite particles. Void growth is accelerated, particularly along grain and subgrain boundaries. In addition, hydrogen charging can cause a transition in fracture mode from ductile hole growth to a combined ductile hole growth and quasi-cleavage mode. Quasi-cleavage areas are generally associated with inclusions which may act as hydrogen sinks while the specimen is being charged and as hydrogen sources while the specimen is being strained. Variations in carbon content, particle size and ferrite grain size, which all contribute to the amount of internal interface, seem to be related to the degree of fracture mode transition. This suggests that grain boundaries and particle-matrix interfaces act as alternative hydrogen sinks.

Acknowledgements

The authors wish to thank A. W. Thompson and R. Garber for their helpful comments. This work was supported by the U. S. Department of Energy through Contract EY-76-S-02-3084 and by the Brown University Metals Research Laboratory, whose facilities were used throughout the course of this study.

References

1.   J. R. Rice: Effect of Hydrogen on Behavior of Materials (A. W. Thompson and I. M. Bernstein, eds.), Chap. 7, p. 455, TMS-AIME, New York, NY, 1976.
2.   K. Yoshino and C.J. McMahon, Jr.: Met. Trans., 1974, Vol. 5, p. 363.
3.   D. P. Smith: Hydrogen in Metals, Chap. 9, p. 145, Univ. Chicago Press, Chicago, ILL, 1948.
4.   A. W. Thompson and I.M. Bernstein: Proc. of the Fourth Int. Conf. on Fracture, Waterloo, Canada (D.M.R. Taplin, ed.), Vol. 2, Chap. 1, p. 249, Univ. Waterloo Press, Ontario, Canada, 1977.
5.   A. R. Rosenfield: Met. Reviews, 1968, Vol. 13, p. 29.
6.   J. R. Rice and D.M. Tracey: J. Mech. Phys. Solids, 1969, Vol. 17, p. 201.
7.   H. Cialone and R.J. Asaro: Met. Trans. A, 1979, Vol. 10A, p. 367.
8.   H. C. Rogers: Trans. TMS-AIME, 1960, Vol. 218, p. 498.
9.   H. C. Rogers: Ductility, Chap. 2, p. 31, ASM, Metals Park, Ohio, 1968.
10.   J. I. Bluhm and R.J. Morrissey: Proc. First Int. Cong. on Fracture, Vol. 3, p. 1739, Japan Soc. Strength and Fracture, Sendai, Japan, 1966.
11.   A. W. Thompson: Effect of Hydrogen on Behavior of Materials (A. W. Thompson and I. M. Bernstein, eds.), Chap. 4, p. 467, TMS-AIME, New York, NY, 1976.
12.   L. Anand and J. Gurland: Acta Met., 1976, Vol. 24, p. 901.
13.   A. S. Argon, J. Im and A. Needleman: Met. Trans. A, 1975, Vol. 6A, p. 815.
14.   A. W. Thompson: Mater. Sci. Eng., 1974, Vol. 14, p. 253.
15.   H. Cialone and R.J. Asaro: unpublished research.
16.   R. Garber: Private communication, 1979.
17.   E. E. Underwood: Quantitative Stereology, Addison-Wesley, Reading, Mass., 1970.
18.   A. W. Thompson: Met. Trans., 1974, Vol. 5, p. 1855.
19.   T. Goldenberg, T.D. Lee and J.P. Hirth: Met. Trans. A, 1979, Vol. 10A, p. 199.

DISCUSSION

R.A. Oriani, University of Minnesota, Minneapolis, MN: The decohesive
action of hydrogen can produce not only a larger number density of microvoids by enhancing nucleation but also a larger mean size by enhancing the
decohesive growth of crack-shaped microvoids. Which effect predominates
probably depends on microstructural details. However, the results of Asaro
and Cialone may exhibit still another contributing effect of hydrogen to the
increased mean size of the microvoids, and that is the internal pressurization by molecular hydrogen consistent with a large input fugacity.

R. Asaro: We have in fact considered the possibility of internal pressurization in our earlier paper (ref. 7) and found this effect to be minor. This
conclusion is based in part on the fact that since the specimens were precharged the hydrogen content is limited. Thus the maximum internal pressure
due to hydrogen should be achieved when the void volume fraction is exceed-
ingly small, since as more and larger voids appear the hydrogen would be
dissipated over a greater void volume. Yet, at a void volume fraction as
low as 0.005 we estimated the maximum internal pressure obtainable to be
approximately 20 MN/m^2, which would have little effect on the void growth
rate. Furthermore, although we did not emphasize this point earlier, the
effect of hydrogen on void growth appears to increase with increasing void
volume fraction. This is more fully addressed in a paper we have submitted
to Metallurgical Transactions A. The point is that, even if our figure of
20 MN/m^2 were a gross underestimate of the hydrogen pressure at a void volume fraction of 0.005, due to the limited supply of hydrogen, the amount of
hydrogen pressure-assisted void growth should drop off with increasing void
volume fraction. Yet we have shown that the amount of hydrogen-assisted
void growth actually increases with increasing void volume fraction. Therefore we feel that the contribution of internal pressurization to void growth
is minimal under the conditions we have employed.

G. Krauss, Colorado School of Mines, Golden, CO: Our results (Craig &
Krauss, this conference) show considerable cleavage related to the substructure of highly tempered 4130 i.e., a ferrite-carbide microstructure, in contrast to the speaker's 1045 results, but similar to the 1017 results. Is
ferrite grain or subgrain size perhaps responsible for these differences?

R. Asaro: Based on the results we presented in Tables I and III and in Figure 7, ferrite grain and subgrain size seem to play some part in determining
whether or not cleavage occurs. However, when one compares steel E, which
has coarser grains but finer carbides, with steel B (coarse grains and carbides) and steel C or D (fine grains and carbides), it appears that the carbide particle size is perhaps more important, since steel E exhibited cleavage with the same sort of low frequency as steels C and D. In addition,
since Table III implies that the amount of interface is important, the carbon content would of course be significant. Finally, Figure 7 indicates
that the charging current density is an important consideration. Our 1045
was not tested after charging at current densities over 4 mA/cm^2 -- judging
from our results a current density of about 8 or 9 mA/cm^2 would probably
cause some cleavage to occur.

L.N. Pussegoda, CANMET, Energy, Mines & Resources, Ottawa, Ontario, Canada:
I presume these tensile tests were done at room temperature. What would
happen if they were done at lower temperatures where hydrogen still assists
fracture? What I am getting at is hydrogen-assisted initiation and propagation of such a crack could probably take place in a cleavage mode if such
a fracture is possible in the microstructure at low temperatures.

<u>R. Asaro</u>: Our tensile tests were conducted at room temperature. In addition, we conducted some 3-point bend tests on Charpy bars made out of our steel A at temperatures ranging from room temperature to -30°C. While the uncharged specimens showed little or no evidence of cleavage, cleavage facets were in fact widespread in the charged specimens even at room temperature. Perhaps still lower temperatures would have caused cleavage to occur in the uncharged specimens as well. In such a case, we believe hydrogen would reduce the local stresses necessary for cleavage to occur, in a manner similar to the way hydrogen assists fracture in a temper embrittled steel.

HYDROGEN EMBRITTLEMENT OF STEELS WITH INTERGRANULAR FRACTURE MODE

T. Inoue, K. Yamamoto and M. Nagumo

Fundamental Research Laboratories, Nippon Steel Corporation
1618 Ida, Nakaharaku, Kawasaki, JAPAN

Steels showing intergranular fracture mode in hydrogen embrittlement
have usually higher susceptibility and it is necessary, therefore, to make
steels free from such fracture mode.

Investigations on hydrogen embrittlement of quenched and tempered low
alloy steels and a plain carbon steel with 600 - 800 MPa tensile strength
have been made with particular reference to the fracture mode. Comparisons
are also made between the fracture mode in hydrogen embrittlement and in
Charpy impact test. Main results are summarized as follows.

1) In low carbon Si-Mn steels, the mode of fracture changes considerably
with the strain rate. Maximum allowable phosphorus concentration at the
grain boundaries for intergranular fracture not to occur is about 2 wt % in
the case of $\dot{\varepsilon} \sim 10^{-4}$ s^{-1}, whereas it is as low as 0.1 wt % in the case of
$\dot{\varepsilon} \sim 10^{-5}$ s^{-1}.

2) In a plain carbon steel with phosphorus concentration at the grain
boundaries of 0.5 - 1.3 wt % the brittle fracture in Charpy test occurs only
transgranularly but the steel exhibited nearly fifty percent of intergranular fracture mode in hydrogen embrittlement.

3) In low carbon Si-Mn steels fine precipitates at the grain boundaries
accelerate the occurence of intergranular fracture mode in hydrogen embrittlement as well as in Charpy test.

Based on the obtained results the condition for intergranular fracture
to occur in hydrogen embrittlement are discussed with various steels.

Fracture mode of hydrogen induced cracking in low alloy steels is either intergranular or transgranular. Steels showing intergranular fracture mode of hydrogen induced cracking have usually higher susceptibility to hydrogen embrittlement and it is necessary to make steels free from such a fracture mode. Although there are many reports (1, 2, 3) that impurity segregation to the grain boundaries increases susceptibility to the hydrogen embrittlement of steels with the tensile strength higher than 1000 MPa, it is not yet clear in steels with strength of 600 - 800 MPa.

The purpose of this study is to make clear the condition for the intergranular fracture in hydrogen embrittlement and to determine the maximum allowable intergranular P concentration. In this study only effect of P was investigated because only P is important in commercial steels, concentration of impurities such as Sb, As or Sn being extremely low.

Experimental Procedure

The chemical compositions of steels used, which were prepared by melting electrolytic iron in a vacuum furnace, are shown in Table I. Ingots (20 - 50 kg in weight) were heated at 1523 K for 1 h in a furnace under an argon atmosphere and were hot rolled to 13 mm thick plates. The heat treatment schedules given to the steels are shown in Table II.

The susceptibility to hydrogen embrittlement was evaluated by the change in reduction of area in tensile test with and without hydrogen charging in experiment 1 and 2. Hydrogen was charged by cathodic electrolysis in 3% NaCl aqeous solution with a current density of 2650 A/m^2. The tensile tests were conducted by an Instron type machine with strain rates of 0.83 x 10^{-4}s^{-1}and 0.83 x 10^{-5} s^{-1} in experiment 1 and 0.83 x 10^{-5} s^{-1} in experiment 2. In experiment 3 the susceptibility was evaluated by a cantilever beam bend test of notched specimens in 3% NaCl aqeous solution with cathodic charging at 2-10 A/m^2(4).

Table I   Chemical Composition of Steels used (wt%)

Steel No.	C	Si	Mn	P	S	B	N	Remark	
1	0.20	0.24	1.48	0.002	0.004	–	0.0020	Si–Mn	
2	0.19	0.25	1.49	0.031	0.004	–	0.0020	Si–Mn–P	Exp.1
3	0.20	0.26	1.49	0.002	0.004	0.0019	0.0020	Si–Mn–B	
4	0.20	0.26	1.49	0.029	0.005	0.0019	0.0020	Si–Mn–P–B	
5	0.46	–	–	0.040	0.003	–	0.0020	Fe–C–P	Exp.2
6	0.15	0.48	1.61	0.008	0.008	0.0009	0.0060	Si–Mn–B–N	
7	0.16	0.49	1.59	0.008	0.007	0.0006	0.0021	Si–Mn–B	Exp.3
8	0.17	0.49	1.58	0.008	0.007	<0.0001	0.0058	Si–Mn–N	

Table II   Heat Treatment Schedule

Steel No.	Quenching	Tempering	Embrittling	Symbol in Fig.
1, 2, 3, 4	1373 K* WQ	873 K 1h WQ	–	◊ ◖ △ ◤
"	"*	"	773 K 1-100 h WQ	○ ● △ ▲
5	1523 K 1h IBQ	923 K 1h WQ	–	
"	"	"	773 K 100 h WQ	
6, 7, 8	1553 K 0.5h IBQ	773 K 1h WQ	–	A1, B, C
6	1553 K 0.5h			
	→1173 K 0.5h IBQ	773 K 1h WQ	–	A2

* Quenching after rolling   WQ (Water Quenched)   IBQ (Ice Brine Quenched)

Fracture surfaces from each specimens were examined by scanning electron microscope. The grain boundary chemistry was analyzed by Auger Electron Spectroscopy (AES) on fracture surface.

Results

<u>Experiment 1</u> The transgranular to intergranular transition with increasing intergranular P concentration changes with the strain rate extensively.

Figure 1 shows the reduction of area with and without hydrogen charging as a function of intergranular P concentration (hereafter denoted by C_B^P which were measured by AES. Typical scanning electron micrographs of the fracture surface from each specimens in Fig. 1 are shown in Fig. 2. It can be seen that the fracture morphology changes with the strain rate.

1) In the case of $\dot{\varepsilon} \sim 10^{-4}$ s^{-1} specimens with C_B^P lower than about 2 wt % are entirely transgranular and the reduction of area is almost constant. This contrasts with the fact that the brittle fracture mode in charpy test contains large amount of intergranular one when C_B^P is about 1 wt % and a significant temper embrittlement occurs for these Si-Mn Steels and other low alloy steels.

For specimens with C_B^P higher than 2 wt % the fracture mode changes to intergranular one and the reduction of area drops much. In this case intergranular hydrogen embrittlement is larger for the specimen containing B (Si-Mn-P-B) than for the specimen without B (Si-Mn-P). This fact will be discussed later.

2) In the case of $\dot{\varepsilon} \sim 10^{-5}$ s^{-1} the fraction of intergranular fracture increases with increasing the intergranular P concentration. This contrasts with the case of $\dot{\varepsilon} \sim 10^{-4}$ s^{-1}. For the case of small $\dot{\varepsilon}$ ($\dot{\varepsilon} < 10^{-5}$ s^{-1}) the effort to decrease C_B^P is effective for reducing the hydrogen embrittlement. Only specimens with C_B^P smaller than 0.1 wt % show completely transgranular fracture.

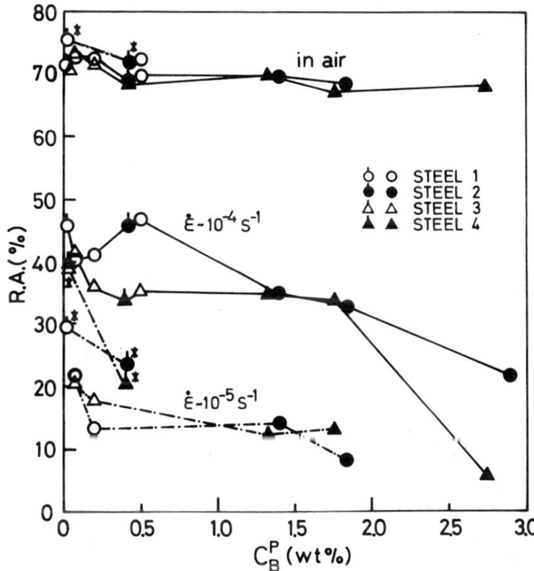

Fig.1 Relation between the reduction of area and the intergranular P concentration for steels 1, 2, 3, 4. Tensile strength of 800MPa except for specimens marked by *(~750 MPa)

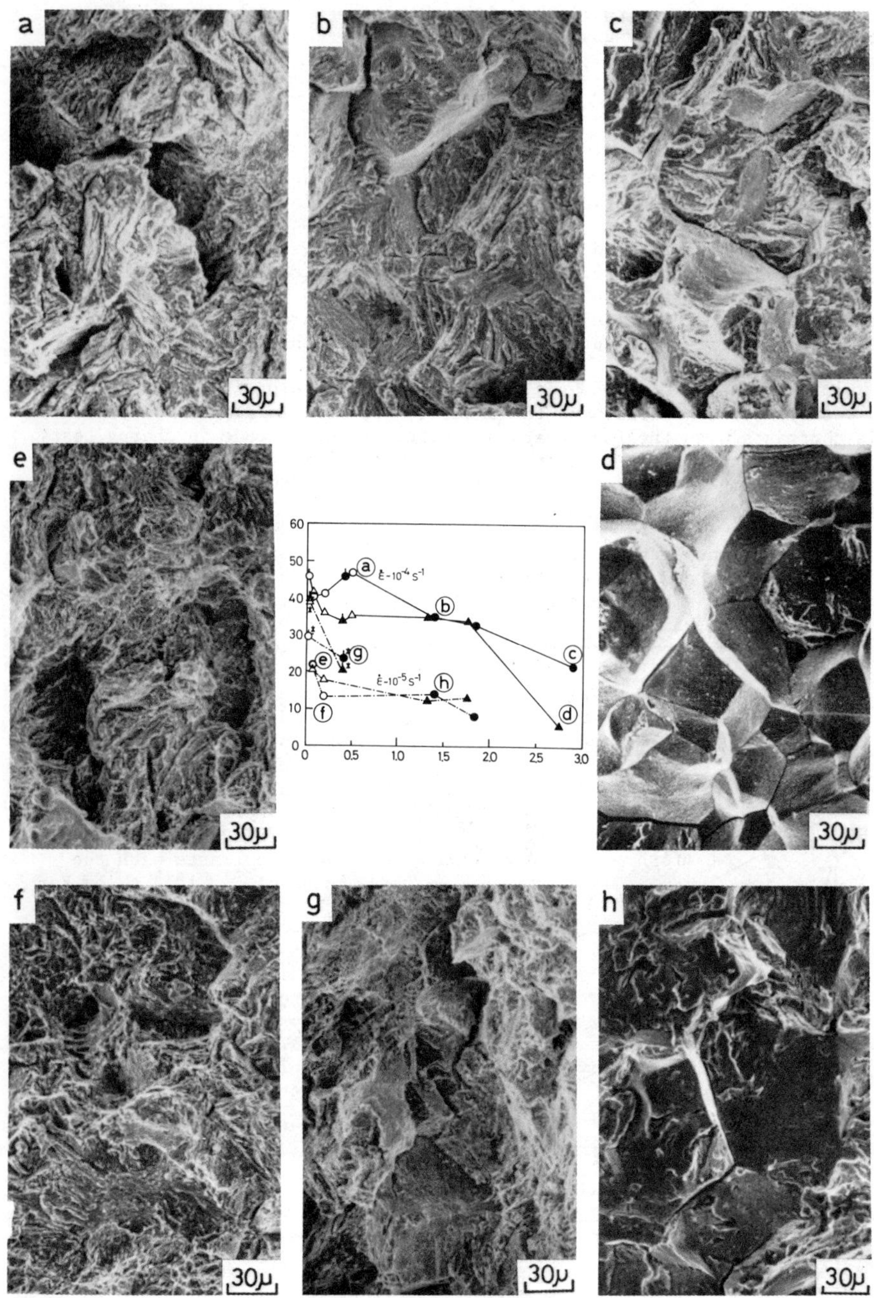

Fig. 2   Typical scanning electron micrographs of the fracture surface from the specimens in Fig. 1.

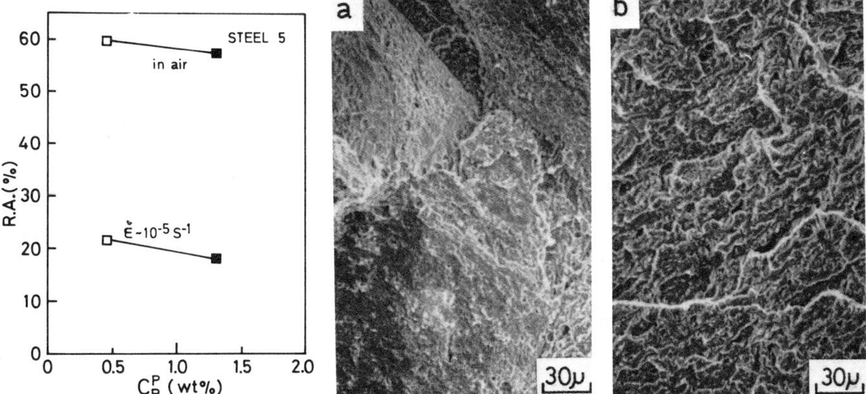

Fig.3 Relation between R.A.
and C_B^P for steel 5.

Fig.4 Typical fracture surface from the
specimen shown in Fig.3 (a) and that
failed in Charpy test (b).

<u>Experiment 2</u>   P segregation to the grain boundaries alone can cause inter-
granular fracture.

Fig. 3 shows the result for the Fe-C-P steel. Typical scanning e-
lectron micrographs of fracture surface from the specimens tested while
undergoing cathodic charging are shown in Fig. 4 as well as that fractured
in Charpy test. It can be clearly seen that the brittle fracture mode in
Charpy test for the plain carbon steel specimen with C_B^P of 1.25 wt % is
entirely transgranular but nearly half of the fracture mode caused by hydrogen embrittlement for the same steel is intergranular.

Tensile strength of these specimen is about 600 MPa which is lower
than those shown in Fig. 1 of 800 MPa. Degree of the drop in reduction of
area for these specimens is large for the strength level.

<u>Experiment 3</u>   Fine precipitates along the grain boundaries increase the
grain boundary fracture.

Curves in Fig. 5 show the time to failure as a function of applied
stress for steels 6, 7 and 8. The threshold stress of delayed failure
for steel 6 (Si-Mn-B-N) is much lower than those for steel 7 (Si-Mn-B)
and steel 8 (Si-Mn-N). From fractographic observation steel 6 showed
intergranular path, whereas transgranular cracking was predominant in
steel 7 and 8. Figure 6 shows an example of AES spectrum taken from the
intergranular fracture surface of steel 6. It was found that C_B^P is same
for steel 6, 7 and 8 but B peak of BN as well as N peak are clearly distinguished for steel 6. These peaks come from small particles of BN along
the grain boundaries. This fact implies that fine precipitates of BN on
the grain boundaries intensify the effect of the intergranular P segregation on the intergranular hydrogen embrittlement.

In experiment 1 it was found that a large amount of B (0.0019 wt %)
also promote intergranular fracture, although N content is very low. In
this case the effect of B can be considered as follows. Figure 7 shows

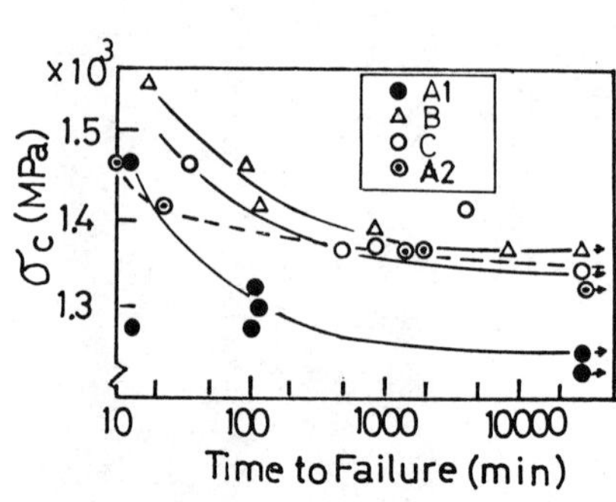

Fig.5   σc-time to failure curve

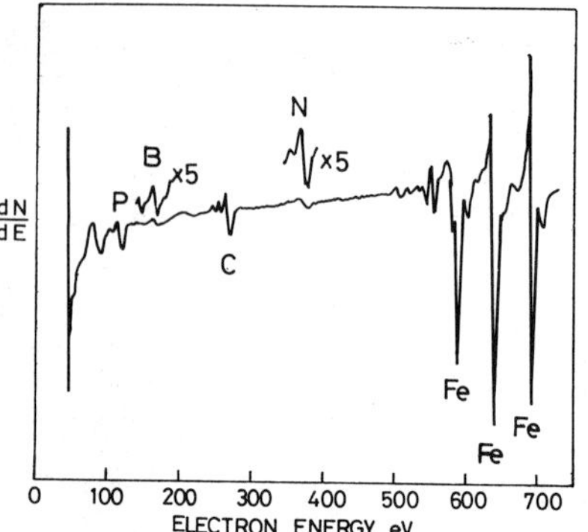

Fig.6   An example of AES spectrum
for A-1.

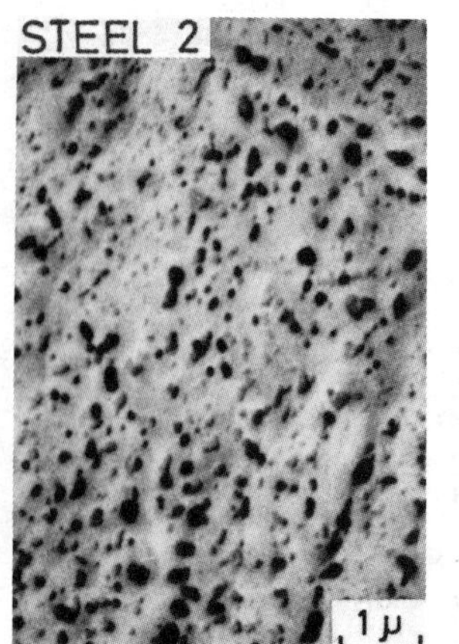

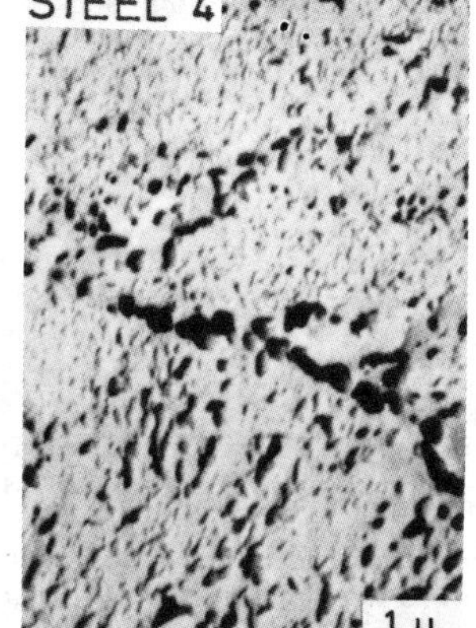

typical examples of
the microstructure of
steel 2 (Si-Mn-P) and
steel 4 (Si-Mn-P-B).
It can be clearly seen
that the grain boundary
carbide is much finer
for steel 4 than for
steel 2. Such fine
precipitates along the
grain boundaries may
promote intergranular
hydrogen embrittlement
caused by the inter-
granular P segregation.

Fig.7   Examples of the microstructure of steel
2 and 4

Discussions

For 600 - 800 MPa tensile strength steels, grain boundary embrittlement caused by impurity segregation is fairly weak and the effect of impurities on the hydrogen embrittlement and the conditions required for the
hydrogen induced intergranular fracture have not yet been made clear.
Conditions that the intergranular fracture occurs in hydrogen embrittlement
can be infered as follows.

Strain Rate Dependence of the Transgranular to Intergranular Transition

On the basis of the results obtained in experiment 1, the degree of
transgranular hydrogen embrittlement and the intergranular one as a function
of C_B^P can be schematically shown as in Fig. 8. The strain rate dependence
of intergranular embrittlement is much larger than that of transgranular one,

transgranular one being dominant in the case of high strain rate (such as $\dot{\varepsilon} \sim 10^{-4}$ s^{-1}). In this case the allowable C_B^P is as high as 2 wt %. In the case of low strain rate ($\dot{\varepsilon} \sim 10^{-5}$ or lower) intergranular fracture is dominant and the allowable C_B^P is as low as 0.1 wt %.

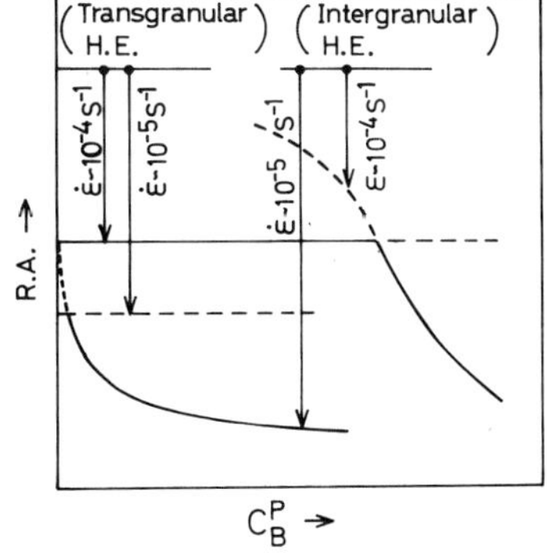

Fig.8 Schematic graph showing the variation of transgranular and intergranular hydrogen embrittlement with C_B^P and $\dot{\varepsilon}$.

The following point is not yet clear. From the study of the fracture morphology and the crystallographic orientation of the fracture surface of transgranular hydrogen induced cracking, it has been found that the inter-lath separation is the dominant one in the transgranular hydrogen embrittlement in these steels. Therefore there is a possibility that impurities can change the degree of transgranular hydrogen embrittlement, although it is assumed in Fig. 8 that transgranular one is independent of C_B^P.

Relation between Temper Embrittlement and Intergranular Hydrogen Embrittlement

It is well known that plain carbon steels even when they contain considerable amount of P are not susceptible to temper embrittlement and no intergranular fracture can be obtained (5). It was found that the P segregation to the grain boundaries can be clearly detected by AES measurement in a Fe-C-P steel and therefore intergranular P segregation alone can not cause temper embrittlement (6). The existence of fine precipitates along the grain boundaries as well as the intergranular P segregation is necessary for temper embrittlement (7). It has been found in experiment 2 that in the case of hydrogen embrittlement intergranular P segregation alone can cause intergranular fracture. This is reasonable because temper embrittlement is related with brittle fracture, whereas hydrogen embrittlement is more related with ductile one.

From the results in experiment 1 and 2 together it can be said that the intergranular P segregation is a necessary and sufficient condition for the intergranular hydrogen embrittlement.

Effect of Fine Precipitates along the Grain Boundaries

The above conclusion does not deny the effect of the precipitates. From the results in experiment 3 and 1 it is infered that either fine precipitates of BN or carbides along the grain boundaries intensify the effect of the intergranular P segregation on the hydrogen embrittlement. Such effect is similar as in the case of temper embrittlement (7). Fine particles closely precipitated along the grain boundaries can block the dislocation movement across the grain boundaries and therefore increases the dislocation density near the boundaries. The increase in the dislocation density resists to the stress relaxation near the crack tip and also promotes hydrogen accumulation.

Conclusions

Investigations were made on the hydrogen embrittlement of low alloy and plain carbon steels in quenched and tempered condition with particular reference to the change in the fracture mode. The results are as follow.

1) In Si-Mn steels the condition for the intergranular fracture changes considerably with the strain rate. Maximum C_B^P (the intergranular P concentration) is about 2 wt % in the case of $\dot{\varepsilon} \sim 10^{-4}$ s^{-1}, whereas C_B^P is as low as 0.1 wt % in the case of $\dot{\varepsilon} \sim 10^{-5}$ s^{-1}.

2) In a plain carbon steel with C_B^P 0.5-1.3 wt % the brittle fracture mode in Charpy test is transgranular but nearly half of the fracture mode caused by hydrogen embrittlement is intergranular.

3) In Si-Mn steels fine particles closely precipitated along the grain boundaries intensify the intergranular fracture in hydrogen embrittlement as well as in Charpy test.

From above results it has been concluded that the intergranular P segregation is a necessary and sufficient condition for the hydrogen embrittlement with intergranular fracture mode and the fine precipitates along the grain boundaries only intensify the effect of P.

Acknowledgement

The authors wish to thank Dr. Hideya Okada, Director of our Laboratories, for his interest in this work and permission to present this paper at the conference.

References

1.   K. Yoshino and C.J. McMahon, Jr.: Met. Trans., 1974, vol 5, pp.363

2.   C.L. Briant, H.C. Feng, and C.T. McMahon, Jr.: Met. Trans., 1978, vol 9A, pp.625

3.   K. Sugino, K. Miyamoto and M. Nagumo: paper presented at 4th International Conference on the Strength of Metals and Alloys, Nacy, France, 1976

4.   T. Inoue, K. Yamamoto, M. Nagumo and K. Miyamoto: Supplement to Trans JIM, 1980, vol 21, pp.433

5.   H.H. Hollomon: Trans. ASM, 1946, vol 36, pp.473

6.   T. Inoue and K. Yamamoto, Tetsu to Hagane, 1979, vol 65, A105

7.   T. Inoue, T. Fuji, K. Yamamoto and T. Horiya: Tetsu to Hagane, '78, vol 46, S 310

HYDROGEN EFFECTS ON BIAXIALLY STRESSED AISI 1015 STEEL

R. P. McNitt, R. D. Sisson, Jr., M. R. Louthan, Jr.,
B. A. Lewis, J. Murali, J. A. Wagner

Environmental Degradation Laboratory
College of Engineering
Virginia Polytechnic Institute and State University
Blacksburg, VA   24061

Disk rupture tests of preflawed spheriodized 1015 steel in hydrogen or
oxygen show that hydrogen reduces the burst pressure. Larger reductions in
burst pressure were observed for samples containing larger preflaws. Pressurization with hydrogen also decreases the low cycle fatigue life and promotes failure in a leak before burst mode. Samples tested in hydrogen
failed along a plane of maximum tensile stress. While, samples tested in
oxygen failed along a plane of maximum shear stress. These results were
attributed to hydrogen enhanced crack growth and hydrogen induced hardening.

Introduction

Mild steels due in part to their availability and good strength to
cost ratio are likely to be used in future energy systems. In these systems
the steel may be in intimate contact with hydrogen gas or compounds of
hydrogen. For reasons of safety and economy it is imperative that the
mechanical behavior of these steels in a hydrogeneous environment be
thoroughly characterized. In particular, the mechanical behavior under
complex stress states should be established. Hydrogen has little effect on
steels loaded in the elastic range; however, effects on the mechanical
behavior in the post yield (plastic) range is of primary interest to actual
engineering applications. Plastic deformation may be expected in regions
of large residual stress, at corners and holes or at other points of stress
concentration.

To establish the effects of hydrogen on the mechanical behavior of
1015 steel (a typical mild steel) in a more complex stress state a series
of monotonic and cyclic biaxial loading experiments were conducted in
oxygen and hydrogen gas. The results of these tests are reported in this
paper.

Experimental Procedure

The materials of construction for vessels that contain high pressure
hydrogen are frequently subject to plane stress biaxial loading where the
principal stress ratio varies for one half (cylindrical geometry) to one

(spherical geometry). When that ratio equals one the in-plane shear stress
is zero and the lattice experiences a maximum dilation. These conditions
are frequently associated with enhanced embrittlement susceptibility;
therefore, this condition was achieved in the specimens used in this study.
To achieve this stress state a disk pressurization system (Figure 1) in
which a thin disk is pressurized from one side was used. During the test
the disk bulges and the bulge increases with pressure until either a leak
develops or the sample bursts. Due to symmetry, the center of the disk is
in a pure biaxial stress state.

Spheroidized AISI 1015 sheet was tested. The microstructure is presented in Figure 2. Disks 7.6 cm in diameter were machined from .4 cm
thick sheet. The uniaxial room temperature mechanical properties of the
steel were:

Yield Stress	236 MPa (34.2 KSI)
Ultimate Stress	325 MPA (47.1 KSI)
Fracture Stress (True)	594 MPA (86.2 KSI)
% Reduction in Area	63%

Test disks were flawed at the center of the pressurized side. Flaw
type A was a circular groove of .6 mm radius cut .15 to .20 mm deep. Flaw
type B was a linear scratch 1 mm long cut approximately .03 mm deep. This
scratch length was either parallel or perpendicular to the rolling direction.

Monotonic or cyclic pressurization tests were conducted. In monotonic
pressurization the sample was pressurized by either hydrogen or oxygen gas
until the sample leaked or burst (Figure 3). The load deformation characteristics of these samples were determined with the aid of electric resistance gauges, an LVDT type transducer and optical devices. The cyclic
pressurization low cycle fatigue tests ($\approx$ 2 cycles per minute) used gas
pressure that varied between two preset limits until the sample failed.
Either high purity hydrogen or oxygen gas was used to pressurize the system.
To increase the purity of the gas in the system the pressure chamber was
filled with high purity gas then evacuated twice prior to the start of each
test. The pressurization rate for the monotonic tests was slow such that 4
to 7 minutes were required to reach 8.3 MPa.

<u>Results</u>

The results of the monotonic tests with samples containing type A
flaws are given in Table I. It is seen that pressurization with hydrogen
reduces the failure pressure by approximately 12%. It should be noted
that, all of the hydrogen pressurized samples failed by leaking, while one
oxygen pressurized sample burst. Microscopic examination of the fracture
surfaces showed only microvoid coalescence in the oxygen tested samples and
quasi cleavage in sampled tested in hydrogen as shown in Figure 4. Previous tests (1) with uniaxial samples and compact tension samples indicated
that hydrogen induced damage occured only in regions of large plastic
deformation such as occured in the necked region or at the tip of a crack
or flaw. Thus it was expected that the gross deformation characteristics
of specimens pressurized by hydrogen and oxygen gas would be similar, as
seen in Figure 5. The bulge height varies linearly with pressure and is
unaffected by the test gas. Larger heights were obtained in oxygen tested
samples because of the larger failure pressures. The observation agrees
with the previous observations that the near yielding behavior of mild

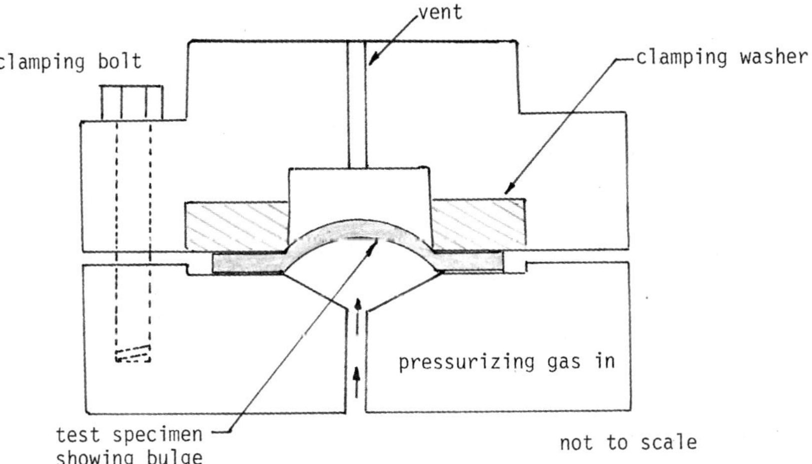

Figure 1.   Cross section schematic of disk pressurizing system with a bulged disk at pressure.

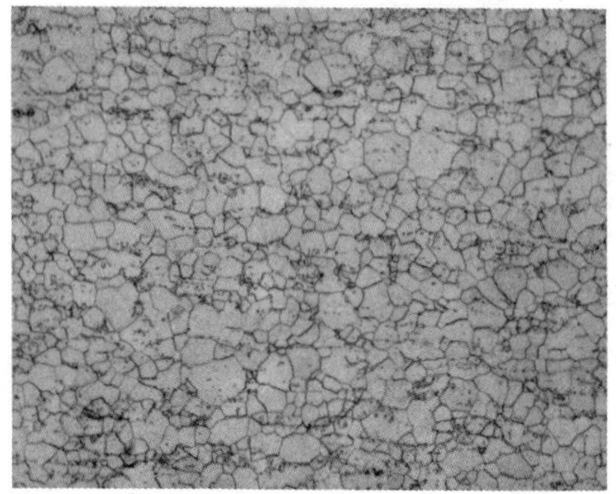

FIGURE 2

Microstructure of AISI 1015 Steel
250X

FIGURE 3

Clockwise from Lower Left Hand Corner
A) Untested Disk, B) Disk Failed in H_2 by Leak Through Crack at Pencil Point,
C) Disk Ruptured in O_2

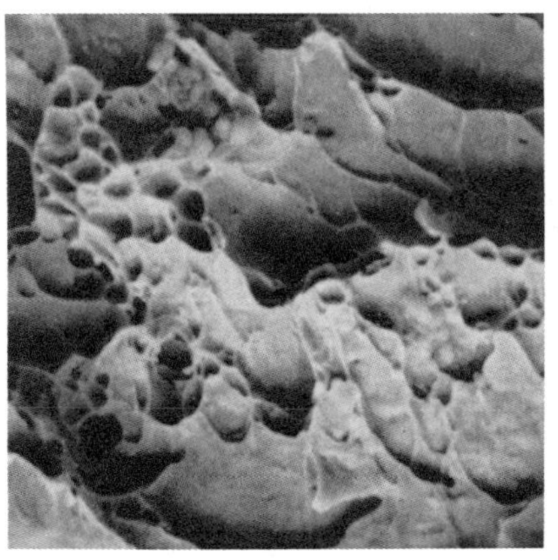

FIGURE 4A

Fractograph of burst fracture surface
of 1015 steel disk fatigued in Oxygen.

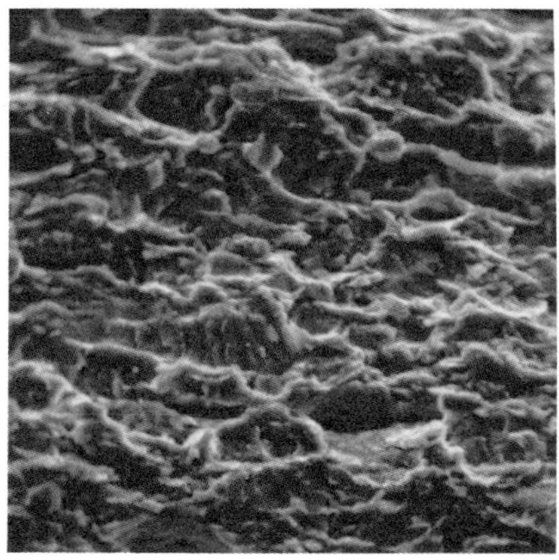

FIGURE 4B

Fractograph of leak fracture surface
of 1015 steel disk fatigued in Hydrogen.

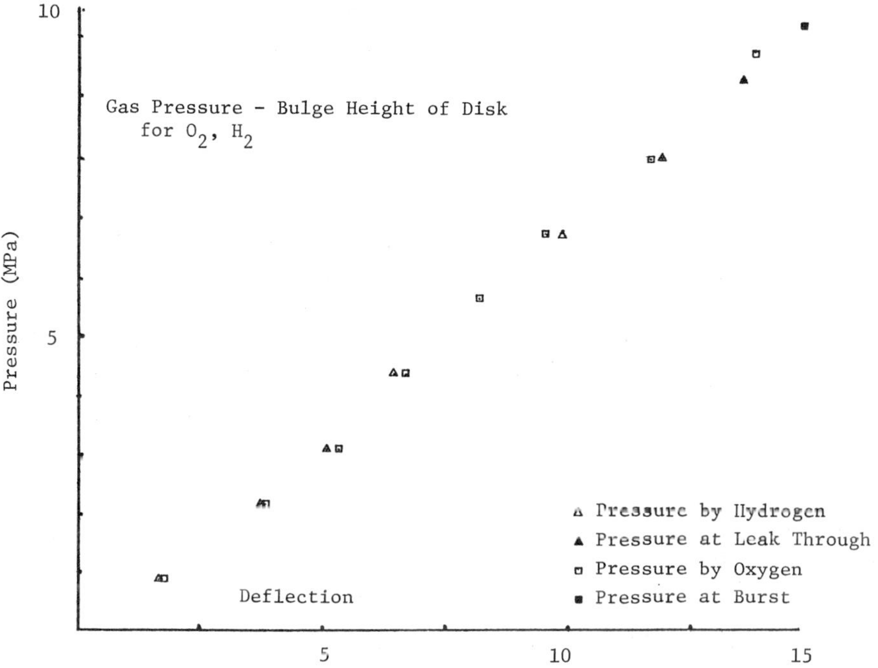

Figure 5 - Bulge Height as Function of Pressure

steels was uneffected by hydrogen when significant hydrogen induced embrittlement was observed (2). This suggests that hydrogen effects in mild steels occured only after large plastic strains. To corroborate this observation several samples were loaded with hydrogen to specific fractions of the oxygen failure pressure, the hydrogen pressure was then removed and the sample was pressurized to failure by oxygen. Pressurizations with hydrogen up to 75% of the oxygen failure pressure caused no damage. However, pressurization by hydrogen to 85% of the oxygen failure pressure caused permanent damage. This result supports the contention that large plastic strains are required for hydrogen induced effects.

Table I
Failure Presure – Static Tests

Pressurizing Gas	H_2	O_2
No. of Samples	5	6
Failure Pressure Range (MPa)	7.6 – 8.1	8.6 – 9
Ave. Failure Pressure	7.7	8.8
Std. Deviation	.04	.03
% Samples Burst	0	16

Several samples were cyclically loaded over the pressure range .7 to 7 MPa in hydrogen or oxygen gas then pressurized to failure. The results of these tests are shown in Table II and indicate that the low cycle fatigue loading in H_2 promotes damage that causes a lowering of the pressure required for failure. The hydrogen damaged samples leaked rather than burst.

Table II
Low Cycle Fatigue – Failure by Overpressure

Specimen	No. of Cycles	Cyclic Pressurizing Gas	Pressure Range	Over-pressurization Gas	Failure Pressure		Fracture Mode
A	50	O_2	.7–7MPa	O_2	9	MPa	Burst
B	50	H_2	"	H_2	8.3	MPa	Leaked
C	100	O_2	"	H_2	9	MPa	Leaked
D	100	H_2	"	O_2	8.1	MPa	Leaked

The type B flaw or scratch could be oriented with respect to the rolling direction. However, the preliminary tests revealed that even when the flaw was perpendicular to the rolling direction these samples tested in hydrogen gas developed leaks with the crack parallel to the rolling direction. Thus the flaw had no effect on the failure process. All flaws were subsequently oriented parallel to the rolling direction and most of the failures for these type B flaws originated in the flaw. The results of these tests are presented in Table III.

Table III
Failure Results of Overpressurized Specimens With Small Flaw

Pressurizing Gas	H_2	O_2
No. of Samples	4	3
Failure Pressure Range (MPa)	8.4 – 8.8	8.4 – 9.1
Ave. Failure Pressure	8.6	8.9
Std. Deviation	.02	.05
% Samples Burst	0	100%

These results do not indicate that hydrogen had a significant influence on the failure pressure. However, it is striking to note that all samples tested in hydrogen failed by leaking while all samples tested in oxygen burst. Clearly, hydrogen affected the failure process even though only small effects on burst pressure were observed.

A number of samples containing type B flaws were cyclically pressurized to near the average oxygen failure pressure. These results are presented in Table IV.

Table IV
Low Cycle Biaxial Fatigue – Small Flaw Specimens

Pressurizing Gas	Pressure Limits	# Cycles to Failure	Failure Mode
O_2	.7 – 8.8 MPa	105	Burst
O_2	"	40	Burst
O_2	"	45	Leak
O_2	"	16	Burst
H_2	"	2	Leak
H_2	"	11	Leak
H_2	"	1	Leak
H_2	.7 – 8.8 MPa	2	Leak
H_2	"	3	Leak
O_2	.7 – 8.1 MPa	No. Failure in ~ 1,100 cycles	
H_2	"	94	Leak
H_2	"	65	Leak
H_2	"	279	Leak

These results indicate that hydrogen reduces the low cycle fatigue life by approximately an order of magnitude. For small flaws hydrogen reduces the low cycle life and changes the failure mode from bursting to leaking.

Discussion

Several observations are apparent from these test results:

• Hydrogen lowers the static failure pressure and the failure stress
  of AISI 1015 steel.

• Hydrogen promotes crack growth such that leaks develop before the
  samples burst.

• The failure surfaces of hydrogen leak-through cracks show quasi
  cleavage, while oxygen tested samples show typically microvoid
  coalesence.

• Hydrogen lowers the low cycle fatigue life of AISI 1015 steel.

• Hydrogen does not effect the elastic nor near yield plastic behavior
  of AISI 1015 steel even when the biaxial loadings are applied to
  the sample.

The observations are consistant with a model which assumes that hydrogen suppresses the plastic processes in mild steels at large plastic strains
and lowers the bond strengths. This lowering of bond strength is seen by
the quasi cleavage crack growth at stress levels lower than the stresses
required for failure of oxygen pressurized samples.

Thus it would appear that for the biaxial stressed disks, hydrogen is
affecting only the deformation processes in the near failure region. In
this region hydrogen enhances crack growth and the cracks preferentially
grow parallel to the rolling direction. In these thin preflawed samples
the subcritical crack growth penetrates the disk prior to reaching a
critical length for unstable extension (burst). However, for specimens
tested in oxygen where subcritical crack growth proceeds by microvoid
coalescence, the critical crack length reached prior to leaking and unstable
failure results. Hydrogen may therefore be suppressing ductile crack formation and enhancing brittle crack growth. The hydrogen enhanced subcritical crack growth occurs at large plastic strains such as might be expected along stringers, surface imperfections, cracks, notches and other
defects or in regions of local plastic instabilities. This conclusion is
supported by the direction of crack growth and the significant reduction in
low cycle fatigue life where local plasticity plays an important role in
low cycle fatigue thus the large scale reduction in cycles to failure are
consistent with the model.

Conclusions

Plastic straining in a hydrogen environment alters the mechanical
behavior of AISI 1015 steel. These changes include a reduction in static
failure pressure and a reduction in low cycle fatigue life. For pressure
vessels these effects will be accompanied by a change in failure mode from
a violent burst to a leak. This type of failure is often considered
favorable because of the leak before burst condition.

However, more testing is required to determine the role of surface
finish and thickness on the failure process before these results can be
extropolated to engineering systems.

These experimental results show that hydrogen suppresses the ductile processes in mild steels and enhances crack growth by a brittle mechanism.

References

1. M. R. Louthan, Jr., R. P. McNitt, J. Murali, N. Sridhar and T. S. Sudarshan, <u>2nd</u> <u>World</u> <u>Hydrogen</u> <u>Energy</u> <u>Conference</u>, Vol. 4, August 1978.

2. J. Murali, Ph.D. Thesis, VPI & SU, August 1980.

Acknowledgments

The authors acknowledge the financial support of NASA and DOE.

The Resistance of Highly Tempered 4130 Steels

to Hydrogen Stress Cracking

Bruce Craig* and George Krauss**

*Research Engineer, Marathon Oil Company, Denver Research Center, Littleton, CO 80160.

**AMAX Foundation Professor, Colorado School of Mines, Golden, CO 80401.

ABSTRACT

The hydrogen stress cracking of highly tempered (700°C) 4130 steels was examined by cathodic charging to failure under three point bending. Standard 4130 steel containing 1% Cr and 0.25% Mo as well as a modification of the base composition containing 0.75% Mo were studied. The time to failure in hydrogen was observed to increase logarithmically as the tempering temperature increased above 600°C. The higher Mo steel had more resistance to hydrogen stress cracking than did the unmodified 4130 at the same temper. Scanning electron microscopy (SEM) showed a transgranular fracture morphology in the presence of hydrogen. The transgranular cracking was found to be initiated at non-metallic inclusions. Crack initiation and propagation around these inclusions resulted in flat fracture zones in which the inclusion was centrally located. These zones spanned several austenite grains, were approximately 50 μm wide by 100 μm long, and were joined by ductile tear ridges. The zones themselves showed a brittle fracture appearance characterized by many fine facets. Correlation of the facets and the associated carbide dispersions with the substructure defined by transmission electron microscopy (TEM) showed that the fine facets correspond to cleavage across the subgrains of the tempered martensite. The beneficial role of Mo is apparently to refine the carbide size, thus retarding the buildup of hydrogen pressure at the carbide-ferrite interfaces.

INTRODUCTION

High strength steels considered to be most resistant to hydrogen stress cracking in sour oil and gas environments have microstructures consisting of dispersed spheriodized carbide particles in a matrix of ferrite. These steels are water quenched and tempered close to the Ac_1 temperature. Several investigators have shown the benefit of a spheriodized structure produced by quenching and tempering (1-4) but even this structure is not completely immune to cracking in hydrogen. Generally the higher the applied stress, or the higher the strength of a given structure the less resistant it is to hydrogen cracking (1). Thus, highly tempered uniform microstructures produce the best hydrogen resistance but only at the expense of strength.

Alloying of low alloy steels such as 4130 with above normal amounts of Mo has been shown to be effective in increasing resistance to hydrogen cracking at higher strength levels (5, 6), presumably because of refinements in austenite grain size and carbide particle size.

Fractographic studies of hydrogen cracking have largely been concentrated on high strength steels tempered at low temperatures where cracking is intergranular (7, 8). However, in more highly tempered steels a transition to transgranular fracture is associated with hydrogen cracking (9). An early study by Farrell and Quarell (10) presents evidence for patches of transgranular cleavage in a hydrogen charged quenched and tempered low alloy steel and related the initiation of the cleavage patches to a central inclusion particle. The replica technique used for the fractography however, did not permit the characterization of the fine details of the fracture on the cleavage patches. More recently, similar fracture patches have been reported in quenched and tempered steels (11) and observed in low carbon steel weld metal (12).

The purpose of this paper is to characterize more fully the nature of the transgranular fracture that develops in highly tempered 4130 type steels. In particular the relationship between the substructure and carbide particle dispersion to the fracture surface features will be discussed.

EXPERIMENTAL PROCEDURE

The composition of the alloys tested is listed below.

	C	Mn	Si	P	S	Cr	Mo	Cb	Al	N
4130A	0.31	0.72	0.35	0.017	0.015	1.02	0.20	-	0.07	0.017
4130C	0.30	0.71	0.35	0.017	0.016	1.01	0.75	0.038	0.07	0.018

Specimens were normalized at 930°C for one half hour and air cooled, followed by austenitizing at 900°C for one half hour and water quenching. The specimens were then tempered for one hour at 700°C. Strips for bent beam testing as described elsewhere (6) were loaded in three point bending and charged with hydrogen in a 2N H_2SO_4 solution with 1000 ppm CS_2 as a poison. The charging current density was 5.0 mA/cm^2. Samples were stressed to 80% of their measured yield strength. Further details on experimental procedure are described in reference (9).

RESULTS AND DISCUSSION

Figure 1 shows the substructure that develops in the 4130A and 4130C steels after tempering at 700°C for one hour. The carbides are spheriodized and found to be concentrated at subboundaries although some are also located within the subgrains. As indicated in Figure 1, the average subgrain size (∿1um) for the 4130A steel was found to be coarser than that (0.7 um) of the 4130C steel. Long intralath carbides that had largely resisted spheriodizing were sometimes observed (9).

Figure 2 shows a typical flat fracture zone characteristic of the transgranular fracture in the 4130A steel tempered at 700°C. The fracture surfaces of the specimens broken in hydrogen consisted of many zones at somewhat different elevations. The zones were joined by ductile tear ridges similar to those shown at the points marked R in Figure 2(a).

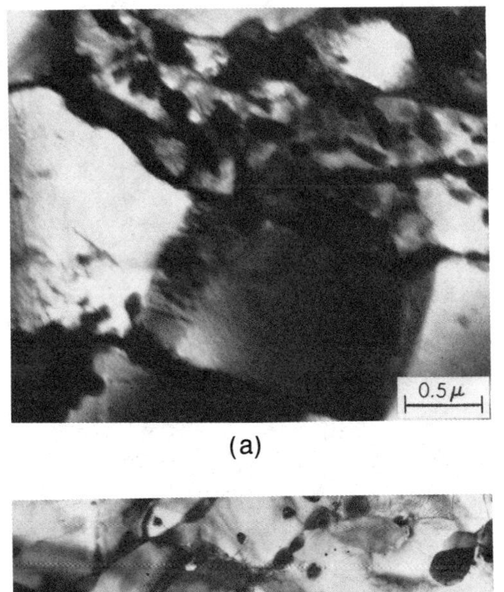

(a)

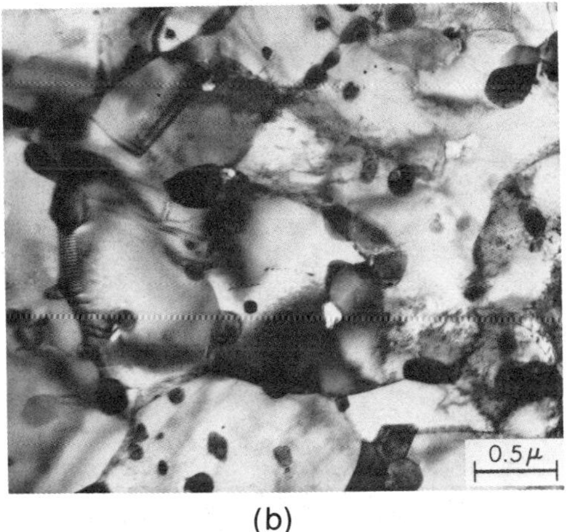

(b)

Fig. 1 Substructures of (a) 4130A and (b) 4130C
        steels, tempered at 700°C for 1 hour
        TEM micrographs.

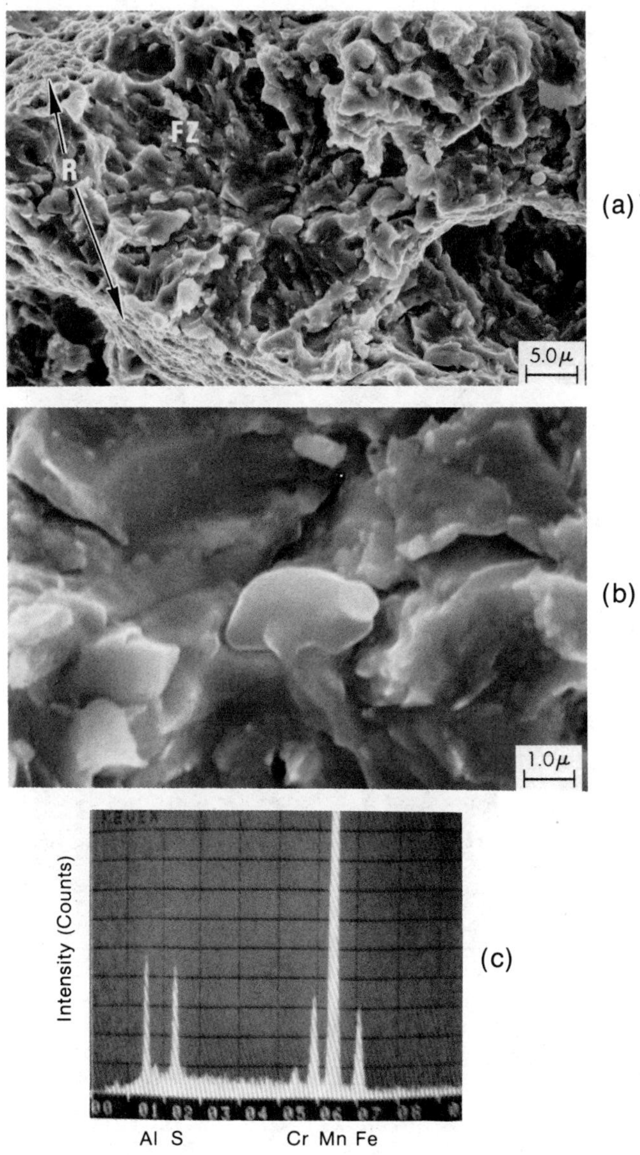

Fig. 2 Fracture in hydrogen of 4130A steel tempered at 700°C after exposure for 110 hours at 437 MPa (63 ksi). (a) flat fracture zone (FZ) and tear ridges (R), (b) initiating inclusion, (c) Energy Dispersive X-ray spectrum for inclusion of (b).

The fracture of each zone apparently had initiated at a large inclusion particle. Figure 2(b) shows the inclusion that initiated the fracture of Figure 2(a) and Figure 2(c) shows an energy dispersive X-ray spectrum that indicates the inclusion is a complex aluminum silicon oxide - manganese sulfide (13). Other inclusions analyzed were similarly complex or single phase oxides or sulfides. The inclusion distribution of both steels was quite similar. In view of the variable composition of the inclusions, size appears to be a more important parameter for initiating fracture than is chemical composition.

Inclusions are well documented traps for hydrogen (14, 15). The high interfacial area of these traps allows sufficient hydrogen to accumulate and build the pressure that eventually initiates the flat fracture zones. However, despite the comparable inclusion content of the 4130A and 4130C steels, fracture was not observed in the 4130C for longer exposures (150 hours) and higher stresses, 572 MPa (83 ksi), than those which caused failure in the 4130A steel. Therefore other factors contribute to hydrogen stress cracking resistance.

The most significant difference between the two tempered steels was the finer substructure and carbide size in the 4130C alloy a result consistent with the increased tempering resistance due to its higher Mo content. Carbides are also well documented hydrogen traps (14, 15). Thus it appears that a finer carbide structure is the key microstructural parameter for improved hydrogen stress cracking resistance. The reason for this may be related to the longer time required to accumulate a critical hydrogen content to drive the brittle fracture.

Figures 3 and 4 show in more detail characteristics of the flat fracture zones. Figure 3(b) shows the fine cleavage facets and the associated carbides of the fracture surface. Many of the facets are quite free of particles and therefore most probably represent transgranular fracture through the subgrains (Figure 1) of the tempered martensite. There are areas of high particle density shown in both Figures 3 and 4. These regions most probably correspond to intergranular fracture between subgrains where the carbides tend to concentrate in a highly tempered structure. The cleavage facets, often bordered by particles, were however the predominate feature of the flat fractures of the hydrogen stress cracking fractographs. Transgranular hydrogen induced cracking has been observed in ferritic microstructures of high purity iron and is promoted by increasing carbon content (16). The subgrains of the highly tempered 4130 steels consist of ferrite which because of the high tempering temperature contains a relatively high carbon content. Thus the observations of the transgranular cleavage fracture is consistent with the previous research findings.

SUMMARY

This paper has emphasized the relationship of the substructure produced by high temperature tempering to the fracture morphology produced by hydrogen stress cracking. The preferred fracture path after initiation at inclusion particles is transgranular cleavage through the ferrite subgrains. Intergranular fracture along subboundaries where carbides have concentrated is sometimes observed. Dispersions of finer carbides and finer substructures appear to promote hydrogen stress cracking resistance.

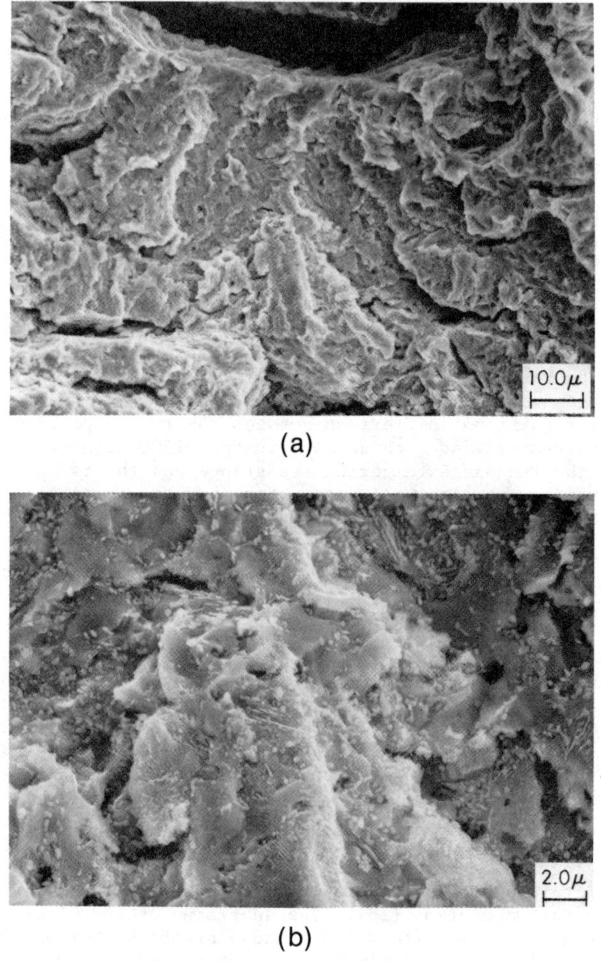

(a)

(b)

Fig. 3 (a) Portion of flat fracture zone in 4130A steel
tempered at 700°C, (b) Detail from part (a).

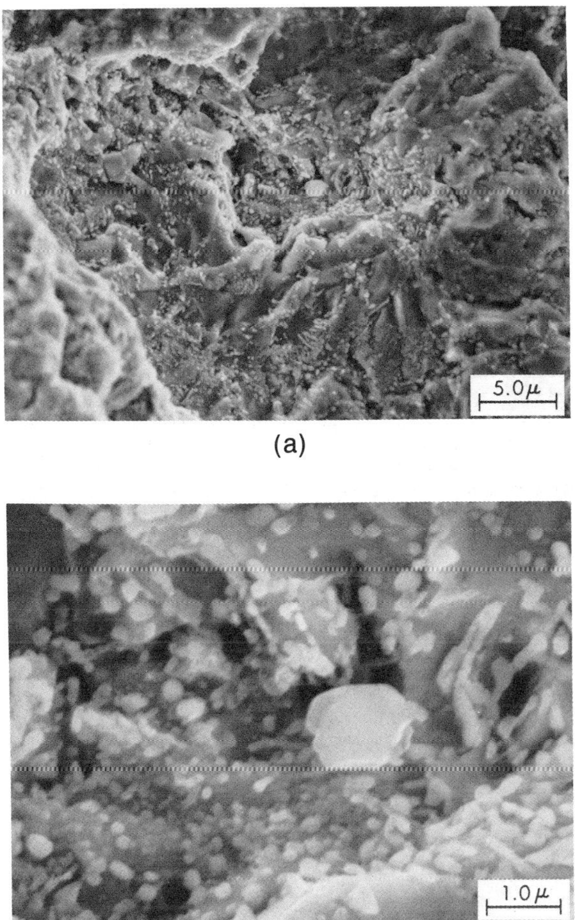

Fig. 4 (a) Portion of flat fracture zone in 4130A
steel tempered at 700°C, (b) detail of
initiation site of zone shown in (a).

ACKNOWLEDGEMENTS

The support of the Marathon Oil Company and the AMAX Foundation for this research is deeply appreciated.

REFERENCES

1. W. M. Cain and A. R. Troiano, Pet Engr, 1965, Vol 37, pp 78-82
2. A. R. Troiano, Trans ASM, 1960, Vol 52, p 54
3. E. Snape, Corrosion, 1967, Vol 23, pp 154-172
4. E. Snape, Corrosion, 1968, Vol 24, pp 261-282
5. P. J. Grobner, W. C. Hagel, and V. Biss, Climax Molybdenum Co., Ann Arbor, Michigan, Unpublished Research
6. P. J. Grobner, D. L. Sponseller, and W. W. Cias, Mat. Perform, 1975, Vol 14, pp 35-43
7. S. K. Banerji, C. J. McMahon Jr., and H. C. Feng, Met Trans A, 1978, Vol 9A, pp 237-247
8. G. W. Simmons, P. S. Pao, and R. P. Wei, Met Trans A, 1979, Vol 9A, pp 1147-1158
9. B. D. Craig and G. Krauss, Met Trans A, to be published
10. K. Farrell and A. G. Quarrell, JISI, 1964, Vol 198, pp 1002-1011
11. A. Ciszewski, T. Radomski, and M. Smialowski, Stress Corr Cracking and Hydrogen Emb of Iron Base Alloys, pp 671-679, NACE, Houston, Texas 1977
12. G. C. Schmid, Mat Perform, 1980, Vol 19, pp 9-18
13. W. Rostoker and J. R. Dvorak, Interp of Metallographic Structures, pp 39-47, Academic Press, N. Y. Second Ed, 1977
14. G. M. Pressouyre and I. M. Bernstein, Met Trans A, 1978, Vol 9A, p 1571
15. J. P. Hirth, Met Trans A, 1980, Vol 11A, pp 861-890
16. I. M. Bernstein, Mat Sci Eng, 1970, Vol 6, pp 1-19

BEHAVIOR OF CLADDED STEELS IN HYDROGEN

G.M. PRESSOUYRE, L. CADIOU and D. CATELIN
Creusot-Loire Research Center
71208 Le Creusot   France

Introduction

Cladded steels are being increasingly used for the protection of equipments to be operated in hydrogen environments ; such is the case for petroleum processing and coal conversion vessels. In service conditions are often characterized by environments containing high proportions of hydrogen or hydrogenated gases (H_2S, NH_3) at elevated temperatures (400 - 450°C, i.e. 750 - 850°F).

In most cases, stainless steel weld metal cladding of the 347 type is used. However, other cladding techniques are now available, such as hot rolling and explosion cladding. Among the differences introduced by these cladding techniques, one concerns the interface between the stainless steel cladding and the base metal. Since the interface represents a structural discontinuity, the presence of hydrogen at this location could give rise to embrittlement problems. This may happen after a long exposure to hydrogen at high temperature, when it has diffused through the cladding, or, for some reason, if the cladding has been damaged. This study has been directed to characterizing possible interface hydrogen embrittlement problems. Several techniques will be used, either to quantify the loss of cohesion at the interface, to analyze hydrogen diffusion along or across the interface, or to visualize cracks morphology.

Experimental

Materials

All test coupons have been taken from industrial heats and claddings. Typical chemical compositions are given in the table below.

(*Code : C : cladding ; BM : base metal ; E : explosion ; EHR : explosion + hot rolling ; HR : hot rolling ; HRNi : hot rolling + nickel foil ; W : weld cladding).

Table I. Typical chemical compositions for cladding and base metals

Material	C	S	P	Si	Mn	Ni	Cr	Mo	other	Code[*]
473 Nb	.024	.005	.02	.50	1.7	9.0	18.5	.35	Nb:.50	C E,EHR,HR
472 T	.02	.008	.024	.45	1.6	10	17.3	.09	-	C HR, HRNI
347	.08	.03	.045	1.0	2.0	11	18	-	-	C W
1 M7	.20	.002	.007	.27	1.0	.20	.20	.05	-	BM E,EHR,HR,W
1.2 F08	.15	.015	.01	.70	.53	.22	1.3	.55	-	BM HR
1.2 Mo6/7	.19	.004	.01	.25	1.0	.15	.20	.50	-	BM HRNi

Five types of claddings were studied, i.e explosion cladding, explosion cladding followed by hot rolling, hot rolled cladding, hot rolled cladding with intermediate nickel foil, and weld metal cladding. Explosion cladding is followed by heat treatment at 570°C for 1/2 hr and air cooled ; weld metal claddings are stress relieved at 675°C for 1 hr. Cladding thicknesses vary between 3 and 5 mm, but for EHR and W plates where it is about 9 mm ; base metal thickness varies between 45 mm and 130 mm. All base metals structures consist of ferrite + pearlite ; all cladding metals structures are of austenite but for 347 that exhibits austenite + ferrite. These structures may best be appreciated on the following pictures on crack morphology. Finally, all regions close to the interfaces show a decarburization of the base metal accompanied by a carburization of the cladding, but for explosion cladding.

Mechanical testing

Mechanical testing was performed on tear, shear and bend specimens (see fig.1 for details). Hydrogen charging was performed either in gas phase at 450°C, under 20 bars hydrogen (300 psi), or cathodically in acid solution (H_2SO_4-N + poison AS_2O_3 ; 30 mA/cm^2) ; charging times were calculated from diffusion data to ensure complete saturation. When gas charging, the whole specimen was exposed to hydrogen ; thus hydrogen was present everywhere, i.e. in the cladding, at the interface, and in the base metal. When cathodically charging, all the specimen besides the interface region was masked by an impermeable lacquer ; thus hydrogen preferentially diffused along the interface. One will note that these charging conditions are much more drastic than usual in service conditions, since here the interface is directly exposed to hydrogen ; more over, the hydrogen equivalent pressure when cathodically charging is very high (10^3 Atm. ; $1.4 \cdot 10^4$ psi).

Diffusion tests

Diffusion tests were performed on disks with the interface as a barrier to diffusion (see fig.1) in gas phase, at 450°C and .07 bars hydrogen (1 psi) on the cladding side ; other tests were also performed at room temperature using an electrochemical permeation technique (ref.1), with the interface parallel to the hydrogen flux (fig.1).

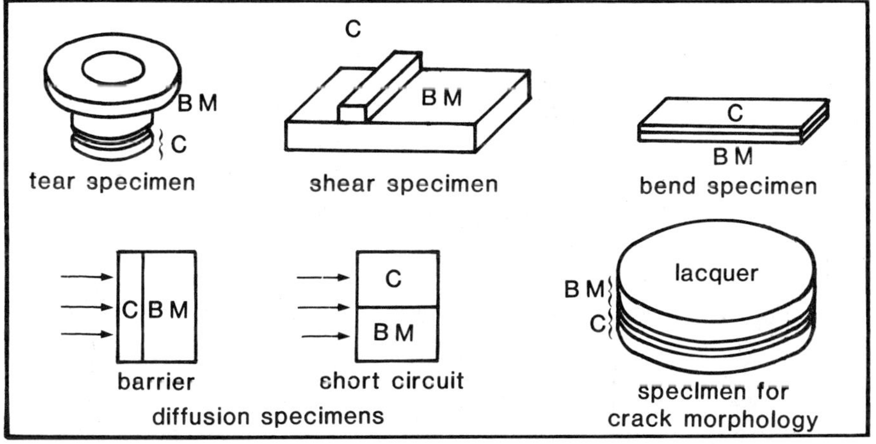

Fig.1. Specimens schematic

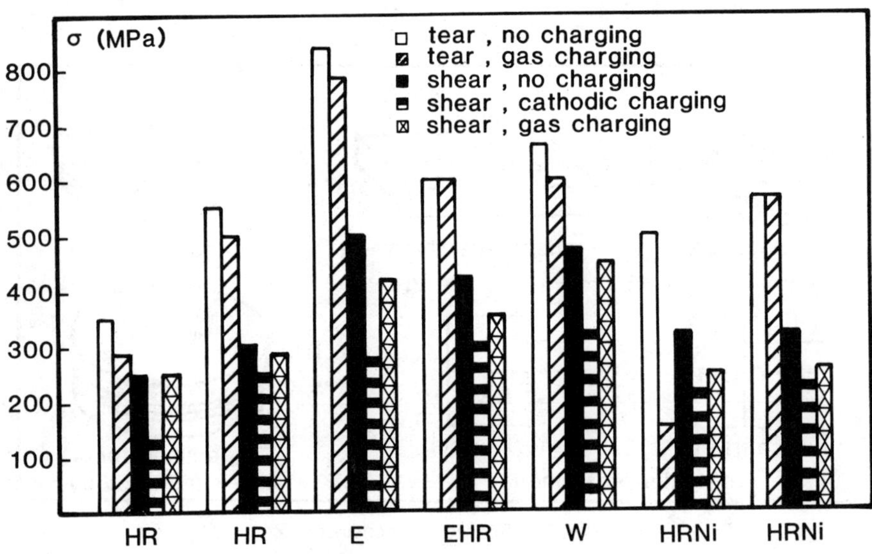

Fig.2.   Mechanical testing results

Crack morphology

Hydrogen induced cracks were obtained in all cases by violently cathodically charging thick disks (current density > 50 mA/cm^2) at the interface (as in mechanical testing).

Results

Mechanical testing

Shear and tear results are shown on fig.2 ; all values are averages over several specimens. In the uncharged state, the explosion cladding interface (E) is the strongest, followed by W, EHR and HR cladding interfaces. When gas charging with our test conditions, little loss of cohesion was observed either in shear or in tear testing ; thus the preceding ranking is conserved. An exception is to be noted in the case of a hot rolled cladding with an intermediate nickel foil, where the average loss of cohesion in tearing after gas charging is important (over 60 % loss). When cathodically charging, the effect is more drastic because in that case, defects are also created at the interface (see crack morphology), contrarily to the gas charging case where only the decrease in cohesion due to hydrogen trapped at the interface is present. Thus, the best resistance is offered, in absolute values, by the W, EHR and E claddings, in that order, followed by hot rolled claddings. One should note however that in this case, most interfaces offer the same resistance to cathodic hydrogen damage, in the range 200 - 300 MPa.

Bend tests were performed using gas charging and an applied stress of 160 MPa. This was done in order to cheek if, in these tests conditions, hydrogen induced cracks were nucleated at the interface. After metallographic examination, none of the interfaces exhibited cracks.

Diffusion tests

Tests with the interface as a barrier to diffusion were performed on a HR and a HRNi cladding to see if the presence of the intermediate nickel foil added a barrier effect. Both interfaces act as a barrier because measured times to reach steady state (t_s) are longer than theoretical times, (130 s) as shown on fig.3 : however, the presence of the nickel foil does not significantly delay hydrogen diffusion (380 s. instead of 300 s.). Moreover, the cladding with an intermediate nickel foil allows more hydrogen to reach the base metal (J_s is higher, see fig.3).

Tests with the interface parallel to the hydrogen flux (short circuit path) were conducted on all types of cladding. When a short circuit path is present, degassing curves exhibit a hump, and the total curve (J total ; fig.3) may approximately be decomposed into 2 curves : one corresponds to the flux along the interface, and the other to the flux through the base metal (the flux through the austenitic cladding is negligible). One may then appreciate the diffusion characteristics of the interface compared to the base metal. Results are summarized below.

Table II. Diffusion results with the interface as a short circuit path

Ratio	HR	HR	E	EHR	W	HRNi	HRNi
D interface / D base metal	67	17	33	17	17	57	36
Flux of hydrogen at steady state along the interface at .H/cm^2 .S x 10^{-12}	3.9	1.7	2,3	2.2	3.6	6	5.6

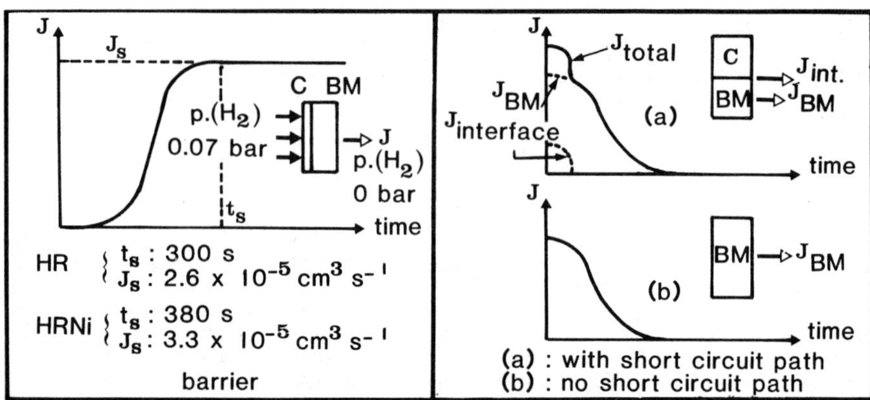

Fig.3. Diffusion results

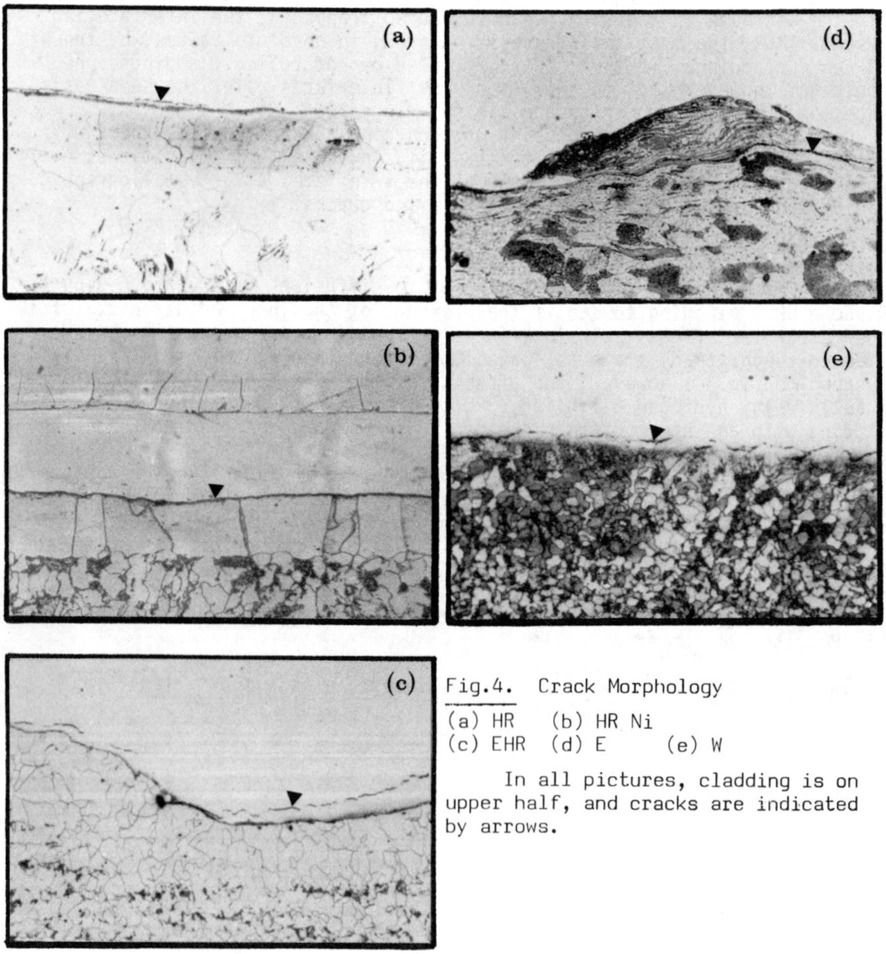

Fig.4. Crack Morphology

(a) HR   (b) HR Ni
(c) EHR (d) E   (e) W

In all pictures, cladding is on upper half, and cracks are indicated by arrows.

The fastest short circuit diffusion occurs in a HR cladding specimen and in the two HRNi cladding specimens : the ratio D interface/D base metal is highest. It seems logical that the highest diffusion coefficient should correspond to the interface with the less cohesion, as shown on fig.2. Also, the diffusion flux is highest in the two HRNi specimens ; this means that more hydrogen diffuses faster along this interface.

<u>Crack morphology</u>

In such drastic conditions as cathodic charging, all cladding interfaces may be cracked. The location of the obtained hydrogen induced cracks gives an indication of the regions of the interface most susceptible to embrittlement. The results are given in figs.4   . In the HR cladding case, cracks grow parallel and close to the interface, in the cladding ; this also is the case for EHR and W specimens. In other words, the crack propagates in that region of the cladding which has been carburized. On the contrary, cracks propagate at the interface in the case of explosion cladding specimens. More peculiar is the location of hydrogen induced cracks in HRNi specimens, at the interface between the nickel foil and the iron foil on the base metal side. Thus, all cladding interfaces are susceptible to hydrogen induced cracking, as long as enough hydrogen is allowed to penetrate this region.

<h2 align="center">Conclusions</h2>

A comparison has been made between different types of cladding, concerning the hydrogen embrittlement susceptibility and diffusion characteristics of the cladding-base metal interface. It can be concluded that :

- when exposed to hydrogen gas (20 bars, i.e. 300 psi ; 450°C), no defects are created at the interface, but there exists a slight loss of cohesion in most cases (but for a hot rolled with an intermediate nickel foil, where the observed decrease is important). The interface offering the highest resistance is that of the explosion cladding.
- when cathodically charging with very high hydrogen input fugacity, hydrogen induced cracks are nucleated at all interfaces. The cohesive shear strength of all interfaces is decreased to comparable levels (150-300 MPa).
- the presence of an intermediate nickel foil does not add any barrier capacity to hot-rolled cladding interfaces (even carbon may diffuse, as evidenced by decarburization). On the contrary, more hydrogen is carried faster along the interface, with obviously possible further diffusion into the base metal. This, and occasional bad behavior in mechanical testing in hydrogen, may stem from the fact that with the nickel foil, the number of interfaces is doubled, as well as the probability of finding a defect susceptible to initiate cracking ; also, the solubility of hydrogen in nickel is relatively high at high temperatures ($\theta > 400°C$).
- if one is to choose a cladding plate for use in a hydrogen environment when the equivalent hydrogen pressure at the interface is about 20 bars at 450°C (this corresponds to an outside hydrogen pressure of about 90 bars, i.e. 1300 psi, for cladding and base metal thicknesses of 5 mm and 50 mm, respectively), any cladding may be taken as long as interface intrinsic cohesive strength is sufficient ; in other words, hydrogen embrittlement of the interfaces is negligible in this case. If in service conditions are much more drastic (cathodic charging being an extreme case), one should take into account hydrogen embrittlement of the interface, and consult the ranking of fig.2.

<h2 align="center">Acknowledgement</h2>

The authors would like to thank P. Tison and J.P. Fidelle at the French CEA, Bruyères le châtel, for performing the high temperature permeation measurements.

References

1. M.A.V. DEVANATHAN and Z.O.J. STACHURSKY, Proc. Roy. Soc., 1962, <u>A 270</u>, p. 90.

DISCUSSION

<u>R. Stoltz, Sandia Laboratories, Livermore, CA</u>: At what binding energy does a defect change from a reversible to an irreversible trap?

<u>G.M. Pressouyre</u>: The changing from a reversible to an irreversible trap is probably not abrupt. Indeed, two traps may have identical binding energies but different reversible characters (this owes to the difference between an attractive and a physical trap, see Ref: G. M. Pressouyre, Met. Trans., 10A, 1979, p. 1571). However, experimental results so far (tritium auto-radiography, e.g.) seem to show that above 0.7 eV, reversibility at room temperature is not probable.

ON THE SYNERGISTIC EFFECTS OF HYDROGEN EMBRITTLEMENT AND
TEMPER EMBRITTLEMENT IN A 9% NI STEEL

S. Tikkamäki and H. Hänninen[x]

Outokumpu Company
Tornio, Finland

[x]Technical Research Centre of Finland
Metals Laboratory
Espoo, Finland

Summary

Tensile tests have been conducted on specimens of a 9% Ni steel
charged with hydrogen through cathodic polarization. The metallurgical
conditions studied were chosen on the basis of impact test results. Two
typical heat treatments were used in hydrogen embrittlement experiments:
severely temper embrittled, and non-embrittled condition. The microstructures of these conditions were characterized by transmission electron
microscopy.

The fracture mode of tensile tested hydrogen charged specimens
changed compared to the uncharged material, in the non-embrittled material, from normal ductile dimpled fracture to a brittle cleavage-like
fracture. In the temper embrittled material, the fracture mode changed
from normal ductile dimpled fracture to a brittle intergranular fracture.
In temper embrittled material the fracture stress was reduced to about one
third of the fracture stress of uncharged specimens. The fracture mode
shift from cleavage to intergranular fracture and the decrease in the
fracture stress are discussed on the basis of possible synergistic effects
between impurity-induced temper embrittlement and hydrogen-induced cracking

Introduction

Quenched and tempered (ASTM A553) or double normalized and tempered (ASTM A353) 9% Ni steels have excellent toughness and strength at temperatures down to about 70 K. Therefore, they are widely used for cryogenic applications, e.g. in the construction of liquified natural gas storage tanks (1,2). The required properties of 9% Ni steel are obtained only by tempering in a relatively narrow temperature range of 830-870 K (3-6). During tempering heat treatment a small amount of austenite (5-10 Vol Pct) is reformed. This phase is stable in cooling down to 70 K and is believed to contribute to good low temperature toughness of this steel (4,5). Previous studies have shown that 9% Ni steel is sensitive to reversible temper embrittlement (5-8). The purpose of the present paper is to demonstrate that temper embrittled 9% Ni steel has also enhanced susceptibility to hydrogen-induced intergranular cracking.

Experimental

The commercial 9% Ni steel used in this study was obtained as 16 mm thick plate with the chemical composition given in Table I. Blanks (200 mm x 70 mm) cut from the plate were austenitized in argon atmosphere either at 1473 K for 10 h or at 1073 K for 1 h following quenching into room temperature water, and tempering at 843 K for 20 h in air. The purpose of such prolonged tempering was to eliminate hardness changes during long time heat treatment to obtain embrittled condition. A portion of the tempered blanks were isothermally aged at 693 K for 500 h to obtain maximum embrittlement (4). The propensity to temper embrittlement was determined by Charpy impact test. Longitudinal standard V-notch specimens were used with the notch machined perpendicular to the rolling direction.

Table I. Chemical Composition, Wt Pct

C	Si	Mn	Ni	Cr	Mo	S	P	As	Sb	Sn
0.06	0.26	0.61	9.3	0.05	0.02	0.014	0.008	0.010	0.002	0.005

Tensile specimens (Fig. 1) for hydrogen embrittlement (HE) testing were prepared by machining to a thickness of about 0.5 mm and further polished by grinding, finally with a 600 grit paper. The loading direction was parallel to the rolling direction of the initial plate.

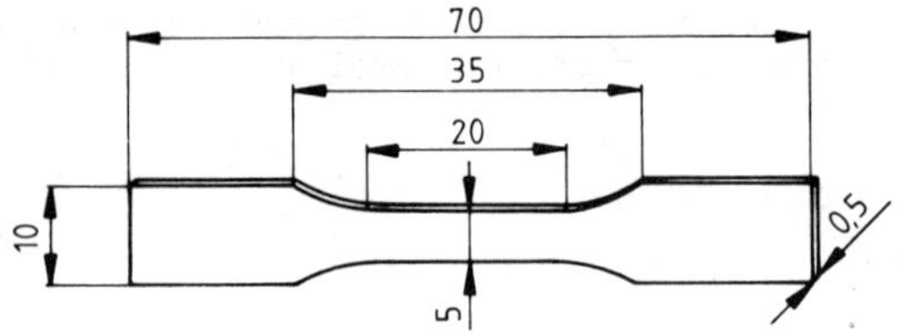

Fig. 1 – The tensile specimen for cathodic charging tests.

The tensile specimens were cathodically charged at 295 K in a 1 N
H_2SO_4 solution containing 0.25 g/l of $NaAsO_2$. The current density was
about 50 mA/cm^2 and a Pt counter electrode was used. The hydrogen charging time was 6 hours. After charging the specimens were tensile tested
within 5 min at room temperature.

Results

Results of hardness and impact tests are summarized in Table II.
Higher austenitization temperature shows a clear effect in impact
properties at low temperatures. The microstructures of tempered and
embrittled conditions are shown in Fig. 2. The TEM study shows no difference in microstructures between the tempered and embrittled materials,
i.e. austenite does not decompose during embrittling heat treatment and
therefore it is not the reason for the observed embrittlement.

Table II. Effect of Heat Treatment on Impact
Properties and Hardness

Austenitized	Tempered				Embrittled			
	Impact Energy, J			Hardness, HV	Impact Energy, J			Hardness, HV
	at 77 K	193 K	293 K		at 77 K	193 K	293 K	
1073 K/1 h	143	190	209	238	40	109	200	243
1473 K/10 h	25	187	219	237	24	190	227	240

Fig. 2 – Microstructures of a 9% Ni steel showing a cellular ferrite
structure and islands of austenite. (a) Tempered condition.
(b) Embrittled condition.

The results of the tensile tests are shown in Table III. Two specimens were used for obtaining reference values and five specimens for obtaining values for hydrogen charged material. Hydrogen charged tempered
large grain material showed large variability in tensile testing results.
Otherwise reproducible results were obtained.

Table III. Tensile Test Results with and without
Cathodic Hydrogen Charging

| Austenitized | Tempered | | | | | | Embrittled | | | | | |
| | Reference | | | Hydrogen charged | | | Reference | | | Hydrogen charged | | |
	$R_{p0.2}$ N/mm^2	R_m N/mm^2	A %	$R_{p0.2}$ N/mm^2	R_m N/mm^2	A %	$R_{p0.2}$ N/mm^2	R_m N/mm^2	A %	$R_{p0.2}$ N/mm^2	R_m N/mm^2	A %
1073 K/1 h	675	730	14	670	705	8	670	720	14	740	750	3
1473 K/10 h	625	660	6	–	600	0	650	690	6	–	230	0

After tensile testing the fracture surfaces were studied by SEM. In
Fig. 3 the normal dimpled fracture morphology of ductile reference specimens is shown. In tempered condition the hydrogen charging shifts the
fracture mode from ductile dimpled fracture to mainly cleavage type fracture. Additionally, smaller intergranular and dimpled areas were observed. In Fig. 4 examples of this hydrogen-induced cracking are presented.

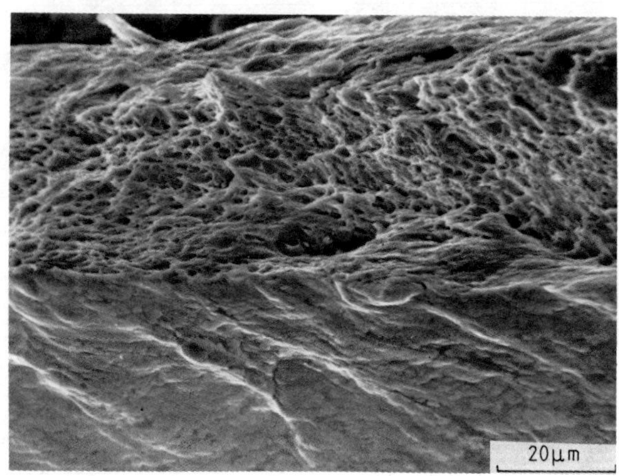

Fig. 3 – Ductile fracture surface of a 9% Ni steel specimen; lower
austenitization temperature, embrittled condition.

In temper embrittled material hydrogen charging shifted the fracture
mode to a brittle intergranular fracture shown in Fig. 5. On the fracture
surface austenite particles are shown. Therefore, the fracture path in
these experiments consists of grain boundaries (ferrite/ferrite) and phase
boundaries (austenite/ferrite). In temper embrittled material the fracture stress was reduced to about one third of the fracture stress of uncharged specimens.

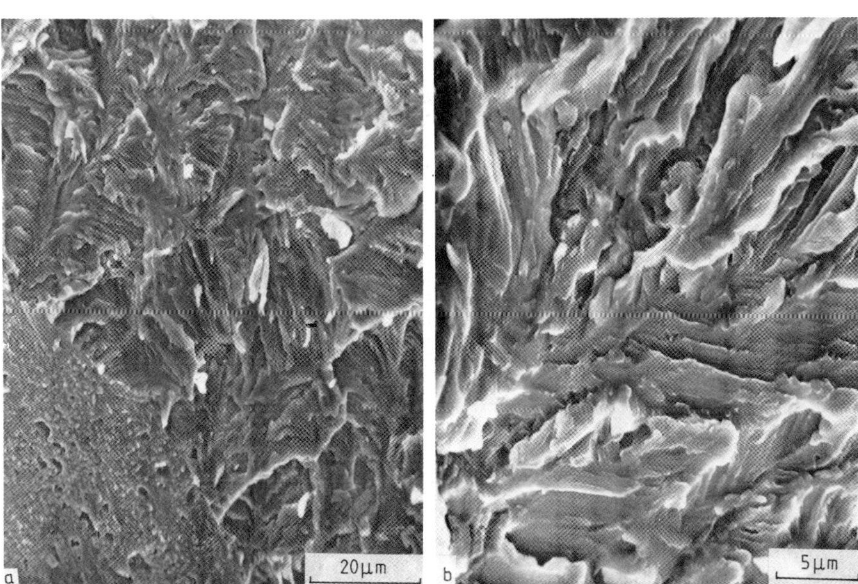

Fig. 4 – Fracture surface morphology of hydrogen-induced cleavage
in cathodically charged tempered 9% Ni steel (high
austenitization temperature). (a) A general view (partly
intergranular fracture). (b) A detail of cleavage fracture.

Discussion

The present study has shown that hydrogen can shift the fracture mode
in a 9% Ni steel from ductile dimpled fracture to cleavage fracture. If
the material is in temper embrittled condition brittle intergranular
cracking occurs, and the fracture stress can be drastically reduced.
Similar phenomena have been observed e.g. in low alloy steels (9) and in
commercial Ni (10) which have been explained by the synergistic influences
of impurity elements and hydrogen. The elements involved are such as P,
S, Sn and Sb, which are known to cause mechanical temper embrittlement, to
poison the hydrogen recombination reaction on the metal surface, to lower
the exchange current density for cathodic reaction in electrolytes and
thereby to increase the amount of hydrogen absorption in metal, to increase the diffusivity of hydrogen along grain boundaries (11), to cooperate with hydrogen may be resulting in hydride formation, and to reduce
the cohesive energy of the grain boundary (9, 12). Impurity segregation
has been shown to be also important in reversible temper embrittlement in
9% Ni steels (8), the role of phosphorus being most prominent.

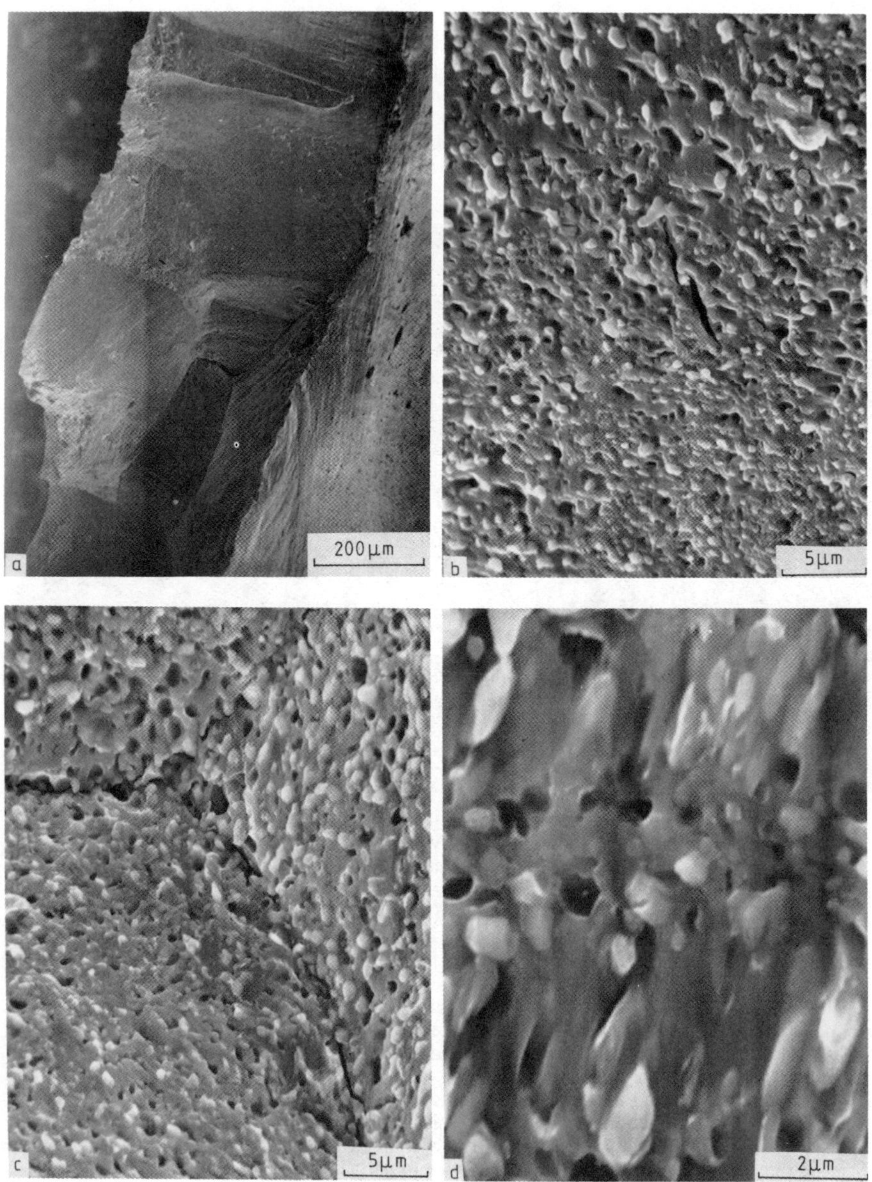

Fig 5 — Intergranular brittle cracking in temper embrittled 9% Ni
steel after cathodic hydrogen charging. (a) A general view
of the fracture surface. (b)–(d) Details: note that the
fracture path follows the grain boundary and phase boundary
between ferrite and austenite (the particles on the fracture
surface).

Hydrogen-induced cracking in the studied steel occurs by cleavage at high stress levels (nearly in the range of uncharged material), when intergranular impurity content is low. As the impurity content in the grain boundaries is increased in prolonged heat treatments, the hydrogen-induced crack path becomes intergranular and the stress level needed for cracking can be drastically lowered as was observed. McMahon et al. (13) have proposed the following model for this kind of crack formation:

(i)   When local stresses in crystal or grain boundary reach the cohesive strength, brittle fracture occurs.

(ii)   The cohesive strength decreases as the local hydrogen concentration increases.

(iii)   Segregated impurities reduce the cohesive strength of a grain boundary (each impurity may reduce the cohesive strength at a different magnitude), probably in proportion to impurity concentration.

(iv)   Brittle cracking occurs when the maximum stress reaches the critical value for cracking ahead of a crack in the region of maximum hydrostatic stress.

Efforts to determine the responsible segregated impurity elements were also done by trying to mechanically fracture the Auger-specimens of studied 9% Ni steel at 77 K intergranularly, but this proved to be extremely difficult. Thus, this steel does not be very prone to pure temper embrittlement. May be the embrittling temperature chosen according to the literature (4) was too low for the studied steel, which had relatively low impurity content. However, the superposition of the effects of segregated impurities and hydrogen can produce brittle intergranular cracking at very low stress levels which can be interpreted according to the above model (13) through the synergistic interaction between segregated impurities and hydrogen. Thus, the levels of impurity segregation needed for HE seem to be much lower than the levels for temper embrittlement in this steel. Immunity to HE or hydrogen-induced SCC cannot therefore be assessed only from temper embrittlement data or from discernable changes in micro-structure.

Acknowledgement

This research was supported in part by Outokumpu Company Foundation (S.T.) and was part of the Reactor Materials Research Project (H.H.) financed by the Ministry of Trade and Industry in Finland. Helpful comments of Prof. J. Forsten are gratefully appreciated.

References

1. R.H. Tharby: Welding and Metal Fabrication, 1973, Feb., pp. 2-12.
2. R. Scott: Met. Technology, 1976, vol. 3, pp. 71-78.
3. G.R. Brophy and A.J. Miller: Trans. ASM, 1949, vol. 41, pp. 1185-1201.
4. C.W. Marschall, R.F. Hehemann and A.R. Troiano: Trans. ASM, 1962, vol. 55, pp. 135-148.
5. S. Nagashima: Toward Improved Ductility and Toughness, Climax Molybdenum Symposium, Kyoto, 1971, pp. 353-355.
6. J. Zvokelj, G. Murry and A. Constant: Rev. de Métallurgie, 1969, vol. 66, pp. 25-35.
7. P.M. Munro: Iron and Steel, 1971, vol. 44, pp. 21-27.

8. C. Gaspard, A. Magnée, C. Coutsouradis and L. Habraken: Metall. Rep. CRM, 1979, vol. 54, pp. 31–39.
9. K. Yoshino and C.J. McMahon, Jr.: Met. Trans., 1974, vol. 5A, pp. 363–570.
10. R.M. Latanision and H. Opperhauser, Jr.: Met. Trans., 1974, vol. 5A, pp. 483–492.
11. M. Kurkela: M.Sc. Thesis. Technical University of Oulu, Finland, 1979.
12. C.L. Briant, H.C. Feng and C.J. McMahon, Jr.: Met. Trans, 1978, vol. 9A, pp. 625–633.
13. C.J. McMahon, Jr., C.L. Briant and S.K. Banerji: Fracture 1977, vol. 1, ICF4, Waterloo, Canada, June 19–24, pp. 363–385.

FRACTOGRAPHY OF ALLOYS TESTED IN HIGH-PRESSURE HYDROGEN*

R. J. Walter, J. D. Frandsen, and R. P. Jewett
Rockwell International/Rocketdyne Division
Canoga Park, California

Scanning electron fractography was performed on specimens tested in high-pressure hydrogen and, for comparison, in air and helium environments. Examination showed that the hydrogen environment produced a brittle fracture mode, either transgranular or intergranular, in the following alloys: Inconel 718, Waspaloy (TMP), Inconel 625, Mar-M-246, and Haynes 188. Cast Inconel 718 and Incoloy 903 contained a large proportion of ductile fracture when da/dN tested at ambient temperature.

Comparison of fracture morphology with crack growth rates showed that the fastest crack growth rates were associated with fracture in boundaries separating the matrix and a precipitate phase. The second fastest crack growth rates were associated with intergranular fracture, and the next fastest rates were associated with transgranular fracture. The specimens with the slowest crack growth rates fractured with considerable microvoid formation.

*This work was sponsored by the National Aeronautics and Space Administration Marshall Space Flight Center, Alabama, under Contract NAS8-27980.

Introduction

This paper summarizes the results of electron fractography examination performed on specimens tested in high-pressure hydrogen. Most of the specimens were cyclic loaded using a 9-minute cycle that simulates an SSME (Space Shuttle Main Engine) mission. The cycle consists of ramping to maximum load in 1.75 seconds, holding at maximum load for 255 seconds; ramping to 92% of maximum load in 1 second and holding at this load for 235 seconds; and ramping to the minimum load (R = 0.01) in 1.2 seconds and holding at this load for 30 seconds.

The ambient temperature crack growth rates for the specimens examined are shown in Fig. 1. All data shown in Fig. 1 were obtained using the 9-minute SSME cycle, except for wrought Inconel 718 in the STA 2 heat treatment condition. The crack growth rates shown for this alloy were calculated from sustained load crack growth data and represent the sustained load crack extension that would be produced during the 9-minute SSME cycle.

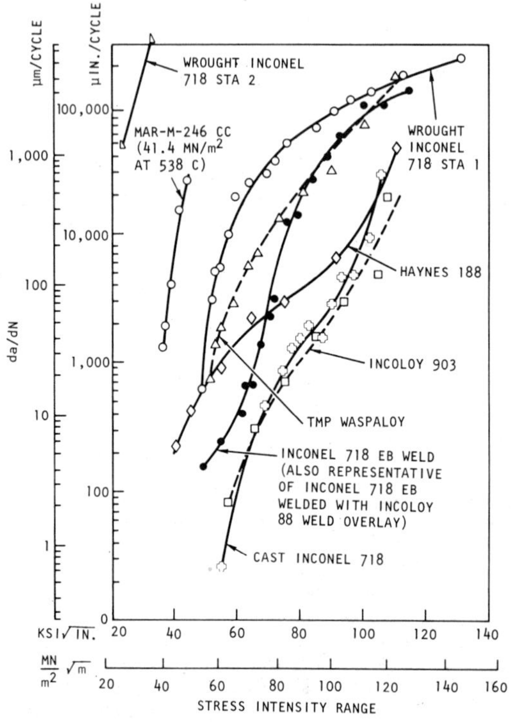

Fig. 1 – Crack Growth Rates (da/dN) Vs Stress Intensity Amplitude (ΔK) for Various Alloys Exposed to 34.5 to 48.3 MN/m^2 (5000 to 7000 psi) Hydrogen at Ambient Temperature

Results and Discussion

The most rapid crack growth occurred during 538 C (1000 F) da/dN tests on Mar-M-246 CC specimens and ambient temperature sustained-load crack growth measurements performed on Inconel 718 in the STA 2 (940, 716-621 C) heat treatment condition. Mar-M-246 and Inconel 718 STA 2 are two-phase alloys of γ (FCC Ni) and γ' (Ni$_3$Al,Ti) in Mar-M-246 and γ and orthorhombic Ni$_3$Nb phases in Inconel 718.

Optical and electron microscopy of Mar-M-246 specimens tested in hydrogen indicated that the fracture was predominantly intergranular. Primary γ' is preferentially present in the grain boundaries, forming the preferred fracture path in hydrogen. Figure 2A is an electron micrograph of a section intersecting the fracture showing crack propagation along the γ/γ' interface. Electron fractography (Fig. 2B) showed the fracture to be dendritic and cellular, and the fracture features were similar to the $\gamma + \gamma'$ cellular structure.

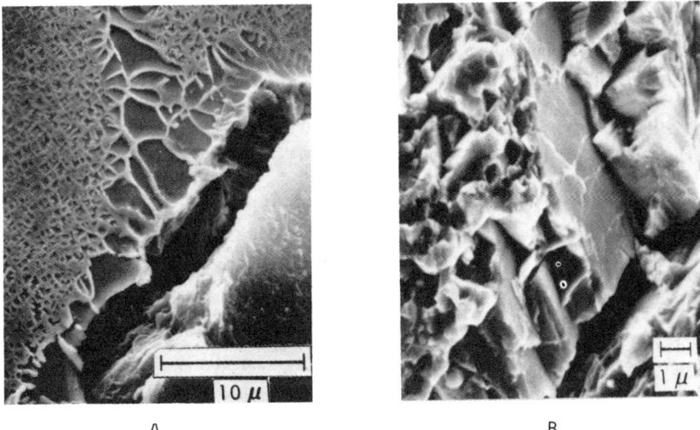

A   B

Fig. 2 – Metallography of Mar-M-246 CC Specimen Cyclic Tested
in Hydrogen at 538 C (1000 F): A. SEM Showing Crack
Growth Along Primary γ'/γ Interface and B. Fractograph of Subcritical Crack Growth

The 940 C (1725 F) STA 2 anneal precipitates a semicontinuous orthorhombic Ni_3Nb phase in Inconel 718. Figure 3A shows the fracture path in
hydrogen followed the matrix $(\gamma)/Ni_3Nb$ interface. Thus, the fractures of
conventionally cast Mar-M-246 and Inconel 718 in the STA 2 condition were
similar. Both alloys are basically two phase, and crack growth in hydrogen
followed the two-phase boundaries.

There are two possible mechanisms for preferred fracture along the interface between the different phases present in these materials. One could

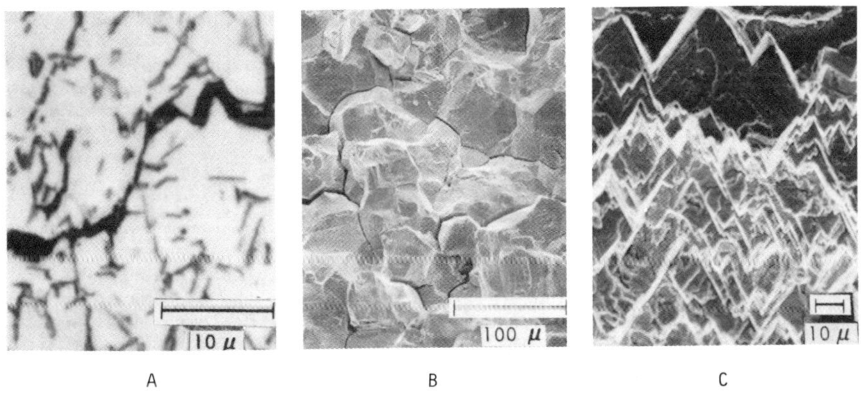

A   B   C

Fig. 3 – Metallography of Inconel 718 Fractured in Hydrogen: A. Photomicrograph of Inconel 718 STA 2 Specimen Tensile Tested at
25 C; B. Fractograph of Inconel 718 STA 1 Specimen Cyclic
Tested at 25 C; and C. Fractograph of Inconel 718 STA 1 Specimen Cyclic Tested at 260 C

be due to an inherently low-strength boundary between the phases. The second
possibility could be the result of different mechanical behavior of the two
phases and the strain concentration required to maintain contiguity at an interface between two materials experiencing a different amount of deformation.
If the deformation of the second-phase particles is sufficiently more difficult than the matrix, the particle becomes a barrier to dislocation movement
and, during dislocation transport, the region surrounding these particles
becomes enriched with hydrogen.

The heat treatment used for the Inconel 718 engine hardware is STA 1
(1038, 760-649 C), which produces a basically single-phase alloy, although
present in the microstructure are γ' and γ'' age-hardening precipitates that
are extremely small and discontinuous. Table I contains the fracture toughness (K_{IC}) and the threshold stress intensity for sustained load crack growth
(K_{TH}) of specimens exposed to 34.5 MN/m^2 (5000 psi) helium and 34.5 MN/m^2
(5000 psi) hydrogen at ambient temperature. Both K_{IC} and K_{TH} were lower in
the STA 2 than in the STA 1 heat treatment conditions; but the most important
factor was the extremely low, 14 MN/m^2 $\sqrt{m}$ (13 ksi $\sqrt{in.}$) K_{TH} of Inconel 718
STA 2 when exposed to 34.5 MN/m^2 (5000 psi) hydrogen.

Table I. K_{IC} and K_{TH} Average Values for Inconel 718
Exposed to 34.5 MN/m^2 (5000 psi) H_2 and He
Environments at Ambient Temperature

HEAT TREATMENT	ENVIRONMENT	K_{IC}, $MN/m^2\sqrt{m}$	K_{TH}, $MN/m^2\sqrt{m}$
STA 2	He	78	48
STA 2	H_2	--	14
STA 1	He	119	112
STA 1	H_2	110	42

Wrought Inconel 718 in the STA 1 heat treatment condition had the third
fastest cyclic crack growth rates in hydrogen, and the threshold stress intensity for significant crack growth during the 9-minute simulated SSME cycles
was about 44 MN/m^2 $\sqrt{m}$ (40 ksi $\sqrt{in.}$) which correlates closely with K_{TH} of
42 MN/m^2 $\sqrt{m}$ (38 ksi $\sqrt{in.}$) for sustained-load crack growth in 34.5 MN/m^2
(5000 psi) hydrogen. Crack growth measurements performed in hydrogen at various cycle durations showed a linear increase of da/dN with cycle duration for
K > 44 MN/m^2 $\sqrt{m}$ (40 ksi $\sqrt{in.}$). Thus, there is considerable evidence that
the 9-minute cyclic crack growth in hydrogen was due mainly (if not entirely)
to sustained-load crack growth.

The fracture of wrought Inconel 718 in the STA 1 heat treatment condition was ductile and transgranular when cyclic loaded in air and mainly intergranular when cyclic loaded in hydrogen. The degree of intergranular fracture produced during the 9-minute cycle was related to the specimen type.
Compact specimens had the least intergranular fracture followed by side-
grooved TDCB specimens and the PTC specimens, which fractured completely intergranularly (Fig. 3B). The stress geometry ahead of the crack was more
nearly plane strain for the PTC specimens than for the compact specimens, and
the crack growth rates obtained on the PTC specimens were about two orders of
magnitude faster than obtained on the compact specimens.

At elevated temperatures, the hydrogen-induced fracture changed from
intergranular to transgranular and, at 260 C (500 F) crystallographic facets
dominated. At 260 C (500 F) the fractography examination indicated a pre-
dominance of chevron-like fracture traces as shown in Fig. 3C. The chevron

angles measured between 60 and 68 degrees. Based on these measurements, the
fracture planes could be either {111} or {110}. Further experimental work
is contemplated to definitively identify the fracture planes.

It can be postulated that hydrogen absorption at ambient temperature is
restricted to the grain boundaries causing a preferred intergranular fracture
mode in this temperature region. The accumulation of hydrogen at the grain
boundaries may be the result of hydrogen transport by dislocation hydrogen
sweeping (Ref. 2) and pileup of dislocations at the grain boundaries. Intergranular hydrogen permeation should also be more rapid than bulk permeation
and the grain boundaries can also be a sink for hydrogen recombination
poisons, such as sulfur, which would accelerate intergranular hydrogen permeation. The change to transgranular fracture at elevated temperatures may be
because hydrogen permeation and diffusion increases with increasing temperature, resulting in a more even dispersion of hydrogen throughout the lattice.

Cast Inconel 718 STA 1 had the slowest crack growth in high-pressure
hydrogen for all of the Inconel 718 conditions tested. Overall, there was a
small accelerating effect of hydrogen on crack extension in cast Inconel 718,
but one of the three specimens tested in hydrogen had the same crack growth
rates as obtained on the specimen tested in air.

The cast Inconel 718 microstructure was large grained with carbide particles outlining the subgrain boundaries. The fractures produced in both air and
in hydrogen contained both ductile and relatively brittle appearing regions
(Fig. 4A), and there was little discernible difference between the fractures
produced in air and in hydrogen.

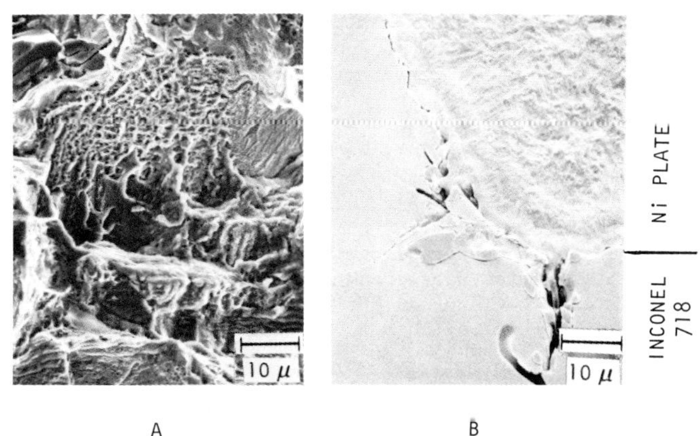

Fig. 4 – Metallography of Cast Inconel 718 Specimen Cyclic Tested
in Hydrogen: A. Fractograph of Typical Region and B. SEM
of Fracture Periphery

To relate the fracture features to the microstructure, a specimen tested
in hydrogen was sectioned, nickel plated to preserve the fracture edge, and
the metallographic mount examined by scanning electron metallography. Figure
4B is an SEM showing the microstructure at the fracture line. It is evident
from Fig. 4B that the fracture path followed the subgrain network, and that
the fracture proceeded both around and through the carbide particles. Correlation with the fractography features (Fig. 4A) suggests that the brittle areas

were associated with propagation around or through the large carbides, and the ductile areas were associated with microvoid initiation by colonies of small carbide particles.

Waspaloy is another age-hardened, high-strength, nickel-base alloy that is embrittled by hydrogen. The cyclic crack growth measurements were performed on TMP (thermal mechanical processed) Waspaloy. TMP Waspaloy is solution-annealed at 1010 C (1850 F) maximum, which is below its recrystallization temperature, and duplex aged at 843 and 760 C (1550 and 1400 F). The microstructure of TMP Waspaloy is duplex (Fig. 5A) consisting of a network of very small grains surrounding large, somewhat elongated grains.

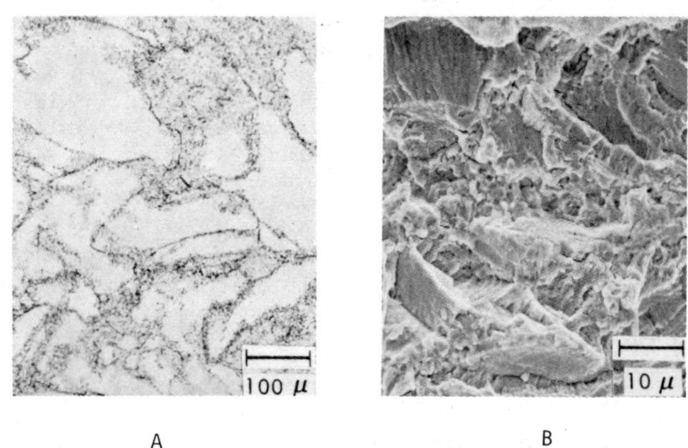

Fig. 5 – Metallography of TMP Waspaloy Specimen Cyclic Tested
in Hydrogen: A. Photomicrograph Showing Microstructure and B. Fractograph

Figure 5B is a typical fractograph of subcritical crack growth of a compact Waspaloy specimen cyclic loaded in hydrogen. The fracture consists of transgranular brittle facets separated by somewhat ductile, fine-featured areas that may be associated with fracture around small carbide particles or through the small grains surrounding the large recrystallized grains. It can be postulated that the tortuous crack growth through the fine-grained areas in the microstructure would produce the slower crack growth rates obtained for TMP Waspaloy compared to Inconel 718, which fractured intergranularly in hydrogen.

Haynes 188 is a non-age-hardened, single-phase, cobalt-base alloy that was tested in high-pressure hydrogen, and the test results showed that the crack growth rates were comparable to wrought Inconel 718 at low-stress intensities, but were slower than Inconel 718 at intermediate-stress intensities. There are a number of areas in the SSME where Haynes 188 is welded to Incoloy 903, and the hardware is then given the Incoloy 903 STA 2 (954, 718-621 C) heat treatment. To determine the likelihood of crack propagation in these areas, a Haynes 188 TDCB specimen in the STA 2 heat treatment condition was cyclic tested in 34.5 MN/m^2 (5000 psi) hydrogen.

Figure 6 is an electron fractograph of subcritical crack growth in Haynes 188. Crack growth throughout the subcritical crack growth range was mainly intergranular with transgranular facets intermixed with intergranular areas and with a total absence of ductility. The fracture was therefore similar to wrought Inconel 718 at ambient temperature.

In actuality, the da/dN data in Fig. 1 suggest that Haynes 188 has a lower threshold stress intensity for significant hydrogen assisted crack growth than wrought Inconel 718. At ambient temperature, Haynes 188 is considerably weaker than Inconel 718, having half of Inconel 718's yield strength. The size of the plastic zone ahead of the crack is determined by the stress intensity and yield strength; and, at the same stress intensities, the plastic zone in the Haynes 188 specimen would be considerably larger than in the Inconel 718 specimens. Thus, the slower crack growth in Haynes 188 at the intermediate stress intensities may be caused by development of a large plastic zone in front of the crack.

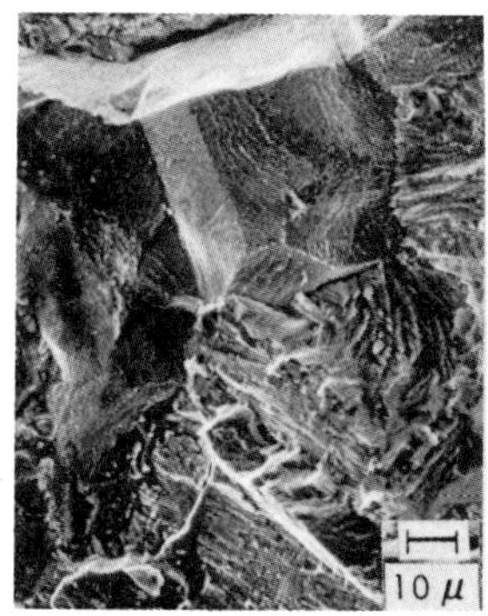

Fig. 6 – Fractograph of Subcritical Crack Growth in Haynes 188 Specimen Cyclic Tested in Hydrogen

Incoloy 903 ranks with cast Inconel 718 as having the slowest crack growth rates obtained in hydrogen, but the rates were, in both cases, somewhat faster than obtained in ambient temperature air. Incoloy 903 is similar to wrought Inconel 718 in that they are both age-hardened iron-nickel alloys and have similar ambient and elevated temperature tensile properties. Incoloy 903 was, however, designed for low-thermal expansion and contains no chromium, and ambient-temperature tensile tests indicate that it is resistant to hydrogen environment embrittlement. The crack growth specimens received the conventional Incoloy 903 heat treatment STA 2, which produced an equiaxed, single-phase microstructure.

The fractography of Incoloy 903 and wrought Inconel 718 specimens were virtually the same when cyclic loaded (9-minute SSME cycle) in air at ambient temperature. That is, the fractures consisted of various sized dimples. The fractures of Incoloy 903 specimens tested in hydrogen were also basically ductile, but comparison with the air fracture suggests a less ductile fracture mode in hydrogen than in air. That is, the large dimples in hydrogen were comparatively shallow, and some of them contained flat brittle-appearing fracture facets. The dimple sizes appeared to be basically the same in hydrogen and in air. Thompson (Ref. 3), on the other hand, showed a 60% reduction of dimple size resulting from tensile testing hydrogen precharged Incoloy 903 specimens.

Fractography examinations also revealed brittle regions not observed during examination of the specimen tested in air. Figure 7 shows one of these regions that consists of transgranular facets separated by colonies of microvoids.

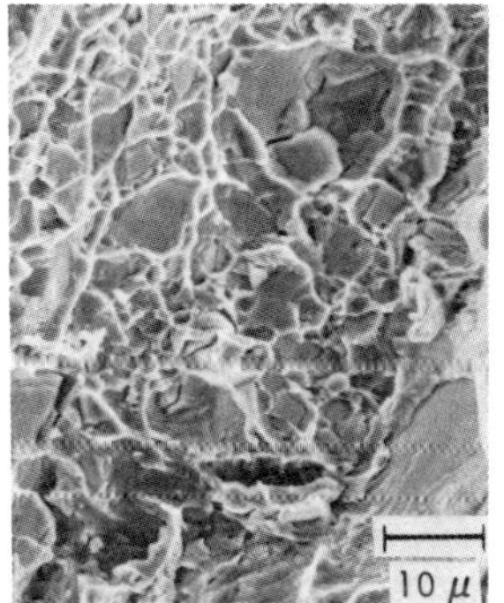

Fig. 7 – Fractograph of Brittle Subcritical Crack Growth Region in Incoloy 903 Specimen Cyclic Tested in Hydrogen

EDAX analysis showed the brittle facets to be high in Ti, Al, and Cb, indicating intermetallic compounds. It appears, therefore, that the hydrogen environment promoted fracture of the relatively large intermetallic particles present in the microstructure.

Despite the lack of hydrogen environment embrittlement of Incoloy 903 specimens that are tensile tested and cyclic loaded at ambient temperatures in high-pressure hydrogen, it is possible to produce hydrogen-induced brittle fracture by high-temperature, high-pressure hydrogen charging. Hollow tensile specimens, pressurized with hydrogen at 538 C (1000 F) and tensile tested at ambient temperature, fractured with a brittle fracture mode around the inside periphery. This would suggest that the resistance of Incoloy 903 to embrittlement is due to lower ambient temperature permeation and diffusion in Incoloy 903 than in Inconel 718, or that a much greater quantity of hydrogen is needed to embrittle Incoloy 903 than Inconel 718. Robertson (Ref. 4) showed very little difference of hydrogen permeation and diffusion between Incoloy 903 and Inconel 718. Thus, the resistance of Incoloy 903 to hydrogen environment embrittlement is evidently related to much greater quantity of hydrogen needed to embrittle Incoloy 903 than Inconel 718, and this hydrogen concentration can most easily be obtained by elevated temperature charging.

Summary and Conclusions

The metallography examination of specimens tested in hydrogen showed that the hydrogen environment produced a brittle fracture mode--either transgranular or intergranular in the following alloys: (1) Inconel 718, (2) TMP Waspaloy, (3) Mar-M-246--conventionally cast, and (4) Haynes 188. The fractures of Incoloy 903 and cast Inconel 718 were basically ductile, although the crack growth data indicated significant increase of crack growth rates due to the hydrogen environment.

Table II lists these alloys in order of their relative crack growth rates, and with each alloy is a brief description of their fracture morphologies. The fastest crack growth rates were associated with fracture in boundaries

Table II. List of Alloys da/dN Tested in Order of Relative
Crack Growth Rate in High-Pressure Hydrogen and Their
Corresponding Fracture Morphology

Relative Crack Growth Rate	Alloy	Fracture Morphology
1	Wrought Inconel 718 STA 2	Transgranular following the γ/Ni₃Cb Interface
2	Mar-M-246 Conventional Cast	Predominantly intergranular fracture that followed the γ/γ' Interface
3	Wrought Inconel 718 STA 1	Predominantly intergranular fracture with some transgranular fracture
4	TMP Waspaloy	Transgranular fracture facets and fine featured regions interpreted as fracture through the continuous small-grained network in the duplex microstructure.
3-5	Haynes 188	Predominantly intergranular fracture with some transgranular fracture facets.
6	Cast Inconel 718	Mixture of brittle transgranular fracture regions and regions containing small microvoids
7	Incoloy 903	Basically ductile fracture with some brittle-appearing areas containing transgranular fracture facets surrounded by very small microvoids

separating the matrix and a precipitate phase. The second fastest crack growth rates were associated with intergranular fracture, and the next fastest rates were associated with transgranular fracture. The specimens with the slowest crack growth rates fractured with considerable microvoid formation.

Acknowledgements

The light metallography was performed by J. A. Gerstmeyer of Rocketdyne and R. A. Spurling of the Rockwell Science Center.

References

1. R. J. Walter and W. T. Chandler: Environmental Degradation of Engineering Materials, pp. 513-22, edited by M. R. Louthan and R. P. McNitt, Virginia Polytechnic Institute Press, Blacksburg, VA, 1977.

2. J. K. Tien, A. W. Thompson, I. M. Bernstein, and R. J. Richards: Met. Trans. A, 1976, Vol. 7A, pp. 821-29.

3. A. W. Thompson: Effect of Hydrogen on Behavior of Materials, edited by A. W. Thompson and I. M. Bernstein, pp. 467-77, TMS-AIME, New York, 1976.

4. W. M. Robertson: Met. Trans. A, 1977, Vol. 8A, pp. 1709-12.

STRESS STATE AND THICKNESS EFFECTS ON DELAYED FAILURE*

M. R. Louthan, Jr., R. D. Sisson, Jr. R. P. McNitt, and P. E. Smith

Environmental Degradation Laboratory
College of Engineering
Virginia Polytechnic Institute
Blacksburg, VA 24061

Tests of quenched and quenched-and-tempered 4340 steel exposed to
6.9 MPa hydrogen showed that the susceptibility to delayed failure is a
function of the stress state at the sample surface and the thickness of
the sample as well as stress level. The minimum principal stress required to cause failure in a 120 hour test of a four point loaded beam
or clamped plate decreased as the specimen thickness increased. For
thin (< 0.25 cm thick) specimens the stress level required to cause
failure was significantly less for "biaxially" loaded (clamped plates)
than for "uniaxially" loaded (four point beams) samples. Analysis of
acoustic emissions during failure indicates that there is a significant
delay time before cracking initiates and that the post initiation failure
process is rapid. Fractographic studies showed large regions of microvoid coalesence, however, regions of intergranular fracture are also
apparent and the amount of intergranular fracture is far greater in
the "biaxially" stressed samples than in the "uniaxially" stressed
samples. These data are discussed in terms of a "critical ligament"
model which assumes that hydrogen induced failure will occur only after
a region of some specific size is embrittled. Stress state and sample
thickness will affect the size of this zone and therefore will affect
delayed failure processes.

*This work is supported by DOE contract E(40-1) 5255 and NASA grant
NAG 2-3.

Introduction

The deleterious effects of hydrogen on the mechanical properties of high strength steels have been well documented (1-5). However, any comprehensive review of hydrogen embrittlement must conclude that test results, although extensive, are contradictary and, to some extent, confusing. Part of the disagreement results from the role of stress state (and hence test technique) in the embrittlement process (6). Because stress state may be important in two very different aspects of degradation, plasticity and dilation, its specific role may vary from system to system. The role of hydrogen in the embrittlement process is also uncertain. Models for embrittlement invoke decohesion, dislocation pinning, reductions in surface energy, internal pressure development, grain boundary effects and slip softening (7). Because of this diversity of potential effects it was recently concluded that "simple mechanisms are not sufficient to rationalize the general degradation --- (and) --- a complex process composed of a number of coupled mechanisms accounts for many of the observed degradation phenonema" (7). During an investigation of the effect of stress state on delayed failure hydrogen embrittlement, specimen thickness was shown to influence the failure processes, including the extent to which plastic deformation accompanied fracture. The observations which are described in this paper are in agreement with the conclusion that hydrogen induced failure generally results from the coupling of several mechanisms.

Materials and Procedures

Disk and beam samples were machined from hot rolled 4340 plate stock which was austenitized at 1115°K for 0.5 hours and oil quenched to produce untempered martensite. The prior austenite grain size was 0.01 mm. The 0.2% offset yield strength of the as-quenched plate was approximately 1200 MPa, the ultimate strength 2100 MPa and the hardness Rc 52-54. The elongation to fracture was $\sim$3% and the room temperature failure mode was typically microvoid coalesence. The surfaces of as-machined samples were hand ground to a 240 grit finish.

Disk specimens were stressed by bending (Figure 1a) as either clamped or unclamped plates so that the tip surfaces of the disk were in biaxial tension or compression. Beam samples were stressed in 4-point bending so that one top surface was in uniaxial tension. Surface stresses were calculated from specimen deflection measurements using plate (disk) and beam theory. Calculations were in excellent agreement with stress levels determined by strain gauge measurements on selected samples. Specimen thickness was controlled by grinding (under coolant) the as-machined samples to predetermined thickness prior to hand grinding the surfaces.

Stressed samples were placed in high pressure hydrogen autoclaves which were evacuated and back-filled with hydrogen several times before being filled with 99.9+% H_2 at 6.9 MPa. Hydrogen purity was controlled by passing the fill gas over an oxygen sieve and through a liquid nitrogen cold trap. The pressurized autoclaves containing the test samples were held at room temperature for 120 hours before the autoclaves were evacuated and the specimens were examined for failure. Selected autoclaves were instrumented for detection of acoustic emissions during the 120 hours exposure.

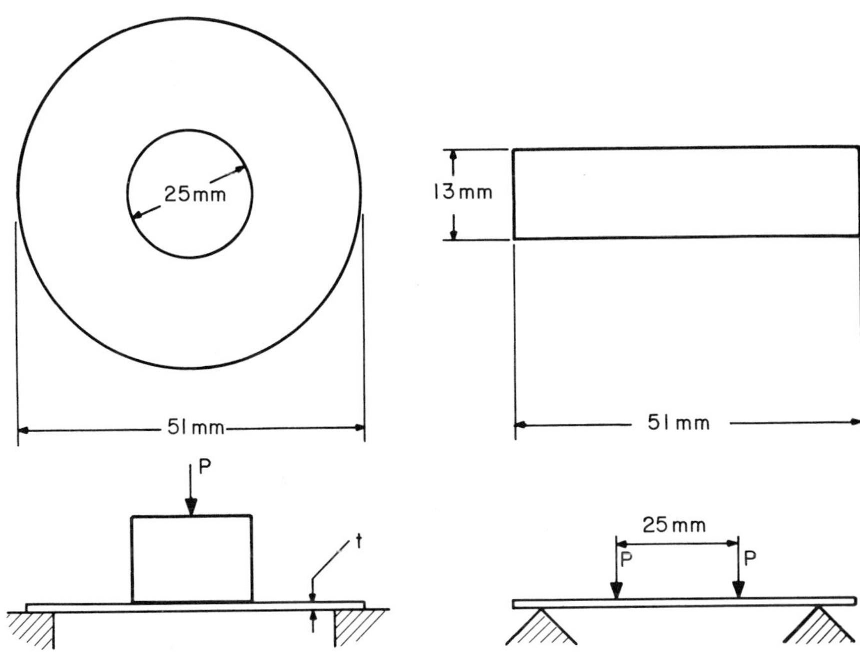

a) Biaxial Loading

b) Uniaxial Loading

Figure 1.   Schematic of Specimens and Loadings.

Failed samples were examined metallographically to determine crack path and fracture mode.

Results

The initial tests with beam and plates samples showed that the susceptibility of the as-quenched 4340 steel to delayed failure was a sensitive function of specimen thickness, macroscopic stress state and maximum stress level (8). The data presented in Reference 8 were extended by the present investigation and both sets of data are summarized in Figure 2. Although there was significant scatter between the maximum survival stress and the minimum failure stress; these results, which represent over 75 individual samples, clearly show that the maximum surface stress necessary to cause failure in 120 hours decreases as the specimen thickness increases. It is also apparent that, for thin samples failure is promoted by a biaxial stress state.

Post failure examination of the beam samples suggested the following failure sequence: crack initiation occurred by intergranular cracking in the maximum stress region. The macroscopic orientation of the cracks was perpendicular to the principal stress axis and intergranular crack growth proceeded until a critical flaw was obtained and the remainder of the specimen failed by ductile rupture processes. This process is shown schematically in Figure 3.

Analysis of acoustic emission data indicates that, once initiated, the failure process was very rapid. The sample was apparently silent throughout most of the exposure sequence and no emissions were detected. After this incubation period one to three periods of large scale emissions were noted. These acoustic emissions lasted for a total of less than one second with each period of emission being about 0.01 sec. Samples which were removed from exposure after the emissions were found to be in two pieces. Examination of samples which were removed from exposure prior to the onset of acoustic emission failed to reveal any evidence of fracture. No partially fractured samples were produced. This metallographic evidence, coupled with the acoustic measurements suggest that failure occurred by a nucleation and growth sequence and that the growth period was less than one second in duration. The nucleation period tended to increase as the maximum stress level decreased; however, sufficient data were not obtained to develop an empirical relationship between stress level and time to crack nucleation. The failure sequence for the biaxially loaded samples was similar to that of the uniaxial samples except for the large region of intergranular fracture which was observed. Acoustic emissions indicated a slower growth process because multiple emission pulses of several seconds in duration were detected. Furthermore, several samples were removed from exposure with the failure process only partially complete. However, examination of samples which were removed from test prior to the onset of acoustic emissions did not reveal any evidence of cracking. A schematic of failure sequence in a biaxially stressed sample is shown in Figure 4.

X-ray analysis of the fracture surfaces showed peak broadening. Such broadening could result from either the rough surface produced by the fracture process or from the localized plastic deformation which accompanied fracture. Microvoid coalescence is a ductile (plastic) fracture process and was observed on all fractured samples. Comparison of the X-ray peak profiles from impact fractured samples and from the hydrogen fractured samples revealed similar peak shapes. Thus, the observed broadening may

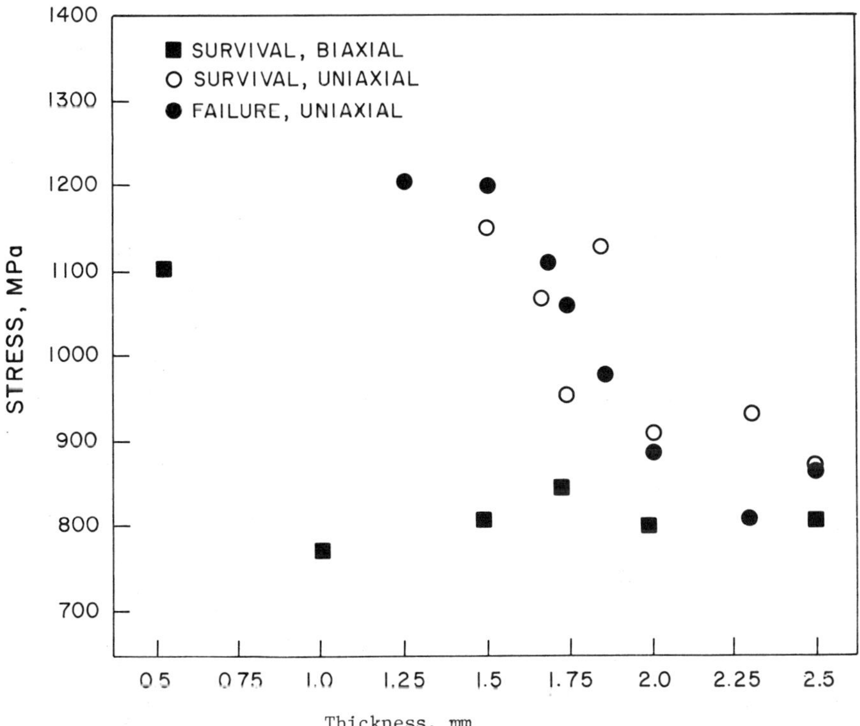

Figure 2. Minimum failure stresses and maximum survival stresses for 4340 samples of various thicknesses.

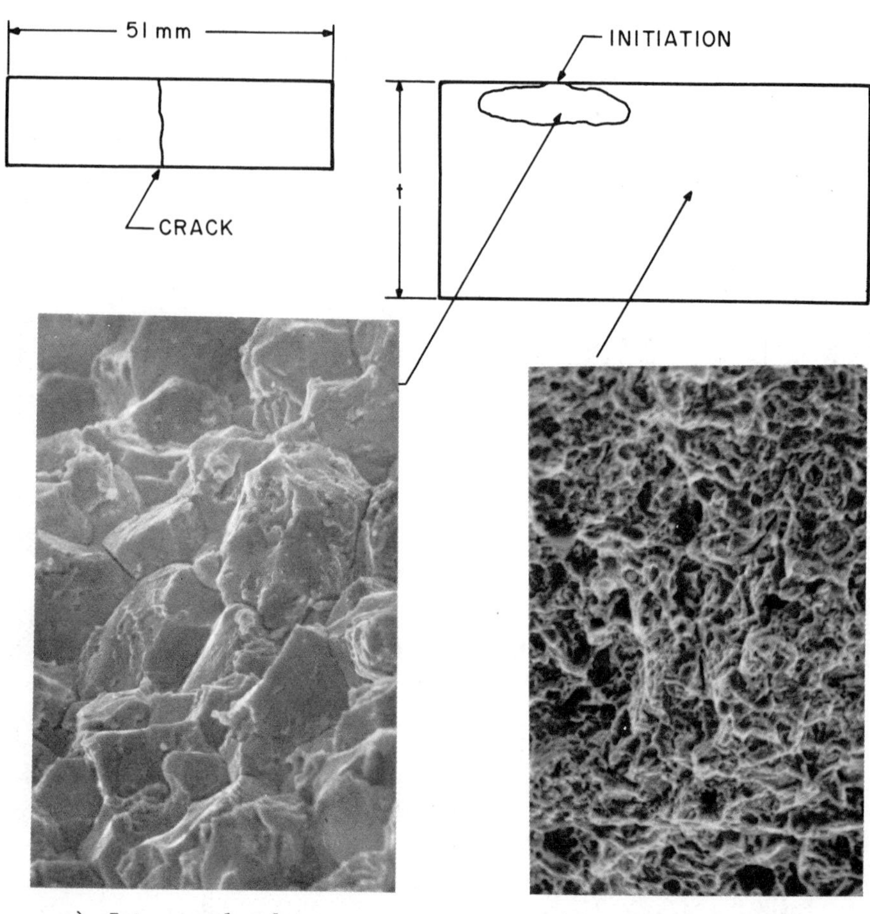

Figure 3.   Fracture Process in Uniaxially Loaded Sample.   Crack initiation began at specimen surface; crack propagated along the grain boundaries until overloaded fracture occurred by microvoid coalescence.

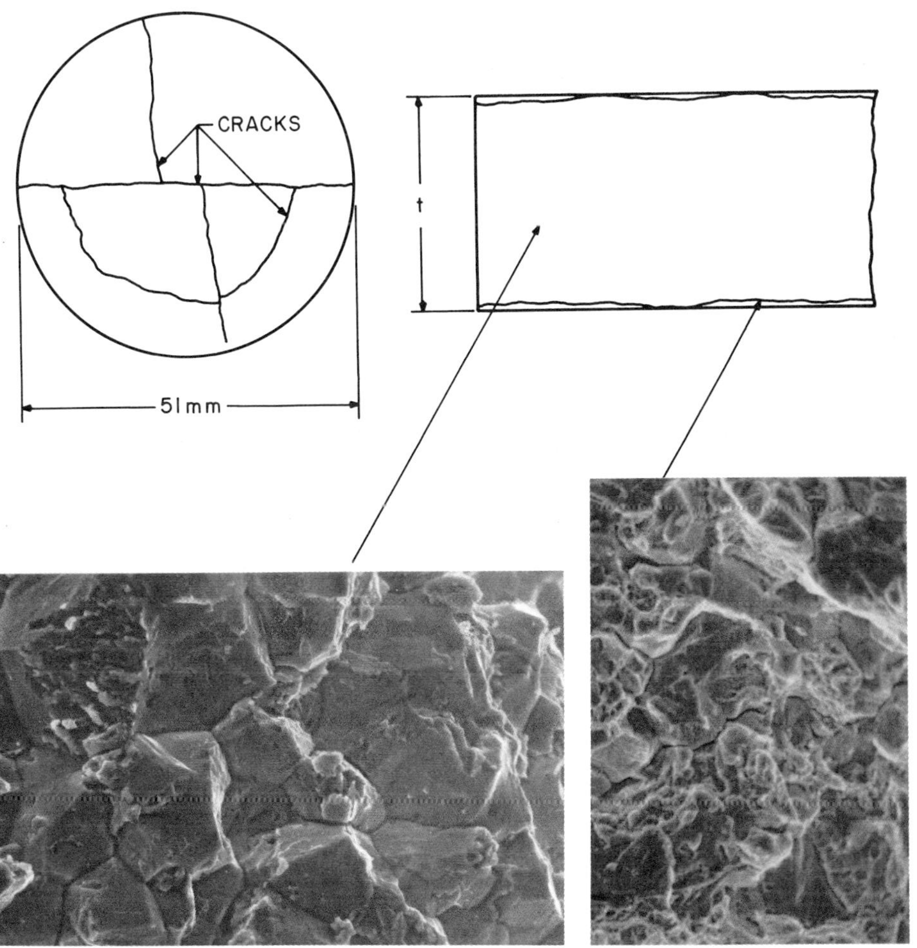

a)   Intergranular cracking   b)   Mixed mode failure

Figure 4.   Fracture Process in Biaxially Loaded Sample.   Crack initiation and propagation occurred by intergranular fracture.   Only very limited regions of microvoid coalescence were found in the near surface area.

simply result from the roughening and plasticity which occurred during overload fracture.

Discussion

The results of this study show that stress state and specimen or test
design will have a significant influence on the apparent susceptibility of
metals and alloys to hydrogen embrittlement. Biaxial stresses were shown
to promote embrittlement. This promotion was apparent both in the stress
level required to cause hydrogen damage and in the extent of intergranular
fracture which accompanied failure. These results, coupled with the
observed incubation period, cannot be explained in terms of dislocation
transport, internal pressure development, slip softening models, or hydrogen
induced reductions in surface energy.

The samples were strained prior to exposure to the hydrogen environment and no additional load was applied. Thus, throughout these tests,
little or no dislocation movement should occur until the specimens began
to fracture unless hydrogen absorption softened the lattice. If hydrogen
absorption induced softening and thereby promoted dislocation motion, such
effects should be enhanced in the uniaxially stressed samples which have
both in plane and out of plane shear stresses. Only out of plane shear
stresses exist in the biaxially loaded specimens. Therefore, the enhanced
embrittlement in the biaxially loaded samples indicate that dislocation
transport and slip softening did not play significant roles in these
hydrogen induced delayed failures.

The long term, and apparently reversible, incubation period which
proceeded the onset of hydrogen damage is not compatible with hydrogen
induced reductions in surface energy. The external surfaces of the samples
contained the highest stresses and were immediately exposed to hydrogen
when the autoclave was pressurized. If hydrogen reduced the surface
energy of the steel and promoted crack growth from surface flaws this
reduction should be almost immediate and long term incubation periods
should not be expected. Embrittlement by internal pressure buildup can
also be discounted by these observations because the samples were stressed
to a constant strain prior to exposure to the high pressure gas. Pressurization of the autoclave placed the sample under hydrostatic pressure
which could conceivably lower localized bulk stresses in the region of
macrovoids and other crack-like defects. The diffusion of hydrogen to
these voids and crack-like defects could raise the stress. However, the
increase in stress due to diffusion of hydrogen will, at a maximum,
simply equal the lowering which results from the external hydrostatic
pressure, thus, no net increase in stress could be achieved. Therefore,
it is concluded that the results of this study are not compatible with
many of the models for hydrogen embrittlement.

Lattice dilation is assumed to be important in the embrittlement because of the enhanced susceptibility in the biaxially loaded samples.
Clearly, the fractographic data emphasizes the importance of grain boundaries
and the incubation period shows that a critical hydrogen concentration must
be reached at some critical depth in the sample before embrittlement will
result. Rationalization of the thickness effect also suggests that embrittlement must occur over some critical depth or ligament. A critical
ligament model has been developed to explain the temperature dependence
of K_{IC} (9). This model predicts that a failure will occur when a critical
failure stress, σ^*, exists over a microstructurally significant depth into
a sample, a. This characteristic microstructural parameter is postulated

to be greater than a grain diameter and has been estimated to be 50 grain
diameters (10).

   The stress through the cross section of an elastically loaded four
point bend sample increases linearly with distance from the neutral axis.
Thus the ratio of the surface stress, σ_m, to the stress at any depth, σ,
must equal the ratio of the specimen thickness, t, to t - a where a is
the depth of which σ is measured. If σ is equated to some critical stress
below which hydrogen embrittlement might not occur, σ^*, the following
equation is obtained:

$$\frac{\sigma_m}{\sigma^*} = \frac{t}{t-a} \tag{1}$$

In this equation σ_m is the minimum surface stress required to provide a
zone depth a which has stresses above the critical stress for hydrogen
embrittlement. If a is related to microstructural features and is a
constant for a given material and loading, then σ_m is clearly a function
of specimen thickness. Empirical fitting of the data in Figure 2 to
Equation 1 gives values of several hundred grain diameters. Although
such values are larger than the 50 grain diameters estimated for overload fracture obtained by considering the probability of favorable grain
orientation for slip, they may be reasonable if one considers that in
this case the probability is not for slip but for grain boundary fracture
because of hydrogen accumulation. If this rationalization is correct, the
critical zone must decrease when biaxial stresses are applied. Such a
decrease is to be expected because the probability of favorable grain
orientations should increase as the stress state goes from uniaxial to
biaxial.

Conclusions

   The results of this study show that delayed failure hydrogen embrittlement is promoted by biaxial stresses. The data are not consistent with
embrittlement models which involve dislocation transport, slip softening,
internal pressure buildup or hydrogen induced reductions in surface energy.
However, the results can be rationalized through a model which invokes
hydrogen accumulation at grain boundaries. This model assumes that specific
boundary-stress axis orientations are more susceptible to embrittlement than
other orientations and that hydrogen induced failure only occurs when
hydrogen has been accumulated to some critical concentration in a region
of some specific (yet undetermined) size. Estimates of the size of this
region suggest that tens to hundreds of grains are involved.

References

1)   Troiano, A. R., _Trans. ASM_, 1960, Vol. 52., p. 54.

2)   Lee, T. D., T. Goldenberg and J. P. Hirth, _Met. Trans. A,_ 1979,
     Vol. 10A, p. 439.

3)   Gerberich, W. W. and Y. T. Chen., _Scripta Met_, 1974, Vol. 8, p. 243.

4)   Oriani, R. A., _Berichte der Bunsen-Geselischalt fur Physikalische
     Chemic_, 1972, Vol. 76, p. 848.

5)   Bernstein, I. M. and A. W. Thompson, <u>Int. Met. Rev.</u>, 1976, Vol. 21, p. 269.

6)   Louthan, M. R., Jr. and R. P. McNitt, <u>Effects of Hydrogen on Behavior of Materials</u>, 1976, TMS-AIME, p. 496.

7)   Hirth, J. P., <u>Met. Trans</u>, 1980, Vol. 11A, p. 861.

8)   Smith, P. W., M. R. Louthan, Jr. and R. P. McNitt, <u>Scripta Met</u>, 1980, Vol. 14, p. 199.

9)   Rice, J. R., <u>Journal of the Mechanics and Physics of Solids</u>, 1972, Vol. 21, p. 395.

10) Auerbach, B. L., Chester, <u>Fracture</u>, Vol. 1, Edited by H. Liebowitz, Academic, Press, New York (1969).

DISCUSSION

<u>R. Stoltz, Sandia Laboratories, Livermore, CA</u>: Do you think that the differences in uniaxial and biaxial fracture stress, is due to differences in the mechanical instability of a uniaxial vs. biaxial sheet loading?

<u>M.R. Louthan, Jr.</u>: Not really. I believe that the difference is due to the associated lattice dilatation and hydrogen interactions with grain boundaries on the regions of dilatation. Mechanical instability may play a role on the overall fracture process, but I feel our results have shown that failure is nucleation-controlled for the samples we studied.

HYDROGEN INDUCED DEGRADATION OF DUCTILE DISPERSED -

PHASE ALLOYS OF IRON AND OF NICKEL

M.W. Joosten
Conoco, Inc.
Ponca City,
Oklahoma 74601

T.D. Lee
Korea Institute of Science
   and Technology
Dong Dae Mun,
Seoul, Korea

T. Goldenberg
Bell Laboratories, Inc.
Allentown, Pennsylvania
          18103

and   J.P. Hirth
Metallurgical Engineering Department
The Ohio State University
Columbus, Ohio 43210

A spheroidized 1090 steel alloy and a thoria-dispersion nickel alloy were
subjected to hydrogen degradation by electrolytic charging at various temperatures and charging current densities. The degradation was revealed in
U-notch bend tests. The qualitative trends in previous work on steels included mode II cracking along characteristic slip traces after precharging
and mode I cracking at drastically reduced strains during dynamic charging.
These trends were followed in the present work. The trends with temperature for the steel help in explaining diverse fracture patterns for such
materials under the influence of hydrogen as found by different workers.
Cracking in the nickel alloy was markedly influenced by directionality
introduced in processing.

I. Introduction

Several groups[1-5] have recently studied the degradation of spheroidized
plain-carbon steels by cathodically precharged hydrogen. All are in agreement that the major effect of hydrogen is to decrease the ductility by
enhancing the void formation-growth-coalescence process leading to ductile
fracture. However, the various studies[1-5] disagree on the details of the
fracture process.

Lee et al[1] performed blunt U-notch plane-strain bend tests on AISI 1095
steels precharged at $100A/m^2$ or less in 1N sulfuric acid with a cathode recombination poison. They found that the major effect of hydrogen was to
promote the onset of plastic instability at the U-notch surface. The
plastic instability led to void formation, and subsequently mode II shear
cracking, along characteristic slip traces at surface strains reduced by a
factor of two relative to the uncharged case. They found a major effect of
hydrogen on void nucleation. The critical strain for void initiation and
for crack initiation was independent on notch curvature, and hence on internal stress concentration, since these events were initiated at the surface.
Cialone and Asaro[2] performed smooth and notched round-bar tensile tests on
1045 steel also charged at a high hydrogen fugacity. They studied void
nucleation by the same methods as Argon et al[6], noted that hydrogen enhanced void nucleation at carbides, and determined that it reduced the
calculated value of the cohesive strength of the carbide interface from 950

MPa to 550 MPa. They found that hydrogen had no direct effect on void growth, the larger voids at a given strain simply arising from earlier void nucleation. In this case, void coalescence occurred by growth and linkage of voids along subboundaries containing the carbides that nucleated the voids, the voids elongating parallel to the tensile axis, a mode of growth in sharp contrast to the cracking along characteristic slip traces found by Lee et al.

Oriani and Josephic[3] studied sheet tensile specimens of spheroidized 1045 steel charged at a hydrogen fugacity of 100 MPa. In agreement with the above works, they found that hydrogen enhanced void nucleation at carbide particles but had little or no effect on void growth. The void configurations resembled those of Cialone and Asaro. Garber et al[4,5] analyzed round-bar tensile tests of 1020 and 1080 steel charged at high fugacity, again using the Argon analysis as well as studying dimple sizes on fracture surfaces. In contrast to the other studies, they found no effect of hydrogen on void nucleation or void growth. Hydrogen did increase void growth in the void coalescence stage. Also in contrast to some of the other work[2,3], but in agreement with Lee et al,[1] void coalescence did not occur along the subboundaries. In their later work these authors performed void counts as in the other studies. In their earlier work[4] they studied dimple size on fracture surfaces. The dimple studies are likely to be less reliable, partly because of uniaxial plastic strain biasing the void shapes and complicating quantitative metallographic analysis[7]. Also since many of the voids form in the final stages of fracture during crack propagation, the state of local plastic strain is even more complex and more difficult to analyze.

The disparate results of these studies illustrate the confusing complexity of the role of hydrogen in ductile fracture and the difficulty in obtaining a unified model for its effects. Accordingly, the present work on bend tests of spheroidized steel at low and elevated temperatures were undertaken to determine whether temperature changes could influence the different crack morphologies observed in the earlier work. Experiments[8] for spheroidized steels indicate a change from ductile to a brittle cleavage fracture with decreasing temperature at $\sim$103K in air. Such a trend is expected theoretically[6] and would be expected at a temperature nearer ambient with hydrogen present.

Also tests were conducted at room temperature on a ductile, dispersed-thoria, nickel-base alloy to study a material with a different size scale of the dispersed second phase for comparison with the steel work. Work on a similar alloy[9] had shown that, in comparison to single-phase nickel alloys, the fine dispersion in a T-D nickel alloy markedly reduced the degradation produced by precharging with hydrogen. In addition to precharged hydrogen, dynamically charged hydrogen tests were performed. The fracture mode of the dynamically charged, nickel and steels changes from a ductile mode II to a brittle mode I with a marked reduction in strain to fracture compared to the precharged case.

II.   <u>Experimental Procedure</u>

Rough-machined samples of hot-rolled bars of AISI 1095 steel were austenitized for 2 h at 750°C, air-cooled, tempered for 20 h at 700°C, and air cooled. The composition and tensile properties of the resulting spheroidized structure are listed in Table I. The structure consisted of a volume fraction of 0.15 of cementite particles of 0.30 μm average diameter. The samples were finish machined by grinding to produce bend specimens of the type illustrated elsewhere[1] with U-notches of 1.19 mm root-radius, the longitudinal axis of the bend specimens corresponding to that of the initial bar.

The notch surfaces were polished to a 600 grit finish. All areas of the specimen except the notch surface were coated with a non-conducting vinyl coating.

Disperson-strengthened (DS) nickel was produced by Sherritt-Gordon by extrusion, cold swaging and annealing. T D nickel is processed similarly but is given further working and heat-treating operations which change the mechanical properties so the behavior of D S nickel should not be associated directly with that of T D nickel. The composition and tensile properties of the D S nickel alloy are listed in Table I. The structure consisted of thoria particles with diameter ranging from 10 to 200 μm, biased toward the smaller size range, in elongated grains with narrow dimensions ranging from 0.2 to 1.0 μm. Bend specimens were prepared in the same manner as for the steels.

The steel samples were preheated or cooled and then immersed in constant temperature solutions of 1N and 7N, respectively, sulphuric acid containing 1g/1 of thiourea, a hydrogen recombination inhibitor. The notch surfaces were charged at a current of 100 A/ m^2 at -25°C or 10 A/ m^2 at 70°C. Samples were precharged with hydrogen for 2.5 h, the current stopped, and immediately strained in-situ or dynamically charged by starting charging and straining simultaneously. A few samples were tested in open-circuit conditions (1 A/ m^2 current density, -258 mV potential, hydrogen scale). The samples were bent in three-point loading at an extension rate of 8.3 μm/s to loadings from the yield point to fracture in 100 kg increments. The plastic deflection angle was measured and converted to notch-root principal strain as in the earlier work[1]. The specimens were cut in half longitudinally and perpendicular to the notch to expose the deformed plane-strain region. The sections and the fracture surfaces were studied by scanning electron microscopy.

The charging conditions correspond to a high hydrogen fugacity[1,10] ($\sim$ 1 GPa) at the surface. As with the room temperature work[1], it was verified that 2.5 h charging at 100 A/ m^2 produced no voids or blisters at -25°C. At 70°C, such charging did lead to blistering, but no voids or blisters were formed in 2.5 h charging at 10 A/ m^2, the current adopted for the subsequent tests.

The D S nickel specimens were tested in the same solution at room temperature at a charging current of 100 A/ m^2. The details of the experimental procedure were the same as for the steels. The open-circuit current density was 1 A/ m^2 at a potential of -123 mV, standard hydrogen scale.

III. <u>Results</u>

A. Steels

The load-deflection values for fracture at -25°C and 70°C in the uncharged condition were in good agreement with the room temperature results for a similar steel[1], so the strains for the uncharged case determined in the earlier study are used as baseline data. The overall characteristics of the hydrogen degradation were the same as for the earlier room-temperature studies[1] at both temperatures. After precharging, the specimens experienced a loss of ductility, and fracture was initiated at the notch-root as a mode II crack. The cracks were preceded by void formation along the logarithmic spiral characteristic slip traces of plasticity theory illustrated for example in ref. 11. These traces also were followed by the mode II cracks, which occasionally changed path from one set of traces to another as they propagated as also observed in earlier work[1, 12].

In the dynamic charging case, a greater loss of ductility was experienced and fracture initiated at the notch root as a mode I crack. These trends are illustrated in Table II which presents the critical strains for void and crack initiation for the various cases. The trends in the load–deflection data[13] were similar to those at room temperature so, since they are less relevant to the discussion than the critical strain values, they are not presented here.

Thus, the overall degradation observed in the room temperature case is reproduced by the studies at –25°C and 70°C. The detailed results at 70°C closely correspond to those at room temperature, while those at –25°C differ in some respects. At 70°C, as at room temperature[1], void initiation occurs in precharged specimens between closely spaced carbide particles, Fig. 1, aligned along the direction of action of the principal normal stress. These voids do not preferentially form on subboundaries or grain-boundaries. The initial portion of the fracture surface is ductile-dimple-rupture as at room temperature. Also in agreement are the observations of quasi-cleavage mode I fracture in the dynamic charging case and the absence of crack branching for either the precharged or dynamically charged case.

For the –25°C case, the precharged samples, while still exhibiting a mode II type crack, exhibited some differences in metallography. The voids in this case were fewer in number and more randomly distributed than in the other cases. Also some carbide cracking was observed near the propagating crack tip together with the more common voids at decohered carbide interfaces. The fractographs revealed a mixed fracture surface composed of ductile tear ridges (with dimple rupture patterns) together with brittle-type facets, which would be associated with separation along boundaries, Fig. 2. The dynamically charged specimens at –25°C exhibited branching cracks, Fig. 3, and a definite tendency for cracking along grain and subboundaries, Figs. 3 and 4, in contrast to all other cases. The latter fractures are the only ones in the present work which resemble those in the work of Cialone and Asaro[2], although the hydrogen fugacity should be much greater in the present case.

B.   D.S. Nickel

In the uncharged condition, the D S nickel failed by formation of a mode II crack which propagated along characteristic shear traces in an analogous manner to the spheroidized steel. Such a crack is illustrated in Fig. 5. Voids formed at the small thoria phases at sites of maxim μm incompatibility stress, i.e. along the longitudinal direction in the samples, as shown in Fig. 6. The fracture surface was a ductile dimple rupture type as shown in Fig. 7, where the small dimples correspond to the smaller thoria particles while the large dimples correspond to the few large thoria particles that reach the size of 200 μm. The void formation and crack initiation occurred virtually at the same critical maximum principal surface strain at the notch root of 0.21. The same critical strain value was observed for notches with notch root radii of 0.59, 1.19 and 1.59 mm, independent of the different internal stresses for the different cases. This behavior is expected since the cracks are surface initiated and is analogous to that of the spheroidized steel[13].

Precharging the D S Ni with hydrogen for 2 h produced no measurable change in mechanical properties. The load-deflection data for the 1.19 mm notch radius specimens were identical to those for the uncharged case as were the fractographic appearances and the fracture mode. Dynamic charging, in contrast, produced a dramatic reduction in ductility, a change to mode I crack propagation, and a change in fractographic appearance to quasi-cleavage. The directionality introduced by working influenced the crack path as shown in Fig. 8. The crack plane followed the longitudinal trace of the elongated grain struc-

TABLE I.   Composition (wt. %) and tensile properties, determined at an 8.5 μm/s elongation rate, for the experimental alloys.

Composition (w/o)	C	Mn	P	S	Si	Fe		
1095 Steel	0.98	0.70	0.010	0.040	0.30	bal.		

D S Nickel	S	C	Cr	Cu	Fe	ThO_2	Ni
	0.0011	0.0040	0.002	0.002	0.022	2.1	bal.

Tensile Properties

	U.Y.P.(MPa)	L.Y.P.(MPa)	U.T.S.(MPa)	El.(%)	R.A.(%)
Steel	387	343	574	30	65
D S Nickel	---	560	820	8	75

TABLE II.   Critical value of the maximum principal plastic surface strain at the notch root at which void formation and crack initiation was observed.

	Void Initiation	Crack Initiation
Uncharged (all temp.)	0.15	0.40
70°C Precharged	0.12	0.17
Dynamically charged	0.09	0.17
-25°C Precharged	0.11	0.15
Open-Circuit	0.15	0.16
Dynamically charged	0.03	0.07
Room temp.[1] Precharged	0.07	0.23
Open-circuit	--	0.12
Dynamically charged	0.05	0.05

Fig. 1. SEM micrograph of voids forming between two carbide particles in close proximity.

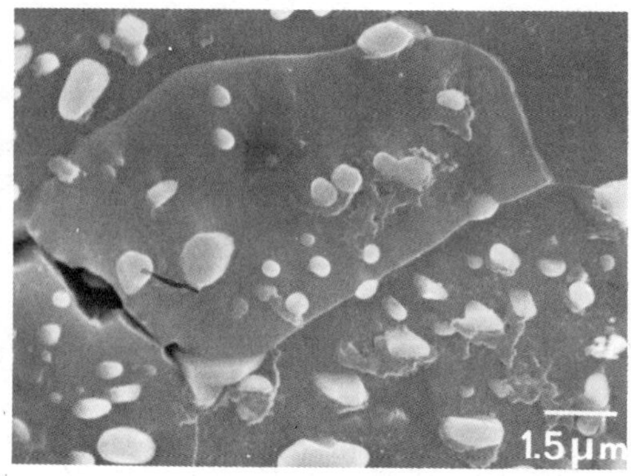

Fig. 3. SEM micrograph of crack branching along subboundaries in a steel sample dynamically charged at -25°C.

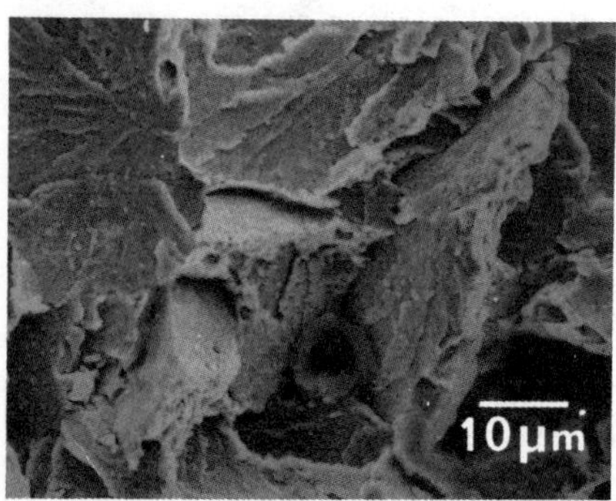

Fig. 2. SEM fractograph of a steel sample precharged at -25°C.

Fig. 4. SEM micrograph of crack progressing along a grain boundary in a steel sample dynamically charged at -25°C.

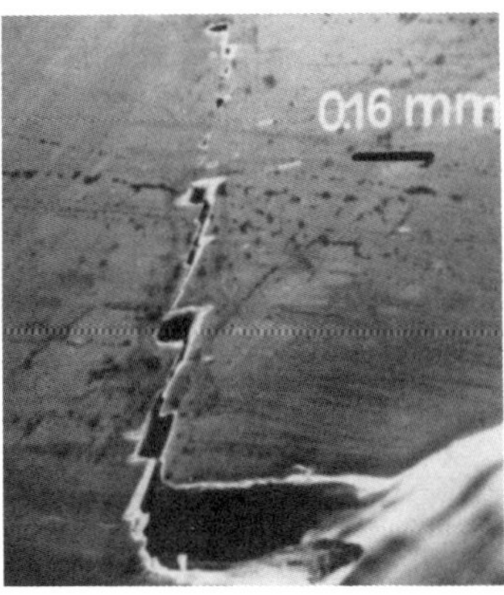

Fig. 5.   SEM micrograph showing crack
propagation along characteristic slip traces in D S
nickel.

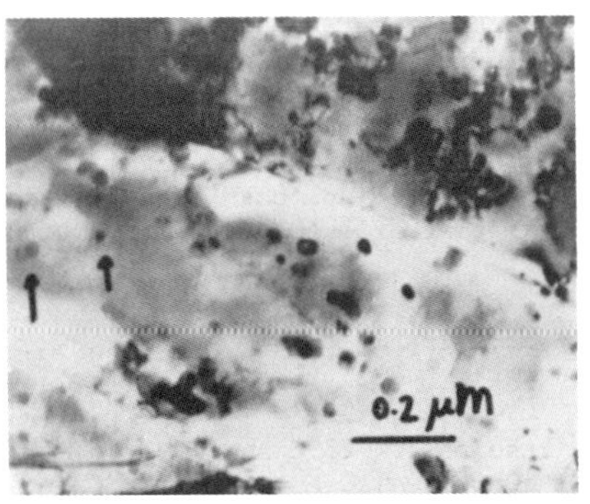

Fig. 6.   TEM micrograph showing void
nucleation at the interface
of thoria particles in D S
nickel.

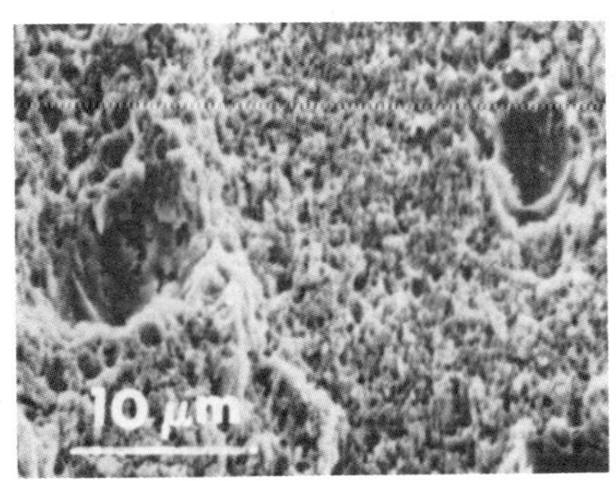

Fig. 7.   SEM fractograph of D S nickel
in mode II region.

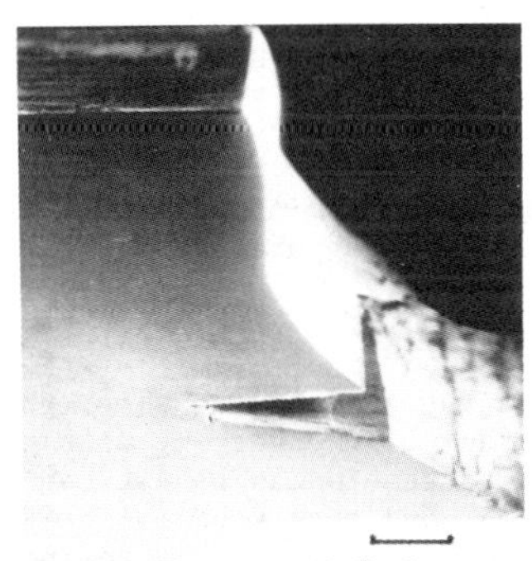

Fig. 8.   SEM micrograph of dynamically
charged D S nickel showing
mode I crack.

ture rather than the trace of the plane on which the maximum principal normal stress acted. Cracking initiated at the yield point, indicating a drop of the critical strain from 0.21 to < 0.01. With increased strain multiple cracks formed as shown to some extent in Fig. 8. Thus, in contrast to the precharged case, the drastic embrittlement associated with dynamic charging does not occur as it did for both spheroidized steels and for a high-strength, quenched-and-tempered AISI 4340 steel[14], where the critical strain for cracking was also < 0.01.

IV.   Discussion

A.   Steels

The results at $-25^{\circ}C$ and $70^{\circ}C$ verify that in the low-strength, spheroidized AISI 1090 steels, a major effect of precharged hydrogen is to promote the on-set of plastic instability. The development of plastic instability is favored[11, 15] by a plane strain state, low work-hardening rate, the presence of a plastic strain gradient and the presence of a free surface. All of these factors are applicable for the present tests, the strain-gradient effect be-ing particularly accentuated in the bend tests. The tendency for plastic instability would be less in the round tensile bars, even if notched, because of a smaller strain gradient, the absence of plane strain conditions and the tendency of localized slip bands to interact and blunt because of the speci-men geometry. This tendency would be less for the sheet-tensile specimens because of the absence of plane strain constraints and a large strain gradi-ent.

The precharged hydrogen results at $-25^{\circ}C$ demonstrate the tendency for a tran-sition to intergranular brittle fracture with decreasing temperature, observed by others[8], and show that hydrogen promotes this transition. On the basis of work by McMahon and coworkers[16, 17] on high-strength steels one would expect the embrittlement to be accentuated by the presence of metalloids such as phosphorus at sub-and grain boundaries. Thus, preferential separation along boundaries would be favored by impurity segregation and the presence of hy-drogen whether the separation were brittle or ductile in the sense of occur-ring by preferential void growth there as in the work of Cialone and Asaro[2]. This suggests that microstructure and heat treatment can play a role in whether boundary separation occurs or not, depending on impurity metalloid content, boundary area, and whether or not carbide particles are present on the boundaries.

In view of the above discussion, the differences among the different studies of the effect of precharging with hydrogen on mechanical properties of spheroidized steels can be rationalized. The major effect is on void nuclea-tion in the bend test studies[1] because of the localization of plastic flow in shear bands which produce large incompatibility stresses at carbide particles. In these studies void growth and coalescence follow because of the shear localization. In the work of Cialone and Asaro[2] and Oriani and Josephic[3], shear instabilities are not favored. Yet we postulate that there is a major influence on void nucleation because of the presence of carbides on sub-boundaries and probably because of segregation of metalloids there. The preferential linking of voids by growth along the boundaries[2,3] would also follow from these two effects. In the work of Garber et al[4,5], neither a tendency for plastic instability nor apparently, preferential promotion of nucleation at boundaries was present. Thus, more randomly distributed voids nucleated and grew with little mutual interaction and with little effect of hydrogen until the late stages of growth.

In the bend tests the mechanistic effect of hydrogen appears to be the pro-motion of plastic instability by influencing the generation of dislocations at the surface by affecting the surface energy of the associated slip step[18,19] and by promoting the planarity of the slip[19,20]. Whether hydrogen influences the carbide decohesion directly, in addition to its indirect effect in generating incompatibility stresses, remains an open question. Hydrogen already trapped at carbide interfaces could contribute to void formation by lowering surface energy. It is unlikely, however, that dislocations could transport much hydrogen to the interfaces to aid in the above process. The binding energy to dislocations is less than that to carbides[19,21] so hydrogen would not readily be drained from carbides. Also fresh dislocations would tend to trap any weakly bound hydrogen, including that in the ferrite lattice, and compete with moving dislocations for the hydrogen.

The dynamic charging results show, as in the room temperature case[1], that the mechanism by which hydrogen produces the severe embrittlement leading to mode I cracking is localized near the crack tip. Previous estimates for room temperature, with a bulk diffusivity of hydrogen of 5×10^{-11} m^2/s, typical for steel[22], were that the effect occurred within a distance $\bar{X} \sim 10\mu$m of the crack tip both for crack initiation and propagation. The compilation of Völkl and Alefeld[23] suggests that at -25°C, the diffusivity and thus the value of $\bar{X}$ would be somewhat smaller. On the other hand, since straining and charging occur simultaneously, it is possible that dislocations transport hydrogen and enhance its apparent diffusivity as suggested by Tien et al[24]. While there is some question about the theoretical prediction of the effect[19], experiment[25] implies that the enhancement could be by a factor of 10^2 to 10^4 which could increase the effective value of $\bar{X}$ to the order of 100 μm to 1 mm. If this were the case, hydrogen would penetrate over a distance large compared to the mean intercept length between carbide particles of 1 μm. For this reason the results are not conclusive in establishing the mechanism for the hydrogen embrittlement in the dynamic charging case. Hydrogen could act at the very crack tip to promote brittle crack propagation. Alternatively it could cause carbide particle decohesion near the crack tip which then could contribute to the crack propagation. Further tests at varying strain rates, temperatures and carbon contents are underway to attempt to further refine the estimate of $\bar{X}$ and hence to provide more discrimination with respect to cracking mechanism.

B.   D S Nickel

In comparison to T D nickel[26], the D S nickel had a finer grain size and substructure, resulting in somewhat larger yield stress as shown in Table I. Transmission electron microscopy showed that the distribution of fine thoria particles in the D S nickel was similar to that in T D nickel. In contrast to T D nickel[26], however, the D S nickel sustained larger strains ($\sim$ .30) prior to void initiation. Thus, differences exist, but both materials are characterized by uniform distributions of extremely fine particles which are potential sites for hydrogen trapping.

The precharging results verify the earlier findings for T D nickel by Thomson and Wilcox[9], and the trapping models for ameliorating the embrittling effect of hydrogen in such alloys[21,27]. Nickel or nickel alloys with widely spaced second-phase particles are highly susceptible to intergranular embrittlement in the presence of hydrogen, the effect being exacerbated by metalloid impurity adsorption to the boundaries, as reviewed by Smith[28]. However, with a finely dispersed second phase as in the T D nickel work[9] and the present D S nickel case, the particles provide a high density of strong trap sites for hydrogen and compete with boundaries for the hydrogen[21,26]. Moreover, if, analogous to the carbide case[21], the binding energy to the particles would exceed that to dislocations, so the particles could both strip dislocations of hydrogen if the particles are unsaturated and resist the flow of hydrogen to dislocations which are free of hydrogen. Both effects would reduce transport of hydrogen to boundaries by dislocations. Hence, the trapping would tend to reduce the flow of hydrogen to boundaries by either diffusion or dislocation transport, and reduce the rate of initiation of cracks at these potentially damaging sites.

In the dynamic charging case, on the other hand, the severe mode I embrittlement occurs just as for the steels. The diffusivity of hydrogen in nickel at $20^\circ C$ is 10^{-13} m^2/s in the absence of trapping[23]. Thus, with bulk diffusion the maximum penetration of hydrogen ahead of the crack tip in the dynamic case would be ~ 0.5 μm, with the likely value being even less because of trapping. Moreover, experiments on nickel[25] indicate an effective diffusivity enhancement by dislocation transport to $\sim 10^{-9}$ m^2/s in the absence of trapping, which would give a maximum value of $\bar{X} \sim 70$ μm.

Just as for the steel case, these values are such that the mechanism of hydrogen degradation in the dynamic charging case is not revealed for D S nickel. However, the distance is smaller in the D S nickel case and the smaller thoria particles are less likely to decohere than the larger carbides in the steel case. Hence, while further work is required to pinpoint the mechanism, the D S nickel results tend more to favor an effect of hydrogen at the brittle crack tip as in decohesion models for hydrogen degradation.

V.   Summary

A.   For the spheroidized 1095 steel, precharging with hydrogen at $70^\circ C$ and $-25^\circ C$ led to a surface initiated mode II crack in agreement with previous room temperature work[1]. The major role of hydrogen degradation was to promote the onset of plastic instability at a reduced critical surface strain. The transition from intragranular void growth at room temperature and $70^\circ C$ to intergranular void growth at $-25^\circ C$ suggests that the disparities among previous results[1-5] at room temperature may have arisen from differing metalloid impurity contents on grain boundaries.

B.   For the spheroidized steel, dynamic charging with hydrogen at $70^\circ C$ and $-25^\circ C$ led to severe degradation in the form of mode I cracking initiated at the surface at a strain near yield. The results confirmed room temperature results, but were inconclusive in establishing the mechanism of degradation because of the range, 10 μm to 1 mm, of possible penetration of hydrogen ahead of the crack tip.

C.   For the precharged case, hydrogen did not degrade the properties of D S nickel, in agreement with earlier work on a similar alloy[9]. The absence of degradation confirms concepts for the amelioration of the embrittling effects of hydrogen in the presence of a high density of trap sites[21, 26], thoria particles in the present case.

D.   Dynamic charging severely degraded the D S nickel, analogous to prior work on several types of steels[1, 14]. The range of penetration of hydrogen, 0.5 to 70 μm, ahead of the crack tip was less than for the spheroidized steel, but was still to large to make possible the establishment of the mechanism of hydrogen degradation.

ACKNOWLEDGMENT

The authors are pleased to acknowledge the support of this research by the National Science Foundation under Grant DMR 7815735.

REFERENCES

1. T.D. Lee, T. Goldenberg and J.P. Hirth: Met. Trans. A, 1979, vol. 10A, pp. 199-208.
2. H. Cialone and R.J. Asaro: Met. Trans. A, 1979, vol. 10A, pp. 367-75.
3. R.A. Oriani and P.H. Josephic: Scripta Met., 1979, vol. 13, pp. 469-72.
4. R.I. Garber, I.M. Bernstein and A.W. Thompson: Scripta Met., 1976, vol. 10, pp. 341-5.
5. R.I. Garber, I.M. Bernstein and A.W. Thompson: Met. Trans. A, 1980, vol. 11A, in press.
6. A.S. Argon, J. Im and R. Safoglu: Met. Trans. A, 1975, vol. 6A, pp. 825-37; A.S. Argon and J. Im: ibid., pp. 839-51.
7. A.W. Thompson: Met. Trans. A, 1979, vol. 10A, pp. 727-31.
8. S.P. Rawal and J. Gurland: Met. Trans. A, 1977, vol. 8A, pp. 691-8.
9. A.W. Thompson and B.A. Wilcox: Scripta Met., 1972, vol. 6 pp. 689-96.
10. P. Wilson, S. Sklarska-Smialowska and M. Smialowski: 1979, Report of the Fontana Corrosion Center, Ohio State University, Columbus, Ohio.
11. C.A. Griffis and J.W. Spretnak: Trans. Iron Steel Inst. Japan, 1969, vol. 9, pp. 372-87.
12. T. Goldenberg, T.D. Lee and J.P. Hirth: Met. Trans. A, 1978, vol. 9A, pp. 1663-71.
13. M.W. Joosten: M.S. Thesis, Ohio State University, Columbus, Ohio, 1978.
14. T.D. Lee, T. Goldenberg and J.P. Hirth: Met. Trans. A, 1979, vol. 10A, pp. 439-48.
15. F.A. McClintock, S.M. Kaplan and C.A. Berg: Int. J. Fract. Mech., 1966, vol. 2, p. 615.
16. S.K. Bauerji, C.J. McMahon, Jr., and H.C. Feng: Met. Trans. A, 1978, vol. 9A, pp. 237-47.
17. C.J. McMahon, Jr., J. Kameda and N. Bandyopadhyay: in Proc. 2nd. Japanese Inst. Met. Symp. on Hydrogen in Metals, Tokyo, November 1979.
18. R.M. Latanision and R.W. Staehle: Acta Met., 1969, vol. 17, pp. 307-20.
19. J.P. Hirth: Met. Trans. A, 1980, vol. 11A, pp. 861-90.
20. S. Moriya, H. Matsui and H. Kimura: Mater. Sci. Eng., 1979, vol. 40, pp. 217-25.
21. G.M. Pressouyre and I.M. Bernstein: Met. Trans. A, 1978, vol. 9A, pp. 1571-80; Acta Met.: 1979, vol. 27, pp. 89-100.
22. R.A. Oriani: Acta Met., 1970, vol. 18, pp. 147-58.
23. J. Völkl and G. Alefeld: in Hydrogen in Metals, Topics in Applied Physics, 1978, vol. 28, pp. 321-48.
24. J.K. Tien, A.W. Thompson, I.M. Bernstein and R.J. Richards: Met. Trans. A, 1976, vol. 7A, pp. 821-9.
25. M. Kurkela and R.M. Latanision: Scripta Met., 1979, vol. 13, pp. 927-32.
26. B.A. Wilcox and A.H. Clauer: in The Superalloys, Wiley, New York, 1971, pp. 197-230.
27. A.W. Thompson: Met. Trans., 1974, vol. 5, pp. 1855-61.
28. G.C. Smith: in Hydrogen in Metals, ASM, Metals Park, Ohio, 1974, pp. 485-513.

DISCUSSION

<u>R. Stoltz, Sandia Laboratories, Livermore, CA</u>:   What was the fracture path
in the dynamically changed DS Nickel and was there evidence of dimples on
the fracture surface?

<u>M.W. Joosten</u>:   The Mode I fracture occurred at greatly reduced strain for the
dynamically charged D.S. Nickel.   The dramatic reduction in ductility was
evidenced by a change in the fractographic appearance to quasi-cleavage.

<u>D.A. Mezzanotte, E.I. du Pont de Nemours, Savannah River Laboratory, Aiken,
SC</u>:   Have you examined the case of a precharged sample undergoing dynamic
charging?   Such an experiment may allow you to separate the effects of internal and external hydrogen and reach some conclusion on the criticality of
dislocation transport to the observed embrittlement.

<u>M.W. Joosten</u>:   The experimental arrangement was such that precharged speci-
mens were exposed to 1 A/m^2 dynamic charging during the test sequence.   It
was thought that this amount of dynamic charging was preferable to hydrogen
degasing.   Some precharged samples were also tested in air at room temperature.   No significant changes in fractography or principal notch root strain
were observed.   At increased charging current density (10-100 A/m^2) the influence of hydrogen at the crack tip should not be altered by the amount of
hydrogen adsorbed during precharging.

HYDROGEN-RELATED EMBRITTLEMENT OF Ni-BASE SUPERALLOYS[*]

by

Nicholas F. Fiore and James A. Kargol

Department of Metallurgical Engineering
and Materials Science
Notre Dame, IN 46556

When single-phase Ni-base superalloys are subjected to certain thermo-
mechanical strengthening treatments, they become highly susceptible to failure by hydrogen embrittlement (HE). There are a number of ways in which the
thermomechanical state of the alloys may influence HE tendency. Micro- and
sub-microstructural changes caused by processing may influence the formation
kinetics and stability of passive films which protect an alloy. The same
changes may alter the alloy surface so as to increase H adsorption and absorption. Increases in bulk H diffusivity may affect H-transport and therefore influence crack initiation and propagation. Thermal exposure may cause
segregation of certain impurities to grain boundaries, and once there, the
impurities may interact with H atoms to promote intergranular failure. The
same thermal exposure may lead to short- and long-range ordering, which may
affect slip mode and cause loss of ductility.

In this paper, the results of a series of experiments on a specific
alloy, aimed at relating the diverse phenomena listed above to HE, are described. Ellipsometric studies relate passive film behavior to the alloy's
thermomechanical state. Electrochemical polarization and surface ultrasonic
studies define the relation between structure and H-generation and H-entry
kinetics. Electrochemical permeability studies establish the relation between structure and bulk H diffusivity. Ductile fracture analysis and fracture mechanics studies relate crack initiation and propagation to ordering
and grain boundary segregation.

The results of the study indicate that HE of the alloy increases in two
distinct stages as a function of aging time. The initial stage may be related to phosphorous segregation and short-range ordering, and the later
stage is related to long-range ordering.

[*]Research sponsored by Division of Materials Research, National Science
Foundation.

Introduction

The most highly alloyed austenitic stainless steels and Ni-base super-alloys were originally developed for high-temperature aerospace applications or for resistance to specific aggressive environments in the chemical processing industry. They were expensive, specialty alloys of limited use. Recently, it has become evident that these materials will be employed in large quantities in the nuclear industry, in sour gas and oil well fields, in coal conversion processes and in geothermal energy recovery processes. It appears that they are the only alloys available for the forseeable future which possess the required combination of strength and resistance to environment-related failure in those aqueous environments, which are characterized by high H fugacity and the presence of S and Cl ions.

Under certain combinations of processing history, operating temperature and environment, these normally strong, ductile and inert alloys fail catastrophically in a brittle fashion in very short times. For example, in sour brines, the alloys fail by a type of Hydrogen Embrittlement (HE) referred to as Hydrogen Stress Cracking (HSC) (1). It is the purpose of this paper to discuss research aimed at establishing the nature of the phenomena which control HSC in certain Ni-base alloys.

The processes which influence embrittlement tendency are diverse. They may originate in the aqueous environment, on the alloy passive film (which protects against general degradation yet catalyzes certain specific embrittlement reactions), or in the alloy itself. The research plan directed toward identifying controlling phenomena is correspondingly diverse, and consists of the 6 sets of experiments depicted in Figure 1.

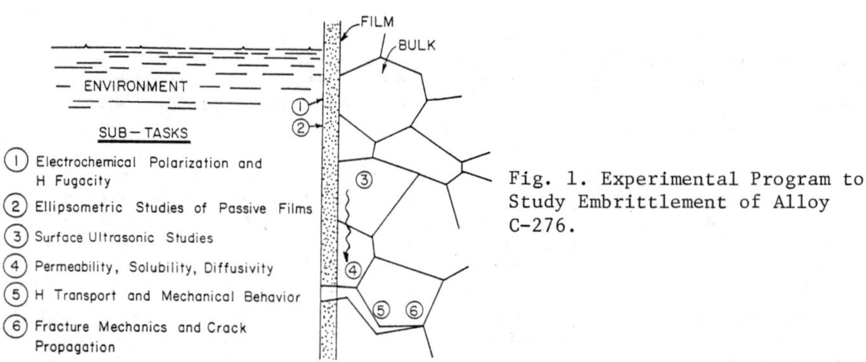

Fig. 1. Experimental Program to Study Embrittlement of Alloy C-276.

Each set is focused on the influence of metallurgical structure and thermo-mechanical treatment (TMT) on the potential embrittlement process, viz.:

1. H Absorption Studies in which the possibility is investigated that bulk structural changes may lead to surface structural changes which in turn affect the electrochemical reaction by which H enters the alloy.

2. Ellipsometry Studies of Passive Film Formation in which the presence and the effectiveness of the passive film as a H-production catalyst or a H-barrier is related to metallurgical structure.

3. Surface Ultrasonic Studies in which changes in the acoustic
    properties of the alloy surface are used to gauge H-surface
    concentration, the pinning of dislocations by H-atoms, and
    the role of H-recombination poisons on embrittlement.

4. H Permeability, Solubility and Diffusivity Measurements in which
    the bulk mobility of H is characterized and related to sub- and
    micro-structure.

5. H Transport and Mechanical Behavior Studies in which the ingress
    and egress of H is related to crack behavior and the micro-
    mechanisms of fracture.

6. Fracture Mechanics Studies in which environment-assisted crack
    propagation is studied as a distinct unit process in embrittle-
    ment.

This series of experiments may be employed in embrittlement studies of
any alloy since the processes involved occur universally in environment-
related failure. The research discussed in this paper has been concentrated
on HSC in a particular material, HASTELLOY* Alloy C-276 (55Ni, 16Cr, 16Mo,
5Fe, 4W, 2Co), which has shown promise as a structural material in deep
sour gas wells. The wells approach 1000m (30,000 ft) in depth and contain
hostile aqueous H_2S-NaCl environemnts at bottom-hole pressures of 140 to
210 MPa (20,000 to 30,000 psi) at temperatures of 200-300°C. The annealed
alloy is extremely resistant to HSC in this environment, but lacks adequate
strength for the structural application. When the alloy is cold-worked to
higher strength levels and then exposed to the elevated temperatures, it
embrittles rapidly, although the aging causes little, if any, modification
in microstructure (2-4).

Two distinct aging-induced phenomena have been proposed as controlling
HSC in the cold worked alloy. Berkowitz and Kane (5) have established that
aging causes P to segregate to grain boundaries. The P, which is a known
H recombination poison, may retard the formation of molecular H_2 from reduced H^+. The increased concentration of H at the surface promotes increased absorption of H at boundaries of the near-surface grains. The excess P
and H may also reduce cohesion and accelerate intergranular failure. Since
Alloy C-276 typically contains 0.04 wt% P (the heat in this study contained
0.042 wt% P), this mechanism may be applicable.

Asphahani (6) in contrast, has attributed the embrittlement to the
enhancement of planar slip caused by aging-induced ordering. Alloy C-276
undergoes both short-range ordering (SRO) at short aging times and long-
range ordering (LRO) at long aging times at temperatures between 200 and
500°C (7). Planar slip may increase crack nucleation at sharp surface
steps, increase internal stress concentrations at dislocation pileups (8),
or promote accelerated transport of H by dislocations (9). All of these
processes promote brittle failure.

Neither model of HSC can account for all aspects of HSC in Alloy C-276.
The mixed-mode (transgranular-intergranular) failure observed under some
test conditions is at variance with the predictions of the segregation
hypothesis (6). The improved embrittlement resistance of low-P heats of
the alloy is not predictable from the ordering hypothesis (3). The present
research therefore is directed toward isolating the factors which dominate
failure in a given situation.

Experimental

Samples for all of the studies were produced from the same heat of Alloy C-276 and ranged in geometry from 3mm-thick plate for the fracture mechanics experiments, to 1 mm-thick sheet for H-absorption, ultrasonic and mechanical tests, to 5 μm-thick foils for the permeability measurements. In all cases, the last processing step in producing the sample consisted of 60% reduction by cold-rolling. The cold-worked (cw) materials were then aged at 500°C for various times.

The experimental procedures used for the permeability (10), H absorption (11), ellipsometry (12), ultrasonic (13), mechanical test (14,15) and fracture mechanics (16) studies have been reported. In general, testing was conducted with samples cathodically polarized in a solution of 5% H_2SO_4 poisoned with 2 g/l thiourea (NH_2CSNH_2). This environment has been shown to simulate sour-well conditions and to provide an accelerated test of HSC resistance (4). In all tests the H content of the samples was measured on a LECO RH-1* Hydrogen Determinator.

Results

Permeability and Diffusivity Studies

Since for HSC to occur, H atoms must move substantial distances in the lattice, it is necessary to have reliable measures of H diffusivity D as a function of alloy TMT. Diffusivity data were not available in the moderate temperature range in which embrittlement is found in Alloy C-276, so electrolytic permeability measurements were performed in annealed, annealed + 100h aged, cw and cw + 100h aged foils (Figure 2). Embrittlement occurs more rapidly in aged material; therefore it might be expected that D would be higher in the aged forms of the alloy. As is seen in Figure 3, the opposite is true. Moreover it will be shown later in this paper that the values of D measured in any form of the alloy are much too low to account for depth of intergranular failure observed in sour-well environments. The results suggest that some accelerated form of transport must be occurring.

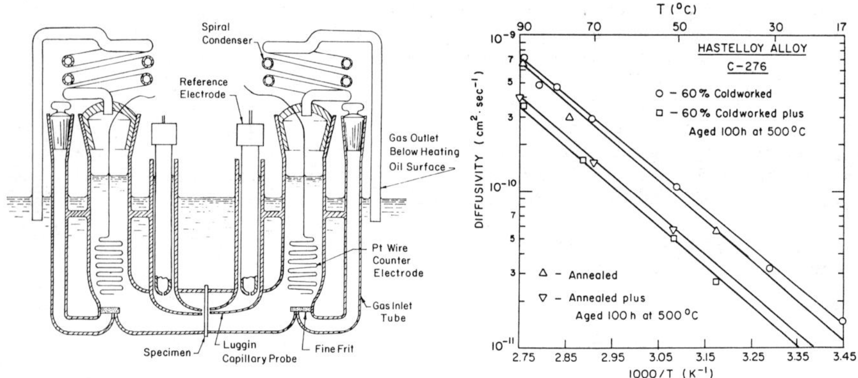

Fig. 2. Electrolytic H Permeability Cell.

Fig. 3. H Diffusivity in Alloy C-276.

*Trademark of Cabot Corporation.

H Absorption Studies

In this set of experiments, the possibility was investigated that aging promotes HSC by causing structural changes near the surface which increase H-entry kinetics by altering the H evolution reaction (h.e.r.). The H-entry kinetics have been analyzed by carefully measuring H ingress and egress, and H overpotential (η), in the cw and cw+aged alloy. These data then allowed a mathematical analysis of surface H concentration and bulk concentration profiles for alloys of various TMT.

The key to the analysis is f_H, the hydrogen fugacity, which drives H from the environment into the alloy. A common expression for the fugacity is

$$f_H = \exp\,[\eta F/zRT],\qquad\qquad(1)$$

where F is Faraday's constant, and z is a constant related to the h.e.r. If TMT changes alloy surface structure so as to increase η or decrease z, then f_H will increase, and the alloy may become more susceptible to embrittlement. The embrittlement susceptibility of an alloy in a given TMT may thus be associated with atomic-scale surface changes rather than bulk structural changes.

Figure 4, in which the ingress measurements are summarized, shows that the overall H content of the 100h-aged alloy is lower than that of the cw alloy. This finding is difficult to account for in terms of the segregation model, since the segregated P is hypothesized to promote H entry. In Figure 5 are displayed plots calculated from the model of H concentration versus distance into the sample after 4h of charging. It is seen that H penetration is greater and surface concentration is higher in the cw material, a result which is not consistent with the segregation model.

As part of this analysis, z was determined for the cw and the cw+a alloys. This parameter is fixed by the rate-controlling step of the h.e.r. and can, in theory, have a large effect on f_H. For these alloys, z was found to be essentially independent of TMT, so that the differences in HSC susceptibility which arise from aging are not associated with structural changes which alter the mechanism of the h.e.r.

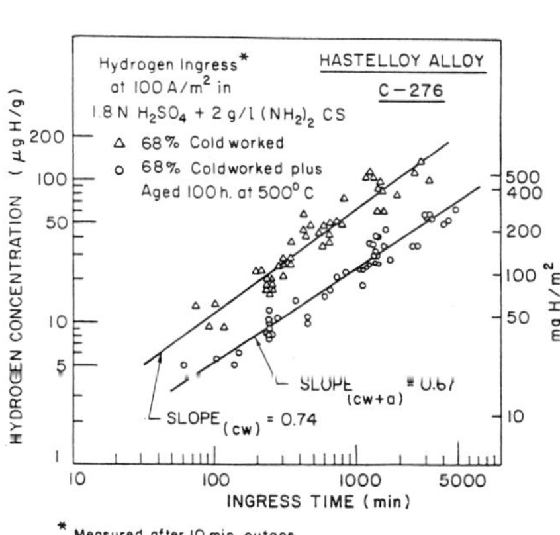

Fig. 4. H Ingress into Alloy C-276.

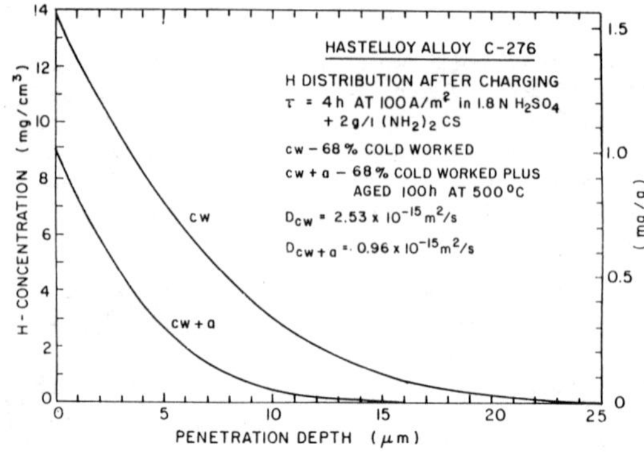

Fig. 5 .
H Concentration Profiles in
Alloy C-276 after 4h charging.

Ellipsometry Studies

It is possible that the increase in cracking tendency caused by aging occurs because the passive film formed on a cw + aged alloy is different than that found on the cw alloy. For example, the film on the cw alloy may be more stable under the cathodic conditions present in HSC than that of the cw+aged alloy.

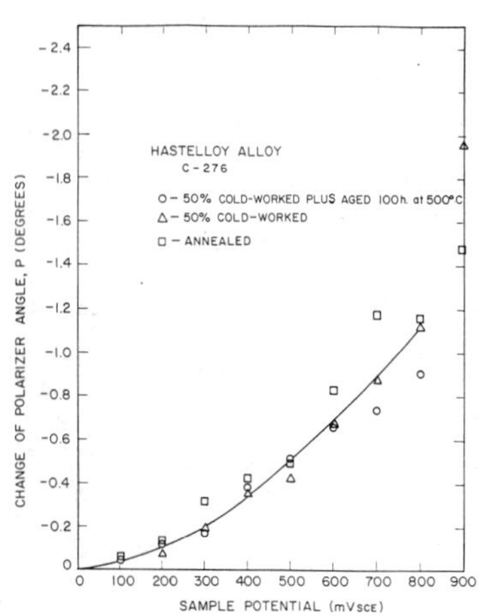

The ellipsometry measurements, which are optical studies of the passive film, show that the rapid film formation which occurs in the anodic range commences at a critical potential, but that this potential is but a mild function of TMT, viz: – 200 mV$_{SCE}$ (Saturated Calomel Electrode) for the annealed alloy, –190 mV$_{SCE}$ for the cw and –170 mV$_{SCE}$ for the cw + 100h aged alloy. The thickness of the film in the anodic range varies parabolically with potential (Figure 6). Once the critical potential has been reached, the film thickness is independent of TMT, as is its rate of removal under cathodic conditions. Passivation behavior therefore provides no insights as to why HSC resistance is such a strong function of TMT or as to whether the segregation or the ordering hypothesis has more merit.

Fig. 6. Change in Polarizer Angle during *in situ* Anodic Polarization of Alloy C-276.

Surface Ultrasonic Wave Studies

It has been established in many studies of HSC that H penetrates much more deeply into a material and causes brittle fracture at depths much larger than can be accounted for by bulk diffusion. One possible explanation for the observed accelerated H-transport is that H atoms are attracted to dislocations and are then rapidly transported to the alloy interior as the dislocations move during plastic deformation. Internal friction techniques are extremely valuable in measuring point defect-dislocation attractions, because the stronger the attraction, the less mobile are the dislocations which are pinned by the defects, and the lower is the dislocation damping component of the over-all internal friction (17).

In the ultrasonic internal friction studies, Rayleigh waves are propagated along the tension surface of a U bend sample (Figure 7), while the sample is undergoing cathodic charging. The attenuation (damping) of the wave may be measured as a function of charging potential and time.

It is evident from Figure 8 that as cathodic current density increases, the attenuation of the wave decreases. This behavior is to be expected if H atoms generated during charging are entering the material and pinning dislocations. The greater the cathodic current density, the greater the H flux into the alloy. As dissolved H atoms migrate to and pin dislocations, the dislocations are less able to dissipate vibrational energy, and attenuation decreases.

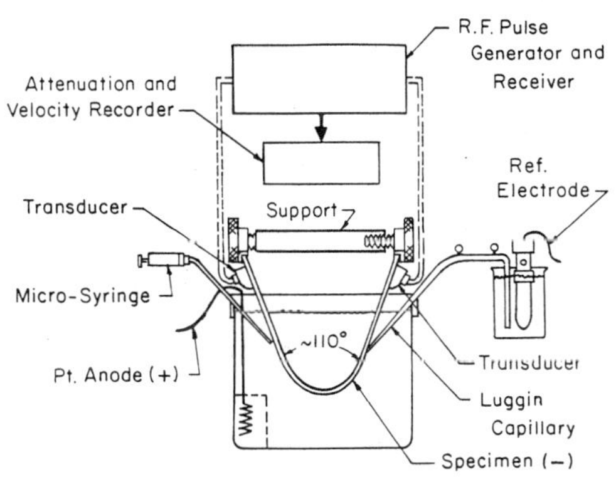

Fig. 7. Surface Ultrasonic Wave Test System.

If the H-dislocation attraction is significant and the cathodic current density removed, then H atoms will egress the sample, dislocations will be de-pinned and attenuation will increase with time. Such behavior is observed (Figure 8); moreover the kinetics of the de-pinning are consistent with predictions of time-dependent dislocation damping theory (18).

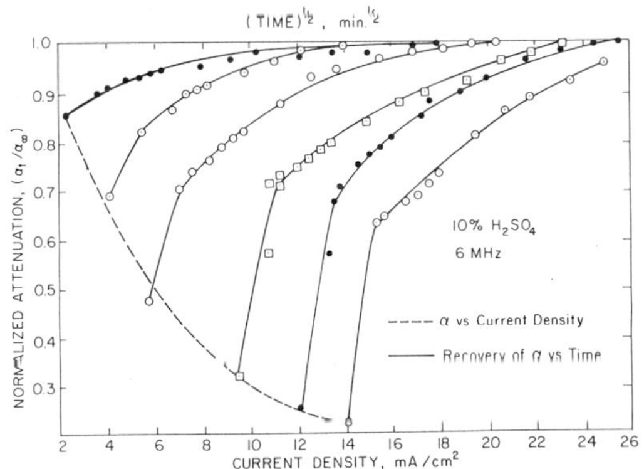

Fig. 8. Surface Ultrasonic Wave Attenuation during in situ H Charging of 60% Cold Worked Alloy C-276.

It is of interest to note that the H-dislocation interactions measured by this technique occur within the first few hundred Å of the surface, so that this method is a highly sensitive gauge of H behavior. The results of these measurements suggest that dislocation transport of H may well play an important role in embrittlement whether the segregation or the ordering models are operative.

Mechanical and Fracture Mechanics Tests

A number of mechanical and fracture mechanics tests have been performed in an effort to isolate mechanisms which control HSC in this alloy. The time-to-failure (ttf) of cw C-bend samples, pre-aged to various times and then stressed during charging was obtained with the procedure developed by Matsushima and Uhlig (20). After short-time aging, ttf decreased markedly (Figure 9) and then reached a plateau. The decrease coincided with times necessary for either P segregation or SRO to occur. (The increase in hardness during this time period is a result of SRO induced hardening). At very long aging times, LRO is substantial, hardness increases once again, and ttf decreases. The short-time embrittlement may be accounted for by either the segregation or the ordering hypothesis; however segregation has gone to completion long before the second decrease in ttf, so the long-time embrittlement is likely to be associated with LRO.

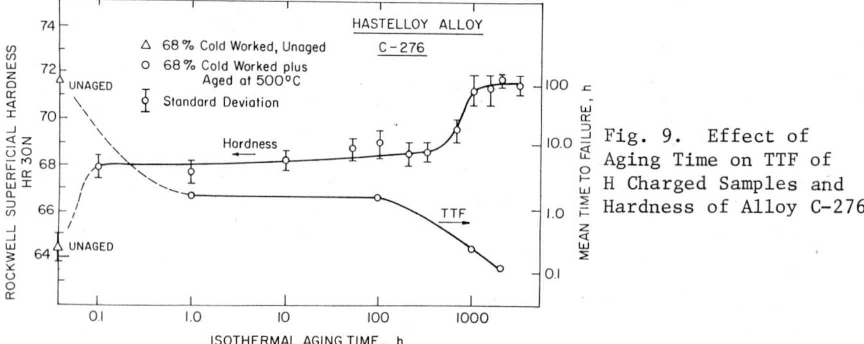

Fig. 9. Effect of Aging Time on TTF of H Charged Samples and Hardness of Alloy C-276.

The conclusions from the C-bend tests about short-time and long-time embrittlement are exactly those reached from fracture mechanics testings. As may be seen from Figure 10, Stage II crack velocity increases rapidly at times coinciding with both P segregation and SRO. The velocity reaches a pleateau until very long aging times pass whereupon it increases rapidly at times coincident with extensive LRO.

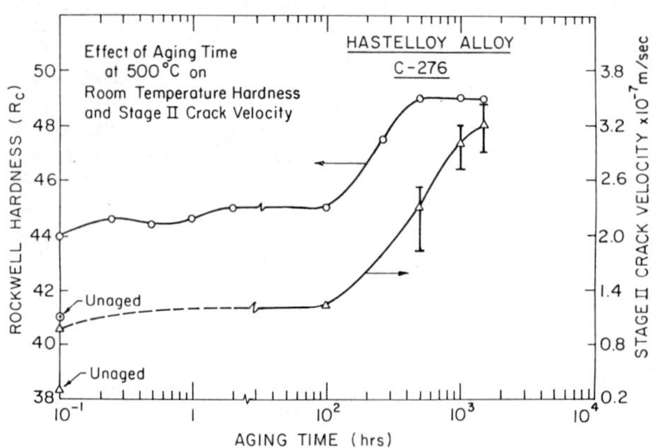

Fig. 10. Effect of Aging Time on Stage II Crack Growth Velocity and Hardness of Alloy C-276

Tensile tests on H-charged samples provide information about accelerated H transport. The fracture surfaces of the samples consist of an intergranular surface region and a ductile inner core (Figure 11). The depth of the region is greater for the cw + 100h aged material (~100μm) than for the cw region (~25μm). Calculations for appropriate values of D however, indicated that H can be transported only 10 μm by bulk diffusion for the observed failure times (Figure 5), and so once again the necessity for some means of accelerated H-transport is apparent. Tensile tests at various strainrates provide still further evidence that this process is dislocation transport. As may be seen from Figure 12, ductility (as measured by reduction in area) is markedly decreased during slow strain-rate testing of samples pulled to fracture during H-charging. For higher strain rates, at which dislocations are likely to move too rapidly to retain their H atmospheres, ductility increases again.

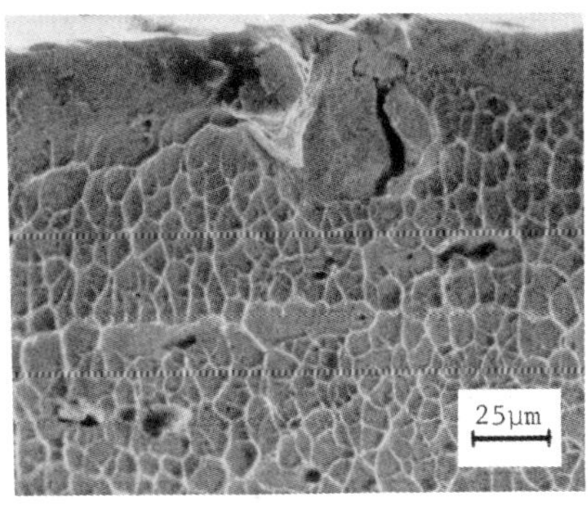

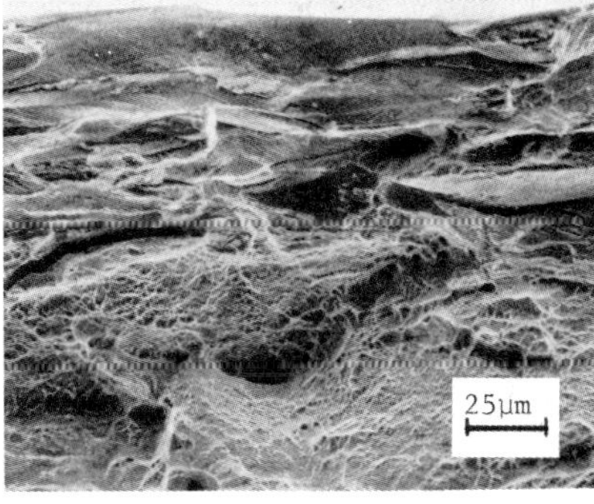

cw   cw + aged

Fig. 11. SEM Fractographs of Alloy C-276 Tensile Test Samples.

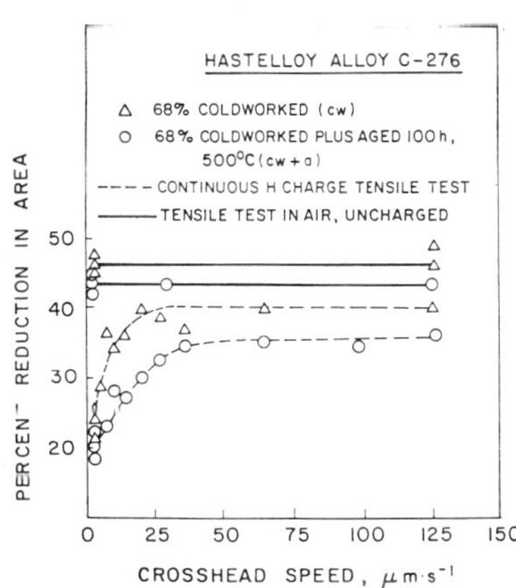

Fig. 12. Effect of Crosshead Speed on Embrittlement of H Charged Alloy C-276.

Kurkela and Latanision (19), in permeability cell measurements which employ _in situ_ plastic deformation of Ni foil which is exposed to H, have shown that straining leads to an increase of a factor of 10^4 in D. These findings, consistent with the surface ultrasonic and mechanical test results of the present study, strongly support the conclusion that dislocation transport of H plays an important role in embrittlement.

Dislocation transport should be enhanced by planar slip which may lead to high local concentrations of H near co-planar dislocation arrays. Figure 13 shows the surface of polished tensile samples of cw and cw + 100h aged alloys. The planarity of slip in the aged material, presumably caused by ordering, is evident.

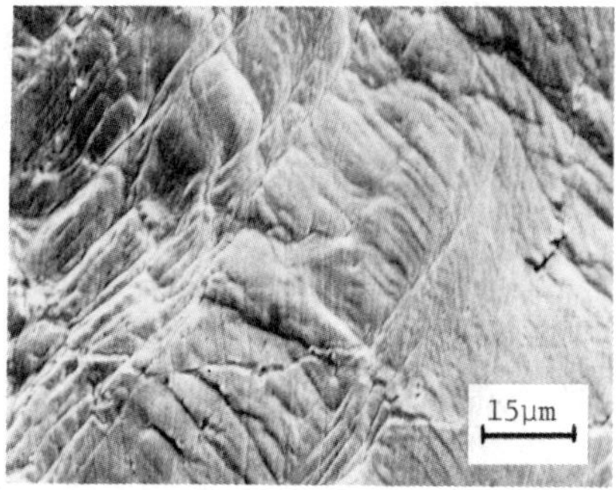

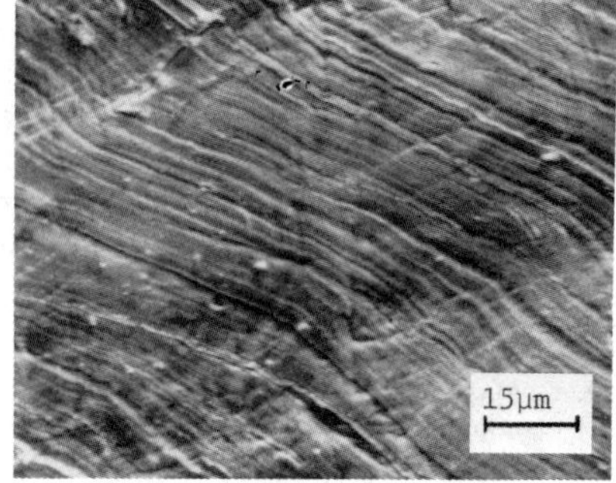

Fig. 13. Surface Slip Lines on Tensile Test Samples of Alloy C-276 in Air. (left: cw; right: cw + aged)

Summary

The permeability, H absorption, ellipsometry and surface ultrasonic studies assist greatly in characterizing HSC in Alloy C-276. Electrochemical and passive film changes induced by TMT are inadequate to explain the radical difference in embrittlement tendency of the cw and cw plus aged forms of the alloy. Bulk diffusivities are much too low to account for the depth of embrittlement in the short times-to-failure observed. This result, along with the surface wave attenuation studies, suggests that dislocation transport plays an important role in embrittlement.

The mechanical tests provide more direct information than the above study about HSC. At short aging times, both P segregation and SRO occur in Alloy C-276; therefore it is not possible to separate the effects of these two phenomena on HSC. However, in a similar Ni alloy, HASTELLOY Alloy G, which does not exhibit either SRO or LRO, 100h aging causes a marked decrease in ttf (21). Ordering therefore is not the only phenomenon contributing to embrittlement. The decrease in ttf in Alloy C-276 is more drastic than that in Alloy G, so SRO has a potent effect on HSC, perhaps by promoting planar slip. In essence then, P segregation and SRO may be acting synergistically to accelerate failure.

A second type of synergism may arise because P, in bulk solution, increases the rate of SRO. The combination of SRO-induced hardening (in a P-containing matrix) and P segregation to grain boundaries, then could lead to accelerated failure. Evidence in support of this hypothesis is found in the segregation study of Berkowitz and Kane (6). Low-P heats of Alloy C-276 resist embrittlement, although the same degree of grain boundary coverage occurs in them as in the high P material. This suggests that matrix as well as grain boundary P composition may be important.

In summary then, short-term embrittlement of this alloy may depend on the interplay of grain boundary segregation of P and SRO, which in turn may be influenced by the presence of bulk P. Long-term embrittlement is almost certainly controlled by LRO. Grain boundary segregation, which would accelerate intergranular fracture alone, cannot account for the observed mixed intergranular-transgranular fracture mode. Moreover, segregation has gone to completion long before the second sharp increase in embrittlement occurs. In either short-term or long-term embrittlement, planar slip and H transport by dislocations contribute to accelerated HSC.

Conclusions

1. Aging at 500°C produces measurable changes in the kinetic parameters
    which influence H entry into cold-worked Alloy C-276. In addition, it
    produces a significant decrease in H diffusivity in the alloy. Neither
    the H absorption nor the H diffusivity changes are adequate to account
    for the marked increase in HSC susceptibility which accompanies aging.

2. Aging causes only minimal changes in the passive film formation on the
    alloy in an H-generating environment.

3. Time-to-failure and fracture mechanics studies show that marked increases
    in embrittlement occur within about one hour of aging and then again
    after 1000 hours of aging. The short-time embrittlement may be associat-
    ed with the segregation of P to grain boundaries, SRO, or to the accel-
    eration of SRO by P in bulk solution. The long-time embrittlement is
    associated with LRO.

4. Planar slip induced by ordering and accelerated transport of H by dis-
    locations are involved in both short-time and long-time embrittlement.

Acknowledgments

The authors are priviledged to acknowledge the cooperative effort of the
following senior investigators: Drs. J. Bellina, R. J. Coyle, Jr., E.
Lunarska, D. Mezzanotte and N. Sridhar. They are also indebted to Dr. A.
Atrens for his help during the early stages of the project and to Drs. A.
Asphahani (Cabot Corporation) and B. Berkowitz (Exxon Research and Develop-
ment Corporation) for their comments throughout the course of the work.
This work has been supported by the Division of Materials Research, National
Science Foundation.

References

1. R. D. Kane and J. B. Greer, J. Pet. Tech., 1977, vol. 29, pp. 1483-1488.

2. R. D. Kane and B. J. Berkowitz, Corrosion, 1980, vol. 36, pp. 29-36.

3. R. D. Kane, M. Watkins, D. F. Jacobs, and G. L. Hancock, Corrosion, 1977,
    vol. 33, pp. 309-320.

4. R. J. Coyle, Jr., J. A. Kargol, and N. F. Fiore, Pro. 11th Offshore
    Tech. Conf., pp. 1325-1330, OTC Committee, 6200 N. Central Expwy,
    Dallas, TX 75206, 1979.

5. B. J. Berkowitz and R. D. Kane, Corrosion, 1980, vol. 36, pp. 24-29.

6. A. I. Asphahani, Proc. Second Int. Cong. on Hydrogen in Metals, Paper
    No. 3C-2, Pergamon Press, Paris, France, 1977.

7. P. Johnson, Ph.D. Thesis, Dept. of Met. Eng. and Mat. Sci., Univ. of
    Notre Dame, Notre Dame, IN 46556, 1980

8. N. S. Stoloff and R. C. Davies, Prog. Mat. Sci., 1966, vol. 13, pp. 1-84.

9. J. K. Tien, A. W. Thompson, I. M. Bernstein, and R. J. Richards, Met.
    Trans., 1976, vol. 7A, pp. 821-829.

10.   D. A. Mezzanotte, J. A. Kargol, and N. F. Fiore, Scripta Met., 1980, vol. 14, pp. 219-223.

11.   J. A. Kargol, N. F. Fiore and R. J. Coyle, Jr., Univ. Notre Dame, Notre Dame, IN 46556, unpublished research, 1980.

12.   J. J. Bellina, St. Mary's College, Notre Dame, IN 46556, unpublished research, 1980.

13.   E. Lunarska-Borowiecka and N. F. Fiore, Univ. Notre Dame, Notre Dame, IN 46556, unpublished research, 1980.

14.   R. J. Coyle, Jr., J. A. Kargol and N. F. Fiore, Univ. Notre Dame, Notre Dame, IN 46556, unpublished research, 1980.

15.   R. J. Coyle, Jr., J. A. Kargol and N. F. Fiore, Scripta Met., in press.

16.   N. Sridhar, J. A. Kargol and N. F. Fiore, Scripta Met., 1980, vol. 14, pp. 225-228.

17.   R. Truell, C. Elbaum and B. Chick, Ultrasonic Methods in Solid State Physics, Academic Press, New York, 1969, Ch. 3.

18.   A. Granato, A. Hikata and K. Lücke, Acta Met., 1958, vol. 6, p. 470.

19.   M. Kurkela and R. M. Latanision, Scripta Met., 1979, vol. 13, pp. 927-932.

20.   I. Matsushima and H. H. Uhlig, J. Electrochem. Soc., 1966, vol. 113, pp. 555-559.

21.   R. J. Coyle, A. Atrens, N. F. Fiore, J. J. Bellina, and M. Jolles, Environment-Sensitive Fracture of Engineering Materials, Z. A. Foroulis, ed., pp. 431-439, AIME Press, Warrendale, PA, 1979.

A COMPARATIVE STUDY OF STRESS-CORROSION CRACKING, HYDROGEN-ASSISTED CRACKING

AND LIQUID-METAL EMBRITTLEMENT IN Al, Ni, Ti and Fe-BASED ALLOYS

S.P. Lynch

Aeronautical Research Laboratories
Defence Science and Technology Organisation
Department of Defence, Australia

Metallographic and fractographic studies of crack growth in an age hardened
aluminium alloy, nickel, a titanium alloy, and a low-alloy, high-strength steel
in a variety of environments, e.g. argon, hydrogen, oxygen, moist air, water, and
liquid metals, are described. Crack growth at high velocities was studied in
particular to limit the number of possible environmental reactions occurring at
crack tips. For the aluminium alloy and nickel, the use of single crystals also
facilitated interpretation of results. Mechanisms of environmentally assisted
cracking (EAC) are discussed in the light of these observations.

Striking similarities between stress-corrosion cracking (SCC),
hydrogen-assisted cracking (HAC) and liquid-metal embrittlement (LME) were
observed in many cases; for example, fracture surfaces of nickel specimens
produced by cracking in liquid mercury and hydrogen gas were indistinguishable
and the degree of embrittlement was about the same in both environments. For LME,
chemisorption is the only reaction which occurs at crack tips. Thus, on the basis
of similarities between SCC, HAC and LME, and other considerations, it is argued
that chemisorption of hydrogen atoms (produced by dissociation of hydrogen or
water molecules) at the tips of cracks is responsible for many cases of HAC and
SCC.

EAC often produced fractures with a brittle, cleavage-like appearance on a
macroscopic   scale; however, crack growth, which did not necessarily occur at
high velocities, was usually associated with quite large (localised) strains, and
small, shallow dimples were usually apparent on fracture surfaces on a
microscopic   scale. Thus, it is proposed that EAC usually occurs by localised
plastic deformation at crack tips and that chemisorption of some (but not all)
species reduces the (shear) strength of interatomic bonds at surfaces and thereby
facilitates the nucleation of dislocations at crack tips.

Introduction

For SCC in aqueous environments, possible material-environment interactions at crack tips include adsorption, dissolution, oxide-film formation, and generation of hydrogen; diffusion of hydrogen ahead of crack tips may produce locally high concentrations of hydrogen in the lattice, precipitation of molecular hydrogen at voids and, in some cases, precipitation of hydrides. Intercrystalline SCC is common and, hence, factors such as grain-boundary segregation then have to be considered in conjunction with the above processes. Since the above phenomena occur on an atomic scale at 'inaccessible' crack tips, it is very difficult to determine, from studies of SCC alone, which process (or processes) are responsible for facilitating crack growth. Thus, it is not really surprising that there is disagreement concerning mechanisms of SCC and that mechanisms based on each of the above processes, and combinations of them, have been proposed for some materials.

The present approach to understanding mechanisms of environmentally assisted cracking (EAC) is based on comparing the characteristics of fractures produced in a wide range of environments at high rates of crack growth in single crystals. Thus, if different environments have common (specific) effects on crack growth, then a process which is common to these environments is probably responsible. Moreover, if rates of EAC are high compared with the maximum rates of dissolution of material and diffusion of hydrogen, then these processes are probably not involved. Furthermore, tests on single crystals avoid the complications associated with intercrystalline cracking and facilitate determination of crystallographic planes and directions of cracking. Having established the characteristics and probable mechanisms of fracture under relatively simple conditions, the mechanism of crack growth for more complex cases, e.g. SCC in commercial alloys at low rates of crack growth, should be more readily deduced.

Previous studies along the above lines by the present author (1–5) have shown that there are very close similarities between SCC, HAC, and LME in some materials and it has been proposed that many cases of EAC can be explained on the basis that chemisorption of environmental species facilitates nucleation of dislocations at crack tips. Further observations of EAC in aluminium, nickel, titanium, and iron-based alloys, which support this proposed mechanism, are presented in the following.

Experimental Details

Materials studied were (i) a high-purity Al 6.27Zn 2.94Mg alloy, strain-annealed to give a grain size of 10–30 mm, and aged to peak-hardness (162 HV10), (ii) nickel (99.5% purity) annealed to give a grain size of 10–20 mm, (iii) a Ti 6Al 4V alloy in the mill-annealed condition with a fine $\alpha + \beta$ structure (Proof stress 830 MNm^{-2}), and (iv) a low-alloy, high-strength steel (D6ac) (0.46 C, 1.2 Mo, 1.1 Cr, 0.12 V, 0.8 Mn, 0.24 Si, 0.004 P, 0.002 S) with a tempered martensitic structure (Proof stress 1200 MNm^{-2}). Compositions are in wt.%.

Environments used were (i) dry gases (argon, helium, hydrogen, oxygen) at ~100 kPa, (ii) moist air (~50% Relative Humidity), (iii) distilled water, (iv) 1000 mol m^{-3} HCl in a 3.5 wt.% NaCl aqueous solution, and (v) liquid metals (mercury, bismuth-lead-tin-indium-cadmium quinary eutectic). Tests were carried out at ~25°C except for those in the liquid alloy (M.P. ~47°C) which were performed at ~60°C. For 'single-crystal' tests on Al–Zn–Mg and Ni, cantilever-beam specimens were notched such that crack growth occurred across a large central grain. Bolt-loaded double-cantilever beam specimens were used for testing the titanium and steel.

Results

Al–Zn–Mg Single Crystals

For monotonically increasing loads (overload), crack growth in dry gaseous environments was associated with very extensive deformation and large, deep dimples (Fig. 1(a)), often with extremely small shallow dimples within them, were observed on fracture surfaces. For aqueous environments, overload fracture surfaces were relatively flat and macroscopically parallel to {100} planes; isolated, large dimples (A) and serrated steps were common (Fig. 1(b)) and, at high magnifications, extremely small, shallow dimples were apparent (Fig.1(c)).

For cyclic loading, changing the environment during testing (with ΔK remaining about the same) showed that, compared with an inert reference environment (helium), hydrogen and oxygen did not influence either fracture-surface appearance or striation spacings. However, fatigue in moist air, distilled water, and the liquid alloy environments resulted in marked changes in fracture plane and fracture-surface appearance, and produced increases in rates of crack growth of about four, six, and over one hundred times, respectively, compared with helium (Fig. 2). Except for the difference in rates, fatigue crack growth in moist air, water, and the liquid alloy was similar, with the same features as described for overload fracture in water. In embrittling environments, crack growth occurred in <110> directions and slip associated with fracture occurred mainly on {111} planes intersecting the crack-tip region (Fig. 2(b)(inset)). Laue X-ray back-reflection patterns from fracture surfaces were very diffuse and it was estimated (by comparison with patterns from specimens given known strains) that strains close to fracture surfaces were ~15% and ~40% for liquid-alloy and water environments, respectively. Changing from 'inert' to 'embrittling' environments (<u>and</u> vice versa) produced abrupt changes in fracture appearance and striation spacings, i.e. transient effects were not observed. Fractures with the 'brittle' characteristics described above were observed for velocities of cracking as high as 25 mm/s during fatigue in water and as low as 0.005 mm/s for sustained-load cracking in water. In the liquid-alloy environment, velocities of crack growth were at least 10 mm/s. Velocities of crack growth were uniform, i.e. crack growth was <u>not</u> discontinuous, even at low velocities.

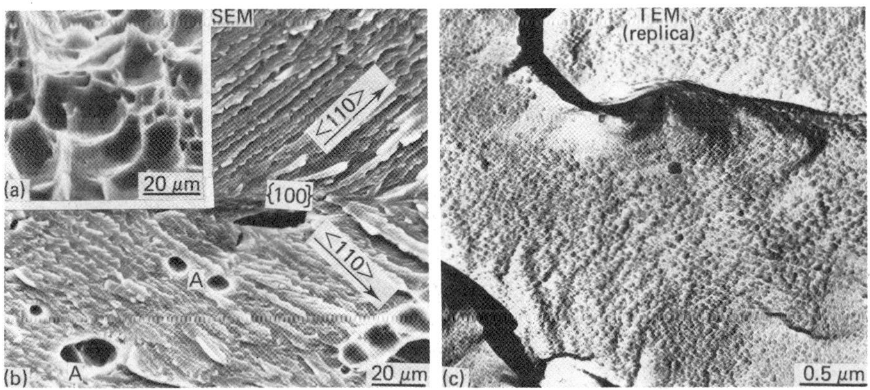

Fig. 1 Fracture-surfaces of Al-Zn-Mg produced by monotonic loading in (a) argon, and (b,c) water.

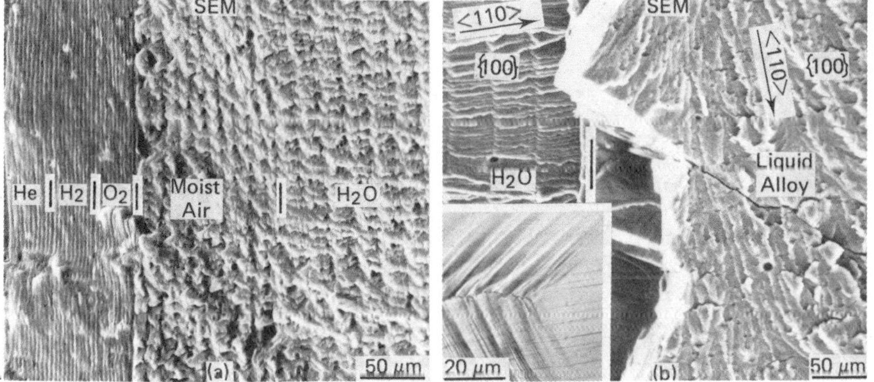

Fig. 2 Fracture-surface of Al-Zn-Mg produced by cyclic loading at 1 Hz in various environments; (inset) slip on side surface after fatigue in water.

Nickel Single Crystals

Liquid mercury and gaseous hydrogen environments produce embrittlement (4) and, for specimens partially cracked in liquid mercury and (after evaporating mercury from cracks) then cracked in hydrogen, it was difficult to distinguish between the respective fracture surfaces unless cracks were closed slightly before cracking in hydrogen to produce a striation mark on the fracture surfaces (Fig. 3(a)). (There was no doubt that mercury was completely removed from cracks since specimens tested separately in mercury and hydrogen confirmed the above results.) Not only were fractures of nickel in mercury and hydrogen environments identical to each other but they also had many features in common with fractures of Al-Zn-Mg in liquid-metal and aqueous environments; for all these cases, fracture surfaces were macroscopically parallel to {100} planes, crack growth occurred in <110> directions, and isolated large dimples (and numerous small dimples) and serrated steps were observed on fractures (Figs. 1-3). For fatigue in nickel, changing from inert to embrittling environments (and vice versa) also produced abrupt changes in striation spacings, striation appearance, and fracture plane. For mercury and hydrogen environments, spacings of 'brittle' striations on {100} planes were approximately the same (at a cyclic frequency of 1 Hz) and about five times those of 'ductile' striations in argon (Fig. 4). The distribution of slip around cracks in nickel for hydrogen and mercury environments was also similar to that for Al-Zn-Mg in embrittling environments, i.e. slip occurred preferentially on {111} planes intersecting crack-tip regions (Fig. 4(inset)), and strains just beneath fracture surfaces for nickel were at least 50%. For nickel in mercury <u>and</u> hydrogen, fractures with 'brittle' characteristics were observed for crack velocities as high as 10 mm/s.

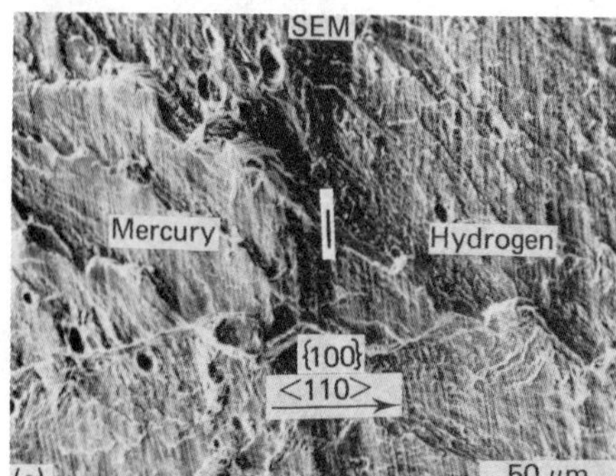

Fig. 3(a) Fracture surface of nickel produced by monotonic loading first in mercury and then in hydrogen; (b) detailed appearance of hydrogen region.

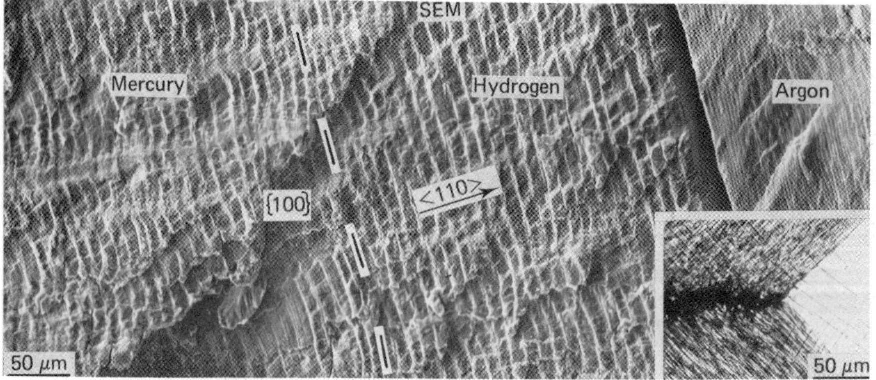

Fig. 4 Fracture surface of nickel produced by cyclic loading at 1 Hz first in mercury, then hydrogen, and then argon; (inset) slip on side surface after fatigue crack growth in hydrogen.

Ti 6Al 4V

For overload fracture in air, equiaxed dimples were observed on fracture surfaces; for SCC in the NaCl + HCl environment, a mixture of 'cleavage' and 'fluted' regions was observed on fracture surfaces (Fig. 5(a)). 'Cleavage' planes were approximately parallel to basal planes of the α-phase (<u>precise</u> orientations were not determined) and 'fluted' regions were deep, elongated dimples (i.e. their profiles were concave on opposing fracture surfaces). Very small, shallow dimples and tear ridges were evident on 'cleavage' regions at high magnifications (Fig. 5(a)(inset)). Sustained-load cracking in mercury produced 'cleavage' and 'fluted' regions (identical to those for SCC) (Fig. 5(b)) and regions of intercrystalline fracture; small, shallow dimples were observed on intercrystalline facets (Fig. 5(c)). For sustained-load cracking in hydrogen, fracture surfaces were completely intercrystalline with small, shallow dimples and crack-arrest markings evident on most facets (Fig. 5(d)). Velocities of sustained-load cracking at high stress-intensity factors (K ~40–60 MN m$^{-3/2}$) for the NaCl + HCl solution, mercury, and hydrogen were ~0.1 mm/s, 10 mm/s and 0.05 mm/s, respectively. Fracture surfaces resulting from cracking at high and low K were fairly similar.

D6ac Steel

Crack growth in air produced dimpled transcrystalline fracture surfaces while sustained-load cracking in both hydrogen and mercury resulted in a mixed intercrystalline/transcrystalline fracture mode (Fig. 6(a,b)). Intercrystalline and transcrystalline facets were completely covered with small, shallow dimples and isolated, large dimples. Velocities of EAC in hydrogen and mercury at high K (~100 MN m$^{-3/2}$) were ~0.001 mm/s and 1 mm/s, respectively; fracture-surface appearance was fairly similar for cracking at high and low K.

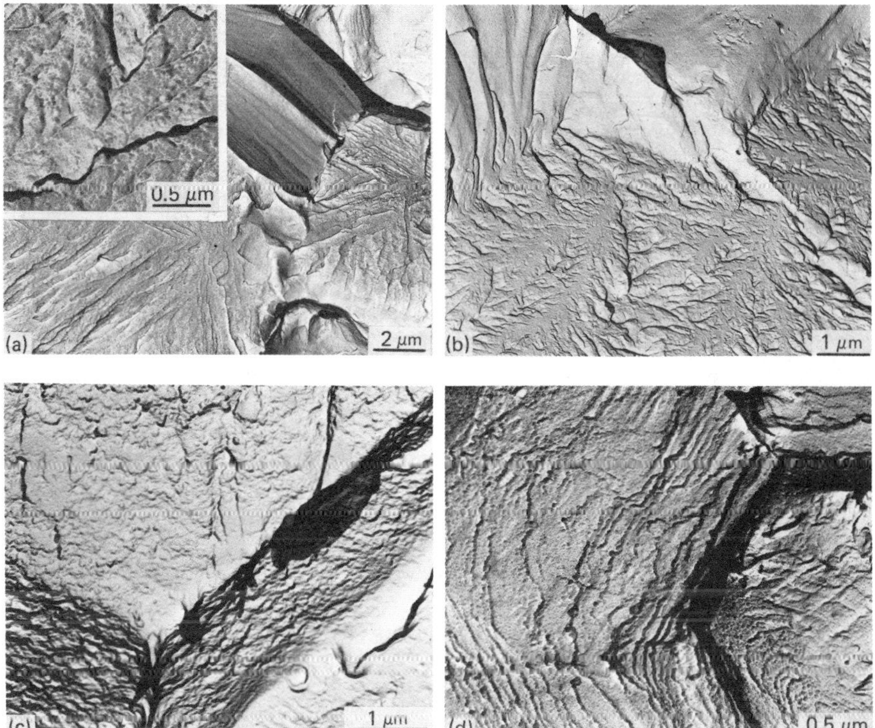

Fig. 5 Fracture-surface replicas of Ti6Al4V produced by sustained-load cracking in (a) NaCl + HCl solution, (b,c) mercury, and (d) hydrogen.

Fig. 6 Fracture surfaces of D6ac steel produced by sustained-load cracking in (a) mercury, and (b,c) hydrogen.

Discussion

Material-Environment Interactions

The remarkably similar characteristics of fractures produced in aqueous, hydrogen, and liquid-metal environments suggest that SCC, HAC, and LME can occur by a common mechanism in many cases. The present results are considered to be particularly significant since crystallographic planes and directions of cracking, distributions of slip around cracks, and microscopic   (as well as macroscopic)   details of fracture surfaces were often the same for fractures in different embrittling environments. Thus, it is proposed that <u>chemisorption</u> is often the most important reaction during SCC, HAC, and LME since chemisorption is the <u>only</u> reaction which generally occurs during LME. For most liquid-metal/solid-metal embrittlement couples (including the present), mutual solubilities are so small and crack growth so rapid that there is insufficient time for other processes, such as dissolution and diffusion, to occur (6).

For high velocities of cracking in aqueous and hydrogen environments, lattice diffusion of hydrogen ahead of cracks cannot occur since rates of diffusion are quite low, e.g. $D\sim10^{-14}m^2/s$ for hydrogen in Al-Zn-Mg. Rapid transport of hydrogen by dislocations into the interior of specimens could possibly occur but would probably be relevant only for cases where crack planes and slip planes approximately coincided; for {100} cracks in Al-Zn-Mg and Ni, dislocations may sweep hydrogen along {111} planes inclined at $\sim54°$ to the line of prolongation of cracks but hydrogen should not accumulate directly ahead of cracks. In aqueous environments, dissolution could not produce the high crack velocities observed since solvated ions would be unable to diffuse away from crack tips sufficiently rapidly to allow continued reaction. Furthermore, moist air or water-vapour environments, where dissolution cannot occur, often have very similar effects to aqueous environments (e.g. Fig. 2(a)) suggesting that dissolution is not responsible for embrittlement. Likewise, the formation of oxide films (and their subsequent rupture) would not be sufficiently rapid to account for the observed high velocities of crack growth. At low crack-growth velocities, oxide-film formation, dissolution, and diffusion of hydrogen ahead of cracks could occur but, since the fracture-surface characteristics (which are generally quite specific) are very similar for high and low rates of cracking, it is considered that mechanisms of crack growth are basically the same at both high and low velocities, and that an explanation based on adsorption best fits the experimental observations. However, other reactions in addition to adsorption could play some role during cracking at low velocities; in particular, rates of repassivation should be an important factor since films would inhibit adsorption at the metal surface.

Hydrogen gas partially dissociates on contact with 'clean' surfaces of Ni, Ti, and Fe (but <u>not</u> Al) resulting in adsorption of atomic hydrogen (7). Similarly, water molecules dissociate on surfaces of Ni, Ti, Fe <u>and</u> Al resulting in adsorption of hydrogen atoms. Thus, it is proposed that adsorption of atomic hydrogen is responsible for HAC and SCC in these materials. On this basis, the absence of any observed effect of molecular hydrogen on crack growth in aluminium alloys is expected since dissociation does not occur. Atomic hydrogen is generated at crack tips in a wide variety of materials and environments and, hence, a mechanism of SCC based on adsorption of atomic hydrogen may be widely applicable. SCC in cases not involving hydrogen (or hydrogen-containing environments) could possibly be due to adsorption of other species such as halide ions, e.g. for SCC of zirconium alloys in iodine vapour (8); however, there are undoubtedly cases of SCC where adsorption is not primarily responsible.

Atomic Mechanisms of Crack Growth

In the absence of dissolution (removal of atoms), crack growth on the atomic scale must occur by either tensile separation of atoms ('tensile decohesion') or shear movement of atoms (dislocation nucleation or egress) at crack tips. For tensile decohesion, it is generally assumed (9,10) that (i) dislocations do not generally intersect crack tips and, hence, strains around cracks should not be large, and (ii) voids do not form ahead of cracks so that fracture surfaces should not be dimpled. Moreover, for FCC metals, crack growth should occur in <100> directions (on {100} planes) to minimise the extent of slip around cracks (11). None of these assumptions were satisfied for the fractures studied in the present work and, hence, it is concluded that crack growth in embrittling environments (as well as in inert environments) occurs by plastic flow. It follows, therefore, that since adsorption can only influence the first few atomic layers at surfaces, an effect on the nucleation or egress of dislocations at crack tips must be responsible for the effect of environment on crack growth. For Ni and Al-Zn-Mg single crystals, it has been proposed (see refs. 4,5 for details) that chemisorption of liquid-metal or hydrogen atoms facilitates the nucleation of dislocations at crack tips, and thereby results in crack growth by an alternate-slip process at crack tips combined with some general matrix slip ahead of cracks which produces nucleation and growth of voids. Such a process would produce dimpled fracture surfaces macroscopically   parallel to {100} planes, with crack growth in <110> directions, as observed. Embrittlement would result simply because crack growth (by slip) is made easier - any process which facilitated fracture would decrease the extent of associated plasticity. An alternate-slip/void-growth mechanism is probably also applicable, with minor modifications, to other cases of EAC including intercrystalline cracking. The latter probably commonly occurs due to either preferential adsorption at grain-boundary/surface intersections or the presence of mechanically weak fracture paths along grain boundaries (e.g. precipitate-free zones, large area fractions of particles) or both. For inert environments, nucleation of dislocations at crack tips is probably difficult relative to general matrix slip and, hence, significant crack growth does not occur until large strains develop ahead of cracks such that extensive egress of dislocations occurs at crack tips.

Effect of Chemisorption on Surfaces

For clean metal surfaces, there is often a small (but significant) contraction in the spacings of atoms between the first and second atomic layers due to the imbalanced atomic forces (12), and it has been suggested (13) that this non-uniformity could hinder the nucleation and egress of dislocations at surfaces. Adsorption often cancels this contraction in the surface layer (or in some cases may produce an expansion or re-construction of the surface layer) (12) and, hence, could facilitate the nucleation of dislocations. More specifically, the transfer of electrons from adsorbed atoms to surface metal atoms may increase the repulsive force between surface atoms and thereby reduce the shear strength of interatomic bonds at surfaces. However, not all chemisorbed species should have such an effect since chemisorption of some species, e.g. oxygen on many metals, can sometimes produce disruption of the surface metal lattice with the incorporation of environmental atoms into the topmost surface layers. Thus, while adsorption of hydrogen may facilitate crack growth, oxygen may not influence crack growth (e.g. Fig. 2) and may inhibit HAC by adsorbing preferentially. The precise behaviour will depend on a number of factors including the type of adsorption, adsorption kinetics, and surface coverage.

'Internal' Embrittlement

Both LME and HAC can result from the presence of embrittling species within the material, viz. low-melting-point metal inclusions, hydrogen in voids, but it is widely accepted that the basic mechanism of embrittlement is the same regardless of the source of embrittling species. Thus, internal embrittlement is probably primarily due to adsorption at tips of internal cracks, which facilitates nucleation of dislocations. However, there will generally be differences between 'external' and 'internal' HAC in the kinetics of crack growth since the latter involves hydrogen diffusion through the lattice to internal voids and, in some cases, the development of high pressures within voids. In some materials, e.g. titanium alloys, the presence of hydrides may also influence crack growth although, unless the hydrides are brittle and form an almost continuous fracture path, their effects may be only small (14).

The nucleation of dislocations could be facilitated by 'dissolved' hydrogen just beneath the first atomic layer at crack tips as well as by adsorbed hydrogen since both could reduce the (shear) strength of interatomic bonds at surfaces. Such 'dissolved' hydrogen may be particularly significant during crack growth in aluminium alloys charged with hydrogen prior to testing, since hydrogen which precipitated in voids would be innocuous. However, hydrogen dissolved in the lattice away from free surfaces probably has no significant effect on fracture.

Closing Remarks

One of the conclusions of the present paper, that adsorption is responsible for many cases of SCC and HAC, is in agreement with many previous studies (e.g. 15-18); however, the present adsorption/dislocation-nucleation theory does not suffer from the drawbacks of previous adsorption theories which were generally based on surface-energy or tensile-decohesion criteria for fracture. Moreover, the present theory goes some way towards reconciling the conflicting views of some other workers. Thus, the present author essentially agrees with Oriani (19) that hydrogen decreases the strength of interatomic bonds in the surface or very near-surface layers, and agrees with Beachem (20) that HAC generally occurs by localised plastic flow. Furthermore, the present theory is <u>exactly</u> the same as that proposed by Clum (21) on the basis of field-ion microscopy studies of iron alloys in the presence of hydrogen. At this stage, the theory is necessarily qualitative since the exact details of the adsorption reaction are not fully understood. However, using a variety of recently developed techniques for studying surfaces, e.g. to determine the common effects of hydrogen and mercury on the surface characteristics of nickel, should lead to further understanding.

References

1. S.P. Lynch: Mechanisms of Environment Sensitive Cracking of Materials, eds. P.R. Swann, F.P. Ford, and A.R.C. Westwood, pp. 201-212, The Metals Society, London, 1977.
2. S.P. Lynch and N.E. Ryan: Hydrogen Damage, ed. C.D. Beachem, pp. 369-376, ASM, Metals Park, Ohio, 1977.
3. S.P. Lynch: Metals Forum, 1979, vol. 2, pp. 189-200.
4. S.P. Lynch: Scripta Met., 1979, vol. 13, pp. 1051-1056.
5. S.P. Lynch: Fatigue Mechanisms, ed. J.T. Fong, pp. 174-213, ASTM STP 675, ASTM, Philadelphia, 1979.
6. M.H. Kamdar: Prog. Mat. Sci., 1973, vol. 15, pp. 289-374.
7. D.O. Hayward: Chemisorption and Reactions on Metallic Films, ed. J.R. Anderson, vol. 1, pp. 225-326, Academic Press, London, 1971.
8. B. Cox and J.C. Wood: in ref. 1, pp. 520-530.
9. A. Kelly, M.R. Tyson and A.H. Cottrell: Phil.Mag., 1976, vol. 15, pp. 567-586.
10. J.R. Rice and R. Thomson: Phil. Mag., 1974, vol. 29, pp. 73-97.
11. S.S. Hecker, D.L. Rohr and D.F. Stein: Met.Trans., 1978, vol. 9A, pp. 481-487.
12. M.A. van Hove: Surface Sci., 1979, vol. 81, pp. 1-7.
13. R.L. Fleischer: Acta Met., 1960, vol. 8, pp. 598-604.
14. J.C. Williams: Effect of Hydrogen on Behavior of Materials, eds. A.W. Thompson and I.M. Bernstein, pp. 367-380, Met.Soc. AIME, New York, 1975.
15. C.M. Ransom and P.J. Ficalora: Met. Trans., 1980, vol. 11A, pp. 801-807.
16. D.P. Williams and H.G. Nelson: Met. Trans., 1970, vol. 1, pp. 63-68.
17. A.S. Tetleman and S. Kunz: Stress Corrosion Cracking and Hydrogen Embrittlement of Iron Based Alloys, eds. R.W. Staehle, J. Hochmann, R.D. McCright and J.E. Slater, pp. 359-375, NACE, Houston, 1977.
18. H.H. Uhlig: ibid., pp. 174-179.
19. R.A. Oriani: ibid., pp. 351-358.
20. C.D. Beachem: ibid., pp. 376-381.
21. J.A. Clum: Scripta Met., 1975, vol. 9, pp. 51-58.

DISCUSSION

D.A. Meyn, U.S. Naval Research Lab, Washington, DC: Was Ti-6-4 fracture in
H_2 gas intergranular? Fractograph was not clear as to this.

S.P. Lynch: Scanning electron micrographs showed that fracture of Ti6Al4V
in hydrogen gas was intercrystalline although some areas of the fracture
surfaces were partially obscured by a film - probably a hydride formed after
fracture. It is also of interest to note, with respect to comparing HAC and
LME that, while fractures in liquid mercury showed only small areas of intercrystalline fracture, fractures in liquid cadmium were predominantly intercrystalline.

J.A. Moskovitz, Kaiser Aluminum, Pleasanton, CA: At what stress intensity
level did you see the rapid crack growth in aqueous solutions in the single
crystal Al-Zn-Mg alloy, i.e. how much was the fracture toughness reduced by
H adsorption?

S.P. Lynch: Velocities of environmentally-assisted cracking in Al-Zn-Mg
crystals in water depended on the rate of increase of crack-opening-displace-
ment and, at high displacement rates, high velocities of cracking (2 mm/s)
were observed. For example, for cyclic loading at high ΔK (values were not
determined) at cyclic frequencies of 24 Hz, spacings of 'brittle' striations
were ~0.5 mm, i.e. crack velocities were ~25 mm/s.

COMPARISON OF HYDROGEN AND OXYGEN SLOW STRAIN-RATE EMBRITTLEMENTS

R. E. Reed-Hill

Department of Materials Science and Engineering
University of Florida
Gainesville, Florida 32611

The embrittlement of Nb and Ta by oxygen is compared with that of
hydrogen in steels and the type Va refractory metals. Hydrogen embrittles
exothermic occluders of H such as Nb and Ta in two ways, one involving
embrittlement due to precipitated hydrides, the other by hydrogen in
solution. Recent work on Nb and Ta with up to 1.6 at. % O has shown that
oxygen can also embrittle these metals in two ways. From roughly 450 to
700 K abnormal work hardening associated with oxygen-induced dynamic
strain-aging promotes brittle fracture by raising the flow stress to where
catastrophic internally nucleated brittle fracture occurs. The low
temperature forms of H and O embrittlements in Nb and Ta are thus not the
same. However, at higher temperatures (700 to 950 K) the O and H embrittlements are analogous with both interstitials in solution. For this type of
embrittlement normally called slow strain embrittlement both hydrogen
and oxygen (a) produce fractures that are preferentially nucleated at the
specimen surface and that can grow in a non-catastrophic fashion, (b) yield
a lower critical stress below which brittle fracture does not occur,
(c) develop a fracture surface morphology that becomes more brittle the
slower the strain-rate or the lower the stress intensity factor. It is
believed that the temperature ranges in which these slow strain-rate
embrittlement phenomena occur can be related to the respective activation
energies for diffusion of H and O if effective activation energies that
account for trapping by impurities, inclusions, grain boundaries and the
dislocation structure can be compared.

It has now been demonstrated that oxygen can induce slow strain-rate embrittlement in niobium[1] and tantalum.[2] An important technical and scientific question is how does the oxygen embrittlement compare with that due to hydrogen observed in both steels[3-4] and non-ferrous metals such as niobium,[5-6] vanadium,[7-8] tantalum,[9] nickel[10] and α-β titanium.[11] In other words, are the oxygen and hydrogen embrittlements different or are they merely different aspects of a more general form of embrittlement?

The Dual Nature of the Oxygen Embrittlement

Recent work on niobium[12] has given a better insight into this matter. This has shown that the loss in ductility in niobium between roughly 450 and 950 K involves two basically different embrittlement phenomena. Evidence for this is presented in Fig. 1 which gives reduction-in-area versus temperature curves for niobium specimens with 0.75 at. % oxygen tested at three strain-rates differing from each other by factors of ten. At this oxygen level in niobium two distinct ductility minima can be observed at the two faster strain-rates. Note that as the strain-rate is decreased the minima broaden in width and become deeper so that at the slowest strain-rate they overlap to form an apparent single ductility minimum. At higher oxygen concentrations (1.6 at. %) the degree of the embrittlement increases[12] and only a single minimum is observed at all of the strain-rates involved in Fig. 1.

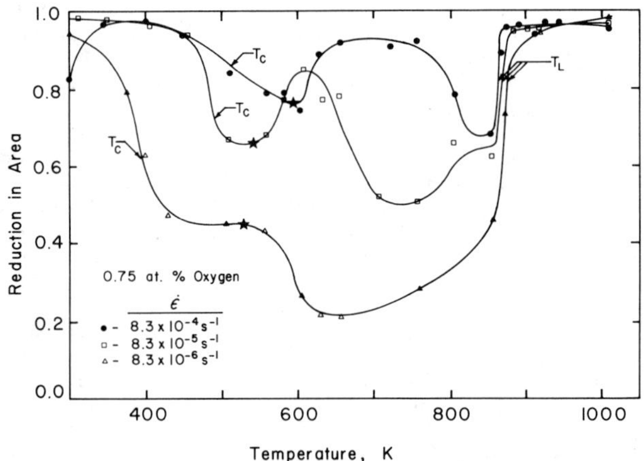

Fig. 1. Reduction-in-area versus temperature curves for Nb-0.75 at. % O specimen.[12]

The Lower Temperature Embrittlement

The loss in ductility on the low temperature side of the embrittlement region from 450 to 950 K has been ascribed[12] to dynamic strain-aging due to oxygen. In this region the stress level during a test rises high enough, due to the increased work hardening rate associated with dynamic strain-aging, that the specimen simply fails catastrophically as a result of an internally nucleated crack. Fracture surfaces in this interval exhibit sharply defined cleavage and intergranular facets. The brittle fracture morphology in this temperature range does not in general change with the strain-rate.

The Higher Temperature Embrittlement

At the high temperature end of the ductility minimum the fracture
mechanism is quite different. Instead of the fractures being internally
nucleated they form at or just under the surface and often along the entire
gage length. Multiple surface cracking of this type is shown in Fig. 2 on
a specimen of tantalum with 0.5 at. % O deformed to failure at 723 K at a
strain-rate of $10^{-3}s^{-1}$. In the niobium specimens of Fig. 1 surface cracking
occurred between the temperatures corresponding to the black stars and the
T_L symbols, designating the upper temperature transitions from brittle to
ductile behavior, on the R.A. vs T curves. These surface cracks become
visible early in a tensile test. Thus, Fig. 3 shows a surface crack in a
tantalum specimen with 0.3 at. % O deformed to only 1.5% strain. This
specimen would normally have an elongation of about 25%.

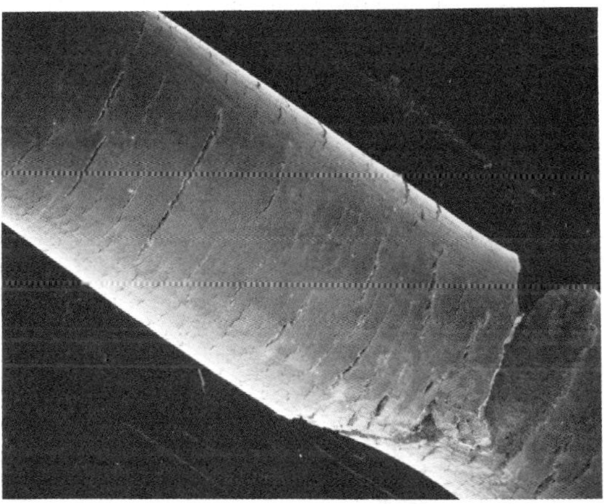

Fig. 2. Multiple surface nucleated cracks on a Ta-0.5
at. % O specimen tested at 723 K and $\dot{\varepsilon} = 10^{-3}s^{-1}$. 15X.

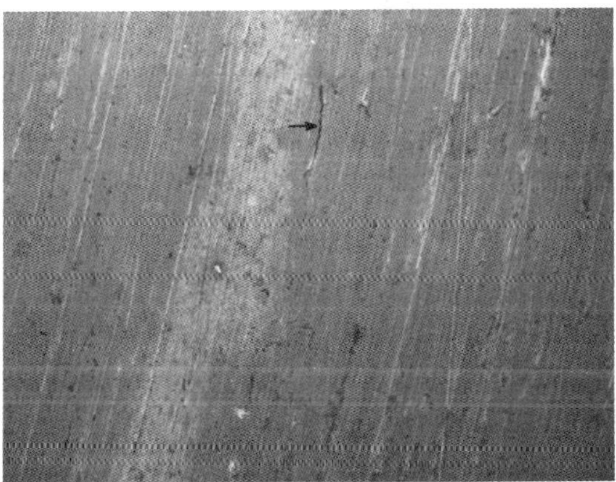

Fig. 3. A surface nucleated crack on a Ta-0.3 at. % O specimen after 1.5% strain. 723 K and $\dot{\varepsilon} = 10^{-3}s^{-1}$. 375X.

The Higher Temperature Fracture Morphology

At very slow strain-rates and at temperatures inside the ductility
minimum close to or at the transition temperature, a surface nucleated crack
tends to pass completely across the cross-section of the specimen leaving a
fracture surface with an appearance like that in Fig. 4. This brittle
transgranular fracture exhibits a number of more or less parallel ridges
and is basically different from the conventional brittle fracture of a
specimen deformed near the lower end of the ductility minimum. At temperatures somewhat below the upper ductility transition temperature, the fracture
morphology shown in Fig. 4 can still be observed but only close to the gage
section surface where the surface cracks nucleated. This is because with
decreasing test temperature the average flow stress level in a tensile test
rises rapidly.[12] A surface crack growing under this higher stress level
eventually tends to become unstable and accelerate as the crack expands. A
shift in the fracture mode then occurs with increasing crack growth velocity.
Evidence for this can be seen in Fig. 7a corresponding to a Nb specimen
tested at 808 K some 80° below the transition temperature.

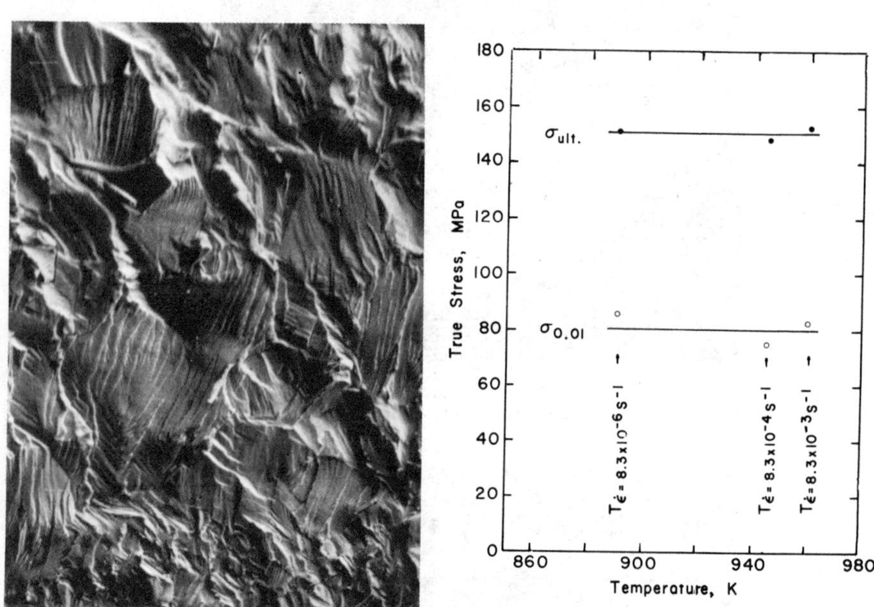

Fig. 4 (Left). A striated brittle fracture surface corresponding to
a higher temperature non-catastrophic failure. Nb-1.6 at. % O tested
at 884 K and $\dot{\varepsilon}$ = 8.3 x 10^{-6}. 300X.

Fig. 5 (Right). At the upper brittle to ductile transition, both the
U.T.S. and the 0.01ε flow stress are independent of the strain-rate
and transition temperature; Nb-1.6 at. % O specimens.

The Higher Temperature Lower Critical Stress

The high temperature shift from brittle fracture to ductile fracture
tends to occur when the average stress level in a tensile test falls below
a fixed value. This is illustrated in Fig. 5 where the ultimate stresses
and the 1% strain flow stresses for specimens tested at the higher temperature transition temperatures are plotted against the transition temperatures

for Nb specimens with 1.6 at. % O deformed at three different strain-rates.
The significant point is that the ultimate stress and the 1.0% strain flow
stress and therefore the average stress are effectively independent of the
strain-rate as well as of the transition temperature which is determined by
the strain-rate. This result is consistent with the concept of an average
lower critical stress that is required for the nucleation and the growth of
surface cracks.

Strain-Rate Dependence of the Higher Temperature Fracture Morphology

In the higher temperature embrittlement region the fracture morphology
is strain-rate dependent. As may be seen in Fig. 6 increasing the strain-
rate at 808 K for niobium specimens with 1.6 at. % O causes both the stress
level in a test (as measured by the U.T.S.) and the ductility (R.A.) to rise.
This increase in ductility with increasing strain-rate is accompanied by a
progressive change in fracture morphology from brittle to ductile. Figure 7
shows representative micrographs corresponding to specimens deformed (a) at
the slowest strain-rate, $8.3 \times 10^{-6}\text{s}^{-1}$, and (b) at the fastest, $8.3 \times 10^{-2}\text{s}^{-1}$
which clearly show the total change in the fracture morphology.

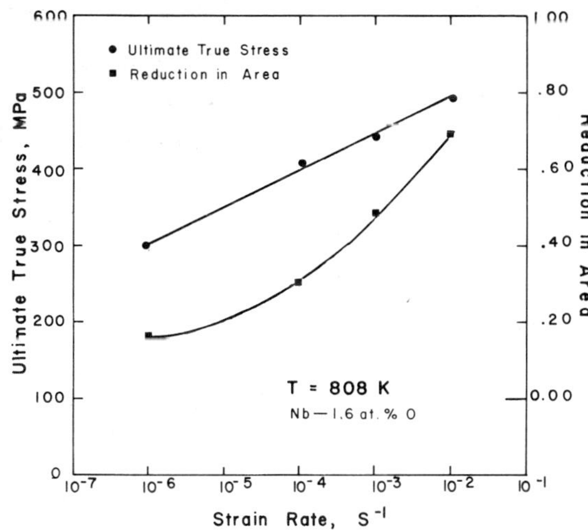

Fig. 6. In the upper temperature embrittlement region
both the ductility (R.A.) and the stress (U.T.S.) increase
with increasing strain-rate.

Relation to Hydrogen Embrittlement

Exothermic occluders of hydrogen such as Nb and Ta also exhibit two
types of embrittlement due to hydrogen depending on whether the hydrogen
concentration does or does not exceed the terminal solubility. Thus,
embrittlement due to hydrogen in solid solution differs from that due to the
presence of hydrides. That due to hydrogen in solid solution normally
increases in severity with decreasing strain-rate while that caused by
hydrides tends to show a reversed strain-rate dependence. For a fixed
concentration of hydrogen, which is in solution at higher temperatures, the
solid solution form of embrittlement will normally predominate at higher
temperatures. However, a shift to the hydride embrittlement form is to be

 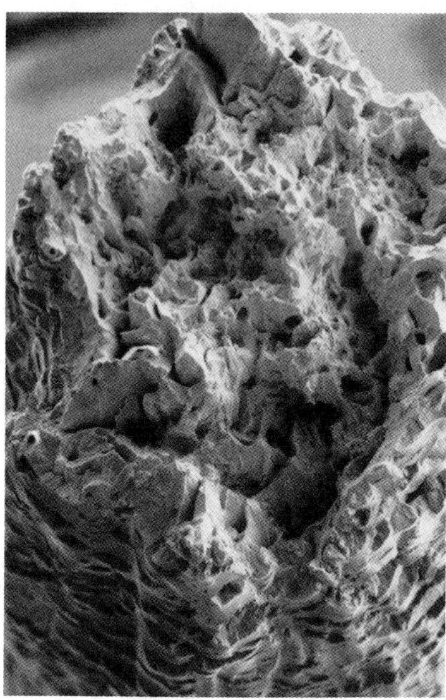

(a)   (b)

Fig. 7.   Fracture surfaces of Nb-1.6 at. % O specimens tested at 808 K
and at (a) 8.3 x $10^{-6}s^{-1}$, and (b) 8.3 x $10^{-2}s^{-1}$.   41X.

expected on decreasing the temperature and thereby causing the hydrogen to
precipitate as hydrides.

In the case of oxygen embrittlement it is possible that oxides should
be a factor to be considered. However, in the work which has been done on
Ta and Nb both phases of the embrittlement can be rationalized in terms of
oxygen in solid solution. This includes the lower temperature embrittlement
(i.e. 450 to 700 K in Nb) which is attributed to dynamic strain-aging and
the higher temperature embrittlement (700 to 950 K in Nb) involving crack
growth that is controlled by the diffusion of oxygen.

With regard to hydrogen, however, it is probable that dynamic strain-
aging is not a significant factor in producing embrittlement. Hydrogen has,
in general, only a minor effect on the work hardening in the metals that it
embrittles. Thus, the lower temperature embrittlements due to both oxygen
and hydrogen in exothermic occluders of hydrogen such as Ta and Nb are
basically different. On the other hand, the higher temperature forms,
involving in both cases the diffusion of interstitials in solid solution,
would appear to have much in common and it is here that one should look for
correlations. This also suggests that the hydrogen embrittlement of steels
which is due to hydrogen in solution should be similar to that due to
oxygen at higher temperatures.

Multiple Surface Cracking

A basic feature of the embrittlement due to oxygen at higher temperatures is multiple cracking that is nucleated at the surface. This is also a

well-documented feature of the embrittlement due to hydrogen in solution. Descriptions of this form of fracture have been given for Nb,[5,13] V,[14] Ni,[10] and various steels.[15-18] In a number of cases direct reference has been made[5,10,13,15,16] to the development of these cracks at an early stage in a tensile test and to their non-catastrophic growth. It would thus appear that this type of fracture is a basic feature of slow strain-rate embrittlement due to either hydrogen or oxygen in solid solution.

The Lower Critical Stress

A basic feature of the slow strain embrittlement of steels is the existence of a lower critical stress below which hydrogen-induced brittle fractures do not occur. This result was deduced using dead loaded notched tensile bars that had been charged with hydrogen and the fracture occurred[4] at a fixed load. This yields a well-defined lower critical (nominal) stress. In a constant cross-head speed tensile test using unnotched specimens the load rises during most of the test. However, if one considers the average stress level, in the rising load part of the stress-strain curve, then one can also deduce a lower critical stress criterion since ductile fracture returned and embrittlement ceased when the average stress, on raising the temperature, fell below the average critical stress. Thus, the oxygen embrittlement, like the hydrogen embrittlement, exhibits a lower critical stress.

Strain-Rate Dependence of the Fracture Morphology

Beachem[19] has demonstrated with the use of wedge loaded and notched steel specimens charged with hydrogen that the fracture morphology in hydrogen embrittlement can be a function of the stress intensity factor with the fracture becoming progressively more brittle the lower the stress intensity factor. The stress intensity factor is proportional to the applied stress. Thus the curve in Fig. 6 showing that the U.T.S. falls with decreasing strain-rate in the higher temperature oxygen embrittlement region of Nb can be interpreted as showing that decreasing the strain-rate lowers the stress intensity factor. The two photomicrographs in Fig. 7 then confirm that in the case of oxygen embrittlement lowering the stress intensity factor (strain-rate) causes the fracture morphology to change from ductile micro-void coalescence to a brittle morphology. Thus, the variation of the fracture morphology with strain-rate or stress intensity factor is basically equivalent in both the hydrogen embrittlement of steels and the oxygen embrittlement of Ta and Nb.

The Activation Energy for Diffusion

According to Troiano[4] the slow strain-rate embrittlement due to an interstitial in solid solution involves a stress induced drift of the interstitial to the region of maximum triaxial stress ahead of a crack. It is logical to assume, therefore, that the temperature of some basic embrittlement feature (i.e. the brittle to ductile transition) is proportional to the activation energy for diffusion of the interstitial in question. In this regard it is necessary to consider the effective activation energy because of trapping of the interstitial by other impurities, precipitates and dislocations. Thus, the temperature of the return of ductility on the high temperature side of the ductility minimum[6] of a Nb specimen with 1 at. % H is about 280 K and for a Nb specimen[12] with 0.75 at. % O it is about 860 K where in both cases the strain-rate is about $10^{-4}s^{-1}$. The ratio of these two temperatures is about 3 to 1. The corresponding activation energies in annealed pure Nb are 10.2 kJ/mole for hydrogen and 110 kJ/mole for oxygen with a ratio of close to 10 to 1. This large difference in the two ratios is probably attributable to a much larger effect of trapping on the diffusion of hydrogen than on the diffusion of oxygen.

Support for the conclusion that the activation energy for the diffusion of hydrogen can be strongly altered by trapping is available from studies made on AISI 4340 steel heat-treated to strength levels above 1,400 MPa. It should first be noted that the activation energy for the diffusion of hydrogen in pure annealed iron is about 8 kJ/mole. Measurements by Steigerwald et al.[20] of the temperature dependence of the time to incubate cracks in static fatigue tests of hydrogen charged 4340 steel gave an activation energy of 38 kJ/mole. Two other papers[16,21] have also reported nearly this same activation energy for other diffusion controlled aspects of the hydrogen embrittlement of 4340 steel. The assumption that this activation energy actually corresponds to the diffusion of hydrogen in this steel is supported by the diffusion study of Beck et al.[22] who measured the diffusion of hydrogen through heat treated 4340 steel diaphragms using a permeation technique and obtained an activation energy of 38.5 kJ/mole.

Acknowledgments

The support of this research by the Army Research Office, Durham, under Contract Number DAAG-79-G-0050 is gratefully acknowledged. Figures 2 and 3 represent work performed by A. A. Garcia. Most of the niobium data quoted in this paper is from the thesis work of P. G. Watson which was supported by the Department of Energy under Contract Number EY-76-S-05-3262.

References

1.   J. R. Donoso and R. E. Reed-Hill, Met. Trans. A, 1976, vol. 7A, pp. 761–5.
2.   A. E. Diaz and R. E. Reed-Hill, Scripta Met., 1979, vol. 13, pp. 491–6.
3.   J. T. Brown and W. M. Baldwin, Jr., Trans. AIME, 1947, vol. 200, pp. 298–304.
4.   A. R. Troiano, Trans. ASM, 1960, vol. 52, pp. 54–80.
5.   T. M. Wood and R. D. Daniels, TMS-AIME, 1965, vol. 233, pp. 898–903.
6.   D. Hardie and P. McIntyre, Met. Trans., 1973, vol. 4, pp. 1247–54.
7.   A. L. Eustice and O. N. Carlson, Trans. AIME, 1961, vol. 221, pp. 238–41.
8.   D. H. Sherman, C. V. Owen and T. E. Scott, TMS-AIME, 1968, vol. 242, pp. 1775–84.
9.   A. G. Ingram, E. S. Bartlett and H. R. Ogden, TMS-AIME, 1963, vol. 227, pp. 131–6.
10. A. H. Windle and G. C. Smith, Met. Sci. J., 1970, vol. 4, pp. 136–44.
11. R. D. Daniels, R. G. Quigg and A. R. Troiano, Trans. ASM, 1959, vol. 51, pp. 843–59.
12. P. G. Watson and R. E. Reed-Hill, submitted to Met. Trans. A.
13. D. Hardie and P. McIntyre, Hydrogene Met. Cong. Intl. 1972, vol. 2, pp. 453–57, Paris, France, 1973.
14. R. H. Van Fossen, Jr., T. E. Scott and O. N. Carlson, J. Less-Com. Met., 1965, vol. 9, pp. 437–51.
15. R. J. Walter and W. T. Chandler, Mat. Sci. Eng., 1971, vol. 8, pp. 90–7.
16. J. G. Morlet, H. H. Johnson and A. R. Troiano, J.I.S.I., 1958, vol. 189, pp. 37–44.
17. D. Hardie and T. I. Murray, Met. Tech., 1978, vol. 5, pp. 145–50.
18. W. Hofmann and W. Rauls, Weld. J., 1965, vol. 44, pp. 226s–30s.
19. C. D. Beachem, Met. Trans., 1972, vol. 3, pp. 437–51.
20. E. A. Steigerwald, F. W. Schaller and A. R. Troiano, TMS-AIME, 1959, vol. 215, pp. 1048–52.
21. C. F. Barth and E. A. Steigerwald, Met. Trans., 1970, vol. 1, pp. 3451–55.
22. W. Beck, J. O'M. Bockris, J. McBreen and L. Nanis, Proc. Roy. Soc., 1966, vol. A290, pp. 220–35.

STRESS CORROSION CRACK PROPAGATION IN ZIRCALOY-2 IN A

CH$_3$OH/HCl MIXTURE

P. Majumdar, M. Golozar and J.C. Scully

Department of Metallurgy
University of Leeds
Leeds LS2 9JT
England.

Stress corrosion crack propagation has been examined in constant extension rate tests at room temperature in CH$_3$OH containing 0.4 Vol% HCl.
Two modes of fracture were observed: an intergranular dissolution mode and
a transgranular cleavage mode which is caused by hydrogen absorbed during
the dissolution process. A reversible embrittlement phenomenon has demonstrated that cleavage is associated with an absorbed species but the phenomenon was observed only in stressed specimens. An additional feature observed during long exposure (>2 days) was the occurrence of intergranular
blisters of hydrogen. These are associated with the intergranular precipitates.

Introduction

Stress corrosion cracking in Zircaloy-2 at room temperature has been
reported by several authors (1-3) who have observed cracking in Hg,
CH$_3$OH/I$_2$, CH$_3$OH/HCl and aqueous NaCl. Two different modes of fracture have
been observed: intergranular and transgranular cleavage, which occur either
together or separately. It is the purpose of this paper to show that the
intergranular cracking is caused by stress-assisted dissolution while the
transgranular cleavage is caused by hydrogen absorbed during the dissolution process.

Experimental Procedure

The Zircaloy-2 alloy was in the form of sheet 1.5mm thick and contained
(wt.%): 1.5Sn, 0.14Fe, 0.1Cr, 0.05Ni, 0.11 O$_2$ and (in ppm) 155C, 7-11H,
43Hf, 25W and 1U. All specimens were annealed at 800°C for 1h in a vacuum
of 10^{-6} torr and such material had a 0.1% Proof Stress of 385 N mm^{-2}, a UTS
of 450 N mm^{-2} and an elongation of ca. 32% for a 25mm gauge length. Two
types of specimens were used for constant extension rate experiments:
(i) specimens 20mm wide with one 6mm deep 60° notch, and (ii) specimens
16mm wide with a 2mm deep 60° notch. The former were used in an Instron
Tensile Testing Machine, while the latter were used on a low crosshead
machine based upon a design by Parkins. Specimens were marked with lines
2mm apart running along the length and crack propagation was followed

optically. Single cracks propagated across the narrowest part of specimens.
The mixture employed consisted of CH_3OH and 0.4 Vol% of concentrated HCl
(35 wt.%) which had been previously determined as the most aggressive proportions. This was placed in a polyethylene beaker cemented about the notch.

Embrittlement experiments of two different types were conducted employing tensile specimens with a 25mm gauge length. In one type of experiment
specimens were strained on the Instron Machine at a crosshead speed of
16.6μm s^{-1} to a stress of 385 N mm^{-2} at which point straining was stopped.
The solution mixture was then added to the beaker cemented about the specimen. Specimens were anodically polarized for 1200s at a current density of
10μA mm^{-2}. The solution was then removed from the beaker. Specimens were
then broken either immediately or removed from the testing machine and replaced after a time interval during which they were maintained at room
temperature or above, which is referred to below as the ageing time. Because the load fell slightly during polarization, this type of experiment
is referred to as a constant strain experiment. In the second type of
experiment specimens were strained at several crosshead speeds within the
range 0.32–83μm s^{-1} while subjected to one of two anodic current densities
(10 and 100μA mm^{-2}) up to a stress of 400 N mm^{-2}. The solution was then
removed and, as in the first type of experiment, specimens were broken at
the same crosshead speed as that used in the first part of the experiment
either immediately or after a time interval during which they were not
under an external stress. Polarization was applied for a period of time.
Typical times were 300s with a crosshead speed of 83μm s^{-1} and 2400s with a
crosshead speed of 16μm s^{-1}, both at the lower current density. These are
referred to as constant crosshead speed experiments.

Experimental Results

Constant Extension Rate Experiments

In these experiments specimens tested under open circuit conditions
exhibited three possible different types of fracture: ductile overload,
transgranular cleavage and intergranular separation, the number and proportion depending upon the crosshead speed as shown in Fig. 1. At relatively high crosshead speeds a two zone intergranular/ductile fracture was
observed, an example of which is shown in Fig. 2. At lower crosshead
speeds a three zone intergranular/cleavage/ductile fracture was observed.
Figure 3 shows a section of such a fracture in which the first two zones
are present.

The crack propagation rate measured varied with crosshead speed.
Typical values were 8.4mm h^{-1} at a crosshead speed of 0.166μm s^{-1} and
0.2mm h^{-1} at 1.6nm s^{-1}. The effect of potential was examined by incorporating a potentiostatic circuit at a single crosshead speed of 0.166μm s^{-1}.
The results are shown in Fig. 4 which also shows the time-to-failure. At
both ends of the potential range examined specimen surfaces darkened during
experiments. Zirconium hydride was identified by X-ray diffraction. Anodic polarization promoted increased intergranular fracture and resulted in
the occurrence of cleavage which was not seen under open circuit conditions.
Cathodic polarization decreased the amount of intergranular fracture as the
employed potential was made less noble and eventually cracking was
completely prevented.

Earlier experiments to determine the most corrosive mixture and to
determine the corrosion rate in the selected mixture under open circuit
conditions indicated intergranular corrosion, the depth of which could be
measured by breaking specimens after a range of times of exposure. The
cause of such fracture appeared to be due only to the selective dissolution
of the grain boundaries. Baking of specimens between exposure and fracture,

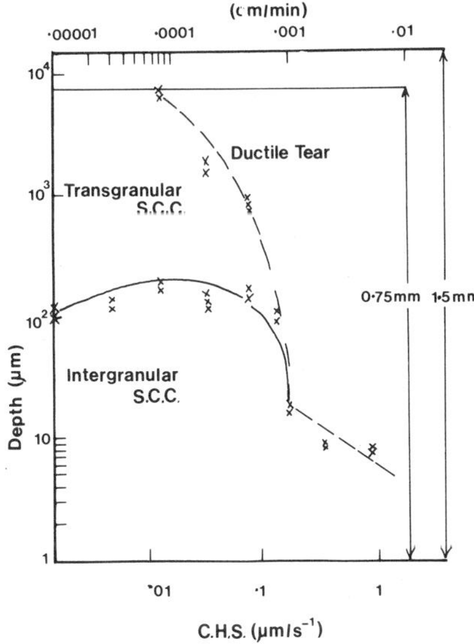

Fig. 1.   The fracture modes observed in constant extension rate experiments.

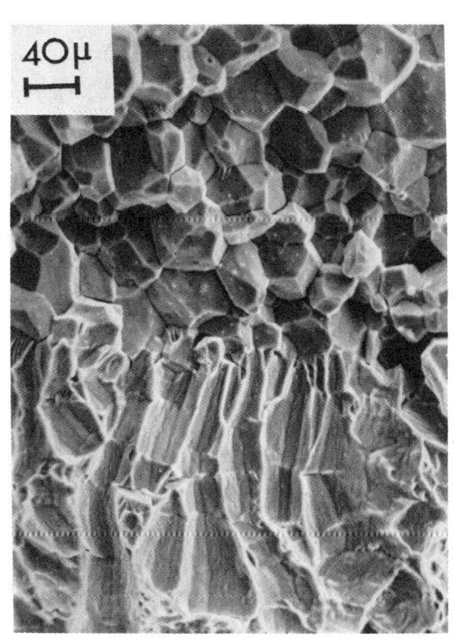

Fig. 2.   An intergranular/ductile two zone fracture: 8.3μm s⁻¹

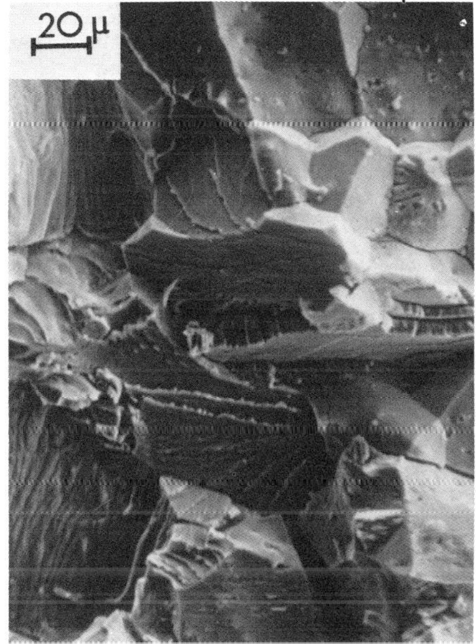

Fig. 3.   An intergranular/cleavage fracture: 3nm s⁻¹.

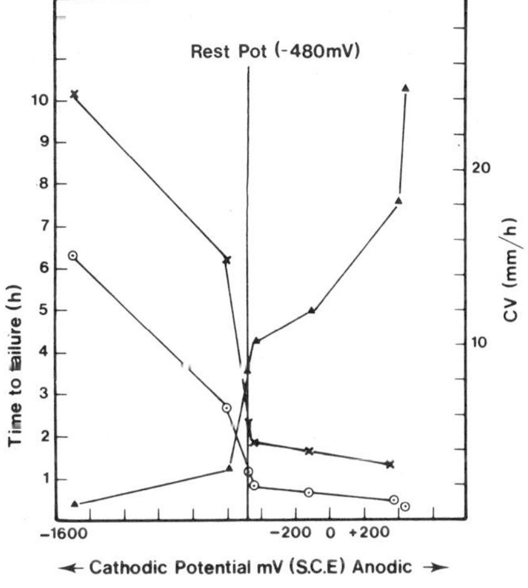

Fig. 4.   The effect of potential upon crack propagation rate and time-to-fracture at a crosshead speed of 0.166μm s⁻¹.

for example, appeared to have no effect and it was not thought that absorbed hydrogen was responsible for any part of the intergranular fracture. The maximum depth of intergranular corrosion was compared with the maximum depth of intergranular stress corrosion observed in the constant extension rate experiments, the comparison being made for identical periods of exposure. The results are shown in Fig. 5. Over a wide range of crosshead speeds the stress corroded specimens exhibited more intergranular separation than the corroded specimens and this is most easily explained as arising from stress (or strain)-assisted dissolution. At the high end of the range the difference might be expected to disappear since such experiments occupy very short periods of time. Complete ductile fracture and no intergranular corrosion are seen. At the low end of the range the disappearance of the stress-assisted intergranular fracture is not so obvious but it can be observed that it is replaced by cleavage.

The explanation for the disappearance of the stress-assisted intergranular fracture is best obtained by consideration of the effect of the applied crosshead speed upon the imposed stress. As the applied crosshead speed is lowered the time taken to initiate the intergranular stress corrosion will increase. During all this time hydrogen is being absorbed. In the low crosshead speed experiments the amount absorbed is great enough to cause transgranular cleavage before the stress is sufficiently high to cause intergranular stress corrosion. Only intergranular corrosion occurs prior to cleavage in such experiments; there is no stress-assisted component.

In the experiments to determine corrosion rate unstressed specimens were subsequently fractured in air. Such specimens did not exhibit any cleavage after being fractured over a wide range of crosshead speeds. This was unexpected since hydrogen is being absorbed and comparison can be made with equivalent experiments on a Ti-O alloy (4) in which a cleavage zone, intermediate to the intergranular and ductile zones, was observed after exposure of unstressed specimens and fracture in air. Such three zone fracture specimens were not seen in Zircaloy-2 specimens. The occurrence of cleavage under such experimental conditions in titanium alloys depends upon the absorption of a sufficient amount of hydrogen and the establishment of a volume of embrittled alloy. The absorbed flux and the diffusion of H atoms are in competition and unless the difference between these is sufficiently great conditions for cleavage will not be met. Anodic polarization of unstressed Zircaloy-2 specimens did not change the result. No cleavage was ever obtained from corroded unstressed specimens.

The difference between specimens tested at a low crosshead speed in Fig. 5 and showing no stress-assisted effect is that the strained specimens were under increasing load during the whole time of the experiment. The occurrence of cleavage under such conditions must have been caused by the slowly increasing stress. This might have been caused by the effect of the stress in breaking surface films and thereby promoting an enhanced rate of H absorption. In addition, there may have also been an effect between the deforming lattice and the absorbed hydrogen. In order to examine this point further and in order to obtain evidence that the observed cleavage was caused by an absorbed species the two subsequent types of experiments were performed (5).

Constant Strain Experiments

Specimens broken immediately after polarization exhibited a three zone fracture. After ageing the intermediate cleavage fracture was less evident and after long ageing times at room temperature or shorter times at higher temperatures it did not occur at all. The gradual reduction of the cleavage zone was clear evidence that it was caused by an absorbed species which would be hydrogen since no other species can be supposed and hydrogen absorption undoubtedly occurs. The reduction was associated, as might be expected, with an increase in ductility, as illustrated in Fig. 6.

Constant Crosshead Speed Experiments

Specimens broken immediately after the end of polarization exhibited a three zone fracture. Upon ageing the subsequent elongation-to-failure increased and the intermediate cleavage zone was reduced and eventually was not observed. The changes in elongation with time are shown in Fig. 7. It can be seen that the degree of embrittlement was greater in experiments at lower crosshead speeds, a result that emphasized that the embrittlement is a time-dependent phenomenon. Figure 7 shows results obtained as described in the experimental section. Specimens were broken after zero ageing, after 24h ageing at room temperature and after 1h at 100°C. Two current densities were employed: 10 and $100\mu A\ mm^{-2}$.

Blisters

During weight loss experiments to determine the corrosion rate under open circuit conditions specimen surfaces developed blisters. They became visible to the naked eye after around a 48h exposure. The number observed was always small and they grew randomly over the type of specimen employed. A blistered specimen is shown in Fig. 8. Such observations were unexpected. Elaborate procedures were adopted to ensure that the blisters were hollow. These included, for example, depressing the surfaces of blisters with needles. A broken blister is shown in Fig. 9. The internal surface of the blister skin and of the blister body were always intergranular and some of this detail is visible in the Figure. In some examples, the intergranular intermetallic precipitates could be seen on the internal blister surface.

Discussion

The results show that stress corrosion crack propagation occurs in two different fracture modes in this sytem. They also provide evidence that there are two different mechanisms of cracking each of which is associated with a specific fracture mode.

The intergranular cracking appears to occur by a form of stress-assisted dissolution, a feature emphasized particularly in the results shown in Fig. 5. This interpretation is also consistent with the effect of potential shown in Fig. 4: anodic polarization increases the amount of intergranular cracking whereas cathodic polarization reduces it. In addition, the open circuit weight loss determination experiments on unstressed specimens had also indicated the occurrence of intergranular corrosion. The reason for this preferential dissolution was not explored. It can only be generally ascribed to the presence in the grain boundaries of unspecified solute elements in solid solution, with the additional possibility of an effect due to the intermetallic precipitates. In the testing solution the alloy appears to form a film or adsorbed layer on the grain surfaces which affords protection to them, leaving only the grain boundaries to be corroded.

The transgranular cleavage cracking appears to be caused by absorbed hydrogen generated at the crack tip. The system therefore is similar to the double fracture mode observed in α-Ti alloys (4) exposed to a similar mixture: intergranular dissolution with absorption of hydrogen leading to transgranular cleavage. Whether the cleavage occurs as the result of a stress- or strain-induced hydride precipitation is not known. By comparison with the fracture mechanism in α-Ti alloys the cleavage mode in Zircaloy-2 is perhaps best described as a form of slow strain-rate hydrogen embrittlement. The principal difference between the behaviour of α-Ti alloys and Zircaloy-2 after exposure to CH_3OH/HCl mixtures is that cleavage in Zircaloy-2 was seen

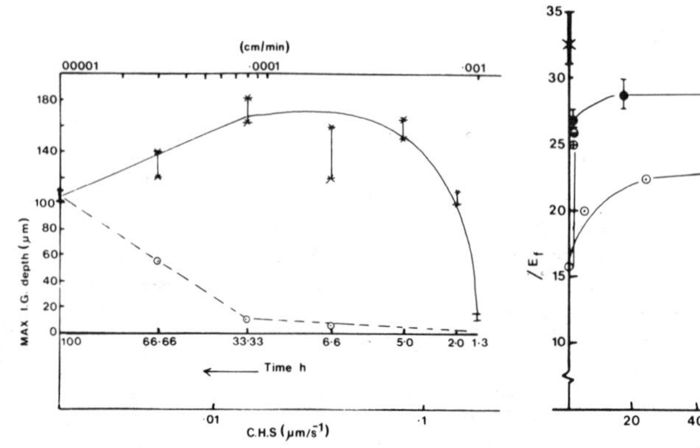

Fig. 5.   A comparison of the amount of intergranular fracture seen in stress corroded specimens and corroded specimens broken in air.

Fig. 6.   Elongation-to-fracture, ε_f, of specimens broken after anodic polarization at 10μA mm^{-2} for 1200s under an initial stress of 385 N mm^{-2} followed by ageing at two temperatures and subsequent fracture in air.

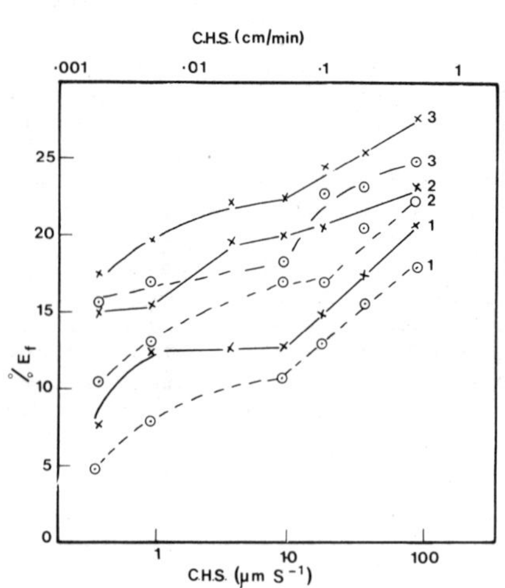

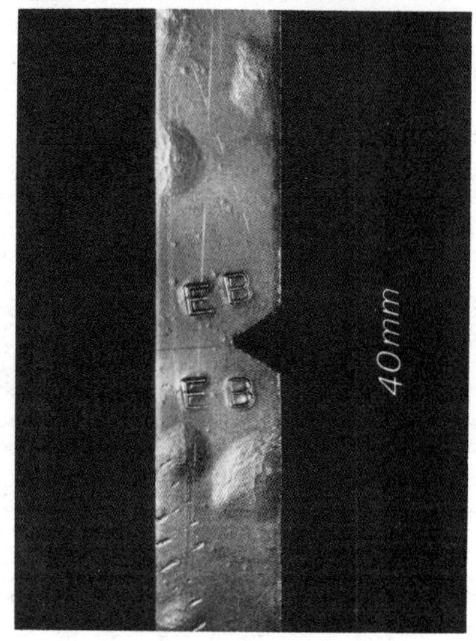

Fig. 7.   Elongation-to-fracture, ε_f, of specimens broken as described in the text: 1. Zero ageing, 2. 24h, 3. 1h at 100°C.

Fig. 8.   A blistered specimen of Zircaloy-2 after 72h exposure.

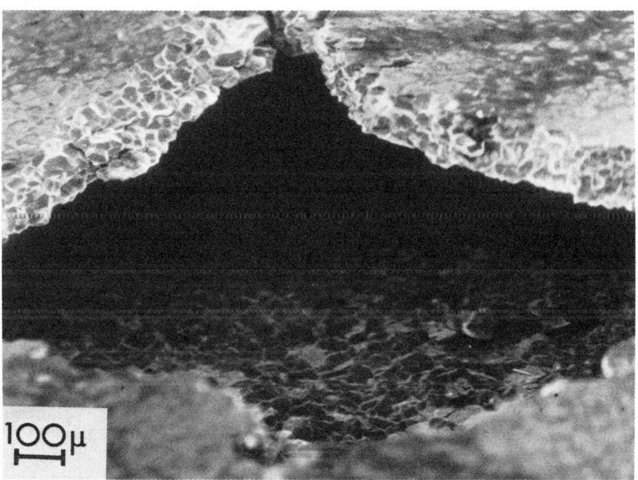

Fig. 9. Part of a blister surface broken mechanically. The intergranular
surface on the inside is clearly visible.

only after the constant strain and constant crosshead speed experiments
described above and not after simple exposure of unstressed specimens. The
function of the stress, or attendant strain or strain-rate, was either to
ensure that an additional amount of hydrogen was absorbed, prior to fracture
in air, sufficient to cause a zone of cleavage fracture, and/or to promote
the necessary interaction between mobile dislocations and absorbed hydrogen
atoms. The exact explanation for this was not examined. The results do
emphasize, however, that this type of transgranular cleavage, when occurring
during stress corrosion crack propagation, is the result of the simultaneous
interaction of the effects of stress and hydrogen absorption.

The observed effect of potential upon the stress corrosion crack propagation rate is consistent with the proposed function of absorbed hydrogen.
Anodic polarization produced additional intergranular fracture and the
higher rates of dissolution associated with this form of fracture could be
expected to promote enhanced hydrogen absorption, consistent with the occurrence of cleavage, which was not seen without anodic polarization at the
crosshead speed employed. The assumption in such an interpretation is the
amount of cleavage observed in specimens broken at the one crosshead speed
employed in the polarization experiments is controlled by the flux of
absorbed hydrogen. From what has already been discussed with regard to the
results shown in Figs. 6 and 7 this assumption is fully consistent: cleavage
occurs as a result of hydrogen absorption at a slowly deforming crack front.
Increasing the amount of hydrogen absorbed by anodic polarization (or by
using a cathodic poison (6)) increases the amount of cleavage observed and
raises the maximum crosshead speed at which cleavage is observed. Thus the
occurrence of cleavage at any one crosshead speed is determined by the absorbed hydrogen flux, and at any one value of absorbed hydrogen flux by the
applied crosshead speed, at least over a certain range of values of each
variable.

The reverse effect of cathodic polarization can be attributed to an increase in the film formation rate, an effect seen also on α-Ti alloys. Such
polarization reduces and can also prevent intergranular dissolution, and
associated with such an effect will be the non-appearance of cleavage fracture.   Film formation does not, however, prevent the entry of discharged
hydrogen into the lattice, it merely reduces the flux to below the level
necessary to cause cleavage. Under cathodic polarization conditions massive
hydride formation does occur on the surface and it cannot therefore be argued that sufficient hydrogen concentration does not occur when the surface is
filmed i.e. that the absorbed hydrogen diffuses too rapidly into the bulk of
the alloy lattice. Since the massive hydride formation does not promote
the cleavage observed in the stress corrosion experiments it must be supposed that intrinsic to the cleavage mechanism are very specific conditions
of absorbed hydrogen flux and deforming unfilmed metal surface.

The unusual occurrence of blisters must be commented upon. The Zr/H
interaction results in an equilibrium such that only very low partial
pressures of hydrogen can exist in contact with a Zr surface. The combination of atoms to form molecular hydrogen in the Zircaloy-2 grain boundaries
may be attributed to a catalytic action of either the intermetallic compounds or of elements in solid solution in that region. Intermetallic compounds
were always visible on the internal blister surfaces and metallographic
examination (6) indicated that they acted as nucleating sites. The exact
conditions under which blisters occur are not known and their occurrence
has not been reported previously. It can be supposed that hydrogen atoms
accumulate preferentially in the grain boundaries. At some stage a relatively small number of blisters are nucleated. These act perhaps as sinks,
possibly because nucleation is difficult, and would account for the small
number of blisters observed in specimens exposed over periods of 2-21 days.

The observation of intergranular blisters inevitably raises the question of whether at a pre-nucleation stage absorbed hydrogen could be responsible for some of the intergranular cracking observed. This was not thought to be the case. The ageing experiments, for example, all failed to demonstrate that the amount of intergranular fracture was at all reduced, a result which would be an expected result if hydrogen was causing any proportion of it. Furthermore, no ageing effect was detected under conditions where cleavage would not be observed, e.g. in weight loss specimens broken afterwards. It could be concluded that any effect of absorbed hydrogen on intergranular stress corrosion cracking is either non-existent or very small. The results of Fig. 5 are consistent with hydrogen having no effect upon intergranular cracking.

Acknowledgements. The authors are grateful to the Sandia Corporation for the provision of the Zircaloy-2.

References

1.   B. Cox, Rev. Coatings and Corrosion, 1975, vol. 1, pp. 366-422.

2.   B. Cox, Corrosion, 1972, vol. 28, pp. 207-18.

3.   K. Elayaperumal, P.K. De and J. Balachandra, Corros. Sci., 1971, vol. 11, pp. 579-89.

4.   J.C. Scully and T.A. Adepoju, Corros. Sci., 1977, vol. 17, pp. 789-812.

5.   P. Majumdar and J.C. Scully, Corros. Sci., 1979, vol. 19, pp. 141-5.

6.   M. Golozar and J.C. Scully, to be published.

DISCUSSION

<u>L. Christodoulou, Carnegie-Mellon University, Pittsburgh, PA</u>: What are the
particles nucleating the blisters you observe?

<u>J.C. Scully</u>:   There are a number of different particles present in the grain
boundaries of Zircaloy-2.   It has not proved possible for us to determine
which of these nucleate the hydrogen blisters.

<u>E.N. Pugh, National Bureau of Standards, Washington, DC</u>: A general question.
The transgranular SC fracture-surfaces you show in this system, which you
attribute to cleavage, are virtually indistinguishable from those produced
in other transgranular SCC cases.   Yet you maintain in your other publications that the latter are produced by anodic dissolution.   Could you comment
on this apparent contradiction?

<u>J.C. Scully</u>:   Zircaloy-2 and α titanium alloys have a history of transgranular
cleavage failures under various circumstances.   It is not altogether surprising that such cleavage should occur under stress corrosion conditions
where hydrogen can be absorbed.   In FCC alloys, however, evidence for such
cleavage is not nearly so apparent.   Your own work (and others) lends support to hydrogen embrittlement of austenitic stainless steels giving fractures similar to those seen after stress corrosion fracture.   For α brass
this has not been shown.   It is also noteworthy that numerous ancillary experiments lend support to the role of hydrogen in the hexagonal alloys, e.g.
reversible embrittlement experiments, whereas no equivalent experiments have
been demonstrated for α brass, and only recently has some evidence been provided for austenitic stainless steels.   At the moment the simplest explanation is being put forward: hexagonal alloys suffer from hydrogen embrittlement under stress corrosion conditions, FCC alloys do not.   Experiments that
demonstrate hydrogen embrittlement as causing stress corrosion fracture have
yet to be done.   It is, perhaps, important to emphasize that transgranular
cracking is being discussed.   The FCC aluminum alloys which exhibit mainly
intergranular fracture do appear to be failing by hydrogen embrittlement
under stress corrosion conditions.

<u>J.P. Fidelle, Commissariat à l'Energie Atomique, Bruyères-le-Châtel, France</u>:
H_2 bubbles are known to cause damage in metals much as Ti, at temperatures
where hydrides no longer exist or when they break down during fast heating.
However, the formation of H_2 blisters you have found in Zr, at temperatures
where Zr hydride is stable is both new and surprising, because H supersaturation would be rather expected to cause hydriding.   Then can it be that the
rate of H generation is so fast that H_2 cannot react as fast with the metal?
Even so, hydriding must take place eventually, as the blisters' internal surfaces, uncontaminated by air, must react with pressurized H_2.   Did you observe
it when opening blisters and looking at their surface?

<u>J.C. Scully</u>:   The observation of the formation of hydrogen blisters during
the corrosion of Zircaloy-2 was a considerable surprise to us.   This paper
reports only their occurrence.   Several different phenomena were observed.
Some blisters broke while exposed to the solution, in some cases due to
stress corrosion because of the stresses in the outer blister skin.   In such
cases the solution entered the blister and caused intergranular corrosion.
Blisters that did not break developed a dark interior surface.   This was
almost certainly due to hydriding for the reasons that the questioner gives.
Positive identification of the hydride has not so far been made.   So far as
the kinetics are concerned, we can only surmise that the recombination reaction, which we suppose is catalysed by one or more of the intermetallic
particles, occurs more rapidly than the hydriding reaction but also, perhaps

at an earlier stage. Since a relatively small number of blisters is observed in any single specimen (2-20) these may act as hydrogen sinks and
cause hydrogen concentration gradients, particularly along the grain boundaries. In this way it can be envisaged that blister nucleation and growth
could occur before general supersaturation had occurred or reached a level
that would give widespread hydride nucleation.

High-Temperature Behavior

THE ROLE OF GAS VENTING IN HYDROGEN ATTACK OF STEEL*

Mark Ransick and Paul Shewmon

Department of Metallurgical Engineering
The Ohio State University
Columbus, Ohio

It is shown that as the hydrogen attack (HA) of steel progresses the
methane bubbles linkup and vent their methane to the surface. This reduces
the internal pressure and stops the swelling. The amount of swelling that
occurs before venting is decreased by banding in the steel, and by prior
cold work. Banding tends to concentrate the bubbles in a plane. Cold work
greatly increases the number of boundaries covered with bubbles. The gross
blistering of some refinery piping under mild conditions of HA reported
by Sorell and Humphries, can also be explained in this way.

*Work supported by the Division of Material Sciences, Office of Basic Energy
Sciences, DOE.

The engineering use of materials is often influenced or inhibited by the internal generation of an insoluble gas during service. In the cases of hydrogen attack (HA) of pressure vessel steel (methane formation)[1], or hydrogen embrittlement of copper (steam formation)[2], the primary concern is the loss of mechanical properties, especially ductility due to the growth of bubbles into fissures. In nuclear fuel elements (Xe and Kr formation) a primary concern is when the gas bubbles will interconnect and vent gas to the fuel element plenum, thus stopping the swelling of the fuel even though the fission process continues to generate noble gases[3]. In the HA of steel pipe, cracking and loss of ductility usually constitutes failure, but in some cases gross blistering or ballooning of the pipe surface constitutes failure[4].

Though the literature on hydrogen attack is appreciable there are virtually no papers discussing what limits the expansion of a steel undergoing HA. Since the complete decarburization of steel is often observed in steel that has undergone HA, one might expect that swelling continues until the carbon is exhausted. However, inspection of Pishko's density data (Fig. 1, Ref. 5) and his photomicrographs (his Fig. 11) shows that in his REM steel the density change is complete before any significant decrease in the volume of pearlite is observed.

This leads to the suggestion that the expansion of a 1020 steel undergoing HA is determined not by the exhaustion of the carbide supply but by the venting of the methane to the surface which relieves the internal pressure driving expansion. This paper describes several experiments aimed at studying this postulate and the occurrence of surface venting. In a related paper on the effect of cold work on HA the existence of venting in cold worked steel was suggested by the great depth of decarburization and very limited swelling of the internal fissures accompanying HA in the strained material, while in the undeformed material HA gave larger fissures (bubbles) but much less dissolution of pearlite[6].

Experimental Procedure

The steel used in these experiments was a silicon killed 1020 steel (0.2% C, 0.7% Mn, 0.3% Si, 0.022% S, 0.01% P). The steel was chosen because it showed a particular propensity to form small (20 μm diam.) blisters on its surface during HA and because of our interest in the effect of cold work on HA in this higher sulfur steel.

The $\frac{1}{4}$ X $\frac{1}{2}$ X $2\frac{1}{2}$ inch samples and three point bending procedure are described elsewhere[6]. For this study the surfaces of the sample to be studied for venting were polished through 0.3 μ alumina or 1.5 μ diamond before bending and exposure to the 6.4 MPa hydrogen at 450°C. The surfaces could thus be studied in the SEM without any surface preparation after HA had occurred. Two samples were studied, one (unbent) was exposed for 88 hr. One (bent) was exposed for 176 hr.

Results

The surface of the sample exposed for 88 hours showed several types of features. The most numerous were small (10 to 25 μm) upraised discs of metal which appeared all across the surfaces (Fig. 1). These were distinct in appearance from blisters. Blisters generally are a factor of ten larger. The upraised discs usually consisted of slip-like terraces (Fig. 2), whereas the surface of the blisters were always smooth. Many of these upraised discs had a relatively smooth side, with the opposite side having definite

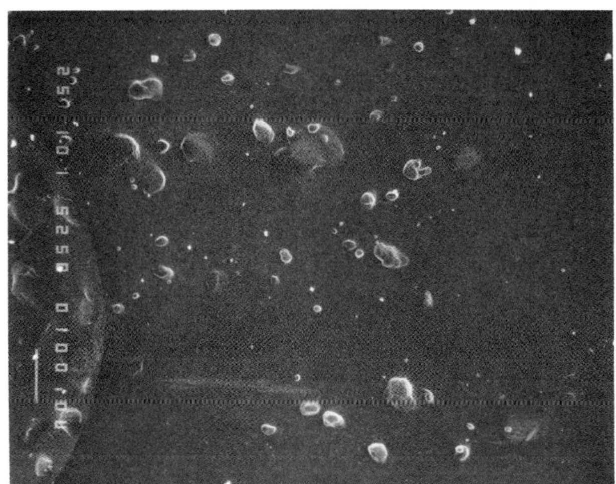

Figure 1:   Copious upraised discs after 88 hr.

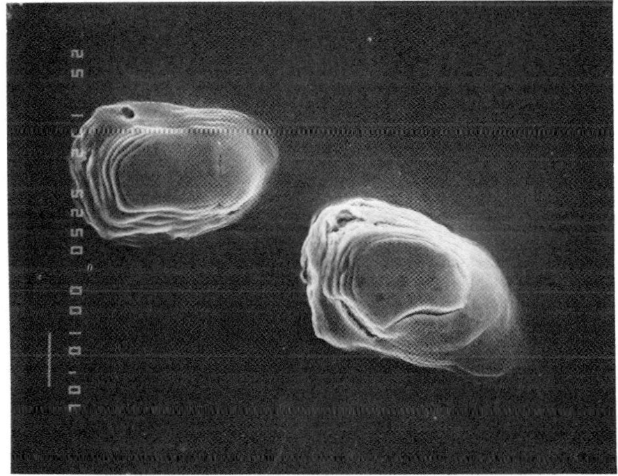

Figure 2:   Higher magnification of raised discs.   Note
terraces on both   and vents on one.

terraces. When this sample was sectioned it was discovered that under these small discs were small inclusions, either MnS or $MnOSiO_2$ type. Adjacent to the inclusion was a void, either in between the inclusion and the surface or on the side of the inclusion opposite the free surface. Fig. 3 shows one such cross section thru a raised disc with a MnS type inclusion beneath the bubble that raised the disc. The inclusion was inside of a pearlite colony which dissolved very little. Several such pictures gave this appearance of a small channel that connected the void to the surface and thus vented the gas. Also, on the surface of these discs there often appear to be small fissures in between some of the terraces. This is clear in the larger disc in Fig. 2.

Another surface feature was the vents or fissures observed on the flat surface. These occurred rarely but when observed several fissures appeared close together. Fig. 4 shows such a region. The contrast at the edges of the fissure makes them stand out. These vents followed along grain boundaries. No vents could be detected that were in the form of a single hole or point on a grain boundary, or at the intersection of three boundaries.

Numerous fissures were located on the surface of the bent bar that was exposed for 176 hours. The fissures that were observed along the surface on the edge of the sample [the side that was ¼ in. thick and normal to the axis around which bending occurred] all followed along what were either clearly defined grain boundaries or along inclusions. The grain boundaries could be readily detected due to the thermal etching that occurred as a result of the 450°C temperature exposure. The fissured vents in the region of tension developed primarily, but not exclusively, parallel to the rolling direction which was also the tensile axis. It was not obvious whether the total number of vented boundaries was greater in the region of tension than in the region that was undeformed. While there exist vents along boundaries in the undeformed region, a higher percentage of the vents occur along inclusions in the undeformed region than in the tensile region, e.g. see Fig. 5. Because of the vents forming along inclusions there exists a degree of vent directionality due to the alignment of the inclusions in the rolling plane. Large vents existed on the sample wherever a large inclusion existed.

The vents that occurred in the side in the region of compression occurred perpendicular to the rolling direction and thus perpendicular to the applied compressive stress. It is more difficult to determine whether these vents occur exclusively along grain boundaries. Due to more numerous slip lines the grain boundaries that are thermally etched are more difficult to observe. Occasionally some of the vents followed a jagged path, unlike the vents in the tension region that clearly follow the smooth grain boundaries. Fig. 6 shows such a jagged vent, which may still follow grain boundaries along its entire length. The vents here in the cold worked material have a constant, narrow width similar to the extended fissures found internally in cold worked material[6].

It is clear that the surface vents in the bent region followed the direction of the primary internal fissures, that is, parallel to the tensile axis on the top (tension) side and normal to the axis of compressive stress on the bottom (compression) side. Examination of a cross-section often showed internal fissures running all the way out to the surface.

Discussion

The principle conclusion of these observations is that the internal

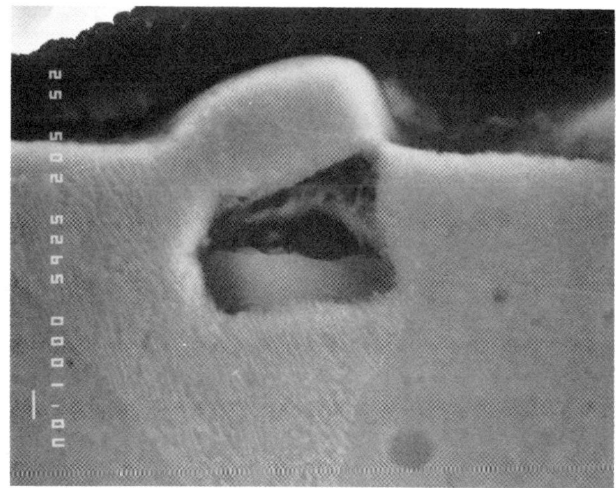

Figure 3:   Section through raised disc. Bottom
of bubble covered with MnS inclusions.

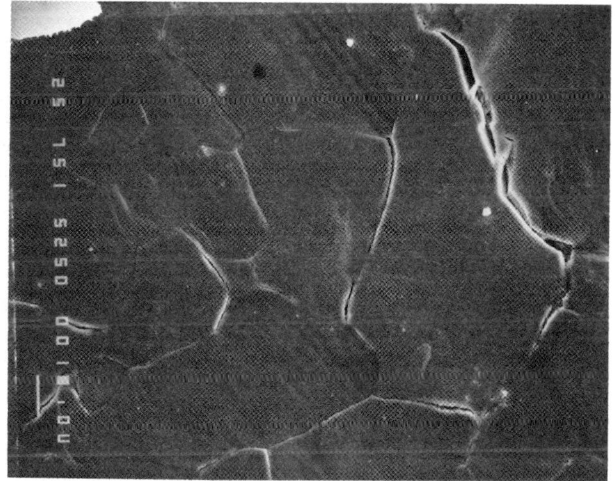

Figure 4:   Cluster of grain boundary vents on polished
surface.

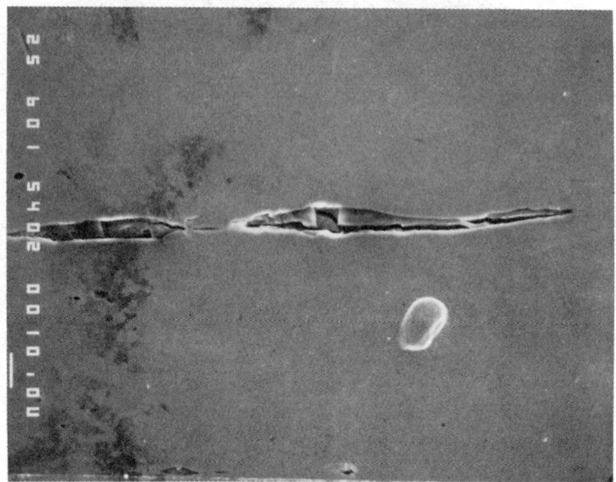

Figure 5: Vents along inclusion on near neutral axis.

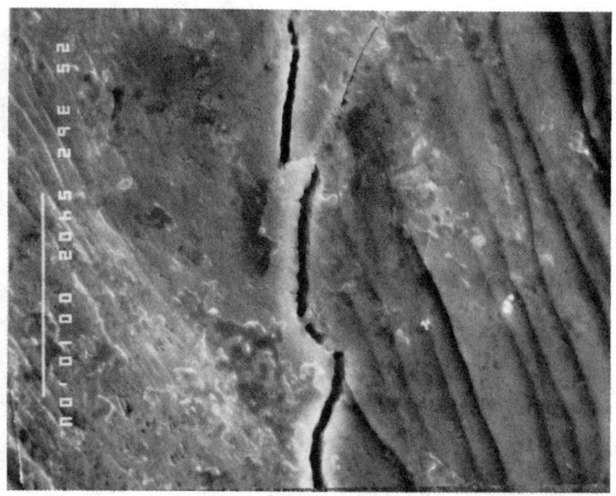

Figure 6: Jagged vent, primarily along grain boundaries
on compression side of bent specimen.

methane filled fissures in steel undergoing HA vent their gas to the surface.
This is shown by the observation of open fissures on the free surface, by
vents formed on upraised discs above small inclusions, and by the observation in samples bent before HA that the surface vents parallel, and often
merge with, the internal fissures. That the fissures can be interconnected
through much or all of the sample fits in with the observations mentioned
in the introduction --the stop of swelling before internal decarburization
had visibly proceeded, and, complete dissolution of pearlite in the cold
worked regions, while the nondeformed region along the neutral axis keeps
its original pearlite colonies.

There has been no published metallographic study of venting in steel
undergoing HA, though some reported observations fit in with it quite well.
Bisaro and Geiger studied the evolution of CH_4 during HA. They reported
occasional bursts which would raise the evolved CH_4 content well above
that which could be accounted for by the diffusion of carbon to the outside
surface and conversion there to CH_4[7]. They suggest that these releases
come from the rupture of blisters filled with Ch_4. The bursts could equally
come from the venting of deeper internal fissures.

The range of swelling values reported for steels undergoing HA has
been substantial (2-10%). It could result from differing times for the
advent of venting. The rare-earth treated steel studied by Pishko was
banded and tended to form planes of bubbles along grain boundaries in the
rolling plane. His samples were cut from plate with cuts normal to the
rolling plane. Thus the planes of bubbles and the resulting fissures
extended right up to a free surface. A minimal $\Delta V/V \simeq 2\%$ resulted before
swelling stopped. Decarburization proceeded without further growth.

A quite different geometry has been reported by Sorell and Humphries[4].
They show examples of large blisters extending several inches along the
circumference of an 8" diam. pipe. Here the rolling plane is in the wall,
and the conditions of service (low hydrogen pressure and temperature for
HA) are such that attack developed slowly and some surface decarburization
occurred. The blisters tend to develop along the rolling plane and often
follow long inclusions. Given decarburization and the absence of any cuts,
normal to the rolling plane, i.e. the circular pipe closes on itself, there
are no easy paths for venting of methane to the surface. The low P_{H_2} and

temperature also keep the methane pressure relatively low so that the
blisters can expand by creep rather than rupture due to high internal
pressure. These are the best conditions to minimize methane venting, and
thus are the conditions that give rise to the growth of large blisters
by creep.

References

1. A Study of the Effects of High-Temperature, High Pressure Hydrogen
    on Low Alloy Steels, _API Publn_ 945, June, 1975.

2. S. Harper, V. A. Callcut, D. W. Townsend, and R. Eborall, _Jnl._ _Inst._
    Metals, _90_ (1961-2) pp. 414 and 423.

3. J. Rest, and C. M. Gehl, _Nucl._ _Engr._ _and Design_ _56_ (1980) pp. 233-56.

4. G. Sorell, and M. J. Humphries, _Matl._ _Performance_ (NACE) _17_ (1978)
    pp. 33-41 (August).

5.   R. Pishko, M. McKimpson, and P. G. Shewmon, <u>Met</u>. <u>Trans</u>. <u>A</u>, <u>10A</u> (1979) pp. 887.

6.   Mark Ransick and P. G. Shewmon, Accepted for publ. in <u>Met</u>. <u>Trans</u>. <u>A</u>.

7.   R. E. Bisaro and G. H. Geiger, <u>Corrosion</u> (NACE) <u>23</u> (1967) pp. 289.

DISCUSSION

<u>M.R. Louthan, Jr., Virginia Polytechnic Institute and State University, Blacksburg, VA</u>: The importance of venting in limiting swelling due to hydrogen attack suggests that specimen size and configuration will play major roles in determining test results. Do you have (or plan) any experiments to investigate such effects?

<u>P.G. Shewmon</u>: I believe the biggest factor in swelling is having no cut surfaces normal to the rolling plane, and the next most effective would be to have a decarburized surface layer before hydrogen attack. This is certainly amenable to experimental verification of the type you suggest but we have no plans to pursue it.

<u>M.W. Joosten, Conoco Inc., Ponca City, OK</u>: What was the difference, if any, in the venting characteristics on the compression side of your bend specimens?

<u>P.G. Shewmon</u>: In the compression region the internal fissures ran primarily normal to the external surface while in the tensile region the fissures tended to parallel the external surface.

EFFECT OF IMPURITY ELEMENTS ON THE FORMATION OF

BUBBLES BY HYDROGEN ATTACK IN 2.25Cr-1Mo STEEL

Tadamichi Sakai and Haruo Kaji[*]

Central Research Laboratory, Kobe Steel, Ltd., Kobe, Japan
* Kakogawa Works, Kobe Steel, Ltd., Kakogawa, Japan

The effects of addition of P, As, Sn, Sb (up to 0.017 pct), and Si
(up to 0.26 pct) on the hydrogen attack behavior of 2.25Cr-1Mo steel were
determined through observation of methane bubbles on grain boundaries using
a scanning electron microscope.

Phosphorus and As were found to restrict the bubble formation.

In contrast to these elements, Sn and Si have the effect to promote
the bubble formation. Antimony also tends to enhance it.

These adverse effects seem to offset each other when the impurities
are added together.

The effect of P and As may be due to the adsorption of these elements
to the bubble surface, which retards the methane formation reaction.

On the other hand, the effect of Sn, Si, and Sb suggests that these
elements do not adsorb to the bubble surface and that these elements in
bulk solution promote the methane formation reaction.

Introduction

Steels for pressure vessels containing high pressure hydrogen at elevated temperature are required to have the resistance to material degradation which is termed hydrogen attack. Hydrogen attack is characterized by methane bubbles which nucleate mainly at carbides on grain boundaries and make growth (1, 2).

In this sense, hydrogen attack is one type of intergranular failure.

Impurity elements P, As, Sn, and Sb, and an alloying element Si (sometimes included in the impurities) are well known to play a significant role in the grain boundary embrittlement termed temper embrittlement. Therefore, the effect of these elements on hydrogen attack is interesting.

Furthermore, with 2.25Cr-1Mo steel, efforts have been made to reduce these elements from the temper embrittlement viewpoint. Therefore, it is necessary to seize the effect of these elements on hydrogen attack for sound material design.

Experimental Procedure

Table 1 lists the chemical compositions of tested steels. High purity 2.25Cr-1Mo steel containing 0.27 pct Si designated as the R steel was used as a base and the impurity elements P, Sn, As, and Sb were added individually and together. Furthermore, in order to clarify the effect of Si, two steels designated as Σ ' and Σ '(Low Si) were used.

These steels were vacuum-induction melted and each ingot of 90 kg in weight was forged to square rods and then was heat treated as is shown in Table 2.

A part of each steel designated as, for example, P(SC) was subjected to the step cooling treatment after the common heat treatment. Step cooling treatment is known to promote the segregation of the impurities to grain boudaries. Therefore, one would expect that the effects of the impurities on the bubble formation are intensified.

Table 1. Chemical Composition of Test Steels, Wt Pct

steel	C	Cr	Mo	Si	P	As	Sn	Sb	Mn	S	Al
R	0.13	2.22	0.97	0.27	0.003	0.003	0.001	0.001	0.49	0.015	0.019
P	0.15	2.26	0.99	0.26	<u>0.016</u>	0.003	0.001	0.001	0.49	0.016	0.020
As	0.14	2.30	0.96	0.26	0.002	<u>0.017</u>	0.001	0.001	0.49	0.008	0.008
Sn	0.14	2.23	0.98	0.26	0.004	0.003	<u>0.016</u>	0.001	0.49	0.016	0.019
Sn´	0.15	2.27	0.99	0.26	0.003	0.003	<u>0.010</u>	0.001	0.49	0.018	0.020
Sb	0.13	2.30	0.97	0.26	0.002	0.001	0.001	<u>0.016</u>	0.46	0.009	0.014
Σ	0.15	2.20	0.99	0.25	<u>0.009</u>	<u>0.008</u>	<u>0.008</u>	<u>0.010</u>	0.49	0.016	0.022
Σ´	0.12	2.19	0.99	0.26	<u>0.013</u>	<u>0.012</u>	<u>0.008</u>	<u>0.008</u>	0.50	0.014	0.016
Σ´(Low Si)	0.11	2.38	0.99	<u>0.06</u>	<u>0.012</u>	<u>0.010</u>	<u>0.010</u>	<u>0.011</u>	0.48	0.012	0.016

N:∿0.005 O:∿0.002 Σ´,Σ´(Low Si): containing 0.2 pct Cu

Table 2. Heat Treatment and Step Cooling Treatment

Heat Treatment	960°C × 3hAC*→ 940°C × 3hAC* →690°C × 3.5hAC→ 690°C × 26.5hFC
Step Cooling Treatment	593°C × 1h$\xrightarrow{5.6{\circ}C/h}$ 538°C × 15h$\xrightarrow{5.6{\circ}C/h}$ 524°C × 24h$\xrightarrow{5.6{\circ}C/h}$ 496°C × 60h$\xrightarrow{2.8{\circ}C/h}$ 468°C × 125h$\xrightarrow{2.8{\circ}C/h}$ 315°C AC

* 42°C/min from 940°C to 400°C

The specimens (5 by 7 by 10 mm) cut from the heat treated steels were exposed to hydrogen at 843 K (570°C) and 31.4 MPa for 320, 400, and 500 hours in a stainless steel pressure vessel surrounded by a resistance-heated furnace.

Hydrogen attack is characterized by methane bubbles which nucleate mainly at carbides and grow on grain boudaries (1, 2). Figure 1 shows the process schematically.

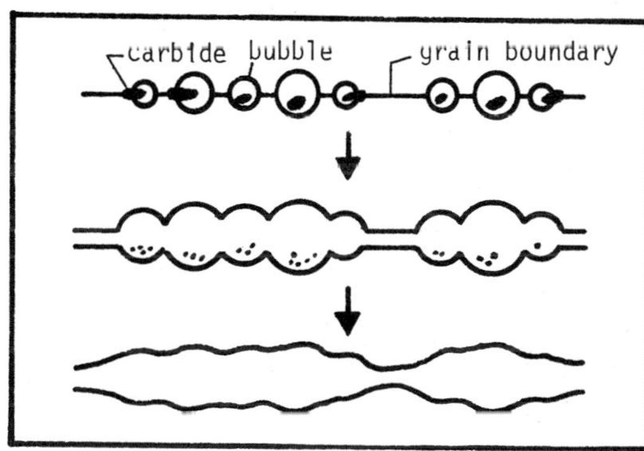

Fig. 1. Schematic illustration showing the process of bubble growth and failure of grain boundary.

Therefore, the total cross section of bubbles per unit area of grain boundaries which was calculated from the bubble density (the number of bubbles per unit area of the grain boundary) and the diameter (bubble growth rate) was used to estimate the extent of hydrogen attack or hydrogen attack susceptibility of the steels.

In order to observe the grain boundaries and take micrographs by a scanning electron microscope, the specimens subjected to hydrogen attack treatments were fractured at low temperature (77 K). The bubble density and diameter were measured on the micrographs. Since the cross section of the bubble resembles an ellipse in shape, the mean value between the lengths of long axis and short axis was used as the diamter of the bubble.

Results

The scanning electron micrographs in Photo. 1 indicate the features of bubbles on grain boundaries in some of the specimens.

It can be seen that the bubbles in the R steel grow to such an extent that they are very close together and the grain boundary is left only in net pattern at the holding time of 500 h.

Compared with the R steel, the P steel has the lower bubble density and the bubbles exist separately even at the 500 h holding time.

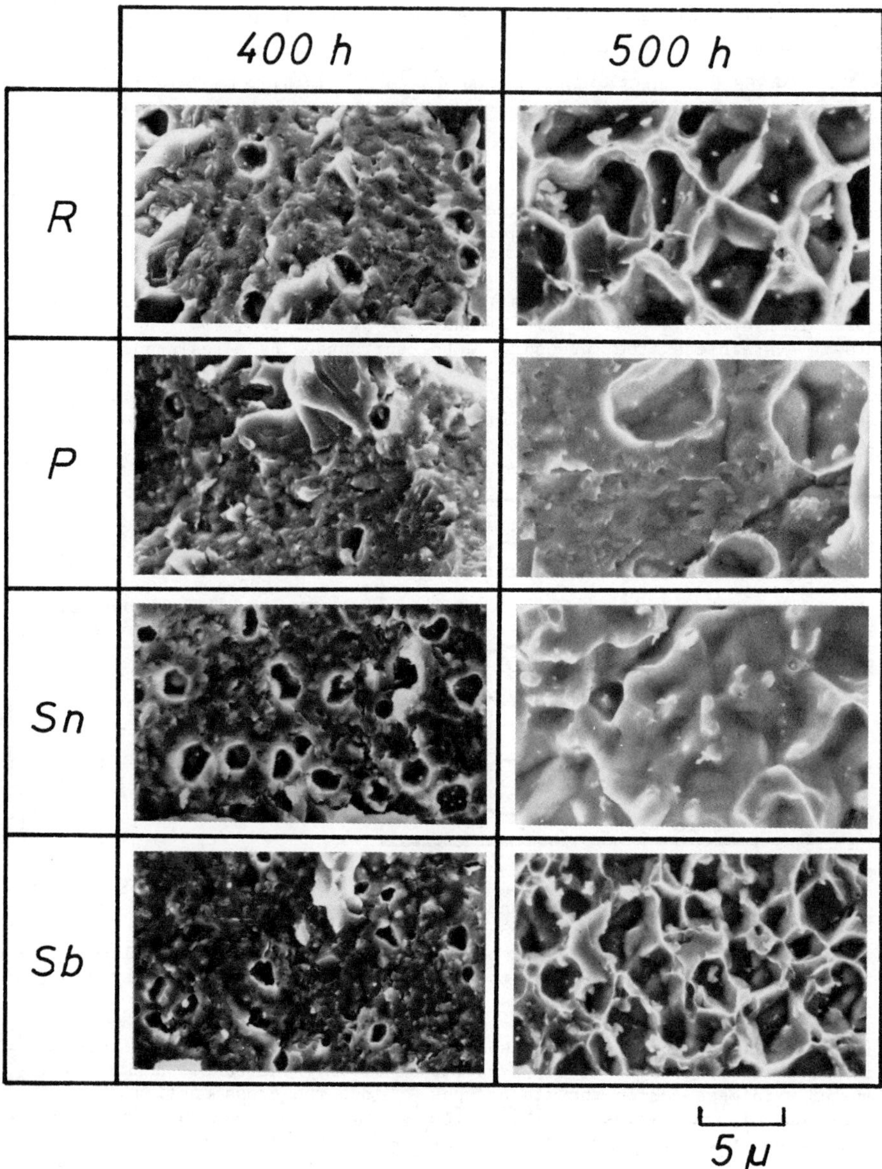

Photo. 1. Scanning electron fractographs showing the bubbles on grain boundaries.

On the other hand, the Sn steel has the higher bubble density and the bubbles have already coalesced at 500 h, indicating the final stage in Fig. 1, where the "grain boundary" surface is melted due to surface diffusion of Fe atoms. The Sb steel also has the higher bubble density, but the bubbles are found to be small.

Figures 2 through 4 indicate the results of measurements.

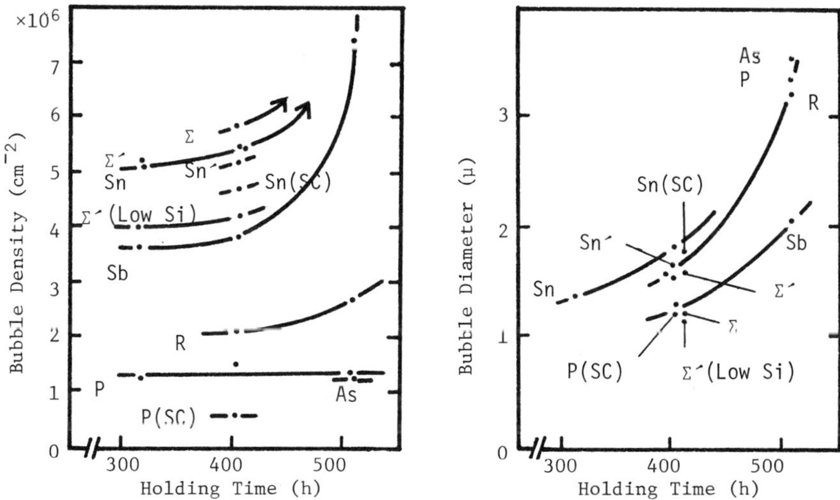

Fig. 2. Bubble density(number of bubbles per unit area of grain boundary) and diameter.

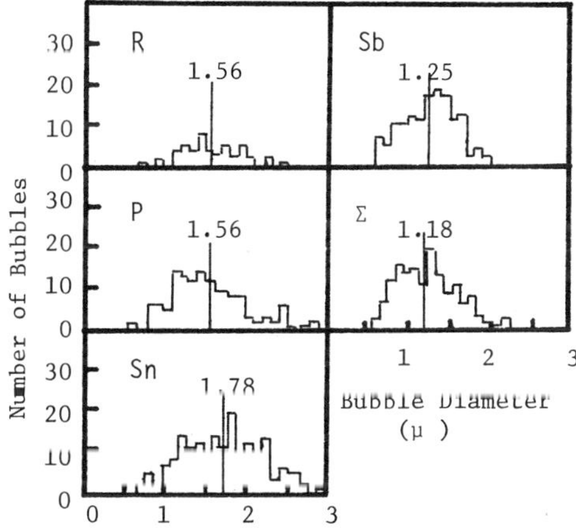

Fig. 3. Histograms of bubble diameter at the holding time of 400h.

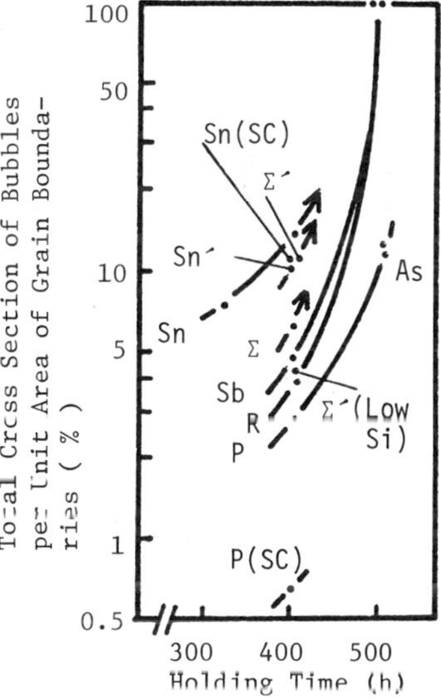

Fig. 4. Total cross section of bubbles per unit area of grain boundary.

Figure 2 shows the bubble density and diameter. Several arrowheads in this figure indicate that the exact estimations were impossible due to the coalescence of bubbles at the 500 h holding time. The bubble diameters were determined by the histograms as shown in Fig. 3.

It can be seen from Fig. 2 that the As and P steels behave similarly and that Si promotes both the bubble density and diamter.

Figure 4 shows the total cross section of bubbles per unit area of grain boudaries in pct estimated from the density and diameter.

The results of measurements including the combined effects of the impurities and the effects of step cooling treatment can be summarized as follows.

(1) Addition of either 0.016 pct P or 0.017 pct As decreases the bubble density by a factor of 0.75 (at 400 h) $\sim$ 0.5 (at 500 h). This effect is intensified by step cooling treatment and the density is further decreased. Step cooling treatment also reduces the bubble diameter.

(2) Addition of 0.016 pct Sn increases both the bubble density and diameter by factors of 2.7 and 1.15· respectively. However, this promoting effect is not intensified by step cooling treatment.

(3) Addition of 0.016 pct Sb also increases the bubble density by 2 (at 400 h) $\sim$ 3 (at 500 h) times, but the bubble diamter is reduced by a factor of 0.8

(4) The bubble density and diameter of the 0.26 pct Si steel (Σ ') are 1.3 and 1.4 times those of the 0.06 pct Si steel (Σ '(Low Si)), respectively.

(5) In the steel (Σ) where all the impurity elements P, As, Sn, and Sb are added together by $\sim$ 0.01 pct each, the bubble density is not decreased, while the bubble diameter is reduced and the total cross section of bubbles takes the medium value.

Discussion

It has been shown that the P steel has low bubble density. Since nucleation sites are mainly carbides on grain boudaries, the number of carbides is an important factor determining the bubble density. However, the carbides in the P steel are much more than the bubbles and, further, do not seem to be less than those in the other steels as shown in Photo. 2.

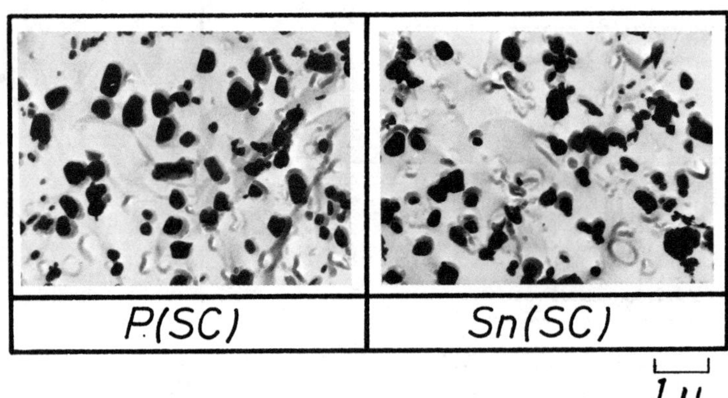

Photo. 2. Carbides on grain boundary before hydrogen attack treatment
(carbon extraction replicas).

Other possible mechanism that may account for the decrease in bubble density is adsorption of P to the bubble surface. Since P is known to segragate strongly to grain boundaries (and to free surface) in 2.25 Cr-1 Mo steel (3), it is expected that the P adsorbs to the surface of the bubble which nucleates on a carbide in a grain boundary. The adsorption, on the one hand, improves the nucleation by lowering the surface energy of the bubble, on the other hand, restricts the lowering of the surface energy by prohibiting hydrogen and carbon from adsorbing to the bubble surface. This prohibition means the retard of the methane formation reaction and causes lowering of the pressure in bubbles which provides the driving force for bubble nucleation (4). In summary, it is probable that P restricts the bubble nucleation and reduces the bubble density through the adsorption to the bubble surface.

In the case where the step cooling treatment is given to the P steel preceding the hydrogen attack treatment, more phosphorus is segregated to the grain boundary and the effect of P must be intensified. This is consistent with the experimental result in the P(SC) steel where the bubble density is further reduced.

Lowering of the pressure in the bubbles also reduces the bubble growth rate (5) or the bubble diameter as shown in the P(SC) steel. On the other hand, the low bubble density has a beneficial effect on the bubble growth because it can keep the carbon activity high during the bubble growth. This may be responsible for the result that the bubbles in the P steel are not small.

The effect of As on the formation of bubbles is essentially the same as that of P. This suggests that, although the segregation of As to grain boundaries has not been reported in 2.25Cr-1Mo steel, the same mechanism would be applied to the effect of As.

The effect of Sn is in contrast to that of P and As, i.e. addition of Sn is beneficial to bubble formation, suggesting the possibilities that, first, Sn does not segregate to the grain boudary and to the bubble surface, secondly, Sn in bulk solution promotes the formation of bubbles. The former agrees to the fact that the effect of Sn is not intensified by step cooling treatment. Another evidence is the fact that Sn does not produce temper embrittlement in 2.25Cr-1Mo steel (3). The latter might be due to a strong repulsive interaction between Sn and C in Fe (6). Tin would reject C from solution and cause it to segregate to grain boundaries and to the bubble surface (7), activating the methane formation reaction.

The same mechanism may be applied to the effects of Si and Sb because it has been reported that Si does not segregate to grain boundaries in 2.25Cr-1Mo steel (8) and Sb does not embrittle this steel (7). However, the effect of Sb is not so strong that the reduction in carbon activity accompanied with the bubble growth could surpass the beneficial effect, leading to the retard of the bubble growth rate.

The effect of Sn, Si, and Sb which promotes the bubble formation seems to be offset by the effect of P and As which retards it: Whereas the steel has the high bubble density not being reduced by P and As, the bubble diameter is small showing the effect of P and As and the total cross section of bubbles per unit area of grain boundaries S is offset. Quantitatively, S of a specimen having a certain composition is roughly approximated at the holding time of 400 h by the following equation.

$$S\ (\%) = 633\ Sn + 46.7\ Sb + 29\ Si - 92.3\ P - 85.7\ As$$

where the element symbols represent wt pct.

Finally, the rapid increase in the bubble density during the holding time of 400 to 500 h is understood by the following mechanism: The pressure in the bubbles is borne by the grain boundary between the bubbles. This tensile stress on the grain boundary rapidly increases as the bubbles grow and come close to each other, causing new bubbles on the grain boundary.

Conclusion

It has been shown that the impurity elements have the effects on the formation of bubbles in different ways.

Phosphorus and As restrict the bubble formation. This may be caused by the adsorption of these elements to the bubble surface which retards the methane formation reaction on the surface.

The effect of Sn and Si is in contrast to that of P and As and promotes the bubble formation. This suggests that Sn and Si do not segregate to grain boundaries and to the bubble surface and that these elements in bulk solution activate the methane formation reaction. Antimony also tends to enhance the bubble formation.

These adverse effects seem to be able to offset each other.

References

1. T. Sakai and H. Kaji : Tetsu-to-Hagané, 1978, vol. 64, p. 430
2. T. Sakai and H. Kaji : Proceedings of Second JIM International Symposium on Hydrogen in Metals, 1979, p. 489
3. J. Yu and C. J. McMahon, Jr. : Met. Trans. A, 1980, vol. 11A, p. 277
4. P. G. Shewmon : Met. Trans. A, 1976, vol. 7A, p. 279
5. R. Raj, H. M. Shih and H. H. Johnson : Scripta Met., 1977, vol. 11, p. 839
6. F. Neumann and H. Schenck : Arch. Eisenhütt., 1959, vol. 30, p. 477
7. C. J. McMahon, Jr. and L. Marchut : J. Vac. Sci. Technol., 1978, vol. 15, p. 450
8. J. Yu and C. J. McMahon, Jr. : Met. Trans. A, 1980, vol. 11A, p. 291

DISCUSSION

<u>P. Shewmon, Ohio State University, Columbus, OH</u>: Did you measure overall volume change or mass density change for your samples? The US-API has published a Nelson curve for 2 1/4 Cr-1 Mo, based apparently on one data point, and concluded that hydrogen attack (HA) doesn't occur but decarburization does. Is there a Nelson curve showing HA for this steel in Japan?

<u>T. Sakai</u>: (1) We did not measure the volume change. (2) There is not such a Nelson curve in Japan. But there are several data including our present experiment showing the hydrogen attack in the steel under severe hydrogen conditions.

<u>S.M. Bruemmer, Battelle Pacific Northwest Labs, Richland, WA</u>: Your explanation for the effect of impurities rests on whether they segregate to grain boundaries and carbide interfaces. Do you plan to analyze grain boundary chemistries for impurity segregation? In reference to your explanation for the effect of Sn and Sb, is it not possible that they also enrich grain and carbide interfaces since they do segregate in iron?

<u>T. Sakai</u>: (1) We are planning to check the impurity segregation using Auger electron microscopy. (2) The extent of segregation of an impurity seems to depend on the alloys. For example, McMahon et al. showed that Sn and Sb segregate in Ni-Cr steel due to strong Ni-Sn, Ni-Sb cosegregation, but not in 2.25 Cr-1 Mo steel. This result is favorable to explain our present result.

<u>S. Stock, University of Illinois, Urbana, IL</u>: What is the spatial distribution of bubbles you observed? Are they randomly distributed, uniformly distributed or clustered?

<u>S. Sakai</u>: In general, bubbles are not clustered. But I do not know whether the bubbles are distributed randomly or uniformly, since we have not made measurement on the spatial distribution.

HIGH TEMPERATURE HYDROGEN ATTACK STUDIES

D. Eliezer
Materials Engineering Department
Ben-Gurion University of the Negev
Beer Sheva, Israel.

Hydrogen attack of SAE 1020 carbon steel was investigated at room temperature after specimen exposure at 525°C to hydrogen and a blend of hydrogen sulfide at pressures of 3.5 MN/m^2 and for different exposure times. Exposure to pure hydrogen resulted in a significant reduction in the mechanical properties. In the hydrogen/hydrogen sulfide blend, hydrogen attack was not observed and the room temperature tensile properties were unchanged. It was observed that bubbles and blisters were not ovonly distributed on the exposed surfaces of the specimens, but instead were very dependent on specimen thickness. During the hydrogen attack, tears form and undergo crack growth. Subcritical crack growth tests were conducted in both a high purity hydrogen and in the hydrogen/ hydrogen sulfide blend, at various temperatures and hydrogen pressures. Slow crack growth was not observed. Data provided from the crack growth studies will be reported.

Introduction

Hydrogen attack occurs as a result of atomic hydrogen diffusing into the steel and combining with carbon in solution to form methane, which remains trapped internally as bubbles. The localized internal decarburization and the build up of methane pressure results in fissures which induce an apparent embrittlement of the macroscopic section of the steel and cause a loss of strength, swelling and ultimately fracture. The temperature range for hydrogen attack of carbon steels extends from about $260^{\circ}C$ to $540^{\circ}C$. This temperature range coincides with the range of service temperatures of steels in petroleum refining and other equipment (1,2,3). The usual means of predicting the behavior of materials in high pressure, high temperature hydrogen is Nelson curves (4). These curves delineate the safe and unsafe hydrogen pressure-temperature regimes for carbon and alloy steel in a purely empirical manner. Understanding of the phenomena responsible for this degradation is far from complete, and therefore the Nelson curves cannot be extrapolated to different environments, service conditions or steels.

The purpose of the present investigation is to understand better the rapid period of hydrogen attack through its influence on the mechanical behavior of materials including hydrogen attack induced subcritical crack growth, under sustained load. The results are applied to anticipate coal gasification systems under conditions of their operation.

Experimental Procedure

The material used in this investigation was commercially produced, 12.5 mm thick SAE 1020 steel plate. Tensile specimens having a gauge length of 28 mm and a thickness of 2.5 mm were cut with their long axis perpendicular to the direction of rolling. Environmental exposures were conducted in an all-metal, sealed tubular stainless-steel chamber surrounded by a resistance-heated tube furnace. Commercially available pure hydrogen and hydrogen sulfide gases were used. Hydrogen attack of SAE 1020 steel was investigated at room temperature after exposure at $525^{\circ}C$ at a pressure of 3.5 MN/m^2 for different periods of time.

The fracture surfaces of all failed specimens were observed by both optical and scanning electron microscopy. Following those exposures where surface scales were observed, the scales were examined using scanning electron microscopy, energy-dispersive X-ray analysis, X-ray diffraction and Mössbauer techniques.

In order to conduct subcritical crack growth tests we used wedge opened loaded (WOL) fracture mechanics type specimens.

Results and Discussion

Figure 1 summarizes the room temperature tensile data observed after a 240 hr exposure at 525°C to vacuum, to hydrogen at 3.5 MN/m^2, and to a blend of 10% hydrogen sulfide in hydrogen at 3.5 MN/m^2 ($P_{S_2} \simeq 10^{-4}$ N/m^2). As can be seen, exposure to hydrogen resulted in a reduction in the yield strength of 57% and in the ultimate strength of 34%, and a very marginal reduction in elongation (less than 1%). Exposure to the hydrogen/hydrogen sulfide blend under the same conditions resulted in no significant reduction in room temperature tensile properties compared with that observed in the control specimen exposed to a vacuum environment. Examination of the fracture surfaces of specimens exposed to all environments indicated massive regions of microvoid coalescence producing ductile rupture (Fig. 2). Surface scale formation was observed on all specimens exposed to the hydrogen/hydrogen sulfide blend. As indicated earlier, this did not lead to a degradation in room temperature mechanical properties (Fig. 1), and there was no evidence of bubble formation on the specimen surfaces. Thus, it appears that the presence of the surface scale significantly hinders the rate of high-temperature hydrogen attack (5).

All scales were found to consist of the FeS phase. The iron sulfide scale was identified by X-ray diffraction and Mössbauer techniques. The Mössbauer spectrum revealed that the isomer shift is 0.697 mm s^{-1}, the quadruple splitting 0.094 mm s^{-1} and the magnetic splitting is 305 KO$_e$. The relatively low values of the isomer shift and quadruple splitting indicate a very strong covalency of the Fe-S bond. Exposure of the samples to the hydrogen/hydrogen sulfide blend for longer times than 240 hours does not change the Mössbauer spectrum.

During this study, it was observed that bubbles and blisters were not evenly distributed on the exposed surfaces of the specimens, but

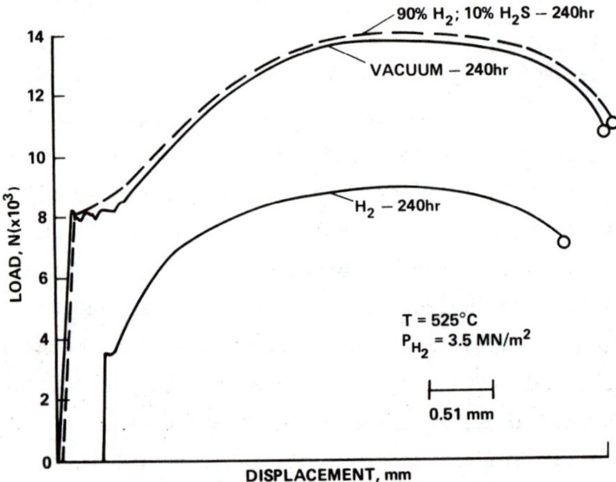

Fig. 1 - Room temperature tensile properties of
SAE 1020 steel after 240 hr exposure at 525°C.

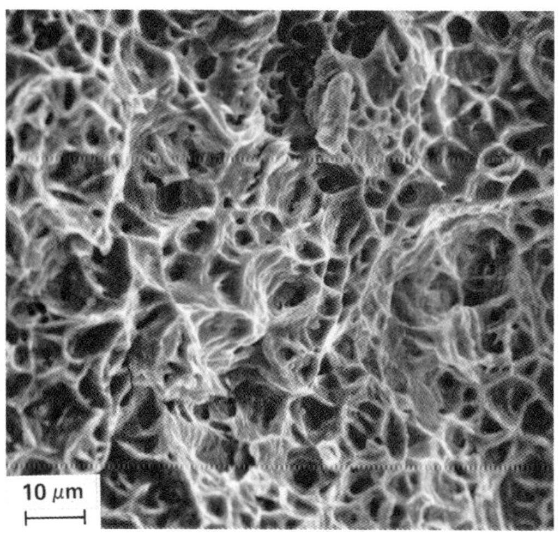

Fig. 2   SEM fractograph of fracture surfaces showing
         massive regions of microvoid coalescence
         producing ductile rupture.

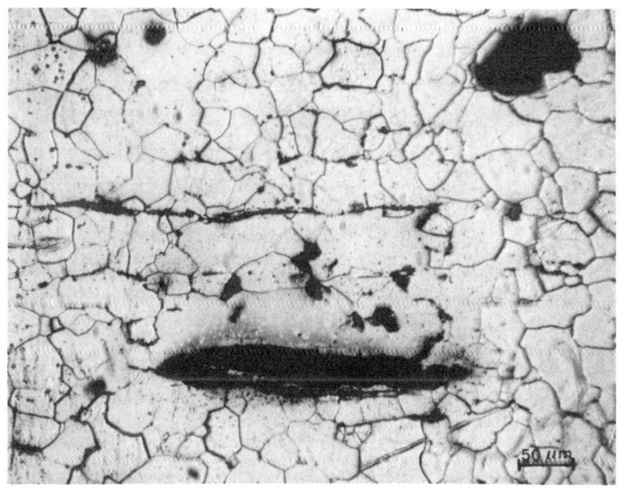

Fig. 3 - Tears observed in a completely decarburized
         specimen, shows preferential grain coarsening
         around the tears.

instead were very dependent on specimen thickness. An examination of the
tensile specimen surface at the gauge length revealed numerous small
bulges or protrusions which appear approximately like hemispherical caps.
These protrusions are termed as blisters. The diameter of these caps are
as large as 2 mm and are considerably bigger than the diameter of the
approximately spherical bubbles which form at the grain boundary in the
interior (6,7). It is interesting to note that the blisters observed on
the gauge length surface are either absent or smaller and fewer on the
shoulders of the tensile specimen. A tear observed in the interior of
the specimen is reported in Figure 3. These tears are extensive at the
gauge length, relatively few at the shoulder, and are not observed in
the WOL specimen. It can be seen (Fig. 3) that the region close to the
tear has very coarse ferrite grain. The preferential recrystallization
and grain coarsening confirms that substantial internal pressure exists
inside the tear. A typical blister observed in a partially hydrogen
attacked specimen is shown in Figure 4. When a tear is located close to
the specimen surface it bulges and produces blisters. It is being
proposed that during the hydrogen attack of a commercial 1020 steel,
tears form and undergo crack growth as a result of the formation and
accumulation of methane under high pressure at the weak interfaces such
as the inclusion/matrix or the ferrite/pearlite band. If a large tear
is located close to the surface, it bulges at 10 to 50% of the equilibrium
pressure of methane and forms into a blister. The tears grow and the
blisters bulge rapidly as compared to the growth of the bubbles and lead
to fissuring and delamination. Therefore, tearing and blistering may
lead to accelerated failure by hydrogen attack. Subcritical crack
growth tests were conducted in both a high purity hydrogen and in the
hydrogen/hydrogen sulfide environments at various temperatures ranging
from $375^{\circ}C$ to $525^{\circ}C$ (time ranging from 10 to 30 days). Wedge opening
loaded specimens (WOL) 1.25 cm thick, fracture mechanics type, were cut
from the SAE 1020 plate so that the crack would propagate in the rolling
direction on a plane normal to the transverse rolling direction. In
these tests, the displacement was held at a known constant value during
a test, load levels and changes in load could be related to specimen
compliance, respectively. In accordance with ASTM recommended procedures,
specimens were fatigue precracked in air and wedge, were driven into the
initial machined notch to produce a fixed crack opening displacement
(COD). After removal from the furnace, the specimens were air cooled
and then reloaded to determine the actual final load. Specimens were

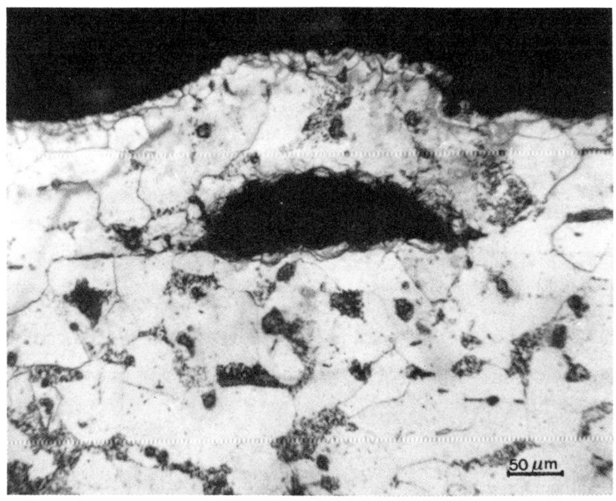

Fig. 4 - Tear bulges into a blister in a partially
decarbourized specimen.

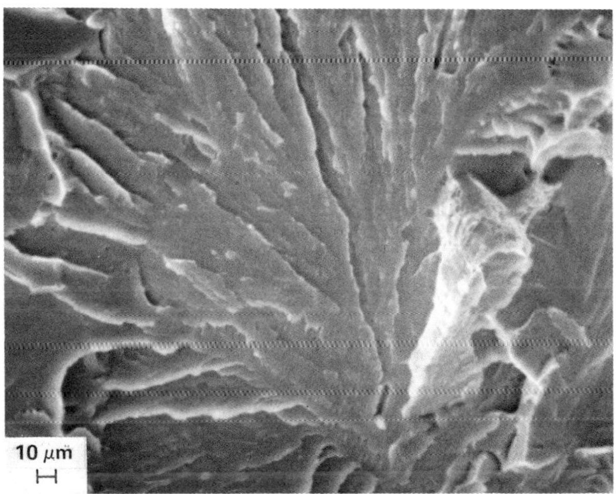

Fig. 5 - SEM fractograph of fracture surfaces of WOL
specimen exposed to hydrogen environment
showing transgranular cleavage.

fatigue cracked post test in order to protect the region that was
hydrogen attacked, and then were failed by overload. The results show
that no slow crack growth takes place under the aforementioned testing
conditions. Detailed fractography was conducted using a scanning
electron microscope (SEM). The primary failure mode in all hydrogen
environments was transgranular cleavage (Fig. 5). It is believed that
this is a result of creep during the elevated temperature exposure. In
the present state of knowledge, subcritical crack growth in SAE 1020
carbon steel is an area of considerable uncertainty.

We believe that this research may provide additional information to
aid in the establishment of design and operational guidelines to ensure
the safe and reliable operation of coal gasification systems.

Conclusions

Hydrogen attack of SAE 1020 carbon steel from a high purity hydrogen
environment was found to be severe with the formation of numerous methane
fissures and bubbles and with a significant reduction in the room
temperature, tensile yield and ultimate strengths. In the hydrogen/
hydrogen sulfide blend, hydrogen attack was not observed and no fissure
or bubble formation occurred and the room temperature tensile properties
were unchanged. The improvement in the mechanical properties in the
hydrogen/hydrogen sulfide blend is due to the formation of an iron
sulfide scale. It was observed that bubbles and blisters were not
evenly distributed on the exposed surfaces of the specimens, but instead
were very dependent on specimen thickness. Subcritical crack growth
studies under conditions of static loading were conducted. Slow crack
growth was not observed.

Acknowledgement

The author acknowledges Dr. H.G. Nelson from NASA Ames Research
Center, Moffett Field, California for his valuable advice and instructive
discussions.

References

1. P.G. Shewmon: Met. Trans. A. 1976, Vol. 7A, pp. 279-86.

2. H.H. Podgurski: Trans. TMS-AIME, 1961, Vol. 221, pp. 389-94.

3. L.C. Wcincr: Corrosion, 1961, Vol. 17, pp. 137-43.

4. Steels for Hydrogen Service at Elevated Temperatures and Pressures
    in Petroleum Refineries and Petrochemical Plants. API Publ. 941,
    June 1977.

5. D. Eliezer and H.G. Nelson: Corrosion, 1979, Vol. 35, pp. 17-21.

6. II.M. Shih and II.II. Johnson: Scripta Mct., 1977, Vol. 11, pp. 151-154.

7. A.A. Sagues, B. Okray Hall and H. Wiedersich: Scripta Met., 1978,
    Vol. 12, pp. 319-326.

FORMATION OF HYDROGEN BUBBLES BY PULSE HEATING

W. Pesch, T. Schober and H. Wenzl
Institut für Festkörperforschung, Kernforschungsanlage Jülich
5170 Jülich, Germany

Hydrogen bubble formation is a well known phenomenon in endothermic occluders of hydrogen (e.g. Cu, Mg) which are quenched and annealed. Hydrogen is introduced here at high temperatures. Quenching then leads to a supersaturation of hydrogen and - diffusivities permitting - to precipitation in form of bubbles. In the present work, the reverse case is investigated. An exothermic occluder (here V) was hydrogen charged to the ($\alpha+\beta$) hydride two phase region at a low temperature. The samples were subsequently pulse heated in vacuum to temperatures not far from the melting point. Under these conditions, a very large hydrogen supersaturation was produced which corresponded to equilibrium pressures of a few thousand bars. Only a minor fraction of the hydrogen could escape to the adjacent surfaces during the pulse. Inspection of the pulse heated samples revealed the morphology of bubbles and cavities arising from the precipitation of hydrogen. Interestingly, the density of bubbles was so high in some regions that actually "metal foam" was formed. The implications of the present results are also discussed.

Introduction

There are two classes of hydrogen occluding metals, the endothermic (examples are Cu, Mg, Ag, Au) and exothermic occluders (Nb, Ta, V, Zr, etc.). In the first group hydrogen bubbles are observed when hydrogen charged samples are quenched from high temperatures (1,2,3). Conversely, cooling of exothermic occluders leads to the precipitation of a solid hydride phase, from the numerous studies we list the TEM work on vanadium as recently reviewed in Ref. 4. In the present work, the opposite case is investigated experimentally which is rapid heating (pulse heating) of an exothermic occluder (here V). Depending on the rate of heating the hydrogen will either leave the specimen by diffusion, or else, will precipitate internally in the form of bubbles provided that nucleation is possible. This latter case of pulse heating of exothermic metals has to our knowledge not been reported so far. There has been however some speculation about "void formation" in the removal of hydrogen from Pd (5). This work demonstrates that pulse heating of vanadium hydrides will generally lead to the formation of internal hydrogen-filled cavities which drastically changes the geometry of the sample. Fig. 1 shows schematically the VH phase diagram with the isobars and the path taken by specimens. A more detailed account of this work will appear elsewhere (6).

Experimental

Vanadium strips 20 and 300 μm thick and about 3 mm wide were hydrogen charged to about 5-20 % using a new Pd charging technique (7). The samples were pulse heated in high vacuum either by the discharge of a capacitor or else by AC heating.

The main requirement of the experiment was that a high temperature with a corresponding high equilibrium pressure was reached before the hydrogen in the central portion had a reasonable chance to escape to the free surface. The peak temperatures were near the melting point and could be reached after a few ms. The heated portions of the samples were then investigated using scanning electron microscopy (SEM), optical microscopy and metallography.

Results

The main observation after heat pulsing is the appearance of blisters and bubbles or irregularly shaped cavities. Most of our topological results on pulse heated V-H alloys may be visualized schematically in Fig. 2; in the following figures some representative examples of the phenomena are given. Fig. 3 shows surface blisters on a pulsed 300 μm foil and Fig. 4 a transverse fracture surface displaying a "metal foam" structure in the interior. A similar effect as in Fig. 4 is depicted in the metallographic section in Fig. 5 where it is seen that a large fraction of the center portions consists of cavities; it also shows that many of the small bubbles appear along grain boundaries. This weakening of the grain boundaries often leads to the ejection of whole grains (Fig. 6) from the surface.

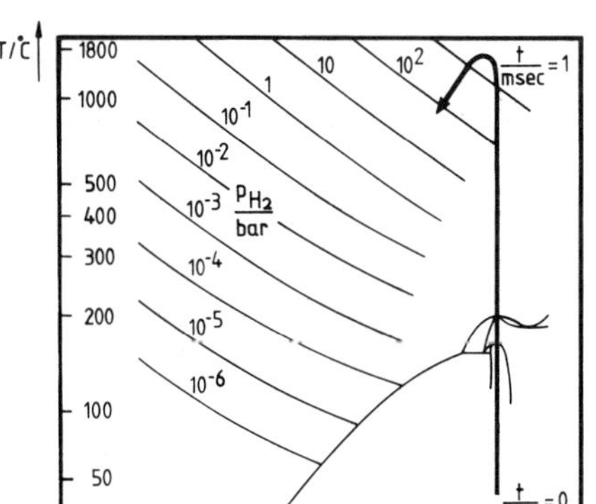

Fig. 1 - Partial V-H phase diagram. The pulse heating of a
VH$_{0.5}$ sample corresponds to a state change along the
thick line, starting (t=0) in the β-phase, crossing
two order-disorder phase boundaries and ending in the
α-phase region. Here, the isobars indicate values of
the equilibrium pressure.

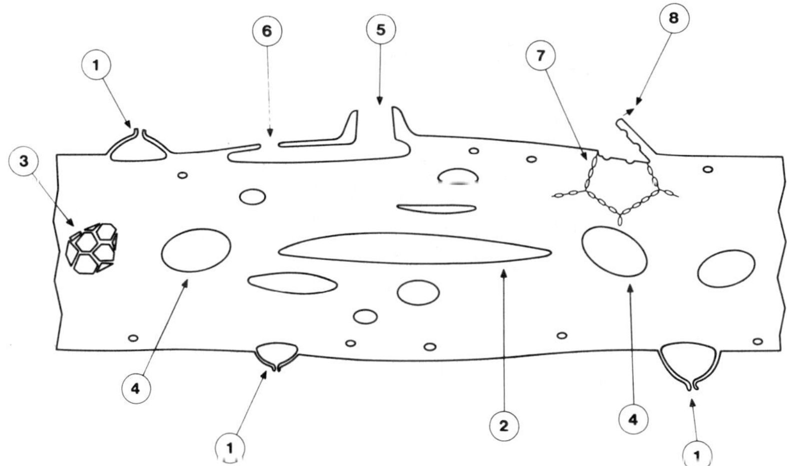

Fig. 2 - Schematic view of the hydrogen precipitation phenomena
observed in pulse-heated VH$_x$ alloys: 1) Surface blisters with perforations. 2) Large flat internal cavities. 3) Metal foam structure. 4) Internal cavities
with spherical shape. 5) Round craters formed by the
ejection of molten material. 6) Surface holes formed
from underlying cavities by surface diffusion. 7) De-
coration of grain boundaries with small bubbles.
8) Whole grains are tilted up or even ejected.

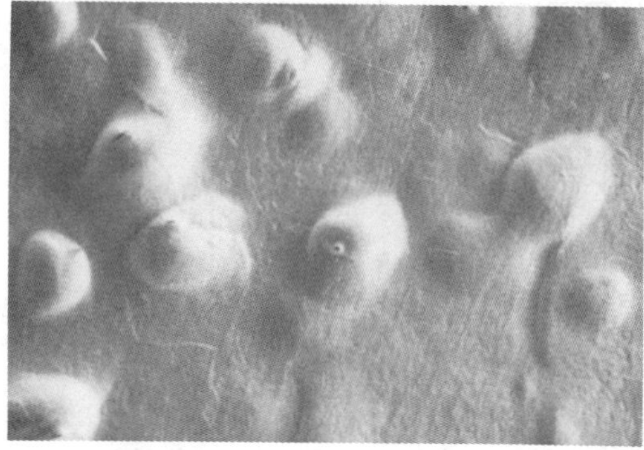

Fig. 3 - SEM micrograph of a pulse heated VH_x foil of a thickness of 300 µm. Visible are dome shaped surface blisters with orifices for hydrogen pressure relief.

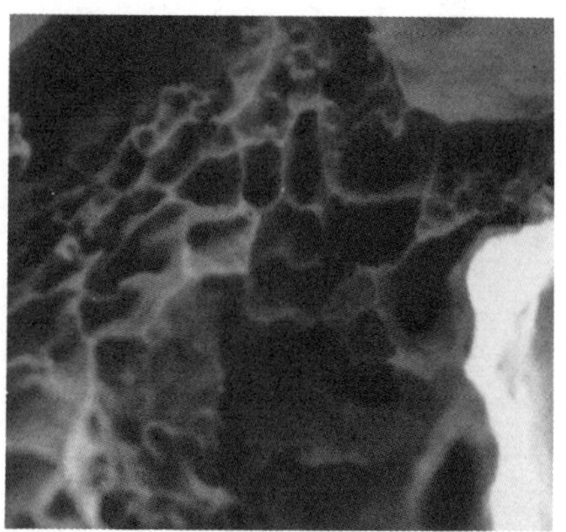

Fig. 4 - SEM micrograph of a pulse heated VH_x-foil. Transverse fracture surface showing the formation of "metal foam" with rather thin cell walls.

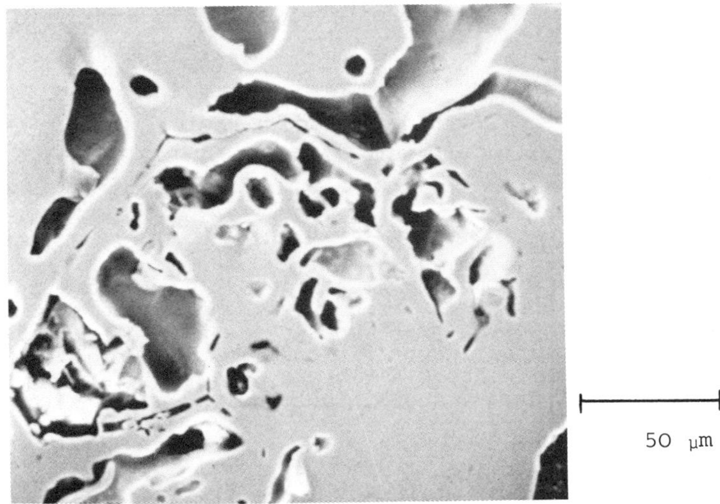

Fig. 5 - SEM micrograph of a pulse heated 20 µm VH$_x$-foil.
Transverse metallographic section showing large
internal cavities. Note precipitation of elongated
bubbles along grain boundaries.

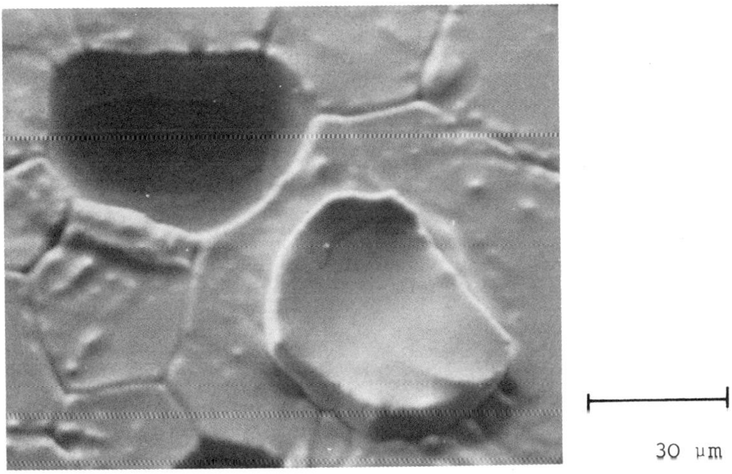

Fig. 6 - SEM micrograph of a pulse heated 20 µm VH$_x$-foil.
After ejection, a whole grain landed in the vicinity of the hole and was bonded to the foil due to
the high temperature. Again, the grain boundaries
had been weakened by H$_2$-bubble formation

Occasionally, these grains were found in the immediate vicinity of their origin.

Discussion

The equilibrium pressure in dilute V-H alloys may be expressed by (after Ref. 8):

$$\frac{1}{2} \log \frac{p_{H_2}}{bar} = 2.83 - \frac{1400 \text{ K}}{T} + \log x \tag{1}$$

Equ. (1) leads to the isobars in Fig. 1. It is readily seen that pressures above 10^3 bars may arise near the melting point of V. At these pressures H_2 bubbles are heterogeneously nucleated at grain boundaries or dislocations. Material is transported away from the bubble sites by interstitial loop punching which is important above a critical pressure p_c (after Ref. 9):

$$p_c = \frac{\mu b}{2\pi r} \ln r/b \tag{2}$$

(μ = shear modulus, b = Burgers vector, r = loop radius). A rough estimate shows that $p_c \approx 10$ kbar at r $\approx$ 0.5 nm.

Bubbles filled with gas of supercritical pressure grow rapidly by loop punching. The growth rate is dominated by the diffusion current of hydrogen into the bubbles. The diffusion coefficient of H in V is about $2 \cdot 10^{-4}$ cm^2/s between 500 $^{\circ}$C and the melting point. Within the heating pulse of about 1 ms the diffusion distance of H (as estimated from $x^2 = 2Dt$) is several μm. Consequently, a sufficient amount of H can reach heterogeneous nucleation sites within the time of the heating pulse. Only a small fraction of H can escape from the foil to the vacuum during the heating pulse. - Hydrogen bubbles in the grain boundaries most probably have similar detrimental effects on the mechanical properties as has been shown for He-bubbles (10).

Finally, we arbitrarily list a few areas where the present effect may be of importance:

- in fusion reactor technology (where certain exothermic occluders may be subject to rapid heating)
- in sputtering studies. Here, ejection of whole grains may lead to very high sputtering yields.
- for the production of a random distribution of bubbles (or voids, after degassing) for flux pinning or diffusion experiments
- in mechanical applications of exothermic H_2-occluders which may be subject to rapid heating
- for the production of metal foam structures.

References

1. J.S. Lally, P.G. Partridge: Phil. Mag.,1966, vol. 13, p. 9

2. W.R. Wampler, T. Schober, B. Lengeler, Phil. Mag., 1976, vol. 34, p. 129

3. W.R. Wampler, T. Schober, B. Lengeler, Proc. 2nd Int. Congr. Hydrog. in Metals, Paris, 1977, 1 B 11

4. T. Schober, H. Wenzl in: "Hydrogen in Metals II", Topics in Appl. Physics Vol. 29, G. Alefeld, J. Völkl Editors, Springer Heidelberg (1978)

5. J.F. Lynch, J.D. Clewley, T.B. Flanagan, Phil. Mag., 1973, vol. 28, p. 1415

6. W. Pesch, T. Schober, H. Wenzl (to be published)

7. T. Schober, A. Carl, J. Less-Common Metals, 1979, vol. 63, p. P53

8. H. Wenzl, Jül-Report N$^{\circ}$ 1640, KFA Jülich, 1980

9. J.H. Evans, J. Nucl. Mat. 1977, vol. 68, p. 129

10. II. Trinkaus, H. Ullmaier, Phil. Mag. A., 1979, vol. 39, p. 563.

PROBLEMS IN ENGINEERING APPLICATIONS

THE EFFECT OF HYDROGEN EXPOSURE ON FRACTURE TOUGHNESS[*]

J. D. Landes and D. E. McCabe
Westinghouse R&D Center

Fracture toughness studies were conducted for a 2¼Cr-1Mo steel exposed six months to a high temperature,high pressure hydrogen gas environment. The material is a normalized and tempered ASTM A387 Class 2, Grade 22 steel; base metal, weld metal,heat affected zone (HAZ) and deliberately temper embrittled weld metal were included. The hydrogen environment was one simulating a coal liquefaction condition with exposure temperature 727 K (850°F) and pressure 21.4 MPa (3500 psi). Companion specimens were exposed to 727 K in air to separate the effect of hydrogen exposure on toughness from that of temperature exposure. Baseline tests were conducted for specimens in the as-received condition. The fracture toughness results were interpreted using a J_{Ic} and elastic-plastic R curve approach. The results showed some temper embrittlement due to the high temperature air exposure with no other effect for this exposure condition. The 727 K hydrogen exposure resulted in a drastic reduction of room temperature toughness by about an order of magnitude on J_{Ic}. This toughness reduction was not observed in a 727 K test. Weld metal specimens displayed a greater reduction in toughness due to the 727 K hydrogen exposure than did the base metal specimens.

This study was part of a larger program which studied the effects of environments simulating coal conversion conditions on the fracture mechanics properties of pressure vessel steels.

[*]This work was supported by the Electric Power Research Institute on Research Project 627-1. Dr. R. Viswanathan was the EPRI Project Manager.

Introduction

Hydrogen is a major constituent of environments produced in the conversion of coal to a synthetic fuel. These environments would be contained at high temperature and pressure in large steel pressure vessels. The aggressive nature of hydrogen may reduce the structural integrity of the pressure vessel steel. To study these potential problems a program was sponsored by the Electric Power Research Institute to experimentally evaluate the effects of high temperature and pressure hydrogen bearing environments on the fracture mechanics type properties of these steels (1). One part of the program was the study of the effect of long exposures to high temperatures and pressure hydrogen on fracture toughness measured subsequent to the exposure. Most of these tests were conducted at room temperature. These results would generally apply to a pressure vessel which had been cooled to room temperature after a period of high temperature operation.

Materials and Test Conditions

The test material was a normalized and tempered grade of $2\frac{1}{4}$Cr-1Mo steel, ASTM A387 Class 2, Grade 22, supplied as a 7-inch thick plate weldment. The fabrication procedure and material properties have been reported previously (1,2). Specimens were taken from material representative of base metal, weld metal and HAZ. A section of weld metal was deliberately temper embrittled (2) to study this condition. Mechanical properties of the material conditions relevant to fracture toughness testing are given in the following table.

Material Condition	Yield Strength MPa (ksi)		UTS MPa (ksi)		FATT K (°F)	
Base	345	(50)	528	(77)	294	(70)
Weld	423	(61)	575	(83)	275	(35)
Weld Temper Embrittled	435	(63)	578	(84)	344	(160)

The specimens were machined as compact specimens, Fig. 1, with B = 25.4 mm (1.0 in.). The specimens were preloaded with self loading devices to relatively low loads, not exceeding the elastic limit of the specimen, so that the crack tip region would be stressed during exposure to hydrogen. The specimens were then placed in an autoclave containing pure hydrogen gas and exposed at 727 K (850°F) under 24.1 MPa (3500 psi) pressure for six months. For each exposed specimen a companion specimen was held in air at 727 K for six months. Following exposure the specimens were rapidly cooled to room temperature and then held in dry ice until tested. This procedure was followed so that the hydrogen would be retained in a manner simulating rapid cooling of a large pressure vessel. Specimens were then tested either at room temperature or 727 K. Included in the testing were specimens exposed to the hydrogen environment, specimens held in air at 727 K and baseline, as received specimens.

The toughness was measured using an elastic-plastic test technique similar to the proposed J_{Ic} test procedure (3,4). (This testing was completed before the procedure was issued and did not conform in all aspects.) This allows the specimens to be loaded into the plasticity regime well beyond the limit on K_{Ic}. The J_{Ic} procedure records stable tearing of a crack on an R curve, Fig. 2, where J is plotted as a function of crack advance, Δa, and J_{Ic} is measured at the beginning of stable tearing. The toughness level can be compared by two criteria; the J_{Ic} value and the slope of the R curve, dJ/da.

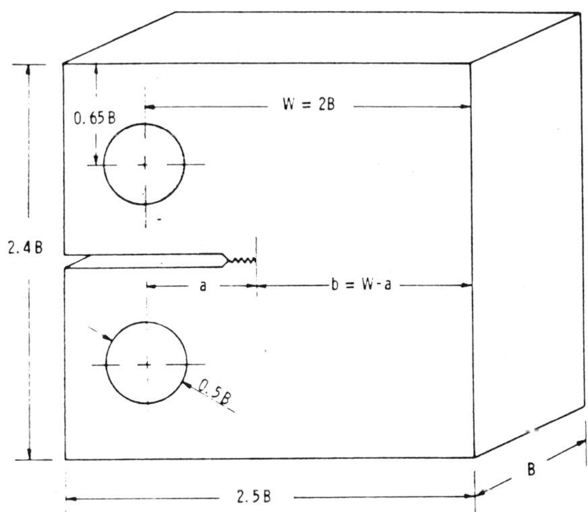

Fig. 1 – Example of a compact specimen with dimensions proportion to thickness.

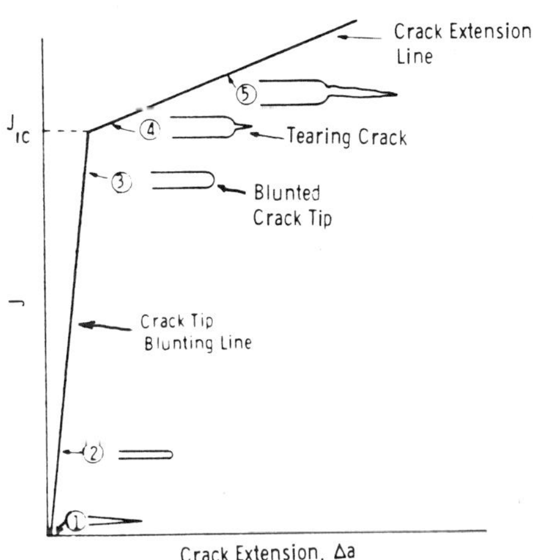

Fig. 2 – R curve schematic of the ductile fracture process.

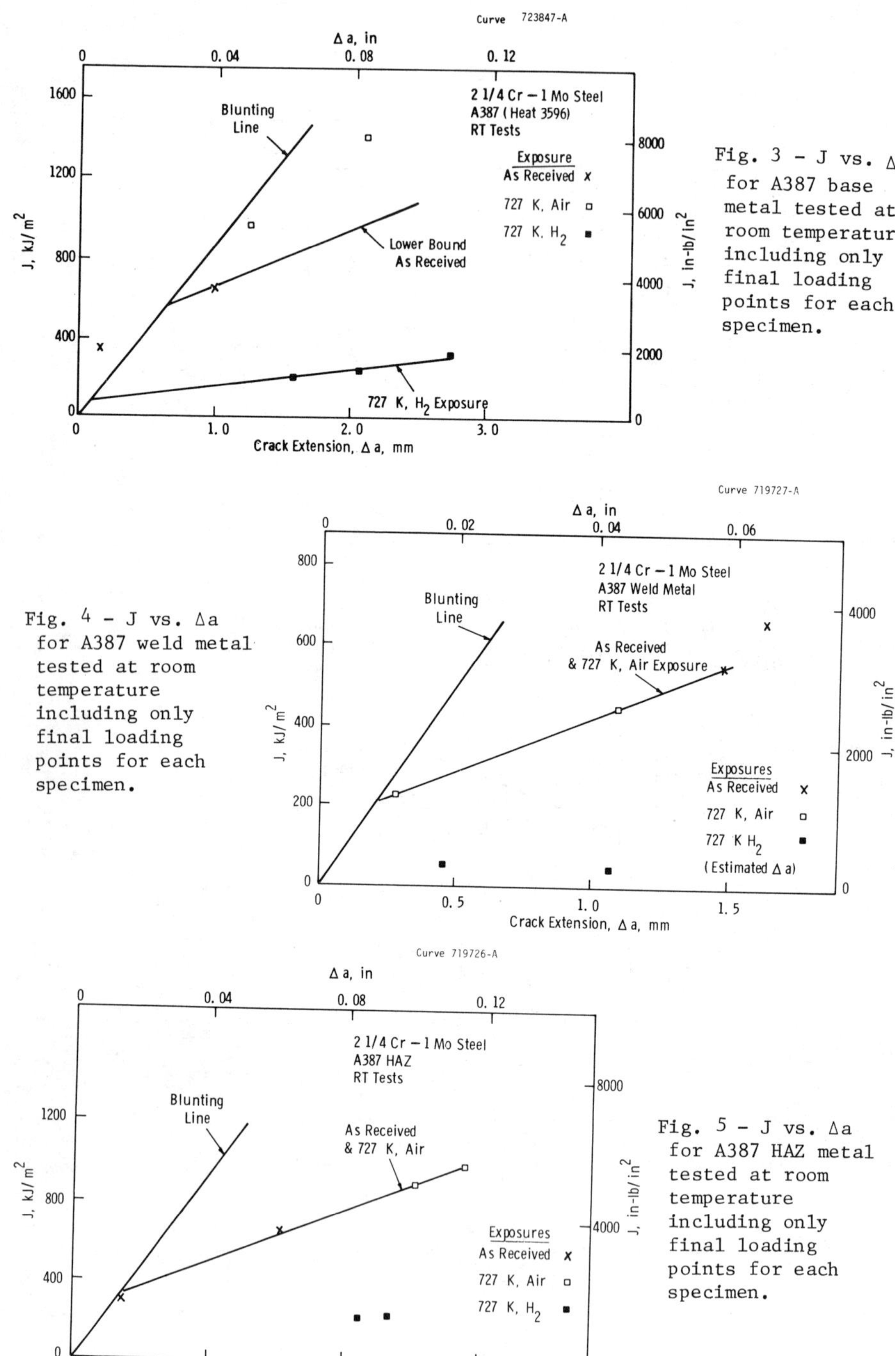

Fig. 3 - J vs. Δa for A387 base metal tested at room temperature including only final loading points for each specimen.

Fig. 4 - J vs. Δa for A387 weld metal tested at room temperature including only final loading points for each specimen.

Fig. 5 - J vs. Δa for A387 HAZ metal tested at room temperature including only final loading points for each specimen.

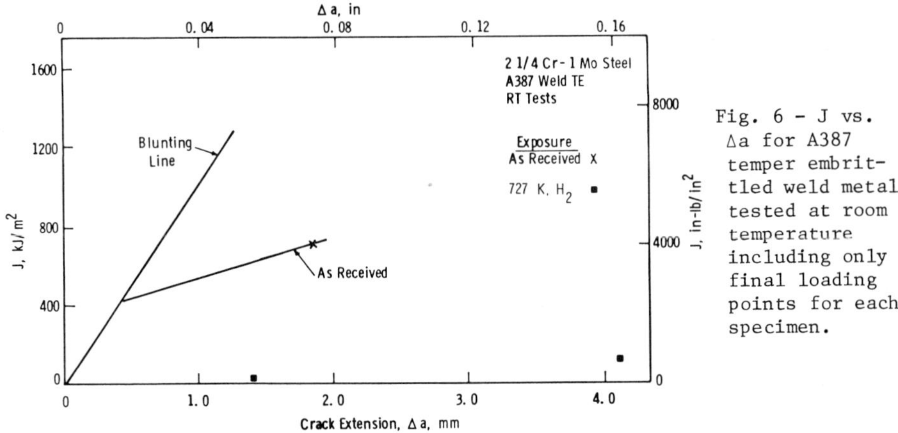

Fig. 6 - J vs. Δa for A387 temper embrittled weld metal tested at room temperature including only final loading points for each specimen.

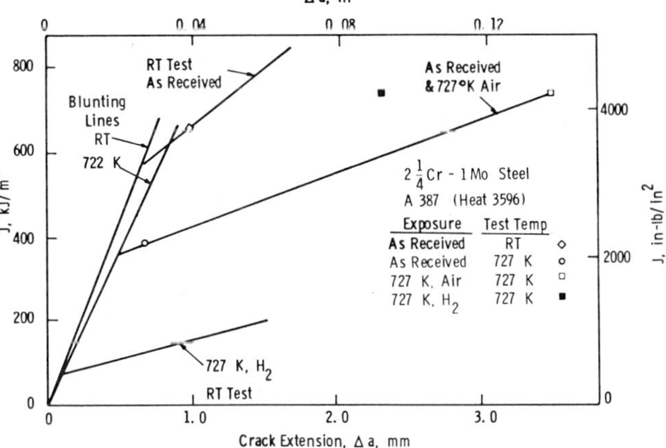

Fig. 7 - J vs. Δa for A387 base metal including only final loading points for each specimen.

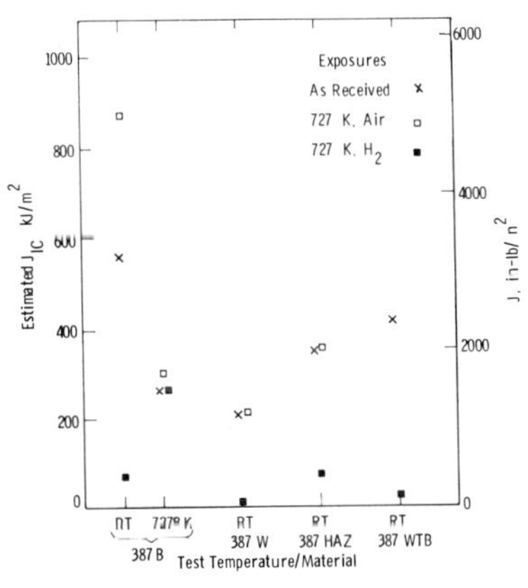

Fig. 8 - Estimated J_{IC} versus temperature and material for various exposure conditions.

The tests were conducted using the unloading compliance test technique (5,6) so that a full R curve could be generated from a single specimen. In many cases the R curve was not properly generated and toughness comparisons were made from crack extensions measured at the final loading point. A complete description of the testing and results is given in a topical report on fracture toughness (7).

Test Results

The test results are all plotted on an R curve format of J versus crack extension, Δa. In cases where R curves were not generated toughness comparison was made on the basis of estimated R curve or relative position on the plots (a single point lying above a given R curve has higher toughness). The data are presented in separate plots for each material condition and comparisons are made between as received, 727 K air exposure and 727 K hydrogen exposure.

Results for base metal tested at room temperature are given in Fig. 3. The as-received specimens and those exposed to 727 K air fall along the same R curve but the specimens exposed to 727 K hydrogen show a marked reduction in toughness. Similar results are observed for room temperature tests on weld metal, Fig. 4, HAZ, Fig. 5 and the deliberately temper embrittled weld metal, Fig. 6.

Tests were conducted at 727°K for the base metal and compared with room temperature test results in Fig. 7. Although the toughness is lower for the as received and 727 K air exposure for the 727 K tests as compared to room temperature, the specimen exposed to 727 K hydrogen shows no loss in toughness at 727 K and has much higher toughness than the room temperature tests of the specimens exposed to hydrogen.

Several of the specimens failed by cleavage after some stable crack extension occurred. Since the value of J at cleavage was always above J_{Ic}, the toughness for these specimens is compared on the R curves presented in the previous figures. Cleavage occurred predominately in the weld metal and deliberately temper embrittled weld metal after exposure to 727 K air and hydrogen. This indicates that the temperature exposure alone may have temper embrittled the material somewhat.

Estimates were made for both J_{Ic} and the R curve slope dJ/da for all material conditions. These are presented in Figs. 8 and 9, respectively. The values are labeled estimates because sufficient data points were not generated to meet the proposed J_{Ic} standard method (4). The results compared both by J_{Ic} and dJ/da show a large reduction in toughness caused by exposure to hydrogen for the room temperature tests. The toughness for the specimen exposed to hydrogen and tested a 727 K shows no loss in toughness.

Discussion

All material conditions of $2\frac{1}{4}$Cr-1Mo steel exposed for six months to hydrogen at 727 K and 24.1 MPa pressure showed a marked loss in toughness when these specimens were tested at room temperature using the J_{Ic} test format. Specimens exposed only to 727 K air for six months did not exhibit any distinguishable loss in toughness except for an increased tendency to fracture by cleavage above J_{Ic}. For a specimen exposed to hydrogen and tested at 727 K no loss in toughness was exhibited. The loss in toughness in the room temperature tests most likely resulted from hydrogen trapped in the crack tip region of the metal when the specimens were cooled after exposure. Hydrogen could have been released from the specimen tested at 727 K during heating to test temperature.

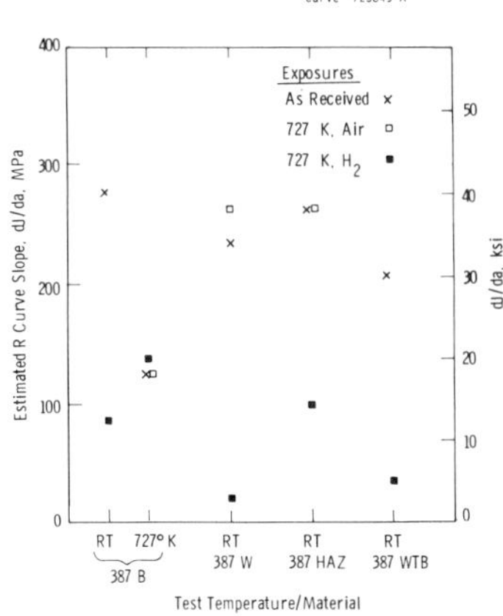

Fig. 9 – Estimated R curve slope, dJ/da, versus test temperature and material for various exposure conditions.

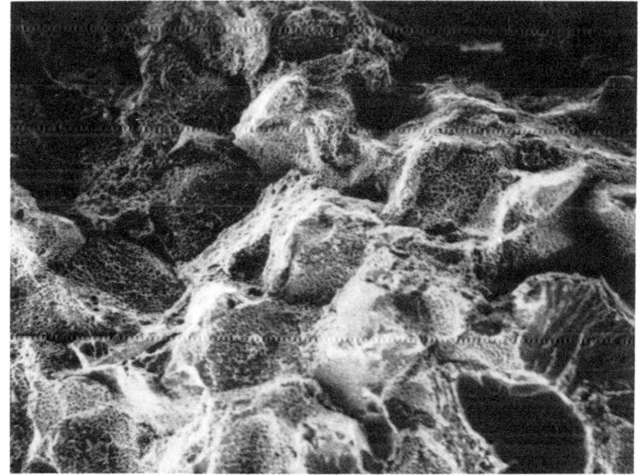

Fig. 10 – Fracture surface for A387 exposed six months at 727 K, 24.1 MPa in H_2 gas.

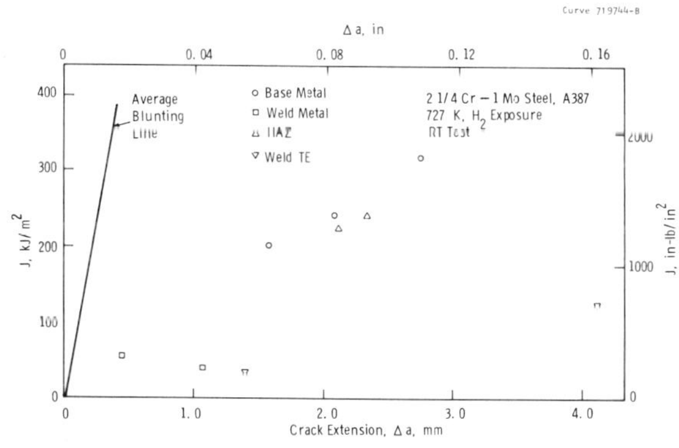

Fig. 11 – J vs. Δa for A387 exposed to 727 K and tested at room temperature including only final loading points for each specimen.

These results show that in an application to an actual pressure vessel containing hydrogen gases failure is more likely to occur if these vessels are cooled too rapidly to room temperature under a pressure loading. Safe operation would require both controlled cooling so that the hydrogen could escape from the metal and reduced loads at room temperature.

The toughness tests are conducted under a monotonically rising load with controlled displacement. The behavior observed as toughness loss could not be distinguished from damage, which might be caused by trapped hydrogen, in the form of subcritical cracking under a sustained load. The fracture surface resulting from the stable cracking during the J_{Ic} test was predominately intergranular, Fig. 10. This mode of cracking could also result from subcritical cracking under a sustained load. Future testing may be directed toward a more detailed study of cracking mechanisms.

Although all of the specimens exposed to the hydrogen showed reduced toughness at room temperature, this toughness loss was more pronounced in the weld metal and deliberately temper embrittled weld metal, Fig. 11. The results plotted in an R curve format in this figure show two distinct toughness levels. Data for the base metal and HAZ fall significantly above data for the weld conditions.

Conclusions

1. Exposure of 2¼Cr-1Mo steel to hydrogen gas at 727 K (850°F) and 24.1 MPa (3500 psi) pressure for six months resulted in a significant loss in fracture toughness for specimens tested at room temperature by the J_{Ic} method.

2. Exposure to 727 K air did not reduce toughness.

3. A specimen exposed to hydrogen and tested at 727 K showed no loss in toughness.

4. The toughness loss which most likely results from hydrogen retained in the crack tip region illustrates the need for proper cooling of a large pressure vessel containing hydrogen gas.

5. The weld metal showed a greater toughness loss than the base metal or HAZ.

References

1. McCabe, D. E. and J. D. Landes, "Design Properties of Steels for Coal Conversion Vessels - Final Report," EPRI Report, March 1, 1980.

2. McCabe, D. E. and J. D. Landes, "Design Properties of Steels for Coal Conversion Vessels - Topical Report on Mechanical Properties of Materials," EPRI AP-1338, Project 627-1, Interim Report, January 1980.

3. Landes, J. D. and J. A. Begley, "Recent Developments in J_{Ic} Testing," in Recent Developments in Fracture Mechanics Test Methods Standardization, ASTM STP 632. American Society for Testing and Materials, 1977, pp. 57-81.

4. Clarke, G. A., et al., "A Procedure for the Determination of Ductile Fracture Toughness Values Using J Integral Techniques," Journal of Testing and Evaluation. JTEVA, Vol. 7, No. 1, January 1979, pp. 49-56.

5. Clarke, G. A., W. R. Andrews, P. C. Paris and D. W. Schmidt, "Single Specimen Tests for J_{Ic} Determination," in Mechanics of Crack Growth, ASTM STP 590, American Society for Testing and Materials, 1976, pp. 27-42.

6. Clarke, G. A. and J. D. Landes, "Toughness Testing of Materials by J
    Integral Techniques," in Toughness Characterization and Specifications
    for HSLA and Structural Steels. Proceedings of a symposium at the 106th
    AIME Annual Meeting, March 1977, pp. 79-111.

7. Landes, J. D. and D. E. McCabe, "Design Properties of Steels for Coal
    Conversion Vessels - Topical Report on Fracture Toughness," EPRI Report
    December 28, 1979.

DISCUSSION

C.G. Interrante, National Bureau of Standards, Washington, DC: Would you
describe the test procedure further and indicate the time lapse that occurred
after exposure to hydrogen and before testing and clarify how you conclude
that the reported effect of hydrogen is a reversible one.

J.D. Landes: The specimens were cooled from the exposure temperature to
room temperature in about 4 hours. They were then placed in dry ice until
they were tested. All tests were conducted within 72 hours of removal from
the exposure chamber. For the room temperature tests, the specimens were
allowed to warm up before testing. For the 727 K tests the specimens were
heated to test temperature in a furnace. The effect of hydrogen on reducing
fracture toughness was considered reversible because there was no effect
of exposure on toughness for the 727 K test. The damage is likely to be
related to an effect of hydrogen trapped in the steel at room temperature
rather than a permanent microstructural damage. The reduction in toughness
must then be related to a hydrogen cracking effect and not to hydrogen
attack.

P.G. Sherman, Ohio State University, Columbus, OH: Did you find any evidence
of methane bubble formation? C. Y. Li at Cornell has shown that exposure of
this alloy to 2000 psi H_2 at 600°C (1112°F) reduces the creep rupture time
substantially even though this wouldn't give hydrogen attack as indicated
by something like the Nelson curve.

J.D. Landes: We did not find evidence of methane bubble formation although
a careful study of microstructural damage was not within the scope of the
program. The observed reversibility of the toughness loss would suggest that
there was not any significant methane bubble formation. In one phase of this
program (not included in this paper) we observed both a higher creep rate and
an enhancement of plastic flow for specimens tested in a mixture of H_2-6% H_2S
at 727 K. Note: 1100°F + 2000 psi is well above the Nelson Curve for
decarburization.

J.K. Tien, Columbia University, New York, NY: On the question of methane
formation, since the phenomenon is reversible, I would think methane didn't
form at the exposure temperature.

J.D. Landes: There was no evidence of methane formation. The reversible
nature of the toughness loss would also suggest no hydrogen attack.

DEVELOPMENT OF A HIGH STRENGTH, HYDROGEN-RESISTANT

AUSTENITiC ALLOY

K. M. Chang, D. H. Klahn and J. W. Morris, Jr.
Lawrence Berkeley Laboratory and
Dept. of Materials Science and Mineral Eng.,
University of California, Berkeley.

Abstract

Research toward high-strength, high toughness non-magnetic steels for
use in the retaining rings of large electrical generators led to the development of a Ta-modified iron-based superalloy which combines high strength
with good toughness after suitable aging. The alloy did, however, show some
degradation in fatigue resistance in gaseous hydrogen. This sensitivity was
associated with a deformation-induced martensitic transformation near the
fracture surface. The addition of a small amount of chromium to the alloy
suppressed the martensite transformation and led to a marked improvement in
hydrogen resistance.

Introduction

The work reported here was done in the course of research toward a new
high-strength non-magnetic steel for use in the retaining rings of large
electrical generators. The constraints on the alloy required that it be
hardenable through thermal processing to ~200 ksi (1410 MPa) yield strength
with a residual fracture toughness, K_{Ic}, greater than 100 ksi $\sqrt{in}$. (110 MPa
$\sqrt{m}$) and that it not embrittle significantly in gaseous hydrogen, which is
used to cool the generator.

Initial work focused on the attainment of the strength, toughness, and
magnetic property objectives. The product of this work (1,2) was a tanta
lum-modified Fe-based superalloy, designated EPRI-E, of nominal composition
Fe-36Ni-3Ti-3Ta-0.5Al-1.3Mo-0.3V-0.01B. If this alloy is upset cross forged
at 1100°C and then given a double-aging treatment of four hours at 750°C
followed by four hours at 670°C, it is paramagnetic and has a yield strength
of 1415 MPa (205 ksi) with 19% elongation and 67% reduction in area. Its
plane strain fracture toughness (K_{Ic}) at this yield strength is ~120 MPa $\sqrt{m}$
(110 ksi $\sqrt{in}$). On the basis of these properties the alloy was subjected to
hydrogen susceptibility tests, whose results, interpretation and consequences are the subject of this paper.

943

Experimental

Laboratory 10kg ingots were induction-melted under partial helium pressure using high purity (99.9+) raw materials. After homogenization the ingots were upset cross-forged from 1150°C into plates, then water quenched. After machining to the appropriate size, the samples were sealed in stainless steel bags filled with argon for thermal treatments and were usually water-quenched afterwards.

The sensitivity of alloy mechanical properties to gaseous hydrogen was determined through tests conducted in a stainless steel chamber which could be evacuated to 10^{-3} torr and then backfilled with hydrogen to 0.8 atm. Evacuation insured that the hydrogen gas in the test chamber had a low concentration of active impurity gases such as O_2, H_2O, and N_2 (<100 ppm). Two types of specimen were tested at room temperature: notched tensile specimens and subsized compact tension specimens tested in a low cycle fatigue mode.

Results

Hydrogen Susceptibility

A series of notched tensile specimens were made from Alloy E to examine the hydrogen susceptibility. The results of testing are listed in Table I. Specimen EH-1 was tested in hydrogen at a crosshead spee of 3×10^{-3} in/min which gave a test time of 10 to 15 minutes. Specimens EH-2 and EH-4 were tested under the same conditions in air. The strength of these three specimens is virtually the same whether measured by the proportional limit or the ultimate load. However, a decrease in the total plastic extension suggested that the ductility of the hydrogen specimen EH-1 was less than that of the specimens tested in air. To investigate the effect of strain rate specimen EH-3 was tested in hydrogen at a lower crosshead speed. The total plastic extension was further reduced to only about one half that of the air specimens. The dependence of notch ductility on the strain rate not only suggests that alloy E is susceptible to hydrogen embrittlement but also indicates that hydrogen transport influences the embrittlement process (3).

Table 1.   <u>Notched Tensile Properties of Alloy E in Air and in Hydrogen.</u>

Specimen	Environment	Cross-Head Speed (mm/sec)	Proportional Limit* MPa (ksi)	Ultimate Stress* MPa (ksi)	Plastic Extension ($\times 10^{-2}$ mm)
EH-1	H_2 (0.8 atm)	1.27×10^{-3}	1898 (275)	2098 (304)	11.4
EH-2	Air	1.27×10^{-3}	1877 (272)	2105 (305)	17.5
EH-3	H_2 (0.8 atm)	1.27×10^{-4}	1932 (280)	2091 (303)	9.1
EH-4	Air	1.27×10^{-3}	1918 (278)	2091 (303)	17.3

* Stress based on actual area at root of notch.

Specimen data: Nominal gauge length   - 12.7 mm
                 Nominal gauge diameter - 3.175 mm
                 Diameter at notch   - 2.54 mm
                 Notch angle   - 45°

Fatigue testing in which crack growth velocities in air and hydrogen are compared has been suggested by Nelson (4) as the most realistic and sensitive test for hydrogen sensitivity. The precracked compact tension specimen EHF-1 was first fatigued in air using an initial cycle $\Delta K=0.5k_{Ic}$, R=6, at a frequency of 1 cycle/sec. The maximum load was automatically adjusted to assure a constant crack opening. A record of load vs. fatigue cycle, taken at intervals of 500 cycles, is shown in Fig. 1. The slope of the curves is proportional to the crack growth velocity. After cycling specimen EHF-1 in air for 4500 cycles, the environment chamber was evacuated and then back-filled with 0.8 atm. hydrogen and the test continued for another 1000 cycles. Immediately following this environmental change, the load-cycle curve changed slope by a factor of 4.3. Specimen EHF-2 was then tested under the same conditions with cyclic loading hydrogen from the beginning. The crack growth velocity of EHF-2 was equal to that of EHF-1 in hydrogen. It was concluded that alloy E in the full-hardened condition is sensitive to hydrogen embrittlement.

Deformation-Induced Martensite

Magnetic measurements of various fracture surfaces of alloy E indicated an increase in the magnetic susceptibility. These and associated x-ray measurements (Table II) showed that the full-hardened alloy E was not completely stable with respect to mechanical deformation and some ferromagnetic α' martensite was induced on the fracture surfaces. The specimens broken in fatigue had a higher martensite fraction on the fracture surface than those broken in fracture toughness tests, despite the low stress intensity employed in fatigue tests. The presence of hydrogen during fatigue resulted in a still higher percentage of martensite, indicating that hydrogen promoted the transformation of austenite to martensite under deformation.

Table II. X-ray Diffraction Analysis on Various Fracture Surfaces of Alloy E.

Phase	Fatigue in H_2	Fatigue in Air	Rapid Fracture
Martensite (%)	86	73	61
Austenite (%)	14	27	39

Figure 2 shows the α' martensite phase formed near the fracture surfaces of fracture toughness and fatigue specimens, respectively. The mechanically-induced martensite takes the form of thin plates which appear in decreasing volume fraction with increasing distance from the fracture surface. The martensite-forming zone is much narrower on the fatigued surfaces than on the surface of the fracture toughness specimen, which presumably reflects the lower level of stress intensity in the fatigue tests. On the other hand, the transformation at the fracture surface is much more extensive in the fatigue tests and nearly covers the fracture surface; backscatter Mössbauer spectroscopic measurements, with a penetration depth of a few tens of angstroms, show 98% transformation on the surface of the speciment fatigued in hydrogen.

These data suggest a causal connection between the induced martensitic transformation and hydrogen susceptibility. Mechanistically, it is well known (5) that martensite is generally more sensitive to hydrogen embrittlement than austenite, hence an extensive martensitic transformation ahead of the propagating crack would be expected to promote crack growth in hydrogen. The data also suggest that the presence of hydrogen promotes the strain-induced transformation.

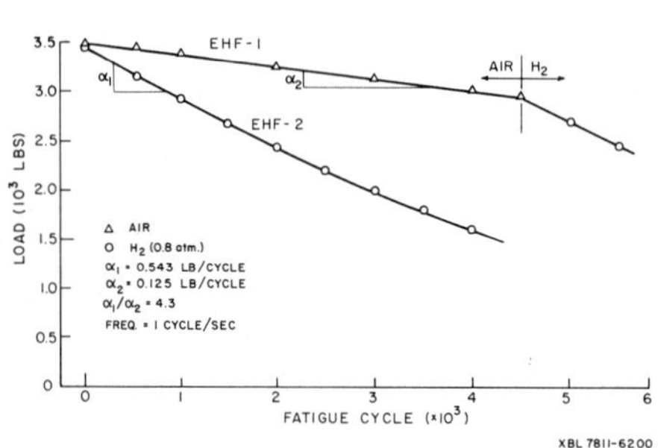

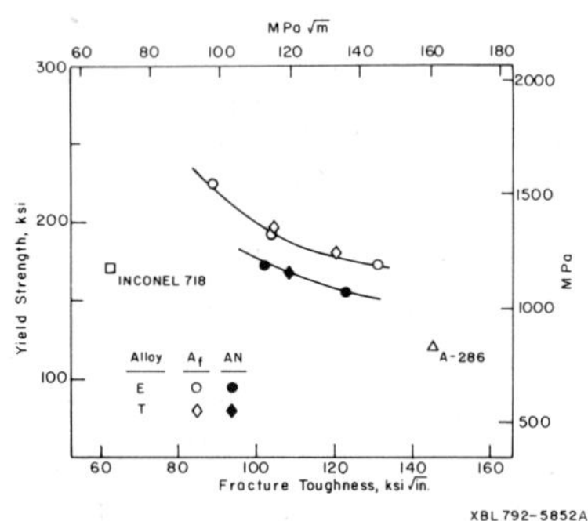

Fig. 1   Hydrogen effect on the fatigue crack growth rate of alloy E.

Fig. 2   Distribution of deformation-induced martensite on (a) fracture, (b) fatigue surface of alloy E.

Fracture

Fatigue   20 μ

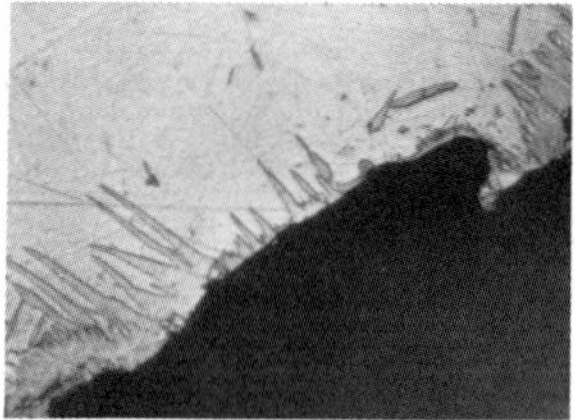

Fig. 3   Yield strength-fracture toughness relations for alloys E and T after processing from the annealed (AN) and forged (A_f) conditions. The two alloys have essentially identical strength/toughness character-istics. Typical data for Inconel 718 and A-286 is also shown for comparison.

Alloy Modification

Given the clear association between the induced martensite transformation and sensitivity to hydrogen, a plausible alloy design approach to improve hydrogen resistance was to modify the alloy chemistry so as to increase stability with respect to martensitic transformation. Given the base composition of the alloy, the most obvious means for increasing austenite stability would be an increase in the nickel content. This approach is, however, precluded by the low temperature ferromagnetism of the alloy at higher nickel contents. Of the possible new element additions, chromium was immediately attractive since it both stabilizes the austenite phase (at least at lower concentrations) and enhances resistance to oxidation and corrosion. The acceptability of Cr modification was, however, dependent on the possibility of introducing it in significant concentration without compromising the desirable mechanical properties of the alloy.

To explore the consequences of Cr addition, a modified alloy was cast with nominal composition Fe-34.5Ni-5Cr-3Ti-3Ta-0.5Al-1.0Mo-0.3V-0.01B, and designated EPRI-T. The alloy differs from EPRI-E only through the Cr addition and a slight concommitant decrease in nickel content to compensate for the decrease in Fe. Age hardening studies revealed that the alloy shows a hardening response resembling that of EPRI-E and has virtually identical strength-toughness properties, as documented in Table III and Fig. 3. In the double-aged and fully hardened condition, however, alloy T is stable with respect to martensite transformation during normal mechanical testing at room temperature. No martensite phase was detected after either tensile or fracture toughness testing. The suppression of the martensitic transformation is also presumably responsible for the slight decrease in the total elongation of Alloy T relative to that of Alloy E.

Table III. Mechanical Properties of Double Aged Alloy E and Alloy T.

Alloy	Yield Strength MPa (ksi)	Tensile Strength MPa (ksi)	Elongation %	Reduction in Area T	Fracture Toughness MPa√m (ksi√in)
E	1332 (193)	1587 (230)	24	60	114 (104)
T	1345 (195)	1539 (223)	18	53	115 (105)

Heat treatments: 750°C/3 hrs + 650°C/9 hrs

To evaluate the hydrogen sensitivity of EPRI-T, notched tensile and fatigue testing was conducted on the alloy after aging to ~200 ksi (1350 ksi) yield strength under conditions identical to those employed for testing of alloy E. Both the notched tensile properties and the fatigue crack growth rate in alloy T in 0.8 atm. hydrogen were essentially identical to those in air; no evident hydrogen sensitivity was found. Post-test examination of the fracture surfaces revealed no evidence for induced martensite.

Following these tests, fatigue pre-cracked compact tension specimens of alloy T were dead-weight loaded to stress intensities as high as 88 MPa√m (80 ksi√in) and immersed in a 3.5% NaCl solution for periods up to one month No evidence of crack growth was found.

Discussion

While the work reported here is not sufficient to establish a causal connection between the induced martensite transformation and the hydrogen sensitivity of alloy E, and while it is possible that the improvement in hydrogen resistance on addition of Cr reflects an influence of alloy chemistry as well as on phase stability, the parallel observations of an increase in the extent of martensitic transformation under fatigue conditions evidencing hydrogen embrittlement and the suppression of hydrogen suscepti- bility on stabilization with respect to strain-induced martensite strongly suggest a causal connection between the two phenomena and offer a promising approach for improving the hydrogen resistance of austenitic steels.

Acknowledgments

The authors are indebted to Dr. Howard G. Nelson, NASA-Ames Research Center, Moffitt Field, California, for assistance in providing the hydrogen test apparatus and helpful discussions. This work was supported by the Electric Power Research Institute under contract #RP-636-2, and made use of experimental facilities at the Lawrence Berkeley Laboratory supported by the Division of Material Science, Office of Basic Energy Sciences, U.S. Department of Energy under contract #W-7405-Eng-48.

References

1. J. W. Morris, Jr., K. M. Chang, D. H. Klahn, G. Thomas, and S. Jin: "Heat Treatment and Mechanical Properties of a Ta-Modified Fe-Base Superalloy," LBL Report No. 9799, Lawrence Berkeley Laboratory, Berkeley, CA., September, 1979.

2. J. W. Morris, Jr. and G. Thomas: EPRI Report FP-1061, Electric Power Research Institute, Palo Alto, CA., April, 1979.

3. J. K. Tien, A. W. Thompson, I. M. Bernstein, and R. J. Richards: Met. Trans. A., 7A (1976) p. 821.

4. H. G. Nelson, "Hydrogen Embrittlement Testing," ASTM STP 543, p. 152.

5. R. B. Benson, Jr., R. K. Dann, and L. W. Roberts, Jr.: Trans. TMS-AIME, 242 (1968) p. 2199.

DISCUSSION

<u>J.P. Fidelle, Commissariat à l'Energie Atomique, Bruyères-le-Châtel, France</u>:
How would you compare the severity of your lab work conditions to service
conditions, where wet H_2 should be less aggressive?

<u>J.W. Morris</u>: Our tests were done in 0.8 atm. dry H_2. To my best knowl-
edge, service conditions are ~5 atm. dry H_2 in US-made machines, while
service temperature is somewhat above the ambient temperature used in our
tests. I understand that manufacturers try to prevent contamination by
water vapor - I do not know how well they succeed. It should be kept in
mind that most retaining ring failures (of which I am aware) have been
caused by exposure to corrosive media during storage or handling, or to
leaks in water-cooled machines causing water contact with the ring. For
this reason, the industry is concerned about stress corrosion cracking resistance in a variety of aggressive media, including distilled and salt
water, H_2S, and industrial acids.

IMMUNITY OF AUSTENITIC RETAINING RING STEEL Fe-18% Mn, 4.5% Cr, 0.5% C

TO GASEOUS HYDROGEN EMBRITTLEMENT

Markus O. Speidel* and Andrejs Atrens

BBC Brown Boveri Research Center, CH-5401 Baden, Switzerland

Abstract

The austenitic, non-magnetic, high-strength generator retaining ring steel X40MnCr18 was examined for gaseous hydrogen embrittlement and for stress corrosion cracking. Both constant strain rate tensile tests and sub-critical crack growth measurements were performed in a variety of environments including air, oil, aqueous solutions and hydrogen. The steel is not affected by hydrogen but does evince intergranular stress corrosion cracking in the presence of water.

1. Introduction

Retaining rings used on Brown Boveri generators are made from the non-magnetic austenitic steel X40MnCr18. With a very large number of Brown Boveri generators in service, not a single failure of a retaining ring has occurred which could be attributed to gaseous hydrogen embrittlement. Therefore the service record implies that gaseous hydrogen embrittlement does not occur. However, the service failure (1) of a retaining ring of different metallurgy in a machine of a different make, prompted the present laboratory investigation.

2. Experimental Procedure

The austenitic non-magnetic steel (of BBC designation St 9100, German standard designation X40MnCr18, standard material number 1.3817) contains 0.3 to 0.6% C, 17 to 19% Mn, and 3 to 5% Cr. Its mechanical strength (ultimate tensile strength of from 1080 to 1380 MPa and a minimum yield strength of 980 MPa) is attained by cold expanding the ring. Specimens were machined from material taken from actual retaining rings, and their orientations could be related to the tangential, radial and axial directions of the ring. The tensile axis was in the tangential direction for the slow strain rate test. For the crack growth measurements, the load axis (which was perpendi-

* now at Eidg. Technische Hochschule Zürich, Switzerland

cular to the crack plane) was tangential and the crack growth direction was radial.

Slow strain rate tensile tests were carried out with strain rates in the range 10^{-3} to 10^{-9} s^{-1}. The slowest strain rate tests required more than 2 years to complete; during this time smooth tensile specimens were exposed to the environment and slowly pulled to fracture.

Subcritical crack growth measurements were performed with DCB (double cantilever beam) type specimens. Each of these was exposed to the environment and then a constant load was applied. Crack velocity was calculated from crack extension divided by the exposure time. Actual exposure times of up to and including 1 year were used to ensure a sufficiently low detection limit for the subcritical crack velocity.

After each test, the fracture surfaces were examined using optical and scanning electron microscopy.

3. Results

3.1. Slow strain rate tests at room temperature

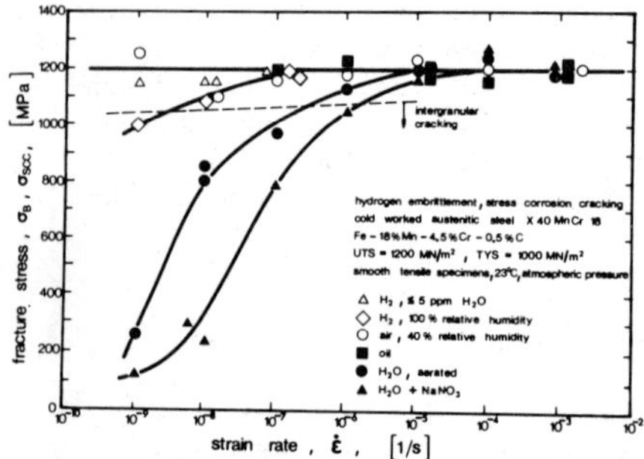

Fig. 1 – Slow strain rate tests. Note that dry hydrogen does not result in a reduced fracture stress. Water reduces the fracture stress, due to intergranular stress corrosion cracking.

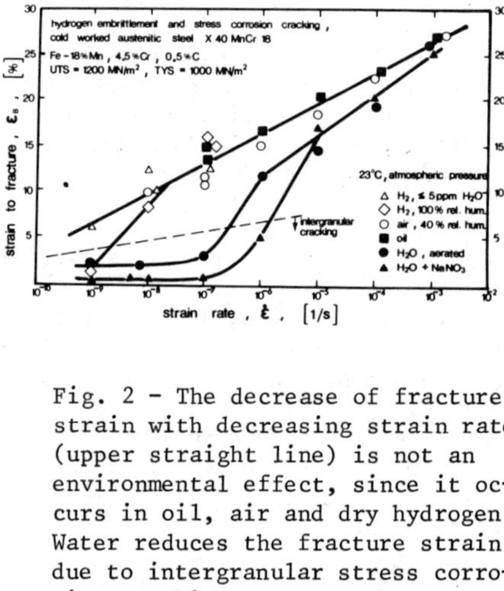

Fig. 2 – The decrease of fracture strain with decreasing strain rate (upper straight line) is not an environmental effect, since it occurs in oil, air and dry hydrogen. Water reduces the fracture strain due to intergranular stress corrosion cracking.

In Fig.s 1 and 2 are shown data obtained in the slow strain rate tensile tests in a variety of environments including wet and dry hydrogen. In these figures, the fracture stress and fracture strain are plotted as functions of the applied strain rate. The most important fact which can be deduced from Fig. 1 is that the fracture stress in dry hydrogen was not reduced as the strain rate was lowered down to 10^{-9} s^{-1}. This indicated that the material was not susceptible to hydrogen embrittlement.

However, in wet hydrogen of 100% relative humidity very slow strain rates resulted in a slightly reduced fracture stress and in intergranular cracking. Such cracking is not attributed to the hydrogen itself but rather to the water which can condense on the specimen surface. This kind of stress corrosion cracking was, of course, much more pronounced if the specimens were immersed in aerated water or nitrate solution. For such conditions a drastic reduction of the fracture stress occurs at slow strain rates, Fig. 1.

The fracture strain (ductility) of the austenitic, non-magnetic retaining ring material is shown in Fig. 2 as a function of strain rate and environment. Obviously, the fracture strain was reduced at lower strain rates. This effect, however, was not due to hydrogen, since it occured also in such innocuous environments as oil and air. However, the presence of water as 100% relative humidity in the hydrogen reduced the fracture strain from 6% to 1% at a strain rate of 10^{-9} s^{-1}, Fig. 2. Immersion in water or nitrate solution caused an even more dramatic reduction of the fracture strain. In all cases the environmental reduction of the fracture strain shown in Fig. 2 was due to intergranular stress corrosion cracking. Gaseous hydrogen embrittlement was not observed.

3.1.1. Fractography

Fractography reinforced the finding of no hydrogen embrittlement in the slow strain rate tensile tests at room temperature; the general decrease in the fracture strain with decreasing strain rate for the specimen fractured in air, dry hydrogen or oil could be attributed to the details of the fracture process, Fig. 3. In all cases there was a ductile intergranular component, e.g. Fig. 4, which corresponded to approximately 10-15% of the fracture surface. The rest of the fracture surface consisted of a ductile transgranular fracture, but the type of transgranular fracture (and the

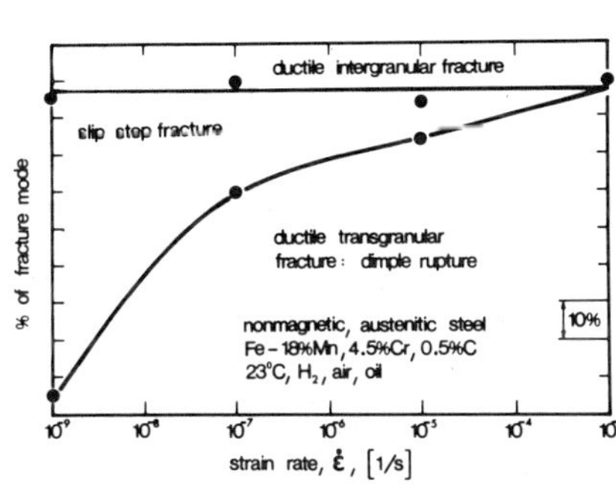

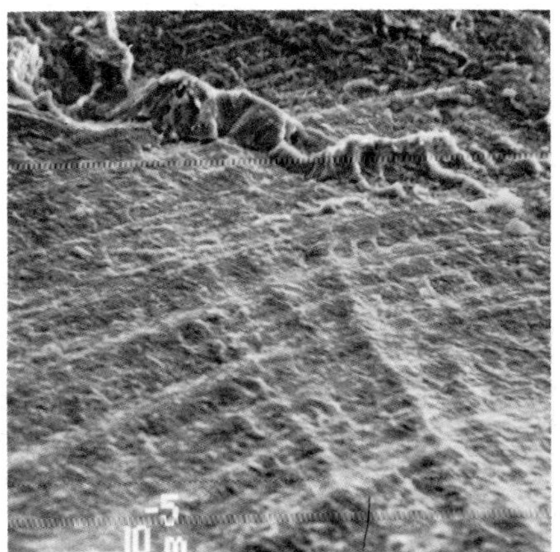

Fig. 3 - Summary of how the fracture modes depend on the strain rate for specimens tested in dry hydrogen or oil or air at 23°C.

Fig. 4 - Ductile intergranular fracture. Note the many small dimples. Specimen tested in hydrogen, $\dot{\varepsilon} = 10^{-9}$ s^{-1}.

amount of each type) depended on the strain rate. At a strain rate of 10^{-3} s^{-1}, the transgranular fracture was all dimple rupture formed by microvoid coalescence, Fig. 5. As the strain rate was decreased from 10^{-3} to 10^{-9} s^{-1} there was a decrease in the fraction of the fracture surface covered by dimple rupture and an increasing amount of the transgranular type of fracture illustrated in Fig. 6. It is thought that this fracture occurs primarily along slip planes and consequently it is called the slip plane fracture mode. However, the existence of many fine dimples indicate that it is a mixed mode of fracture. This change in transgranular fracture mode was observed in dry hydrogen, air and oil and corresponded to the decrease in ductility shown in Fig. 2.

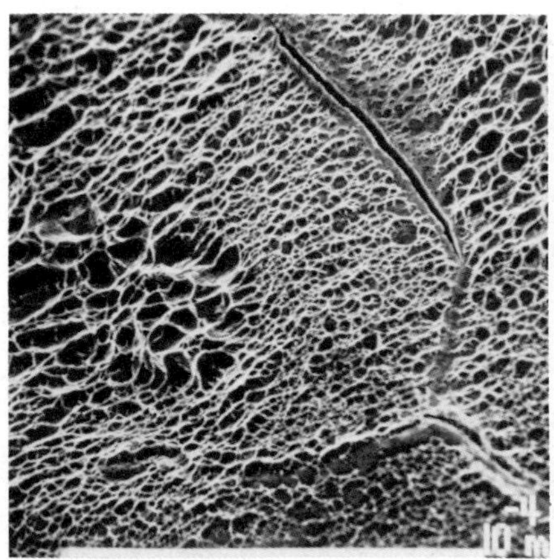

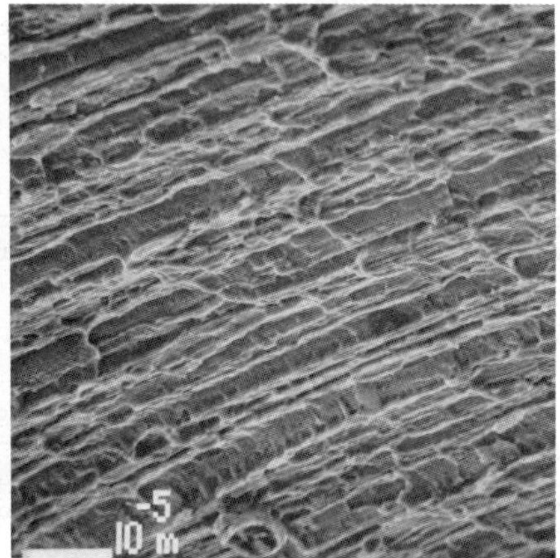

Fig. 5 – Dimple rupture. Note that grain boundaries are often visible. Here a microcrack is seen along part of the grain boundary. Specimen tested in air, $\dot{\varepsilon} = 10^{-3}$ s^{-1}.

Fig. 6 – Slip plane fracture mode. The very many small dimples indicate this fracture mode was microscopically ductile. Specimen tested in air, $\dot{\varepsilon} = 10^{-7}$ s^{-1}.

In contrast to the ductile fracture modes discussed above, the specimens tested at slow strain rates in moist hydrogen or in water had brittle intergranular fractures, Fig.s 7 and 8.

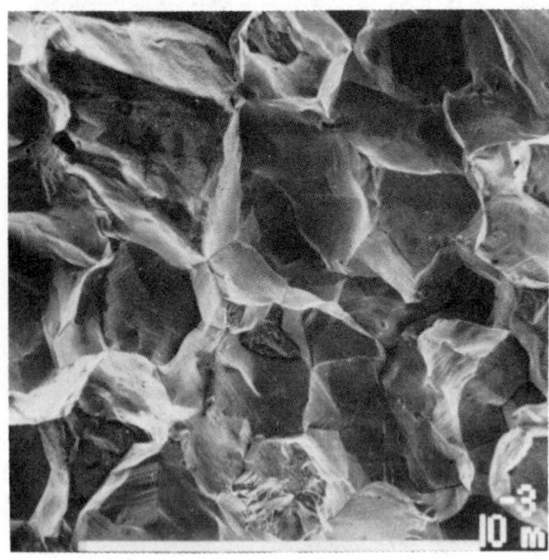

Fig. 7 – Brittle intergranular fracture. Note that some of the grain boundaries are covered with slip steps. Specimen tested in water, $\dot{\varepsilon}=10^{-9}$ s^{-1}.

Fig. 8 – Magnified view of part of figure 7 showing the slip steps on the grain boundaries.

Clearly visible in Fig. 8 on the grain boundary are the slip steps associated with the mechanical deformation of the material prior to testing. There
are also secondary cracks which appear to be along slip planes. In contrast
to the ductile intergranular fracture illustrated in Fig. 4, there are not
many features in Fig.s 7 and 8 which could be interpreted as being associated with ductile fracture.

The dividing line between specimens with ductile fracture modes and
those with the brittle intergranular fractures is shown in Fig. 2. A further
interesting point concerns the specimen tested at a strain rate of 10^{-8} s^{-1}
in hydrogen of 100% relative humidity. This specimen was slightly less ductile than the other specimens tested at this strain rate. Moreover, the
fracture surface of this specimen had a component of the brittle intergranular fracture. Thus this specimen was a transition between the ductile fracture in dry hydrogen (or oil or air) and the totally intergranular fracture
in water.

3.2. Slow strain rate tests at 100°C

The previous slow strain rate tests showed that dry hydrogen at room
temperature did not cause hydrogen embrittlement. Now, it is well known
that steels are most prone to hydrogen embrittlement at around room temperature and thus one would expect even less tendency of the steel to embrittle
in the elevated temperature service environment. To verify that this was
indeed the case, slow strain rate tests were conducted at 100°C in dry H$_2$
and N$_2$.

As expected there was no effect of hydrogen on the fracture stress or
the strain to fracture. For example, there was no difference between the
ductility measured in hydrogen or the ductility measured in nitrogen, Fig. 9.
However, in contrast to the room temperature results, the strain to fracture
was a constant 30% irrespective of the strain rate.

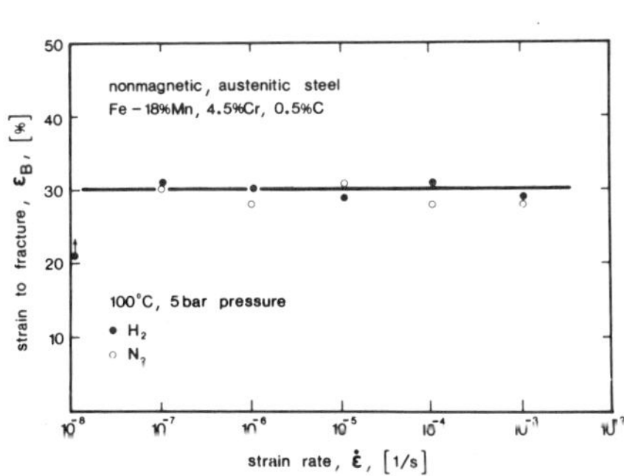

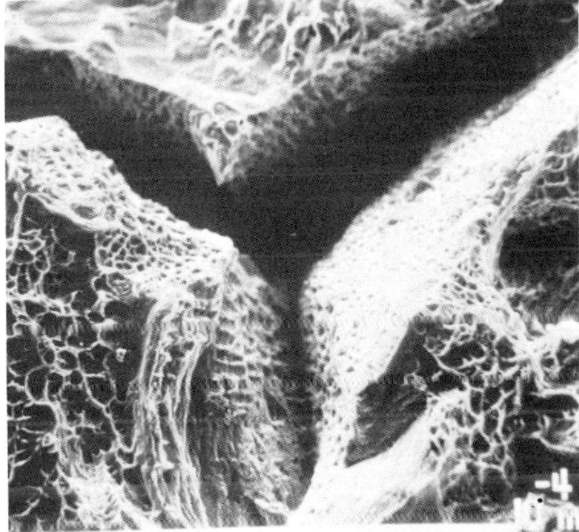

Fig. 9 - The strain to fracture of
austenitic ring steel as a function
of strain rate in 5 bar H$_2$ and N$_2$ at
100°C. Hydrogen has no effect.

Fig. 10 - Dimple rupture on the grain
boundaries. Specimen tested in hydro-
gen at 100°C, $\dot{\varepsilon} = 10^{-7}$ s^{-1}.

3.2.1. Fractography

Fractography again reinforced the findings of the mechanical tests: there was no observable difference between specimens tested in hydrogen and specimens tested in nitrogen. Moreover, corresponding to the constant strain to fracture irrespective of strain rate was a constancy of fracture mode: in all cases approximately 90% of the fracture surface was due to dimple rupture. Furthermore, the remaining 10%, the intergranular component, appeared more ductile. For example, grain boundaries of the specimen tested in hydrogen at a strain rate of 10^{-7} s^{-1} were mostly covered with dimples as shown in Fig. 10.

3.3. Fracture mechanics tests in hydrogen

The critical stress intensity at room temperature of the retaining ring steel X40MnCr18 is about 120 MPa.m$^{1/2}$. In order to investigate subcritical crack growth, fatigue precracked DCB specimens were loaded in hydrogen (at 23°C and atmospheric pressure) to 80 and 100 MPa.m$^{1/2}$. No crack growth occurred in 1 year. As the accuracy of the crack length measurement was 0.1 mm, the crack velocity was less than 4×10^{-12} m.s^{-1}, Fig. 11. Thus the threshold stress intensity exceeds 100 MPa.m$^{1/2}$ for gaseous hydrogen embrittlement at 23°C and atmospheric pressure. To verify that elevated temperature and pressure have no deleterious effect, a fatigue precracked DCB was loaded to 100 MPa.m$^{1/2}$ in 5 bar hydrogen at 100°C. No crack growth occurred. Thus, hydrogen embrittlement is not a problem with the non-magnetic, austenitic retaining ring steel provided that the hydrogen is dry. In hydrogen of high humidity, however, intergranular subcritical crack growth occurred, Fig. 11. The crack growth rate in hydrogen with 100% relative humidity was lower than the stress corrosion crack growth rates in water. However, at low stress intensities the crack growth rates in both environments tended to converge, Fig. 11. This is understandable since water can condense in the fatigue precrack. Therefore, a high humidity in the hydrogen environment must be avoided to prevent stress corrosion cracking of steel. The intergranular subcritical crack growth which occurred in humid hydrogen is attributed to the water and not to the hydrogen itself. Such cracking would occur in all kinds of gases containing 100% relative humidity.

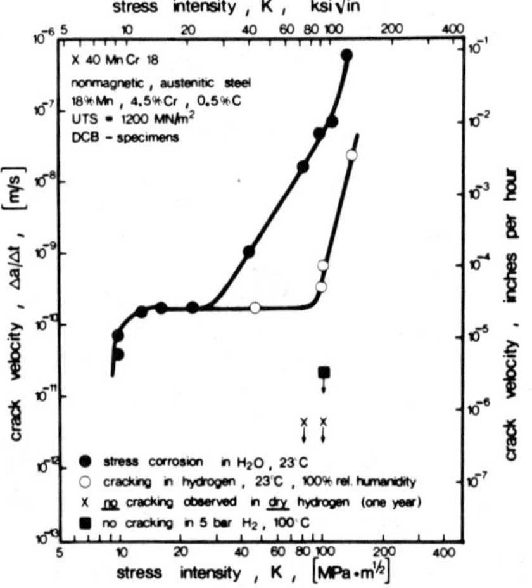

Fig. 11 – Fatigue-precracked, fracture-mechanics specimens cut from a retaining ring show no crack growth in dry hydrogen when loaded to a stress intensity of 100 MPa.m$^{1/2}$ for a full year. The presence of water however results in intergranular stress corrosion cracking.

4. Discussion

The present study has shown that there is no gaseous hydrogen embrittlement for the austenitic non-magnetic retaining ring steel X40MnCr18. This complete absence of hydrogen embrittlement contrasts with work (2-5) on another austenitic non-magnetic retaining ring material whose composition 18Mn4Cr and mechanical properties are similar to that of the material used in this study. The material 18Mn4Cr has grain boundary precipitates, virtually zero radial ductility, low K_{IC} values, an appreciable crack velocity in vacuum and a slightly higher crack velocity in hydrogen (2-5). All these observations contrast starkly with comparative results for the steel X40MnCr18 and point to improper processing of the material 18Mn4Cr.

There are probably a combination of factors responsible for the absence of gaseous hydrogen embrittlement of X40MnCr18. Firstly, it is well known that gaseous hydrogen is not a very severe environment compared with free corrosion (6) or cathodic charging (7), both of which can be equated to gaseous hydrogen at very high pressures (8). Secondly, austenitic steels are, in general, usually fairly resistant to hydrogen embrittlement, with a general correlation of increased resistance to hydrogen with increasing austenitic stability and increasing stacking fault energy (SFE) (9,10). In particular, there are several other reports (11-13) of austenitic manganese steels resistant to gaseous hydrogen embrittlement. The resistance of the austenitic steels to hydrogen can be partly attributed to the low hydrogen diffusity in the matrix, which limits the hydrogen access to the material. Then the austenitic stability and SFE have been associated (10) with hydrogen transport by dislocation drag: high austenitic stability is associated with high SFE and fine slip which in turn is less likely to lead to local high hydrogen concentrations. Moreover, the martensite which can form in alloys of low austenitic stability like 304 could provide short circuit diffusions paths, as well as providing in itself a phase susceptible to hydrogen cracking.

The alloy X40MnCr18 is a stable austenite with no magnetic or martensite phases present (14-16). Furthermore, although slip is coarse during the manufacture of the retaining ring, the slip bands are saturated during this process and subsequent slip occurs by dislocation motion in between the slip bands (15), thus minimizing the possibility of hydrogen transport by dislocation drag.

5. Conclusions

The present study has shown that there is no gaseous hydrogen embrittlement for the austenitic non-magnetic retaining ring steel X40MnCr18. This can be attributed in part to the relative mildness of the environment and to a limited possibility for hydrogen to penetrate into the steel.

Acknowledgements

Acknowledgement is given to J. Albrecht for many fruitful discussions, to G. Keser and R. Häring for help with the experiments and R. Baumann for help with the scanning electron microscopy.

References

1. C.B. Jolly, M.C. Murphy, A.N. Paterson and D.J. Petty, <u>CEA Thermal and Nuclear Power Section</u>, Halifax, Nova Scotia 1975.

2. B. Mukherjee, <u>ASTM STP 642</u>, p. 264.

3. I.M. Austen, P. McIntyre and E.F. Walker, <u>Proceedings of the Metallurgists Conference on Materials in Power Plant, Edinburg</u>, 1975, p. 120.

4. P. McIntyre and E.F. Walker, <u>The Second International Congress on Hydrogen in Metals</u>, Paris 1977, paper 3B8.

5. K. Akhurst, <u>Ph.D.Thesis</u>, Imperial College, University of London, 1979.

6. L.S. Darken and R.P. Smith, <u>Corrosion</u> 1949, vol 5, pp. 1-16.

7. A. Atrens, J.J. Bellina, N.F. Fiore and R.J. Coyle: <u>The Metal Science of Stainless Steels</u>, E.W. Collings and H.W. Kings, eds., pp. 54-69. The Met. Soc. AIME, 1978.

8. A. Atrens, D. Mezzanotte, N.F. Fiore and M.A. Genshaw, <u>Corrosion Sci.</u>, in press.

9. A.W. Thompson, <u>Handbook of Stainless Steels</u>, D. Peckner and I.M. Bernstein, eds., McGraw Hill, 1977, chapter 46.

10. A.W. Thompson and I.M. Bernstein, <u>Advances in Corrosion Science and Technology</u>, M.G. Fontana and R.W. Staehle, eds., pp. 53-175, Plenum Press, 1980.

11. R.A. McCoy, <u>Hydrogen in Metals</u>, I.M. Bernstein and A.W. Thompson, eds., pp. 169-178, Amer. Soc. Met., 1974.

12. A.W. Thompson, <u>ibid.</u>, pp. 179-182.

13. R.B. Benson, <u>ibid.</u>, pp. 183-197.

14. C. Maggi, BBC Brown Boveri and Co. Baden, unpublished work.

15. J. Albrecht, BBC Brown Boveri and Co. Baden, unpublished work.

16. K. Sipos, L. Remy and A. Pineau, <u>Met. Trans.</u>, 1976, vol 7A, pp 857-864.

DISCUSSION

D.A. Mezzanotte, Jr., E.I. du Pont de Nemours & Co. Aiken, SC: How do you
account for the lack of a strain rate dependence for the total strain to
fracture in your 100°C tests?

A. Atrens: It is clear that the total strain to failure depends critically
on the details of the fracture process. If the fracture mode illustrated in
Fig. 6 is at least partly caused by fracture along slip planes, then one would
expect localized slip to promote this slip plane fracture mode whereas increased cross slip would favor dimple rupture. This leads one to conclude
that at 23°C, slip becomes more localized as the strain rate is decreased,
perhaps due to some sort of strain ageing or dislocation rearrangement. On
the other hand, at 100°C, the slip morphology appears to be independent of
strain rate. In this case the extra thermal energy should promote easier
cross slip and consequently the dimple rupture fracture mode.

J.W. Morris, Jr., University of California, Berkeley, CA: I have two questions: (1) Your tests refer to an alloy at the 1000 MPa strength level
(~140 ksi). Much of the concern about the hydrogen sensitivity of 18Mn-5Cr
of which I am aware refers to a somewhat higher strength alloy (170 ksi or
~1200 MPa). Have your investigations included these higher alloy strengths?
(2) There have been rumors of hydrogen-induced failures in 18Mn-5Cr rings
at the 140 ksi level. You mentioned a retaining ring failure as the impetus
for this work. How do you account for such failures in light of your results?

A. Atrens: 1. We have not investigated higher strength versions. 2. The retaining ring failure mentioned in the introduction was an alloy of a completely
different metallurgy made by a different company. Our experimental results
support our service experience that gaseous hydrogen embrittlement is not a
problem for our steel.

R.H. Bellows, General Electric Company, Schenectady, NY: Regarding the fractograph showing dimpled intergranular failure, what were the conditions of
test leading to that fracture?

A. Atrens: In general, the intergranular component (which is about 10% of
the total) of the fracture of specimens strained at 100°C was more ductile
than that of specimens strained at 23°C. The particular micrograph shown
was of a specimen tested in 5 bar H_2, 100°C, $\dot{\xi} = 10^{-7}$ s^{-1}.

A STUDY OF HYDROGEN PERMEATION AND SUSCEPTIBILITY TO CRACKING OF A TURBINE

DISC STEEL SUBJECTED TO HYDROGEN CHARGING FROM A STRONGLY ALKALINE SOLUTION [+]

P T Wilson[*], Z Szklarska-Smialowska and M Smialowski[**]

The Ohio State University, Fontana Corrosion Center, Columbus OH

Hydrogen permeation transients were measured for a Ni-Cr-Mo-V turbine-disc steel, ASTM A471, electrochemically charged with hydrogen from a 12 mol kg^{-1} NaOH solution at T = 30°C, 60°C and 85°C. The apparent diffusion coefficient of atomic hydrogen was found to depend linearly on the steady-state permeation current density. The results indicated that a proportion of the hydrogen dissolved in the steel was associated with traps and that the trapping was a reversible process. The trapping rate of hydrogen was lower than the rate of untrapping; the latter inc reased with increasing temperature more rapidly than did the rate of trapping.

Slow strain-rate tests were completed at room temperature and at T = 90°C in a 12 mol kg^{-1} NaOH solution, at the corrosion potential and also at various cathodic charging potentials. At room temperature the tensile-stressed steel cracked readily when subject to cathodic charging. At charging current densities exceeding some very low threshold value the time to failure was almost independent of the charging current density. At 90°C cracking occurred at the corrosion potential and the cracking susceptibility increased with increasing cathodic current density. At identical charging currents, the cracking susceptibility was significantly less at T = 90°C than at room temperature. The main cause of the reduction in the susceptibility of the steel to hydrogen-induced cracking with increase in temperature is thought to be related to the decrease in the proportion of trapped hydrogen atoms with increase in temperature, which arises from the difference in activation energies of the trapping and untrapping processes.

* Present address: Chemistry Division, DSIR, Private Bag, Petone,
                   New Zealand.

** Present address: Institute of Physical Chemistry of the Polish Academy
                    of Sciences, 01-224 Warsaw, Poland.

+ This work was supported by the Electric Power Research Institute, Palo Alto.,
  Calif.

Introduction

The environmentally-assisted cracking of medium-strength low-alloy
steels in hot concentrated caustic solutions has been reported and reviewed(1).
The possibility exists that hydrogen embrittlement contributes to this type
of failure, the hydrogen entering the steel as a result of a corrosion
reaction of the steel with an alkaline environment. The electrochemical
hydrogen permeation technique (2) is sufficiently sensitive to allow the
estimation of the small amounts of hydrogen which dissolve in a steel as a
result of a corrosion reaction, and also provides a means of comparing the
mobility and trapping characteristics of hydrogen atoms in a metal lattice
at different temperatures. By combining results from the above technique
with slow strain-rate testing (3), it is possible to compare the cracking
susceptibility of a steel having different concentrations of hydrogen and at
different temperatures. This paper reports the results of such a study.

Experimental

Hydrogen Permeation

Steady-state and transient hydrogen permeation was measured using the
cell shown in Figure 1, at temperatures of 30, 60 and 85°C. The cell was
machined from teflon because the corrosivity of the test environment,
12 mol kg^{-1} NaOH, did not permit the use of glass. The steel, type ASTM

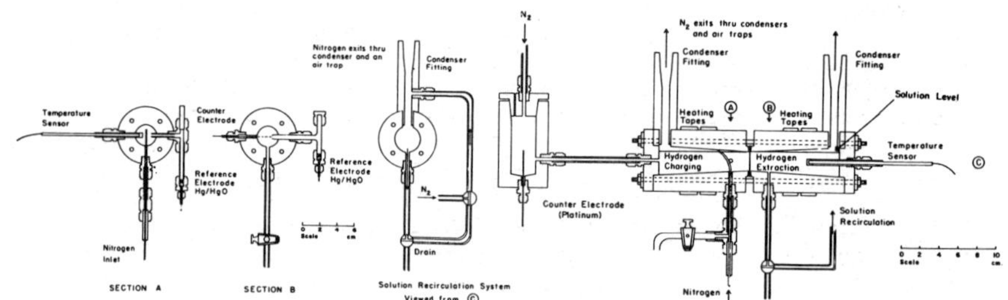

Fig.1. The electrochemical hydrogen permeation cell.

A471 had the composition and mechanical properties listed in Table I. Steel
membranes were ground to a thickness of 1.0 mm, polished to a 600# finish,

Table I. Composition and Mechanical Properties of ASTM A471 Turbine Disc
Steel.

Element	C	Mn	P	S	Si	Ni	Cr	Mo	V	Cu	Fe
Weight%	0.26	0.24	0.008	0.009	0.013	3.54	1.63	0.36	0.12	0.13	Bal

0.2% Yield Strength	1034 MN m^{-2}
Ultimate Tensile Strength	1152 MN m^{-2}
Percent Elongation	16
Percent Reduction in Area	55

and heated in Argon at T=400°C for 30 minutes; one side of the membrane was
palladized (4).

Hg/HgO/NaOH reference electrodes (5) were used for all potential measure-

ments. They contained the same solution as that in the compartment to which
they were connected. Solutions were deoxygenated using nitrogen which had
been purified by passing it through chromium solutions in contact with a zinc
amalgam (6), and humidified by passing through the test solution at the test
temperature. A gas-pumping arrangement in the anodic compartment, Figure 1
both deoxygenated and circulated the solution; N_2 bubbling in the cathodic
compartment also provided solution agitation necessary for good temperature
control which was maintained to $\pm1^oC$ at T=30 and 60^oC, and $\pm2^oC$ at T=85°.
The electrolyte in the cathodic compartment was tested at intervals to confirm its concentration. The possibility of oxide films and, or N_2 or H_2
bubbles interfering with the cathodic charging was recognized. The first
possibility was prevented by having a minimum cathodic charging current
which protected the surface. The second possibility was minimized because
of the sloping contour of the cell. A series of both increasing and decreasing transients were measured by changing the charging current in a step-wise
manner; a steady-state permeation current was observed before the next
current step. An experiment at a particular temperature was concluded by
measuring the permeation current at the corrosion potential.

<u>Slow Strain-Rate Tensile Tests</u>

Tensile test specimens of the steel were strained at a constant rate of
10^{-6} s^{-1} using a multi-specimen strain-rate machine (7), in air, and in
12 mol kg^{-1} NaOH at T=25^oC and 90^oC. In the NaOH solutions measurements
were made at the corrosion potential and also at various hydrogen charging
currents. The test cell was of all teflon construction. The reference
electrodes and N_2 purification system were the same as that used for the
permeation experiments, similar precautions being taken to exclude O_2 and
prevent changes in solution concentration. The tensile test-specimen was a
6.35 mm diameter rod with a 12.5 mm reduced section of diameter 2.54 mm.
Time to failure was used as a measure of cracking susceptibility, although
the reduction in area at fracture was also measured and the fracture surface
was examined using scanning electron microscopy.

Results

<u>Hydrogen Permeation Measurements</u>

The steady-state permeation currents are listed in Table II   together

Table II.   Dependence of Hydrogen Permeation on Charging Current

Charging Current Density, $A\ m^{-2}$	Permeation Current Density x 10^{-2}, $A\ m^{-2}$		
	at T = 30^oC	at T = 60^oC	at T = 85^oC
0.95	1.12	2.0	3.0
.3.18	1.8	3.25	4.9
9.5	2.23	4.5	8.4
31.8	3.6	6.5	12.9
i = i$_{corr}$	not detected	0.5	1.0
Background Current	0.2	0.6	1.0

with the background currents and the permeation currents which were measured
when the cathodic side of the membrane was at the corrosion potential. The
relationship between steady-state permeation current and charging current

was linear on a log-log graph. The temperature dependence of the steady-state permeation currents is shown in Figure 2.

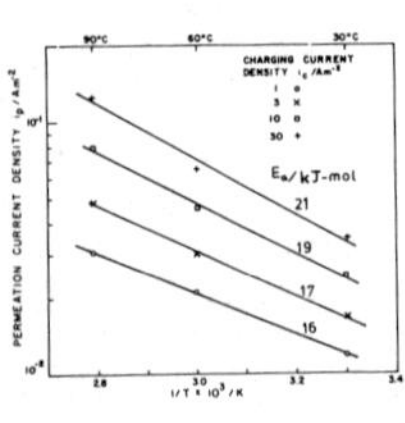

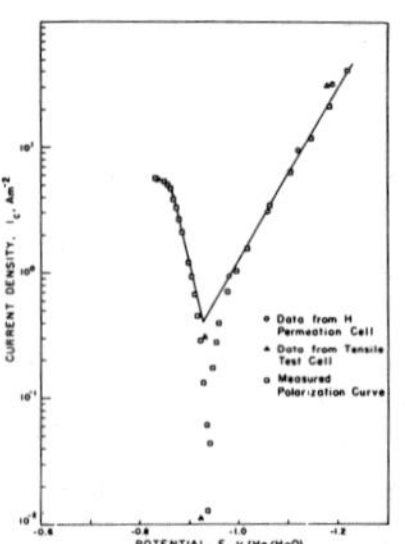

Fig.2. Temperature dependence of hydrogen permeation

Fig.3. Polarization curve for ASTM A471 steel in 12 mol kg^{-1} NaOH, T=90°C

The potential-current measurements from both the permeation cell and the slow strain-rate cell are plotted on a polarization curve for the test steel, Figure 3, at T=90°C, and are in reasonable agreement. Potential measurements in the cathodic side of the permeation cell provide a sensitive indication of the presence of any interfering bubbles, O_2 leakage or surface films, hence the significance of this agreement. At room temperature similar agreement was found although the corrosion potential tended to drift anodically.

The mathematical treatment for the hydrogen permeation technique has been given by McBreen et.al. (8) and Nanis and Namboodhiri (9). For a constant diffusion coefficient a rising transient is described by equation 1, where J_n is a normalized

$$J_n = \frac{J - J_1}{J_2 - J_1} = \frac{2}{\sqrt{\pi \tau}} \sum_{n=o}^{\infty} \exp\left[\frac{-(2n+1)^2}{4\tau}\right] \qquad 1$$

flux and J_1 and J_2 are steady-state fluxes at the beginning and end of the experiment; the other parameters and boundary conditions are defined by McBreen et. al. (8) For the decreasing transient two possible equations apply (9), depending on the boundary condition at the inlet surface. For the decreasing transient experiments described in this paper, the charging current was reduced in steps, with the lower charging current always sufficiently cathodic to prevent any oxide film formation. Consequently it is reasonable to assume that the input surface remained permeable to hydrogen and the "fast" (9) decay equation applied.

$$J_n = \frac{J - J_1}{J_1 - J_2} = \frac{2}{\sqrt{\pi \tau}} \sum_{n=o}^{\infty} \exp\left[\frac{-(2n+1)^2}{4\tau}\right] \qquad 2$$

The permeation transients were reasonably reproducible for a given charging current and showed no dependence on charging history, or peaks, which would have indicated irreversible damage to the steel. Diffusion coefficients calculated at fractional values of the normalized flux showed a small variation within a single transient but there was a more significant variation between transients of differing permeation current. These results are shown in Figures 4; similar results were observed at T=60°C; the temperature dependence is shown in Figure 5.

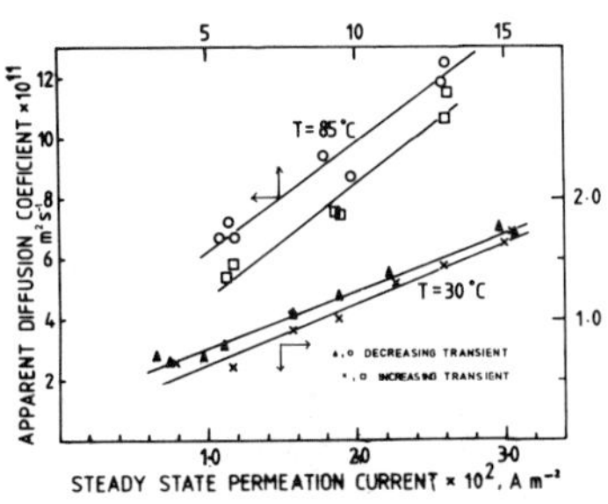

Fig.4. Variation of apparent dif-fusion coefficient (at Jn = 0.5) with steady-state permeation current.

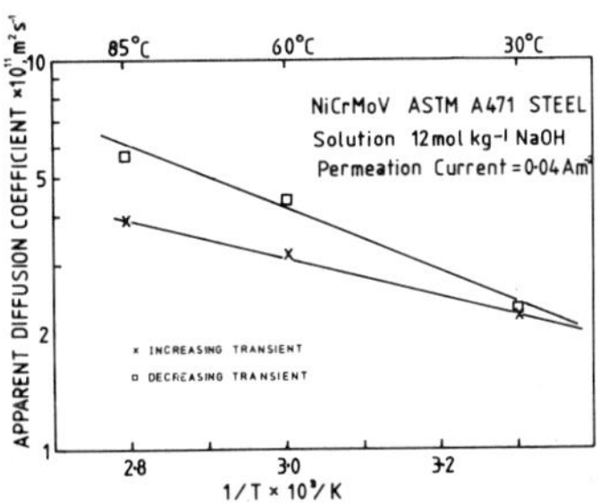

Fig.5. Temperature dependence of apparent diffusion coefficient.

Slow Strain-Rate Tensile Tests

Figure 6 shows a typical load-elongation curve together with the method for estimating the time to failure. No localized corrosion was observed at

Table III. Results of Slow Strain Rate Tests on ASTM A-471 Turbine Disc Steel.

Environment	Charging Current Density i_c, A m^{-2}	Times to Failure At room temperature	Average	t_f, hrs at T=90°C	Average
Air	--	53.8,58.8,52.4,54.0	54.8	--	--
		54.5,56.4	55.4	40.3,41.3	40.8
12 mol kg^{-1}	i_{corr}				
NaOH	0.318	27.5,35.9,36.0	33.1	36.4,36.9	36.6
	3.18	26.8,28.6,26.7,26.0	27.0	34.0,35.0	34.5
	31.8	26.1, 26.7	26.4	30.2,30.3	30.25

both test temperatures. The results are listed in Table III. Figure 7 relates the time to failure, to the steady-state permeation rates found in the permeation experiment, for the charging currents employed in the slow-strain rate test. The microstructure of the test steel was that of course grained lower bainite with relatively large needles of cementite in a matrix of untempered martensite. Specimens that failed in tests conducted at room temperature in air, exhibited a ductile fracture of the double cup and cone type with an average reduction of area at feacture, RA = 60%. A similar type

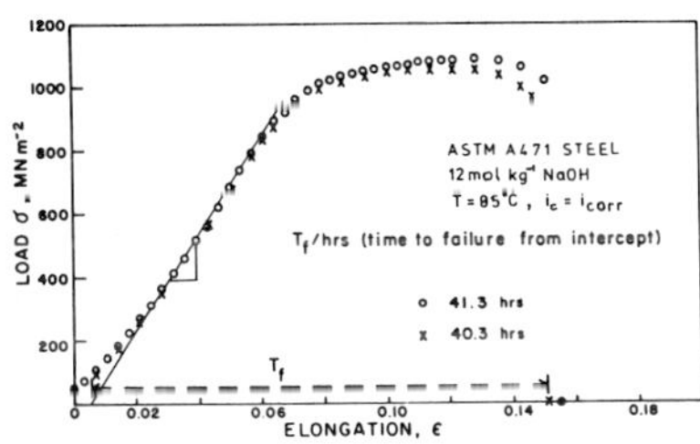

Fig.6. Typical load-elongation curve.

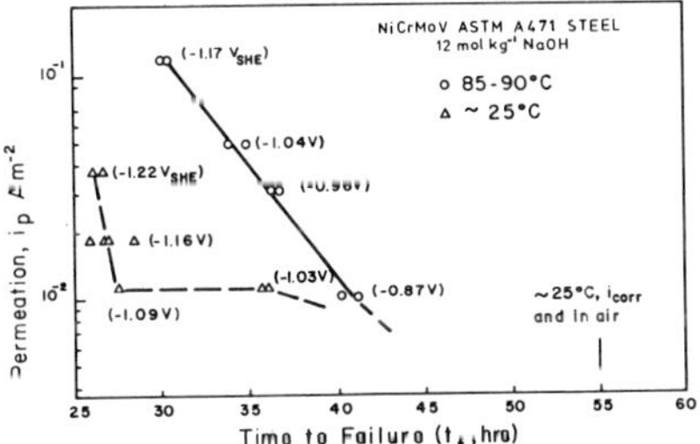

Fig.7. Relationship between time-to-failure and the corresponding steady-state hydrogen permeation current.

of ductile fracture with RA = 58% was observed for specimens that failed at
room temperature at the corrosion potential. The reduction in area for the
specimen tested at T = 90°C at the corrosion potential was approximately 40%.
All remaining specimens tested in 12 mol kg^{-1} NaOH with the application of
cathodic charging currents failed in a brittle mode with practically no
reduction in area at fracture. Typical SEM fractographs are shown in Figure 8.
On the reduced section of all specimens tested under cathodic polarization
conditions, numerous cracks were apparent on both sides of the fracture.

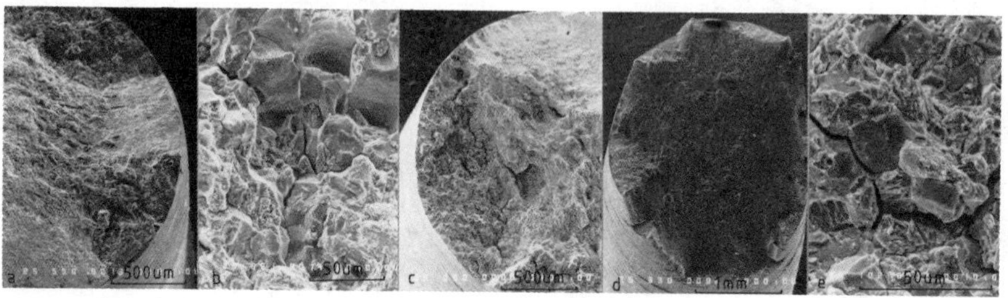

Fig.8. Fractographs of specimens that failed in the slow strain-rate test.
a. T=25°C, i_c=0.318Am^{-2}, t_f=36hrs, RA≈0. Intergranular rupture at upper and
bottom right; quasi-cleavage with needles of cementite elsewhere.

b. T=25°C, i_c=0.318Am^{-2}, t_f=27.5hrs. RA≈0. A large portion of the fracture
surface showed intergranular rupture.

c. T=25°C, i_c=3.18Am^{-2}, t_f=26.8hrs. RA≈0. Shows intergranular rupture and
dimples at left, cracks on external surface and at upper right a portion
broken by mechanical overload.

d. T=90°C, i_c=0.318Am^{-2}, t_f=36.9hrs. RA≈0. Shows dimples and a few intergranular cracks, plus numerous cleavage facets.

e. T=90°C, i_c=31.8Am^{-2}, t_f=30.3hrs. RA≈0. Surface shows large intergranular
cracks, dimples and cleavage facets.

Discussion

Hydrogen Permeation

The absence of peaks in the permeation-time plots recorded at constant
charging current, and the reasonable reproducibility of the results of duplicate measurements, indicated that no irreversible damage occurred to the steel
membrane during the recording of the transients. The steady-state permeation
current, i_p depends on both the lattice diffusion coefficient D_l and the concentration at the charging surface, C_o through equation 3, where the other

$$i_p = F D_l C_o / L \qquad (3)$$

parameters have their usual meanings. Consequently the temperature dependence
of the steady-state permeation current measures the effect of temperature on
both of the above variables. The activation energy for hydrogen lattice diffusion in low alloy steels has been determined (10,11) although a common
feature of these results is a difference in activation energies between data
recorded at high temperatures, T>150°C and that recorded at lower temperatures.
It is generally accepted that the high temperature result is the true lattice
diffusion activation energy. Comparing these activation energies which fall
in the range, 8–12kJ mol^{-1}, with that for steady-state permeation in Figure 4,
shows clearly that at a constant charging current the amount of hydrogen entering the steel increases with temperature.

The calculated diffusion coefficients were plotted as a function of steady-state permeation current density because this parameter is directly proportional to the concentration of hydrogen at the input surface, C_0 through equation 3. Following Johnson et al (12) and Oriani (13), it is reasonable to explain the observed variation of "apparent" diffusion coefficient with hydrogen concentration as a consequence of hydrogen trapping. Assuming that there are a finite number of traps, then the "apparent" diffusion coefficient should "approach" a maximum value as the traps become saturated. The reproducibility of these results also indicated that the trapping occurring during the recording of the transients, was completely reversible although it should be noted that irreversible traps would have been filled with hydrogen after the first application of a cathodic charging current, and would not have had any observable effect on subsequent transients. If local equilibrium existed between the mobile and trapped hydrogen populations during the recording of the transients (13), then the apparent diffusion coefficient for an increasing transient should equal that for the decreasing transient, provided an equal number of traps are being filled or emptied respectively. Consequently diffusion coefficients calculated for different transient directions must be compared at equivalent lattice hydrogen concentrations. The apparent diffusion coefficients for the different transient directions are compared in Figure 4,in which the initial steady-state current for a decreasing transient and the final steady-state current for an increasing transient were used to represent equivalent lattice hydrogen concentrations. The diffusion coefficient for the two types of transients are approximately equal at T=30° but diverge at the higher temperature. The diffusion coefficient for the decreasing transient becomes larger,indicating a more rapid emptying of traps. This trend is shown more clearly in Figure 5 where the temperature dependence of the apparent diffusion coefficients are compared.

Slow Strain-Rate Tensile Tests
<u>Slow Strain-Rate Tensile Tests</u>

The results indicate that at room temperature at the corrosion potential there was no environmental effect on the mechanical properties of the steel, which differed from that found at T=90°C. This occurred because at T=90°C the oxygen solubility in the solution was much lower than at room temperature and therefore it was easier to deoxygenate, and the corrosion current was much higher. Consequently, at room temperature the specimen surface was passivated, preventing any hydrogen from dissolving in the steel whereas at T=90°C the specimen surface remained active allowing hydrogen to dissolve in the steel. This conclusion was supported by the results obtained in the electrochemical permeation experiment where a hydrogen permeation current was observed at T=60°C and 90°C, but not at 30°C. The lack of reproducibility in the time to failure observed at a charging current of i_c = 0.318 A m^{-2} at room temperature was probably also due to the fact that at low cathodic charging currents and also at the corrosion potential even traces of dissolved oxygen would effect the experimental result as indicated by the drift in corrosion potential.

The results indicate that under the conditions of the slow strain-rate tensile test carried out at room temperature, this low alloy steel was extremely sensitive to initiation and propagation of hydrogen-induced cracks. Once initiated, the crack propagated until the tensile strength of the remaining cross-sectional area was attained; this produced the final rupture by tensile overload. In the fractograph, Figure 8c, the area damaged by hydrogen-induced cracking can be clearly distinguished from that ruptured by mechanical tensile overload. It is evident from Figure 7 that between room temperature and T=90°C an essential change in the intensity of the damaging process takes place. This can also be concluded from the different appearances of the fracture surfaces shown in Figures 8a and 8b, and 8d respectively. A one hundred times larger charging current density was necessary to produce at T=90°C similar effects to those observed at i_c = 0.318 A m^{-2} at room temperature; compare Figure 8b with Figure 8e;   this difference in charging current density corresponds to at least a 10 times difference in input-hydrogen concentration.

It is generally acknowledged that the susceptibility of steels to hydrogen embrittlement has a maximum at room temperature and decreases markedly with both increasing and decreasing temperatures. In gaseous hydrogen this behaviour has been explained in terms of chemisorption kinetics (14) controlling the transport of hydrogen to the crack initiation site. With the results in this paper, and also those of Townsend (15), it is clear that the amount of dissolved hydrogen increased with increasing temperature. A possible explanation for this lower susceptibility to hydrogen embrittlement with increasing temperature is suggested by the results of the hydrogen permeation experiments. At room temperature the rates of hydrogen trapping and untrapping did not differ substantially, while at $T=90^{o}C$ the rate of untrapping was higher than the rate of trapping which suggests that under the conditions of the present experiments the atoms of hydrogen which were weakly associated with some internal features acting as reversible traps in the steel, played a significant part in the damaging process.

Conclusions

Hydrogen permeation transients measured for a low alloy steel, type ASTM A471 in 12 mol kg^{-1} NaOH indicated that the amount of hydrogen dissolved in the steel at a given charging current increased with increasing temperature. A portion of the hydrogen dissolved in the steel was associated with traps; the trapping was a reversible process, which did not produce irreversible damage in the unloaded steel. The rate of untrapping increased with increasing temperature more rapidly than did the rate of trapping. With the charging side of the steel membrane at the corrosion potential, a finite hydrogen permeation current was observed at $T=60^{o}C$ and $90^{o}C$ but not at room temperature, where passivation probably prevented any hydrogen from entering the steel. Slow strain-rate tensile tests of the low alloy steel in the same solution with cathodic charging, showed that the steel was extremely susceptible to hydrogen embrittlement at room temperature with the susceptibility decreasing with increasing temperature. The main cause of the reduction in the susceptibility to hydrogen-induced cracking with increase in temperature was thought to be related to the decrease in the proportion of trapped hydrogen atoms with increase in temperature, which arises from the difference in activation energies of the trapping and untrapping processes.

References

1.   C.S. Carter and M.V. Hyatt : Int. Conf. on Stress Corrosion Cracking and Hydrogen Embrittlement of Iron Base Alloys, Firminy, France, June, 1973. p. 583.
2.   M.A.V. Devanathan and Z. Stachurski : Proc. Roy. Soc., 1962, A270, p.90.
3.   G.M. Ugiansky and J.H. Payer, eds. : Stress Corrosion Cracking - The Slow Strain-rate Technique, ASTM STP 665. 1977.
4.   N. Hall ed. : Metal Finishing - Guidebook and Directory, 1979, p.298
5.   D.J.G. Ives and G.J. Janz : Reference Electrodes, Academic Press, New York 1961.
6.   L. Meites, Polarographic Techniques, 2nd ed., p.89. J. Wiley, New York,1965.
7.   W.T. Nutter, A.K. Agrawal and R.W. Staehle, ASTM STP, 665, 1977, p. 375.
8.   J. McBreen, L. Nanis and W. Beck, J. Electrochem. Soc. 1966, 113, p. 1218.
9.   L. Nanis and T.K.G. Namboodhiri, J. Electochem. Soc., 1972, 119, p. 691.
10.   O.F. Angeles, R.J. Stueber and G.H. Geiger, Corrosion 1976, 32, p. 179.
11.   F.R. Coe and J. Moreton, J. Iron and Steel Inst. 1966, 204, p. 366.
12.   H.H. Johnson, N. Quick and A.J. Kumnick, Scripta. Met., 1979, 13, p. 67.
13.   R.A. Oriani, Acta. Met., 1970, 18, p. 147.
14.   D.P. Williams and H.G. Nelson, Met. Trans. 1970, 1, p. 63.
15.   H.E. Townsend, Corrosion 1972, 28, p. 39.

DISCUSSION

N.J.H. Holroyd, University of Newcastle-Upon-Tyne, Newcastle-Upon-Tyne, Great Britain: Is the embrittlement promoted at the free corrosion potential in hydroxide solutions at 90°C, solely due to hydrogen?

S. Smialowska: At T = 90°C some participation of a SCC process of the anodic type cannot be excluded, but there is no direct evidence for it.

G.M. Pressouyre, Creusot-Loire Research Center, Le Creusot, France: When performing first permeation transients, both irreversible and reversible traps are active. When degassing or performing second permeation transients, only reversible traps are active (irreversible traps are filled). Hence the difference between first permeation (the curve here) and degassing curve may not only be due to different reversible trapping kinetics, but also to irreversible trapping occurrence. See Ref.: G.M. Pressouyre and I. M. Bernstein, Met. Trans., 1978, 9A, p. 1571.

S. Smialowska: In our experiments we took into consideration only the subsequent, not the first, permeation transients, because the first recorded transient is likely to be affected by surface oxides formed on the cathodic charging current.

R. Jones, Battelle Northwest, Richland, WA: Some hydrogen traps are beneficial while others are not; could you elaborate on the type of hydrogen traps you think are active in this material?

S. Smialowska: Unfortunately our experiments did not provide information about the type of hydrogen traps which are active in this steel.

M.Elices*, V.Sánchez*, I. Bernstein**, A.Thompson**,J.Piñero*.

*E.T.S.I. Caminos. Universidad Politécnica. Madrid.

**Carnegie-Mellon University. Pittsburgh.

There are several basic forms of high-strength steel that are used currently in prestressed concrete construction. In general, these can be divided into three groups: wires, stress-relieved strands and high-tensile alloy bars. This paper refers to eutectoid, patented, cold drawn and stress-relieved wires, by far the most widely used type of wires.

The paper summarizes the first results of an extensive research program to determine hydrogen embrittlement susceptibility of prestressing steels. It comprises a theoretical approach to find expressions relating time to fracture to stress intensity factor and hydrogen concentration provided by the environment and an experimental approach to check the validity of the model developed and to show the influence on hydrogen susceptibility of surface condition and degree of drawing.

Introduction

Stress corrosion cracking of prestressing steel is customarily attributed to hydrogen embrittlement. Although this fact has not yet been proved,
it is widely accepted that hydrogen embrittlement plays an important role
in stress corrosion cracking of these high strength steel in actual
environments. The International Federation of Prestressing (F.I.P.) has
therefore trended towards a hydrogen embrittlement test to study stress
corrosion susceptibility of prestressing steels. Among various candidates,
the ammonium thiocyanate test has been chosen. Testing procedure is
described in an F.I.P. report (1).

This test is very sensitive to the steel surface condition and testing
smooth samples leads to great scattering in the time to fracture. As it
will be shown, steels which show a very different behavior (two orders of
magnitude in fracture times) when testing smooth samples may have however
the same bulk hydrogen susceptibility.

It seems reasonable therefore to study hydrogen susceptibility of
prestressing steels by testing precracked specimens and to apply Fracture
Mechanics concepts to evaluate $K_{I,HE}$ for this particular environment.
This should enable us to quantify tests with smooth samples and evaluate
threshold stress intensities in actual environments, needed in design.

Model of Hydrogen Induced Cracking

This model for the initiation of hydrogen embrittlement cracking in
aqueous environments is based on the work of Doig and Jones (2) and of Gerberich and Chen (3). It is assumed for these steels that the controlling
step in cracking is hydrogen diffusion towards the region of highest hydrostatic stress. No micromechanism (4) of cracking is assumed.

Crack initiation time is assumed to be the time required to build up
a critical hydrogen concentration, C_c, anywhere in the notch tip region.
Moreover, we will assume that time to fracture is for practical purposes
the same as initiation time. This hypothesis is reasonable if we take
into account the velocities of crack propagation reported (4) for
hydrogen embrittled high strength steels, about 10^{-5} m/sec. and the
experimental facts that acoustic emission measurements have detected crack
propagation at the very end of the sample life-time. (1, Appendix).
This hypothesis has been checked experimentally by the authors by a twofold
procedure: Testing degasified samples dismounted after more than ninety per
cent of the assumed fracture time, no damage has been observed; Compliance
measurements during testing show no changes up to the fracture time.

The hydrogen concentration built up under equilibrium conditions
$C(x,\infty)$ at any point, x, within the stress field is given (3) by

$$C(x,\infty) = C_o \exp\left[\sigma_H(x)V_H/RT\right] \tag{1}$$

where C_o is the equilibrium concentration in the absence of stress provided
by the NH_4SCN test solution, σ_H the hydrostatic component of the applied
stress, V_H is the partial molar volume R is the gas constant and T is
the absolute temperature.

Under transient conditions the hydrogen concentration at time t, $C(x,t)$, is given (2) by

$$C(x,t) = C_o \{1 - erf \left[x/2(Dt)^{1/2}\right]\} \exp\left[\sigma_H(x)V_H/RT\right] \qquad [2]$$

where D is the diffusion coefficient of hydrogen in metal and

$$erf\ z = 2\pi^{-1/2} \int_o^z \exp(-y^2)\,dy\ ; \qquad [3]$$

this assumes that D is independent of stress and concentration, which has been verified for hydrogen in iron over a range of stresses by Beck and coworkers (5). The diffusion problem has been simplified by assuming a plane infinite source of hydrogen diffusing into a plane infinite specimen with a fixed concentration. More refined calculations performed by Van Leeuwen (6) for round notched specimens or taking into account the finite dimensions of our samples does not seem to improve very much the final results due to the degree of approximation involved in our model.

The stress field at the tip of an elastoplastic crack is a subject of active research and, moreover, if we add the difficulties found in our problem (i.e. the three dimensional character of a thumb nail crack in a round bar plus the possible effects of anisotropy due to cold drawing during wire manufacture) it does not seem appropriate to use the last refinements of a two dimensional isotropic theory.

We will assume a hydrostatic stress distribution in the symmetry plan as sketched in Fig. 1. It is agreed that σ_H increases a bit away from the crack tip in the plastic region and decrease like $r^{-1/2}$ in the elastic zone near a crack with small yielding zone (7).

In the plastic region σ_H maximum can be given (3) by

$$\sigma_H \simeq \sigma_y \left(pcf - \frac{1}{2}\right) \simeq \sigma_y \left(\frac{1}{2} + \alpha K/\sigma_y\right) \qquad [4]$$

where σ_y is the yield stress, K the stress intensity factor and the plastic constraint factor pcf is given by $pcf \simeq 1 + \alpha K/\sigma_y$ as proposed by Hahn and Rosenfield (8) and based on considerable data by Krafft (9), where $\alpha = 12.58\ m^{-1/2}$.

In the elastic region, near the crack tip, σ_H is given by

$$\sigma_H = \beta \frac{K_I}{\sqrt{2\pi x}} \qquad [5]$$

where β ranges between 2/3 for plane stress and $2(1+\nu)/3$ for plane strain and the stress intensity factor can be obtained approximately from the results of Daoud et al. for a straight crack in a circular bar in tension (10):

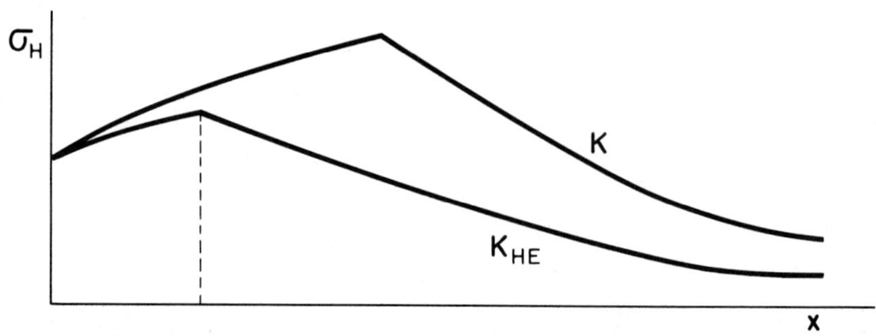

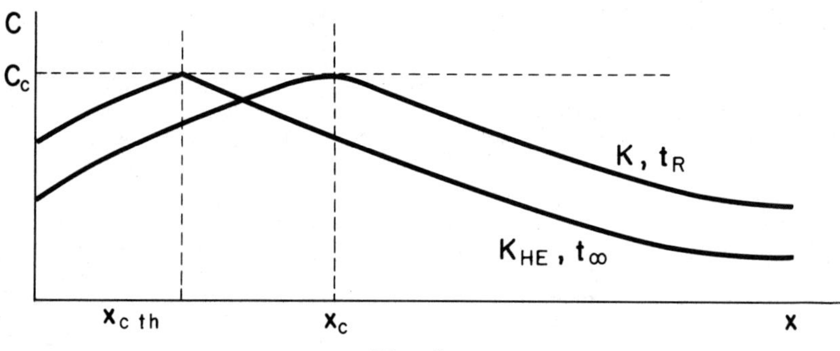

Fig. 1.

$$K_I = \left[1.11 - 3.59\left(\frac{a}{D}\right) + 24.87\left(\frac{a}{D}\right)^2 - 53.39\left(\frac{a}{D}\right)^3 + 57.23\left(\frac{a}{D}\right)^4\right]\sigma\sqrt{\pi a} \qquad [6]$$

valid in the range $(0.06 < a/D < 0.7)$

As stated before, the crack initiation time is assumed to be the time required to build up a critical hydrogen concentration, C_c, anywhere in the notch tip region. From the foregoing analysis, the time to fracture t_R, assumed to be equal to crack initiation time, is therefore derived by equating $C(x_c, t_R)$ from eq. [2] to C_c:

$$C_c = C(x_c, t_R) = C_o\{1 - \text{erf}\left[x_c/2(Dt_R)^{1/2}\right]\}\exp\left[\sigma_y(1/2 + \alpha K)V_H/RT\right] \qquad [7]$$

The threshold stress intensity factor for hydrogen embrittlement K_{HE}, is related to the critical concentration by:

$$C_c = C(x_{cth}, \infty) = C_o \exp\left[\sigma_y(1/2 + \alpha K_{HE})V_H/RT\right] \qquad [8]$$

A link between time to fracture t_R and stress intensity factor K, can be obtained from eqs. [7] and [8]:

$$1 = \{1 - \text{erf}\left[x_c/2(Dt_R)^{1/2}\right]\}\,\exp\left[\sigma_y(K - K_{HE})\,\alpha V_H/RT\right] \qquad [9]$$

If initial and critical concentration values C_o and C_c were known, one experimental value of t_R for a well defined stress intensity K would be sufficient to determine the threshold stress intensity K_{HE}, by eliminating from eqs. [7] and [9] the unknown x_c.

Experimental Results and Model Checking

The above model has been checked through the hydrogen embrittlement test proposed by the F.I.P.; the testing procedure is described elsewhere (1). Our experimental program has been carried out with precracked samples, to allow us to monitor K values. In this way it differs from the standard test for smooth samples in which the wires are stressed at 80 percent of the u.t.s. This test offers the additional advantage that the potential of the wire with respect to s.c.e. remains fairly constant, about -760mV. pH value for the solution at 50°C is 3.9.

Two couples of eutectoid cold drawn steels have been tested. The first one, steels 1 and 2, are both commercial and with almost the same mechanical properties but extremely different behavior in the NH_4SCN test for unnotched samples (6 and 500 hours for fracture time for steel 1 and 2 respectively). The second one, steels 3 and 4 are both obtained from the same batch, with very different degree of drawing, from 13 to 7 mm diameter for steel 3 (71% R.A.) and from 10 to 7 mm diameter for steel 4 (51% R.A.). Chemical composition and mechanical properties are shown in Tables I and II.

Table I. Chemical Composition

Steel N°	C	Mn	Si	P	S	heat treatment
1	.81	.60	.27	.02	.03	stress relieved
2	.80	.65	.20	.02	.03	stabilised
3 and 4	.79	.82	.23	.02	.03	stress relieved

Table II. Mechanical Properties

Steel N°	Yield Stress σ_y (MPa)	u.t.s. σ_R (MPa)	ε_u (%)	R.A. (%)	K_c ($MNm^{-3/2}$)	Fracture time t_R (h)
1	1495	1715	5.5	24	94	6.2
2	1481	1721	6.3	39	108	500.
3	1775	1881	4.6	32	112	4.8
4	1438	1641	6.3	39	92	31.

All specimens were precracked. Transverse precracking was performed in 3 point bending, using starter notches produced by a triangular jeweller's file having a corner radius of about 0.2 mm. To ensure the sharpest possible cracks, the stress intensity factor range ΔK was such that the last 1 mm of crack took not less than 40,000 cycles to form, equivalent

90
80
70
60
50
40
30
20
10
0
K_I (% K_IC)
THEORETICAL RESULTS. STEELS 1, 2
EXPERIMENTAL RESULTS. STEEL 1
EXPERIMENTAL RESULTS. STEEL 2
t_R(1) UNNOTCHED ≃ 6 h.
t_R(2) UNNOTCHED ≃ 500 h.
K_IHE = .27 K_IC
(>1.000 h.)
(>400 h.)
(558 h.)
(>1.000 h.)
0   10   20   30   40   50   60
TIME TO FAILURE (hours)

Fig. 2.

90
80
70
60
50
40
30
20
10
0
K_I (% K_IC)
THEORETICAL RESULTS. STEELS 3, 4
EXPERIMENTAL RESULTS. STEEL 3
EXPERIMENTAL RESULTS. STEEL 4
t_R(3) UNNOTCHED ≃ 5 h.
t_R(4) UNNOTCHED ≃ 31 h.
(>500 h.)
K_IHE = .32 K_IC
(>500 h.)
(>500 h.)
(127 h.)
(>500 h.)
K_IHE = .21 K_IC
0   10   20   30   40   50   60
TIME TO FAILURE (hours)

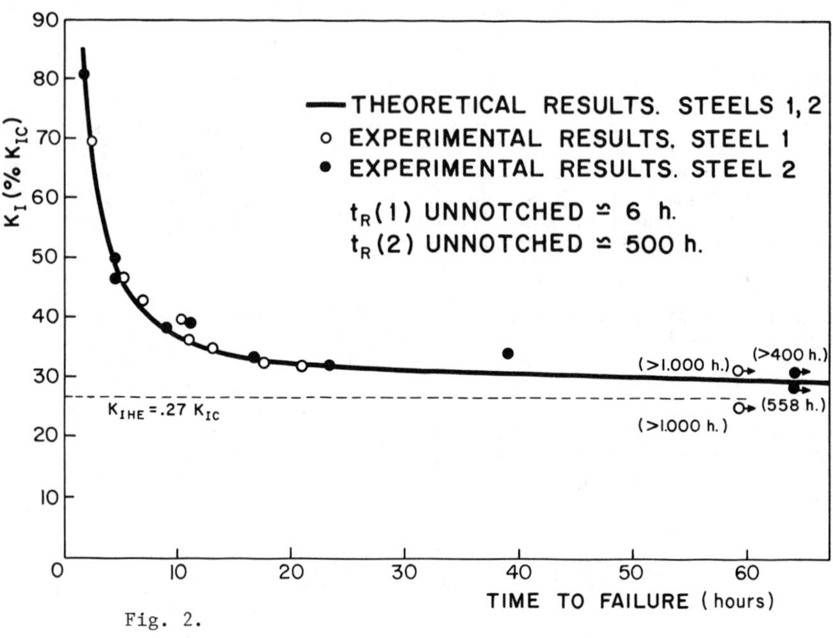

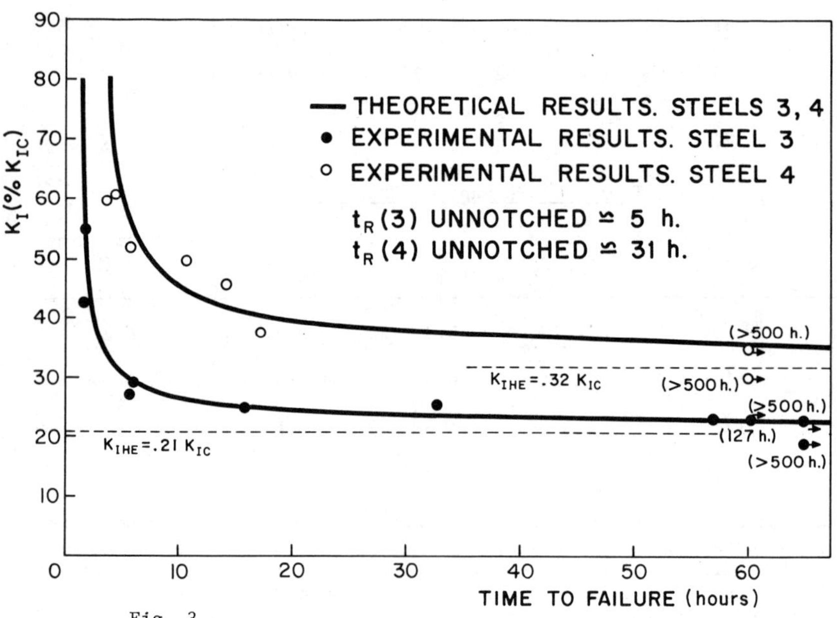

Fig. 3.

to the requirements for fatigue precracking of ASTM E399.

Results of K versus time to rupture t_R are drawn in Fig. 2 (steels 1 and 2) and in Fig. 3 (steels 3 and 4).

The prediction of t_R from the model needs a knowledge of C_o and C_c as stated before. Instead of undertaking the measurement of hydrogen concentrations the problem can be solved if a relationship between x_c, K and σ_y is found. From Fig. 1 it is reasonable to suppose that x_c will increase with K and that the relationship will be independent of material constants other than σ_y. The procedure followed has thus been to calculate for each test the best set (x_c, K_{HE}) which fits eq. [9], K_{HE} being the same for each steel. Afterwards, a plot of $\ln x_c$ vs. $\ln (K/\sigma_y)$, Fig. 4, shows a linear relationship in the testing range $(-4 < \ln (K/\sigma_y) < -3)$ for all steels tested, suggesting the following correlation

$$x_c = 126.8 \ (K/\sigma_y)^{1.12} \tag{10}$$

where the variable dimensions are: K (MN m$^{-3/2}$), σ_y (MN/m^2), and x_c (10^{-4} m).

Theoretical curves K-t_R can be drawn from eq. [9] by substituting the unknown x_c from eq. [10]. In Figure 2 and 3 such theoretical curves have been plotted, and a good agreement with the experimental values is observed.

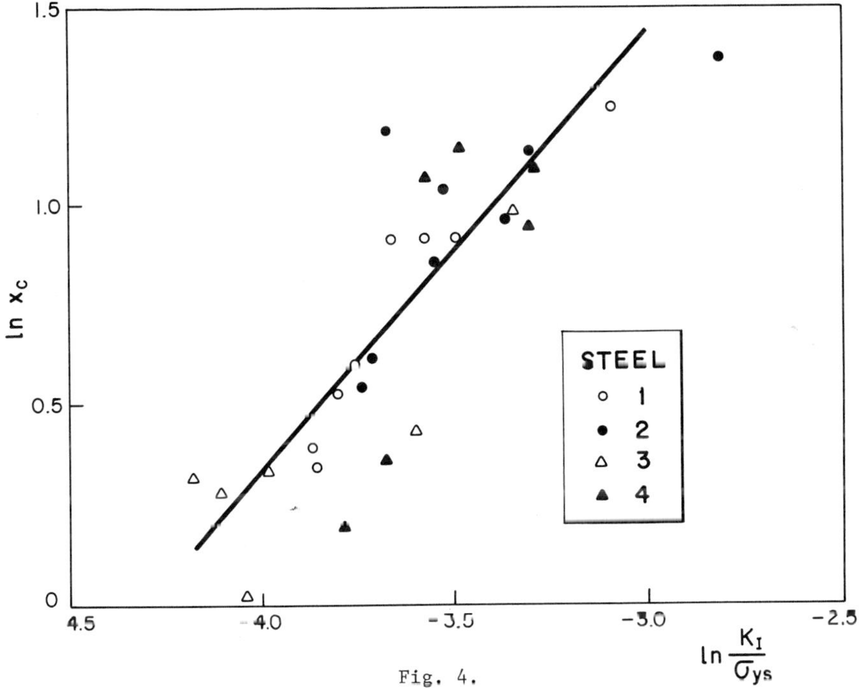

Fig. 4.

Conclusions

1. A model for hydrogen embrittlement kinetics in eutectoid cold drawn steels has been developed which is capable of predicting the threshold stress intensity and a relation between K_I and time to fracture.

2. A correlation between x_c and K/σ_y has been found valid for a wide range of cold drawn steels having different mechanical properties, and several degrees of cold work.

3. The test proposed by the F.I.P. is a measure of surface susceptibility to hydrogen embrittlement, as shown by testing two commercial steel batches; they exhibit very different behavior when testing smooth samples and nevertheless they have almost the same K_{HE}.

4. If stress corrosion cracking of prestressing steels in actual environments is proved to be due to hydrogen embrittlement, a knowledge of C_o would permit one to calculate the threshold stress intensity for this particular environment, and to predict time to fracture as a function of the stress intensity applied, allowing an accurate design for this kind of steel.

References

1. F.I.P. Report on Prestressing Steel: 5. Stress Corrosion Cracking Resistance. Test on Prestressing Tendons. F.I.P., Wexham Spring, Slough SL3 6PL, England, September 1979.

2. P. Doig and G. T. Jones: Met. Trans. A, 1977, vol. 8A, pp. 1993-98.

3. W. W. Gerberich and Y. T. Chen: Met. Trans. A, 1975, vol. 6A, pp. 271-78.

4. I. M. Bernstein and A. W. Thompson: Advances in Corrosion Science and Technology, vol. 7, pp. 53-175, M. Fontana, R. Staehle, Eds., Plenum Press, New York, 1980.

5. W. Beck, J. O. M. Bockris, J. McBreen and L. Nanis: Proc. Roy. Soc. 1965, vol. A290, pp. 220-26.

6. H. P. Van Leeuwen: "A Quantitative Analysis of Hydrogen Induced Cracking" Report NLR TR 74084 U, National Aerospace Laboratory, NLR, The Netherlands 1972.

7. D. Broek: Elementary Engineering Fracture Mechanics, 2nd. Ed., p. 105, Sijthoff and Nordhoff, Alphen aan den Rijn, 1978.

8. G.T. Hahn and A. R. Rosenfield: Trans. ASM, 1966, vol. 59, pp. 909-19.

9. J. M. Krafft: Report of Naval Research Lab., Jan. 1966, cited in (8).

10. O. Daoud, D. J. Cartwright and M. Carney: Strain Analysis, 1978, vol. 13, pp. 83-89.

SULFIDE STRESS CRACKING RESISTANCE OF

SUPERALLOYS FOR SOUR GAS WELL APPLICATIONS

E. P. Whelan

Climax Molybdenum Company
Ann Arbor, Michigan

The sulfide stress cracking fracture toughness K_{issc} of the nickel-base alloys Hastelloy C-276, Hastelloy X(LC) (low carbon) and Hastelloy G, and of the cobalt-base Haynes Alloy No. 188, has been determined in a NACE aqueous solution containing 5% NaCl, 0.5% CH_3COOH and saturated with H_2S. Double cantilever beam fracture toughness specimens were used in galvanic contact with steel. The alloys were processed to produce four levels of yield strength, with maximum levels of 1347 MPa (195.3 ksi), 1158 MPa (167.9 ksi), 1161 MPa (168.4 ksi), and 1426 MPa (206.8 ksi) being achieved for Hastelloy C-276, Hastelloy X(LC), Hastelloy G, and Haynes Alloy No. 188, respectively. Prior to testing, alloys were aged for 200 hours at 205 C (400 F) in order to simulate down-hole exposure. Cracking occurred intergranularly. Mean fracture toughness values of 58.6 MPa $m^{1/2}$ (53.3 ksi in.$^{1/2}$), 90.7 MPa $m^{1/2}$ (82.5 ksi in.$^{1/2}$), 59.3 MPa $m^{1/2}$ (54.0 ksi in.$^{1/2}$) and 54.4 MPa $m^{1/2}$ (49.6 ksi in.$^{1/2}$) were obtained for Hastelloy C-276, Hastelloy X(LC), Hastelloy G, and Haynes Alloy No. 188, respectively, at their maximum yield strength levels.

Introduction

Gas wells are being drilled to depths as great as 7300 m (24,000 ft) in the Mississippi fields. The gas present in these deep wells is contaminated by up to 50% hydrogen sulfide, up to 10% carbon dioxide and frequently contains chlorides (1). The combination of these contaminants with down-hole temperatures close to 200 C (390 F) or more and pressures as high as 150 MPa (22 ksi) produces an extremely demanding environment (1), with hydrogen embrittlement being potentially a major source of well casing and tubing failure. One particular form of embrittlement occurring in these environments is known as sulfide stress corrosion cracking (SSC), and fracture toughness methods have been developed to measure the SSC cracking resistance of several casing and tubing alloys (2,3,4). As the yield strengths of suitable alloys are increased, their SSC cracking resistance decreases, and currently steels with yield strengths in the range 690-760 MPa (100-110 ksi) represent the upper limit for ferrous alloys with acceptable SSC resistance.

In the search for further increases in SSC resistance, attention has been directed towards Ni-base and Co-base superalloys (5,6,7,8,9). Solid solution and precipitation strengthened superalloys can be processed to high yield strength levels, and these alloys are being evaluated in both the laboratory and the field for resistance to sulfide stress cracking under sour gas environmental conditions. C-ring and bent beam tests have been used most frequently in laboratory studies on these alloys. However, more quantitative data on resistance to cracking can be obtained using fracture mechanics test methods. Consequently, in the present study, the Ni-base alloys Hastelloy C-276, Hastelloy X(LC) (a low carbon version of Hastelloy X), and Hastelloy G, and the Co-base Haynes Alloy No. 188, have been evaluated at various yield strength levels to determine their resistance to sulfide stress cracking, using fracture toughness specimens in a standard NACE test solution (2).

Experimental Procedures

The compositions of the alloys studied are listed in Table 1. Hastelloy C-276, Hastelloy X(LC) and Hastelloy G were tested after cold working 0%, 20%, 40% and 60%, while Haynes Alloy No. 188 was cold worked 40% then aged for 1 hour at 705 C (1300 F), 760 C (1400 F), 815 C (1500 F) and 870 C (1600 F) to produce four yield strength levels for each alloy. All alloys were given a low temperature aging treatment at 205 C (400 F) for 200 hours after processing. This treatment was designed to simulate gas well down-hole exposure, and had been reported previously as causing a decrease in sulfide stress cracking resistance (6,7).

Table 1. Nominal Alloy Compositions (wt %)

Alloy	C	Mn	Si	Cr	Ni	Co	Mo	W	V	Cb+Ta	Fe
Hastelloy C-276	0.003	0.45	0.03	15.0	Bal	2.0	16.0	3.5	0.2	−	6.0
Hastelloy X(LC)	0.02	0.5	0.5	22.0	Bal	2.0	9.0	0.5	0.05	−	19.0
Hastelloy G[a]	0.03	1.3	0.35	22.2	Bal	−	6.5	0.5	−	2.1	19.5
Haynes Alloy No. 188[b]	0.1	1.0	0.4	22.0	22.0	Bal	−	14.0	−	−	3.0

[a]Includes 2.0% Cu

[b]Includes 0.004% B, 0.05% La

Room temperature tensile properties were measured parallel to the longitudinal or rolling direction of each plate and were also measured in a transverse direction for Hastelloy C-276, Hastelloy G and Haynes Alloy No. 188.

Four double cantilever beam (DCB) specimens were machined from each alloy plate to the dimensions shown in Figure 1, where B = 6.35 mm (0.25 in.) and B_n = 4.32 mm (0.17 in.). Specimens were taken from both the longitudinal and transverse directions for alloys Hastelloy C-276, Hastelloy G and Haynes Alloy No. 188 and from the longitudinal direction only for Hastelloy X(LC).

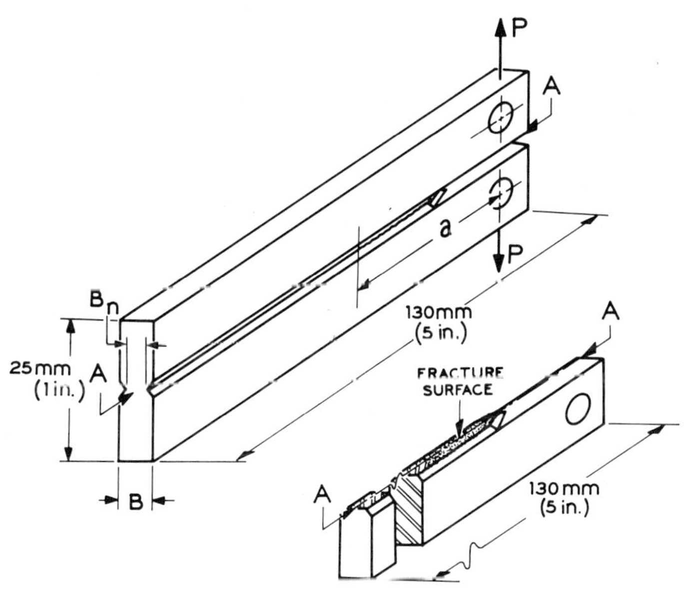

Fig. 1 - Double Cantilever Beam (DCB) Specimen Used to Determine K_{Issc} Values

Sulfide stress cracking tests were performed using the procedure described by Heady (2). Double cantilever beam specimens were stressed using tool steel wedges, hardness HRC 55, and were tested with carbon steel plates attached to each side. The galvanic contact between such steel plates and DCB specimens has been observed to intensify the degree of sulfide stress cracking for corrosion resistant alloys (10). Initial tests were performed using a wedge thickness of 2.41 mm (0.095 in.). The stress provided by this wedge thickness was generally insufficient to produce cracking during a six-week exposure period and therefore further testing was performed on the same specimens using thicker wedges, with many of the specimens being fatigue precracked prior to testing. The thickness of the wedges used was calculated using the procedure described in the Appendix to Heady's paper (2), using estimated K_{Issc} values of between 77 and 104 MPa m$^{1/2}$ (70 and 92 ksi in.$^{1/2}$). Wedge thicknesses ranged from 2.41 mm (0.095 in.) to 2.84 mm (0.112 in.). As will be seen later, these wedge thicknesses were sufficient to initiate and propagate cracks in many higher strength specimens, but at the same time many of the lower strength longitudinal specimens and most of the transverse specimens remained uncracked.

The DCB specimens were exposed for a total period of six weeks to a standard NACE H$_2$O-5% sodium chloride-0.5% acetic acid solution continuously saturated with H_2S. Control DCB specimens were inspected after two weeks, four weeks, and six weeks exposure to determine when crack propagation had ceased. In general, cracking occurred between the second and fourth weeks of test, with little cracking occurring during the first two-week period and negligible crack propagation during the final two weeks of testing. When testing was complete, the fracture toughness for each fully cracked specimen was calculated using the equation (2):

$$K_{Issc} = Pa \left[2 \times 3^{1/2} + 2.38 \frac{h}{a}\right] \left[\frac{B}{B_n}\right]3^{-1/2} [Bh]^{-3/2}$$

where

P = load on specimen load line
a = crack length from load line
B = specimen thickness
B_n = specimen web thickness
h = specimen half height

For specimens that did not crack, or for which a crack was initiated in the chevron region but did not propagate into the parallel-sided reduced section, a minimum fracture toughness value was calculated that was based on the observation that the critical crack length, a, for these specimens would be less than the length from wedge to chevron or wedge to end of chevron crack.

Results

The yield strengths of the alloys in the longitudinal direction are listed in Table 2. Yield strengths ranged up to 1346 MPa (195.3 ksi) for Hastelloy C-276, to 1158 MPa (167.9 ksi) for Hastelloy X(LC), to 1161 MPa (168.4 ksi) for Hastelloy G and to 1426 MPa (206.8 ksi) for Haynes Alloy No. 188. Longitudinal and transverse tensile properties were similar.

Table 2. Longitudinal Yield Strength and Fracture Toughness

Alloy	Condition	0.2% Offset Yield Strength, MPa (ksi)		Fracture Toughness K_{Issc}, MPa m$^{1/2}$ (ksi in.$^{1/2}$)	
Hastelloy C-276	0% CW	384	(55.7)	35.9[a]	(32.6)
	20% CW	889	(129.0)	76.0[a]	(69.3)
	40% CW	1158	(167.9)	76.0	(69.3)
	60% CW	1346	(195.3)	58.6	(53.3)
Hastelloy X(LC)	0% CW	296	(43.0)	35.8[a]	(32.5)
	20% CW	748	(108.5)	80.4[a]	(73.2)
	40% CW	1029	(149.3)	89.6[a]	(81.5)
	60% CW	1158	(167.9)	90.7	(82.5)
Hastelloy G	0% CW	310	(45.0)	31.3[a]	(28.5)
	20% CW	791	(114.7)	81.8	(74.4)
	40% CW	980	(142.2)	66.5	(60.5)
	60% CW	1161	(168.4)	59.3	(54.0)
Haynes Alloy No. 188	870 C/1 h	1134	(164.5)	79.2	(72.0)
	815 C/1 h	1260	(182.7)	66.8	(60.8)
	760 C/1 h	1335	(193.6)	54.0	(49.1)
	705 C/1 h	1426	(206.8)	54.5	(49.6)

[a]Minimum K_{Issc} value

The mean fracture toughness values calculated for each strength level of
each alloy are also listed in Table 2. Adequate crack propagation for K_{Issc}
calculation occurred for the three highest yield strength levels of Hastelloy
G and Haynes Alloy No. 188, for the two highest yield strength levels of
Hastelloy C-276 and for the highest yield strength level only for Hastelloy
X(LC). Specimens at other yield strength levels produced only minimum K_{Issc}
values because of insufficient crack propagation. Figure 2 shows a plot of
the variation of mean fracture toughness values with yield strength level for
adequately cracked specimens of the four alloys. As expected, fracture toughness decreases as yield strength increases.

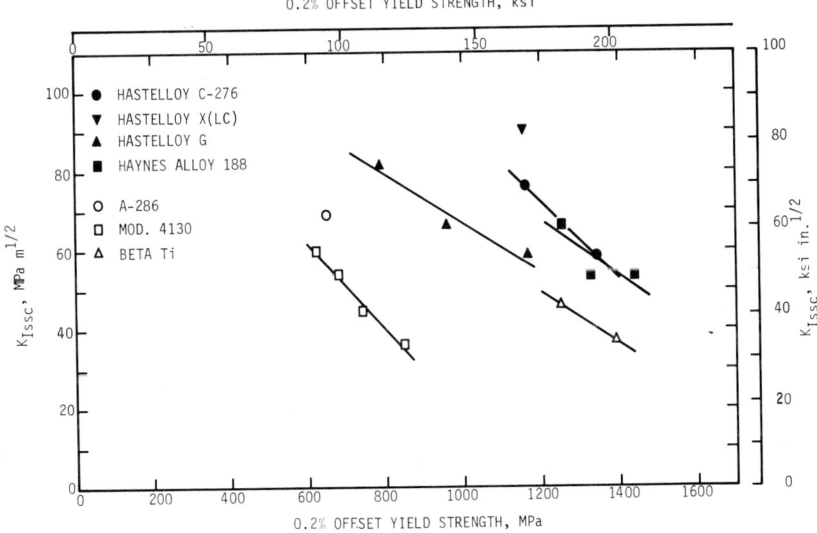

Fig. 2 - Variation of Fracture Toughness with Yield Strength

Transverse specimens from the four alloys exhibited greater resistance to
sulfide stress cracking than longitudinal specimens to the extent that only
minor crack propagation occurred. In general, cracks in the transverse specimens propagated out of the plane of the reduced web section of the DCB specimen as they followed the elongated grain boundaries. Haynes Alloy No. 188,
with a more equiaxed grain structure than the three cold worked alloys, exhibited the greatest degree of transverse specimen cracking. Due to its incomplete nature, fracture toughness data for the transverse specimens is not
reported here.

Scanning electron microscopic examination of the fracture surfaces from
the longitudinal DCB specimens indicated that fracture was predominantly
intergranular, Figure 3. This fracture mode was more evident in specimens
from heavily cold worked solid solution strengthened alloys such as Hastelloy
C-276, Figure 3a, than in specimens from the precipitation strengthened Haynes
Alloy No. 188, Figure 3c. Metallographic sections through these fracture surfaces, Figures 3b and 3d, illustrate the intergranular and twin boundary
features of the fracture surface. One characteristic observed for many transverse specimens was crack initiation followed by propagation of the crack
front out of the plane of the reduced web section and along the grain boundaries of the elongated grains that lay normal to the reduced section. Thus
the fracture mode was predominantly intergranular for both longitudinal and
transverse specimens.

a

b

c

d

Fig. 3 - Microstructure of Fracture Surfaces. a) and b) Hastelloy C-276, c) and d) Haynes Alloy No. 188.

Discussion

Cold worked Hastelloy C-276 and precipitation strengthened Haynes Alloy No. 188 exhibit equivalent K_{Issc} values for equivalent yield strength levels, with Haynes Alloy No. 188 being tested to higher yield strength levels than Hastelloy C-276. There appears to be no clear advantage for the cobalt-base alloy over the nickel-base alloy in this particular sulfide stress corrosion test. These two alloys exhibit the best combination of high yield strength and high fracture toughness. Hastelloy X(LC) exhibits a high K_{Issc} value at its highest yield strength level, which is, however, only equivalent to the yield strength level of 40% cold worked Hastelloy C-276. Hastelloy G exhibits a somewhat lower resistance to sulfide stress cracking than the other three alloys and, moreover, displays susceptibility to sulfide stress cracking at lower yield strength levels than the other three alloys.

Included in Figure 2 are K_{Issc} data on A-286 (11), a modified version of AISI 4130 (12) and a Beta Titanium alloy (13). The superiority of the four alloys studied here over A-286 and modified AISI 4130, in terms of their significantly higher fracture toughness at equivalent yield strength levels, is clearly evident. The fracture toughness of the Ni-base alloys is also greater than that of the Beta Titanium alloy, although to a signficantly lesser extent. The higher ratio of yield strength to density for the Beta Titanium alloy, as compared with the Ni-base and Co-base alloys studied here, is stimulating further interest in β-titanium alloys for deep sour well applications.

The higher fracture toughness of the transverse specimens is associated with the orientation of their deformation-induced elongated grain structure with respect to the plane of the reduced web section of the DCB specimen. A similar result has been observed and discussed in a study of the embrittlement of cold worked superalloys (6) in the NACE solution used here. In that study, no fracture was observed for Hastelloy C-276 specimens stressed in such a manner that the plane of crack propagation would lie transverse to the rolling direction, and so to the longitudinal axis of elongated grains. Cracking failures were observed only in specimens stressed in a manner equivalent to that of the present longitudinal specimens, the crack propagation plane lying parallel to the rolling direction. This sharp distinction was not observed in the present work, in which limited cracking, confined to the chevron and propagating out of the reduced web section, was observed for the transverse Hastelloy C-276 specimens. As indicated earlier, Haynes Alloy No. 188 exhibited the greatest degree of transverse specimen cracking in the present study, as compared with the other alloys, although the sulfide fracture toughness values obtained for the transverse specimens of Haynes Alloy No. 188 were higher than those for the longitudinal specimens.

In assessing the applicability of the alloys studied here to sour well use, due account must be taken of sources of failure other than H_2S. Resistance to chloride-induced attack is essential, and studies elsewhere have indicated that Hastelloy C-276 has superior resistance to cracking in chloride-containing sour gas well environments than Hastelloy X(LC) (14). Thus the apparent attractiveness of Hastelloy X(LC) in the current study does not necessarily indicate the potential of this alloy for sour gas well use.

The experimental conditions used in this study provide a moderately severe test of the sulfide stress cracking resistance of the alloys under study, both by the use of a low temperature aging treatment to simulate downhole exposure and by the use of galvanic contact to promote hydrogen attack. As mentioned previously, this galvanic contact, simulating localized contact between a Ni-base tube and a steel casing, has been shown to intensify sulfide stress cracking in the Co-base alloy MP35N (9). In addition, low temperature aging of Ni-base alloys has been shown to decrease their resistance to sulfide stress cracking in C-ring tests (6). At the same time, actual gas well conditions will involve exposure to the hydrogen-embrittling environment at temperatures and pressures in the vicinity of 200 C (395 F) and 150 MPa (22 ksi), respectively (1). These down-hole conditions may possibly lead to a change in SSC resistance in all alloys due to increased hydrogen solubility and diffusivity.

Summary

In this evaluation of the sulfide stress cracking resistance of four superalloys, the ambient temperature fracture toughness values for Hastelloy C-276 and Haynes Alloy No. 188 are equivalent at equivalent yield strength levels, and these alloys exhibit a superior combination of high yield strength with high fracture toughness. Hastelloy X(LC) exhibits high fracture toughness at a lower yield strength level, while Hastelloy G exhibits a somewhat lower resistance to sulfide stress cracking.

Acknowledgements

The alloys used in this study were kindly provided by the High Technology Materials Division of Cabot Corporation.

The author would like to thank Drs. P. J. Grobner, W. C. Hagel, D. L. Sponseller and R. I. Garber for useful discussions, and M. A. Zeoli and R. W. McConnell for technical assistance.

References

1.   R. N. Tuttle and J. W. Kochera, "Effect of Hydrogen on Behavior of Materials," A. W. Thompson and I. M. Bernstein, eds. p. 531, The Metallurgical Society of AIME, New York, 1976.
2.   R. B. Heady, Corrosion - NACE, 1977, vol. 33, pp. 98-107.
3.   P. J. Grobner, D. L. Sponseller and D. E. Diesburg, Trans. ASME, J. Eng. Ind., 1976, vol. 98, pp. 708-716.
4.   P. J. Grobner, D. L. Sponseller and D. E. Diesburg, Corrosion - NACE, 1979, vol. 35, pp. 175-185.
5.   M. Watkins and J. B. Greer, J. Petroleum Tech., 1976, vol. 28, pp. 698-704.
6.   R. D. Kane, M. Watkins, D. F. Jacobs and G. L. Hancock, Corrosion - NACE, 1977, vol. 33, pp. 309-320.
7.   R. D. Kane and B. J. Berkowitz, Corrosion - NACE, 1980, vol. 36, pp. 29-36.
8.   A. I. Asphahani, Hydrogen Met., Proc. Int. Congr. 2nd, 1977, Paper 3C2, Pergamon, Oxford, U.K. 1978.
9.   P. W. Rice, Materials Performance, 1978, vol. 17, pp. 16-25.
10.   P. J. Grobner, Private communication.
11.   E. P. Whelan, Unpublished research.
12.   J. A. Straatmann, P. J. Grobner and D. L. Sponseller, Paper 77-Pet-48, ASME, New York, 1977.
13.   S. Ohtani, T. Nishimura and Y. Moriguchi, "Alloys for the 80's," Paper 21, Climax Molybdenum Company of Michigan, 1980.
14.   R. D. Kane, J. B. Greer, D. F. Jacobs, H. R. Hanson, B. J. Berkowitz and G. A. Vaughn, Corrosion '79, Paper Number 174, 1979.

TOUGHNESS LOSSES AND FRACTURE BEHAVIOR

OF LOW STRENGTH CARBON-MANGANESE STEELS IN HYDROGEN

S. L. Robinson and R. E. Stoltz

Sandia National Laboratories
Livermore, CA 94550

The effect of hydrogen on the toughness of low strength carbon-manganese steels has been measured using both J-integral and burst testing methods. J_{IC}/K_{IC} measurements of A516 steel plate were made in air and in hydrogen at pressures from 3.45 to 34.5 MPa. Cracks were oriented in the LI orientation. Burst tests were performed on pre-flawed A106 steel pipe sections, using nitrogen and nitrogen-hydrogen gas mixtures. Internal pre-flaws were oriented in the ST direction and K_{IC} values calculated from the measured burst pressures. The two test methods gave similar results if crack orientation differences are considered. A reduction in K_{IC} of 30% in 6.9 MPa hydrogen and 55% in 34.5 MPa hydrogen was observed when compared to air or nitrogen. Hydrogen induces a quasi-cleavage fracture mode, characterized by flat fracture through the ferrite and plastic tearing in the pearlite. Considerable stable crack growth resistance remains in hydrogen, and the fracture process is considered to be a plasticity driven rather than a cleavage process. To maintain a similar factor of safety, pipelines carrying hydrogen at 6.9 MPa would require a 30% decrease in operating pressure due to the decrease in K_{IC}, which leads to a 42% decrease in rate of energy transmission when compared to natural gas.

Introduction

Recently, proposals[1,2] for the use of renewable energy sources have suggested the use of hydrogen as an energy transmission medium. The costs and safety aspects of hydrogen transmission through pipelines are partly dependent upon the effects of hydrogen on the mechanical properties of the pipeline steels. Systems cost studies[1,3] have assumed that under pipeline conditions the hydrogen-caused property degradation is negligible. This assumption is based upon the rule of thumb that below 700 MPa yield strength, steels are unaffected by hydrogen.

The degree of hydrogen degradation is now known to be a strong function of the testing procedure[4], the microstructure of the steel[5], purity of the gas[6,7], and pressure[8]. For example, cyclic loading of low strength carbon steels in low pressure[7] and high pressure[9] hydrogen causes accelerated fatigue crack growth. In contrast, carbon steel specimens do

Figure 1.
Crack Growth Resistance
Data for A516-70 Steel
in Hydrogen.

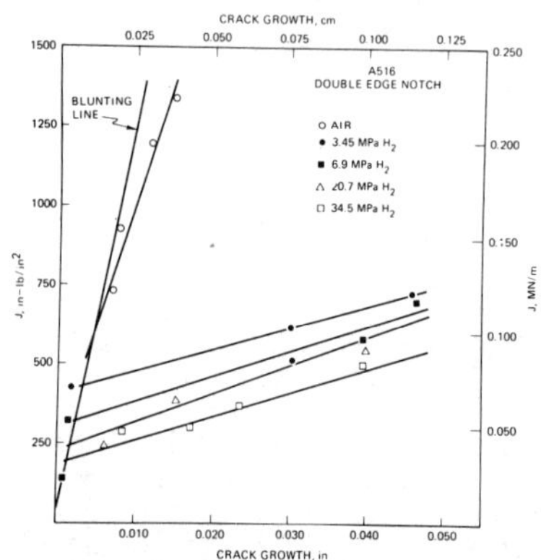

Figure 2a.
Minimal Stretch Zone
Formation in A516-70
Steel Tested in Hydrogen.

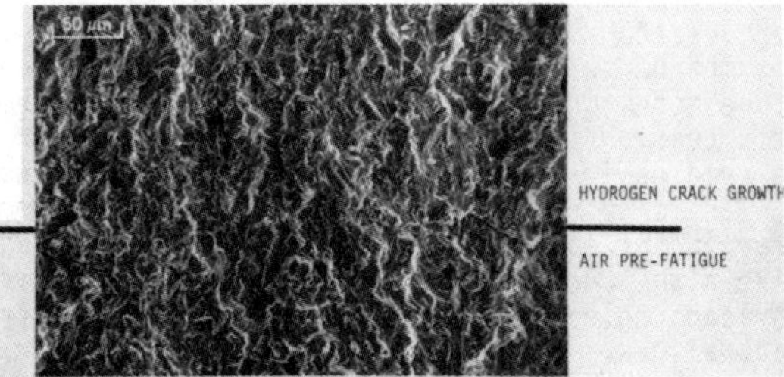

not show sustained load crack growth under nonfluctuating loads, at hydrogen pressures approaching 100 MPa[10,11]. In the low strength class of steels which are important to pipeline applications, a rising load appears to be necessary for hydrogen embrittlement effects to be observed. This is in contrast with effects observed in high strength steels in which time dependent crack extension is seen[8] under static loading conditions.

Pipeline failures often result from the mechanical deformation of the pipe[12], and the tougness of the steel becomes critical. Two methods were used to measure the toughness of pipeline steels in hydrogen and inert environments: the J-integral concept[13] and burst testing of internally flawed pipes. In the J-integral toughness test, J_{IC} and the resistance curve slope of A516 grade 70 steel were measured as a function of hydrogen pressure. In the burst tests, intentionally flawed A106 grade B pipes were burst using nitrogen, and a hydrogen-nitrogen mixture. Apparent fracture toughnesses were calculated from the burst pressures and compared to J-integral test results. Fractographic analysis was performed in order to relate the fracture features to the ferrite-pearlite microstructure.

Experimental Procedures

The materials tested were, A516 grade 70 and A106 grade B, both silicon-killed pearlitic carbon-manganese steels. The A516 was obtained in a 2.5 cm thick plate, and the A106 in nominal 10 cm diameter schedule 40 seamless pipe. The chemical analyses are listed in Table I. Both materials exhibit pearlite colonies of similar size and volume fraction, and elongated manganese sulfide inclusions. In the A106 pipe these inclusions are predominantly in the midthickness region and oriented along the pipe axis.

All J_{IC} tests were performed using double-edged notched samples. The experimental procedures are similar to those in Ref. 14. The J-values are considered accurate to ±0.004 MN/m and the crack lengths to ±0.005 cm.

Burst tests were conducted on nominal 10 cm diameter pipes of A106 grade B steel. The longitudinal internal flaws had flank angles of 105°, a root radius of less than 0.08 mm, and penetrated 20% of the wall thickness. Inert gas tests were conducted by continuous pressurization with nitrogen until burst. Hydrogen tests were conducted at a constant 6.9 MPa hydrogen pressure, and overpressure with nitrogen to burst. The pressurization rate was approximately 0.7 MPa per minute. All tests were conducted at a temperature of 288 $\pm$ 3 K. The applied stress intensity at burst was calculated by using the analysis in Ref. 15, corrected for shell curvature effects on hoop stress[16,17,18].

Results

Figure 1 summarizes the results of crack extension vs applied J tests for A516 in air and in hydrogen at pressures from 3.45 to 34.5 MPa (500 to 5000 psi). Measurements of J_{IC}, K_{IC} equivalents, and the crack resistance slope are given in Table II. As shown, the J_{IC} values decrease continuously with increasing hydrogen pressure. The resistance slope, however, decreases from air to 3.45 MPa hydrogen, but remains essentially constant with increasing hydrogen pressure up to 34.5 MPa hydrogen.

Scanning electron fractography showed that fracture in air was by void growth and coalesence. A large stretch zone and change in height of the fracture surface occured in the transition from the prefatigue zone to the rising load crack extension region. The contrary was true in the hydrogen tests, regardless of pressure. Figure 2a shows the pre-fatigue to hydrogen

Figure 2b.
Fracture Mode in A516-70
Steel in 3.45 MPa Hydrogen.

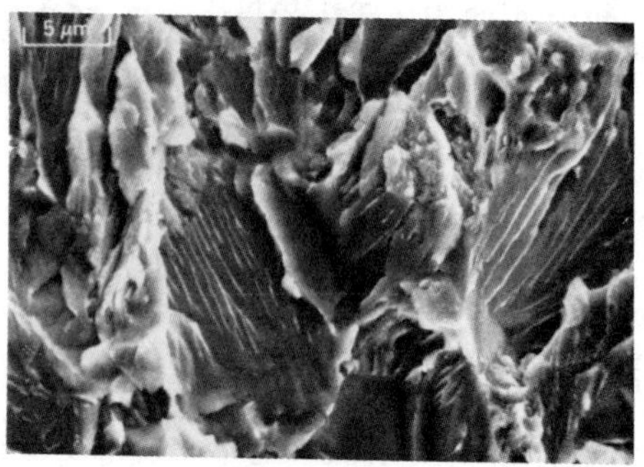

Figure 3.
Burst Test Results
for Internally Flawed
A106-B Steel Pipe

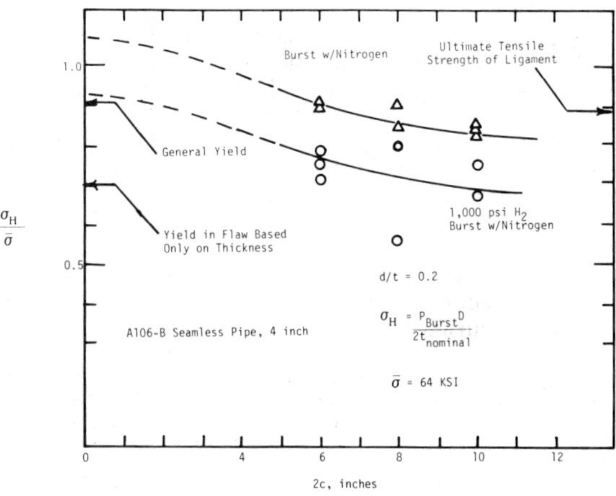

Figure 4.
Cross Section of
Pipe Wall after
Hydrogen Burst Test.

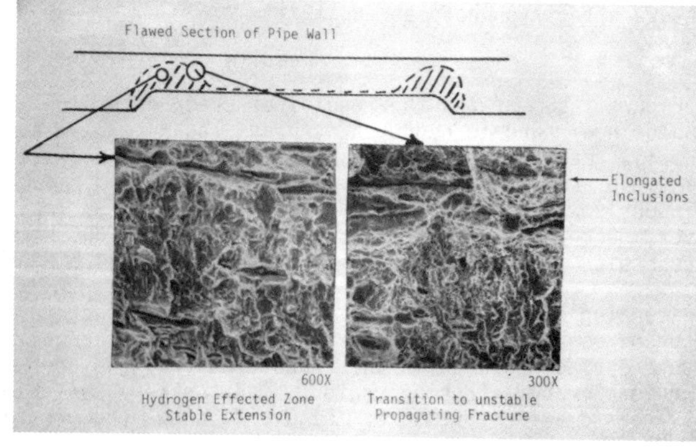

crack growth transition. No stretch zone is present and the overall
fracture surface morphology is very similar in the two regions. The
fracture mode in hydrogen was by "quasi-cleavage", as seen in Figure 2b.
Fracture through the free ferrite is relatively flat and crystallographic,
with some evidence of river markings. These regions are separated by tear
ridges associated with the pearlite colonies. No evidence for void
nucleation and growth was observed. Examination of samples from all the
hydrogen tests showed no difference in fracture morphology from 3.45 to
34.5 MPa test pressure.

Similarly, hydrogen induced losses in toughness were observed in the
burst tests. The normalized hoop stress at burst for internally flawed
pipes is shown in Figure 3. The failure strength decreases with increasing
flaw length, and approaches a limiting value. The introduction of a 6.9
MPa hydrogen pressure causes a loss of approximately 15% in the normalized
hoop stress at burst, in the range of flaw lengths tested. The calculated
stress intensities (K_{IC}) at burst are reported in Table II for the
nitrogen and 6.9 MPa hydrogen plus nitrogen tests.

The region of hydrogen affected crack growth in the burst test was
nonuniform, occurring principally at the crack ends, as seen in Figure 4.
The stress state triaxiality is highest at that location. At higher
magnification, the quasicleavage mode is seen to be composed of relatively
flat, crystallographic zones separated by tear ridges. The flat faceted
zones are similar in size to the free ferrite grain size. The ductile
fracture zone is associated with overload failure, evidenced by elongated
shear dimples in the transition region. Early fracture of sulfide-matrix
interfaces occurred, and because of the short-transverse crack orientation,
possibly contributed significantly to the crack advance. Significant
stretch zone formation was observed in most hydrogen burst tests, probably
due to the relative bluntness of the precrack.

Discussion

Hydrogen Induced Toughness Losses and Fracture Mode Changes

The measured equivalent K_{IC} from J-measurements in air is 150 MPa
$\sqrt{m}$ for A516-70, and 114 MPa $\sqrt{m}$ from burst tests of A106-B steel. This
24% difference between the two values can be explained by orientation
effects, which have been shown to cause a 17 to 30% difference in K_{IC}[20].
The absolute values of K_{IC} and dJ/da compare well with published
values[19,20]. At 6.9 MPa H_2 pressure, a 34% loss in K_{IC} is obtained
from J measurements, compared to a 25% loss from burst testing. The
absolute differences are again attributable to orientation effects. The
toughness values at 6.9 MPa hydrogen of 85 and 104 MPa $\sqrt{m}$ respectively
indicate significant fracture resistance remains. The crack propagation
resistance has decreased by nearly tenfold, but remains at a relatively
high level, indicating that large amounts of plastic work are required to
propagate the crack once initiated.

As discussed in the introduction, a rising load or imposed plastic
deformation is necessary before hydrogen effects are observed in low
strength carbon and carbon-manganese steels. Two points emerge from the
data of Figure 1. First, the energy to initiate fracture, J_{IC}, decreases
with increasing hydrogen pressure and approaches a saturation level. The
largest change occurs in going from air to 3.45 MPa hydrogen, while the
effect of hydrogen is near saturation at 34.5 MPa. This variation is very
similar to the effect of hydrogen on fatigue crack growth[7], if one
plots crack growth at a fixed ΔK vs hydrogen pressure.

The large change in J_{IC} from air to 3.45 MPa hydrogen is a result of the change in fracture mode. Ductile crack growth in air is by link-up of subsurface voids with the blunted crack tip. This process involves large plastic strains over a region comparable to the crack opening displacement[21]. In the air test this distance is ~ 200 μm (COD ≈ 0.5 J_{IC}/σ_y). In hydrogen the amount of plastic deformation is smaller, as reflected in the quasi-cleavage mode through the ferrite. Additionally, the fracture process occurs over a much smaller distance ahead of the crack, as can be deduced from Figure 2a. The overall fracture appearance does not change between the fatigue and rising load regions. The fatigue striation spacing is ~1 μm, which defines the process zone over which each fatigue fracture step occurs. Since this is much less than the ferrite grain size (~30 μm), the ferrite grains are visible on the fracture surface. During hydrogen crack growth the ferrite grains are also visible, indicating that the fracture process similarly occurs in a region less than the ferrite grain size.

The second feature of Figure 1 is the near saturation of J_{IC} at 35 MPa hydrogen. Figure 2b shows that the primary fracture feature in hydrogen is a quasi-cleavage mode through the ferrite. (Tear ridges in the pearlite are formed after the ferrite has failed as can be seen in the transition region between hydrogen fracture and overload). While Beacham originally suggested that this quasi-cleavage is a true cleavage process and fracture occurs on (100) cleavage planes,[22] recent work by Hinotani and co-workers[23] shows that the fracture plane in iron can be the (110) slip plane. Fracture in hydrogen has been proposed to occur by accumulation of dislocations on the slip plane, through accelerated injection as discussed by Hirth.[24] Saturation effects on J_{IC} would then be related to the saturation of dislocations, limiting their ability to move hydrogen along the slip plane. Hirth[24] calculates that dislocation cores saturate at pressures between 0.1 MPa (15 psi) and 60 MPa (8500 psi) while the elastic field around the dislocations saturate at pressures greater than 60 MPa. These values are dependent on the choice of binding energies of hydrogen to the various defects, but they do bracket the pressures where J_{IC} appears to saturate (Table II).

The works by Tahayama[25] and Hinotani[23] both suggest that the mechanism of hydrogen embrittlement in iron and low strength steels is through a change in the distribution of plastic deformation rather than decohesion leading to true cleavage. This concept is supported by both the J-tests and burst test results. No sustained load cracking was observed either in A516 or A106; continued plastic deformation is required for stable crack growth. In addition, no regions of true cleavage, as would be seen in low temperature fracture of these steels, were observed. While the measurements do not support Beacham's[26] idea that a lowering of the flow stress occurs, the hydrogen fracture behavior of these C-Mn steels is consistent with Lynch's suggestion[27] that hydrogen causes a localization of flow in the material ahead of the crack tip.

Finally, while the J_{IC} values decrease with increasing hydrogen test pressure, the slope of the dJ/da resistance curve is independent of hydrogen pressure. The J_{IC} value is a measure of initial crack growth and since the ferrite fractures first, the J_{IC} value is controlled by the hydrogen-dislocation interaction described above. Once crack growth has initiated, however, the resistance is determined by the strongest region of the microstructure, i.e., the pearlite colonies. Increasing hydrogen pressure does not affect the ductile tearing through the pearlite, and thus the resistance slope is independent of test pressure. Spherodizing the cementite to increase the free ferrite volume fraction should act to reduce the crack growth resistance in hydrogen.

<u>Engineering Implications</u>

The J-integral test measurement and the burst test measurement give similar results for the calculated equivalent K_{IC} when orientation effects are considered. J_{IC} techniques are less cumbersome and less expensive than burst test methods, and can be adapted to test cracks oriented similarly to naturally occurring cracks. Consequently, the J-measurement technique is being developed more fully, for the purpose of testing other relevant pipeline and pressure vessel materials.

Toughness decreases of 25 to 34% are observed in comparing inert environments (air or nitrogen) with 6.9 MPa hydrogen. This decrease would necessitate pressure derating of approximately 30% in natural gas pipelines in order to maintain a similar factor of safety. An additional penalty is paid in energy efficiency of a hydrogen pipeline at 30% reduced pumping pressure due to the differences in compressibility of hydrogen and natural gas[1]. In terms of energy throughput, a hydrogen pipeline at 70% full pressure delivers 42% less energy than a fully rated natural gas system.

The effect of the tenfold loss in crack growth resistance in hydrogen is more difficult to quantify. Operating pressure reductions of 30% due to hydrogen associated K_{IC} losses would reduce the occurence of crack initiation. Once initiation occurs, crack propagation is then controlled by the resistance term. The absolute level of dJ/da in high pressure hydrogen is still high compared to many technologically useful alloys (see for example the summary by Paris[28]). It is probably feasible to use the existing or newly built pipeline systems, with this level of degradation. However, an accelerated failure rate will probably be experienced, because the tenfold decrease in dJ/da would bias preexisting flaws towards mechanical instability leading to rupture. This problem deserves further study, to ensure safe hydrogen handling in current pipelines.

Summary

1. In A516 and A106, K_{IC} decreases by 25 to 30% at 6.9 MPa H_2 and by
    55% at 34.5 MPa.
2. Satisfactory agreement in measured fracture toughness is obtained
    between J-integral methods and burst test results when orientation
    differences are considered.
3. Hydrogen assisted fracture occurs by a "quasi-cleavage" mode involving
    flat faceted fracture of free ferrite grains.
4. The hydrogen-induced fracture mode is considered to be a plasticity
    driven, rather than a true cleavage process.
5. The toughness losses of 30% implies a decrease of 42% in
    the rate of energy transmission through a hydrogen pipeline. The
    decrease in dJ/da may bias pre-existing flaws towards mechanical
    instability leading to failure.

Acknowledgments

This work is supported by DOE Contract DE-AC04-76DP00789. The assistance of A. Salmi and R. Habekoss has made this work possible.

TABLE I
CHEMISTRIES AND PROPERTIES OF STEELS

	C	Mn	Si	P	S	Fe
A516-70	0.21	1.04	0.21	0.012	0.020	BAL
A106-B	0.15	0.41	0.30	0.015	0.015	BAL

	Yield, MPa	Ultimate, MPa	% Elongation
A516-70	345	528	30
A106-B	400	483	40

TABLE II
TOUGHNESS MEASUREMENTS FROM J-TESTS AND BURST TESTS

Condition	J_{IC}, MN/m (in-lb/in^2)		K_{IC}, MPa $\sqrt{m}$ (ksi $\sqrt{in}$)		dJ/da MPa (lb/in^2)	
J-TESTS						
AIR	0.121	(697)	150	(137)	516	(7.5 x 10^4)
3.5 MPa H2	0.076	(438)	119	(108)	47	(6.9 x 10^3)
6.9 MPa H2	0.056	(322)	102	(93)	55	(8.1 x 10^3)
20.7 MPa H2	0.042	(243)	89	(81)	54	(8.9 x 10^3)
34.5 MPa H2	0.036	(207)	82	(75)	57	(8.3 x 10^3)
BURST TESTS						
NITROGEN			114	(104)		
6.9 MPa H$_2$ plus NITROGEN to burst			85	(77)		

REFERENCES

1. D. P. Gregory: A Hydrogen Energy System, Catalog No. L21173, American Gas Association, Arlington, VA, 1973.
2. J. O'M. Bockris: Energy, The Solar Hydrogen Alternative, J. Wiley and Sons, 1975.
3. A. Konopka, and J. Wurm: Transmission of Gaseous Hydrogen, Procedings, 9th Intersociety Energy Conversion Engineering Conf., 1974, p. 405.
4. M. R. Louthan, and R. P. McNitt: In Effect of Hydrogen on Behavior of Metals, p. 496. Eds. A. W. Thompson, I. M. Bernstein, AIME New York, 1976.
5. A. W. Thompson: Transmission and Storage, Vol. II of Hydrogen: Its Technology and Implications, p. 89, CRC Press, Cleveland, 1978.
6. W. Hoffman, W. Rawls: Welding Journal, 1965, Vol. 44, Research Suppl, p. 225.
7. H. G. Nelson, Effect of Hydrogen on Behavior of Metals, Eds. A. W. Thompson, I. M. Bernstein, p. 602, AIME New York, 1976.
8. D. P. Williams, and H. G. Nelson: Met. Trans. 1970, Vol. 1, p. 63.
9. R. J. Walter, R. J. and W. T. Chandler: p. 273, Loc. Cit.
10. A. W. Loginow, E. H. Phelps: Corrosion, 1975, Vol 31 p. 404.
11. S. L. Robinson, A. J. West: Hydrogen Compatibility of Structural Materials for Pressure Vessels and Pipelines, SAND77-8613, Sandia National Laboratories report.
12. R. B. Smith: 5th Symposium on Line Pipe Research, pg. F-1, American Gas Association, Arlington, VA. Cat. L30174, 1974.
13. J. A. Begley and J. D. Landes: Fracture Toughness, pg. 1-39, ASTM STP 514, ASTM 1972.
14. W. R. Hoover: Hydrogen Compatibility of Structural Materials for Energy Storage and Transmission, SAND79-8200, Sandia National Labs Report; also SAND79-8202.

15. A. Green, I, Sneddon: Proc. Cambridge Phil. Soc., 1950, Vol 146 p. 159.
16. F. Erdogan, and M. Ratwani: Nuclear Engineering and Design, 1972, Vol 20 p. 265.
17. A. R. Duffy, R. J. Eiber, W. A. Maxey: Practical Fracture Mechanics for Structural Steel, page M-1, Ed. M. O. Dobson, U.K. Atomic Energy Authority, 1969.
18. G. T. Hahn, M. Sarrate, and A. R. Rosenfield: International Journal of Fracture Mechanics 1969, Vol 6 p. 197.
19. P. Nguyen-Day and G. Phelippeau: J. Testing and Evaluating, 1979, Vol 7, pp. 310-316.
20. A. D. Wilson: Journal of Engineering materials and Technology, 1979, Vol. 101, 265.
21. R. N. McMeeking: J. Mech. Phys. Solids, 1977, Vol. 25, pp. 357-381.
22. C. D. Beachem: Met. Trans., 1973, Vol. 4, pp. 1999-2000.
23. S. Hinotani, S. Terasaki and F. Nakasoto: Proceedings of the Second JIM International Symposium, Hydrogen in Metals, p. 421, 1980, Vol. 21.
24. J. P. Hirth: Met. Trans., 1980, Vol. 11A, p. 861.
25. T. Takeyama and H. Takahashi: Proceedings of the Second JIM International Symposium, Hydrogen in Metals, 1980, Vol. 21, p. 409.
26. C. D. Beacham: Met. Trans., 1972, Vol. 3A, p. 439.
27. S. P. Lynch: Scripta Metallurgica, 1979, Vol. 13, p. 1051.
28. P. C. Paris, H. Tada, Z. Zahour and H. Ernst, Elastic-Plastic Fracture, pg. 5-36, ASTM STP 668, ASTM, 1979, pp. 5-36.

DISCUSSION

I.M. Bernstein, Carnegie-Mellon University, Pittsburgh, PA: Could you assess from your results the material aspects of the reliability of using existing natural gas pipelines to transport hydrogen?

S.L. Robinson: I would anticipate a greatly accelerated rate of failure if operating pressures are not reduced i.e. reduction of the stress intensities at existing flaws. Because of the low BTU content of hydrogen, Gregory of IGT has suggested that 2 or 3 parallel pipelines should be employed rather than larger diameters or higher pressures. This course of action would also be indicated by this work.

M. Urednicek, Foothills Pipe Lines (Yukon) Ltd., Calgary, Alberta, Canada: Assuming, as you also pointed out, that possible damage is externally induced, can you indicate why you have selected a model with internal rather than external defects when analysing changes in stress intensity factor?

S.L. Robinson: I have presented here only a part of the total program we are conducting. Another part of the study involves burst testing of externally-flawed pipes after thermal charging which simulates long-term containment of hydrogen in the pipe.

HYDROGEN COMPATIBILITY OF A LINE PIPE STEEL

D. J. Christenson[*], I. M. Bernstein[*], A. W. Thompson[*],
E. J. Danielson[*], M. Elices[+], and F. Gutierrez-Solana[+]

[*]Carnegie-Mellon University, Pittsburgh, PA
[+]Escuela Técnica Superior, Madrid, Spain

The effect of hydrogen on the mechanical properties of line pipe steel
with an equiaxed ferrite-pearlite microstructure was studied. Hydrogen was
introduced into the steel using either electrochemical or external hydrogen
gas exposure. Tensile tests were performed on both the uncharged and hydrogen charged steel, and the results analyzed for strength, ductility, fracture behavior, and differences between the effects of internal and external
hydrogen.

It was established that hydrogen had no significant effect on the
strength of the steel, while causing a severe reduction in ductility, characterized by reduction of area losses up to 80%. Examination of fracture
surfaces showed the hydrogen-free steel exhibited a dimpled failure while
the embrittled specimens showed extensive evidence of quasi-cleavage. There
were no observed qualitative differences found between the two methods of
hydrogen introduction, either in reduction of area loss or fracture surface
appearance.

Introduction

The addition of hydrogen as a feed-stock to natural gas pipelines would help to extend supplies as well as making use of an abundant by-product, produced, for example, by electrolysis of water using off-peak electricity(1,2). However, by introducing such hydrogen into existing natural gas pipelines, there is a risk of embrittling the pipe, particularly for moderate to high hydrogen pressures.

This study was undertaken to examine the effects of both internal hydrogen and an external hydrogen atmosphere on the subsequent mechanical properties of line pipe steel. It has been found, for some materials, that both internal and external hydrogen produce similar results(3-6). If so, an electrochemical method of testing hydrogen effects can be used to test pipes to predict performance when they are exposed to external hydrogen atmospheres. This was done using tensile tests on pipe sections subjected to either external hydrogen atmospheres of varying pressure or to hydrogen introduced internally by high fugacity electrochemical cathodic charging.

Materials

The materials studied were three different samples of Spanish line pipe steel in either plate or pipe form. Compositions are given in Table I. Figures 1-3 are three-dimensional constructs showing the microstructure in the rolling direction, longitudinal transverse, and short transverse directions. The plate material has a heavily banded ferrite-pearlite structure while both pipes exhibit an equiaxed ferrite-pearlite structure(7). The microstructures in both pipes are similar except the grain size of Pipe No. 1 is smaller than that of Pipe No. 2.

Experimental Procedure

The yield and ultimate tensile strengths as well as the reduction of area (RA) associated with the fracture were determined from tensile tests for hydrogen-free and hydrogen charged samples. For the plate material, typical values were, respectively, 160 MPa, 240 MPa, and 70 pct(7).

In one set of tests, the hydrogen was generated electrochemically by cathodic charging in a IN H_2SO_4 solution, which was poisoned with CS_2 and As_2O_3 to reduce hydrogen recombination(7-9). The charging current was varied in order to vary the external fugacity and resultant internal hydrogen concentration of the steel. These tensile specimens were tested immediately, at an initial strain rate of $3.3 \times 10^{-4}s^{-1}$, in order to minimize hydrogen loss. The average hydrogen concentration was determined by vacuum extraction techniques on similarly charged pieces of the pipe material.

A second set of tests was performed to study the effect of an external hydrogen atmosphere. Tensile specimens were tested at the same cross-head speed as for charged specimens, in a chamber containing various hydrogen pressures. This method of hydrogen introduction most resembles the actual in-use conditions of the pipe.

The fracture morphologies of both types of test specimens, as well as those of the hydrogen-free samples, were examined by scanning electron microscopy. These results were then compared to find the correspondence of the external hydrogen with the amount of internal hydrogen. This comparison

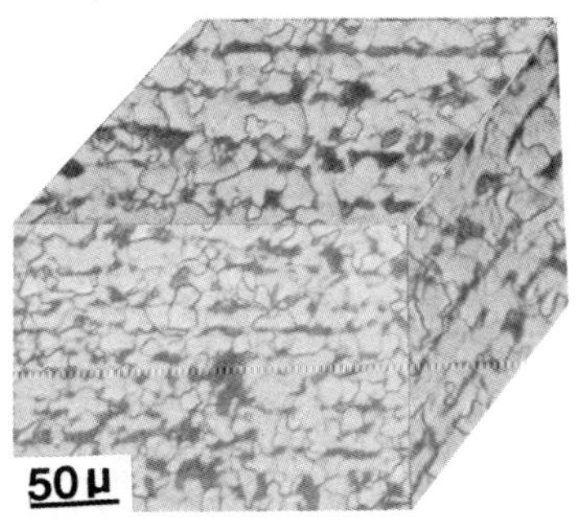

Fig. 1. Microstructure of plate material.

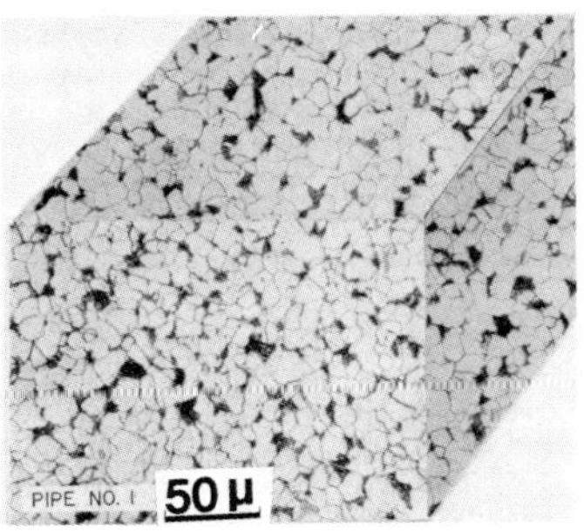

Fig. 2. Microstructure of Pipe No. 1

Fig. 3. Microstructure of Pipe No. 2.

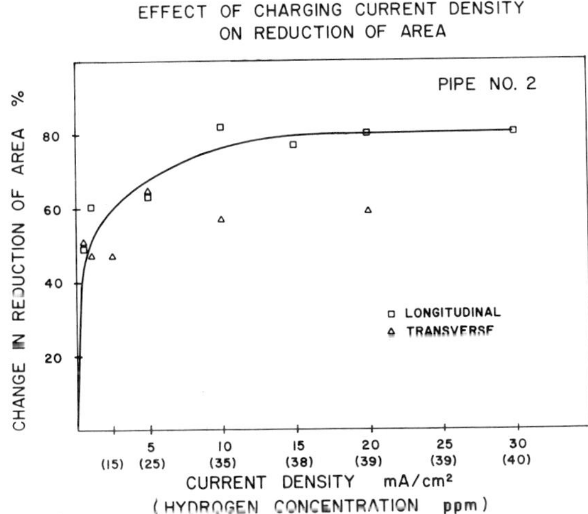

Fig. 4. Effect of charging current density on reduction of area.

was done on the basis of similar fracture behavior and morphology and similar changes in reduction of area.

Results

Internal Hydrogen

The internal hydrogen behavior of the Pipe No. 2 material was tested in both the rolling direction and the longitudinal transverse (LT) direction, to establish the role, if any, of mechanical and crystallographic texture. The uncharged specimens had 0.2% offset yield strengths of 266 MPa and 286 MPa, and ultimate tensile strengths of 414 and 417 MPa, in the two respective directions. The accompanying RA values were 61% for the rolling direction and 51% for the LT direction. Tensile specimens, charged with current densities ranging from 0.5 to 30 mA/cm^2, were also tested. Measurements of the yield and ultimate strengths averaged 294 MPa and 424 MPa, independent of current density, showing no significant effect of hydrogen on the material's strength properties in either testing direction. However, the RA decreased with increased charging current density, with a maximum loss of 80%. Figure 4 shows a plot of the reduction of area loss as a function of the charging current density. The numbers in parentheses show the effective average hydrogen concentration in weight parts per million for each of the current densities.

The uncharged fracture surfaces were characterized by a dimpled surface, typical of the ductile fracture of low strength steels, as seen in Figure 5. On the other hand, the hydrogen charged samples showed evidence of quasi-cleavage (10), whose presence was established by a study of matching surface stereo pairs. Figure 5 shows several examples of such fracture surfaces for varying hydrogen concentrations and for both the rolling direction and the LT orientations.

External Hydrogen

The effect of an external hydrogen atmosphere on the steel was studied on the plate material as well as on both Pipe No. 1 and Pipe No. 2. These tensile specimens were notched before testing to facilitate localized hydrogen absorption in the region of expected fracture (11). Notched ultimate tensile strengths were 530 MPa for the plate material and an average of 450 MPa for the pipes, while the notched yield strength for both types of material was 300 MPa. These strengths did not vary significantly with the magnitude of the external hydrogen pressure. However, the reduction of area associated with the fracture decreased with increasing hydrogen pressure and had a maximum loss of about 52% for the pipe and 70% for the plate material, as seen in Figure 6, which is a plot of the dependence of reduction of area loss on the external hydrogen pressure. The behavior of the two pipes is very similar, suggesting that the grain size of the ferrite (over the range investigated) had no effect on the degree of embrittlement. The plate material, however, had a much higher reduction of area loss than the pipe, suggesting that its banded microstructure makes it more susceptible to hydrogen embrittlement. Since neither Pipe No. 1 nor the plate material were tested with internal hydrogen, this has not been confirmed.

Fracture surfaces of these tensile specimens were examined by scanning electron microscopy, several examples of which are shown in Figure 7. All of the fracture surfaces exhibited two modes of fracture behavior; tearing near the notch and a combination of quasi-cleavage and dimpled failure over

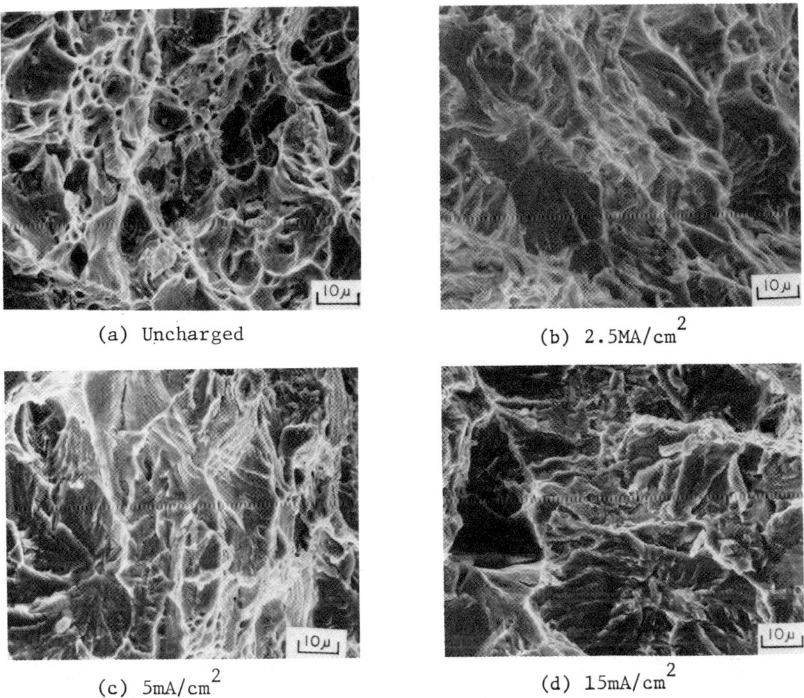

(a) Uncharged

(b) $2.5MA/cm^2$

(c) $5mA/cm^2$

(d) $15mA/cm^2$

Fig. 5.   Fracture surfaces of cathodically charged specimens.

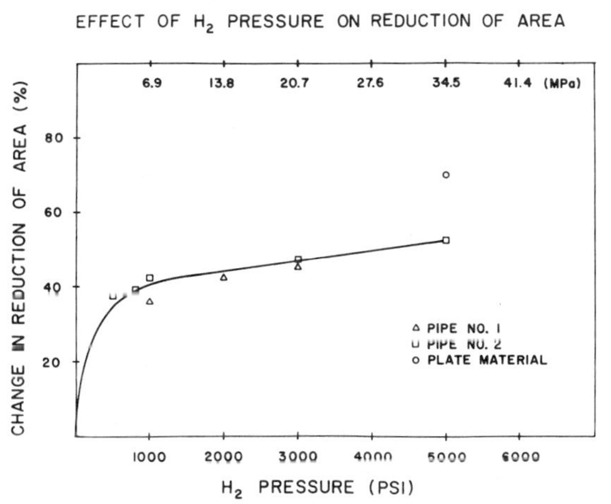

Fig. 6.   Effect of external hydrogen atmosphere on reduction of area.

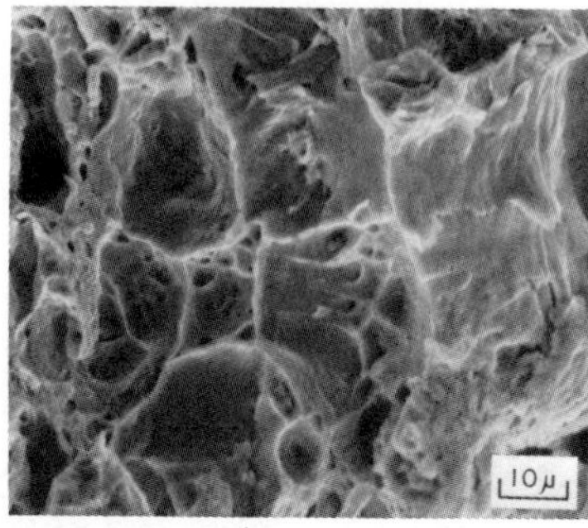

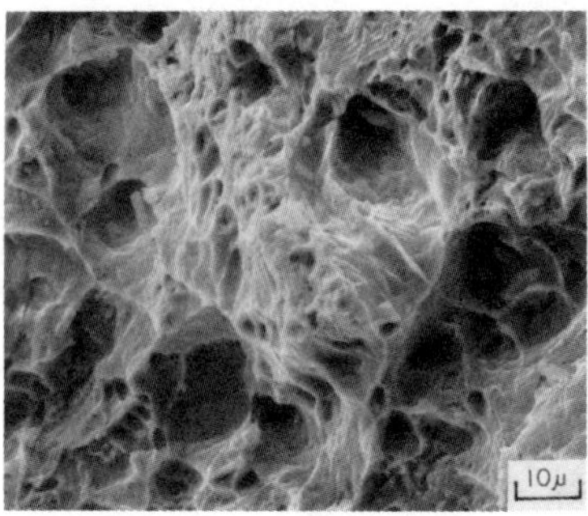

(a) 500 psi hydrogen pressure

(b) 3000 psi hydrogen pressure
(near notch)

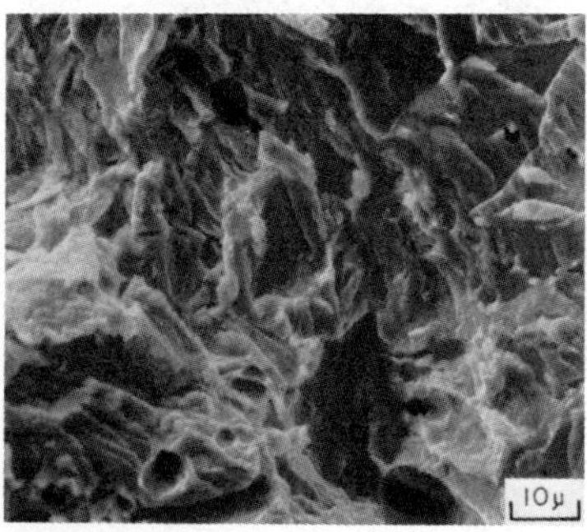

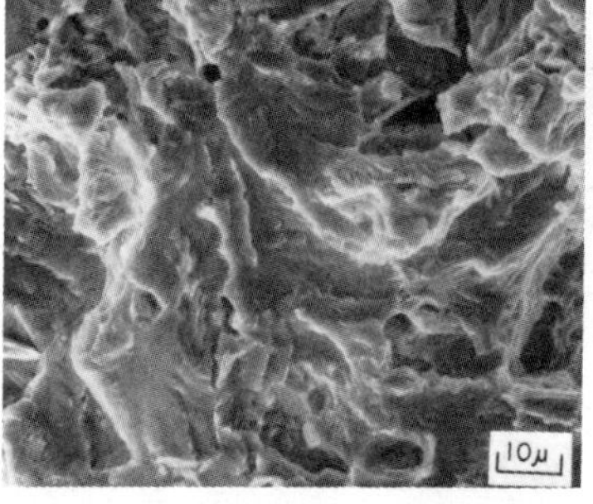

(c) 3000 psi (away from notch)

(d) 5000 psi hydrogen pressure

Fig. 7.   Fracture surfaces of specimens tested in external hydrogen
atmospheres.

the rest of the surface. As the external hydrogen was increased, the concentration of dimples decreased until, at the highest pressures, the fracture surfaces away from the notch closely resembled those of the cathodically charged specimens.

Discussion

A comparison of results from both types of testing shows that both methods of hydrogen introduction produced severe hydrogen embrittlement. However, in neither case did the hydrogen concentration have an effect on the subsequent strength of the steel, as is typically observed (5). By comparing the fracture surfaces and the reduction of area losses, a correlation between the effects of external hydrogen pressure and internal hydrogen concentration induced by cathodic charging can be found for this material. Hydrogen pressure of 3000 psi (21 MPa), a value which has been suggested for use in pipelines (12,13), produced very similar results to those for a specimen charged with a current density of $2.5mA/cm^2$, which in turn corresponds to an average hydrogen content of 15 ppm. Thus the more convenient electrochemical technique can be used in the laboratory to simulate elevated gas pressure exposure.

In general, the fracture behavior of the specimens charged with the lowest current densities corresponded quite well with those of the highest hydrogen pressures used. However, testing on samples charged with current densities less than $2.5mA/cm^2$ will be necessary in order to find a precise relation between external hydrogen pressure and internal hydrogen concentration.

To further examine the question of whether hydrogen can be safely transmitted through this pipe, a series of bend-bar tests (9) is planned in order to study more directly the effect of hydrogen on the toughness of this steel. From the viewpoints both of general mechanical performance, and particularly of pipeline inspection requirements, there are indications(13-15) that toughness in the presence of hydrogen will play a pivotal role in the design of pipelines to carry hydrogen gas.

Conclusions

1. Hydrogen introduction by either electrochemical or external gaseous exposure has no significant effect on the strength of this steel.

2. The introduction of hydrogen results in a decrease in ductility for both internal and external hydrogen.

3. Reduced ductility fracture caused by external or internal hydrogen is qualitatively similar and a correspondence between the two can be found, on the basis of fracture behavior and morphology. For example, charging at $2.5mA/cm^2$ gave results similar to tests in 3000 psi (21 MPa) of external hydrogen.

Acknowledgments

We are grateful for the high-pressure hydrogen testing, performed by R. E. Stoltz at Sandia National Laboratories, and for the financial support of Butano S.A., Spain.

References

1. D. P. Gregory, Scientific American, 1973, vol. 228 (No. 1), pp 13-21.
2. A. L. Hammond, W. D. Metz, and T. H. Maugh, Energy and the Future ,
    pp 117-23, A.A.A.S., Washington, 1973.
3. W. W. Gerberich, J. Garry, and J. F. Lessar, Effect of Hydrogen on Be-
    havior of Materials (A. W. Thompson and I. M. Bernstein, eds.), pp 70-82,
    TMS-AIME, New York, 1976.
4. R. P. Gangloff and R. P. Wei, Met. Trans., (1977), vol 8A, pp 1043-53.
5. J. P. Hirth, Met. Trans., (1980), vol 11A, pp 861-90.
6. A. W. Thompson and I. M. Bernstein, Advances in Corrosion Science and
    Technology (M. G. Fontana and R. W. Staehle, eds.), vol. 7, pp 53-175,
    Plenum, New York, 1980.
7. I. M. Bernstein and A. W. Thompson, "Report of an Investigation on the
    Hydrogen Compatibility of Spanish Pipeline Steel," Carnegie-Mellon
    University, Pittsburgh, PA, 1980.
8. B. B. Rath and I. M. Bernstein, Met. Trans., (1971), vol. 2, pp 2845-51.
9. J. E. Costa and A. W. Thompson, Met. Trans. A, in press.
10. C. D. Beachem, J. Basic Eng. (Trans. ASME, Series D), 1965, vol. 87,
    pp 299-306.
11. D. Hardie and P. Bowker, Effect of Hydrogen on Behavior of Materials
    (A. W. Thompson and I. M. Bernstein, eds.), pp 251-261, TMS-AIME,
    New York, 1976.
12. D. P. Gregory, A Hydrogen-Energy System, Amer. Gas Assoc., Chicago, 1972.
13. A. W. Thompson, Transmission and Storage (Vol. 2 of Hydrogen: Its
    Technology and Implications), K. E. Cox and K. D. Williamson, eds.,
    pp 83-124, CRC Press, Cleveland, 1977.
14. R. B. Thompson, A. W. Thompson, and D. O. Thompson, Effect of Hydrogen
    on Behavior of Materials (A. W. Thompson and I. M. Bernstein, eds.),
    pp 612-22, TMS-AIME, New York, 1976.
15. A. W. Thompson and I. M. Bernstein, Int. J.1 Hydrogen Energy, 1977,
    vol. 2, pp 163-73.

Table I. Composition of Line Pipe Steel

Material	C	Mn	Si	P	S	Other
Plate	0.16	1.45	0.27	0.020	0.040	--
Pipe No. 1	0.20	1.48	0.44	0.030	0.030	--
Pipe No. 2	0.10	0.66	0.23	0.020	0.025	0.15 Cr, 0.17 Co

THE EFFECTS OF CATHODIC PROTECTION AND OVER
PROTECTION ON THE TENSILE DUCTILITY AND CORROSION
FATIGUE BEHAVIOUR OF X-65 PIPELINE STEEL

B.R.W. Hinton and R.P.M. Procter

Corrosion and Protection Centre,
University of Manchester Institute
of Science and Technology, England.

The environmental cracking of X-65 pipeline steel in chloride
solutions has been investigated under tensile loading conditions at
various applied potentials and strain rates, and under cyclic loading
conditions at -0.8 V (S.C.E.) and various frequencies. Potentials more
negative than the -0.8 V and slow strain rates promote loss of ductility
while fatigue crack propagation rates are enhanced by low cyclic
frequencies.

Extensive electron fractography has indicated that these effects
are due to the formation of brittle modes of fracture which are considered to result from the presence of cathodically evolved hydrogen.

Although an applied potential of -0.8 V in NaCl solution prevents
corrosion and does not affect the tensile ductility of plain specimens,
it has no such beneficial effect on the fatigue behaviour of pre-cracked
specimens. This is thought to be the result of the differences in
chemistry between the bulk solution and that within a fatigue crack.

Introduction

Many offshore natural gas transmission pipelines are constructed from low strength steels such as A.P.I. 5LX-65. These steels, with the correct heat treatment, are considered immune to failure by hydrogen embrittlement under static loading conditions (1).   However, under cyclic loading in hydrogen gas (2) and in aqueous solutions, (3) these steels fail by fatigue cracking at rates in excess of those observed in air.   There is also evidence that such steels may be embrittled by slow strain rate tensile loading in aqueous solutions of high hydrogen fugacity (4).

Pipelines are fabricated by welding, and welded joints may contain defects which could act as sites for crack nucleation. Furthermore, fluctuating gas demands, and the use of pipelines as reservoirs, can produce pressure variations which could subject the pipeline to cyclic, or steadily increasing tensile, loads. Finally, the sea water environment is an aggressive one, and cathodic protection to potentials of around -0.8 V (S.C.E.) is usually employed to prevent general corrosion of the pipeline; however, any potential more negative than -0.8 V would result in undesirable hydrogen evolution.   Under certain conditions, therefore, combinations of these factors could lead to environmental cracking of gas pipelines.

In this paper, the effect of slow tensile loading on the ductility of X-65 steel in chloride solutions, and the effect of low frequency cyclic loading on the fatigue cracking rate in chloride solutions are examined. In particular, the behaviour under both loading conditions is compared at a potential of -0.8 V.

Experimental Details.

All specimens were taken from one length of 1m diameter pipe made from X-65 steel with the following composition: C 0.08%, Mn 1.15%, S 0.011% and P 0.013%, and mechanical properties: σy 450 MPa, U.T.S. 550 MPa, %El. 17% and % RA 77%. The steel had a banded ferritic – pearlitic microstructure.

Tensile specimens with a gauge length of 25 mm and 2.5 mm diameter were tested in a Parkins type constant strain rate tensile machine. Fatigue experiments were carried out with pin-loaded, pre-cracked single edge-notched tension specimens of dimensions 280 x 55 x 10 mm. These specimens were tested in a servo-hydraulic machine under load control, with a load ratio of 0.15 and with both sinusoidal and triangular wave-forms. It was established that the degree of environmental enhancement of fatigue cracking is the same for both waveforms. Crack length was monitored by measuring compliance with a clip gauge.

Both tensile and fatigue tests were conducted in an aerated 3.5% NaCl solution of initial pH = 8.   The potential was controlled with a potentiostat.

Results.

Tensile Test Data

The variations in ductility, as indicated by the percentage reduction in area, with potential for different strain rates are shown

in Figure 1.   As the potential is reduced from the open circuit corrosion
potential to -0.8 V (S.C.E.), the ductility remains constant, similar to
that obtained in air and independent of strain rate. However, at potentials
more negative than -0.8 V, there is a sharp decrease in ductility with decreasing potential.   The effect is more pronounced as the strain rate is
reduced. At potentials of around -1.6 V, an R.A. as low as 15% is obtained
at the slowest strain rate, compared with 75% at -0.8 V. These losses in
ductility are considered to be the result of cracking due to cathodically
evolved hydrogen.

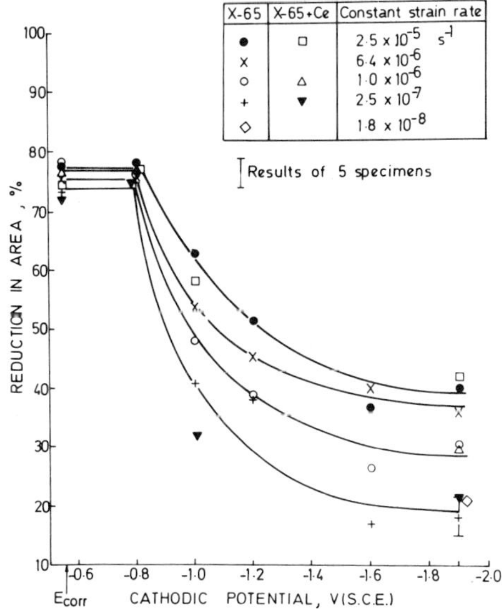

Fig. 1. The variation in % R.A. with po-
         tential for various strain rates.

<u>Fractography of Tensile Specimens.</u>

     The appearance of the specimen fracture surfaces varies according to
the combination of potential and strain rate at which the test is conducted.
The different features are summarised schematically in Figure 2.   In
Region 1 of Figure 2 the fracture is ductile and the fracture surface
morphology consists of dimples resulting from microvoid formation and
coalescence Figure 3(a).   In Region 2, the fracture surface consists
of surface cracks which are normal to the tensile axis, and microvoid
coalescence Figure 3(b).   These cracks are produced by the coalescence of
many transgranular cracks, all with an appearance typical of cleavage,
that is, macroscopically smooth facets with river markings, Figure 3(c).
These cracks are also observed within the necked region of the specimen,
adjacent to the fracture surface.   Metallographic examination of un-
broken specimens indicated that these cracks do not form until after
the specimen has necked during testing.

     Figure 3(d) shows that considerable lack of ductility is associated
with the fracture of specimens in Region 3.   Here the fracture surfaces
consist predominantly of areas resulting from cleavage fracture, as

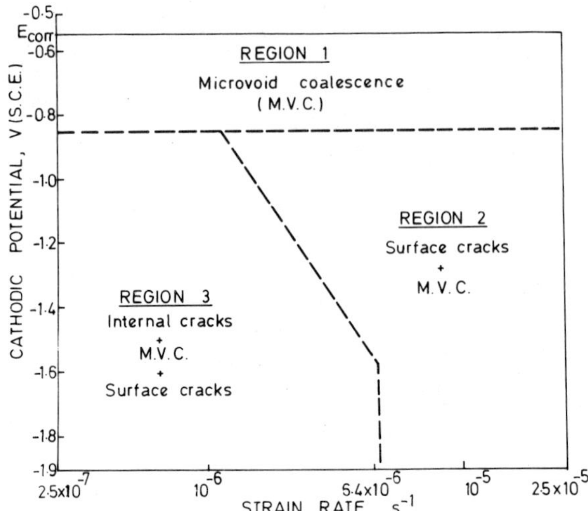

Fig. 2. A schematic diagram showing the
distribution of fractographic
features for combinations of
potential and strain rate.

illustrated in Figure 3(c).   However, metallographic sectioning confirmed that these cleavage cracks form internally and without contact
with the test solution. They are associated with large inclusions,
pearlite colonies and planes of segregated MnS inclusions.

Some specimens were pre-charged with hydrogen immediately prior to
testing at the most negative potential. No additional loss in ductility
was observed. This suggests that straining, and hence freshly created
surface, is necessary for sufficient hydrogen to enter the specimen.

Fatigue Test Data.

The variation in fatigue crack propagation rate (f.c.p.r.) with the
stress intensity range, ΔK, at a potential of -0.8 V is shown in Figure 4
for various frequencies. It is evident that for all frequencies considered some increase in the f.c.p.r. occurs relative to the rate in air.
This is in marked contrast to the effect on tensile ductility of a
potential of -0.8 V. At low levels of ΔK the f.c.p.r. is independent of
frequency; however, with increasing ΔK, a dependence of f.c.p.r. on
frequency develops and a point is reached, for each frequency, where the
data change slope and depart from the frequency independent region. This
point occurs at higher ΔK and higher f.c.p.r. values as the frequency is
decreased. The new slope of the data decreases with decreasing frequency
until at 0.01 Hz it becomes zero. At higher ΔK levels the data again
change slope and become parallel to the air data.

Fractography of Fatigue Specimens.

Detailed examinations were carried out at intervals of 1 mm along the
fracture surfaces of fatigue specimens, using both the scanning electron

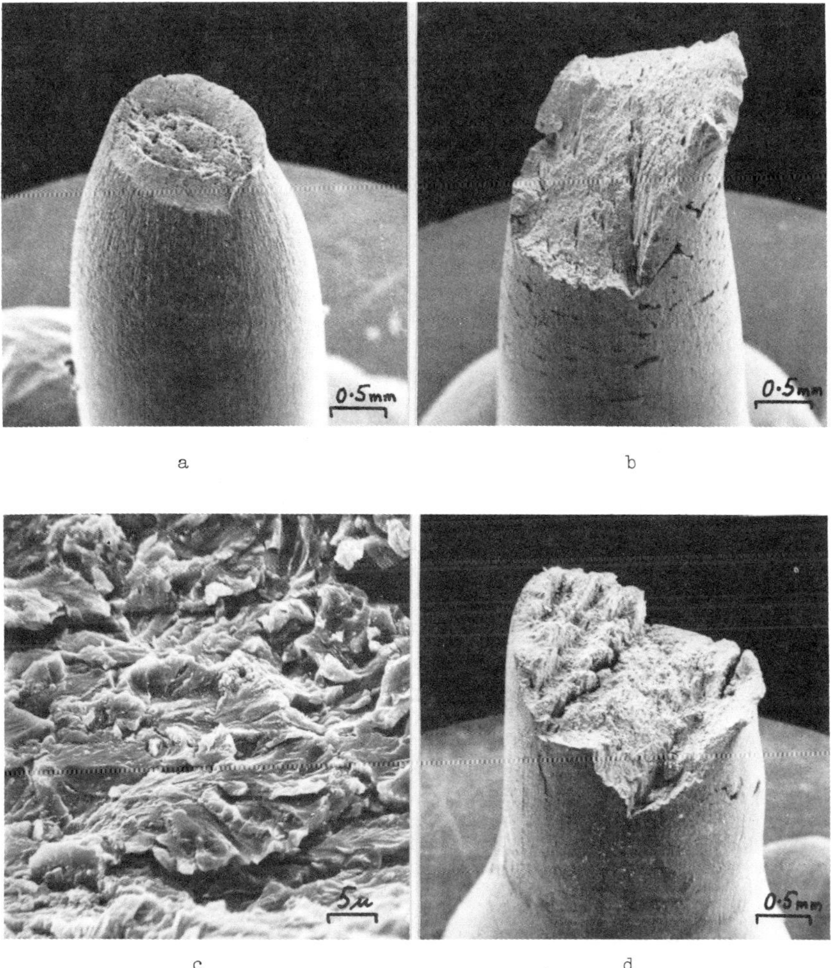

Fig. 3.   Fractured Specimens (a) −0.8 V, 6.4 x 10^{-6} s^{-1} (Region 1)
(b) 1.6 V, 2.5 x 10^{-5} s^{-1} (Region 2)
(c) An area of cleavage fracture
(d) −1.0 V, 2.5 x 10^{-7} s^{-1} (Region 3)

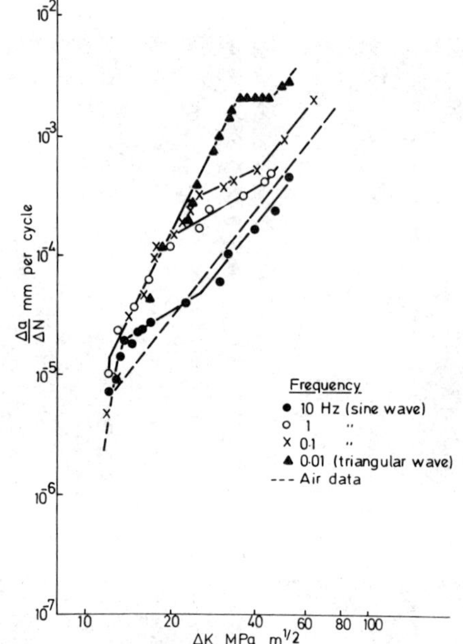

Fig. 4. Fatigue crack propagation rate as a function of ΔK for various frequencies at −0.8 V.

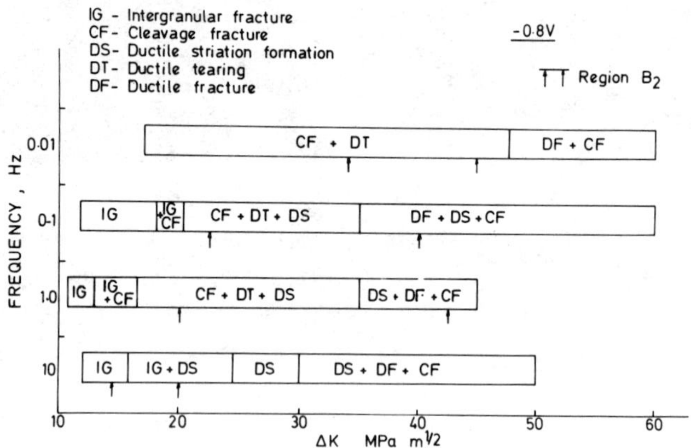

Fig. 5. A schematic diagram showing the variation in fracture mode with ΔK for various frequencies at −0.8 V.

microscope and two stage plastic carbon replicas with the transmission electron microscope. In air, crack propagation occurs predominantly by ductile striation formation. However, the fracture modes present during cracking in chloride solution are complex and vary greatly with both ΔK and frequency; this variation is summarised schematically in Figure 5. Changes in mode with increasing ΔK are gradual rather than sharply defined; the first symbol in each region represents the major fracture mode present in that region.

At low ΔK values intergranular cracking occurs (Figure 6). With increasing ΔK this is gradually replaced by cleavage facets, interspersed with ductile striations and/or ductile tearing (Figure 7).

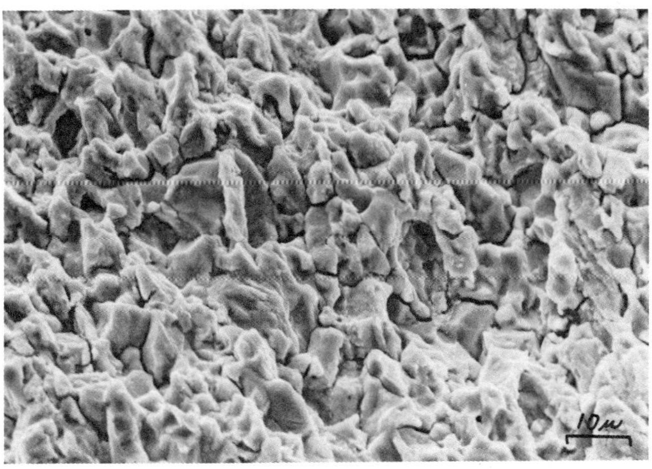

Fig. 6. Intergranular cracking. -0.8 V, 0.1 Hz, $\Delta K = 17$ MPam$^{\frac{1}{2}}$.

At still higher ΔK levels, depending on frequency, there is a change to more predominantly ductile fracture modes.   In these regions the fracture surfaces consist of areas of ductile fracture, resulting from shearing between cracks on different planes, microvoid coalescence and ductile tearing, interspersed with cleavage facets and/or ductile striations (Figure 8). The vertical arrows on Figure 5 indicate the transition region between the frequency independent regions of the data in Figure 4 and the region where the data is parallel to the air data. It is significant that the change from generally brittle to generally ductile modes of crack extension occurs in the vicinity of this transition region.

Discussion

The occurrence of brittle modes of fracture such as cleavage and intergranular cracking is responsible for the decreased tensile ductility and the increased f.c.p.r., compared with air data. These modes of fracture are clearly the result of the interaction of the applied stress and cathodically evolved hydrogen.

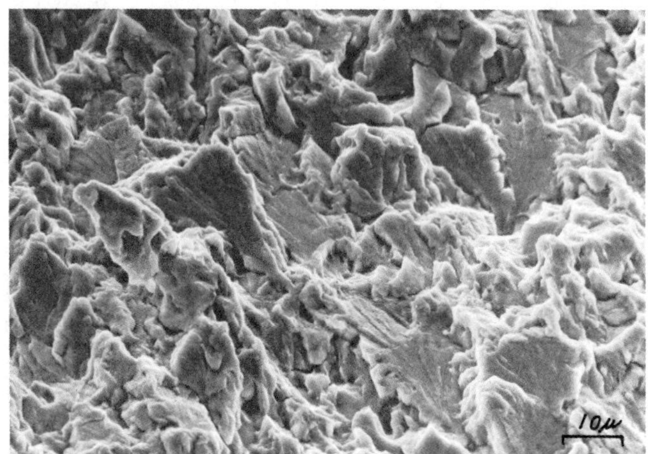

Fig. 7. Cleavage facets and ductile tear
       ridges. -0.8 V, 0.01 Hz, $\Delta K = 32$ MPam$^{\frac{1}{2}}$.

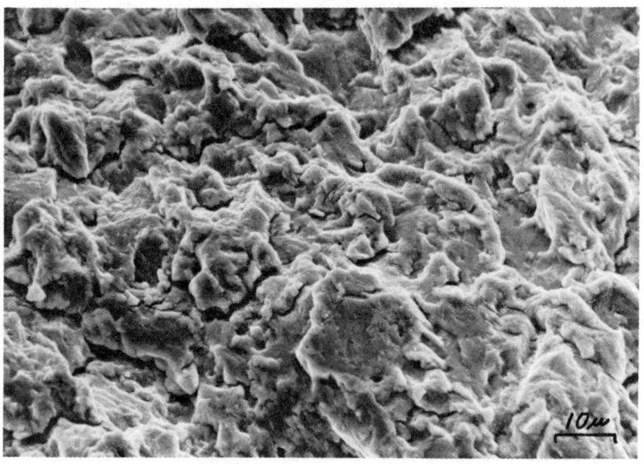

Fig. 8. Ductile tear ridges, M.V.C, ductile
       striations and cleavage facets.
       -0.8 V, 0.1 Hz, $\Delta K = 40$ MPam$^{\frac{1}{2}}$.

   At the faster strain rates, under tensile loading, hydrogen cracks
initiate at, and propagate in from, the surface thereby reducing the
specimen cross section and promoting earlier failure with reduced ductility.
At lower potentials more hydrogen is available and hence lower ductilities
are observed. At slower strain rates, and particularly at lower potentials,
more internal cracking is observed as there is more time available for more
hydrogen to diffuse to traps such as inclusions and pearlite colonies.

Very little surface cracking occurs under these conditions because such cracking appears to be associated with necking and, as Figure 5 illustrates, very little necking occurs under these test conditions.

At low ΔK levels fatigue crack propagation occurs by brittle modes of fracture and the rate is independent of frequency. This evidence, together with further results (to be published) (5) which show that the slope of the data in this region increases with decreasing potential, suggest that at low ΔK levels cracking is controlled predominantly by the environment. At high ΔK levels ductile modes of fracture are predominant and the data are parallel to the air data; however, the data are also somewhat dependent on frequency. This suggests that cracking in this region is controlled mainly by the mechanics of the fatigue process with less environmental influence. The transition between the two regions is thought to occur when the applied crack opening displacement can no longer be accommodated by largely brittle crack extension. This is probably due to the diffusion limitation on the distance ahead of the crack tip over which the concentration of hydrogen, necessary for cracking, can be achieved during each cycle.

Slow strain rate tensile loading is thought to resemble closely the loading conditions occurring ahead of a fatigue crack during the tensile half cycle. The analogous observations of decreased ductility, with more cleavage cracking, occurring with slower straining rates and the faster f.c.p.r., with more cleavage cracking, that is observed with decreasing cyclic frequency, reinforce this conclusion. However, the environmental conditions during the two tests must be different because an enhanced f.c.p.r. is observed at -0.8 V, whereas at this potential the ductility is unchanged from that observed in air. This difference is thought to arise from the presence of the fatigue pre-crack, at the tip of which the environment is different from the outside bulk environment. At the smooth tensile specimen surface the rate of hydrogen evolution is low at -0.8 V and at near neutral pH values. Within the fatigue crack, the potential is undoubtedly more positive than -0.8 V; this permits some dissolution of Fe to Fe^{2+}. The subsequent hydrolysis of Fe^{2+} ions results in the formation of H^+ ions and a lowering of the pH at the crack tip which effectively increases the amount of hydrogen available; this leads to the enhanced f.c.p.r. Furthermore, oxygen is known to adsorb preferentially at active sites on freshly created surfaces. However, the environment in a fatigue crack will soon become deaerated, so the hydrogen evolution reaction, and hydrogen entry, at active sites near the crack tip will suffer less competition from oxygen adsorption, and the oxygen reduction reaction then would be the case on smooth surfaces exposed to the bulk, aerated environment. It is therefore suggested that some loss of ductility would occur if pre-cracked tensile specimens were tested under slow strain rate conditions; such a test would possibly represent more closely fatigue cracking under low cyclic frequencies.

The observation that, at -0.8 V, the tensile ductility is unaffected at all strain rates, whereas the f.c.p.r. is enhanced to a certain degree at all frequencies, has important practical implications for pipeline operation. It suggests that cathodic protection to the normal working levels is only safe provided there are no crack-like defects present! However, the normal working stresses on a pipeline are fairly low and Figure 6 shows that environmental effects are reduced at low ΔK levels.

Conclusions

1.   The tensile ductility of X-65 is reduced if loading is carried out in 3.5% NaCl solutions at potentials less than -0.8 V. The ductility decreases as the potential and strain rate are decreased.

2.   The rate of fatigue crack propagation in X-65 is increased if cyclic loading occurs in 3.5% NaCl solutions at -0.8 V. In general, the f.c.p.r. increases with decreasing frequency.

3.   The reduced ductility and enhanced f.c.p.rs. are due to the occurrence of brittle modes of fracture that result from the presence of cathodically evolved hydrogen.

4.   The enhanced f.c.p.r. at -0.8 V as opposed to the absence of any effect under tensile loading at -0.8 V is thought to be due to differences in chemistry between the solution at the crack tip and the bulk solution.

References

1.   T.P. Greoneveld and A.R. Elsea, "Hydrogen Stress Cracking in Natural Gas Transmission Pipelines" pp 727-737 in <u>Hydrogen in Metals</u>, I.M. Bernstein, A.W. Thompson ed. A.S.M. Metals Park, Ohio, 1974.

2.   H.G. Nelson, "Hydrogen-Induced Slow Crack Growth of a Plain Carbon Pipeline Steel under Conditions of Cyclic Loading". pp 602-611 <u>Effect of Hydrogen on Behaviour of Materials</u>, A.W. Thompson, I.M. Bernstein ed. A.I.M.E. N.Y., 1976.

3.   O. Vosikovsky, "Fatigue Crack Growth in an X-65 Line-Pipe Steel at Low Cyclic Frequencies in Aqueous Environments," Trans. A.S.M.E. <u>Journal of Engineering Materials and Technology</u>. <u>97</u> (4) 1975 pp 298-304.

4.   A. Kawashima, K. Hashimoto, S. Shimodaira, "Hydrogen Electrode Reaction and Hydrogen Embrittlement of Mild Steel in Hydrogen Sulphide Solutions", <u>Corrosion - NACE</u> <u>32</u> (8) 1976 pp 321-331.

5.   B.R.W. Hinton, Ph.D. Thesis, University of Manchester, U.K. 1980.

DISCUSSION

N.J.H. Holroyd, University of Newcastle-Upon-Tyne, Newcastle-upon-Tyne, Great Britain: The fracture mode of embrittlement given by slow strain rate testing at nominal strain rates of about $10^{-6}s^{-1}$ and that given by low frequency corrosion fatigue testing (<0.1Hz) are usually similar. In the case of X65 linepipe steel in chloride solutions this was not observed; slow strain rate testing gave transgranular fracture whereas corrosion fatigue testing gave intergranular fracture. Can you explain this observation?

B.R.W. Hinton: The fracture mode resulting from slow strain rate tests and low frequency fatigue tests is generally the same i.e. cleavage facet formation. However, at low ΔK values the fracture mode during cyclic loading is intergranular. We believe that this is due to the distance, over which the crack tip stresses are high, being of the order of the grain size. With the tensile tests, general yielding has occurred over the entire gauge length diameter when cracking commences, and this dimension is very much larger than the grain size; thus intergranular cracking is not observed.

M. Meshii, Northwestern University, Evanston, IL: Why did you carry out the monotonic tests at strain rates considerably lower than those for the fatigue tests?

B.R.W. Hinton: The strain rates begin to overlap at a frequency of 0.0001 Hz. You will appreciate that testing at such frequencies is time consuming, and results at this frequency are yet to be obtained. However, when they are available, we consider that they will only strengthen the similarities between the two types of loading data, which we have, concerning the effects of potential and loading rate. Furthermore, we believe that they will reinforce the proposal that constant slow strain rate tensile loading conditions closely resemble the loading patterns experienced by elements at the tip of a slowly cycled fatigue crack.

HYDROGEN-ASSISTED FATIGUE OF PERIODICALLY

PRESSURIZED STEEL CYLINDERS *

M. Kesten and K.-F. Windgassen
Messer Griesheim GmbH
Industriegase
P. O. B. 4709
4000 Düsseldorf 1
West Germany

Service failures of various hydrogen pressure vessels experienced in
Europe within the last 10 years have been the cause of considerable
concern.

As a contribution to the solution of the problem of hydrogen compatibility
with pressure vessel steels long-term fatigue tests have been carried out
with commercially produced Cr-Mo-steel cylinders being periodically
pressurized with hydrogen gas in the pressure range between 20 and 195 bar.

In keeping the experimental conditions as close to the actual filling
practice as possible the test frequency was set to 5 cycles per hour. For
each cycle the rate of increasing the pressure in the interval between 120
and 195 bar was maintained at 0.2 bar/sec. By means of local reduction of
the wall thickness the amplitude of the hoop stress of the test cylinders
was individually adjusted to values covering the range between the stress
at yield and the normal service stress.

The results clearly demonstrate the detrimental influence of hydrogen. The
fatigue life of the hydrogen cylinders is reduced by a factor of approximately 6 compared to similar tests with oil. As a consequence, this
indicates the need to generally control the service practice of hydrogen
transport vessels.

Comparison of the experimental results with the case histories of real
service failures reveals the decisive role which is played by internal
surface defects. Based on a thorough fractographic analysis of the
fracture surfaces and on the hydrogen-assisted fatigue crack growth data
being available for the type of steel investigated a general model is
being presented describing the conditions necessary for hydrogen failures
to occur in pressure vessels.

---

* This work has been supported by the German Bundesministerium für
  Forschung und Technologie, Contract No. FEKKs, 1.1/10

Introduction

The majority of all "merchant" hydrogen which is produced and consumed in Europe is stored and distributed in the gaseous state as pressurized gas. Nearly all of this is distributed to dispersed customers and transported to the customer's facility by trucks equipped with either steel cylinders (Fig. 1) or tube vessels (Fig. 2). The usual service pressure of this transport equipment being 200 bars. This system of storing and transporting hydrogen in steel vessels did not change significantly since the beginning of this century when large scale consumption of hydrogen began (Fig. 3).

Within the last decade a number of service failures have been experienced because of brittle crack formation.

In most incidents the vessels failed by leaking. But at a few occasions there also occured catastrophic failures where cylinders or vessels ruptured or even disintegrated resulting in more or less severe property damage.

Although reports on failed hydrogen cylinders are going back to the end of the last century (1), it may be just 10 or 15 years ago that people became aware of the fact that these failures are to be attributed to the specific degradating effect that gaseous hydrogen has on pressure vessel materials.

Since that time a lot of valuable research work has demonstrated how the hydrogen susceptibility of a material is correlated to certain features of its microstructure (2). This microstructural approach so far has been very

Fig. 1: Trailer truck with 330 50-litre steel cylinders to carry   3,000 m³ H_2.

Fig. 2: Tube trailer carrying 9 forged steel vessels, H_2-capacity   3,500 m³

successful. It offers the chance that in the future much more resistant materials may be developed by strictly controlling their microstructure (3).

However, as far as the safety of the present hydrogen storage and transport equipment is concerned there is still another way to look at the problem (4). Taking into account that nearly all service failures are

Fig. 3: Early horse-cart for hydrogen distribution

somehow associated with some kind of surface defects on the vessel wall it seems to be promising to approach the problem from the design point of view. That means to concentrate on those factors of influence which follow from the particular service conditions and the surface state of a commercially produced vessel.

Assuming that the experienced failures are basically the result of an hydrogen induced and assisted fatigue action, long term low cycle periodic pressurization tests were carried out with actual steel cylinders.

Experimental procedure and results

Since it was considered most important that the experimental conditions matched as close as possible the actual service conditions the pressurization cycle as shown in Fig. 4 was used.

Compared with real transport vessels the test frequency had to be somewhat accelerated. But in order to realize a loading rate comparable with the actual filling practice the pressure built up was in two steps.

In the second interval which was considered most important for the embrittling process the rate of increasing the hydrogen pressure was 0.23 bar/sec. This corresponds to a strain rate in the cylindrical wall in the order of 10^{-6} sec^{-1} depending on the actual wall thickness of the test cylinders. A 4 minutes hold time at the maximum pressure of 190 bar was followed by rapid discharging.

This pressurization cycle allowed for a test frequency of 5 cycles per hour.

Because all cylinders were subjected to the same maximum internal pressure the stress amplitude in the cylindrical wall was varied and individually adjusted for each test cylinder by locally reducing its wall thickness according to the required hoop stress, by machining the cylinder from outside.

The test cylinders used were commercially produced out of seamless tubes from a chrome-moly steel, the most common grade for gas cylinders and forged pressure vessels. The chemical composition, heat treatment and

resulting mechanical properties of the cylinder batch
is given in the Table.

The internal surface of the
test cylinders was as manufactured. A 100 % ultra-
sonic inspection guaranteed
absence of manufacturing
surface defects like laps,
folds or scratches with a
depth in excess of 0.5 mm.

Prior to the testing the
cylinders had to be filled
with lead in order to
reduce their free hydrogen
volume.

Fig. 5 summarizes the
results of the fatigue
tests. The hoop stress
in the cylindrical sec-
tion at maximum hydrogen
pressure is plotted versus
the number of cycles to
failure. The data ob-
tained for hydrogen are
compared with the results
of fatigue tests per-
formed with identical
cylinders but using
hydraulic oil instead of
hydrogen. The latter
compare well with other
available data of hydraulic fatigue tests with ordinary steel cylinders.

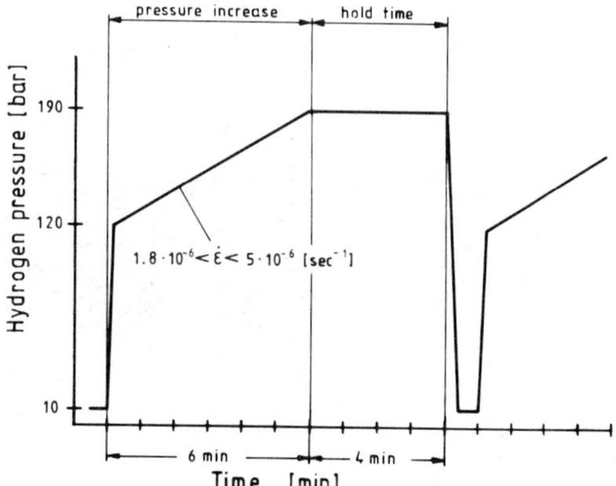

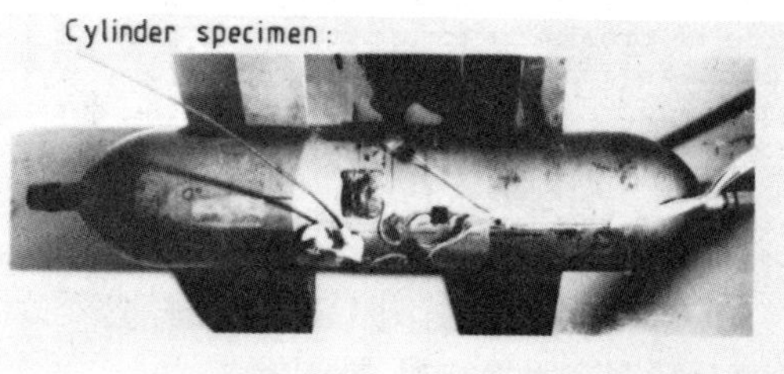

Fig. 4: Pressurization cycle used in the
fatigue experiments

Steel 34 Cr Mo 4 oil quenched and tempered (40 min/630°C)

Composition (Wt %)

C	Si	Mn	P	S	Cr	Mo
0.36	0.22	0.67	0.021	0.009	0.99	0.14

Yield Strength:	847 MPa
Ultimate Tensile Strength:	959 MPa
Elongation:	16 %

Table: Chemical composition and mechanical properties
       of the cylinder material

Discussion

The results obtained clearly demonstrate the detrimental effect which
hydrogen has on the low cycle fatigue life of commercial steel cylinders.
The possible consequences for the endurance limit are not yet clear. Tests

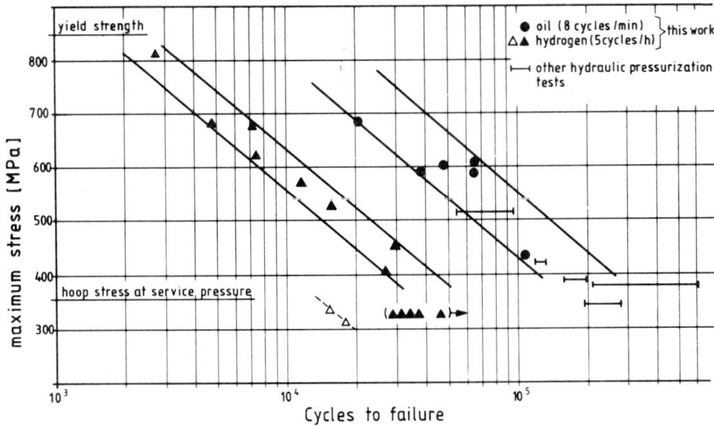

Fig. 5: Summary of the fatigue results

at cyclic stress levels corresponding to the lower end of the curve are
still going on.

Life expectancy calculations based on the established data give rather
promising results. Keeping in mind that the maximum allowable service
stress in cylinders and transport vessels of this steel grade is about
360 MPa, this should enable roughly 30,000 filling cycles which from the
practical point of view would mean a normally unlimited service life.

This, however, is in obvious contrast to the considerable number of service
failures and the fact that all the known failures actually did occur after
a much shorter service life.

The explanation for this apparent contradiction is the predominant role
which is played by the state of the internal surface coming in contact
with the hydrogen gas.

Even minor surface defects or irregularities may drastically reduce the fatigue life. This is demonstrated by the two points which fall to the left
of the general curve. They are the results of two test cylinders with
hardly visible manufacturing defects which would have been considered insignificant by normal inspection standards.

In Fig. 6 the attempt was made to correlate the actual service life of
failed vessels with the size of the preexisting surface defect which initiated the failure. The respective K-values of the starter defects were
roughly estimated from the available details of the case histories and
failure investigations.

Although showing a considerable scatter the plotted data are in line with
the fatigue test results. They indicate that the presence of a certain
critical surface defect is a necessary precondition for hydrogen failures
to occur.

When further comparing the established experimental results and the additional information from various service failures with the normal fatigue be-

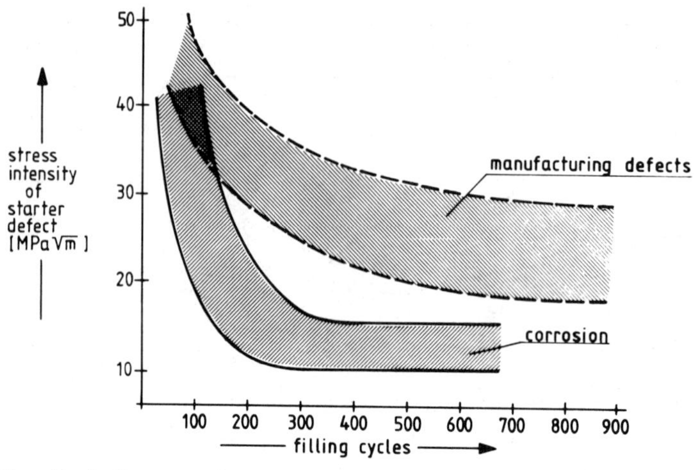

Fig. 6: Influence of surface defects on pressure vessel life

haviour of the investigated material and the available literature data on hydrogen-assisted fatigue and hydrogen sustained crack growth, a general model can be formulated which describes the conditions necessary for hydrogen failures to occur in pressure vessels (Fig. 7).

Basically, a vessel failure will proceed in four individual steps:

- The initiation of a fatigue crack above a critical stress intensity, ΔK_0^H.

- The environmentally assisted fatigue crack growth at stress intensities below K_{IH}. Where K_{IH} is the threshold stress intensity for sustained crack growth in the presence of hydrogen (5).

- Hydrogen crack growth under sustained load at stress intensities above K_{IH}, with a perhaps additive fatigue component.

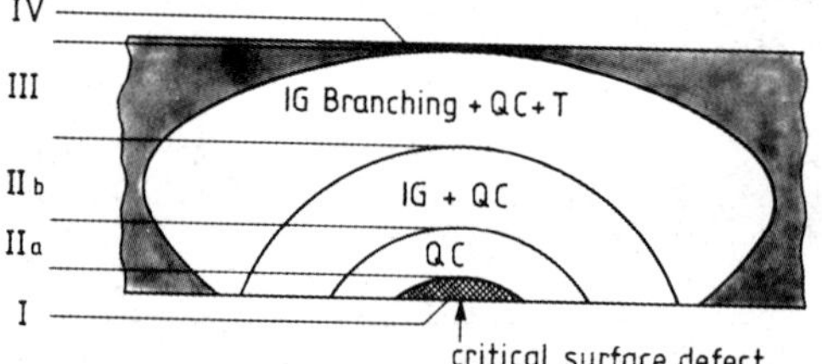

Phase I	Fatigue crack initiation $\Delta K > \Delta K_0^H$
Phase II	Hydrogen-assisted fatigue crack growth $K_{max} < K_{IH}$ II a : $K_{max} < K_{max}^T$; II b : $K_{max} > K_{max}^T$
Phase III	Hydrogen sustained crack growth $K_{max} > K_{IH}$
Phase IV	Critical crack growth $K_{max} > K_{IC}$

Fig. 7: Development of a hydrogen-assisted fatigue failure

- The final fracture with no necessary assistance from hydrogen.

For conservative estimates of the service life of a hydrogen vessel the
first step is the decisive one. If the conditions for crack initiation are
met failure will inevitably occur. The contributions of the three other
steps to the total service life have to be neglected because of the
following reasons:

According to the currently known results of hydrogen fatigue crack growth
experiments it is safe to assume that hydrogen accelerates the normal
growth rates by at least a factor of 50, strongly depending on the frequency of loading and on the hydrogen pressure (6).

As an example the estimated number of cycles necessary to grow an assumed
0.5 mm deep, semielliptical surface crack to a depth of 2 mm would require
roughly 15,000 cycles in an inert environment, but only 300 cycles in the
presence of compressed hydrogen.

Such a short period of crack extention is insignificant compared to the
total projected service life and the inspection intervals of an ordinary
hydrogen vessel.

The same reasoning holds for any sustained hydrogen crack growth.

The individual steps in the course of the development of a vessel failure
are always accompanied by respective changes in mechanism which finds its
reflection in specific fracture modes.

By a thorough fractographic inspection of the fracture surfaces produced
with the fatigue tested cylinders it was possible to distinguish between
succeeding areas each having characteristic fracture appearance.

Fig. 8 gives an example of a typical fracture surface and also shows some
microscopical details which are specific for the respective fracture zones.

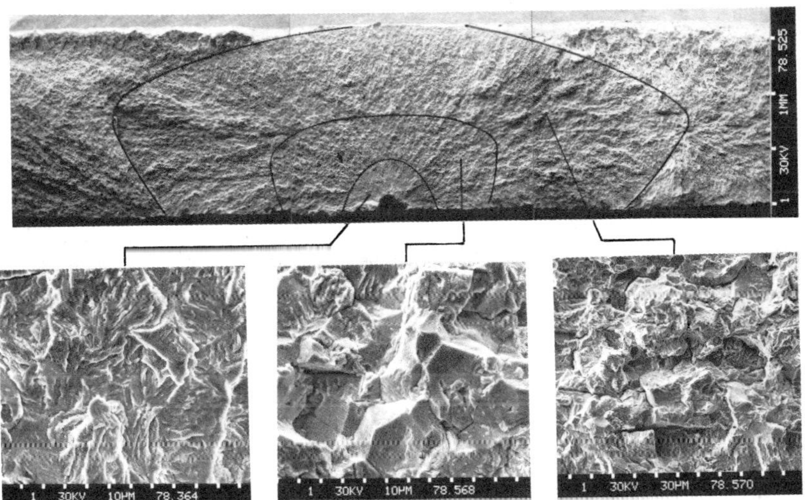

Fig. 8: Fracture morphology of the fatigue-tested cylinders

Immediately adjacent to the starter defect there is a smooth area where the fracture path is completely transgranular. The shape of this zone is clearly identifiable because of a rather abrupt change in fracture morphology to an area where the fracture is largely intergranular. Both areas correspond to hydrogen-assisted fatigue crack growth where the stress intensity is still below the threshold fracture toughness, K_{IH}. In all test specimens the change is observed to take place at a rather constant stress intensity value of about 23 MPa $\sqrt{m}$.

These findings are in agreement with recent results published by Ritchie and co-workers (7). During hydrogen-assisted crack growth experiments with quenched and tempered chrome-moly steels they observed a sudden crack acceleration above a certain constant maximum stress intensity K^T_{max} which also was accompanied by a respective change from a transgranular to a predominately intergranular fracture mode.

The expected shift from the hydrogen-assisted fatigue crack growth to hydrogen sustained crack growth, generally, is not easy to detect. It is not always accompanied by a sudden change in fracture mode. For this third fracture zone, however, crack branching and secondary cracking is a characteristic feature. The crack path may be either inter- or transgranular. Very often the crack front tends to follow non-metallic inclusions.

All these morphological features are identical to what is found when actual service failures are evaluated. This again proves that the failure of hydrogen vessels, in fact, is a fatigue problem. More than in inert environments the dominant process is crack initiation at critical defects. But as has been shown through the fatigue experiments, hydrogen vessels can be operated safely if the state of their internal surface is closely controlled.

References

1. A. Martens, "Untersuchung eiserner Behälter zur Aufbewahrung von Wasserstoffgas", _Zeitschrift VDI, 40_ (26) (1896) pp 717 - 723

2. A. W. Thompson and I. M. Bernstein, "The Role of Metallurgical Variables in Hydrogen-Assisted Environmental Fracture", _Advances in Corrosion Science and Technology_, Vol. 7, pp. 53-175, Plenum, New York, 1980.

3. G. M. Pressouyre, C. Zmudzinski and L. Bretin, "Optimization of Steels for Use in Hydrogen Environment" presented at CEC-Seminar "Hydrogen as an Energy Vector", Brussels, Feb. 1980.

4. M. Kesten, "Die Entstehung von Sprödbrüchen bei Wasserstoff-Druckgasbehältern und Maßnahmen zu deren Verhinderung", _Gas Aktuell, 18_ (1979) pp 9 - 15.

5. A. W. Loginow and E. H. Phelps, "Steels for Seamless Hydrogen Pressure Vessels", _Trans. ASME, J. Eng. Ind._ (1975) pp 274 - 282.

6. Private Communication, P. McIntyre and D. Harris, British Steel Corp., Swinden Labs.

7. S. Suresh, C. M. Moss and R. O. Ritchie, "Hydrogen-Assisted Fatigue Crack Growth in 2 1/4 Cr-1 Mo Low Strength Steels" pp 481 - 484, presented at 2. JIM Symposium, "Hydrogen in Metals", Minakami, Japan, Nov.1979

DISCUSSION

<u>M.R. Louthan,Jr., Virginia Polytechnic Institute, Blacksburg, VA 24061</u>: Our experience has shown that defects which are far too small to be controlled in commercial practice are sufficient to serve as crack initiation sites. How do you propose to prevent (or limit) hydrogen enhanced low cycle fatigue when such small defects are effective initiators?

<u>M. Kesten</u>: The results of our fatigue experiments fully support your experience since they also show that surface defects which, from the practical point of view, one would consider to be very minor and difficult to detect, nevertheless, may considerably reduce the fatigue life of a hydrogen vessel. One obvious consequence would be to limit the service life of actual transport vessels based on experimentally established data. According to our test results, however, such a measure would have no practical consequences for the average transport vessel. If the vessels are free of those surface defects which can be properly detected by established inspection standards the vessels can be safely operated for many thousand fillings.

<u>P. Azou, Ecole Centrale des Arts et Manufactures, Chatenay-Malabry, France</u>: Have you tried to realize a mechanical treatment of the inner wall to prevent hydrogen damage?

<u>M. Kesten</u>: Shot peening of the internal surface of a vessel is generally considered to be beneficial because it creates compressive stresses in a surface layer of the vessel wall. In fact, shot peening is applied by most vessel manufacturers as a means of cleaning the inner surface to make visual inspection more meaningful. In our experiments we were not able to study systematically the effect of such a treatment because of the long duration of the tests..

<u>L. Christodoulou, Carnegie-Mellon University, Pittsburgh, PA</u>: What was the purity level of hydrogen used in your experiments and what do you think is the influence of impurities such as water vapor on the failure you observe?

<u>M. Kesten</u>: A basic objective of our tests was to adapt the experimental conditions as close as possible to the actual service situation. Therefore, hydrogen of the most common grade was used coming directly from a commercial purification unit. The maximum concentration of the main impurities was 5 vpm O_2, 10 vpm H_2O, and 100 vpm N_2. According to recent investigations water vapor is thought to act as an inhibitor in a similar manner as oxygen does. A certain water content in the hydrogen gas, therefore, may be of advantage as long as there is no precipitation of liquid water on the vessel wall.

<u>C. Altstetter, University of Illinois, Urbana, IL</u>: Is there a beneficial effect of overpressurizing to create favorable surface stresses?

<u>M. Kesten</u>: It is common practice that every pressure vessel is subjected to a hydraulic test where the test pressure exceeds the normal service pressure by 50%. In addition to this an even further overpressurization was never taken into consideration so far. In fact, it seems doubtful whether such a treatment could be carried out with the necessary control it would require. Since it is the stress concentration at the internal surface defect which we are concerned with one would have to be sure just to create favourable surface stresses at these locations and not to generate new cracks or extend already existing crack-like defects.

EFFECT OF HYDROGEN ON THE BEHAVIOR OF

EUROPEAN STEELS USED FOR HYDROGEN PRESSURE VESSELS

H.Barthélémy-J. Bryselbout -G. Galéa-C. Barbe

L'AIR LIQUIDE - C.R.C.D. 78350 Jouy en Josas

France

We have investigated the behavior of low alloyed steels, with chromiun
and molybdenum, with manganese and with chromiun, nickel and mobybdenum.
These are quenched and tempered steels, with tensile strength between 700
and 1000 MPa.

This investigation has been carried out in using as test samples
clamped disks deformed under hydrogen pressure according to the test introduced by FIDELLE. This test has been adapted to take service conditions of
vessels of our Company into account. We have exercised particular care to
study the influence of oxygen and water vapor pollution of hydrogen and the
influence of the deformation velocity. Different embrittlement indexes have
been defined to compare the behavior of steels in the presence of different
hydrogen qualities. The accuracy obtained enables us to distinguish steels
having closely related behaviors : an increase of tempering temperature of
10° Celsius involves an appreciable decrease of embrittlement index. The
inhibitory influence of oxygen and water vapor is also discussed, this inh-
ibition much depends on the velocity of pressure increase.

Finally a new fatigue apparatus which reproduces actual conditions of
vessels fillings has been developped. This kind of test is very important
for vessels submitted to very numerous fillings during their life, sometimes
one per day. First results are commented, clearly showing the detrimental
effect of repetitions of hydrogen fillings.

I. INTRODUCTION

The European Companies distributing hydrogen generally use steel pressure vessels of capacity from 10 to 250 Nm3 of gas at 200 bar.

In recent years some failures of such vessels have been reported, that are due to hydrogen embrittlement of steels.

It is now well known that factors which can influence the failure of these vessels in the presence of hydrogen are those depending on, steel quality (chemical composition, mechanical properties, heat treatment, inclusions). service conditions (purity of hydrogen, frequency of vessels filling) and internal stress conditions (due to design, surface roughness, quenching defects, corrosion).

This paper presents some results of different tests, the aim of which is to investigate the effect of steel quality and service conditions of hydrogen pressure vessels.

II. EXPERIMENTAL PROCEDURE

Tested steels are low alloyed steels which are the most commonly used for hydrogen pressure vessels in Europe, three chromium-molybdenum steels A, B and C and one manganese steel D. As a reference, we have also introduced a low alloyed steel with chromium nickel and molybdenum E, widely used in the presence of gaseous hydrogen. Chemical compositions of these steels (table I), heat treatments and mechanical tensile values (table II) are given.

TABLE I - chemical compositions of steels

Steels %	C	Si	Mn	P	S	Cr	Mo	Ni	Al	Ti
A	0,35	0,27	0,90	0,026	0,016	1,00	0,29	0,08	0,013	0,02
B	0,30	0,33	0,57	0,015	0,01	2,50	0,49	0,25	0,03	0,00
C	0,33	0,25	0,73	0,024	0,014	1,06	0,16	–	0,005	–
D	0,25	0,22	1,5	0,014	0,015	0,17	0,12	0,16	0,001	–
E	0,22	0,27	0,71	–	–	2,73	0,35	0,79	–	–

Rupture test of clamped disks introduced by FIDELLE (1,2) have seemed to us to be the most adapted to carry out this kind of investigation.

First tests (3,4,5) have confirmed that hydrogen effects is mainly important during plastic deformation of steel and that the adjunction of a constant loading time (no deformation) seems to have no sensible influence. So for these investigations we have chosen only testing methods which led to variable deformation of disks from the beginning of the test to the rupture.

These two testing methods are "dynamic rupture test" and "fatigue test"

TABLE II - mechanical tensile values and heat treatments of steels

Steels	Tensile Strength (MPa)	Yield Stress (MPa)	Elongation %	Austenisation	Quenching	Tempering
A_1	935	805	21,2	875°C-30 mn	oil	675°C-30 mn
A_2	963	852	19,4	875°C-30 mn	oil	665°C-30 mn
B_1	900	728	18,4	975°C-30 mn	air	650°C-30 mn
B_2	981	843	16,3	975°C-30 mn	air	615°C-30 mn
C_1	937	832	18,3	850°C-25 mn	oil	620°C-50 mn
C_2	976	882	18,6	850°C-25 mn	oil	600°C-50 mn
D	918	902	18,5	850°C-30 mn	water	540°-30 mn
E	690	530	28,2	900°C-300 mn	oil	675°-300 mn

- Dynamic rupture test : In this test disks specimens are loaded by monotonic pressure increase of gas at constant velocity from the atmospheric pressure to the rupture pressure.

We have already discribed in detail the test cell and the testing methods (3), we only point out that tests are carried out with hydrogen of different purities and helium as inert reference, and higher ratio $\dfrac{PH_e^{*}}{PH_2}$ is, more important the hydrogen embrittlement of the concerned material is. To compare different steels, the chosen criterion is the embrittlement index "I" equal to maximum value of the ratio $\dfrac{PH_e}{PH_2}$ as a function of the pressure increase rate.

- Fatigue test : Dynamic rupture test is a very useful test which allows to compare behaviors of different materials in the presence of different hydrogen qualities in defining an embrittlement index. Nevertheless this test conducted with monotonic loading does not take into account micro-damages produced by cyclic stresses due to the very numerous fillings of pressure vessels, sometimes one per day. This is the reason why we developped a new fatigue apparatus which reproduces actual conditions of vessels fillings.

For this apparatus (fig. 1 and 2), we used the same kind of cell with disk specimen. Two electromagnetic valves regulated by two different timers command the inlet and the outlet of gas. A pressure gauge located in the upper part of the cell detects the rupture of the disk, stops the cyclemeter and the whole apparatus. An example of a pressure cycle obtained is given (fig. 3).

* PH_e and PH_2 : rupture pressures caused by pure helium and hydrogen of different purity at given pressure increase rate.

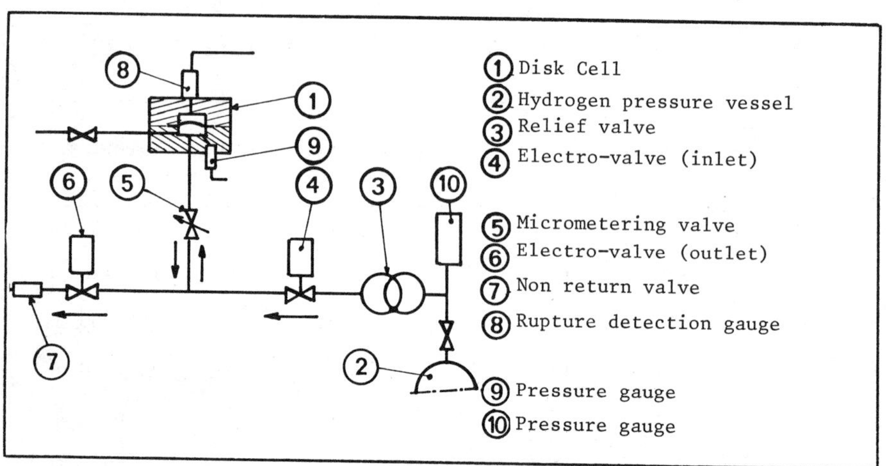

Fig. 1 Principe of the Fatigue Apparatus

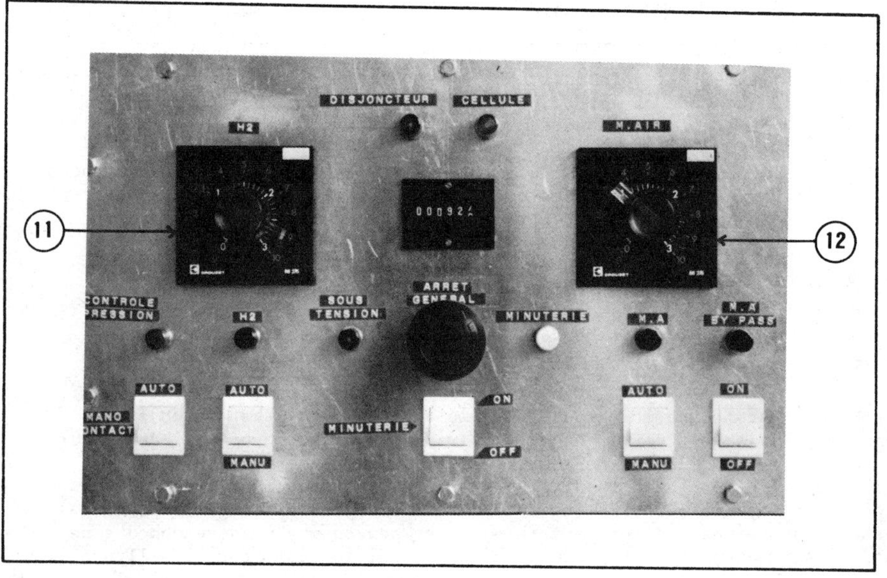

Fig. 2 Regulating system of the Fatigue Apparatus

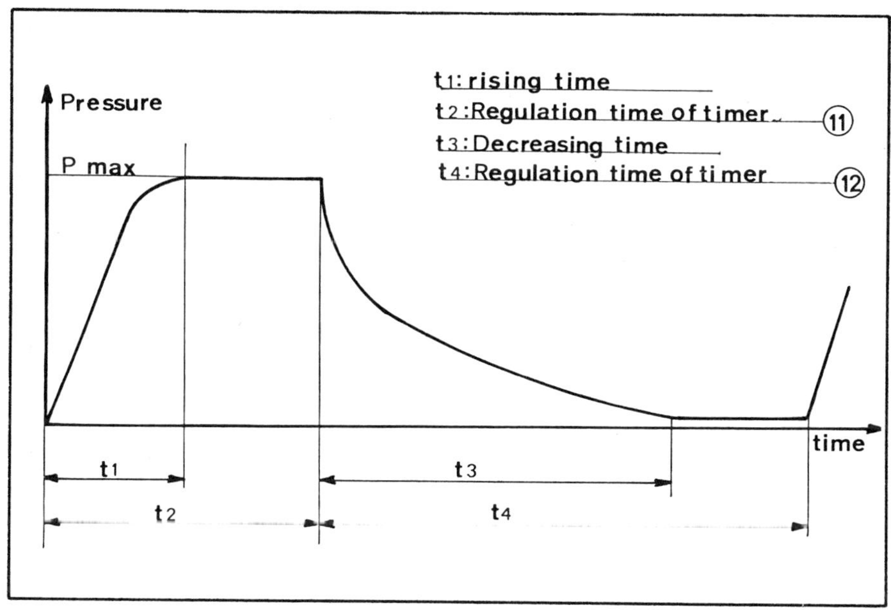

Fig. 3 Fatigue Apparatus - Cycle of pressure
obtained

III RESULTS AND DISCUSSION

A. Dynamic rupture test

This kind of test allowed us to bring out the influence of different
factors.

1) <u>Pressure increase rate</u> : We have reported (fig. 4) the values of ratio
$\frac{PH_e}{PH_2}$ obtained with each steel with pure hydrogen as a function of pressure
increase rate ; it appears that maximum value of this ratio (embrittlement
index) is obtained for a pressure increase rate less than 1 bar/mn except
for "steel "A_1".

2) <u>Heat treatment</u> : We can see that for the three steels A,B,C, hydrogen
embrittlement decreases when tempering temperature increases (table II and
III), an increase of tempering temperature of 10° Celsius involves an appreciable decrease of embrittlement index. On the other hand, an increase of
tempering temperature leads to a decrease of tensile strength (table II).

3) <u>Hydrogen impurity</u> : For all tested steels I_2 is higher than I_1 and I_3 is
higher than I_2 (table III). This show that the adjuction of oxygen or water
vapor in hydrogen involves a decrease of embrittlement. Nevertheless this
inhibitory effect is not very important for a pressure increase rate less
than 1 bar/mn : difference between I_2 et I_1 is only about 5%.

4) Tensile strength : Hydrogen embrittlement of a steel is not directly
connectable to its tensile strength. Some different steels can have the
same tensile strength and different embrittlement index and reciprocally.

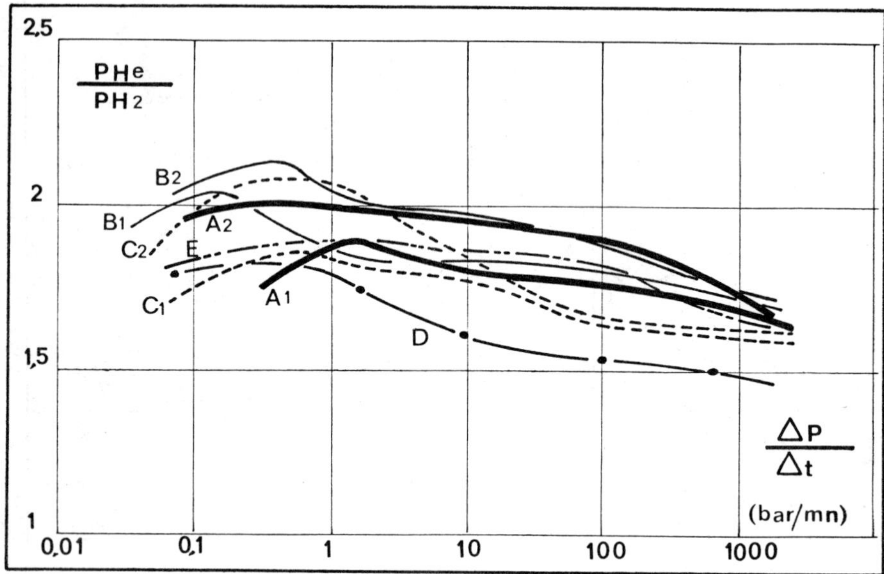

Fig. 4 Embrittlement of steels in pure hydrogen

TABLE III - embrittlement index

Index : purity	steels	A_1	A_2	B_1	B_2	C_1	C_2	D	E
I_1 : O_2 et $H_2O <$ 1 ppmV		1,90	2,03	2,03	2,13	1,85	2,10	1,80	1,90
I_2 : O_2 = 120 ppmV		1,85	1,95	1,83	2,06	1,73	1,92	1,70	1,70
I_3 : O_2 = 140 ppmV		-	-	-	-	-	1,85	-	-
H_2O = 80 ppmV									

B. Fatigue test

At the present time these tests have been only carried out with two different heat treatments C_1 and C_2 of the same steel grade C, in using hydrogen of high purity, industrial hydrogen and nitrogen as inert reference. Pressure cycles were chosen in such a way that : t1 = t2 = 4,5 s and t3 = t4 = 10 s (fig. 3).

Reported results of the maximum pressure of cycles as a function of the number of cycles leading to rupture (fig. 5) show that :

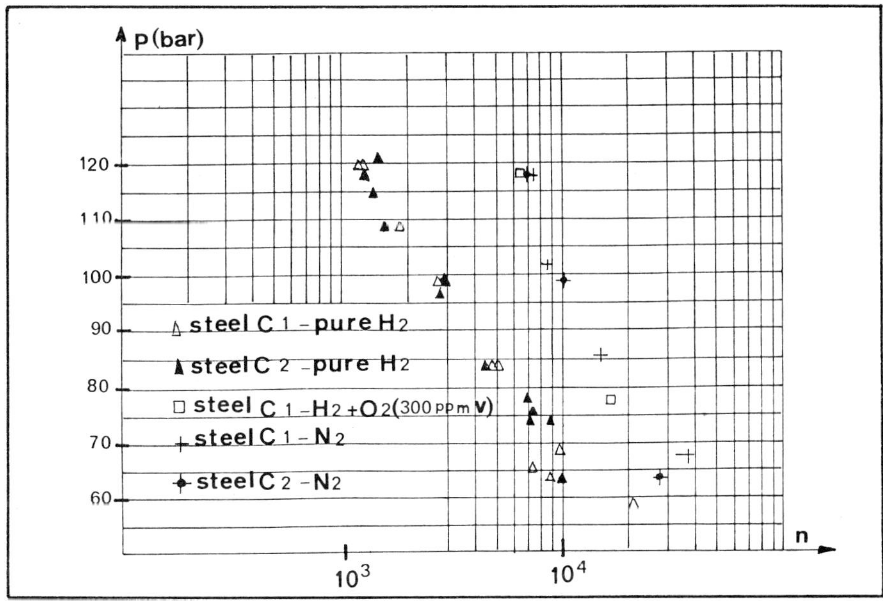

Fig. 5 Fatigue tests results

- it is possible to obtain by this technique results with very low dispersion. This is probably due to the fact that the surface quality of disks is
good enough and that defect introduced to produce plastic area, is not due
to mechanical notch but, it is the consequence of the clamping which is always the same for each cell ;

- maximum pressure value of cycles with pure hydrogen (PH_2) can be correlated to the number of cycles leading to rupture with hydrogen (nH_2) according to :

$$P_{H2} = A \log nH_2 + B \qquad (1)$$

- the value of B obtained by extrapolation (n = 1) is practically equal to
the rupture pressure value obtained, in the same range of pressure increase
rate by dynamic rupture test ;

- the number of cycles obtained at rupture with nitrogen (nN_2) is higher
than the one obtained with pure hydrogen. Ratio $\dfrac{nN_2}{nH_2}$ for a fixed type of
cycle, gives a practical idea of the detrimental effect of repetitions of
hydrogen fillings ; we obtained :

$$3 < \frac{nN_2}{nN_2} < 6 \qquad (2)$$

- the number of cycles obtained at rupture with industrial hydrogen containing about 300 ppm volume of oxygen ($nH_2 + O_2$) is practically equal to the
one obtained with nitrogen. Industrial hydrogen of very poor quality does
not seem to have any embrittlement effect during test carried on with <u>high</u>

<u>pressure increase rate</u> selected :

- it is not possible to differenciate the hydrogen behavior of both heat treatment chosen for steel C. This is in agreement with results obtained by dynamic rupture test (fig. 4) ; for chosen pressure increase rate (between 800 and 1600 bar/mn) it is also not possible to differenciate both heat treatments. Our next fatigue tests will be performed with lower pressure increase rate, and will certainly lead to a more important detrimental effect of hydrogen.

CONCLUSIONS

1. Rupture dynamic tests and fatigue tests are very well adapted to investigate hydrogen embrittlement of pressure vessels steels and particulary to bring out the influence of factors which affect steel quality and hydrogen purity. Rupture dynamic tests enable to define an embritllement index and then to classify the behaviors of different material in the presence of hydrogen of different qualities, fatigue tests allow to see the eventual additional effect of repetitive micro-damages and with the ratio $\frac{nN_2}{nH_2}$ to give a concrete idea of detrimental effect of hydrogen on the life of an hydrogen pressure vessel.

2. Results confirm the influence of different factors, benefic effect of high tempering temperature, inhibiting effect of oxygen and water vapor as impurity into hydrogen. But they mainly show the important influence of deformation velocity of steels. Plastic deformation is not only necessary to have a detrimental effect of hydrogen, but it seems also that steels are more sensible when the deformation rate corresponds to pressure increase rate of dynamic rupture test lower than 1 bar/mn. Some steels can have the same hydrogen behavior when they are deformed at high rate and different behaviors at low rate. This results must be confirmed by fatigue test at low deformation rate.

ACKNOWLEDGMENT

We wish to thank the Directorate General for Research, Science and Education of the Commission of the European Communities to have supported a part of this research.

REFERENCES

(1) J.P. Fidelle, L.R. Allemand, C. Roux, M. Rapin : "Hydrogen in Metals" Valduc, Fr. Ed. by J.P. Fidelle, M. Rapin, Publ. CEA, 91680 Bruyères le Chatel Fr. (1969) pp 137-172.

(2) J.P. Fidelle ASTM STP 543 (1974) pp 34-50 and pp 221-253

(3) H. Barthélémy, A.M. Brass, J. Bryselbout, J. Koenig : "Hydrogen as an Energy Vector". Brussels, Belgium - 12-14/02/80, Ed. D. Nicolay CCE - DG XIII - A/3 - Bat. J. Monnet B4-072 Lux.(1980).

(4) H. Barthélémy, R. Guenot, M. Aucouturier, A.M. Brass - IIIe Congrès National sur la Technologie des Appareils à Pression Paris, Fr. (oct. 1980) to be published.

(5) R. Clermont, M. Aucouturier, A.M. Brass, J.Y. Faudou, R. Guenot : "Hydrogen in metals" Chatenay-Malabry, Fr. Publ. by Pergamon Press, Paris (1977).

DISCUSSION

A.W. Thompson, Carnegie-Mellon University, Pittsburgh, PA: Let me ask both you and Dr. Kesten: American distributors of hydrogen gas (in both trucks and laboratory cylinders) claim not to have any service failures which can be attributed to hydrogen. Can you comment on why this might be, provided you think their reports are accurate?

H. Barthelemy and M. Kesten: First of all, we do not question the reports of American distributors of hydrogen gas. As regards the European situation, it is necessary to remember that there are different regulations, different gas distribution techniques and different pressure vessel manufacturers in Europe. Therefore I shall restrict my statement to our experience. In France we have added to the national regulations our own specifications regarding the mechanical properties of steels for hydrogen vessels. This is probably the reason why we have had no service failure during the last 15 years. Nevertheless if we try to find some technical reasons which could explain that the hydrogen embrittlement of hydrogen transport vessels might be less important in U.S., it is possible to notice that the maximum values for hoop stress, filling pressure, U.T.S. of steels and the average frequency of fillings are lower in the U.S. than in Europe, but very often these differences between all these factors of influence in U.S. and in Europe are not very significant.

B.C. Odegard, Sandia National Labs, Livermore, CA: Observations on the inner surface of a hydrogen pressure vessel made in the U.S. showed a heavy oxide film and cracks evident through the film. This vessel had been in the laboratory for an unknown period of time in the empty state.

H. Barthelemy: Maybe this was due to an admission of water vapor during the empty state and to the condensation of free water during the following pressure increase (next gas filling).

EFFECT OF THERMO-MECHANICAL TREATMENT ON FRACTURE

TOUGHNESS OF A HYDRIDE FORMING MATERIAL*

A. Arora
Rockwell International Science Center
Thousand Oaks, CA 91360

Zirconium and its alloys are candidate materials for many nuclear components. The presence of hydrogen (in excess of terminal solid solubility) in these alloys leads to the loss in ductility and causes component failures. In Zr-2.5% Nb, the available hydrogen precipitates in the regions of high stress concentration and gives rise to local embrittlement. In this study, a thermo-mechanical treatment has been employed to trap the diffusing hydrogen. An advantage of this treatment is that it can be applied to in-service components. Results to date are beneficial. Slow crack growth (SCG) tests on specimens show that the crack growth rates are reduced by a factor of 10 by this treatment. The duration of the thermo-mechanical treatment (up to 60 hours) is also found to have strong effects on the crack growth rates. SEM studies performed on specimens after SCG tests, indicate increased amount of cracking in secondary planes. Snoek ordering of hydrogen-interstitials is suggested to be responsible for improved fracture toughness.

*This work was performed at Metallurgical Sciences Laboratories, Faculty of Engineering, University of Manitoba, Winnipeg, Canada.

Introduction

Zr-2.5% Nb is a high strength material commercially used for pressure reactor tubes and reactor core components. Presence of hydrogen in this alloy has been observed to result in considerable loss in ductility due to the formation of hydrides under elevated stress and temperature conditions (1-5). Despite a large number of studies in this area, very little is understood about the processes of transporting hydrogen to the elevated stress sites, trapping of hydrogen at impurities, hydrogen accumulation, phase transformation of α-Zr to zirconium hydride and the fracture of these hydrides under high applied stress. However, observations of the existence of brittle hydride phase in Zr-2.5% Nb have been made (4) and sufficient data exists on the fracture toughness of as-received and hydrided Zr-2.5% Nb (6, 7). In this study, attempts were made to improve the fracture resistance of this material by employing a thermo-mechanical treatment (TMT), this paper reports the results of that study.

Experimental Procedure

The specimens tested here were cut from the pressure tubes in a circumferential-longitudinal (C-L) direction.* Details on specimen geometry and dimensions are given in Ref. 8 along with a description of the experimental test set-up. In brief, the specimens (compact tension type) contained a notch and a sharp fatigue pre-crack to serve as a crack starter. Specimens were brought to the test temperature (i.e., the temperature at which cracking is observed to occur) and were loaded under dead weight to obtain a desired crack tip stress intensity. The incremental crack growth was measured continuously during the test by using acoustic emission techniques (8-10).

A thermo-mechanical treatment was employed on the specimens prior to the test. One advantage of this treatment is that it could be employed in situ, that is, a specimen can be loaded to a stress level below the threshold stress intensity, K_{th}, and aged for a desired length of time at temperatures at which the static strain aging phenomenon is known to occur in that material. In the tests reported here, a specimen, typically, was loaded to a crack tip stress intensity of 2.5 MNm$^{-3/2}$ (i.e., approximately 50% of the reported threshold stress intensity) and then aged at 100°C for desired length of time, up to 60 hours, to determine the fracture kinetics dependence on the duration of the treatment. Subsequently, the slow crack growth (SCG) test was resumed on the specimen merely by increasing the load to obtain the test stress intensity and adjusting the temperature to the test temperature, which in our tests remained unchanged. The test stress intensity was selected to be 19 MNm$^{-3/2}$ and belonged to stage II of the velocity-stress intensity curve for this material, at which the crack velocity is only weakly dependent on the stress intensity (3, 4).

Crack growth data were obtained from thermo-mechanically treated as well as untreated specimens. In order to observe the effect of hydrogen content on TMT, the tests were performed on both as-received as well as hydrided specimens. The hydrided specimens contained approximately 200 ppm hydrogen as compared to as-received specimen which contained

---

*The specimens were provided by Atomic Energy of Canada, Ltd., WNRE, Pinawa, Canada.

10 ppm hydrogen. All the specimens were vacuum annealed at 400°C for
2 hours to relieve any residual stresses incurred during cold working and
fatigue precracking. In addition, hydrided specimens were given a homogenization treatment following hydriding in order to evenly distribute the
hydrogen in the specimen.

On termination of the test, the specimens were fast fractured and
the fracture surface was prepared for SEM (scanning electron microscope)
observation. The surface was cleaned ultrasonically, dried and coated
with gold (by deposition technique) with 30 deg shadow angle. The fracture surface was then observed in SEM at 60 deg to the incident electron
beam in order to determine relevant features of the surface which would
indicate the fracture mechanism involved.

Results

Beneficial results were obtained from the tests conducted in this
study. The resistance to fracture increased considerably with the application of the TMT. An example of this is shown in Fig. 1, where the
incremental crack growth is plotted with time on a log-log scale for an
as-received specimen with and without the treatment. A decrease in crack
growth rates from 2.9×10^{-9} to 7.3×10^{-10} m/s occurred when the specimen was subjected to the TMT for 24 hours. In addition, the time to
incubate the cracking was observed to decrease with this treatment.
Cracking due to pre-existing hydrides also decreased considerably during
the initial stages of loading (9).

The tests on hydrided specimens showed similar effects of TMT. A
consistent increase in fracture toughness occurred with increasing duration of the treatment. Results of SCG tests on hydrided specimens are
shown in Fig. 2. The TMT and the test conditions are given in the
figure. Cracking in the specimen with 60 hours of treatment was observed
to vertually cease after 220 hours with no further crack growth until
470 hours. In another test, where the crack growth was observed to cease
after 130 hours in an as-received specimen with 40-hour treatment, the
crack tip stress intensity was increased in steps to further resume
cracking. An incremental rise in the crack tip stress intensity from
19 $MNm^{-3/2}$ to 20 and 21 $MNm^{-3/2}$, respectively, did not induce any cracking for the periods of 24 hours. A further increase in K_I to 22 $MNm^{-3/2}$,
however, resulted in subsequent crack growth; this suggests a definite
increase in fracture toughness of the treated Zr-2.5% Nb.

Results on the crack growth rates in specimens with various TMT
durations are reported in Fig. 3. A constant decrease in crack growth
rate with increasing treatment duration is evident from this figure. The
accuracy in the measurement of crack velocity for specimens with longer
aging times (more than 50 hours) is limited by the sensitivity of the
technique; however, results clearly indicate that a better resistance to
fracture can be obtained with longer durations of the TMT. From the
obtained crack growth rates, the time to failure was estimated and is
plotted in Fig. 4 with the duration of treatment for hydrided specimens.
Since the specimen with 60 hours duration did not show significant crack
growth between 220 to 470 hours, it may be estimated not to have failed
in 10^6 s. Results on time to failure in untreated specimens were in
agreement with the observations made on delayed hydrogen cracking in
round notched bars cut from Zr-2.5% Nb pressure tubes (11). The data
from this study are summarized in Table I.

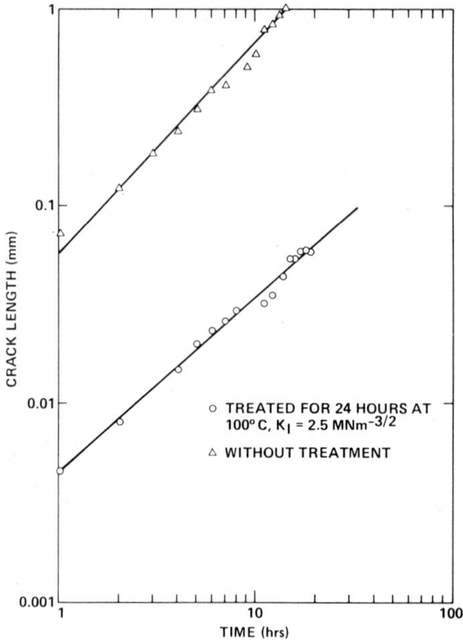

Fig. 1 - Effect of thermo-mechanical treatment on slow crack growth. Test temperautre = 100°C; K_I = 19 MNm$^{-3/2}$; hydrogen content = 10 ppm.

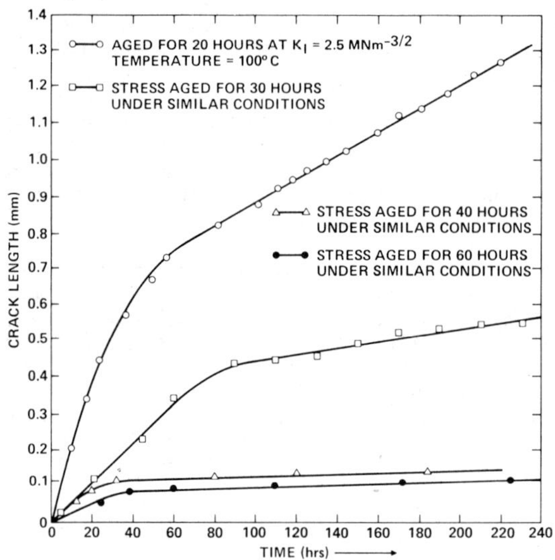

Fig. 2 - Effect of TMT duration on slow crack growth. Test Temperature = 100°C; K_I = 19 MNm$^{-3/2}$; hydrogen content = 200 ppm.

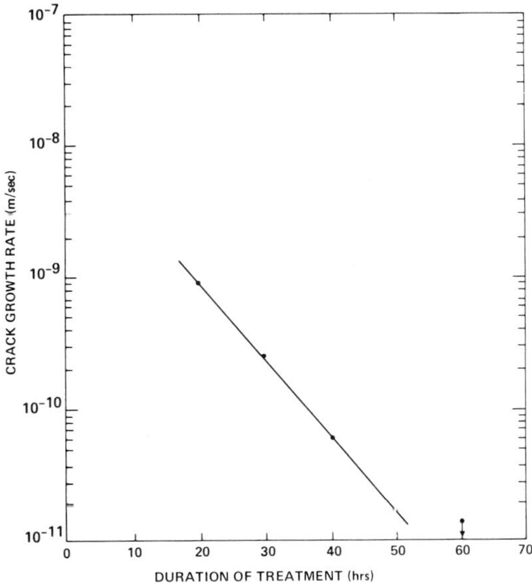

Fig. 3 - Effect of TMT duration on crack growth kinetics.

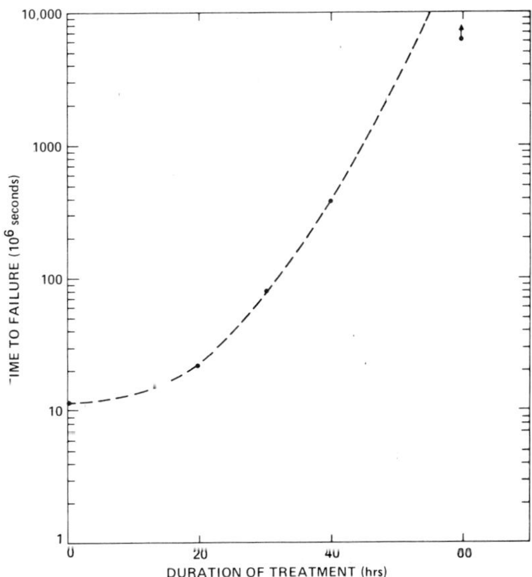

Fig. 4 - Time to failure vs TMT duration for specimens under static loads, K_I Initial = 20 MNm$^{-3/2}$

Table I. Summary of Results

Duration of Treatment	Hydrogen Content	Crack Growth Rates		Reduction Ratio
		Without Treatment	With Treatment	
hrs	ppm	m/sec	m/sec	
24	10	2.9×10^{-9}	7.3×10^{-10}	0.25
40	10	2.1×10^{-9}	2.6×10^{-10}	0.12
20	200	2.0×10^{-9}	9.0×10^{-10}	0.45
30	200	2.0×10^{-9}	4.0×10^{-10}	0.20
40	200	2.0×10^{-9}	6.0×10^{-11}	0.03
60	200	2.0×10^{-9}	$\sim 10^{-11}$	<0.01

Fracture surface of the specimens tested in this study were observed in SEM. Figure 5 shows the fractographs obtained from two specimens, one without the treatment, (a), and the other with 60 hours of TMT, (b) and (c). The fracture surface in Fig. 5a shows the slow crack growth region with features of brittle fracture indicating delayed hydride cracking, and is similar to the previous observations (4, 5). A limited amount of secondary cracking is seen (indicated by arrow). Figures 5b and c show the features of the fast fracture region of specimen with 60 hours treatment. A large number of secondary cracks are clearly evident, even at the low magnification (Fig. 5b). Figure 5b suggests that the large sized cracks are induced by fracturing the hydride particles in C-R (circumferential-radial) orientation, i.e., the particles with their length along the axis of the pressure tube and the width in parallel with the loading axis. Figure 5c shows the section view of the fractured hydride as a deep crack on the secondary plane.

Discussion and Conclusion

These results indicate that application of this thermo-mechanical treatment yields high resistance to crack growth on planes normal to the loading axis by hydride coarsening on planes parallel to the loading axis. In view of the existing theories on the mechanisms of strengthening, the following possibilities may be considered to provide high strength and toughness to Zr-2.5% Nb: 1) stress reorientation of hydride particles (12), 2) strain hardening due to cell substructure formation (13, 14), 3) hydrogen trapping and particle coarsening due to impurity effects (15, 16), and 4) static strain aging to induce Snoek ordering (17, 18). The data, to date, are insufficient to determine conclusively the prime mechanism or mechanisms involved in the process of strengthening, but on the basis of existing data it is possible to evaluate the role of each mechanism. Stress reorientation of hydride particles can be disregarded because 1) the stress levels involved during TMT are too low ($\sim$0.5 K_{th}), and 2) only the hydrides taken into solution can become reoriented. Formation of a dense cellular substructure is also unlikely and has previously been ruled out, based on yield drop (18), as a necessary prerequisite for strain aging in Zr-Nb systems; this also does not conform with the observation of large sized hydride particles on the secondary planes. Impurity effects on hydrogen embrittlement behavior (resulting from TMT) may be considered to provide improved fracture properties. Trapping of hydrogen at pre-existing hydrides may occur

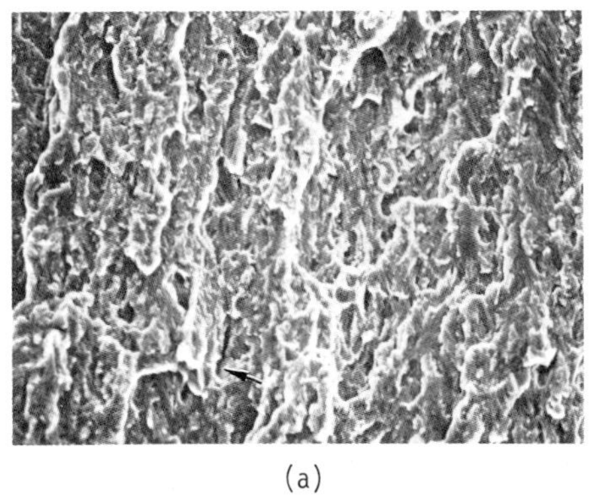

(a)

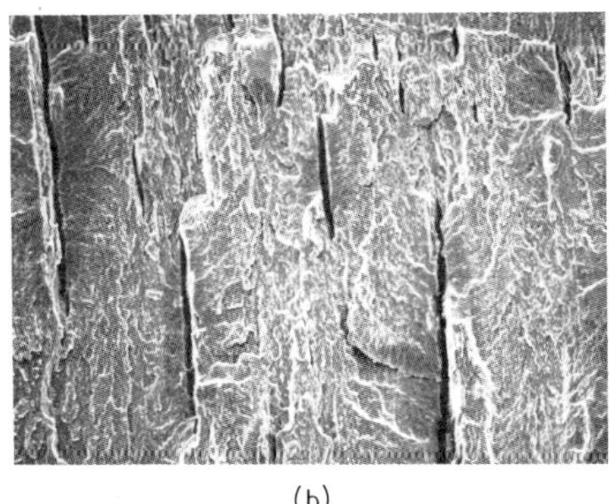

(b)

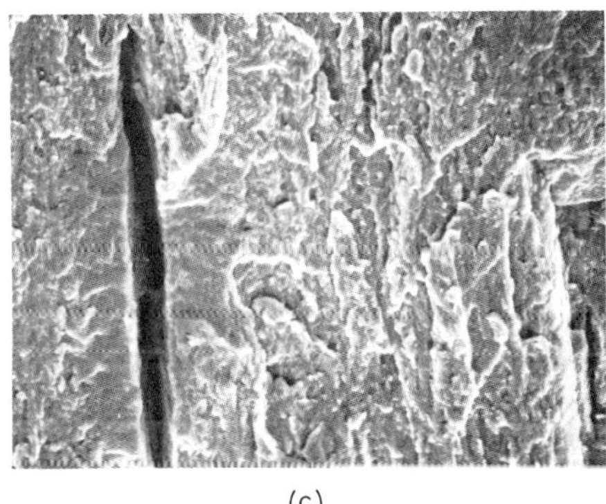

(c)

Fig. 5 - SEM fractographs of specimen. (a) x672 without treatment, (b) x164, and (c) x574 with treatment, 60 hours. Direction of crack propagation is from top to bottom.

followed by particle coarsening; however, an increased amount of activation energy would be required to coarsen the particles (19) to the
observed size and, thus, is less probable in our case. Strain aging
characteristics of Zr-1% Nb and Zr-2.5% Nb have been invesigated in
detail (17, 18), and the role of the interstitials, e.g., hydrogen and
oxygen, in strain aging has been determined. Thorpe and Smith (18) have
shown the existence of major peaks, at 350 and 573K, in static strain
aging response of Zr-1% Nb with temperature. They argue that while the
573K peak (with few other minor peaks) is due to oxygen interstitials,
the increase in stress, $\Delta\sigma$, in strain aging response at 350K can be
attributed to interstitials hydrogen which diffuses more rapidly than
oxygen. An interstitial-interstitial (i-i) type of pair formation due to
Snoek ordering may be considered to result from the static strain aging
phenomenon. The fractographic observations in this study seem to agree
with this hypothesis which will result in hydrogen accumulation due to
H-H pair formation and hydride precipitation to form large size hydride
particles. Another consideration in Snoek ordering is that of the axis
of alignment of i-i pairs. As expected, the H-H pairs will align themselves with the loading axis to provide strengthening of the material
and, consequently, the growth of hydrides on secondary planes will occur;
this is in agreement with our observations. Further work is required to
determine conclusively the mechanism involved in strengthening of this
material.

In conclusion, the thermo-mechanical treatment employed in this
study on as-received and hydrided specimens has clearly demonstrated an
increase in fracture resistance of Zr-2.5% Nb to hydrogen induced delayed
cracking. The increase in the fracture toughness is observed to be
directly dependent on the duration of the treatment. Static strain aging
which involves Snoek ordering of hydrogen interstitials is suggested to
be responsible for the improved strength of this material.

References

1. C. J. Simpson and J. Moerman: *Effect of Hydrogen on Behavior of
    Materials* (A. W. Thompson and I. M. Bernstein, eds) pp. 428-39,
    AIME, New York, 1976.
2. L. A. Simpson and M. P. Puls: *Met. Trans.*, 1979, Vol. 10A, pp. 1093-
    105..
3. L. A. Simpson and K. Nuttal: *ASTM STP 633*, pp. 608-29, 1977.
4. R. Dutton, K. Nuttal, M. P. Puls and L. A. Simpson, *Met. Trans.*,
    1977, Vol. 8A, pp. 1553-62.
5. K. Nuttal, D. P. McCooeye, A. J. Rogowski and F. J. Havelock: *Scr.
    Met.*, 1976, vol. 10, pp. 979-82.
6. L. A. Simpson: *Proc. ICM-3*, Cambridge, Aug. 1979.
7. L. A. Simpson and C. D. Cann: *J. Nucl. Mater.*, 1979, Vol. 87,
    pp. 303-16.
8. A. Arora and K. Tangri: *Expt. Mechanics*, in press.
9. A. Arora and K. Tangri: *Intl. Advances in NDT*, Gorden Breach Science
    Publ., New York, 1980.
10. A. Arora: *Proc. 3rd Intl. Conf. on NDE in Nuclear Industry*, ASM,
    Metals Park, 1980.
11. C. E. Coleman and J. F. R. Ambler: *ASTM STP 633*, pp. 589-607, 1977.
12. D. Hardie and M. W. Shanahan: *J. Nucl. Mater.*, 1975, Vol. 55,
    pp.1-13.
13. D. Kalish and H. J. Rack: *Met. Trans.*, 1972, Vol. 3, pp. 1885-92.

14.   A. J. Bedford, P. G. Fuller and D. R. Miller: _J. Nucl. Mater._, 1972, Vol. 43, pp. 164-74.
15.   T. K. Sinha and M. K. Asundi: _J. Nucl. Mater._, 1977, Vol. 67, pp. 315-21.
16.   R. D. Kane and B. J. Berkowitz: _Corrosion-NACE_, 1980, Vol. 36, pp. 29-36.
17.   P. Delobelle and C. Oytana: _Met. Trans._, 1978, Vol. 9A, pp. 740-41.
18.   W. R. Thorpe and I. O. Smith: _J. Nucl. Mater._, 1979, Vol. 80, pp. 35-42.
19.   S. K.Bhattacharya and K. C. Russell: _Met. Trans._, 1972, Vol. 3, pp. 2195-99.

SUBJECT INDEX

The letter "T" or "F" following an entry
indicates that a table or figure appears on that page.

O

Ordering
　　effect on hydrogen embrittle-
　　　ment and permeability in
　　　nickel, 412 (F)

Oxide films
　　as a diffusion barrier, 61, 111,
　　　187
　　as an adsorption barrier, 136
　　effect on environmental fatigue,
　　　483 (F), 489
　　effect on stress corrosion
　　　cracking, 517 (F)
　　enrichment by magnesium, in
　　　aluminum alloys, 515

Oxygen
　　effect on; adsorption of hydro-
　　　gen, 136
　　　corrosion of aluminum alloys,
　　　488 (T)
　　　crack rates, 58, 486
　　　hydrogen trapping in
　　　titanium, 185

P

Palladium
　　as a surface coating, in per-
　　　meation, 5, 100, 151, 172,
　　　414

Peierls stress
　　effect of hydrogen on, 200, 226,
　　　250

Permeation transients
　　by electrochemical charging,
　　　22 (F), 105–112
　　by gas phase, 20–25, 20 (F)
　　effect of, cathodic poisons,
　　　154
　　　implanted species, in iron,
　　　143–152
　　　stress, 173–175
　　electrical analogue model for,
　　　106
　　first and second transients, 346
　　time lag measurements, 22, 345
　　values derived from, 5–25,
　　　412 (T)

Potential
　　effect on fracture mode in iron
　　　and nickel, 371 (F)
　　effect on stress corrosion
　　　cracking of aluminum alloys,
　　　456 (F,T)

Pre-exposure embrittlement
　　in aluminum alloys, 451, 473,
　　　504, 514

R

Refractory Metals
　　niobium-oxygen alloy, 874
　　tantalum-oxygen alloy, 874
　　vanadium, 924

S

Slow strain rate tests, 450
　　in NaNO$_3$, 953
　　in NaOH solutions, 961
　　in water and moist H$_2$, 953

Stacking fault energy
　　effect on hydrogen suscep-
　　　tibility, 537, 545, 591
　　in austenitic stainless steel,
　　　effect of hydrogen, 582
　　values of, 545 (F)

Stainless steels
　　17Cr-14Ni (pure), 556
　　18Cr-6Ni (pure), 556
　　18Cr-10Ni (pure), 556
　　21-6-9, 542, 607–618, 637, 655
　　304, 527, 576, 583, 666
　　304L, 527, 565, 583, 607, 619
　　　630, 637, 655
　　310, 576, 630
　　316, 527, 576, 637
　　316L, 565, 583
　　A286, 542, 597, 645
　　JBK-75, 645, 655
　　cathodic charging of, 164, 261,
　　　527, 619
　　diffusion of hydrogen and
　　　deuterium in, 165 (T)
　　effect of α' martensite on
　　　hydrogen cracking of, 527,
　　　560, 571, 619, 629
　　effect of, Σ martensite on
　　　hydrogen cracking of, 537,
　　　629
　　　sensitization on hydrogen
　　　behavior, 527–540, 571, 629,
　　　665
　　　strain rate on hydrogen
　　　embrittlement, 655
　　　temperature on hydrogen
　　　embrittlement, 655
　　hydrogen-induced ductility
　　　losses in, 541–553
　　hydride formation, 261
　　lattice parameter changes, due
　　　to hydrogen, 262 (F)
　　stress dependence of diffusi-
　　　vity, 173–175